Calculus in Context

Calculus in Context

Background, Basics, and Applications

Alexander J. Hahn
University of Notre Dame

Johns Hopkins University Press
Baltimore

Printed in the United States of America on acid-free paper
9 8 7 6 5 4 3 2 1

Johns Hopkins University Press
2715 North Charles Street
Baltimore, Maryland 21218-4363

Library of Congress Control Number: 2016948163

ISBN 978-1-4214-2230-5 (hardback; alk)
ISBN 1-4214-2230-1 (hardback: alk)
ISBN 978-1-4214-2231-2 (ebook)
ISBN 1-4214-2231-X (ebook)
www.press.jhu.edu

Composed using the LaTeX Document Preparation System. TeX is a trademark of the American Mathematical Society.

Contents

Preface

Overview of the Content of the Text

The first four chapters present the basics of standard geometry, trigonometry, algebra, and coordinate geometry. This discussion is framed by the contributions to geometry and astronomy from ancient Greece (Archimedes and Ptolemy, for instance) to the scientific revolution (including those of Copernicus, Galileo,

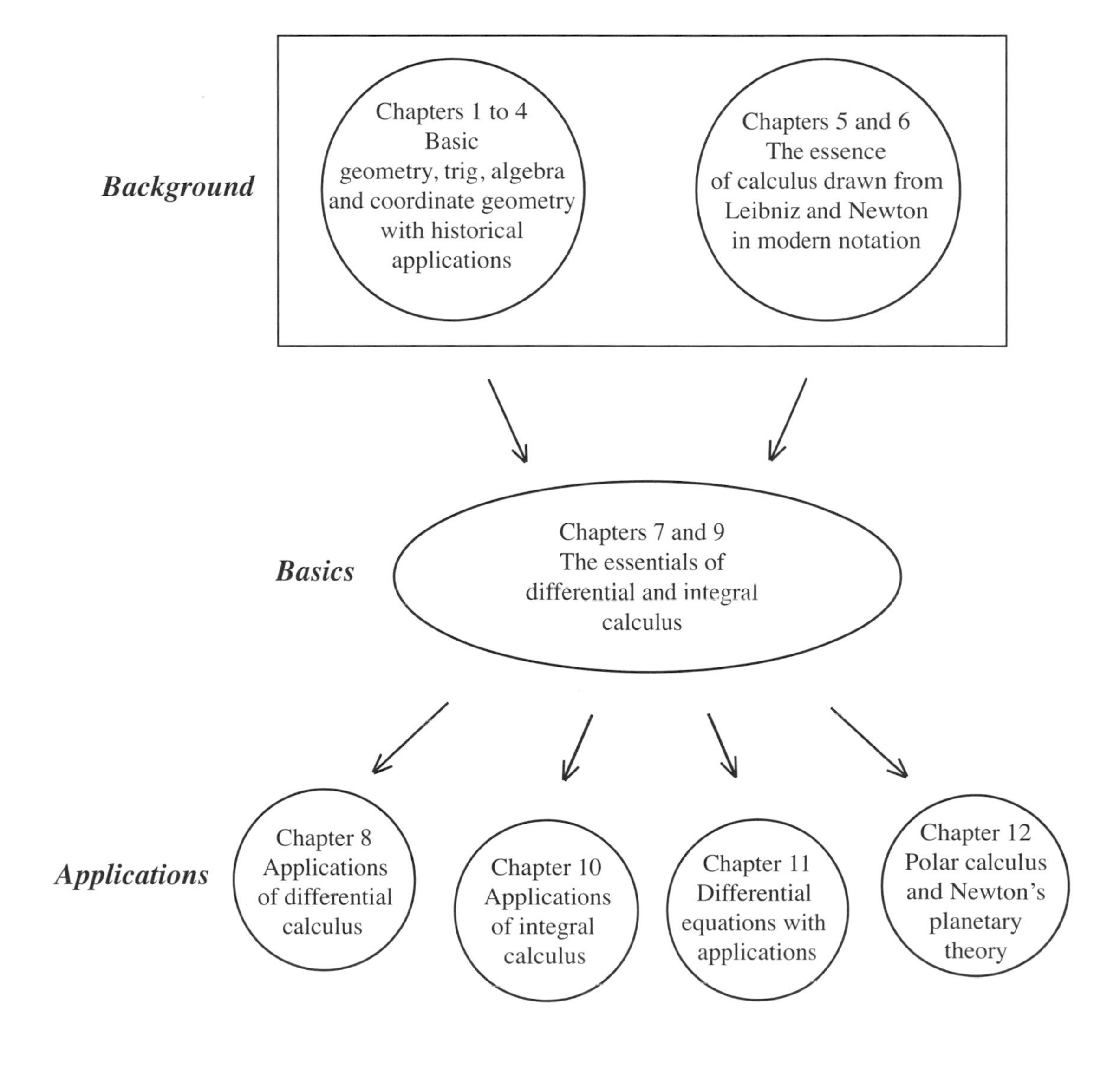

Kepler, and Descartes). Chapter 5 draws a brief, basic calculus course from the pages of Leibniz and Chapter 6 adds fundamental insights and applications from the work of Newton. With its extensive Problems and Projects sections, Part I of the text is a complete essential calculus and the precalculus preliminaries in 270 pages. The subject is framed by its historical context, but the notation and the selection of topics is current. If calculus and precalculus are the primary concerns of a course, then Chapter 2 with its focus on area computations of Archimedes and Chapter 3 with its goal of developing the Sun-centered model of the motion of the planets can be skipped. If included, these two chapters show Greek geometry and trigonometry in action. The scientific revolution reached its culmination in the latter part of the 17th century with the publication of Newton's *Principia* and its explanation of the planetary orbits with a combination of fundamental laws of motion and force and the strategies of calculus. A discussion of the central aspects of this challenging work—it too can be omitted if the target is calculus—concludes the first part of the text.

Part II of the text starts by engaging calculus again, this time much more fully and deeply. Chapter 7 consists of a comprehensive treatment of the essentials of differential calculus, and Chapter 9 does the same for integral calculus. All basic aspects of calculus are dealt with in 107 pages plus 32 pages of problems. A course may well start with Part II and use the discussions of Part I—especially Chapters 1 and 4—for reference and review. After the presentation of the basics of calculus, Part II turns its attention to applications. At this point, an instructor's strategy vis-à-vis the topics presented in this text should shift more into a "pick and choose" mode. If the power of calculus to inform is to be illustrated convincingly and beyond such trivial (or contrived) situations as "the farmer and the fence," then explanations of some of the very basic aspects of science and engineering that calculus informs need to be included. The fact is that the explanation of such central concepts as acceleration, force, momentum, torque, moments of inertia, and the properties of lenses require core ideas of calculus. Chapters 8 and 10 apply calculus (differential in the first case and integral in the second) to mechanics (in particular to statics and dynamics), optics, architecture, the geometry of the pseudosphere, planetary motion, and ballistics. Chapter 11 provides the fundamentals of differential equations and then turns to applications. The focus is on first- and second-order differential equations, complex numbers, power series, and applications to motion with air resistance and the action of shock absorbers. Chapter 12 starts with a study of polar functions and then completes the agenda of the text by returning to Newton's orbital mechanics and applying the calculus of polar functions to reach its important conclusions. The various applied segments of the text do not build on each other but are independent. Whether they are selected or not is entirely up to the intentions and preferences of the instructor. The number of pages involved in Part II gives some sense of the way the material is emphasized. The applications of calculus in Chapters 8, 10, 11, and 12 take up 200 pages and their Problems and Projects sections—many of which are study segments—about 70 pages more.

Expressed with the metaphor of language, this book combines the vocabulary and basic grammar of calculus with much of the poetry and the stories that have been written with it. The notation used in the text is modern throughout, the mathematics is current and relevant, and the wide-ranging Problem and Projects segments provide the student with skill-sharpening opportunities. An instructor should exercise careful judgment when assigning the problems, as the level of difficulty varies.

How This Text Can Be Used

In the hands of an engaged instructor, these materials and their organization are suitable for two-semester calculus courses of different degrees of difficulty and intensity and a wide range of student involvement. A course can be more elementary—for example, by starting with and incorporating discussions from Part I—or more advanced—for instance, by beginning with the more theoretical approach to calculus of Part II and then turning to the applications in the later chapters.

I have taught the topics of this text at the University of Notre Dame for many years (to a large extent from my earlier book *Basic Calculus: From Archimedes to Newton to Its Role in Science* [Springer-Verlag, 1998]). My approach was to build the material from the ground up in standard lecture format and to provide students with lots of opportunities to engage questions and to solve (mostly routine) problems on the board. The students responded well to this approach and learned a lot. The fact that students are increasingly exposed to calculus in high school moved me recently to use a much different strategy. In the spirit of the *flipped classroom* and *inquiry-based learning*, I split the class into teams (of three, sometimes two) and handed each team a "packet" with an applied topic from Chapters 8 and 10. The assignment for each team was to "reverse engineer" the material of the packet (mostly outside of class). The students were to begin by identifying, reviewing, studying, and learning the fundamental aspects of calculus that their topic involved from earlier versions of Chapters 7 and 9. They would then expand their investigations to relevant background materials and to the mathematical particulars. I assisted their efforts during office hours and with clarifying lectures in class. After a final "go through" with me, the student teams presented their topics to the class (typically taking two or three class periods). The students were first-year Notre Dame honors program students and hence among the best at Notre Dame, but their interests and intended majors were primarily in arts and letters disciplines. The results were positive for this small class of about 30. Students gained an understanding and a sense of ownership of strategies of calculus and an appreciation of its applications within meaningful contexts. They stepped to the board confidently and, often with PowerPoint assistance, presented cogent explanations. Briefly put, this was a flipped classroom based on the text (rather than on instructor-selected mini-videos).

Letting students dive in, select a topic, wrestle with it, and prepare it for presentation—with the instructor providing both guidance during office hours and clarifying comments in class—can be a very profitable approach. Does it not have to be one of the primary aims of a college education to enable students to engage sophisticated written materials, to extract the essence of the matter, and, finally, to present their findings articulately in front of an audience?

The Role of Technology

This text's emphasis is on the mathematics and not on computation and related technology. In this regard, the text makes use of a number of the websites with powerful and wide ranging computational capacity. Internet searches will display many more examples of such websites.

http://web2.0calc.com (a good basic calculator with memory)

https://www.desmos.com/calculator (a good graphing calculator)

http://www.integral-calculator.com (computes definite and indefinite integrals)

http://www.emathhelp.net/calculators/calculus-2/trapezoidal-rule-calculator/
(computes trapezoidal approximations)

http://www.emathhelp.net/calculators/calculus-2/simpsons-rule-calculator/
(computes Simpson approximations)

https://www.desmos.com/calculator/hmmutc9ija (computes Taylor polynomials)

https://www.symbolab.com/solver/ordinary-differential-equation-calculator
(solves differential equations)

http://www.math-cs.gordon.edu/~senning/desolver/ (applies Euler's method)

https://www.desmos.com/calculator/ms3eghkkgz (polar graphing calculator)

With a few exceptions, the units used throughout the book are the standard metric units—meter, kilogram, and second (MKS)—and units, such as the newton, that are derived from them. The most important exception involves the discussion about the American suspension bridges, including the Brooklyn Bridge, the George Washington Bridge, and the Golden Gate Bridge. They are discussed as they were built—in feet and pounds.

Acknowledgments

A warm and emphatic "thank you" goes to a number of my colleagues and friends at the University of Notre Dame. Those in the Department of Mathematics deserve particular mention. A special word of thanks to Timothy O'Meara, former provost, for his support and inspirational example over decades. A warm word of gratitude to my friends in Notre Dame's Kaneb Center for Teaching and Learning, especially Kevin Barry, Chris Clark, and Kevin Abbott, for always emphasizing that teaching must be measured not by the quality of the exposition, but by the learning experience of the students. Many thanks to Kevin Barry for developing the simulation of the elliptical orbits of the planets on the website

http://learning.nd.edu/orbital/orbital-info.html.

It relies on the mathematics of Kepler as Section 10.4 presents it.

Much gratitude is due the many hundreds of bright students of the Glynn Family Honors Program at Notre Dame. Your engaged participation was a great and constant source of satisfaction. I would like to thank them all in the name of the small last group of Glynn students that I taught: Mary Bernard, Brennan Buhr, Emily Daly, Eileen DiPofi, Lauren Jhin, Stephen Kelly, David Korzeniowski, Gilberto Marxuach, Connor Reilly, Terese Schomogyi, Fabiola Shipley, and Mary Szromba. Many talented undergraduates have assisted me over the years as instructors of the weekly problem sessions for my course and as graders. The last two—Alexis Doyle and Erin Thomassen—deserve special mention as two of the best.

Finally, I wish to thank the people at Johns Hopkins University Press, especially Vince Burke, who was supportive from the first conversations about this project through to its completion, and Ashleigh McKown for her thorough and constructive copyediting. A big thank-you to John Thoo, professor of mathematics at Yuba College in Marysville, California, for setting up the LaTeX format for the text that was compatible with the requirements of Johns Hopkins University Press. John came to my assistance again and again when glitches arose and fine tuning was needed.

This book is dedicated to my wife

Marianne Hahn

Thank you so much for your constant encouragement over the years. Not one page of this book could have been written without your devoted support.

Part I

Chapter 1

The Astronomy and Geometry of the Greeks

Chapter 1 presents basic facts and central concepts of geometry and trigonometry. These subjects had earlier roots, but the Greeks formulated them in a comprehensive and structured way. The Greek explanation of the motion of the Sun, Moon, and planets of the universe provides a motivating context for this study.

By the beginning of the last millennium B.C., the city-states of Greece had started to flourish. These trading towns along the Aegean and southern Italian coasts were ruled by an independent, politically aware merchant class. Their growing trade made them wealthy and connected them with other lands on the shores of the Mediterranean and beyond. Their outward-looking orientation spawned a new rational point of view. Rather than the acceptance of *mythos,* the account that gods and demigods controlled nature and unleashed its forces on a whim, there was a realization that observed phenomena operated in accordance with a rational *logos* that reason could begin to sort out and comprehend. Greek statesmen, philosophers, architects, sculptors, dramatists, and mathematicians approached and shaped reality with this mind-set.

From 600 B.C. until about 200 A.D., Greek geniuses, working in Greece and its colonies along the coast of the Mediterranean, laid the foundations of mathematics and science. Many of their answers, such as "all matter is made up of the four basic constituents earth, air, water and fire," were wrong or incomplete, but the important fact is that when they asked "are there basic elements that combine to make up all matter?" they posed the right questions. To grasp the movement of the myriad points of light that is the visible universe was of particular interest. Mathematics was regarded to be central to both the understanding of the natural world and the design of the universe. From about 300 B.C. onward, Alexandria (in today's Egypt) with its great library was the hub of this activity. Working in the museum, a state-supported institute for advanced studies (a "house of the muses" for the arts and sciences), scholars investigated astronomy, mathematics, and medicine. Around 300 B.C., Euclid put together the *Elements*, a comprehensive treatise that built on the mathematics of the Babylonians and Euclid's Greek predecessors, Thales and Pythagoras, among others. It is a tightly structured exposition, driven by logic and organized into 13 books. Ten fundamental statements are given a defining position as axioms or postulates. They are placed at the beginning and could be examined critically at the outset. All else is derived from them in the form of several hundred propositions about various aspects of plane geometry and properties of numbers. Today's mathematical theories still adhere to this basic structure.

Hipparchus, working in the 2nd century B.C., is the Euclid of astronomy. He catalogued the stars, observed the planets, and determined the lengths of the seasons. Relying on the perspective of a *geocentric* universe, one that has the Earth fixed at its center, he developed the work of his predecessors into more definitive geometric models of the motion of the planets. Hipparchus also studied triangles and advanced a trigonometry based on chords of circular arcs. Building on this work, Claudius Ptolemy (around 150 A.D.) refined the trigonometry of chords and devised an elaborate scheme of circles that describes how the Sun, Moon, and planets move in the geocentric context. His explanation was the accepted theory of the motion of the heavens until the discoveries of Copernicus, Galileo, and Kepler placed the Sun at the center of the universe and told us that the planets move around it in elliptical orbits.

Chapter 1 begins with a brief overview of the Greek understanding of the universe. It then introduces basic concepts of abstract geometry and trigonometry and shows how the Greeks put them to practical use. Most of the material presented has its origins in two works, the *Elements* of Euclid and the *Almagest* of Claudius Ptolemy, two of the most influential volumes in the history of Western civilization. The emphasis of the *Elements* is on abstract mathematics, and in particular on geometry. The importance of this work is undiminished today. It is still an essential component of today's mathematics curriculum. The emphasis of the *Almagest* is on astronomy. It would be the accepted explanation of the mechanics of the solar system for about 1500 years, until it was displaced in the 17th century by the insights of Kepler and Newton. The focus of this chapter is only on some of the most important aspects. These are presented from a modern perspective, in terms of both concepts and notation. The achievements of the Greeks—you will soon be in a position to judge for yourself—are remarkable. "Without the concepts, methods and results found and developed by previous generations right down to Greek antiquity," Hermann Weyl (1885–1955), one of the leading mathematicians of the 20th century, would write, "one cannot understand either the aims or the achievements of mathematics in the last fifty years."

1.1 THE GREEKS EXPLAIN THE UNIVERSE

Today we know that the Earth rotates once a day about its north-south axis and that it moves in a yearlong orbit about the Sun. However, from the vantage point of an observer on the surface of the Earth, the Earth is fixed and the Sun is seen to move in the sky and (on a clear night) the stars to rotate in the heavens along circular paths. It is not surprising that this was the point of view of the Greek philosophers. (After all, as you sit with this book in your hands, does it seem to you that you're moving? Do not the Sun and stars move when you look out at them?) In their daily rotation, the stars retain their position relative to each other, so that the Greeks regarded them as fixed in place on a rotating *celestial sphere*. They positioned the spherical Earth at the center of this sphere and tracked the path of the Sun against the celestial sphere as a great circle. The Moon, Mercury, Venus, Mars, Jupiter, and Saturn are the planets. (Neptune, Uranus,

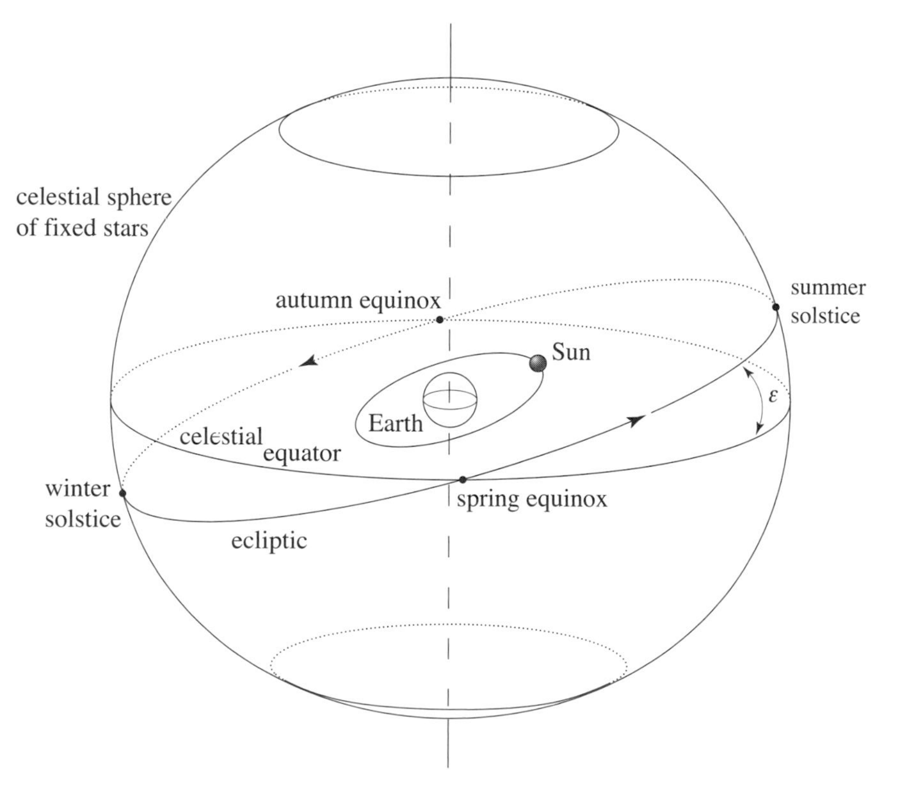

Fig. 1.1. The Greek picture of the universe

and Pluto had as yet not been detected.) They look like stars in the night sky, but they were observed to move or wander (in Greek, *planetes* = wanderers) relative to the background of the fixed stars. Each of the Greek thinkers provided additional details. For example, according to Aristotle, each of the planets and the Sun is attached to its own crystal sphere whose rotation moves it along and according to Pythagoras, each of the heavenly bodies contributes a certain tone to an orchestrated "music of the spheres."

The essence of the structure of the Greek universe is illustrated in Figure 1.1. The axis is the continuation of the north-south axis of the Earth. The *celestial equator* is the projection of the Earth's equator against the celestial sphere. The *ecliptic* is the projection of the path of the Sun against the celestial sphere. Both the celestial equator and the ecliptic are great circles on the celestial sphere. The angle ε between the planes that these circles determine is the *obliquity of the ecliptic*. The two points where the circles intersect are the *spring equinox* and *autumn equinox*. The word "equinox" has the Latin root *aequinoctium* from *aequi* = equal and *noctium* = of nights. On the days when the Sun is at these two locations, the time from sunrise to sunset is the same as the time from sunset to sunrise for any point on Earth. The point on the ecliptic where the Sun reaches its highest position above the plane of the celestial equator is the *summer solstice*, and the point of lowest position below this plane is the *winter solstice*. The word "solstice" is derived from the Latin *sol* = sun and *sistere* = to stand still. At the summer solstice, the Sun slows its ascent, stops, and begins its descent. At winter solstice, this is reversed. The Sun slows its descent, stops, and begins to ascend. These four points divide the ecliptic into four equal circular arcs.

The ancient astronomers knew the pattern of stars in the night sky well. They grouped the myriad points of light into several dozen recognizable clusters that were later called *constellations* (from the Latin word *stella* = star). Twelve of these clusters were arranged in a belt around the 360° of the ecliptic (the influence of the Babylonian base 60 number system is apparent here), about 10° on either side of the ecliptic. The names of these clusters

> Aries (Ram), Taurus (Bull), Gemini (Twins), Cancer (Crab), Leo (Lion),
> Virgo (Maiden), Libra (Scales), Scorpio (Scorpion), Sagittarius (Archer),
> Capricorn (Water Goat), Aquarius (Water-Bearer), and Pisces (Fish)

tell us that were derived from the shapes that they suggested (and a good deal of imagination). Since several had names of animals, this belt of constellations became known as the *zodiac* (in Greek, *zoidion* = the diminutive for animal). See Figure 1.2. This form of the zodiac entered Greek astronomy in the 4th century B.C. by way of the Babylonians. At the time of the Greek astronomers, spring equinox marked the start of Aries, summer solstice that of Cancer, autumn equinox of Libra, and winter solstice of Capricorn. Each of these constellations took up about 30° of the 360° degrees along the ecliptic. The zodiac gains its impor-

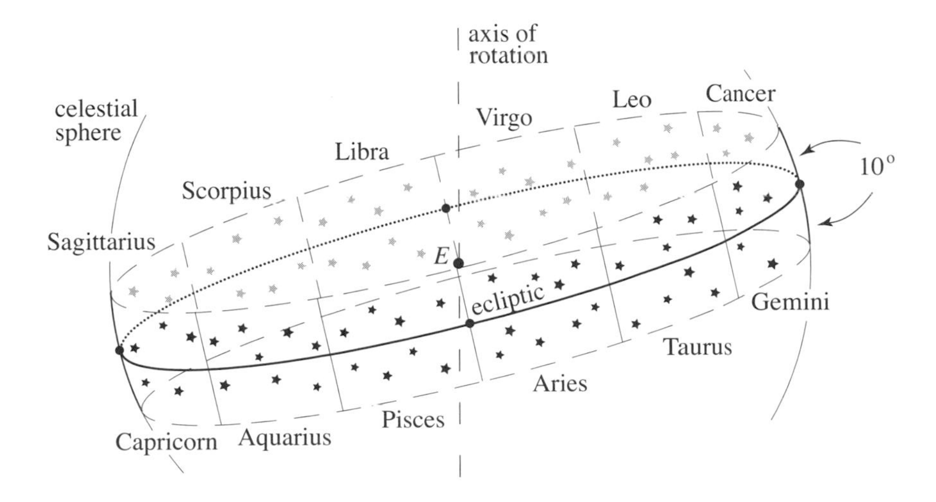

Fig. 1.2

tance from the fact that the planets appear as points of light that always move within the constellations of the zodiac. Only the Moon drifted beyond the band of these constellations. Astronomers tracked the annual path of the planets as they moved through the constellations from one to the next. Because the stars cannot be seen when the Sun is in the sky, the path of the Sun against the celestial sphere (this is the ecliptic) was observed and tracked soon after sunset and before sunrise when the constellations are visible. The science of astronomy and the practice of astrology developed side by side. Charting the planets in the zodiac is central to both. It is the first step in our understanding the planetary orbits, and it is the basis of the superstition of horoscopes.

The following ritual must have been performed by the Greek astronomers thousands of times. Plant a *gnomon*—this is a straight pole—at a point A in the middle of a flat horizontal stretch of ground. (The word "gnomon" comes from an ancient Greek word meaning "knower.") Use a plumb line to set this pole in a vertical position. This is the segment GA of Figure 1.3a. At some time in the morning, after the Sun has risen in the east, mark the endpoint W of the shadow (it lies west of A) and measure the length AW. As the Sun moves up in the sky, the shadow will shorten and rotate. When the Sun starts its descent, the shadow will begin to lengthen. At some time before the Sun sets in the west, the tip of the shadow will reach a point E (toward the east) on the ground where its length AE will equal AW. By the symmetry of the process, the bisector of the angle WAE lies on the north-south line through A. This line is called the *meridian* through A. At the moment in the day when the shadow falls on the meridian, it has reached its shortest length. The Sun is now at its highest point in the sky. This moment is *noon* on that day. A *day* is

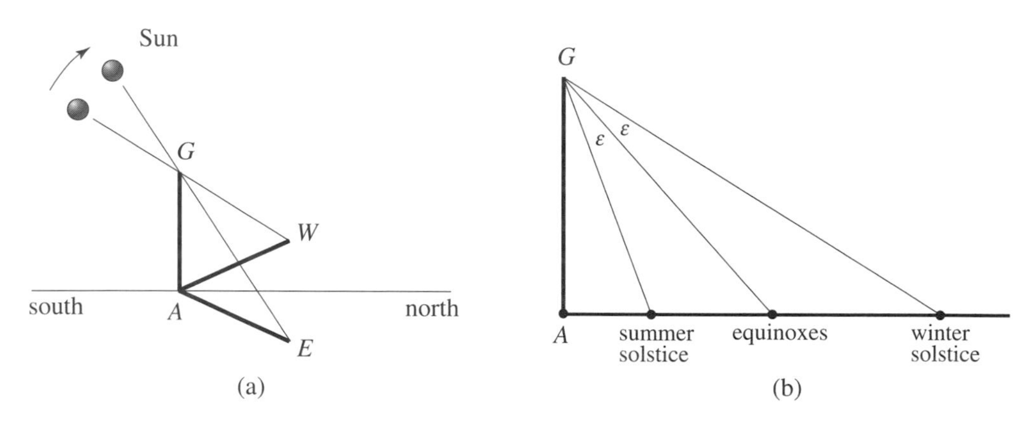

Fig. 1.3

defined to be the duration of time between two successive noons. The division of a day into 24 equal units defines the *hour*. (With this definition, the hour changes slightly from day to day. Today's hour is standardized and fixed.)

Since the focus is on the movement of the Sun against the celestial sphere, the daily rotation of this sphere (in other words, the daily rotation of the Earth) must be discounted. To achieve this, gnomon measurements are taken exactly at noon each day. The day on which the noon shadow of the gnomon is shortest, in other words, the day of the year on which the Sun is at its highest point in the sky, is summer solstice. Winter solstice is the day on which the noon shadow of the gnomon is longest. This is the day of the year on which the noon Sun is at its lowest point. Summer solstice is the day of longest daylight, and winter solstice is the day of shortest daylight. Figure 1.1 tells us that the angle at G between the summer and winter solstice positions is 2ε and that the bisection of this angle provides the two equinox positions. This is illustrated in Figure 1.3b. On the days of spring and autumn equinox, the number of hours during which the Sun is above the horizon is the same as the number of hours during which it is below it.

The seasons of the year are determined as follows: *spring* is the period from spring equinox to summer solstice; *summer* runs from summer solstice to autumn equinox; *autumn* from autumn equinox to winter solstice; and *winter* from winter solstice to spring equinox. The *year* is defined to be the time period between two successive summer solstices (or winter solstices, autumn equinoxes, or spring equinoxes).

The astronomer-mathematician Hipparchus of Nicaea (near today's Istanbul), who lived from about 190 to 120 B.C. and is regarded to be the greatest astronomer of antiquity, made shadow measurements that gave the following results:

spring: $94\frac{1}{2}$ days *summer:* $92\frac{1}{2}$ days *autumn:* $88\frac{1}{8}$ days *winter:* $90\frac{1}{8}$ days

After adding these numbers, Hipparchus knew that the year was $365\frac{1}{4}$ days long. Shadow measurements also told him that the obliquity of the ecliptic is approximately $\varepsilon = 24°$.

Refer back to Figure 1.1. The spring equinox, summer solstice, autumn equinox, and winter solstice positions divide the ecliptic into four equal circular arcs. The fact that the lengths of the seasons are different means that the Sun traces out these four arcs in different lengths of time. In spite of this irregularity, the great Greek philosophers were all in agreement about the following

1. The Earth is fixed and positioned at the center of a spherical universe.

2. All other bodies in the universe move along circular paths with constant speed.

After all, our experience tells us that the Earth stands still and that spheres and circles are perfect geometric shapes. There was simply no other other possibility for the structure of the universe or the motion of the stars, Sun, and planets within it.

Another observation stood in conflict with these two principles. The motion of the planets (and the Sun as well) has two aspects: they are carried along by the daily rotation of the celestial sphere of stars, but they also have their own independent motion. While the planets generally moved from east to west against the backdrop of the constellations of the zodiac, their movement was irregular. They were observed to intermittently slow down, stop, reverse direction for some time only to stop once more, and then continue

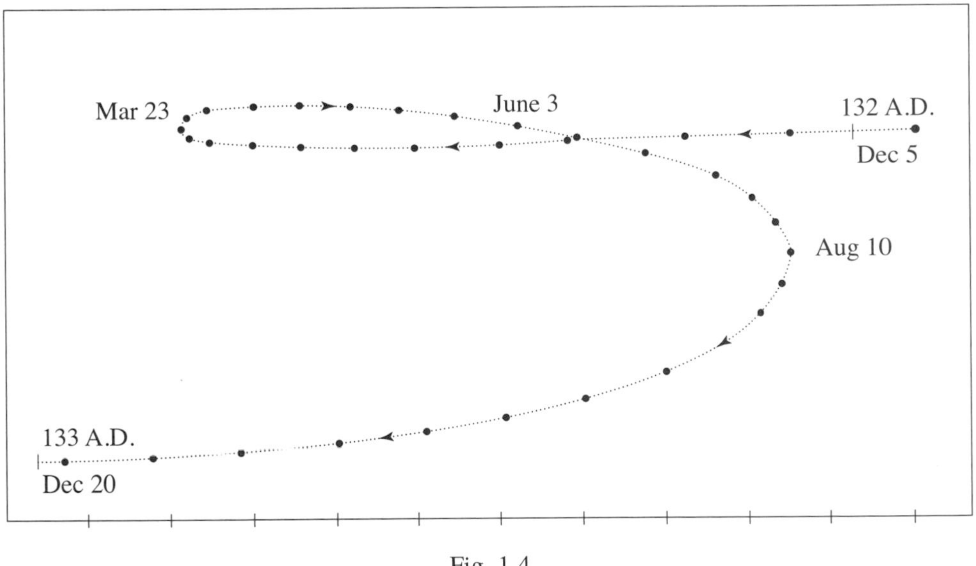

Fig. 1.4

in the original direction. For example, the path that Saturn traced against the night sky as the great Greek astronomer and mathematician Claudius Ptolemy (from about 90 to 170 A.D.) observed it in the years 132–133 A.D. is pictured in Figure 1.4. Ptolemy had documented the phenomenon of *retrograde* motion.

Clearly there was a problem. The observed motion of the Sun and the planets appeared to contradict the fundamental principles that Greek philosophy had declared to be nonnegotiable. The challenge that confronted the Greek astronomers seemed to be embedded in a contradiction: to develop a mathematical description of the universe that conformed to the two principles *and* was consistent with the data that Greek (and Babylonian) observers had accumulated over a number of centuries. They were to wrestle with this challenge for 500 years before Claudius Ptolemy finally completed the task.

1.2 ACHIEVING THE IMPOSSIBLE?

The problem of the Sun's motion was solved by Hipparchus as follows. He kept the Earth fixed at the middle of the universe, but he moved the center of the Sun's circular orbit off to the side. Figure 1.5 illustrates what he did. Focus on the circle centered at O and the four arcs into which the Sun's orbit is divided. Observe that during spring, when the projection S^* of the Sun against the ecliptic moves from spring equinox to summer solstice, the circular arc traced out by the Sun is the longest of the four. The arc for summer is next, then winter, and, finally, the arc for autumn is shortest. With the understanding that the

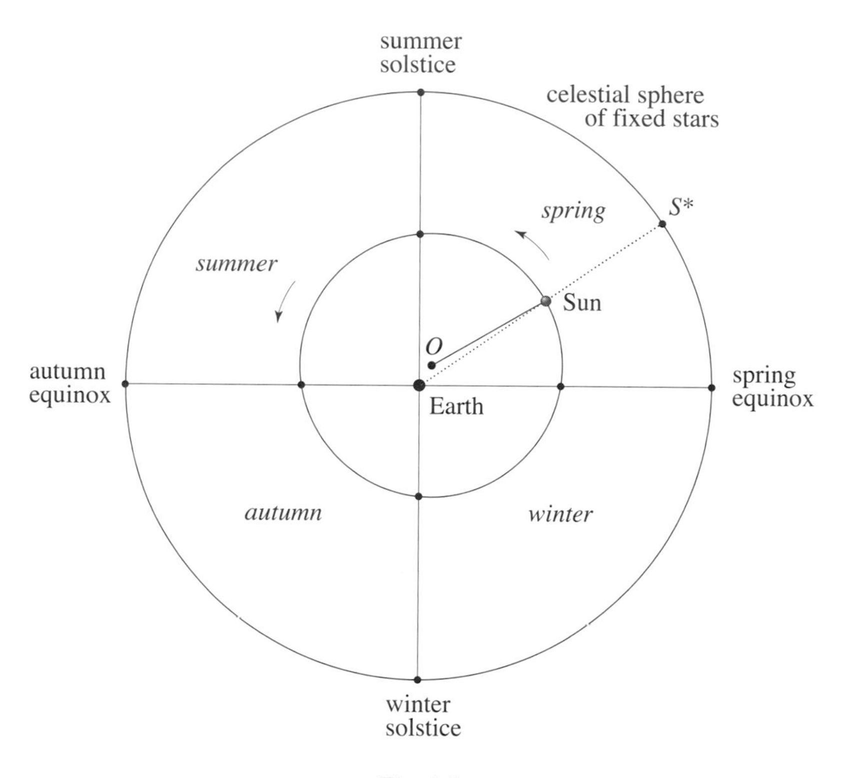

Fig. 1.5

Sun's speed is constant, spring is the longest season, and summer, winter, and autumn follow in that order. Notice that this is consistent with what Hipparchus had observed: spring, summer, autumn, and winter were $94\frac{1}{2}$, $92\frac{1}{2}$, $88\frac{1}{8}$, and $90\frac{1}{8}$ days long, respectively. By using this data, Hipparchus was able to determine the location of O relative to that of the Earth. Notice also that the solution of Hipparchus meets the requirement that the Earth is at the center of the universe and the Sun orbits in a circle.

This took care of the Sun, but what about the irregular motion of the planets? Hipparchus takes another off-centered circle—such a circle was called *deferent* by the ancient astronomers—its center O at some distance from the Earth at E. Hipparchus adds a smaller circle—an *epicycle*—placing its center D on the deferent. The motion of the planet P is a combination of two simultaneous circular motions. As the center D of the epicycle moves along the deferent, the planet P rotates along the epicycle. See Figure 1.6. In the figure, the two rotations are both counterclockwise around O and D, respectively. The two motions take place at constant speeds with the speed of P around the epicycle greater than that of D around the deferent. Since D moves counterclockwise around O, the planet P moves counterclockwise relative to E, most of the time. However, when P is in the part of its orbit closest to O, the motion of P on its epicycle

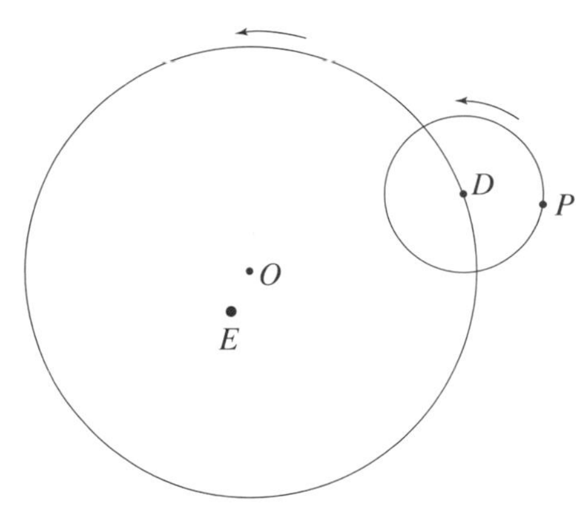

Fig. 1.6

will predominate over that of D on the deferent, so that during this part of its orbit, P will be seen from E to have changed direction. The two planes that the deferent circle and the epicycle determine are close to, but not the same as, the plane of the ecliptic. This means that P will move slightly, but noticeably, up and down relative to the ecliptic. If the various parameters—namely, the ratio of the radii of the two circles, the planes that they determine, as well as the speeds of the points D and P around them—are carefully chosen, then the motion of P in Hipparchus's model will, as seen from E, exhibit the type of behavior recorded in Figure 1.4.

To develop mathematical models that were in agreement with their observations, Greek astronomers needed to carefully identify the location of the stars and planets. Using the spring equinox position and the celestial equator as a frame of reference, Hipparchus specified the location $\times$ of a star or planet by recording the angle α—the *right ascension*—where $0 \leq \alpha \leq 360°$ and the angle δ—the *declination*—where $-90° \leq \delta \leq 90°$ from the point of the observation. See Figure 1.7. (This is still a common approach for listing planetary and stellar positions.) The star catalog that Hipparchus drew up listed more than 850

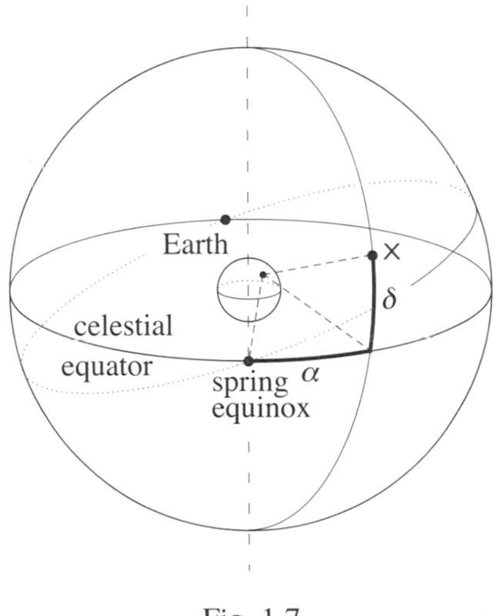

Fig. 1.7

stars. Hipparchus was also a great geographer. His map of the world as the Greeks knew it at the time (about 190–125 B.C.) is shown in Figure 1.8. Notice the coordinate system that organizes the geographic information it provides. A careful look shows that the primary horizontal and vertical axes intersect at the island of

Rhodes, the place where Hipparchus lived and worked. The city of Alexandria, founded by Alexander the Great near the Nile Delta about 150 years earlier, is seen to lie on the same meridian. The system of horizontal lines of latitude and slanting lines of longitude bends so as to give a sense of the curving shape of the

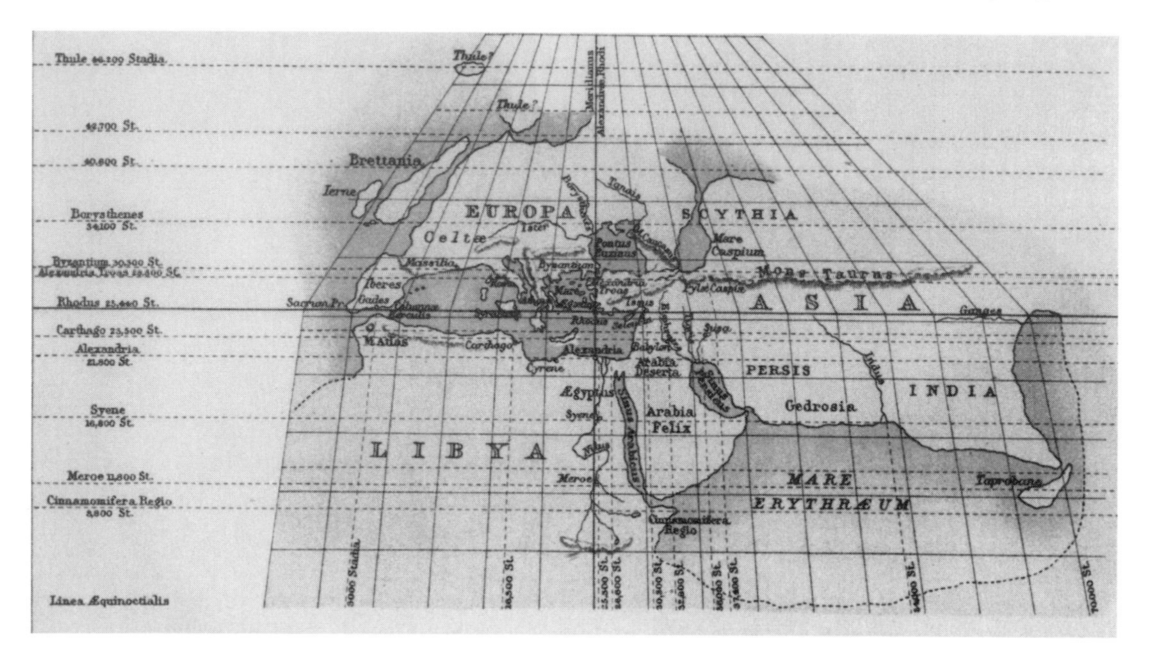

Fig. 1.8. The map of Hipparchus. Image provided by the Hesburgh Library, University of Notre Dame.

Earth. In the construction of his map, Hipparchus incorporated reports from travelers and made use of the fact that celestial observations, such as the angles between stars and vertical gnomons, and the lengths of the Sun's shadow cast by gnomons on the summer and winter solstices, provide information about the location of the observation.

The Greek Earth-centered explanation of the universe was put into final form around 150 A.D. by Claudius Ptolemy working in Alexandria (by then a part of the Roman Empire). His *H Megale Syntaxis,* or "The Great Treatise" is an encyclopedic compilation—a grand synthesis—of Greek mathematics and astronomy. The superlative of the Greek *megale* (great) is *megiste* (greatest). This is the word that gave the treatise its Arabic name *al-majisti* and, in turn, the name *Almagest*. The *Almagest* lays out what is known as the Ptolemaic system of the universe today. Important components build on the contributions of Hipparchus. It fine-tunes the deferent-epicycle model so that it is better aligned with what was observed. In order to give a more accurate account of a planet's speed as it moves around the Earth, Ptolemy adds another point Q, the *equant*, to the model. It lies on the diameter of the planet's orbit through O and E, on the side opposite E but at the same distance from O. It determines the speed of the point D as follows. Think of a beam of light that rotates at Q with constant speed (like the one of a lighthouse) in such a way that D, moving on the deferent, is always illuminated by the beam. See Figure 1.9. When even this more elaborate scheme does not mesh with observations, Ptolemy introduces additional epicycles. The required configurations of circles are more complicated for the inner planets (Mercury and Venus) than for the outer planets (Mars, Jupiter, and Saturn) known at the time. The configuration of circles for the Moon is the most elaborate. In fact, *Almagest* presents a progression of three different lunar theories of increasing complexity, a complexity forced by persistent discrepancies with observations. (This is hardly surprising. After all, the Moon in its orbit around Earth is also pulled by the Sun, and Earth's bulge at the equator adds gravitational forces that introduce further complications.)

Ptolemy's system was an extremely intricate clockwork of circles upon circles. A total of 40 epicycles and deferents were needed to describe the motions of the planets, Sun, and Moon. The Ptolemaic system should not be viewed as a description of the physical universe, but rather as an abstract mathematical

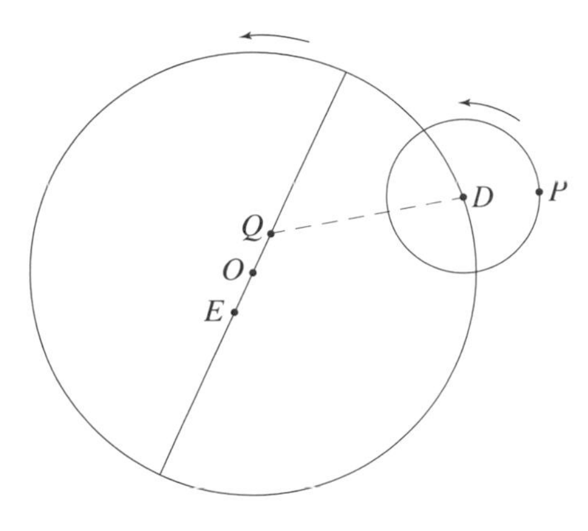

Fig. 1.9

scheme designed to predict the positions of the objects moving within it. Given the requirements that this clockwork needed to conform to—an Earth-centered system in which celestial objects move in orbits defined by circles—it succeeded in a remarkable way. In the way it describes a large and important class of natural phenomena in mathematical terms, *Almagest* is unrivaled in antiquity. As was already observed, it would be the accepted theory of planetary motion until the 17th century, when Galileo confirmed the Sun-centered view, Kepler introduced his elliptical orbits, and Newton's laws and mathematics provided the explanations.

Ptolemy also improved Hipparchus's maps. His *Geographia* provides geographical coordinates of 8000 localities and lays the foundation of the science of cartography. Its map of the known world captures Europe and North Africa remarkably well. See Figure 1.37. This map and the extensive catalogue of stars of the *Almagest* proved invaluable to the early Spanish and Portuguese navigators of the 15th century.

What the Greek astronomer-mathematicians accomplished was in fact astonishing. Can you blame them for looking at things from the perspective of a fixed Earth? Our own intuition—shaped as it is by our daily existence on our planet—confirms this point of view. From the perspective of a fixed Earth, the planets do in fact move along epicycles (but elliptic epicycles) carried by the deferent of the Sun's orbit about the Earth. Today we are able to think outside our earthly box and understand the solar system more simply as Sun centered, and we know that the approach of the Greeks was too complicated. The Greeks would not be able to provide definitive answers, but their geometry, trigonometry, and early progress with calculus would later be developed into the mathematical tools that did. The fact is that their mathematics and especially their geometry are as relevant today as they were then.

1.3 GREEK GEOMETRY

Given its importance within their evolving description of the universe, it is not surprising that the Greeks pursued geometry for its own sake. The mathematician-philosophers Thales (from about 620 to 550 B.C.) and Pythagoras (from about 570 to 495 B.C.) contributed substantially to this effort. Pythagoras founded a secretive fraternity of "Pythagoreans" that pursued mathematics within a belief system that included the idea that numbers explained everything. Euclid's *Elements* took the insights of the Babylonians, Thales, Pythagoras, and others and shaped them into a comprehensive, tightly structured study of geometry and numbers.

We turn to some basic concepts and facts from the *Elements*. Let's start with a look at angles. Two line segments with a common endpoint form an *angle*. It is labeled by θ in Figure 1.10a. A common way to measure or quantify the extent of the opening of an angle is the *degree*, denoted by $^\circ$. Place the angle into a circle so that the common endpoint is at the center. See Figure 1.10b. Declare the entire circle to have 360 degrees (see Figure 1.10c) and apportion degrees proportionally. So an angle consisting of half the circle is equal to 180°, one consisting of a quarter circle measures 90° and is called a *right* angle, and so on. The degree is subdivided into minutes and seconds. The *minute* is equal to $\frac{1}{60}^{\circ}$, and the *second* is equal to $\frac{1}{60}$ of a minute and hence $\frac{1}{3600}^{\circ}$. The terms "minute" and "second" have Latin roots. The minute

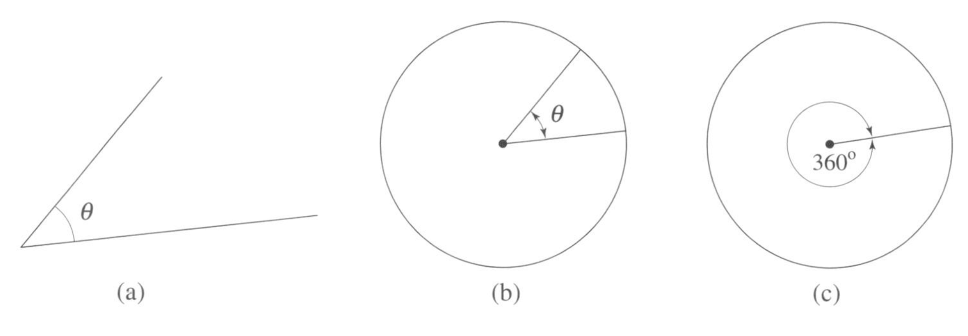

Fig. 1.10

comes from *pars minuta prima* meaning "first small part" and the second from *pars minuta secunda* meaning "second small part." Today, the degree, minute, and second are designated by $^{\circ}$, $'$, and $''$, respectively. So 15° 23′ 47" is shorthand for the angle of $15 + \frac{23}{60} + \frac{47}{3600}$ degrees. This scheme for measuring angles is consistent with the Babylonian base 60 number system and is believed to have originated with the Babylonians. This approach was later adopted by the Greeks, used by Copernicus in the 16th century, by Newton in the 17th, and is still used today.

The *Elements* devotes much space to the study of triangles. We'll consider a few aspects of this next. But first, some terminology. A corner of a triangle is called a *vertex*. When a triangle is drawn in such a way that one of its sides is horizontal (or nearly so), then that side is often referred to as the *base* of the triangle. One of the most fundamental facts tells us that the interior angles of a triangle add up to 180°. In Figure 1.11, a typical triangle is considered and α, β, and γ denote its three angles. The line at the top is

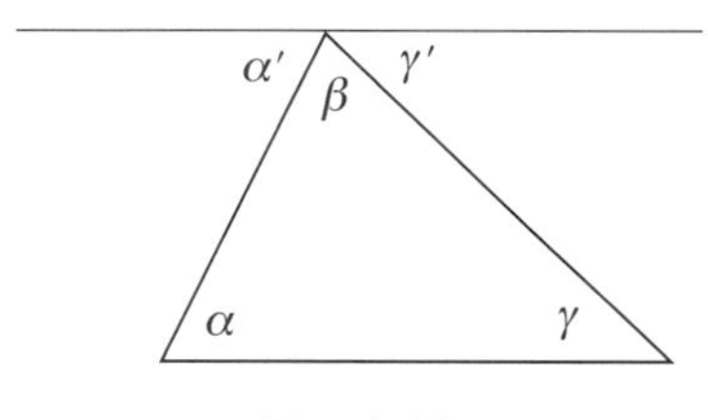

Fig. 1.11

parallel to the base, and the angles that this line makes with the sides of the triangle are α' and γ', respectively. It follows from the figure that

$$\alpha + \beta + \gamma = \alpha' + \beta + \gamma' = 180^{\circ}.$$

A triangle is called *isosceles* if two of its sides are equal. An *equilateral* triangle is a special case. It has all three sides equal to each other. Consider an isosceles triangle. Let A, B, and C be its vertices, AB its base, and AC and BC the two equal sides. See Figure 1.12a. The fifth proposition in Book I of Euclid's *Elements* asserts that the angle at A is equal to the angle at B. To verify this, Euclid proceeds as follows. He extends Figure 1.12a to Figure 1.12b in such a way that $CD = CE$ and draws segments from A to E and B to D. Study each of the two triangles ΔCAE and ΔCBD with a focus on the angle they share at C and the two pairs of sides CA, CE, and CB, CD, each of which form the angle at C. Since $CA = CB$ and $CE = CD$, it follows that the triangle ΔCAE can be flipped around so that it is identical to ΔCBD. This means that $\angle CAE = \angle CBD$. It also means that the angle at E is equal to the angle at D and that $AE = BD$. Because $BE = AD$, the triangle ΔABE can be flipped so that it coincides with ΔBAD. It follows that $\angle BAE = \angle ABD$. This fact, along with the equality $\angle CAE = \angle CBD$, implies that $\angle CAB = \angle CBA$, precisely what needed to be verified.

Euclid's fifth proposition became known as the *pons asinorum*, Latin for "bridge of asses." There are at least two explanations for the name. One is that a part of the diagram of Figure 1.12b resembles a bridge.

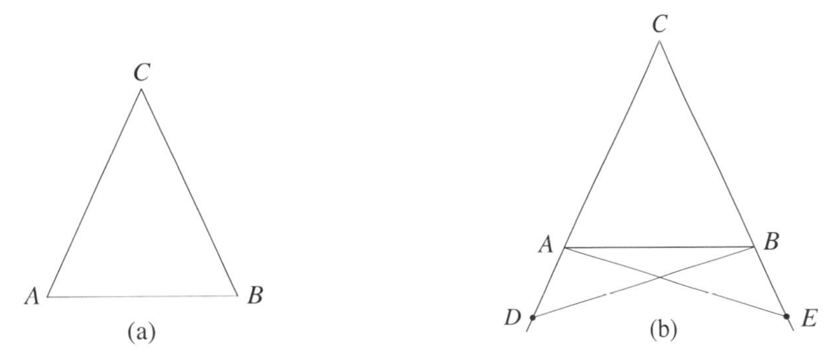

Fig. 1.12

The more common explanation is that this proposition is the first real test for a reader of the *Elements* and that it serves as a bridge to the harder propositions that follow. The phrase *pons asinorum* is used today to refer to any critical test of ability or understanding that separates the quick mind from the slow.

An important consequence of the *pons asinorum* is that for any triangle inscribed in a circle in such a way that one of its sides is a diameter of the circle, the angle opposite that side is a *right* angle. Any triangle with such an angle is a *right* triangle. Thus a triangle constructed inside a circle in this way is a right triangle. In Figure 1.13a, O is the center of the circle, AB is a diameter, and C is the third vertex of

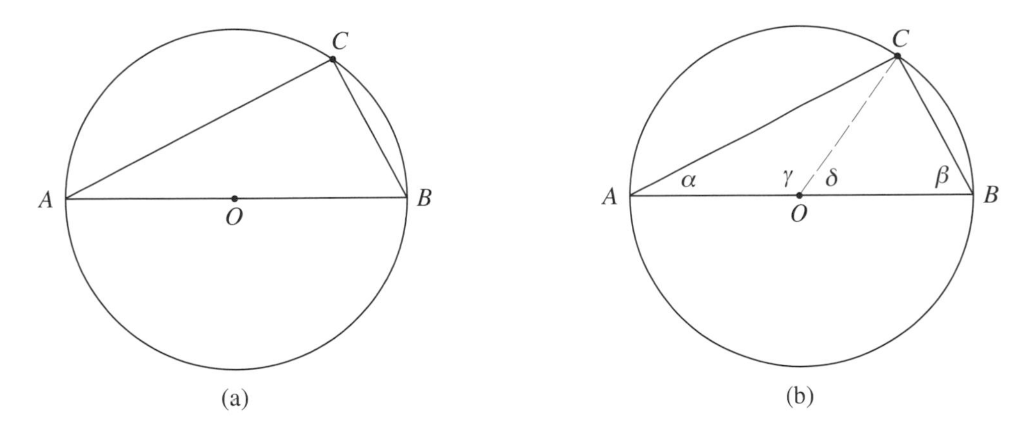

Fig. 1.13

the triangle. To verify the assertion, connect O to C and notice that both ΔAOC and ΔCOB are isosceles triangles. See Figure 1.13b. Let α and β be the angles at A and B, respectively. The application of the *pons asinorum* tells us that $\angle ACO = \alpha$ and $\angle OCB = \beta$. It follows that $2\alpha + \gamma = 180°$ and $2\beta + \delta = 180°$. So

$$2\alpha + (\gamma + \delta) + 2\beta = 360°.$$

Because $\gamma + \delta = 180°$, we get $2\alpha + 2\beta = 180°$. Therefore $\angle ACO + \angle OCB = 90°$. So the triangle ΔABC has a right angle at C.

Two triangles T and T' are *similar* if their angles match up. What this means is that there is a correspondence between the angles of T and T' in such a way that matching or corresponding pairs of angles are equal. The two triangles ΔABC and $\Delta A'B'C'$ in Figure 1.14 illustrate the point. They are similar because the angles at A and A', B and B', and C and C' are respectively equal. (Why is it the case that if two pairs of angles match up, then the remaining pair does also?) Two triangles are *congruent* if they are the same in the sense that one of them can be moved or flipped and repositioned so as to coincide

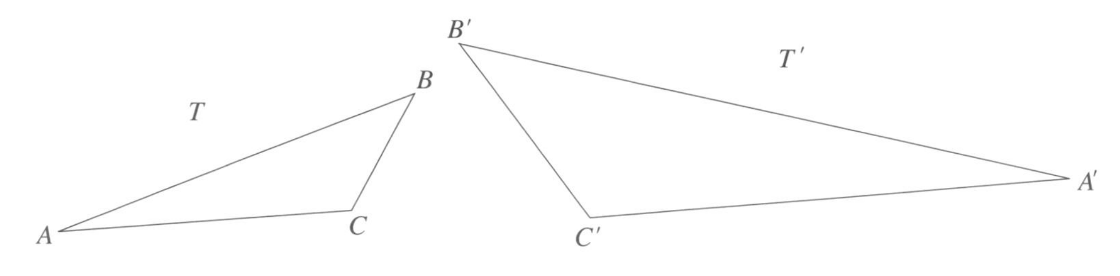

Fig. 1.14

with the other. Return to the proof of the *pons asinorum* and notice that the triangles ΔCAE and ΔCBD are congruent, and that ΔABE and ΔBAD are congruent as well. For two congruent triangles, the repositioning maneuver establishes a correspondence between its respective angles. So congruent triangles are similar. Let T and T' be any two similar triangles. As in Figure 1.14, label their vertices A, B, C and A', B', C', respectively, in such a way that the pairs of angles at A and A', B and B', and C and C' correspond to each other and are equal. The correspondence between the angles sets up a correspondence $AB \to A'B'$, $AC \to A'C'$, and $BC \to B'C'$ between their sides. One of the most important and useful facts in all of geometry says that the ratios of corresponding sides of similar triangles are equal or, more explicitly, that

$$\frac{AB}{A'B'} = \frac{AC}{A'C'} = \frac{BC}{B'C'}.$$

Since Thales was an early Greek genius who understood this relationship, it is often referred to as Thales's theorem. The proof in Euclid's *Elements* is provided in part 1B of the Problems and Projects section at the end of this chapter.

Here and elsewhere we'll do the following. For two points P and Q in the plane, PQ or QP will label the segment that connects them. However, if PQ appears in an equation, it will be understood to be the length of the segment PQ or, equivalently, the distance between P and Q.

Let's suppose that an inquisitive Greek traveler to Egypt comes upon a pyramid. He is interested in its size and determines the dimensions of its base by pacing off its sides. He then turns his attention to the height of the pyramid. On this bright sunny afternoon, he paces off the distance from the side of the pyramid to the tip of the shadow that it casts and estimates that the tip of the shadow is 310 paces from the center of the pyramid's base. Our traveler knows that he is 3 paces tall and measures the length of his shadow as 5 paces. Organizing the information he has, he draws the diagram of Figure 1.15 in the sand. The triangle ABC represents the pyramid and h its height. The two slanted lines depict the light rays from

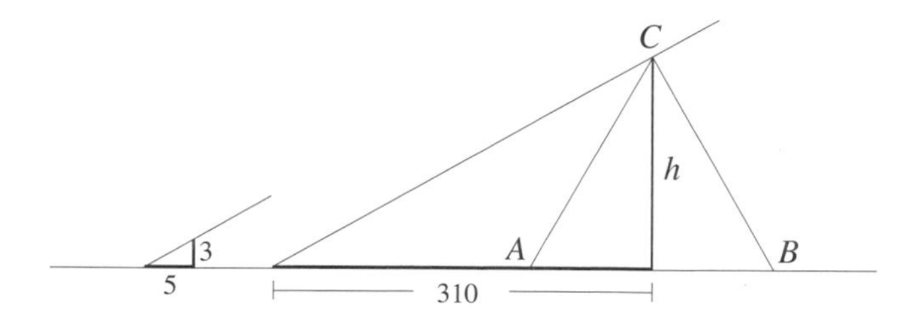

Fig. 1.15

the sun that determine the shadows. Our Greek traveler understands similar triangles and concludes that $\frac{h}{310} = \frac{3}{5}$. So he knows that the height of the pyramid is approximately equal to $h = \frac{3 \cdot 310}{5} = 186$ paces. This traveler could have been the Greek historian Herodotos, whose chronicles recall his visit to the great pyramids of Egypt (near today's Giza) in the 5th century B.C. and report some of its dimensions. Herodotos may well have been familiar with the studies that Thales undertook about one century earlier.

1.4 THE PYTHAGOREAN THEOREM

It turns out that the Pythagorean theorem, a designation that honors the great Greek mathematician, was already known to the Babylonians more than 1000 years before the time of Pythagoras. Consider any right triangle. The side opposite the right angle is called the *hypotenuse*. Let a, b, and c be the lengths of the sides of the triangle with c the length of the hypotenuse. The theorem of Pythagoras asserts that $a^2 + b^2 = c^2$.

History does not tell us exactly how the Babylonians or the Pythagoreans verified this theorem. One of the simplest verifications is visual and is provided by the sequence of diagrams of Figure 1.16. Start with the right triangle on the left. Now take a square of side length $a + b$. Go to each corner of the square

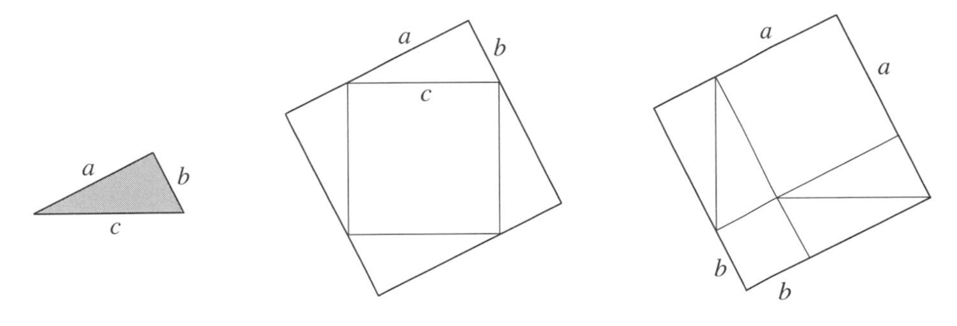

Fig. 1.16

and mark off the lengths a and b, as shown in the middle of the figure. Complete this to a copy of the triangle by drawing in the hypotenuse. After thinking for a moment, you will observe that the quadrilateral formed by the four hypotenuses is a square of side length c. Therefore the part of the area of the surrounding square that falls outside the four triangles is c^2. Now rearrange the positions of the upper and lower right triangles inside the larger square, as indicated by the figure on the right. This time the part of the area of the larger square that falls outside the four triangles is $a^2 + b^2$. It follows that $a^2 + b^2 = c^2$. There is a proof of the Pythagorean theorem in Euclid's *Elements*, but it is more complicated.

The Pythagorean theorem works "in reverse" also. Namely, if in some triangle the lengths of the sides satisfy the relationship $a^2 + b^2 = c^2$, then the triangle is a right triangle with c the length of the hypotenuse and a and b the lengths of the other two sides. It appears that ancient "rope stretchers" used this fact in the construction of buildings. Take, for example, a rope that has a length of 12 units. If the rope is stretched out horizontally on the ground in the triangular pattern indicated in Figure 1.17, then the lengths of the

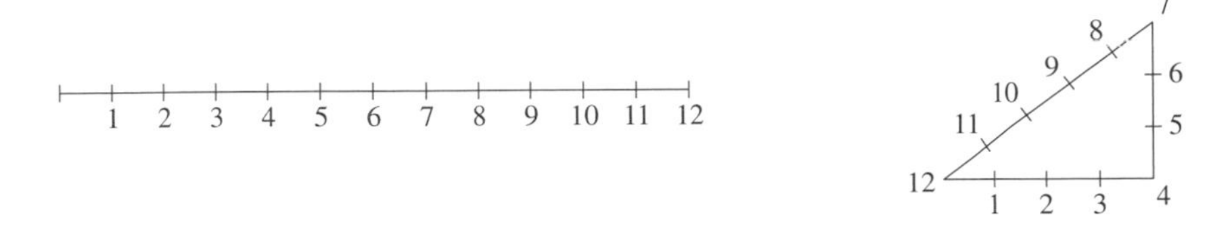

Fig. 1.17

sides satisfy the relationship $4^2 + 3^2 = 16 + 9 = 25 = 5^2$, so that the angle at 4 is a right angle. The construction of two perpendicular walls can now begin!

The primary purpose of a number system is to count and measure things. Let's assume that some unit of length is given. Nowadays it would be the centimeter, meter, kilometer, inch, foot, or mile, for example.

We will say that a line segment is *measurable* in our unit of length if its length is a rational number $\frac{m}{n}$ with m and n both positive integers. So the segment is measurable if its length can be expressed precisely in whole or fractional units. For instance, if the unit is the centimeter, then the segments of lengths $26\frac{4}{7} = \frac{186}{7}$ centimeters, as well as $53\frac{14}{29} = \frac{1551}{29}$ and $1726\frac{951}{3657} = \frac{6{,}312{,}933}{3657}$ centimeters, are all measurable.

The question presents itself as to whether all line segments are measurable "on the nose." Assume that the segment pictured in Figure 1.18 has length 1. Take another line segment of length 1, place the two segments together at right angles, and form the right triangle pictured in the figure. It turns out (surprise?) that the hypotenuse of this triangle is *not* measurable. Here is the argument. Let d be the length of the hypotenuse and suppose $d = \frac{r}{s}$ with r and s positive integers. After canceling all factors common to both r and s, $d = \frac{n}{m}$ where n and m have no common factors. By the Pythagorean theorem,

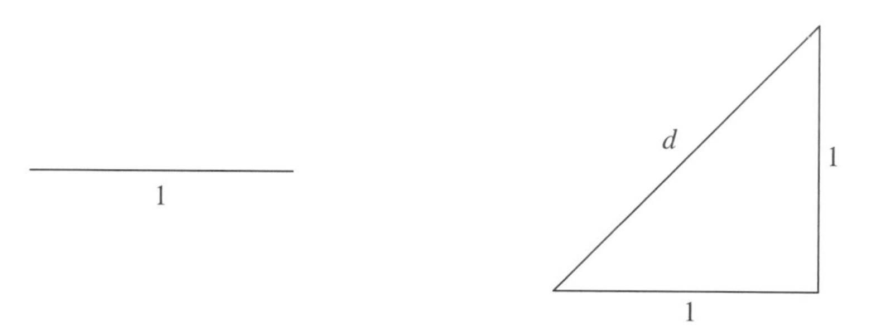

Fig. 1.18

$d^2 = 1^2 + 1^2 = 2$. So $\frac{n^2}{m^2} = 2$ and $n^2 = 2m^2$. It follows that n^2 is an even number. If n were to be odd, then $n = 2k + 1$ for some positive integer k, but then $n^2 = 4k^2 + 4k + 1$ and n^2 would be odd as well. Since this is not so, n is even. So $n = 2k$ for some positive integer k. It follows that $4k^2 = 2m^2$, so that $m^2 = 2k^2$. But this means that m^2 and hence m are both even. Therefore both n and m have 2 as a common factor, contradicting what we know about n and m. Our argument started with the supposition that d is measurable and moved in a strictly logical way to an impossible consequence. The inescapable conclusion is that d cannot be measurable.

This argument can be found in Euclid's *Elements* (Proposition 117 of Book X). There is evidence that the great philosopher Aristotle also knew this argument, and it is likely that it was already known to the Pythagoreans. It was the deeply held belief of this secret fraternity that mathematics is the underlying explanation of all things and that all reality finds its ultimate explanation in numbers and mathematics. As we have just seen, however, the numbers of the Pythagoreans were unable to come to grips with the very basic matter of measuring lengths. Legend has it that when one of Pythagoras's disciples pointed this out, the Pythagoreans, at sea at the time, threw the bearer of this message overboard and swore everyone else to secrecy.

The point that we need to take away is this. While the hypotenuse discussed above arises as a perfectly valid geometric construction, its length d cannot be measured with the number system of the Greeks. In particular, their numbers ran into limitations that their geometry did not. Indeed, Greek geometry and trigonometry flourished in a way that their numerical studies did not. In retrospect, it seems ironic that the Pythagorean theorem—an assertion about the lengths of the sides of a triangle—derives its name from a person or school that did not possess a number system with which length could always be measured. Today we can put it this way. The system of numbers of the form $\frac{m}{n}$— in other words, the rational numbers—is too small. The preceding demonstration of the nonmeasurability of d shows that $\sqrt{2}$ is a number that is not rational. It was almost 2000 years after the appearance of the *Elements* before the larger real number system was developed. The base ten positional system of Indian-Arabic origin with its symbols $1, 2, \ldots,$ and (very importantly) 0 provided the start. Positional means that the meaning of the number 7, for example, depends on its position in the larger number in which it appears. For instance, 7 has one meaning in the number 75, a different meaning in the number 763, and yet another meaning in 7826. Its meaning depends on its position in the number. Only after the introduction of the decimal point early in the 17th century was the system complete. For instance, 7826.5329 is the number

$$7826.5329 = 7 \cdot 1000 + 8 \cdot 100 + 2 \cdot 10 + 6 \cdot 1 + 5\,\tfrac{1}{10} + 3\,\tfrac{1}{100} + 2\,\tfrac{1}{1000} + 9\,\tfrac{1}{10000}.$$

This number is a rational number because it is a sum of rational numbers. On the other hand, as we just saw, $\sqrt{2}$ is not rational, it is *irrational*. But it does have an infinite decimal expansion that begins with

$$\sqrt{2} \;=\; 1.4142... \;=\; 1 + 4\,\tfrac{1}{10} + 1\,\tfrac{1}{100} + 4\,\tfrac{1}{1000} + 2\,\tfrac{1}{10000} + \cdots.$$

Consider a straight line that runs infinitely in both directions. Fix a point and label it 0. Now take a unit of length and mark off a point one unit to the right of 0 and label it 1. Continue in this way to get $2, 3, \ldots$. Repeat this on the left of 0, but use the negative numbers $-1, -2, \ldots$ to label the points. Do a similar thing with tenths, hundredths, thousandths, and so forth, and their negatives, and so on. This process is illustrated for $\sqrt{2}$ in Figure 1.19. It gives rise to the real number line: every point on the line corresponds to a real number—in other words, a number given by a decimal expansion—and every number given by

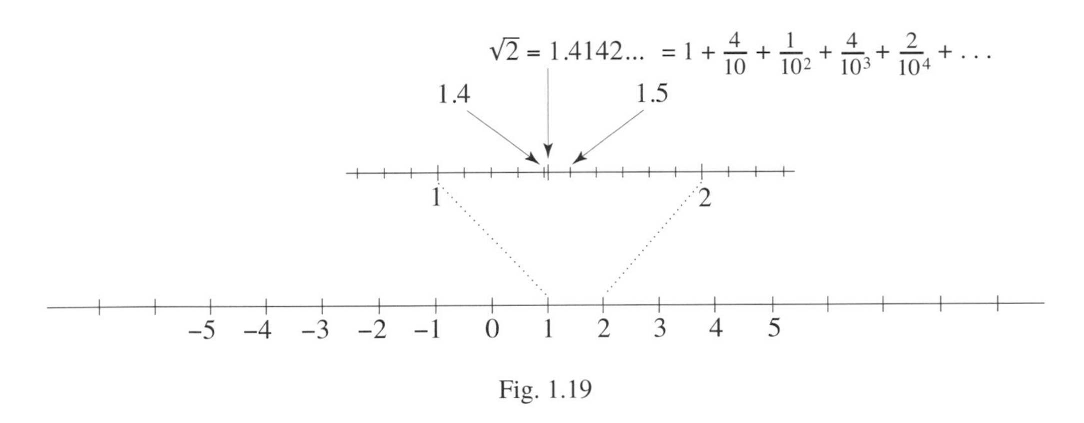

Fig. 1.19

a decimal expansion corresponds to a point on the line. The operations of addition, subtraction, multiplication, and division of real numbers also have Indian-Arabic origins.

Since the notation of the number system of the Greeks (see part 2J of the Problems and Projects section at the end of Chapter 2) is not appropriately sophisticated, and since their mathematical methods did not include infinite processes (such as the decimal expansion for $\sqrt{2}$), the construction of the real number system was beyond their reach. They did discuss quantities that we call irrational today, but only in geometric terms (see part 4F of the Problems and Projects section at the end of Chapter 4). In this text we have already used real numbers in important ways, in particular, in the statement of the Pythagorean theorem (as this involves the lengths of the sides of a triangle), and we will continue to do so.

1.5 THE RADIAN MEASURE OF AN ANGLE

A more useful numerical measure of an angle than the degree is the radian measure. Consider any angle θ, and place it in a circle so that the two segments that define it meet at the circle's center. Let r be the radius of the circle, and let s be the length of the arc that θ cuts out from the circumference. See Figure 1.20a. The *radian* measure of the angle θ is defined to be the ratio $\frac{s}{r}$. We will write $\theta = \frac{s}{r}$. The concept of radian measure has Arabic roots, but it was fully recognized as a natural unit of angular measure only in the early 18th century in England.

The immediate question that arises is this: Does the radian measure of an angle depend on the size of the circle into which the angle is placed? If this measure of an angle is to be a meaningful concept, then it should not. Let's investigate. Place θ into another circle. Let R be its radius and S the length of the arc that θ cuts out. See Figure 1.20b. The question that must be addressed is this: Is $\frac{s}{r} = \frac{S}{R}$?

Let n be a positive integer, and divide the angle into n equal sections. Each section determines a pair of points on each of the two circles. Connect consecutive points on the smaller circle with straight line segments. Since the lengths of these segments are all equal, we let d_n be this common length. Do the same thing for the larger circle, and let D_n be the common length of these longer segments. Note that the lengths d_n and D_n depend on the n that you have picked. These segments together with consecutive radii

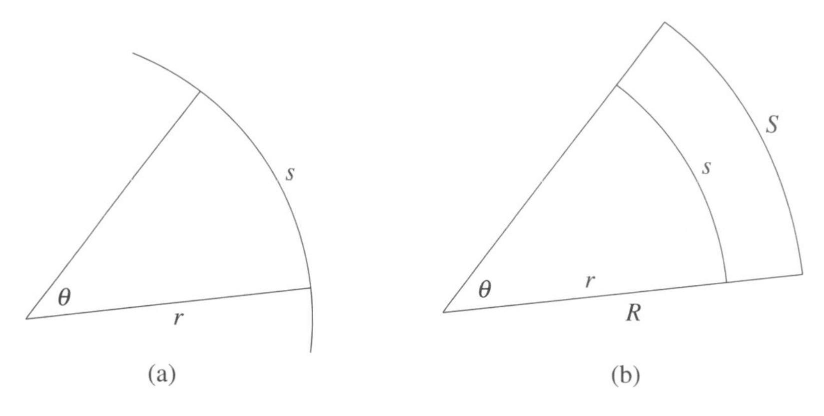

Fig. 1.20

create isosceles triangles inside both the smaller and larger circular sectors. It follows from the *pons asinorum* that each of the smaller triangles (with sides of length r) is similar to each of the larger ones (with sides of length R). Since ratios of corresponding sides are equal, it follows that $\frac{d_n}{r} = \frac{D_n}{R}$. The case $n = 4$ is shown in Figure 1.21. Observe next that if n is taken to be large, then the lengths of the n segments of length d_n add up to s approximately. So the number nd_n is nearly equal to s, and therefore $\frac{nd_n}{r}$ is nearly

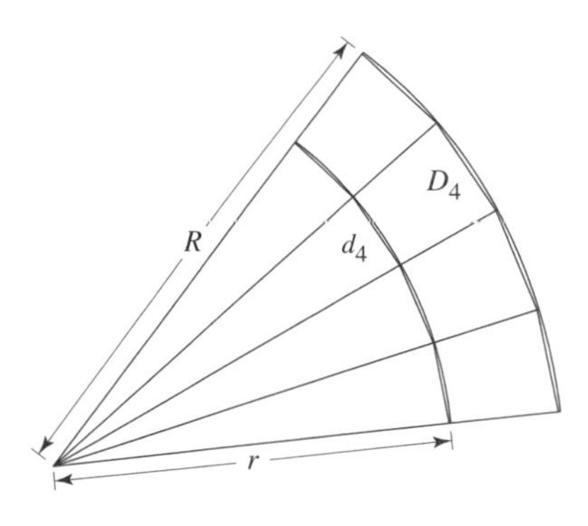

Fig. 1.21

equal to $\frac{s}{r}$. If n is taken sequentially larger and larger, the numbers $\frac{nd_n}{r}$ close in on $\frac{s}{r}$. We abbreviate this by writing

$$\lim_{n\to\infty} \frac{nd_n}{r} = \frac{s}{r}.$$

The symbol lim is short for "limit" and refers to the closing-in process just described. The symbol ∞ represents "infinity," and $\lim\limits_{n\to\infty}$ means that n, by being taken larger, is being "pushed to infinity." Proceeding in exactly the same way within the larger circle, we get

$$\lim_{n\to\infty} \frac{nD_n}{R} = \frac{S}{R}.$$

Since $\frac{d_n}{r} = \frac{D_n}{R}$ for any n, we see that

$$\frac{s}{r} = \lim_{n\to\infty} \frac{nd_n}{r} = \lim_{n\to\infty} \frac{nD_n}{R} = \frac{S}{R}.$$

So we have established the required equality $\frac{s}{r} = \frac{S}{R}$. Observe that the radian measure of an angle is a ratio of lengths. It is therefore a real number.

We will see in later chapters that such limit arguments—not used by the Greeks because of their infinite aspect—are the cornerstone of calculus.

Now take a circle of radius 1 as in Figure 1.22a. The length of one-half of its circumference is a number that is denoted by π. Consider the angle 180° and observe that its radian measure is $\frac{\pi}{1} = \pi$. Now take a circle of radius r as in Figure 1.22b, and let c be its circumference. Using this circle, we get that the

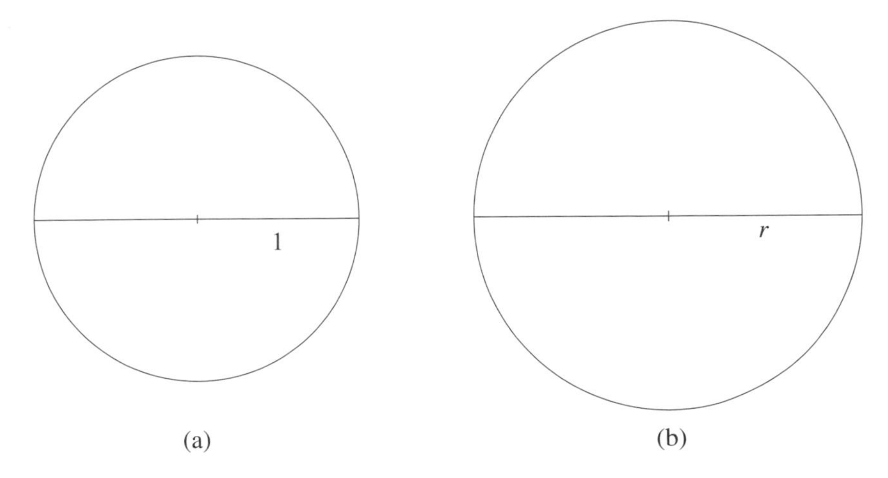

Fig. 1.22

radian measure of 180° is equal to $\frac{c/2}{r}$. Since the radian measure of an angle is the same regardless of which circle is used, it follows that $\frac{c/2}{r} = \pi$. This simplifies to the well-known formula

$$c = 2\pi r$$

for the circumference of a circle of radius r.

Because 180° corresponds to π radians, it is easy to fill in the information of Table 1.1. Notice that 1°

Degrees	Radians
360°	2π
180°	π
90°	$\frac{\pi}{2}$
60°	$\frac{\pi}{3}$
45°	$\frac{\pi}{4}$
30°	$\frac{\pi}{6}$
1°	$\frac{\pi}{180}$

Table 1.1

$= \frac{\pi}{180} = 0.017453\ldots$ radians and that 1 radian corresponds to $(\frac{180}{\pi})^\circ = 57.295779\ldots$ degrees. Since π represents a length, it is a real number. The question of what it is equal to occupied the mathematicians of antiquity and is still relevant today. Archimedes, who lived from 287 to 212 B.C. and was the most remarkable mathematical thinker of ancient Greece (we will encounter his exploits soon), showed that

$$3.1408 < 3\tfrac{10}{71} < \pi < 3\tfrac{1}{7} < 3.1429$$

in developing a scheme that provides estimates for π that are as accurate as one might wish. It leads to the decimal expansion $\pi = 3.141592\ldots$. See Problem 1.15 of the Problems and Projects section of this chapter. It turns out that π, like $\sqrt{2}$, is an irrational number (but this is difficult to establish).

The Greek Eratosthenes (from 276 to 194 B.C.) was director, as Euclid had been before him, of the museum of Alexandria, the research institute sponsored by the pharaohs of Egypt's last dynasty. It was the commonly accepted view among Greek philosophers that the Earth is round. Eratosthenes set out to mea-

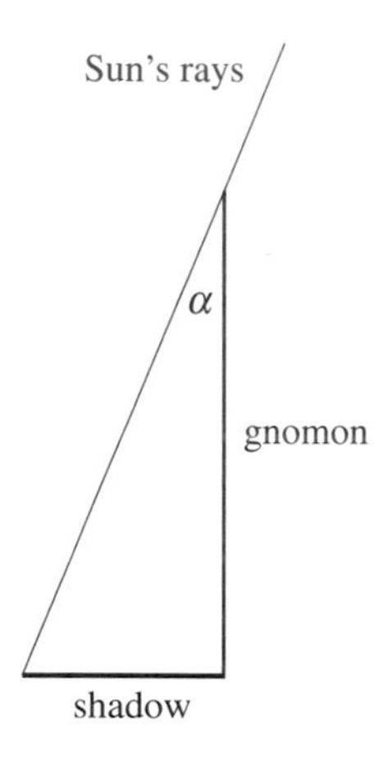

Fig. 1.23

sure its size. He knew that at noon on summer solstice of each year in the town of Syene (near today's Aswan in Egypt) the Sun shines into the bottom of a deep well. This means that at precisely that time, the

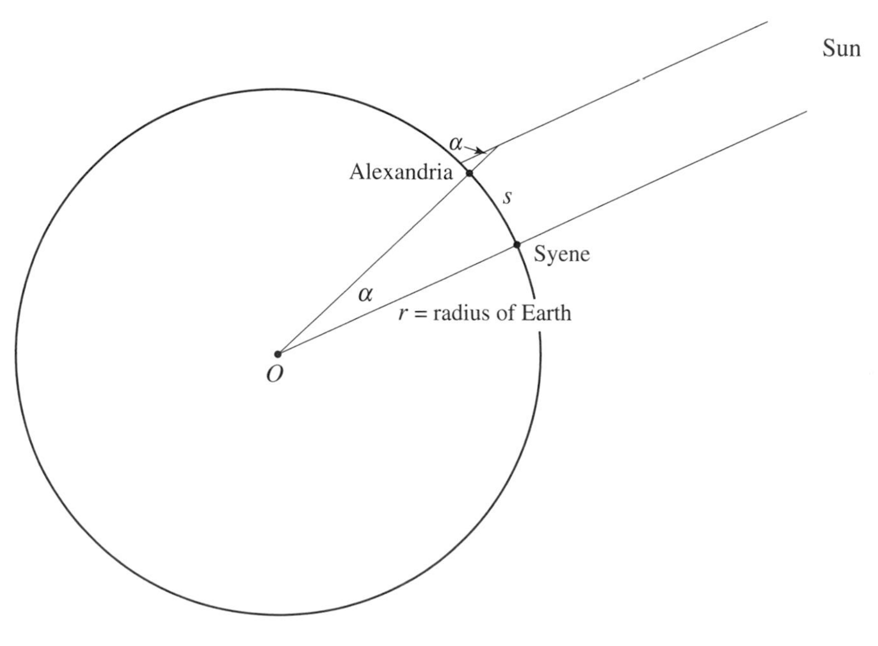

Fig. 1.24

Sun is directly overhead at this location. In Alexandria, again precisely at noon on summer solstice, Eratosthenes measured the length of the shadow of a gnomon (recall that a gnomon is simply a straight rod in vertical position) and determined the angle α of Figure 1.23 to be 7.5°. A careful look at Hipparchus's map of Figure 1.8 tells us that Syene is due south of Alexandria and (see the left margin of the map) that the distance from Syene to Alexandria is 21800 − 16800 = 5000 Greek stadia. Hipparchus drew his map about 50 years after Eratosthenes worked, but Eratosthenes had this information also. The rest was easy for Eratosthenes. Today we would argue from Figure 1.24 as follows. On the one hand,

$$\alpha = (7\tfrac{1}{2})\tfrac{\pi}{180} = \tfrac{15(3.14)}{360} = 0.13 \text{ radians},$$

and on the other hand, $\alpha = \frac{s}{r} = \frac{5000}{r}$. Therefore

$$r = \tfrac{5000}{\alpha} = \tfrac{5000}{0.13} = 38{,}500 \text{ stadia}.$$

So Eratosthenes—quite literally with a stick, some geography, and some geometry, cemented together by pure thought—had measured the size of the Earth! Since 100 Greek stadia are the equivalent of about 16 kilometers, Eratosthenes's estimate of Earth's radius corresponds to about $385 \cdot 16 \approx 6200$ kilometers. This estimate is rather good! Today's accurate value for the radius of the Earth is 6317 kilometers.

1.6 GREEK TRIGONOMETRY

Trigonometry is the study of triangles and the application of this study. The word "trigonometry" is Greek for "measuring the triangle" (in Greek, *trigono* = triangle, and *metrein* = to measure). History gives Hipparchus credit not only for transforming Greek astronomy from a purely theoretical into a predictive science, but also for his contributions to trigonometry. We have already seen how Ptolemy's *Almagest* refined Greek astronomy. The fact is that this treatise advanced geometry and trigonometry as well. This section and the Problems and Projects section at the end of this chapter recalls basic and important aspects of trigonometry that the Greeks were familiar with within their own conceptual approach.

Figure 1.25a shows a typical right triangle. For the given angle θ, define the "sine," "cosine," and "tangent" to be the following ratios of lengths:

$$\sin\theta = \frac{a}{h}, \quad \cos\theta = \frac{b}{h}, \quad \text{and} \quad \tan\theta = \frac{\sin\theta}{\cos\theta} = \frac{\frac{a}{h}}{\frac{b}{h}} = \frac{a}{b}.$$

In these definitions only the size of the angle θ matters. The particular right triangle into which it is placed is not relevant. This is easily seen as follows. Suppose that the right triangle of Figure 1.25b is used instead. Because the two triangles have the angle θ and their right angles in common, their respective

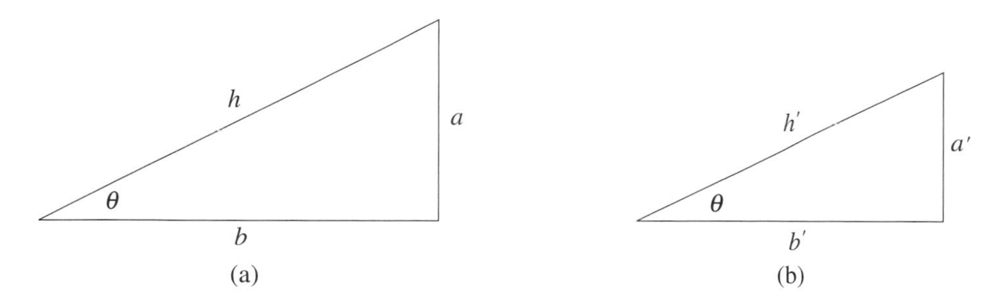

Fig. 1.25

third angles are equal as well. It follows that the triangles of Figures 1.25a and 1.25b are similar. Therefore

$$\frac{a'}{h'} = \frac{a}{h}, \quad \frac{b'}{h'} = \frac{b}{h}, \quad \text{and} \quad \frac{a'}{b'} = \frac{a}{b},$$

so it doesn't matter which right triangle is used to compute the ratios $\sin\theta$, $\cos\theta$, and $\tan\theta$.

Figure 1.26a depicts a circle with center O, an angle θ with $\theta < 180°$, and the points B and C on the circle that it determines. Choose A so that the segment AOB is a diameter of the circle and put in the

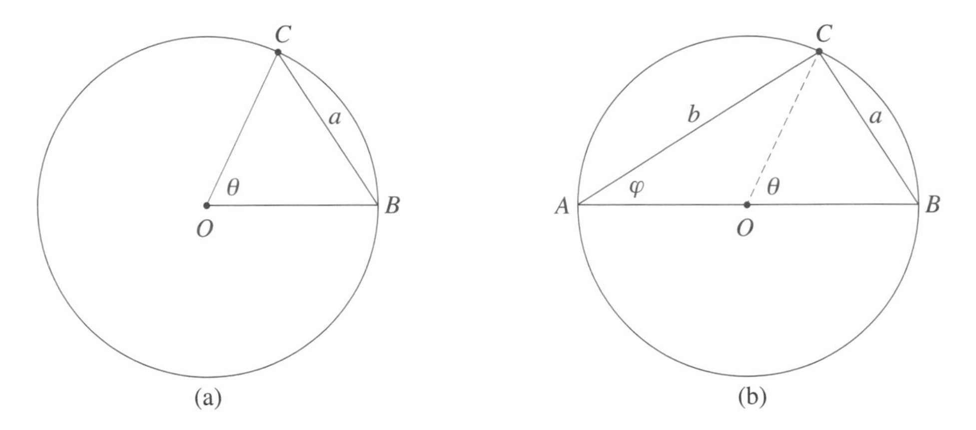

Fig. 1.26

segment AC. Refer to Figure 1.26b. Let $\angle OAC = \varphi$. Because the triangle ΔAOC is isosceles, $\angle ACO = \varphi$. It follows that $2\varphi + (180° - \theta) = 180°$. So $2\varphi = \theta$ and hence

$$\varphi = \tfrac{1}{2}\theta.$$

We point out as an aside that Figure 1.26 connects today's trigonometry with the original formulation of the Greeks. The segment CB of Figure 1.26a is the *chord* of the angle θ. Let a be its length and define

$$\text{crd}\,\theta = a.$$

This is the basic concept of Greek trigonometry. Let $AC = b$ and notice that $\text{crd}\,(180° - \theta) = b$. Since AB is a diameter, $\angle ACB$ is a right angle. So with $d = AB$, we get $\sin\varphi = \frac{a}{d}$ and $\cos\varphi = \frac{b}{d}$. Therefore

$$\sin\varphi = \tfrac{1}{d}\,\text{crd}\,2\varphi \quad\text{and}\quad \cos\varphi = \tfrac{1}{d}\,\text{crd}\,(180° - 2\varphi).$$

So both the sine and the cosine can be expressed in terms of the chord of Greek trigonometry.

We will now compute $\sin\theta$, $\cos\theta$, and $\tan\theta$ for some standard values of θ. Consider the right triangle

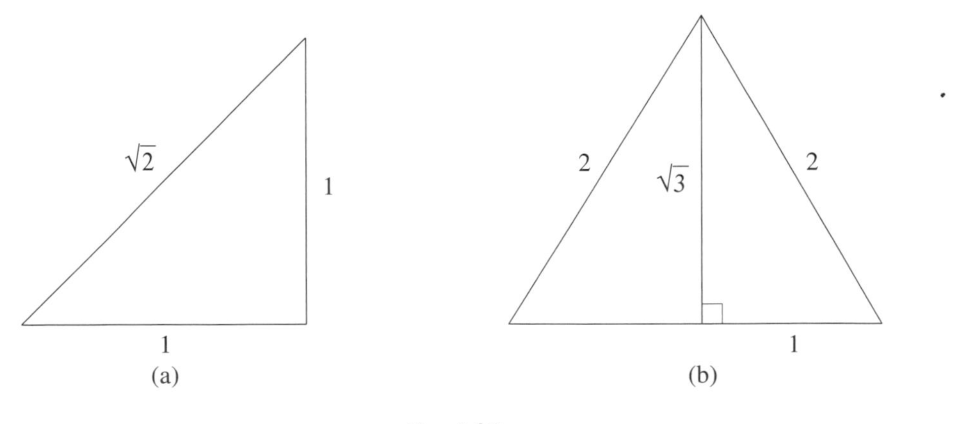

Fig. 1.27

of Figure 1.27a. Since it is isosceles, the acute angles are each 45° or $\frac{\pi}{4}$ radians. Therefore

$$\sin\frac{\pi}{4} = \frac{1}{\sqrt{2}},\ \cos\frac{\pi}{4} = \frac{1}{\sqrt{2}},\ \text{and}\ \tan\frac{\pi}{4} = 1.$$

Next, take the equilateral triangle of Figure 1.27b. It has sides of length 2 and height h. By the Pythagorean theorem, $h^2 + 1^2 = 2^2$. So $h^2 = 3$ and $h = \sqrt{3}$. It follows from the *pons asinorum* that each angle is equal to $60°$ or $\frac{\pi}{3}$ radians. Therefore

$$\sin\frac{\pi}{3} = \frac{\sqrt{3}}{2},\ \cos\frac{\pi}{3} = \frac{1}{2},\ \text{and}\ \tan\frac{\pi}{3} = \frac{\sqrt{3}}{1} = \sqrt{3}.$$

Since each of the smaller angles at the top is $30°$ or $\frac{\pi}{6}$, it follows that

$$\sin\frac{\pi}{6} = \frac{1}{2},\ \cos\frac{\pi}{6} = \frac{\sqrt{3}}{2},\ \text{and}\ \tan\frac{\pi}{6} = \frac{1}{\sqrt{3}}.$$

Observe that $\sin 0$, $\cos 0$, and $\tan 0$ do not make sense for the simple reason that there is no right triangle with angles of $0°$. The same thing is true for any angle $\theta \geq 90°$. The definition of the sine, cosine, and tangent that Figure 1.25 provides does not apply to such angles. To give meaning to the sine, cosine, and tangent of any angle θ with $0° \leq \theta \leq 180°$, use a circle of radius 1 and proceed as follows. If $0° \leq \theta < 90°$, complete θ to the diagram shown in Figure 1.28a and define $\sin\theta = a$ and $\cos\theta = b$. This

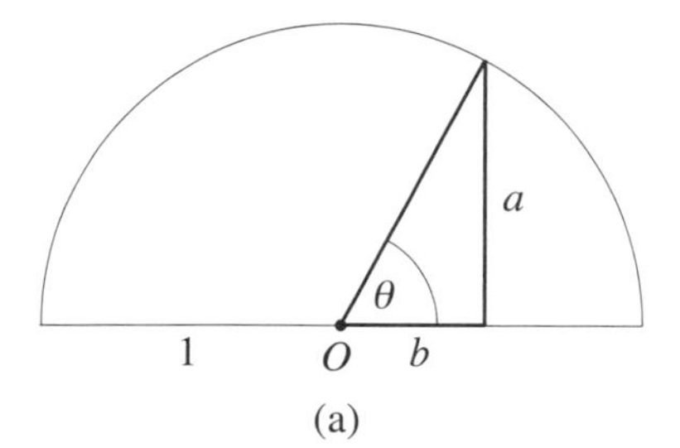

(a)

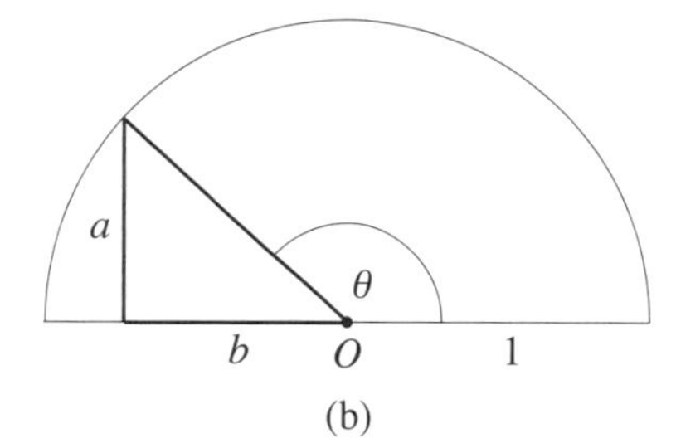

(b)

Fig. 1.28

agrees with the previous definition. If $90° \leq \theta \leq 180°$, complete θ as shown in Figure 1.28b and define $\sin\theta$ and $\cos\theta$ by

$$\sin\theta = a \quad \text{and} \quad \cos\theta = -b.$$

The tangent is given by $\tan\theta = \frac{\sin\theta}{\cos\theta}$ in either case.

There are many identities that relate $\sin\theta$, $\cos\theta$, and $\tan\theta$. We'll develop several of them next. For example, by applying Pythagoras's theorem to Figure 1.28, we get immediately that

$$\sin^2\theta + \cos^2\theta = 1$$

for any θ satisfying $0 \leq \theta \leq 180°$. Here and later in this text, we will write $\sin^2\theta$ in place of the less efficient $(\sin\theta)^2$. We'll do the same thing for the other trigonometric functions.

Let θ satisfy $0° \leq \theta < 90°$. So $180° - \theta > 90°$. Consider Figure 1.28a for θ and Figure 1.28b for $180° - \theta$. Study the two diagrams simultaneously. Convince yourself that the two right triangles are similar and hence that

$$\sin(180° - \theta) = \sin\theta \quad \text{and} \quad \cos(180° - \theta) = -\cos\theta.$$

It follows, for example, that $\sin 120° = \sin(180° - 60°) = \sin 60° = \frac{\sqrt{3}}{2}$ and $\cos 120° = \cos(180° - 60°) = -\cos 60° = -\frac{1}{2}$.

Consider any triangle. Let α, β, and γ be its angles, and denote the lengths of the sides opposite these angles by a, b, and c, respectively. The law of sines and the law of cosines are two of the most fundamental facts of trigonometry. The *law of sines* asserts that

$$\frac{\sin\alpha}{a} = \frac{\sin\beta}{b} = \frac{\sin\gamma}{c},$$

and the *law of cosines* tells us that

$$c^2 = a^2 + b^2 - 2ab\cos\gamma.$$

The law of cosines—stated here for the angle γ—holds analogously for the angles α and β as well.

In the verification of each of the two laws, two different kinds of triangles need to be considered. The answer to the question "Are all of its angles less than 90° or not?" distinguishes the two types. An angle θ is *acute* if $0 \le \theta \le \frac{\pi}{2}$, and it is *obtuse* if $\frac{\pi}{2} < \theta \le \pi$. A triangle is *acute* if all of its three angles are acute, and *obtuse* if it has an obtuse angle.

If the triangle that we are now considering is obtuse, we'll let γ be the obtuse angle. Notice that in

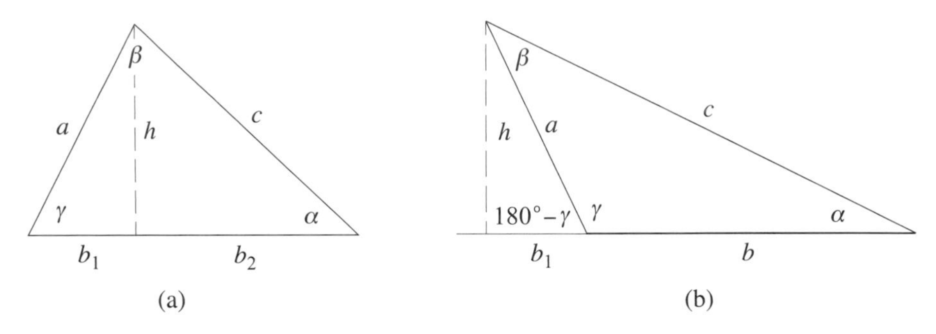

Fig. 1.29

Figure 1.29, (a) is the acute case and (b) is the obtuse case. In each case, h is the height of the triangle. We'll first take up the verification of the law of cosines.

Case (a). Let $b = b_1 + b_2$ and consider Figure 1.29a. Since $\cos\gamma = \frac{b_1}{a}$, we see that

$$a^2 + b^2 - 2ab\cos\gamma = a^2 + b^2 - 2bb_1 = a^2 + (b_1 + b_2)^2 - 2(b_1 + b_2)b_1 = a^2 + b_2^2 - b_1^2.$$

Applying the Pythagorean theorem twice, we get $a^2 = b_1^2 + h^2$ and $h^2 + b_2^2 = c^2$. Therefore

$$a^2 + b_2^2 - b_1^2 = b_1^2 + h^2 + b_2^2 - b_1^2 = h^2 + b_2^2 = c^2.$$

The proof of Case (a) is complete.

Case (b). By a basic trig identity, $\cos\gamma = -\cos(180° - \gamma) = -\frac{b_1}{a}$. Since $a^2 = h^2 + b_1^2$,

$$a^2 + b^2 - 2ab\cos\gamma = a^2 + b^2 + 2bb_1 = h^2 + b_1^2 + b^2 + 2bb_1^2 = h^2 + (b_1 + b)^2.$$

Since $c^2 = h^2 + (b_1 + b)^2$, the proof is done in Case (b).

We'll start the proof of the law of sines with Figure 1.29a. Since $\sin\gamma = \frac{h}{a}$ and $\sin\alpha = \frac{h}{c}$, we get $a\sin\gamma = c\sin\alpha$. Therefore

$$\frac{\sin\alpha}{a} = \frac{\sin\gamma}{c}.$$

This observation tells us more generally that for any two acute angles of a triangle, the ratios of the sine of the angle divided by the length of the side opposite it are the same. This implies that the proof of the law of sines is complete in the case of Figure 1.29a. It also implies that

$$\frac{\sin\alpha}{a} = \frac{\sin\beta}{b}$$

in the case of Figure 1.29b. To finish the proof in the case of Figure 1.29b, we show that $\frac{\sin\gamma}{c} = \frac{\sin\alpha}{a}$. Notice that $\sin\alpha = \frac{h}{c}$ and, using the identity $\sin\gamma = \sin(180° - \gamma)$, that $\sin\gamma = \frac{h}{a}$. So $a\sin\gamma = c\sin\alpha$. Therefore $\frac{\sin\gamma}{c} = \frac{\sin\alpha}{a}$, and hence

$$\frac{\sin\alpha}{a} = \frac{\sin\beta}{b} = \frac{\sin\gamma}{c}.$$

So the proof in the obtuse case of Figure 1.29b is also complete.

Example 1.1. A triangle has sides $a = 6$ and $b = 7$ units. The angle between them is $\gamma = 118°$. Can the third side and the other two angles of the triangle be determined? By the law of cosines, the third side c of the triangle (opposite the given angle) satisfies $c^2 = 6^2+7^2-2(6\cdot 7)\cos 118° = 85-84\cos 115° \approx 124.4356$. So $c = 11.1551$ (accurate up to four decimal places). By the law of sines, the angles α and β, opposite the sids a and b, respectively, satisfy

$$\frac{\sin\alpha}{a} = \frac{\sin\beta}{b} = \frac{\sin 118°}{11.1551} = 0.0792.$$

It follows that $\sin\alpha = 6(0.0792) = 0.4749$ and $\sin\beta = 7(0.0792) = 0.5541$. By pushing the inverse sine button of a calculator, we can conclude that $\alpha = 28.3537°$ and $\beta = 33.6463°$.

Have a look at the right triangle of Figure 1.30. The vertex of θ is at the center of a circle of radius 1. Notice that $\theta = \frac{s}{1} = s$ in radians, that $\sin\theta = \frac{a}{1} = a$, and by comparing lengths, that $\sin\theta < \theta$. Now sup-

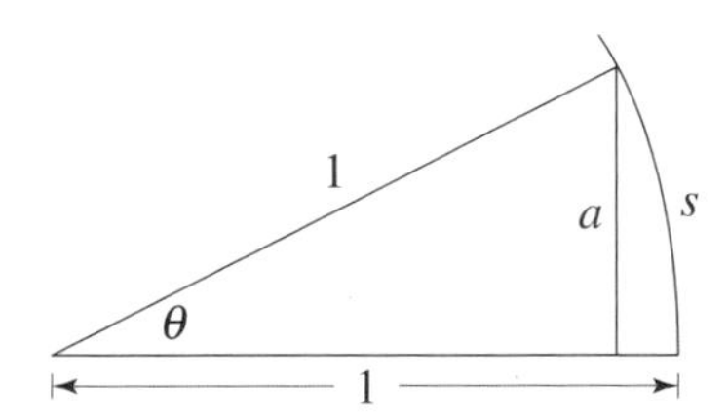

Fig. 1.30

pose that $\theta = s$ is small. The figure shows that $\sin\theta = a$ is also small, and that the lengths s and a are close to each other. So for small θ with θ in radians, there is close agreement between $\sin\theta$ and θ. If θ is taken sequentially smaller and smaller, $\sin\theta$ and θ get sequentially closer. As a consequence, the ratio $\frac{\sin\theta}{\theta}$ closes in on 1. Table 1.2 provides evidence for this. Check the entries of the table (making sure that

θ (degrees)	θ (radians)	$\sin\theta$	$\frac{\sin\theta}{\theta}$
15°	$\frac{\pi}{12} = 0.261799$	0.258819	0.988616
10°	$\frac{\pi}{18} = 0.174533$	0.173648	0.994931
5°	$\frac{\pi}{36} = 0.087266$	0.087156	0.998731
4°	$\frac{\pi}{45} = 0.069813$	0.069756	0.999188
3°	$\frac{\pi}{60} = 0.052360$	0.052336	0.999543
2°	$\frac{\pi}{90} = 0.034907$	0.034899	0.999797
1°	$\frac{\pi}{180} = 0.017453$	0.017452	0.999949

Table 1.2

your calculator is in "radian mode" and not "degree mode" when you compute $\sin\theta$). In terms of limit notation, we have observed that

$$\lim_{\theta\to 0}\frac{\sin\theta}{\theta} = 1.$$

We will see later that this limit is crucial for the development of the calculus of the trigonometric functions.

1.7 ARISTARCHUS SIZES UP THE UNIVERSE

Aristarchus of Samos (from about 310 to 230 B.C.) was another of the great Greek astronomers. Samos is the Greek island off the west coast of today's Turkey from which Pythagoras came as well. Little is known about Aristarchus, but what is known is this: he believed that the universe has the Sun fixed at its center, that Earth revolves around the Sun, and that it rotates about its own axis in the process. History has given Copernicus credit for these insights, but the fact is that Aristarchus had arrived at this *heliocentric* model about 1800 years earlier.

What will interest us is Aristarchus's treatise *On the Magnitudes and Distances of the Sun and Moon* and its use of "cosmic" trigonometry. His analysis rests on the following hypotheses and observations:

A. The Moon receives its light from the Sun.

B. The Moon revolves in a circle about the Earth with the Earth at the center.

C. When an observer on Earth looks out at a precise half Moon, so that the angle $\angle EMS$ in Figure 1.31 is 90°, the angle $\angle MES$ can be measured to equal 87°.

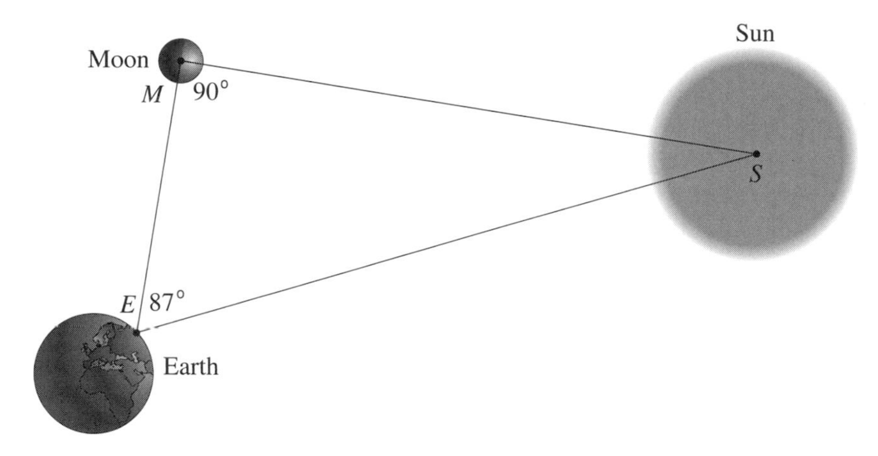

Fig. 1.31

D. At the instant of a total eclipse of the Sun, the Moon and the Sun (as viewed from Earth) subtend the same angle, and this angle can be measured to be 2°. Refer to Figure 1.32.

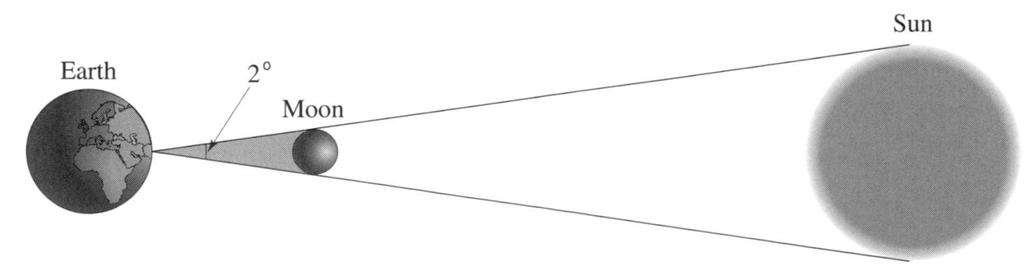

Fig. 1.32

E. During a lunar eclipse, the shadow indicated in Figure 1.33 has a width four times the radius r_M of the Moon. (This was based on how long the Moon was observed to be in Earth's shadow.)

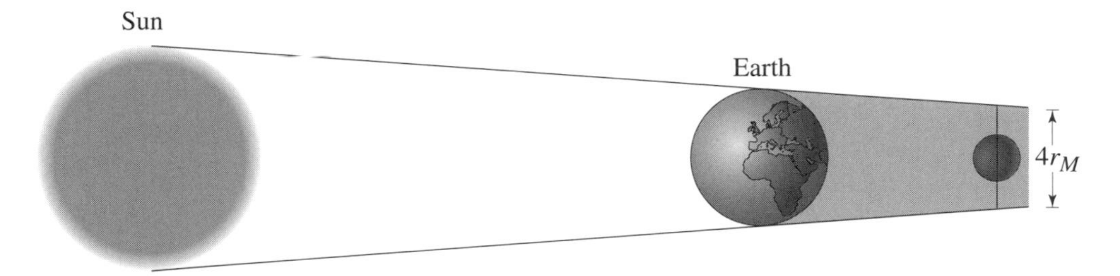

Fig. 1.33

It should be noted that Aristarchus knew that the diagrams he sketched were not at all to scale. (Observe also that northern Europe and Africa in these figures are depicted more accurately than in the maps of Hipparchus and Ptolemy in Figures 1.8 and 1.37. A more accurate picture of the continental land masses and the separating oceans would not begin to be known until the voyages of discovery in the 15th century.)

Aristarchus studies these observations and combines the information that they provide to derive estimates for

$$\begin{aligned} r_E &= \text{the radius of the Earth,} \\ r_M &= \text{the radius of the Moon,} \\ r_S &= \text{the radius of the Sun,} \\ D_M &= \text{the distance from the Earth to the Moon, and} \\ D_S &= \text{the distance from the Earth to the Sun.} \end{aligned}$$

Realizing that he would only obtain approximations, he could take π equal to 3 even though he would have been aware of much better estimates (possibly those of Archimedes). Figure 1.34 is taken directly

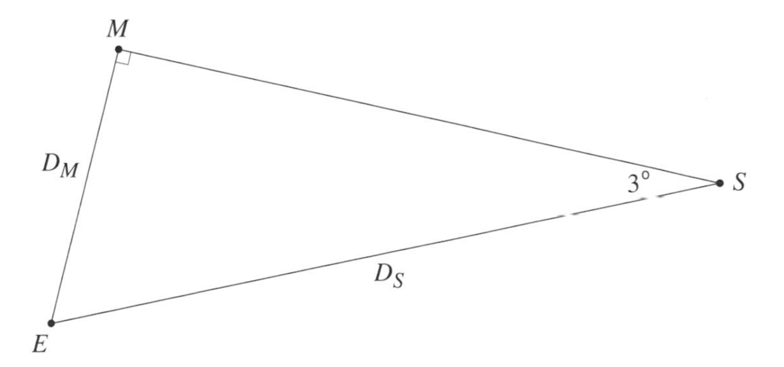

Fig. 1.34

from Observation C. Notice that $3° = \frac{\pi}{60}$. Therefore $\frac{D_M}{D_S} = \sin 3° = \sin \frac{\pi}{60}$. Since $3° = \frac{\pi}{60}$ radians is a small angle, $\sin \frac{\pi}{60}$ is approximately equal to $\frac{\pi}{60}$. See Table 1.2. Aristarchus could take $\frac{\pi}{60}$ to be $\frac{3}{60} = \frac{1}{20}$. In this way, Aristarchus obtains the approximation $\frac{D_M}{D_S} = \frac{1}{20}$ and arrives at the estimate

$$\frac{D_S}{D_M} = 20.$$

Aristarchus has established in particular that the Sun is much farther from Earth than the Moon.

Refer next to Observation D, the situation of the solar eclipse. Figure 1.35 is an elaboration of Figure 1.32. Radii of the Moon and the Sun have been inserted, and some distances have been labeled. The angle at E, indicated as being equal to $1°$, is obtained by bisecting the $2°$ angle of Figure 1.32. By similar triangles, $\frac{r_S}{r_M} = \frac{D_S}{D_M}$, and therefore

$$\frac{r_S}{r_M} = 20.$$

Observe also that $\frac{r_M}{D_M} = \sin 1°$. Since $1° = \frac{\pi}{180}$ radians, $\frac{r_M}{D_M} = \sin \frac{\pi}{180}$. Since $\frac{\pi}{180}$ is a small angle, Aristarchus could use the information that Table 1.2 provides and take $\sin \frac{\pi}{180}$ equal to $\frac{\pi}{180}$. With π equal

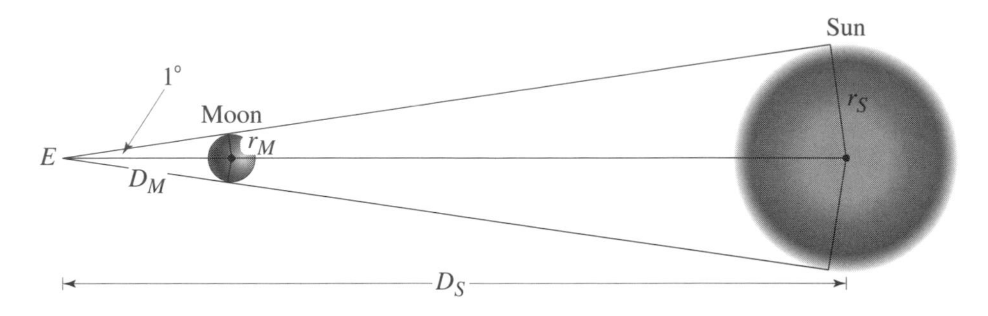

Fig. 1.35

to 3, Aristarchus gets the approximation $\frac{r_M}{D_M} = \frac{1}{60}$ and therefore the estimate

$$\frac{D_M}{r_M} = 60.$$

From Observation E and Figure 1.33, Aristarchus obtains Figure 1.36. This figure shows a light ray that is tangent to both the Sun and the Earth. The radii of the Sun and the Earth drawn into the figure are

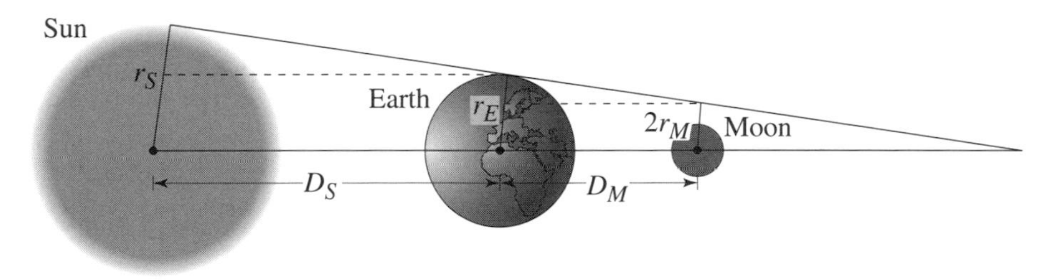

Fig. 1.36

both perpendicular to this light ray. The extension of the radius of the Moon indicated in the figure is perpendicular to this light ray as well. Because the two triangles with the "dotted" bases are similar,

$$\frac{r_E - 2r_M}{r_S - r_E} = \frac{D_M}{D_S}.$$

Recalling that

$$\frac{D_M}{D_S} = \frac{r_M}{r_S},$$

Aristarchus gets

$$\frac{r_E - 2r_M}{r_S - r_E} = \frac{r_M}{r_S}.$$

After cross-multiplying, $r_S r_E - 2r_S r_M = r_M r_S - r_M r_E$. So $r_S r_E + r_M r_E = 3r_S r_M$. Dividing this last equation through by $r_S r_M$ gives us $\frac{r_E}{r_M} + \frac{r_E}{r_S} = 3$. Since $\frac{r_S}{r_M} = 20$, we have $r_S = 20r_M$. By a substitution,

$$3 = \frac{r_E}{r_M} + \frac{r_E}{20r_M} = \frac{20r_E + r_E}{20r_M} = \frac{21r_E}{20r_M}.$$

Therefore $\frac{r_E}{r_M} = \frac{60}{21} = \frac{20}{7}$, and hence $r_M = \frac{7}{20}r_E$. Since $\frac{r_S}{r_M} = 20$, it follows that $r_S = 20r_M$, and therefore that

$$r_S = 7r_E.$$

After inserting $r_E = 38500$ stadia, the value that Eratosthenes derived, Aristarchus has achieved the estimates

$$r_M = 13500 \text{ stadia} \quad \text{and} \quad r_S = 270{,}000 \text{ stadia}.$$

Since $\frac{D_M}{r_M} = 60$, he also has the approximations $D_M = 60r_M = 810{,}000$ stadia, and by inserting this value into $\frac{D_S}{D_M} = 20$,

$$D_S = 16{,}200{,}000 \text{ stadia}.$$

We have presented a streamlined version of Aristarchus's argument. The latter was more elaborate and complicated. He made use of inequalities and got slightly different answers. For example, instead of $\frac{r_M}{r_S} = \frac{1}{20}$, he obtained $\frac{1}{20} < \frac{r_M}{r_S} < \frac{1}{18}$, and instead of $\frac{r_M}{D_M} = \frac{1}{60}$, he had $\frac{1}{60} < \frac{r_M}{D_M} < \frac{1}{45}$. However, the essence of his analysis has been retained. Table 1.3 compares Aristarchus's estimates to modern values. Notice

	Aristarchus (kilometers)	Actual (kilometers)
r_E radius of Earth	6200[a]	6370
r_M radius of Moon	2200	1740
r_S radius of Sun	43000	695,500[b]
D_M Earth to Moon	130,000	384,570[c]
D_S Earth to Sun	2,600,000	150,000,000

[a] As was pointed out, this is taken from Eratosthenes. While Eratosthenes lived about 40 years after Aristarchus, similar estimates had been achieved earlier (and were known to Aristotle).

[b] The Sun consists of gas. The radius given is that of the photosphere, the illuminated part. The part from the center to 30% of its radius has in essence all the shining power and 60% of the mass. The part from the center to 60% of its radius has 95% in of the mass.

[c] Radar and laser measurements. The distance varies from about 350,000 to 410,000 kilometers because of the elliptical nature of the Moon's orbit.

Table 1.3

that Aristarchus's value for the radius of the Moon is reasonably accurate. However, his estimates for the distance to the Moon, the radius of the Sun, and its distance from the Earth are off by factors of 3, 16, and 50, respectively. In any case, using only some basic observations and pure thought in the form of trigonometry, Aristarchus provided at least some idea of the vast distances in the solar system and began to unravel some of its mystery. Aristarchus's strategy is correct in principle. With more accurate measurements of the angles involved, he would have done much better. See Problem 1.33.

Little attention was paid to Aristarchus's Sun-centered understanding of the universe. The Greek astronomers and philosophers dismissed it as being in clear contradiction with the facts. The Earth-centered view of Ptolemy's *Almagest* prevailed until Copernicus revived Aristarchus's idea in the 16th century. The computation of the distance D_S from Earth to the Sun turned out to be very challenging. It was not computed with suitable accuracy until the 17th century, when the astronomers Cassini and Flamsteed came within $7\frac{1}{2}\%$ of the correct value. Their approach to the determination of this distance is taken up in Section 3.7 of Chapter 3.

Numbers are used to count and to determine the size of things. When they are used to count, such as the number of apples in a display in a supermarket, the result can be accurate "on the nose." When measuring things such as the dimensions of a room, or an interval of time, this is rarely the case. When a room is measured with a tape measure to be 5.52 meters by 7.36 meters, this is most certainly only an approximation. When a number a approximates a number b, we will write $a \approx b$ from now on. This notation is fundamentally inadequate, because it does not distinguish between tight approximations such as $\pi \approx 3.1415927$ or those of Example 1.1 and Table 1.2, Eratosthenes's loose estimate of 6200 kilometers for the radius of Earth, and Aristarchus's very rough approximation of 2200 kilometers for the radius of the Moon. But in a given situation, the context of the discussion will give a sense of the accuracy of the symbol $\approx$. With this understanding, this symbol will see frequent use throughout this text.

1.8 PROBLEMS AND PROJECTS

1A. Ptolemy's Maps. Ptolemy's work *Geographike Syntaxis* laid the foundation of the science of cartography. It reached western Europe from Constantinople as a Greek manuscript and was translated into Latin early in the 15th century. Remarkable maps were reconstructed by medieval cartographers from the precise positional information that the text supplied. Figure 1.37 depicts an early printed version of Ptolemy's map of the world. Important areas of Europe around the Mediterranean Sea are easily recognized. North Africa as well as the Arabian peninsula are also easily made out. India and the Far East, on the other hand, are off target. Central and southern Africa are unknown territory. Ptolemy devised a carefully spaced grid

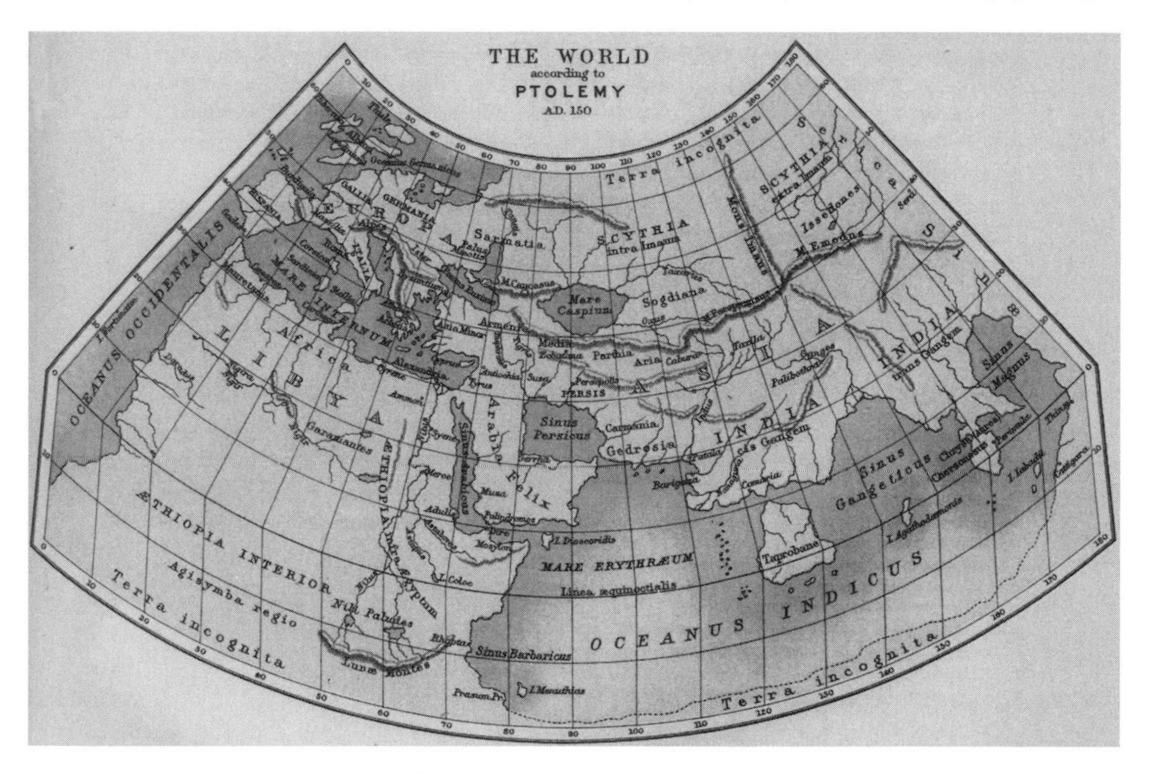

Fig. 1.37. Claudius Ptolemy's map of the world. Image provided by the Hesburgh Library, University of Notre Dame.

to organize his map and to provide precise positions of its key features. He used his grid like a coordinate system. The coordinates of the 8000 locations that the *Geographike* contains enabled the reconstruction of the maps. The strip of numbers at the lower boundary of the map tells us that Ptolemy's lines of latitude divide the known part of the globe into 180 degrees. Ptolemy knew about earlier Greek estimates of the circumference of the Earth and had a sense that he had mapped about half the globe (of course, he had no knowledge whatever about the missing half). Christopher Columbus used Ptolemy's map to make the case to Queen Isabella and King Ferdinand of Spain that he could reach Asia by sailing west. The fact that

Ptolemy's map underestimated the size of the globe, and hence the distance that needed to be traveled, might have contributed to the success of Columbus's argument.

1B. Similar Triangles. Let ΔABC and $\Delta A'B'C'$ be two similar triangles, and suppose that $A \to A'$, $B \to B'$, and $C \to C'$ is the correspondence between the equal angles. The discussion that follows presents a proof of the fact that the ratios of corresponding sides are equal. Since their angles match up, the smaller triangle ΔABC can be positioned inside the larger $\Delta A'B'C'$, as shown in both Figures 1.38a and 1.38b. The correspondence between the angles sets up the correspondence $AB \to A'B'$, $AC \to A'C'$, and $BC \to B'C'$ between the sides. Notice that the area of $\Delta BCB'$ in Figure 1.38a is equal to the area of $\Delta BCC'$ in Fig-

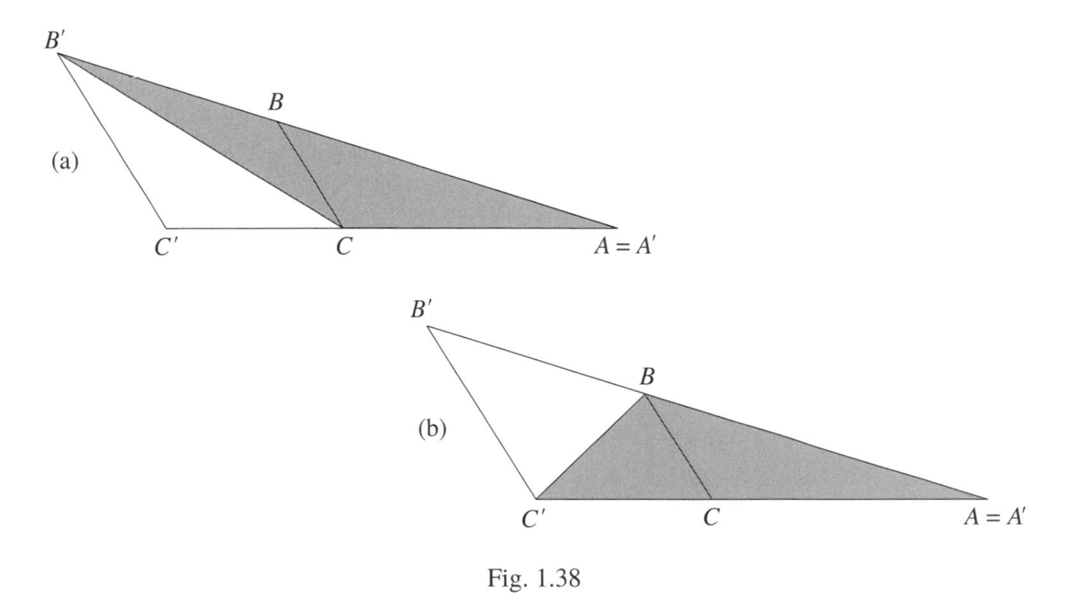

Fig. 1.38

ure 1.38b because the two triangles have the same base BC and, with respect to this base, equal heights. It follows that the areas of the triangles $\Delta A'B'C$ and $\Delta A'C'B$ highlighted in Figures 1.38a and 1.38b are equal as well. Study Figure 1.38a and observe that with respect to their bases $A'B'$ and AB, the two triangles $\Delta A'B'C$ and ΔABC have the same height h. In the same way, the two triangles $\Delta A'C'B$ and ΔACB in

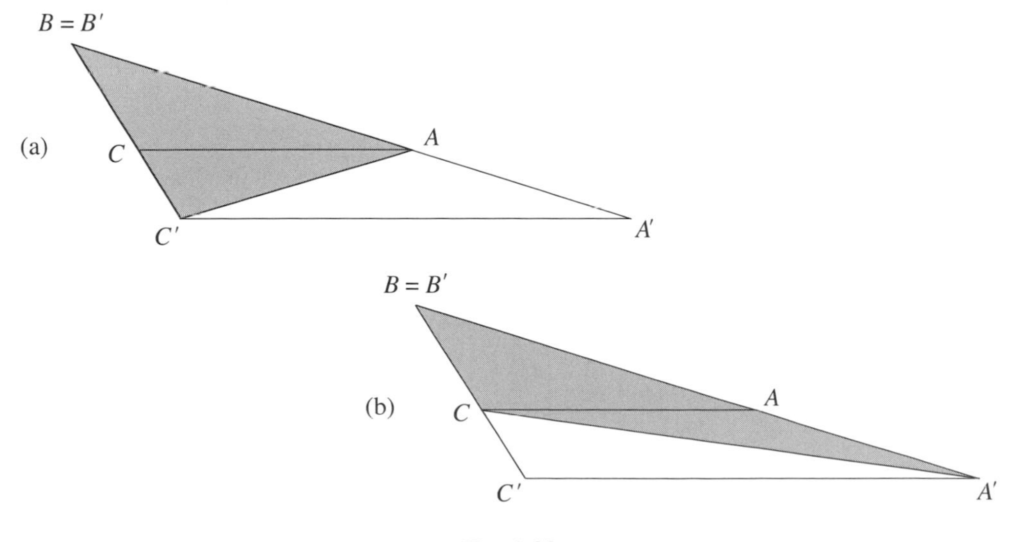

Fig. 1.39

Figure 1.38b have the same height h' relative to the bases $A'C'$ and AC. It follows directly from this that

$$\frac{\text{Area }\Delta A'B'C}{\text{Area }\Delta ABC} = \frac{\frac{1}{2}A'B'\cdot h}{\frac{1}{2}AB\cdot h} = \frac{A'B'}{AB} \quad \text{and} \quad \frac{\text{Area }\Delta A'C'B}{\text{Area }\Delta ACB} = \frac{\frac{1}{2}A'C'\cdot h'}{\frac{1}{2}AC\cdot h'} = \frac{A'C'}{AC}.$$

Because the areas of the triangles $\Delta A'B'C$ and $\Delta A'C'B$ are equal, $\frac{A'B'}{AB} = \frac{A'C'}{AC}$. Use a similar argument and Figures 1.39a and 1.39b to show that $\frac{B'C'}{BC} = \frac{A'B'}{AB}$ and conclude that $\frac{A'B'}{AB} = \frac{A'C'}{AC} = \frac{B'C'}{BC}$. This is the assertion of Thales's theorem.

1.1. Consider two similar triangles. Show that the property of the equality of the ratios of corresponding sides follows from the law of sines. [Comment: This argument is much shorter and simpler than the one presented above. But there is a fundamental flaw, because the definitions of the radian measure of an angle and the sine of an angle both rely on the "corresponding sides" property of similar triangles.]

1C. The Pythagorean Theorem. The next two problems present proofs of the Pythagorean theorem. One is attributed to a celebrated scientist of the 20th century; the other has its origins in ancient China.

1.2. Let ΔABC be a right triangle, and label the vertices so that AB is the hypotenuse. Let c be the length of the hypotenuse, and let a and b be the lengths of the other two sides. Draw in the perpendicular segment PC from C to AB. See Figure 1.40 and note that $c = c_1 + c_2$. Why are the two triangles ΔPAC and ΔPBC both similar to ΔABC? Show that c is equal to both $\frac{a^2}{c_2}$ and $\frac{b^2}{c_1}$. Conclude that

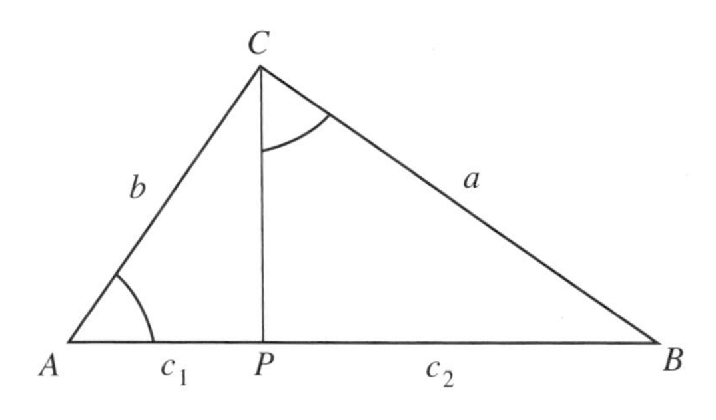

Fig. 1.40

$a^2 + b^2 = c^2$. [Credit for this proof of the Pythagorean theorem has been given to the famous physicist Albert Einstein.]

1.3. The Pythagorean theorem was also known to the Chinese. Their old visual proof for the (3, 4, 5)

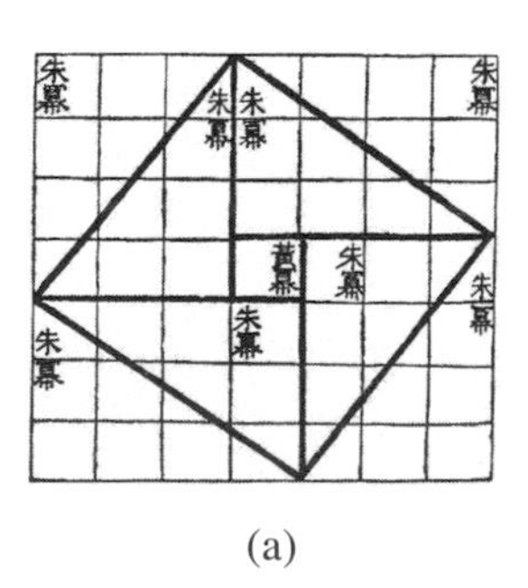

(a)

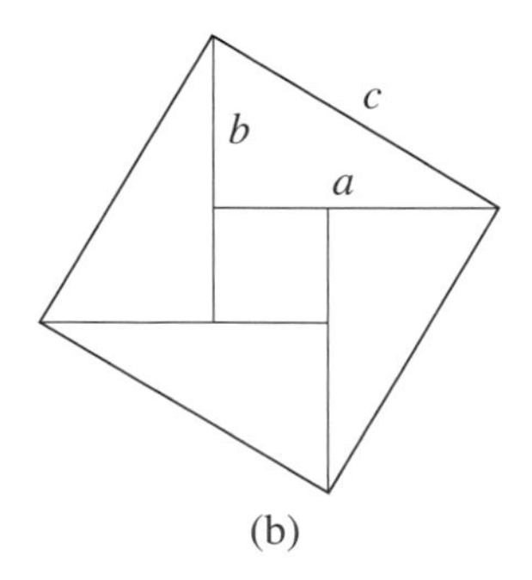

(b)

Fig. 1.41. The figure on the left comes from an old Chinese text composed before A.D. 200.

triangle of Figure 1.41a is easily adapted to the general case of Figure 1.41b. It depicts four identical right triangles (each with sides of lengths a, b, and c) arranged inside a square. Determine the size of the inner square, and use the diagram to verify the Pythagorean theorem.

1D. More about $\sqrt{2}$. The proof presented in the text that $\sqrt{2}$ is irrational is over 2000 years old. It was already mentioned that Aristotle was aware of it, but that it was probably already known to the Pythagoreans.

A recent geometric proof follows. Assume, if possible, that $\sqrt{2} = \frac{n}{m}$, where n and m are positive integers. So $n^2 = 2m^2$. It follows that there is a square with side length a positive integer n that is a sum of the areas of two squares both with side length a positive integer m. If this square is not the smallest one with this property, pick a smaller one. If that one is not the smallest one, pick a smaller one yet. This process must stop at some point to provide a smallest square with this property. This smallest one and the two squares that are together equal to it in area are depicted in Figure 1.42. Let this smallest one have side

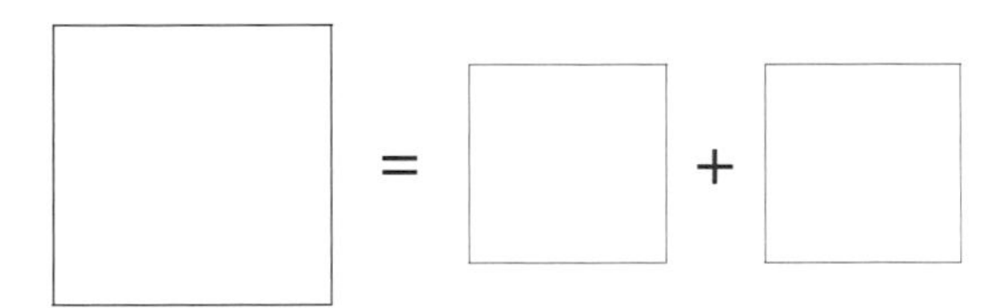

Fig. 1.42

length j, and let i be the side length of each of the two other squares. Now slide the two smaller squares onto the larger one, one of them into the upper right corner, and the other into the lower left corner, as shown in Figure 1.43. This move produces three smaller squares along the diagonal. Label the middle

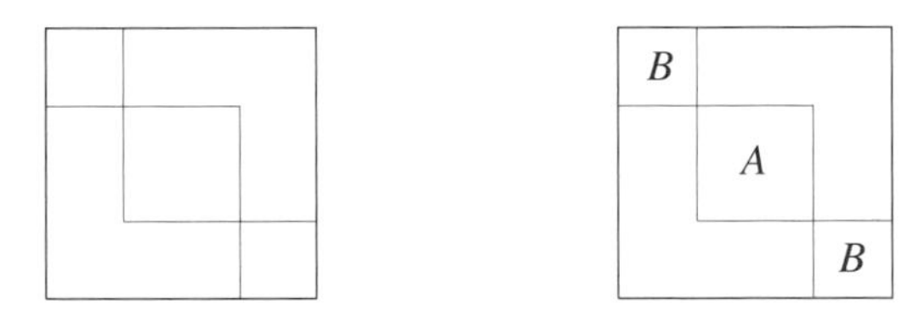

Fig. 1.43

one by A and the two in the corners by B.

1.4. After thinking about the sliding move that Figure 1.43 depicts, show that the area of A is twice the area of B. Then compute the side lengths of A and B in terms of j and i, and show that both squares have integer sides. Why can we now conclude that $\sqrt{2}$ must be irrational?

1E. Angles, Arcs, and Circles. The next several problems examine the measurement of angles and arcs.

1.5. Fill in the following blanks with numbers that are accurate up to the second decimal place:

i. 1 radian = ________ degrees

ii. 1 degree = ________ radians

iii. 78.5° = ________ radians

iv. 1.238 radians = ________ degrees

1.6. In the circular sector of Figure 1.44, $\theta = 57.3°$ and $r = 3$. What is the length of the arc AB?

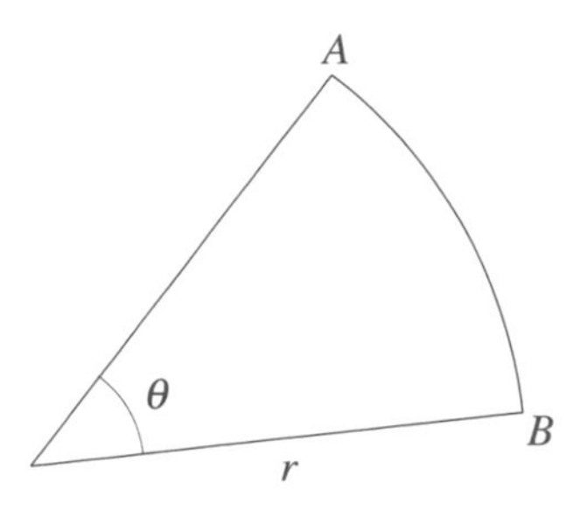

Fig. 1.44

1.7. Place an angle θ at the center of a circle of radius 2. Let the endpoints of the arc that it cuts from the circle be A and B. See Figure 1.44. What is θ equal to in degrees if the length of the arc from A to B is $1\frac{1}{2}$?

1.8. An arc on a circle is 4 centimeters long. If the radius of the circle is 5 centimeters, find the angle determined by the endpoints of the arc and the center of the circle. If the angle is 21°, find the radius of the circle.

1.9. Let O be the center of a circle, and let B and B' be two points on it. Let C be the midpoint of BB' and construct DE perpendicular to BB' and through C. This is illustrated in Figure 1.45. Verify that O lies on DE. [Hint: Euclid proceeds as follows. The segments OB and OB' are equal. So are BC and $B'C$. Because $\Delta BOB'$ is isosceles, the angle at B is equal to the angle at B'. It follows that the

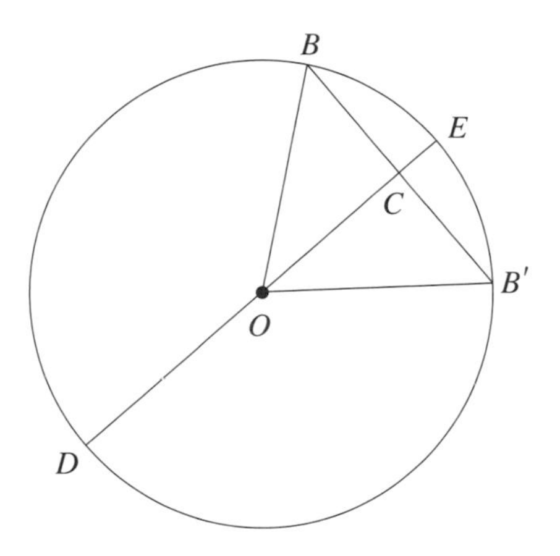

Fig. 1.45

triangles ΔOCB and $\Delta OCB'$ are congruent. So one can be repositioned so that it coincides with the other. Therefore $\angle BCO = \angle B'CO$ and hence both are right angles. So O lies on the perpendicular bisector of BB'.]

1.10. Given two distinct points in the plane, describe how you can obtain all the circles that go through them. [Hint: Use the information provided by Problem 1.9.]

1.11. Given three points in the plane that do not lie on the same line, show that there is exactly one circle that goes through them.

1F. Celestial Navigation. Written records of navigation using stars, or celestial navigation, go back to Homer's *Odyssey*, when, for example, Calypso advises Odysseus to "keep the Bear" (a constellation) on his left-hand side to sail from Calypso's island to his next destination. Elementary celestial navigation was common in the 16th century. The considerations of Eratosthenes and a star catalogue were used to determine the position—at least, the latitude—of a ship at sea.

Suppose a ship is positioned on the globe as shown in Figure 1.46a. Note that the angle β determines its north-south position. This is the ship's *latitude*. As Figure 1.46b shows, β can be determined if one

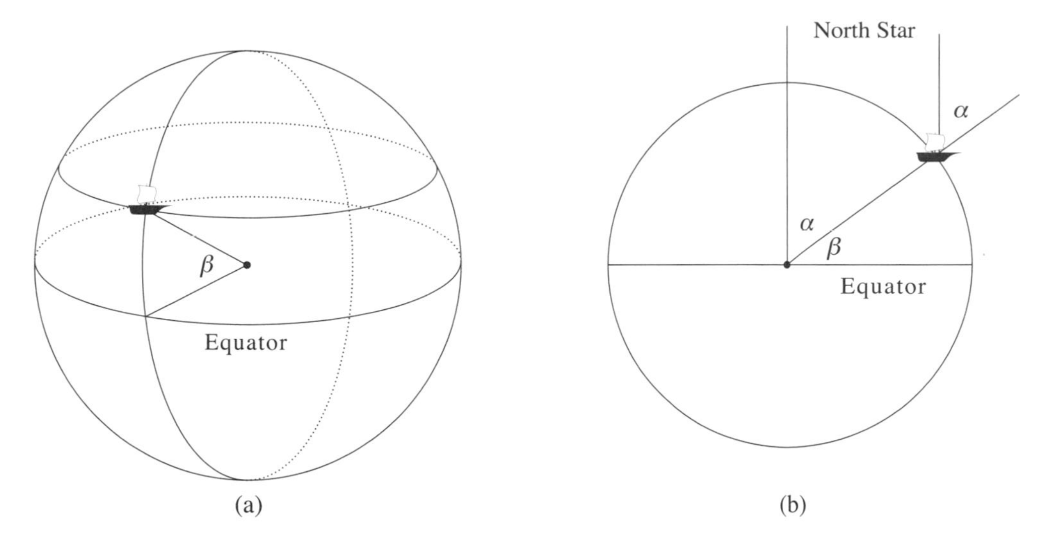

Fig. 1.46

knows α. But α can be measured: it is the angle between a plumb line and the line of sight to the North Star.

1.12. Express the distance from the ship to the equator (along the arc) in terms of α and the radius of the Earth.

Why doesn't this strategy work for determining the *longitude*—i.e., the east-west position—of a ship? Without the solution of this "problem of longitude," a ship's position cannot be fully determined. This problem turned out to be very elusive and was not solved until the 18th century. Large cash prizes were offered for a solution by the governments of seafaring states like Spain, Portugal, Venice, Holland, and England. The British "Board of Longitude" offered the largest amount, a sum of 20,000 pounds. Note that as the Earth turns, the stars travel across the sky at a rate of 360° per day, or 15° per hour. Thus, if time can be measured accurately aboard a ship, the sky will provide its position. A ship's rolling motion rendered pendulum clocks useless. The issue therefore centered on the construction of a clock accurate under the severe conditions on board. The matter was not settled until 1761–1762. A precision clock (referred to as marine chronometer) devised by the Englishman John Harrison (it had taken him 19 years to complete) aboard one of His Majesty's ships lost only about 5 seconds in 80 days at sea.

1.13. Suppose that the angle α is measured to be 53°. How many kilometers is the ship north of the equator? (Use the modern value of 6370 kilometers for Earth's radius.)

1G. Approximating π. Take a circle of radius 1 and let O be its center. Let n be some positive integer and place n points on the circle in such a way that any two consecutive points are the same distance apart. The figure obtained by connecting consecutive points with line segments is called a regular polygon of n sides. Denote the length of any one of the equal sides by s_n. Placing a point equidistant between each two consecutive points of this polygon gives us n more points, for a total of $2n$ points. Connecting every two consecutive points of this larger set of points gives us a regular polygon of $2n$ sides. Denote the length of one of its sides by s_{2n}. Figure 1.47 shows one side of the n-gon and several consecutive sides of the $2n$-gon.

1.14. Show the following:

i. $OT^2 = 1 - \frac{s_n^2}{4}$.

ii. $QT^2 = \left(1 - \frac{\sqrt{4-s_n^2}}{2}\right)^2$.

iii. $s_{2n}^2 = 2 - \sqrt{4 - s_n^2}$.

iv. Why ns_{2n} is an approximation for π that gets more and more accurate with increasing n.

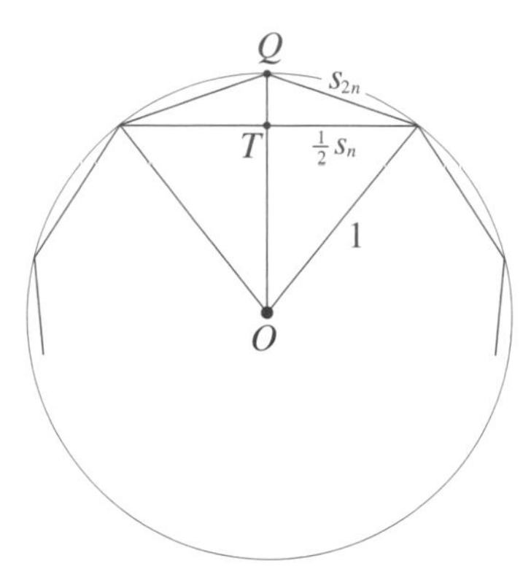

Fig. 1.47

1.15. **i.** Inscribe a regular hexagon into the circle, and show that $s_6 = 1$.

ii. Use the conclusion of Problem 1.14iii to verify that $s_{12} = \sqrt{2 - \sqrt{3}}$, $s_{24} = \sqrt{2 - \sqrt{2 + \sqrt{3}}}$,

$s_{48} = \sqrt{2 - \sqrt{2 + \sqrt{2 + \sqrt{3}}}}$, and $s_{96} = \sqrt{2 - \sqrt{2 + \sqrt{2 + \sqrt{2 + \sqrt{3}}}}}$.

Show that $\pi \approx 48s_{96} \approx 3.141032$. Go on to the next step to get the better approximation $\pi \approx 96s_{192} \approx 3.141452$.

iii. Consider a square and show that $s_4 = \sqrt{2}$.

iv. Show that $s_8 = \sqrt{2 - \sqrt{2}}$, $s_{16} = \sqrt{2 - \sqrt{2 + \sqrt{2}}}$, $s_{32} = \sqrt{2 - \sqrt{2 + \sqrt{2 + \sqrt{2}}}}$, and

$s_{64} = \sqrt{2 - \sqrt{2 + \sqrt{2 + \sqrt{2 + \sqrt{2}}}}}$. Go to the next step to show that $\pi \approx 64s_{128} \approx 3.141277$.

v. The decimal expansion of π starts with $\pi \approx 3.141592$.

1.16. The concept of measurability (see Section 1.4) is perhaps most concretely illustrated by the consideration of the numbers that underlie our monetary system. All dollar amounts are expressed in the form $\$x\frac{yz}{100}$. So only rational numbers, indeed only certain rational numbers, are allowed. In a supermarket, one will occasionally find, say, three items for a dollar. A single item is not measurable within the system. Why not?

1H. Basic Trigonometry. The next set of problems has a focus on elementary trigonometry.

1.17. Use the appropriate triangles to fill in the following blanks.

i. $\cos\frac{\pi}{6} =$ ________ **ii.** $\cos\frac{\pi}{4} =$ ________

iii. $\cos\frac{\pi}{3} =$ ________ **iv.** $\tan\frac{\pi}{6} =$ ________

v. $\tan\frac{\pi}{4} =$ ________ **vi.** $\tan\frac{\pi}{3} =$ ________

1.18. Compare (use a calculator) the values of α, $\sin\alpha$, $\tan\alpha$ for

i. $\alpha = 0.1$ radians:
$\sin\alpha =$ ________, $\tan\alpha =$ ________.

ii. $\alpha = 0.01$ radians:
$\sin\alpha =$ ________, $\tan\alpha =$ ________.

iii. $\alpha = 0.001$ radians:
$\sin\alpha =$ ________, $\tan\alpha =$ ________.

1.19. Illustrate with a diagram similar to that of Figure 1.30 that if $\theta' > \theta > 0$, then $\cos\theta' < \cos\theta$.

1.20. Show that if θ is given in degrees, then $\lim_{\theta\to 0}\frac{\sin\theta}{\theta} = \frac{\pi}{180}$. [Hint: If θ is given in degrees, then $\theta\cdot\frac{\pi}{180}$ is the same angle expressed in radians. So for θ in degrees, $\frac{180}{\pi}\cdot\frac{\sin\theta}{\theta} = \frac{180}{\pi}\cdot\frac{\sin(\theta\cdot\frac{\pi}{180})}{\theta} = \frac{\sin(\theta\cdot\frac{\pi}{180})}{\theta\cdot\frac{\pi}{180}}$.]

1.21. Explain why $\lim_{\theta\to 0}\dfrac{\tan\theta}{\theta} = 1$ (with θ in radians).

1.22. The secant of θ is defined by $\sec\theta = \frac{1}{\cos\theta}$. Verify the identity $\sec^2\theta = \tan^2\theta + 1$. The cosecant and cotangent of an angle θ are defined by $\csc\theta = \frac{1}{\sin\theta}$ and $\cot\theta = \frac{1}{\tan\theta} = \csc\theta = \frac{\cos\theta}{\sin\theta}$. Show that $\csc^2\theta = \cot^2\theta + 1$.

1I. More Basic Trigonometry. The problems of this segment present outlines of both the half-angle and addition formulas for the sine, cosine, and tangent. These trig identities are verified below with some restrictions on the angles involved. They are valid in general, however.

1.23. Consider a circle with center O, radius 1, and diagonal AB. Let C be any point on the upper semicircle, and let $\theta = \angle COB$. Choose P on AB so that CP is perpendicular to AB. By results in

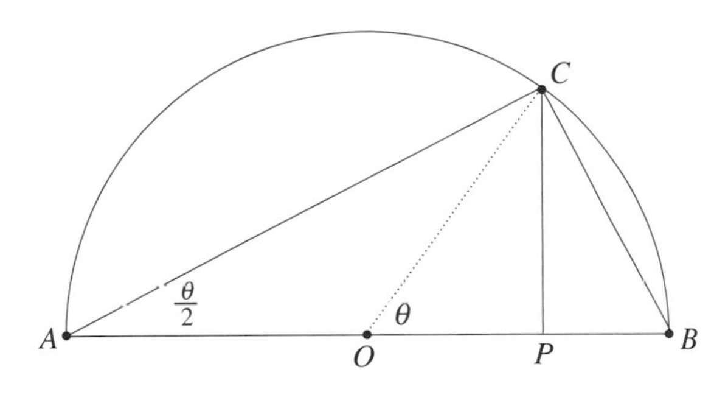

Fig. 1.48

Sections 1.3 and 1.6, $\angle ACB$ is a right angle and $\angle BAC = \frac{\theta}{2}$. Observe that $\sin\theta = CP$ and $\cos\theta = OP$. Use the Pythagorean theorem to show that $AC^2 = 2(1 + \cos\theta)$.

i. Verify the half-angle formulas

$$\sin^2\frac{\theta}{2} = \frac{1}{2}(1 - \cos\theta),\ \cos^2\frac{\theta}{2} = \frac{1}{2}(1 + \cos\theta),\text{ and }\tan^2\frac{\theta}{2} = \frac{1-\cos\theta}{1+\cos\theta}.$$

[Hint: For the first formula, multiply $\sin^2\frac{\theta}{2} = \frac{CP^2}{AC^2}$ by $\frac{1-\cos\theta}{1-\cos\theta}$ and simplify.]

ii. Verify the equalities $\tan\frac{\theta}{2} = \frac{\sin\theta}{1+\cos\theta}$ and $1 - \tan^2\frac{\theta}{2} = \frac{2\cos\theta}{1+\cos\theta}$, and use them to show that $\tan\theta = \frac{2\tan\frac{\theta}{2}}{1-\tan^2\frac{\theta}{2}}$.

Since it relies on Figure 1.48, the above verification of these half-angle formulas assumed that the angle θ is acute. The same argument also works in the case of Figure 1.49a and hence in any situation where $\angle CAB = \frac{1}{2}\theta$ is acute. We will now see that the first step in this argument works for an obtuse angle $\angle CAB$ as well.

1.24. In Figure 1.49b, $\angle CAB < 180°$ is any obtuse angle embedded in a circle with center O and radius 1. Connect B and C to O and consider the angle θ. Show that $\angle CAB = \frac{1}{2}\theta$. [Hint: Draw the diameter AD and the segments BD and CD.]

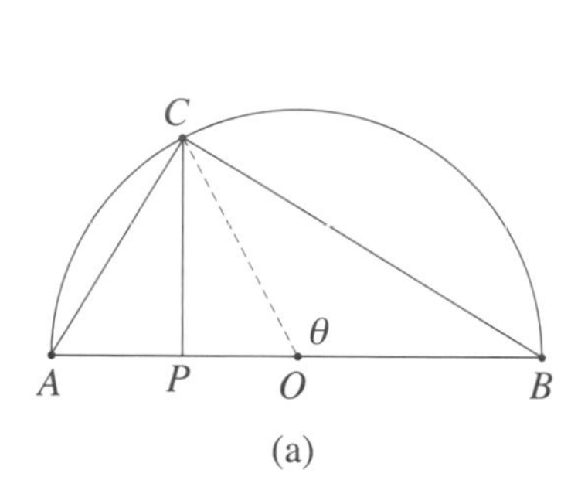

(a)

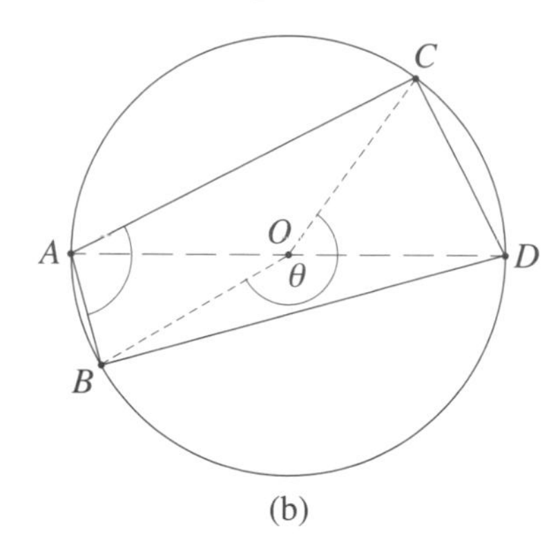

(b)

Fig. 1.49

1.25. Let α and β be two angles with sum less than $\frac{\pi}{2}$, and place them into the two right triangles ΔABC and ΔBDC shown in Figure 1.50. Complete the diagram to the rectangle $RSDC$ and draw in the parallel AT to SD. Show that $\angle SBA = \beta$.

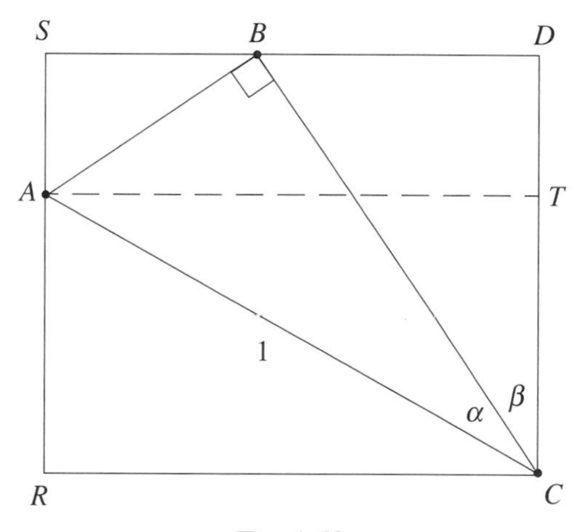

Fig. 1.50

i. Show that $AT = SB + BD = AB\cos\beta + BC\sin\beta$, and conclude that

$$\sin(\alpha + \beta) = \sin\alpha\cos\beta + \cos\alpha\sin\beta.$$

ii. Show that $TC = DC - SA = BC\cos\beta - AB\sin\beta$, and hence that

$$\cos(\alpha + \beta) = \cos\alpha\cos\beta - \sin\alpha\sin\beta.$$

iii. Divide the top and bottom of $\dfrac{\sin\alpha\cos\beta + \cos\alpha\sin\beta}{\cos\alpha\cos\beta - \sin\alpha\sin\beta}$ by $\cos\alpha\cos\beta$, and conclude that

$$\tan(\alpha + \beta) = \frac{\tan\alpha + \tan\beta}{1 - (\tan\alpha)(\tan\beta)}.$$

1.26. Use conclusions from Problem 1.25 to show that

i. $\sin 2\alpha = 2(\sin\alpha)(\cos\alpha)$,

ii. $\cos 2\alpha = \cos^2\alpha - \sin^2\alpha = 1 - 2\sin^2\alpha = 2\cos^2\alpha - 1$, and

iii. $\tan 2\alpha = \dfrac{2\tan\alpha}{1-\tan^2\alpha}$.

The solutions of the next three problems rely on the law of sines or the law of cosines.

1.27. Consider a triangle ΔABC. The length of the side AB is 8, and the angles at A and B are $\frac{\pi}{5}$ and $\frac{\pi}{7}$, respectively. Determine the lengths of the other two sides.

1.28. Two sides of a triangle have lengths 7 and 11. The angle between these two sides is $\frac{\pi}{5}$. What is the length of the third side?

1.29. Show that if the three sides a, b, and c of a triangle satisfy the relationship $a^2 + b^2 = c^2$, then the triangle is a right triangle.

1J. Some "Inverse" Trigonometry. The solutions of some of the problems that follow require a calculator with inverse trig feature. (Use three-decimal-place accuracy.)

1.30. What angles α, β, γ, and φ in degrees and radians satisfy $\sin\alpha = \frac{1}{2}$, $\tan\beta = 1$, $\sin\gamma = \frac{\sqrt{3}}{2}$, and $\tan\varphi = \sqrt{3}$, respectively?

1.31. Fill in the blanks:

i. sin ____° = 0.219

ii. sin ____° = 0.834

iii. sin ____ (in radians) = 0.002

iv. sin ____ (in radians) = 0.664

v. tan ____° = 0.774

vi. tan ____° = 1.478

vii. tan ____ (in radians) = 10.473

viii. tan ____ (in radians) = 27.664

1.32. For each of the three triangles in Figure 1.51 find the length of the sides as well as the angles that have not been specified.

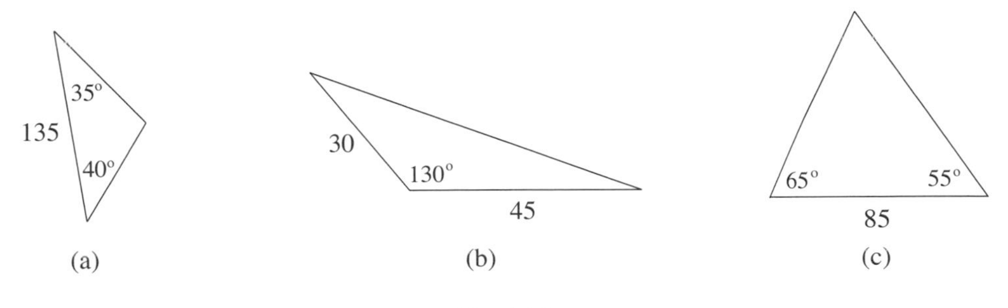

Fig. 1.51

1.33. You are given a triangle with sides of lengths 11, 7, and 5. Determine its three angles.

The next two discussions specify angles in terms of minutes. Recall that one minute, denoted by $1'$, is equal to $\frac{1}{60}^{\circ}$.

1K. Aristarchus's Argument Revisited. Go back to Section 1.7 and change the assumptions as follows. In Observation C take $89°50'$ instead of $87°$; in Observation D take $\frac{1}{2}^{\circ}$ instead of $2°$; and in Observation E take $5r_M$ instead of $4r_M$. Keep the estimate $r_E = 6200$ kilometers for Earth's radius.

1.34. Use these assumptions and repeat Aristarchus's computation of the distances r_M, r_S, D_M, and D_S. (Use a calculator and compute with an accuracy of four decimal places.) Compare your conclusions with the modern values of Table 1.3. [Comment: Your conclusions should be close to these values. However, in the context of Figure 1.31, a precise determination of a half Moon and an accurate measurement of the angle $89°50'$ would be difficult to achieve even today. Since the Sun would have to be low in the sky, careful corrections would have to be made for the fact that light is bent by the atmosphere (the phenomenon of refraction).]

1L. Ptolemy and the Moon. Ptolemy computed the distance D_M between the Earth and the Moon as follows. In Figure 1.52, E and M are the centers of the Earth and Moon, respectively, and r_E is the radius

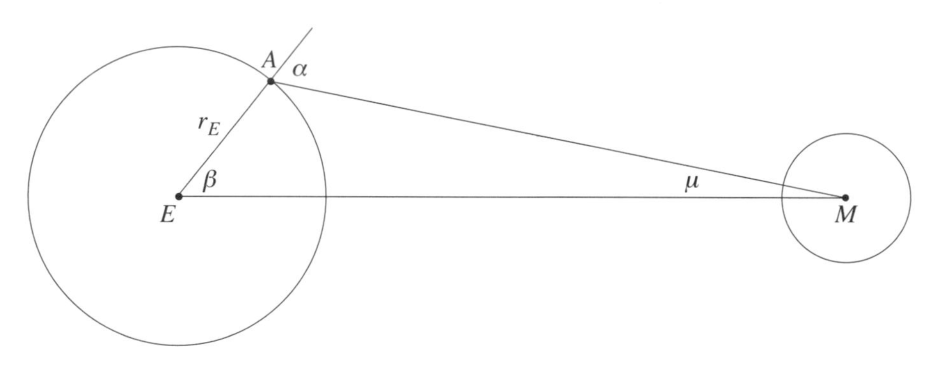

Fig. 1.52

of the Earth. The point A represents Alexandria. One day, Ptolemy measured the angle α to be $50°55'$ and computed β to be $49°48'$.

1.35. Take $r_E = 6200$ kilometers from Eratosthenes. Supply the details to Ptolemy's argument as outlined below.

i. Show that $\alpha = \beta + \mu$.

ii. Show that $D_M = r_E \cdot \frac{\sin\alpha}{\sin\mu}$ by using the law of sines and a trig formula from Section 1.6.

iii. Obtain $D_M \approx 247{,}000$ kilometers and compare this with today's value.

1M. Ptolemy's Theorem. The *Almagest* contains a comprehensive exposition of Greek trigonometry. Presupposing a thorough knowledge of the *Elements* of Euclid, Ptolemy develops Greek trigonometry along with a table of chords that allowed him to explain and compute the motions of the Sun, Moon, planets, and stars. The trigonometry in the *Almagest* consists primarily of numerical calculations—often very intricate—in the context of triangles in the plane and on the sphere. His emphasis on numerical methods stands in contrast to the common opinion that Greek mathematics deliberately avoided such methods in favor of purely geometrical arguments. From the large body of mathematics in the *Almagest*, we'll consider only a theorem of geometry. It is known as Ptolemy's theorem today.

Proposition. Let θ be any angle with $0° < \theta < 360°$, and place it into a circle with center O. Consider the arc on the circle that θ determines, and let A and B be its endpoints. Refer to Figure 1.53. Let P be any

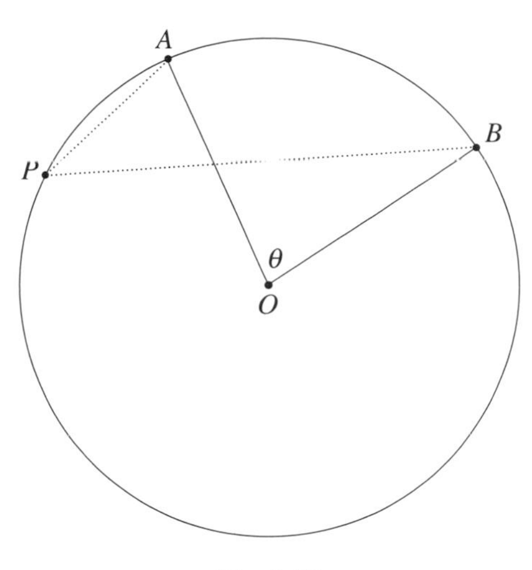

Fig. 1.53

point on the circle that falls outside this arc. Then $\angle APB = \frac{\theta}{2}$.

To verify the proposition, put in the diameter POC and let $\phi = \angle BOC$. See Figure 1.54. By applying a result of Section 1.6, we get

$$\angle APC = \tfrac{1}{2}(\theta + \phi) \quad \text{and} \quad \angle BPC = \tfrac{1}{2}\phi.$$

So $\angle APB = \angle APC - \angle BPC = \frac{1}{2}(\theta + \phi) - \frac{1}{2}\phi = \frac{\theta}{2}$, as required by the proposition.

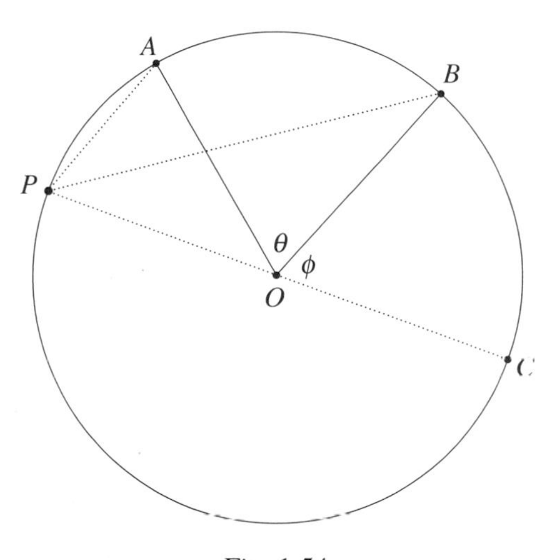

Fig. 1.54

1.36. In the argument above, it was assumed that $\theta \leq 180°$. Use the conclusion of Problem 1.24 and provide the details of the verification in the case $\theta > 180°$.

Make note of the following consequence of the proposition.

Corollary. Fix an arc on a circle with endpoints A and B, and let P be any point on the circle and outside the arc. Then the angle $\angle APB$ is the same, no matter where (on the circle and outside the arc) P is taken.

We now turn to Ptolemy's theorem and its proof.

Ptolemy's Theorem. Let A, B, C, and D be any four points on a circle. Then

$$AC \cdot BD = AB \cdot CD + AD \cdot BC.$$

The proof relies on the corollary that was just discussed and basic properties of similar triangles. Figure 1.55 illustrates the typical situation. Let E be the point of intersection of the two diagonals, and draw AF such that $\angle DAF = \angle BAE$. See Figure 1.56.

By the corollary, $\angle ADB = \angle ACB$. Therefore $\angle DAF = \angle BAE = \angle BAC$ and $\angle ADF = \angle ADB = \angle ACB$. Figure 1.57 extracts the triangles ΔADF and ΔACB from Figure 1.56. These triangles have two

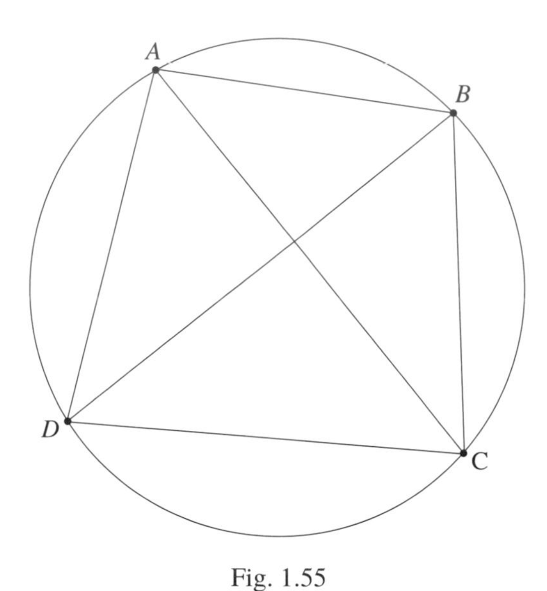

Fig. 1.55

angles in common. They are labeled 1 and 2. Since the sum of the angles of any triangle is 180°, the angles labeled 3 are also equal, so that ΔADF and ΔACB are similar. Take one more look at Figure 1.57.

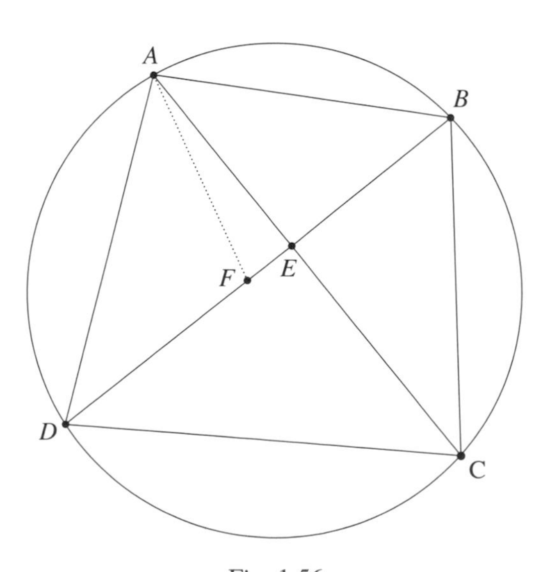

Fig. 1.56

Since AD and AC, and DF and CB, are corresponding pairs of sides,

(a)
$$\frac{AD}{AC} = \frac{DF}{BC}.$$

By another application of the corollary to Figure 1.56, $\angle ACD = \angle ABD = \angle ABF$. Since

$$\angle DAC = \angle DAF + \angle FAE = \angle BAE + \angle EAF = \angle BAF,$$

the triangles ΔACD and ΔABF have two angles in common. They are the angles labeled 1 and 2 in

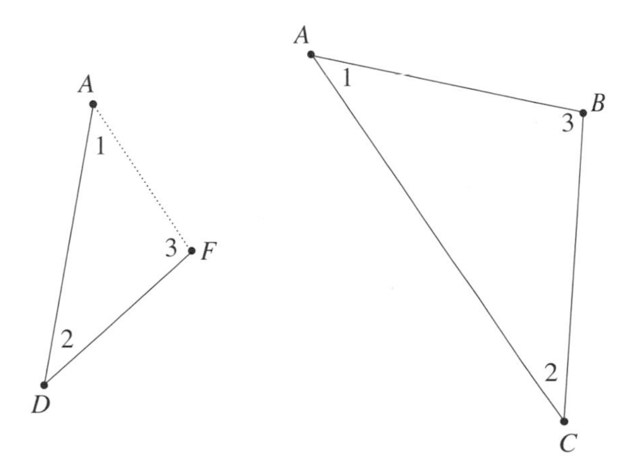

Fig. 1.57

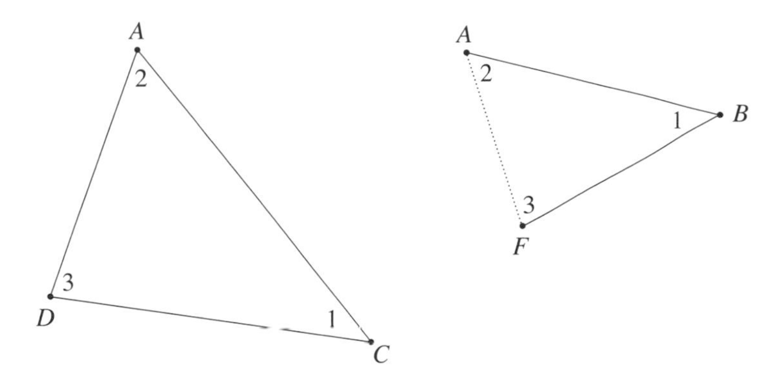

Fig. 1.58

Figure 1.58. Therefore they too are similar. This implies the equality of the ratios

(b)
$$\frac{AB}{AC} = \frac{BF}{CD}.$$

By equation (a), $DF = \frac{AD}{AC}BC$, and by equation (b), $BF = \frac{AB}{AC}CD$. A look at Figure 1.56 tells us that

$$BD = BF + DF = \frac{AB}{AC}CD + \frac{AD}{AC}BC.$$

Multiplying through by AC finally shows that

$$(AC) \cdot (BD) = (AB) \cdot (CD) + (AD) \cdot (BC).$$

The proof of Ptolemy's theorem is now complete.

1.37. Repeat the proof of Ptolemy's theorem in the situation of Figure 1.59 rather than Figure 1.56.

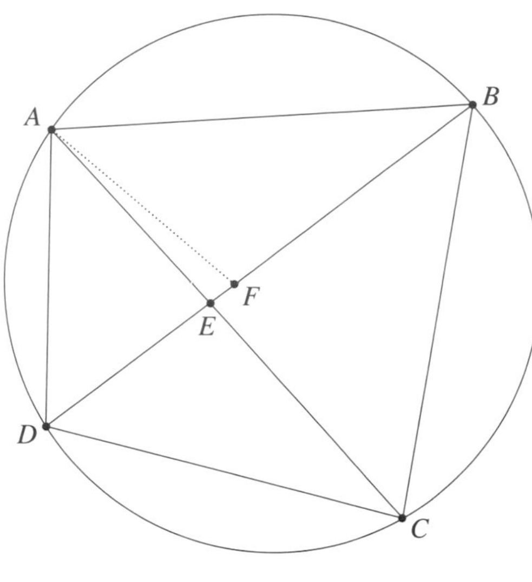

Fig. 1.59

Ptolemy's theorem can be used to verify a number of the basic facts of geometry and trigonometry. These include Pythagoras's theorem, the addition formula for the sine and the law of cosines, among others.

1.38. Use Ptolemy's theorem to verify Pythagoras's theorem.

1.39. Let CD be a diameter of a circle of diameter equal to 1. Let α and β be two acute angles, and place them into Figure 1.60 at the point C as shown. By a result in Section 1.3, the triangles ΔCAD and ΔCBD are both right triangles. Now put in the diameter through B and O, and let E be the point

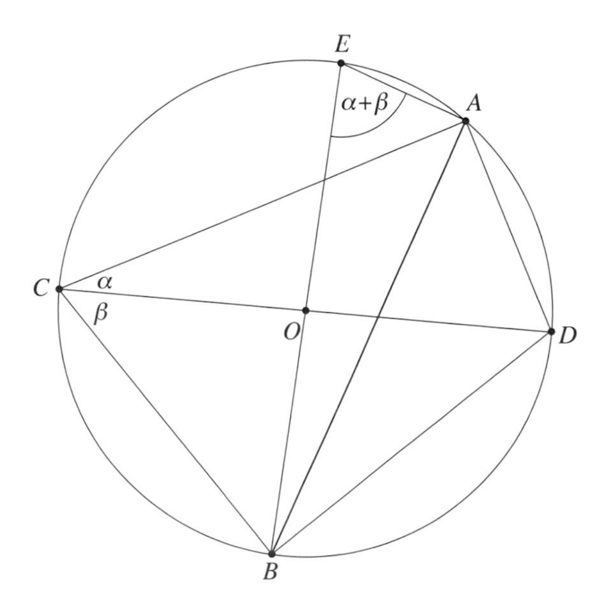

Fig. 1.60

that it determines. The triangle ΔBAE is also right. Show that

$$\sin(\alpha + \beta) = \sin \alpha \cos \beta + \cos \alpha \sin \beta.$$

[Hint: Use Ptolemy's theorem as well as the corollary.]

1.40. Assume that the four vertices of a quadrilateral fall on a circle. The four sides and the two diagonals determine four triangles. Consider a pair of these triangles that are not adjacent (so they are opposite) and show that they are similar. [Hint: Use Ptolemy's corollary.]

1.41. Consider a circle and any arc on it. Let A and B be the endpoints of the arc, and let P be a point on the circle outside the arc. The radius of the circle is 3, and the length of the arc is 4. Determine the angle $\angle APB$ first in radians and then in degrees.

Let A and B be two points on a circle. The two points determine two arcs. Moving in the same direction around the circle, there is the arc from A to B, and then there is the arc from B to A. When one of the two arcs is under consideration, the other arc will be referred to as the *complementary* arc.

1.42. Consider any circle and any arc on it. Let A and B be the endpoints of the arc, and let θ be the angle that the arc and the center O of the circle determine. Show that $\angle APB = \pi - \frac{\theta}{2}$ for any point P on the arc except for the points A and B. [Hint: Apply Ptolemy's proposition to P and the complementary arc.]

1N. More about Circles and Triangles.

1.43. Let A and B be two distinct points in a plane. Let P be the midpoint of the segment AB. Let L be the line through P that is perpendicular to the segment AB. This L is the perpendicular bisector of the segment AB. Show that the points on L are precisely the points that are equidistant from A and B.

1.44. Let A, B, and C be points in a plane that do not all fall on the same line. Show that there is a circle through A, B, and C. It follows that any triangle can be inscribed in a circle. [Hint: Let the point O be the intersection of the perpendicular bisectors of the sides AB and BC. Why is O the center of the circle in question? Use Problem 1.43.] Show that there is only one such circle.

1.45. Let a triangle ΔACB with sides a, b, and c be given, and place it into a circle as shown in Figure 1.61. Complete the triangle to the trapezoid $ACBD$. Let $\gamma = \angle ACB$, and let d be the distance from C to

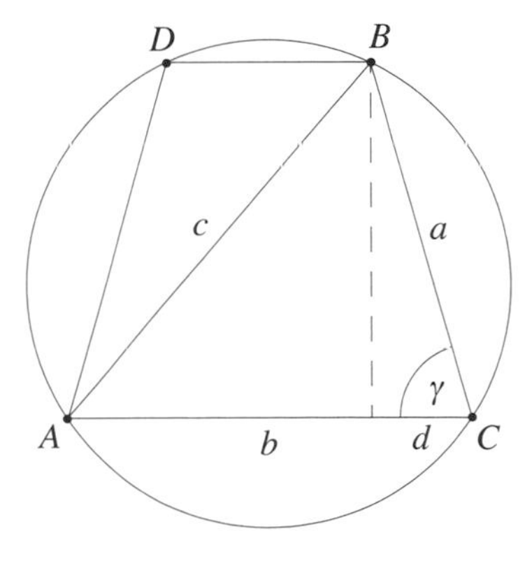

Fig. 1.61

the point of intersection of AC and the perpendicular from B. Apply Ptolemy's theorem to verify the law of cosines

$$c^2 = a^2 + b^2 - 2ab\cos\gamma.$$

1.46. Consider a triangle inscribed in a circle. We saw in Section 1.3 that if a side of the triangle is a diameter of the circle, then the angle opposite to that side is a right angle. Show conversely, that if one of its angles is a right angle, then the side opposite that angle is a diameter. [Hint: Label the triangle ΔABC and let the right angle be at C. Let O be the center of the circle and extend CO to a diameter COD of the circle. Observe that the triangles ΔCAD and ΔCBD have right angles at A and B respectively. Conclude that $ACBD$ is a rectangle.]

1.47. Consider any arc on a circle of radius r, and let A and B be its endpoints. Let P be any point on the circle but not on the arc, and let $\angle APB = \alpha$ (in radians). Verify that arc $AB = 2r\alpha$ and $AB = 2r(\sin\alpha)$. [Hint: Apply Ptolemy's proposition, the law of cosines, and a trig identity.]

1.48. A triangle has sides a, b, and c. With b as base, its height is h. Show that the radius r of the circle on which the three vertices of the triangle lie is $r = \frac{ac}{2h}$. (By Problems 1.44 or 1.11, such a circle always exists.) [Hint: See Figure 1.62 and compute the sine of the angle α in two ways. Use

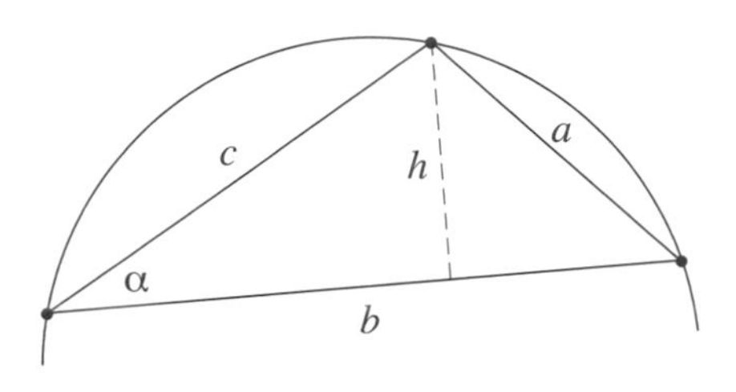

Fig. 1.62

Problem 1.47 for one of the computations.]

1.49. Let a, b, and c be any three positive numbers. Is there a triangle that has a, b, and c as the lengths of its sides? Show that this is so if $a + b > c$. Start by taking segments of lengths a and b and aligning them in a straight line, as shown in Figure 1.63. Conclude that if a, b and c are positive numbers

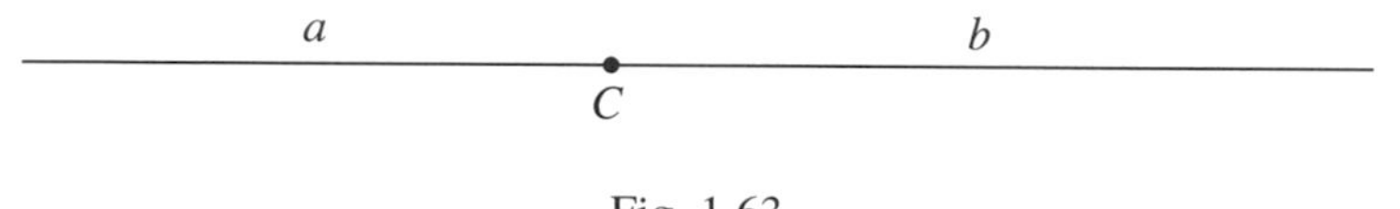

Fig. 1.63

such that $a + b > c$, then also $a + c > b$ and $b + c > a$.

1.50. A triangle has sides of lengths 7 and 11 and (with the third side as base) height 4. Show that the length of the third side is approximately 16. Show that the radius r of the circle on which the three vertices of the triangle lie is $r = \frac{77}{8}$.

1.51. Why is there a triangle with side lengths 7, 11, and 17? Compute the angle between the sides of lengths 7 and 11. Compute the height of the triangle with respect to the base 17. What is the radius of the circle on which the three vertices lie? [Hint: Use the law of sines.]

10. Highways and Billboards. A car moves along a straight stretch of highway S. A passenger at P looks out at an approaching billboard. We'll assume that S as well as the bottom edge AB of the billboard lie in the same plane. Figure 1.64 provides a view from above. Concentrate on the angle $\angle APB$ between the passenger's lines of sight to the left and right edges of the billboard. A look at the figure confirms that $\angle APB$ first increases, but then decreases again. Therefore at some point before the car passes the billboard, $\angle APB$ attains a maximum value. Can the location of P at which this maximum value occurs be determined in some way? (Surprise: The intuitively so obvious answer that this location must be the intersection of S with the perpendicular bisector of AB is wrong.)

The problem that follows provides a basic fact that is the key to the answer.

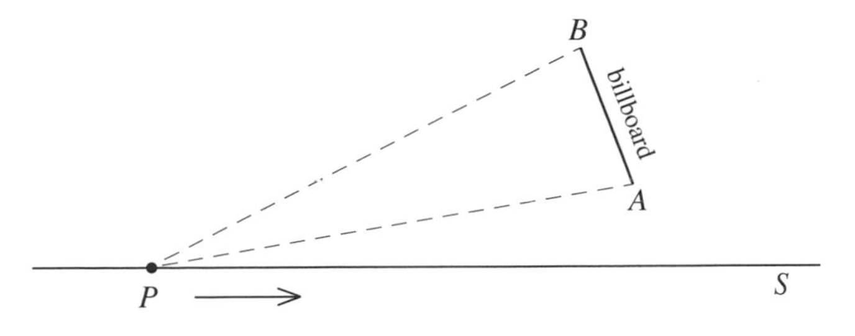

Fig. 1.64

1.52. Consider a circle with center O and any arc on it. Let A and B be the endpoints of the arc, and let $\angle AOB = \varphi$. Consider the line L determined by A and B. The line L divides the plane into two parts. Suppose that Q is any point in the plane on the side of L opposite the arc AB, and let $\alpha = \angle AQB$. Show that if Q is outside the circle, then $\alpha < \frac{\varphi}{2}$; if Q is on the circle, then

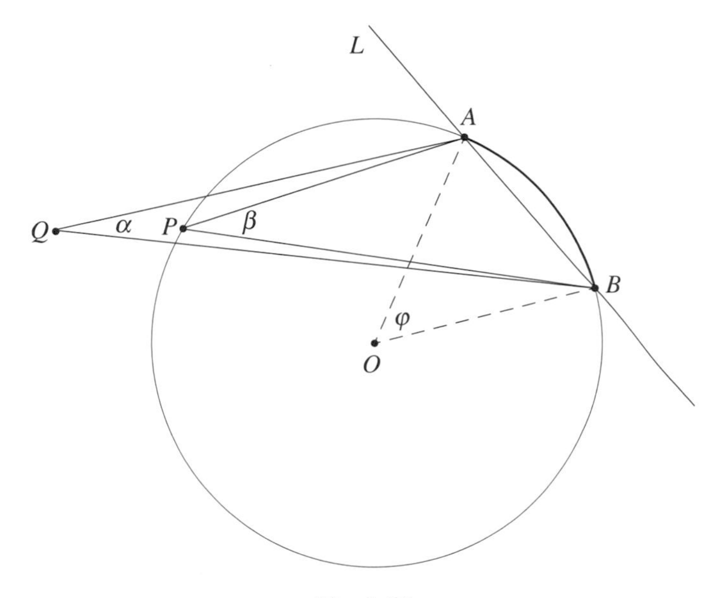

Fig. 1.65

$\alpha = \frac{\varphi}{2}$; and if Q is inside the circle, then $\alpha > \frac{\varphi}{2}$. [If Q is on the circle, apply Ptolemy's proposition. If Q is outside the circle, study Figure 1.65 carefully.]

We are now in a position to answer the billboard question.

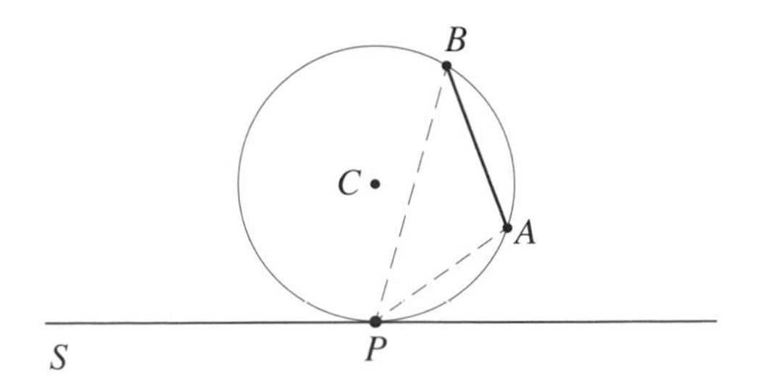

Fig. 1.66

1.53. Consider a circle through both A and B that has the line S as a tangent. See Figure 1.66. Then the point of tangency P is the location at which the angle $\angle APB$ is a maximum.

That this is so is a direct consequence of the conclusion of Problem 1.52. The issue that remains is this: given a line in the plane and two points that are both on one side of the line, is there a circle through the two points that has the line as a tangent? The answer is yes, but the verification is tricky. See construction 38 on page 350 of Hartshorne's *Geometry: Euclid and Beyond* of the References at the end of the text.

Chapter 2

The Genius of Archimedes*

This chapter starts with a brief classical account of the conic sections. It then explores the infinite process with which Archimedes computes areas and describes his approach to the area of a parabolic section.

Archimedes—the most famous scientist of antiquity—was born in 287 B.C. in the Greek city-state of Syracuse, a port on the Mediterranean island of Sicily. There is historical reference to the fact that he spent considerable time in Alexandria, and there seems little doubt that he studied with the successors of Euclid. After his studies, he returned to Syracuse and lived there in complete absorption with his mathematical investigations.

He became a legend in his own time. Late in the 3rd century B.C., Syracuse became embroiled in the struggle between Rome and Carthage for control of the western Mediterranean. In his *Parallel Lives*, the historian Plutarch (about 46 to 126 A.D.) recounts Archimedes's efforts in the defense of the city against the Romans:

> When, therefore, the Romans assaulted the walls in two places at once, fear and consternation stupefied the Syracusans, believing that nothing was able to resist that violence and those forces. But when Archimedes began to ply his engines, he at once shot against the land forces all sorts of missile weapons, and immense masses of stone that came down with incredible noise and violence, against which no man could stand; for they knocked down those upon whom they fell, in heaps, breaking all their ranks and files. In the meantime huge poles thrust out from the walls over the ships, sunk some by the great weights which they let down from on high upon them; others they lifted up into the air by an iron hand [and soon] such terror had seized upon the Romans, that, if they did but see a little rope or a piece of wood from the wall, instantly crying out, that there it was again, Archimedes was about to let fly some engine at them, they turned their backs and fled.

The Roman attack on Syracuse was repelled. A lengthy siege was later successful, and Syracuse was conquered and destroyed in 212 B.C. Archimedes perished during the destruction. Plutarch relates several versions of his death. The one most widely cited finds Archimedes, oblivious to the city's capture, absorbed in the study of a particular diagram that he had sketched in the sand. When a Roman soldier confronted him, Archimedes requested time to complete his deliberations. The impatient soldier, however, ran him through with his sword.

The work of Archimedes is impressive—as we shall soon see. Plutarch speaks of Archimedes's

> purer speculations [and] studies, the superiority of which to all others is unquestioned, and in which the only doubt can be, whether the beauty and grandeur of the subjects examined, or the precision and cogency of the methods and means of proof, most deserve our admiration. It is not possible to find in all geometry more difficult and intricate questions, or more simple

* This chapter can be skipped, if the focus is on basic mathematics, and in particular the mathematics required for the pursuit of calculus. However, some of the concepts and results of the first three sections will be referred to later.

> and lucid explanation. No amount of investigation of yours would succeed in attaining the proof, and yet once seen, you immediately believe you would have discovered it; by so smooth and so rapid a path he leads you to the conclusion required.

Archimedes was also the quintessential eccentric scientist. His deep absorption in thought

> made him forget his food and neglect his person, to that degree that when he was occasionally carried by absolute violence to bathe, or have his body anointed, he used to trace geometrical figures in the ashes of the fire, and diagrams in the oil on his body.

A famous episode recounts how, after a particularly satisfying discovery, Archimedes ran through the streets of Syracuse in naked celebration shouting "Eureka, Eureka!" (Eureka, or $\varepsilon\upsilon\rho\varepsilon\kappa\alpha$, is Greek for "I have found it.")

It is, of course, difficult to separate fact from fiction and reality from legend in Plutarch's account of Archimedes's remarkable talents as inventor of machines of war. However, the brilliance of Archimedes's investigations into mathematics and physics can be confirmed. Much of his work has survived in transmitted form. Archimedes understood the law of the lever, namely, that the rotational effect of a force is given as the product of its magnitude times the distance to the axis of rotation. His law of hydrostatics (in Greek, *hydro* = water, *statikos* = at rest) told him that the force with which a liquid pushes up against a floating object is equal to the weight of the volume of the liquid that the object displaces. He calculated the centers of mass (or gravity) of planar regions and solids with methods that anticipated modern integral calculus.

This chapter starts by introducing three remarkable families of curves: the parabola, the ellipse, and the hyperbola that the Greeks discovered. These *conic sections*—obtained by intersecting a cone with a plane—were well known to Euclid. They were analyzed much more extensively in the comprehensive treatise *Conics* by Archimedes's contemporary Apollonius (from about 260 to 190 B.C.) of Perga (an ancient city that was located on the south coast of today's Turkey). The parabola and the ellipse—as we will see in subsequent chapters—form the basis for our understanding of the motion of projectiles, the orbits of the planets around the Sun, and the geometry of lenses and mirrors of modern optical instruments.

2.1 THE CONIC SECTIONS

The *conic sections* are the curves that can be obtained by intersecting or cutting a double cone with a plane. See Figure 2.1. The word "section" comes from the Latin word for "cut." We will have a look at

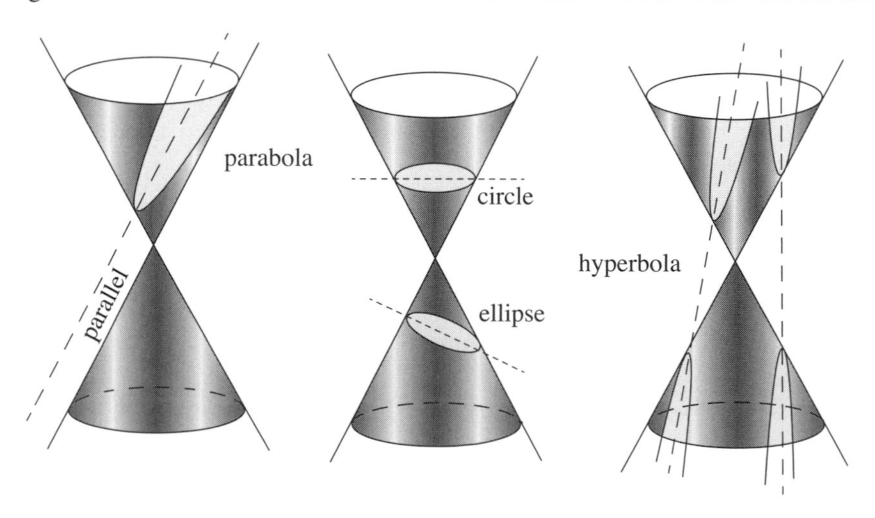

Fig. 2.1

Apollonius's analysis of the parabola and ellipse (but consider the hyperbola only briefly). We'll start by providing definitions of these curves that reside in the plane and do not require us to "step out" into three-dimensional space (as Figure 2.1 clearly does). We will also list—without proofs—a few of the basic properties of the conic sections that we will need later.

Let a plane—a perfectly flat mathematical plane—be given. So that distances can be measured—say, between two points in the plane—a unit of length—say, the centimeter or meter—is provided as well. For two points P and Q in the plane, PQ or QP will label the segment that connects them. If PQ appears in an equation, it will be understood to be the length of the segment PQ, or equivalently, the distance between P and Q.

Let's begin with the parabola. Fix both a line D and a point F in the plane with F not on D. The *parabola* determined by D and F is the collection of all points P in the plane such that the distance from P to F is equal to the (perpendicular) distance from P to D. Figure 2.2 illustrates what has been described. It locates several points P on the parabola. In each case, the length of the segment from P to F is equal to the length of the dashed segment from P to D. The line D is called the *directrix* of the parabola, and the point F is called the *focal point* or *focus* of the parabola. The line through the focal point perpendicular to

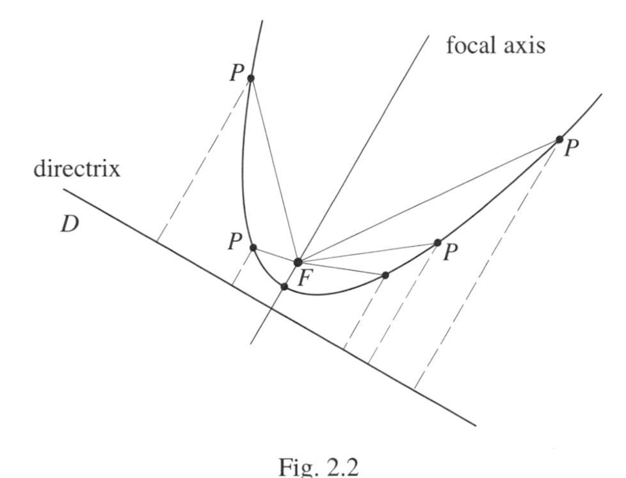

Fig. 2.2

the directrix is the *focal axis* of the parabola. The figure shows the focal point, directrix, and focal axis. We'll recall—without proofs—just two basic propositions about the parabola from the work of Apollonius.

Proposition P1. Let P be any point on a parabola, and consider the tangent line at P. The angle at P that the tangent makes with the line from P to the focal point is equal to the angle at P that the tangent makes

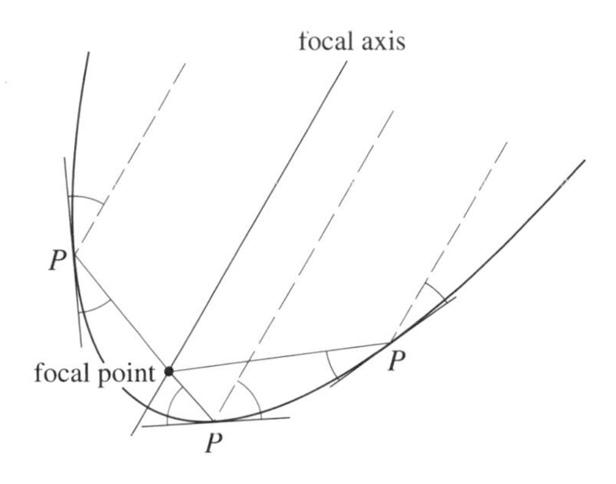

Fig. 2.3

with the line from P parallel to the focal axis.

Figure 2.3 illustrates this "reflection property" of the parabola for three different choices of the point P. We point out in passing that when light is reflected by a mirror, the angle of an incoming beam of light is equal to the angle of the reflected beam. It follows from the proposition that if a mirror has the shape obtained by rotating a parabolic arc one revolution around its axis, then any light ray parallel to the axis will be reflected through the focal point. This is the reason for the use of parabolic mirrors both in telescopes (for collecting light) and in the headlights of cars (for projecting light).

Consider any parabola, cut it with any straight line, and let S and S' be the points of intersection. See Figure 2.4a. There is a point V on the parabola, such that the tangent line at V is parallel to the cut SS'. The parabolic region SVS' is called a *parabolic section*, and V is the *vertex* of the parabolic section. As

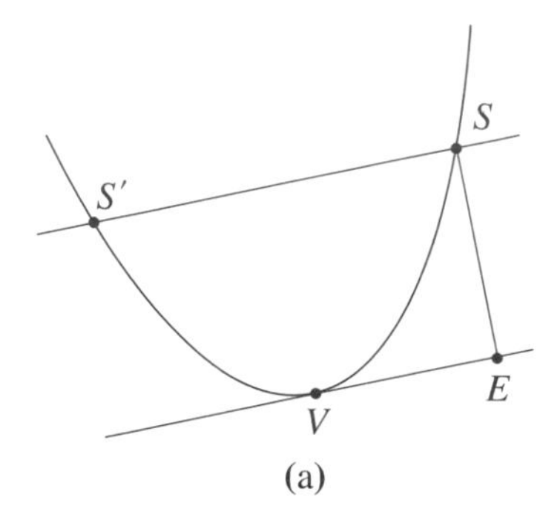

(a)

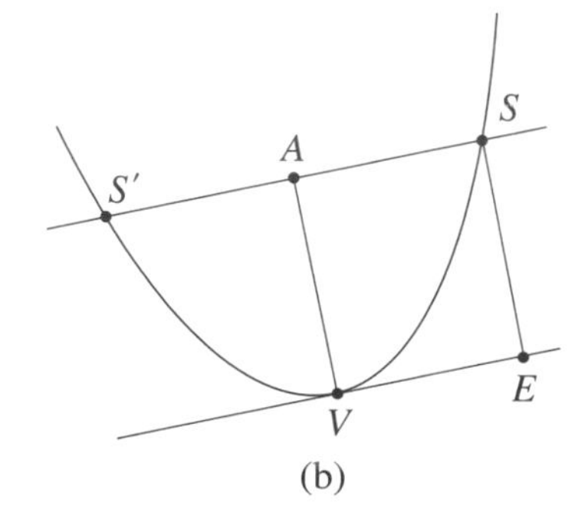

(b)

Fig. 2.4

already mentioned, a *section* is a region obtained from another by a cut. Now let A be the midpoint of the segment SS', connect A with V, and take E on the tangent such that ES is parallel to VA. Refer to Figure 2.4b.

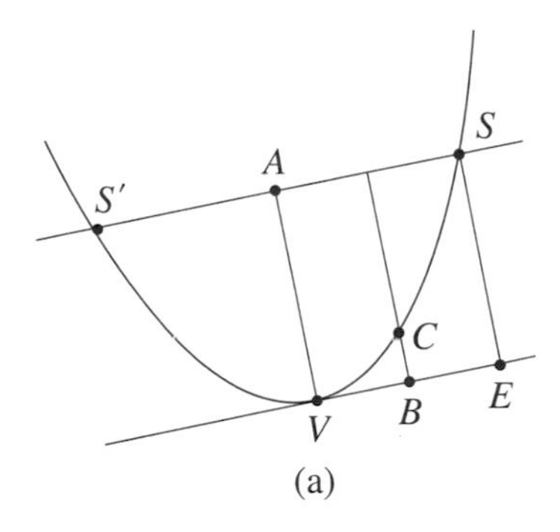

(a)

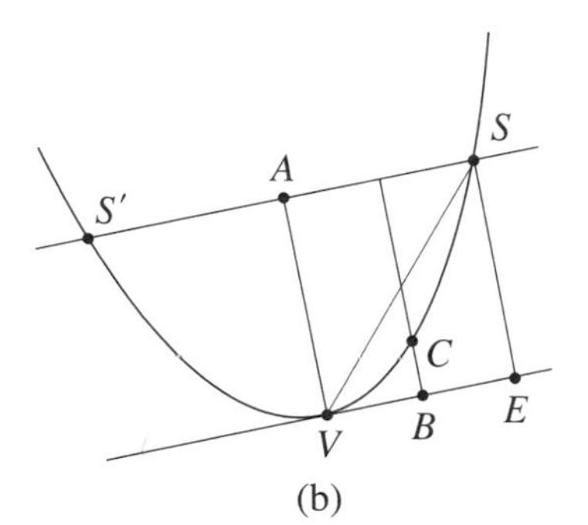

(b)

Fig. 2.5

Let B be any point on VE, and let C be the point on the parabola such that BC is parallel to VA. See Figure 2.5a. Finally, in Figure 2.5b, the segment VS is drawn in.

Proposition P2. For any point B on EV,

$$\frac{BC}{ES} = \frac{VB^2}{VE^2}.$$

Proposition P3. If B is the midpoint of VE, then C is the vertex of the parabolic section SVC.

We will see in a moment that the connection between the distances that Proposition P2 provides is a property of the parabola that is still central today. Proposition P3, on the other hand, is not. We mention it only because it is a key ingredient in Archimedes's study of the area of a parabolic section.

We turn to the ellipse next. Fix any two points F_1 and F_2 in the plane and a constant k that is greater than the distance between F_1 and F_2. Consider the collection of all points P such that the distances from P to F_1 and from P to F_2 add up to k or, in terms of an equation, such that $PF_1 + PF_2 = k$. This collection of points is the *ellipse* determined by the points F_1 and F_2 and the constant k. In the context of Figure 2.6, notice that P is on the ellipse precisely if the lengths of the solid segment from P and the dashed segment from P add to k. The points F_1 and F_2 are the *focal points* of the ellipse. The line through the focal points

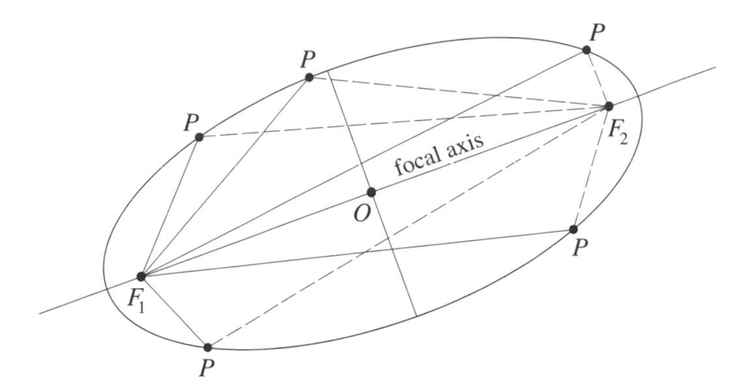

Fig. 2.6

is the *focal axis* of the ellipse. The midpoint O of the segment F_1F_2 is the *center* of the ellipse. Any segment from a point on the ellipse, through the center, to the point on the ellipse on the other side is a *diameter* of the ellipse. One-half the length of the diameter through the two focal points is the *semimajor axis* of the ellipse, and one-half the length of the perpendicular diameter is the *semiminor axis*. Let a be the semimajor axis and $c = OF_1 = OF_2$. The ration $\frac{c}{a}$ is the *eccentricity* ε of the ellipse. Observe that $\varepsilon < 1$. The eccentricity is a measure of how "elliptical" an ellipse is. Consider a circle with center O and radius r. Why is it an ellipse with focal points $O = F_1 = F_2$, and $k = 2r$? Its semimajor and semiminor axes both equal to r, and its eccentricity is 0.

Two basic propositions about the ellipse—again without proofs—and again from the work of Apollonius follow next.

Proposition E1. Let P be any point on an ellipse, and let the segment AB be tangent to the ellipse at P. Then $\angle F_1PA = \angle F_2PB$.

This "reflection property" of the ellipse is illustrated in Figure 2.7. Suppose that the ellipse is a circle.

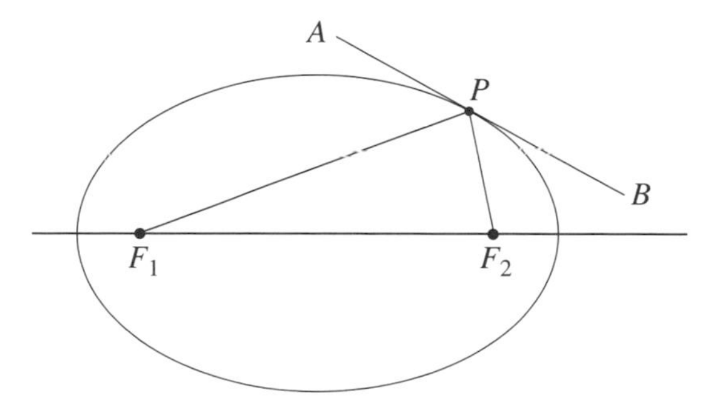

Fig. 2.7

Then the two focal points are the same point (the center of the circle), and the segments from F_1 and F_2 to P are the same radius of the circle. Figure 2.7 and Proposition E1 inform us that the angle that this radius makes with the tangent is equal to 90°. We can conclude for any circle and any radius of the circle, that the tangent to the circle at the point where the radius meets the circle is perpendicular to the radius. A second basic property of the ellipse follows.

Proposition E2. Let PQ be any diameter of an ellipse, and let B be any point on PQ. Choose C such that

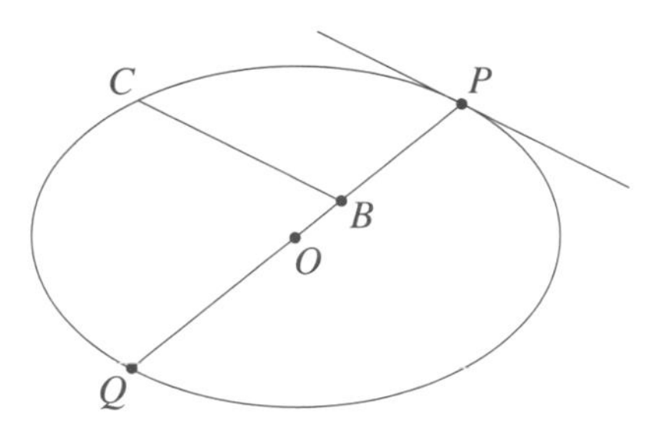

Fig. 2.8

CB is parallel to the tangent at P. Then the ratio $\frac{QB \cdot BP}{CB^2}$ is the same no matter where B is chosen on PQ.

See Figure 2.8. Consider Proposition E2 in the case of a circle. Because OP is a radius, the tangent at P is perpendicular to OP and hence to the diameter QP. Since CB is parallel to the tangent, CB and QP are perpendicular. See Figure 2.9. By an application of a fact from Section 1.3, the triangle ΔPCQ has a

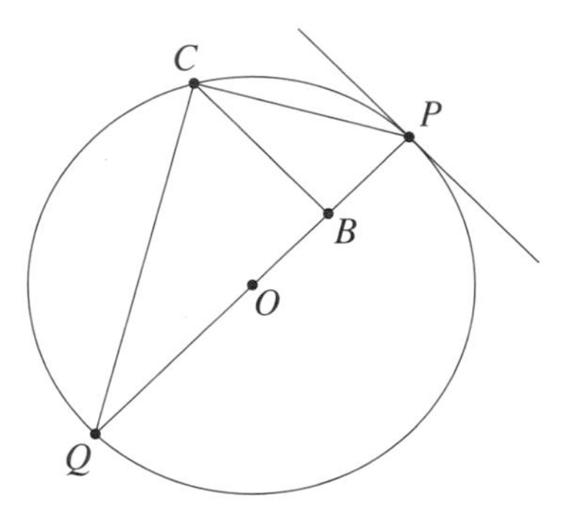

Fig. 2.9

right angle at C. Compare the triangles ΔPCB and ΔPCQ. They both have right angles, and they have $\angle CPB$ in common. It follows that they are similar. By the same argument, ΔQBC is also similar to ΔPCQ. Therefore ΔPCB is similar to ΔQBC, and hence $\frac{PB}{CB} = \frac{CB}{QB}$. So $\frac{QB \cdot PB}{CB^2} = 1$. We have shown that in the case of the circle, the ratio of Proposition E2 is always equal to 1.

Example 2.1. Let's have another look at Propositions P2 and E2. Turn to Figure 2.10a. Choose S so that it is one unit from the focal axis, and let $ES = c$. Now let $VB = x$ and $BC = y$. By an application of Proposition P2, $\frac{y}{c} = \frac{x^2}{1^2}$. Therefore $y = cx^2$. Go next to Figure 2.10b. Let a be the length of PO. Consider the segment from O to the ellipse that is perpendicular to PQ, and let b be its length. With $B = O$, we get that $\frac{QB \cdot BP}{CB^2} = \frac{a \cdot a}{b^2} = \frac{a^2}{b^2}$. Now let B be any point on PQ. Let $x = OB$ and $y = BC$. If B is to the right of O, then $QB = a - x$, $BP = a + x$, and we get that $\frac{QB \cdot BP}{CB^2} = \frac{(a+x)(a-x)}{y^2}$. Check that this equality also holds if B is to the left of O. Proposition E2 tells us that the ratio $\frac{QB \cdot BP}{CB^2}$ is the same for any B, and therefore that $\frac{(a+x)(a-x)}{y^2} = \frac{a^2}{b^2}$. After a little algebra, $a^2 - x^2 = a^2 \cdot \frac{y^2}{b^2}$. So $1 - \frac{x^2}{a^2} = \frac{y^2}{b^2}$, and hence $\frac{x^2}{a^2} + \frac{y^2}{b^2} = 1$.

In the case of the parabola, the equation $y = cx^2$ relates the distance x from the point B to a fixed vertical axis to the distance y from B to a fixed horizontal axis. A similar thing is true for the ellipse. The equation $\frac{x^2}{a^2} + \frac{y^2}{b^2} = 1$ relates the distance x from the point B to a fixed vertical axis to the distance y

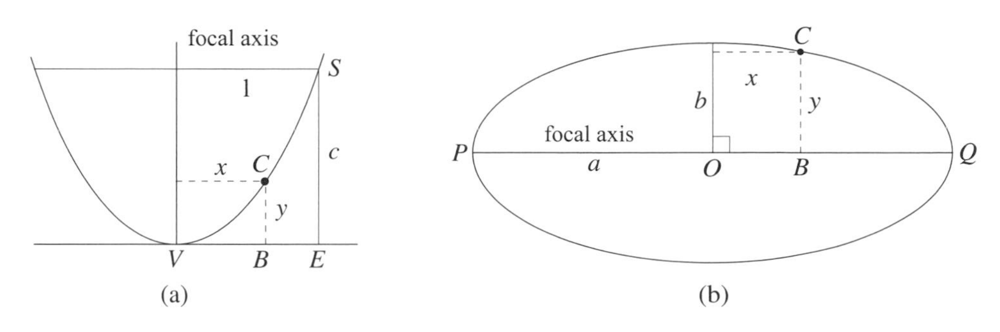

Fig. 2.10

from B to a fixed horizontal axis. If you have seen an xy-coordinate system in action, you will realize that Apollonius had such a system in his grasp! Note, however, that he did so only for the special cases that he considered: the parabola and the ellipse (and the hyperbola).

The hyperbola is defined next. Again, fix two points F_1 and F_2 in the plane and a constant $k > 0$. Suppose that k is less than the distance between F_1 and F_2. The collection of points P such that the absolute value of the difference between PF_1 and PF_2 is equal to k is a *hyperbola*. Refer to Figure 2.11. The points F_1 and F_2 are the *focal points* of the hyperbola. The line determined by the two focal points is

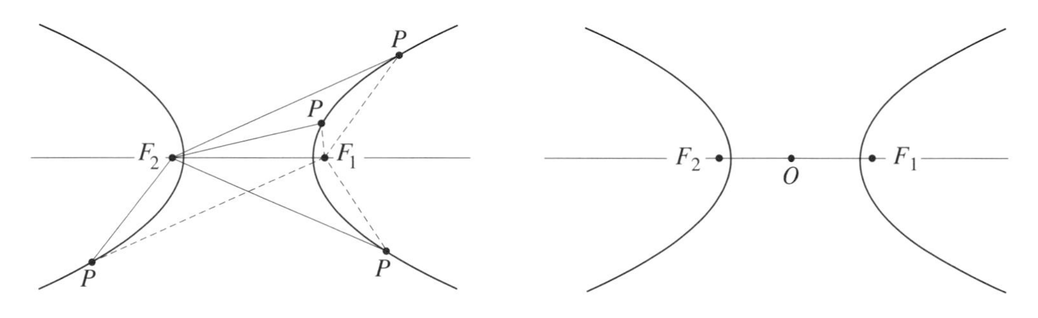

Fig. 2.11

called the *focal axis*. The midpoint O between the two focal points is the *center* of the hyperbola. Hyperbolas will play only a minor role in this text.

The conic sections, studied by the Greeks in the abstract, turned out later to be of remarkable relevance. We will see that they explain the motion of projectiles, the orbits of the planets, and those of natural and artificial satellites. They also determine the geometries of the lenses and mirrors of sophisticated optical instruments. Hyperbolas are relevant as well. Space probes designed to investigate an outer planet are often redirected by allowing them to pass near an inner planet. The gravitational field of the inner planet accelerates the probe and bends its path along a hyperbolic curve before it escapes and continues its mission.

2.2 THE QUESTION OF AREA

We now turn to the concept of area for regions in the plane. If b is the base of a rectangle and h the height, then its area A is defined by A $= bh$. See Figure 2.12. This is the basic situation, and the determination of the area of any other region is ultimately reduced to this case. Consider next a parallelogram with base b and height h. By cutting off the indicated triangle on the left and reattaching it on the right, as shown in Figure 2.13, the given parallelogram is transformed—without change in area—into a rectangle with the same base b and height h. It follows that the area A of the parallelogram is A $= bh$.

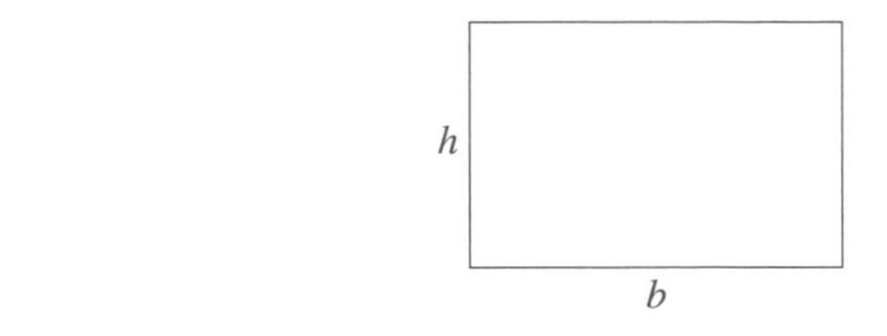

Fig. 2.12

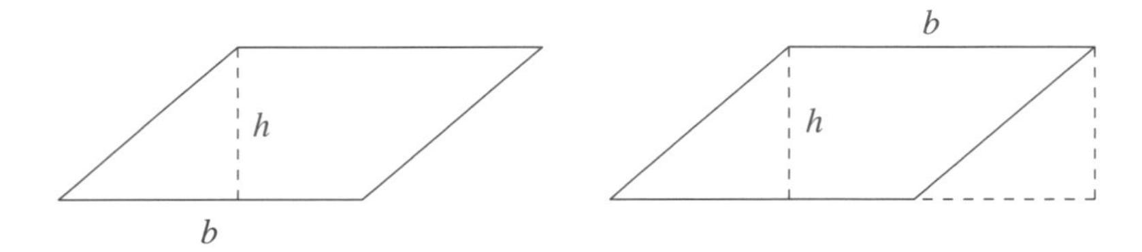

Fig. 2.13

Turn next to the case of a triangle with base b and height h. Take a copy of the same triangle, flip it, and attach it to the original as shown in Figure 2.14. The resulting larger figure is a parallelogram with

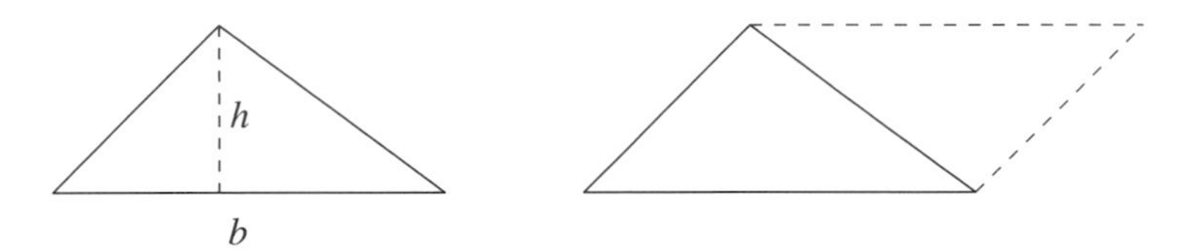

Fig. 2.14

base b and height h. Since this larger area is equal to bh, it follows that the area A of the original triangle is $A = \frac{1}{2}bh$.

We next consider the more subtle situation of the area of a sector of a circle with radius r. A *sector* is a pie-shaped region. Consider the sector determined by the center O and the segments OB and OC, and let θ be the angle of the sector. See Figure 2.15. We will determine the area A of the sector by thinking of it as consisting of lots of very thin triangular wedges. Partition the given sector into, say, n smaller sectors

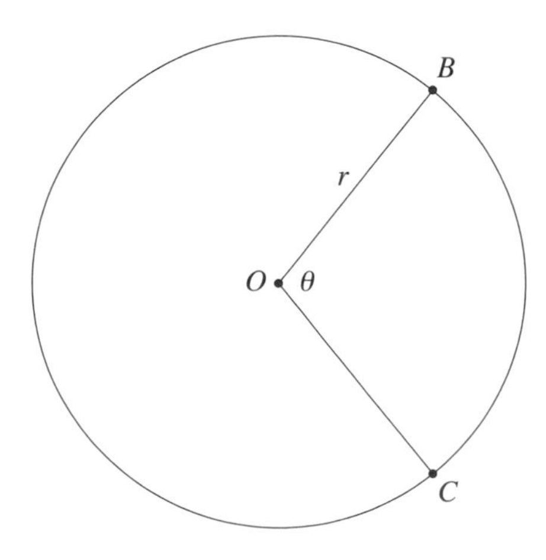

Fig. 2.15

all of the same size. This determines $n - 1$ points between B and C. After consecutively connecting all of these points, we have inscribed an isosceles triangle into each of the n sectors. All of these triangles have the same shape and size. The base as well as the height of the isosceles triangles certainly depend on n, so we denote them by b_n and h_n, respectively. The case $n = 5$ is shown in Figure 2.16. The bases of the triangles—taken together—approximate the arc BC. So $b_n n$ approximates the length of arc BC. The area

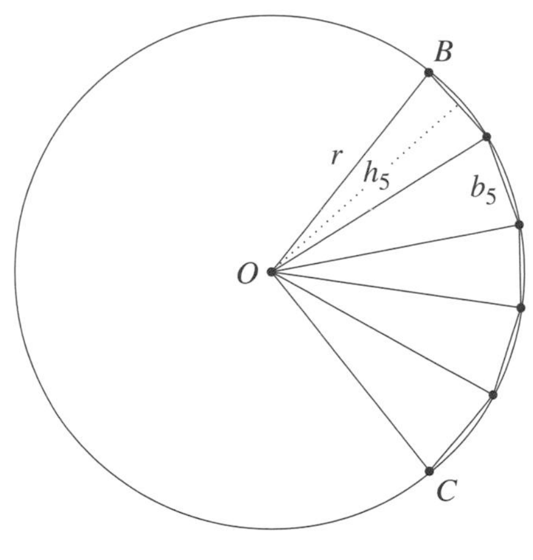

Fig. 2.16

of each inscribed triangle is $\frac{1}{2}h_n b_n$. Taken together, these n triangles approximate the sector BOC. Let $\text{A}_n = (\frac{1}{2}h_n b_n)n = \frac{1}{2}h_n(b_n n)$ be the sum of the areas of all n triangles. So A_n approximates the area of the sector BOC. When n is taken to be sequentially larger and larger, two things happen simultaneously:

(1) The numbers h_n close in on the radius r, and the numbers $b_n n$ on arc BC. As a result, the numbers $\frac{1}{2}h_n(b_n n)$ close in on $\frac{1}{2}r(\text{arc } BC)$.

(2) The numbers A_n close in on the area of the sector BOC.

Because the numbers $\text{A}_n = \frac{1}{2}h_n(b_n n)$ can close in on one thing only, it follows that the area of the sector BOC must be equal to $\frac{1}{2}r(\text{arc } BD)$. In the notation of limits, we can abbreviate what was just asserted by writing

$$\text{Area of sector } BOC = \lim_{n\to\infty} \text{A}_n = \lim_{n\to\infty} \tfrac{1}{2}h_n(b_n n) = \tfrac{1}{2}r(\text{arc } BC).$$

Since we know from Section 1.5 that $\theta = \frac{\text{arc } BD}{r}$ radians, we have now established the fact that the area A of a sector of radius r and angle θ (given in radians) is equal to

$$\text{A} = \tfrac{1}{2}r^2\theta.$$

With $\theta = 2\pi$, this gives us the familiar expression πr^2 for the area of a circle of radius r.

2.3 PLAYING WITH SQUARES

Consider a square with side of length a. Of course, the area A of the square is $\text{A} = a^2$. Nothing else needs to be said. Nevertheless, we will now consider a completely different approach to this area. It may seem completely silly, but it will pay dividends later.

Subdivide the square into four equal squares and color three of them black as shown in Figure 2.17. Denote the area of the resulting black "L" by **B**. (Note that $\mathbf{B} = \frac{3}{4}\text{A}$, but this is irrelevant for now.) Do the same thing with the remaining white square in the upper right corner. Since each of these subdividing

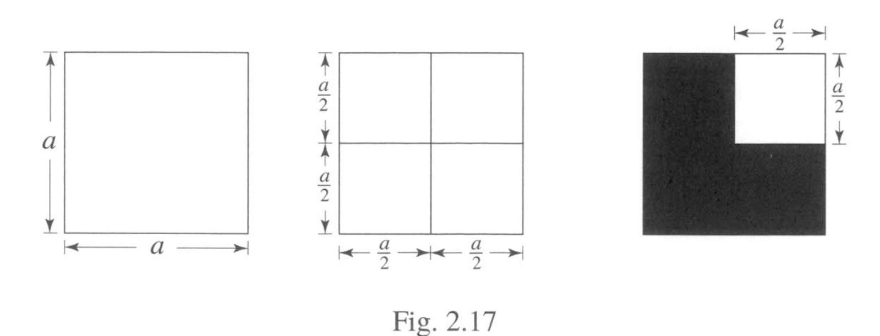

Fig. 2.17

squares has area one-fourth that of the original white square, it follows that this smaller black L has area $\frac{1}{4}\mathbf{B}$. Repeating this construction again produces a black L, this time of area $\frac{1}{4}(\frac{1}{4}\mathbf{B}) = (\frac{1}{4^2})\mathbf{B}$, and so on. See Figure 2.18. Now consider the following steps. Step 0 is the creation of the black area **B**. Step 1 adds the black L of area $\frac{1}{4}\mathbf{B}$ to **B** and produces the black area $\mathbf{B} + \frac{1}{4}\mathbf{B}$. Step 2 adds the next black L and produces the black area $\mathbf{B} + \frac{1}{4}\mathbf{B} + \frac{1}{4^2}\mathbf{B}$. Step 3 adds the next black L of area $\frac{1}{4}(\frac{1}{4^2}\mathbf{B}) = \frac{1}{4^3}\mathbf{B}$, and so on. The results

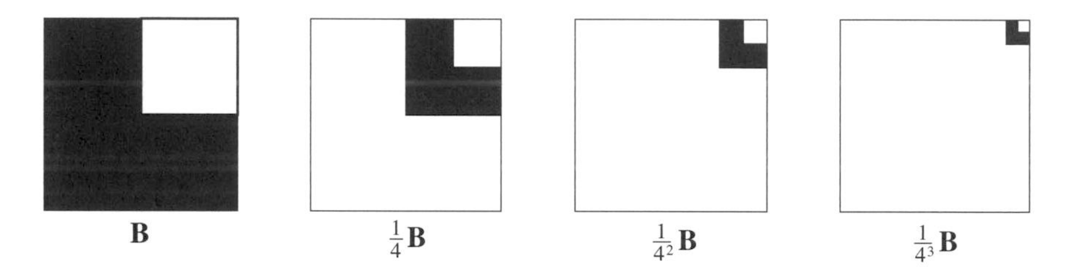

Fig. 2.18

of Steps 0 through 3 are illustrated in Figure 2.19. They produce the black areas

$$\mathbf{B}, \quad \mathbf{B} + \tfrac{1}{4}\mathbf{B}, \quad \mathbf{B} + \tfrac{1}{4}\mathbf{B} + \tfrac{1}{4^2}\mathbf{B}, \quad \mathbf{B} + \tfrac{1}{4}\mathbf{B} + \tfrac{1}{4^2}\mathbf{B} + \tfrac{1}{4^3}\mathbf{B}.$$

This construction can be continued indefinitely. Make note of the emerging pattern and observe that for any number n (possibly very large), Step n creates the black area

$$\mathbf{B} + \tfrac{1}{4}\mathbf{B} + \tfrac{1}{4^2}\mathbf{B} + \tfrac{1}{4^3}\mathbf{B} + \cdots + \tfrac{1}{4^n}\mathbf{B}.$$

Each step produces an approximation of the area A of the original square. The area of the small white square that remains in the upper right corner is the error of the approximation. Let's compute these errors. After Step 0, the area of the small white square is $\frac{1}{4}$A. After Step 1, it is one-fourth of this, or $\frac{1}{4^2}$A. After Step 2, it is one-fourth that of Step 1, or $\frac{1}{4^3}$A. And after Step 3, it is $\frac{1}{4^4}$A. Continue the pattern, and observe that the area of the white square after Step n is $\frac{1}{4^{n+1}}$A. For example, the error after Step 14 is $\frac{1}{4^{15}}$ (and hence

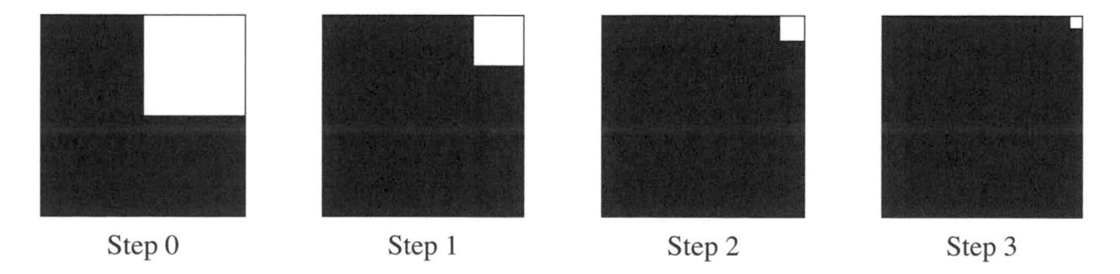

Fig. 2.19

less than 0.000000001) of the area A. Since $\frac{1}{4^{n+2}}$ is less than $\frac{1}{4^{n+1}}$, it follows that the error of each step is less than the preceding error. In summary, for any n, Step n produces the partition

$$\mathrm{A} = (\mathbf{B} + \tfrac{1}{4}\mathbf{B} + \tfrac{1}{4^2}\mathbf{B} + \cdots + \tfrac{1}{4^n}\mathbf{B}) + \tfrac{1}{4^{n+1}}\mathrm{A}$$

of the area of the original square into a black area and a white area. The black area is an approximation of A, and the white area $\frac{1}{4^{n+1}}$A is the error of the approximation. The approximation can be made as accurate as we need to have it. Simply choose n large enough so that the error $\frac{1}{4^{n+1}}$A is small enough!

In the present case of the square, all this is rather superfluous because we already know that its area is $\mathrm{A} = a^2$. The point is, however, that this strategy can be applied to compute the areas of regions that are not known beforehand. We will see shortly how Archimedes used it to compute the area of a parabolic section.

We conclude with a consequence of our analysis. By Figure 2.18, $\mathbf{B} = \frac{3}{4}\mathrm{A}$, so that $\mathrm{A} = \frac{4}{3}\mathbf{B}$. Substituting this into the preceding equation for A gives

$$\tfrac{4}{3}\mathbf{B} = (\mathbf{B} + \tfrac{1}{4}\mathbf{B} + \tfrac{1}{4^2}\mathbf{B} + \cdots + \tfrac{1}{4^n}\mathbf{B}) + (\tfrac{1}{4^{n+1}} \cdot \tfrac{4}{3}\mathbf{B}).$$

After canceling $\mathbf{B}$, we get $\frac{4}{3} = 1 + \frac{1}{4} + \frac{1}{4^2} + \cdots + \frac{1}{4^n} + \frac{1}{4^n} \cdot \frac{1}{3}$. So for any positive integer n,

$$1 + \tfrac{1}{4} + (\tfrac{1}{4})^2 + \cdots + (\tfrac{1}{4})^n + \tfrac{1}{3}(\tfrac{1}{4})^n = \tfrac{4}{3}.$$

This is an identity established by Archimedes. Note that $1 + \frac{1}{4} + (\frac{1}{4})^2 + \cdots + (\frac{1}{4})^n = \frac{4}{3} - \frac{1}{3}(\frac{1}{4})^n$. Note also that for larger and larger n, the term $\frac{1}{3}(\frac{1}{4})^n$ gets closer and closer to 0. Therefore the numbers

$$\begin{gathered} 1 + \tfrac{1}{4} \\ 1 + \tfrac{1}{4} + (\tfrac{1}{4})^2 \\ 1 + \tfrac{1}{4} + (\tfrac{1}{4})^2 + (\tfrac{1}{4})^3 \\ \vdots \\ 1 + \tfrac{1}{4} + (\tfrac{1}{4})^2 + \cdots + (\tfrac{1}{4})^n \\ \vdots \end{gathered}$$

close in on $\frac{4}{3}$. We abbreviate this observation by writing

$$\lim_{n\to\infty}\left(1 + \tfrac{1}{4} + (\tfrac{1}{4})^2 + \cdots + (\tfrac{1}{4})^n\right) = \tfrac{4}{3}.$$

Let's rewrite the preceding sum more compactly as

$$1 + \tfrac{1}{4} + (\tfrac{1}{4})^2 + \cdots + (\tfrac{1}{4})^n = \sum_{i=0}^{n} (\tfrac{1}{4})^i.$$

This "sigma notation" (the letter Σ is the capital Greek sigma, the equivalent of "*S*" suggesting "sum") was not used until the Swiss mathematician Leonhard Euler introduced it in the 18th century. The notation works from the bottom up. So $i = 0$ is the first i. Plug it into $(\frac{1}{4})^i$ to get $(\frac{1}{4})^0 = 1$ and put in a +. Next, put in what you get by letting $i = 1$—namely, $(\frac{1}{4})^1$—and put in another +. Continue the pattern until the last i—namely, $i = n$—is reached. The resulting sum is what the sigma notation stands for. With this shorthand notation, the earlier limit equation can be rewritten $\lim_{n\to\infty} \sum_{i=0}^{n} (\frac{1}{4})^i = \frac{4}{3}$.

Example 2.2. Following this "sigma" summation strategy for $\sum_{i=3}^{6} 2^i$, we get

$$\sum_{i=3}^{6} 2^i = 2^3 + 2^4 + 2^5 + 2^6 = 8 + 16 + 32 + 64 = 120.$$

2.4 THE AREA OF A PARABOLIC SECTION

We now turn to Archimedes's work *Quadrature of the Parabola*, in which he computes the area of a section of a parabola. Quadrature refers to "squaring," and squaring a given region bounded by one or more curves refers to the classical problem of constructing (in the sense of Euclid with straightedge and compass) a square with the same area as the given region.

Take any parabolic section SVS', where V is the vertex. (Refer back to Figure 2.4a.) Inscribe the triangle $\Delta SVS'$ as shown in Figure 2.20. Archimedes's objective is the proof of the fact that the area of the parabolic section is equal to four-thirds the area of the inscribed triangle. His argument consists of a

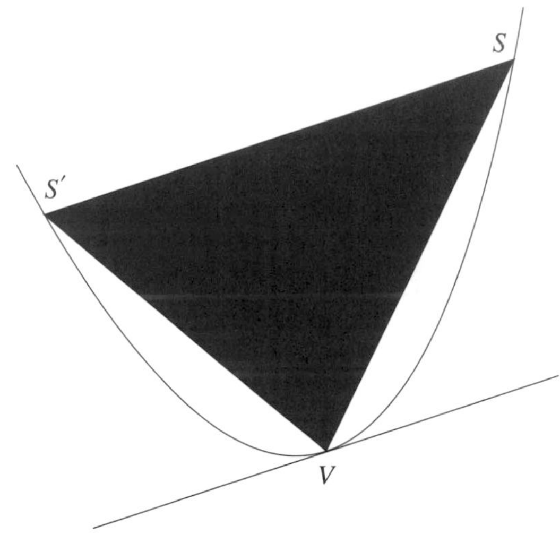

Fig. 2.20

number of steps, each carefully laid out and each rather routine. The argument as a whole, however, is formidable. Let's follow along.

Refer to Figure 2.21. Let A be the midpoint of the segment $S'S$. Draw a line through S parallel to VA,

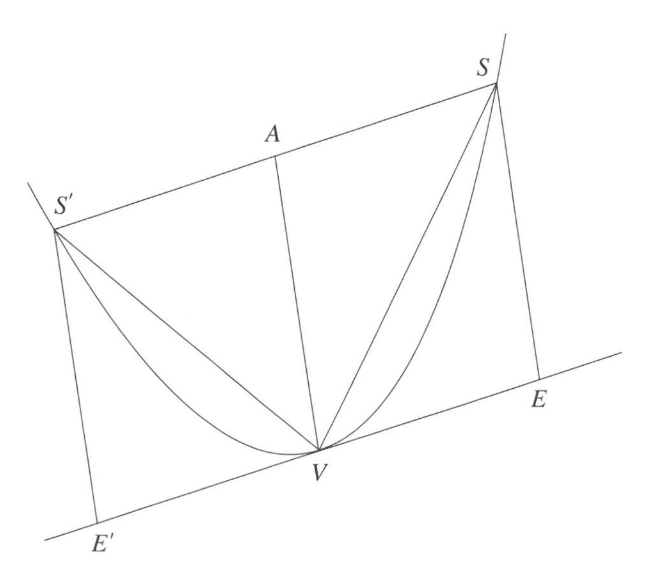

Fig. 2.21

and let its point of intersection with the tangent at V be E. Do a similar thing on the other side of VA. Next, Archimedes takes B to be the midpoint of the segment VE and draws the segment BD parallel to VA. See Figure 2.22. The point C is the intersection of BD with the parabola, and G is the intersection of

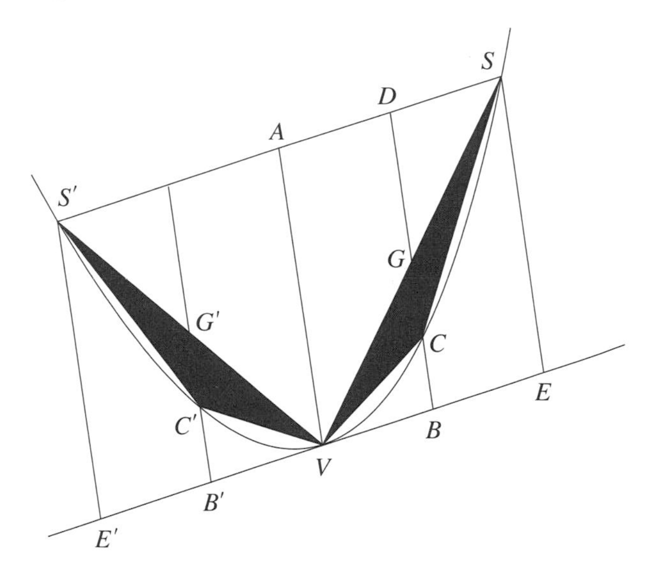

Fig. 2.22

BD and VS. He does the same thing on the other side. The key fact in the argument—concentrate on the right side of Figure 2.22—is as follows.

Proposition A. The area of the triangle ΔVAS is equal to four times the area of the triangle ΔVCS.

When applied to the left side of Figure 2.22, Proposition A tells us that

$$\text{Area}\,\Delta VAS' = 4\,\text{Area}\,\Delta VC'S'.$$

Therefore

$$\text{Area}\,\Delta SVS' = 4(\text{Area}\,\Delta VCS + \text{Area}\,\Delta VC'S').$$

Again in reference to Figure 2.22, Archimedes has verified the following.

Proposition B. $\text{Area}\,\Delta VCS + \text{Area}\,\Delta VC'S' = \frac{1}{4}\,\text{Area}\,\Delta SVS'$.

The brilliance of Archimedes's argument becomes apparent now: since B and B' are the midpoints of VE and VE', respectively, Proposition P3 of Section 2.1 applies to assert that C and C' are the respective vertices of the parabolic sections VCS and $VC'S'$. Consider now the two parabolic sections VCS and $VC'S'$ with their circumscribed parallelograms. See Figure 2.23. Applying Proposition B to each of them shows that the sum of the areas of the two black triangular regions on the right is $\frac{1}{4}\text{Area}\,\Delta VCS$, and that of the two black triangles on the left is $\frac{1}{4}\text{Area}\,\Delta VC'S'$. Since we have already seen that

$$\text{Area}\,\Delta VCS + \text{Area}\,\Delta VC'S' = \tfrac{1}{4}\text{Area}\,\Delta SVS',$$

the four black triangles together have area

$$(\tfrac{1}{4})^2\text{Area}\,\Delta SVS'.$$

Let $\mathbf{B} = \text{Area}\,\Delta SVS'$. Recall that the black area of Figure 2.22 is $\frac{1}{4}\mathbf{B}$, and that of Figure 2.23 is $(\frac{1}{4})^2\mathbf{B}$. Archimedes's procedure of forming triangles can be repeated again and again. The principle is now the

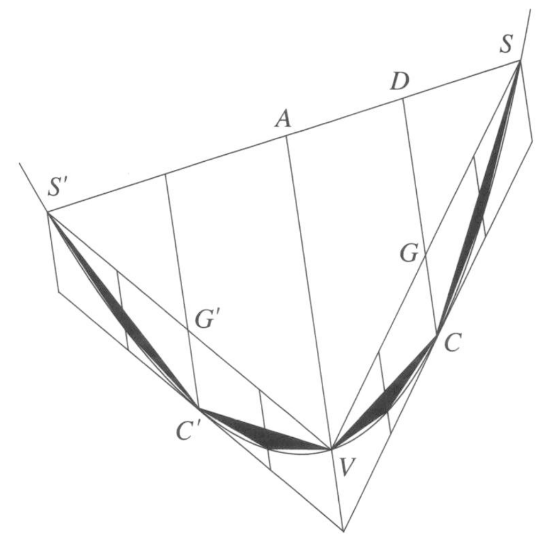

Fig. 2.23

same as that in the approximation of the area of the square in Section 2.3. When the black areas produced by each step are taken together, they provide an approximation of the area of the parabolic section. The first three approximations, obtained by adding the black regions already determined (those of Figures 2.20, 2.22, and 2.23), are shown in Figure 2.24. Observe the pattern, and notice that when the areas of the triangles pro-

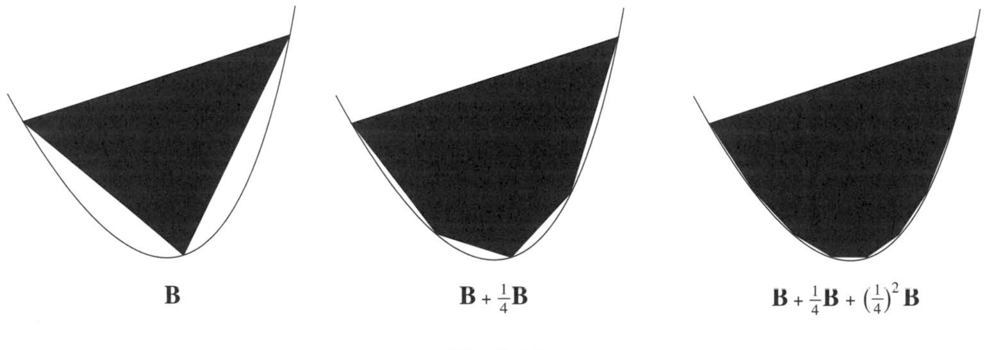

Fig. 2.24

duced after n steps are added together, an approximating inscribed region of area

$$\mathrm{A}_n = \left(1 + \tfrac{1}{4} + (\tfrac{1}{4})^2 + (\tfrac{1}{4})^3 + \cdots + (\tfrac{1}{4})^n\right)\mathbf{B}$$

is obtained. Now let n get larger and larger, and recall from Section 2.3 that

$$\lim_{n\to\infty}\left(1 + \tfrac{1}{4} + (\tfrac{1}{4})^2 + (\tfrac{1}{4})^3 + \cdots + (\tfrac{1}{4})^n\right) = \tfrac{4}{3}.$$

Because the numbers A_n close in on

$$\tfrac{4}{3}\mathbf{B} = \tfrac{4}{3}\text{Area}\,\Delta SVS',$$

it follows that this is the area of the parabolic section SVS'. This fact is known as

Archimedes's theorem. The area of the parabolic section SVS' is equal to four-thirds the area of the inscribed triangle.

To complete the proof of Archimedes's theorem, we still need to verify Proposition A. Figure 2.25 is taken from Figure 2.22. The key step is to show that the area of ΔGCV is equal to one-half of the area of ΔDGV and that the area of ΔGCS is equal to one-half of the area of ΔDGS. By putting the pieces together, it follows that the area of ΔVCS is equal to one-half of the area of ΔDVS. But the area of ΔDVA is equal to the area of ΔDVS (because the triangles have equal bases and heights). Therefore the area of

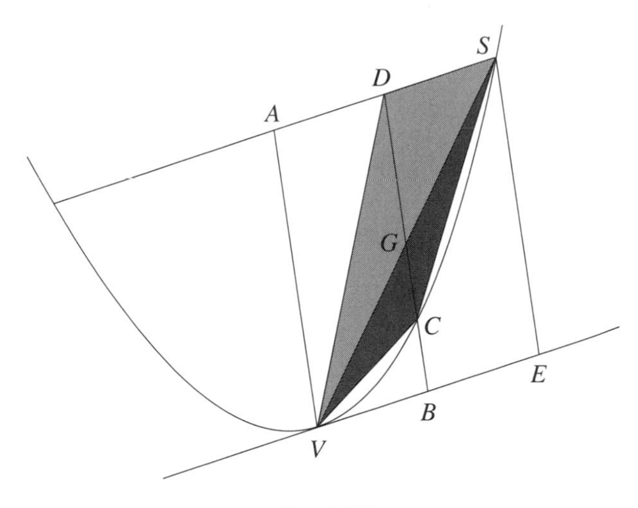

Fig. 2.25

ΔVCS is equal to one-fourth that of ΔVAS. So Area ΔVAS = 4Area ΔVCS, as Proposition A asserts.

It remains to demonstrate that the area of ΔGCV is equal to one-half of the area of ΔDGV and that the area of ΔGCS is equal to one-half of the area of ΔDGS. To show this, Archimedes focuses on the segment $BCGD$. Because B is the midpoint of VE, it follows that $\frac{VB}{VE} = \frac{1}{2}$. So by Proposition P2 of Section 2.1, $\frac{BC}{ES} = \frac{VB^2}{VE^2} = \frac{1}{4}$. Since $ES = BD$, he gets

$$BC = \tfrac{1}{4}BD.$$

Because the triangles ΔVBG and ΔVES are similar, $\frac{BG}{ES} = \frac{VB}{VE} = \frac{1}{2}$. Since $ES = BD$, it follows that $BG = \frac{1}{2}BD$. So G is the midpoint of BD, and hence $BG = GD$. Because $BG = \frac{1}{2}BD$, it follows that $\frac{1}{2}BG = \frac{1}{4}BD$. But $\frac{1}{4}BD = BC$, so $BC = \frac{1}{2}BG$. Therefore C is the midpoint of BG, and $CG = BC$. By a substitution,

$$CG = \tfrac{1}{2}GD.$$

Figure 2.26 consists of two figures extracted from Figure 2.25. Concentrate on Figure 2.26a. Note that

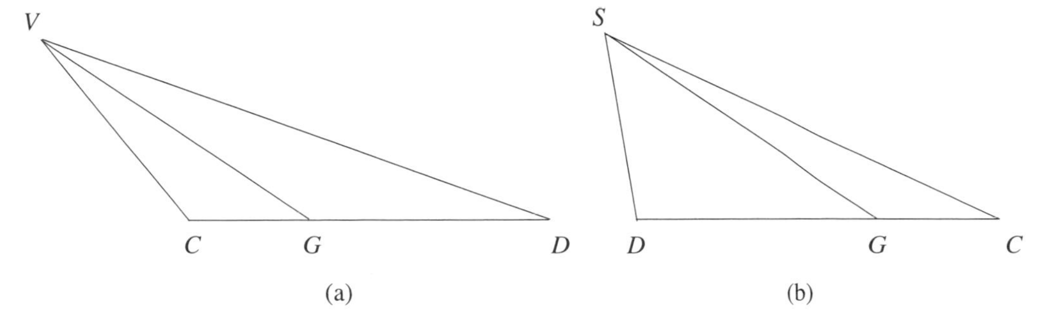

Fig. 2.26

the two triangles ΔGCV and ΔDGV have the same height. Since $CG = \frac{1}{2}GD$, it follows that

$$\text{Area}\,\Delta DGV = 2\text{Area}\,\Delta GCV.$$

Concentrating on Figure 2.26b, we get in the same way that Area $\Delta DGS = 2$Area ΔGCS.

We have witnessed a remarkable argument. Its mathematical precision is decidedly superior to most of what would follow over the next 1800 years! In its last step (the "limit" step), the proof you saw differs from Archimedes's original. The Greeks did not consider (or perhaps, more accurately, did not accept as legitimate) arguments that used "infinite" processes. Instead, Archimedes lets A be the area of the parabolic section SVS' and $A' = \frac{4}{3}$Area $\Delta SVS'$, and demonstrates the equality $A = A'$ by showing that $A > A'$ and $A < A'$ are both impossible. About 2000 years later, arguments involving infinite processes become the heart and soul of integral calculus, but mathematicians wrestled with their legitimacy until the 19th century.

2.5 THE METHOD OF ARCHIMEDES

Historians of science believe that Archimedes knew the area formula for the parabolic section before he devised the rigorous proof described above. They raise the question as to how Archimedes might first have discovered this result. It was thought that a certain treatise of Archimedes called *Method* might have provided the answer. There were brief references to the *Method* in the transmitted literature, but the work itself seemed not to have survived. In the early 1900s, a Danish historian came across a catalogue of the holdings of the libraries of the Greek Orthodox Church. The catalogue gave a brief description of a document containing—in two superimposed layers—an older manuscript overwritten by a later one. The partially washed out older layer was described to be mathematical, the newer as religious. With a heightened sense of interest, the historian undertook the long journey to Constantinople (now known by its Turkish name Istanbul) in the summer of 1906 and found a volume consisting of about 200 leaves of parchment. It contained Greek Orthodox prayers from the 13th century, and indeed most of the leaves showed, at right angles to it, a mathematical manuscript in 10th-century notation underneath! Some of it was readable. When he studied it, he realized that one of the texts it included could only be the *Method* of Archimedes. He took photographs and produced a transcription. This had been an important find. It gave insights into the method that Archimedes had used to discover facts about various areas and volumes, which he later proved in other ways.

We turn to the *Method* and its discussion of the area of a parabolic section. Let's begin by recalling Archimedes's law of the lever. A *lever* consists of a rigid arm and a fixed point around which the arm can pivot freely. This point is the *fulcrum*. In Figure 2.27, the arm is represented by the segment and the fulcrum by the tip of the triangle positioned at F. Archimedes's law of the lever says the following: if two weights w_1 and w_2 are placed on the lever on opposite sides of the fulcrum, at the respective distances d_1 and d_2 from the fulcrum, in such a way that $w_1d_1 = w_2d_2$, then the system balances; i.e., it is in *stable equilibrium* with the lever in any position, not only the horizontal. (It is assumed that the system is balanced before the weights are placed on it.) Notice that a great weight w_1 can be balanced by a small

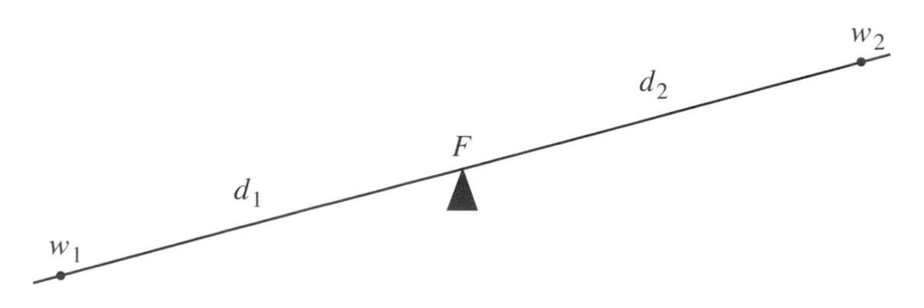

Fig. 2.27

weight w_2 provided that d_2 is large enough. This is why Archimedes is to have said in reference to the lever, "Give me a place to stand on and I will move the Earth." Suppose, for example, that $d_1 = 1$, $w_1 = 3000$, and that $w_2 = 20$. What does d_2 have to be equal to so that the system is in balance? For equilibrium, $w_1 d_1 = w_2 d_2$. So $3000 = 20 \cdot d_2$ and $d_2 = 150$.

In his work *On the Equilibrium of Planes*, Archimedes applies the law of the lever to determine the *centroid* of a triangle. Let's begin by explaining what the centroid of a region is in intuitive terms. Consider a planar region without holes, and assume it to be made of a thin, flat sheet of a rigid material (of uniform thickness). If you wanted to balance this region on the tip of a finger or—to make this more precise, on the tip of a pin, in such a way that the region remains in stable position—then the point at which the pin needs to be placed is called the *center of mass of the region*. If the region is also homogeneous (no lumps), then the center of mass is the geometric center, or *centroid*, of the region.

Suppose that the region is the triangle ΔABC. Let A' be the midpoint of BC and put in the segment AA'. Refer to Figure 2.28. Use a set of lines parallel to BC and another set of lines parallel to AA' to divide the triangle into a grid of identical parallelograms as shown in the figure. So that we can see what is going on, there are relatively few parallelograms, and they are relatively large. But in the argument of Archimedes that follows, suppose that there are lots of them, all tiny, and that they essentially fill out the

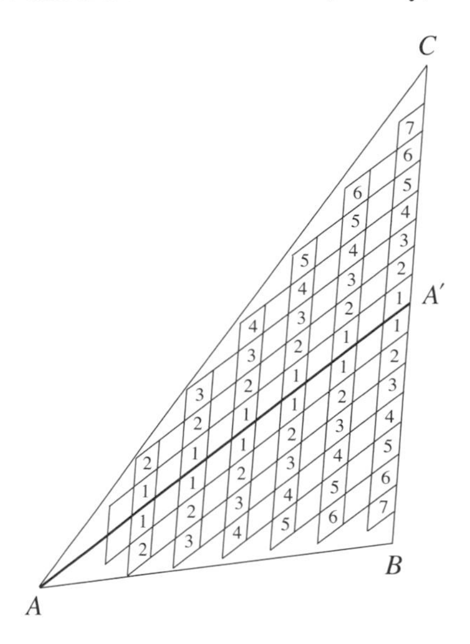

Fig. 2.28

two halves of the triangle ΔABC. Consider the 12 columns of parallelograms of the figure and number the parallelograms in each column in the indicated way. Refer to any of the columns and notice that each parallelogram above the line AA' is matched by one below the line. Matching pairs are given the same number. (For clarity, only the parallelograms of every other column are numbered.) Run a lever through the middle of any of the columns, put the fulcrum on AA', and notice that the parallelograms above

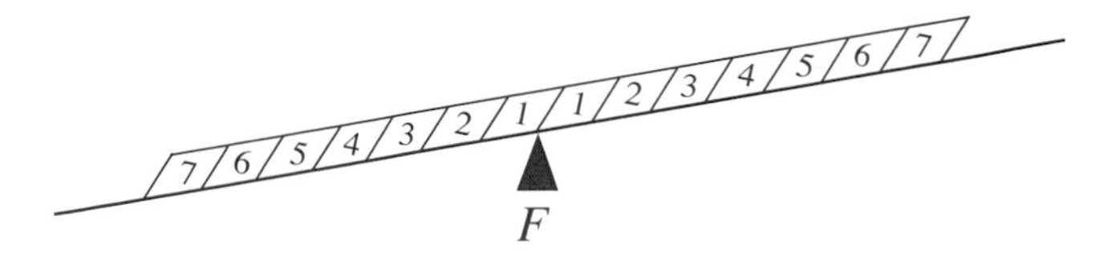

Fig. 2.29

the line AA' balance those below the line. Refer to Figure 2.29. This is so for each of the columns. Consider a knife with a straight cutting edge. Support the triangle ΔABC by placing the knife's cutting edge under it along the segment AA'. It follows from our discussion that the triangle balances on the knife's edge. Therefore its centroid M lies on the segment AA'.

Now let B' be the midpoint of the side AC. The discussion above tells us that M must also lie on BB'. So M is the intersection of AA' and BB'. See Figure 2.30a. We'll now determine precisely where the point M falls along AA' (or BB'). Extend the segment BB' to a point D in such a way that AD is parallel to BC, as shown in Figure 2.30b. The angles of $\Delta BMA'$ and ΔAMD match up (two matching pairs are

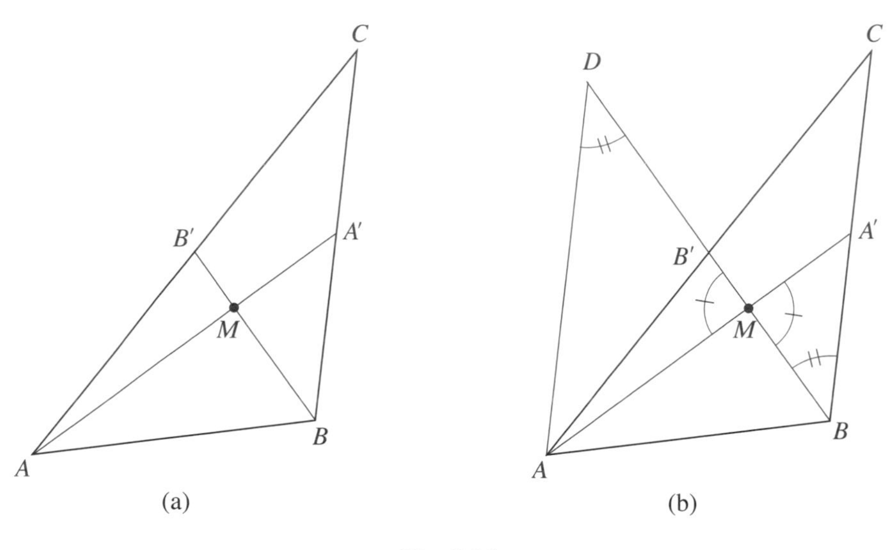

Fig. 2.30

pointed out in the figure), and it follows that $\Delta BMA'$ and ΔAMD are similar. So $\frac{AM}{AD} = \frac{A'M}{A'B}$. The triangles $\Delta BB'C$ and $\Delta DB'A$ are also similar, so that $\frac{BC}{CB'} = \frac{AD}{AB'}$. Since $CB' = AB'$, this means that $AD = BC$. Therefore $\frac{A'M}{AM} = \frac{A'B}{AD} = \frac{A'B}{BC} = \frac{1}{2}$. So $MA' = \frac{1}{2}AM$. It follows that $3MA' = AA'$, and therefore that $MA' = \frac{1}{3}AA'$. Archimedes's determination of the location of the centroid M is now complete.

In order to get to the essence of the *Method* efficiently, we will deal with the parabolic section VCS of Figure 2.22 and prove that its area is equal to $\frac{4}{3}$ times that of the area of the inscribed triangle ΔVCS

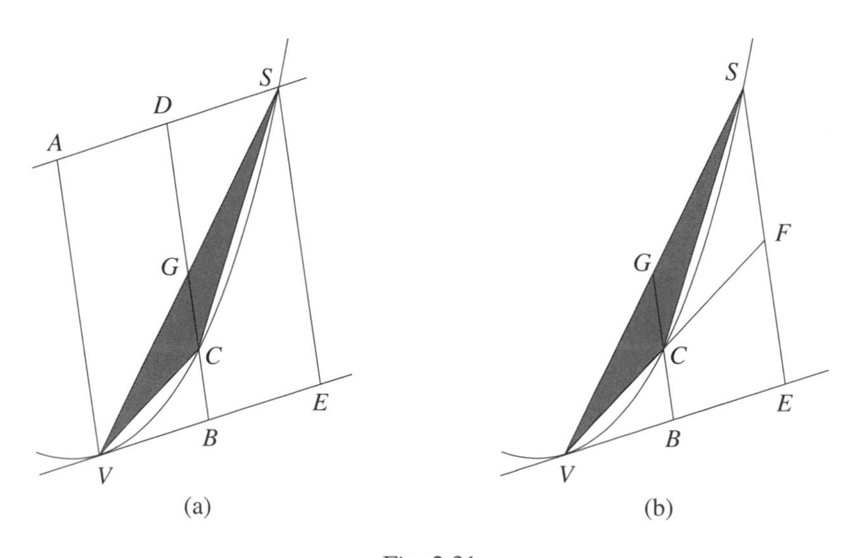

Fig. 2.31

(rather than proving that the area of the parabolic section SVS' of Figure 2.20 is $\frac{4}{3}$ times that of the inscribed triangle $\Delta SVS'$).

To this end, we'll start with Figure 2.31a (taken from Figure 2.22) and some of the facts about this figure already developed. The vertex C of the parabolic section VCS is the intersection of BD with the parabola, and G is the intersection of BD and VS. By Proposition A, Area $\Delta VAS = 4$Area ΔVCS. Since the triangles VAS and VES are identical, their areas are the same. Therefore

$$\text{Area}\,\Delta VES = 4\text{Area}\,\Delta VCS.$$

Next, extend the segment VC to a point F on ES. See Figure 2.31b. Recall that B is the midpoint of VE and C is the midpoint of BG. Since ΔVBC is similar to ΔVEF, we have $\frac{BC}{EF} = \frac{VB}{VE}$, and since ΔVBG is similar to ΔVES, $\frac{BG}{ES} = \frac{VB}{VE}$. It follows that $\frac{BC}{EF} = \frac{BG}{ES}$, and hence that $\frac{EF}{ES} = \frac{BC}{BG} = \frac{1}{2}$. Therefore F is the midpoint of ES.

Let X be an arbitrary point on VE, and draw XZ parallel to ES. See Figure 2.32. Let N and Y be the indicated points of intersection. Archimedes next proceeded to the important equality

$$XZ \cdot NF = YZ \cdot VF.$$

By similar triangles, $\frac{VN}{VX} = \frac{VF}{VE}$, and therefore $VN = \frac{VX \cdot VF}{VE}$. Also by similar triangles, $\frac{XZ}{VX} = \frac{ES}{VE}$, and hence

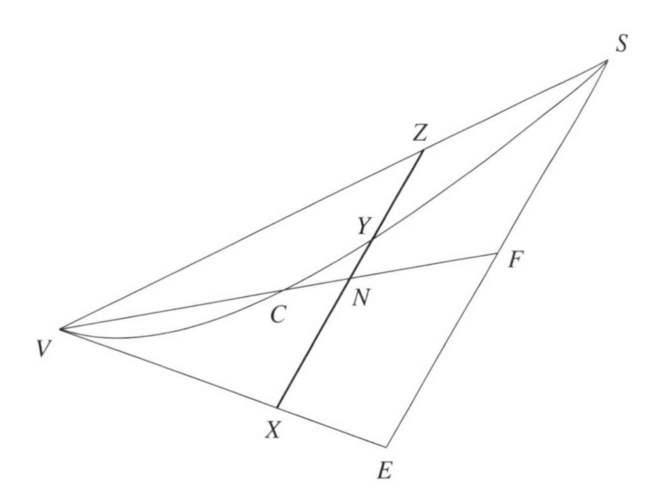

Fig. 2.32

$XZ = \frac{VX \cdot ES}{VE}$. By an application of Proposition P2 of Section 2.1, $\frac{XY}{ES} = \frac{VX^2}{VE^2}$. By combining these equalities,

$$\begin{aligned} VN \cdot XZ &= \left(\frac{VX \cdot VF}{VE}\right)\left(\frac{VX \cdot ES}{VE}\right) \\ &= \frac{VX^2}{VE^2} \cdot VF \cdot ES = \frac{XY}{ES} \cdot VF \cdot ES \\ &= XY \cdot VF. \end{aligned}$$

So by Figure 2.32, $(VF - NF) \cdot XZ = VN \cdot XZ = XY \cdot VF = (XZ - YZ) \cdot VF$. After multiplying out and subtracting, Archimedes gets $-XZ \cdot NF = -YZ \cdot VF$. The equality $XZ \cdot NF = YZ \cdot VF$ is therefore verified.

We have arrived at the ingenious aspect of the *Method*. This is the "mechanical" comparison of the area of the parabolic section VCS with that of the triangle ΔVES, by regarding both to be made of a uniform, rigid material, slicing each into very thin strips, and comparing their weights by suspending them at opposite ends of a lever.

Extend the segment VF of Figure 2.32 to a point K such that $FK = VF$, and consider the segment VFK to be a lever with fulcrum at F. Refer to Figure 2.33. Now think of the triangle ΔVES as sliced up

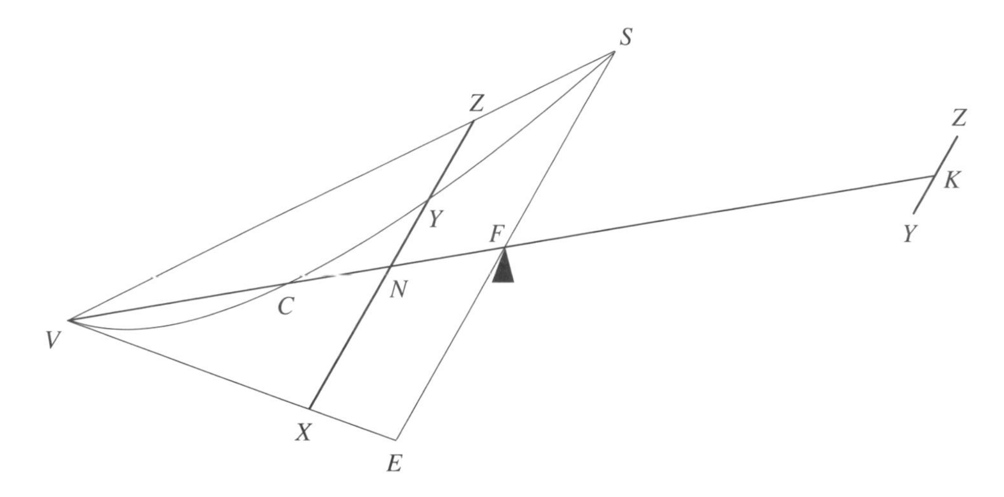

Fig. 2.33

into very thin strips, all of the same width d as shown in Figure 2.34a. For every strip XZ of ΔVES, there is the corresponding strip YZ of the parabolic section VCS. See Figure 2.33. In this way, the parabolic section is also sliced into strips. Refer to Figure 2.34b. Choose a unit of length and a unit of weight in such

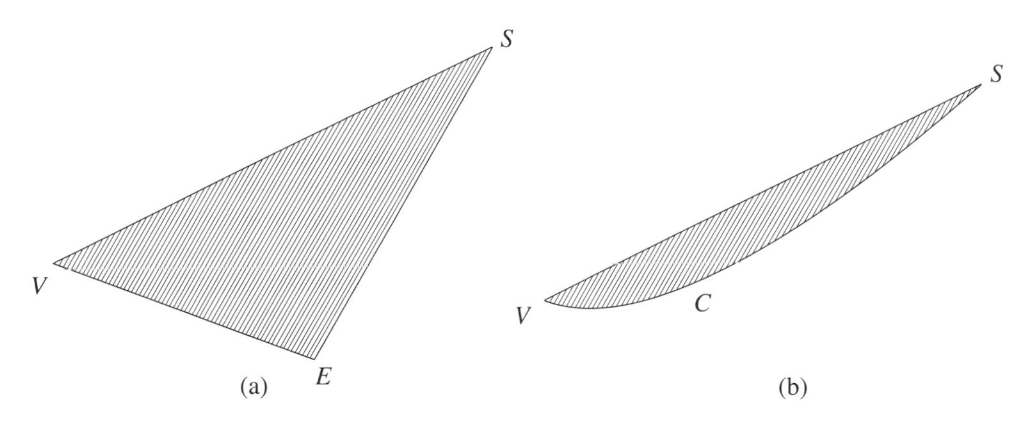

Fig. 2.34

a way that 1 unit of area of the homogeneous material weighs 1 unit. So the weight of any strip is numerically equal to d times its length. The equality $XZ \cdot NF = YZ \cdot FK$, Figure 2.33, and the law of the lever together tell us that each strip XZ of ΔVES is balanced by the corresponding strip YZ of the parabolic sector VCS placed at the point K. Therefore the entire triangle ΔVES in its location in Figure 2.33 is balanced by all the strips YZ of the parabolic sector VCS placed at K. It follows that the entire weight of ΔVES placed at its centroid M balanced the entire weight of the parabolic sector VCS placed at K. Recall that $MF = \frac{1}{3}VF$. The fact that the units were chosen in such a way that area is equal to weight tells us that

$$(\text{Area } \Delta VES) \times \tfrac{1}{3}VF = (\text{Area of parabolic section } VCS) \times FK,$$

as illustrated in Figure 2.35. Recall that Area ΔVES = 4Area ΔVCS and that $FK = VF$. So

$$\tfrac{4}{3} \text{ Area } \Delta VCS = \text{Area of parabolic section } VCS.$$

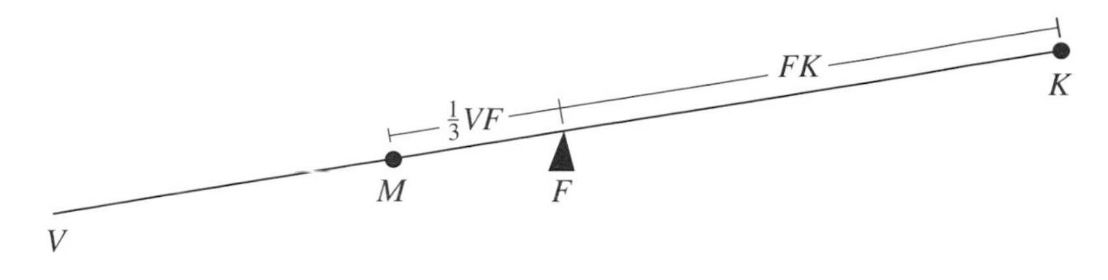

Fig. 2.35

This completes the proof of the area formula that Archimedes presents in the *Method*.

The story of the *Method* has an interesting final chapter. We'll pursue it in the upcoming Problems and Projects section.

We have had a glimpse of the mathematical genius of Archimedes. The Problem and Projects section will tell us more. We can now understand what the historian Plutarch meant when he praised "the precision and cogency of the methods and means of proof" of Archimedes's studies. Several of the basic concepts and strategies of calculus—such as tangent lines to curves, the determination of areas as sums of infinite aggregates, and computations by slicing regions into thin strips—are already found in the works of Archimedes.

2.6 PROBLEMS AND PROJECTS

All the problems that follow are related to the concerns of this chapter. As the comments will indicate, only some of them were explicitly considered by Archimedes. There is no question, however, that he could have solved them all.

2A. About Parabolas and Ellipses. To get a sense of the conic sections of the Greeks, get a flashlight. Direct it against a flat wall. The wall is a physical plane that cuts the cone of light that the flashlight emits. If the flashlight is directed so that the tube housing the batteries is perpendicular to the wall, a circular area is illuminated. If the flashlight is good, then the circles will be clean and distinct. Tilt the flashlight away from the perpendicular position, and an ellipse will appear. Tilt more, but carefully, until the beam farthest from the wall no longer hits the wall. At the moment this occurs, you will get a parabola. (What curves appear as the flashlight begins to be directed away from the wall?)

2.1. Figure 2.36a shows a parabola. The point V is the intersection of the parabola with its focal axis,

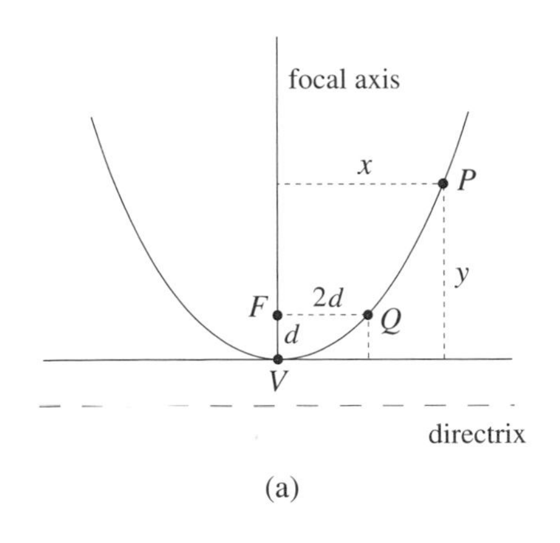

(a)

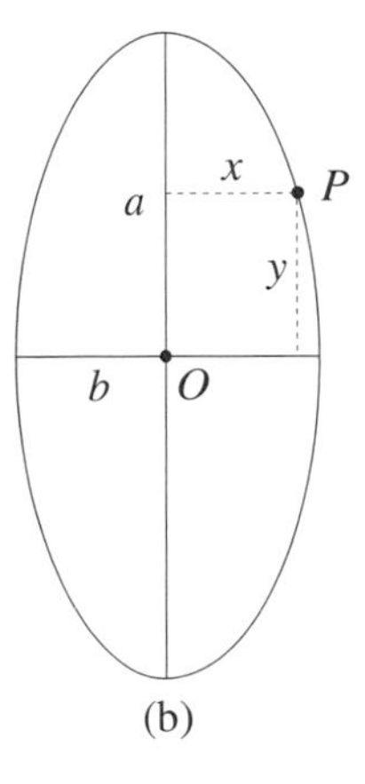

(b)

Fig. 2.36

and d is the distance from V to the focal point F.

i. Show that the point Q specified in the figure is on the parabola.

ii. Let P be any point on the parabola, let x be the distance from P to the axis, and let y be the distance from P to the tangent at V. Use one of the propositions of Section 2.1 to show that $y = \frac{1}{4d}x^2$.

2.2. The ellipse in Figure 2.36b has semimajor axis a and semiminor axis b. Let x and y be the indicated distances. Use one of the propositions of Section 2.1 to show that $\frac{x^2}{b^2} + \frac{y^2}{a^2} = 1$.

2.3. Consider an elliptical mirror. Suppose that there is a light source at one of the focal points F. What do all the light rays that emanate from F and strike the mirror have in common? The same principle with a source of high-intensity sound waves in place of a light source often plays an important role in the medical procedure to remove kidney stones. Describe what this role might be.

2B. Trapezoids and the Pythagorean Theorem. The next two problems consider trapezoids and a presidential proof of the Pythagorean theorem.

2.4. A *trapezoid* is a quadrilateral with two parallel sides. Let a and b be the lengths of the two parallel sides of a trapezoid, and let h be the distance between them. Show that the area of the trapezoid is $\frac{1}{2}(a + b)h$ by making use of Figure 2.37a.

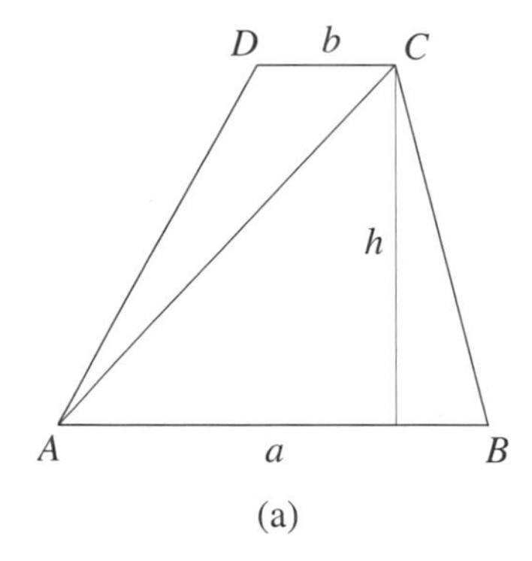

(a)

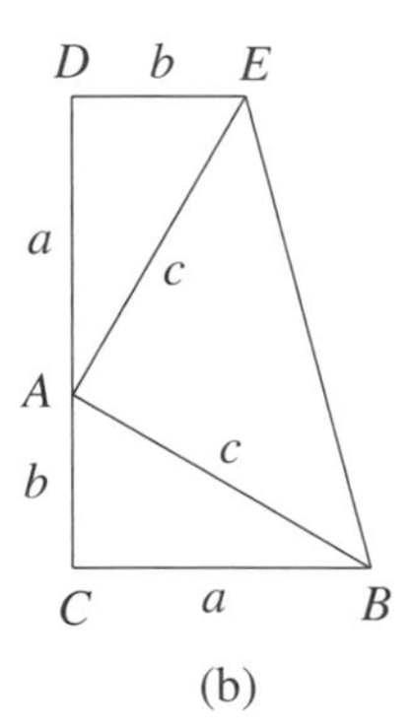

(b)

Fig. 2.37

2.5. Let ΔABC be a right triangle with sides a, b, and c and right angle at C as shown in Figure 2.37b. Extend the side CA to a straight line, and attach a copy ΔADE of the given triangle to the line as indicated. Why is the angle $\angle EAB$ equal to 90°? Compute the area of the quadrilateral $CBED$ in two ways to show that $a^2 + b^2 = c^2$. [This proof of the Pythagorean theorem was devised by President James Garfield. Garfield was a college professor, a major general in the Civil War, and a Republican senator from Ohio before he became president in 1881. He was assassinated in the same year. His proof appeared in the *New England Journal of Education* in 1876.]

2C. Heron and the Area of a Triangle. The mathematician Heron lived and worked in Alexandria in the 1st century A.D. An interesting formula for the area of a triangle bears his name, but it was later determined that it was Archimedes who discovered it 300 years earlier. The formula asserts that if a, b, and c are the sides of a triangle, then its area is equal to

$$\sqrt{s(s-a)(s-b)(s-c)}, \quad \text{where } s = \tfrac{1}{2}(a+b+c).$$

The next several problems outline a later proof of Heron's formula. It combines some of the trigonometry of Section 1.6 with some algebraic maneuvering. Let $T = \Delta ABC$ be a triangle. As it is depicted in Figure 2.38, the angle at A is acute. However, the verification of Heron's formula outlined below is valid without this assumption.

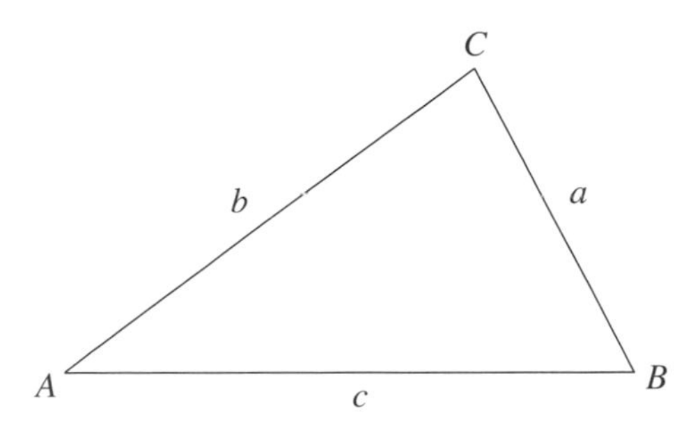

Fig. 2.38

2.6. Let $\alpha = \angle CAB$. Show that area of T is $\frac{1}{2}bc\sin\alpha$.

2.7. Use the law of cosines to verify that area $T = \frac{1}{4}\sqrt{(2bc)^2 - (b^2 + c^2 - a^2)^2}$.

2.8. Apply the identity $x^2 - y^2 = (x + y)(x - y)$ several times to check that

$$\begin{aligned}
\text{area } T &= \tfrac{1}{4}\sqrt{[2bc + (b^2 + c^2 - a^2)][2bc - (b^2 + c^2 - a^2)]} \\
&= \tfrac{1}{4}\sqrt{[(b + c)^2 - a^2)][a^2 - (b - c)^2]} \\
&= \tfrac{1}{4}\sqrt{[(b + c) + a)][(b + c) - a)][a - (b - c)][a + (b - c)]} \\
&= \sqrt{\tfrac{1}{2}[(b + c) + a)] \cdot \tfrac{1}{2}[(b + c) - a)] \cdot \tfrac{1}{2}[a - (b - c)] \cdot \tfrac{1}{2}[a + (b - c)]} \\
&= \sqrt{s(s - a)(s - b)(s - c)}.
\end{aligned}$$

2.9. Find the area of the triangle with sides 5, 7, and 10.

2D. Circles and Related Areas. The next set of problems studies areas defined by circular arcs.

2.10. Depicted in Figure 2.39 are an equilateral triangle and a square. In each case the sides have length 2. All circular arcs all have radius 1 and are centered at the vertices. Find the areas of the two highlighted

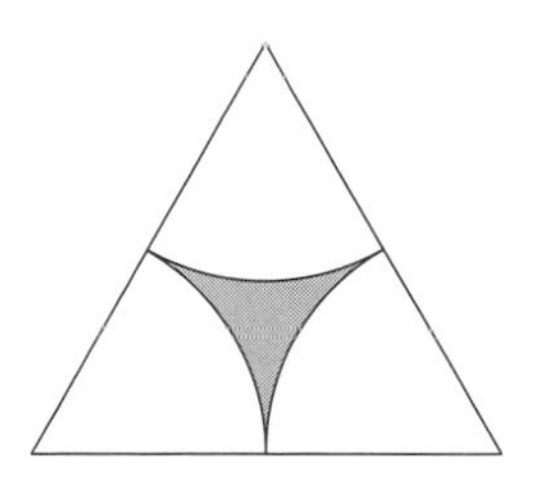

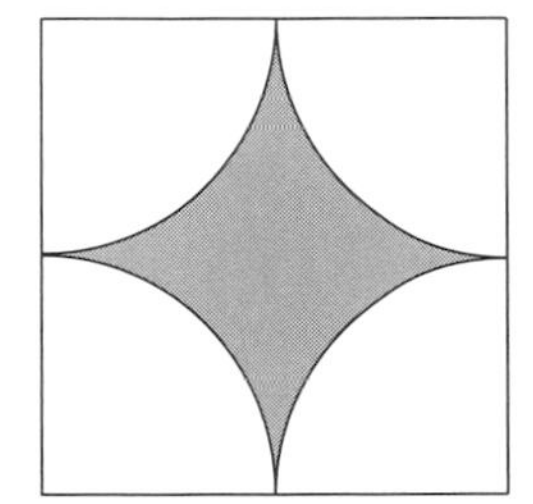

Fig. 2.39

star-shaped regions.

2.11. Show that the shaded section of the circle in Figure 2.40 has area $\frac{1}{2}r^2\theta - r^2 \sin\frac{\theta}{2}\cos\frac{\theta}{2}$. Then use the half-angle formula of Problem 1.26i to show that the area is equal to $\frac{1}{2}r^2(\theta - \sin\theta)$.

2.12. Compute the area of the shaded circular section of Figure 2.40 in the situation where $r = 7$ and $\theta = 50°$. Then compute this area again in the case $r = 5$ and arc AB has length 8.

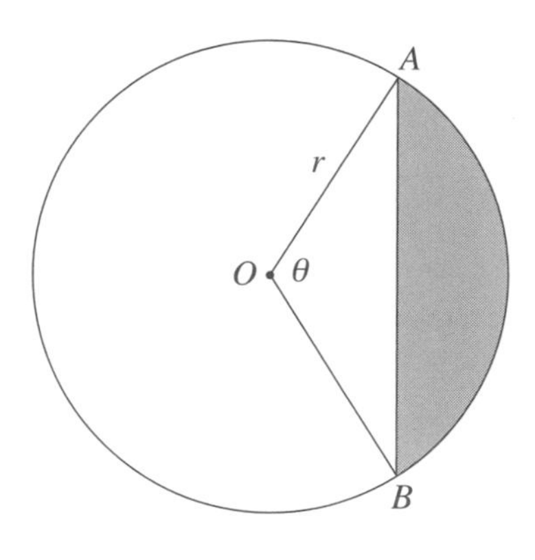

Fig. 2.40

2.13. In Figure 2.41, O is the center of a circle and the radii AO and OB meet at right angles. The second arc connecting A and B is half of the circle with diameter AB and center C. Show that the area of the shaded moon-shaped region bounded by the two circular arcs is equal to the area of the triangle ΔAOB. [Hint: Show that the area of a quarter circle of radius r is equal to one-half the area of a

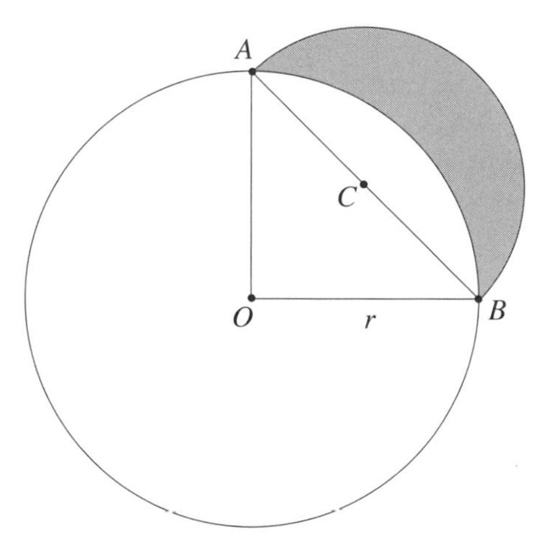

Fig. 2.41

circle of diameter AB.]

Moon-shaped regions such regions such as those of Figures 2.41 and 2.42 are called *lunes* ("luna" is Latin for "moon"). They were studied by the ancient Greeks and later Islamic mathematicians. The famous man of the Renaissance Leonardo Da Vinci was also fascinated by them.

2.14. Verify the conclusion of Problem 2.13 by using the area formula of Problem 2.11.

2.15. Figure 2.42 shows three semicircles with diameters a, b, and c arranged so that their diameters form a triangle. By a result of Section 1.3, the triangle is a right triangle with hypotenuse c. The two lunes that the configuration determines are shaded.

i. Show that the sum of the areas of the two lunes is equal to the area of the right triangle.

ii. Why is the area of the triangle and hence that of lunes greatest when $a = b$?

iii. For what (if any) values of a and b is the area of the triangle smallest?

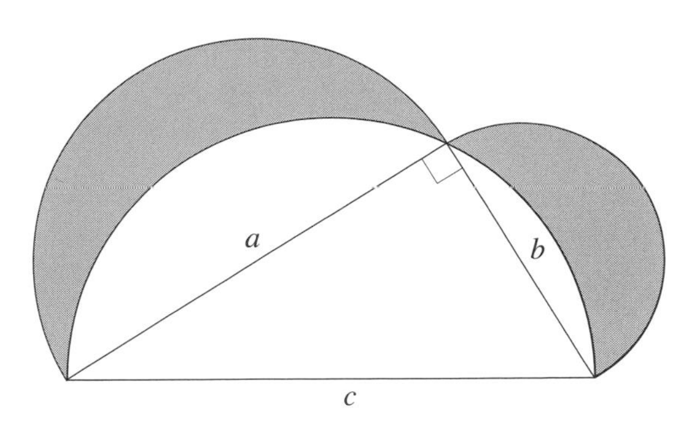

Fig. 2.42

2.16. Consider a point P on the diameter AB of a circle, and let a and b be the lengths AP and PB. Let $y = PC$ be the height of the triangle ΔABC. Refer to Figure 2.43a.

i. Show that $y = \sqrt{ab}$. Do so three times. First, by applying a conclusion of Example 2.1 to the circle, then by using the similarity of the triangles ΔACP and ΔCBP, and finally by applying the Pythagorean theorem to each of the triangles ΔABC, ΔACP, and ΔCBP.

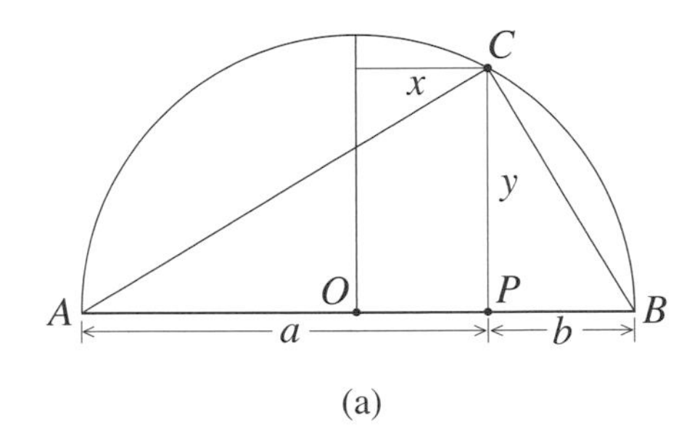

(a)

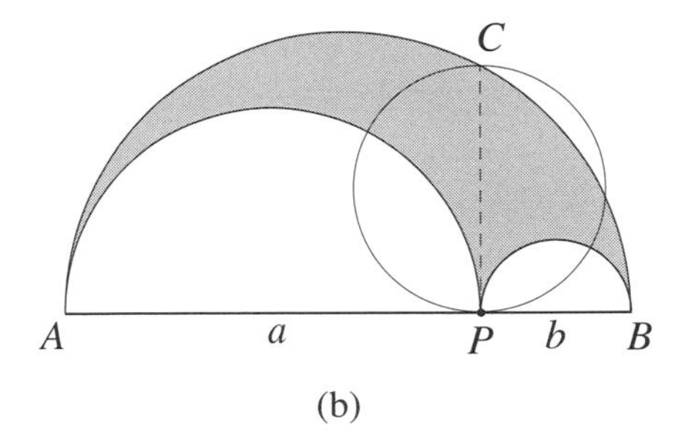

(b)

Fig. 2.43

ii. Conclude that the area of the triangle ΔABC is $\frac{1}{2}\sqrt{ab}(a+b)$.

iii. Refer to Figure 2.43b and show that the shaded area is equal to the area of the circle with diameter CP.

The *Book of Lemmas* is a collection of 13 propositions about the geometry of circles. The work came down to us in transcription from the Arabic and it is often attributed to Archimedes. One of the figures it studies is the shaded region of Figure 2.43b called the *arbelos*. The word comes from the Greek word meaning "shoemaker's knife." (Apparently, the blade of a knife used by ancient Greek shoemakers resembled this shape.)

2.17. Another problem that the *Book of Lemmas* deals with concerns the following geometric figure. Consider the upper half of a circle with diameter AB. See Figure 2.44. The points C and D divide the diameter in such a way that $AC = DB$. Two semicircles with diameters AC and DB are placed above the segment AB, and another semicircle with diameter CD is placed below the segment AB. Let O be the center of the segment AB, and let PQ be the perpendicular to AB through O. Verify the observation of Archimedes that the area of the geometrical shape with boundary determined by the four semicircles is equal to the area of the circle with diameter PQ.

The shape of Figure 2.44 is commonly referred to as a *salinon*. In the English mathematical literature, the word is said to come from the Greek for "salt cellar." In the German literature, it is said to be Greek for "salt barrel." As to what precisely this shape has to do with the storage of salt does not seem to be obvious.

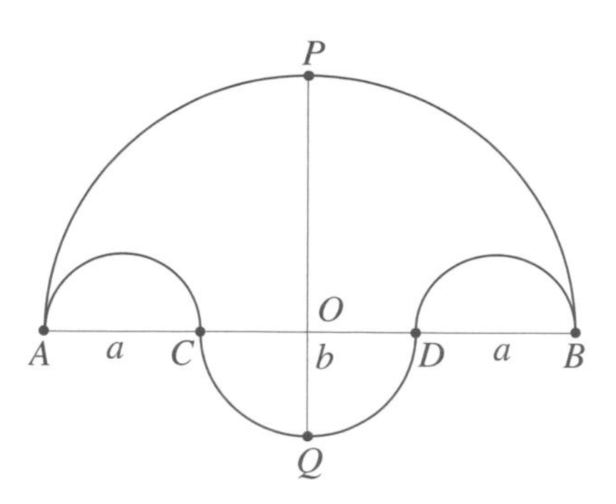

Fig. 2.44

2.18. Consider a semicircle of radius r. Place any square into the semicircle so that the base is on the diameter and one corner is on the circle. Let a be the length of the side of this square. The square is shown on the left in Figure 2.45a. Now place a second square to the right of the first one as shown in the figure, and let b be the length of its side. (To see that such a square exists, start with a smaller square and increase its diagonal until the upper right corner touches the circle). Show that the sum

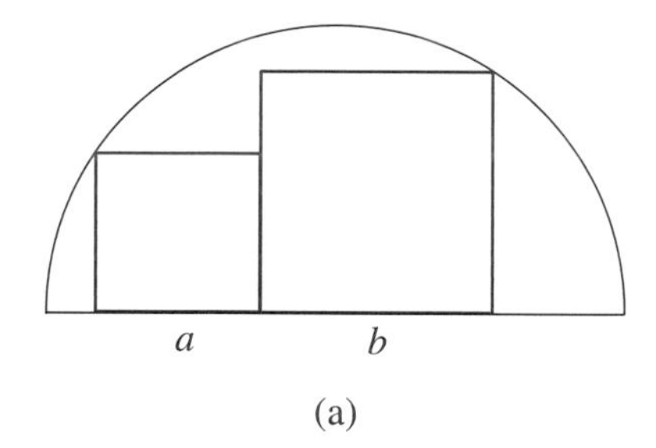

(a)

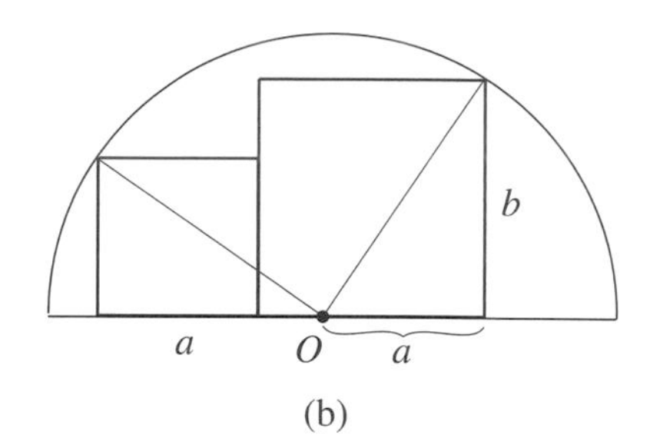

(b)

Fig. 2.45

of the areas $a^2 + b^2$ of the two squares is equal to r^2. [Hint: Place the point O on the diameter by marking off a distance a from the lower right vertex of the square on the right and draw in the two segments from O to the semicircle, as shown in Figure 2.45b. Show that both of these segments have length $\sqrt{a^2 + b^2}$. Conclude that O must be the center of the circle. To do this, suppose that some other point on the diameter is the center.]

2E. Sigma Notation and Areas. The problems in this section focus on areas given by infinite processes.

2.19. Write the sums $1 + 2 + 3 + 4$, $2^2 + 3^2 + 4^2 + 5^2$, $3^3 + 4^4 + 5^5 + 6^6$, and $4^8 + 5^{10} + 6^{12}$ using sigma notation.

2.20. The squares in Figure 2.46 are each of side 1. Continue the indicated pattern to show that

Fig. 2.46

$$\lim_{n\to\infty} \sum_{i=1}^{n} \left(\tfrac{1}{2}\right)^i = 1.$$

2.21. Consider a triangle with base b and height h. Inscribe into the triangle the parallelograms R_1, R_2, and R_2', where the base of R_1 is $\frac{1}{2}b$, and the bases of R_2 and R_2' are $\frac{1}{2}\left(\frac{1}{2}b\right) = \frac{1}{4}b$. Continue the pattern illustrated by Figure 2.47, and fill the triangle with the indicated sequence of parallelograms. Show

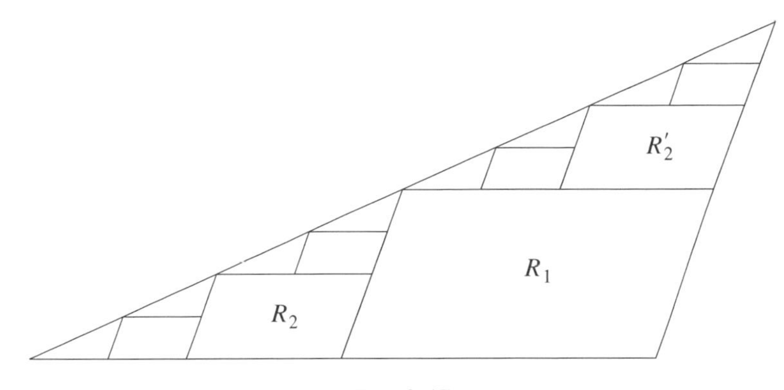

Fig. 2.47

that the area of this triangle is $\frac{1}{2}bh$ by adding up the areas of all the parallelograms. In other words, show that

$$\tfrac{1}{2}\, bh = \text{Area } R_1 + (\text{Area } R_2 + \text{Area } R_2') + \ldots .$$

Draw a similar figure for an acute triangle, and carry out the argument in this case.

2.22. Explain why Figure 2.48 shows that $\frac{1}{4} + \frac{1}{4^2} + \frac{1}{4^3} + \cdots = \frac{1}{3}$. This result is equivalent to Archimedes's

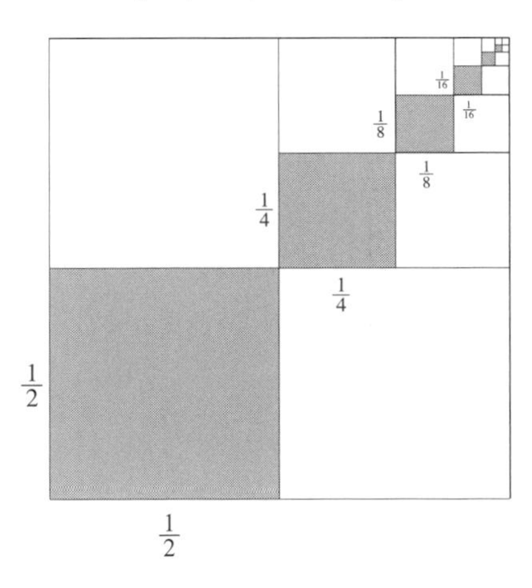

Fig. 2.48

equality $1 + \frac{1}{4} + \frac{1}{4^2} + \frac{1}{4^3} + \cdots = \frac{4}{3}$.

2.23. Let x be any number, and consider the sum $S_n = 1 + x + x^2 + \cdots + x^{n-1}$, where $n \geq 1$ is an integer. Since $xS_n = x + x^2 + \cdots + x^n$, we see that

$$(1 - x)S_n = S_n - xS_n = (1 + x + x^2 + \cdots + x^{n-1}) - (x + x^2 + \cdots + x^{n-1} + x^n) = 1 - x^n.$$

Therefore

$$S_n = \frac{1 - x^n}{1 - x}$$

for any $x \neq 1$. Suppose that $|x| < 1$ and push n to infinity. Observe that x^n goes to 0, so that $\lim_{n\to\infty} S_n = \frac{1}{1-x}$. What happens when $x = 1, -1$, or when $x > 1$?

2.24. Let a and r be any fixed numbers, and consider the infinite sum $a + ar + ar^2 + \cdots$. If $|r| < 1$, then show that this infinite sum is equal to $\frac{a}{1-r}$. [Hint: Apply the result of Problem 2.23.]

2.25. Use the conclusion of Problem 2.23 to show that $\frac{1}{2} + \frac{1}{2^2} + \frac{1}{2^3} + \cdots = 1$ and that $1 + \frac{1}{4} + \frac{1}{4^2} + \frac{1}{4^3} + \cdots = \frac{4}{3}$.

2F. Archimedes's Theorem. The next several problems illustrate Archimedes's formula for the area of the parabolic section.

2.26. Consider a parabolic section where the cut is 7 units long and the perpendicular distance from the cut to the vertex is 4 units. What is the area of the parabolic section?

2.27. A parabola is cut with a straight line. The length of the cut is 5, and the area of the resulting parabolic section is 16. What is the distance from the vertex of the parabolic section to the cut?

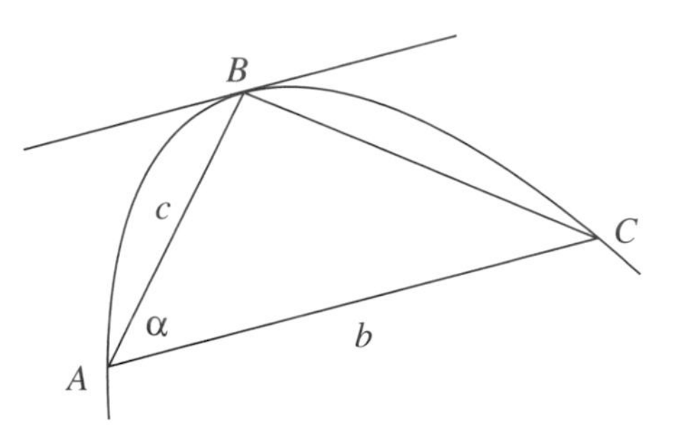

Fig. 2.49

2.28. Figure 2.49 shows a parabola cut by the segment AC. The tangent to the parabola at B is parallel to AC. Show that the area of the parabolic section ABC is $\frac{2}{3}bc\sin\alpha$.

2.29. The distance from the focal point F of a parabola to its directrix is 3. The parabola is cut parallel to the directrix at a distance of 7 units from the directrix. Determine the area of the parabolic section. [Hint: Refer to Figure 2.50, analyze the triangle SFS', and find the length of the cut.]

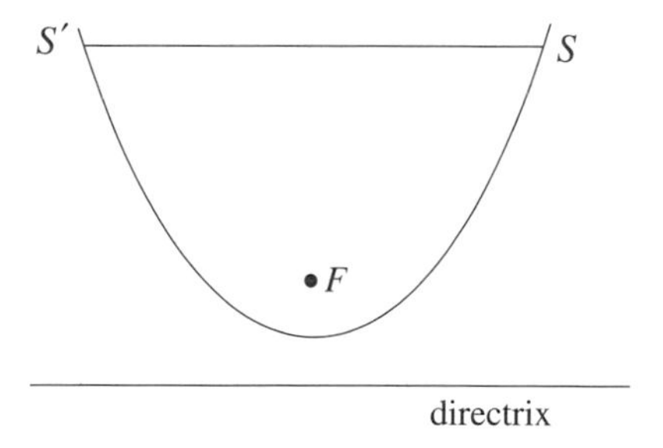

Fig. 2.50

2.30. Figure 2.51 shows a circle with center O and radius r that is cut by the segment SS'. The segments OS and OS' determine the angle θ, and the segment OV bisects this angle. Show that the tangent to the circle at V is parallel to the cut SS'. (Apply the conclusions of Problem 1.9 and Proposition E1 to show that both are perpendicular to OV.) Now use the half-angle formula of Problem 1.26i to show that

i. The area of the circular section SVS' is $\mathrm{A} = \frac{1}{2}\theta r^2 - r^2\sin\frac{\theta}{2}\cos\frac{\theta}{2} = \frac{1}{2}r^2(\theta - \sin\theta)$ and the area of the triangle SVS' is $\mathrm{B} = (r\sin\frac{\theta}{2})(r - r\cos\frac{\theta}{2}) = \frac{1}{2}r^2(2\sin\frac{\theta}{2} - \sin\theta)$.

ii. It follows that $\frac{\mathrm{A}}{\mathrm{B}} = \frac{\theta - \sin\theta}{2\sin\frac{\theta}{2} - \sin\theta}$. Show that $\frac{\mathrm{A}}{\mathrm{B}}$ is equal to $\frac{\pi}{2}$ when $\theta = \pi$ and $\frac{\frac{\pi}{2}-1}{\sqrt{2}-1}$ when $\theta = \frac{\pi}{2}$.

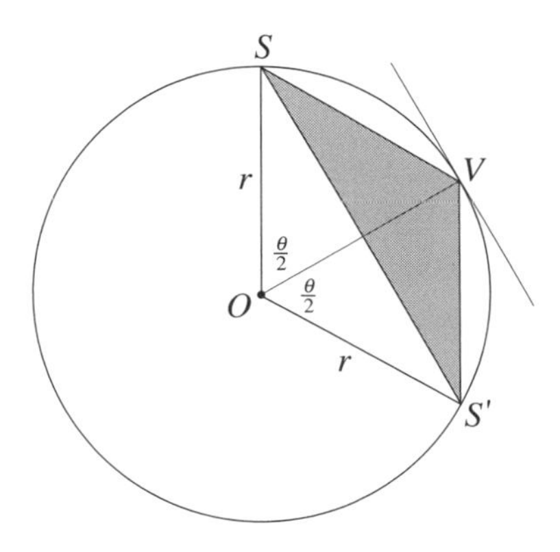

Fig. 2.51

Archimedes's theorem tells us for any parabolic section and its inscribed triangle, that the ratio of their areas is always equal to $\frac{4}{3}$ regardless of the shape and size of the parabolic section. The conclusion of Problem 2.30 informs us in the context of a circular section and its inscribed triangle that the ratio of these areas does depend on the shape—namely, the defining angle θ—of the circular section.

2G. The Lever and Its Use in the Method. The next two problems refer to the lever of Figure 2.27. Some unit of length and some unit of weight are given.

2.31. Let $w_1 = 5$, $d_1 = 3$, and $w_2 = 7$. What does d_2 have to be so that the lever is in balance?

2.32. Let $w_1 = 80, w_2 = 15$, and assume that the lever is balanced. It is known that the two weights are 9 units apart. Find d_1 and d_2.

2.33. Show that the verification of the equality $XZ \cdot NF = YZ \cdot VF$ applies not only in the case of Figure 2.32, but also in the case of Figure 2.52.

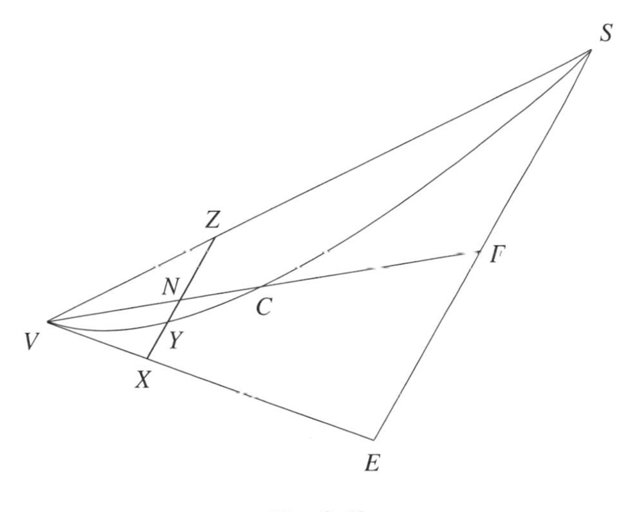

Fig. 2.52

2H. Archimedes's Law of Hydrostatics. When an object floats in a fluid, there is an upward force that acts on it to counteract its weight. Archimedes discovered that this upward force, known as *buoyant force* today, is equal to the weight of the amount of fluid that the immersed object displaces.

2.34. In Figure 2.53a, a block of wood is floating motionlessly in a pool of still water. The vertical forces that act on it are considered in Figure 2.53b. The force of gravity, denoted by F_1, acts downward

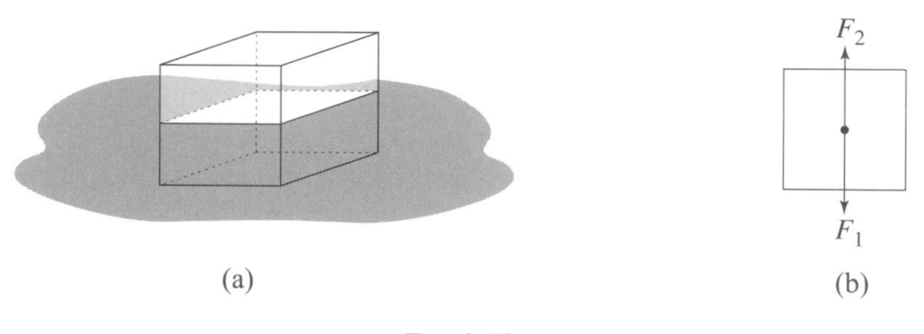

Fig. 2.53

and has magnitude equal to the weight of the block. The upward force F_2 on the block as given by Archimedes's law of hydrostatics is equal to the weight of the water that it displaces. Since the block is not moving, the net vertical force on it is zero. Hence $F_1 = F_2$: the weight of the block is equal to the weight of water that it displaces. Suppose that the block weighs 10 pounds. So the block displaces 10 pounds of water. Since water weighs 62.5 pounds per cubic foot, the block displaces $\frac{10}{62.5} = 0.16$ cubic feet of water. Suppose that the block is a cube with side length 0.6 feet and that it floats with its top surface parallel to the surface of the water. How high along the 1-foot height of the block is the water level?

2.35 The hull of a boat has a volume of 5000 cubic feet. The hull weighs 100,000 pounds. What is the maximum cargo it can carry?

2I. Archimedes and the Crown. The ruler of Syracuse commissioned a goldsmith to make a crown and provided an appropriate quantity of gold for the purpose. The finished crown was identical in weight to the amount of gold that he had given, but the ruler suspected that the goldsmith retained some of the gold for himself and that he substituted an amount of silver. He asked Archimedes to analyze the crown and verify whether the goldsmith had stolen some of the gold. When Archimedes hit on the answer (it is based on his law of hydrostatics), he was elated and is said to have run naked through the streets of Syracuse shouting "eureka" (I have found it).

2.36. Let the crown weigh a total of w pounds, and let it be made up of w_1 pounds of gold and w_2 pounds of silver. Suppose that w pounds of pure gold loses f_1 pounds when weighed in water, that w pounds of pure silver loses f_2 pounds when weighed in water, and that the crown loses f pounds when weighed in water. Show that

$$\frac{w_1}{w_2} = \frac{f_2 - f}{f - f_1}.$$

[Hint: Suppose first that $w_1 = \frac{1}{3}w$. How much weight will the gold in the crown lose in terms of f_1, and how much will the silver lose in terms of f_2? Then, more generally, since w_1 is some fraction of w, suppose that $w_1 = cw$ for some constant c.]

2.37. Suppose the crown of Problem 2.36 displaces a volume of v cubic units when immersed in water, and that lumps of pure gold and pure silver, of the same weight as the crown, displace v_1 and v_2 cubic units, respectively, when immersed in water. Show that

$$\frac{w_1}{w_2} = \frac{v_2 - v}{v - v_1}.$$

Given this formula, all Archimedes has to do is immerse the crown in water and then do the same for quantities of pure gold and silver equal in weight to the crown. He could measure the amount of water that is displaced in each case. He then knew the exact ratio of the weight of gold to the weight of silver in the crown.

2.38. Pure gold has a density of 0.698 pounds per cubic inch, and silver has a density of 0.379 pounds per cubic inch. A crown made of an amalgam of gold and silver weighs 3 pounds and displaces 5 cubic inches when immersed in water. What percentage of the crown is gold?

2J. The Number Scheme of Archimedes. One version (there are variations) of the traditional number system of the Greeks is listed in Figure 2.54. The digamma, koppa, and sampi were taken from an older alphabet of the Phoenicians.

Other numbers are formed by juxtaposition, using the rule that larger numbers go on the left and smaller ones on the right. For example: $\kappa\varepsilon = 25$, $\lambda\zeta = 37$, $\upsilon\pi\eta = 488$. To designate thousands, the units symbols were used with a stroke before the letter to avoid confusion. For example: $,\gamma = 3000$ and $,\beta\tau\pi\delta = 2384$. (To distinguish between numerals and letters, the Greeks sometimes put a bar over the

Units	Tens	Hundreds
1 = α (alpha)	10 = ι (iota)	100 = ρ (rho)
2 = β (beta)	20 = κ (kappa)	200 = σ (sigma)
3 = γ (gamma)	30 = λ (lambda)	300 = τ (tau)
4 = δ (delta)	40 = μ (mu)	400 = υ (upsilon)
5 = ε (epsilon)	50 = ν (nu)	500 = ϕ (phi)
6 = ϛ (digamma)	60 = ξ (xi)	600 = χ (chi)
7 = ζ (zeta)	70 = ο (omicron)	700 = ψ (psi)
8 = η (eta)	80 = π (pi)	800 = ω (omega)
9 = θ (theta)	90 = ϟ (koppa)	900 = ϡ (sampi)

Fig. 2.54

numerals: $\overline{,\varepsilon\chi o} = 5670$.) For 10000, an M (for $\mu\upsilon\rho\iota\alpha\sigma$ = myriad) was used. This was combined with other symbols as follows.

$$\beta\mathrm{M} = 20000$$

$$\iota\delta\mathrm{M},\eta\phi\xi\zeta = 14(10{,}000) + 8567 = 148{,}567$$

$$\upsilon\mathrm{M} = (400)(10000) = 4{,}000{,}000$$

$$\omega\mu\varepsilon\mathrm{M},\beta\tau\pi\delta = 845(10{,}000) + 2384 = 8{,}452{,}384$$

$$,\delta\tau\pi\delta M,\overline{\varepsilon\sigma\xi\gamma} = 4384(10000) + 5263 = 43{,}845{,}263$$

This scheme is similar in principle to the one that the Romans used. Addition and multiplication in this system was cumbersome. This was surely one reason why Greek algebra and its applications lagged behind Greek geometry.

2.39. Write the numbers 85; 842; 34547; 2,875,739; and 99,999,999 using the Greek system.

Notice that the largest number in the Greek system is the number (in Hindu-Arabic),

$$99{,}999{,}999 = 9999(10000) + 9999.$$

It is one short of $(10000)(10000) = 100{,}000{,}000 = 10^8$.

We turn now to Archimedes's scheme for writing large numbers. It is best comprehended "in the abstract." Take any positive integer n and consider the sequence

$$n, n^2, \ldots, n^n, n^{n+1}, \ldots, n^{2n}, n^{2n+1}, \ldots, n^{3n}, \ldots, n^{n\cdot n} = n^{n^2}$$

of consecutive powers of n. Since the exponents provide a count, we see that there are n^2 numbers in the sequence. For $n = 2$, the sequence is 2, 4, 8, and $2^4 = 16$. Observe that the gaps between the numbers are 2, 4, and 8, and that they grow by a factor of 2 from step to step. For $n = 3$, the sequence is

$$3, 9, 27, 81, \dots, 3^9 = 19{,}683.$$

Now the gaps are $6, 18, 54, \dots$. They grow by a factor of 3 from step to step. In the general case, the gaps grow by the factor n from step to step.

2.40. What is the last term of the sequence of Archimedes for $n = 4, n = 5$, and $n = 6$?

For $n = 10$, there are $10^2 = 100$ numbers in the sequence, with the last one equal to $10^{10\cdot 10} = 10^{100}$. The number scheme that Archimedes introduces takes n to be the very largest number of the Greek system plus 1. In other words, Archimedes takes $n = 100{,}000{,}000 = 10^8$. So the first few numbers that Archimedes considers are

$$10^8,\ 10^{16},\ 10^{24},\ 10^{32},\ 10^{40},\ 10^{48}, \dots .$$

The twelfth number is $10^{8\cdot 12} = 10^{96}$. There are $(10^8)^2 = 10^{16}$ numbers in all, and the last one is the enormous number $10^{10^{16}}$.

Having produced this huge array of numbers, Archimedes looks for a context in which to apply it. When he finds it, he is pleased to address his manuscript *The Sandreckoner* to his benefactor, the king of Syracuse. By telling the king about the work of Aristarchus, his manuscript documents Aristarchus's achievement for the historical record:

> Aristarchus brought out a book consisting of some of the hypotheses, in which the premises lead to the result that the universe is many times greater than that now so called. His hypotheses are that the fixed stars and the Sun remain unmoved, that the Earth revolves about the Sun in the circumference of a circle, the Sun lying in the middle of the orbit, and that the sphere of the fixed stars, situated about the same center as the Sun, is so great ...

Then he states his purpose:

> I say then that, even if a sphere were made up of sand, as great as Aristarchus supposes the sphere of the fixed stars to be, I shall still prove that, of the numbers named by me, some exceed in multitude the number of grains of sand in a mass which is equal in magnitude to the sphere referred to, provided that the following assumptions are made ...

In other words, Archimedes imagines the entire cosmos to be packed with sand and proposes to count the number of grains of sand in question! By a strange combination of wild speculations (that have

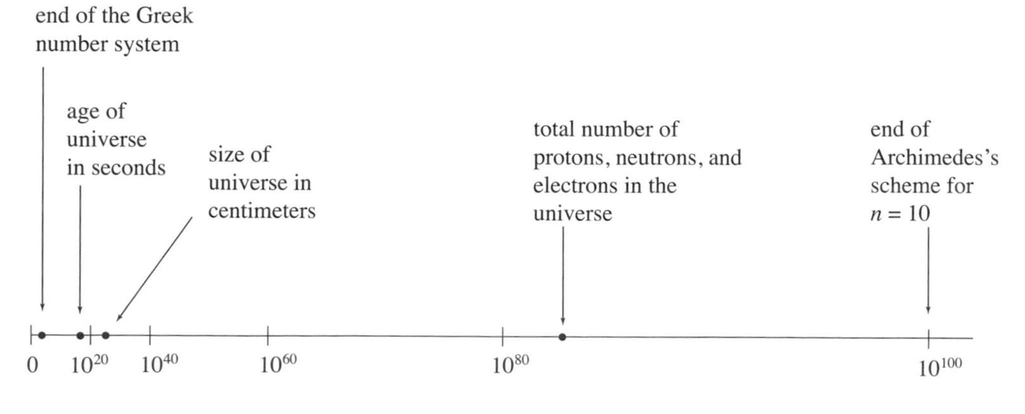

Fig. 2.55

Aristarchus's estimates as starting point), a careful measurement of the angular diameter of the Sun, and lots of delicate geometric analysis, Archimedes concludes that the distance from the Earth to the sphere of fixed stars is no larger than about 5×10^{12} kilometers. By counting the number of grains of sand in spheres of larger and larger sizes—he starts with a sphere the size of a poppy seed, moves to one of diameter of one finger breadth, and so on—he announces that 10^{63} grains of sand are more than enough to fill the sphere of fixed stars.

Because Archimedes's number system handles 10^{63} with ease, his goal was accomplished! The point for us is not so much the usefulness of Archimedes's scheme, but rather the grandiose nature of his speculations. Archimedes, in other words, thought big! A comparison with the numbers of Figure 2.55 tells us how big. His scheme $n = 10$ already handles the largest numbers that arise in our physical universe.

2K. The Palimpsest Reappears. The document that the Danish historian discovered in Istanbul in 1906 is the oldest (by 400 years) surviving transcription of any work of Archimedes. It is a *palimpsest* (from the Greek *palimpsestos*, a composite of *palin* = "again" and *psestos* = "rubbed smooth"), namely, a document containing—in two superimposed layers—an older script overwritten by a later one. In the case of the manuscript discovered by the Danish historian, it seems that a Greek priest in the 13th century needed parchment for a prayer book. He came upon a collection of Archimedes's works written about 250 years earlier and cleaned off the Archimedean text as best he could. He then wrote on these pages the text of a collection of Greek prayers at right angles to the Archimedean text still faintly visible underneath. On a visit to Istanbul in the 1840s, a German biblical scholar came across this bound collection. Intrigued by it,

Fig. 2.56. Image from the *Archimedes Palimpsest Project* (http://www.archimedespalimpsest.org) released for use by the private owner of the manuscript. The manuscript is held by the Walters Art Museum, Baltimore, MD.

he took a page with him (it is now in the Cambridge University Library). But he had no clue about the significance of what he had come across.

In October of 1998, the auction house Christie's announced that the very manuscript that our Danish historian had studied would be put up for sale in New York City by an unidentified French family. The ownership of the palimpsest was immediately challenged in a federal court by the Greek government and the Greek Orthodox Patriarchate of Jerusalem. The manuscript had been the property of the Greek Orthodox Church, belonging to a monastery of the Patriarchate of Jerusalem before it was moved to a monastery in Constantinople. It was alleged that the manuscript was stolen from there during the years 1922–24. How the palimpsest came to France is not known. In the 1920s it was the property of a collector in Paris. The French family that put the manuscript up for auction came into its possession in the 1930s. Just hours before the auction, the court ruled that the French family did have legal title, and the sale went ahead. An American buyer outbid a representative from the Greek government and purchased the manuscript for (gasp) 2.2 million dollars. (The lawyer representing the buyer stated that he was "a private American" who worked in "the high-tech industry," but that it was not Bill Gates. A German magazine later reported that the likely buyer was Jeff Bezos, founder of Amazon.com.)

The palimpsest was brought to the Walters Art Museum in Baltimore for analysis and restoration (it had suffered considerably from mold). An extensive study from 1999 to 2008, undertaken by scientists from universities and technology labs, processed digital images in various wavelengths, including ultraviolet and visible light, and uncovered most of the mathematical text of the bottom layer. This analysis included the use of highly focused X-rays (of the Stanford Linear Accelerator Center), ideal for revealing the intricate structure and composition of all kinds of matter. The result of this effort was satisfactory on the whole. Figure 2.56 depicts one of the pages of the palimpsest. The "before" on the left shows the text of the prayer, and the "after" on the right some of Archimedes's diagrams. The palimpsest is a unique source for the diagrams that Archimedes drew in the sand in Syracuse in the 3rd century B.C.

Historians have pointed out that Archimedes does not explain how he initially arrived at some of his important results. This explanation is contained in the *Method*. It is now confirmed that this million-dollar manuscript is an exposition of Archimedes's central strategy to use principles and concepts from the physics of mechanics to solve problems that are purely mathematical. He determines the areas, volumes, and centroids of regions in the plane and in space by slicing and dicing them into small parallel fragments and by applying his law of the lever to them. The proofs, correct and rigorous by today's standards, are brilliant applications of geometry. Our study of the parabolic section is but one example. Archimedes's slicing methods would become a central strategy of calculus about 2000 years later. The palimpsest contains other works, but none more important than the *Method*. It continues to be our only source for this remarkable treatise.

Chapter 3

A New Astronomy*

The two previous chapters illustrate the remarkable quality of the mathematics of the Greeks. The Greeks made decisive contributions to geometry not only with their ingenious insights, but by presenting geometry as an axiomatic mathematical structure that builds the propositions and theorems of the subject in a cohesive and logical way from a foundation of central definitions and postulates. They applied their geometry and trigonometry in spectacular ways to estimate the sizes of and distances between Earth, Moon, and Sun, and created geocentric mathematical models that simulate the motion of the Sun, Moon, and planets. They introduced the ellipse, parabola, and hyperbola, and developed their basic properties. And finally, they developed subtle mathematical methods for computing areas and volumes that anticipate essential aspects of modern calculus.

When the work of Euclid, Aristarchus, Apollonius, Archimedes, Ptolemy, and that of the Greek philosophers arrived in Europe from the storehouse of transcriptions gathered by Islamic civilizations, light began to displace the darkness of the Middle Ages. In time, the scholars of Europe studied and absorbed the learning of Greece. When they started to question and challenge it, a scientific revolution got underway. The Greek geocentric picture of the universe began to collapse in the 16th century with Copernicus's publication of his *De Revolutionibus Orbium Coelestium* (*On the Revolutions of the Heavenly Spheres*). Going against conventional wisdom, he moved the Sun to the center of the universe and placed the planets into circular orbits around it. (The introduction of an early unpublished version of the *De Revolutionibus* mentions "that Aristarchus of Samos was of the same opinion," but the published editions dropped this reference.) Kepler refined this heliocentric explanation. He spent years in the painstaking analysis of the orbital data that the astronomer Tycho Brahe had amassed with his large and accurate new instruments. Kepler's effort to align Copernicus's heliocentric model with this data was finally rewarded when he hit on his laws of planetary motion. In several treatises early in the 17th century, Kepler observed that the orbits of the planets are ellipses, each with the fixed Sun at a focus, described the speed with which they move, and found the numerical relationship that all orbits share. At around the time Kepler developed his elliptical theory, Galileo pointed his telescope to the observe the night sky. He discovered four moons circling the planet Jupiter and he studied the changing shape of the sunlit portion of the planet Venus. The four moons were inconsistent with the Greek Earth-centered view, and the phases of Venus exhibited a pattern that confirmed Copernicus's Sun-centered explanation. Galileo also knew that the heliocentric theory provides a simple explanation for the looping reversals of direction in the planets' observed path (that the Greeks had already noted). As a consequence, Galileo became a fervent and outspoken advocate of Copernicus's Sun-centered system. At the same time, he developed his newly conceptualized physics of motion. While Greek astronomy and physics are superseded, the mathematics of this scientific revolution was predominantly Greek geometry and trigonometry (until the appearance of the new mathematical approaches of Newton and Leibniz late in the 17th century).

---

* This chapter can be skipped, if the focus is on basic mathematics, particularly the mathematics required for the pursuit of calculus. However, a few matters will be referred to in a later chapter.

The reception of the ideas of Copernicus, Kepler, and Galileo by the Church was decidedly negative. Its theologians pointed to passages in the Bible asserting that, according to God's command, "the sun stopped in the middle of the sky and delayed going down about a full day." The teachings of the Church are based on the Bible, and it is the exclusive role of the Church to interpret scripture. Therefore the new astronomy—with its fixed Sun and moving Earth—presented a direct challenge to its authority. Copernicus, aware of the potential difficulties in this regard, published his treatise in the year of his death. Kepler was excommunicated by the Lutheran Church (but primarily for his theological and not scientific views). Galileo was instructed by the Catholic Church that he could discuss the Copernican system as a mathematical hypothesis, but not as the description of the physical universe. In spite of this, his treatise *Dialogo sopra i due massimi sistemi del mondo* (*Dialogue Concerning the Two Chief World Systems*) did not hide its support for the Copernican point of view. What made things worse was that the *Dialogo* flew off the printing presses in Italian, the language of the people on the street, rather than in Latin, the highbrow language of scholarly discourse and scientific communication. After a trial by the Roman Inquisition, Galileo was silenced and placed under house arrest for the remainder of his life. It was only much later in 17th century, when the astronomers Cassini and Flamsteed used their improved telescopes to gain an accurate sense of the vast distances of the planets from the Sun and when Newton's genius combined his laws of motion with completely new mathematical methods to explain what Kepler had observed, that the heliocentric point of view began to find acceptance.

3.1 A FIXED SUN AT THE CENTER

Copernicus (from 1473 to 1543) was born in Poland. He studied astronomy in Bologna and medicine in Padua. After returning to Poland, he settled into his position as canon of the Church. He found ample time to pursue astronomy in his little study, said to have been in the tower of the cathedral. He was not satisfied with the complicated astronomy of Ptolemy and published his own comprehensive study *De Revolutionibus* in 1543.

Let's turn to a description of the basic structure of Copernicus's Sun-centered universe. The Sun is motionless at the center of an immense, unmoving sphere of fixed stars. The six planets Mercury, Venus, Earth, Mars, Jupiter, and Saturn (known since their discovery by the ancient astronomers) orbit the Sun in circles. The radii of these circles increase in the same order. The Moon is in circular orbit around the Earth. The motion of the Earth has two primary aspects, both illustrated in Figure 3.1. One is the Earth's daily rotation around an axis through its poles. In the figure, this axis is represented by the arrow N (defining the direction north). The speed of the rotation is constant. The other motion is Earth's circular orbit around the Sun. The Earth's axis of rotation is perpendicular to the plane of its equator, but not to

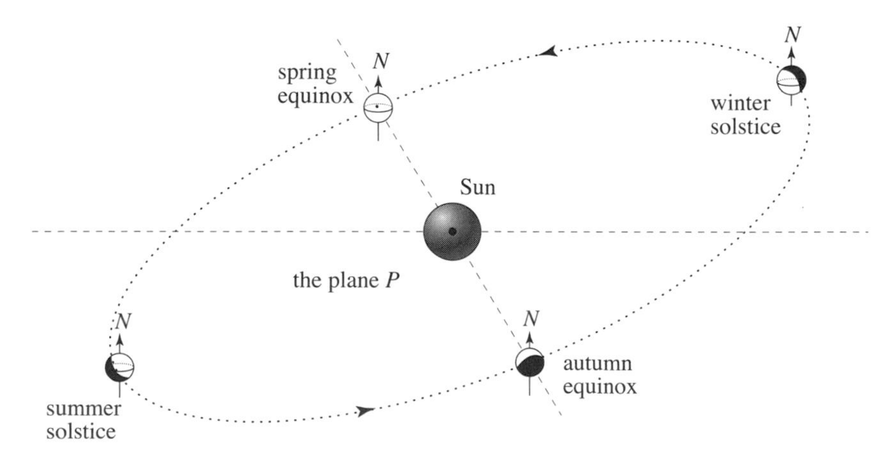

Fig. 3.1

the plane of its orbit.

Let P be the plane through the center of the Sun parallel to the plane of Earth's equator, or, equivalently, perpendicular to the Earth's axis of rotation. By considering the position of the Earth in its orbit relative to this plane, we gain a better understanding of what was already observed in Section 1.1. From the perspective of Figure 3.1, the Earth moves alternately above and below the plane P. There are two occasions when the Earth's center and hence its equator lie in the plane P. On either of these two days, consider a point on the Earth's surface at sunrise, follow it around, and notice that one-half a rotation of the Earth later, the Sun will set over this point after 12 hours. These are the days of *spring equinox* and *autumn equinox*. Since the center of the Earth lies in the plane P at these two positions, the line that joins them—the line of equinoxes—lies in P. This is one of the two dotted lines in the figure. The other is the line in the plane P that is perpendicular to the line of equinoxes. When the Earth is at its lowest point below the plane P, the Sun is highest in the sky at noon in the Northern Hemisphere and shines down on it most directly. This is *summer solstice*. On this day, the Sun is above the horizon for the longest period of time at any location in the Northern Hemisphere. It is the day of longest daylight. From the perspective of Figure 3.1, when the Earth is at its highest point above the plane P, the Sun is lowest in the sky at noon in the Northern Hemisphere, and the smallest portion of this hemisphere is exposed to the Sun. This is *winter solstice*. It is the day on which the Sun is above the horizon for the shortest period of time in the Northern Hemisphere. It is the day of shortest daylight. The figure indicates the dark and sunlit regions of the Earth at each of the four positions we have singled out. These four positions define the seasons. The time the Earth moves from spring equinox to summer solstice is *spring*, the time it moves from summer solstice to autumn equinox is *summer*, the time from autumn equinox to winter solstice is *autumn* or *fall*, and the time from winter solstice to spring equinox is *winter*.

Figure 3.2 is a close-up of the summer solstice position of Figure 3.1. When interpreted from within the Sun centered perspective, "the obliquity of the ecliptic" that the ancient astronomers had measured is the angle between the plane of the Earth's equator and the plane of its orbit. This angle, approximately $23\frac{1}{2}^\circ$, is shown in the figure. The circle that this angle determines on the surface of Earth is known as the Tropic of Cancer. The Tropic of Capricorn is the corresponding circle in the Southern Hemisphere. Con-

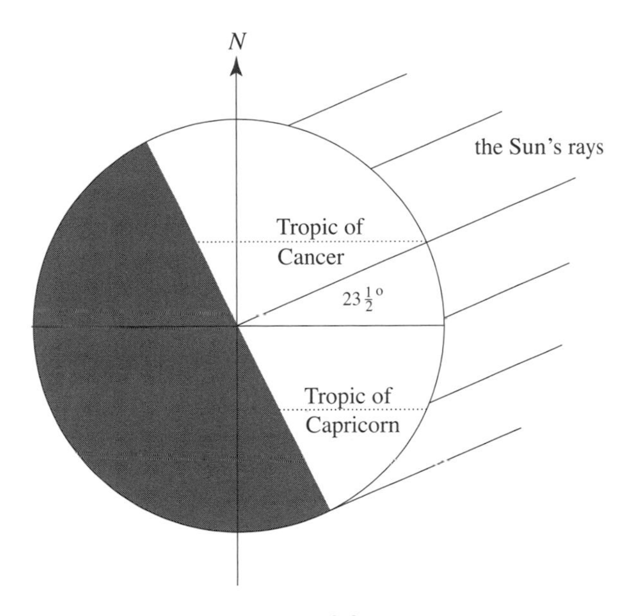

Fig. 3.2

sult an atlas and confirm that Aswan, Egypt, lies near the Tropic of Cancer. We now understand why the Sun shone into the very bottom of the well in Syene (the ancient city near today's Aswan) at summer solstice. Recall that this fact was instrumental in Eratosthenes's computation of the radius of Earth. (Refer to Section 1.5.)

From Copernicus's Sun-centered perspective, the explanation of the mysterious retrograde motion of the planets that the ancient astronomers observed is straightforward. For example, consider the observations of Saturn by Ptolemy that are captured in Figure 1.4. Because Saturn takes about 30 Earth years to complete one orbit, it follows that Saturn completes $\frac{1}{30}$ of its orbit in one year. Figure 3.3 tells us why Ptolemy saw what he did when he looked out on the heavens in the years 132 and 133 A.D. It shows the (dotted) lines of sight of an observer on Earth looking out on Saturn against the background of the fixed stars. Observe that the general pattern of the positions of Saturn on the sphere of fixed stars is consistent with Ptolemy's

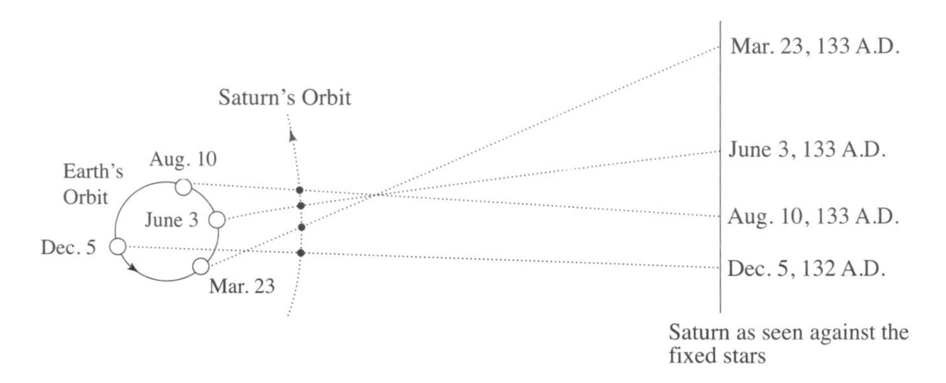

Fig. 3.3

record. The fact that the plane of the orbit of Saturn is slightly different from the plane of Earth's orbit explains the fact that Ptolemy observed the retrograde motion of Saturn to rise and fall slightly against the backdrop of the fixed stars.

Copernicus's Sun-centered perspective also explains the phenomenon of the "precession of the equinoxes" that the ancient astronomers had noticed. This is the observation that the line through the two equinox positions revolves slowly. In his Earth-centered context, the Greek astronomer Hipparchus had assessed it to be about 1° per century, or one complete revolution every 36,000 years. Figure 3.1 tells us that the Earth's axis of rotation is tilted with respect to the plane of its orbit. Figure 3.4 makes this explicit. Copernicus correctly understood the precession of the equinoxes to be a consequence of the fact that the

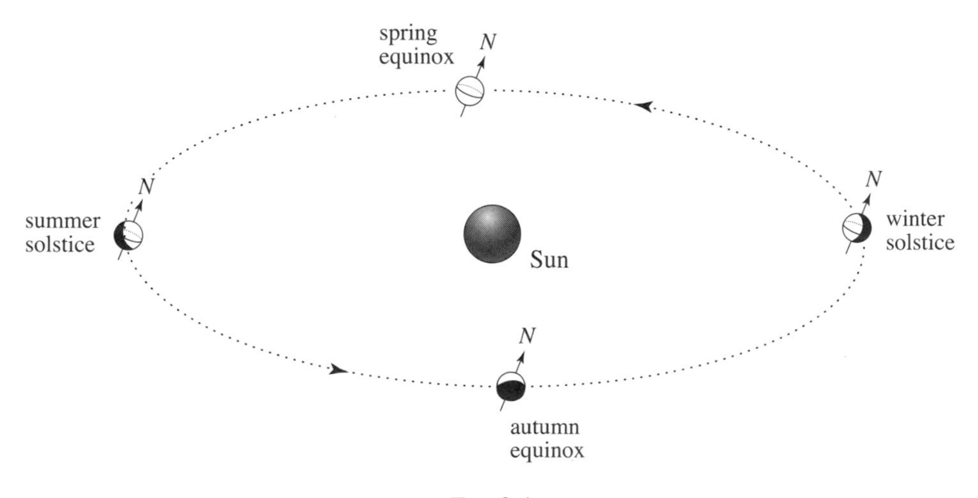

Fig. 3.4

Earth's axis of rotation revolves. What happens is captured in Figure 3.5. This revolution of the axis is extremely slow. By relying on observations of the ancients, Copernicus's *De Revolutionibus* records that the Earth's axis moves through one cycle in 25,816 years. Today's more accurate value is 25,772 years.

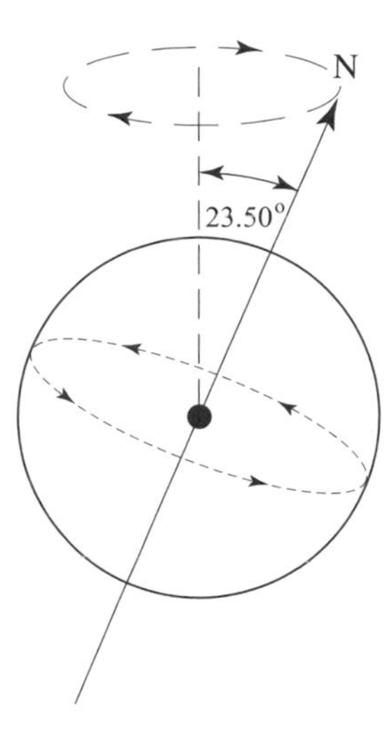

Fig. 3.5

Now comes the all-important question. Was Copernicus's model that has the Sun at the center and the planets in circular orbits around it in sync with what was observed in the heavens? With an accuracy that measured up to the standards of the time? The answer is no! Certainly not if the orbits are taken to be circles with center the Sun (or Earth in the case of the Moon). After all, Kepler would inform us later that these orbits are ellipses. Copernicus was aware of inaccuracies, and the *De Revolutionibus* responded by modifying the circular orbits with the same geometric constructions that Hipparchus and Ptolemy had used. See Sections 1.1 and 1.2. The off-centered deferent circle (with an abstract point the center of the circle rather than the Sun) became the basis of Copernicus's orbital geometry. But attached to it were smaller circles, most prominently epicycles, and ultimately Copernicus's scheme of circles was nearly as complicated as Ptolemy's had been. However, the central purpose of his additional circles was completely different. Instead of having to explain the complications that resulted from the Earth-centered point of view (the phenomenon of retrograde motion, for instance), Copernicus's supplementary circles were small and were built in to account for the differences between his basic model of the orbit (the deferent circle) and the actual orbit (the ellipse).

3.2 COPERNICUS'S MODEL OF EARTH'S ORBIT

For his description of the orbit of the Earth around the Sun, Copernicus chose the same off-centered circle that Hipparchus had used (to describe the orbit of the Sun around the Earth). Neither Hipparchus nor Copernicus required the addition of epicycles. We'll begin Copernicus's study with Figure 3.6. The larger circle represents the sphere of the fixed stars. The Sun S is positioned at its center. The positions of the Earth—projected against the sphere of the stars—at spring equinox, summer solstice, autumn equinox, and winter solstice are indicated. The smaller circle, centered at the point O, represents the path of the Earth. The Earth traces it out at constant speed. The positions of the Earth in its orbit at spring equinox, summer solstice, autumn equinox, and winter solstice are marked. As the Earth E moves along its circular path, its projection E^* moves on the sphere of stars. Given that E moves along its orbit at constant speed, we see by looking at the four segments of Earth's orbit that spring (from spring equinox to summer solstice) is the longest season, followed by summer, winter, and autumn. Section 1.1 informs us that this was in fact the state of things when Hipparchus looked skyward in the 2nd century B.C. This model reconciles the observed differences in the lengths of the seasons with the requirement that the orbit be a perfect circle that is traversed at constant speed.

When Copernicus looked at the heavens in the year 1515 and—no doubt—made shadow measurements like the Greeks had done before him (it was nearly 100 years before Galileo studied the skies with his telescope), he determined that spring was $92\frac{51}{60}$ days long and that summer with a length of $93\frac{14.5}{60}$

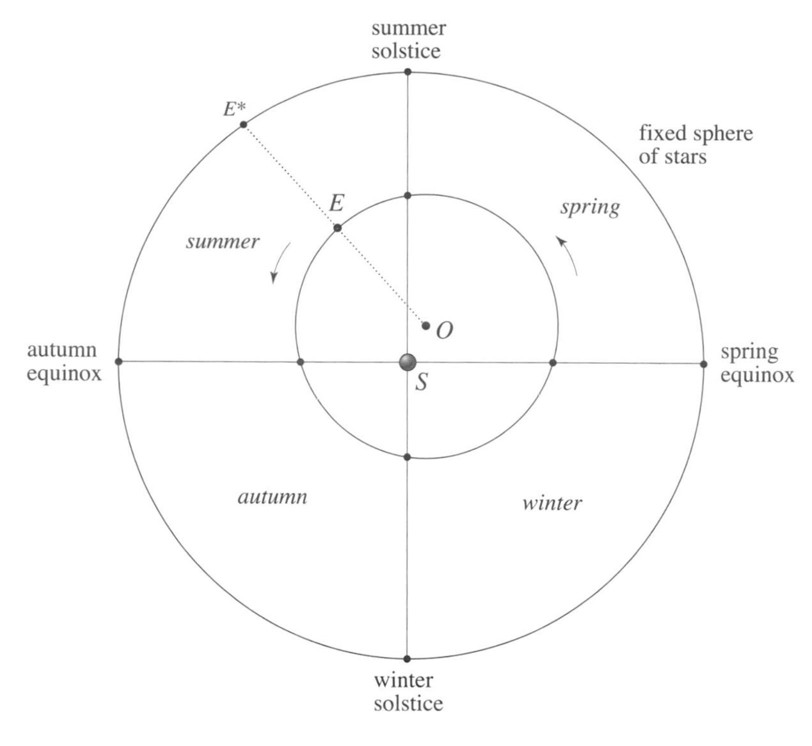

Fig. 3.6

days was the longest season of the year. For Earth to move from the spring equinox to its autumn equinox position required $92\frac{51}{60} + 93\frac{14.5}{60} = 186\frac{5.5}{60}$ days. Copernicus used this information to construct his model for the orbit of the Earth around the Sun.

To understand Copernicus's answer, start with Figure 3.7. In the figure, S is again the position of the Sun at the center of the universe and O is the center of Earth's circular orbit. Because summer and not spring was the longest season then, Copernicus needed to place O in the summer quadrant as in the figure (and not in the spring quadrant as in Figure 3.6). But where exactly did he have to place O so that his model reflects what he observed about the lengths of the seasons in the year 1515? The notation in Figure 3.7 is as follows:

A is the position of the Earth at spring equinox

B is the position of the Earth at summer solstice

C is the position of the Earth at autumn equinox

D is the position of the Earth at winter solstice

E is a typical position of the Earth in its orbit

E^* is the projection of E against the sphere of fixed stars

F is the position of the Earth, called *aphelion* (from Greek-Latin roots meaning "from the Sun"), when it is farthest from the Sun

λ is the angle $\angle FSB$

In addition, we'll let $r = OE$ be the radius of Earth's orbit and $c = SO$.

Copernicus solved the problem of the placement of O relative to S by using his observations about the lengths of the seasons to determine the distance c (in terms of r) and the angle λ. Following what Hipparchus had done, Copernicus assumed that the axes AC and BD intersect perpendicularly at S. The points A', B', and C' on Earth's circular orbit are determined by the lines through O parallel to AC and BD, respectively.

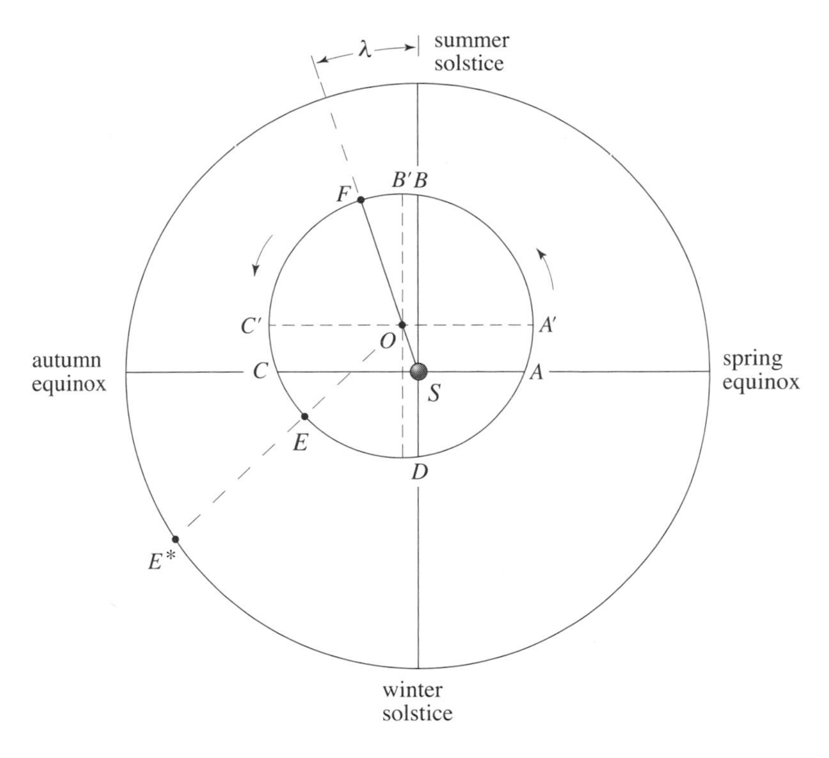

Fig. 3.7

The lengths of the arcs AA' and BB' are the key. They determine the right triangle with hypotenuse SO and hence c and λ.

The details follow (using current notation and terminology). Consider the motion of the Earth E on its circular path, and let ω be the angle traced out by the rotating segment OE in one day. Since the Earth completes the full circle in $365\frac{1}{4}$ days, it follows that $\omega = \frac{360°}{365\frac{1}{4}} = 0.986°$. Therefore

$$\omega = (0.986°)\frac{\pi}{180°} = 0.0172 \text{ radians.}$$

As Copernicus had observed, it took Earth $186\frac{5.5}{60} = 186.0917$ days to move from the spring equinox to the autumn equinox position. Since the Earth traces out 0.0172 radians per day, it follows that $\angle AOC = 186.0917 \times 0.0172 = 3.2010$ radians. Since r is the radius of Earth's orbit, it follows that arc $AC = 3.2010r$. Figure 3.7 tells us that

$$\text{arc } AC = \text{arc } AA' + \text{arc } A'C' + \text{arc } C'C,$$

and hence that $2(\text{arc } AA') = 3.2010r - \pi r$. So

$$\text{arc } AA' = \tfrac{1}{2}(3.2010r - 3.1416r) = 0.0297r.$$

During spring, the Earth moves from A to B. Since Copernicus had observed that spring is $92\frac{51}{60} = 92.8488$ days long, $\angle AOB = 92.8488 \times 0.0172 = 1.5970$ radians. It follows that arc $AB = 1.5970r$. Turning to Figure 3.7 again, we get

$$\text{arc } AB = \text{arc } AA' + \text{arc } A'B' - \text{arc } B'B.$$

Therefore arc $BB' = \text{arc } AA' + \text{arc } A'B' - \text{arc } AB = 0.0297r + \frac{\pi}{2}r - 1.5970r$, so that

$$\text{arc } BB' = 0.0297r + 1.5708r - 1.5970r = 0.0035r.$$

Figure 3.8 extracts important features from Figure 3.7 including the right triangle with hypotenuse $c = SO$. In the figure, a and b are the lengths of the other two sides of this right triangle. Turn to the angle $\alpha = \angle AOA'$. Since $\alpha = \frac{\text{arc } AA'}{r} = 0.0297$ radians, $\alpha = 0.0297\frac{180}{\pi} = 1.70°$. The fact that $\sin\alpha = \frac{a}{r}$, $\alpha = \frac{\text{arc } AA'}{r}$, and that this angle is small told Copernicus that therefore a and arc $AA' = 0.0297r$ are indis-

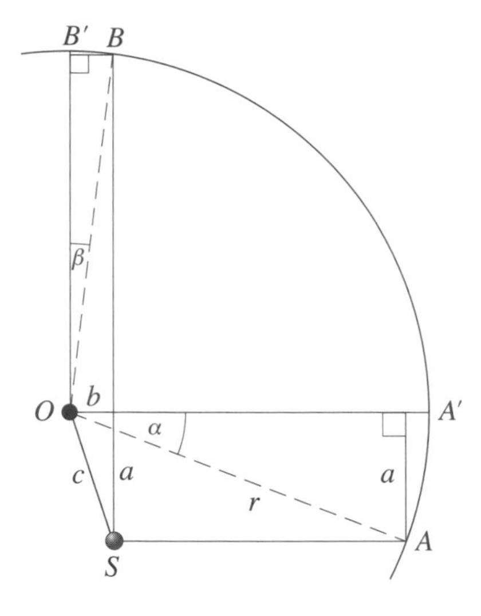

Fig. 3.8

guishable. Table 1.2 provides the evidence. A similar thing is true for b and arc $BB' = 0.0035r$. These observations provided Copernicus with the information of Figure 3.9. Copernicus could now apply the

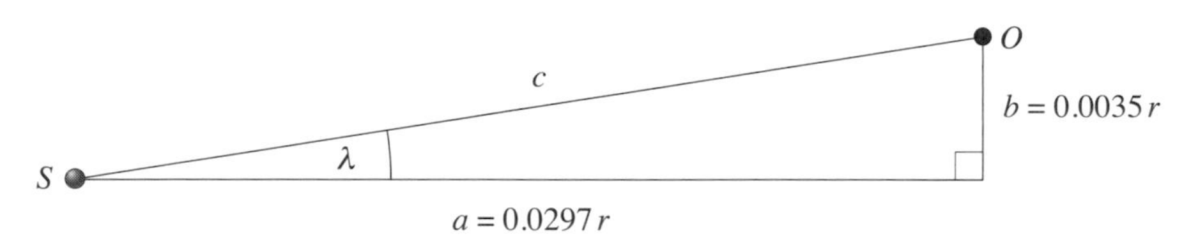

Fig. 3.9

the Pythagorean theorem to get

$$\begin{aligned} c &\approx \sqrt{(0.0297r)^2 + (0.0035r)^2} = r\sqrt{(297 \times 10^{-4})^2 + (35 \times 10^{-4})^2} \\ &= r(10^{-4}\sqrt{297^2 + 35^2}) \\ &\approx 0.0299r, \end{aligned}$$

so that that $c \approx 0.0299r$. Figure 3.9 shows that the angle λ satisfies $\tan\lambda \approx \frac{0.0035r}{0.0297r} = \frac{0.0035}{0.0297} \approx 0.1178$. From this, Copernicus could conclude that $\lambda \approx 6.72° \approx 6\frac{2}{3}°$. (Today, we can check this by pushing the inverse tangent button of a calculator.) To complete his model of the Earth's orbit, Copernicus needed to know its radius r. Since c is small relative to r, Copernicus knew that this radius is approximately equal to the distance from Earth to the Sun. But the determination of this distance was elusive. Table 1.3 tells us that the value of about 8 million kilometers (or 5 million miles) that Copernicus had obtained was too small by a factor of about 20 (and only a little better than the value Ptolemy achieved). It was not until about 150 years later that the French-Italian astronomer Cassini and the English astronomer Flamsteed came close, when they measured this distance to within 6% of its correct value.

Copernicus's descriptions of the orbits of the other planets and the Moon are much too complicated to consider here. For instance, to resolve the challenges that the Moon's orbit presents, Copernicus needs a deferent, an epicycle, and another epicycle on top of that. We turn instead to Copernicus's analysis of the distances of the other planets from the Sun (in terms of the distance from the Earth to the Sun).

3.3 ABOUT THE DISTANCES OF THE PLANETS FROM THE SUN

We will start with a fact from Euclid's *Elements*. Consider a circle with center O, a point P on the circle, and the tangent line to the circle at P. Choose the point A so that AP is a diameter of the circle. Let Q be a point on the circle different from P. Since AP is a diameter, $\angle AQP = 90°$. (See Section 1.3, in particular, Figure 1.13.) Let $\theta = \angle QAP$, and let L be the line through Q and P. All this is depicted in Figure 3.10. Keeping A and P fixed, push Q to P along the circle. The angle $\angle AQP$ remains a right angle throughout the process. The line L rotates toward the tangent at P, and θ goes to 0. The angle $\varphi = \frac{\pi}{2} - \theta$ goes to $90°$ but also to $\angle APT$. It follows that the angle between AP and the tangent at P must be $90°$. Therefore a

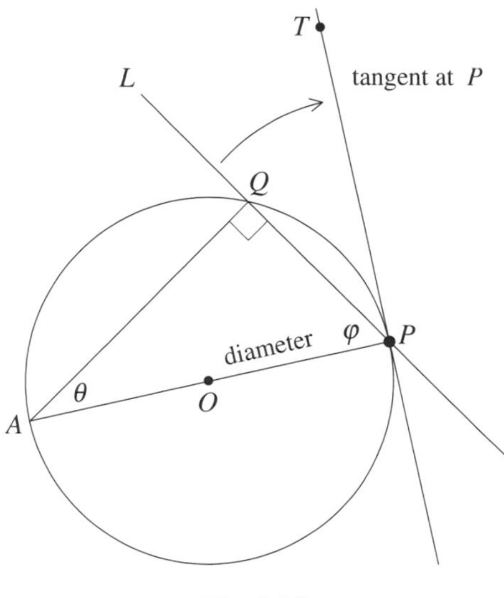

Fig. 3.10

radius of a circle to any point on it is perpendicular to the tangent of the circle at that point. The verification just provided has used the idea of limit and is very much in the spirit of differential calculus. Euclid's verification, a proof by contradiction, is taken up in Problem 3.10.

Copernicus used this fact in an ingenious way to gain a sense of the distances of the planets from the Sun. Consider the inner planet Venus, for example. Venus orbits the Sun inside the orbit of the Earth. Refer to Figure 3.11. In the figure, S represents the Sun, E is a typical position of the Earth in its orbit, and

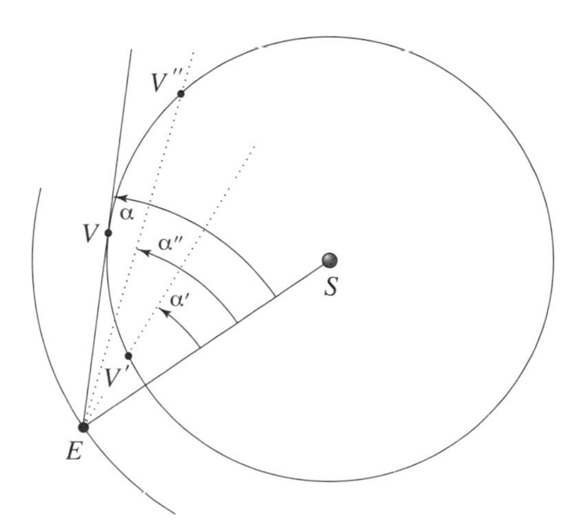

Fig. 3.11

V, V', and V'' are positions of Venus. Observing both Earth E and Venus V again and again over time, and measuring the angle $\alpha = \angle VES$, Copernicus found that $\alpha \approx 46°$ is the maximum value of this angle. A look at the figure tells us that α is greatest when the line of sight to Venus is tangent to Venus's circular orbit around S. But then, by the fact from Euclid just described, the angle between the line of sight to Venus and the radius VS of the orbit is a right angle. It follows that

$$\sin\alpha = \frac{VS}{ES}.$$

Since $\alpha \approx 46°$ and $\sin 46° \approx 0.72$, Copernicus obtained

$$VS \approx 0.72\, ES.$$

(Copernicus's argument has made some simplifications. For instance, it takes the orbit of Venus to be a circle and ignores the second, smaller epicycle that Copernicus's model calls for. It also assumed that the Sun S is the center of the orbits of both Earth and Venus. In Copernicus's approach, S and centers of the orbits are distinct.)

Copernicus's calculations of this ratio for the outer planets, namely those that orbit the Sun outside the orbit of Earth, are more complicated. Consider Mars, for instance. In Figure 3.12, M and E as well as M' and E' show Mars and Earth at two different times in their respective orbits. In the situation M' and E' they are aligned with the Sun S, and Mars is said to be *in opposition*. In the other situation, $\angle MES$ is a right angle, and M is said to be *at quadrature*. When Mars is in opposition, the side of Mars that the Sun illuminates faces Earth. At that time, Mars appears brighter than at any other time. Copernicus waited for

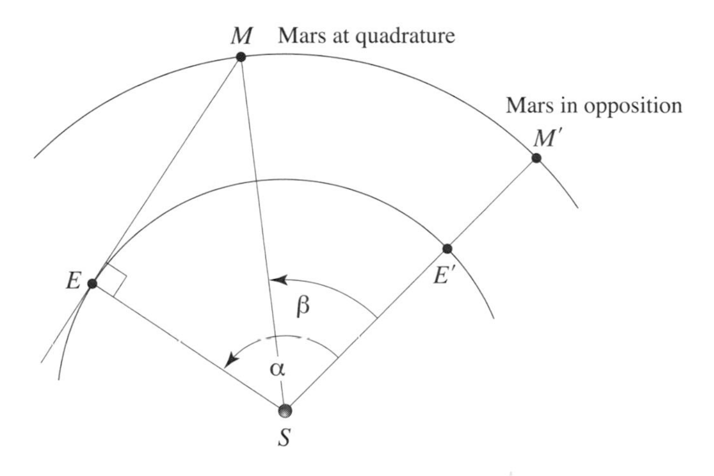

Fig. 3.12

opposition to occur and then measured the time it took for Mars to be at quadrature. Since he knew the orbital periods (the time required for a single orbit) of the two planets, he could estimate the angles α and β that they traced out during this time. These estimates told him that $\alpha - \beta \approx 49°$. In view of Figure 3.12,

$$\cos(\alpha - \beta) = \frac{ES}{MS}.$$

Since $\cos 49° \approx 0.656$, Copernicus obtained

$$MS \approx \frac{ES}{0.656} \approx 1.52\, ES.$$

Table 3.1 provides the ratio of the distance from the planet to the Sun over the distance from Earth to the Sun that Copernicusderived for each of the known planets. The table also lists the ratios obtained by

Planet	Copernicus	Kepler	Modern
Mercury	0.38	0.389	0.387
Venus	0.72	0.724	0.723
Earth	1.00	1.000	1.000
Mars	1.52	1.523	1.524
Jupiter	5.22	5.200	5.202
Saturn	9.17	9.510	9.539

Table 3.1

Kepler as well as the modern values. (We have seen that Copernicus already realized that the distances of the planets from the Sun vary as they orbit. Thus the distance ratios in the table vary as well.)

3.4 TYCHO BRAHE AND PARALLAX

The Danish nobleman Tycho Brahe (from 1546 to 1601) was a colorful and volatile character (his nose was partially cut off in a duel when he was a student, and he lived with a nose of gold and silver thereafter) who developed an early passion for astronomy. In 1572 he observed a bright star in the sky in a location where there had been no star before. This "nova stella" (in Latin *nova* = "new" and *stella* = "star") was visible for about 18 months and then disappeared entirely. Five years later, Tycho observed a comet moving through the sky and coming closer. Tycho did not know what we know today: that a new star is the explosive death of a massive, distant star (a supernova) and that comets come from the far reaches of the solar system, round the Sun, and return to their distant realms. Tycho was convinced that what he saw took place far from the Earth, beyond the orbit of the Moon. But how could this be? No lesser authorities than the Greek philosophers had asserted that the sphere of fixed stars was unchanging and that the planets were held in place by impenetrable spheres of crystal. Tycho knew that this had to be wrong.

Tycho gained such a reputation as astronomer that the king of Denmark provided him with an island (in the sound between Denmark and Sweden), the resources to set up an astronomical observatory, and a large annual grant to support it. There, Tycho designed and crafted large precision instruments that allowed him to observe celestial phenomena with a much higher degree of accuracy than was previously possible. (It was about 30 years before Galileo studied the heavens with his telescope.) The positions of the planets and stars that Tycho charted were on average accurate to about 1 minute, or $\frac{1}{60}$ of a degree of arc. This was 10 times more precise than the accuracy of 10 minutes, or $\frac{1}{6}$ of a degree that Hipparchus, Ptolemy, and Copernicus had achieved. The fact that the diameter of a full Moon has an angular width of about $\frac{1}{2}^\circ = 30$ minutes provides a sense of this improvement. With these instruments Tycho and his assistants systematically observed and recorded the positions of stars and the movements of the planets. The data that they collected over about 20 years were on an massive scale.

Tycho was interested in measuring the distances of the objects he observed. Were they close to the orbiting Moon, or were they far beyond its orbit? His measurements relied on the phenomenon of parallax. The principle involved is simple and easily illustrated. Stretch out your arm in front of you and look at your upturned thumb. Look at it first with one eye closed, then with the other eye closed. As you switch from one eye to the other, observe that the position of your thumb as seen against a wall or other surface in the background will shift. If you were to measure the difference between the angles of the two lines of sight, you could estimate the distance between your eyes and your thumb. Such an estimate is of little to no interest, but the fact is that the same strategy can be applied to the vast expanse of the night sky to estimate how far some distant objects in the heavens are from Earth. This is the *method of parallax*. (The word "parallax" comes from the Greek *parallaxis*, meaning "a change.")

Turn to Figure 3.13. Suppose that an astronomer observes an object A in the heavens with Earth in position E. Under the assumption that Copernicus's explanations are correct, the Earth will be in a different position E' some months later. When A is observed again, now from location E', the astronomer should be able to detect a shift in the position A^* of the object as seen against the fixed pattern of distant stars. The figure illustrates what is involved. The star that appeared in 1572 provided Tycho with an excellent opportunity to apply this strategy. According to the Greek philosophers, changes in the universe

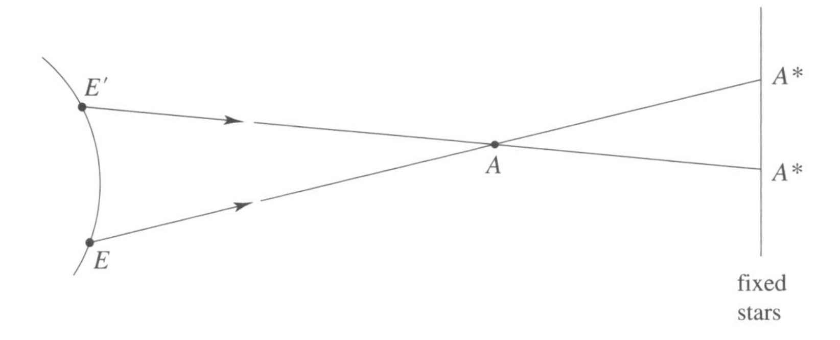

Fig. 3.13

could only occur in realms close to the Earth. So this new star was thought to be relatively near. But when Tycho pointed his precision instruments skyward to observe the new star, he found no such shift. There was *no* parallax. He knew then that either the Earth did not move, or that his new star was so far away that its parallax was too small to measure. Brahe did not believe that his star could be that far away and concluded that the Earth must be fixed and that Copernicus had to be wrong. Tycho developed his own "Tychonic" system of the universe: the Moon and the Sun are in orbit about a fixed Earth, but all other planets are in orbit about the Sun. For over a century, the Ptolemaic, Copernican, and Tychonic descriptions of the universe existed side by side.

It would take more than another 250 years before powerful split-image telescopes were developed that could measure the very tiny angles of parallax of the stars. Only then was the orbital motion of the Earth around the Sun confirmed. These measurements also meant that the universe is unimaginably large. Even the nearest stars are enormous distances from our Sun. The distance that light travels in one year, referred to as the *light-year*, is the unit of length with which distances between objects in the universe are usually measured. In the 1800s, the speed of light was known with an accuracy close to its modern value of 300,000 kilometers per second, and its was known that it takes light from the Sun about 8 minutes to travel to Earth. The parallax in the position of 61 Cygni, a relatively near system of two stars, was measured to be 0.00001742 degrees. This meant that the light from these two stars takes an almost incomprehensible 11 years to make the trip to Earth. So 61 Cygni is 11 light-years from us. It was established more recently that the outer reaches of the universe are an astonishing 14 billion light-years away.

But let's get back to Tycho. When he fell out of favor with the new king of Denmark, he accepted an appointment as astronomer to the royal court of the Austrian emperor in Prague. He brought with him some of his instruments as well as the wealth of accurate data that he had collected. In Prague, Tycho returned to the problem of determining which of the planetary systems provided the most accurate picture of the universe. He turned his attention to Mars. In 1593, he noticed with a sense of disbelief that over a period of several weeks both the Ptolemaic and Copernican predictions for the position of Mars were off-target by around 5 degrees. This error—ten times the angular width of the Sun or a full Moon—was huge. The planet was nowhere near its predicted place in either system. Table 3.2 tells us what he recorded

	Right Ascension	Difference
Tycho's observation	$346^\circ\ 7\frac{1}{2}'$	
Ptolemaic (Alphonsine) tables	$351^\circ\ 26'$	$+5^\circ\ 18\frac{1}{2}'$
Copernican tables	$342^\circ\ 0'$	$-4^\circ\ 7\frac{1}{2}'$

Table 3.2

in his log book. "Right ascension" refers to the angle between the observed position of Mars and the position of spring equinox. See Section 1.2 and Figure 1.7. Refer to Tycho was never able to draw definitive conclusions from the large discrepancies that he had detected. But he realized that both the Ptolemaic and the Copernican explanations had to be wrong.

3.5 KEPLER'S ELLIPTICAL ORBITS

Johannes Kepler (from 1571 to 1630) was born near Stuttgart, in southern Germany. His father was a mercenary whose only pleasures seem to have been drink and war. He deserted his family soon after Johannes's birth. Young Kepler was sickly and poor, but he was also a brilliant student. Being deeply religious, he enrolled at the University of Tübingen to study theology. While there, he learned about the principles of the Copernican system from a professor of mathematics (privately, since the university administration frowned upon such heretical studies). After he graduated, he took the dual posts in the Austrian provincial capital Graz of mathematician of the province and teacher at the Evangelical Seminary. He continued to pursue his interests in astronomy, especially in the relationship between the distances and speeds of the planets, as well as the driving forces that move them in their orbits.

Kepler's work came to Tycho's attention, and he arranged for Kepler to come to Prague in 1600 to become his assistant. Soon thereafter, Tycho died (a drinking-related bladder problem appears to have been a contributing factor), and Kepler was appointed to succeed him as imperial astronomer to the Austrian court. He was now free to use Tycho's massive and accurate data set in the pursuit of his own investigations. Kepler would have found reference in Tycho's log book to the discrepancies that he noticed between the predicted and the observed positions of Mars in 1593. Not surprisingly, therefore, Kepler turned his attention to the orbit of Mars.

Kepler soon discovered that underlying the problem with Mars was not Mars at all, but the orbit that had been assigned to the Earth. Kepler knew that Mars returns to the same position in its orbit every 687 days while the Earth makes a complete revolution every $365\frac{1}{4}$ days. So during the time Mars completes one revolution around the Sun, the Earth completes one revolution and then continues on for another $321\frac{3}{4}$ days. Kepler turned to the precise observations that Tycho had made in his campaign to measure the distance to Mars. He started with March 5, 1590, and considered the positions E_1 and M of Earth and Mars in their orbits at that time. For three consecutive returns of Mars to M (on January 21, 1592; Decem-

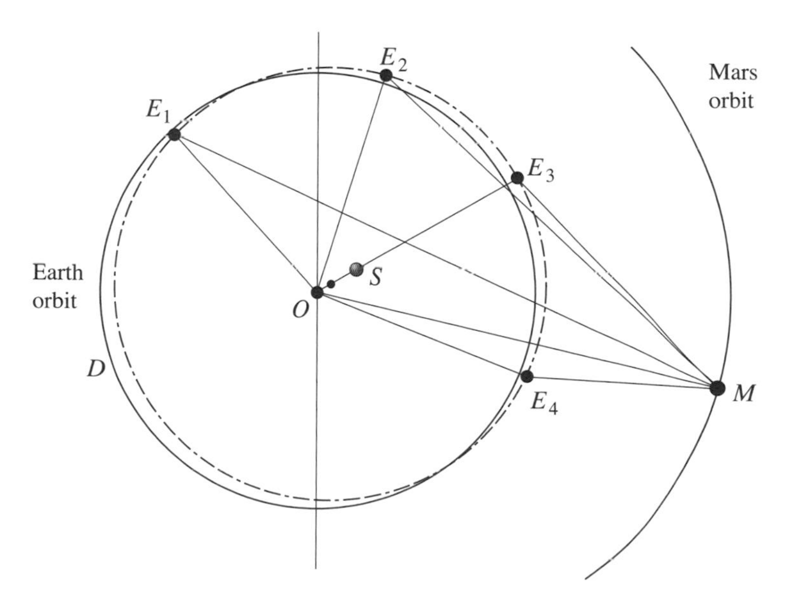

Fig. 3.14. Adapted from the first figure in Chapter 24 of Kepler's *Astronomia Nova*, 1609.

ber 8, 1593; and October 26, 1595), he marked the positions E_2, E_3, and E_4 of Earth in its orbit. Figure 3.14 captures what he recorded. The point O is the center of Earth's orbit, and S marks the position of the Sun. Kepler found the relevant angles in Tycho's data (listed there in degrees, minutes, and seconds):

For position E_1: $\angle E_1OM = 127.0836°, \angle E_1MO = 20.7958°$, and $\angle OE_1M = 32.1206°$

For position E_2: $\angle E_2OM = 84.1761°, \angle E_2MO = 35.7731°$, and $\angle OE_2M = 60.0508°$

For position E_3: $\angle E_3OM = 41.2711°, \angle E_3MO = 42.3583°$, and $\angle OE_3M = 96.3706°$

For position E_4: $\angle E_4OM = 1.6347°, \angle E_4MO = 3.3847°$, and $\angle OE_4M = 174.9806°$

Letting d be the distance OM, he turns to the law of sines: From ΔE_1OM, he gets:

$$\frac{\sin 32.1206°}{d} = \frac{\sin 20.7958°}{OE_1}, \text{ and therefore } OE_1 = 0.66774\,d.$$

From ΔE_2OM, he gets:

$$\frac{\sin 60.0508°}{d} = \frac{\sin 35.7731°}{OE_2}, \text{ and therefore } OE_2 = 0.67467\,d.$$

From ΔE_3OM, he gets:

$$\frac{\sin 96.3706°}{d} = \frac{\sin 42.3583°}{OE_3}, \text{ and therefore } OE_3 = 0.67794\,d.$$

Finally, from ΔE_4OM, he gets:

$$\frac{\sin 174.9806°}{d} = \frac{\sin 3.3847°}{OE_4}, \text{ and therefore } OE_4 = 0.67478\,d.$$

Since the radius of a circle is the same no matter where it is measured, Kepler concluded that "the circle that Copernicus described about the point O [the solid circle D in Figure 3.14] ... is not Earth's path. There is instead some other circle $E_1E_2E_3E_4$ on which the Earth is found, whose center lies in the direction of the Sun." So the circle that Copernicus had determined (see Figure 3.7) was off. Since Kepler knew that any three of the four points E_1, E_2, E_3, E_4 determine a circle, he could draw the corrected orbit of the Earth into his diagram (as the dashed curve in Figure 3.14). He could also determine where the center of this orbit is and draw it in (as the point in the figure between O and S). The location of the center of the revised orbit of Earth told Kepler that the distance OS is less than that of the Copernicus's original orbit. In reference to Figure 3.8 (with r the radius of the revised orbit), he determined that the center of the corrected orbit is a distance of about $c = 0.0153\,r$ from the Sun S. Notice that 0.0153 is roughly one-half of the value 0.0299 that Copernicus had achieved. (Albert Einstein referred to the approach that Kepler devised to correct Copernicus's circular orbit of the Earth as "an idea of true genius.")

When Kepler turns to the orbit of Mars, Figure 3.14 is again critical. This time, with the revised circular orbit of the Earth in place, he inserts Tycho's observations about the position M of Mars, and checks whether the recorded locations of M are consistent with the numerical data that he has for the positions of the points E_1, E_2, E_3, and E_4. His computations show that there are discrepancies. Kepler adjusts the position of M by moving it closer to S, and the differences decrease. But they persisted, and Kepler concluded his study by noting that the agreement would be precise "if we suppose that Earth's path is not a perfect circle, but is narrower at the sides ... ." The simple but ingenious idea to study the orbits of Earth and Mars simultaneously had allowed Kepler to correct Copernicus's circular orbit of the Earth and the location of its center from the Sun. It now suggested to him that these orbits might not be circles after all. They appeared to be curves that are pushed in at the sides. After three more years of grueling work, Kepler concluded that the orbit of Mars had to be an oval. Ruling out the possibility that the orbit is an ellipse—"if our figure were a perfect ellipse, the job would have been done by Archimedes"—he decided that it is "truly oval ... [like] an egg ... [with] two vertices, one more blunt, the other sharper, and

...visibly inclined at the sides." More hard work followed.

Finally, Kepler noticed something "quite by chance" when he compared the geometry of his orbit of Mars against a circle. Figure 3.15 considers a circle with center O, the position S of the Sun, and the narrower orbit of Mars. What Kepler noticed was that $\frac{CM}{MO} = 0.00429$ and also that $\frac{SM}{MO} = 1.00429$. That .00429 occurred twice in this way could not be a coincidence. It meant that $\frac{CO}{MO} = \frac{CM+MO}{MO} = 0.00429 + 1 = 1.00429$. Therefore $SM = CO = AO$. Kepler knew well that if the dashed curve through A and M were to be an ellipse with S at a focus, then the equality $SM = AO$ would follow as a direct consequence of its

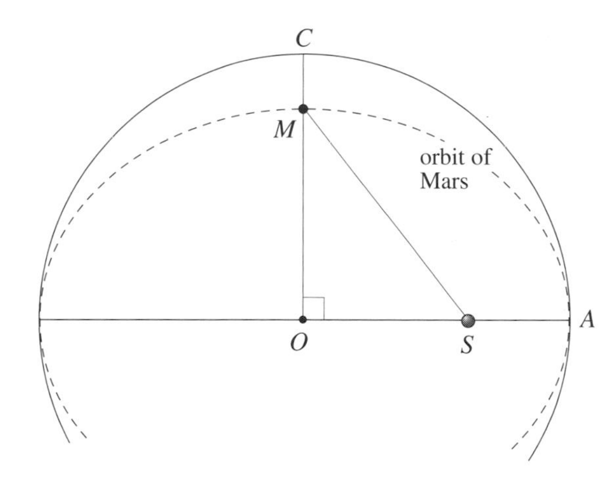

Fig. 3.15

geometry. So could it be ...? Kepler realized that this was a breakthrough and wrote, "I felt as if I were awakened from sleep to see a new light." From $\frac{CO}{MO} = 1.00429$, it follows that $\frac{MO}{CO} = 0.99573$. When Kepler noticed that this equality also applied to other points on the orbit of Mars, namely that

$$\frac{M_1O_1}{C_1O_1} = 0.99573, \ \frac{M_2O_2}{C_2O_2} = 0.99573, \ \frac{M_3O_3}{C_3O_3} = 0.99573,$$

and so on, as illustrated in Figure 3.16, he knew that the orbit of Mars satisfies the basic property of the

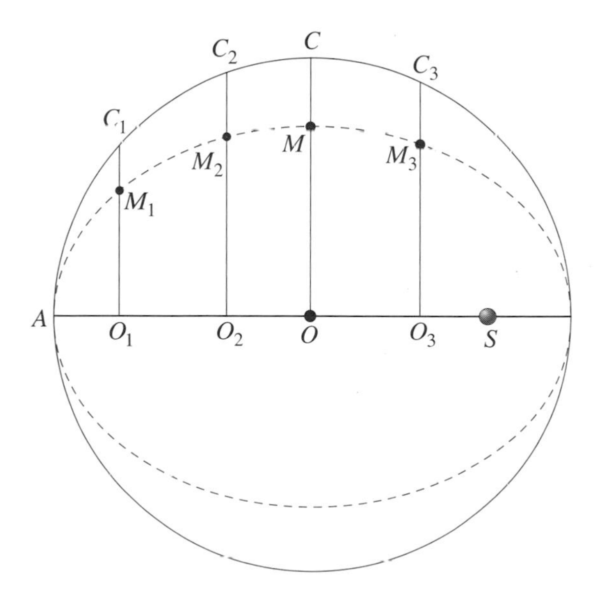

Fig. 3.16

ellipse as a curve obtained by starting with a circle, fixing a diameter, and shrinking the distances from the circle to the diameter by the same fixed coefficient. Now he was convinced. The orbit of Mars could only be an ellipse with the Sun at a focal point. What Kepler had discovered also told him what the eccentricity of this ellipse would be. By Figure 3.15 and the Pythagorean theorem,

$$OS^2 = SM^2 - MO^2 = AO^2 - (0.99573^2)CO^2 = (1 - 0.99148)AO^2 = 0.00852\,AO^2.$$

So $OS = 0.09230\,AO$, and it follows that $\frac{OS}{AO} = 0.09230$ is the eccentricity of the ellipse. Refer to Section 2.1 for some basic facts about the ellipse. (Sections 4.4 and 4.5 take up a more definitive algebraic analysis of this curve.)

By 1605, Kepler's battle with Mars was over. After disagreements with Tycho's heirs over the use of Tycho's data were resolved, Kepler published the story of his discoveries in 1609 in his book *Astronomia Nova* (*The New Astronomy*). This famous work tells the story of his many years of hard work as well as the evolution in his thinking. Rather than a polished account of his final, definitive conclusions, it is a story of the difficult process of scientific discovery. Kepler's scientific writing also deals with strange (by the standards of today's scientific literature) theological and mystical elements. The *Astronomia Nova* contains what are now called Kepler's first two laws of planetary motion.

1. Each planet P moves in an elliptical orbit with the Sun S at one of the focal points of the ellipse. See Figure 3.17. Incidentally, the word "focus" for each of the two defining points of the ellipse was

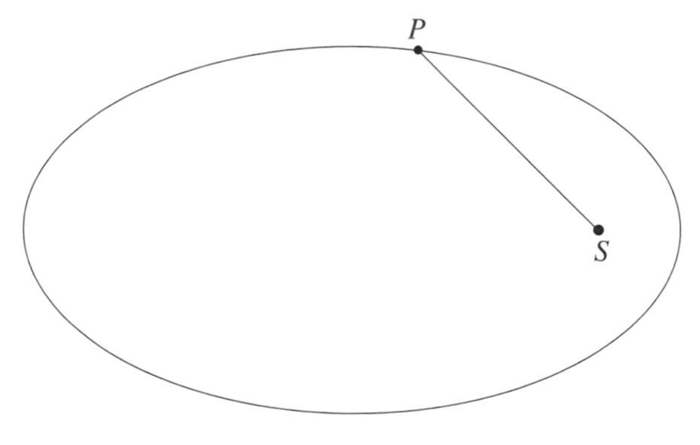

Fig. 3.17

provided by Kepler. It comes from the Latin word meaning "hearth."

2. A given planet sweeps out equal areas in equal times. Stated more precisely, this second law says the following: suppose that it takes a planet time t_1 to move from position P_1 to Q_1 in its orbit, and

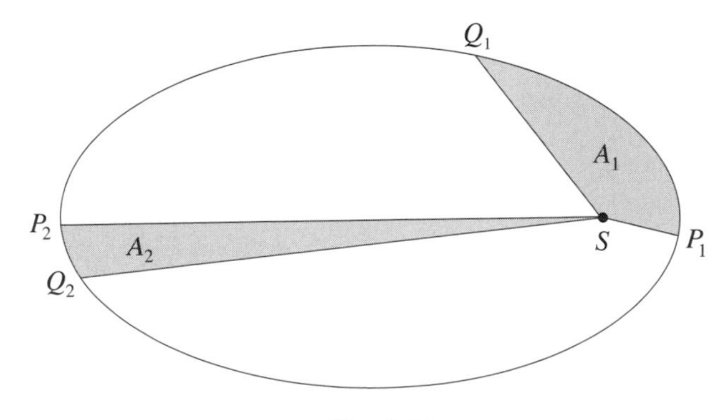

Fig. 3.18

that the segment from the Sun S to the planet sweeps out the area A_1 in the process. See Figure 3.18. Suppose that the planet later moves from position P_2 to Q_2 in time t_2 and that it sweeps out the area A_2. According to Kepler's second law,

$$\text{if } t_1 = t_2, \text{ then } A_1 = A_2.$$

A moment's reflection about Figure 3.18 shows that the second law implies that a planet moves faster when it is closer to the Sun than when it is farther away.

Let P be any planet, and consider its elliptical orbit. In Figure 3.19, the point O is the center of the ellipse and Sun S is at one of the focal points. The distance a is the semimajor axis of the ellipse, and c is

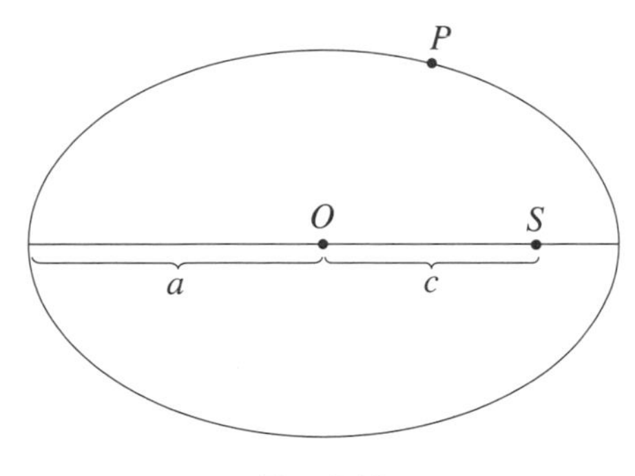

Fig. 3.19

the distance between O and S. The point of the orbit of P closest to S is called *perihelion*, and the point in the orbit of P farthest from S is called *aphelion*. Notice that when P is at perihelion, its distance from S is $a - c$. Similarly, at aphelion the distance of P from S is $a + c$. Since

$$a = \frac{(a + c) + (a - c)}{2},$$

the semimajor axis is the average of the greatest and least distances of the planet from the Sun. The ratio $\varepsilon = \frac{c}{a}$ is the eccentricity of the elliptical orbit. Notice that the closer ε is to zero, the closer the aphelion distance $a(1 + \varepsilon)$ is to the perihelion distance $a(1 - \varepsilon)$, and hence the closer the ellipse is to a circle. The length of time T it takes for P to complete one revolution is the *period* of its orbit.

Kepler published his third and final law in 1619 in his *Harmonices Mundi* (*The Harmony of the World*). It provides a connection between the orbits of any two planets around the Sun.

3. Let M and J be any two planets, and consider their elliptical orbits. Let a_M and a_J be the respective semimajor axes, and let T_M and T_J be their periods. Kepler's third law states that

$$\frac{a_M^3}{T_M^2} = \frac{a_J^3}{T_J^2}.$$

With his laws of planetary motion, Kepler had hit upon the three insights that are fundamental to the understanding of the workings of the solar system. The ellipses sketched above to illustrate Kepler's laws are much more "elliptical" or flatter than those of the planets. These are all close to being circles, a fact that makes Kepler's discoveries even more remarkable. The convoluted collection of circles with which Ptolemy and Copernicus described the motions of the planets was now superseded by a beautifully simple system of elliptical orbits. The speeds with which the planets trace out their orbits are also simply explained. His laws suggested to Kepler that there must be some force emanating from the Sun that keeps the planets in their orbits. Kepler suspected that such a force decreases with increased distance from the Sun. It is in fact the case that the average velocities of the planets fall off as their distances from the Sun increase. However, Kepler was unable to determine the precise quantitative measure of this force. This and related matters remain—as we shall see—for Isaac Newton to explain. It would be another 70 years before Newton identified gravitational force, established the inverse square law of gravitational attraction, and verified mathematically all that Kepler had observed.

In 1627, Kepler published the *Rudolphine Tables*. Dedicated to his patron, Emperor Rudolf II of Austria, in whose employ he was during his years in Prague, they were the first set of astronomical tables that made it possible to calculate the position of the planets with an accuracy sufficient to make reliable predictions. Earlier tables were much in use by navigators, astronomers, and astrologers, but their predictions gave errors of up to 5° of arc. The fact that the angular diameter of the Moon is about $\frac{1}{2}^\circ$ of arc gives a sense of the magnitude of such errors. The Rudolphine Tables provided an accuracy of $\frac{1}{6}^\circ$ of arc (and better). In putting together his table, Kepler made extensive use of the newly developed tool of logarithms (invented by the Scotsman John Napier only about 10 years before).

The few years that remained in Kepler's life were difficult. His religious beliefs—which saw no contradiction between his astronomy and the Bible—were attacked, and he was labeled a heretic. Europe had become embroiled in the Thirty Years' War. This conflict between Catholic and Protestant forces for the control of Europe devastated the population and left many cities in ruins.

The new astronomy did not find immediate and general acceptance. Ptolemy's description of the motion of the planets was difficult to dislodge, because it rested on the comprehensive system of Aristotelian thought, a tight weave of philosophical discourse, explanations, and observations, deeply rooted and difficult to refute. Aristotle had laid out a comprehensive approach to the investigation of all natural phenomena. He was the unchallenged authority at the time Copernicus and Kepler were proposing their new order for the universe. Central to the Aristotelian understanding was the distinction between the perfect, permanent celestial realm on the one hand, and the changeable, often volatile nature of earthly phenomena on the other. Mathematics, especially geometry, could inform the perfect, predictable celestial domain, and in particular the pursuit of astronomy. It could not, however, shed light on the ever-fluctuating, supposedly unpredictable happenings on or near Earth's surface. According to this view, mathematics could say nothing about the motion of cannonballs and bullets that are fired, or rocks and balls that are thrown. It was not until Aristotelian learning was drawn into question and overthrown that the new astronomy could make any serious headway. The challenge that the founders of the new astronomy faced was not to correct certain faulty theories and to replace them by better ones. Rather, they had to destroy one world order and replace it by another. They had to restructure the mind-set and intellectual framework within which science was approached. This difficult task fell in large measure to Galileo.

3.6 THE STUDIES OF GALILEO

Galileo Galilei (from 1564 to 1642) was born in the city of Pisa near Florence. He exhibited an independence of outlook at an early age. He was sent to the University of Pisa, but left four years later when his father could no longer afford the considerable fees. Guided by a family friend who was a distinguished mathematician, Galileo learned from the works of Plato, Aristotle, Euclid, Archimedes, and Ptolemy. He would soon master the geometry and physics of the Greeks. In time, however, he began to have questions, and he developed experiments to test what he was learning. He devised both "thought" experiments of penetrating reflections as well as actual experiments that generated numerical data that he could study.

Aristotle's theory of motion proclaimed that the heavier a body is, the faster it falls. Galileo had problems with this assertion, and engaged in the following thought experiment. He contemplated a cannonball in free fall. Next, he considered it to be cut into two pieces, and he thought of both pieces falling side by side. See Figure 3.20. From the point of view of the fall, nothing had changed. Therefore the cannonballs in the two situations must drop with the same speed. According to Aristotle, however, each of the two separate halves of the ball—being lighter—must drop more slowly than the ball as a whole. Aristotle had to be wrong. Galileo had discovered that all bodies fall in the same way, regardless of their weight (assuming that they are heavy enough and move slowly enough for air resistance to be negligible).

Fig. 3.20

In 1592, Galileo moved to Padua, near Venice, to accept the position of professor of mathematics at the university there. His attention turned to quantitative aspects of falling objects. Instead of considering a ball in free fall, he let it roll down an inclined plane. This slowed things down and made it possible to quantify observations. Galileo placed a spherical bronze ball, round and smooth, on the inclined plane and let it descend from rest. He fixed a unit of time and a unit of length. (In some of his experiments he used the *tempo* ≈ 0.01 seconds and the *punto* ≈ 0.94 millimeters.) He measured and recorded the distance c the ball rolled from time $t = 0$ to $t = 1$. The ball continued to roll and pick up speed. More careful measurements of times and distances revealed that the ball rolled a distance $3c$ from time $t = 1$ to $t = 2$, a distance $5c$ from time $t = 2$ to $t = 3$, and a distance $7c$ from time $t = 3$ to $t = 4$, and so on. See Figure 3.21. So the ball traveled the distances

$$\begin{aligned} &c \text{ from the start at } t = 0 \text{ to time } t = 1,\\ &c + 3c = 4c = 2^2c \text{ from the start to time } t = 2,\\ &c + 3c + 5c = 9d = 3^2c \text{ from the start to time } t = 3,\\ &c + 3c + 5c + 7c = 16c = 4^2c \text{ from the start to time } t = 4, \end{aligned}$$

and so on. The evident pattern is that the distance rolled by the ball is proportional to the square of the

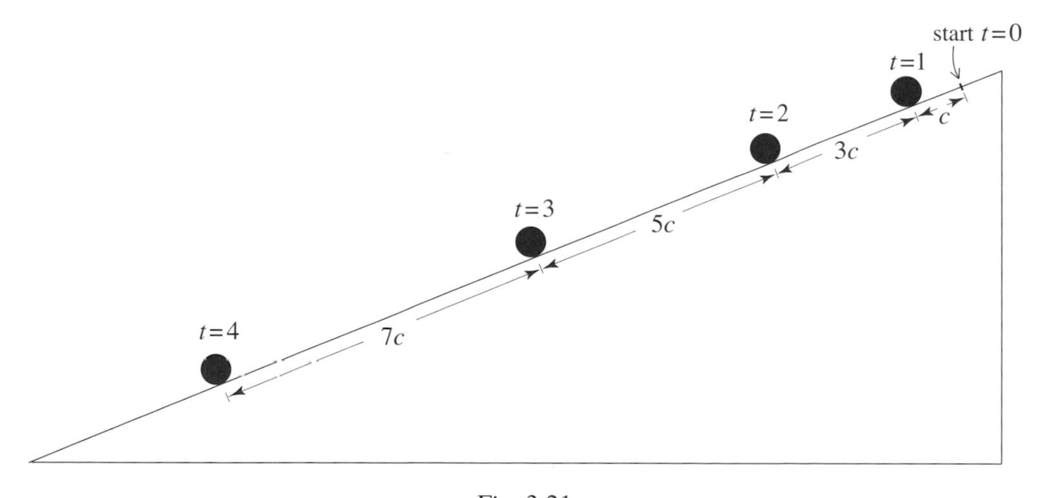

Fig. 3.21

time that the motion takes. So if the ball rolls down an inclined plane starting from rest, then after an elapsed time t, the distance that the ball has rolled from its start is given by

$$ct^2,$$

where c is a constant that depends on the inclination of the plane. Galileo undertook such experiments in the years 1600–1608 in Padua. Much later, in his treatise *Discorsi e dimostrazioni matematiche, intorno à due nuove scienze* (*Discourses and Mathematical Demonstrations Relating to Two New Sciences*) of 1638, he recalled that "in such experiments, repeated a full hundred times, we always found that the spaces traversed were to each other as the squares of the times, and this was true for all inclinations of the plane ..."

Because the expression ct^2 is true for any incline plane, Galileo reasoned that the same equation is also valid in the case of free fall. So if a ball is dropped from rest and falls for a time t, then (again ignoring the effects of air resistance) it will fall a distance of ct^2, where c is some constant. This is Galileo's famous *times squared law of fall*. It appears that Galileo did not (or could not) determine an explicit value for the constant c. (After Galileo's time, g was thought to be about 7.3 meters/second^2, or 24 feet/second^2 in today's units. Later in the 17th century, the Dutch scientist Huygens found a value close to today's $c = \frac{g}{2}$ where g is the gravitational constant 9.8 meters/second^2, or 32 feet/second^2.)

Aristotle's physics also asserts that all objects have a natural motion toward the center of the Earth and that the motion of an object in any other direction is possible only so long as a mover is in contact with the object to drive it onward. The moving agent in the case of a thrown object (once it leaves the hand of the thrower) is the disturbance produced in the air. As it moves, a temporary vacuum is formed behind the object. This vacuum is filled by the compressed air that rushes around the object from the front. This in turn creates a void in front of the object, which is filled as the object continues to move forward. For Aristotle, no motion is possible without air, so in a vacuum, motion is impossible. In brief, the fundamental principle of Aristotle's theory of motion is this: an object that is in motion will stop unless an external force continues to propel it.

Galileo, however, was skeptical and returned to his experiments. He released a ball at a certain elevation on an inclined plane. He let it roll up another inclined plane and observed that it reached its original height. See Figure 3.22. He took a second inclined plane with a smaller inclination and noticed that the same was true. Galileo continued this experiment in his mind, imagining inclined planes of smaller and smaller inclinations. Carrying his thinking to its logical end, he concluded that on a flat surface (with

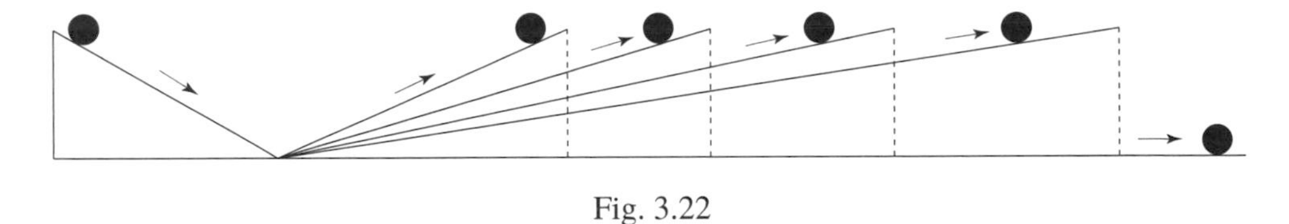

Fig. 3.22

no friction other than the cause of the roll), a perfectly spherical ball would roll forever. Galileo had thus conceived of a horizontal version of the *law of inertia*: "we may remark that any velocity once imparted to a moving body will be rigidly maintained as long as the external causes of acceleration or retardation are removed, a condition which is found only on horizontal planes ... ." Put another way, a body that is in a state of either rest or horizontal motion will continue in this same state, unless it is acted upon by some external force. This stands in sharp contrast to Aristotle's principle.

During his years in Pisa, Galileo let his bronze ball roll off a horizontal table and observed its curving flight to the ground carefully. He thought as follows about the trajectory that the ball describes:

> Imagine any particle projected along a horizontal plane without friction, then we know ... that this particle will move along this same plane with a motion that is uniform and perpetual ... but if the plane is limited and elevated, then the moving particle, which we imagine to be a heavy one, will on passing over the edge of the plane acquire, in addition to its previous and uniform and perpetual motion, a downward propensity ... so that the resulting motion ... is compounded of one which is uniform and horizontal and of another which is vertical and naturally accelerated.

So it was Galileo's insight that the flight of the ball—or that of any projectile—could be regarded as the *simultaneous composite of a horizontal motion and a vertical motion.* By applying his earlier observations, Galileo concluded that the ball's trajectory is parabolic. Today, we can formulate Galileo's thinking as follows. Suppose that d is the height of the table, and let the ball start its flight to the ground at time $t = 0$. See Figure 3.23. Consider its position at any time t later. Let x be the horizontal distance that the ball has moved during time t. By Galileo's law of horizontal inertia, its horizontal motion is "uniform" and "rigidly maintained," so that the speed $\frac{x}{t}$ of the ball in the horizontal direction is equal to some constant s.

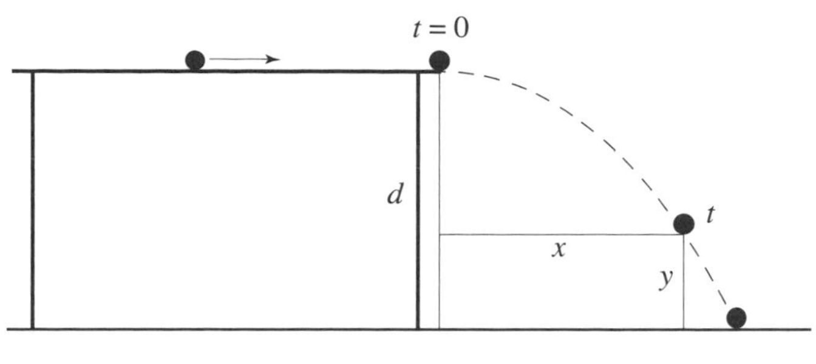

Fig. 3.23

If y is the distance of the ball above the ground at time t, then by Galileo's times squared law of fall, $d - y = ct^2$ for the constant c. Since $t = \frac{x}{s}$, we get that

$$y = -c(\tfrac{x}{s})^2 + d = -\tfrac{c}{s^2}x^2 + d.$$

Elementary coordinate geometry (as explained in Section 4.3) will tell us that this is the equation of a parabola. Remarkably, at around the time that Kepler determined that planetary orbits are ellipses, Galileo discovered that the curving trajectory described by thrown object is a parabola. When Galileo carried out his investigations, the real number system was as yet not established. Consequently, Galileo did not express his conclusions in terms of algebraic equations, and worked instead with the proportions of the Greeks.

Of course, some of Aristotle's thinking is consistent with ordinary common sense. After all, a real spherical ball would not roll forever on a horizontal plane. As a consequence, a correct explanation of the phenomenon of motion and the laws that govern it could not have been achieved by better observations from within the older system of ideas. A new mind-set and framework were required. To correctly understand what was going on, it was necessary to think not about real bodies as they are actually observed, but about idealized bodies moving in idealized space, one without friction and resistance. In ordinary experience, there are no perfectly spherical balls moving on perfectly smooth and straight inclined planes and rolling away to infinity. But by imagining them, Galileo stripped away the inessentials and revealed the laws at the core. He knew that there would be discrepancies between the conclusions of an experiment and the assertions of the basic laws. But he also knew that the interfering factors must be separated from the fundamental principles. Upon arriving at the basic principles in this abstracted context, the variables of friction and resistance can be reinserted. Greek geometry had done a similar thing in a much simpler context. Its focus was not on the wheel and the ball, but on their abstractions, the circle and the sphere.

Galileo also had a considerable impact on astronomy. He heard about the invention of the telescope, fashioned his own, and turned it toward the sky in the years 1609 and 1610 with spectacular results. When

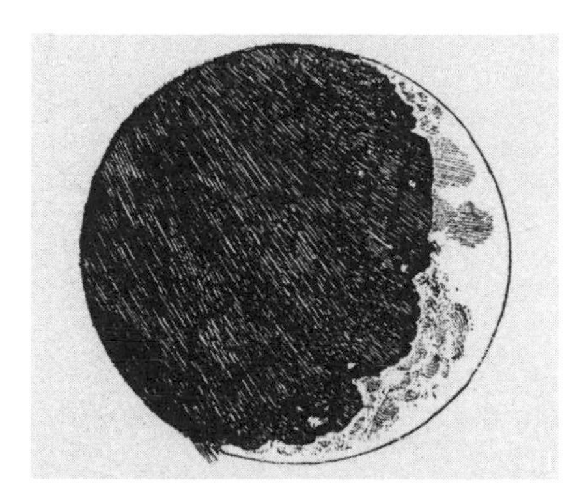

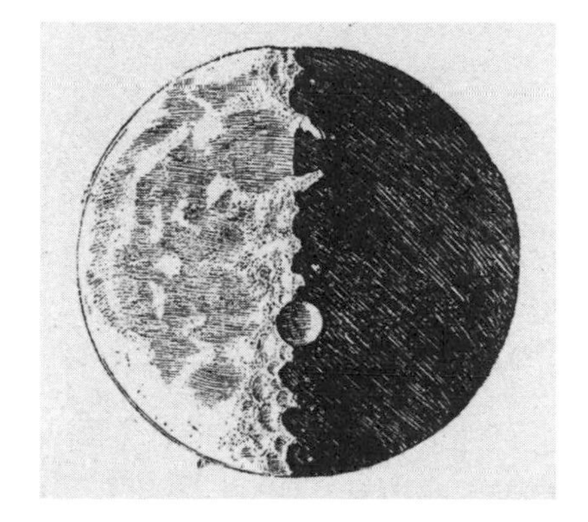

Fig. 3.24. Galileo's engravings of the lunar surface from the *Sidereus Nuncius*, 1610.

he looked at the Moon, he saw clearly that is was not at all a perfect sphere, as the Greek philosophers had asserted, but that it was deeply scarred and cratered. Figure 3.24 shows two sketches from Galileo's book the *Sidereus Nuncius* (*Starry Messenger*) that describe what he observed.

In January of 1610, he trained his telescope on Jupiter. Diagrams in the *Siderius Nuncius* tell us what he saw. See Figure 3.25. He discovered—sometimes on the left, sometimes on the right, and at times

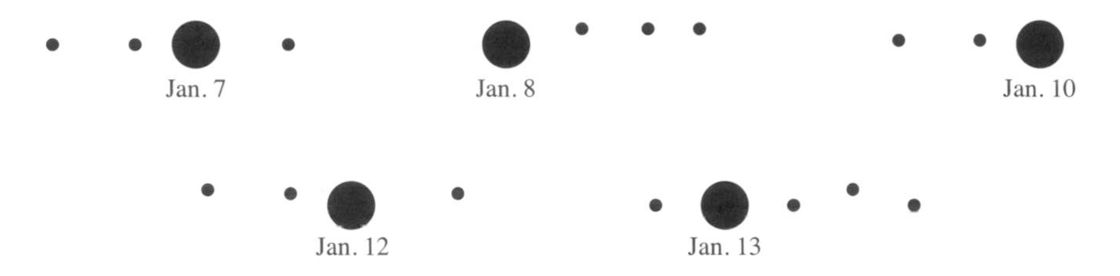

Fig. 3.25

hidden from view altogether—what were most certainly four moons changing position as they orbited the planet. The existence of this system, which certainly does not have the Earth at its center, gave support to the viewpoint of Copernicus. Galileo studied the phases of Venus and concluded that the pattern they exhibit is consistent with the heliocentric hypothesis. In the Ptolemaic system of Figure 3.26, Venus (as seen from Earth) always appears more or less crescent shaped. In the Copernican system of Figure 3.27,

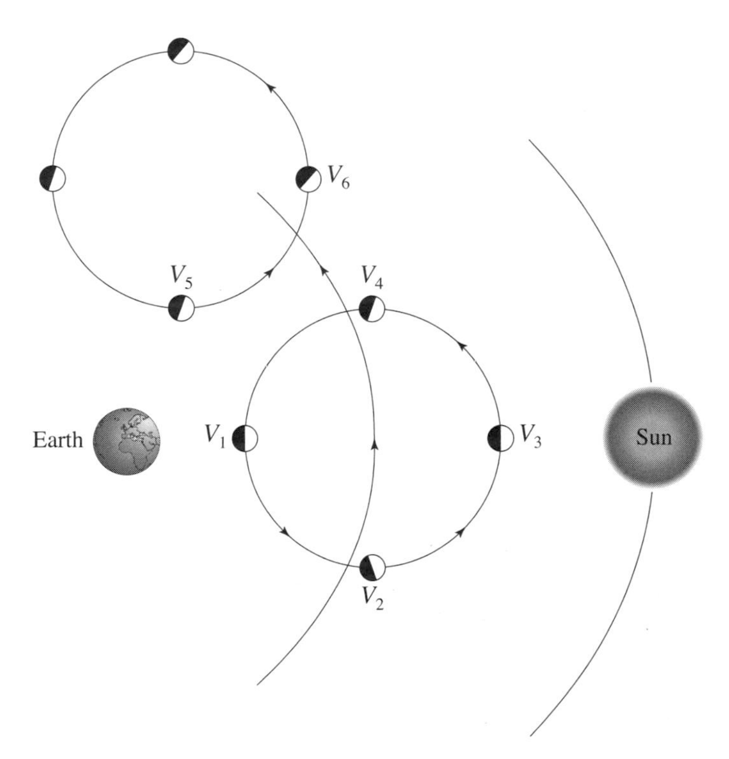

Fig. 3.26

however, Venus has the full range of phases that Galileo observed. Therefore the basic configuration of the Ptolemaic model had to be in error and that of Copernicus correct. He became a vocal advocate of the Copernican arrangement of the universe. Galileo's eloquent treatise *Dialogo sopra i due massimi sistemi del mondo* (*Dialogue Concerning the Two Chief World Systems*) of 1632, on its face an impartial comparison of the merits of the Ptolemaic and Copernican systems, did not hide his support of the latter.

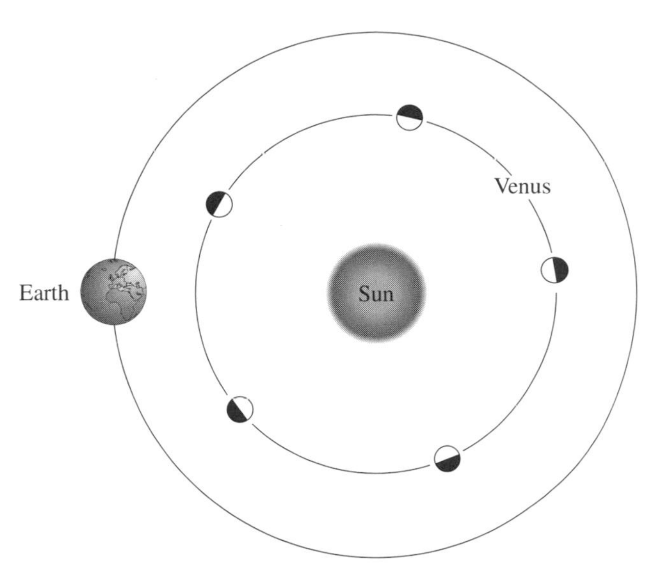

Fig. 3.27

As we have seen, Galileo's contributions to the development of science were significant. His new experimental approach to the understanding of the phenomenon of motion began the transition from medieval to modern science, and his discoveries with the telescope and writings about astronomy gave validity to the Copernican heliocentric description of the universe. Galileo was an exceptional teacher and eloquent speaker, and his fame soon spread. When a great hall with a capacity of more than 1000 proved too small, his astronomy lecture would be moved outside. He decided to publish his findings in pamphlets, dialogues, letters, and books. He wrote not in the Latin of the academy, but in Italian, the language of the street. The points of view that Galileo promoted would lead to difficulties with the Church. Challenging Aristotle's physics or regarding the Copernican system as a mathematical theory was one thing, but proposing that the Copernican point of view explained the workings of God's heaven was quite another. This was contradictory to scripture, undermined the authority of the Church, and had to be fought. After 20 years of hearings, misunderstandings, charges, revisions, and injunctions, Galileo was forced to recant, and the Inquisition sentenced him to house arrest for life.

The general acceptance of Keplerian heliocentric astronomy by the scientific community came in the latter part of the 17th century, when Newton provided a mathematical explanation of the orbits of the Earth and the planets based on the basic laws of motion and the concept of centripetal force. He showed that the equations of gravitational force applied to provide a unified explanation of motion on Earth and in the heavens from Galileo's parabolas to Kepler's elliptical orbits.

3.7 THE SIZE OF THE SOLAR SYSTEM

Ptolemy's value of around 8,000,000 kilometers (5,000,000 miles) for the distance from Earth to the Sun was accepted by Copernicus, and neither Kepler nor Galileo had a much better grasp of it. But in the latter part of the 17th century, this distance was reassessed when two of the star astronomers of the day set about the task of calculating it anew. The Italian Giovanni Cassini (1625–1712) was already famous for his observations of Jupiter and Saturn when he was called by the French king to direct the new astronomical observatory in Paris. The other was John Flamsteed (1646–1719), a young astronomer working in central England, who would soon become the first Astronomer Royal of the observatory in Greenwich that the king of England was establishing.

Both Cassini and Flamsteed knew that the fall of the year 1672 would provide a great opportunity for such a calculation. For a week or two, Mars would not only be in opposition with Sun, Earth, and Mars aligned (in the positions S, E', and M' depicted in Figure 3.12), but Mars would also be near its perihelion. Both Cassini and Flamsteed knew the eccentricity $\varepsilon_E = 0.017$ of Earth's orbit and with a_E its semimajor axis, that the Earth's distance from the Sun varies from $a_E(1-0.017) = 0.983a_E$ (at perihelion) to $a_E(1+0.017) = 1.017a_E$ (at aphelion) or from about 98% of a_E to about 102% of a_E. This variation fell within their tolerance for error, so they assumed that the distance from the Earth to the Sun is equal to a_E. With Mars they had to be more careful. With a_M the semimajor axis of its orbit, and Kepler's value $\varepsilon_M = 0.0926$ for its eccentricity, they knew that the distance from Mars to the Sun varies from $a_M(1-\varepsilon_M) = a_M(0.9174)$ at perihelion to $a_M(1+\varepsilon_M) = a_M(1.0926)$ at aphelion, or from about 90% of a_M to about 110% of a_M. However, since Mars would be near its perihelion, they could assume that its distance from the Sun was close to $a_M(1-\varepsilon_M)$. Therefore they could let d be the distance between Earth

Fig. 3.28

and Mars, work with the diagram in Figure 3.28, and use the approximations

$$a_E + d \approx a_M(1-\varepsilon_M),$$

and hence $a_M \approx \dfrac{a_E + d}{1-\varepsilon_M}$. Letting T_E and T_M be the periods of the orbits of Earth and Mars, Flamsteed and Cassini could apply Kepler's third law to get

$$\frac{a_E^3}{T_E^2} = \frac{a_M^3}{T_M^2} \approx \frac{(a_E+d)^3}{(1-\varepsilon_M)^3 T_M^2},$$

hence

$$\frac{a_E}{T_E^{2/3}} \approx \frac{a_E+d}{(1-\varepsilon_M)T_M^{2/3}},$$

and therefore

$$a_E \cdot (1-\varepsilon_M)\left(\tfrac{T_M}{T_E}\right)^{2/3} \approx a_E + d.$$

So they got $d \approx [(1-\varepsilon_M)\left(\tfrac{T_M}{T_E}\right)^{2/3} - 1]a_E$ and

$$a_E \approx \frac{d}{(1-\varepsilon_M)\left(\tfrac{T_M}{T_E}\right)^{2/3} - 1}.$$

Knowing that the periods of the orbits of Earth and Mars were $T_E = 365.25$ days and $T_M = 686.95$ days, Cassini and Flamsteed got $(1-\varepsilon_M)\left(\tfrac{T_M}{T_E}\right)^{2/3} - 1 \approx (0.9074)(1.5237) - 1 = 0.3826$, and they could conclude that

$$a_E \approx 2.61d.$$

It therefore remained for Cassini and Flamsteed to measure the distance d. They both thought that Mars would be near enough and its angle of parallax from an appropriate baseline on Earth large enough to make a reasonably accurate estimate of the distance d possible. With the distance d and hence a_E in hand, the distances between all other planets and the Sun could be computed by the trigonometric methods that Copernicus had already developed. See Section 3.3. This would answer the central question in astronomy of the time: What is the size of the solar system?

A lot had changed since Tycho Brahe pointed his instruments to the sky and noted that the new star that he observed had no parallax and since Galileo first looked through a telescope to study the heavens. The telescopes that Cassini and Flamsteed had available were more powerful and precise than those of

Galileo. They were equipped with micrometer eyepieces and telescopic sights that made it possible to measure angular separations to within small fractions of degrees. Cassini and Flamsteed were confident that with their instruments they would be able to measure the parallax of Mars, and hence determine its distance from Earth. But they also knew that this would be a delicate task that faced a number of serious difficulties: the angles of parallax for Mars would be small, in spite of the fact that Mars is relatively close to Earth. They knew that assessing the parallax of planets and stars is like assessing the thickness of a human hair held at arm's length. In addition, Mars would change its observed position not only as a consequence of parallax, but also because of its continuing motion in its orbit. Finally, there was the problem of refraction. Light rays bend in the Earth's atmosphere when they travel from an object in the skies to an observer on Earth. So the actual and observed positions of the object in the sky are slightly different. But Cassini and Flamsteed pressed ahead. Since Mars in opposition would be close to Earth and under full sunlight at night, the conditions for detecting and measuring the parallax of the planet were optimal.

The strategy that Cassini and Flamsteed needed to pursue is simple in principle. From an observation point B, Mars is sighted as a point of light in the night sky at a location M^* within a cluster of the fixed stars of a familiar constellation. This is repeated from a different location B' far from B, with the result that the observed position of M^* within the same star cluster will have shifted slightly. See Figure 3.29. If our astronomers measure this shift accurately, they can derive an estimate of the angle θ, and they can

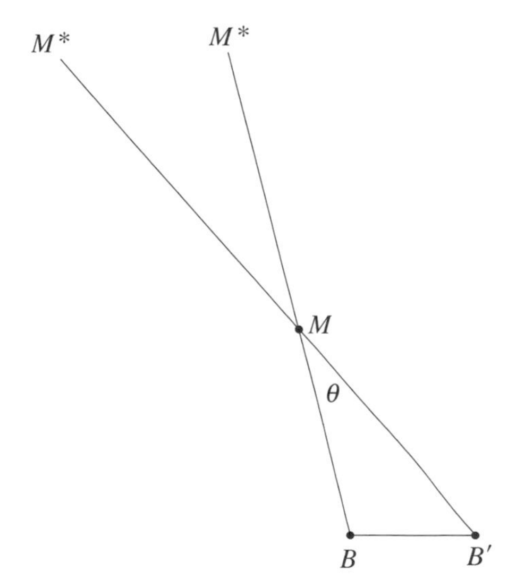

Fig. 3.29

then approximate the distance between Mars and Earth.

The determination of the angle θ proceeds as follows. Refer to Figure 3.30a. From the vantage point B, the astronomer fixes a star A in the cluster and carefully measures the angle $\alpha = \angle ABM^*$. After the location of his observation post has changed to B', he locates the star A once more and measures the angle $\alpha' = \angle AB'M^*$. Because A is far from Earth, the lines of sight BA and $B'A$ are essentially parallel. It follows from Figure 3.30b that

$$(\alpha + \beta) + (\alpha' + \beta') = \pi \text{ and } \beta + \beta' + \theta = \pi.$$

After a substitution, $\alpha + \alpha' - \theta + \pi = \pi$, so that $\theta = \alpha + \alpha'$ is determined. The *angle of parallax* $p(M)$ of M relative to the baseline BB' is defined by $p(M) = \frac{1}{2}\theta$. Notice that $p(M) = \frac{1}{2}\theta = \frac{1}{2}(\alpha + \alpha')$ is the average of the measured angles α and α'. Because the angles involved are small, the angle of parallax $p(M)$ and related angles are measured in seconds. With 1 degree = 60 minutes and 1 minute = 60 seconds,

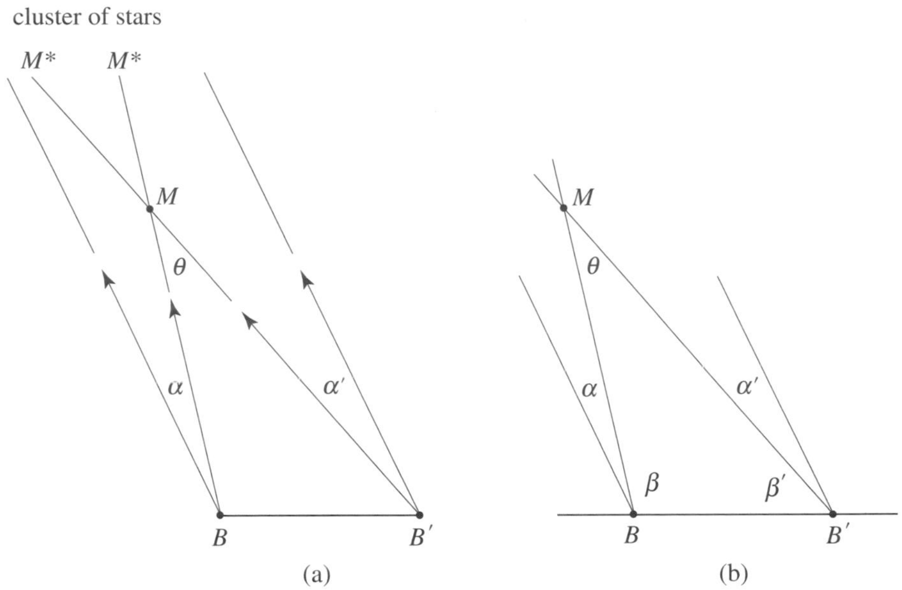

(a) (b)

Fig. 3.30

$1° = 3600''$. Since $180°$ is equal to π radians, the angle of parallax expressed in radians is

$$p(M) \times \frac{1}{3600} \times \frac{\pi}{180} = p(M) \times \frac{\pi}{648{,}000} \approx p(M) \cdot 4.85 \times 10^{-6} \text{ radians.}$$

Since the distance from Earth to Mars is huge relative to the distance between the observation points B and B', the distances from B to M and B' to M can be taken to be equal. We will set this distance equal to $d(B,M)$ and assume that the triangle $\Delta BMB'$ is closely approximated by the circular sector BMB' of

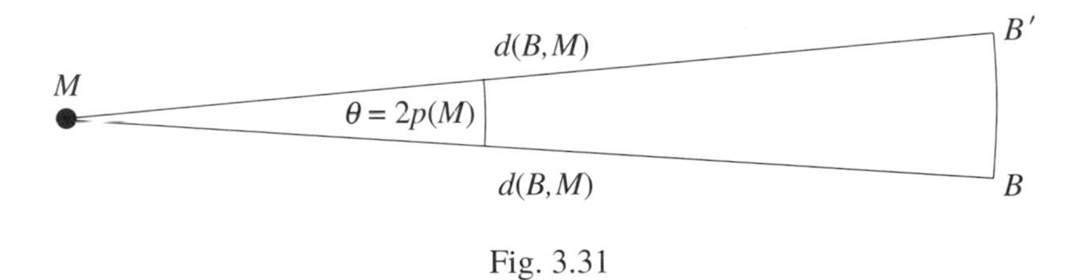

Fig. 3.31

Figure 3.31. Using the definition of radian measure, we get $2p(M)(4.85 \times 10^{-6}) \approx \frac{BB'}{d(B,M)}$. Therefore

$$d = d(B,M) \approx \frac{BB'}{9.7p(M)} \times 10^6,$$

where $p(M)$ is the angle of parallax of Mars in seconds. It follows that the angle $p(M)$ together with the distance BB' provide the approximate distance between Earth and Mars.

Flamsteed made all the necessary measurements during a single night. Late in the evening of October 6, 1672, he observed Mars from his observation post in Derby, his hometown in central England. The point B of Figure 3.32a marks his location on the surface of our planet. Precisely 6 hours and 10 minutes later, he observed Mars again. The Earth had turned by slightly more than a fourth of a complete rotation, and Flamsteed's observation post had rotated to the point B' in the figure. Flamsteed knew the latitude of his location (the angle between the segment OB from Earth's center O and the equatorial plane) to be about $53°$, so that he could compute the length BB' of his baseline to be the equivalent of about 5300 kilometers

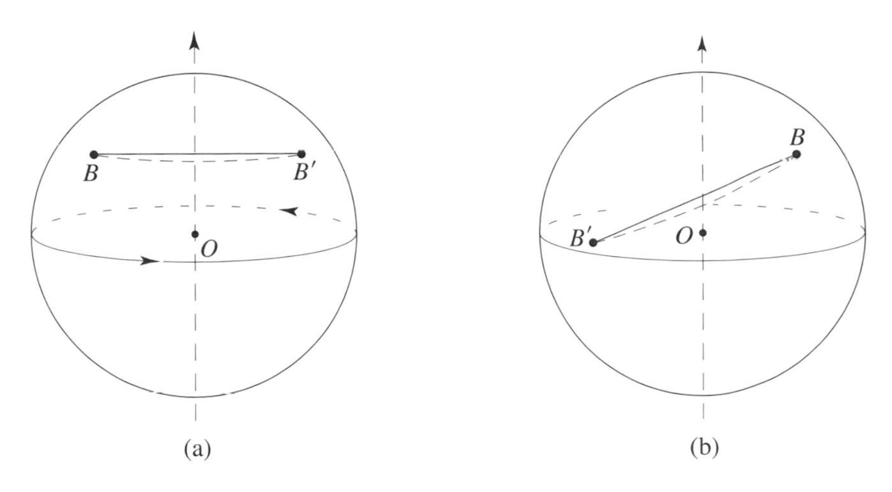

Fig. 3.32

(or about 3300 miles). As Figure 3.30a illustrates it, Flamsteed peered into the night sky and found M^* near three bright stars in the constellation Aquarius. Letting A be one of them, he measured the angle $\alpha = \angle ABM^*$ and later the angle $\alpha' = \angle AB'M^*$. He found the angle $\theta = \alpha + \alpha'$ to be about 21 seconds and concluded that the angle $p(M) = \frac{1}{2}\theta$ of parallax (with respect to the base line BB') was about 10.5 seconds. Doing the equivalent of plugging his information into the formula derived above gave Flamsteed the estimate

$$d = d(B,M) \approx \frac{BB'}{9.7p(M)} \times 10^6 \approx \frac{5{,}300}{(9.7)(10.5)} \times 10^6 \approx 52{,}000{,}000 \text{ kilometers.}$$

Therefore Flamsteed arrived at the value $a_E \approx 2.61d \approx 136{,}000{,}000$ kilometers (or 85,000,000 miles) for the semimajor axis a_E of the Earth's orbit, and hence the average distance from Earth to the Sun.

As director of the Paris observatory, Domenico Cassini had considerable resources at his disposal. It was his idea to approach the measurement of the parallax of Mars from two different vantage points B and B' on Earth at the same time! This would eliminate the difficulty of having to quantify the motion of the planet during the time between measurements. While Cassini remained in Paris, his colleague Jean Richer was sent on an expedition to Cayenne in French Guiana, a French colony in South America just north of the equator (on the Atlantic coast near the northernmost tip of Brazil). Knowing the latitude and longitude of Cayenne (today this is specified as about 5° north of the equator and 52° west of Greenwich), Cassini could estimate the distance between Paris and Cayenne to be the equivalent of about 6700 kilometers (or 4200 miles). Refer to Figures 3.30 and 3.32b, taking B to be Paris and B' to be Cayenne. In September and October of 1672, much as Flamsteed had done, both men observed Mars at M^* close to a star A in the constellation Aquarius. While Cassini and his assistants measured the angle $\alpha = \angle ABM^*$ in Paris, Richer measured the angle $\alpha' = \angle AB'M^*$ in Cayenne. By comparing solar information (such as the times of sunrise and the times the Sun reached its highest position in the sky) at the two locations, they were able to make repeated pairs of measurements at close to the same time. Cassini then waited nearly a year for his colleague to return to Paris with his data! After evaluating their data carefully, they concluded that the angle $\theta = \alpha + \alpha'$ was about 26 seconds. So the corresponding angle of parallax $p(M) = \frac{1}{2}\theta$ with respect to their baseline BB' was approximately 13 seconds. This provided Cassini with the estimate

$$d = d(B, M) \approx \frac{BB'}{9.7\, p(M)} \times 10^6 \approx \frac{6{,}700}{(9.7)(13)} \times 10^6 \approx 53{,}000{,}000 \text{ kilometers,}$$

and therefore the value $a_E \approx 2.61d \approx 140{,}000{,}000$ kilometers (or 87,000,000 miles) for the semimajor axis of Earth's orbit. This was slightly better than what Flamsteed had achieved and only about 7% less than today's value of 150,000,000 kilometers. In view of what Copernicus already knew (see Table 3.1), it was now understood that Saturn, the farthest of the known planets, was about $9.51 \times 140{,}000{,}000 \approx 1{,}300{,}000{,}000$ kilometers (or 830,000,000 miles) from the Sun. For the first time, astronomers had a good sense of the actual size of the solar system! (Incidentally, at the time Cassini and Flamsteed made their measurements, the *Earth radius* was the primary unit of distance that astronomers used. The meter and kilometer did not as yet exist—they would be introduced in France after the French Revolution.)

What Cassini and Flamsteed observed in the summer and fall months of the year 1672 is depicted in Figure 3.33. It shows the dotted sequence of the positions of Mars against the stars of Aquarius. On the night of Flamsteed's observations, Mars was at its turnaround from retrograde to normal orbital motion. This meant that the shift in the observed position of Mars between the two measurements was

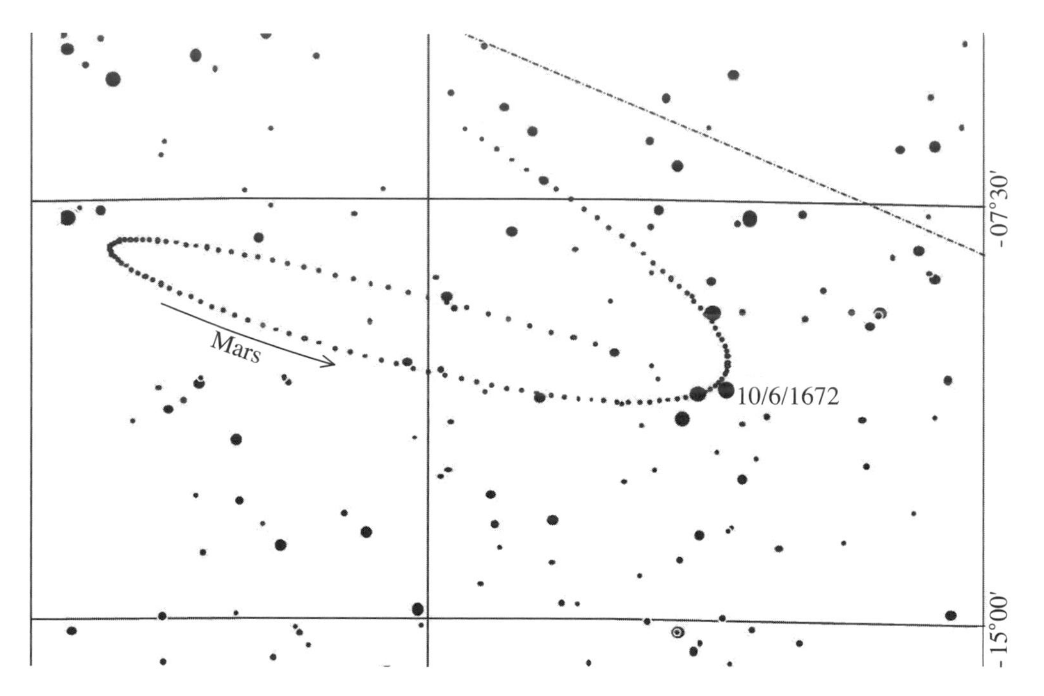

Fig. 3.33. Image reproduced with permission from Parker Moreland, who created it from a screenshot made possible by TheSky Astronomy Software © Software Bisquc, Inc.

almost exclusively due to parallax rather than the motion of the planet itself.

And yet, in spite of what Flamsteed and Cassini had achieved, uncertainty and questions remained. In fact, Flamsteed and Cassini expressed doubts. In their published accounts, both astronomers stood by their results, but in public lectures they suggested that the only conclusion to be drawn from their observations of 1672 was that the parallax that they had measured was smaller than the accuracy of their instruments. An astronomer who tried to replicate these findings later arrived at inconsistent results and could only conclude that the parallax of Mars was hiding behind instrumental uncertainties. It would take another 200 years before there was unquestioned success. In 1877, the a Scottish astronomer David Gill went off to Ascension Island in the South Atlantic (in the middle of nowhere between Brazil and Africa), waited for Mars to be in opposition, accurately measured the parallax of Mars, and determined the semimajor axis of Earth's orbit to be 149,800,000 kilometers (or 93,080,000 miles). The telescope he used was highly precise. It had an objective lens that was split into two halves. With one half fixed, the other half could slide along the separating diameter to produce a double image of a celestial object. An accurate measurement of the angular separation between two nearby objects A and A' could be made as follows. Start by superimposing the two images of the objects. Then keep the image of A fixed and adjust the calibrated sliding mechanism in such a way that the superimposed image moves to coincide with the observed position of A'. In this way, the angular separation between A and A' could be measured to within a fraction of a second.

In the 20th century, after the invention of radar and the precise determination of the speed of light, distances in the solar system could be determined accurately by timing (with atomic clocks) how long it takes a radar beam traveling at the speed of light to travel to an object and bounce back. This approach resulted in the value of 149,597,892 kilometers for the semimajor axis of Earth's orbit. This measurement has since become the basis of the official definition of the *astronomical unit* au as

$$1 \text{ au} = 149{,}597{,}892 \text{ kilometers}.$$

The au is the unit with which distances in the solar system are usually listed. Table 3.3 provides current values for the basic orbital data for Mercury, Venus, Earth, Mars, Jupiter, and Saturn, the six planets that

	Mercury	Venus	Earth	Mars	Jupiter	Saturn
Semimajor axis (in au)	0.387	0.723	1.000	1.524	5.203	9.537
Orbital eccentricity	0.206	0.007	0.017	0.093	0.048	0.054
Orbit period (in years)	0.241	0.615	1.000	1.881	11.863	29.447

Table 3.3

were known at the time of the heliocentric revolution in astronomy. The seventh planet—Uranus—would not be discovered for another 100 years (in 1781) and the eight planet—Neptune—not for another 150 years (in 1842).

3.8 PROBLEMS AND PROJECTS

3A. A Shadow Measurement. We begin by putting one of the shadow measurements of the Greeks discussed in Section 1.1 into the heliocentric context. Refer to Figure 3.1 and rotate the spring equinox position of the Earth in the plane P so that it is in the same "vertical" plane as its summer solstice position. This is done in Figure 3.34. The points C and C' denote the center of the Earth, and S is the center of the Sun. (The figure is not drawn to scale.) Notice that $\angle CSC'$ is the angle between the Earth's equatorial plane and the plane of its orbit around the Sun. This angle was called the obliquity of the ecliptic by the ancients. It was denoted by ε in Section 1.1, but it will now be denoted by γ. (The symbol ε will from now on be reserved for the eccentricity of an elliptical orbit.) The Earth's axis of rotation and the regions of

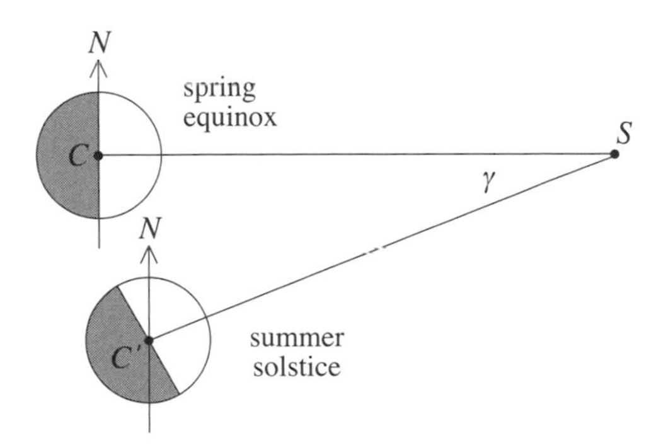

Fig. 3.34

sunlight and shadow on the Earth are indicated in the figure. The points A and A' in Figure 3.35a represent the same point on the surface of the Earth, and the points G and G' are the tips of a vertical gnomon placed at A and A'. (Recall that a gnomon is simply a straight stick with which angles of shadows are measured. The gnomons are drawn larger so as to make them and the shadows that they cast visible in the diagram.)

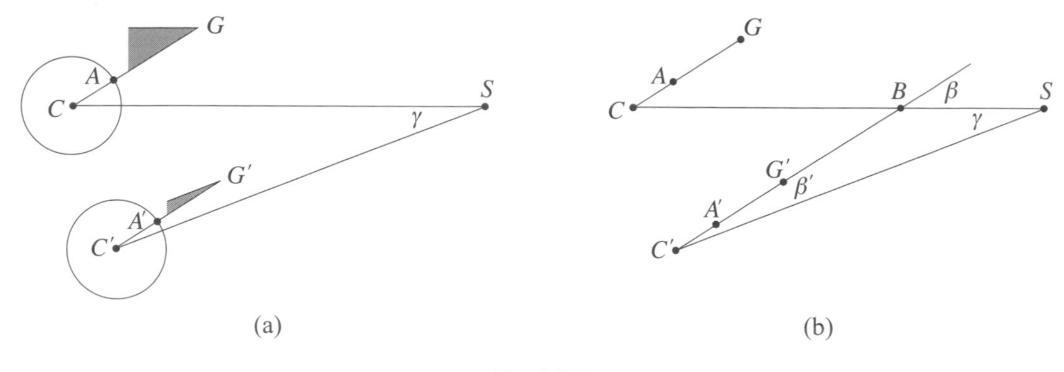

Fig. 3.35

Denote the angles at G and G' by α and α', respectively. Figure 3.35b is derived from Figure 3.35a. The segment $C'A'G'$ is extended to a point B on CS, and β and β' are the indicated angles.

3.1. **i.** Why are the segments CAG and $C'A'G'$ parallel?

ii. Why are the angles β and β' equal to α and α', respectively?

iii. Show that $\gamma = \beta - \beta' = \alpha - \alpha'$.

The measurement of the angles α and α' provides the estimate $\gamma = \alpha - \alpha'$ for the angle between the plane of the Earth's equator and the plane of the Earth's orbit about the Sun. Recall that the Greeks measured γ to be 24° in their Earth-centered context. A more accurate value for this angle is $23\frac{1}{2}°$.

3B. About Clocks, Arcs, and Copernicus's Model of Earth's Orbit.

3.2. Figure 3.36 depicts the clock face of Big Ben in London. Its minute hand is 14 feet long, and its hour hand is 9 feet long. Through what respective distances have the tips of the minute and hour

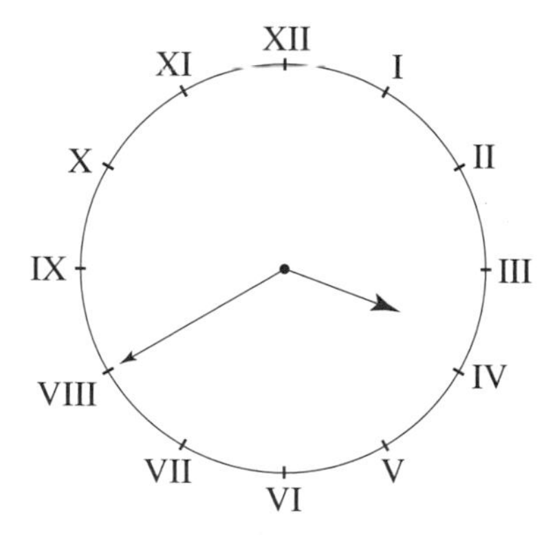

Fig. 3.36

hands moved since the hands were in the one o'clock position?

3.3. The circle in Figure 3.37 has center O and a radius of 1 meter. Like the hour hand of a clock, the arrow rotates clockwise at a rate of one revolution in 12 hours. The points A and D are positioned in such a way that the segment connecting them is parallel to the diameter BC of the circle. Determine the length of the arc CD, given that the arrow requires 7.5 hours to rotate from A to D.

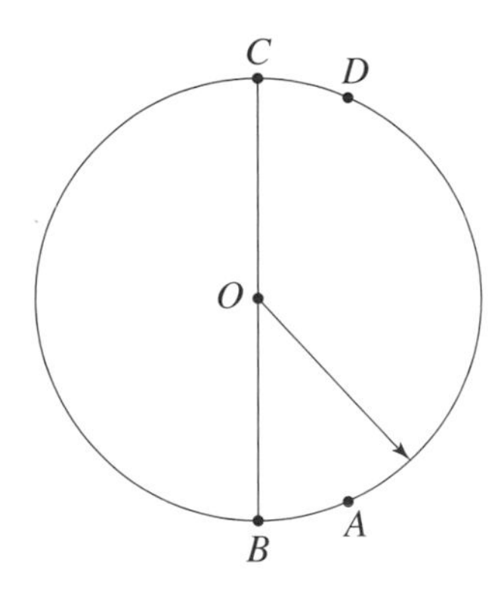

Fig. 3.37

3.4. The hour hand of the clock on a clock tower is 2 feet long, and the minute hand is 2.5 feet long. The clock loses one minute per hour to an accurate clock. How many feet does the tip of the hour hand travel in 72 hours? How many feet does the tip of the minute hand travel during this time?

3.5. Copernicus did not arrive at definitive model of Earth's orbit around the Sun. Nonetheless, his model is consistent with phenomena that were confirmed. One of them is related to an observation of Kepler made decades later. Figure 3.38 is taken from Figure 3.7. It shows the Sun S and the Earth E in its circular orbit with O at the center. Let $\mathbf{r}$ be the arrow from O to E, and note that $\mathbf{r}$ rotates with constant angular speed. Let $\mathbf{v}$ be the arrow from S to E. During autumn, E moves from C to D (autumn equinox to winter solstice). In the process, $\mathbf{v}$ rotates from C to D and hence through exact-

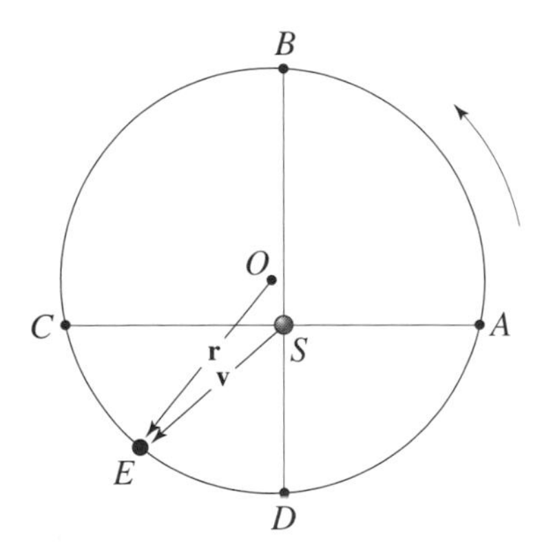

Fig. 3.38

ly 90°. At the same time, $\mathbf{r}$ also rotates from C to D, and hence through an angle less than 90°. Therefore during autumn, $\mathbf{v}$ rotates, on average faster than $\mathbf{r}$. During summer, both $\mathbf{r}$ and $\mathbf{v}$ rotate from B to C. Show that during summer, $\mathbf{v}$ rotates more slowly than $\mathbf{r}$. It follows that $\mathbf{v}$ has, on average, greater angular velocity during winter, when the Earth is closer to the Sun, than during summer, when it is farther away. This observation is consistent with Kepler's second law.

3.6. The distance from the planets to the Sun is known to vary. One way to measure this variation is to take the average of the maximum and minimum distances, namely $\frac{1}{2}(\max + \min)$, and then measure

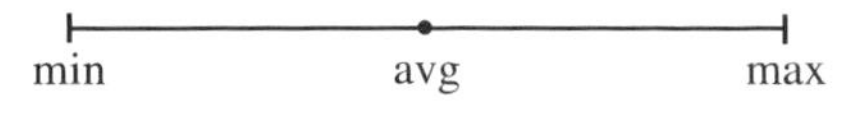

Fig. 3.39

the deviation of the distance from this average. The maximum deviation from the average is

$$\max - \text{avg} = \text{avg} - \min = \tfrac{1}{2}(\max - \min).$$

This means (see Figure 3.39) that the ratio (max – avg)/avg = (avg – min)/avg is an indicator of the variation of the distance. Let's now return to Copernicus's model of Earth's orbit around the Sun. Refer to Figure 3.7. Since $SO = c$, the maximum distance of Earth from the Sun is $r + c$ and the minimum distance is $r - c$. The average of the two is $\frac{1}{2}[(r+c)+(r-c)] = r$ and $\frac{1}{2}[(r+c)-(r-c)] = c$. Therefore, in the case of Earth's orbit,

$$\frac{\max - \text{avg}}{\text{avg}} = \frac{\text{avg} - \min}{\text{avg}} = \frac{c}{r} \approx 0.0299,$$

or about 3%. This ratio is larger than the actual value of 1.7%, but again the model reflects the phenomenon.

3.7. Compute the angle β of Figure 3.8 first in radians and then in degrees. Discuss the accuracy of using the length of arc $B'B$ as an estimate for the side b of the central triangle depicted in Figure 3.9.

3.8. Develop Copernicus's model for the Earth's orbit again using the data of Hipparchus's time: spring, $94\frac{1}{2}$ days; summer, $92\frac{1}{2}$ days; autumn, $88\frac{1}{8}$ days; and winter, $90\frac{1}{8}$ days.

- **i.** Adapt Figures 3.7 and 3.8 to this situation.
- **ii.** Show that $c = 0.0412r$ and that the angle λ between aphelion and the summer solstice position falls to the right of summer solstice and is equal to 24.70°.

3.9. Approximate values for the current lengths of the seasons are: spring, 92 days, 18 hours, 20 minutes, or 92.764 days; summer, 93 days, 15 hours, 31 minutes, or 93.647 days; autumn, 89 days, 20 hours, 4 minutes, or 89.836 days; and winter, 88 days, 23 hours, 56 minutes, or 88.997 days. This and more recent information can be derived from data provided by the United States Naval Observatory (USNO) on the website

http://aa.usno.navy.mil/data/docs/EarthSeasons.php

Develop Copernicus's model of the Earth's orbit around the Sun using these values for the lengths of the seasons. Use your model to answer the following: how many days after summer solstice is Earth farthest from the Sun (aphelion) and how many days after winter solstice is Earth closest to the Sun (perihelion). Compare this with the information given by the USNO.

3C. Tangents to Circles and Copernicus's Distance Computations. The fact that a line tangent to a circle is perpendicular to the radius to the point of tangency was important in Copernicus's computations.

3.10. The proof of this fact in Euclid's *Elements* goes by contradiction. In Figure 3.40, L is the tangent to the circle at the point A, and OA is a radius. Suppose that the angle between OA and L is not 90°, and assume instead that OB is perpendicular to L. So ΔOBA is a right triangle and OA is its hypotenuse. Why is $OA > OB > OC$, and why does this complete the proof?

3.11. A *Parade Magazine* column of June 15, 2003, asked "If a person were standing at sea level looking out over the ocean, how far away would the horizon be?" The answer was "The horizon—in this case, where the sky meets the sea—isn't so far away: only about 2.5 miles." Assess the accuracy of this reply. [Let the distance of the person's eyes above the ground to be 6 feet. Take 3950 miles for the radius of the Earth. Use 1 mile = 5280 feet to convert the distances to miles. Draw a sketch of what is going on and apply the Pythagorean theorem.]

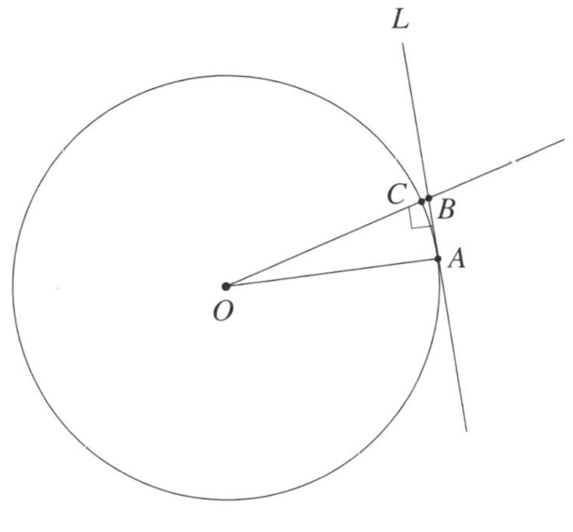

Fig. 3.40

3.12. We are in Islamic North Africa in the later Middle Ages. From a vantage point M on the slope of a mountain 1.5 miles (or 7920 feet) above sea level, a clear view is had of the Mediterranean Sea. From this site, a geometer from the court of the Emir measures the angle that a plumb line makes with his line of sight to the sea at the horizon. He first measures this angle to be $88\frac{1}{6}^\circ$, then $88\frac{1}{2}^\circ$, but finally concludes that $88\frac{1}{3}^\circ$ is most accurate. He uses this measurement to estimate the radius of the Earth. How does he do it, and what estimate does he achieve? What estimates did his first two measurements given him? How susceptible is this method to error? [Use Figure 3.41. In the figure, O is the center of the Earth and H is the horizon as seen from M. Just in case you're wondering ... this is fiction. However, Islamic scholars did make estimates of the size of the Earth in precisely

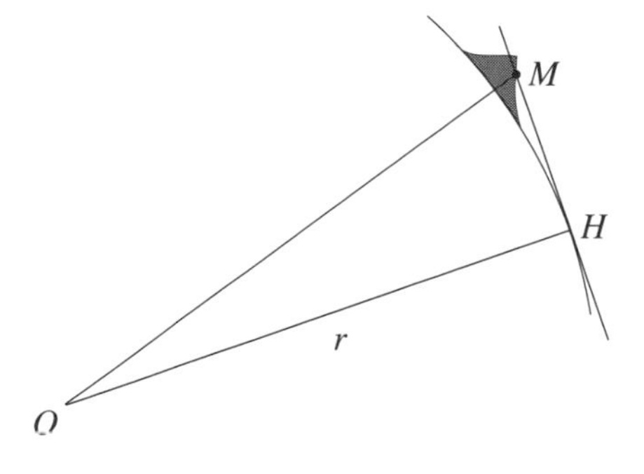

Fig. 3.41

this way. They perfected the astrolabe, an astronomical instrument of the Greeks, before the 10th century and used it to measure such angles with the accuracy mentioned.]

3.13. Consider a quadrilateral and verify that the four interior angles add up to 360°. [Hint: Add a line segment between two of the vertices of the quadrilateral to split it into two triangles.]

3.14. You are standing in front of a large cylindrical gasoline tank. You have access to only a small part of its perimeter, as the rest of the tank is fenced in. You are interested in determining the radius of the tank. You can measure distances by walking them off, and you have a device for measuring angles. Is this enough to get at the radius? You think for a while. Then you draw Figure 3.42. You know that the interior angles of a quadrilateral add to 360°, and ... aha! What is your solution?

3.15. Copernicus considered Figure 3.11 in the case of Mercury and calculated the angle α to be 22°. What estimate for the distance of Mercury from the Sun could he derive from this measurement?

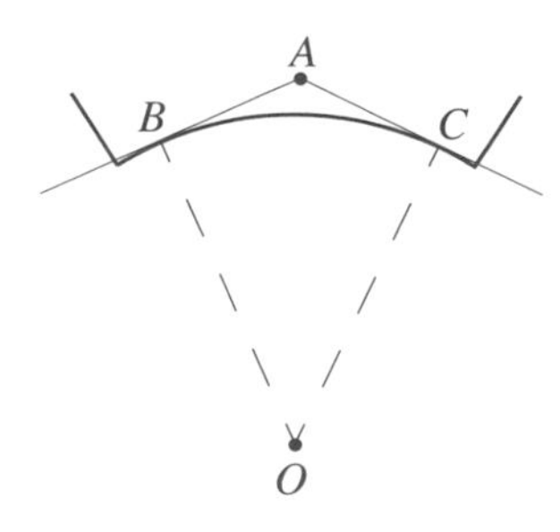

Fig. 3.42

3.16. Copernicus studies the orbit of Jupiter at opposition and then again at quadrature. See Figure 3.43. The data that he collects tell him that $\alpha - \beta \approx 79°$. How does he derive the distance estimate $JS \approx 5.24ES$ from this?

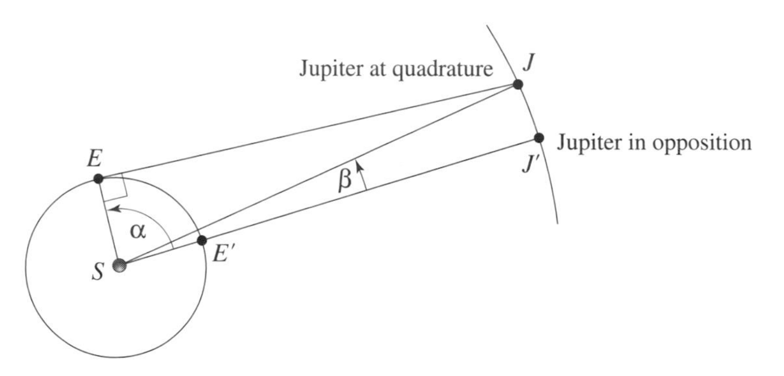

Fig. 3.43

3.17. We'll discuss an approach to the distance of any outer planet (relative to the Earth to Sun distance) that does not make use of the location of the planet at quadrature, but does conform to the spirit of Copernicus's work. We'll apply it to Mars, under the assumption that both Earth and Mars have circular orbits with center the Sun S. Start with the situation in which Mars at M' is in opposition. See Figure 3.44. After a certain time, say, a few months, Earth and Mars will have moved to positions

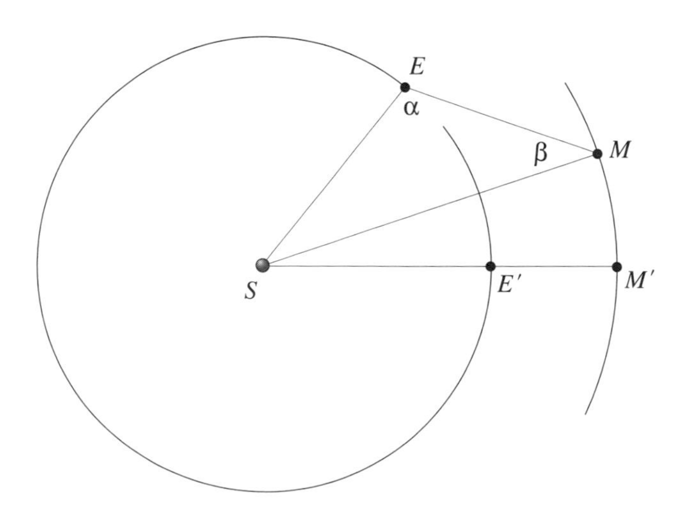

Fig. 3.44

E and M, respectively. Copernicus could estimate the angles $\angle MSM'$ and $\angle ESE'$ because he knew the elapsed time since opposition and the periods of Earth and Mars. Knowing these two angles would have provided him with an estimate for the angle $\angle MSE$. Since he was able to measure α, he could now estimate β. By applying the law of sines to the triangle ΔMSE, Copernicus could get

$$\frac{\sin\beta}{ES} = \frac{\sin\alpha}{MS},$$

and he could conclude that

$$\frac{MS}{ES} = \frac{\sin\alpha}{\sin\beta}.$$

3.18. Refer to the orbits of Earth and Mars as depicted in Figure 3.44. Suppose that Copernicus knew that the period of the orbit of Mars is about 1.9 years. Let's suppose that he observed Mars for two months after the planet was in opposition and that he then measured the angle α to be $114.9°$. What estimate for the distance of Mars from the Sun could he have derived from this measurement? Compare this answer with data from Table 3.1.

An important question about the famous diagram—Figure 3.14—that Kepler devised to conclude that Copernicus's circular model of Earth's orbit did not measure up to Kepler's standards of accuracy concerns the data on which this conclusion was based. How did Tycho determine the angles of the four important triangles in the years 1590 to 1595?

3.19. **i.** Why is Earth's angular speed along Copernicus's circular orbit $\frac{360}{365\frac{1}{4}} \approx 0.9856$ degrees per day? Does the point representing Earth move clockwise or counterclockwise around this circle?

ii. Why would Mars have been in opposition shortly after Earth was at the point E_4? How could Tycho have used information about when this opposition occurred to determine the angles $\angle E_1OM$ to $\angle E_4OM$?

iii. Could Tycho have obtained information about the angles $\angle OE_1M$ to $\angle OE_4M$ by measuring $\angle SE_1M$ to $\angle SE_4M$ instead?

iv. How would Tycho have determined the remaining angles $\angle E_1MO$ to $\angle E_4MO$?

3.20. *Some Fiction about Kepler.* During a time when Kepler still thought that the orbit of Mars was a circle, he used Tycho's data to plot the position of the center C of the circle as well as three positions of Mars on a sheet of paper. He knew that the period of the orbit of Mars was 687 days. The difference in time between position M_1 and M_2 was 57 days and that between positions M_2 and M_3 was 72 days. After some thought and a few calculations, he realized that while the positions of Mars that he had plotted were correct, the place on the paper where he had put the center C of the circle was off. But Kepler quickly corrected the error and repositioned C. How did he do this?

3D. Consequences of Kepler's Laws. Consider a planet P (or an asteroid or comet) in its elliptical orbit around the sun S. See Figure 3.19. Let a be its semimajor axis and c the distance from either focal point to the center O of the ellipse. The ratio $\varepsilon = \frac{c}{a}$ is the eccentricity of the ellipse. We have already observed that $a + c = a + \varepsilon a = a(1 + \varepsilon)$ is the maximum distance from P to S, and that $a - c = a - \varepsilon a = a(1 - \varepsilon)$ is the minimum distance from P to S. We first turn to the variation of the distance of the planet P as discussed in Problem 3.6.

3.21. Show that $\text{avg} = \frac{1}{2}(\max + \min) = a$ and that

$$\frac{\max - \text{avg}}{\text{avg}} = \frac{\text{avg} - \min}{\text{avg}} = \frac{c}{a} = \varepsilon,$$

the eccentricity of the orbit. This provides an interpretation of the eccentricity ε of a planet's elliptical orbit in terms of the variation of its distance from the Sun. Kepler knew that $\varepsilon = 0.017$ for the Earth and $\varepsilon = 0.092$ for Mars.

3.22. Figure 3.45 depicts the focal axis of an ellipse with center O and focal points F_1 and F_2. The point A is on the ellipse with $AO = a$ equal to the semimajor axis. The point B is a point on the ellipse with $BO = b$ equal to the semiminor axis. Refer back to the definition of the ellipse in Section 2.1 and use the figure to show that the constant k is equal to $(a - c) + (a + c) = 2a$. Then verify that

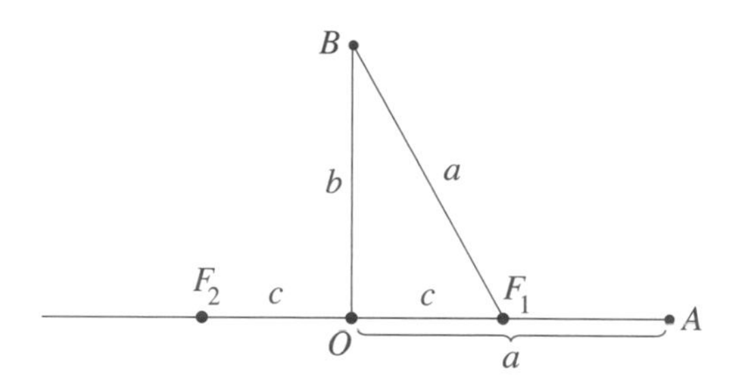

Fig. 3.45

$BF_1 = a$, that $a^2 = b^2 + c^2$, and that $b = \sqrt{a^2 - c^2} = a\sqrt{1 - \varepsilon^2}$. Conclude from this that the closer the eccentricity ε of the ellipse is to 0 the more circular it is, and the closer the eccentricity is to 1 the flatter it is.

3.23. Three ellipses E_1, E_2, and E_3 are given. All have the same semimajor axis a. Their semiminor axes are b_1, b_2, and b_3 with $b_1 < b_2 < b_3 < a$. Refer to Figure 3.46 and explain why the respective right focal points of these ellipses labeled 1 for E_1, 2 for E_2, and 3 for E_3 are in the positions shown in

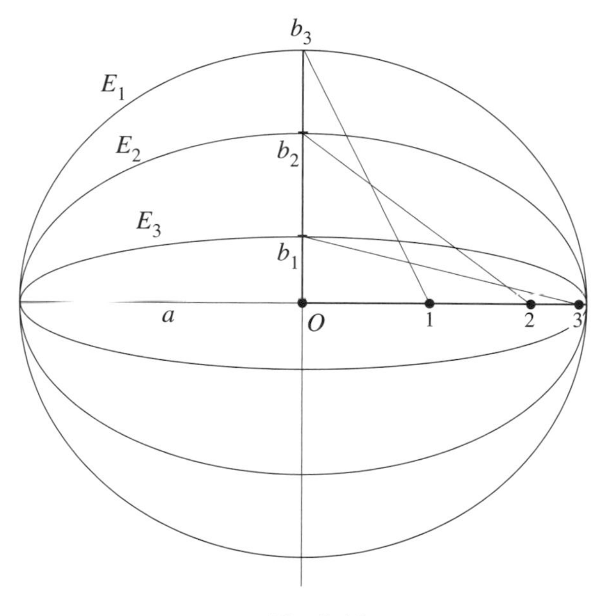

Fig. 3.46

the figure. The figure gives a sense of the connection between the geometry of the ellipse and the location of its focal points.

3.24. Use data from Table 3.3 to compute the semiminor axis $b = a\sqrt{1 - \varepsilon^2}$ for Mercury. If you were to draw the elliptical orbit of Mercury at a scale of 1 au = 25 centimeters would you be able to visually distinguish it from a circle? To answer the last question, compare Mercury's ellipse to its surrounding circle. How far from the center of your ellipse would the point representing the Sun be?

Kepler's second law—study Figure 3.18—implies that a planet's speed is greater when it is nearer the Sun than when it is farther away. It follows that a planet's speed is greatest at perihelion and least when it is at aphelion. We'll denote these two speeds by $v_{\max}$ and $v_{\min}$, respectively. Let P and A be the perihelion and aphelion positions of the planet, respectively. Consider two small and equal intervals of time I_P

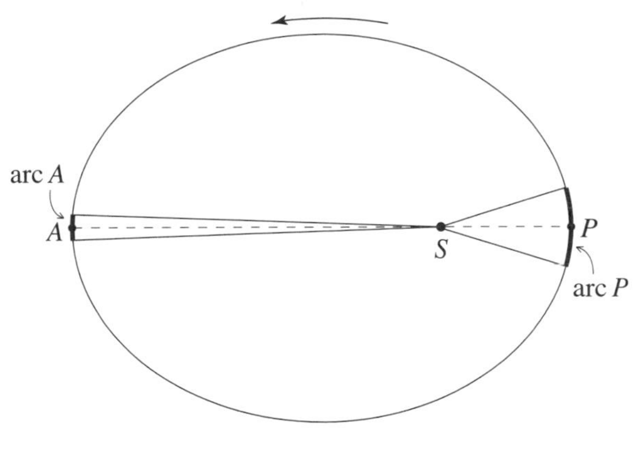

Fig. 3.47

and I_A, and suppose that the planet is at P at the midpoint of the time interval I_P and at A at the midpoint of the time interval I_A. During these equal time intervals the planet traces out two small arcs along its elliptical orbit. One has center the point P and the other is centered at the point A. Refer to Figure 3.47.

3.25. Let arc P and arc A designate the lengths of the two arcs depicted in Figure 3.47.

i. Use an area formula from Section 2.2 and Kepler's second law to show that $\frac{1}{2}a(1-\varepsilon)(\text{arc } P) \approx \frac{1}{2}a(1+\varepsilon)(\text{arc } A)$.

ii. The fact that average speed is equal to distance traveled divided by the time it takes tells us that the average speeds of the planets in traversing the two arcs are $\frac{\text{arc } P}{I_P}$ and $\frac{\text{arc } A}{I_A}$, respectively. Since the arcs are both small, these averages are approximately equal to $v_{\max}$ and $v_{\min}$, respectively.

iii. Show that the observations in parts (i) and (ii) imply that

$$\frac{v_{\max}}{v_{\min}} \approx \frac{\text{arc } P}{\text{arc } A} \approx \frac{1+\varepsilon}{1-\varepsilon}.$$

iv. When the time interval $I_P = I_A$ is pushed to zero, the approximations involved get tighter and tighter so that $\frac{v_{\max}}{v_{\min}} = \frac{1+\varepsilon}{1-\varepsilon}$.

v. Check that $\frac{v_{\max}}{v_{\min}}$ is approximately 1.03 for Earth, 1.20 for Mars, and 1.52 for Mercury.

3.26. Table 3.3 tells us that all planets (except Mercury) have eccentricities $\varepsilon \approx 0$, so that their orbits are nearly circular. Explain why $v_{\max} \approx v_{\min}$ for these planets and why their orbital speeds are nearly constant. Many comets have elliptical orbits with eccentricities $\varepsilon \approx 1$. For such flat orbits $v_{\max}$ is much greater than $v_{\min}$. Show for a comet with orbital eccentricity ε satisfying $0.999 < \varepsilon < 1$, that $v_{\max} > 1999 v_{\min}$.

3.27. Consider planets A and B with orbital eccentricities ε_A and ε_B, respectively. Show that if $\varepsilon_A > \varepsilon_B$, then the speed ratio $\frac{v_{\max}}{v_{\min}}$ for planet A is greater than that for planet B.

3.28. The speed of a planet satisfies the condition that $v_{\max}$ is almost exactly 10% greater than $v_{\min}$. Find the ratio $\frac{v_{\max}}{v_{\min}}$ for this planet, and this information to detrmine the eccentricity of its orbit. Then use Table 3.3 to identify the planet.

3.29. Suppose that an astronomer who understands Kepler's laws, knows the periods of all the planets and the semimajor axis of Earth's orbit. How could he compute the semimajor axes of the orbits of all the planets?

3E. Galileo's Numerical and Experimental Insights. The next two problems consider matters relevant to Galileo's experimental efforts.

3.30. Consider the sum $1 + 2 + 3 + \cdots + (k - 1) + k$ of the first k consecutive positive integers. There is an easy way to obtain a formula for this sum. Write the sum twice to get

$$(1 + 2 + 3 + \cdots + (k - 1) + k) + (1 + 2 + 3 + \cdots + (k - 1) + k).$$

By adding from the inside out, we get that this is equal to $k + 1$ added k times, and hence $k(k + 1)$. It follows that $1 + 2 + 3 + \cdots + (k - 1) + k = \frac{k(k+1)}{2}$.

3.31. In his experiments with inclined planes, Galileo considered sums of consecutive odd numbers of the form $1 + 3 + 5 + \ldots$. Notice that $1 + 3 = 2^2$, $1 + 3 + 5 = 3^2$, $1 + 3 + 5 + 7 = 4^2$, and that $1 + 3 + 5 + 7 + 9 = 5^2$. This suggests that the sum of the first k odd numbers should be equal to k^2. Consider the term $2i - 1$. Plugging $i = 1, 2, 3, \ldots, (k - 1), k$ into it provides the list $1, 3, 5, \ldots, 2k - 3, 2k - 1$ of the first k odd integers. Verify the equality

$$1 + 3 + 5 + \cdots + (2k - 3) + (2k - 1) = k^2$$

by using the strategy of the solution of Problem 3.30.

The summation formulas of Problems 3.30 and 3.31 can also be verified by applying the *principle of induction*. Fix an integer $m \geq 0$, and let a statement S_n be given for each $n \geq m$.

Principle of Induction. Suppose that

(1) statement S_m is true, and

(2) for every $k \geq m$, the truth of statement S_k implies the truth of statement S_{k+1}.

Then statement S_k is true for all $k \geq m$.

The validity of this principle can be seen as follows. Suppose k is any integer with $k \geq m$, and let $i = k - m$. Since S_m is true, it follows from (2), that S_{m+1} is true, and by repeatedly applying (2), that $S_{m+2}, S_{m+3}, \ldots$, and finally S_{m+i}, are all true. Since $m + i = k$, S_k is true, and the verification is complete.

Return to Problem 3.30. Let $m = 1$, and let S_k be the statement $1 + 2 + 3 + \cdots + (k - 1) + k = \frac{k(k+1)}{2}$. Since $1 = \frac{1 \cdot 2}{2}$, statement S_1 is true. Suppose $k \geq 1$ and that S_k is true. So $1+2+3+\cdots+(k-1)+k = \frac{k(k+1)}{2}$. It follows that $[1+2+3+\cdots(k-1)+k]+(k+1) = \frac{k(k+1)}{2}+(k+1) = \frac{k(k+1)+2(k+1)}{2} = \frac{(k+1)(k+2)}{2}$, and therefore that statement S_{k+1} is true. Parts (1) and (2) of the principle of inductions are now established. So S_k is true for all $k \geq 1$. Use the principle of induction in a similar way to verify the formula of Problem 3.31.

Experts in the history of mathematics have determined that the principle of induction was first discovered by the Sicilian mathematician Franciscus Maurolycus in the year 1557 about 50 years before Galileo's studies of astronomy and the physics of motion.

3.32. *Galileo and some Fiction.* Galileo begins an experiment by dropping his metal ball from the edge of a table straight down to the floor. With the ball starting from rest in repeated trials, Galileo concludes that it takes the ball $\frac{2}{5}$ of a second to fall to the floor. Continuing his experiment, Galileo rolls the ball on the table. He measures its speed to be a constant 3 meters per second. The ball soon reaches the edge of the table and falls to the falls in a parabolic arc. Before measuring anything, Galileo knows that the ball will hit the floor $\frac{6}{5}$ meters from the base of the table. Explain how Galileo reaches this conclusion. Your explanation should identify the laws and principles of projectile motion that Galileo relies on and specify how he applies them.

During his years in Padua, Galileo put an inclined plane on a table and recorded the following experiment in his working papers. In repeated trials, he released a ball from rest at various vertical distances y above the table and measured the corresponding distances x between the foot of the table and the point at which the ball hit the horizontal floor. The inclination of his plane was fixed, and the ball rolled horizontally for a small stretch before it flew off to the ground. Figure 3.48 illustrates his experimental setup. In the figure, d is the distance from the beginning of the roll to the bottom of the plane, t is the time this takes, and v is the speed that the ball attains at the bottom of the plane. In an earlier study, Galileo found that the average speed $\frac{d}{t}$ of the ball down the plane is equal to $\frac{1}{2}v$. Galileo considered the ball's fall from the edge of the table to the ground to consist of a combination of a horizontal component and a vertical component. By

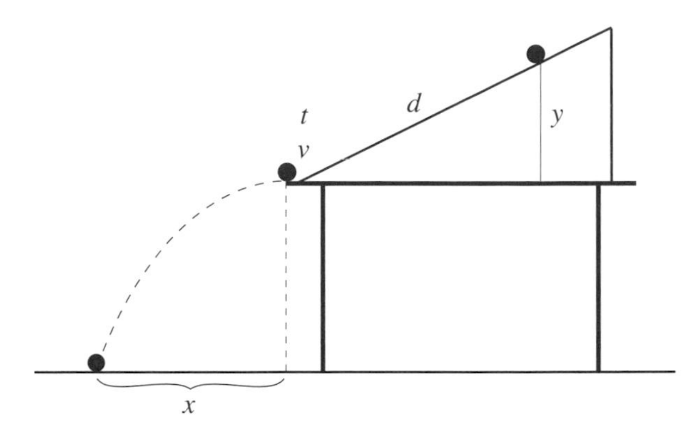

Fig. 3.48

his law of horizontal inertia, the horizontal component has constant speed v. The vertical component of the flight of the ball is vertical free fall with zero initial speed. Since this is the same for any horizontal speed v, the time of the ball's fall from the edge of the table to the ground is the same for any starting height y.

3.33. In Galileo's time, mathematical relationships were not written in terms of algebraic equations. Galileo himself expressed quantitative relationships in terms of proportions and in words. We will write the relationship "proportional to"—meaning that one variable is equal to a constant multiple times the other—by placing the symbol $\propto$ between the two variables. For example, if α is the fixed angle of the inclined plane, then $\sin\alpha = \frac{y}{d}$, so that $y = cd$ with $c = \sin\alpha$. Therefore $y \propto d$.

i. How did Galileo know that $d \propto t^2$? How did he conclude that $y \propto t^2$?

ii. Galileo knew that $\frac{1}{2}v \cdot t = d$. How did he deduce from this that $v \propto t$ and hence that $v \propto \sqrt{y}$.

iii. Let t_0 be the time it takes for an object to fall from the edge of the table straight down to the floor. Why does it follow that $x = v \cdot t_0$ and hence that $x \propto v$.

iv. By putting (ii) and (iii) together, Galileo concluded that $x \propto \sqrt{y}$.

It is the relationship $x \propto \sqrt{y}$ that Galileo's experiment tested and successfully confirmed. So Galileo's experiment tested several aspects of his account of motion at once. Notice that the two parts of the experiment (the one above the table and the other below) together bypass the time variable t. As a result, Galileo needed to attend only to distance measurements and not the much more elusive measurement of elapsed time. The mathematics underlying this experiment is studied in detail in Chapter 8.

3F. More Parallax. We'll start with two simple examples of parallax, then consider diurnal and stellar parallax, and finally, supply a detail that Flamsteed would have needed for his analysis.

3.34. The muse of astronomy Urania sees a Greek column off in the distance. Behind it on the horizon, she recognizes the familiar outline of Mount Olympus and notices that the top of the column is in line with a certain distinctive feature of the mountain. Urania moves 12 strides roughly parallel to the face of the mountain and observes a shift in the position of the top of column against its background. It is now aligned with a second distinctive feature of the mountain. Returning to the original spot, she estimates the angle between her lines of sight to the two features of the mountain to be about 5 degrees. After concentrated thought, Urania concludes that the column is about 140 strides away. How did she reach this conclusion? [Hint: The fact that Urania's baseline is short makes it possible to estimate the parallax of the column from the measurement of the one angle.]

3.35. *A Parallax Project.* Several students go to a spot on campus from which they have a clear view of a building that is a few hundred yards (say, 200 yards) away. It should have a long facade that is roughly perpendicular to the line of sight. A student is selected and takes a position $*$ a number of paces—say, 20 paces away—from the spot (in the direction of the building). The "pace" is the operative unit of length. The other students mark out a "baseline" (a few paces long) through the spot and roughly parallel to the facade of the building. The students then use the method of stellar parallax to estimate the number of paces from the baseline to $*$. Finally, they pace off the distance to $*$ and compare it with the predicted value. [The approach that Urania used to determine the distance of the column can be adapted to this situation.]

In the parallax measurement of Flamsteed, the baseline BB' involved two points B and B' on the surface of the Earth, where B' is obtained from B by the Earth's daily rotation. See Figure 3.32a. Such parallax is referred to as *diurnal* parallax . (The word "diurnal" comes from the Latin *diurnalis*, meaning "daily.") The parallax measurements of both Flamsteed and Cassini involve relatively short baselines and are suitable for estimating the distances of objects within the solar system. Recall that the appropriate measure of such distances is the astronomical unit au. All the stars are far from Earth—much, much farther than any object in our solar system—but they are not all the same distance away from us as the Greek astronomers had thought. Even within the same constellation, some are relatively near and others are imaginable far. are not all lie far beyond the solar system, The distances of the nearer stars can be estimated by measuring the *stellar parallax* of such stars. This makes use of the baseline determined by locations E and E' of Earth on opposing sides of its orbit around the Sun. The standard measure of the distances to the stars is the light-year ly. This units is related to the au by 1 ly = 63,241 aus.

3.36. Figure 3.49 shows the Earth on opposite ends E and E' of its orbit around the Sun S as well as a nearer star A. The angle $p(A)$ defined by $p(A) = \frac{1}{2}\angle EAE'$ is the *stellar parallax* of the star A. As in earlier situations, the distance $d(A,E)$ of A from E satisfies $d(A,E) \approx \frac{EE'}{9.7p(A)} \times 10^6$, where $p(A)$ is given in seconds. The first successful computations of stellar parallax were carried out for the stars 61 Cygni, Vega, and Alpha Centauri in the late 1830s by three different astronomers working independently.

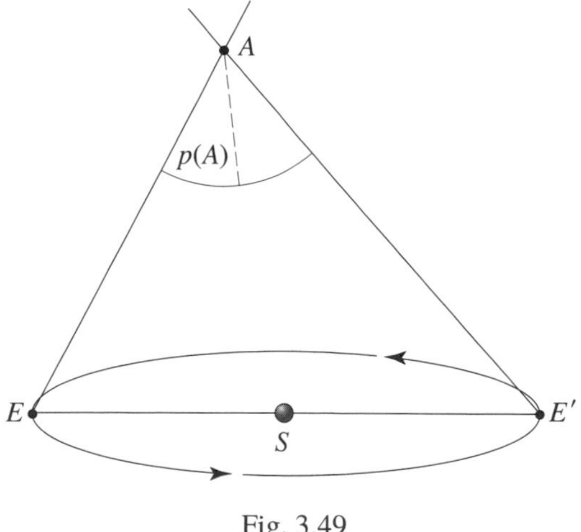

Fig. 3.49

The modern values of their angles of parallax are 0.29, 0.13, and 0.75 seconds, respectively. History has credited Friedrich Bessel with his parallax measurement for 61 Cygni as having been first. Take the Sun-Earth distance ES as 1 au, and then estimate the distances to these stars in light-years.

3.37. Tycho Brahe was able to pinpoint the angular position of a star with an accuracy of 1 minute, or $\frac{1}{60}^\circ$. This suggests that he would have been able to detect a stellar parallax of 60 seconds. Since he found no parallax when he observed the bright new star that had suddenly appeared, this would have meant that the parallax of his "stella nova" was less than this amount. What might he have been able to conclude from this about the distance of this star from Earth?

3.38. Figure 3.50a shows Earth with center O rotating around its axis. A location in the Northern Hemisphere with latitude φ is shown to have rotated from position B to B'. Figure 3.50b shows a "top view" of the circle of rotation as well as the two positions B and B'. The angle θ measures the difference between the positions B and B'. Let r_E be Earth's radius.

i. Show that the radius of the circle of Figure 3.50b is $r_E \cos\varphi$.

ii. Verify that the distance from B to B' is equal to $(r_E \cos\varphi)\sqrt{2(1-\cos\theta)}$.

iii. Turn to Flamsteed's calculation of the parallax of Mars. The latitude of Derby is known to

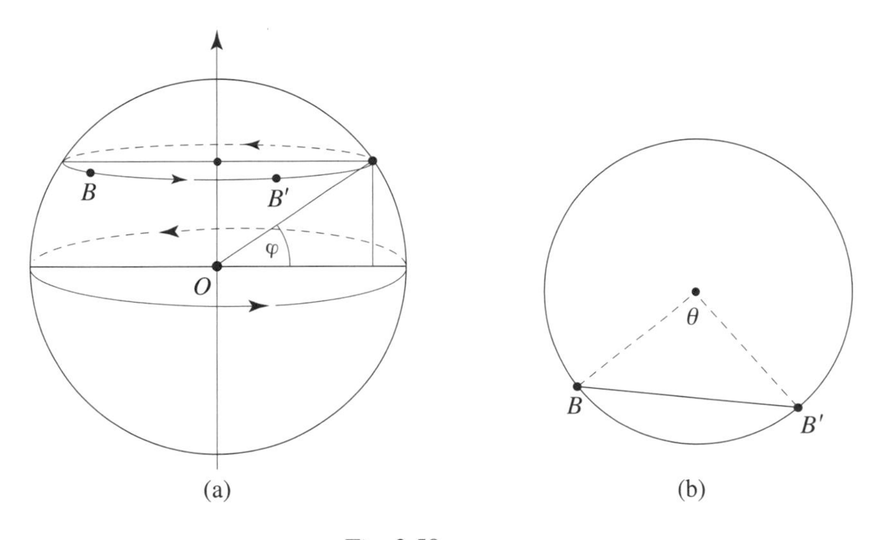

Fig. 3.50

be $\varphi = 52.92°$, and the elapsed time between his measurements was 6 hours and 10 minutes. Use this information and the value $r_E = 6370$ kilometers to show that Flamsteed's baseline BB' was in fact 5550 kilometers (rather than 5300 kilometers). This estimate has relied on the assumption that Earth's surface is a sphere. (This is in fact not quite true. Instead, its surface as a sphere that is flattened at the poles.)

3G. Mars and Earth on Close Approach.

3.39. It is known today that the perihelion and aphelion distances for Earth are 147,098,291 kilometers and 152,098,233 kilometers, respectively, and that the perihelion and aphelion distances for Mars are 206,655,216 kilometers and 249,232,432 kilometers, respectively. Assuming that Mars is in opposition and at perihelion, show that the distance d between Earth and Mars lies in the range $54{,}556{,}983 \leq d \leq 59{,}556{,}925$ kilometers. Notice in particular that the estimates for the distance d between Earth and Mars (with Mars in opposition as in Figure 3.28)—both Flamsteed's $d \approx 52{,}000{,}000$ kilometers and Cassini's $d \approx 53{,}000{,}000$ kilometers—came up short. What does this imply about their measurements of the parallax of Mars?

Given the conclusion of Problem 3.39, that the closest possible approach of Mars to Earth is 54,556,983 kilometers, the question that arises is "What is the closest that Mars has actually gotten to Earth?" The following information appeared in the article "Up Close and Personal with Mars" in the Science Times Section of the *New York Times* on Tuesday, August 19, 2003. Minutes before 6 o'clock Eastern time on August 27, the separation between Mars and Earth was precisely 55,758,006 kilometers (or 34,646,418 miles). It was asserted that this was the closest that Mars had been in about 60,000 years. There is no Mars approach on the record that was closer. But in the year 2287, the article went on to say "Mars and Earth will cruise even closer." The 10 closest Mars approaches to Earth from 3000 B.C. to 3000 A.D. are listed below. They were computed by scientists at the Jet Propulsion Laboratory in Pasadena, California, in response to questions about Mars's close approach in 2003.

aus	kilometers	date
0.37200418	55,651,033	Sept. 8, 2729
0.37200785	55,651,582	Sept. 3, 2650
0.37217270	55,676,243	Sept. 5, 2934
0.37225400	55,688,405	Aug. 28, 2287
0.37230224	55,695,623	Sept. 11, 2808
0.37238224	55,707,590	Aug. 30, 2571
0.37238878	55,708,568	Sept. 2, 2366
0.37271925	55,758,006	Aug. 27, 2003
0.37279352	55,769,117	Aug. 24, 2208
0.37284581	55,776,939	Aug. 22, 1924

Chapter 4

The Coordinate Geometry of Descartes

The first three chapters of this text have put the geometry of the Greeks and the ingenious way with which they used it on extensive display. But in spite of their ingenuity, they missed some important elements. Since they did not develop the symbolism of algebra, they could not express quantitative relationships in terms of algebraic equations. They did not have the coordinatized line and the correspondence between points and real numbers that it expresses. So there was no coordinate system and no coordinate plane. Without algebraic equations and coordinate planes, there were no graphs and no exploitations of the powerful interplay between algebra and geometry. While Greek mathematics is fundamental, substantial, and brilliant, it is fundamentally incomplete.

The construction of a coordinate system and the exploration of equations in two variables requires three ingredients that were developed toward the beginning of the 17th century. One of them is the real number system and its visual realization as the number line. Another is the development of algebra as a symbolic language of variables (such as x, y, and z), constants (such as a, b, and c), and operations (such as $+, -, \times$, and $\sqrt{\ }$). The third is the recognition that by placing two number lines in a plane at right angles to each other, each pair of numbers representing a solution of an algebraic equation in two variables can be visualized as a point in the plane, and the locus of all solutions as a curve in the plane. The mathematical study of curves and related structures would soon blossom into the discipline of calculus. We saw in Chapter 3 how the basic principles underlying Greek physics and astronomy were rethought and reformulated. When calculus (as well as its three- and higher dimensional versions) played a powerful informative role when it was applied to the newly conceptualized astronomy and physics, modern science was born and took off. It is not much of an exaggeration to say that the fusion of algebra and geometry—made possible by the coordinate system—has had a transformative influence.

The thinkers who contributed most to the creation of coordinate geometry were the two Frenchmen René Descartes (1596–1650) and Pierre de Fermat (1601–1665). Descartes was a renowned philosopher, and Fermat was a lawyer and high-ranking government official (he served as counselor to the court of law of Toulouse). Both achieved lasting fame as mathematicians, and both are regarded—along with Kepler, Galileo, Leibniz, and Newton—as belonging to the greats of 17th-century mathematics. Descartes and Fermat recast and extended the geometry of the classical Greek treatises, most notably those of the Greek Apollonius (see Section 2.1), with a new symbolic algebraic language. They are also given credit for their early contributions to differential and integral calculus.

While Fermat's work was circulated in manuscript form in 1636, it was not published until 1679, long after his death. Thus history has given the primary credit for the development of coordinate, or analytic geometry to Descartes. Descartes received his early education from the Jesuits, who recognized both his physical weakness and his mental alertness. They allowed him to read and meditate in bed well past traditional rising hours. This became a lifelong habit that Descartes found conducive to intellectual output. At the age of 22, amazingly, he became a professional soldier. The lulls in military activity during the winter months allowed him to pursue mathematics and philosophy. One day in 1619, while in service in Bavaria, he escaped the cold by shutting himself into a "stove." He recounted that, during his meditations in this warm room, a divine spirit revealed a new philosophy to him. When he emerged from the stove, he had applied the mathematical method to philosophy and conceived the principles of analytic geometry. In his famous work *Discours de la Méthode*, written in 1629 and published in 1637, he recalls that

> Archimedes, in order that he might draw the terrestrial globe out of its place and transport it elsewhere [recall that Archimedes supposedly asserted, in reference to the lever, that if he had a place to stand he could move the world], demanded that only one point should be fixed and immovable; in the same way I shall have the right to conceive high hopes if I am happy enough to conceive one thing only which is certain and indisputable.

He then makes the most famous announcement in all of philosophy: "Je pense, donc je suis." "Cogito ergo sum." "I think, therefore I am."

Descartes wrote in support of Copernicus's heliocentric planetary system, but—afraid that he would suffer the same fate as Galileo—decided against its publication. After having lived and worked in Holland for 20 years, he accepted the offer of Queen Christina of Sweden to move to Stockholm and become her philosophy tutor. His remark "men's thoughts are frozen here, like the water" took aim at the intellectual climate there. He soon contracted pneumonia and could not recover. Descartes died in Stockholm in 1650, less than a year after he arrived.

We now turn to *La Géométrie*, the 100-page appendix of the *Discours de la Méthode* that sets out his analytic geometry. We'll discuss Descartes's new and surprisingly modern algebraic notation and develop a modern version of his coordinate geometry. Turning to the conic sections (see Section 2.1), we'll derive the equations of the circle, parabola, ellipse, and hyperbola. Finally, we'll consider the circle of radius 1 to recast the fundamental concepts of trigonometry.

4.1 THE REAL NUMBERS

The real number system was briefly described in Section 1.4. The point was made that in order to be able to apply geometry in computational contexts, a number system was needed with which any length could be measured. Such a system needed to come with a notation for numbers so devised that efficient numerical procedures for addition, multiplication, and division could be developed. The real numbers is such a system. It establishes a powerful link between geometry on the one hand, and arithmetic and algebra on the other. Even though its construction had been completed before Descartes composed his *Géométrie*, it is probable that it was not as yet well enough established. In any case, Descartes did not make use of the real numbers in this treatise. When he says early in Book I,

> to add the lines BD and GH, I call one a and the other b, and write $a + b$. Then $a - b$ will indicate that b is subtracted from a; ab that a is multiplied by b; $\frac{a}{b}$ that a is divided by b; aa or a^2 that a is multiplied by itself; a^3 that this result is multiplied by a, and so on ...

he is not writing about the addition, subtraction, and multiplication, and division of the *lengths* of line segments expressed as real numbers. Instead, he is referring to the addition, subtraction, and so on of the line segments themselves, just as the Greeks had done in their geometry and as Euclid's *Elements* had presented it. However, given the centrality of the real number system in today's mathematics, this text will use the real numbers system in describing the interplay between algebra and geometry that Descartes discovered. This is therefore an opportune time within our historical narrative to discuss the development of this system.

Consider the Roman numbers

I, II, III, ..., IX, X, ..., XXIV, ..., LXVI, ..., XCIII, ..., CCLIX, ..., MMDCCXV, ...

If you were asked to multiply and then divide the last two Roman numbers of this list within the Roman scheme, you would find this cumbersome in the first case and completely baffling in the second. This would also be true in the case of the Greek system of Figure 2.54. Mathematicians from the Islamic world made important progress toward a new number system. By the 9th century, the scholar al-Khwarizmi working in Baghdad had adopted the ten symbols $1, 2, 3, 4, 5, 6, 7, 8, 9$, and 0 as well as the strategy for writing larger numbers from Indian mathematics. In this scheme, the Roman numerals above are written $1, 2, 3, \dots, 9, 10, \dots, 24, \dots 66, \dots, 93, \dots, 259, \dots, 2715, \dots$. This *Hindu-Arabic* number system is *positional*. The symbol 2 means a different thing in each of the numbers 2, 24, 259, and 2715—namely, two, twenty, two hundred, and two thousand—in accordance with its position in each of these numbers. A treatise of al-Khwarizmi pointed to the importance of zero by instructing his readers, "when nothing remains, put down a small circle so that the place be not empty, but the circle must occupy it." The Latin title *Algoritmi de Numero Indorum* of this work tells us that the term *algorithm* is derived from the name of its author. Another of al-Khwarizmi's treatises deals with the problem of solving equations. Its Arabic title *Hisab al-jabr*... (it refers to the procedure of rearranging and combining terms) informs us that the term *algebra* also has Arabic origins. Islamic scholars developed effective and sequential procedures, or algorithms, for carrying out addition, subtraction, multiplication, and division for their numbers. Their number system also enabled them to express lengths numerically, including the length of the diagonal of the 1 by 1 square as well as the circumference of the circle with diameter 1, the number now written as π.

It took time for the Islamic system of numbers to reach Western Europe and gain acceptance. Even a pope was involved in the promotion of this effort. In the 11th century, when Pope Sylvester II was the young monk Gerbert, his abbot sent him from France to Islamic Spain to study mathematics. Great institutions of learning, such as the University of Cordoba, were making advanced education available to thousands of students. Gerbert not only learned the new number system but also absorbed the questioning and probing spirit of Islamic scholarship. This is the spirit that would lead to the establishment of the first European universities about 100 years later.

Another promoter of Islamic mathematics in Western Europe was Leonardo of Pisa (about 1175-1250), known today as *Fibonacci*. (This name, conferred by a mathematician of the 18th century, is a contraction of the Latin *filius Bonacci*, meaning "son of Bonacci" or possibly "son of good nature.") Leonardo's father was a trade official who facilitated the commercial dealings of the merchants of Pisa in what is today the north African country of Algeria. Young Leonardo joined him there and became acquainted with the mathematics of Euclid, Apollonius, and Archimedes. Taught methods of accounting by Islamic scholars, he learned how to calculate "by a marvelous method through the nine figures" together with the figure 0, "called zephirum in Arabic." On his return, he published the *Liber Abaci* in 1202. This historic book (contrary to the suggestion of the title, it has nothing to do with the abacus) introduced Western Europe to Islamic arithmetic and algebra, as well as the practice of using letters instead of numbers to generalize and abbreviate algebraic equations. It provided Western Europe with the first thorough exposition of the Hindu-Arabic numerals and the methods of calculating with them.

The acceptance of the new system—not yet equipped with the decimal point—was slow. As late as 1299, the merchants of Florence were forbidden to use it. They were told instead to either use Roman numerals or to write out numbers in words. The historical record informs us that Hindu-Arabic numbers appeared on gravestones in German states in 1371, and that they were imprinted on coins of Switzerland in 1424, of Austria in 1484, of France in 1485, of German states in 1489, of Scotland in 1539, and of England in 1551. They can be seen on early versions of the historic map of Ptolemy printed in a German state in 1482 (see the rim of numbers at the bottom of the later depiction in Figure 1.37). They were used by Leonardo da Vinci in 1487 within his design of the central vault for the Cathedral of Milan (still under construction at the time).

Decimal fractions began to gain use in Europe after they were developed in a 1585 publication by the Flemish mathematician and engineer Simon Stevin (1548-1620). *De Theinde*, or *The Tenth*, is a 29-page booklet that presents an elementary and thorough account of them. It was written for the benefit of "...stargazers, surveyors, carpet-makers, wine-gaugers, mint-masters and all kind of merchants." The French version *La disme* appeared in the same year. An English translation, *Dime: The Art of Tenths, or Decimal Arithmetic*, was published in 1608 in London. (This translation inspired Thomas Jefferson to propose a decimal currency for the United States and most likely the name "dime" for the 10-cent coin.) A little later, John Napier (1550-1617), a Scotsman and one of the inventors of logarithms, put the "dot on the *i*" by adding the decimal point. What is now our modern number system was notationally and computationally complete.

With the *decimal* or *base* 10 number system (and a given unit of length) it is possible—and this is the important point—to express *any length* in terms of a number. Whether the length is *rational*, that is, of the form $\frac{n}{m}$ for integers n and m, or *irrational*, such as the d in Figure 1.18, does not matter. For instance, the lengths $5\frac{1}{4}$ and $71\frac{2}{3}$ are expressed as 5.25 and 71.6666..., meaning that $5\frac{1}{4} = 5 + \frac{2}{10} + \frac{5}{10^2}$ and $71\frac{2}{3} = 71 + \frac{6}{10} + \frac{6}{10^2} + \frac{6}{10^3} + \frac{6}{10^4} + \dots$. In the same way, the lengths $\sqrt{2}, \sqrt{3}, \sqrt{5}, \sqrt{6}, \sqrt{7}, \dots$ that arise in the evolving spiral of Figure 4.1 by successive application of the Pythagorean theorem, as well as those given by other square roots, cube roots, and higher roots, can all be written with this system. For example, $\sqrt{2} = 1.414213562...$, $\sqrt{3} = 1.732050808...$, $\sqrt{5} = 2.236067977...$, and $\pi = 3.141592654...$. We will see in the first part of the Problems and Projects section that, with the exception of $\sqrt{4} = 2$, $\sqrt{9} = 3$, $\sqrt{16} = 4, \dots$, all the numbers $\sqrt{n}$ that arise in this spiral (and its continuation) are irrational. So for each

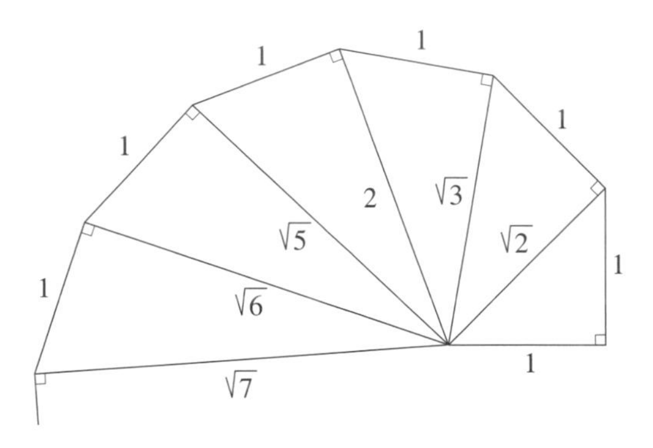

Fig. 4.1

of them, the full infinite progression of the decimal expansion is needed to achieve equality. Stopping at any point gives only an approximation. For example, $\sqrt{2} \approx 1.41421$ and $\sqrt{2} \approx 1.41421356$ are approximations.

The collection of all numbers, positive and negative, that are expressible in this decimal notation is called the *set of real numbers*. Take a unit of length and a straight line that extends infinitely in both directions. Fix a point on the line and label it with the number 0. This point is the *origin*. Using distance as measure, mark off the numbers $1, 2, 3, 4, \dots$ on the right of 0 and $-1, -2, -3, -4, \dots$ on the left of 0. In the same way, every real number corresponds to a point on the line, and in turn every point on the line

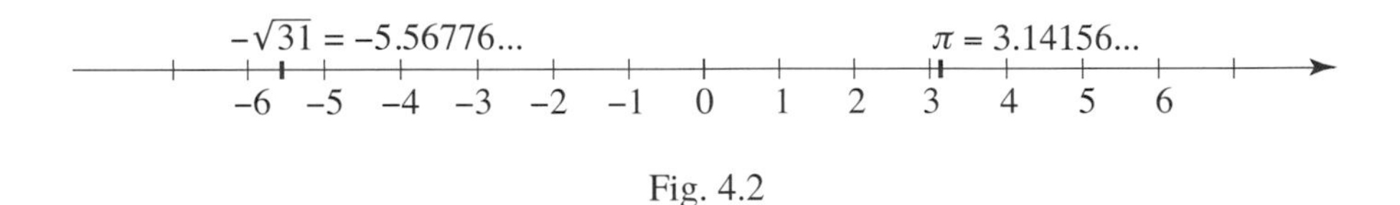

Fig. 4.2

corresponds to a real number. Positive numbers are on the right of 0 and negative numbers are on the left. Figure 4.2 illustrates what has been described. It also shows the locations of the numbers $-\sqrt{31} = -5.56776...$ and $\pi = 3.14156...$. A line with this numerical structure is a *number line*.

The number line and its unit of length make it possible to assign a length to any line segment. In fact, in the three previous chapters, we often labeled line segments with their lengths, and we will do so going forward.

The symbolic algebraic notation we use today to express mathematical problems and the operational strategies with which we solve for unknowns took a long time to develop. Until the 16th century, it had been common practice to express such problems entirely with words and abbreviations. This includes the work of Islamic scholars briefly discussed above. This practice began to change in the 1560s when a copy of the *Arithmetica* of the Greek Diophantus was discovered in Rome. Diophantus, one of the last great Greek mathematicians, solved a variety of algebraic problems in this work and made use of some symbolic notation in the process. In the 1590s, the French mathematician François Viète (1540–1603), in part influenced by the *Arithmetica*, introduced new symbolic elements and began to develop a general method of problem solving. When he wrote

$$B5 \text{ in } A \text{ quad } -C \text{ plano } 2 \text{ in } A+A \text{ cubum aequatum } D \text{ solido}$$

in Latin, he meant what we would express today by the equation $5bx^2 - 2cx + x^3 = d$. So Viète uses the capital letter A to express a variable, the capitals B and C to express constants, and he uses + and − for addition and subtraction. The use of the word "aequatum" tells us that he does not have a symbol for equality. As was the mind-set of the time, Viète retains the connection with geometry. In the example, all terms represent volume. The term "$B5$ in quad A" represents the one-dimensional magnitude $B5$ times the two dimensional unknown A^2. In "C plano 2 in A," the term C plano 2 is a two-dimensional magnitude multiplied by the one-dimensional unknown A. Since "A cubum" is A^3 and the label "D solido" refers to volume, this means that all the terms in Viète's equation are three-dimensional. The connection with geometry required that all terms of an equation needed to have the same dimension. Viète goes on to provide the following illustration of his method.

> Given the difference of two sides (think of "side" as an unknown) and their sum, to find the sides. Let the difference B of the two sides be given, and also let their sum D be given. It is required to find the sides. Let the lesser side be A. Then the greater will be $A + B$. Therefore, the sum of the sides will be $A2 + B$. But the same sum is given as D. Wherefore, $A2 + B$ is equal to D. And, by antithesis, $A2$ will be equal to $D - B$, and if they are all halved, A will be equal to $D1/2 - B1/2$. Or, let the greater side be E. Then the lesser will be $E - B$. Therefore, the sum of the sides will be $E2 - B$. But the same sum is given as D. Therefore, $E2 - B$ will be equal to D, and by antithesis, $E2$ will be equal to $D + B$, and if they are all halved, E will he equal to $D1/2 + B1/2$. Therefore, with the difference of two sides given and their sum, the sides are found. For, indeed, half the sum of the sides minus half their difference is equal to the lesser side, and half their sum plus half their difference is equal to the greater. Which very thing the inquiry shows. Let B be 40 and D 100. Then A becomes 30 and E becomes 70.

Today, we would simply let x and y be the respective lengths of the two sides, and set $x - y = b$ and $x + y = d$. After two additions, $2x = b + d$ and $2y = d - b$. So $x = \frac{1}{2}(b + d)$ and $y = \frac{1}{2}(d - b)$. Finished! With $b = 40$ and $d = 100$, we get $x = 70$ and $y = 30$.

We now have a sense of the state of algebra at the time Descartes began to compose his *Géométrie* in the 1620s. Early in this treatise he lets a and b be positive constants (within his framework these are line segments) and looks for solutions of the equation $y^2 = -ay + b^2$ (again in terms of segments). Descartes used $\propto$ in place of our $=$ and writes $yy \propto -ay + bb$, but also $y^2 \propto -ay + b^2$. His solution starts with the right triangle ΔMLN of Figure 4.3a and applies the Pythagorean theorem to get $MN^2 \propto \frac{1}{4}a^2 + b^2$, and hence $MN \propto \sqrt{\frac{1}{4}a^2 + b^2}$. In Figure 4.3b, he marks off the point P so that $PN \propto LN$, and lets $y \propto MP$. So

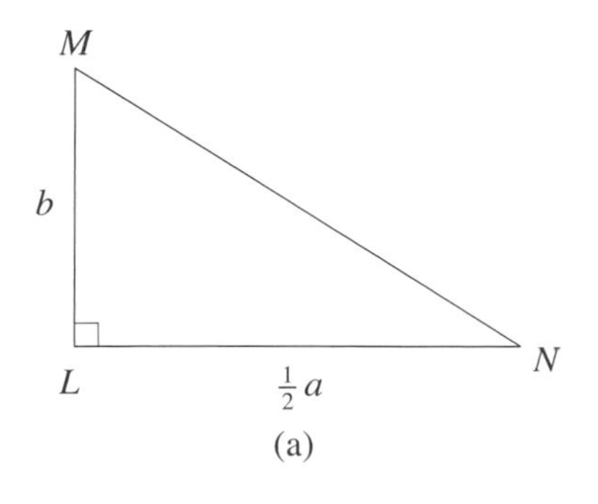

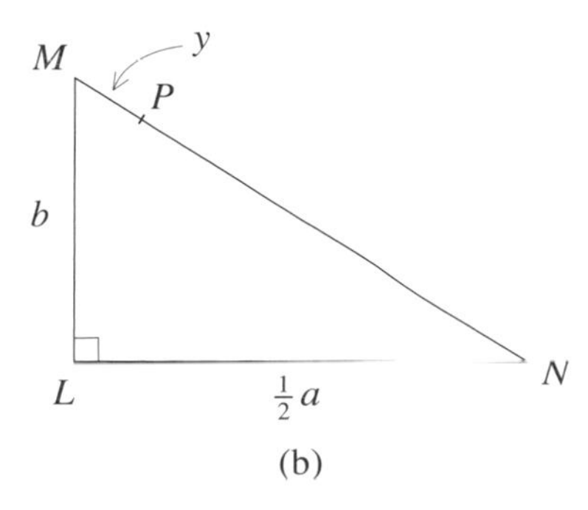

Fig. 4.3

$y \propto MN - LN \propto \sqrt{\frac{1}{4}a^2 + b^2} - \frac{1}{2}a$. Since $(y + \frac{1}{2}a)^2 \propto \frac{1}{4}a^2 + b^2$, he gets $y^2 + ay + \frac{1}{4}a^2 \propto \frac{1}{4}a^2 + b^2$, and hence $y^2 \propto -ay + b^2$. The negative solution $y \propto -\sqrt{\frac{1}{4}a^2 + b^2} - \frac{1}{2}a$ is ignored.

Instead of writing the equation

$$B5 \text{ in } A \text{ quad } -C \text{ plano } 2 \text{ in } A{+}A \text{ cubum aequatum } D \text{ solido},$$

as Viète had done, Descartes would have written it as either

$$5bxx - 2cx + xxx \propto d \quad \text{or} \quad 5bx^2 - 2cx + x^3 \propto d,$$

not seeming to prefer the one form over the other. The second version is precisely (with the exception of $\propto$ in place of =) the way we would write it today. A short passage from the third book of his *Géométrie* illustrates Descartes's algebra:

> given $x^3 - \sqrt{3}x^2 + \frac{26}{27}x - \frac{8}{27\sqrt{3}} \propto 0$, let there be required another equation in which all the terms are expressed in rational numbers. Let $y \propto \sqrt{3}x$ and multiply the second term by $\sqrt{3}$, the third by 3 and the last by $3\sqrt{3}$. The resulting equation is $y^3 - 3y^2 + \frac{26}{9}y - \frac{8}{9} \propto 0$. Next let it be required to replace this equation by another in which the known quantities are expressed only by whole numbers. Let $z \propto 3y$. Multiplying 3 by 3, $\frac{26}{9}$ by 9, and $\frac{8}{9}$ by 27, we have
>
> $$z^3 - 9z^2 + 26z - 24 \propto 0.$$
>
> The roots of this equation are 2, 3, 4; and hence the roots of the preceding equation are $\frac{2}{3}$, 1, $\frac{4}{3}$, and those of the first equation are
>
> $$\tfrac{2}{9}\sqrt{3},\ \tfrac{1}{3}\sqrt{3}, \text{ and } \tfrac{4}{9}\sqrt{3}.$$

Example 4.1. It was Descartes's aim to deal with the equation $x^3 - \sqrt{3}x^2 + \frac{26}{27}x - \frac{8}{27\sqrt{3}} = 0$ by transforming it into the simpler one $z^3 - 9z^2 + 26z - 24 = 0$ that he knew how to solve. Carry out how we would do this today: substitute $x = \frac{1}{\sqrt{3}}y$ into Descartes's original equation, multiply through by $3\sqrt{3}$, then substitute $y = \frac{1}{3}z$ and, finally multiply through by 27 to obtain the last equation.

4.2 THE COORDINATE PLANE

The discussion that follows explores in today's mathematical language what Descartes achieved. Just as the points on a line can be identified with the real numbers by using a coordinate line, so the points in a plane can be identified with pairs of real numbers. Start by drawing two perpendicular coordinate lines that intersect at their origins. One line is placed horizontally with the positive direction to the right. The other is placed vertically with the positive direction upward. These are the *coordinate axes*. The horizontal axis is often called the *x-axis*, and the vertical axis the *y-axis*. However, these designations depend on the variables used. For example, in a context where a variable time t or angle θ is considered, an axis might be labeled by t or θ.

Any point P in the plane determines a unique ordered pair of numbers as follows. Draw the line through P parallel to the y-axis. This line intersects the x-axis at some point, and this point has a coordinate, say a, on this number line. Similarly, draw a line through P parallel to the x-axis to get a coordinate, say b, on the y-axis. Refer to Figure 4.4. In this way, Descartes assigns to the point P the pair of numbers (a, b). The first number, a, is called the *x-coordinate* of P, and the second number, b, is called the *y-coordinate* of P. We say that P is the point with coordinates (a, b), and we will often write (a, b) in place of P. For example, the point S in Figure 4.4 has coordinates $(3, -3.4)$, so $S = (3, -3.4)$. By reversing the process,

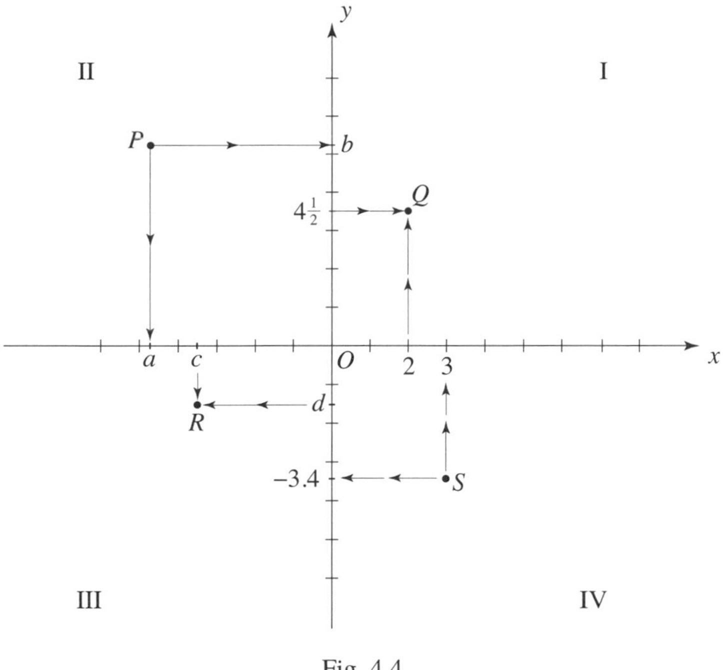

Fig. 4.4

we can start with an ordered pair, say (c, d), and arrive at the corresponding point R. For example, $(2, 4\frac{1}{2})$ corresponds to the point Q. The point $(0, 0)$ is called the *origin* and is often denoted by O. The notation 0 (zero) will be used for this point if the attention is on the x-axis or the y-axis.

This setup is called the *rectangular* or *Cartesian* (for Descartes) *coordinate system*. It provides the connection between geometric objects—points, collections of points, and curves—and algebra—pairs of numbers, collections of such pairs, and equations. The plane supplied with a coordinate system is called the *Cartesian plane*. The coordinate axes divide the Cartesian plane into four *quadrants*, labeled I, II, III, and IV in Figure 4.4. Notice that the first quadrant consists of the points whose x- and y-coordinates are both positive.

Check that the distance between the points -5 and 3 on the number line is 8. Observe that this is equal to the absolute value of $-5 - 3$ or, equivalently, that of $3 - (-5)$. Similarly, the distance between 6.4 and 4.6 is 1.8. This is the absolute value of either $6.4 - 4.6$ or $4.6 - 6.4$. In general, the distance between the points a and b on a number line is equal to $a - b$ if $b \leq a$ and $b - a$ if $b > a$. See Figure 4.5. In the first

Fig. 4.5

case, $a - b \geq 0$ and hence $|a - b| = a - b$. In the second case, $a - b < 0$ and hence $|a - b| = -(a - b) = b - a$. Observe, therefore, that the distance between a and b is equal to the absolute value $|a - b| = |b - a|$ in either case.

Consider any two points $P_1 = (x_1, y_1)$ and $P_2 = (x_2, y_2)$ in the plane, and refer to Figure 4.6. By the remarks just made, the distance between the two x-coordinates is $|x_1 - x_2|$, and the distance between the two y-coordinates is $|y_1 - y_2|$. It follows that the segment P_1P_2 is the hypotenuse of the right triangle of Figure 4.7. Therefore, by the Pythagorean theorem,

$$P_1P_2 = \sqrt{|x_1 - x_2|^2 + |y_1 - y_2|^2}.$$

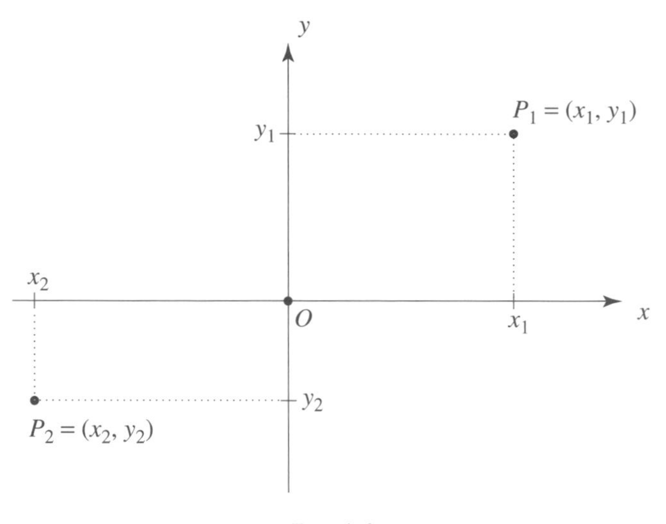

Fig. 4.6

Since $|x_1 - x_2|^2 = (x_1 - x_2)^2$ and $|y_1 - y_2|^2 = (y_1 - y_2)^2$, the distance between the points $P_1(x_1, y_1)$ and $P_2(x_2, y_2)$ is equal to

$$P_1P_2 = \sqrt{(x_1 - x_2)^2 + (y_1 - y_2)^2}.$$

Observe that we are continuing an earlier practice: when a segment, here P_1P_2, occurs in a mathematical

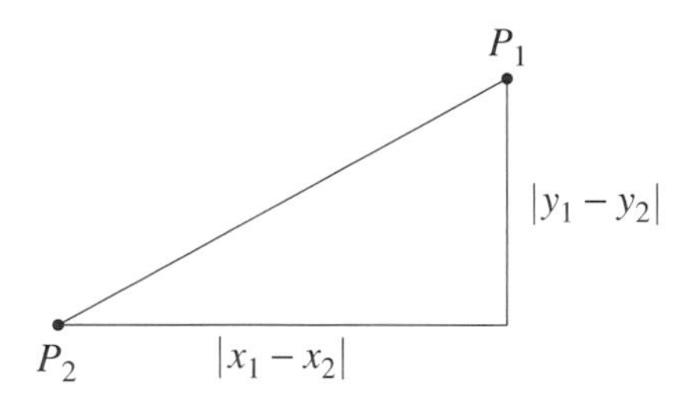

Fig. 4.7

expression, the reference will always be to the length of the segment.

Example 4.2. The distance between $(1, -2)$ and $(5, 3)$ is

$$\sqrt{(1 - 5)^2 + (-2 - 3)^2} = \sqrt{4^2 + 5^2} = \sqrt{41}.$$

Since $\sqrt{(5 - 1)^2 + (3 - (-2))^2}$ is also equal to $\sqrt{4^2 + 5^2} = \sqrt{41}$, the order in which the points are considered has no effect on the result.

The *graph* of an equation in x and y is the set of all points (a, b) in the plane with the property that the values $x = a$ and $y = b$ satisfy the equation. For example, the point $(1, -3)$ is a point on the graph of the equation $3x^2 + y^2 = 12$, since $3(1)^2 + (-3)^2 = 12$. In the same way, the point $(5, 2)$ is on the graph of the equation $2x^2 - 5y^3 = 10$, because $2(5)^2 - 5(2)^3 = 50 - 40 = 10$.

The *y-intercepts* of a graph are the y-coordinates of the points of intersection of the graph with the y-axis. They are found by setting $x = 0$ and solving for y. When this is done for the graph of $xy^3 + y^2 - x^3 - 3 = 0$, for example, we get $y^2 = 3$, and therefore $y = \pm\sqrt{3}$. In the same way, the *x-intercepts* of a graph are the x-coordinates of the points of intersection of the graph with the x-axis. They are found by setting $y = 0$ and solving for x. For $xy^3 + y^2 - x^3 - 3 = 0$, we get $x^3 = -3$, so that $x = -\sqrt[3]{3}$.

A graph is a pictorial representation of an equation. It provides a picture of the algebra and clarifies what is going on. In the other direction, when a curve has to be understood, it is often of great advantage to have an equation of the curve, that is to say, an equation whose graph is the curve in question. This translates the geometric concern into algebra. The algebra allows for precise computations that can reveal exact numerical information. In other words, if a curve can be represented by an algebraic equation, then the rules of algebra can be used to analyze it. This complementary duality between geometry on the one hand and numbers and algebra on the other is the essence of analytic geometry. This interplay is of crucial importance in the solution of many mathematical problems.

Consider a circle with center $C = (3, 2)$ and radius 4. By definition, a point $P = (x, y)$ in the plane is

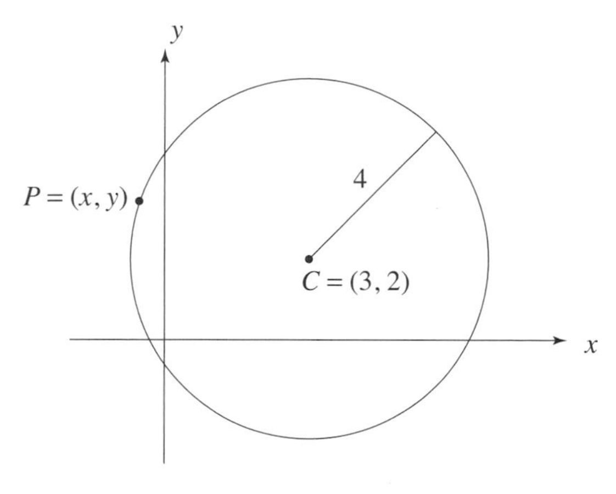

Fig. 4.8

on the circle precisely if the distance from P to C is equal to 4. See Figure 4.8. Put another way, $P = (x, y)$ is on the circle exactly when

$$PC = \sqrt{(x-3)^2 + (y-2)^2} = 4.$$

Squaring both sides, we see that this is so precisely when

$$(x-3)^2 + (y-2)^2 = 16.$$

Proceeding in the same way, we see that a point $P = (x, y)$ is on the circle with radius r and center (h, k) exactly when its coordinates satisfy

$$(x-h)^2 + (y-k)^2 = r^2.$$

This is the *standard form* of the equation of the circle with radius r and center (h, k). If the center is the origin $O = (0, 0)$, the equation is $x^2 + y^2 = r^2$.

Example 4.3. Find an equation of the circle with radius 3 and center $(2, -5)$. Putting $r = 3$, $h = 2$, and $k = -5$, we obtain $(x-2)^2 + (y+5)^2 = 9$.

Figure 4.9 shows several circles of radius 2 along with their standard equations. The following consequence can be drawn from the figure. Let h and k be constants. If in the equation $x^2 + y^2 = 4$, x is replaced by $x - h$, then the graph of the new equation is obtained by shifting the original graph by h units to the right if h is positive and h units to the left if h is negative. The circle $(x+6)^2 + y^2 = 4$ is an example with $h = -6$. In the same way, if y is replaced by $y - k$, then the graph is shifted k units up if k is positive and k units down if k is negative. The circle $x^2 + (y-5)^2 = 4$ is an example with $k = 5$. Replacing both x and y by $x - h$ and $y - k$, respectively, shifts the graph both horizontally and vertically. The circles $(x-7)^2 + (y-3)^2 = 4$ and $(x+9)^2 + (y+3)^2 = 4$ provide examples. In the first case, $h = 7$ and $k = 4$, so

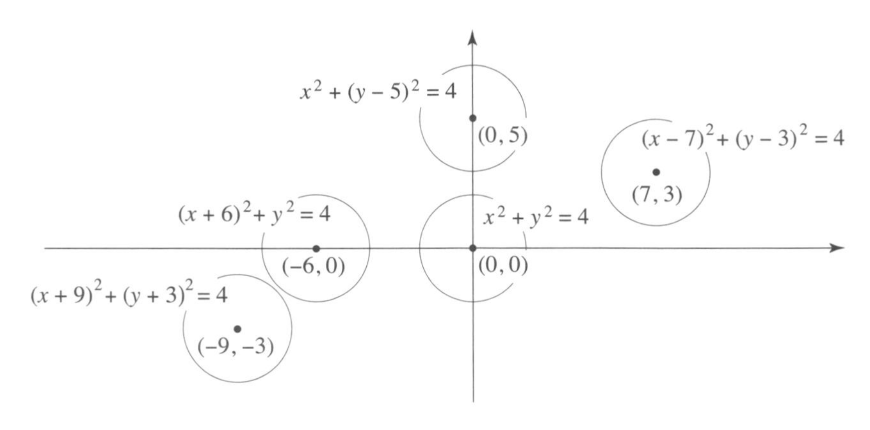

Fig. 4.9

that the circle $x^2 + y^2 = 4$ is shifted 7 units to the right and 3 units up, and in the second case, $h = -9$ and $k = -3$, so that the circle $x^2 + y^2 = 4$ is shifted 9 units to the left and 3 units down.

The fact is that what has just been observed is true in general, not only for circles. Whatever the graph of the equation is, replacing x by $x - h$ shifts the graph h units to the right for a positive h and to the left for a negative h, and analogously for y and $y - k$. What is not apparent from the case of the circle is the fact that such left/right or up/down *shifts* or *translations* do *not rotate the graph.* It follows that if the graph of an equation is *shifted* or *translated* in the xy-plane vertically and/or horizontally in such a way that it is not rotated in the process, then the equation of the new graph is obtained by replacing x by $x - h$ and y by $y - k$, where h and k are determined by the distances and directions of the shifts. For example, the graph of $y - 2 = (x + 3)^2$ has the same shape and orientation as the graph of $y = x^2$. It is obtained by translating the graph of $y = x^2$ in such a way that the point $(0, 0)$ on the original graph ends up at the point $(-3, 2)$ of the shifted graph.

Example 4.4. Show that the graph of the equation $x^2 + y^2 + 2x - 6y + 7 = 0$ is a circle and find its center and radius. Rewrite the equation as $(x^2 + 2x) + (y^2 - 6y) = -7$. We'll now complete the square—this procedure is discussed in detail in the Problems and Projects section of this chapter—for each variable. This means that we take

$$(x^2 + 2x \qquad) + (y^2 - 6y \qquad) = -7,$$

divide the coefficient of x by 2, square the result, insert this into the first blank, then divide the coefficient of y by 2, square this result, and add it into the second blank. So we add a 1 to the x terms and 9 to the y terms. To keep the $=$ in tact, we also add 1 and 9 to the right side. We therefore get $(x^2 + 2x + 1) + (y^2 - 6y + 9) = -7 + 1 + 9$. Notice that the x terms within the first pair of parentheses and the y terms within the second pair have both been *completed to squares*, and that the equation has been transformed to $(x + 1)^2 + (y - 3)^2 = 3$. This is the equation of the circle with center $(-1, 3)$ and radius $\sqrt{3}$.

Any equation of the form $ax + by + c = 0$ where a, b, and c can be any constants (but not all equal to 0) is a straight line. Conversely, any line has an equation of this form. For now, we'll only provide a few examples. Their graphs are sketched in Figure 4.10. The graph of the equation $x - 4 = 0$, or $x = 4$, is the set of all points (x, y) in the plane with $x = 4$. This is a vertical line through the point $x = 4$ on the x-axis. In a similar way, $y + 5 = 0$ is a horizontal line through the point $y = -5$ on the y-axis. Notice that the points $(-6, 0)$ and $(0, 4)$ are both on the graph of the equation $2x - 3y + 12 = 0$. Therefore the graph of this equation is the line that these two points determine. A similar thing is true for the points $(-1, -5)$, $(4, -6)$ and the equation $x + 5y + 26 = 0$. Finally, consider the line L through $(0, 0)$ and (h, k). Notice that both $(0, 0)$ and (h, k) satisfy the equation $\frac{k}{h}x - y = 0$, so that L is the graph of this equation, or equivalently of $y = \frac{k}{h}x$. Lines—the word will always mean *straight* lines—will be studied in detail in Section 5.1.

Coordinate geometry made mathematics a double-edged tool. Geometric concepts could be formulated

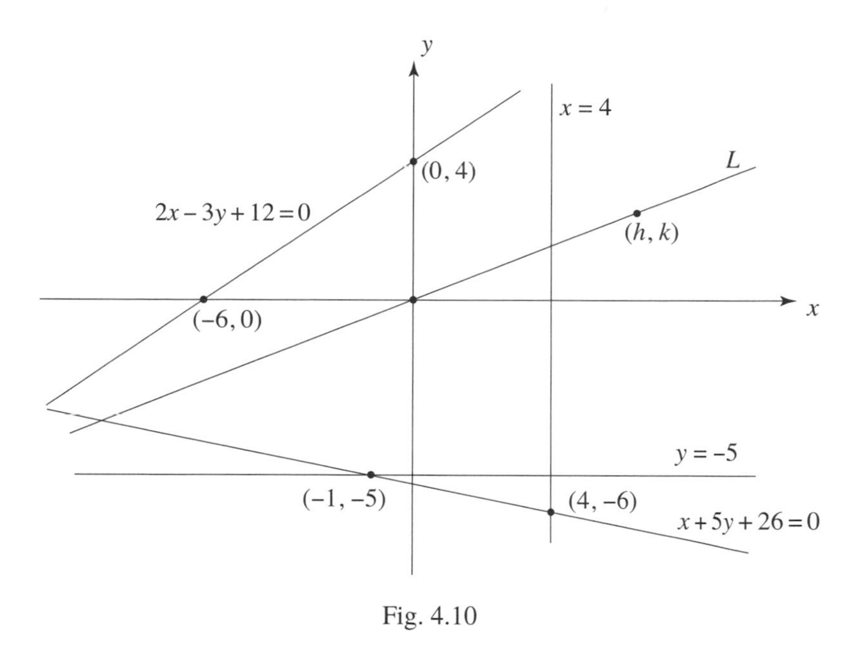

Fig. 4.10

and solved algebraically and conversely, by interpreting algebraic statements geometrically one could gain an intuitive understanding of the algebra. The Italian-French mathematician Luigi Lagrange (1736–1813) put it this way "as long as algebra and geometry traveled separate paths their advance was slow and their applications limited. But when they joined company they drew from each other fresh vitality and from then on marched at a rapid pace towards perfection."

4.3 ABOUT THE PARABOLA

Let D be a line and F a point in the plane with F not on D. The line and the point determine a parabola as the set of all points P in the plane with the property that the distance from P to F is equal to the distance from P to D. The point F is the *focal point*, and the line D is the *directrix* of the parabola. The line through the focal point perpendicular to the directrix is the *focal axis* of the parabola. The parabola has already been introduced in Section 2.1. Move the parabola so that its directrix D is horizontal. See Figure 4.11.

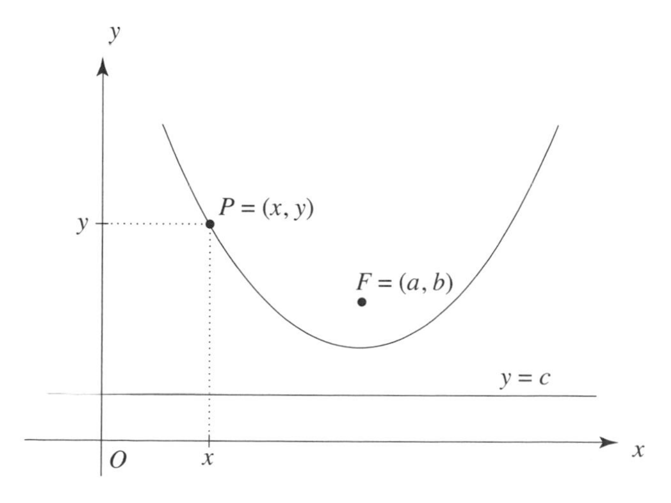

Fig. 4.11

Suppose D crosses the y-axis at the point c, and observe that $y = c$ is the equation of D. Let $F = (a, b)$ be the focus. Since F is not on D, it follows that $b \neq c$.

Let $P = (x, y)$ be any point in the plane. Then the distance from P to D is $|y - c|$, and the distance from P to F is

$$PF = \sqrt{(x-a)^2 + (y-b)^2}.$$

So P is on the parabola precisely when $|y - c| = \sqrt{(x-a)^2 + (y-b)^2}$. Squaring both sides gives $(y-c)^2 = (x-a)^2 + (y-b)^2$. After multiplying things out and simplifying, we have

$$\begin{aligned} y^2 - 2cy + c^2 &= x^2 - 2ax + a^2 + y^2 - 2by + b^2 \\ 2by - 2cy &= x^2 - 2ax + a^2 + b^2 - c^2 \\ 2(b-c)y &= x^2 - 2ax + (a^2 + b^2 - c^2). \end{aligned}$$

Since $b \neq c$, we can divide through by $2(b - c)$ to get

$$(*) \quad y = \Big(\frac{1}{2(b-c)}\Big)x^2 - \Big(\frac{a}{b-c}\Big)x + \Big(\frac{a^2 + b^2 - c^2}{2(b-c)}\Big).$$

By letting $A = \frac{1}{2(b-c)}$, $B = -\frac{a}{b-c}$, and $C = \frac{a^2+b^2-c^2}{2(b-c)}$, we have shown that any parabola with a horizontal directrix has an equation of the form

$$y = Ax^2 + Bx + C.$$

Example 4.5. Find an equation of the parabola with directrix $y = 5$ and focus $(2, -3)$. This is the special case $c = 5$, $a = 2$, and $b = -3$ of the parabola just discussed. Since $\frac{1}{2(b-c)} = \frac{1}{2(-3-5)} = -\frac{1}{16}$, $-\frac{a}{b-c} = -\frac{2}{-3-5} = \frac{1}{4}$, and $\frac{a^2+b^2-c^2}{2(b-c)} = \frac{4+9-25}{-16} = \frac{3}{4}$, we see that $y = -\frac{1}{16}x^2 + \frac{1}{4}x + \frac{3}{4}$ is an equation of this parabola.

We have just established the fact that *if a parabola has a horizontal directrix*, then it has an equation of the form $y = Ax^2 + Bx + C$ (with $A \neq 0$). This argument can be reversed to show that the graph of any equation of the form $y = Ax^2 + Bx + C$ (with $A \neq 0$) is a parabola with horizontal directrix. The next example illustrates what is involved.

Example 4.6. Show that the graph of the equation $y = 2x^2 + 5x - 4$ is a parabola, and find its directrix and focus. Work backward using equation $(*)$. Since $\frac{1}{2(b-c)} = 2$, it follows that $b - c = \frac{1}{4}$. Because $-\frac{a}{b-c} = 5$, we now get that $a = -\frac{5}{4}$. From the equality $-4 = \frac{a^2+b^2-c^2}{2(b-c)}$ we get $a^2 + b^2 - c^2 = -4 \cdot 2(b-c)$. So $b^2 - c^2 = -a^2 - 4 \cdot 2(b-c) = -\frac{25}{16} - \frac{8}{4} = -\frac{57}{16}$. Use of the factorization $b^2 - c^2 = (b+c)(b-c)$ shows that $b + c = -\frac{57}{16} \cdot 4 = -\frac{57}{4}$. Since $b - c = \frac{1}{4}$, it is now easy to solve for b and c to get $b = -7$ and $c = -7\frac{1}{4}$. It follows from equation $(*)$ that $y = 2x^2 + 5x - 4$ is the equation of the parabola with directrix $y = -7\frac{1}{4}$ and focus $(-\frac{5}{4}, 7)$.

Example 4.7. Consider the parabola $y = x^2 + 3x + 4$. The completing of the square procedure provides relevant information about the parabola. Accordingly, rewrite the equation as

$$\begin{aligned} y &= x^2 + 3x + 4 = x^2 + 3x + (\tfrac{3}{2})^2 - (\tfrac{3}{2})^2 + 4 \\ &= (x + \tfrac{3}{2})^2 + 4 - \tfrac{9}{4}. \end{aligned}$$

So $y = (x + \frac{3}{2})^2 + \frac{7}{4}$. Notice that the smallest possible value of y is $\frac{7}{4}$, and it occurs for $x = -\frac{3}{2}$. Since $y - \frac{7}{4} = (x + \frac{3}{2})^2$, the graph of this parabola is obtained by taking the graph of $y = x^2$ and shifting it without rotating it $\frac{3}{2}$ to the left and $\frac{7}{4}$ units up.

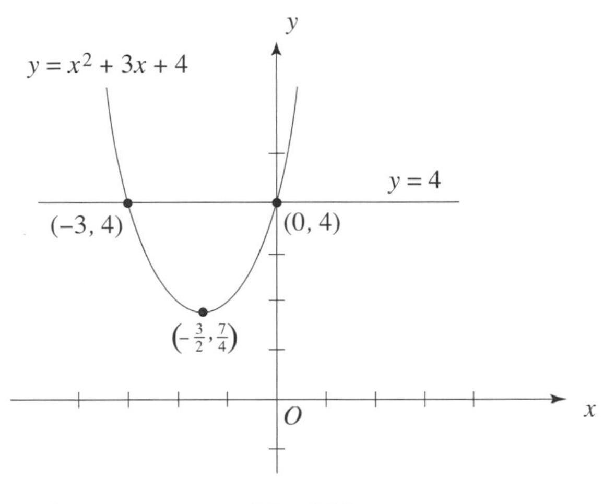

Fig. 4.12

Example 4.8. In this example, we'll cut the parabola $y = x^2 + 3x + 4$ with the horizontal line $y = 4$ and compute the area of the resulting parabolic section. See Figure 4.12. This can be done by applying the theorem of Archimedes of Section 2.4. We saw in the previous example that $\frac{7}{4}$ is the smallest value of $y = x^2 + 3x + 4$ and that it is achieved for $x = -\frac{3}{2}$. Therefore $(-\frac{3}{2}, \frac{7}{4})$ is the lowest point on the parabola. Since the tangent line hugs the curve near the point of contact, the tangent to the parabola at $(-\frac{3}{2}, \frac{7}{4})$ must be horizontal. So this tangent is parallel to the cut $y = 4$. It follows from Section 2.1 that $(-\frac{3}{2}, \frac{7}{4})$ is the vertex of the parabolic section. The points of intersection of the parabola and the line $y = 4$ are needed next. If (x, y) is on both the parabola and $y = 4$, then $y = x^2 + 3x + 4$ and $y = 4$. So $x^2 + 3x + 4 = 4$, and $x^2 + 3x = 0$. By factoring, $x^2 + 3x = x(x + 3) = 0$, so $x = 0$ or $x = -3$. Therefore the points of intersection are $(-3, 4)$ and $(0, 4)$. By Archimedes's theorem, the area of the parabolic section is $\frac{4}{3}$ times that of the inscribed triangle. Refer to Figure 4.12 again. The inscribed triangle has base 3 and height $4 - \frac{7}{4} = \frac{9}{4}$, so the triangle has area $\frac{1}{2} \cdot 3 \cdot \frac{9}{4} = \frac{27}{8}$. Thus the area of the parabolic section is $\frac{4}{3} \cdot \frac{27}{8} = \frac{9}{2}$.

Example 4.9. Draw the graph of the parabola $y = 2x^2$. Set up a table of values and plot the corresponding points, and then join them by a smooth curve to obtain the graph. Figure 4.13 shows this graph as well as

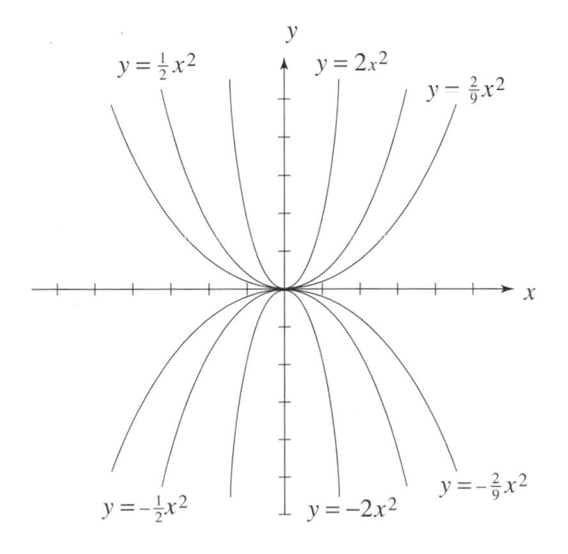

Fig. 4.13

the graphs of several other parabolas of the form $y = Ax^2$.

Consider the equation $y = Ax^2 + Bx + C$, with $A \neq 0$. For large positive or negative x, the term Ax^2 will dominate the others. So if $A > 0$, then y is positive when $|x|$ is large, and if $A < 0$, then y is negative when $|x|$ is large. It follows that the parabola $y = Ax^2 + Bx + C$ opens upward if $A > 0$ and downward if $A < 0$. The argument that demonstrated that any parabola with a horizontal directrix has an equation of the form $y = Ax^2 + Bx + C$ can be used to show that any parabola with a vertical directrix has an equation of the form $x = Ay^2 + By + C$ with $A \neq 0$ and that the graph of any equation of this form is a parabola with a vertical directrix. Such a parabola opens to the right if $A > 0$ and to the left if $A < 0$.

4.4 ABOUT THE ELLIPSE

Let F_1 and F_2 be two points in the plane, and let k be a positive real number such that k is greater than the distance F_1F_2. The two points and k specify an *ellipse* as the set of all points P in the plane such that the distance from P to F_1 plus the distance from P to F_2 add up to k, or, put another way, such that $PF_1 + PF_2 = k$. The points F_1 and F_2 are the *focal points* of the ellipse, and the line that they determine is the *focal axis*. Section 2.1 has already introduced the ellipse and illustrated these concepts.

Let P_1 and P_2 be the two points on the ellipse that lie on the focal axis, and let P_3 on the ellipse lie on the perpendicular to the focal axis through the *center* C of the ellipse. Let $a = CP_1 = CP_2$, $c = CF_1 = CF_2$, and $b = CP_3$. Refer to Figure 4.14. In the figure, P_1 cannot be at F_1 or to the left of F_1, because this would mean that $a \leq c$ and lead to the contradiction $k = P_1F_1 + P_1F_2 = (c - a) + (c + a) = 2c = F_1F_2$. So $a > c$, and

$$P_1F_1 + P_1F_2 = (a - c) + (a + c) = 2a,$$

so that $k = 2a$. Since $2a = k > F_1F_2 = 2c$, it follows that $a > c$ (as already noted in the figure). The lengths a and b are the *semimajor axis* and *semiminor axis* of the ellipse, respectively. By the Pythagorean

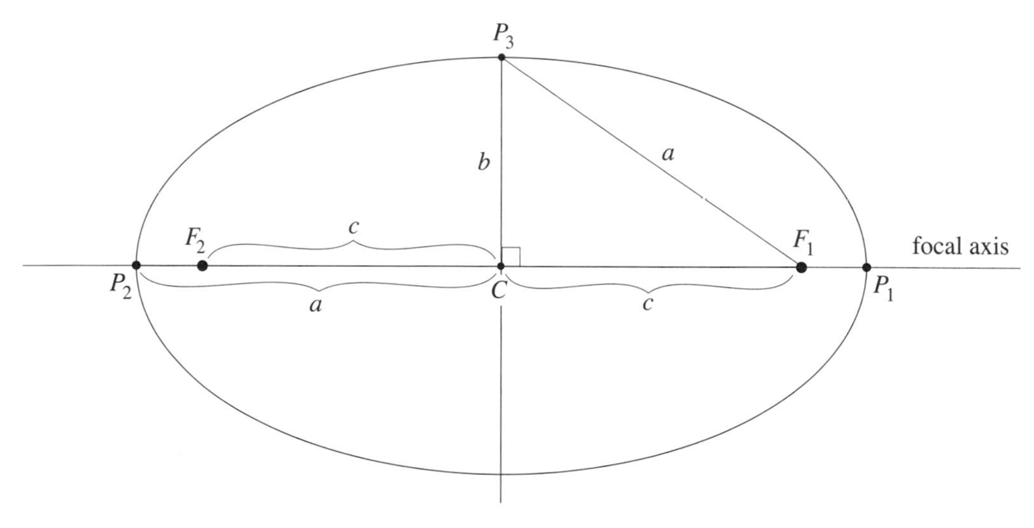

Fig. 4.14

theorem and the fact that P_3 is on the ellipse, $2a = k = P_3F_1 + P_3F_2 = 2P_3F_1 = 2\sqrt{b^2 + c^2}$. So $a = \sqrt{b^2 + c^2}$, and hence

$$a^2 = b^2 + c^2.$$

The *eccentricity* ε of the ellipse is defined by $\varepsilon = \frac{c}{a}$. Notice that $\varepsilon < 1$. If $\varepsilon = 0$, then $c = 0$. In this case, $F_1 = F_2 = C$. So for any point P on the ellipse, $2PC = PF_1 + PF_2 = 2a$, and it follows that the ellipse is a circle with radius a. In general, the closer ε is to 0, the closer a and b are to each other, and the more "circular" the ellipse is. On the other hand, the closer ε is to 1, the closer c is to a, the smaller b is relative to a, and the more "elliptical" the ellipse.

Place the ellipse in such a way that the focal axis coincides with the x-axis and the center C with the origin O. See Figure 4.15. Note that

$$F_1 = (c, 0) \quad \text{and} \quad F_2 = (-c, 0).$$

Now let $P = (x, y)$ be any point on the ellipse. The concern will be the precise conditions on the coordinates x and y of P that will guarantee that $P = (x, y)$ falls on the ellipse. The task will be to rewrite

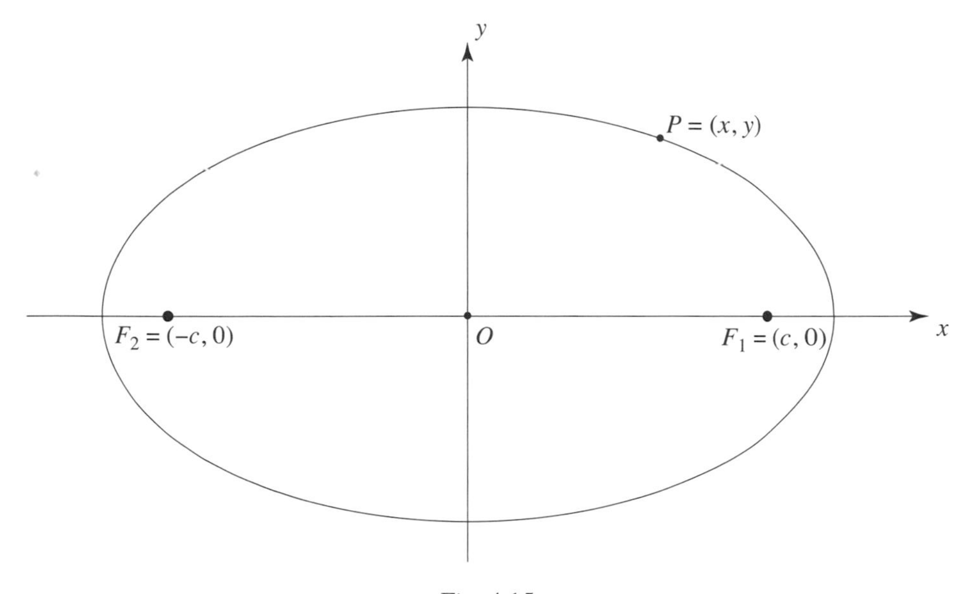

Fig. 4.15

the equality $PF_1 + PF_2 = 2a$ in terms of the coordinates of the points involved. The use of the distance formula of Section 4.2 informs us that the point $P = (x, y)$ is on the ellipse precisely if

$$\sqrt{(x-c)^2 + (y-0)^2} + \sqrt{(x-(-c))^2 + (y-0)^2} = 2a.$$

By simplifying, squaring both sides, canceling, moving things around, and simplifying once more, this equation is transformed in successive steps to

$$\begin{aligned}
\sqrt{(x-c)^2 + y^2} + \sqrt{(x+c)^2 + y^2} &= 2a, \\
\sqrt{(x-c)^2 + y^2} &= 2a - \sqrt{(x+c)^2 + y^2}, \\
(x-c)^2 + y^2 &= 4a^2 - 4a\sqrt{(x+c)^2 + y^2} + (x+c)^2 + y^2, \\
(x-c)^2 &= 4a^2 - 4a\sqrt{(x+c)^2 + y^2} + (x+c)^2, \\
x^2 - 2cx + c^2 &= 4a^2 - 4a\sqrt{(x+c)^2 + y^2} + x^2 + 2cx + c^2, \\
a\sqrt{(x+c)^2 + y^2} &= a^2 + cx, \\
a^2((x+c)^2 + y^2) &= a^4 + 2a^2cx + c^2x^2, \\
a^2(x^2 + 2cx + c^2 + y^2) &= a^4 + 2a^2cx + c^2x^2, \\
a^2x^2 + 2a^2cx + a^2c^2 + a^2y^2 &= a^4 + 2a^2cx + c^2x^2, \\
a^2x^2 - c^2x^2 + a^2y^2 &= a^4 - a^2c^2, \\
(a^2 - c^2)x^2 + a^2y^2 &= a^2(a^2 - c^2).
\end{aligned}$$

Since $b^2 = a^2 - c^2$, this last equation becomes $b^2x^2 + a^2y^2 = a^2b^2$. Because $a > c \geq 0$ and $b > 0$, we can divide $b^2x^2 + a^2y^2 = a^2b^2$ by a^2b^2 to get

$$\frac{x^2}{a^2} + \frac{y^2}{b^2} = 1.$$

This is the *standard equation* of the ellipse. Observe that if $a = b$, then this equation is equivalent to $x^2 + y^2 = a^2$, so that we are dealing with a circle with center O and radius a.

Example 4.10. Show that the graph of $9x^2 + 16y^2 = 144$ is an ellipse. Determine its semimajor and semiminor axes, its two focal points, and its eccentricity. Divide both sides of the equation by 144 to get

$$\frac{x^2}{16} + \frac{y^2}{9} = 1.$$

The equation is now in standard form. Since $a^2 = 16$ and $b^2 = 9$, we get $a = 4$ and $b = 3$ for the semimajor and semiminor axes, respectively. Since $c^2 = a^2 - b^2 = 4^2 - 3^2 = 7$, $c = \sqrt{7}$, and the focal points are $(-\sqrt{7}, 0)$ and $(\sqrt{7}, 0)$. The eccentricity is $\varepsilon = \frac{\sqrt{7}}{4} \approx 0.66$.

With the ellipse placed so that its focal axis lies on the x-axis and its center is at the origin, then—as we have seen—its equation is of the form

$$\frac{x^2}{a^2} + \frac{y^2}{b^2} = 1$$

with $a \geq b$. If the ellipse of Figure 4.14 is is placed so that the focal axis is on the y-axis with the center again at the origin, then the ellipse has an equation of similar form, but this time the semimajor axis a is aligned with the y term and the semiminor axis b with the x term, and the equation is

$$\frac{x^2}{b^2} + \frac{y^2}{a^2} = 1.$$

Example 4.11. Consider the equation $25x^2 + 100x + 9y^2 - 54y = 44$. Show that it is an ellipse, and determine its semimajor and semiminor axes and its eccentricity. Since an ellipse involves squares, let's see what happens if we complete the squares for the terms $25x^2 + 100x$ and $9y^2 - 54y$. Doing this, we get

$$25x^2 + 100x = 25(x^2 + 4x) = 25((x^2 + 4x + 4) - 4) = 25(x + 2)^2 - 100, \text{ and}$$
$$9y^2 - 54y = 9(y^2 - 6y) = 9((y^2 - 6y + 9) - 9) = 9(y - 3)^2 - 81.$$

It follows that $25x^2 + 100x + 9y^2 - 54y = 25(x + 2)^2 - 100 + 9(y - 3)^2 - 81 = 44$, and therefore that $25(x + 2)^2 + 9(y - 3)^2 = 44 + 100 + 81 = 225$. Dividing both sides by $225 = 25 \cdot 9$ brings the original equation into the form $\frac{(x+2)^2}{3^2} + \frac{(y-3)^2}{5^2} = 1$. This is the ellipse obtained from $\frac{x^2}{3^2} + \frac{y^2}{5^2} = 1$ by shifting it 2 units to the left and 3 units up. Therefore its focal axis is vertical, and the semimajor and semiminor axes are $a = 5$ and $b = 3$, respectively. Since $c^2 = a^2 - b^2 = 25 - 9 = 16$, it follows that $c = 4$. So the focal points are $(0, \pm c) = (0, \pm 4)$, and the eccentricity is $\varepsilon = \frac{4}{5} = 0.8$.

Example 4.12. The ellipse of Figure 4.15 has equation $\frac{x^2}{a^2} + \frac{y^2}{b^2} = 1$. Solving this equation for y, we get

$$y^2 = b^2(1 - \tfrac{x^2}{a^2}) = b^2(\tfrac{a^2-x^2}{a^2}) = \tfrac{b^2}{a^2}(a^2 - x^2),$$

and therefore $y = \pm\frac{b}{a}\sqrt{a^2 - x^2}$. Since $y \geq 0$ for the upper half of the ellipse, the upper half is the graph of $y = \frac{b}{a}\sqrt{a^2 - x^2}$. In the same way, the lower half of the ellipse is the graph of $y = -\frac{b}{a}\sqrt{a^2 - x^2}$. Surround the ellipse with a circle of radius a also centered at the origin O. Solving its equation $x^2 + y^2 = a^2$ for y gives us $y = \pm\sqrt{a^2 - x^2}$. As in the case of the ellipse, the upper half of the circle is the graph of

$y = \sqrt{a^2 - x^2}$, and the lower half is the graph of $y = -\sqrt{a^2 - x^2}$. Let (x_0, y_0) be a typical point on the upper half of the circle. Since $y_0 = \sqrt{a^2 - x_0^2}$, it follows that $\frac{b}{a}y_0 = \frac{b}{a}\sqrt{a^2 - x_0^2}$, and hence that the point $(x_0, \frac{b}{a}y_0)$ is on the upper half of the ellipse. In terms of the graphs, this says that if a point on the upper half of the surrounding circle is pushed parallel to the y-axis toward the x-axis in such a way that its distance from the x-axis shrinks by a factor of $\frac{b}{a}$, then the point lands on the upper half of the ellipse. This is exactly what Kepler observed about the orbit of Mars and its surrounding circle. See Figure 3.16. For each pair of positions C_i and M_i, he found that $(0.99573)\frac{C_i}{O_i} = \frac{M_i}{O_i}$. This observation confirmed his belief that the orbit of Mars must be an ellipse.

4.5 QUADRATIC EQUATIONS IN x AND y

It turns out that there is a tight relationship between the conic sections and quadratic equations in the variables x and y. Before we can take this topic on, we'll need to consider the hyperbola. This is the one conic section that has received little attention up to now. (Section 2.1 only provided the definition.)

Fix two distinct points F_1 and F_2 in the plane as well as a positive number k less than the distance between F_1 and F_2. The points and k define a *hyperbola* as the set of all points P such that the absolute value of the difference in the lengths of the segments PF_1 and PF_2 is equal to k. The two points F_1 and F_2 are the *focal points* of the hyperbola. The line that the focal points determine is the *focal axis*. For no point P on the hyperbola can its distance to F_1 be equal to its distance to F_2, because if this were so, then $|PF_1 - PF_2| = 0$. But this is not possible since $k > 0$. Therefore for a given point P on the hyperbola, either F_1 or F_2 is closer. The points P closer to F_1 form one component, or *branch*, of the hyperbola, and those closer to F_2 form the other.

Let C be the midpoint of the segment F_1F_2 and let $c = CF_1 = CF_2$. Let P_1 be the point that is both on the focal axis and on the right branch of the hyperbola and let $a = CP_1$. Refer to Figure 4.16. In the figure, P_1 cannot be at F_1 or to the right of F_1, because this would mean that $c \leq a$ and lead to the contradiction $k = P_1F_2 - P_1F_1 = |(a+c)-(a-c)| = 2c = F_1F_2$. Therefore $c > a$ and $P_1F_2 - P_1F_1 = (a+c)-(c-a) = 2a$, so that $k = 2a$. Form the right triangle with base $a = CP_1$ and hypotenuse c, and let b be its third side. Since $c^2 = a^2 + b^2$, we know that $b = \sqrt{c^2 - a^2}$. In analogy with the ellipse, the lengths a and b are known

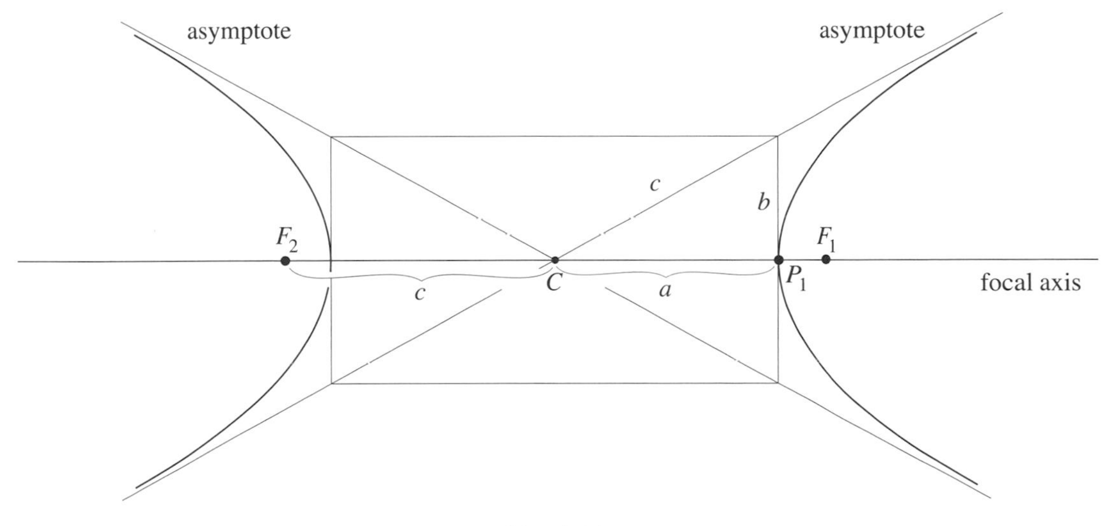

Fig. 4.16

as the *semimajor* and *semiminor axes* of the hyperbola. The *eccentricity* ε of the hyperbola is defined by $\varepsilon = \frac{c}{a}$. Since $c > a$, it follows that $\varepsilon > 1$. Notice that if ε is close to 1, then a is close to c, so b is small relative to a, and the opening of the hyperbola is narrow. Alternatively, if ε is much larger than 1, then b is large relative to a and the opening is wide.

Extend the right triangle to the $2a \times 2b$ rectangle with center C shown in Figure 4.16. We will soon see that the two extended diagonals of this rectangle are both *asymptotes* of the hyperbola, meaning that the hyperbola converges to these two lines, as shown in the figure. The word comes from the Greek *asumptotos*, meaning "not falling together."

Place the hyperbola so that the focal axis coincides with the x-axis and the point C with the origin. This is done in Figure 4.17. Notice that $F_1 = (c, 0)$ and $F_2 = (-c, 0)$. Let $P = (x, y)$ be any point on the parabola. The question arises, as in the situation of the ellipse, as to what the exact conditions on the coordinates x and y are that guarantee that the point $P = (x, y)$ is on the hyperbola. It follows from the

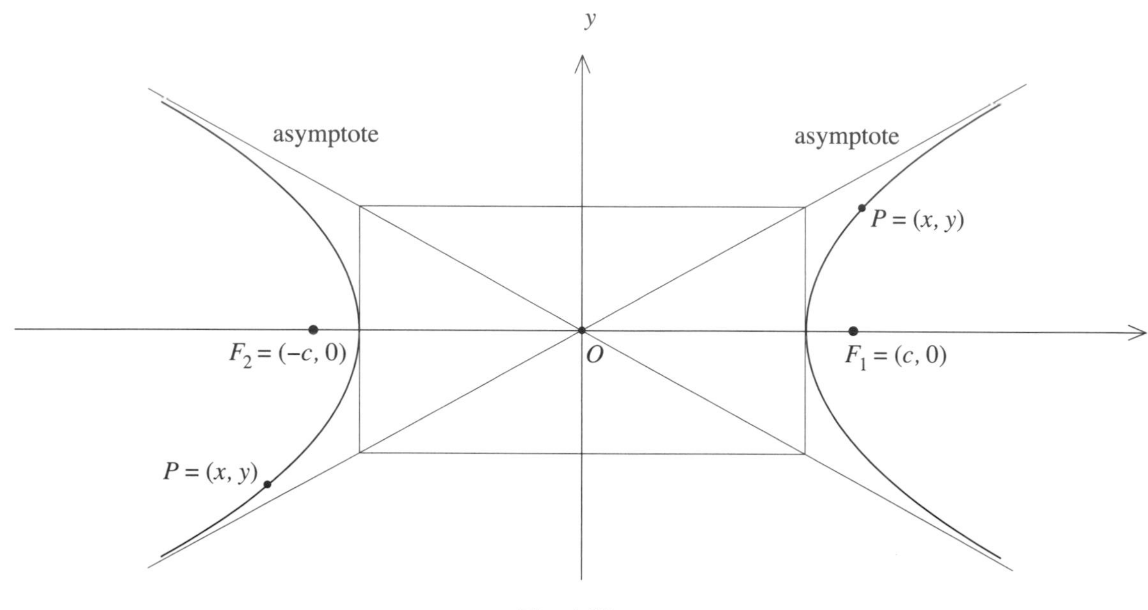

Fig. 4.17

distance formula of Section 4.2 that if P is on the right branch of the hyperbola, then

$$|PF_2 - PF_1| = PF_2 - PF_1 = \sqrt{(x+c)^2 + y^2} - \sqrt{(x-c)^2 + y^2} = 2a.$$

In the same way, if P is on the left branch, then

$$|PF_1 - PF_2| = PF_1 - PF_2 = \sqrt{(x-c)^2 + y^2} - \sqrt{(x+c)^2 + y^2} = 2a.$$

By playing a lengthy game of algebra similar to the one played for the ellipse, it can be shown in either case that the condition that the coordinates of the point $P = (x, y)$ must satisfy is

$$\frac{x^2}{a^2} - \frac{y^2}{b^2} = 1.$$

This is the *standard equation* of the hyperbola. It follows that $\frac{x^2}{a^2} = 1 + \frac{y^2}{b^2}$, so that $\frac{x^2}{a^2} \geq 1$. Therefore $x^2 \geq a^2$, and hence $x \geq a$ or $x \leq -a$. Since $\frac{y^2}{b^2} = \frac{x^2}{a^2} - 1 = \frac{x^2 - a^2}{a^2}$, we see that $y^2 = \frac{b^2}{a^2}(x^2 - a^2)$ and, after taking square roots, that

$$y = \pm\tfrac{b}{a}\sqrt{x^2 - a^2}.$$

Observe that if x^2 is large relative to a^2, then the effect of a^2 is minor and

$$y = \pm\tfrac{b}{a}\sqrt{x^2 - a^2} \approx \pm\tfrac{b}{a}\sqrt{x^2}.$$

The larger the x^2 term is, the better this approximation. Let's focus on the right branch $x \geq a$ of the hyperbola, and consider a point $P = (x, y)$ on the hyperbola with x much larger that a. If $y > 0$, then, by the observation just made, $y \approx \frac{b}{a}x$, so that the point is close to the line $y = \frac{b}{a}x$. If $y < 0$, then in the same way, the point is close to the line $y = -\frac{b}{a}x$. The same analysis shows that the hyperbola is close to the lines

$$y = \tfrac{b}{a}x \quad \text{and} \quad y = -\tfrac{b}{a}x$$

for large negative x as well. So these lines are asymptotes for both branches of the hyperbola.

Example 4.13. Consider the equation $9x^2 - 4y^2 = 36$. Divide both sides by 36 to get the standard form $\frac{x^2}{2^2} - \frac{y^2}{3^2} = 1$ of the equation of a hyperbola. Since $a = 2$ and $b = 3$, $c = \sqrt{2^2 + 3^2} = \sqrt{13}$. The eccentricity of the hyperbola is $\varepsilon = \frac{c}{a} = \frac{\sqrt{13}}{2} \approx 1.80$. Since the asymptotes are the lines $y = \pm\frac{3}{2}x$, it is now possible to provide at least a rough sketch of the graph.

By interchanging the roles of x and y, we get the equation

$$\frac{y^2}{a^2} - \frac{x^2}{b^2} = 1.$$

This also represents a hyperbola. Its graph is obtained by rotating that of Figure 4.17 by 90°.

Example 4.14. Sketch the curve $4x^2 - 4y^2 = 8$. After dividing both sides by 8, we obtain

$$\frac{x^2}{(\sqrt{2})^2} - \frac{y^2}{(\sqrt{2})^2} = 1.$$

This equation has the form discussed above with $a = \sqrt{2}$ and $b = \sqrt{2}$. So its graph is a hyperbola, and the lines $y = \frac{\sqrt{2}}{\sqrt{2}} = x$ and $y = -\frac{\sqrt{2}}{\sqrt{2}} = -x$ are asymptotes. The graph is sketched in Figure 4.18a. Turn next to

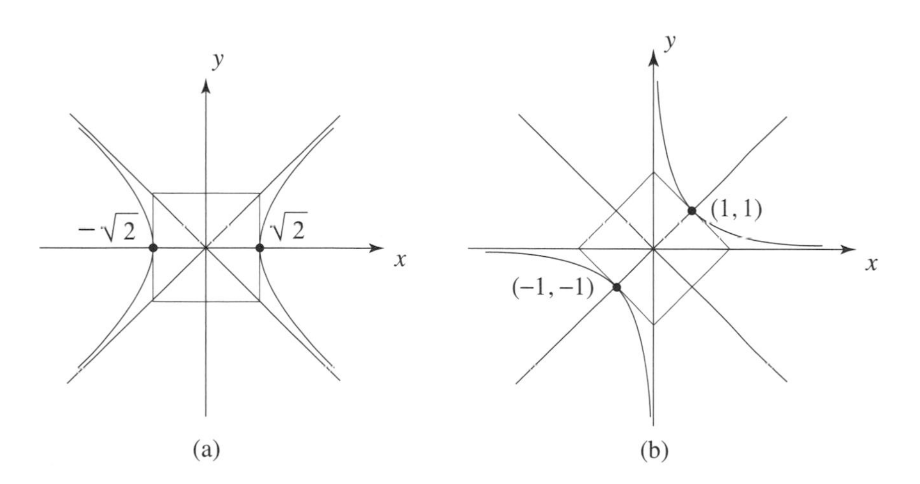

Fig. 4.18

the equation $xy = 1$. Since $y = \frac{1}{x}$, its graph has the shape shown in Figure 4.18b. It can be verified that it is a hyperbola with asymptotes $y = x$ and $y = -x$. In fact, it can be shown that by rotating the graph of $xy = 1$ clockwise by 45°, the hyperbola of Figure 4.18a is obtained. Since the distance between $(1, 1)$ and $(-1, -1)$ is $\sqrt{(1-(-1))^2 + (1-(-1))^2} = 2\sqrt{2}$, it follows that this rotation moves the square and hence the two asymptotes of Figure 4.18b to the square and the two asymptotes of Figure 4.18a.

It is not difficult to see that the graph of any equation of the form $Ay + Bx + C = 0$, where A, B, and C are constants with at least one of A or B not zero, is a line and that any line has an equation of this form. (The upcoming Section 5.1 will verify this). We'll now turn to the quadratic version of this fact. Consider any quadratic equation in the variables x and y of the form

$$(*) \qquad Ax^2 + Bxy + Cy^2 + Dx + Ey + F = 0,$$

where A, B, C, D, E, and F are constants (not all of which are zero). Go back over them and notice that the equations of the circles, parabolas, ellipses, and hyperbolas considered in this chapter can all be written in this form. We will call the equation $(*)$ *degenerate* if its graph has either no points on it, consists of a single point, or is a line or a combination of two lines. The equation $x^2 + y^2 + 1 = 0$ is degenerate because it has no solutions. The equation $x^2 + y^2 = 0$ is degenerate because the only solution is $x = 0$, $y = 0$. The equation $x^2 - y^2 = 0$ is degenerate because it can be factored as $(x - y)(x + y) = 0$, so that its graph is the combination of the two lines $x - y = 0$ and $x + y = 0$. Notice that if $A = B = C = 0$, then equation $(*)$ is always degenerate.

Example 4.15. Consider the equation $4x^2 - 16x + 3y^2 + 36y + F = 0$. By completing the square, show that it is degenerate for $F \geq 124$ and that it represents an ellipse for $F < 124$.

Example 4.16. The equation $2x^2 - 7xy + 6y^2 + 7x - 11y + 3 = 0$ is another example of a degenerate equation. Check by multiplying that

$$2x^2 - 7xy + 6y^2 + 7x - 11y + 3 = (x - 2y + 3)(2x - 3y + 1).$$

Since $(x - 2y + 3)(2x - 3y + 1) = 0$ means that either $x - 2y + 3 = 0$ or $2x - 3y + 1 = 0$, the graph of the equation is a combination of the graphs of the two lines $x - 2y + 3 = 0$ and $2x - 3y + 1 = 0$.

We will summarize—without proofs—the basic connections between conic sections and quadratic equations in two variables. Descartes was aware of them. (The intention of our summary is largely historical. It will play no role role later in this text.) For instance, in his *Géométrie*, he studies the equation

$$y^2 = cy - \tfrac{c}{b}xy + ay - ac$$

and asserts that it is a hyperbola. While saying more generally, "If this equation contains no term of higher degree than the rectangle of two unknown quantities, or the square of one, the curve belongs to the first and simplest class, which contains only the circle, the parabola, the hyperbola, and the ellipse, . . . ," Descartes does not present the supporting details. Later in the 17th century, the Englishman John Wallis (1616–1703) and the Dutchman Jan de Witt (1625–1672) did go into details. Wallis did so in his comprehensive treatise on the conic sections. This work also introduced the notation $\leq$ and $\geq$ for "less than or equal to" and "greater than or equal to," as well as ∞ for infinity. Jan de Witt studied the equation

$$\tfrac{b^2}{a^2}x^2 + \tfrac{2b}{a}xy + y^2 - bx + 2cy + c^2 = 0$$

and identified it as a parabola. He too pointed out that any quadratic equation in two variables (that is not degenerate) can be transformed into one of standard form, and therefore that it represents a conic section. De Witt provided all the mathematical details that verify this assertion. Today, the close relationship between conic sections and quadratic equations is standard fare in today's expansive calculus books.

Basic Fact 1. Any conic section in the xy-plane is the graph of an equation of the form

$$Ax^2 + Bxy + Cy^2 + Dx + Ey + F = 0.$$

Consider a quadratic equation $ax^2 + bx + c = 0$, where a, b, and c are constants with $a \neq 0$, and recall that its solutions are given by the quadratic formula

$$x = \frac{-b \pm \sqrt{b^2 - 4ac}}{2a}.$$

Observe that the term $b^2 - 4ac$ controls what happens. If $b^2 - 4ac = 0$, then there is only one solution, namely $x = -\frac{b}{2a}$. If $b^2 - 4ac > 0$, then there are two solutions. Finally, if $b^2 - 4ac < 0$, then there are no solutions. Notice that if $b = 0$ and $\frac{-c}{a} \geq 0$, then the solutions have the simple form $x = \pm\sqrt{\frac{-c}{a}}$.

A criterion that is similar in flavor to what was just described provides information about the equations of form $(*)$ that are not degenerate.

Basic Fact 2. If $Ax^2 + Bxy + Cy^2 + Dx + Ey + F = 0$ is not degenerate, then its graph is a conic section. The graph is a parabola if $B^2 - 4AC = 0$, an ellipse if $B^2 - 4AC < 0$, and a hyperbola if $B^2 - 4AC > 0$.

Because the proofs of Basic Fact 1 and Basic Fact 2 are beyond the intentions of this text, we'll only make a brief comment about them. Observe that the conic sections depicted in Figures 4.11, 4.15, and 4.17 have equations that can be rearranged to satisfy Basic Fact 1. It is easy to check that these rearranged equations conform to Basic Fact 2. The fact that any conic section in the xy-plane can be moved (by a combination of a translation and a rotation) to coincide with one of the conic sections depicted in these figures can be exploited to supply the proofs in general.

Basic Fact 3. If $Ax^2 + Bxy + Cy^2 + Dx + Ey + F = 0$ is not degenerate, then $B = 0$ precisely if the graph of the equation is a conic section with focal axis parallel to either the x-axis or the y-axis.

We'll illustrate the importance of completing the square in connection with Basic Fact 3. Let's have a look at the equation $3x^2 - 6y^2 + 42x - 24y + 107 = 0$. Notice that $B^2 - 4AC = 0 + 72 = 72$. So by Basic Fact 2, the graph of the equation is a hyperbola (unless the equation is degenerate). Because $B = 0$, we can complete the squares to get

$$\begin{aligned}3x^2 + 42x - 6y^2 - 24y + 107 &= 3(x^2 + 14x) - 6(y^2 + 4y) + 107\\ &= 3(x^2 + 14x + 7^2 - 7^2) - 6(y^2 + 4y + 4 - 4) + 107\\ &= 3(x^2 + 14x + 7^2) - 6(y^2 + 4y + 4) - 3\cdot 7^2 + 6\cdot 4 + 107\\ &= 3(x + 7)^2 - 6(y + 2)^2 - 16.\end{aligned}$$

After dividing both sides of $3(x + 7)^2 - 6(y + 2)^2 = 16$ by 16, we get $\frac{(x+7)^2}{\frac{16}{3}} - \frac{(y+2)^2}{\frac{16}{6}} = 1$, and hence $\frac{(x+7)^2}{(\frac{4}{\sqrt{3}})^2} - \frac{(y+2)^2}{(\frac{4}{\sqrt{6}})^2} = 1$. It follows that the graph of $3x^2 + 42x - 6y^2 - 24y + 107 = 0$ is the hyperbola $\frac{x^2}{a^2} - \frac{y^2}{b^2} = 1$ of Figure 4.17 (with $a = \frac{4}{\sqrt{3}}$ and $b = \frac{4}{\sqrt{6}}$), shifted without rotation in such a way that its center at the origin $(0, 0)$ is shifted to the point $(-7, -2)$.

4.6 CIRCLES AND TRIGONOMETRY

This section will use the xy-coordinate plane to extend the definitions of $\sin\theta$, $\cos\theta$, and $\tan\theta$ to the situation where θ can be any real number. Consider a circle of radius r with center the origin. Let $P = (x, y)$ be a point on the circle in the first quadrant. Draw a radius to P, and let θ be its angle with the positive x-axis. Denote by s the length of the arc that this angle cuts from the circumference. Refer to Figure 4.19 and notice that $\theta = \frac{s}{r}$ in radians. It follows from the figure that $\sin\theta = \frac{y}{r}$ and $\cos\theta = \frac{x}{r}$. Therefore $x = r\cos\theta$ and $y = r\sin\theta$.

When $r = 1$, these relationships have the simpler form

$$\theta = s,\ x = \cos\theta, \text{ and } y = \sin\theta.$$

We will therefore take $r = 1$. The circle with center the origin and radius 1 is called the *unit circle*. Its equation is $x^2 + y^2 = 1$, and its circumference is 2π. Until now—see Sections 1.5 and 1.6—only angles θ with radian measure $0 \leq \theta \leq \pi$ have been considered and the definitions of $\sin\theta$ and $\cos\theta$ required this restriction as well. We will now expand the concepts of angle, sine, and cosine, show how any real

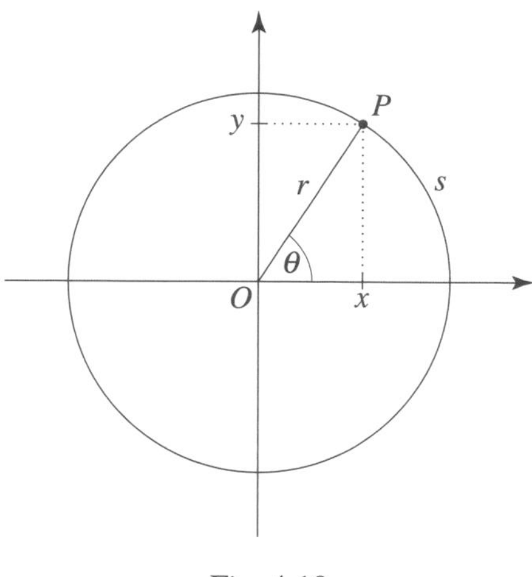

Fig. 4.19

number θ can be interpreted as an angle, and how $\sin\theta$ and $\cos\theta$ make sense for any real number θ.

So let θ be any real number. Assume first that $\theta \geq 0$ and consider a segment of length θ, as depicted in Figure 4.20. Let A be the right endpoint of the segment and P_θ the left endpoint. The notation emphasizes that by starting at A, the left endpoint depends on the length θ. Measure off the distance θ along the perimeter of the unit circle: start at the point $(1, 0)$ and proceed in a *counterclockwise* direction until the

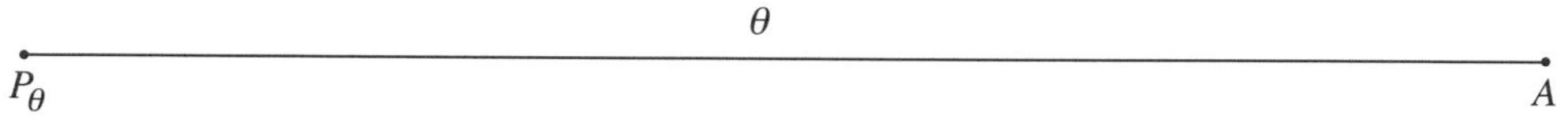

θ

P_θ A

Fig. 4.20

entire distance θ is measured off and the left endpoint P_θ is placed on the circle. Think of the segment as a string of length θ, and wind it around the circle. Place its right end A at the point $(1, 0)$. Then go around the circle counterclockwise (possibly many times) until P_θ lands on the circle. See Figure 4.21. Recall that

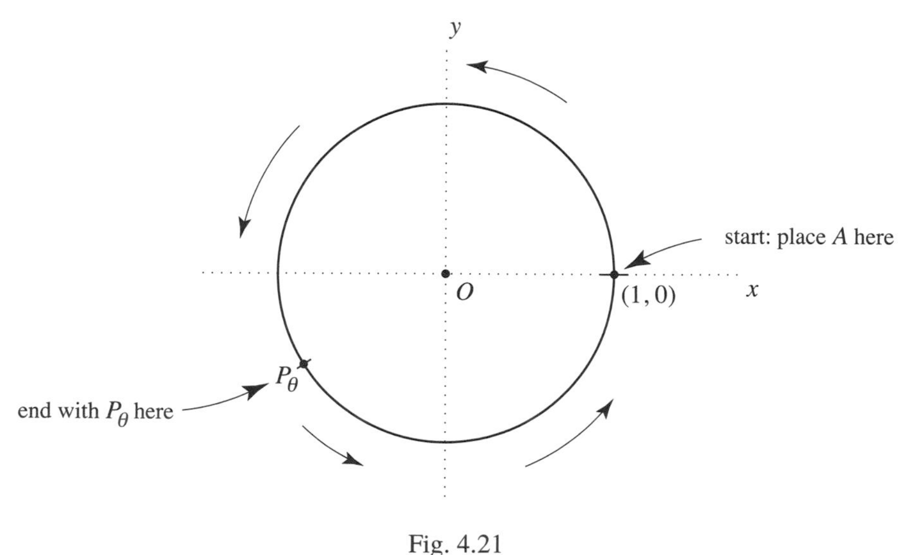

Fig. 4.21

the circumference of the unit circle is 2π. So for example, to measure off $\theta = \frac{\pi}{2}$, start at $(1, 0)$, go around (counterclockwise) the first quarter of the circle, and stop at the point $P_{\frac{\pi}{2}} = (0, 1)$. To measure off $\theta = \pi$, go around the first two quarters of the circle and stop at $P_\pi = (-1, 0)$. For $\theta = \frac{3\pi}{2}$, go around the first three quarters to $P_{\frac{3\pi}{2}} = (0, -1)$. For $\theta = 2\pi = \frac{4\pi}{2}$, go around all four quarters to $P_\theta = A = (1, 0)$.

What if θ is a negative real number? Then $-\theta$ is positive, and we'll let the segment of Figure 4.20 have length $-\theta$. To place the point P_θ in this case, measure off the length $-\theta$ on the perimeter of the circle. As before, start at $A = (1, 0)$, but this time go in the *clockwise* direction to measure off the segment and to locate the point P_θ. For example, with $\theta = -\frac{\pi}{2}$, start at $(1, 0)$, go around a quarter circle clockwise and stop at the point $P_{-\frac{\pi}{2}} = (0, -1)$. To measure off $\theta = -\pi$, go around one-half the circle clockwise and stop at $P_\pi = (0, -1)$, and so on.

In this way, any real number θ can be interpreted as an angle in radians: it is the opening generated by the segment from O to $(1, 0)$ as it rotates (while fixed at O) from $(1, 0)$ to the point P_θ. If the number is positive, the rotation is counterclockwise and the angle is positive. If the number is negative, the rotation is clockwise and the angle is negative. In either case, this is how the angle of any radian measure θ is defined.

What about degrees? The fact that an angle of π radians corresponds to $180°$ means that an angle of 1 radian is equal to $(\frac{180}{\pi})^\circ \approx 57.30°$. (Refer back to Section 1.5.) Therefore an angle of θ radians can be assigned $\theta \cdot \frac{180}{\pi}$ degrees. So for example, an angle of 100π radians is equal to 18,000°, an angle of 10π radians is equal to 1800°, an angle of 0.1π radians is equal to 18°, and an angle of 0.01π radians is equal to 1.8°. In the other direction, an angle of 1 degree is equal to $\frac{\pi}{180}$ radians. So an angle of θ degrees is equal to $\theta \cdot \frac{\pi}{180}$ radians. For example, the angle 10° is equal to $10 \cdot \frac{\pi}{180} \approx 0.175$ radians and 100° is equal to $100 \cdot \frac{\pi}{180} \approx 1.745$ radians.

We now turn to the definitions of $\sin\theta$ and $\cos\theta$ for any real number θ. Continue to consider the unit circle. For any real number θ, locate the point P_θ. Let the coordinates of P_θ be x and y and define

$$\cos\theta = x \quad \text{and} \quad \sin\theta = y$$

as illustrated in Figure 4.22. Since $P_{\frac{\pi}{2}} = (0, 1)$, we see that $\cos\frac{\pi}{2} = 0$ and $\sin\frac{\pi}{2} = 1$. Since $P_\pi = (-1, 0)$,

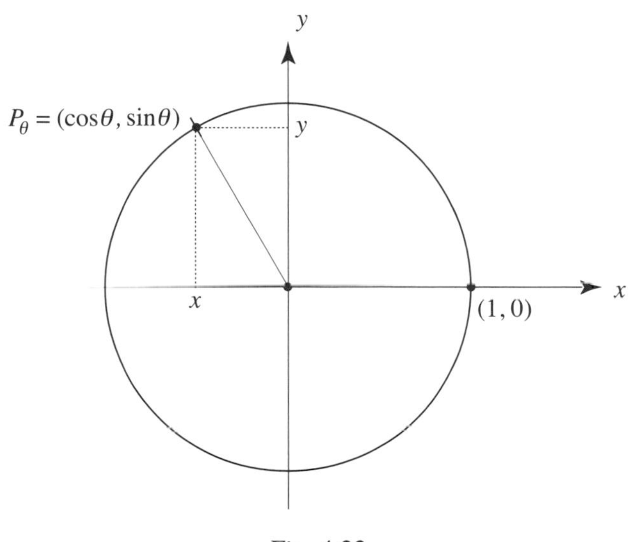

Fig. 4.22

we get $\cos\pi = -1$ and $\sin\pi = 0$. Finally, $P_{\frac{3\pi}{2}} = (0, -1)$ implies that $\cos\frac{3\pi}{2} = 0$ and $\sin\frac{3\pi}{2} = -1$. A look at Section 1.6—in particular, Figure 1.28—confirms that for any angle θ with $0 \leq \theta \leq \pi$, the definition of the sine and cosine that has just been presented agrees with what is specified there. This tells us, for example, that

$$\cos\tfrac{\pi}{6} = \tfrac{\sqrt{3}}{2} \text{ and } \sin\tfrac{\pi}{6} = \tfrac{1}{2},\ \cos\tfrac{\pi}{4} = \tfrac{1}{\sqrt{2}} \text{ and } \sin\tfrac{\pi}{4} = \tfrac{1}{\sqrt{2}}, \text{ and that } \cos\tfrac{\pi}{3} = \tfrac{1}{2} \text{ and } \sin\tfrac{\pi}{3} = \tfrac{\sqrt{3}}{2}.$$

Let θ be an angle given in degrees. Then the sine and cosine of θ are equal to the sine and cosine of the radian equivalent $\theta \cdot \frac{\pi}{180}$ of the angle θ. So $\sin\theta = \sin(\theta \cdot \frac{\pi}{180})$ and $\cos\theta = \cos(\theta \cdot \frac{\pi}{180})$.

We'll illustrate the above with some examples. For several θ, the locations of P_θ as well as its coordinates $\cos\theta$ and $\sin\theta$ are first approximated, and then determined precisely with a calculator.

Example 4.17. i. Let's consider $\theta = 23$ radians. Since $\frac{23}{2\pi} \approx 3.66$, we see that

$$23 \approx 3.66(2\pi) = 3(2\pi) + 0.66(2\pi) = 3(2\pi) + 1.32\pi = 3(2\pi) + \pi + 0.32\pi \approx 3(2\pi) + \pi + \tfrac{\pi}{3}.$$

So an approximate location of the point P_{23} is obtained by starting at the point $(1, 0)$ of the unit circle and going around it counterclockwise for two complete revolutions, another half revolution, and stopping after another sixth of a revolution. This means that the point P_{23} lies in the third quadrant and that the angle the segment from the origin O to P_{23} makes with the negative y-axis is approximately equal to $\frac{\pi}{6}$. Study Figure 4.21 for this situation and conclude that $\cos 23 \approx -\sin\frac{\pi}{6} = -0.5$ and $\sin 23 \approx -\cos\frac{\pi}{6} = -\frac{\sqrt{3}}{2} \approx -0.87$. A calculator provides the more accurate values $\cos 23 \approx -0.532$ and $\sin 23 \approx -0.846$.

ii. Let's take $\theta = 100$ radians. Since $\frac{100}{2\pi} \approx 15.92$, we get

$$100 \approx 15.92(2\pi) = 15(2\pi) + 0.92(2\pi) = 15(2\pi) + 1.84\pi = 15(2\pi) + \pi + 0.84\pi = 15(2\pi) + \pi + 1.68(\tfrac{\pi}{2})$$
$$= 15(2\pi) + \pi + \tfrac{\pi}{2} + 0.68(\tfrac{\pi}{2}) \approx 15(2\pi) + \pi + \tfrac{\pi}{2} + \tfrac{2}{3}(\tfrac{\pi}{2}) \approx 15(2\pi) + \pi + \tfrac{\pi}{2} + \tfrac{\pi}{3}.$$

This implies that an approximate location of the point P_θ can be obtained by starting at the point $(1, 0)$ of the unit circle, going around it counterclockwise for 15 complete revolutions, then another half revolution, then a quarter revolution and then another sixth revolution. So P_{100} lies in the fourth quadrant and the angle the segment from O to P_{100} makes with the negative y-axis is approximately $\frac{\pi}{3}$. A look at Figure 4.21 provides the conclusion that $\cos 100 \approx \sin\frac{\pi}{3} = 0.87$ and $\sin 100 \approx -\cos\frac{\pi}{3} \approx -0.50$. A calculator tells us more accurately that $\cos 100 \approx 0.862$ and $\sin 100 \approx -0.506$.

iii. Let's take $\theta = -34$ radians. Since $\frac{34}{2\pi} \approx 5.41$,

$$-34 = -5.41(2\pi) = -5(2\pi) - 0.41(2\pi) = -5(2\pi) - 1.64(\tfrac{\pi}{2}) = -5(2\pi) - \tfrac{\pi}{2} - 0.64(\tfrac{\pi}{2}) = -5(2\pi) - \tfrac{\pi}{2} - 1.28(\tfrac{\pi}{4})$$
$$= -5(2\pi) - \tfrac{\pi}{2} - \tfrac{\pi}{4} - 0.28(\tfrac{\pi}{4}) = -5(2\pi) - \tfrac{\pi}{2} - \tfrac{\pi}{4} - 1.12(\tfrac{\pi}{16}) = -5(2\pi) - \tfrac{\pi}{2} - \tfrac{\pi}{4} - \tfrac{\pi}{16} - 0.12(\tfrac{\pi}{16}).$$

Since $\frac{\pi}{2}$, $\frac{\pi}{4}$, $\frac{\pi}{16}$, and $0.12\frac{\pi}{16}$ correspond to 90°, 45°, 11.25°, and 1.35°, we get the approximate location of P_{-34} by starting at $(1, 0)$, going around the unit circle clockwise for five complete revolutions, then another quarter of a revolution or 90°, and then for another 60°. Return to Figure 4.21 and observe that this means that $P{-34}$ is in the second quadrant and that the angle that the segment from the origin O to P_{-34} makes with the negative x-axis is about 30°. It follows that $\cos(-34) \approx -\cos 30° \approx -0.87$ and $\sin(-34) \approx -\sin 30° = -0.50$. A calculator provides the values $\cos(-34) \approx -0.849$ and $\sin(-34) \approx -0.529$.

Refer back to Figure 4.22. By observing the moving point P_θ, notice that as θ varies from 0 to $\frac{\pi}{2}$, $\sin\theta$ varies from 0 to 1, and as θ moves from $\frac{\pi}{2}$ to π, $\sin\theta$ varies from 1 to 0, and so on. Plotting the various points $(\theta, \sin\theta)$ provides the graph of Figure 4.23. Doing the same thing for the cosine, we get the graph of

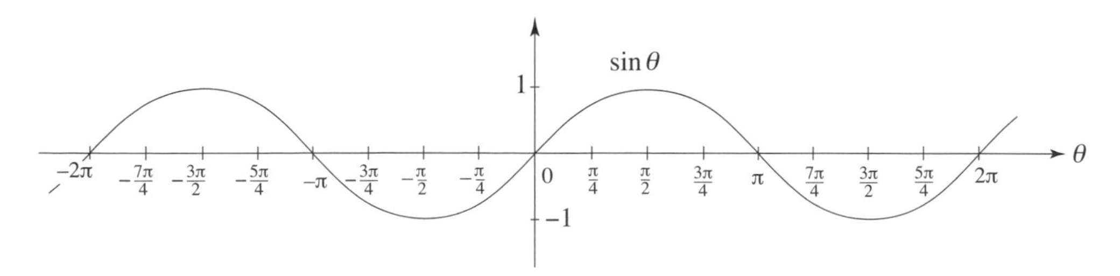

Fig. 4.23

Figure 4.24. The sine and cosine satisfy a number of basic identities. These can often be "read off" from Figure 4.22. Let θ be any real number. Since the point $P_\theta = (\cos\theta, \sin\theta)$ is on the unit circle, it satisfies

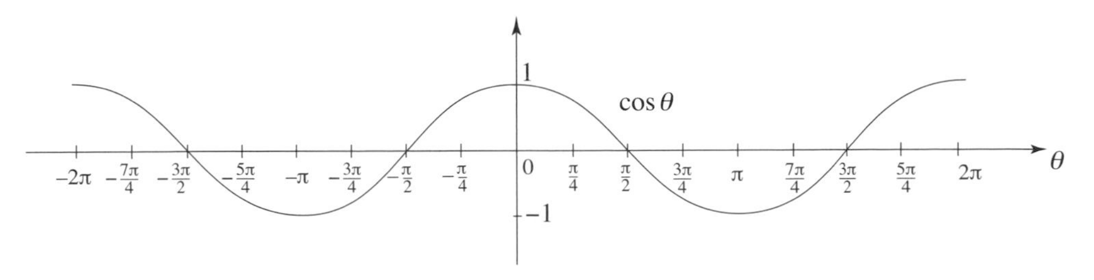

Fig. 4.24

the equation $x^2 + y^2 = 1$. Therefore

$$\sin^2\theta + \cos^2\theta = 1.$$

Let θ be any real number. Observe that P_θ and $P_{(\theta+2\pi)}$ are the same point. Therefore

$$\sin(\theta + 2\pi) = \sin\theta \;\text{ and }\; \cos(\theta + 2\pi) = \cos\theta.$$

Consider any θ as well as $-\theta$. A comparison of the points P_θ and $P_{-\theta}$ (see Figure 4.25a, for example) tells us that they have the same x-coordinates, and that the y-coordinate of one is the negative of the y-coordi-

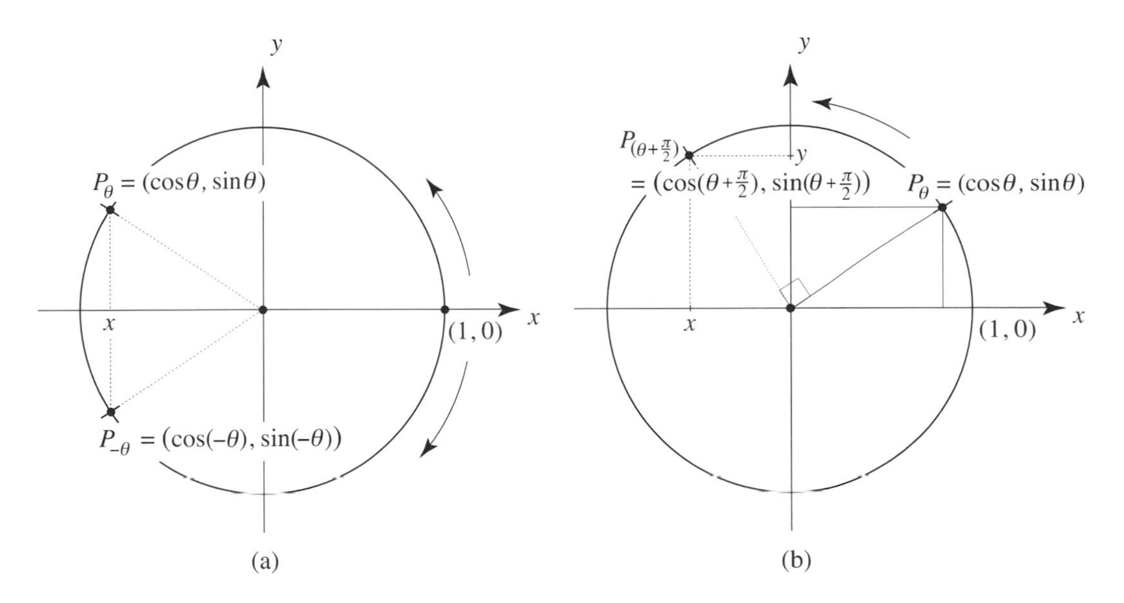

Fig. 4.25

nate of the other. It follows that

$$\cos(-\theta) = \cos\theta \;\text{ and }\; \sin(-\theta) = -\sin\theta.$$

Next, consider any θ and also $\theta + \frac{\pi}{2}$ as well as the corresponding points P_θ and $P_{(\theta+\frac{\pi}{2})}$. Figure 4.25b illustrates a typical situation. A careful study of the figure tells us that the triangle determined by P_θ and its coordinates is similar to the triangle determined by $P_{(\theta+\frac{\pi}{2})}$ and its coordinates x and y. It follows from this and the figure that $\cos(\theta + \frac{\pi}{2}) = x = -\sin\theta$ and that $\sin(\theta + \frac{\pi}{2}) = y = \cos\theta$. Thus the identities

$$\cos(\theta + \tfrac{\pi}{2}) = -\sin\theta \;\text{ and }\; \sin(\theta + \tfrac{\pi}{2}) = \cos\theta$$

are another consequence of the study of Figure 4.22.

Example 4.18. Verify the identities $\cos(\pi - \theta) = -\cos\theta$ and $\sin(\pi - \theta) = \sin\theta$ in two ways. First by applying identities already derived. Then again by making use of a diagram similar those in Figure 4.25.

Why are these formulas valid with $-\pi$ in place of π? Consider applying the identities $\sin(-\theta) = -\sin\theta$ and $\cos(-\theta) = \cos\theta$.

Let's turn to the study of the tangent

$$\tan\theta = \frac{\sin\theta}{\cos\theta}.$$

Suppose first that $0 \le \theta \le \frac{\pi}{2}$. If $\theta = 0$, then $\tan\theta = 0$. Again go to Figure 4.22. As θ increases, the sine increases and the cosine decreases. Since both are positive, $\tan\theta$ increases. When θ is close to $\frac{\pi}{2}$, the sine is close to 1 and the cosine is close to 0, so $\tan\theta$ is very large. If $\theta = \frac{\pi}{2}$, then $\cos\theta = 0$, so that $\tan\theta$ is not defined. When $-\frac{\pi}{2} \le \theta \le 0$, the situation is similar. Since the sine is negative and the cosine is positive,

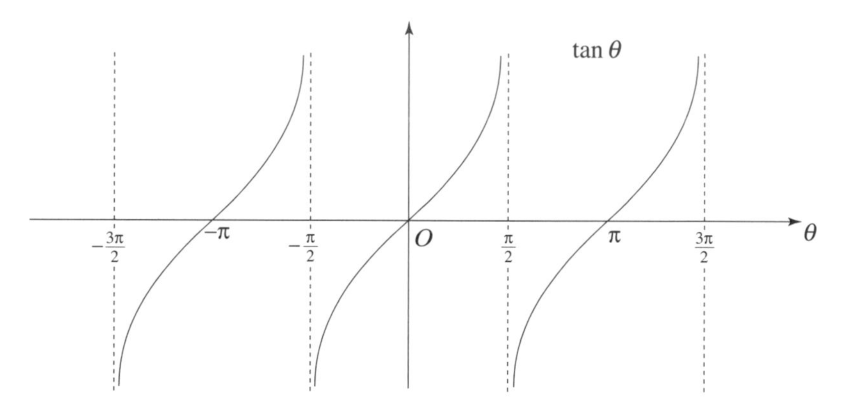

Fig. 4.26

$\tan\theta$ is now negative. The rest of the graph of the tangent is simply a repetition of the pattern for $-\frac{\pi}{2} \le \theta \le \frac{\pi}{2}$. It is sketched in Figure 4.26. Let θ be any angle. Since $\sin(\theta + \pi) = -\sin\theta$ and $\cos(\theta + \pi) = -\cos\theta$, it follows that $\tan(\theta + \pi) = \tan\theta$. Similarly, $\tan(-\theta) = -\tan\theta$.

The remaining trig expressions—the secant, cosecant, and cotangent—are defined by $\sec\theta = \dfrac{1}{\cos\theta}$, $\csc\theta = \dfrac{1}{\sin\theta}$, and $\cot\theta = \dfrac{1}{\tan\theta}$, respectively. Of these three, only the secant will be relevant later.

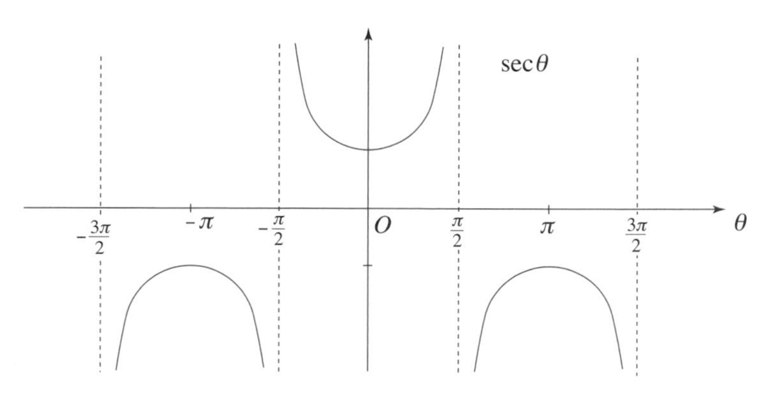

Fig. 4.27

Studying the graph of $\cos\theta$ and considering $\frac{1}{\cos\theta}$ tells us that the graph of the secant has the form depicted in Figure 4.27.

Suppose that we are given any point Q in the plane. Let r be its distance from the origin O. So Q lies on the circle of radius r and center O. Let θ be the angle between the radius OQ and the positive x-axis.

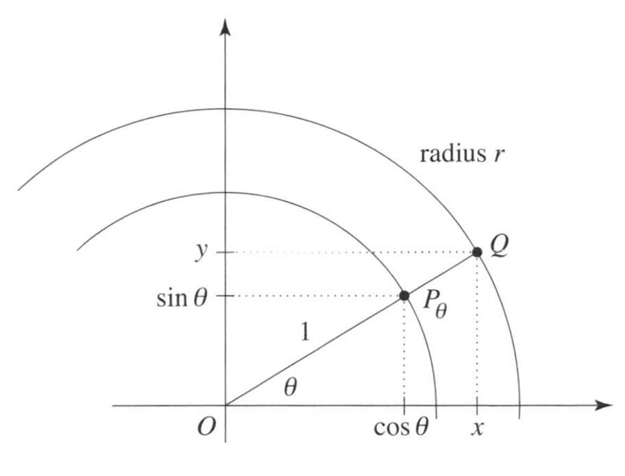

Fig. 4.28

See Figure 4.28. Consider the corresponding point $P_\theta = (\cos\theta, \sin\theta)$ on the unit circle. By similar triangles,

$$\frac{x}{\cos\theta} = \frac{y}{\sin\theta} = \frac{r}{1} = r.$$

Therefore the coordinates of Q are $x = r\cos\theta$ and $y = r\sin\theta$. Notice that the location of the point Q can be specified by its distance r from the origin and the angle θ. This observation is the basis of the *polar coordinate system* that will play an important role in Chapter 12. The equations $x = r\cos\theta$ and $y = r\sin\theta$ tell us how the polar coordinate system is related to the Cartesian coordinate system.

Example 4.19. In Figure 4.28, let Q be the point with $\theta = \frac{\pi}{4}$ and $r = 2.5$. Then the coordinates of Q are $x = (2.5)\cos\frac{\pi}{4} = 2.5 \cdot \frac{\sqrt{2}}{2} = \frac{5}{4}\sqrt{2}$ and $y = (2.5)\sin\frac{\pi}{4} = 2.5 \cdot \frac{\sqrt{2}}{2} = \frac{5}{4}\sqrt{2}$.

The "measure off the distance on the perimeter" strategy illustrated in Figure 4.21 could have been applied to a circle with center O and any radius r to define the radian measure of an angle as well as its sine and cosine for any real number. However, for $r \neq 1$, a division by r would have been required for all three definitions. This is illustrated by Figure 4.28 and the formulas $\cos\theta = \frac{x}{r}$ and $\cos\theta = \frac{x}{r}$ that arise from it.

It seems to be difficult to pinpoint when and by whom the trigonometric ratios were first expressed in terms of coordinates of points on a circle centered at the origin. But the Swiss mathematician Leonhard Euler (1707–1783) certainly made use of this connection. Euler's mathematical output, astonishing in terms of its volume, range, as well as depth, established him as the greatest mathematician of the 18th century. He was born and educated in Basel, Switzerland, but spent most of his career in St. Petersburg and Berlin.

4.7 PROBLEMS AND PROJECTS

The first segment of these Problems and Projects presents basic facts about numbers, and the second illustrates the completing of the square procedure. This is followed by classical problems designed to test basic skills in algebra. Information about factoring polynomials is provided next, and then there is a look at the geometric "numbers" of the Greeks that Descartes used in his analytic geometry. Problems involving the coordinate plane, various aspects of circles, parabolas, ellipses, and hyperbolas, and, finally, basic trigonometry conclude the section.

4A. About Prime, Rational, and Irrational Numbers. The facts discussed in this segment were known to the Greeks and can be found in Euclid's *Elements*. An integer greater than 1 is a *prime number* if it can be divided only by 1 and itself. The numbers 2, 3, 5, 7, 11, 13, and 17 are examples. It is a fact that if a prime p divides a product nm of integers, then p divides either n or m. The fundamental theorem of arithmetic asserts that any positive integer greater than 1 can be written as a product of powers of prime numbers, and that this can be done in only one way (except for the order of the factors). Check, for example, that $324 = 2^2 \cdot 3^4$, $825 = 3 \cdot 5^2 \cdot 11$, and $4536 = 2^3 \cdot 3^4 \cdot 7$, and there is no other way of doing this (other than $324 = 3^4 \cdot 2^2$ or $4536 = 7 \cdot 2^3 \cdot 3^4$, for instance).

4.1. Determine the prime factorizations of 28, 143, 192, and 720.

If a number is a square, then all primes in its prime factorization occur to an even power. For instance, $4536^2 = (2^3)^2 \cdot (3^4)^2 \cdot 7^2 = 2^6 \cdot 3^8 \cdot 7^2$. It is also the case that if all the primes in the product occur to an even power, then the number is a square. For example, $324 = 2^2 \cdot 3^4 = (2 \cdot 3^2)^2 = 18^2$.

4.2. Let m be a positive integer. Show that if $\sqrt{m}$ is a rational number, then m must be a square. Conclude that if m is not a square, then $\sqrt{m}$ is irrational. [Hint: If $\sqrt{m} = \frac{s}{t}$ for positive integers s and t, then $m \cdot t^2 = s^2$. If one of the primes in the factorization of m occurs to an odd power, then the same thing is true for $m \cdot t^2$.]

So with the exception of $\sqrt{4} = 2$, $\sqrt{9} = 3$, $\sqrt{16} = 4$, and so on, the numbers $\sqrt{2}, \sqrt{3}, \sqrt{5}, \sqrt{6}, \sqrt{7}, \sqrt{8}, \ldots$ of the spiral of Figure 4.1 are all irrational.

4.3. Consider the numbers $\sqrt{n}$ for $n = 1, 2, \ldots, 100$. How many of these 100 numbers are rational? How many are irrational? Consider $\sqrt{n}$ for $n = 1, 2, \ldots, 1{,}000{,}000$. How many of these numbers are rational? [Hint: To answer the last question, the fact that $1000^2 = 1{,}000{,}000$ could be helpful.]

Let r be any real number. Is there a pattern in its decimal expansion from which it is possible to tell whether r is rational or irrational? Suppose that in the decimal expansion of r there is a repeating block of numbers from some point on. For example, let's suppose that $r = 23.74865865865...$ and that the block 865 keeps repeating. Then $100r = 2374.865865...$ and $100{,}000r = 2374865.865865...$. Notice that $100{,}000r - 100r = 2{,}374{,}865 - 2374 = 2{,}372{,}491$. So $99{,}900r = 2{,}372{,}491$. It follows that $r = \frac{2{,}372{,}491}{99{,}900}$ is a rational number. In the logically opposite direction, is it also true that the decimal expansion of any rational number has a block that keeps repeating? From the (longhand) division algorithm applied to, say, t divided into s, we know that there is, at every step, a remainder that is less than t. So there are only a finite number of possibilities for such a remainder. If 0 is a remainder, then the division process stops and 0 is the repeating block. For instance, $\frac{581}{25} = 23.24 = 23.24000...$. If 0 is not a remainder, the process keeps going, so that there must be a point when a remainder occurs again. But at that point the cycle repeats. Divide 132 into 1124, for instance. A look at Figure 4.29 tells us that the first remainder is 68, the next is 20, and then 68 recurs, so there is a repetition after two steps. In other examples of the division process, many steps may be needed before the same remainder reappears. But there will be a repeating

```
       8.515...
132 | 1124
      1056
       680
       660
       200
       132
        680
         20
```

Fig. 4.29

cycle, so there is a repeating block in the decimal expansion. It follows that the rational numbers are precisely those real numbers that have a decimal expansion with a repeating block of numbers.

4.4. Consider the numbers $1.\underline{7}777...$, $2.\underline{67}67...$, $4.\underline{728}728...$, and $35.34672\underline{638}638...$. Each is a rational number. (The repeating blocks are underlined.) Express each of them as a fraction of positive integers. [Hint: Let r be one of the numbers. For the first one, consider $r - 10r$.]

4.5. Determine the decimal expansions of the rational numbers $\frac{5}{4}$ and $\frac{468}{198}$.

On the other extreme of the repetitive decimal expansions of rational numbers are the decimal expansions of some irrational numbers that are completely unpredictable. The number π is an example. The expansion $\pi = 3.1415926535898732...$ has no pattern. Its numbers flow in a completely random way. One more point. The numbers obtained by cutting the expansion off—say, 3.141, 3.14159, 3.141592653, and so on—are all rational because $3.141 = \frac{3141}{1000}$, $3.14159 = \frac{314{,}159}{100{,}000}$, and $3.141592653 = \frac{3{,}141{,}592{,}653}{1{,}000{,}000{,}000}$, and so on. This illustrates the fact that any real number can be approximated by rational numbers to any desired degree of accuracy.

4.6. Show that there is a rational number between any two distinct real numbers.

4.7. An important irrational number e (we will encounter it in Chapter 10) has the decimal expansion 2.718281828... . Find integers n and m both less than 100,000 such that the rational number $2\frac{n}{m}$ approximates e to an accuracy of nine decimal places.

Can any rectangle be subdivided into a finite array of nonoverlapping (except at the boundaries) identical squares if they are taken small enough? This is certainly possible if the rectangle is a square. In this case, the subdividing squares can be made as small as one may wish. But are such subdivisions possible for any rectangle? A rectangle of size 2×3 can be subdivided into six 1×1 squares and each of these can be subdivided into smaller identical squares in many ways. What about a rectangle of size $1 \times \sqrt{2}$? This time it's not obvious whether this can be done.

4.8. Can a rectangle with sides 3 and 5 be subdivided into a finite array of nonoverlapping (except at the boundaries) identical squares? In many different ways? What about a rectangle with sides $\sqrt{2}$ and $\sqrt{8}$?

4.9. Consider a rectangle R, and let the lengths of its sides be a and b, respectively. Show that R can be subdivided into a finite array of nonoverlapping (except at their boundaries) identical squares if and only if $\frac{a}{b}$ is a rational number. In the current situation, the phrase "if and only if" means that the two statements "if R can be subdivided, then $\frac{a}{b}$ is a rational number" *and* "if $\frac{a}{b}$ is a rational number, then R can be subdivided" both hold. [Hint: If $\frac{a}{b} = \frac{m}{n}$ for positive integers m and n, let $\frac{a}{m} = \frac{b}{n} = s$ be the side length of the subdividing square.]

4.10. Which of the following rectangles can be subdivided into identical nonoverlapping (except at the boundaries) squares: The rectangle with sides 1 and $\sqrt{2}$? The rectangle with sides $\sqrt{3}$ and $\sqrt{6}$? The rectangle with sides $\sqrt{20}$ and $\sqrt{45}$? [Hint: Use the result of Problem 4.9.]

4B. Completing the Square and the Quadratic Formula. Quadratic equations were already solved by the Pythagoreans and before them by the Babylonians. The solutions of the general quadratic equation $ax^2 + bx + c = 0$ (with $a \neq 0$) are given by the quadratic formula $x = \frac{-b \pm \sqrt{b^2 - 4ac}}{2a}$. Today's verification of this formula uses a procedure known as *completing the square*. The procedure consists of several algebraic steps that are illustrated below in the case of the quadratic polynomial $6x^2 + 28x - 80$.

As first important step, factor out the coefficient of the x^2 term. So $6x^2 + 28x - 80 = 6(x^2 + \frac{28}{6}x - \frac{80}{6})$. Focus on $\frac{28}{6}x = \frac{14}{3}x$, divide the coefficient $\frac{14}{3}$ by 2 to get $\frac{14}{6} = \frac{7}{3}$, and square this to get $\frac{49}{9}$. Now rewrite $6(x^2 + \frac{14}{3}x - \frac{80}{6})$ as $6(x^2 + \frac{14}{3}x + \frac{49}{9} - \frac{49}{9} - \frac{80}{6})$. Regroup to get $6[(x^2 + \frac{14}{3}x + \frac{49}{9}) - \frac{49}{9} - \frac{80}{6}]$. Because $(x^2 + \frac{14}{3}x + \frac{49}{9}) = (x + \frac{7}{3})^2$, you now have

$$6x^2 + 28x - 80 = 6(x^2 + \tfrac{28}{6}x - \tfrac{80}{6}) = 6[(x + \tfrac{7}{3})^2 - \tfrac{49}{9} - \tfrac{80}{6}] = 6[(x + \tfrac{7}{3})^2 - \tfrac{169}{9}].$$

Having rewritten $6x^2 + 28x - 80$ as $6[(x + \frac{7}{3})^2 - \frac{169}{9}]$, you have completed the square for the quadratic polynomial $6x^2 + 28x - 80$. Notice that it is now easy to solve $6x^2 + 28x - 80 = 0$ for x. Divide $6[(x + \frac{7}{3})^2 - \frac{169}{9}] = 0$ by 6 to get $(x + \frac{7}{3})^2 - \frac{169}{9} = 0$. So $(x + \frac{7}{3})^2 = \frac{169}{9}$, and hence $x + \frac{7}{3} = \pm\sqrt{\frac{169}{9}} = \pm\frac{13}{3}$. Therefore $x = -\frac{7}{3} \pm \frac{13}{3}$. So $x = 2$ or $x = -\frac{20}{3}$.

4.11. Repeat these steps to complete the square for $4x^2 - 8x - 12$. What is the smallest value that $4x^2 - 8x - 12$ can have? For what value of x does it occur? Solve $4x^2 - 8x - 12 = 0$ for x.

4.12. Complete the square for the polynomial $-5x^2 + 3x + 4$. Then use the result to solve $-5x^2 + 3x + 4 = 0$ for x. Try the same thing for $-5x^2 + 3x - 4$. [Note: The solution of the equation $-5x^2 + 3x - 4$ requires square roots of negative numbers. Such *complex numbers* will be considered Chapter 10.]

4.13. Solve the equation $3x^2 + 21x + 12 = 0$ by completing the square for the polynomial $x^2 + 7x + 4$.

4.14. Verify by completing the square that the solutions of $ax^2 + bx + c = 0$ (with $a \neq 0$) are given by the quadratic formula $x = \frac{-b \pm \sqrt{b^2 - 4ac}}{2a}$. What happens when $a = 0$?

4.15. Let x and d be any two positive numbers. Study the diagrams in Figure 4.30, and write a paragraph that discusses their connection with the completing the square procedure.

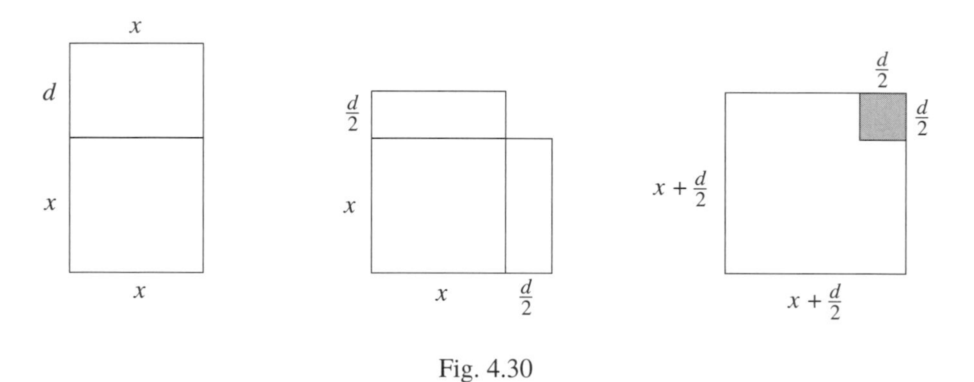

Fig. 4.30

4C. ***The Greek Anthology*** is a collection of about 3700 Greek sayings, songs, rhetorical exercises, and mathematical puzzles, many written in verse. Some of them may have been compiled as early as the 7th century B.C. and others as late as the 10th century A.D. The problems below come from *The Greek Anthology*. In some of them it is not explicit what the question is. In others, the information that is given is not sufficient. You will need to decide what the questions are and to supply additional information as and if needed. One of the problems has infinitely many answers. Determine them all.

4.16. The Muses stole and divided among themselves, in different proportions, the apples I was bringing from Helicon. Clio got the fifth part, and Euterpe the twelfth, but divine Thalia the eighth. Melpomene carried off the twentieth part, and Terpsichore the fourth, and Erato the seventh; Polyhymnia robbed me of 30 apples, and Urania of 120, and Calliope went off with a load of 300 apples. So I come to thee with lighter hands, bringing these 50 apples that the goddesses left me. How many apples did I bring from Helicon?

4.17. Make me a crown weighing 60 minae, mixing gold and brass, and with them tin and much wrought iron. Let the gold and brass together form two-thirds, the gold and tin together three-fourths, and the gold and iron three-fifths. Tell me how much gold you must put in, how much brass, how much tin, and how much iron, so as to make the whole crown weigh 60 minae. [A number of references describe the mina as a unit of weight roughly equal to 1 pound. So it seems that this weighty crown was intended for no ordinary mortal.]

4.18. Brick-maker, I am in a great hurry to erect this house. Today is cloudless, and I do not require many more bricks, but I have all I want but 300. Thou alone in one day couldst make as many, but thy son left off working when he had finished 200, and thy son-in-law when he had made 250. Working all together, in how many hours can you make these?

4.19. The Sun, the Moon, and the planets of the revolving zodiac spun such a nativity for thee; for a sixth part of thy life to remain an orphan with thy dear mother, for an eighth part to perform forced labor for thy enemies. For a third part the gods shall grant thee homecoming, and likewise a wife and a later-born son by her. Then thy son and wife shall perish by the spears of the Scythians, and then having shed tears for them thou shalt reach the end after 27 years.

4.20. Of the four spouts one filled the whole tank in a day, the second in two days, the third in three days, and the fourth in four days. What time will all four take to fill it?

4.21. The Graces were carrying baskets of apples, and in each was the same number. The nine Muses met them and asked them for apples, and they gave the same number to each Muse, and the nine and the three had each of them the same number. [Let x be the number of apples each of the Graces had initially. If each of the three Graces gave y apples, then each of the muses received $\frac{y}{3}$ of them. Now ask yourself what the question might be and show that it has infinitely many answers.]

4D. The Algebra of Diophantus. One of the last great Greek mathematicians, Diophantus lived around 250 A.D. Working in Alexandria, he pursued arithmetic and algebra. He devised a new algebraic symbolism to express equations in shorter and more comprehensible ways. By using abbreviations for frequently occurring concepts and operations—"equals" for instance—Diophantus took an important step from verbal algebra toward symbolic algebra. The following passage from *The Greek Anthology* provides some information about Diophantus's life. Use it to determine how old he became.

4.22. Diophantus passed one-sixth of his life in childhood, one-twelfth in youth, and one-seventh more as a bachelor. Five years after his marriage was born a son who died four years before his father, at half of his father's [final] age.

The next several problems give a flavor (in today's notation) of the problems that Diophantus took on. In the first two, a and b are positive constants.

4.23. Find numbers x and y such that $\frac{x}{y} = a$ and $\frac{x^2-y^2}{x+y} = b$.

4.24. Find numbers x and y such that $xy = a$ and $x + y = b$.

4.25. Find four numbers such that the sums obtained by adding any three at a time are 20, 22, 24, and 27.

4.26. In the right triangle ABC with right angle at C, AD bisects the angle at A. Find the set of smallest integers for the lengths CD, AC, AD, DB, AB, such that the ratios $CD : AC : AD$ are equal to $3:4:5$. [Hint: Let x be any number such that $CD = 3x$, $AC = 4x$, and $AD = 5x$. Refer to Figure 4.31. Since CD and AC are integers, $AC - CD = x$ is an integer. Use the triangle to obtain values for $\tan\frac{\theta}{2}$ and $\tan\theta$. An application of the formula $\tan\theta = \frac{2\tan\frac{\theta}{2}}{1-\tan^2\frac{\theta}{2}}$ (see Problem 1.23ii) tells us that $75x = 7y$. A fact about prime numbers tells us that $x = 7$ is the smallest possible x. With $x = 7$, we get $y = 75$ and $z = 100$, so that all the dimensions are now determined. Finally, use the law of sines to check that the segment AD does in fact bisect the angle at A.]

4.27. For any rational number c, express c^2 as the sum of two nonzero rational squares. [Hint: Let x be any rational number with $x \neq 0$, and consider $2x - c$. Set $c^2 = x^2 + (2x - c)^2$ and solve for x.]

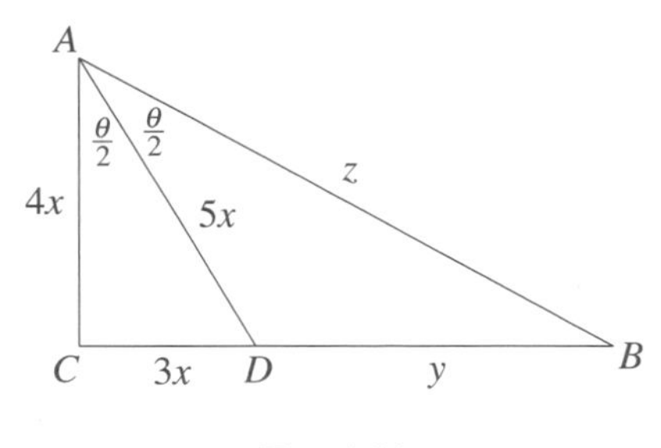

Fig. 4.31

While reading his Latin version of Diophantus's *Arithmetica,* Fermat came across the problem above and Diophantus's solution (as outlined), saw how Diophantus expressed any square as a sum of two squares, and jotted the following note into the margin of his book,

> On the other hand, it is impossible to separate a cube into two cubes, or any fourth power into two fourth powers, or generally any power except a square into two powers with the same exponent. I have discovered a truly marvelous proof of this, which however the margin is not large enough to contain.

The margin was not the only place the proof could not be found. None of Fermat's published or working papers included it. Soon the mathematical chase was on. What had been Fermat's proof? How could this simple statement be verified (with or without Fermat)? Some of the brightest mathematical minds, the great Leonhard Euler among them, were drawn to this question. Was there a solution of the equation $x^n + y^n = z^n$ for $n \geq 3$ with x, y, and z all rational and all nonzero? If such a rational solution exists, then by clearing denominators, a solution involving nonzero positive integers x, y, and z exists as well. "Does the equation $x^n + y^n = z^n$ have solutions with x, y, and z nonzero positive integers for $n \geq 3$?" is how the problem, known later as *Fermat's last theorem*, was passed from one generation of mathematicians to the next. Their pursuits expanded the theory of numbers in substantial ways, but no one was able to crack the problem until the end of the 20th century, when the English mathematician Andrew Wiles supplied the proof in 1993. Fermat's hunch proved to be correct: there were no solutions. The proof involved lots of recent, advanced, and highly sophisticated mathematics. The wildest dreams of either Fermat or Euler could not have expressed images of the mathematical concepts required in the proof.

The next few problems are examples of early Islamic algebra. The quadratic formula solves each of them.

4.28. I multiply a third of a quantity plus a unit by a fourth of a quantity plus a unit, and it becomes 20 units. Determine the quantity.

4.29. I multiply a third of a quantity by a fourth of the quantity in such a way as to give the quantity itself plus 24 units. Determine the quantity.

4.30. Find a number such that if 7 is added to it and the sum multiplied by the root of 3 times the number, then the result is 10 times the number.

The mathematician Gerbert (later Pope Sylvester II, 999–1003) solved the following problems. They were considered to be difficult at the time.

4.31. Gerbert expressed the area of an equilateral triangle of side a as $\frac{a}{2}(a - \frac{a}{7})$. Show that this is an approximation that is equivalent to the approximation $\sqrt{3} \approx \frac{12}{7}$. Use a calculator to check its accuracy.

4.32. Let x and y be the legs of a right triangle, and express x and y in terms of the height h and the area A of the triangle.

4E. Polynomials: Roots and Factoring. Let $p(x)$ be a polynomial. A number d such that $p(d) = 0$ is a *root* of $p(x)$. Note that if $p(x)$ has a factorization of the form $p(x) = (x - d)q(x)$, then d is a root of $p(x)$. This fact works also in reverse. If d is a root of $p(x)$, then $x - d$ divides $p(x)$. Let's return to the polynomial $z^3 - 9z^2 + 26z - 24$ considered by Descartes. (See the discussion toward the end of Section 4.1.) Assume that $z^3 - 9z^2 + 26z - 24$ has a factorization of the form $(z - a)(z - b)(z - c)$ with a, b, and c integers. Then $abc = 24$, and it follows that a, b, and c are all divisors of 24. Notice also that $-a - b - c = -9$, so that $a + b + c = 9$. Since the divisors of 24 are $1, 2, 3, 4, 6, 8, 12$, and 24, each of a, b, and c must come from these numbers (or there negatives). Descartes checked that $z - d$ divides $z^3 - 9z^2 + 26z - 24$ precisely for $d = 2, 3$, and 4. He then confirmed that

$$z^3 - 9z^2 + 26z - 24 = (z - 2)(z - 3)(z - 4).$$

4.33. Use the quadratic formula to find the roots of the polynomial $-2x^2 + 7x - 5$. Then factor it.

4.34. Use the quadratic formula to find the roots of $3x^2 - 9x + 8$. Factor the polynomial.

4.35. Consider the generic quadratic polynomial $ax^2 + bx + c$. By the quadratic formula, $x = \frac{-b+\sqrt{b^2-4ac}}{2a}$ and $x = \frac{-b-\sqrt{b^2-4ac}}{2a}$ are both roots of the polynomial. Therefore both

$$x + \frac{b - \sqrt{b^2 - 4ac}}{2a} \quad \text{and} \quad x + \frac{b + \sqrt{b^2 - 4ac}}{2a}$$

divide $ax^2 + bx + c$. Confirm this by computing $(x + \frac{b-\sqrt{b^2-4ac}}{2a}) \cdot (x + \frac{b+\sqrt{b^2-4ac}}{2a})$.

4.36. Find the roots of the polynomial $x^3 + 6x^2 - 9x - 14$ and factor it.

4.37. Find the roots of $x^3 - 4x^2 - 4x - 5$ and factor the polynomial.

4F. The "Numbers" of Descartes. For Descartes, a positive number is a line segment and two line segments are equal if they have the same length. Even though he cannot express the length of every segment with a number (this is what the real number system does), he can assess whether two segments have the same length by placing them next to each other. Descartes describes his numbers and his notation for them early in his *Géométrie*, when he says "in geometry, to find required lines it is merely necessary to add or subtract other lines; or else, taking one line which I call unity in order to relate it as closely as possible to numbers" and again later "it is not necessary thus to draw the lines on paper, but it is sufficient to designate each by a single letter. Thus to add the lines ... I call one a and the other b, and write $a + b$. Then $a - b$ will indicate that b is subtracted from a." Figure 4.32 illustrates what Descartes describes. The

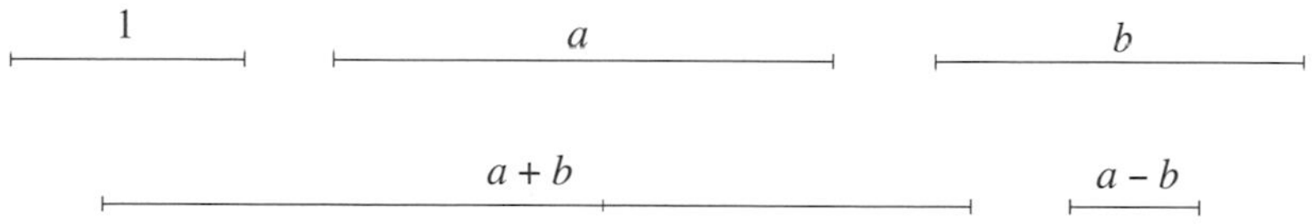

Fig. 4.32

line segment "unity" labeled 1 can have any length, but once chosen, it is fixed. This segment has the multiplicative property (by definition) that for any segment a, $1 \cdot a = a \cdot 1 = a$. The addition and subtraction of segments are done by attaching or superimposing them as in the figure.

4.38. The multiplication of the segments a and b is also geometric. In Figure 4.33a, O is a point from which two line segments emerge. The points O, A, C, and B are placed in such a way that $OA = 1$, $AC = a$, and $OB = b$. Figure 4.33b has AB drawn in and has placed the segment CE parallel to AB. Use the similarity of the triangles ΔOAB and ΔOCE to show that $\frac{BE}{a}$ is equal to $\frac{b}{1}$. It follows that $\frac{b}{1} = \frac{BE}{a}$. This equality provides the definition multiplication. The product of a and b is the segment BE.

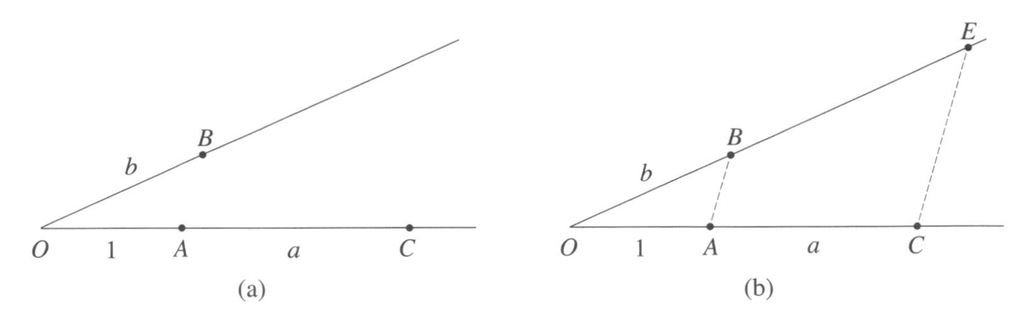

Fig. 4.33

Descartes as well as the Greeks understood the product and addition that arises in the equality $a^2 + b^2 = c^2$ of the Pythagorean theorem in this way. If the segments a and b have lengths expressible in positive rational numbers, then the product ab defined in Problem 4.38 reduces to the product of rational numbers. The quotient $\frac{a}{b}$ of two segments a and b is defined in a similar way (see Problem 4.40). The ratios in Apollonius's original versions of Propositions P2 and E2 of Section 2.1 have this meaning. Square roots of segments can also be defined. Problem 2.16i and Figure 2.43a demonstrate the construction of the segment $\sqrt{ab}$.

4.39. Use a relabeled version of Figure 4.33 to define the ratio $\frac{a}{b}$.

4G. The Coordinate Plane

4.40. Find the distance between the points (1,1) and (4,5), and then between $(1,-6)$ and $(-1,-3)$.

4.41. Place the two points $A = (-3,-7)$ and $B = (6,8)$ into the coordinate plane. Find the length of the segment AB.

4.42. Let a and b be two points on a number line. Show that $c = \frac{a+b}{2}$ is the midpoint of the segment that they determine. Use this fact and similar triangles to show that the midpoint of the segment determined by the points $P_1 = (x_1, y_1)$ and $P_2 = (x_2, y_2)$ is the point $\left(\frac{x_1+x_2}{2}, \frac{y_1+y_2}{2}\right)$.

4.43. Find the midpoints of the following segments.

i. $(1,3)$ and $(7,15)$ **ii.** $(-1,6)$ and $(8,-12)$

4.44. Consider the triangle with vertices $A = (6,-7)$, $B = (11,-3)$, and $C = (2,-2)$. Show that it is a right triangle by using the Pythagorean theorem.

4.45. Show that the points $A = (-1,3)$, $B = (3,11)$, and $C = (5,15)$ lie on the same line by using the distance formula.

4.46. Sketch the graphs of the following equations.

i. $x = 3$ **ii.** $y = -2$

iii. $|y| = 1$ **iv.** $xy = 0$

In the same way that curves in the Cartesian plane are represented by algebraic equations, regions in the plane are given by algebraic expressions involving inequalities.

4.47. Describe and sketch the regions given by the following sets of points in the plane.

i. The set $\{(x,y) \mid x \geq 1\}$. The notation means that this is "the set of all points (x, y) with the property that $x \geq 1$."

ii. The set $\{(x, y) \mid |x| < 1 \text{ and } |y| < 1\}$. This is "the set of all points (x, y) with the property that $|x| < 1$ and $|y| < 1$."

4.48. Sketch the given region in the xy-plane.

i. $\{(x, y) \mid xy < 0\}$

ii. $\{(x, y) \mid 0 \leq y \leq 4 \text{ and } x \leq 2\}$

iii. $\{(x, y) \mid |x| < 3 \text{ and } |y| < 2\}$

4H. About Circles and Parabolas

4.49. Sketch the graph of the circle $(x-3)^2 + (y+5)^2 = 7$.

4.50. Find an equation of the circle with center $(3, -1)$ and radius 5.

4.51. Consider the circle $x^2 + y^2 = 9$ in the xy-plane. Translate the circle 7 units down and 3 units to the right. What is the equation of the circle in this new position?

4.52. By completing squares, show that the graph of the equation $x^2 + y^2 - 4x + 10y + 13 = 0$ is a circle. What are its center and radius?

4.53. Under what condition on a, b, and c does the equation $x^2 + y^2 + ax + by + c = 0$ represent a circle? [Hint: Complete squares.] When this condition is satisfied, find expressions for the center and radius of the circle.

4.54. Consider the parabola $y = x^2 + 4x + 7$, and find the coordinates of the lowest point. Sketch the graph of the parabola.

4.55. Determine the directrix and focus of the parabola $y = 3x^2 - 2x + 5$.

4.56. Consider the parabola in Figure 4.34. Its focus is the point $F = (a, b)$ and its directrix the line $x = c$. Determine an equation of the parabola.

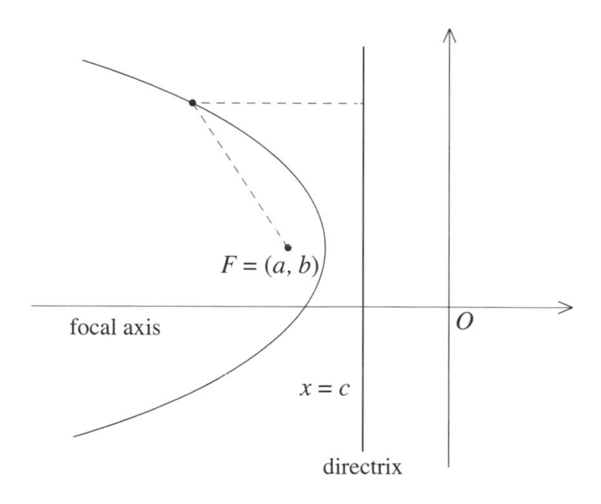

Fig. 4.34

4.57. Sketch the graphs of the parabolas $y = -x^2$ and $y = x^2 + 1$. Then shade in the region R in the plane given by $R = \{(x, y) \mid -x^2 \leq y \leq x^2 + 1\}$.

4.58. Let $P = (x, y)$ be a point in the plane. What conditions on x and y place P into the parabolic section of Figure 4.12?

4.59. Sketch the parabolic section determined by the parabola $y = 3x^2 + 6x + 7$ and the line $y = 8$. Let $P = (x, y)$ be any point in the plane. What are the conditions on the coordinates x and y to guarantee that P lies in the parabolic section?

4.60. Consider the parabolas $y = x^2 + x - 11$ and $y = 2x^2 - 4x - 7$. Both of them open upward. Make use of "completing the square" to find the lowest point and the x-intercepts for each of the parabolas. What are their y-intercepts? Determine the points where the two parabolas intersect, and then sketch both of their graphs. Express the region that the two parabolas enclose in set theoretic notation.

4I. The Trajectory of a Projectile. A projectile flies from the barrel of a gun with an initial speed of v_0. Figure 4.35 shows an xy-coordinate system placed in such a way that the path of the projectile lies in the coordinate plane. The origin O is the point at which the projectile leaves the barrel. The line L is determined by the path of the center of the projectile while it is in the barrel, and θ is the angle between L and the horizontal axis. In the part of the trajectory that is shown, $B = (x, y)$ indicates the location of the center of the projectile in a typical position. We'll let t the projectile's time of travel from O to B. (Air resistance is ignored.) If gravity had not acted, the projectile would have proceeded (by the principle of

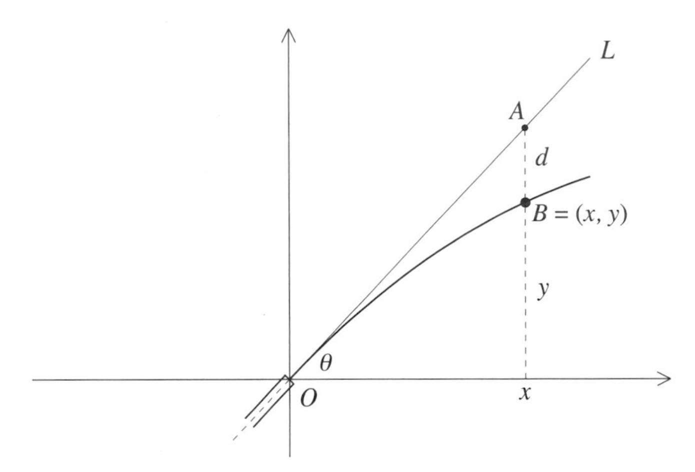

Fig. 4.35

inertia) along the line L to a point A. This inertial component of the flight proceeds from O to A with constant speed v_0 for the duration of time t. It follows that $\cos\theta = \frac{x}{OA} = \frac{x}{v_0 t}$. The action of gravity deflects the projectile downward from L. This gravitational component of the motion starts from rest at A and proceeds vertically to B taking the time t. If d is the distance from A to B, then by Galileo's times squared law of fall, $d = bt^2$ for some constant b. (See Section 3.6.) The two components of the motion have been considered independently, but they occur simultaneously to produce the trajectory. Notice that $\tan\theta = \frac{y+d}{x}$.

4.61. Put all this together to show that

$$y = \frac{-b}{(v_0 \cos\theta)^2}\, x^2 + (\tan\theta)\, x,$$

and conclude that the trajectory is a parabola. (Galileo's principle of inertia is limited to horizontal motion, so that his analysis for the parabolic trajectory only considered the case $\theta = 0$.)

4J. More about Conic Sections

4.62. The graphs of the equations $y = \frac{1}{2}x^2$, $3x^2 + 4y^2 = 6$, and $3x^2 - 4y^2 = 12$ are a parabola, an ellipse, and a hyperbola, respectively. By using equations developed in Sections 4.3, 4.4, and 4.5 and studying Figures 4.11, 4.15, and 4.17, determine the locations of the focus and the directrix of the parabola, the semiminor and semimajor axes of the ellipse, and the equations of the two asymptotes that guide the shape of the hyperbola. Sketch the three graphs.

4.63. Find the semimajor axis, semiminor axis, and eccentricity of the ellipse $\frac{x^2}{25} + \frac{y^2}{4} = 1$.

4.64. Consider the ellipse $\frac{x^2}{5^2} + \frac{y^2}{3^2} = 1$. Show that $(4,0)$ is a focal point. Consider the circle with center $(4,0)$ and radius 2. Calculate the coordinates of the points of intersection of the circle and the ellipse.

4.65. Draw the graphs of the equations $\frac{x^2}{6^2} + \frac{y^2}{4^2} = 1$ and $\frac{(x-2)^2}{6^2} + \frac{(y-4)^2}{4^2} = 1$.

4.66. Identify the graph of the equation. Is it a parabola, circle, or ellipse? Sketch the graph in each case.

i. $y = -x^2 + 3x + 4$ **ii.** $x = 2y^2$

iii. $x^2 + 4y^2 = 16$ **iv.** $9x^2 + 2y^2 = 12$

4.67. Figure 4.36 depicts an ellipse with semimajor axis a and semiminpr axis b. Let $c = \sqrt{a^2 - b^2}$, and show that the points $F_1 = (0, c)$ and $F_2 = (0, -c)$ are the focal points of the ellipse and that the defining constant k is equal to $2a$. Show that $\frac{x^2}{b^2} + \frac{y^2}{a^2} = 1$ is an equation of the ellipse.

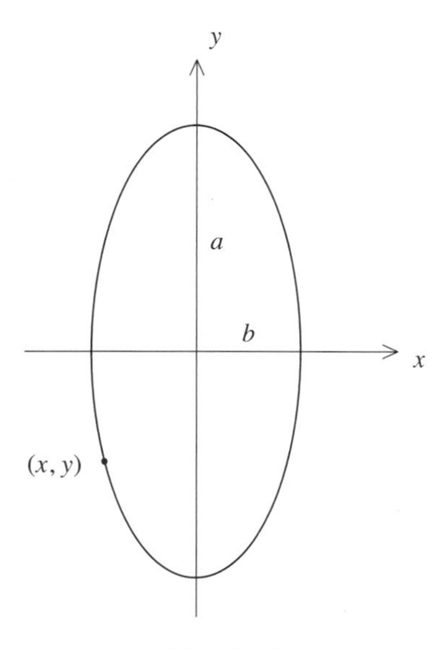

Fig. 4.36

4.68. The graph of the equation $x^2 + 4y^2 - 6x + 8y + 9 = 0$ is a conic section. Apply the $B^2 - 4AC$ criterion to see that it is an ellipse. By completing squares, rewrite the equation in such a way that you can identify the center of the ellipse as well as its semimajor and semiminor axes. Sketch the graph of the ellipse.

4.69. To draw an ellipse with focal points F_1 and F_2, semimajor axis a, and semiminor axis b, proceed as follows. Notice that the distance between the focal points is $2\sqrt{a^2 - b^2}$. Tie a string into a loop of length $2a + 2\sqrt{a^2 - b^2}$. Take a board, place thumbtacks at F_1 and F_2, and put the loop around the tacks. Take a pencil and stretch the string taut. Keep the string stretched, the pencil in contact with the surface, and go around for a complete revolution. Explain why you have drawn an ellipse and why it has semimajor axis a and semiminor axis b.

4.70. Fix an xy-coordinate system. Take a line segment of fixed length, and let P be a fixed point on it. Slide the segment around in the plane over all positions with the property that one endpoint of the segment is always on the x-axis and the other on the y-axis. Show that the locus of all positions of P is an ellipse.

4.71. Show that $16x^2 - 96x + 25y^2 - 100y - 356 = 0$ is a hyperbola. Determine its focal points and its eccentricity.

4.72. The hyperbolas $x^2 - y^2 = 1$ and $\frac{x^2}{2} - \frac{y^2}{2} = 1$ have the same focal axis and are shaped by the same pair of intersecting asymptotes. Determine these asymptotes and sketch the graphs of the two hyperbolas in the same coordinate plane.

4.73. Carry out the details of the verification of the fact the hyperbola of Figure 4.17 has equation $\frac{x^2}{a^2} - \frac{y^2}{b^2} = 1$. Consider both branches of the hyperbola.

4.74. By factoring the left side of the equation $x^2 + 4xy + 4y^2 + 6x + 12y + 9 = 0$, show that the equation is degenerate and that its graph is a single line. [Hint: The conclusion "single line" suggests a strategy.]

4.75. Show that the points $P = (x, y)$ for which x and y satisfy $3x^2 + 19x - 2xy - y^2 + 9y + 14 = 0$ are the points that lie on one of the lines $y = x + 7$ or $y = -3x + 2$.

4.76. Let $Ax^2 + Bxy + Cy^2 + Dx + Ey + F = 0$ be the equation of a conic section. It is a fact that under a translation or a rotation of the conic section, the equation of the conic section changes but the term $B^2 - 4AC$ remains the same. On the other hand, two very different-looking conic sections can have the same $B^2 - 4AC$. Why do parabolas provide examples of this? Turning to ellipses, write down the equation of a flat ellipse that has the same $B^2 - 4AC$ as the circle $x^2 + y^2 - 1 = 0$.

4.77. Let $Ax^2 + Cy^2 + Dx + Ey + F = 0$ be the equation of a conic section (where $B = 0$). Show that after rearranging constants, the equation can be brought into a shifted version of one of the basic forms listed in Sections 4.3, 4.4, and 4.5. Conclude that its focal axis is parallel to either the x-axis or the y-axis. [Hint: Assume first that both A and C are nonzero, complete the squares, and eliminate the degenerate cases.]

4K. About Angles and Trigonometry

4.78. Consider the angles $\theta = 50$ and $\theta = -25$ (both in radians). In each case, follow the strategy of Example 4.17 to find the approximate location of the point P_θ on the unit circle. Obtain approximations for $\cos\theta$ and $\sin\theta$ in each case. Find more precise values with a calculator.

4.79. Let $\theta = 17.52$ radians. Find the approximate location of the point P_θ on the unit circle with the strategy of Example 4.17. Use your conclusion to find approximations for $\cos\theta$ and $\sin\theta$. Check your estimates against the values a calculator provides. [Hint: Evaluate $\cos 15°$ and $\sin 15°$ with the half-angle formulas $\sin^2\frac{\theta}{2} = \frac{1}{2}(1 - \cos\theta)$ and $\cos^2\frac{\theta}{2} = \frac{1}{2}(1 + \cos\theta)$. See Problem 1.23.]

4.80. Refer to Figure 4.25b. Consider an angle θ such that the point P_θ falls into the second quadrant, and verify the identities $\cos(\theta + \frac{\pi}{2}) = -\sin\theta$ and $\sin(\theta + \frac{\pi}{2}) = \cos\theta$ in this situation.

4.81. Let θ be a typical angle. Use Figure 4.22 to compare the points P_θ and $P_{(\theta - \frac{\pi}{2})}$ and conclude that $\sin(\theta - \frac{\pi}{2}) = -\cos\theta$ and $\cos(\theta - \frac{\pi}{2}) = \sin\theta$.

4.82. Verify the identities $\sec(\theta + 2\pi) = \sec\theta$, $\sec(-\theta) = \sec\theta$, $\sec(\theta + \pi) = -\sec\theta$, and the formula $\sec^2\theta = \tan^2\theta + 1$.

4.83. Let r be a positive number and consider the point (x, y) with $x = r\cos\theta$ and $y = r\sin\theta$. Show that this point is on the circle of radius r centered at O. Show that every point on this circle has coordinates $(r\cos\theta, r\sin\theta)$ for some angle θ. Let θ increase from $\theta = 0$ to $\theta = 4\pi$, and check that in the process the circle is traced out exactly twice in a counterclockwise way.

4.84. Let a and b be a positive numbers, and consider the ellipse $\frac{x^2}{a^2} + \frac{y^2}{b^2} = 1$. Show that the point $(a\cos\theta, b\sin\theta)$ is on the ellipse for any angle θ and that the point traces out the entire ellipse, if θ is allowed to vary from 0 to 2π.

Chapter 5

The Calculus of Leibniz

This chapter begins with the study of lines in the coordinate plane. It then discusses Leibniz's method for computing tangent lines of curves. It makes explicit use of the function concept and limits (both are only implicit in Leibniz) to illustrate Leibniz's differential and integral calculus and his notation. The chapter also tells the mathematical story of Kepler's wine barrels.

The stage was set by the Greeks Apollonius and Archimedes; the Italians Cavalieri and Torricelli; the German Kepler; the Frenchmen Descartes, Fermat, and Robeval; the Dutchman Huygens; the Englishmen Wallis and Barrow; Gregory from Scotland; and a number of other mathematicians who put many of the components into place. It remained for Newton and Leibniz to add some decisive touches but, more importantly, to synthesize special cases and examples into a general theory and integrated set of methods with applications to a wide variety of mathematical problems: the differential and integral calculus. Historians of science recognized long ago that Leibniz and Newton developed calculus independently. Newton did so in the years 1665 and 1666, but he did not publish his findings until much later. Leibniz made his initial discoveries in the 1670s and published his discoveries. The realm of calculus has two conceptual components: differential calculus and integral calculus. The slope of a straight line is a measure of its steepness. The ratio "vertical rise per unit horizontal run" or "rise over run" quantifies the steepness of the line. The question "What about the steepness of a curve?" is the beginning of differential calculus. The area of a rectangle is the product of its width and height. The question "What is the area of a region that is enclosed by curves?" is the beginning of integral calculus. Remarkably, as we will see, there is a fundamental connection between differential and integral calculus. This chapter will focus on the work of Leibniz.

The German Wilhelm Gottfried Leibniz (1646–1716) was in his late 20s when on a diplomatic mission in Paris on behalf of his patron, the duke of a Germanic state. Inspired by some of the intellectuals of this city, he developed calculus from 1673 to 1676. Leibniz's treatment of the subject was more algebraic and clearer than Newton's more geometric approach. This is the reason this text starts with Leibniz's calculus instead of Newton's. The notation or symbolism with which mathematics is expressed is important. Notation can capture the process and guide the solution of a problem. It can advance understanding and facilitate applications. The notation of Leibniz's calculus does this very well. It therefore prevailed over Newton's and is still in use today. The work of Leibniz had great impact on the development of mathematics. The Jesuit priest Pierre Varignon (1654–1722) learned Leibniz's calculus and used it to rework the geometric arguments of Newton's *Principia*. The two Swiss brothers Jakob Bernoulli (1654–1705) and Johann Bernoulli (1667–1748) also learned calculus from Leibniz's publications and expanded it in new directions. Johann Bernoulli used it to solve the problem of determining the mathematical shape of the hanging chain. A French nobleman, the Marquis de L'Hospital, hired Johann Bernoulli to teach him the new mathematics and then published what he was taught as his own calculus text, *Analysis of the Infinitely Small*, in 1696. (L'Hospital's rule was the work of Bernoulli.) Newton's *Principia* was groundbreaking and formidable in the way it combined calculus and the laws of force to explain the orbits of the planets, but L'Hospital's was an elegant calculus text from which the subject could be learned. Another student of the Bernoulli brothers, the Swiss Leonhard Euler (1707–1783), became the most prolific and influential mathematician of the 18th century. His work—in dozens of published volumes—advanced calculus, developed new fields of mathematics, and applied mathematics to the study mechanics, artillery, music, and ships (and a number of other subjects).

Leibniz's fame as a philosopher and scientist spread throughout Europe. He was in correspondence with most of the important European scholars of the day. He was named to the French Academy of Sciences, and the Habsburg emperor bestowed on him the title of baron. But Leibniz's last years were unhappy ones. His long dispute over the priority of the creation of the calculus with Newton and his English colleagues became increasingly bitter. When the son of his patron became King George I of England in 1714 and moved his court to London, Leibniz remained behind in Hanover in isolation. His work was neither understood nor appreciated during his lifetime. He suffered increasingly from illness and died in 1716.

5.1 STRAIGHT LINES

Before considering Leibniz's mathematics, we turn to the quantitative analysis of lines that received only brief attention in Section 4.2. When a line is considered, it is understood to be a straight line. Lines that are not straight are referred to as curves. We will be working with a plane that is equipped with an xy-coordinate system.

Consider the equation $2x - 3y - 4 = 0$. It can be rewritten in many algebraically equivalent ways, for example, $2x - 3y = 4$, or $3y = 2x - 4$, or $y = \frac{2}{3}x - \frac{4}{3}$. If the values $x = a$ and $y = b$ satisfy one of these equations, they satisfy all of them. Therefore all of these equations have the same graph. We will work with

$$y = \tfrac{2}{3}x - \tfrac{4}{3}.$$

If $x = 5$, then $y = \frac{10}{3} - \frac{4}{3} = \frac{6}{3} = 2$. So $x = 5$ and $y = 2$ satisfy the equation, and the point $(5, 2)$ is on the graph. In the same way, $(4, \frac{4}{3})$, $(3, \frac{2}{3})$, $(2, 0)$, $(1, -\frac{2}{3})$, $(0, -\frac{4}{3})$, $(-2, -\frac{8}{3})$, and so on are all on the graph. Plot these points, and observe that all fall on the single line L pictured in Figure 5.1. Move along L from left to right and notice—compare the points $(2, 0)$ and $(5, 2)$ for example—that for every 3 units of increase in the direction of the positive x-axis, there is a corresponding increase of 2 units in the direction of the positive y-axis. Hence there is an increase of $\frac{2}{3}$ units in the y-coordinate for every increase of 1 unit in the

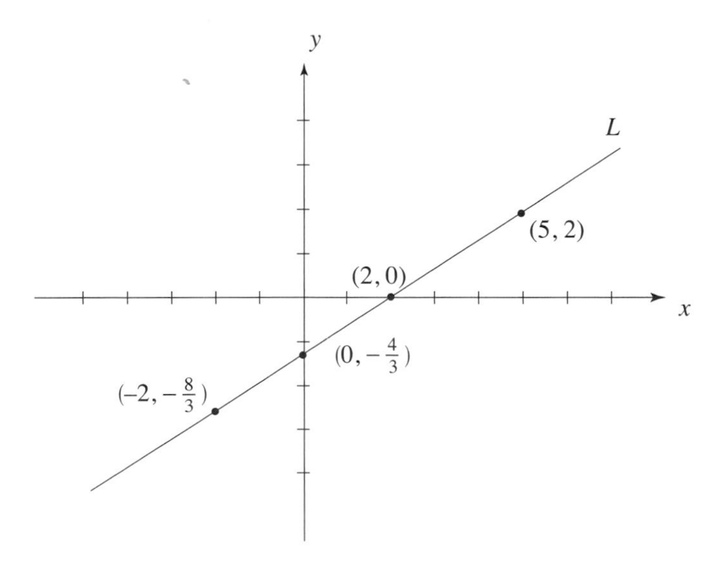

Fig. 5.1

x-coordinate. So the number $\frac{2}{3}$ measures the pitch, or steepness, of the line. It is called the *slope* of the line and is often denoted by m. In the case under consideration now, $m = \frac{2}{3}$.

The slope m can be determined as follows: consider any two points on L, say, $(3, \frac{2}{3})$ and $(-2, -\frac{8}{3})$. Form the differences $\frac{2}{3} - \frac{-8}{3} = \frac{10}{3}$ between the y-coordinate of the first point and the y-coordinate of the second, as well as the difference $3 - (-2) = 5$ between the x-coordinate of the first point and the x-coordinate of the second. The ratio $m = \frac{\frac{10}{3}}{5} = \frac{2}{3}$ of these differences is the slope of the line. Any two distinct points $P_1 = (x_1, y_1)$ and $P_2 = (x_2, y_2)$ on the line can be used to compute the slope in this way.

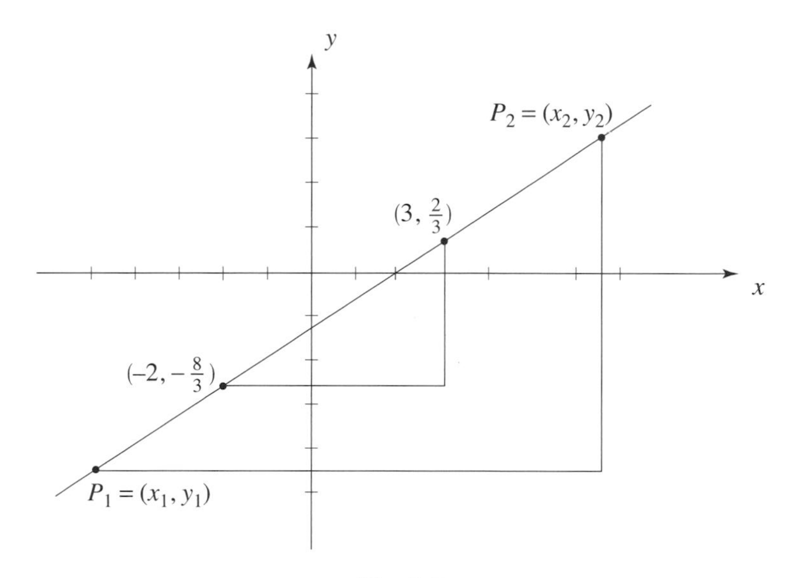

Fig. 5.2

This is a consequence of the basic property of similar triangles. Refer to Figure 5.2 and notice that the right triangle with hypotenuse given by $(3, \frac{2}{3})$ and $(-2, -\frac{8}{3})$ is similar to the right triangle with hypotenuse P_1P_2. It follows that the ratios of the lengths of the vertical over the horizontal legs of the two triangles are equal. Therefore $m = \frac{2}{3} = \frac{y_2 - y_1}{x_2 - x_1}$.

In this way, the slope m of any nonvertical line L can be computed. Let $P_1 = (x_1, y_1)$ and $P_2 = (x_2, y_2)$ be any two distinct points on L. Since the points are distinct, $x_2 \neq x_1$, and

$$m = \frac{y_2 - y_1}{x_2 - x_1}.$$

Because $\frac{y_2 - y_1}{x_2 - x_1} = \frac{y_1 - y_2}{x_1 - x_2}$, the order in which the points P_1 and P_2 are taken in this computation does not matter.

If L is a vertical line, the x-coordinates of all points are the same, so $x_2 = x_1$ and the slope of L is not defined. Consider the equation $2x + 10 = 0$ or, equivalently, $x = -5$. The points (x, y) that satisfy it are precisely the points $(-5, y)$, where y can be anything. So the graph of the equation $x = -5$ is the vertical line through $(-5, 0)$. It is an example of a line that has no slope.

Let L be any nonvertical line. Let $P_1 = (x_1, y_1)$ and $P_2 = (x_2, y_2)$ be two points on L, and take $x_2 = x_1 + 1$. So the slope of L is

$$m = \frac{y_2 - y_1}{x_2 - x_1} = y_2 - y_1,$$

equal to the difference in the two y-coordinates. If this difference is positive, the line rises from left to right; if it is negative, it falls; and if it is 0, then the line is horizontal. Now let L' be any other nonvertical line. Keep x_1 and x_2 the same, and let $P'_1(x_1, y'_1)$ and $P'_2 = (x_2, y'_2)$ be the corresponding points on L'. Then the slope of L' is

$$m = y_2' - y_1'.$$

So for 1 unit of increase in the x-coordinates, the changes in the y-coordinates are $y_2 - y_1$ and $y_2' - y_1'$. It follows that L' is parallel to L precisely if their slopes m and m' are equal.

Figure 5.3 shows several lines through the origin and their slopes. Going from left to right, the lines with a positive slope slant upward and those with a negative slope slant downward. Notice also that the

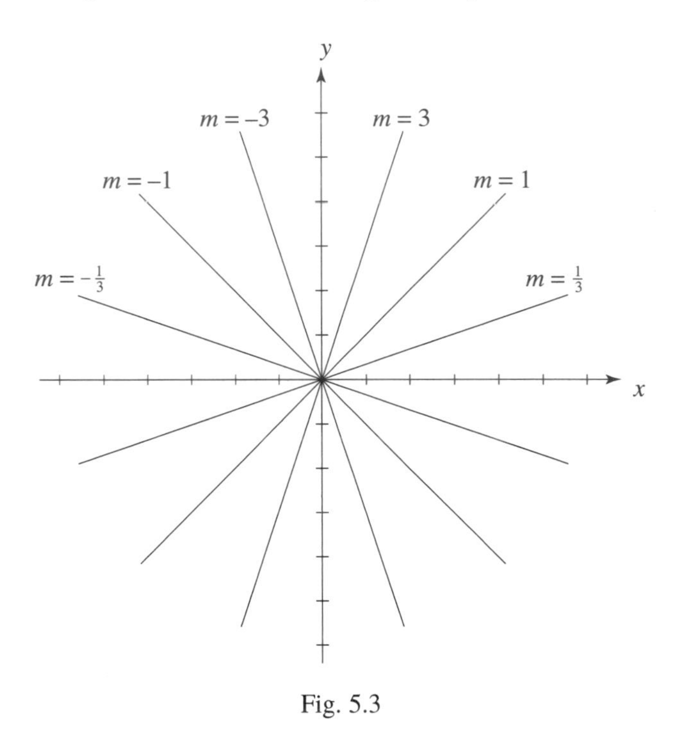

Fig. 5.3

steepest lines are those for which the absolute value of the slope is largest. Consider a line with slope $m \neq 0$. Similar triangles can be used to verify that the line perpendicular to it has slope $-\frac{1}{m}$. Figure 5.3 provides examples of pairs of lines that are perpendicular.

Example 5.1. Sketch the graph of the equation $5x + 3y = 15$ and determine its slope. It suffices to find

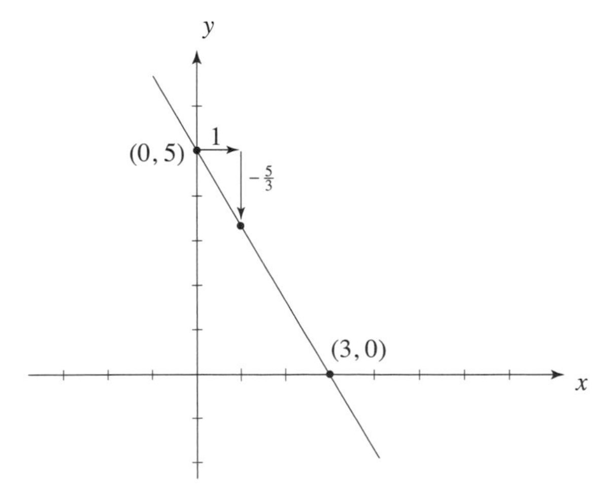

Fig. 5.4

two points on the line and to draw the line through both. It is easiest to find the intercepts. Substituting $y = 0$ into the equation, we get $5x = 15$. So $x = 3$, and $(3, 0)$ is on the graph. Substituting $x = 0$, we see that $(0, 5)$ is on the graph. The graph is shown in Figure 5.4. Using these two points to compute the slope, we get

$$m = \tfrac{(5-0)}{(0-3)} = -\tfrac{5}{3}.$$

So a change of $+3$ in x results in a change of -5 in y. Equivalently, a change of 1 unit in x produces a change of $-\frac{5}{3}$ units in y.

Suppose that L is a nonvertical line. Select any two distinct points $P_1 = (x_1, y_1)$ and $P_2 = (x_2, y_2)$ on L. As before, its slope is $m = \frac{y_2-y_1}{x_2-x_1}$. Now let $P(x, y)$ be any point in the plane. A look at Figure 5.5 shows that $P = (x, y)$ is on L if (and only if) the segment P_1P is parallel to L. So $P = (x, y)$ is on L precisely if

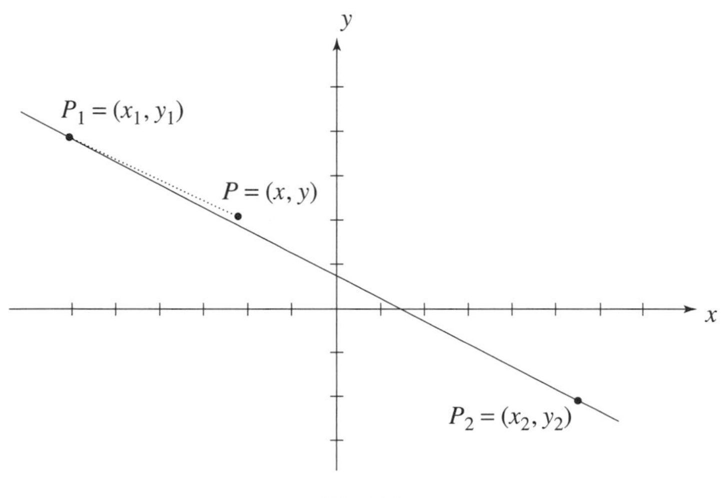

Fig. 5.5

the slope of the segment P_1P is equal to m. Since the slope of P_1P is $\frac{y-y_1}{x-x_1}$, the point $P = (x, y)$ is on L precisely if $\frac{y-y_1}{x-x_1} = m$. It follows that

$$\boxed{y - y_1 = m(x - x_1)}$$

is an equation that all points (x, y) on the line L (but no points off the line) satisfy. An equation of a line arranged in this way is said to be in *point-slope* form.

Example 5.2. Find an equation of the line through $(1, -7)$ with slope $-\frac{1}{2}$. Using the point-slope form of the equation with $m = -\frac{1}{2}$, $x_1 = 1$, and $y_1 = -7$, we get

$$y + 7 = -\tfrac{1}{2}(x - 1).$$

This, as well as $2y + 14 = -x + 1$ or $x + 2y + 13 = 0$, are equations of the line.

Example 5.3. Find an equation of the line through the points $(-1, 2)$, and $(3, -4)$. The slope of the line is

$$m = \tfrac{-4-2}{3-(-1)} = -\tfrac{3}{2}.$$

Using the point-slope form with $x_1 = -1$ and $y_1 = 2$, we obtain

$$y - 2 = -\tfrac{3}{2}(x + 1).$$

This equation can also be written as $3x + 2y = 1$.

Suppose a nonvertical line has slope m and y-intercept b. The y-intercept is the point where the line intersects the y-axis. So $(0, b)$ is on the line. With $x_1 = 0$ and $y_1 = b$, the point-slope form of the equation of the line is $y - b = m(x - 0)$. Therefore

$$\boxed{y = mx + b}$$

is an equation of the line with slope m and y-intercept b. An equation of a line arranged in this way is said to be in *slope-intercept* form. If a line is horizontal, its slope is $m = 0$. So the equation of such a line has the form $y = b$, where b is the y-intercept.

Example 5.4. Put each of the lines of Figure 4.10 (except $x = 4$) into slope-intercept form. Does the information about the slope that it provides in each agree with what the graph of the line suggests? Why is $x = 4$ excluded?

Take any equation of the form $Ax + By + C = 0$, where A, B, and C are constants. If $A = B = 0$, then $C = 0$. Disregard this case, and suppose that either $A \neq 0$ or $B \neq 0$. Suppose that $B \neq 0$. Then $Ax+By+C = 0$ is equivalent to $y = -\frac{A}{B}x - \frac{C}{B}$. A look at the point-slope form shows that this is an equation of the line with slope $-\frac{A}{B}$ and y-intercept $-\frac{C}{B}$. If $B = 0$, then $A \neq 0$. Now the equation is $Ax + C = 0$ or, equivalently, $x = -\frac{C}{A}$. As in the earlier cases, $x = 4$ and $x = -5$; this is an equation of the vertical line through $(-\frac{C}{A}, 0)$. It follows from this discussion that the graph of any equation of the form $Ax + By + C$ is a line (unless $A = B = C = 0$). Any such equation is therefore called a *linear* equation.

Now that we know a few things about lines, we can turn to Leibniz's study of the steepness of curves.

5.2 TANGENT LINES TO CURVES

Leibniz's first publication of his work on calculus appeared in 1684. His article *Nova methodus pro maximis et minimis, itemque tangentibus calculi genus* (*A new method for maxima and minima as well as tangents, ...*) introduces the term "calculus differentialis" and contains the general rules of differentiation. Unfortunately, it was nearly impossible to read. Even Leibniz's friends and students, the brothers Bernoulli (who would become famous mathematicians themselves), commented that this article is "an enigma rather than an explication."

The discussion of the work of Leibniz that follows in this and subsequent sections is an expanded version of his original manuscripts. It also adds some concepts and notational elements that were not introduced until later. It is the goal to clarify some of the "enigmas" while retaining the essence of Leibniz's analysis.

Let an equation in the variables x and y be given. (We'll look at specific examples in a moment.) Its graph is some curve in the xy-coordinate plane, and this curve will be the focus of study in this section. Fix a point P on the curve and consider the tangent line to the curve at P. Assume that it is not vertical, and let m_P be its slope. See Figure 5.6a. Leibniz describes the following procedure for finding m_P. Take some other point $Q \neq P$ on the curve. While keeping P fixed, push Q toward P. Refer to Figure 5.6b and notice that as Q closes in on P, the line through P and Q closes in on the tangent line at P. So as Q is pushed to P, the slope of the line through P and Q homes in on the slope m_P of the tangent. Let's have an algebraic look at what is going on. Let x and y be the coordinates of P, and let x' and y' be the coordinates of Q. Let $\Delta x = x' - x$ and $\Delta y = y' - y$ be the differences between the x-coordinates and y-coordinates of P and Q. So the coordinates of Q are $x' = x + \Delta x$ and $y' = y + \Delta y$. With this notation, the slope of the line through P and Q is equal to the ratio

$$\frac{(y + \Delta y) - y}{(x + \Delta x) - x} = \frac{\Delta y}{\Delta x}.$$

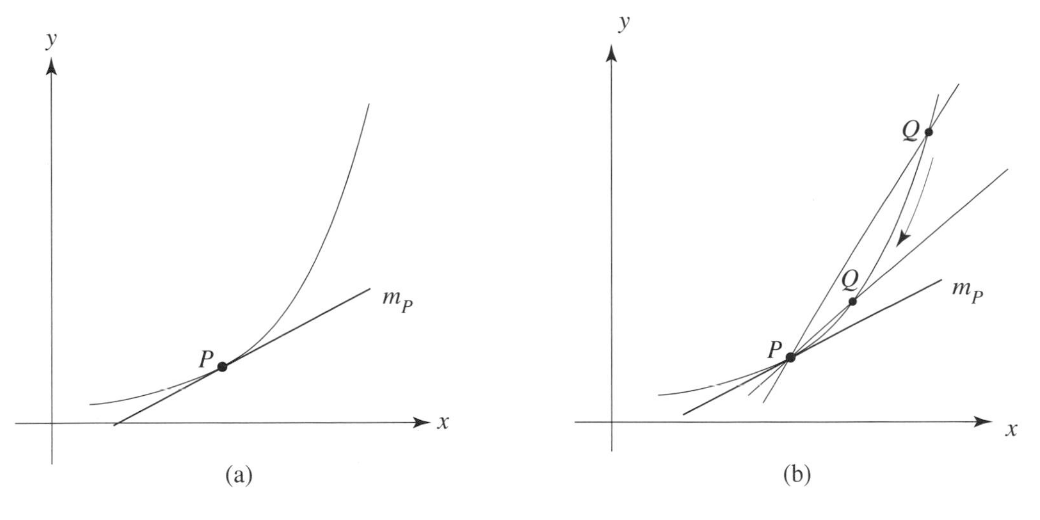

Fig. 5.6

Turn next to Figure 5.7 and note that pushing Q to P is the same thing as pushing Δx to zero. Therefore, when Δx is pushed to zero, the slope $\frac{\Delta y}{\Delta x}$ of the line through P and Q closes in on the slope of the tangent.

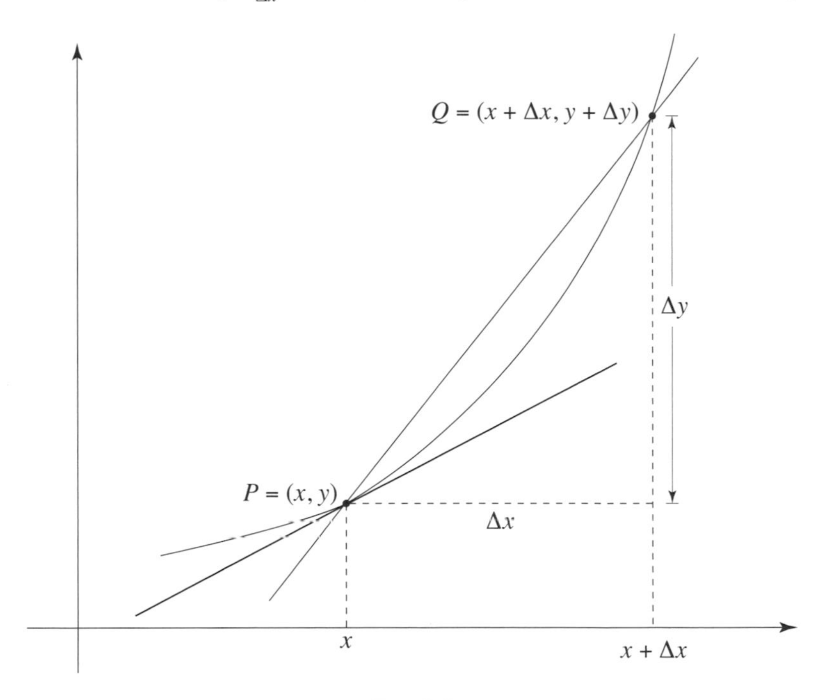

Fig. 5.7

Rewritten in the abbreviated notation of limits—which Leibniz did not use—what Leibniz has verified is that

$$\lim_{\Delta x \to 0} \frac{\Delta y}{\Delta x} = m_P.$$

Leibniz describes his limit maneuver by saying "in which the ordinate [this refers to the term $x + \Delta x$] is moved nearer and nearer to the fixed ordinate [referring to Δx] until it ultimately coincides with it." Notice that he cannot simply set $\Delta x = 0$, for then Δy is also equal to 0, and $\frac{\Delta y}{\Delta x}$ is the term $\frac{0}{0}$. This ratio provides no information, in fact, this ratio does not make sense.

Leibniz turns to the parabola to illustrate his method. He takes $y = Ax^2$, where A is any constant. As above, $P = (x, y)$ is any point on the graph regarded to be fixed. Since

$$Q = (x + \Delta x, y + \Delta y)$$

is also on the graph, $y + \Delta y = A(x + \Delta x)^2$. We continue in Leibniz's own words (in translation from his Latin):

> Then, since $y = Ax^2$, by the same law, we have
>
> $$y + \Delta y = A(x + \Delta x)^2 = Ax^2 + 2Ax\Delta x + A(\Delta x)^2;$$
>
> and taking away the y from the one side and the Ax^2 from the other, we have left
>
> $$\frac{\Delta y}{\Delta x} = 2Ax + A\Delta x$$
>
> provided that the difference [this refers to Δx] is not assumed to be zero until the calculation is purged as far as possible by legitimate omissions [the cancellation of Δx], and reduced to ratios of non-evanescent quantities, and we finally come to the point where we apply our result to the ultimate case.

Expressed in limit notation, Leibniz's ultimate case is

$$\lim_{\Delta x \to 0} \frac{\Delta y}{\Delta x} = \lim_{\Delta x \to 0} (2Ax + A\Delta x) = 2Ax.$$

So the slope of the tangent line at any point $P = (x, y)$ on the parabola $y = Ax^2$ is $m_P = 2Ax$.

Let's apply Leibniz's result to the special case $y = x^2$. So $A = 1$, and the slope of the tangent at any point (x, y) on the graph is $2x$.

Example 5.5. For the point $P = (3, 9)$, $x = 3$, so that the slope of the tangent at P is $m_P = 2 \cdot 3 = 6$. At the point $P = (-2, 4)$, $x = -2$, and the slope of the tangent at P is $m_P = 2 \cdot (-2) = -4$. Figure 5.8 illustrates what has just been observed.

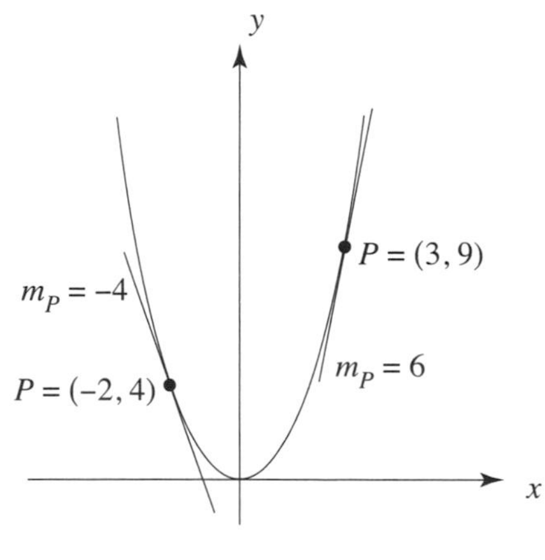

Fig. 5.8

Leibniz next turns to the graph of the equation $y = Ax^3$, where A is a constant. As before, $P = (x, y)$ is a fixed point and $Q = (x + \Delta x, y + \Delta y)$ any other point on the graph. Another look at Figure 5.7 shows what is going on. Since both P and Q are on the graph, $y = Ax^3$ and

$$\begin{aligned} y + \Delta y &= A(x + \Delta x)^3 \\ &= A\left(x^3 + 3x^2\Delta x + 3x(\Delta x)^2 + (\Delta x)^3\right). \end{aligned}$$

After subtracting $y = Ax^3$ from both sides, Leibniz gets $\Delta y = 3Ax^2\Delta x + 3Ax(\Delta x)^2 + A(\Delta x)^3$. Dividing both sides by Δx, he has

$$\frac{\Delta y}{\Delta x} = 3Ax^2 + 3xA\Delta x + A(\Delta x)^2.$$

Pushing Δx to zero, Leibniz can conclude that

$$\lim_{\Delta x \to 0} \frac{\Delta y}{\Delta x} = \lim_{\Delta x \to 0} \left(3Ax^2 + 3xA\Delta x + A(\Delta x)^2\right) = 3Ax^2.$$

He has shown that the slope of the tangent line to the curve $y = Ax^3$ at any point $P = (x, y)$ is equal to $m_P = 3Ax^2$.

Consider the curve $y = x^3$ and the point $P = (2, 8)$ on it. Taking $A = 1$ and $x = 2$ in the discussion just completed, we see that the slope of the tangent at P is $m_P = 3 \cdot 2^2 = 12$. Since $P = (2, 8)$ is on the tangent, we find by using the point-slope form of the equation of a line that $y - 8 = 12(x - 2)$, or $y = 12x - 16$, is an equation of the tangent.

Let's take the parabola $x = y^2$ next. Repeating once more what Figure 5.7 describes, fix a point $P = (x, y)$ on its graph, and let $Q = (x + \Delta x, y + \Delta y)$ be any other point on it. Since the coordinates of Q satisfy the equation of the parabola,

$$x + \Delta x = (y + \Delta y)^2 = y^2 + 2y\Delta y + (\Delta y)^2.$$

Since P does also,

$$\Delta x = 2y\Delta y + (\Delta y)^2 = \Delta y(2y + \Delta y)$$

by subtracting and factoring. Therefore $\frac{\Delta y}{\Delta x} = \frac{1}{2y+\Delta y}$. Push Δx to zero and refer to Figure 5.7. As Q approaches P, the quantity Δy goes to zero. It follows that the slope of the tangent to the parabola at $P = (x, y)$ is

$$m_P = \lim_{\Delta x \to 0} \frac{\Delta y}{\Delta x} = \frac{1}{2y}.$$

Example 5.6. What is the slope of the tangent of the parabola $x = y^2$ at $P = (3, \sqrt{3})$? Since $y = \sqrt{3}$, it follows from the formula just established that $m_P = \frac{1}{2\sqrt{3}}$. What is the slope of the tangent at the origin $P = (0, 0)$? When $y = 0$, the formula $m_P = \frac{1}{2y}$ does not make sense. This is explained by the fact that the tangent to the parabola $x = y^2$ at the origin is vertical. So it has no slope. Confirm this by sketching a graph of $x = y^2$.

What happens when Leibniz's method is applied to a line? Let's take $y = \frac{1}{2}x + 1$, for example. See Figure 5.9. Since P and Q both satisfy this equation, we get

$$y + \Delta y = \tfrac{1}{2}(x + \Delta x) + 1 = \tfrac{1}{2}\Delta x + \tfrac{1}{2}x + 1,$$

and hence $\Delta y = \frac{1}{2}\Delta x$. So $\frac{\Delta y}{\Delta x} = \frac{1}{2}$. Pushing Δx to zero has no effect on $\frac{1}{2}$, and therefore

$$\lim_{\Delta x \to 0} \frac{\Delta y}{\Delta x} = \tfrac{1}{2}.$$

We have confirmed the obvious: the tangent at any point P on the line is the line itself, and it has slope $\frac{1}{2}$.

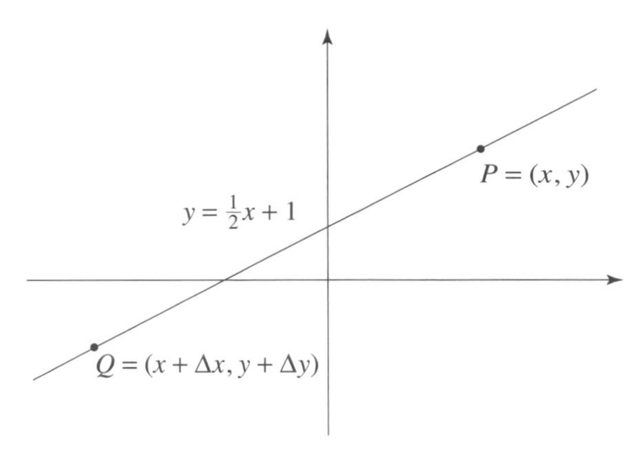

Fig. 5.9

Example 5.7. For our last example, let's have a look at the circle $x^2 + y^2 = 5^2$. Once more, let $P = (x, y)$ be a fixed point and $Q = (x + \Delta x, y + \Delta y)$ any another point on the circle. Since Q is on the circle, $(x + \Delta x)^2 + (y + \Delta y)^2 = 5^2$, so

$$x^2 + 2x\Delta x + (\Delta x)^2 + y^2 + 2y\Delta y + (\Delta y)^2 = 5^2.$$

Since $x^2 + y^2 = 5^2$, we get that

$$2x\Delta x + (\Delta x)^2 + 2y\Delta y + (\Delta y)^2 = 0.$$

After collecting the Δy terms on the left and the Δx terms on the right and factoring, we see that

$$\Delta y(2y + \Delta y) = -\Delta x(2x + \Delta x).$$

Concentrate on the ratio

$$\frac{\Delta y}{\Delta x} = -\frac{2x + \Delta x}{2y + \Delta y},$$

and push Δx to zero. Refer back to Figure 5.7 once more. Since Q goes to P, note that Δy goes to zero. So it follows that the slope of the tangent to the circle at the point $P = (x, y)$ is

$$m_P = \lim_{\Delta x \to 0} \frac{\Delta y}{\Delta x} = -\frac{2x}{2y} = -\frac{x}{y}.$$

Notice that the point $P = (3, 4)$ is on the circle $x^2 + y^2 = 5^2$. From the fact just derived, the slope of the tangent to the circle at P is $-\frac{3}{4}$.

Find the two points (x, y) on the circle that have tangents with slope 1. To get $-\frac{x}{y} = 1$, we need to have $y = -x$. Therefore $x^2 + (-x)^2 = 2x^2 = 5$, so that $x = \pm\sqrt{\frac{5}{2}}$. Therefore the points are $\left(\sqrt{\frac{5}{2}}, -\sqrt{\frac{5}{2}}\right)$ and $\left(-\sqrt{\frac{5}{2}}, \sqrt{\frac{5}{2}}\right)$. What happens at the points $(5, 0)$ and $(-5, 0)$?

5.3 THE FUNCTION CONCEPT

It was Leibniz who introduced the word "function" into the dictionary of mathematics in the 1670s. He used it as a general term for various geometric quantities associated to a point on a curve. Still, functions are implicit rather than explicit in Leibniz's work. About 50 years later, Johann Bernoulli gave the first definition of a function of one variable as any quantity composed of the variable and constants. He used the notation φx for the value of a function φ at x. For Leonhard Euler in the 18th century, a function was an "analytic expression composed in any way" from variables and constants. After Euler proclaimed mathematical analysis to be the study of functions in the 1740s, the function concept began to assume a central position in mathematics. The limit concept was emphasized by the French mathematician-physicist-philosopher Jean-Baptiste D'Alembert in the 1750s. Later, in the work of the French-Italian mathematician Luigi Lagrange, the slope in the role of the derivative was treated as a function. The logically tight reformulation of calculus provided in the 19th century by the Frenchman Augustin-Louis Cauchy and others also included the facilitating notation $\lim\limits_{\Delta x \to 0} \frac{\Delta y}{\Delta x}$.

In order to add clarity to our discussion of Leibniz's calculus, we will introduce the concept of a function and make use of it throughout the remainder of the chapter.

A *function* is a rule that assigns exactly one real number to each number from a given set of real numbers. Such a rule is often denoted by f. If x is a typical number from the given set, then $y = f(x)$ is the number that f assigns to x. In this way, the equation $y = f(x)$ also becomes a notation for the function. The rule $f(x) = 3x - 7$ is an example. Note that $f(2) = 3 \cdot 2 - 7 = -1$ and $f(4) = 3 \cdot 4 - 7 = 5$, and so on. So f assigns -1 to the number 2, and 5 to the number 4, and $3x - 7$ to a typical real number x. Functions of the form $f(x) = 6$, or $g(x) = -7$, or $h(x) = b$ with b a constant, are called *constant functions*. Functions of the form $f(x) = 3x - 7$, $g(x) = -\frac{1}{2}x + 6$, or more generally $h(x) = mx + b$ with m and b constants, are *linear functions*. Those of the form $f(x) = ax^2 + bx + c$, where a, b, c are constants are the *quadratic functions*. The rules specified so far are *defined*, meaning that they make sense, for all real numbers x. The rule $f(x) = \sqrt{x}$ is a function that is defined only for $x \geq 0$.

The *graph* of a function f is the set of all points (x, y) in the xy-plane that satisfy the equation $y = f(x)$. The graph of a constant function $y = f(x) = a$ is a horizontal line, the graph of a linear function $y = f(x) = mx + b$ is a line with slope m and y-intercept b, and the graph of a quadratic function $y = ax^2 + bx + c$ is a parabola that opens upward if $a > 0$ and downward if $a < 0$.

We'll now list examples of some basic functions. Figures 5.10, 5.11, and 5.12 depict their graphs.

(a) The rule $f(x) = x^2$ is a function. This rule f is defined for all x. For example, $f(2) = 4$, $f(3) = 9$, and $f(-2) = 4$. So the points $(2, 4)$, $(3, 9)$, and $(-2, 4)$ are all on the graph.

(b) The rule $f(x) = \sqrt{x} = x^{\frac{1}{2}}$ is a function. Observe that $f(0) = 0$, $f(1) = 1$, $f(3) = \sqrt{3}$, and $f(4) = 2$. So the points $(0, 0)$, $(1, 1)$, $(3, \sqrt{3})$, and $(4, 2)$ are on the graph. This function is defined only for $x \geq 0$. For example, $f(-2) = \sqrt{-2}$ is not defined.

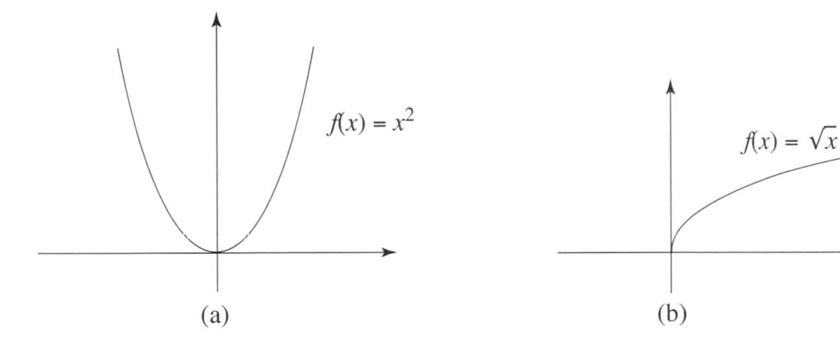

Fig. 5.10

(c) The rule $g(x) = x^3$ is a function. It is defined for all x. Note that $(1, 1)$, $(-1, -1)$, and $(-2, -8)$ are on the graph.

(d) The rule $g(x) = \sqrt[3]{x} = x^{\frac{1}{3}}$ is a function defined for all x. Check that the points $(-1, -1)$, $(8, 2)$, and $(-27, -3)$ are on its graph.

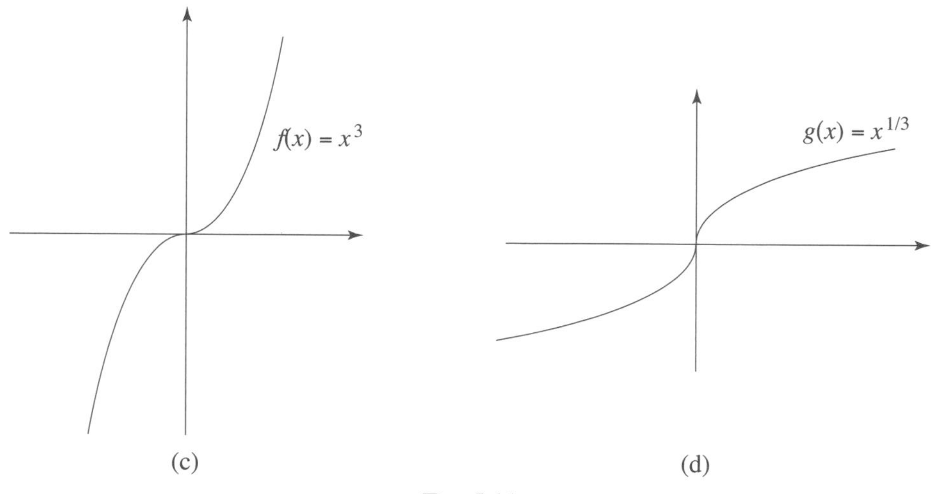

Fig. 5.11

(e) The rule h given by $h(x) = \frac{1}{x} = x^{-1}$ is a function. It is defined for all x except $x = 0$. The points $(1, 1)$, $(2, \frac{1}{2})$, and $(\frac{1}{2}, 2)$ are on the graph.

(f) The rule $h(x) = \frac{1}{x^2} = x^{-2}$ is a function defined for all $x \neq 0$. Since $h(2) = \frac{1}{4}$, $h(-1) = 1$, and $h(3) = \frac{1}{3}$, the points $(2, \frac{1}{4})$, $(-1, 1)$, and $(3, \frac{1}{3})$ are on the graph.

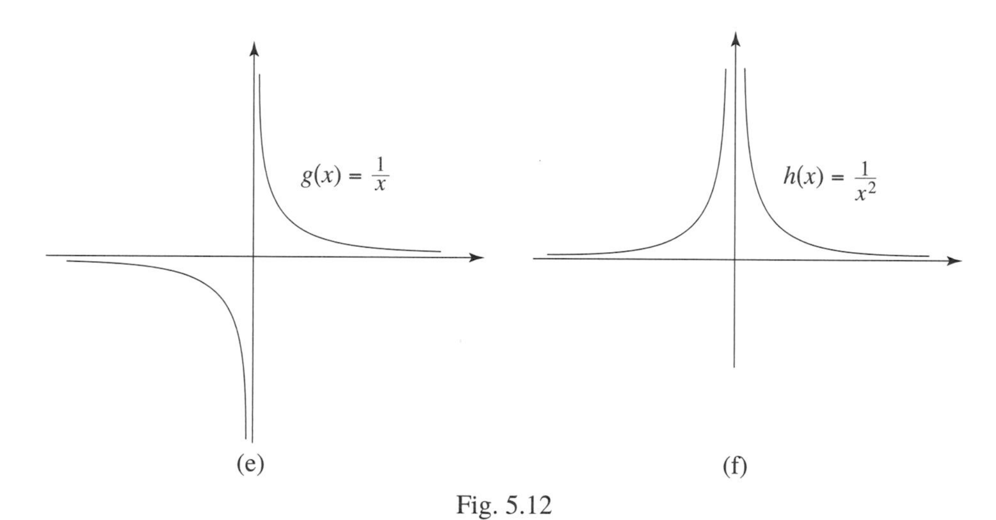

Fig. 5.12

Section 4.6 extended the definitions of the sine, cosine, tangent, and secant from angles satisfying $0 \leq \theta \leq 2\pi$ to angles θ that can be any real number. Use the variable x in place of θ, and observe that the discussion in Section 4.6 defines the functions

$$y = \sin x,\ y = \cos x,\ y = \tan x, \text{and}\ \ y = \sec x.$$

The functions $y = \sin x$ and $y = \cos x$ are defined for all x, and both $y = \tan x$ and $y = \sec x$ are defined for all x except $x = \pm\frac{\pi}{2}, \pm\frac{3\pi}{2}, \pm\frac{5\pi}{2}, \ldots$. Figures 4.23, 4.24, 4.26, and 4.27 provide their graphs.

Let f be any function. For any x there is at most one y such that the point (x, y) is on its graph of f. If f is defined at x, then $y = f(x)$ is the only such y, and if f is not defined at x, then there is no such y. It follows that any vertical line can cross the graph of f at most once. This property singles out the graphs of functions $y = f(x)$ within the larger category of the graphs of equations in x and y. For example, consider the equation $y^2 = x$. Since the pair of numbers $y = \sqrt{x}$ and $y = -\sqrt{x}$ with $x \geq 0$ are both solutions, the equation $y^2 = x$ does not determine y uniquely in terms of x. Similarly, for a given x, the equation $y^2 = x^2$ provides two possibilities for y, namely, $y = x$ and $y = -x$. Therefore neither of the graphs of Figure 5.13 is the graph of a function of x. However, the equations $y = f_+(x) = \sqrt{x}$ and $y = f_-(x) = -\sqrt{x}$ define a pair of functions for $x \geq 0$ with graphs the upper and the lower halves of the parabola of Figure 5.13a, respec-

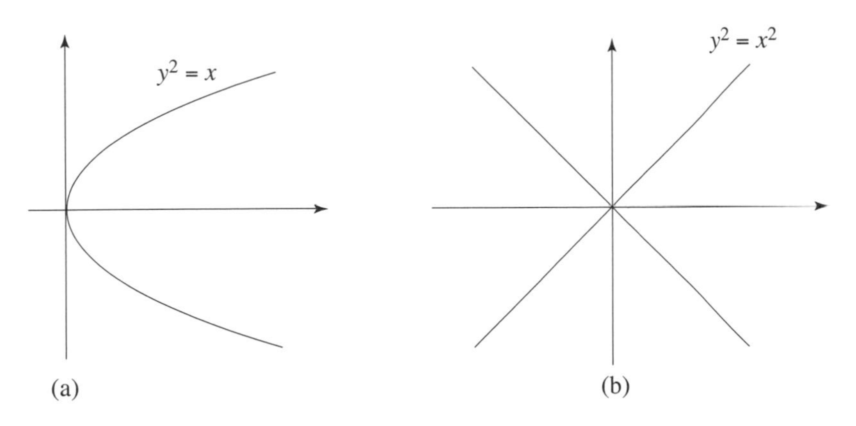

Fig. 5.13

tively. Similarly, the equations $y = g_+(x) = |x|$ and $y = g_-(x) = -|x|$, where $|x|$ is the absolute value of x, define functions with graphs the respective upper and lower halves of Figure 5.13b.

We will now use the function concept to reformulate a matter that was already discussed. Let's have another look at the computation of the slope of a tangent line.

5.4 THE DERIVATIVE OF A FUNCTION

Let f be a function, and fix a point $P = (x, y)$ on its graph. Since $y = f(x)$, it follows that $P = (x, f(x))$. Suppose the graph has a nonvertical tangent at P. Since P is the only point on the graph with this x-coordinate, we will designate the slope of this tangent by m_x, instead of m_P, as was done in Section 5.2. Now let $Q = (x', y')$ be another point on the graph. So $y' = f(x')$. Set $x' - x = \Delta x$ and $y' - y = \Delta y$. Note that $x' = x + \Delta x$, $y' = f(x') = f(x + \Delta x)$, and that $Q = (x + \Delta x, f(x + \Delta x))$. See Figure 5.14. Observe that

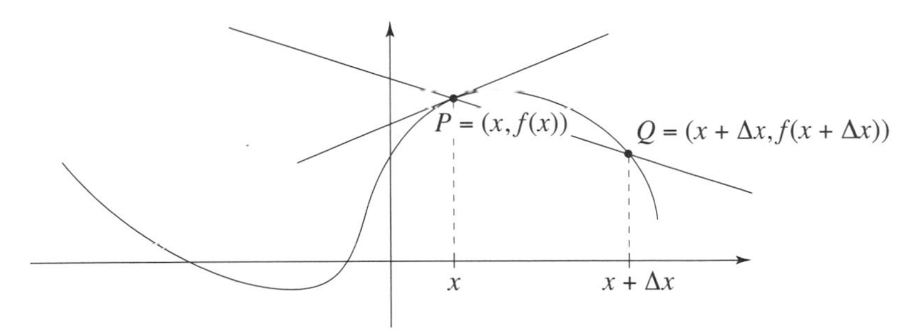

Fig. 5.14

the slope of the line through P and Q is

$$\frac{f(x + \Delta x) - f(x)}{\Delta x}.$$

Since $\Delta y = y' - y = f(x + \Delta x) - f(x)$, this is the ratio $\frac{\Delta y}{\Delta x}$ that was studied in Section 5.2. When Δx is pushed to zero, Q is pushed to P along the graph, and the line through P and Q rotates into the tangent line at P. So the slope of the line through P and Q closes in on the slope m_x of the tangent to the graph at $P = (x, f(x))$. In limit notation, this is expressed as

$$m_x = \lim_{\Delta x \to 0} \frac{f(x + \Delta x) - f(x)}{\Delta x}.$$

In Figure 5.14, the point Q is depicted to the right of P. This implies that Δx is positive throughout the push of Q to P. The point Q and its push to P can also be done from the left, so that Δx is negative throughout the push. The limit $\lim\limits_{\Delta x \to 0} \frac{f(x+\Delta x)-f(x)}{\Delta x}$ is equal to the same m_x. If the limit $\lim\limits_{\Delta x \to 0} \frac{f(x+\Delta x)-f(x)}{\Delta x}$ is some finite number for a particular x, then we say that the function f is *differentiable* at x. The discussion above has informed us that if the graph of f has a nonvertical tangent at the point $(x, f(x))$, then f is differentiable at x and the limit is the slope m_x of this tangent. So the derivative f' is defined at x and has value m_x. If the graph of a function is smooth near a point P, then a tangent line to the graph can be placed at P. This invites the intuitive or visual interpretation: a function f is differentiable at the number x if the graph of f is smooth at the point $(x, f(x))$ and has a nonvertical tangent there.

Now consider the rule that assigns the number m_x to a given number x. This rule defines a function

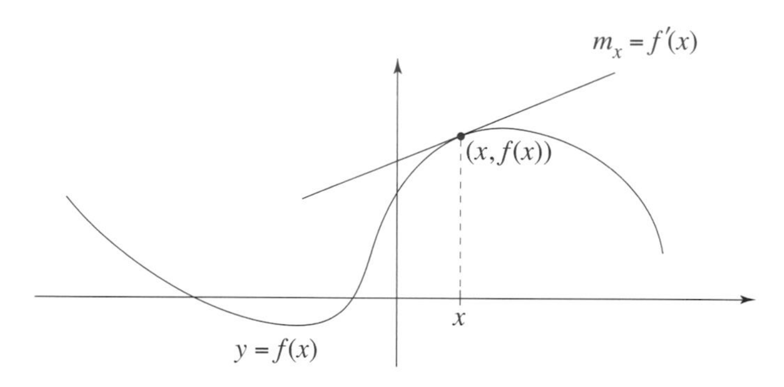

Fig. 5.15

called the *derivative of* f. It is denoted by f'. Once more: the derivative f' of a function f is also a function with rule given by

$$f'(x) = m_x,$$

where m_x is the slope of the tangent to the graph of f at the point $(x, f(x))$. See Figure 5.15. The notation f' for the derivative of a function f was the later contribution of the French-Italian mathematician Luigi Lagrange.

Suppose that f is a linear function. Then it has the form $f(x) = mx + b$, with m and b constants. Its graph is a line with slope m. Take any point (x, y) on the graph. The tangent to the line is the line itself, so it follows that $m_x = m$. Therefore $f'(x) = m$ for all x. Suppose that $m = 0$. Then $f(x) = b$, and $f'(x) = 0$ for all x. For example, the derivative of $f(x) = \frac{1}{2}x + 1$ is $f'(x) = \frac{1}{2}$, the derivative of $f(x) = -3x + 6$ is $f'(x) = -3$, and the derivative of $g(x) = 5$ is $g'(x) = 0$.

We now turn to the derivatives of the basic functions listed in Section 5.3. The labels (a), (b), and so on, parallel those of Section 5.3.

(a) Consider $f(x) = x^2$. Since

$$\begin{aligned}\frac{f(x + \Delta x) - f(x)}{\Delta x} &= \frac{(x + \Delta x)^2 - x^2}{\Delta x} = \frac{x^2 + 2x\Delta x + (\Delta x)^2 - x^2}{\Delta x} \\ &= \frac{2x\Delta x + (\Delta x)^2}{\Delta x} = \frac{\Delta x(2x + \Delta x)}{\Delta x} \\ &= 2x + \Delta x,\end{aligned}$$

it follows that

$$f'(x) = \lim_{\Delta x \to 0} \frac{f(x+\Delta x) - f(x)}{\Delta x} = 2x.$$

(b) Let $y = f(x) = \sqrt{x} = x^{\frac{1}{2}}$. We will not compute $f'(x) = \lim_{\Delta x \to 0} \frac{f(x+\Delta x)-f(x)}{\Delta x}$ directly (but try if you know how to use to the conjugate of a radical expression), but instead refer to Section 5.2. By squaring both sides, we get $y^2 = x$. The graph of $f(x) = \sqrt{x} = x^{\frac{1}{2}}$ is the upper half of the graph of the parabola $x = y^2$. See Figure 5.13a. It was shown in Section 5.2, using Leibniz's tangent method, that the slope of the tangent to this parabola at any point $P = (x, y)$ is $m_P = \frac{1}{2y}$. If the point P is on the graph of f, then $y = x^{\frac{1}{2}}$ and hence $m_x = m_P = \frac{1}{2}x^{-\frac{1}{2}}$. Therefore $f'(x) = \frac{1}{2}x^{-\frac{1}{2}}$. What is the slope of the tangent to the graph of $f(x) = \sqrt{x} = x^{\frac{1}{2}}$ at the point $(2, \sqrt{2})$? Substituting $x = 2$ into $f'(x) = \frac{1}{2}x^{-\frac{1}{2}}$ tells us that this slope is $f'(2) = \frac{1}{2}2^{-\frac{1}{2}} = \frac{1}{2\sqrt{2}}$.

(c) Let $g(x) = x^3$. It was shown in Section 5.2, using Leibniz's tangent method, that the slope of the tangent to the curve $y = x^3$ at any point $P = (x, y)$ is $m_P = 3x^2$. It follows that $g'(x) = 3x^2$.

(d) The derivative of $g(x) = x^{\frac{1}{3}}$ is $g'(x) = \frac{1}{3}x^{-\frac{2}{3}}$. This follows as in some of the earlier cases by applying Leibniz's tangent method to the equation $y^3 = x$.

(e) The derivative of $h(x) = \frac{1}{x} = x^{-1}$ is $h'(x) = -\frac{1}{x^2} = -x^{-2}$. This can be done with Leibniz's tangent method, but it also follows directly from the definition of $h'(x)$. By use of some algebra,

$$\frac{h(x+\Delta x) - h(x)}{\Delta x} = \frac{\frac{1}{x+\Delta x} - \frac{1}{x}}{\Delta x} = \frac{\frac{x-(x+\Delta x)}{(x+\Delta x)x}}{\Delta x} = \frac{1}{\Delta x}\left(\frac{-\Delta x}{(x+\Delta x)x}\right) = \frac{-1}{(x+\Delta x)x}.$$

By pushing Δx to zero, $h'(x) = \lim_{\Delta x \to 0} \frac{h(x+\Delta x)-h(x)}{\Delta x} = -\frac{1}{x^2}$.

(f) The derivative of $h(x) = \frac{1}{x^2} = x^{-2}$ is $h'(x) = -2\frac{1}{x^3} = -2x^{-3}$. Do this by computing the limit $h'(x) = \lim_{\Delta x \to 0} \frac{h(x+\Delta x)-h(x)}{\Delta x}$ directly. What about the slope of the tangent to the graph of $h(x) = \frac{1}{x^2}$ at $(-1, 1)$? This slope is computed by evaluating $h'(x) = -2x^{-3}$ at $x = -1$. Hence it is $h'(-1) = (-2)(-1) = 2$.

Leibniz introduces the notation $\frac{dy}{dx}$ and $\frac{d}{dx}f$ for the derivative of $y = f(x)$. He also develops many of the basic properties of the derivative, including the constant, sum, and difference rules. The following example illustrates all three. Let $f(x) = \sqrt{x} = x^{\frac{1}{2}}$, $g(x) = x^3$, and $h(x) = \frac{1}{x} = x^{-1}$. By Examples (b), (c), and (e) above, the derivative of the function $f(x) - 2g(x) + \frac{1}{3}h(x)$ is equal to

$$\begin{aligned}\frac{d}{dx}(f(x) - 2g(x) + \tfrac{1}{3}h(x)) &= f'(x) - 2g'(x) + \tfrac{1}{3}h'(x) = \tfrac{1}{2}x^{-\frac{1}{2}} - 2\cdot 3x^2 + \tfrac{1}{3}\cdot(-1)x^{-2} \\ &= \tfrac{1}{2}x^{-\frac{1}{2}} - 6x^2 - \tfrac{1}{3}x^{-2}.\end{aligned}$$

Consider a function f and its graph. Suppose that the graph is in one connected piece, without any breaks or gaps. A function with such a graph is called *continuous*. Figure 5.16 provides an example. Notice that the graph of f is smooth with nonvertical tangents from its beginning on the left up to $x = b_1$. So the function is differentiable for $x < b_1$. For the x-coordinates a_1, a_2, and a_3, the tangents to the graph are horizontal so that $f'(a_1) = f'(a_2) = f'(a_3) = 0$. For the coordinates b_1, b_2, and b_3, the graph is not smooth, so that f is not differentiable at any of these values. At $x = b_4$ the graph is smooth and has a tangent. But the tangent is vertical, so that f is not differentiable at b_4. The function f is differentiable everywhere, except at $x = b_1, b_2, b_3$, and b_4. Observe that the graph has "maximal point" for $x = a_2, a_4$, and $x = b_3$, and "minimal points" for $x = a_3$ and b_1. Notice that if a maximal point or a minimal point occurs for a value x, then either the graph of f has a horizontal tangent at $(x, f(x))$, or it is not smooth at $(x, f(x))$. Put another way, if f has a maximum or minimum value at the point $(x, f(x))$, then either $f'(x) = 0$ or $f'(x)$ is not defined.

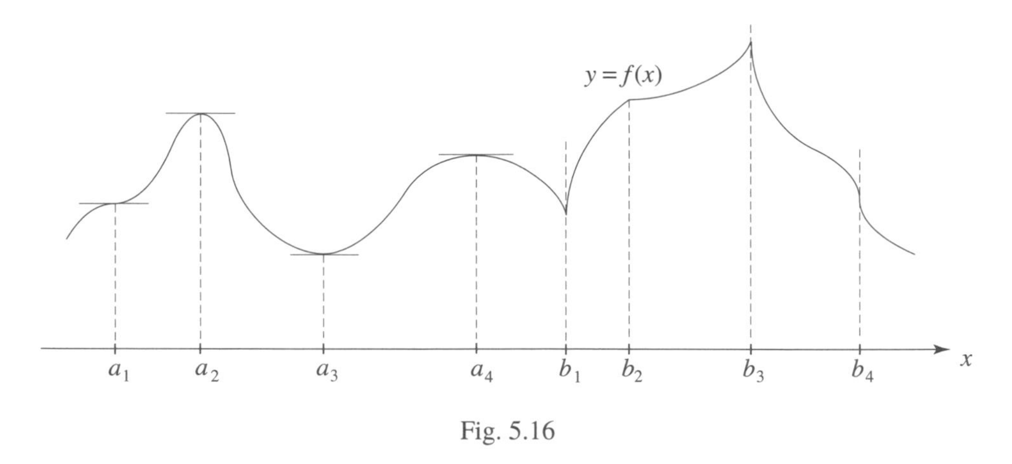

Fig. 5.16

Let $y = f(x)$ be a function that is differentiable for all x satisfying $a < x < b$, where $a < b$ are constants. Suppose that $f'(x) > 0$ for all such x. Then for each x with $a < x < b$, the tangent line to the graph of f at $(x, f(x))$ has a positive slope. This means that the graph rises through any such x—Figure 5.17 illustrates

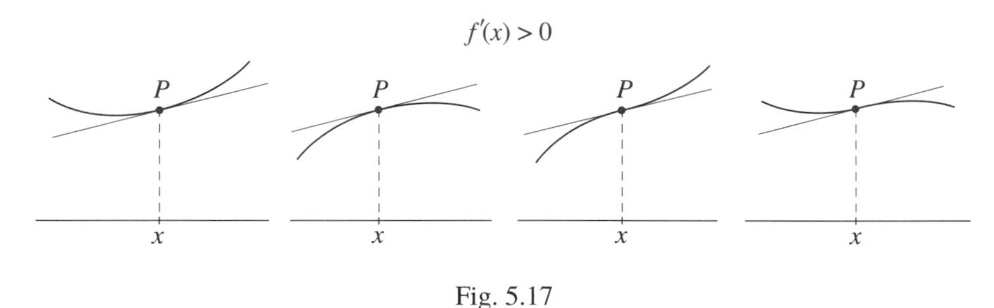

Fig. 5.17

the different possibilities—and therefore that the graph of f rises over the interval $a < x < b$. Similarly, if $f'(x) < 0$ for all x with $a < x < b$, then the graph of f falls over the interval $a < x < b$. The following criterion is a direct consequence of what we have observed.

Max-Min Criterion. (1) If there is a number c with $a < c < b$ such that $f'(x) > 0$ for all x with $a < x < c$ and $f'(x) < 0$ for all x with $c < x < b$, then the graph of f rises to the left of c and falls to the right of c, so that f reaches a maximum value at c.

(2) Alternatively, if there is a number c with $a < c < b$ such that $f'(x) < 0$ for all x with $a < x < c$ and $f'(x) > 0$ for all x with $c < x < b$, then the graph of f falls to the left of c and rises to the right of c, and f has a minimum value at c.

The point $c = a_2$ of Figure 5.16 illustrates the first case of the criterion, and the point $c = a_3$ illustrates the second.

The next section applies the analysis just described to problems that Fermat and Kepler solved long before Newton and Leibniz developed their calculus. Fermat's approach anticipated a strategy of calculus. Kepler's lengthy and verbal geometric methods did this too.

5.5 FERMAT, KEPLER, AND WINE BARRELS

We have already met Pierre de Fermat (1601-1665) as one of the important contributors to coordinate geometry. We will see in a moment that he also used some of the principles of calculus in the problems that he tackled (and we will see later in Chapter 8 that he articulated one of the basic principles of the science of optics). Today, he is perhaps most famous for keeping other mathematicians busy and scratching their heads. We saw (in 4D of the Problem and Projects segment of the previous chapter) that Fermat confounded generations of mathematicians with his "last theorem." He also challenged his contemporaries with mathematical questions. One of them was the Italian scientist Evangelista Torricelli (1608-1647), best known for the invention of the barometer, but also a brilliant contributor to mathematics and physics. He assisted Galileo's scientific efforts during the last months of the old master's life. The Grand Duke of Tuscany appointed Torricelli to succeed Galileo as professor of mathematics at the Florentine Academy. Fermat's problem for Torricelli? Determine for any triangle ΔABC the point P with the property that the sum of the distances $PA + PB + PC$ is a minimum. How Torricelli solved this difficult problem is described in Problem 5.45 of the Problems and Projects section.

We'll now turn to calculus and Fermat's early use of the fact that at a maximum or minimum value of a smooth curve, the slope of the tangent to the curve is zero. In the 1630s, he considers a rectangle with sides x and y. The perimeter p of the rectangle is $p = 2x + 2y$. Fermat asks the question: Of all the rectangles with the same perimeter p, which has the largest area? Figure 5.18 shows three of the many (infinitely many) rectangles with the same perimeter p. Fermat's answer proceeds as follows. The area of

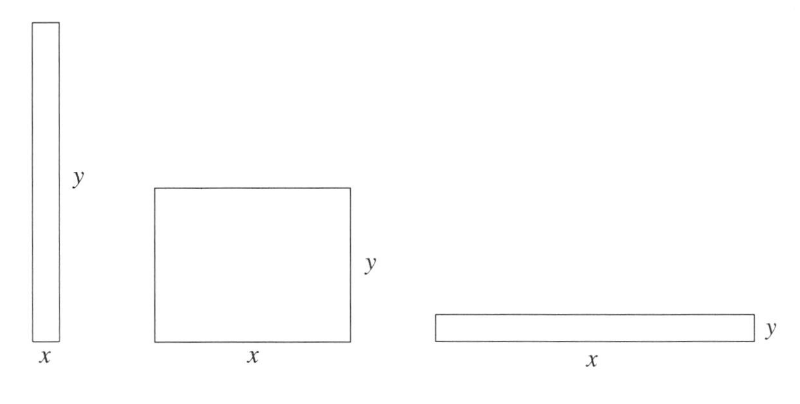

Fig. 5.18

the rectangle is $A = xy$. Since $2x + 2y = p$, we know that $y = \frac{1}{2}(p - 2x)$, and therefore that

$$A = xy = x\tfrac{1}{2}(p - 2x) = \tfrac{1}{2}px - x^2.$$

Since $A(x)$ is a function of the one variable x, the considerations of the previous section apply. By properties established there, $A'(x) = \frac{1}{2}p - 2x = 2(\frac{p}{4} - x)$. The value $x = \frac{p}{4}$ is the only one for which $A'(x) = 0$. Note that $A'(x) > 0$ for $x < \frac{p}{4}$ and $A'(x) < 0$ whenever $x > \frac{p}{4}$. So the graph of $A(x)$ rises for $x < \frac{p}{4}$ and falls for $x > \frac{p}{4}$, and it follows that the function $A(x)$ attains its maximum value for $x = \frac{p}{4}$. The side y of the rectangle of maximal area is $y = \frac{1}{2}(p - 2x) = \frac{1}{2}(p - \frac{p}{2}) = \frac{p}{4}$. It follows that the rectangle of maximal area with a given perimeter is the square.

Kepler considered related questions already in 1613. He sets the stage for one of his studies by recalling the occasion of his second marriage.

> At the time I celebrated my second wedding in November 1613, there were wine barrels piled up on the banks of the Danube near Linz [in Austria]. It had been a rich harvest and the barrels could be bought at a reasonable price. It was the duty of the new husband and caring father of the family to supply his house with ample drink. After several barrels were already in my cellar, the wine merchant came to measure the volume of the barrels. He measured all the barrels in the same way without any regard for their shape and without any additional reflection or calculation. He inserted a rod diagonally through the central hole [through which the barrel is filled or emptied] until its tip reached the far inner edge of the barrel. He then read off a number from the markings on the rod at the central hole that represented the volume of the barrel. I was puzzled that this method could provide an accurate measure of the

contents. It seemed to me that as a newlywed it was reasonable that I should test this method with geometric principles and to bring to light any relevant laws.

Kepler clearly thought it to be consistent with his status as newly married man to let himself be absorbed by mathematical studies. But now to his wine barrels. The way wine merchants in Austria (from about a year before his second marriage to just about the end of his life, Kepler resided in the city of Linz, the regional capital of Upper Austria) measured the volume of the barrels is illustrated in Figure 5.19. As Kepler described, they would position the barrel on its side with its central opening at the top, insert a measuring rod into it, and read off the distance from there to the farthest reaches at the bottom edge of the

Fig. 5.19. Measuring the volume of a wine barrel in 1485 (*left*) and in 1531 (*right*).

barrel. In Figure 5.20b, this is the distance s. For a measured distance s, there is a formula that then provides the official assessment of the volume of the barrel. In this way, the distance s determined the cost of a full barrel of wine. Kepler's publication *A New Stereometry of Barrels* of 1615 studied (among other things) whether this method of assessing the volume of a barrel was fair. Kepler's arguments consisted of lengthy, verbal discussions about the geometries involved. What we present here is how Leibniz might have analyzed the matter, and we'll reformulate Kepler's analysis accordingly.

Kepler devoted a good part of his study to cylindrical barrels. Suppose that a cylinder has height h, that its circular cross section has radius r, and that the diagonal through the entire barrel has length d. See Figure 5.20a. A question that Kepler considered is the following. For a fixed d, what is the shape of the cylindrical barrel—that is to say, what is the relationship between r and h—that provides it with the largest possible volume? Put another way, of all cylinders that have the same diagonal d, what is the shape of the one that has the largest volume? He answered as follows: of all cylinders with equal diagonals d, the one that has the largest volume has its height h over its circular diameter $2r$ in the ratio of 1 to $\sqrt{2}$.

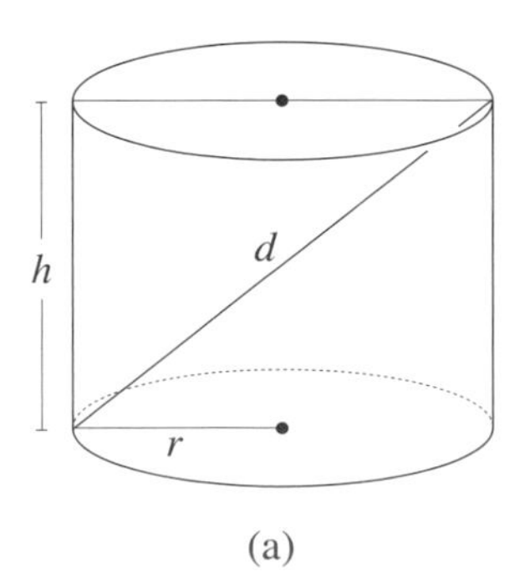

(a)

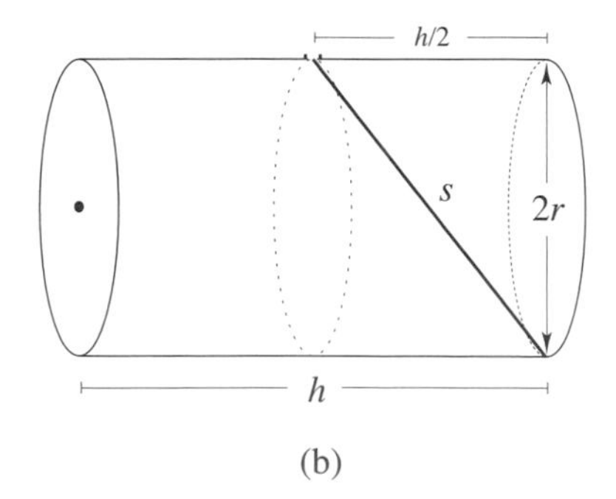

(b)

Fig. 5.20

We'll establish this fact with a much more efficient argument than Kepler's. Applying Pythagoras's theorem to Figure 5.20a tells us that $d^2 = (2r)^2 + h^2$. So $4r^2 = d^2 - h^2$, and hence $r^2 = \frac{1}{4}(d^2 - h^2)$. The volume V of the cylinder is equal to the area of its base πr^2 times its height h. Therefore

$$V = \pi r^2 h = \tfrac{\pi}{4}(d^2 - h^2)h = \tfrac{\pi}{4}d^2 h - \tfrac{\pi}{4}h^3.$$

Since we are taking d as fixed, the volume V is a function of h. By applying what we learned in the previous section,

$$V'(h) = \tfrac{\pi}{4}d^2 - \tfrac{3\pi}{4}h^2 = \tfrac{\pi}{4}(d^2 - 3h^2).$$

The derivative $V'(h)$ is equal to zero when $d^2 = 3h^2$, or $h^2 = \frac{1}{3}d^2$. Because $h > 0$, this means that $h = \frac{1}{\sqrt{3}}d$. Observe that $V'(h) = \frac{\pi}{4}(d^2 - 3h^2)$ is positive when $h < \frac{d}{\sqrt{3}}$ and negative when $h > \frac{d}{\sqrt{3}}$. It follows that the graph of $V(h)$ rises for $h < \frac{d}{\sqrt{3}}$ and falls for $h > \frac{d}{\sqrt{3}}$, as summarized in Figure 5.21. Therefore $V(h)$ attains

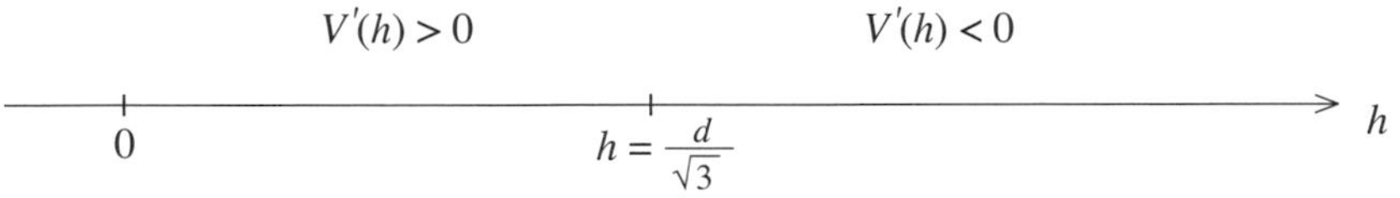

Fig. 5.21

its maximum value at $h = \frac{1}{\sqrt{3}}d$. For this h, $r^2 = \frac{1}{4}(d^2 - h^2) = \frac{1}{4}(3h^2 - h^2) = \frac{1}{2}h^2$. So $4r^2 = 2h^2$, hence $2r = \sqrt{2}h$, and therefore

$$\frac{h}{2r} = \frac{1}{\sqrt{2}}.$$

Precisely as Kepler had asserted, for a fixed diagonal d, the cylindrical barrel that has its height h over the diameter $2r$ of its circular cross section in the ratio of 1 to $\sqrt{2}$ is the one that has the maximum volume.

Kepler Cylinders. Let's turn to Kepler's primary concern: the wine merchant's method for measuring the volume of a barrel. The focus now is on Figure 5.20b and the length s. We'll start by asking Kepler's question again, but this time for s rather than d. For a given s, what is the shape of the cylindrical barrel that has the largest volume?

By the Pythagorean theorem, $s^2 = (2r)^2 + (\frac{h}{2})^2$. So $4r^2 = s^2 - \frac{h^2}{4}$, and hence $r^2 = \frac{1}{4}(s^2 - \frac{h^2}{4})$. Therefore the volume of the barrel is

$$V = \pi r^2 h = \pi\tfrac{1}{4}(s^2 - \tfrac{h^2}{4})h = \tfrac{\pi}{4}s^2 h - \tfrac{\pi}{16}h^3.$$

Since s is taken to be fixed, V is a function of h. Using facts from the previous section,

$$V'(h) = \tfrac{\pi}{4}s^2 - \tfrac{3\pi}{16}h^2 = \tfrac{\pi}{16}(4s^2 - 3h^2).$$

So $V'(h) = 0$ when $3h^2 = 4s^2$. Since $h > 0$, this implies that $h = \frac{2}{\sqrt{3}}s$. For h smaller than $\frac{2s}{\sqrt{3}}$, $V'(h) > 0$ and for h larger than $\frac{2s}{\sqrt{3}}$, $V'(h) < 0$. It follows that the volume $V(h)$ attains its largest value when $h = \frac{2}{\sqrt{3}}s$. This maximum value for the given s is

$$V = \pi r^2 h = \pi\tfrac{1}{4}(s^2 - \tfrac{h^2}{4})h = \tfrac{\pi}{4}(s^2 - \tfrac{1}{3}s^2)\tfrac{2}{\sqrt{3}}s = \tfrac{\pi}{4}(\tfrac{2}{3}s^2)\tfrac{2}{\sqrt{3}}s = \tfrac{\pi}{3\sqrt{3}}s^3.$$

For $h = \frac{2}{\sqrt{3}}s$, $r^2 = \frac{1}{4}(s^2 - \frac{h^2}{4}) = \frac{1}{4}(\frac{3}{4}h^2 - \frac{1}{4}h^2) = \frac{1}{8}h^2$. So $h^2 = 2 \cdot 4r^2$ and therefore

$$\frac{h}{2r} = \sqrt{2}.$$

So for a fixed s, the cylindrical barrel that has its height to the diameter of its circular cross section in the ratio of $\sqrt{2}$ to 1 is the one that has the maximum volume. Notice that this barrel shape is substantially different from the one before. In the first case, the diameter was larger than the height by a factor of $\sqrt{2}$. See Figure 5.22a. In the second situation, this is reversed. The height is larger than the diameter by a factor of $\sqrt{2}$. See Figure 5.22b. We'll refer to a cylinder of this shape as a *Kepler cylinder.*

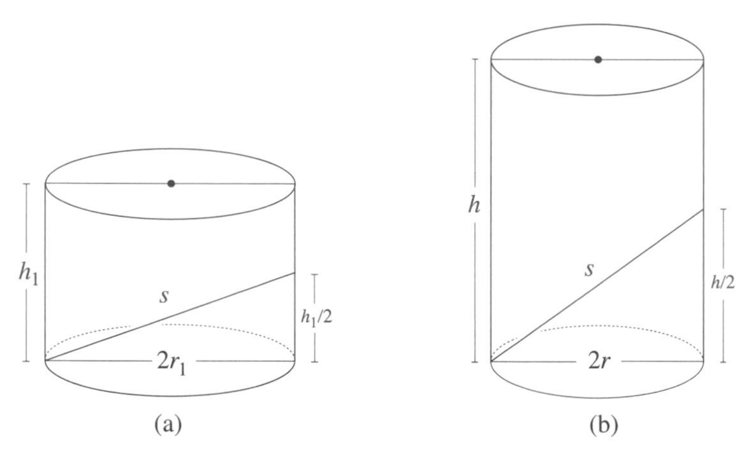

Fig. 5.22

Let's return to the method that the Austrian wine merchants used to determine the volume of *any* wine barrel. After they measured the diagonal s, they assessed the volume of the barrel with the rule

$$V_{\text{rule}} = 0.6s^3.$$

Where might this 0.6 have come from? Let's look at a barrel in the shape of a Kepler cylinder. We saw that its volume is equal to $V = \frac{\pi}{3\sqrt{3}}s^3$ in terms of s and that this is the maximum volume that a cylindrical barrel can have for a given s. Since $\frac{\pi}{3\sqrt{3}} \approx 0.6046 \approx 0.60$, the assessed $V_{\text{rule}} = 0.6s^3$ is a close approximation of this maximum volume. We know from our earlier analysis that a Kepler cylinder satisfies the condition that its height over the diameter of its circular cross section is in the ratio of $\sqrt{2}$ to 1. To repeat, for such barrels, the value $V_{\text{rule}} = 0.6s^3$ provides a very accurate assessment of their volumes. But what about the official assessment $V_{\text{rule}} = 0.6s^3$ for the volumes of barrels of different shapes? For these, the assessed volume can exceed the actual volume of the barrel by significant amounts. Since the price of the wine in the barrel is determined by this official estimate of its volume, this translates to a higher price for the customer and hence a larger profit for the merchant.

Let's apply the rule $V = 0.6s^3$ to the cylindrical barrel of Kepler's first analysis, as shown in Figure 5.22a. Its radius and height are r_1 and h_1, respectively, and its diagonal measure is s. We know that for this barrel, $\frac{\text{height}}{\text{diameter}} = \frac{h_1}{2r_1} = \frac{1}{\sqrt{2}}$. Combining $4r_1^2 = 2h_1^2$ with $4r_1^2 = s^2 - \frac{h_1^2}{4}$ (from Figure 5.22a), we get $\frac{9}{4}h_1^2 = s^2$. Therefore $h_1^2 = \frac{4}{9}s^2$, so that $h_1 = \frac{2}{3}s$. Using the volume formula $V_1 = \pi r_1^2 h_1$, we now get

$$V_1 = \pi \cdot \tfrac{1}{2}h_1^2 \cdot h_1 = \tfrac{\pi}{2}h_1^3 = \tfrac{\pi}{2}(h_1^2)^{\frac{3}{2}} = \tfrac{\pi}{2}(\tfrac{2^2}{3^2}s^2)^{\frac{3}{2}} = \tfrac{\pi}{2}(\tfrac{8}{27})s^3 = \tfrac{4\pi}{27}s^3 \approx 0.47s^3$$

for the actual volume of this barrel in terms of the measure s. The difference between the assessed and actual volumes is $V_{\text{rule}} - V_1 \approx 0.6s^3 - 0.47s^3 = 0.13s^3$. Since $\frac{0.13s^3}{0.6s^3} \approx 0.22$, the assessed volume is about 22% greater than the actual volume. The profit from this 22% lands in the merchant's pocket.

Kepler was fully aware that cylinders are only rough approximations of the wine barrels that were in use. See Figure 5.19. He therefore went on to refine his study. We'll return to Kepler's analysis and his model of the Austrian wine barrel later in the chapter. But we go first to develop the essential aspects of Leibniz's integral calculus.

5.6 THE DEFINITE INTEGRAL

Let's turn to the question of area. We certainly know what the area of a rectangle is and how to compute it, and we know how to compute the area of a parallelogram and a triangle. Making sense out of the meaning of area in the case of a region in the plane with curving boundaries, such as the one of Figure 5.23, is much

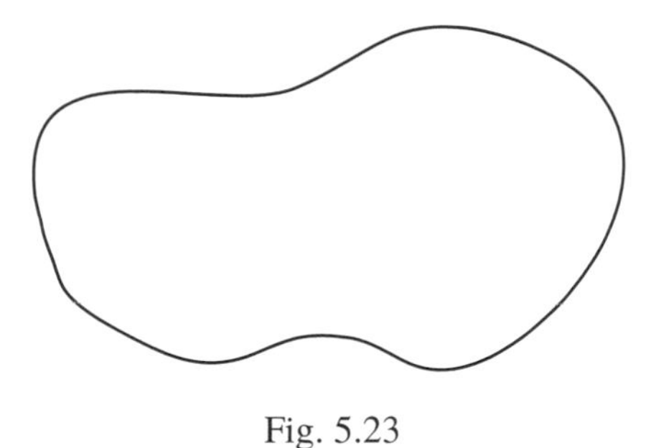

Fig. 5.23

more of a challenge. How does one go about assigning a number to it?

We have already seen how this question was approached in special cases. They are illustrated in Figures 5.24a and 5.24b. Pick a point inside the region, draw from it a web of lines going outward, and complete the scheme to a collection of triangles. If this is done with lots of extremely thin triangles, then the sum of the areas of all the triangles will essentially equal the area of the curved region. This is the way

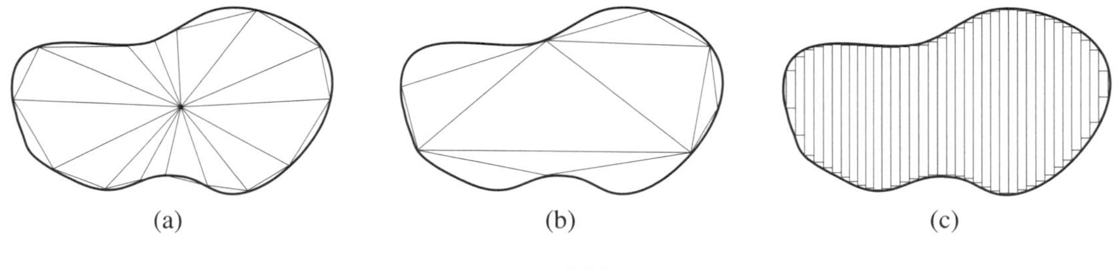

Fig. 5.24

the area of a circle was computed in Section 2.2. Another way is to inscribe a large triangle in the middle and then add more and more triangles to fill in the region. This is the method that Archimedes used to compute the area of the parabolic section. See Section 2.4. The rectangular Cartesian coordinate system suggests that the region can be filled out with fine vertical rectangles. The sum of their areas will also closely approximate the area of the region. See Figure 5.24c. The thinner the rectangular slices, the better the approximation will be. The idea of the definite integral begins with this approach to the computation of area.

In a manuscript written in October 1675, Leibniz considers a coordinate system and a curve $\boldsymbol{C}$, as shown in Figure 5.25a. We'll suppose that the curve is the graph of a function $y = f(x)$ defined for all x with $a \leq x \leq b$. We'll assume that the function is continuous, so that its graph flows in one piece over this interval without breaks or gaps. Leibniz selects a few points on the x-axis between a and b. In Figure 5.25b, seven such points are selected. They are evenly spaced and divide the interval into eight equal pieces each of length $dx = \frac{b-a}{8}$. If x is one of these points, then $x + dx$ is the next one. Along with the graph they determine a rectangle of area $f(x)\,dx$.

Figure 5.26a shows all of the rectangles that are obtained in this way. Observe that these rectangles, when taken together, fill out—in an approximate way—the region under the curve $\boldsymbol{C}$ over the segment from a to b. So the sum of their areas approximates the area under the graph (over the interval from a to b). Leibniz realizes that if more points are taken, then this sum of rectangles will represent the area under the graph more accurately. See Figure 5.26b. (In the figure the rectangles alternate in color to make them distinctly visible.) And he knows that if lots of points are selected between a and b and tightly packed so

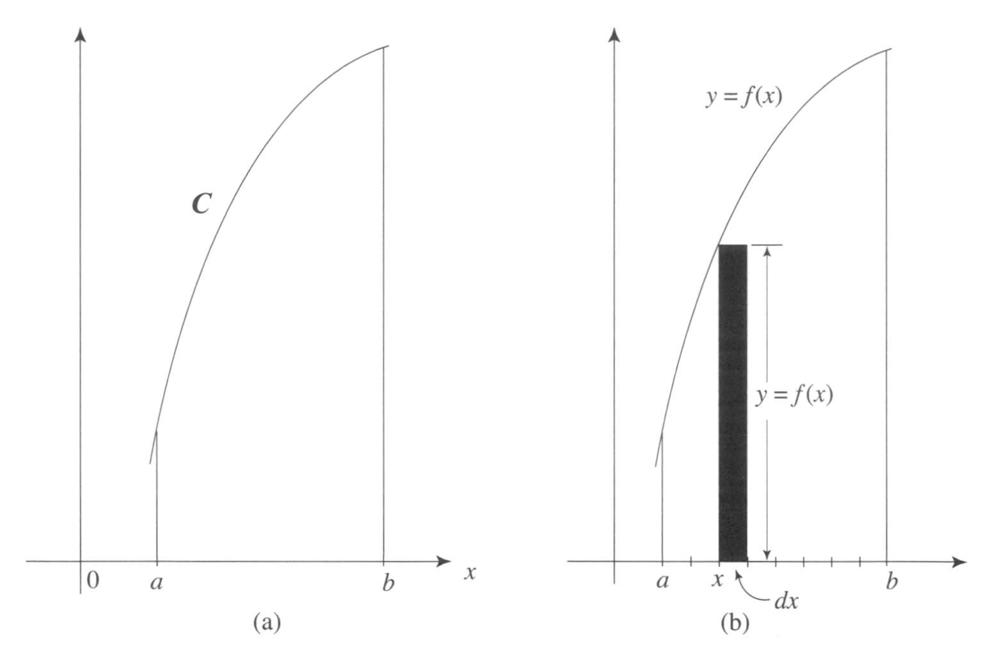

Fig. 5.25

that the gaps between successive points are all very small, then this sum of a large number of terms will very tightly approximate the area under the curve. So if n is a very large positive integer (large relative to

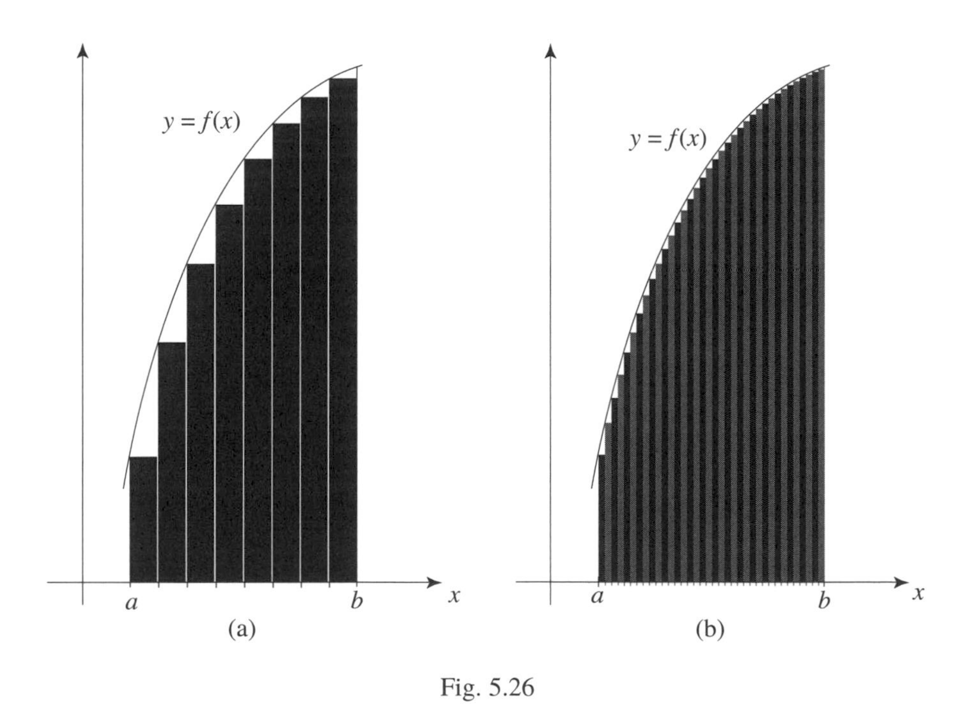

Fig. 5.26

the length $b - a$ of the interval), and if the interval from a to b is divided into n equal intervals by the $n - 1$ points

$$x_1 < x_2 < \cdots < x_{n-1},$$

then with $dx = \frac{b-a}{n}$, the sum

$$f(a)\,dx + f(x_1)\,dx + f(x_2)\,dx + \cdots + f(x_{n-1})\,dx$$

of the areas of the resulting rectangles is a close approximation of the area under the graph of $y = f(x)$. (Following the convention of Western cultures, we have proceeded from left to right.)

After various experiments with notation, Leibniz finally uses an elongated S (Latin is the scientific language of his day, and the Latin word for "Sum" is *Summa*) and writes $\int f(x)\,dx$ to denote the sum of all the $f(x) \cdot dx$ determined by the very tightly packed set of points between a and b. In 1822, the French mathematician Fourier inserted a and b into the notation—to indicate that the terms run from $x = a$ to $x = b$. The resulting

$$\int_a^b f(x)\,dx$$

is the modern shorthand notation for this sum. Like no other mathematician before him, Leibniz realized the fundamental importance of notation:

> With the notation one has to consider the fact that it should be convenient as regards the process of invention. This is especially the case if it expresses the innermost nature of the thing with economy. In this way, the effort involved in the thought process is minimized in a wonderful way.

To this end, he engaged in notational "experiments." For instance, he had originally written $\frac{d}{x}$ for dx and used "omn." (from the Latin *omnes*, meaning "all") to denote the sum. (You will appreciate the importance of notation if you try to multiply two large numbers written in Greek or Roman numerals.) The symbol $\int_a^b f(x)\,dx$ has a name. It is called *the definite integral of the function $f(x)$ from a to b.*

It is important to think of the symbol $\int_a^b f(x)\,dx$ as more than just a number that provides the area under the curve. Include in your thinking the process that produces this number. This starts with a very large n relative to the distance $b - a$, forms the very small $dx = \frac{b-a}{n}$, and includes the points $a < x_1 < x_2 < \cdots < x_{n-1} < b$ that subdivide the interval $a \le x \le b$ and determine the sum

$$f(a)\,dx + f(x_1)\,dx + f(x_2)\,dx + \cdots + f(x_{n-1})\,dx$$

(of the very large number n) of terms $f(x)\,dx$. This sum makes sense independently of its interpretation as area. Such sums arise in many disciplines of mathematics, science, and engineering. They can represent volume, the length of a curve, or the area of a surface. In physics and engineering, they can represent fundamental concepts such as force, energy, momentum, and moment of force.

One final comment needs to be made. No matter how large the positive integer n that has been chosen, a larger n can always be taken, and the summation process can be repeated again, and again, each time with a larger n and hence with a smaller dx. The definite integral

$$\int_a^b f(x)\,dx$$

is in fact the limit of a sequence of numbers obtained in this way. This completes the definition of the definite integral. A comparison of Figures 26a and 26b (and our imagination) tells us that in the context of areas, this limit is no longer an approximation, but it is *equal to the area* under the graph of the function $y = f(x)$. So the definition of the definite integral as a long sum of terms of the form $f(x)dx$ is only a "working" definition. However, this working definition is the way to think, and the value it provides approximates the limiting value in the same way that, say, 11.999999999 approximates 12.

It's time to consider a concrete example. Consider the circle with center $(8, 0)$ and radius 6. Its equation is $(x-8)^2 + y^2 = 6^2 = 36$. Solving for y, we get $y^2 = 36 - (x-8)^2$ and hence $y = \pm\sqrt{36-(x-8)^2}$. For a positive y, we get $y = \sqrt{36-(x-8)^2}$. It follows that $f(x) = \sqrt{36-(x-8)^2} = \sqrt{-x^2+16x-28}$ defines a function that has the upper half of the circle as its graph. Consider this function for x restricted to $2 \le x \le 8$. This part of the graph of the circle is depicted in Figure 5.27.

Let's take $n = 12$ and divide the interval from $x = 2$ to $x = 8$ into 12 equal pieces each of length

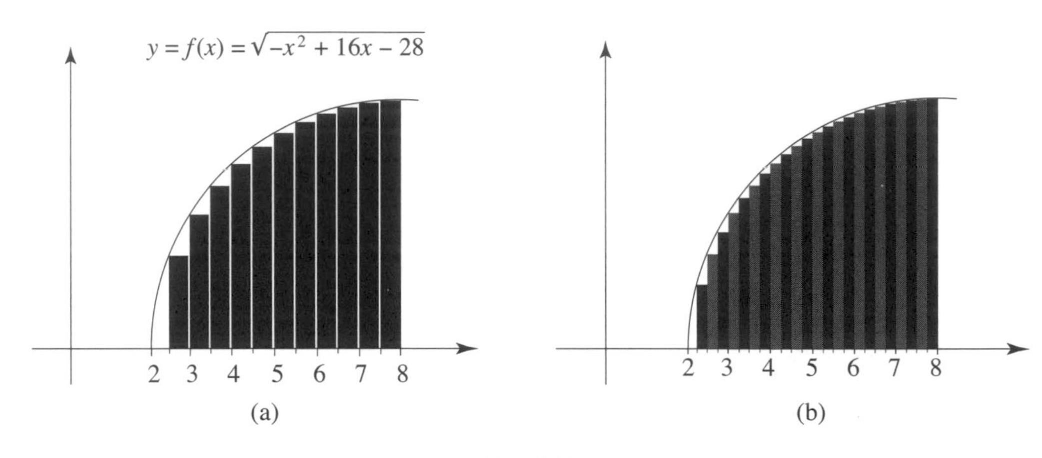

Fig. 5.27

$dx = \frac{8-2}{12} = 0.5$ and turn to Figure 5.27a. The points that divide the interval are

$$2,\ 2.5,\ 3,\ 3.5,\ 4,\ 4.5,\ 5,\ 5.5,\ 6,\ 6.5,\ 7,\ 7.5,\ 8.$$

The sum of the areas of the 12 rectangles (the first has height equal to 0) depicted in the figure is

$$\begin{aligned}
&f(2)\cdot 0.5 + f(2.5)\cdot 0.5 + f(3)\cdot 0.5 + f(3.5)\cdot 0.5 + f(4)\cdot 0.5 + f(4.5)\cdot 0.5 \\
&\qquad + f(5)\cdot 0.5 + f(5.5)\cdot 0.5 + f(6)\cdot 0.5 + f(6.5)\cdot 0.5 + f(7)\cdot 0.5 + f(7.5)\cdot 0.5 \\
&= [f(2) + f(2.5) + f(3) + f(3.5) + f(4) + f(4.5) + f(5) + f(5.5) + f(6) + f(6.5) + f(7) + f(7.5)]\cdot 0.5 \\
&\approx [0 + 2.40 + 3.32 + 3.97 + 4.47 + 4.87 + 5.20 + 5.45 + 5.66 + 5.81 + 5.92 + 5.98]0.5 \\
&= (53.05)(0.5) \approx 26.53.
\end{aligned}$$

Let's next take $n = 24$ and $dx = \frac{8-2}{24} = 0.25$ and look at Figure 5.27b. To make the 24 rectangles distinctly visible, they alternate in color between black and gray. The sum of their areas is

$$\begin{aligned}
&f(2)\cdot 0.25 + f(2.25)\cdot 0.25 + f(2.50)\cdot 0.25 + f(2.75)\cdot 0.25 + f(3)\cdot 0.25 + f(3.25)\cdot 0.25 + \cdots \\
&\qquad + f(6.75)\cdot 0.25 + f(7)\cdot 0.25 + f(7.25)\cdot 0.25 + f(7.5)\cdot 0.25 + f(7.75)\cdot 0.25.
\end{aligned}$$

The approximate numerical values of these terms need to be computed next. The inclusion of all of them does not add clarity, so that most of these computations are omitted in the sum that follows:

$$\approx [0 + 1.71 + 2.40 + 2.90 + 3.32 + 3.67 + \cdots + 5.87 + 5.92 + 5.95 + 5.98 + 5.99]0.25 \approx 27.44.$$

Neither of the two sums that have been calculated is equal to

$$\int_a^b f(x)\,dx = \int_2^8 \sqrt{-x^2+16x-28}\,dx$$

because in each case the number n that has been selected to illustrate the process is far to small and the resulting dx far too large. The definite integral

$$\int_2^8 \sqrt{-x^2 + 16x - 28}\, dx$$

is obtained by selecting a huge n so that the points between 2 and 8 are very tightly packed and all the dx are very small. The precise value of the integral is the area under the upper half of the circle from 2 to 8. Since this is a quarter of a circle of radius 6, we can conclude that

$$\int_2^8 \sqrt{-x^2 + 16x - 28}\, dx = \tfrac{1}{4}\pi \cdot 6^2 = 9\pi \approx 28.27.$$

The Italian Bonaventura Cavalieri (1598–1644) contributed to the study of geometry, trigonometry, optics, and astronomy. He was a student of Galileo, was drawn into the study of the motion of projectiles, and became professor of mathematics at the University of Bologna. He discovered a comparative method of computing areas that is directly related to our discussion. We will take it up next.

5.7 CAVALIERI'S PRINCIPLE

Cavalieri considered two areas A_1 and A_2 that are positioned and aligned as shown in Figure 5.28a. For every x between the two numbers a and b, the lengths $d_1(x)$ and $d_2(x)$ of the vertical cross-sectional cuts

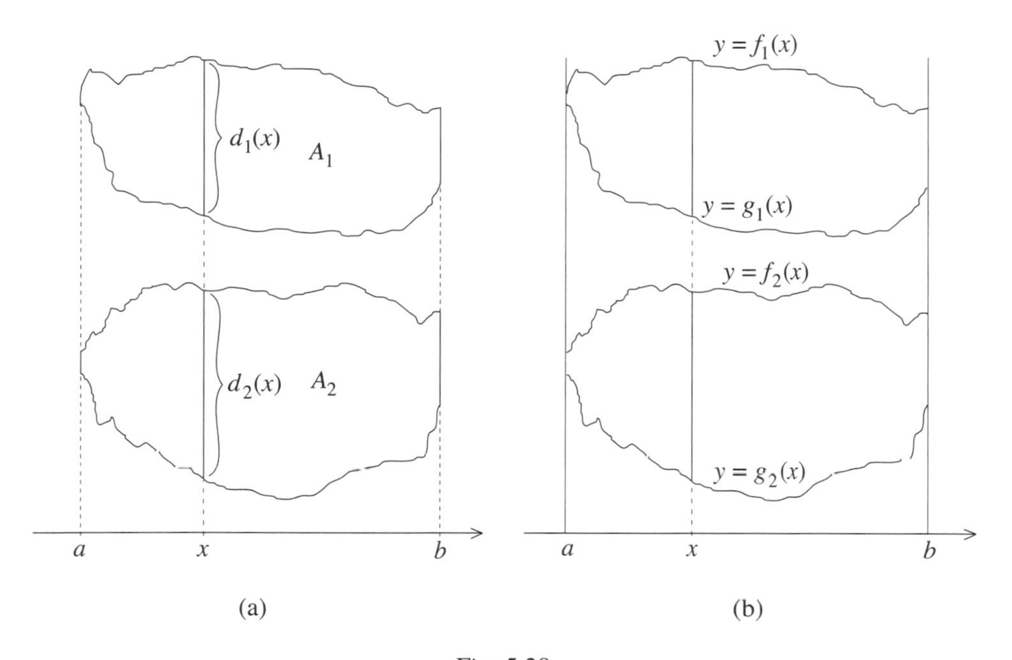

Fig. 5.28

through A_1 and A_2 are related by the assumption that $d_2(x) = cd_1(x)$ for all x and some positive constant c. (In the figure, $c = 1.25$.) Cavalieri recognized the following principle as self-evident.

Cavalieri's Principle. If $d_2(x) = cd_1(x)$ for all x with $a \leq x \leq b$ and a fixed positive constant c, then $A_2 = cA_1$.

Observe, in particular, that if the area A_1 and the constant c are both known, then the area A_2 can be computed.

We will now see that Cavalieri's principle is a consequence of Leibniz's calculus. Let $f(x)$ and $g(x)$ be two functions defined on an interval of the from $a \le x \le b$, and let c be any constant with $c \ge 0$. From the definition of the definite integral as a sum, in combination with the equalities

$$(cf(x))dx = c \cdot f(x)dx \quad \text{and} \quad ((f(x) \pm g(x))dx = f(x)dx \pm g(x)dx$$

that hold for any x and any dx, it follows that

$$\int_a^b cf(x)\,dx = c\int_a^b f(x)\,dx \quad \text{and} \quad \int_a^b (f(x) \pm g(x))\,dx = \int_a^b f(x)\,dx \pm \int_a^b g(x)\,dx.$$

For example,

$$\int_3^7 \sqrt{7}(x^3 + 2x^2)\,dx = \sqrt{7}\int_3^7 (x^3 + 2x^2)\,dx = \sqrt{7}\int_3^7 x^3\,dx + 2\sqrt{7}\int_3^7 x^2\,dx.$$

Let's return to Figure 5.28a. The fact that for each of the two regions any vertical line through x with $a \le x \le b$ goes through the upper boundary as well as the lower boundary at exactly one point means that both upper boundaries and both lower boundaries are graphs of functions. This conclusion is highlighted in Figure 5.28b. It follows from this fact and the figure that

$$A_1 = \int_a^b f_1(x)\,dx \;-\; \int_a^b g_1(x)\,dx = \int_a^b (f_1(x)\,dx \;-\; g_1(x))\,dx = \int_a^b d_1(x)\,dx \quad \text{and}$$

$$A_2 = \int_a^b f_2(x)\,dx \;-\; \int_a^b g_2(x)\,dx = \int_a^b (f_2(x)\,dx \;-\; g_2(x))\,dx = \int_a^b d_2(x)\,dx.$$

Since $d_2(x) = cd_1(x)$ for all x,

$$A_2 = \int_a^b d_2(x)\,dx = \int_a^b cd_1(x)\,dx = c\int_a^b d_1(x)\,dx = cA_1,$$

as asserted by Cavalieri's principle. As an application of Cavalieri's principle, we will compute the area of the ellipse

$$\frac{x^2}{a^2} + \frac{y^2}{b^2} = 1$$

with semimajor axis $a > 0$ and semiminor axis $b > 0$. Solving this equation for y, we get $\frac{y^2}{b^2} = 1 - \frac{x^2}{a^2} = \frac{a^2 - x^2}{a^2}$. So $y^2 = \frac{b^2(a^2-x^2)}{a^2} = \frac{b^2}{a^2}(a^2 - x^2)$, and $y = \pm\frac{b}{a}\sqrt{a^2 - x^2}$. Therefore $y = \frac{b}{a}\sqrt{a^2 - x^2}$ is an equation of the upper half of the ellipse. In the same way, $y = \sqrt{a^2 - x^2}$ is an equation of the upper half of the circle $x^2 + y^2 = a^2$. Figure 5.29 illustrates these observations. Since the radius of the circle is a, the area of the

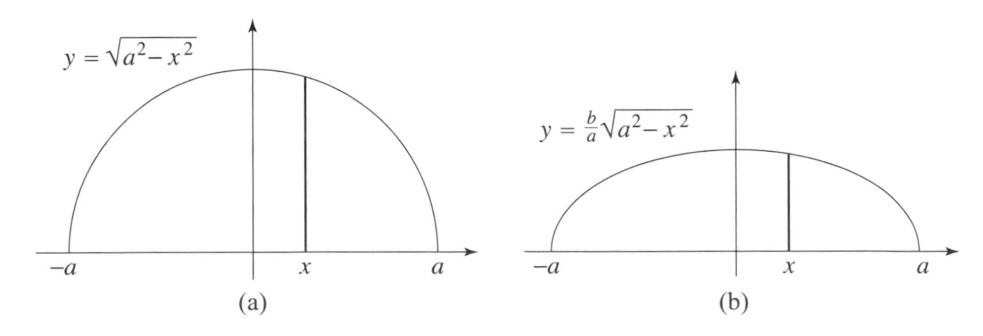

Fig. 5.29

upper half of the circle is $A_1 = \frac{1}{2}\pi a^2$. Let $d_1(x) = \sqrt{a^2 - x^2}$ and $d_2(x) = \frac{b}{a}\sqrt{a^2 - x^2}$. With $c = \frac{b}{a}$, we see that $d_2(x) = cd_1(x)$ for all x with $-a \le x \le a$. So by Cavalieri's principle the area A_2 of the upper half of the ellipse is $A_2 = \frac{b}{a}A_1 = \frac{b}{a} \cdot \frac{1}{2}\pi a^2 = \frac{1}{2}\pi ab$. Therefore

$$ab\pi$$

is the area of the full ellipse with semimajor axis a and semiminor axis b.

5.8 DIFFERENTIALS AND THE FUNDAMENTAL THEOREM

Now that we know what $\displaystyle\int_a^b f(x)\,dx$ means, we'll develop a strategy for calculating it. Going through the process of adding myriad tiny terms would be laborious at best and impossible at worst. So the question is: Is there a viable strategy for computing this number? It turns out that differential calculus plays a central role.

Suppose that two functions $y = f(x)$ and $y = g(x)$ have the same derivative. So $g'(x) - f'(x) = 0$. This means that the slope of the tangent to the graph of $y = g(x) - f(x)$ at any point is zero. So all its tangents are horizontal. If the graph of $g(x) - f(x)$ were to curve upward or downward at any point, the tangent line would either rise or fall. See Figure 5.17, for example. Since this is not the case, the only possibility is that the graph of $g(x) - f(x)$ is a horizontal line. Because every horizontal line has the form $y = C$ for some constant C, it follows that $g(x) - f(x) = C$ for all x. Therefore

$$g(x) = f(x) + C.$$

Let a function f be given. A function F that has the property that its derivative F' is equal to f is called an *antiderivative* of f. It follows from what has just been verified that if $y = F(x)$ is an antiderivative of a function, then any other antiderivative of the function has the form $F(x) + C$ for some constant C.

Leibniz's original notation $\displaystyle\int f(x)\,dx$ is used for the generic antiderivative—also called *indefinite integral*—of f. The fact that any antiderivative of f is given by a fixed antiderivative F plus a constant is expressed by the equality

$$\boxed{\int f(x)\,dx = F(x) + C}$$

Example 5.8. We learned in Section 5.4 that $\frac{d}{dx}x^2 = 2x$, $\frac{d}{dx}x^3 = 3x^2$, $\frac{d}{dx}x^{\frac{3}{2}} = \frac{3}{2}x^{\frac{1}{2}}$, and $\frac{d}{dx}x^{-1} = -x^{-2}$. It follows therefore that

$$\int x\,dx = \tfrac{1}{2}x^2 + C,\ \int x^2\,dx = \tfrac{1}{3}x^3 + C,\ \int x^{\frac{1}{2}}\,dx = \tfrac{2}{3}x^{\frac{3}{2}} + C, \text{ and } \int x^{-2}\,dx = -x^{-1} + C,$$

with C a constant in each case.

Let a function $y = f(x)$ be given. Let $P = (x, f(x))$ be a point on the graph. Suppose that the graph has a nonvertical tangent at P. We know that the slope of this tangent is $f'(x)$. Refer to Figure 5.30. Take a horizontal segment of any length dx. Use it and the tangent to form the right triangle with base dx shown in the figure. Let dy be its height. Leibniz calls this triangle a *characteristic triangle*. Since the slope of the tangent is also equal to $\frac{dy}{dx}$, we know that $\frac{dy}{dx} = f'(x)$. Therefore

$$dy = f'(x) \cdot dx.$$

Leibniz calls dx the *differential of* x and the related quantity $dy = f'(x) \cdot dx$ the *differential of* y. For example, in the case of the parabola $y = x^2$, the slope of the tangent is $f'(x) = 2x$, so that $dy = 2x \cdot dx$. In the case of the function $y = x^3$, the slope of the tangent is $f'(x) = 3x^2$ and $dy = 3x^2 \cdot dx$.

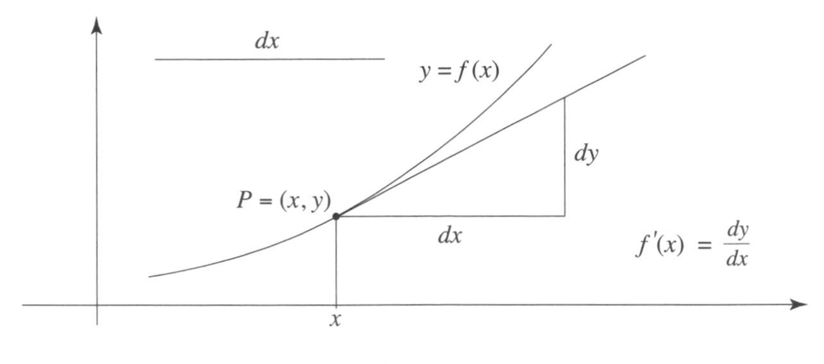

Fig. 5.30

Suppose that the function $y = f(x)$ is differentiable for all x in an interval $a \leq x \leq b$, and let $\boldsymbol{C}$ be its graph. Now let $F(x)$ be any antiderivative of f. So $F'(x) = f(x)$ for all x with $a \leq x \leq b$. Let $\boldsymbol{A}$ be the graph of $y = F(x)$. Set $c = F(a)$ and $d = F(b)$. We'll consider $y = f(x)$ and $y = F(x)$ and their graphs

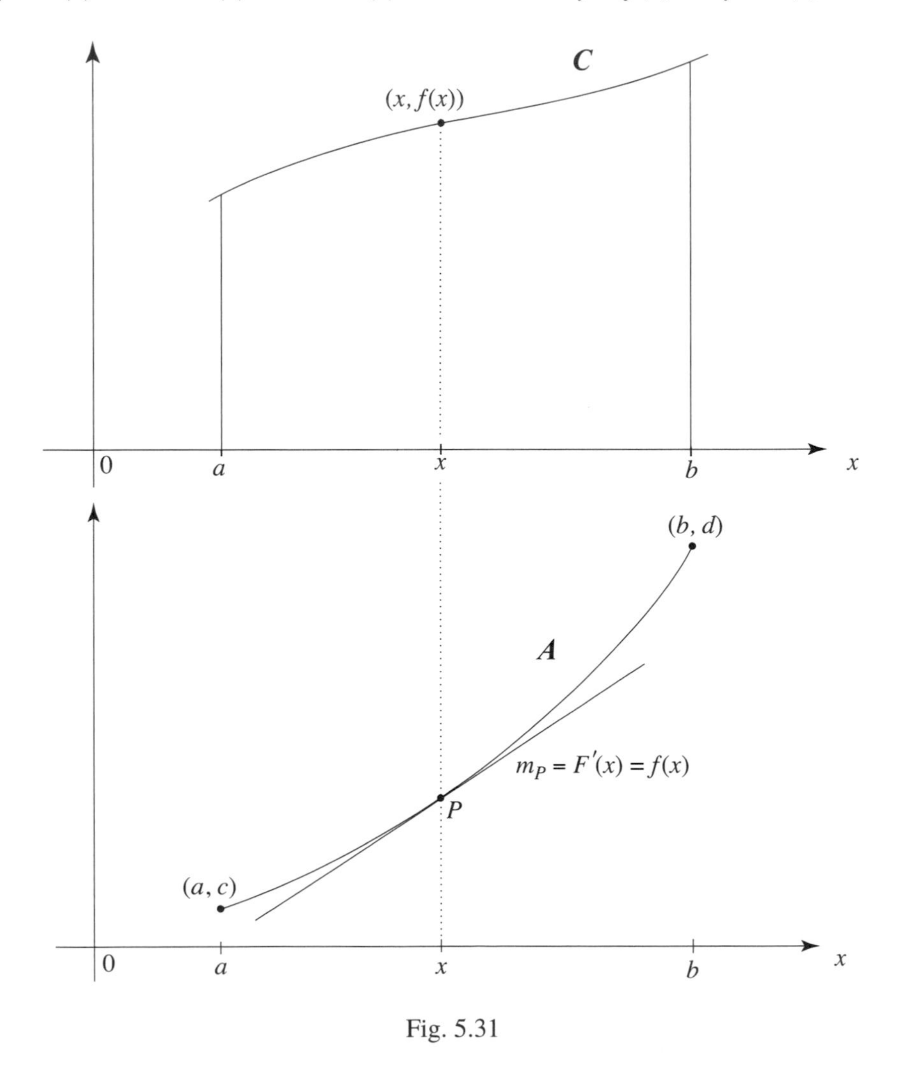

Fig. 5.31

simultaneously, as shown in Figure 5.31. The point $(x, f(x))$ is a typical point on the curve $\boldsymbol{C}$, and $P = (x, F(x))$ is the corresponding point on $\boldsymbol{A}$. Since $y = F(x)$ is an antiderivative of $y = f(x)$, we know that the slope m_P of the tangent to $\boldsymbol{A}$ at P is equal to $m_P = F'(x) = f(x)$.

Next, Leibniz takes any dx, considers a segment of length dx, and places a characteristic triangle with sides dx and dy at the point P of the curve $\boldsymbol{A}$. Because $\frac{dy}{dx} = m_P = F'(x) = f(x)$, we know that

$$dy = F'(x) \cdot dx = f(x) \cdot dx.$$

This equality, interpreted geometrically in Figure 5.32, tells us that the length dy of the vertical segment

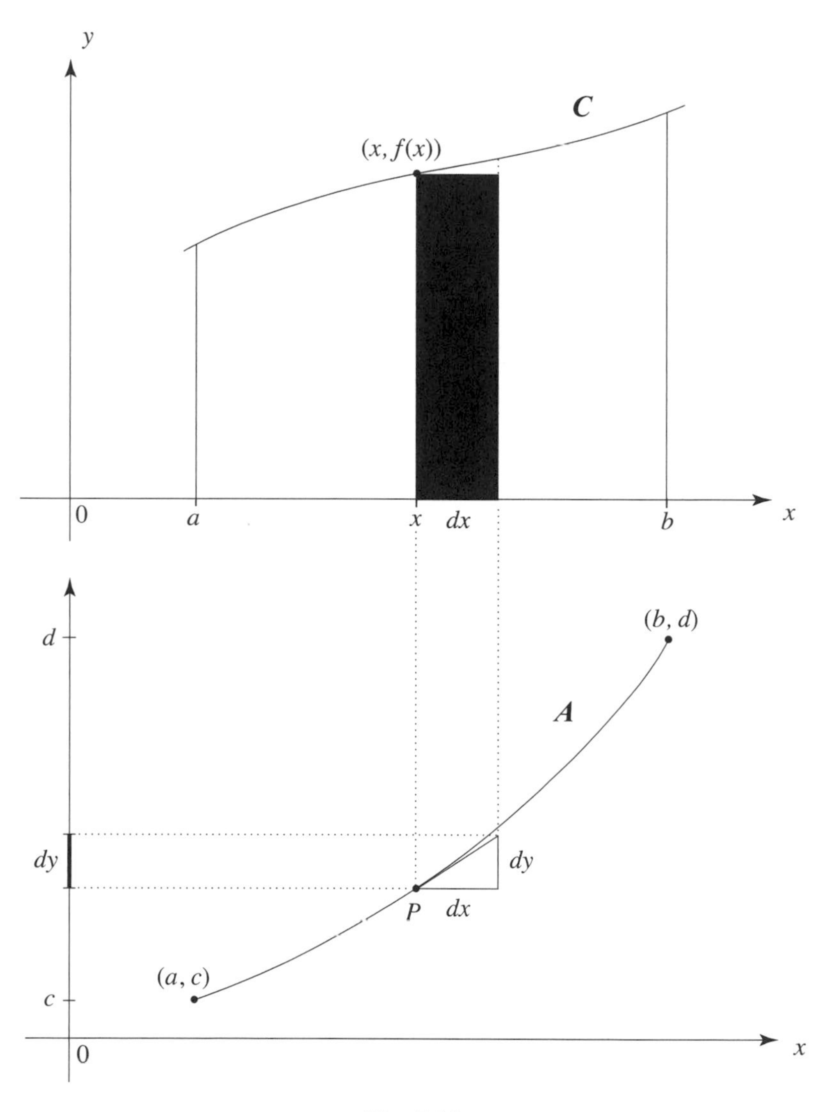

Fig. 5.32

of the characteristic triangle at P is equal to the area $f(x) \cdot dx$ of the corresponding rectangle (shown in black) under the graph of the the function $y = f(x)$.

Now Leibniz takes a very large number n relative to the length $b - a$ and divides the interval from a to b into n very small equal intervals, each of length $dx = \frac{b-a}{n}$. The division points determine, exactly as described in Section 5.6, many very thin rectangles with base dx under the graph $\boldsymbol{C}$ of the function $y = f(x)$. They are depicted in the upper part of Figure 5.33 (with a relatively small n to keep things visible). To distinguish the rectangles, they are shown in alternating colors of black and gray. As we saw in Figure 5.32, the area $f(x) \cdot dx$ of each rectangle is numerically equal to the length of the corresponding segment dy. Considering his rectangles from left to right, Leibniz concludes from his construction as depicted in Figure 5.33 that the sum of the areas $f(x) \cdot dx$ of all of the rectangles is an approximation of the area under the graph $\boldsymbol{C}$ of the function f and that the equal sum of the lengths of the segments dy along the y-axis is an approximation of the distance from $c = F(a)$ to $d = F(b)$. Taking larger and larger n,

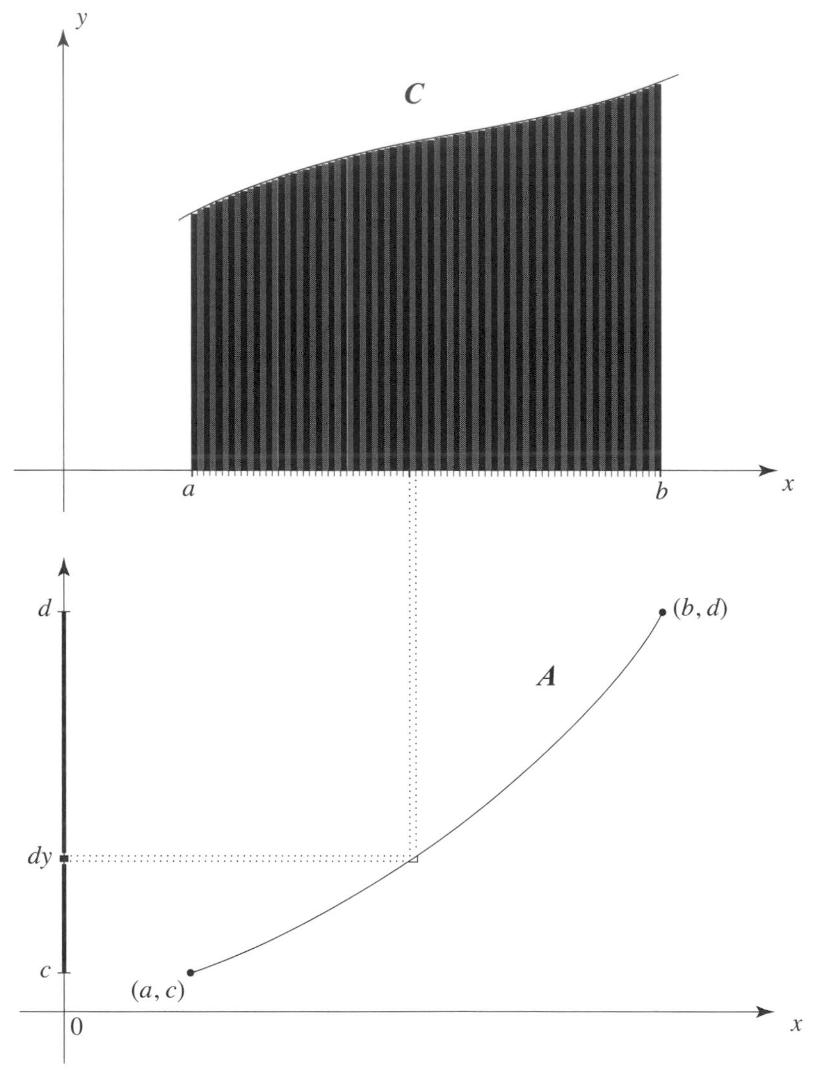

Fig. 5.33

it follows that in the limit, the area under the graph $\boldsymbol{C}$ of the function $y = f(x)$ above the interval from a to b is equal to $d - c = F(b) - F(a)$. Leibniz has therefore demonstrated that

$$\boxed{\int_a^b f(x)\,dx = F(b) - F(a)}$$

where $y = F(x)$ is an antiderivative of the function $y = f(x)$. This formula is known as the *fundamental theorem of calculus*. The difference $F(b) - F(a)$ is often denoted by $F(x)\big|_a^b$.

What the fundamental theorem of calculus tells us this: to evaluate a definite integral $\int_a^b f(x)\,dx$, find an antiderivative F of f and compute the difference

$$F(x)\Big|_a^b = F(b) - F(a).$$

Let's consider some specific examples to get a sense of what Leibniz has accomplished.

Example 5.9. Consider the function $f(x) = x^3$, and let $a = 2$ and $b = 6$. Take $n = 40{,}000$ and $dx = \frac{6-2}{40{,}000} = \frac{1}{10{,}000}$. Since n is large relative to $b - a = 4$, the sum

$$2^3 \cdot \tfrac{1}{10{,}000} + (2 + \tfrac{1}{10{,}000})^3 \cdot \tfrac{1}{1000} + (2 + \tfrac{2}{10{,}000})^3 \cdot \tfrac{1}{10{,}000} + (2 + \tfrac{3}{10{,}000})^3 \cdot \tfrac{1}{10{,}000} + \cdots + (5 + \tfrac{9{,}999}{10{,}000})^3 \cdot \tfrac{1}{10{,}000}$$

is very nearly equal to $\int_2^6 x^3\,dx$. Since $y = \frac{1}{4}x^4$ is an antiderivative of $y = x^3$, this integral is equal to $\frac{1}{4}x^4\Big|_2^6 = \frac{1}{4}(6^4 - 2^4) = \frac{1}{4}(1296 - 16) = 320.$

Example 5.10. Turn to Figure 5.34a. Consider the graph of $y = \sqrt{x} = x^{\frac{1}{2}}$ over the interval $0 \le x \le 4$. An

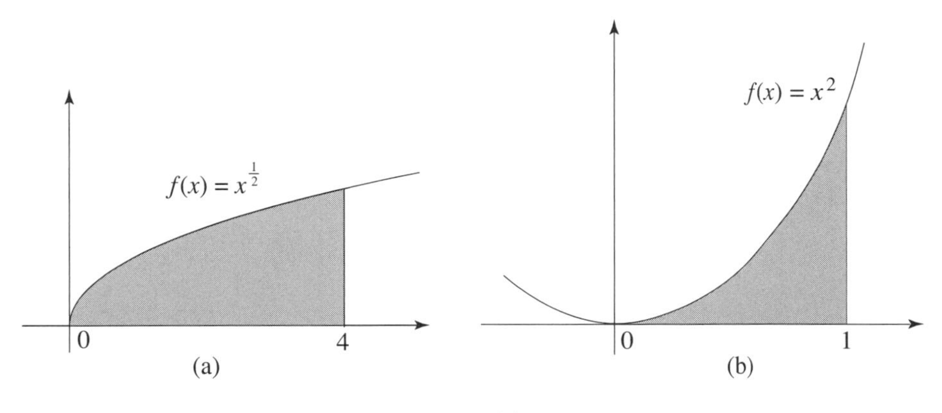

Fig. 5.34

application of the fundamental theorem tells us that the area under it is equal to

$$\int_0^4 x^{\frac{1}{2}}\,dx = \tfrac{2}{3}x^{\frac{3}{2}}\Big|_0^4 = \tfrac{2}{3}4^{\frac{3}{2}} = \tfrac{2}{3}2^3 = \tfrac{16}{3}.$$

By one more application of this theorem, the area under the graph of $y = x^2$ and over the interval $0 \le x \le 1$ as depicted in Figure 5.34b is equal to

$$\int_0^1 x^2\,dx = \tfrac{1}{3}x^3\Big|_0^1 = \tfrac{1}{3}.$$

5.9 VOLUMES OF REVOLUTION

We conclude our discussion of Leibniz's calculus by showing that the summation process that gives rise to the definite integral can be applied to problems other than the computation of area. We will illustrate this by computing volumes of regions obtained by revolving curves. Once this has been done, we'll return to Kepler's wine barrels.

Let's begin with a fact about volumes. Suppose that we are given a region in a horizontal the plane with straight or curving boundaries. Leibniz's summation process tells us how to compute its area A.

Form a solid S by taking an identical copy of the region and move it vertically for a distance h in such a way that it remains parallel to the original region and oriented in exactly the same way. Figure 5.35a provides an example of what has just been described. It is a fact that the volume of S with its two identical flat sides is equal to $A \cdot h$. The fact that the volume of a cylinder of radius r and height h is equal to $\pi r^2 h$

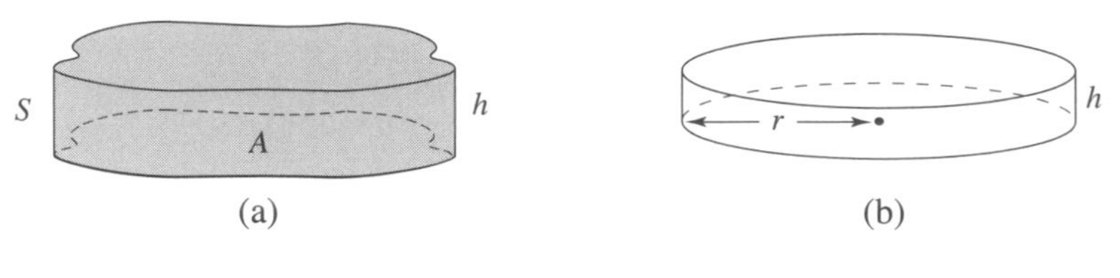

Fig. 5.35

is a special case of this more general fact. See Figure 5.35b. We will now investigate the volumes of more complicated solids.

Consider a semicircle of radius r, and place a coordinate system as shown in Figure 5.36. As was already done several times in earlier sections of this chapter, select a large number n (large relative to the radius r), and divide the interval from 0 to $2r$ into n equal segments each of length $dx = \frac{2r}{n}$. Let

$$0 < x_1 < x_2 < \cdots < x_{n-2} < x_{n-1} < 2r$$

be the $n - 1$ subdivision points. Consecutive points are separated by the distance dx. Let x be a typical subdivision point, and consider the rectangle under the semicircle that x, the corresponding y-coordinate, and the distance dx from x to the next point determine. This very thin rectangle is shown in the figure. Now revolve the semicircular region and the thin rectangle along with it one complete revolution about the x-axis. The semicircle sweeps out a sphere S of radius r, and the thin rectangle sweeps out a thin cylindrical disc of radius y and thickness or height dx. So the circular base of this thin cylinder has area

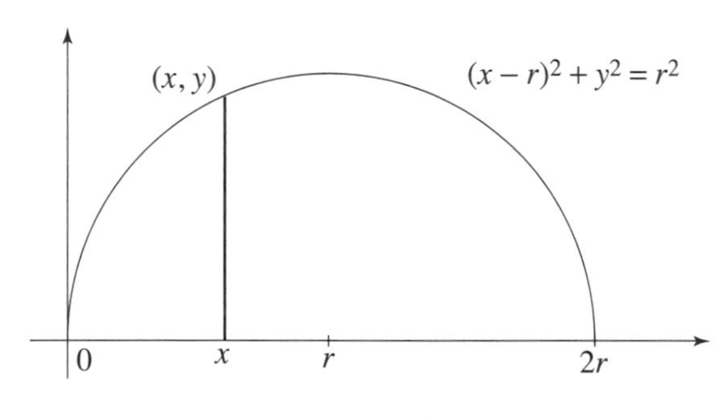

Fig. 5.36

πy^2, so that its volume is $\pi y^2 dx$. Since the point (x, y) satisfies $(x - r)^2 + y^2 = r^2$, observe that $y^2 = r^2 - (x - r)^2 = 2rx - x^2$. Therefore the volume of the thin cylinder is

$$\pi y^2\, dx = (2\pi rx - \pi x^2)\, dx.$$

Doing this for all n rectangles that the subdivision points determine, we see that the long sum

$$(2\pi ra - \pi a^2)\, dx + (2\pi rx_1 - \pi x_1^2)\, dx + (2\pi rx_2 - \pi x_2^2)\, dx + \cdots + (2\pi rx_{n-1} - \pi x_{n-1}^2)\, dx$$

is a very tight approximation of the volume V of the entire *solid of revolution*, the sphere S of radius r. The larger the n that is taken, the thinner each of the cylinders and the tighter this approximation. It follows that with n growing larger and larger, we get in the limit that

$$V = \int_0^{2r} (2\pi r x - \pi x^2)\,dx = \pi \int_0^{2r} (2rx - x^2)\,dx.$$

Because the function rx^2 is an antiderivative of $2rx$, and $\frac{1}{3}x^3$ is an antiderivative of x^2, it follows that $G(x) = rx^2 - \frac{1}{3}x^3$ is an antiderivative of $g(x) = 2rx - x^2$. By the fundamental theorem of calculus,

$$\pi \int_0^{2r} (2rx - x^2)\,dx = \pi\Big[G(x)\Big|_0^{2r}\Big] = \pi[r(2r)^2 - \tfrac{1}{3}(2r)^3] = \pi[r(2r)^2 - \tfrac{2r}{3}(2r)^2] = \pi[\tfrac{r}{3}(2r)^2] = \tfrac{4}{3}\pi r^3.$$

With Leibniz's summation method combined with the fundamental theorem of calculus, we derived the formula

$$V = \tfrac{4}{3}\pi r^3$$

for the volume of a sphere of radius r.

With this argument many volumes of revolution can be computed. Let f be a function that satisfies $f(x) \geq 0$ for all x between a and b on the x-axis. A typical situation is sketched in Figure 5.37. Proceed exactly as in the situation of the semicircle. The interval from a to b is subdivided into a large number n segments each of length $dx = \frac{b-a}{n}$. Each of these segments determines a rectangle under the curve of thickness dx. A typical one is shown in the figure. It is placed at x and has height $f(x)$. Now rotate the region bounded by the graph, the x-axis, and the lines $x = a$ and $x = b$ for one complete revolution about

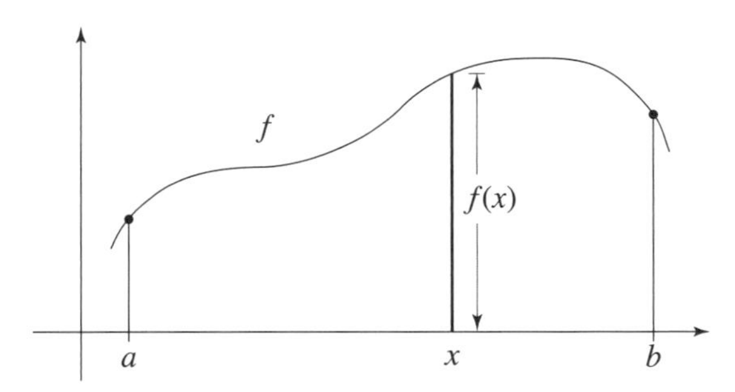

Fig. 5.37

the x-axis. As in the previous case of the sphere, the volume V of the resulting solid is approximated by the sum of all the volumes of all of the thin cylindrical discs determined by the rotation of the rectangles. The typical disc shown has circular area $\pi \cdot (f(x))^2$ and height dx, so it has volume $\pi(f(x))^2 dx$. By pushing n to infinity, we get simultaneously the two sides of the equality

$$\boxed{V = \int_a^b \pi f(x)^2\,dx}$$

As an aside, notice that this definite integral is also equal to the area under the graph of the function $\pi(f(x))^2$ from a to b.

In Figure 5.38, both h and r are positive numbers. The triangular region is determined by the x-axis, the vertical line through h, and the line through the origin with slope $\frac{r}{h}$. Observe that the solid of revolution is a cone of height h and circular base of radius r. The relevant function is $f(x) = \frac{r}{h}x$. By substituting into the formula just developed, we see that the volume V of the cone is

$$V = \int_0^h \pi(f(x))^2\,dx = \int_0^h \pi(\tfrac{r}{h})^2 x^2\,dx = \pi(\tfrac{r}{h})^2 \int_0^h x^2\,dx.$$

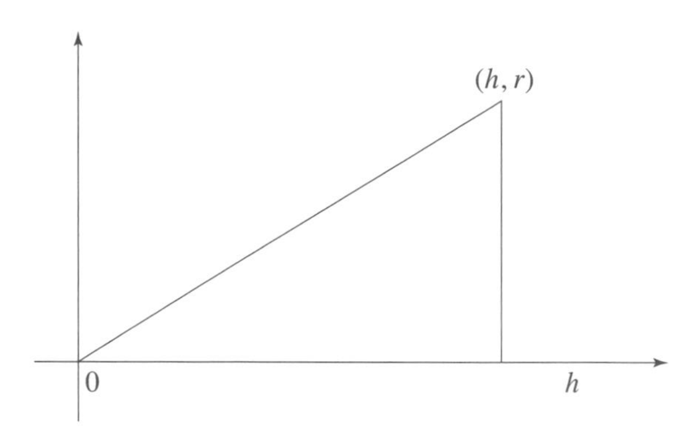

Fig. 5.38

Since $\frac{1}{3}x^3$ is an antiderivative of x^2, it follows that the volume of a cone with height h and circular base of radius r is

$$V = \pi(\tfrac{r}{h})^2\big[\tfrac{1}{3}x^3\big|_0^h\big] = \pi(\tfrac{r}{h})^2\tfrac{h^3}{3} = \tfrac{1}{3}\pi r^2 h.$$

Let's return to the newlywed Kepler and the analysis of wine barrels that absorbed him during his honeymoon. Understandably, he takes particular interest in the standard *Austrian barrel*. He models the Austrian barrel as follows. He takes two identical cutoff cones and attaches them at their larger circular

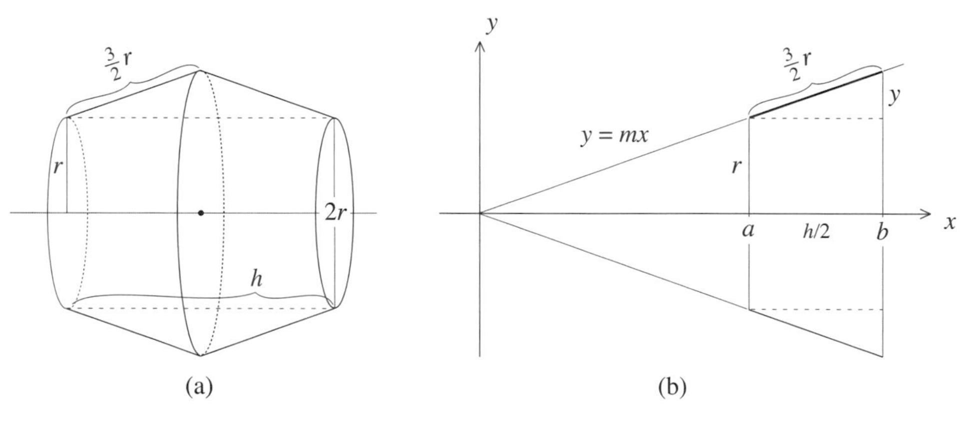

Fig. 5.39

ends to form the barrel shape depicted in Figure 5.39a. Let h be the height of this barrel shape, and let r be the radius of the circular cross sections at its ends. Kepler imposes two conditions. The first is that the height h and the diameter $2r$ of the circular end satisfy the condition $\frac{h}{2r} = \sqrt{2}$ or $\frac{h}{2} = \sqrt{2}r$. This means that the cylindrical core, as outlined by the dotted lines of Figure 5.39a, is a Kepler cylinder. The second condition is that the length of the slanting segment of each cutoff cone is $\frac{3}{2}r$.

We'll now set out to determine the volume V of Kepler's barrel design as a volume of revolution. Turn to Figure 5.39b. Let $y = mx$ be the line through the origin with slope m, and consider the segment of this line over an interval $0 < a \leq x \leq b$. The first question is this: For what m, a, and b does this segment, when revolved around the x-axis, generate the left half of the barrel shape depicted in Figure 5.39a? Notice that with h the height of the barrel, $b - a = \frac{h}{2} = \sqrt{2}\,r$. So by the Pythagorean theorem, $\left(\frac{3}{2}r\right)^2 = y^2 + (\sqrt{2}r)^2$. Therefore $y^2 = (\frac{9}{4} - 2)r^2 = \frac{1}{4}r^2$. So $y = \frac{1}{2}r$ and hence $m = \frac{\frac{1}{2}r}{\sqrt{2}r} = \frac{1}{2\sqrt{2}}$. Since $m = \frac{r}{a}$, we have determined that

$$a = \tfrac{r}{m} = 2\sqrt{2}\,r \quad\text{and}\quad b = \sqrt{2}\,r + a = 3\sqrt{2}\,r.$$

Feeding these data into the volume of revolution formula and multiplying by 2 (to pick up both cutoff cones), we get that the volume V of Kepler's model of the Austrian barrel satisfies

$$\begin{aligned} V &= \pi\int_a^b (mx)^2dx = 2\pi m^2\Big[\tfrac{x^3}{3}\Big|_{2\sqrt{2}r}^{3\sqrt{2}r}\Big] = \tfrac{2\pi m^2}{3}[27\cdot 2\sqrt{2}\,r^3 - 8\cdot 2\sqrt{2}\,r^3] \\ &= \tfrac{2\pi m^2}{3}38\sqrt{2}\,r^3 = \tfrac{2\pi}{3}\cdot\tfrac{38}{8}\sqrt{2}\,r^3 = \tfrac{19}{6}\sqrt{2}\pi r^3. \end{aligned}$$

Now that Kepler has designed his model of the Austrian barrel and computed its volume, his interest turns to the wine merchant's assessment $V_{\text{rule}} = 0.6s^3$ of the volume. Figure 5.40 combines elements from

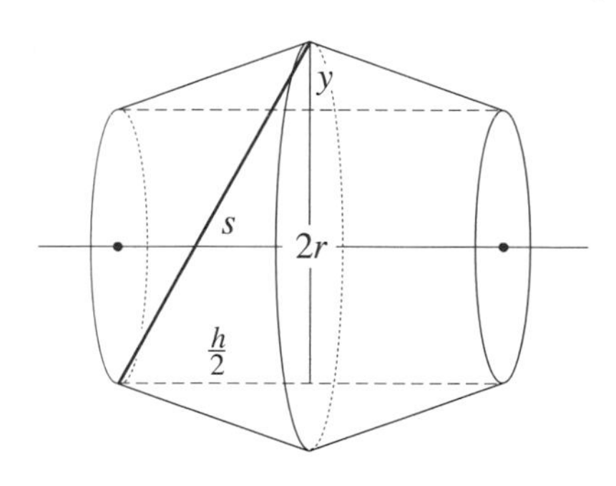

Fig. 5.40

Figures 5.39a and 5.39b to tell us that

$$s^2 = (2r+y)^2 + (\tfrac{h}{2})^2 = (2r+\tfrac{1}{2}r)^2 + (\sqrt{2}r)^2 = (\tfrac{5}{2}r)^2 + 2r^2 = (2+\tfrac{25}{4})r^2 = \tfrac{33}{4}r^2.$$

So $r = \frac{2}{\sqrt{33}}s$ and hence $r^3 = \frac{8}{33\sqrt{33}}s^3$. Therefore the volume of Kepler's model of the Austrian barrel in terms of the measure s is

$$V = \tfrac{19}{6}\sqrt{2}\pi r^3 = \tfrac{19}{6}\sqrt{2}\pi\cdot\tfrac{8}{33\sqrt{33}}s^3 \approx 0.59s^3.$$

Since $V_{\text{rule}} = 0.6s^3$, the official measure of the volume of Kepler's model of an Austrian barrel was only a little more than its actual volume. Kepler knew that he had bought his supply of wine at a fair price. His second marriage was off to a good start. History would tell us that it was much happier than his first.

5.10 PROBLEMS AND PROJECTS

5A. Lines and Their Equations. The first block of problems deals with basic properties of lines.

5.1. Determine the slope of an and an equation for the line through the points $(2,-3)$ and $(-6,2)$.

5.2. Find an equation for the line through the two points $(-2,7)$ and $(5,2)$.

5.3. Write an equation for the line with slope -3 and y-intercept 4.

5.4. Write an equation for the line that has slope $\frac{1}{2}$ and the point $(3,-2)$ on its graph.

5.5. Find an equation of the line with slope -3 through $(-6,-7)$.

5.6. Sketch the line that has equation $y = \frac{1}{3}x - 4$.

5.7. Sketch the line with slope -2 and y-intercept -1.

5.8. What are the slope and y-intercept of the line that has equation $2x + 7y + 2 = 0$?

5.9. What is the slope-intercept form of the equation of the line given by $3x + 5y + 2 = 0$?

5.10. Determine the equation of the line that crosses the x-axis the point $(3,0)$, has positive slope, and makes an angle of 45° with the positive part of the x-axis. Repeat this for the angles 30° and 25°.

5.11. Consider the line in the xy-plane that the two points $(5,1)$ and $(-4,7)$ determine. Sketch the line. Find equations for this line in two-point form, in point-slope form, and in slope-intercept form.

5.12. What is going on in Figure 5.41? The two triangular regions have the same base of 13 units and the same height of 5 units, and yet their areas are different (as the subdivisions of the areas show).

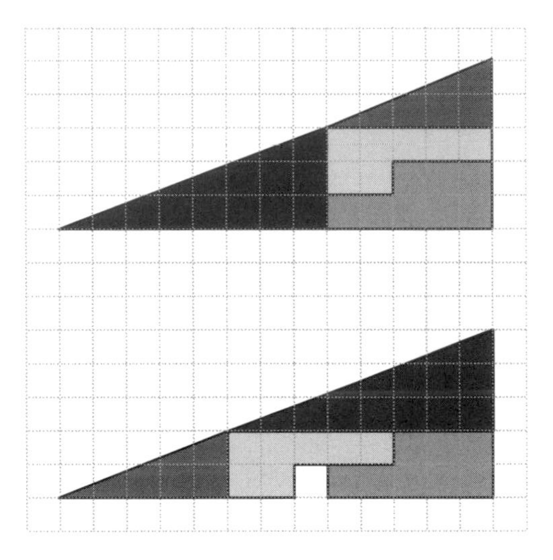

Fig. 5.41

5.13. Consider the circle $(x-3)^2 + (y+2)^2 = 20$ and the line $y = x - 3$. Sketch both on an xy-plane. Then find the points of intersection of the circle and the line.

5.14. Find the points of intersection of the circle with center $(2,3)$ and radius 5 and the line through its center with slope $\frac{1}{2}$.

5.15. Two lines L_1 and L_2, neither of them vertical nor horizontal, are given. Let their slopes be m_1 and m_2, respectively. Show that if the lines are perpendicular, then $m_2 = -\frac{1}{m_1}$. Conversely, show that if

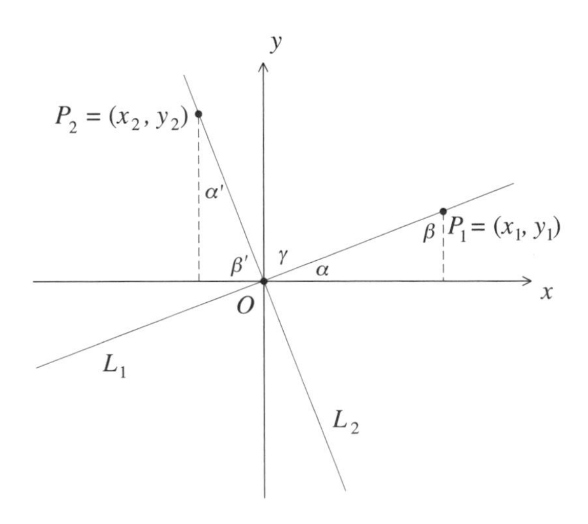

Fig. 5.42

$m_2 = -\frac{1}{m_1}$, then the lines are perpendicular. [Hint: Move the lines without changing their slopes so that their point of intersection lies at the origin. Choose a point $P_1 = (x_1, y_1)$ on L_1 and a point $P_2 = (x_2, y_2)$ on L_2. If the lines are perpendicular, then $\beta = \beta'$, so also $\alpha = \alpha'$. Use the "ratio of the corresponding sides" property of similar triangles to conclude that $m_2 = -\frac{1}{m_1}$. If $m_2 = -\frac{1}{m_1}$, then $\tan\beta = \tan\beta'$. It follows from Figure 4.26, that $\beta = \beta'$. So the lines are perpendicular.]

5B. Computing Slopes of Tangents. For the next set of problems, use Leibniz's tangent method "push Δx to zero" to determine the slope m_P of the tangent to the given curve at the given point.

5.16. The parabola $y = x^2$ at the point $P = (2, 4)$.

5.17. The curve $y = x^3$ at the point $P = (2, 8)$.

5.18. The curve $y = \frac{1}{x^2}$ at any point $P = (x, y)$.

5.19. The graph of the equation $y^2 = 2x + 7$ at any point $P = (x, y)$ with $y \neq 0$. Find the slope of the tangent line at the point $(1, -3)$, and determine an equation of this line.

5.20. The curve $x = y^3$ at any point $P = (x, y)$. [Hint: What happens to $\frac{\Delta y}{\Delta x}\Delta y$ when Δx is pushed to zero? Refer to Figure 5.7 in Section 5.2 and the surrounding discussion.]

5.21. The ellipse $\frac{x^2}{5^2} + \frac{y^2}{4^2} = 1$ at $P = (x, y)$. [If you get $m_P = -\frac{4^2}{5^2} \cdot \frac{x}{y}$, you are right.]

5.22. The curve $y^3 = 3x^2 + 7$ at a point $P = (x, y)$. Show that $m_P = \frac{2x}{(3x^2+7)^{\frac{2}{3}}}$. [Hint: Use the formula $(a + b)^3 = a^3 + 3a^2b + 3ab^2 + b^3$.]

5.23. Consider the circle $x^2 + y^2 = r^2$. Let (x_0, y_0) be any point on the circle with $y_0 \neq 0$, and use Leibniz's tangent method to show that the slope of the tangent line to the circle at (x_0, y_0) is equal to $-\frac{x_0}{y_0}$.

i. Consider the radius of the circle from the origin to the point (x_0, y_0) and show that it is perpendicular to this tangent.

ii. Suppose that $r = 1$. Consider a line $y = -\frac{1}{3}x + b$ with slope $-\frac{1}{3}$ and y-intercept b. For which constants b is the line tangent to the circle?

5C. Functions and Derivatives. The next set of problems studies basic functions and their derivatives.

5.24. Return to Figures 5.13a and 5.13b. Check the assertions that the text makes about the pairs of functions $y = f_+(x)$ and $y = f_-(x)$ as well as $y - g_+(x)$ and $y - g_-(x)$.

5.25. Consider the circle $x^2 + y^2 = 9$. See Figure 5.43. Define explicitly two functions with the property that their two graphs lie on the circle, and the entire circle is the composite of the two graphs.

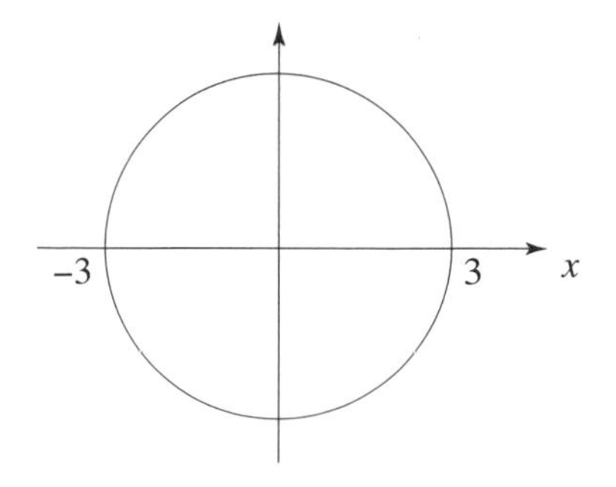

Fig. 5.43

5.26. Water is poured *at a constant rate* into each of the three drinking glasses and the three vases shown in Figure 5.44. Each vessel is empty at the start. The graph under the glass labeled (a) represents the height of the water in the glass as a function of time. Think carefully what is going on in the other five cases, and draw the time/height graphs for each. [This problem was adapted from a 2006 advertisement in a Tokyo subway for a tutorial service for middle school students.]

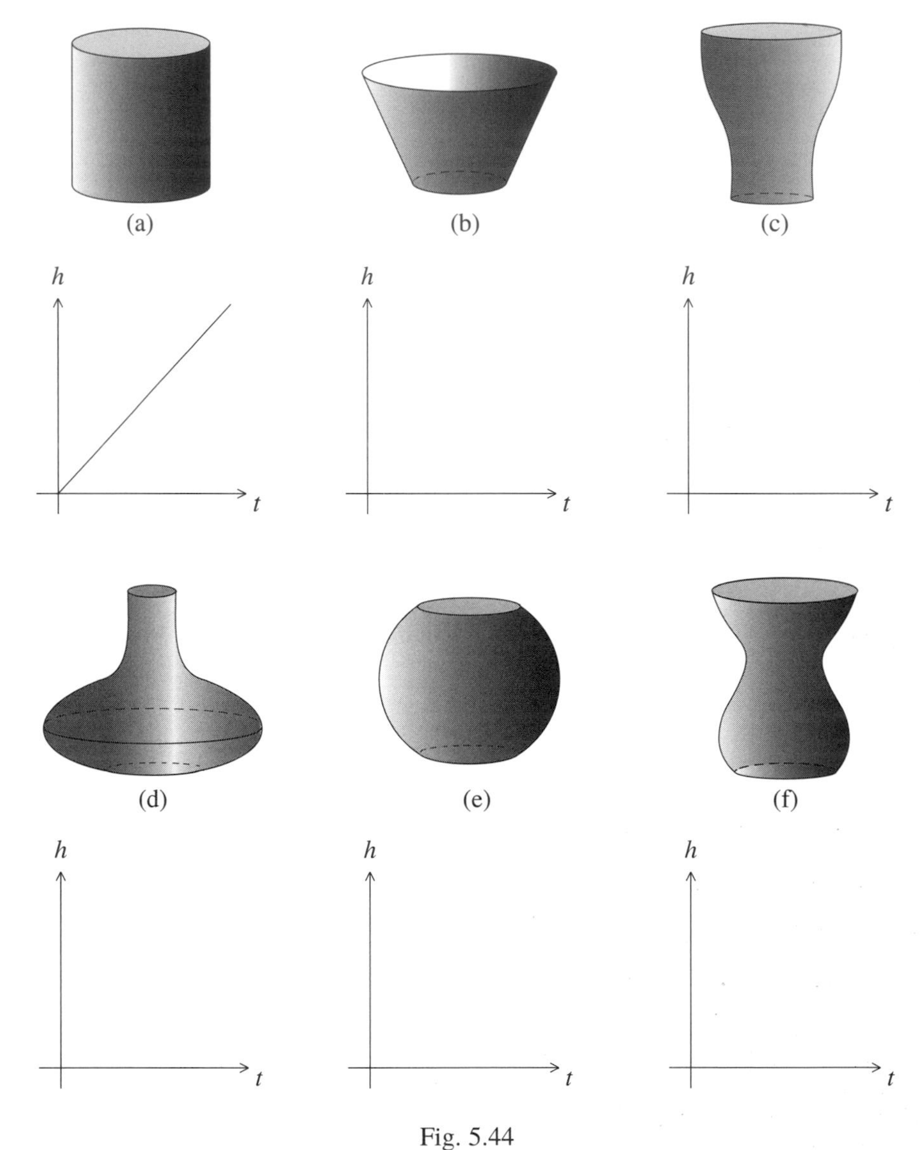

Fig. 5.44

5.27. Let $f(x) = x^3$. Determine $f'(x)$ by evaluating

$$\lim_{\Delta x \to 0} \frac{f(x+\Delta x) - f(x)}{\Delta x}.$$

[Hint: Use the formula $(a+b)^3 = a^3 + 3a^2b + 3ab^2 + b^3$.]

5.28. Consider $h(x) = \frac{1}{x^2}$. Show that $h'(x) = \frac{-2}{x^3}$ by evaluating $\lim_{\Delta x \to 0} \frac{h(x+\Delta x) - h(x)}{\Delta x}$. [Hint: Take common denominators and cancel.]

5.29. Use the derivative formulas already developed to compute the slopes of the tangents to the graphs of the given function at the given points.

i. $f(x) = x^3$ at $(-2, -8)$
ii. $g(x) = \sqrt[3]{x} = x^{\frac{1}{3}}$ at $(-3, \sqrt[3]{-3})$
iii. $f(x) = \frac{1}{x}$ at $(-\frac{1}{3}, -3)$
iv. $f(x) = \frac{1}{x^2}$ at $(-2, \frac{1}{4})$

5.30. By using facts from the text, compute the derivatives of each of the following functions.

i. $f(x) = -10$
ii. $y = 4x + 7$
iii. $f(x) = 7x^2 - 5x + 2$
iv. $y = 2x^{\frac{1}{3}} + \pi x^3$
v. $g(x) = \frac{3}{x} + 3x - 6$
vi. $f(x) = 2x^3 + 3x + 4 - \frac{1}{x^2}$
vii. $h(x) = 4x^{\frac{1}{2}} + 5x^{-1}$

5.31. Consider the function $f(x) = 5x^2 - 2x^3$, and let $P = (x, y)$ be a point on its graph. Determine an equation for the tangent line to the graph at P. [Suggestion: Change the notation for the coordinates of P to, say, $P = (x_0, y_0)$.]

5.32. Consider the parabola $y = x^2$ and the line $y = 3x - 4$. Show that they do not intersect. Move the line toward the parabola without changing its slope. At what point will it first touch the parabola?

5.33. Figure 5.17 illustrates all possible behaviors of a function $y = f(x)$ at and near a point $P = (x, f(x))$ on its graph where $f'(x) > 0$. Sketch the analogous figure in the case $f'(x) < 0$.

5.34. Show that the derivative of $f(x) = \sqrt{5^2 - x^2}$ is $f'(x) = \frac{-x}{\sqrt{5^2-x^2}}$ and that the derivative of $f(x) = (3x^2 + 7)^{\frac{1}{3}}$ is $f'(x) = \frac{2x}{(3x^2+7)^{\frac{2}{3}}}$. [Use the conclusions of Example 5.7 and Problem 5.22.]

5.35. Find antiderivatives for the functions $f(x) = 1 - 3x^3 + 2x^{-\frac{1}{2}}$, $g(x) = -\frac{1}{3}x^{-2} + 8x^{\frac{1}{3}}$, and $h(x) = -4 + 3x^{-2} + 7x^{\frac{1}{2}}$.

5.36. Find antiderivatives for $f(x) = \frac{x}{\sqrt{1-x^2}}$ and $g(x) = \frac{x}{(3x^2+7)^{\frac{2}{3}}}$. [Hint: Use the conclusions of Problem 5.34.]

5D. Maximum and Minimum Values. The next several problems make use of the strategy developed in Sections 5.4 and 5.5 for finding the maximum and minimum values of a function. In each case, justify whether the answer is in fact the minimum or maximum that you are looking for.

5.37. Consider a rectangle with fixed area A and variable sides x and y. Let p be the perimeter of the rectangle. Determine the values of x and y that provide the smallest possible p.

5.38. As part of his analysis of wine barrels, Kepler considers a rectangle, lets d be the length of its diagonal, and shows that of all rectangles with diagonal d, the square has the largest area. Verify Kepler's assertion. [Hint: Let x be a side of the rectangle, and express its area A as a function of x. Notice that if A has a maximum value for some x, then A^2 has its maximum value for the same x.]

5.39. What is the largest area that the rectangle inscribed under the graph of the function $g(x) = \frac{1}{x}$ as shown in Figure 5.45 can have?

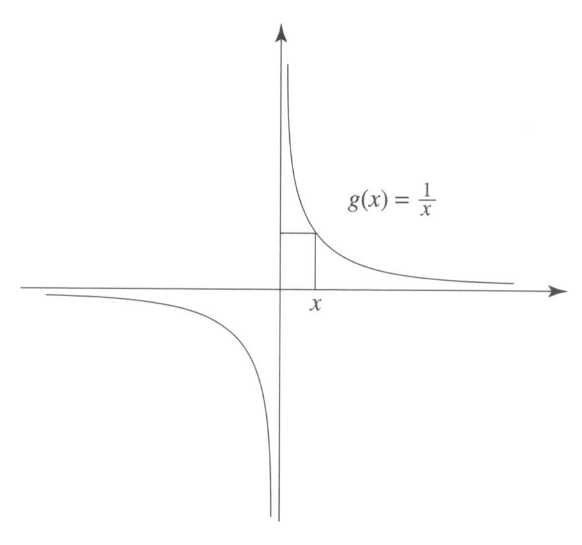

Fig. 5.45

5.40. Find that point on the parabola $y = x^2 + 1$ that is closest to the point (3, 1). [Hints: Find that point on the parabola such that the square of its distance to (3, 1) is minimal. Also use the fact that if a is a root of a polynomial, then $x - a$ divides it.]

5.41. A segment S is divided as shown in Figure 5.46. The product of the lengths of the two smaller segments is known to be 225 units. What is the shortest segment S for which this is possible? What are the lengths of the two smaller segments for the shortest S?

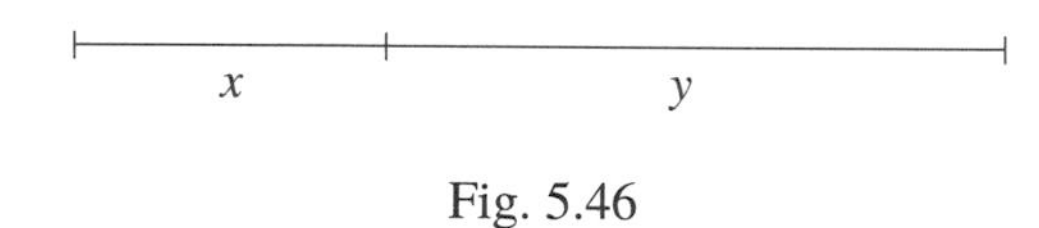

Fig. 5.46

5.42. If the segment of Figure 5.46 is 1200 units long and the product of the squares of the lengths of the two pieces is to be as large as possible, how long should the two pieces be?

5.43. Consider the triangular region determined by the graph of the line $y = 3 - x$. Inscribe a rectangle as shown in Figure 5.47. When both the triangle and the rectangle are revolved once around the x-axis, a cone and an inscribed cylinder result. Determine the largest volume this cylinder can have.

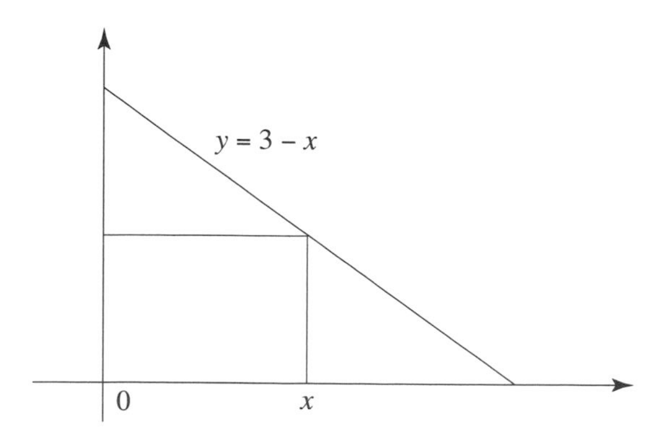

Fig. 5.47

5.44. Use derivatives to compute the distance between the line $y = -\frac{1}{2}x + 5$ and the point $(-4, 3)$. [Hint: Let (x, y) be any point on the line. Express the distance between $(-4, 3)$ and (x, y) as a function of x. Then determine the smallest value of the square of this distance.]

5.45. Let's turn to the problem that Fermat posed to Torricelli. Let's take a reasonably random triangle ΔABC and assume that each of the three angles is less than $120°$.

i. Torricelli solved Fermat's problem for such triangles as follows. He completed the sides of the triangle ΔABC to equilateral triangles ΔABL, ΔACM, and ΔBCN (in each case taking the triangle that falls outside the given ΔABC). Then Torricelli constructed the circles that each of the three sets of points A, B, L; A, C, M; and B, C, N determine. He then he verified that these three circles intersect at a single point P inside ΔABC. This is the point P that minimizes the sum $PA + PB + PC$. Finally, he proved with a geometric argument that this point P minimizes the sum $PA + PB + PC$. Take a sheet of paper and a compass, and carry out the construction of P. Today this point is known as the *Fermat-Torricelli point* of the triangle. The proof that P solves the problem is difficult. Let it be. (If your construction runs into difficulties, have a look at Problem 1.9.)

ii. Reformulate Fermat's problem by setting up an xy-coordinate system and by making use of the coordinates of the points A, B, C, and P. Why is the solution of this reformulated problem not a max-min problem of the type that was discussed in Sections 5.4 and 5.5?

5E. The Definite Integral. The next several problems illustrate the definition of the definite integral as well as the fundamental theorem of calculus.

5.46. Consider the graph of the equation $y = \frac{1}{x^2}$ over the interval from 2 to 4. Select the points

$$2 < 2\tfrac{1}{3} < 2\tfrac{2}{3} < 3 < 3\tfrac{1}{3} < 3\tfrac{2}{3} < 4.$$

Identify dx, compute the sum of the terms $f(x) \cdot dx$, and sketch the corresponding rectangles under the graph. Estimate the value of $\int_2^4 \frac{1}{x^2}\,dx$. Evaluate the integral. How good was your estimate?

5.47. Sketch the graph of the equation $y = \sqrt{x}$ over the interval from 0 to 3. Select the points

$$0 < \tfrac{1}{4} < \tfrac{1}{2} < \tfrac{3}{4} < 1 < \tfrac{5}{4} < \tfrac{3}{2} < \tfrac{7}{4} < 2 < \tfrac{9}{4} < \tfrac{5}{2} < \tfrac{11}{4} < 3$$

and identify the dx. Compute the sum of the terms $f(x) \cdot dx$. Sketch the corresponding rectangles under the graph, and use them to estimate the value of $\int_0^3 \sqrt{x}\,dx$. Evaluate the integral. How good was your estimate?

5.48. Consider the function $g(x) = 9 - x^2$ with $0 \le x \le 1$. Use the points

$$0 < 0.2 < 0.4 < 0.6 < 0.8 < 1$$

on the x-axis to subdivide the interval $[0, 1]$ and compute the sum of the areas $g(x) \cdot dx$ of all the rectangles that these points determine. Do so with an accuracy to three decimal places. This sum is an approximation of the area under the graph of $g(x) = 9 - x^2$ over $0 \le x \le 1$. Repeat this computation (again with accuracy to three decimal places) with the points

$$0 < 0.1 < 0.2 < 0.3 < 0.4 < 0.5 < 0.6 < 0.7 < 0.8 < 0.9 < 1$$

to get another approximation of the area under this graph. Before you do either computation, answer the question: Which of these two approximations would you expect to be better? Finally, find the area exactly.

5.49. Consider the function $f(x) = 16 - x^2$ with $-2 \leq x \leq 2$. Select the points

$$-2 < -1.5 < -1 < -0.5 < 0 < 0.5 < 1 < 1.5 < 2$$

in the interval $-2 \leq x \leq 2$. Compute the sum of the areas $f(x) \cdot dx$ of all the rectangles that these points determine. Do so with accuracy to three decimal places. This sum is an approximation of the area under the graph of $f(x) = 16 - x^2$ over $-2 \leq x \leq 2$. Compare this approximation with the value of $\int_{-2}^{2} (16 - x^2)\, dx$.

5.50. Consider the function $y = f(x) = \sqrt{4 - x^2}$. Its graph is the upper half of the circle of radius 2 with center the origin. Take the points

$$0 < 0.2 < 0.4 < 0.6 < 0.8 < 1 < 1.2 < 1.4 < 1.6 < 1.8 < 2$$

on the x-axis to subdivide the interval $0 \leq x \leq 2$ and compute the sum of the areas $f(x) \cdot dx$ of all the resulting rectangles (again with accuracy to three decimal places) under the graph. Observe that this sum is an estimate of the area under the upper half of the circle and over the segment from 0 to 2 on the x-axis. Use the geometry of the situation to determine the precise area.

5.51. Continue with the function $y = f(x) = \sqrt{4 - x^2}$. Use the points

$$-\sqrt{2} < -\tfrac{1}{2}\sqrt{2} < 0 < \tfrac{1}{2}\sqrt{2} < \sqrt{2}$$

on the x-axis to subdivide the interval $-\sqrt{2} \leq x \leq \sqrt{2}$, and compute the sum of the areas $f(x) \cdot dx$ of all the rectangles that they determines (again with accuracy to three decimal places). Observe that this sum is an estimate of the definite integral $\int_{-\sqrt{2}}^{\sqrt{2}} \sqrt{4 - x^2}\, dx$. Use geometry to determine the precise value of this integral.

5.52. Consider the sum

$$3^2 \cdot \tfrac{1}{1000} + (3 + \tfrac{1}{1000})^2 \cdot \tfrac{1}{1000} + (3 + \tfrac{2}{1000})^2 \cdot \tfrac{1}{1000} + (3 + \tfrac{3}{1000})^2 \cdot \tfrac{1}{1000} + \cdots + (5 + \tfrac{999}{1000})^2 \cdot \tfrac{1}{1000}\,.$$

Choose a function $y = f(x)$ and constants a and b such that this sum is very nearly equal to $\int_a^b f(x)\, dx$. Determine an approximate value of the sum.

5.53. Consider the sum

$$\sqrt{4} \cdot \tfrac{1}{10{,}000} + \sqrt{4 + \tfrac{1}{10{,}000}} \cdot \tfrac{1}{10{,}000} + \sqrt{4 + \tfrac{2}{10{,}000}} \cdot \tfrac{1}{10{,}000} + \cdots + \sqrt{7 + \tfrac{9{,}999}{10{,}000}} \cdot \tfrac{1}{10{,}000}\,.$$

Choose a function $y = f(x)$ and constants a and b such that this sum is closely approximated by $\int_a^b f(x)\, dx$. Determine this approximate value.

5F. Areas, Volumes, and the Fundamental Theorem. The problems that follow are basic explorations of the definite integral.

5.54. Use the fundamental theorem of calculus to evaluate the definite integrals

$$\int_0^3 x^2\, dx, \quad \int_{-8}^{-2} \frac{1}{x^2}\, dx, \quad \text{and} \quad \int_3^{12} \sqrt{x}\, dx.$$

Sketch the areas that you have determined.

5.55. Determine the area between the parabola $y = \frac{1}{6}x^2$ and the line $y = \frac{2}{3}x$.

5.56. Consider the parabolic section obtained by cutting the parabola $y = -3x^2 + 2x + 1$ with the x-axis. Express the area of the parabolic section as a definite integral. Compute the area by applying the fundamental theorem of calculus.

5.57. Consider the parabolic section obtained by cutting the parabola $y = -x^2 + 7x - 6$ with the line $y = 2$. Express the area of the parabolic section as a definite integral. Go to Section 2.4 and look up Archimedes's theorem. To compute the area, consider the application of both Archimedes's theorem and the fundamental theorem of calculus. Choose the simpler of the two methods.

5.58. Figure 5.48 shows the graph of the function $f(x) = \sqrt{x}$. The point $Q(x_0, y_0)$ is any point on the graph and $P = (x_0, 0)$. Let A be the area of the region under the graph of f and over the segment from the origin O to P. Let B be the area of the triangle determined by the tangent to the graph at

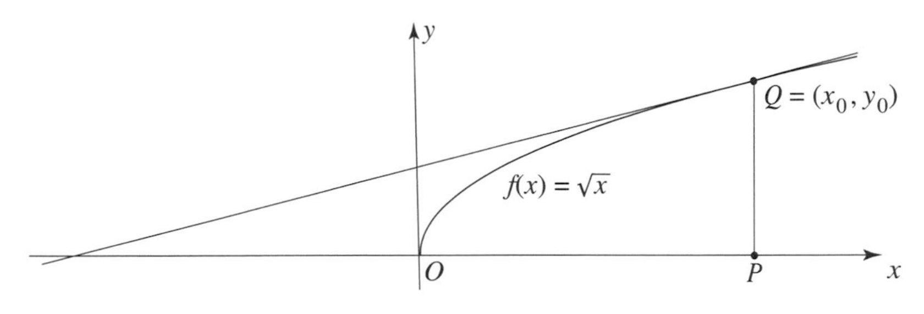

Fig. 5.48

Q, the segment PQ, and the x-axis. Show that $A = \frac{2}{3}B$ no matter where Q is taken.

5.59. Determine the volume of the solid obtained by rotating the region above the x-axis and under the graph of the function $y = \sqrt{x}$, $0 \le x \le 4$, one complete revolution about the x-axis.

5.60. Consider the upper half of the ellipse $\frac{x^2}{5^2} + \frac{y^2}{4^2} = 1$ in Figure 5.49. Express the following as definite integrals.

i. The area of the upper half of the ellipse.
ii. The volume of the solid obtained by rotating it one revolution about the x-axis.

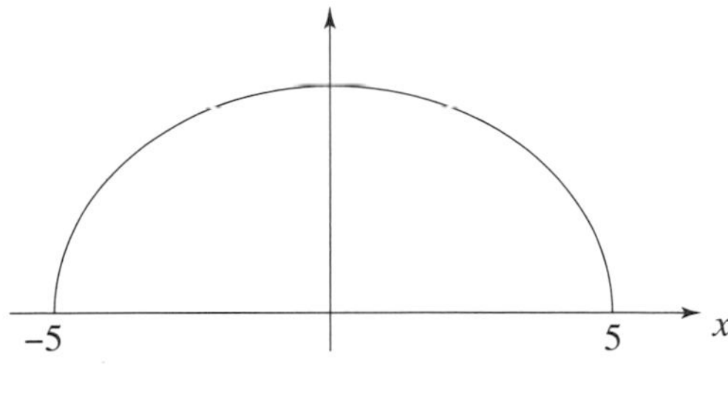

Fig. 5.49

5.61. The equation of the circle with radius 2 and centered at the origin is $x^2 + y^2 = 4$. Evaluate the definite integral

$$\int_0^2 \sqrt{4 - x^2}\,dx$$

after drawing a picture of the area that it represents. Let a be a positive number and evaluate $\int_0^a \sqrt{a^2 - x^2}\,dx$ in the same way.

5.62. Evaluate the definite integral $\int_0^5 \frac{2}{5}\sqrt{5^2 - x^2}\, dx$ and sketch the area that it represents.

5.63. Use the equality $\int_{-a}^{a} \frac{b}{a}\sqrt{a^2 - x^2}\, dx = \frac{b}{a}\int_{-a}^{a} \sqrt{a^2 - x^2}\, dx$ to show that the area of the ellipse with semimajor axis a and semiminor axis b is equal to $ab\pi$.

5.64. Figure 5.50 depicts a semicircle of radius 5 and inscribed into it one-half of the ellipse with semimajor

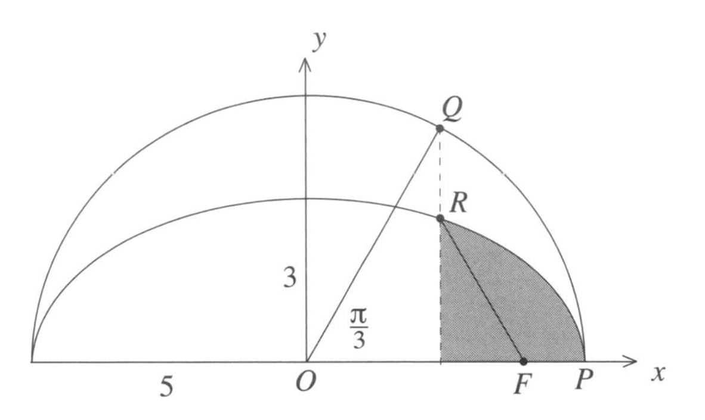

Fig. 5.50

axis 5 and semiminor axis 3.

i. Given that $\angle POQ = \frac{\pi}{3}$, check that $Q = (\frac{5}{2}, \frac{5}{2}\sqrt{3})$.

ii. Show that the graph of the function $y = \frac{3}{5}\sqrt{5^2 - x^2}$ is the half of the ellipse shown in the figure.

iii. Use the formula for the area of the sector of a circle (from Section 2.2) and Cavalieri's principle to compute the area of the shaded region of the figure.

iv. Compute the coordinates of the focal point F of the ellipse. Determine the area of the elliptical sector FPR.

5.65. Use the fundamental theorem of calculus to verify the equalities

$$\int_a^b cf(x)\, dx = c\int_a^b f(x)\, dx \quad \text{and} \quad \int_a^b (f(x) \pm g(x))\, dx = \int_a^b f(x)\, dx \pm \int_a^b g(x)\, dx.$$

5.66. Turn to Kepler's model of the Austrian barrel as depicted in Figure 5.39. Instead of taking the length of the slanting segment shown in the figure as $\frac{3}{2}r$, assume it to be $\frac{4}{3}r$. Under this assumption, the barrel's girth around its middle will be less. Calculate the volume of this "leaner" version of the Austrian barrel. Is the merchant's method for assessing the volume for this model more or less accurate than for Kepler's model of the Austrian barrel? Before answering this question, study the analysis of Kepler's cylinder in Section 5.5 and take a guess.

Chapter 6

The Calculus of Newton

This chapter begins with Newton's calculus of simple functions and describes how he uses power series in more complicated situations. The discussion then turns to the calculus of moving points on the line and in the plane and applies it to derive Galileo's parabolic trajectories. The chapter concludes with a look at essential aspects of Newton's *Principia*. This last segment begins with a study of force in general and centripetal force in particular, and finishes with an analysis of the essential aspects of Newton's planetary theory.

The circumstances surrounding the two geniuses who developed calculus into a powerful and widely applicable new mathematics could not have been more different. We saw that Leibniz was on a diplomatic mission to Paris when, stimulated by the intellectual life that this city offered, he turned to the serious pursuit of mathematics. Isaac Newton (1642–1727), on the other hand, was an English university student at Cambridge barely in his 20s. He had gone back to the family farm during the years 1665 and 1666 when the Great Plague ravaged London and forced the university to close. With his own extraordinary insights and great powers of concentration, he formulated the basic principles of the physics of motion, realized that they applied throughout the universe, and developed the mathematics that allowed him to extract the information that the basic laws provided. He demonstrated that the parabolic trajectories that Galileo had described and the elliptical orbits of the planets that Kepler had documented are both much more than observed realities—they are mathematical consequences of the fundamental laws of motion. Newton first informed his colleagues about his mathematical discoveries by circulating the manuscript *De Analysi per Aequationes Numero Terminorum Infinitas* in 1669. Isaac Barrow, Newton's teacher at Cambridge, was so impressed by the powerful mathematical display in "The Analysis by Means of Equations with an Infinite Number of Terms" that he resigned his professorship at Cambridge and arranged for Newton to succeed him. Newton delayed the publication of some of his discoveries (this manuscript would not be published until 1711) because some of his earlier published accounts had embroiled him in time-consuming disputes with contemporary scientists.

The astronomer Edmund Halley visited Cambridge in 1684 and asked Newton what the shape of the orbit of a planet would be under the action of a force directed to the Sun with a strength that is inversely proportional to the square of its distance from the Sun. Newton immediately answered that the orbit would be an ellipse and indicated that he had proved this long ago. Halley at once realized the importance of this work and persuaded Newton to publish a treatise that would set out his planetary theory in detail. The treatise appeared in the summer of 1687. Titled *Philosophiae Naturalis Principia Mathematica*, and known today simply as the *Principia*, it introduces the basic strategies of calculus and demonstrates their conceptual and computational power by applying it to the planetary orbits. The *Principia* along with Darwin's *Origin of Species* of 1859 are regarded to be the greatest books of science ever written.

After the publication of the *Principia*, Newton's scientific output declined. In 1689 he was elected as Cambridge University's member of Parliament. (Isaac Newton spoke in the House of Commons only once. As he rose, a hush descended on the House in anticipation of what he would have to say. Newton observed that there was an open window causing a draft and asked if it could be closed.) In 1696 he moved to London and assumed the post of warden and later that of master of the Mint. He took his duties there very seriously and oversaw the replacement of the coinage of England. In 1703 he was elected president of the Royal Society, and in 1704 his work *Optiks* appeared. By that time he was famous throughout the scientific world. In 1705, the queen conferred knighthood on Newton. The reviews of Newton the man are mixed. A look at his personal library reveals substantial interests in religion as well as alchemy. A look at his character finds him to be noble and sensitive, but also petty, suspicious, and withdrawn. He was a man who was admired and revered, but he aroused little affection.

Newton developed calculus earlier, but Leibniz published his version first. This became the basis for an ongoing bitter dispute about priority. Had Leibniz invented his calculus of differentials on his own, or had he "borrowed" from Newton? The controversy involved not only the two scientists themselves, but also their supporters. Today's scientific community is in agreement that the two men had invented their versions of the calculus independently. A comparison of their work leads to the conclusion that Leibniz had the better algebraic notation and the more penetrating presentation of the basic concepts, but that Newton's applications within his more geometric language were more powerful.

We'll explain some of Newton's important insights, such as power series and the calculus of moving points, by building on the discussion of Chapter 5. In order to so more effectively, we'll use today's notation and terminology, the language of functions, and the symbolism of limits (all already done in the explanation of the calculus of Leibniz). Only toward the end of the chapter in the discussion of Newton's theory of the motion of the planets, will we closely follow his arguments in order to gain a more concrete sense of how he thought.

6.1 SIMPLE FUNCTIONS AND AREAS

An early English translation (the scientific language of the 17th century was Latin) of Newton's *De Analysi* begins as follows:

> The General Method which I had devised some considerable time ago for measuring the quantity of curves by an infinite series of terms you have, in the following, rather briefly explained than narrowly demonstrated.
>
> To the base AB of some curve AD let the coordinate BD be perpendicular [see Figure 6.1]

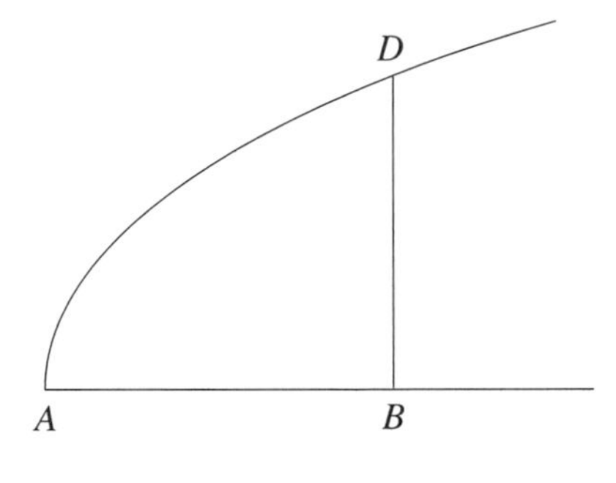

Fig. 6.1

> and let AB be called x and BD y.

Before he turns to the study of "curves by an infinite series of terms," Newton presents his calculus of simple functions, namely, functions of the form

$$f(x) = cx^r,$$

where c is a positive constant and r is a positive rational number. For any positive rational number $r = \frac{m}{n}$, with m and n positive integers, $x^r = x^{\frac{m}{n}}$ is either the nth root of the mth power $\sqrt[n]{x^m} = (x^m)^{\frac{1}{n}}$ of x, or, equivalently, the mth power of the nth root $(\sqrt[n]{x})^m = (x^{\frac{1}{n}})^m$ of x. If n is even, then x needs to satisfy $x \geq 0$. For instance, $x^{\frac{1}{2}} = \sqrt{x}$ does not make sense as a real number unless $x \geq 0$. The functions

$$f(x) = 4x^2,\ f(x) = 3x^{\frac{1}{2}} = 3\sqrt{x},\ f(x) = 6x^{\frac{3}{4}},\ f(x) = 9x^{\frac{7}{5}},\ \text{and}\ f(x) = 7x^3$$

are all examples of the simple functions that Newton examined.

The shape of the graph of $f(x) = cx^r$ depends on both r and c, but the three basic cases to consider are

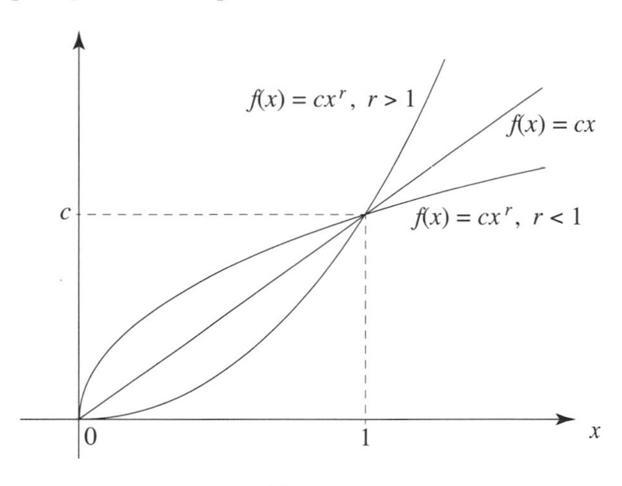

Fig. 6.2

$r < 1$, $r = 1$, and $r > 1$. They are captured in Figure 6.2.

The fundamental theorem of calculus—see Section 5.8—depends critically on the answer to the following question: Does any function have an antiderivative? Or, better, under what conditions does a function have an antiderivative? Newton answered this question by showing that the area function defined by the graph of a simple function provides an antiderivative for it. (In this regard, Newton approached the matter of areas of curving shapes as Leibniz did in Section 5.6.)

Here is what Newton does. Figure 6.3 presents a typical example of a simple function. (The fact that the figure considers the case $r < 1$ is not relevant.) For any $x \geq 0$, Newton lets $A(x)$ be the area under the graph of $y = cx^r$ over the interval from 0 to x. Add a Δx to the x of Figure 6.3. We'll take $\Delta x > 0$, but

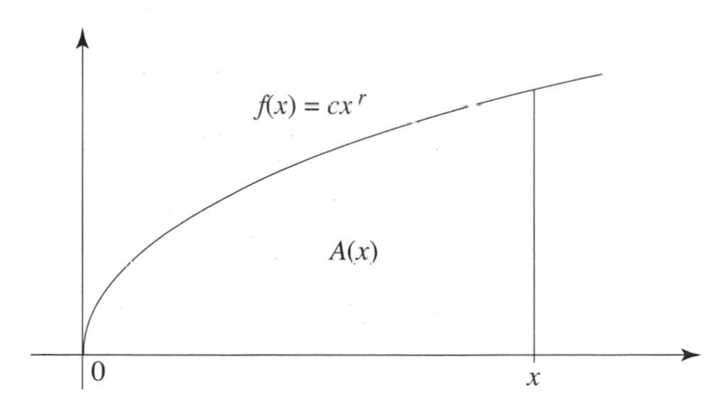

Fig. 6.3

the argument that follows is easily be adapted to the case $\Delta x < 0$. Figure 6.4 shows x and $x + \Delta x$ as well as the two vertical lines that x and $x + \Delta x$ determine. The difference $A(x + \Delta x) - A(x)$ is the area under the graph of $y = f(x)$ above the x-axis and between the two vertical lines. Focus on the horizontal dotted segment between the two lines. When this segment is below the graph, the area of the rectangle that the interval from x to $x + \Delta x$ and the dotted line determine is less than $A(x + \Delta x) - A(x)$. Moving the dashed

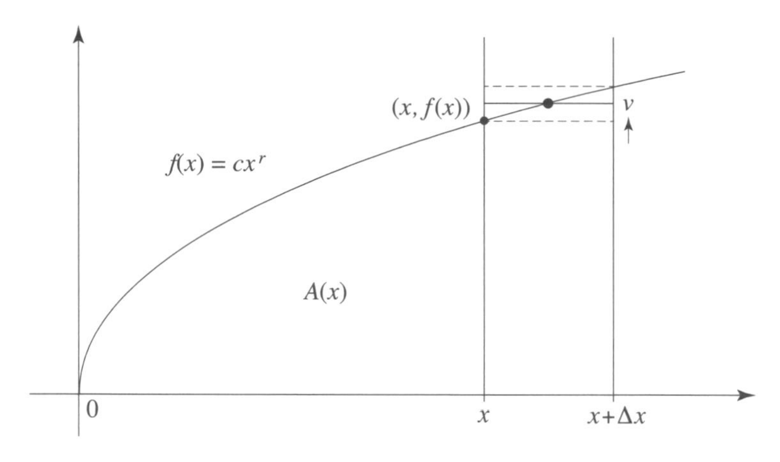

Fig. 6.4

horizontal segment up increases the area of the rectangle. When the dotted line is above the graph, the area of this rectangle is greater than $A(x + \Delta x) - A(x)$. It follows that there is some height v such that the area $v\Delta x$ of the rectangle is equal to $A(x + \Delta x) - A(x)$. Therefore

$$\frac{A(x + \Delta x) - A(x)}{\Delta x} = v.$$

Keeping x fixed, push Δx to zero, and observe that v gets pushed to the y-coordinate $f(x)$ in the process. It follows that

$$\lim_{\Delta x \to 0} \frac{A(x + \Delta x) - A(x)}{\Delta x} = f(x).$$

The left side is the derivative $A'(x)$ of $A(x)$. So Newton's argument has shown that the derivative $A'(x)$ of the area function $A(x)$ of the function $f(x)$ is the function $f(x)$. To repeat:

The area function $A(x)$ of a given simple function $y = f(x)$ is an antiderivative of the function.

What Newton has established for simple functions is closely related to what has already been discussed more generally in Section 5.8. For example, the understanding that $\int f(x)\,dx$ is the generic antiderivative or indefinite integral of $y = f(x)$ tells us that

$$\int f(x)\,dx = A(x) + C.$$

Combining Newton's fact with the fundamental theorem of calculus informs us that for a simple function $y = f(x)$,

$$\int_a^b f(x)\,dx = A(b) - A(a).$$

The definition of the area function $A(x)$ confirms that $\int_a^b f(x)\,dx$ is the area under the graph of $y = f(x)$ and over the interval $a \leq x \leq b$.

6.2 THE DERIVATIVE OF A SIMPLE FUNCTION

Newton considers any simple function $f(x) = cx^r$, where r is any positive rational number, and c is any constant (c can now be negative as well), and develops the formula

$$f'(x) = crx^{r-1}$$

for the derivative of the function. His approach to the antiderivative had been entirely geometric. His derivation of the formula for the derivative is algebraic.

The way Newton proceeds was already illustrated in Leibniz's context of Section 5.2. He fixes any point $P = (x, y)$ on the graph of $y = f(x)$. He takes another point $Q = (x', y')$ on the graph and considers the line that the two points determine. Newton then pushes Q to P and makes use of the fact that, in the process, the slope of the line through P and Q closes in on the slope of the tangent to the graph at $P = (x, y)$. The slope of the tangent at P is the derivative $f'(x)$ that Newton is after.

Let's carry out the details. Let $r = \frac{m}{n}$ with m and n positive integers. Put $x' - x = \Delta x$ and $y' - y = \Delta y$. So $x' = x + \Delta x$ and $y' = y + \Delta y$. It follows that $Q = (x + \Delta x, y + \Delta y)$ and that the slope of the line through

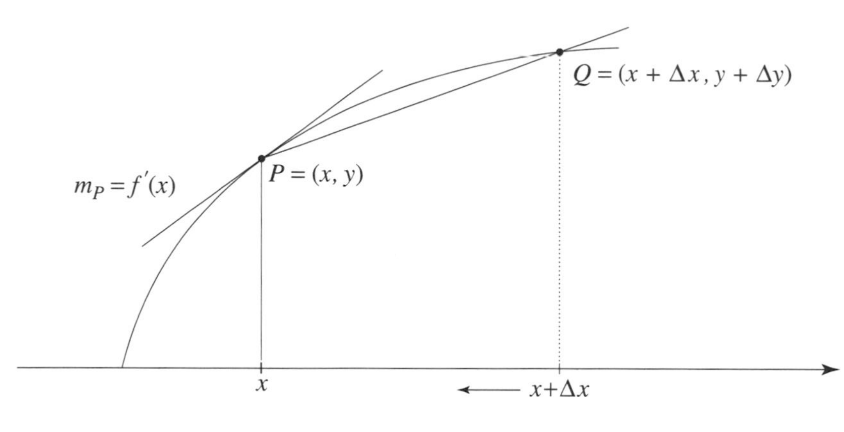

Fig. 6.5

P and Q is equal to $\frac{\Delta y}{\Delta x}$. Pushing Q to P amounts to the same thing as pushing Δx to zero. See Figure 6.5. So it is the limit

$$\lim_{\Delta x \to 0} \frac{\Delta y}{\Delta x}$$

that Newton needs to determine. As was the case with Leibniz, Newton wrestles with the articulation of his ideas. He does not make explicit use of the limit concept, so that he describes the limit involved in the definition of the derivative in this way:

> by the ultimate ratio of evanescent quantities is to be understood the ratio of the quantities not before they vanish, not afterwards, but with which they vanish.

Newton continues his analysis by raising both sides of $y = cx^{\frac{m}{n}}$ to the nth power to get $y^n = c^n x^m$. Because $Q = (x + \Delta x, y + \Delta y)$ is on the graph,

$$(y + \Delta y)^n = c^n (x + \Delta x)^m.$$

To multiply out the n factors

$$(y + \Delta y)^n = (y + \Delta y)(y + \Delta y) \cdots (y + \Delta y),$$

the y and Δy from any one group $(y + \Delta y)$ must be multiplied by the y and Δy from each of the other $n-1$ groups. The product $y \cdot y \cdots y = y^n$ is one term that arises in this way. Fixing any Δy and multiplying it by the y from each of the other $n - 1$ groups gives the product $y^{n-1}\Delta y$. Since n different Δy's can be picked to do this, $y^{n+1}\Delta y$ will occur n times. The result of the multiplication will therefore be

$$(y + \Delta y)^n \;=\; y^n + ny^{n-1}\Delta y \;+\; \text{more terms.}$$

Notice that each of these additional terms must contain at least two factors of Δy (because the terms that contain no or one Δy are already accounted for). Doing the same thing with $(x + \Delta x)^m$, Newton arrives at an equation of the form

$$y^n + ny^{n-1}\Delta y + \text{terms with } (\Delta y)^2 \text{ as factor} = c^n\Big(x^m + mx^{m-1}\Delta x + \text{terms with } (\Delta x)^2 \text{ as factor}\Big).$$

Because $P = (x, y)$ is on the graph, $y^n = c^n x^m$. Therefore after a subtraction,

$$ny^{n-1}\Delta y + \text{ terms with } (\Delta y)^2 \text{ as factor} = c^n\Big(mx^{m-1}\Delta x + \text{terms with } (\Delta x)^2 \text{ as factor}\Big).$$

Now divide both sides of this equation by Δx to get

$$ny^{n-1}\tfrac{\Delta y}{\Delta x} + \text{ terms with } \Delta y \cdot \tfrac{\Delta y}{\Delta x} \text{ as factor} = c^n\Big(mx^{m-1} + \text{terms with } \Delta x \text{ as factor}\Big).$$

Next, push Δx to zero. Since Δy also goes to 0 in the process (see Figure 6.5), all terms with $\Delta y \cdot \frac{\Delta y}{\Delta x}$ as factor go to zero, and because $\displaystyle\lim_{\Delta x \to 0} \frac{\Delta y}{\Delta x} = m_P = f'(x)$, it follows that only

$$(*) \qquad ny^{n-1}f'(x) = c^n mx^{m-1}$$

remains. Recalling that $y = cx^{\frac{m}{n}}$, and raising both sides to the $n-1$ power, we get by simple algebra that

$$y^{n-1} = c^{n-1}x^{\frac{m}{n}(n-1)} = c^{n-1}x^{m-\frac{m}{n}} = c^{n-1}x^m x^{-\frac{m}{n}}.$$

By substituting this for y^{n-1} in equation $(*)$ above, we obtain

$$nc^{n-1}x^m x^{-\frac{m}{n}} f'(x) = c^n mx^{m-1}.$$

After canceling c^{n-1} and x^{m-1} from both sides, we have $nx \cdot x^{-\frac{m}{n}} f'(x) = cm$. Since $\frac{m}{n} = r$, we get $nx^{1-r}f'(x) = cm$ and, finally,

$$f'(x) = c \cdot \tfrac{m}{n} \cdot \frac{1}{x^{1-r}} = crx^{r-1}.$$

Newton has derived the formula

$$\boxed{f'(x) = crx^{r-1}}$$

for the derivative of the simple function $f(x) = cx^r$, where r is any positive rational number, c any constant. We will see in Sections 7.5 and 7.11 that this formula holds for any real constant r.

Example 6.1. For $f(x) = 3x^{100}$, $f'(x) = 300x^{99}$, and for $g(x) = 2x^5 - 4x^{\frac{3}{5}}$, $g'(x) = 10x^4 - \frac{12}{5}x^{-\frac{2}{5}}$.

With $c = 1$, the formula asserts that the derivative of the function $f(x) = x^r$ is $f'(x) = rx^{r-1}$. Turn to Section 5.4 and observe that the examples (a)–(d) follow from this fact. By applying Newton's derivative result for $r > -1$, we see that the derivative of $\frac{1}{r+1}x^{r+1}$ is x^r. So $\frac{1}{r+1}x^{r+1}$ is an antiderivative of x^r. In restated form, this is the formula

$$\boxed{\int x^r\,dx = \tfrac{1}{r+1}\,x^{r+1} + C, \text{ whenever } r > -1.}$$

Example 6.2. Find the area function $A(x)$ for $f(x) = x^3$. Since $F(x) = \frac{1}{4}x^4$ is an antiderivative of f and $A(x)$ is as well, we know that $A(x) = \frac{1}{4}x^4 + C$. Since $A(0) = 0$, we see that $C = 0$, and therefore $A(x) = \frac{1}{4}x^4$.

6.3 FROM SIMPLE FUNCTIONS TO POWER SERIES

Having dealt with simple functions and sums of simple functions, the *De Analysi* turns to the problem of determining areas under more complicated curves. Newton begins this discussion as follows:

> But if the value of y, or any of its terms be more compounded than the foregoing, it must be reduced into more simple terms; by performing the operation in letters, after the same manner as arithmeticians divide in decimal numbers, extract the square root, or resolve affected equations; and afterwards by the preceding rules you will discover the Superficies of the curve sought.

What Newton realized is that is that a more "compounded" or complicated function can be reduced to simpler terms by representing it as an infinite series. Such a representation makes it possible to solve some problems, including area problems, that are not solvable by other means. The central idea is that

Fig. 6.6

some infinite sums can add up to a finite number. The progression of 1×1 squares and a pattern of black regions inside them provide an example. Continuing the pattern and adding the area of the black region that is drawn in at each step—see Figure 6.6—tells us that

$$\tfrac{1}{2} + \tfrac{1}{4} + \tfrac{1}{8} + \tfrac{1}{16} + \cdots = 1.$$

How did Newton make use of this idea? One of his examples considered the function $f(x) = \frac{1}{1+x}$. The "operation in letters" that he applied to it is polynomial division. Dividing $1 + x$ into 1, Newton gets

$$\begin{array}{rl} & \;1 \\ 1 + x & \overline{)\;1} \\ \text{subtract} \longrightarrow & \;\underline{1 + x} \\ & \;\;\; - x \end{array}$$

Step 1

after the first step, and

$$\begin{array}{rl} & \;1 - x \\ 1 + x & \overline{)\;1} \\ & \;\underline{1 + x} \\ & \;\;\; - x \\ \text{subtract} \longrightarrow & \;\underline{- x - x^2} \\ & \;\;\;\;\;\;\; x^2 \end{array}$$

Step 2

after the second. This division process can be continued indefinitely. Below is the state of the computations after Step 5. The evolving pattern should now be clear. If $|x| < 1$, then the successive remainders

$$-x, x^2, -x^3, x^4, -x^5, \ldots$$

$$
\begin{array}{r|l}
 & 1 - x + x^2 - x^3 + x^4 \\
\hline
1 + x & 1 \\
 & \underline{1 + x} \\
 & \quad - x \\
 & \quad \underline{- x - x^2} \\
 & \qquad\quad x^2 \\
 & \qquad\quad \underline{x^2 + x^3} \\
 & \qquad\qquad - x^3 \\
 & \qquad\qquad \underline{- x^3 - x^4} \\
 & \qquad\qquad\qquad x^4 \\
 & \qquad\qquad\qquad \underline{x^4 + x^5} \\
 & \qquad\qquad\qquad\quad - x^5
\end{array}
$$

Step 5

become smaller and smaller, so that $\frac{1}{1+x}$ is approximated by $1 - x + x^2 - x^3 + x^4 - x^5 + \ldots$. The more terms that are included, the smaller the $\pm x^i$, and the better the approximation. For $x = -\frac{1}{2}$, for instance, we get

$$1 + \tfrac{1}{2} + \tfrac{1}{4} + \tfrac{1}{8} + \tfrac{1}{16} + \ldots = \frac{1}{\frac{1}{2}} = 2,$$

a result that follows immediately from the sequence of 1×1 squares just discussed. Taking $x = -\frac{1}{4}$ tells us that

$$1 + \tfrac{1}{4} + (\tfrac{1}{4})^2 + (\tfrac{1}{4})^3 + (\tfrac{1}{4})^4 + \ldots = \tfrac{4}{3}.$$

This equality had already been observed by Archimedes. See Section 2.3.

If only the first few terms of the infinite sum are used, then approximations are achieved. These become more and more accurate as more and more terms are added in. For example,

$$1 + \tfrac{1}{4} + (\tfrac{1}{4})^2 = 1.3125,\ 1 + \tfrac{1}{4} + (\tfrac{1}{4})^2 + (\tfrac{1}{4})^3 \approx 1.3281, \text{ and } 1 + \tfrac{1}{4} + (\tfrac{1}{4})^2 + (\tfrac{1}{4})^3 + (\tfrac{1}{4})^4 \approx 1.3320$$

get closer and closer to $\frac{4}{3} \approx 1.33333333\ldots$.

We'll now let $x = 0.50$. Then $\frac{1}{1+x} = \frac{1}{1.5} = \frac{2}{3}$. Working with accuracy to four decimal places (for instance), we get the following pyramid.

$$
\begin{gathered}
1{-}x = 1.000{-}0.5000 = 0.5000 \\
1{-}x{+}x^2 = 0.5000{+}0.2500 = 0.7500 \\
1{-}x{+}x^2{-}x^3 = 0.7500{-}0.1250 = 0.6250 \\
1{-}x{+}x^2{-}x^3{+}x^4 = 0.6250{+}0.0625 = 0.6875 \\
1{-}x{+}x^2{-}x^3{+}x^4{-}x^5 = 0.6875{-}0.0313 = 0.6562 \\
1{-}x{+}x^2{-}x^3{+}x^4{-}x^5{+}x^6 = 0.6562{+}0.0156 = 0.6718 \\
1{-}x{+}x^2{-}x^3{+}x^4{-}x^5{+}x^6{-}x^7 = 0.6718{-}0.0078 = 0.6640 \\
1{-}x{+}x^2{-}x^3{+}x^4{-}x^5{+}x^6{-}x^7{+}x^8 = 0.6640{+}0.0039 = 0.6679 \\
1{-}x{+}x^2{-}x^3{+}x^4{-}x^5{+}x^6{-}x^7{+}x^8{-}x^9 = 0.6679{-}0.0020 = 0.6659 \\
1{-}x{+}x^2{-}x^3{+}x^4{-}x^5{+}x^6{-}x^7{+}x^8{-}x^9{+}x^{10} = 0.6659{+}0.0010 = 0.6669
\end{gathered}
$$

As expected, the numbers running down the pyramid on the right close in on $\frac{1}{1+x} = \frac{2}{3} = 0.6666\ldots$. For a positive number x smaller than $0.5 = \frac{1}{2}$, the remainder terms $\pm x^i$ will get smaller faster, and the approximations of $\frac{1}{1+x}$ by

$$1 - x + x^2 - x^3 + x^4 - x^5 + x^6 - \ldots$$

will be better. In particular, it follows from the fifth layer of the pyramid that the function $f(x) = \frac{1}{1+x}$ is approximated by the polynomial function $1 - x + x^2 - x^3 + x^4 - x^5$ to within $0.6667 - 0.6562 \approx 0.0105$ over the entire interval $0 \leq x \leq \frac{1}{2}$.

By exploiting this fact, Newton arrived at the approximation

$$\int_0^{\frac{1}{2}} \frac{1}{1+x}\,dx \approx \int_0^{\frac{1}{2}} (1 - x + x^2 - x^3 + x^4 - x^5)\,dx.$$

Observe that this integral represents the area under the graph of $f(x) = \frac{1}{1+x}$ between $x = 0$ and $x = \frac{1}{2}$. See Figure 6.7. The definite integral on the right is easily evaluated. Taking antiderivatives term by term

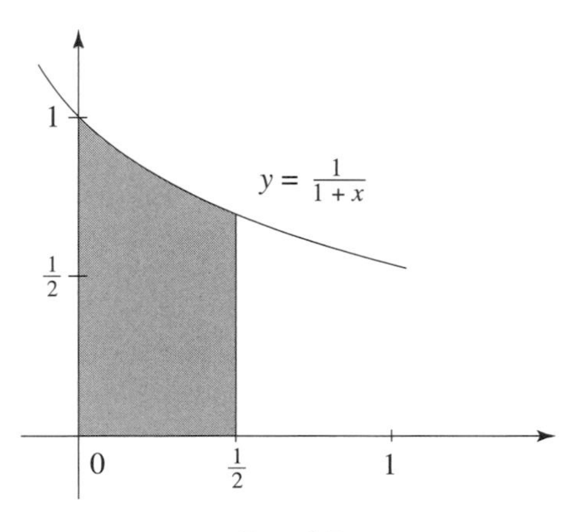

Fig. 6.7

shows that

$$F(x) = x - \tfrac{1}{2}x^2 + \tfrac{1}{3}x^3 - \tfrac{1}{4}x^4 + \tfrac{1}{5}x^5 - \tfrac{1}{6}x^6$$

is an antiderivative of $1 - x + x^2 - x^3 + x^4 - x^5$. Therefore, by the fundamental theorem of calculus,

$$\begin{aligned}
\int_0^{\frac{1}{2}} \frac{1}{1+x}\,dx &\approx \int_0^{\frac{1}{2}} (1 - x + x^2 - x^3 + x^4 - x^5)\,dx \\
&= F(\tfrac{1}{2}) - F(0) = F(\tfrac{1}{2}) = \tfrac{1}{2} - \tfrac{1}{2}(\tfrac{1}{2})^2 + \tfrac{1}{3}(\tfrac{1}{2})^3 - \tfrac{1}{4}(\tfrac{1}{2})^4 + \tfrac{1}{5}(\tfrac{1}{2})^5 - \tfrac{1}{6}(\tfrac{1}{2})^6 \\
&\approx 0.5 - 0.125 + 0.042 - 0.016 + 0.007 - 0.003 \\
&= 0.405.
\end{aligned}$$

If greater accuracy is required, we can use more terms in the approximation for $\frac{1}{1+x}$ and carry more decimals. Take

$$1 - x + x^2 - x^3 + x^4 - x^5 + x^6 - x^7 + x^8 - x^9 + x^{10},$$

for instance. Continuing the pattern of the preceding computation for five more steps and rounding off at the sixth decimal place shows that

$$\begin{aligned}
\int_0^{\frac{1}{2}} \frac{1}{1+x}\,dx &\approx \tfrac{1}{2} - \tfrac{1}{2}\left(\tfrac{1}{2}\right)^2 + \tfrac{1}{3}\left(\tfrac{1}{2}\right)^3 - \ldots + \tfrac{1}{11}\left(\tfrac{1}{2}\right)^{11} \\
&\approx 0.500000 - 0.125000 + 0.041667 - 0.015625 + 0.006250 - 0.002604 \\
&\quad + 0.001116 - 0.000488 + 0.000217 - 0.000098 + 0.000044 \\
&= 0.405479.
\end{aligned}$$

How accurate are the approximations that this procedure achieves? To decide, use the following rule of thumb. To evaluate $\int_0^{\frac{1}{2}} \frac{1}{1+x}\,dx$ with an accuracy of up to, say, four decimal places, keep computing terms (and adding and subtracting them) until the term reached is zero when rounded to four decimal places. The approximation obtained at this point should be accurate up to four decimal places. In the example, the term that rounds to zero is 0.000044. So the approximation 0.405479 should be accurate up to four decimals. This is correct since the value of the integral $\int_0^{\frac{1}{2}} \frac{1}{1+x}\,dx$ up to six-decimal-place accuracy is 0.405465. The determination of the precise value of this integral makes use of logarithms and will be discussed later in the text (in Section 7.11). A word of caution is in order. This rule of thumb calls for numbers to be rounded off. This procedure can lead to errors that are small in the current context but that can be large in others. Such errors are known as *roundoff errors*.

Any infinite sum of the form

$$a + ax + ax^2 + \cdots + ax^{k-1} + ax^k + \dots,$$

where a is a nonzero constant, is called a *geometric series*. Newton knew that any such series is quickly analyzed as follows. For any positive integer n, let

$$S_n = a + ax + ax^2 + \cdots + ax^{n-2} + ax^{n-1}.$$

Since $xS_n = ax + ax^2 + \cdots + ax^{n-1} + ax^n$, we get that $S_n(1-x) = S_n - xS_n = a - ax^n$. Therefore

$$S_n = \frac{a - ax^n}{1-x}.$$

Suppose that $|x| < 1$, and push n to infinity. Since x^n goes to zero, we have established that

$$a + ax + ax^2 + \cdots + ax^{k-1} + ax^k + \dots = \frac{a}{1-x}.$$

In sigma notation, the geometric series in expressed as

$$\sum_{k=0}^{\infty} ax^k = \frac{a}{1-x}.$$

This sigma notation was invented to ease the notational effort of writing down long sums. The notation works this way in this example (and similarly in all others): first plug $k = 0$ into ax^k to get $ax^0 = a$, then plug $k = 1$ into ax^k to get ax, and keep going and keep adding.

For no x with $|x| \geq 1$, does the geometric series add to a finite number. For instance, for $x = 1$, the sum $a + a + a + \cdots +$ keeps getting larger and larger. For $x = -1$, the sum $a - a + a - a + \dots$ jumps back and forth between a and 0 and hence does not add up to a fixed finite number. Taking $a = 1$ and replacing x by $-x$ leads to the infinite series that Newton got with his division process. Use of the geometric series determined by $a = x = \frac{1}{2}$ confirms that

$$\tfrac{1}{2} + \tfrac{1}{4} + \tfrac{1}{8} + \tfrac{1}{16} + \cdots = \frac{\frac{1}{2}}{1-\frac{1}{2}} = 1.$$

With $a = 1$ and $x = \frac{1}{4}$, we get the earlier result

$$1 + \tfrac{1}{4} + \tfrac{1}{8} + \tfrac{1}{16} + \cdots = \frac{1}{1-\frac{1}{4}} = \tfrac{4}{3}.$$

Infinite sums involving powers of a variable are called *power series* in today's mathematics. If x is the variable, we will refer to it as a *power series in x*. If a power series in x has a finite sum for a specific value of x, we say that the power series *converges* for that value. If it does not converge, it is said to *diverge*. For example, we have seen that the power series $1 - x + x^2 - x^3 + x^4 - x^5 + \dots$ converges for $x = -\frac{1}{4}$ and that the sum

$$1 + \tfrac{1}{4} + \left(\tfrac{1}{4}\right)^2 + \left(\tfrac{1}{4}\right)^3 + \left(\tfrac{1}{4}\right)^4 + \dots.$$

is equal to $\frac{4}{3}$. If only the first three, four, one hundred, or one thousand terms are added, then only an approximation of $\frac{4}{3}$ is achieved.

Useful power series can be obtained from other power series by substitution. For example, replacing x by $x - 1$ in the power series

$$\frac{1}{1+x} = 1 - x + x^2 - x^3 + x^4 - x^5 + \dots$$

leads to the series

$$\frac{1}{x} = 1 - (x-1) + (x-1)^2 - (x-1)^3 + (x-1)^4 - (x-1)^5 + \dots.$$

Since the power series for $\frac{1}{1+x}$ converges for all x with $|x| < 1$, this power series for $\frac{1}{x}$ converges for all x with $|x - 1| < 1$. This is the set of x satisfying $0 < x < 2$. Check that in sigma notation,

$$\frac{1}{x} = \sum_{k=0}^{\infty}(-1)^k(x-1)^k.$$

Over the interval $0 < x < 2$, the series can be antidifferentiated term by term to obtain an answer to the question

$$\int \frac{1}{x}\,dx = ?$$

that escaped Newton's integral formula of Section 6.2. We will see in Section 7.11 that the natural logarithm function provides the definitive answer to this question.

For another example that demonstrates the relevance of substitution, replace x with $\frac{1}{9}x^3$ in the power series for $\frac{1}{1+x} = 1 - x + x^2 - x^3 + x^4 - x^5 + \dots$ to get

$$\frac{1}{1+\frac{1}{9}x^3} = 1 - \tfrac{1}{9}x^3 + \tfrac{1}{9^2}x^6 - \tfrac{1}{9^3}x^9 + \tfrac{1}{9^4}x^{12} - \tfrac{1}{9^5}x^{15} + \dots.$$

This power series converges for $|\frac{1}{9}x^3| < 1$, so for $|x^3| < 9$, and hence for $|x| < 9^{\frac{1}{3}}$. Because $\frac{1}{9+x^3} = \frac{1}{9(1+\frac{1}{9}x^3)} = \frac{1}{9}\,\frac{1}{1+\frac{1}{9}x^3}$, we find that

$$\frac{1}{9+x^3} = \tfrac{1}{9} - \tfrac{1}{9^2}x^3 + \tfrac{1}{9^3}x^6 - \tfrac{1}{9^4}x^9 + \tfrac{1}{9^5}x^{12} - \tfrac{1}{9^6}x^{15} + \dots.$$

Because $2 = 8^{\frac{1}{3}} < 9^{\frac{1}{3}}$, we get by antidifferentiating term by term and applying the fundamental theorem of calculus that

$$\begin{aligned}
\int_0^2 \frac{1}{9+x^3}\,dx &= \left(\tfrac{1}{9}x - \tfrac{1}{4\cdot9^2}x^4 + \tfrac{1}{7\cdot9^3}x^7 - \tfrac{1}{10\cdot9^4}x^{10} + \tfrac{1}{13\cdot9^5}x^{13} - \tfrac{1}{16\cdot9^6}x^{16} + \dots\right)\Big|_0^2 \\
&= \tfrac{1}{9}2 - \tfrac{1}{4\cdot9^2}2^4 + \tfrac{1}{7\cdot9^3}2^7 - \tfrac{1}{10\cdot9^4}2^{10} + \tfrac{1}{13\cdot9^5}2^{13} - \tfrac{1}{16\cdot9^6}2^{16} + \dots \\
&\approx 0.222222 - 0.049383 + 0.025083 - 0.015607 + 0.010672 - 0.007707 + \dots \\
&\approx 0.185280.
\end{aligned}$$

A much tighter value of this integral— accurate up to six decimal places—is 0.188556. But this is hard to achieve. The fact that the antiderivative of the function $f(x) = \frac{1}{9+x^3}$ is complicated—it involves concepts from Sections 7.11 and 9.9 as well as computational trickery—means that it is difficlut to apply the fundamental theorem. Turning to the power series solution above, notice that the first six terms of the infinite series that provide the approximation 0.185280 get smaller, but they do so slowly. How many terms of this series are needed to obtain the value 0.188556? Consider the terms $\frac{1}{9}2$, $-\frac{1}{4\cdot 9^2}2^4$, $+\frac{1}{7\cdot 9^3}2^7$, and so on in the sum above. Observe that the exponent of the 9 in the denominator counts the number of the term, and if k is the number of the term, then $3k-2$ provides the other factor of the denominator as well as the exponent of 2. So the tenth term is $\frac{1}{28\cdot 9^{10}}2^{28} \approx 0.002750$, the twentieth is $\frac{1}{58\cdot 9^{20}}2^{58} \approx 0.000409$, and the thirtieth is $\frac{1}{88\cdot 9^{30}}2^{88} \approx 0.000083$. Since this last term rounds to 0.0001 at the fourth decimal, we need to continue the process to obtain the sought-after six-decimal-place accuracy. The fortieth term is $\frac{1}{118\cdot 9^{40}}2^{118} \approx 0.000019$, the fiftieth is $\frac{1}{148\cdot 9^{50}}2^{148} \approx 0.000005$, the sixtieth is $\frac{1}{178\cdot 9^{60}}2^{178} \approx 0.000001$, and the sixty-fifth term is $\frac{1}{193\cdot 9^{65}}2^{193} \approx 0.0000006$. Finally, the sixty-seventh term $\frac{1}{199\cdot 9^{67}}2^{199} \approx 0.00000046$ rounds to 0 in the sixth decimal. So according to our rule of thumb, 67 terms are required for the tight approximation 0.188556 of the integral. These computations would have been labor intensive for Newton. However, today's calculators and computers handle them with ease. Incidentally, our calculators make use of power series when computing trigonometric and logarithmic quantities (among others).

With his power series, Newton introduced a potent method of calculation that is applicable to a wide span of mathematical problems. Power series have remained in a central position in mathematics ever since Newton first made use of them. Section 11.7 takes up this subject in more definitive terms.

6.4 THE MATHEMATICS OF A MOVING POINT

Two years after his *De Analysi* of 1669, Newton presented a second, more extensive exposition of his ideas in the treatise *Methodus Fluxionum et Series Infinitarum* (*Method of Fluxions and Infinite Series*). It would not be published until 1736. After discussing the role of power series as a tool for computing areas under curves, Newton went on to say,

> So much for computational methods of which in the sequel I shall make frequent use. It now remains, in illustration of this analytic art, to deliver some typical problems and such especially as the nature of curves represent. But first of all I would observe that difficulties of this sort may all be reduced to these two problems alone, which I may be permitted to propose with regard to the space traversed by any local motion however accelerated or retarded:
>
> (1) Given the length of the space continuously (that is at every time), to find the speed of motion at any time proposed.
>
> (2) Given the speed of motion continuously, to find the length of the space described at any time proposed.

What is Newton saying here? Consider a point or a particle moving in the Cartesian plane. Take a stopwatch and start it at time $t = 0$. After any time $t \geq 0$, it will have some position in the plane. Let the x-coordinate of this position be $x(t)$ and the y-coordinate $y(t)$. The position of the point and hence the coordinates $x(t)$ and $y(t)$ will vary with t. Both $x(t)$ and $y(t)$ are functions of t. Newton calls such functions *fluents*. Figure 6.8 shows a typical situation. At $(x(t), y(t))$, the point will have a certain velocity in the x-direction and also in the y-direction. They are also functions of t. Newton calls them the *fluxions* of $x(t)$ and $y(t)$. The two problems that Newton singled out are these:

(1) Given the functions $x(t)$ and $y(t)$ that determine the position of the point at any time t, find the velocities in the x- and y-directions that together determine the velocity of the point for any t.

(2) Given the velocities of the point in the x- and y-directions at any time t, find the functions $x(t)$ and $y(t)$ that determine its position for any t.

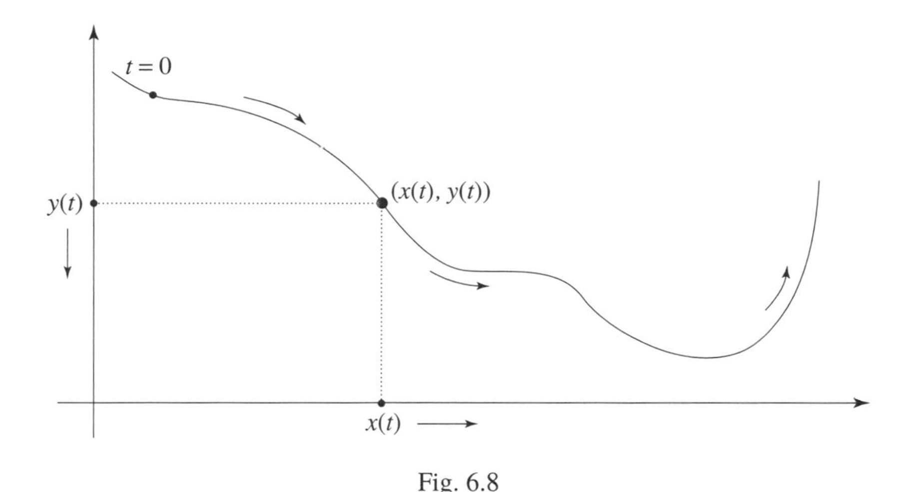

Fig. 6.8

Briefly put, to illustrate "this analytic art," Newton proposed to develop the calculus of motion. He presented precisely what is necessary for a fundamental understanding of the basic observations of Kepler about the orbits of the planets in the solar system (in Section 3.5) as well as those of Galileo about the paths of projectiles (in Section 3.6). Newton's mathematics of motion and its application will be the focus of the rest of the chapter. In reference to Figure 6.8, observe that if the functions $x(t)$ and $y(t)$ are both understood, then the motion of the point is understood. Therefore the mathematics of the motion of a point in the plane (and also in space) reduces to that of a point moving along a coordinate line.

Suppose a point is moving on a coordinate axis. We'll assume that it moves smoothly without any sudden jumps, fits, and restarts. Begin to observe it at time $t = 0$. Let $p(t)$ be the function that specifies its coordinate at any time $t \geq 0$. It is the *position function* of the point. See Figure 6.9. The position $p(0)$ is the *initial position* of the point. The units of distance and time can be meters and seconds, miles and hours,

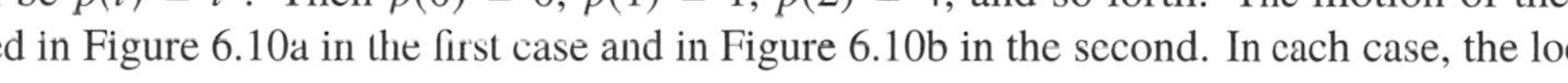

$p(0)$ $p(t)$

Fig. 6.9

and so forth, as a specific problem would provide them.

Suppose that p is given explicitly by $p(t) = t$. So the position of the point on the number line is $p(0) = 0$ at time $t = 0$, $p(1) = 1$ at time $t = 1$, $p(2) = 2$ at time $t = 2$, and so on. Next let the position function be $p(t) = t^2$. Then $p(0) = 0$, $p(1) = 1$, $p(2) = 4$, and so forth. The motion of the point is illustrated in Figure 6.10a in the first case and in Figure 6.10b in the second. In each case, the location of

0 1 2 3
$t = 0$ $t = 1$ $t = 2$ $t = 3$
(a)

0 1 4
$t = 0$ $t = 1$ $t = 2$
(b)

Fig. 6.10

the point is listed above the coordinate axis, and the time at which the point is at a given location is indicated below it.

Suppose next that $p(t) = t^2 - 4t + 3$. Let's begin by investigating the graph of p with horizontal axis the t-axis of time. By Section 4.3, the graph is a parabola that opens upward. The derivative $p'(t) = 2t - 4$ is 0 when $t = 2$. Since $p(2) = 4 - 8 + 3 = -1$, it follows that the parabola has a horizontal tangent at $(2, -1)$. This point is at the bottom of the parabola, so the smallest value that $p(t)$ can have is -1, and it occurs when $t = 2$. Since $t^2 - 4t + 3 = (t - 1)(t - 3)$, it follows that $p(t) = 0$ precisely when $t = 1$ or $t = 3$. The graph of p is sketched in Figure 6.11. Keep in mind that this is the graph of the position function, not

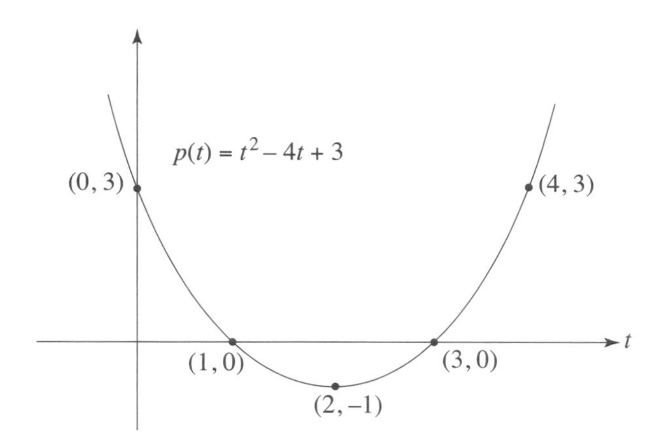

Fig. 6.11

the path of the moving point. We'll now turn to the study of the motion of the point determined by the position function

$$p(t) = t^2 - 4t + 3 = (t-1)(t-3)$$

for $t \geq 0$. Since $p(0) = 3$, the initial position is the point 3 on the axis. Since $p(t)$ decreases initially (the graph in Figure 6.11 falls), the point moves to the left from its initial position and reaches the point 0 at $t = 1$ and the point -1 at $t = 2$. There it stops, changes direction (note that the graph begins to rise), and at $t = 3$ it is back at 0. At $t = 4$ it is at 3 and continues moving to the right. The motion of the point is

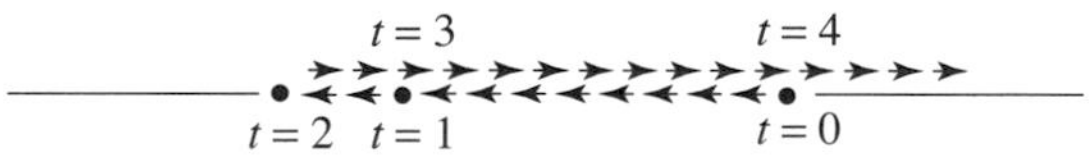

Fig. 6.12

described schematically in Figure 6.12. The three relevant points $-1, 0$, and 3 are singled out on the axis.

Let's study the general case of a point moving on a coordinate axis with position function $p(t)$. Fix some instant $t_1 \geq 0$. Let an additional amount of time Δt elapse, and consider the instant $t_1 + \Delta t$. We have

$$\begin{aligned}
p(t_1) &= \text{the position at time } t_1\\
p(t_1+\Delta t) &= \text{the position at time } t_1 + \Delta t\\
p(t_1+\Delta t) - p(t_1) &= \text{the change in position over the time interval from } t_1 \text{ to } t_1 + \Delta t\\
&= \text{the net distance (possibly negative) covered by the point in the time interval from } t_1 \text{ to } t_1 + \Delta t\\
\frac{p(t_1+\Delta t) - p(t_1)}{\Delta t} &= \text{the net distance covered divided by the time taken to cover the distance}\\
&= \text{the definition of the } \textit{average velocity} \text{ over the time interval from } t_1 \text{ to } t_1 + \Delta t.
\end{aligned}$$

To make things more concrete, return to the point with $p(t) = t^2 - 4t + 3$ for $t \geq 0$, and suppose that distance is measured in meters and time in seconds. Take $t_1 = 1$ second. Now the point is exactly 1 second into its trip. Let another $\Delta t = 3$ seconds elapse. So $t_1 + \Delta t = 4$ seconds. Since $p(t_1 + \Delta t) - p(t_1) = p(4) - p(1) = 3 - 0 = 3$, the point has covered a net distance of 3 meters in these 3 seconds. Because

$$\frac{p(t_1 + \Delta t) - p(t_1)}{\Delta t} = \frac{p(4) - p(1)}{3} = \frac{3 - 0}{3} = 1,$$

the point has an average velocity of 1 meter per second over the time interval from $t_1 = 1$ to $t_1 + \Delta t = 4$. Next take $\Delta t = 2$. Now $t_1 + \Delta t = 3$, and

$$\frac{p(t_1 + \Delta t) - p(t_1)}{\Delta t} = \frac{p(3) - p(1)}{2} = \frac{0 - 0}{2} = 0.$$

The net distance covered by the point is zero, and it has an average velocity of zero. With $\Delta t = 1$, we have $t_1 + \Delta t = 2$, and the average velocity of

$$\frac{p(2) - p(1)}{1} = \frac{-1 - 0}{1} = -1 \text{ meters per second.}$$

The minus sign indicates that the point has moved in the negative direction (from $p(1) = 0$ to $p(2) = -1$) over the time interval from $t_1 = 1$ to $t_1 + \Delta t = 2$.

Turn to any moving point, and consider the ratio

$$\frac{p(t_1 + \Delta t) - p(t_1)}{\Delta t}$$

for progressively smaller elapsed times Δt. This is the average velocity of the moving point computed over smaller and smaller time frames near t_1, namely, the average velocity from t_1 to $t_1 + \Delta t$ for a shrinking Δt. In the process, the average velocity $\frac{p(t_1+\Delta t)-p(t_1)}{\Delta t}$ closes in on a number called the *velocity at the instant*t_1,

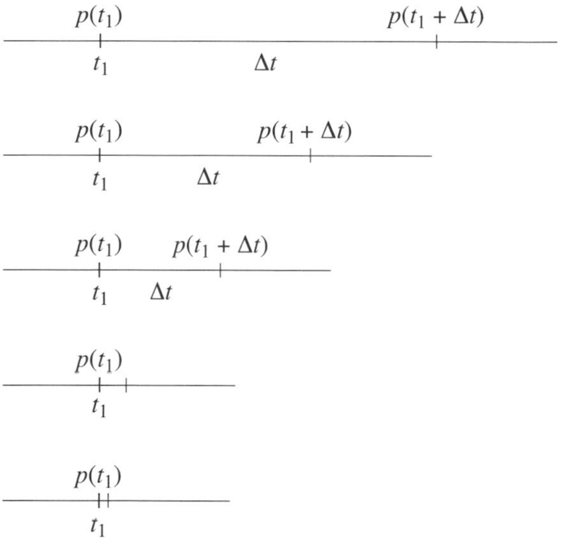

$p(t_1)$ $p(t_1 + \Delta t)$
t_1 Δt

$p(t_1)$ $p(t_1 + \Delta t)$
t_1 Δt

$p(t_1)$ $p(t_1 + \Delta t)$
t_1 Δt

$p(t_1)$
t_1

$p(t_1)$
t_1

denoted by $v(t_1)$. This is the rate at which the position changes at the instant t_1. It is obtained by pushing Δt to zero and observing what happens to the ratio $\frac{p(t_1+\Delta t)-p(t_1)}{\Delta t}$. Therefore

$$v(t_1) = \lim_{\Delta t \to 0} \frac{p(t_1 + \Delta t) - p(t_1)}{\Delta t}.$$

The velocity $v(0)$ at $t = 0$ is called the *initial velocity* of the point. In the diagram, the position $p(t_1 + \Delta t)$ is to the right of position $p(t_1)$ throughout the contraction of Δt to zero. If Δt is positive, then $t + \Delta t$ occurs after t, the point of the diagram moves from left to right, and the limit $v(t_1)$ is positive (or zero). If this diagram were to illustrate a situation of negative Δt, then $t + \Delta t$ occurs before t, and the point of the diagram moves from right to left. In this case, the numerator $p(t_1 + \Delta t) - p(t_1)$ is positive and Δt is negative, so that the limit $v(t_1)$ is negative (or zero). A diagram in which the positions $p(t_1 + \Delta t)$ are to the left of $p(t_1)$ would illustrate the other possible approaches of the point to position $p(t_1)$.

Let's see what happens when this is done with $t_1 = 1$ second for the point with position $p(t) = t^2 - 4t + 3 = (t-1)(t-3)$ for $t \geq 0$. Taking $\Delta t = \frac{1}{2}, \frac{1}{4}, \frac{1}{8}, \ldots, \frac{1}{64}$, we get

$$t_1 + \Delta t = \tfrac{3}{2}, \tfrac{5}{4}, \tfrac{9}{8}, \ldots, \tfrac{65}{64}.$$

The successive average velocities are

$$\begin{aligned}
\frac{p(\frac{3}{2}) - p(1)}{\frac{1}{2}} &= \frac{\left(\frac{3}{2}-1\right)\cdot\left(\frac{3}{2}-3\right)-0}{\frac{1}{2}} = \tfrac{1}{2}\left(-\tfrac{3}{2}\right)2 \\
&= -1.5 \text{ meters per second,} \\
\frac{p(\frac{5}{4}) - p(1)}{\frac{1}{4}} &= \frac{\left(\frac{5}{4}-1\right)\cdot\left(\frac{5}{4}-3\right)-0}{\frac{1}{4}} = \tfrac{1}{4}\left(-\tfrac{7}{4}\right)4 \\
&= -1.75 \text{ meters per second,} \\
\frac{p(\frac{9}{8}) - p(1)}{\frac{1}{8}} &= \frac{\left(\frac{9}{8}-1\right)\cdot\left(\frac{9}{8}-3\right)-0}{\frac{1}{8}} = \tfrac{1}{8}\left(-\tfrac{15}{8}\right)8 \\
&= -1.88 \text{ meters per second,} \ldots, \\
\frac{p(\frac{65}{64}) - p(1)}{\frac{1}{8}} &= \frac{\left(\frac{65}{64}-1\right)\cdot\left(\frac{65}{64}-3\right)-0}{\frac{1}{64}} = \tfrac{1}{64}\left(-\tfrac{127}{64}\right)64 \\
&= -1.98 \text{ meters per second.}
\end{aligned}$$

Continuing to push Δt to zero, we see that the velocity at the instant $t_1 = 1$ is

$$\begin{aligned}
v(1) &= \lim_{\Delta t \to 0} \frac{p(1+\Delta t) - p(1)}{\Delta t} = \lim_{\Delta t \to 0} \frac{(1+\Delta t-1)(1+\Delta t-3)-0}{\Delta t} = \lim_{\Delta t \to 0} \frac{\Delta t(\Delta t - 2) - 0}{\Delta t} \\
&= \lim_{\Delta t \to 0} (\Delta t - 2) = -2 \text{ meters per second.}
\end{aligned}$$

Again, the minus sign means that the point is moving in the negative direction at time $t_1 = 1$.

Example 6.3. Check that a repetition of this calculation with negative Δt, in particular with $\Delta t = -\frac{1}{2}, -\frac{1}{4}, -\frac{1}{8}, \ldots, -\frac{1}{64}, \ldots$ and hence with $t_1 + \Delta t = 1 + \Delta t = \frac{1}{2}, \frac{3}{4}, \frac{7}{8}, \ldots, \frac{63}{64}, \ldots$, results in the same value $v(1) = -2$ meters per second for the velocity at time $t_1 = 1$.

Change notation from t_1 to t for a typical instant of time. After referring back to the definition of the derivative of a function in Section 5.4, observe that the limit

$$\lim_{\Delta t \to 0} \frac{p(t+\Delta t) - p(t)}{\Delta t}$$

has simultaneously three different meanings.

(1) The velocity $v(t)$ at time t of the point that has position function p.

(2) The value $p'(t)$ of the derivative of the position function p at time t.

(3) The slope of the tangent line of the graph of the function p at the point $(t, p(t))$.

To repeat, if $p = p(t)$ is the position function of a moving point, then its derivative

$$p'(t) = \lim_{\Delta t \to 0} \frac{p(t + \Delta t) - p(t)}{\Delta t}$$

is the velocity $v(t)$ of the point at time t. The assumption that the point moves smoothly tells us—as noted in Section 5.4—that the function $p(t)$ is differentiable and hence that this limit exists. If $v(t)$ is positive, then the point is moving to the right at time t, and if $v(t)$ is negative, the point is moving to the left at time t. The *speed* of the point at time t is the absolute value $|v(t)|$ of the velocity. Note that velocity is the combination of speed and direction. The fact that $v(t) = p'(t)$ is positive when the point moves to the right and negative when it moves to the left means that the function $p'(t)$ incorporates information about both speed and direction.

The point with position function $p(t) = t^2 - 4t + 3$ has a velocity of $v(t) = p'(t) = 2t - 4 = 2(t - 2)$ and a speed of $|2(t - 2)|$ at any time t. Since $v(t)$ is negative for $0 \leq t < 2$, the point moves to the left during this time. When $t = 1$, the point's velocity is $v(1) = -2$, and its speed is $|-2| = 2$. Since $v(2) = 0$, the point stops at $t = 2$, and since $v(t)$ is positive for $t > 2$, it moves to the right thereafter. A more complicated example follows.

Example 6.4. Suppose that the position function of a moving point is $p(t) = \frac{3}{4}t^4 - 13t^3 + 75t^2 - 168t - 5$ for any time $t \geq 0$. So its velocity is $v(t) = p'(t) = 3t^3 - 39t^2 + 150t - 168 = 3(t^3 - 13t^2 + 50t - 56)$. The initial position and initial velocity of the point are $p(0) = -5$ and $v(0) = -168$. A substitution tells us that $v(2) = 3(8 - 52 + 100 - 56) = 0$. So the point stops at time $t = 2$. Since 2 is a root of the polynomial $v(t)$, we know that $(t - 2)$ divides $v(t)$. Check the factorization $v(t) = 3(t - 2)(t^2 - 11t + 28)$ to confirm this. From the quadratic formula applied to $t^2 - 11t + 28 = 0$, we know that

$$t = \tfrac{11 \pm \sqrt{11^2 - 4 \cdot 28}}{2} = \tfrac{11 \pm \sqrt{121 - 112}}{2} = 4 \text{ or } 7,$$

and hence that $v(4) = 0$ and $v(7) = 0$. This information in turn leads to the factorization $v(t) = 3(t - 2)(t - 4)(t - 7)$. It tells us how the point moves. For $0 \leq t < 2$, observe that $v(t) < 0$, so that the point moves to the left. After it stops at $t = 2$, we see that $v(t) > 0$ for $2 < t < 4$, so that the point moves to the right over this time interval. At $t = 4$ it stops again, and since $v(t) < 0$ for $4 < t < 7$, it then moves to the left. After its final stop at $t = 7$, $v(t) > 0$ so that it moves to the right thereafter.

Now that we understand the mathematics of a point moving on a coordinate line, we can turn to the motion of a point in the plane. Suppose that a point moves smoothly along a curve in the xy-plane. The motion starts at time $t = 0$. For any time $t \geq 0$, we let $x(t)$ and $y(t)$ be the x- and y-coordinates of the position of the point. Figure 6.8 provides one example, and Figure 6.13 provides another. Since the motion

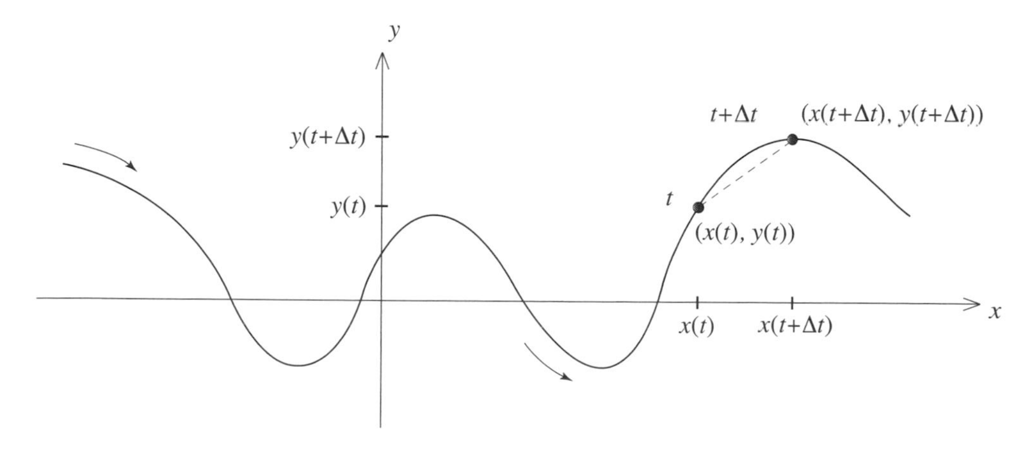

Fig. 6.13

is smooth, both $x(t)$ and $y(t)$ are differentiable functions of t. By applying what we already learned, we know that $x'(t)$ and $y'(t)$ are the respective velocities of the x- and y-coordinates of the point. Think of the derivative $x'(t)$ as the velocity of the shadow of the point on the x-axis that a light source parallel to the y-axis produces. In the same way, we can regard the derivative $y'(t)$ to be the velocity of the shadow of the point on the y-axis.

Assuming that the velocities $x'(t)$ and $y'(t)$ are known, it should be possible to determine the velocity of the moving point in the plane. And it is! Fix any moment t in time, and let an additional (positive) time Δt elapse thereafter. Figure 6.13 shows the point at the two instances t and $t + \Delta t$. For a small elapsed time Δt, the two positions will be near each other, so that by the Pythagorean theorem, the distance traveled by the point is approximately equal to

$$\sqrt{[x(t+\Delta t) - x(t)]^2 + [y(t+\Delta t) - y(t)]^2}.$$

The smaller the Δt, the closer the points are to each other, and the better this approximation. It follows that the average speed of the point during the time interval $[t, t + \Delta t]$ is approximately equal to

$$\frac{\sqrt{[x(t+\Delta t) - x(t)]^2 + [y(t+\Delta t) - y(t)]^2}}{\Delta t}.$$

This is average speed rather than average velocity, because there is no information about the direction of the motion. After some algebra (pull the Δt under the radical as $(\Delta t)^2$, and distribute it over the two parts of the sum), this is equal to

$$\sqrt{\left[\frac{x(t+\Delta t) - x(t)}{\Delta t}\right]^2 + \left[\frac{y(t+\Delta t) - y(t)}{\Delta t}\right]^2}.$$

Taking the limit $\lim\limits_{\Delta t \to 0}$ of this term and using the fact that

$$\lim_{\Delta t \to 0} \frac{x(t+\Delta t) - x(t)}{\Delta t} = x'(t) \quad \text{and} \quad \lim_{\Delta t \to 0} \frac{y(t+\Delta t) - y(t)}{\Delta t} = y'(t),$$

we obtain that the speed of the point at time t is equal to

$$v(t) = \sqrt{x'(t)^2 + y'(t)^2}.$$

Example 6.5. i. Let $x(t) = t$ and $y(t) = t^2$ with $t \geq 0$. Since $y(t) = x(t)^2$, the point moves on the parabola $y = x^2$. Since $t \geq 0$, it starts at $(0, 0)$. The speed of the point at any time t is $v(t) = \sqrt{x'(t)^2 + y'(t)^2} = \sqrt{1 + 4t^2}$. So the point starts with an initial speed of 1 and speeds up without limit from there.

ii. Let $x(t) = \sqrt{t} = t^{\frac{1}{2}}$ and $y(t) = t$ with $t \geq 0$. Again $y(t) = x(t)^2$, and this point too moves on the parabola $y = x^2$. Its speed is $v(t) = \sqrt{x'(t)^2 + y'(t)^2} = \sqrt{(\frac{1}{2}t^{-\frac{1}{2}})^2 + 1} = \sqrt{\frac{1}{4}t^{-1} + 1} = \sqrt{\frac{1+4t}{4t}}$. So the point starts with infinite speed and then slows down with $v(t)$ getting closer and closer to 1.

The fact is that $x'(t)$ is positive when the shadow on the x-axis moves to the right and negative when it moves to the left. In the same way, the shadow on the y-axis moves up if $y'(t)$ is positive and down if $y'(t)$ is negative. This means that both $x'(t)$ and $y'(t)$ contain information about the direction of the motion and that both represent velocity. The expression $\sqrt{x'(t)^2 + y'(t)^2}$ is always positive (or zero) and includes no information about direction. It is therefore "pure" speed. Subsequent discussions will generally distinguish between velocity and speed, but will use the notation v or $v(t)$ for both.

A look at Figure 6.13 confirms that the direction of the motion of the point at time t is determined by the tangent line to the curve at the point $(x(t), y(t))$ and the sign of the terms $x'(t)$ and $y'(t)$ (since they tell us about left/right and up/down). One more look at Figure 6.13 shows that the slope of the dashed segment is $\frac{y(t+\Delta t)-y(t)}{x(t+\Delta t)-x(t)}$, so that the slope of the tangent at $(x(t), y(t))$ is

$$\lim_{\Delta t \to 0} \frac{y(t+\Delta t)-y(t)}{x(t+\Delta t)-x(t)} = \lim_{\Delta t \to 0} \frac{\frac{y(t+\Delta t)-y(t)}{\Delta t}}{\frac{x(t+\Delta t)-x(t)}{\Delta t}} = \frac{\lim\limits_{\Delta t \to 0} \frac{y(t+\Delta t)-y(t)}{\Delta t}}{\lim\limits_{\Delta t \to 0} \frac{x(t+\Delta t)-x(t)}{\Delta t}} = \frac{y'(t)}{x'(t)}.$$

We close this discussion with a technical point. In the last step of the derivation of the formula $v(t) = \sqrt{x'(t)^2 + y'(t)^2}$, the operation $\lim\limits_{\Delta t \to 0}$ needed to be maneuvered past both the square root and the square. We will see why is this possible when the calculus of derivatives is revisited in Chapter 7 (in particular in Problem 7.15 of the Problems and Projects section).

6.5 GALILEO AND ACCELERATION

The most fundamental concept underlying the phenomenon of motion is acceleration. In brief, acceleration is change in velocity. It is closely related to force, and it is best understood in this context. The downward pull of gravity is a force that acts on any object. Let's suppose that you are a tennis player getting ready to serve. After holding the tennis ball still, you toss it vertically upward (as your body coils and the arm with the racket moves to hit the ball). As long as you hold the ball still, its velocity is zero. When your hand moves the ball upward, the force of your hand on the ball exceeds the downward gravitational force. At the instant your hand releases the ball, this excess force will have provided it with an upward velocity. But at this instant and thereafter, gravity—as the only force acting on the ball (other than a negligible amount of air resistance)—will slow the ball's upward progress until it stops at the top of its flight. (Shortly thereafter, with the ball on its way down, your racket makes contact and propels it forward.) Notice that changes in the velocity of the ball were caused by changes in the net force acting on it.

We'll now study acceleration in the abstract. Let the velocity function $v(t)$ of a point moving smoothly on a coordinate axis be given. Fix some instant $t_1 \geq 0$. Let an additional amount of time Δt elapse, and consider the instant $t_1 + \Delta t$. Then

$$\begin{aligned}
v(t_1) &= \text{the velocity at time } t_1 \\
v(t_1 + \Delta t) &= \text{the velocity at time } t_1 + \Delta t \\
v(t_1 + \Delta t) - v(t_1) &= \text{the change in velocity over the time interval from } t_1 \text{ to } t_1 + \Delta t \\
\frac{v(t_1 + \Delta t) - v(t_1)}{\Delta t} &= \text{the average change in velocity per unit time over the time interval from } t_1 \text{ to } t_1 + \Delta t \\
&= \text{the definition of the } \textit{average acceleration} \text{ over the time interval from } t_1 \text{ to } t_1 + \Delta t.
\end{aligned}$$

Consider the average acceleration $\frac{v(t_1+\Delta t)-v(t_1)}{\Delta t}$ over the time interval from t_1 to $t_1 + \Delta t$ for progressively smaller elapsed times Δt that shrink to zero. As this occurs, the ratio $\frac{v(t_1+\Delta t)-v(t_1)}{\Delta t}$ closes in on a number that measures the rate at which the velocity is changing at the instant t_1. This the point's *acceleration at the instant* t_1. It is denoted by $a(t_1)$. In the mathematical shorthand of limits, the acceleration at time t_1 is

$$a(t_1) = \lim_{\Delta t \to 0} \frac{v(t_1 + \Delta t) - v(t_1)}{\Delta t}.$$

Changing notation from t_1 to t, we see that $a(t)$ is the derivative $v'(t)$ of the velocity function $v(t)$. Let $p = p(t)$ be the position function of the moving point, and recall that $v(t) = p'(t)$. So $a(t)$ is the derivative of $p'(t)$. Denoting this *second derivative* of p by $p''(t)$, it follows that $a(t) = p''(t)$. What unit is acceleration measured in? For example, if distance is in meters and time in seconds, then velocity is given in meters per second, so that acceleration is expressed in the unit (meters per second) per second, or meters per second squared.

We have analyzed the motion of a point on a line and observed how its position function determines its velocity and, it turn, how the velocity determines its acceleration. We will now see how to move in the other direction from acceleration to velocity to position. This will make important use of the analysis at the beginning of Section 5.8. If two functions have the same derivative, then one of the functions is equal to the other function plus a constant. In particular, if the derivative of a given function is always zero, then it has the same derivative as a constant function. It follows that the given function must be a constant.

Let a point move on a coordinate axis with acceleration $a(t) = v'(t) = 0$ for all $t \geq 0$. By the fact just mentioned, this means that $v(t) = C$ for some constant C. Taking $t = 0$ tells us that $C = v(0)$, the initial velocity of the point. Notice that the derivative of $p(t)$ is $p'(t) = v(t) = v(0)$ and that the derivative of $v(0)t$ is also $v(0)$. Therefore $p(t) = v(0)t + D$ for some constant D. Taking $t = 0$, we get that $D = p(0)$. So the position function of the moving point is given by

$$p(t) = v(0)t + p(0).$$

It is determined by the initial position and the initial velocity of the point.

Suppose next that a point moves along a coordinate axis with constant acceleration $a(t)$ for $t \geq 0$. Since the constant is equal to the acceleration $a(0)$ at $t = 0$, we know that $a(t) = a(0)$ for all $t \geq 0$. The fact that the functions $v(t)$ and $a(0)t$ have the same derivative $v'(t) = a(t) = a(0)$ means that $v(t) = a(0)t + C$ for some constant C. Setting $t = 0$, we see that $C = v(0)$ is the initial velocity. Hence

$$v(t) = a(0)t + v(0).$$

The derivatives of the position function $p(t)$ and the function $\frac{1}{2}a(0)t^2 + v(0)t$ are both equal to $v(t) = a(0)t + v(0)$, and therefore $p(t) = \frac{1}{2}a(0)t^2 + v(0)t + D$ for some constant D. Plugging $t = 0$ into this equation shows that $D = p(0)$. Therefore the position function

$$p(t) = \tfrac{1}{2}a(0)t^2 + v(0)t + p(0)$$

of the moving point is determined by the constant acceleration $a(0)$, the initial velocity $v(0)$, and the initial position $p(0)$ of the point. If the initial velocity $v(0) = 0$, then the constant acceleration $a(0)$ leads to a displacement of the point of

$$p(t) - p(0) = \tfrac{1}{2}a(0)t^2$$

during the time from 0 to t.

For a point moving along a coordinatized line, the functions $p(t)$, $v(t) = p'(t)$, and $a(t) = v'(t)$ for the position, velocity, and acceleration of a moving point all have a directional aspect. When, for a given t, $p(t)$, $v(t)$, or $a(t)$ is positive, then the position of the point is on the positive side of the axis, the point moves in the direction of the positive axis, or its positive acceleration implies an increase in the velocity. Analogously, when $p(t)$, or $v(t)$, or $a(t)$ is negative for a given t, then the point is positioned on the negative axis, moves in the negative direction, or its negative acceleration implies a decrease in its velocity. The speed $|v(t)|$, on the other hand, always satisfies $|v(t)| \geq 0$ and does not have a directional aspect.

Galileo observed in the early 1600s that a ball rolling down an inclined plane is "uniformly accelerated." He meant by this that if the ball has velocity $v(t)$ at time t and velocity $v(t + \Delta t)$ at a later time $t + \Delta$, then the ratio

$$\frac{v(t + \Delta t) - v(t)}{(t + \Delta t) - t} = \frac{v(t + \Delta t) - v(t)}{\Delta t}$$

is equal to the *same constant* regardless of the time t of the first observation and the elapsed time Δt. (The constant depends only on the inclination of the plane.) It is easy to see that Galileo's uniformly accelerated motion is equivalent to a motion undergoing constant acceleration. To see this, fix an instant t and push the Δt of Galileo's ratio to zero. During this process, the ratio remains constant and this constant does not depend on t. Therefore the acceleration

$$a(t) = v'(t) = \lim_{\Delta t \to 0} \frac{v(t + \Delta t) - v(t)}{\Delta t}$$

of the rolling ball is equal to this same constant for any time $t \geq 0$. Conversely, if the acceleration $a(t)$ is equal to the constant $a(0)$ for all t, then $v(t) = a(0)t + C$. So Galileo's ratio is equal to $\frac{v(t+\Delta t)-v(t)}{\Delta t} = \frac{a(0)(t+\Delta t)+C-(a(0)t+C)}{\Delta t} = a(0)$ for any t and Δt. So the motion is uniformly accelerated in Galileo's sense.

Galileo's experiments (refer to Section 3.6) told him that the acceleration of a ball rolling down an inclined plane is constant and does not depend on the mass or weight of the ball. He found this to be so for many different inclinations of the plane, and concluded that the acceleration of an object in free fall near Earth's surface should also be constant and not depend on the mass or weight of the object. (He did realize that light or fast-moving objects are slowed by air resistance.) Today we know that this constant—denoted by $-g$ (the g for *gravity* and the – because the acceleration is directed downward)—has the value

$$g = 9.81 \text{ meters/second}^2 \text{ or } 32.17 \text{ feet/second}^2.$$

Suppose an object is in vertical free fall near the Earth's surface. Place a y-axis along the path of the fall with the origin at ground level. Let $y(t)$ be the position of the object at any time $t \geq 0$. The initial position of the object is $y(0)$, and its initial velocity is $v(0)$. Substituting $a(t) = -g$ into one of the equations developed above tells us that for for any time $t \geq 0$, the velocity and position functions of the object are

$$v(t) = -gt + v(0) \quad \text{and} \quad y(t) = -\tfrac{g}{2}t^2 + v(0)t + y(0).$$

Let's assume (contrary to the view of most historians of science) that Galileo did actually drop a cannonball from the Leaning Tower of Pisa. See Figure 6.14. We'll take distance in meters and time in

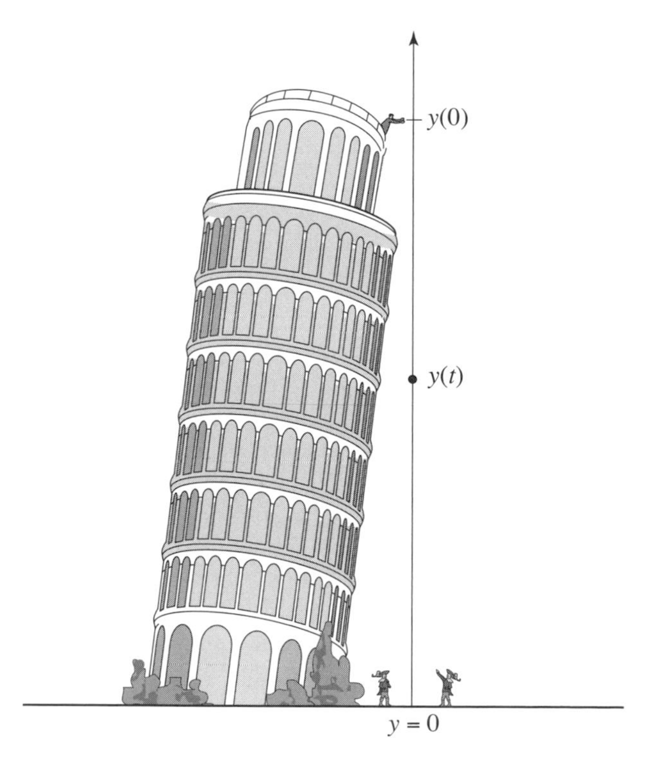

Fig. 6.14

seconds. Since the tower is 56 meters high, the initial position of the cannonball would have been $y(0) = 56$ with an initial velocity $v(0) = 0$ (assuming that it was dropped from rest). This meant that

$$y(t) = -\tfrac{g}{2}t^2 + 56 \text{ meters}.$$

Suppose the cannonball hit the ground at the instant t_1. So $y(t_1) = 0$ and hence $\frac{g}{2}t_1^2 = 56$. Since $g = 9.81$ meters/sec^2, it follows that $t_1^2 = \frac{112}{9.81}$ and $t_1 = \sqrt{\frac{112}{9.81}} \approx 3.38$ seconds. The velocity of the cannonball at impact would have been $v(t_1) = -gt_1 + v(0) \approx -9.81(3.38) \approx -33.15$ meters per second.

The question arises as to how significantly air resistance would have affected the fall of a cannonball from the Leaning Tower of Pisa. This matter is taken up in Project 11H of the Problems and Projects section of Chapter 11. The short answer? Very little.

6.6 DEALING WITH FORCES

Having discussed acceleration, it is time to study force. Think of a force as the kind of push or pull encountered in our everyday experience. The force of gravity—in the form of the weight of an object—is a force that we experience everywhere, literally with every step we take. It was one of Newton's major achievements that he laid bare the fundamental concepts and laws that underlie the basic physics of motion. After defining force as an "action exerted upon a body, in order to change its state, either at rest, or of uniform motion, in a right line," he states the basic laws:

> **Law I.** That every body perseveres in its state of resting, or of moving uniformly in a right line, as far as it is not compelled to change that state by an external force impressed upon it.
>
> **Law II.** That the change of motion is proportional to the moving force impressed; and is produced in the direction of the right line, in which that force was impressed.
>
> **Law III.** That reaction is always contrary and equal to action; or, that the mutual actions of two bodies upon each other are always equal, and directed to contrary parts.

Let's elaborate on what Newton is saying. A force has a magnitude that—in the context of our discussion—can be measured. For example, an amount of push or pull can be measured by the amount of the displacement it produces on some standardized steel spring. A force also has a direction. Its magnitude and direction together determine the force. Let's first consider Newton's second law. It says that if a force acts on an object, then the magnitude F of the force is proportional to the acceleration a that it produces in the motion of the object, and the mass m of the object is the constant of proportionality. In other words,

$$\text{force} = \text{mass} \times \text{acceleration}, \quad \text{or in symbols,} \quad F = ma.$$

Measure F and measure a, and the mass m of the body is determined by $m = \frac{F}{a}$. An increase (or decrease) in F produces a corresponding increase (or decrease) in a. The second law also says that the direction of the acceleration is the same as that of the force. Newton's first law is a direct consequence of his second law. If the magnitude of a force on an object is zero, then the acceleration it produces in the motion must be zero, and hence the object moves with constant velocity, possibly zero. The third law is the assertion that for every force there is always an equal and opposite force. If you push against a wall, the wall will push back on you with an equal and opposite force. If this force did not exist, you would push the wall over. So forces always occur in pairs. It is important to note that the two forces never act on the same object. Your push against a wall is a force on the wall. The wall pushing back is a force on you.

Let's have another look at the point with position function $p(t) = t^2 - 4t + 3$ from Section 6.4. Since $v(t) = 2t - 4$, the point's acceleration is $a(t) = v'(t) = 2$. Suppose that the point is a particle of mass m. The force on this particle is $F = ma = 2m$. Since it is positive, it pushes the particle in the positive direction. The particle starts at time $t = 0$ at $p(0) = 3$ with an initial velocity of $v(0) = -4$. So at the start it moves to the left. Since the force acts to the right, the particle slows down and comes to a stop. Since

$v(2) = 0$, this happens at time $t = 2$. Since the force continues to push to the right, the particle accelerates to the right, gaining in speed in the process. Compare what has been described with Figure 6.12.

Galileo's discovery that the acceleration of an object in free fall is constant is explained by Newton in terms of the equation

$$F = -mg,$$

where F is the force of gravity on the object, g the gravitational constant, and m is the mass of the object. The direction of the force is downward (toward the center of the Earth), and its magnitude mg is called the *weight* of the body. The force of gravity on an object depends on its mass m. On the other hand, its acceleration $-g$ is the same, regardless of the object's mass.

In the metric system, mass is often given in kilograms, abbreviated by kg; length in meters, abbreviated by m; and time in seconds, abbreviated by s. So acceleration is in meters/second^2—abbreviated by m/s^2—and force in kg $\cdot$ m/s^2, a unit that is appropriately called the *newton* and abbreviated by N. The development of the metric system of units (in common use internationally) and its conversion to the American system (closer to the intuitive grasp of an American reader) will be discussed in Section 6.10.

Forces are represented by arrows called *vectors*. A vector representing a force points in the direction of the force. To represent the magnitude of a force, both a unit of length (the centimeter, meter, inch, or foot, for example) and a unit of force (the newton or pound, for instance) need to be given. A force of x units in magnitude is represented by a vector of x units in length. So the magnitude of the force is *numerically equal to* the length of the vector representing it. For example, in Figure 6.15a, the three vectors represent forces of magnitudes 40 N, 75 N, and 85 N. The upcoming discussion focuses on forces but applies to vectors in general. For example, the position, velocity, and acceleration of a moving point can each be represented by a vector, an arrow with direction and length (both functions of time), where the direction is the relevant direction and the length represents magnitude.

Consider two forces with magnitudes F_1 and F_2 acting on the same point. What can be said about their combined effect? If the two forces act in the same direction, then this is a force with magnitude $F_1 + F_2$ in that direction. If they act in directly opposed directions, then their combined effect is a force acting in the direction of the larger force. If, say, F_1 is larger, then the magnitude of the combined force will be $F_1 - F_2$. In general, the combination of two forces is determined as follows. Position the vectors

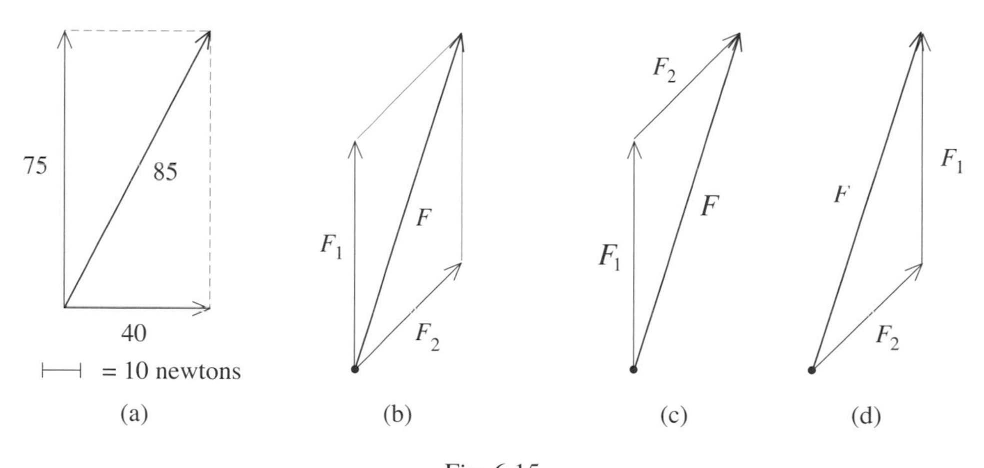

Fig. 6.15

that represent the two forces in such a way that their initial points coincide. As Figure 6.15b shows, this configuration determines a parallelogram, and the diagonal of the parallelogram from the common initial point determines a vector. This vector represents a force, and this force—both in direction and in magnitude—is the combined effect, or *resultant*, of the two given forces. This fact is the *parallelogram law of forces*. Figures 6.15c and 6.15d show that the resultant is also obtained by placing the two vectors end to tip. Figure 6.15a provides a numerical example. The fact that the vectors of magnitudes 40 and 75 are perpendicular means that the magnitude of the resultant is given by the Pythagorean theorem as $\sqrt{40^2 + 75^2} = \sqrt{7225} = 85$ N.

If several forces act at the same point, then their resultant is determined by applying the parallelogram law to two forces at a time. Take the four forces F_1, F_2 and G_1, G_2 considered in Figures 6.16a and 6.16b. The resultant of the first pair is the vector F, and that of the second pair is the vector G. By the parallelogram law, the resultant of F and G is the vector Q in Figure 6.16c. The resultant of any number of vectors can also be obtained by placing them end to tip in any order. The perimeter of the diagram of

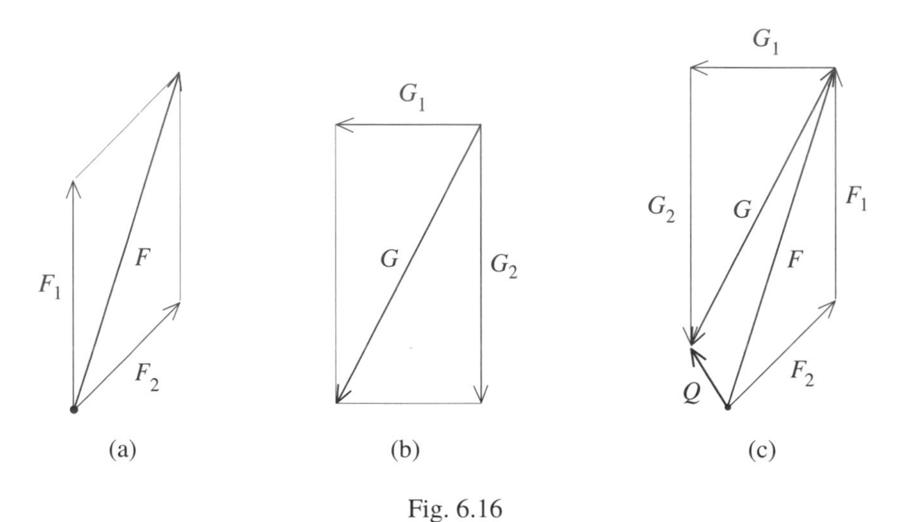

Fig. 6.16

Figure 6.16c shows how the resultant Q of F_1, F_2, G_1, and G_2 is obtained in this way.

The important fact is that *the magnitude of the resultant is numerically equal to the length of the vector that represents it*. This means that the representation of forces by vectors is much more than a convenient way to think about forces: it provides a fundamental insight into the way forces act. Our discussion about forces makes use of the following convention. When referring to a force as F, it will usually be understood that the symbol F also represents the magnitude of the force. The direction is given by the direction of the corresponding vector.

Let's look at the parallelogram law quantitatively. In Figure 6.17a, φ is the angle between the two forces F_1 and F_2. An application of the law of cosines (see Section 1.6) to Figure 6.17b, tells us that the

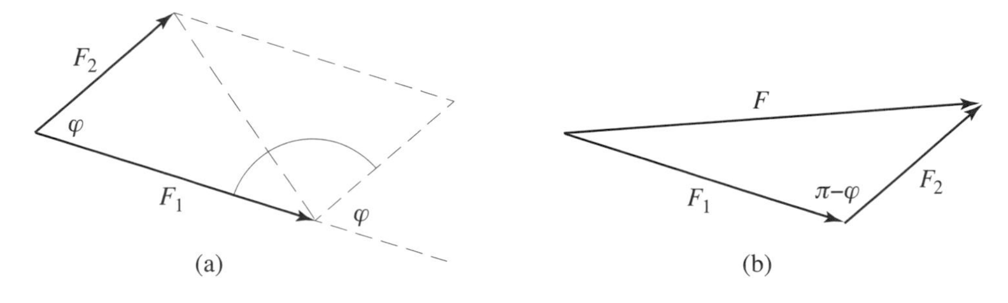

Fig. 6.17

resultant of the two forces F_1 and F_2 has magnitude $F = \sqrt{F_1^2 + F_2^2 - 2F_1F_2\cos(\pi - \varphi)}$. Turn to Section 4.6, and use the trig identity $\cos(\pi - \theta) = -\cos\theta$ of Example 4.18 to conclude that

$$F = \sqrt{F_1^2 + F_2^2 + 2F_1F_2\cos\varphi}\,.$$

Example 6.4. Let F_1 and F_2 be two forces acting on a point. Suppose that their magnitudes are 3 N and 4 N, respectively, and that the angle between them is 50°. The magnitude of the resultant of the two forces is

$$F = \sqrt{3^2 + 4^2 + 2 \cdot 3 \cdot 4 \cos 50^\circ} \approx \sqrt{25 + 24 \cos 50^\circ} \approx 6.36 \text{ N}.$$

For a given force F, there are many ways of finding two forces F_1 and F_2 such that F is the resultant of F_1 and F_2. A general way for doing this is to start with any parallelogram that has the vector representing F as a diagonal. See Figure 6.18a. We'll let φ_1 and φ_2 be the respective angles between the diagonal of

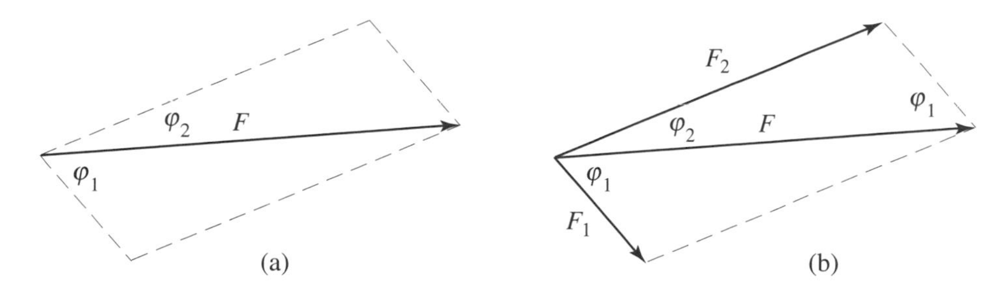

Fig. 6.18

the parallelogram and its two sides. Figure 6.18b shows how the two sides provide vectors F_1 and F_2 that have F as their resultant. Applying the law of sines (see Section 1.6) to Figure 6.18b, we get

$$\frac{\sin \varphi_2}{F_1} = \frac{\sin \varphi_1}{F_2} = \frac{\sin(\pi - (\varphi_1 + \varphi_2))}{F} = \frac{\sin(\varphi_1 + \varphi_2)}{F}.$$

For the last equality apply the identity $\sin(\pi - \theta) = \sin \theta$ of Example 4.18 in Section 4.6. It follows that the magnitudes of F_1 and F_2 are

$$F_1 = F \frac{\sin \varphi_2}{\sin(\varphi_1 + \varphi_2)} \quad \text{and} \quad F_2 = F \frac{\sin \varphi_1}{\sin(\varphi_1 + \varphi_2)}.$$

Example 6.5. Let F be a force and let F_1 and F_2 be two forces that have F as their resultant. Suppose that $F = 15$ N and that the angles that F_1 and F_2 make with F are $\varphi_1 = 20^\circ$ and $\varphi_2 = 40^\circ$, respectively. Check that the formulas for the magnitudes of F_1 and F_2 provide the approximations $F_1 \approx 11.13$ N and $F_2 \approx 5.92$ N, respectively.

Let a force F be given. The dashed line of Figure 6.19a specifies a direction relative to the direction of the force F with an angle θ, where $0^\circ \leq \theta \leq 90^\circ$. Figure 6.19b drops a perpendicular from the tip of the

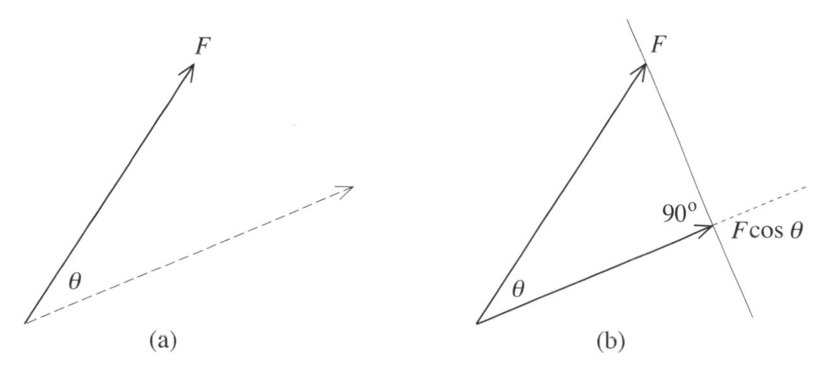

Fig. 6.19

vector F to the dashed line and puts in the vector that this perpendicular determines. This vector represents a force called the *component of F in the direction θ*. If l is the length of this vector, then $\cos\theta = \frac{l}{F}$. It follows that the magnitude of the component of F in the direction θ is $F\cos\theta$. (To remember that the magnitude of the component is given by the cosine rather than the sine, consider the component to be, in a sense, "adjacent" to the original force.) In the same way, the *component of the force in the direction* $90° - \theta$ has magnitude $F\cos(90° - \theta)$. Figure 6.20 draws this component in and tells us that

$$F\cos(90° - \theta) = F\sin\theta.$$

By an application of the parallelogram law, the resultant of the two components is equal to the original

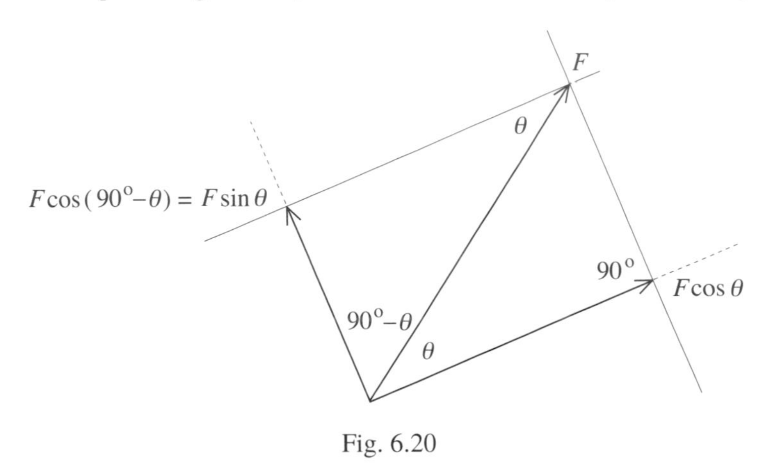

Fig. 6.20

force F. We have described the *decomposition of the force F* into two components. One component in the direction θ and the other component perpendicular to it. They have the respective magnitudes

$$F\cos\theta \quad \text{and} \quad F\cos(90° - \theta) = F\sin\theta.$$

This is the most important example of the more general construction illustrated in Figure 6.18. Taking $\varphi_1 = \theta$ and $\varphi_2 = 90° - \theta$ in the more general situation, we get that the magnitudes of the two components are $F_1 = F\frac{\sin\varphi_2}{\sin(\varphi_1+\varphi_2)} = F\frac{\sin(90°-\theta)}{\sin\frac{\pi}{2}} = F\cos\theta$ and that $F_2 = F\frac{\sin\varphi_1}{\sin(\varphi_1+\varphi_2)} = F\sin\theta$. As expected, this agrees with what was observed above.

Let's return to Galileo's inclined plane and assume that its angle of elevation is β. We'll let an ice cube slide down the plane. The surface of the plane is warm and the bottom surface of the cube has melted, so that the cube slides frictionlessly down the plane. If m is the mass of the cube, then the gravitational force on it has magnitude mg. Separate this force into a component F along the plane and a component G perpendicular to it. See Figure 6.21a. By Newton's third law, the plane pushes on the cube with a force of magnitude G in the opposite direction. It follows that F is the resultant of the forces acting on the ice cube. Notice that the right triangle determined by the inclined plane shares (in addition to the right angle) an angle with the right triangle determined by the force vectors mg and F. Therefore the two triangles share the angle β as well. Because $\sin\beta = \frac{F}{mg}$, it follows that

$$F = mg\sin\beta.$$

Using $F = ma$, we see that the acceleration of the cube is $a = g\sin\beta$. Place an axis with origin 0 at the top of the plane so that its positive direction is down the plane. Let the ice cube start at time $t = 0$, and let $p(t)$ be the coordinate of the center of the bottom surface of the ice cube at any elapsed time t. Assume that the initial position and the initial velocity of the ice cube are both zero. Applying the discussion of Section 6.5, we obtain that the velocity $v(t)$ and position $p(t)$ of the ice cube at any time $t \geq 0$ to be

$$v(t) = a(0)t + v(0) = (g\sin\beta)t \quad \text{and} \quad p(t) = \tfrac{1}{2}a(0)t^2 + v(0)t + p(0) = \tfrac{1}{2}(g\sin\beta)t^2.$$

Now turn to Galileo's rolling ball depicted in Figure 6.21b. Are the forces acting on it and hence its

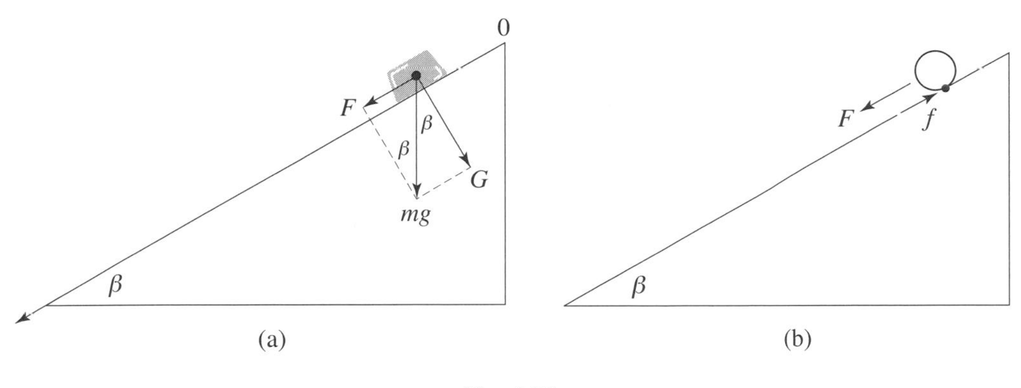

Fig. 6.21

motion the same as that of the sliding ice cube? Or not? A moment's thought tells us that the answer is no! After all, what other than the friction of the plane causes the ball to roll? If f is this frictional force (that might depend on time), then the force that pushes the ball down the plane is $F - f$ rather than the previous $F = mg\sin\beta$. (Historians of science have debated whether Galileo was aware of this difference.) The analysis of the rolling ball is more complicated than that of the sliding ice cube. It will be taken up in Chapter 8.

As we have seen, force is a physical reality with both a magnitude and a direction that can be analyzed with vectors. Velocity and acceleration are two others. Vectors representing velocity and acceleration can be separated into components and added by forming resultants in the same way that this is done for vectors representing forces. With these ideas, Galileo's early grasp of a number of the key elements of the physics of motion gave way to Newton's more comprehensive understanding. Equally important was the development of calculus, a mathematics powerful enough to successfully extract the consequences of the order that the laws of physics impose. The next section makes this concrete.

6.7 THE TRAJECTORY OF A PROJECTILE

We will now turn to the analysis of the trajectory of a projectile. The basic understanding is that a projectile is an object that is launched with some initial velocity, but that is "left on its own" thereafter. We will assume that only gravity acts on the object after its launch and ignore complicating factors such as air resistance, spin, and the shape of the object. These complicating factors are negligible for projectiles that are relatively heavy and move at relatively slow speeds (such as a basketball jump shot for example).

Let an xy-coordinate system be given, and consider Figure 6.22. It depicts a projectile that starts its flight at time $t = 0$ from its initial position at the point $(0, y_0)$. At any time $t \geq 0$, $x(t)$ and $y(t)$ are the respective coordinates of the position of the projectile. Note that $x(0) = 0$ and $y(0) = y_0$. The derivatives $x'(t)$ and $y'(t)$ are the respective velocities of the projectile in the x- and y-directions. As illustrated in Figure 6.23a, the initial velocities $x'(0)$ and $y'(0)$ and the parallelogram law determine the initial velocity vector. It is represented by a vector with direction specified by the angle φ_0 (with $0 \leq \varphi_0 \leq \frac{\pi}{2}$) known as *the angle of elevation*. By the Pythagorean theorem or the discussion that concludes Section 6.4, its magnitude $v_0 = \sqrt{x'(0)^2 + y'(0)^2}$ is the *initial speed* of the projectile. It follows from Figure 6.23a that $\cos\varphi_0 = \frac{x'(0)}{v_0}$ and $\sin\varphi_0 = \frac{y'(0)}{v_0}$, and therefore that

$$x'(0) = v_0\cos\varphi_0 \quad \text{and} \quad y'(0) = v_0\sin\varphi_0.$$

Figure 6.23b informs us that at any time t, the velocity vector is the resultant of the vectors $x'(t)$ and $y'(t)$, and, by appealing to Section 6.4 once more, that $\sqrt{x'(t)^2 + y'(t)^2}$ is the speed of the projectile. The only force on the projectile during its flight is gravity acting in the negative y-direction. In particular, the force in the x-direction is zero. Therefore, by Newton's second law, the acceleration in the x-direction is zero.

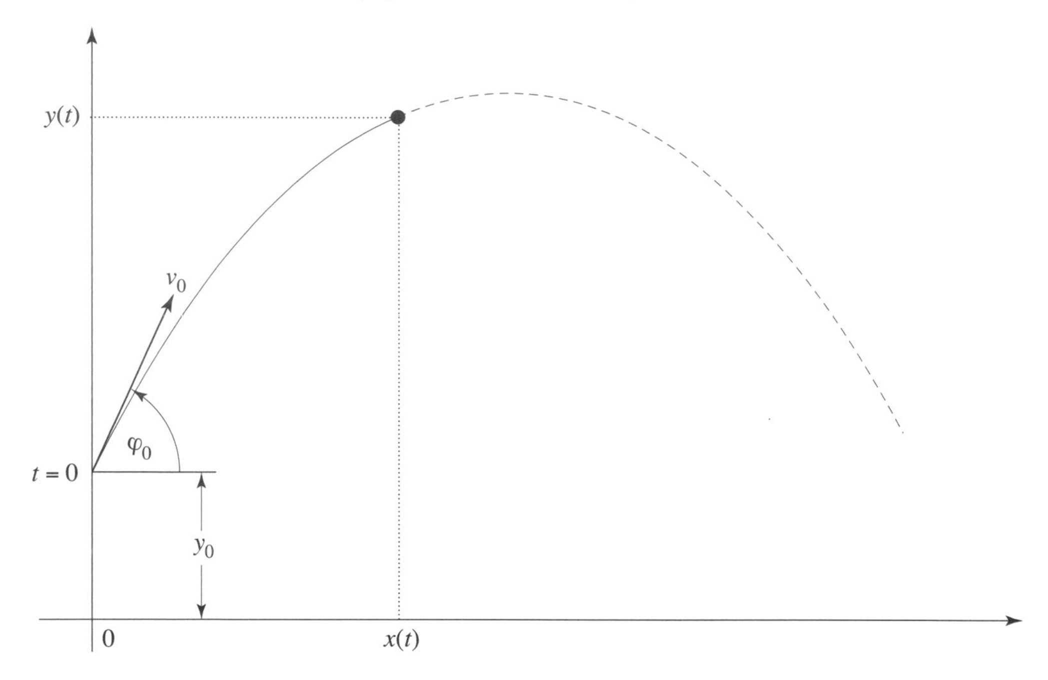

Fig. 6.22

Thus $x''(t) = 0$. Since $x(0) = 0$ and the initial velocity in the x-direction is $x'(0)$, it follows from a discussion in Section 6.5 that

(6a) $$x'(t) = x'(0) = v_0 \cos\varphi_0 \quad \text{and} \quad x(t) = (v_0 \cos\varphi_0)t.$$

Because gravity produces an acceleration of $-g$ in the y-direction, $y''(t) = -g$. Since $y'(0) = v_0 \sin\varphi_0$

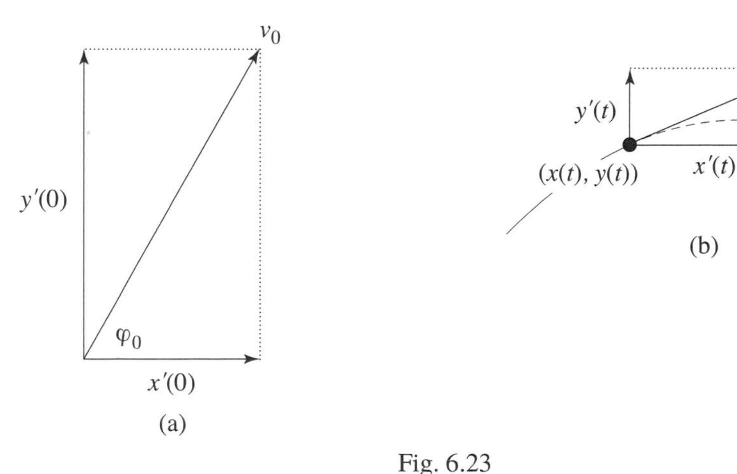

Fig. 6.23

and $y(0) = y_0$, we find by the discussion in Section 6.5 that

(6b) $$y'(t) = -gt + v_0 \sin\varphi_0 \quad \text{and} \quad y(t) = -\tfrac{g}{2}t^2 + (v_0 \sin\varphi_0)t + y_0.$$

By the second equation of (6a), $t = \frac{x(t)}{v_0 \cos\varphi_0}$. So $t^2 = \frac{(x(t))^2}{v_0^2 \cos^2\varphi_0}$, and by a substitution into the expression for $y(t)$,

$$\begin{aligned} y(t) &= \frac{-g}{2v_0^2\cos^2\varphi_0}\,x(t)^2 + \frac{v_0\sin\varphi_0}{v_0\cos\varphi_0}\,x(t) + y_0 \\ &= \frac{-g}{2v_0^2\cos^2\varphi_0}\,x(t)^2 + (\tan\varphi_0)x(t) + y_0. \end{aligned}$$

Observe, therefore, that at any time in its flight, the position (x, y) of the projectile satisfies the equation

$$\text{(6c)} \qquad y = \Big(\frac{-g}{2v_0^2\cos^2\varphi_0}\Big)x^2 + (\tan\varphi_0)x + y_0.$$

By Section 4.3, this is the equation of a parabola. Therefore the trajectory of the projectile is a parabola, a specific parabola determined by the constants v_0, y_0, and φ_0.

What is the maximal height reached by the projectile? This maximum occurs at the instant at which the vertical component of the velocity is zero, so at the instant t_1 at which $y'(t_1) = 0$. By the first equation of (6b), $0 = y'(t_1) = -gt_1 + v_0\sin\varphi_0$, or $t_1 = \frac{v_0\sin\varphi_0}{g}$. Inserting this t_1 into the second equation of (6b) informs us that

$$\begin{aligned} y(t_1) &= -\tfrac{g}{2}t_1^2 + (v_0\sin\varphi_0)t_1 + y_0 = -\tfrac{1}{2g}v_0^2\sin^2\varphi_0 + \tfrac{1}{g}v_0^2\sin^2\varphi_0 + y_0 \\ &= \tfrac{1}{2g}v_0^2\sin^2\varphi_0 + y_0. \end{aligned}$$

Therefore the maximal height that the projectile attains is

$$\text{(6d)} \qquad \tfrac{1}{2g}v_0^2\sin^2\varphi_0 + y_0.$$

For the rest of our discussion, we will assume that the terrain is flat and horizontal and that the x-axis represents the ground. At what time and how far downrange will the projectile hit the ground? Notice that impact occurs precisely at the time t_{imp} for which $y(t_{\text{imp}}) = 0$. Refer to the second equation of (6b), and observe that the time of impact can be found by solving

$$\tfrac{g}{2}(t_{\text{imp}})^2 - (v_0\sin\varphi_0)t_{\text{imp}} - y_0 = 0$$

for t_{imp}. By the quadratic formula and the fact that $t_{\text{imp}} \geq 0$,

$$\text{(6e)} \qquad t_{\text{imp}} = \tfrac{1}{g}\Big(v_0\sin\varphi_0 + \sqrt{v_0^2\sin^2\varphi_0 + 2gy_0}\Big).$$

The *range* of the projectile is the horizontal distance from the initial position of the projectile to the point of impact. So the range is equal to the x-coordinate R of the projectile at the time of impact t_{imp}. It follows from the second formula of (6a) that

$$\text{(6f)} \qquad R = \tfrac{v_0}{g}\cos\varphi_0\Big(v_0\sin\varphi_0 + \sqrt{v_0^2\sin^2\varphi_0 + 2gy_0}\Big).$$

An easy algebraic maneuver and the trig formula $\sin 2\varphi_0 = 2\sin\varphi_0\cos\varphi_0$ of Problem 1.26 provides the alternative expression

$$\text{(6g)} \qquad R = \tfrac{v_0^2}{2g}\sin(2\varphi_0) + \tfrac{v_0}{g}\sqrt{\tfrac{v_0^2}{4}\sin^2(2\varphi_0) + 2gy_0\cos^2\varphi_0}.$$

What angle φ_0 gives the maximal range, that is, the largest possible R? Suppose $y_0 = 0$. A look at the expression (6g) in this case shows that R is maximal when $\sin(2\varphi_0)$ is largest. Since $0 \leq \varphi_0 \leq \frac{\pi}{2}$, it follows that $0 \leq 2\varphi_0 \leq \pi$. The largest value of $\sin(2\varphi_0)$ in this interval is 1, and it occurs for $2\varphi_0 = \frac{\pi}{2}$, that is, for

$\varphi_0 = \frac{\pi}{4}$. So if $y_0 = 0$, the maximal range is

$$R_{\max} = \tfrac{v_0^2}{2g} + \tfrac{v_0}{g}\sqrt{\tfrac{v_0^2}{4}} = \tfrac{v_0^2}{2g} + \tfrac{v_0^2}{2g} = \tfrac{v_0^2}{g}. \tag{6h}$$

If $y_0 \neq 0$, then $\varphi_0 = \frac{\pi}{4}$ does not provide the greatest range R. However, the determination of the angle φ_0 that does so is beyond our current mathematical firepower.

The speed of the projectile at any time t is

$$v(t) = \sqrt{(x'(t))^2 + (y'(t))^2} = \sqrt{v_0^2 \cos^2 \varphi_0 + g^2t^2 - 2(gv_0 \sin \varphi_0)t + v_0^2 \sin^2 \varphi_0}.$$

Since $v_0^2 \cos^2 \varphi_0 + v_0^2 \sin^2 \varphi_0 = v_0^2(\cos^2 \varphi_0 + \sin^2 \varphi_0) = v_0^2$, this is equal to

$$v(t) = \sqrt{v_0^2 + g^2t^2 - 2g(v_0 \sin \varphi_0)t}. \tag{6i}$$

The speed at impact is obtained by plugging the moment of impact t_{imp} into this formula. The formulas for $x'(t)$ and $y'(t)$ (see (6a) and (6b)) together with the discussion that concludes Section 6.4 tell us that the slope of the trajectory at any time is

$$\frac{y'(t)}{x'(t)} = \frac{-gt + v_0 \sin \varphi_0}{v_0 \cos \varphi_0} = \tfrac{-g}{v_0}(\sec \varphi_0)t + \tan \varphi_0.$$

With $t = t_{\text{imp}}$, this provides the slope of the trajectory at the point of impact.

As already remarked, the study above applies to an object that is relatively heavy and tossed with a slow speed. In such a situation the impact of air resistance, the shape of the object, and any path-altering rotation are negligible. So our analysis applies to an apple that a schoolboy tosses, or to a basketball shot, or to the soft shot of a volleyball setter (but not the kill shot of a spiker). However, anyone who has ever stuck a hand out of the window of a car moving at a speed of, say, 100 kilometers (or about 62 miles) per hour and felt the stiff resistance of the air has experienced the considerable impact that air resistance can have.

The hammer throw is an Olympic event. The "hammer" is a iron ball with an attached wire and handle. It has a mass of 7.260 kg (the equivalent of 16 pounds). A thrower spins more and more quickly while holding the handle of the hammer with arms extended, propelling it in large circles at ever-increasing speeds before releasing it and letting it fly off. As of October 2016, the world record for the men's hammer throw is held by Yuriy Sedykh, a Ukranian, who threw it 86.74 m (284 feet 7 inches) at the European Track and Field Championships in Stuttgart, Germany, on August 30th, 1986. Having now stood for over 30 years, it is one of the oldest records in track and field. Yuriy came into the championship meet as the world record holder with 86.66 m. The four best throws in his sequence of six at this championship meet were 85.28 m, 85.46 m, 86.74 m, and 86.68 m.

The record-setting throw was recorded with a camera at 200 frames per second. The frames were analyzed carefully, the phases of his throw were dissected, and much data about body positioning, timing, and technique were collected. Of current relevance is the following information:

The speed of the ball at release: $v_0 = 30.7$ m/s (or about 105 kilometers per hour).

The height of the ball at release: $y_0 = 1.66$ m.

The angle of elevation of the ball at release: $\varphi_0 = 39.9°$.

Given Stuttgart's latitude and elevation above sea level, the g for Stuttgart is about 9.81 m/s^2. Inserting the various constants into the range formula

$$R = \tfrac{v_0}{g} \cos \varphi_0 \left(v_0 \sin \varphi_0 + \sqrt{v_0^2 \sin^2 \varphi_0 + 2gy_0}\right),$$

we get that $R \approx 96.50$ m. This is the distance that Yuriy's throw would have reached in a vacuum. Therefore air resistance reduced Yuriy's record throw by $96.50 - 86.74 = 9.76$ m, or by a little more that 10%.

Air resistance has only a minor effect on relatively heavy projectiles moving at velocities that are relatively small. Its effects on lighter objects or those that are propelled at higher speeds are more substantial. In the context of classical artillery, for instance, we will see in the Problems and Projects section that the formulas developed above are of little or no value. Mathematical models and computer simulations of high-speed trajectories need to take the aerodynamics of the projectile, changes in air density, and humidity conditions into account. Such models and simulations are understandably complex.

6.8 NEWTON STUDIES THE MOTION OF THE PLANETS

After discussing aspects of Newton's differential calculus—the method of "prime and ultimate ratios"—the *Principia* turned to the study of centripetal force. A *centripetal force* is one "by which bodies are drawn, impelled, or any way tend towards a point, as to a centre." So a centripetal force is one that always acts in the direction of a single fixed point. This fixed point is called the *center of force*.

Figure 6.24 provides an example. Take a string and tie a ball P of weight W to one end. Hold the other end of the string at H, and twirl it so that the ball moves along a circle in a fixed horizontal plane. Figure 6.24a shows the vector F that represents the pull of the string on the ball as well as its horizontal and vertical components. Because P does not move vertically, the vertical component F_2 balances the downward pull of the weight W. In terms of magnitudes, $F_2 = W$. It follows that the resultant of F and the weight W is the horizontal component F_1 pointing in the direction of the fixed center O of the circle. See Figure 6.24b. This force is therefore an example of what Newton calls a *centripetal force*. Figure 6.24a

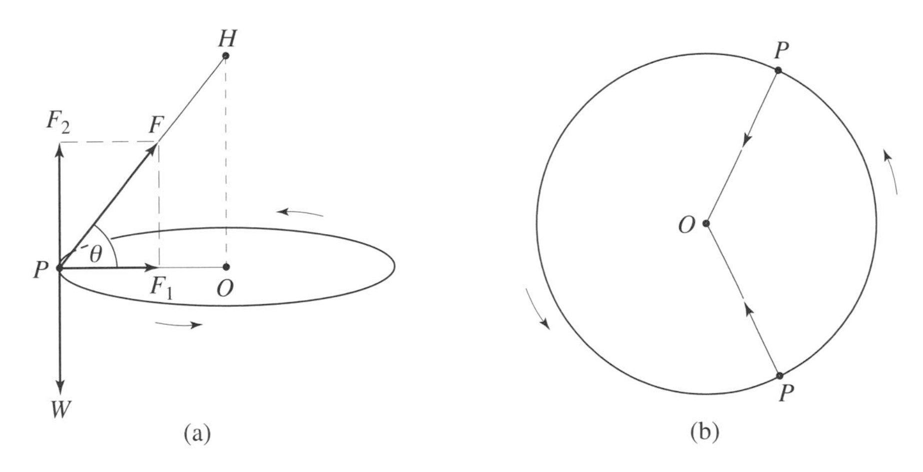

Fig. 6.24

also illustrates Newton's third law. As the ball is twirled, there is a force pulling on the hand at H in the direction of P. It is equal in magnitude to the force F pulling on P, but it acts in the opposite direction.

Example 6.6. Suppose that the string PH has length 1.6 m and that it pulls with a force of $F = 9$ N. Let the radius OP of the circle be 0.7 m. From the triangle ΔHPO, we see that $\cos\theta = \frac{0.7}{1.6} = 0.4375$. It follows that $F_1 = F\cos\theta \approx 3.94$ N. By the Pythagorean theorem,

$$F^2 = F_1^2 + F_2^2 \approx 15.5039 + F_2^2 \approx 15.5039 + W^2.$$

Since $F^2 = 81$, we get $W \approx 8.09$ N. If m is the mass of W, then $W = mg$, and hence $m \approx \frac{8.09}{9.81} \approx 0.8255$ kg.

Book I of the *Principia* develops an abstract theory of motion for objects that are pulled by a centripetal force. It makes use of the idea of the *point-mass*, a theoretical construct of a mathematical point with nonzero mass. The high point of the *Principia* comes in Book III, the *System of the World*. It applies the abstract theory of centripetal force to gravitational force and the study of the orbits of the planets. In this regard, Newton realized, after wrestling with the issue for some time, that the analysis of the Sun's gravitational force on a planet can assume that the mass of the planet is located at a point, its center of mass. Combined with his universal law of gravitation, Newton's analysis has remarkable consequences that include the fundamental explanation for what Kepler had merely observed: the three laws of planetary motion. The *Principia*'s arguments are subtle, difficult, and often terse and opaque. It is, in short, quite a challenge to understand them. One reason for this is that Newton often used delicate Greek geometry known to the trained mathematicians of his time but not to today's students of mathematics. It is the goal of the final three sections of this chapter to make the essential aspects of Newton's arguments and conclusions transparent.

Let's consider a point-mass in motion under the action of a single force, a continuously acting centripetal force. Its magnitude can vary, but it is directed toward a fixed point. With the later application to the motion of the planets around the Sun in mind, we'll follow Newton and label the point-mass by P and the

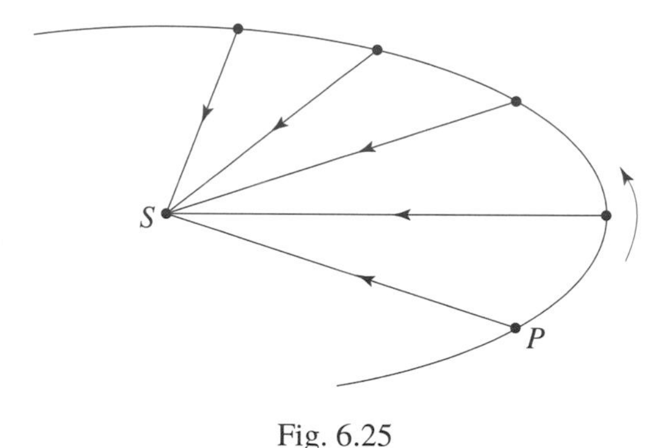

Fig. 6.25

fixed center of the centripetal force by S. See Figure 6.25. For a similar reason, we will refer to the path traced out by P as the *orbit* of P.

Newton begins his analysis by showing that the motion of the point satisfies Kepler's second law. The single assumption that P moves under the action of a centripetal force with center S is enough to guarantee that the segment PS sweeps out equal areas in equal times. This single assumption also implies that the orbit of P lies in a plane that also contains S. Think of it this way. At any time in its motion, consider the velocity vector of P and the plane that S and this vector determine. The point P moves in the direction of this vector, and the centripetal force pulling P toward S lies in this plane. Since this force has no component that can move P outside this plane, P must remain in it.

Now to Kepler's second law. Let Q_1 and Q_2 be any two points of the orbit. Suppose that P traces out

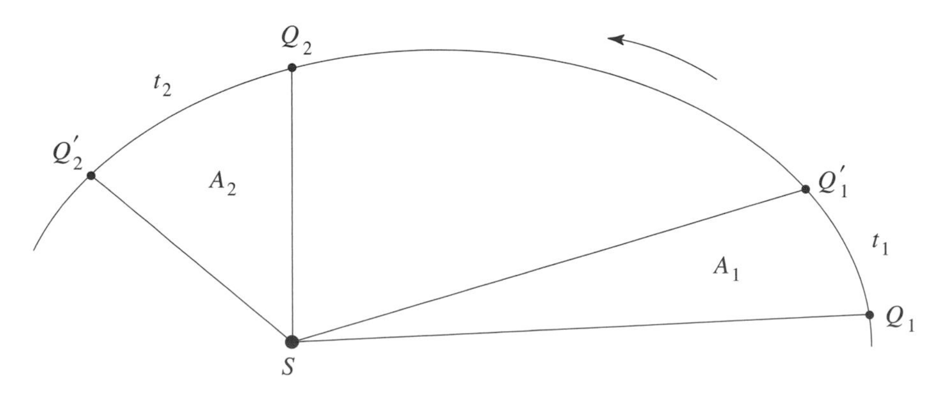

Fig. 6.26

the wedge Q_1SQ_1' during time t_1 and the wedge Q_2SQ_2' during time t_2. Let the areas of the two wedges be A_1 and A_2, respectively. See Figure 6.26. If Newton can show that $\frac{A_1}{t_1} = \frac{A_2}{t_2}$, then $t_1 = t_2$ implies that $A_1 = A_2$, and he has verified Kepler's second law. To show that $\frac{A_1}{t_1} = \frac{A_2}{t_2}$, Newton proceeds as follows. He fixes an interval of time Δt and assumes that Δt is very small compared to both t_1 and t_2. Starting at Q_1, Newton regards the centripetal force to be acting only at the beginning of each interval of duration Δt but not at any other time. So the force acts "machine gun style" as a sequence of pops or impulses. Each pop deflects the point-mass P in the same way that a billiard ball is deflected by the cushion of the table. Between two successive pops, the magnitude of the force is zero, so that by Newton's first law of motion, the path of the point-mass is a straight line. In this way, the actual curving orbit of the point from Q_1 to Q_1' to Q_2 to Q_2' is approximated by a sequence of line segments. In turn, the wedges Q_1SQ_1' and Q_2SQ_2' are approximated by aggregates of thin triangles.

Since Δt is very small, the time t_1 is equal to (at least approximately) the sum of a huge number, let's say n, Δts, and similarly the time t_2 is the sum of another huge number, let's say m, Δts. So $t_1 \approx n \cdot \Delta t$ and $t_2 \approx m \cdot \Delta t$. Starting at Q_1, let the pops of the force occur at P_1, P_2, P_3, and so on. Refer to Figure 6.27 and observe

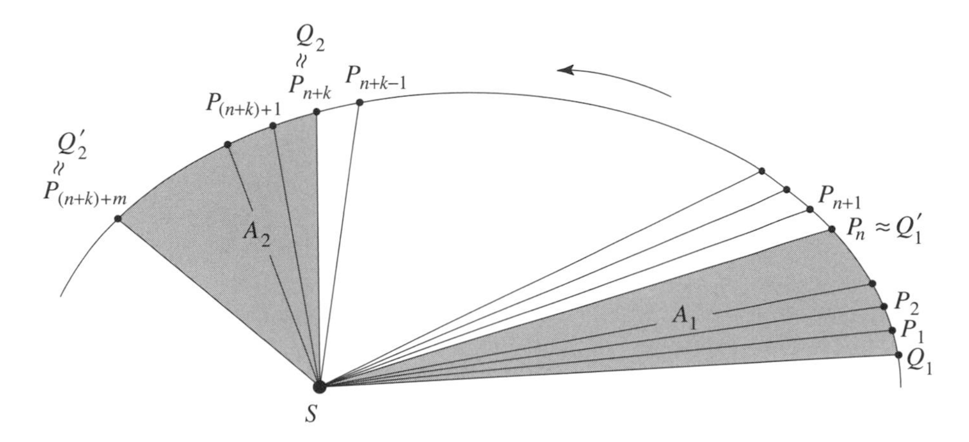

Fig. 6.27

that the region Q_1SQ_1' is approximated by n triangles and Q_2SQ_2' is approximated by m more triangles.

Newton's diagrams are notationally different but conceptually the same. He has the force acting at points A, then Δt later at B, then another Δt later at C, another Δt later at D, and so on, but not in between—see Figure 6.28a. Newton then draws the historic diagram of Figure 6.28b and goes about the task of verifying that the areas of the consecutive triangles ΔASB and ΔBSC are equal. He proceeds as follows. He assumes that the action of the force at B is delayed for another Δt until P reaches position c. There, the force acts parallel to SB to push P to position C (to which P would have been deflected had the force acted at B). Since the time intervals of the motion from A to B and B to c are the same and since P continues to c at the speed it has at B, it follows that the segments AB and Bc have the same lengths. Newton then extracts the two triangular figures of Figure 6.29 from his diagram. Since their bases AB and Bc are equal and their heights are both h, the triangles

$$\Delta ASB \text{ and } \Delta BSc$$

of Figure 6.29a have the same area. Turning to Figure 6.29b, Newton sees that the two triangles ΔBSc and ΔBSC have the same base SB and (since cC is parallel to SB) the same height. So he knows that

$$\Delta BSc \text{ and } \Delta BSC$$

have the same area. He can now conclude that the areas of ΔASB and ΔBSC are the same. By a repetition of his argument, that any two successive triangles in Figure 6.28b that the points S along with A, B, C and

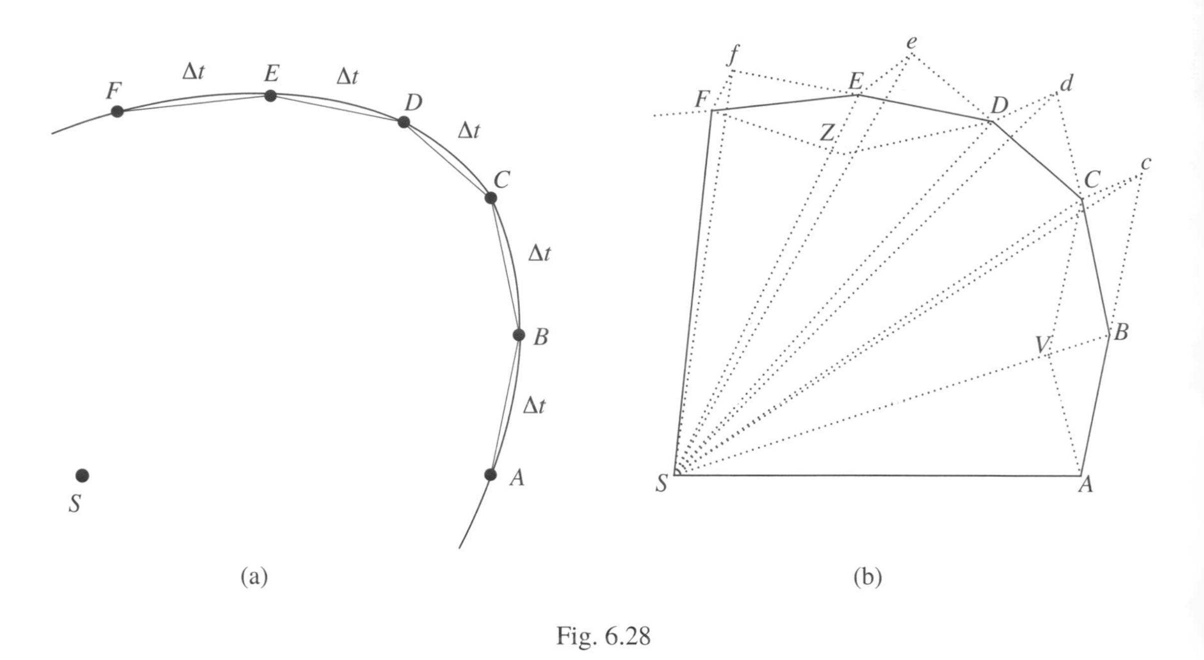

(a) (b)

Fig. 6.28

so on determine have the same area. Therefore, all of the triangles $\Delta ASB, \Delta BSC, \Delta CSD$ and so on have

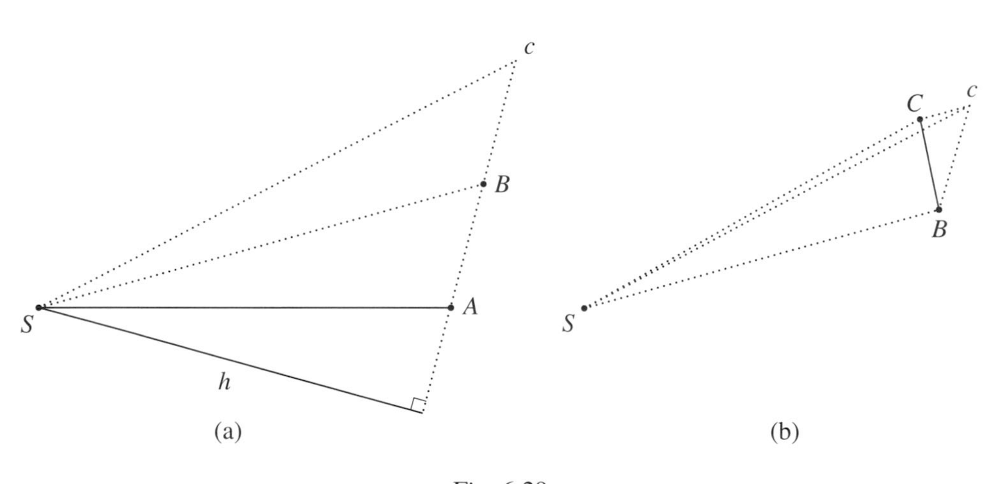

(a) (b)

Fig. 6.29

the same area.

Let's return to Figure 6.27. Applying what has just been described, we know that all the triangular slivers from Q_1 to P_1 to P_n and so on that the time intervals Δt determine have the same area. Let's denote this area by ΔA. Since the area A_1 is approximated by n such triangles and A_2 by m, we know that $A_1 \approx n \cdot \Delta A$ and $A_2 \approx m \cdot \Delta A$. Since $t_1 \approx n \cdot \Delta t$ and $t_2 \approx m \cdot \Delta t$, it follows that

$$\frac{A_1}{t_1} \approx \frac{n \cdot \Delta A}{n \cdot \Delta t} = \frac{\Delta A}{\Delta t} = \frac{m \cdot \Delta A}{m \cdot \Delta t} \approx \frac{A_2}{t_2}.$$

These are only approximations because the Δts do not exactly add up to t_1 and t_2, and the triangular slivers for Q_1SQ_1' and the m slivers for Q_2SQ_2' do not fill out these regions "on the nose." However, if the above construction is repeated, each time for smaller and smaller Δt (and therefore larger and larger n and m) the approximations $t_1 \approx n \cdot \Delta$ and $t_2 \approx m \cdot \Delta$ get better and better, and in turn: the rapid-fire sequence of impulses will approximate the continuously acting force, myriad connected line segments will approximate the curving orbit, and the approximation $\frac{A_1}{t_1} \approx \frac{A_2}{t_2}$ will tighten to an equality. Newton, referring to the diagram in Figure 6.28b, put it this way,

> Now let the number of those triangles be augmented, and their breadth diminished in infinitum; and their ultimate perimeter ADF will be a curve line: and therefore the centripetal force, by which the body is perpetually drawn back from the tangent of this curve, will act continually; and any describ'd areas $SADS$, $SAFS$, which are always proportional to the times of description, will, in this case also, be proportional to those times.

So Newton has concluded that if t is any time interval and A_t the area of the sector that the segment SP sweeps out during the time t, then the ratio $\frac{A_t}{t}$ is the same, regardless of what t is taken and where in the orbit the area is traced out. With a bow to Kepler, we will call the constant

$$\kappa = \frac{A_t}{t}$$

the *Kepler constant* of the orbit and denote it by the Greek letter κ. Having verified Kepler's second law, Newton turns to the connection between the magnitude of the centripetal force on P and the geometry of its orbit.

6.9 CONNECTING FORCE AND GEOMETRY

Newton continues to consider a point-mass propelled by a single centripetal force. In order to study the centripetal force at a typical position of the point-mass, he studies the dynamics of the force *near* this position. These dynamics bring the geometry of the orbit into play. In principle, his approach is closely related to the study of the tangent line at a point of a curve. Its slope can only be assessed by studying the "flow" of the curve near the point.

In Figure 6.30, Newton lets S be the center of force and considers the point-mass in a typical position P. A very small interval of time later, the point has moved from P to Q. Since the elapsed time is very small, Q is very close to P, and the centripetal force will not vary much during this motion. So during the motion from P to Q, Newton takes the centripetal force to be constant both in direction—parallel to SP—and magnitude. He puts in the tangent RPZ to the orbit at P, choosing R such that QR is parallel to SP, and completes QR and RP to the parallelogram shown in the figure. As he had done in his study of Kepler's

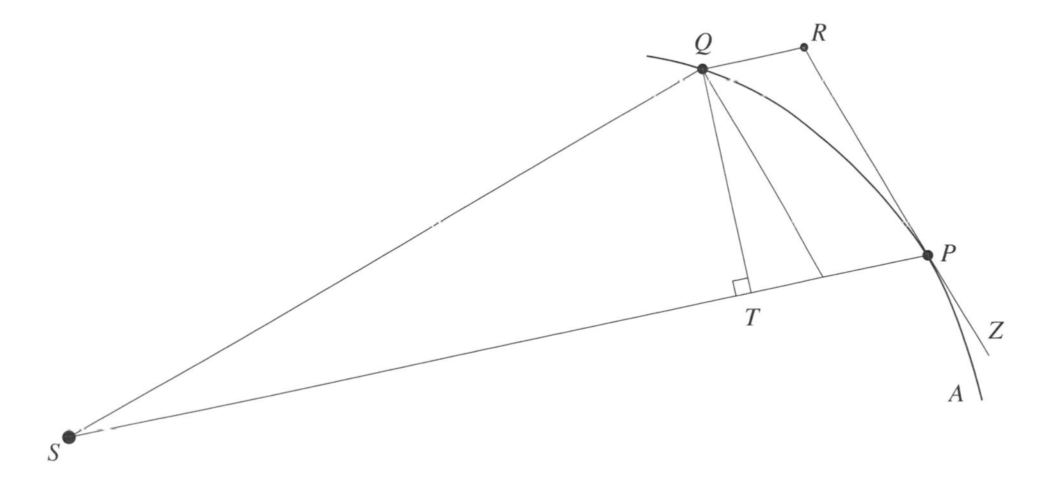

Fig. 6.30

area law, Newton thinks of the motion of the point-mass from P to Q to be separated into two components:

(1) The motion from P to R. This part takes place along the tangent line at P. The velocity is constant and equal to the velocity of the point-mass at P. This part of the motion has zero acceleration and is free of the action of the centripetal force.

(2) The motion from R to Q. This part takes place in the direction from P to S and has zero initial velocity. This is the accelerated part of the motion. It is entirely determined by the centripetal force.

As the point-mass proceeds from P to Q, these two motions take place simultaneously. But by regarding the motion from P to Q as occurring separately in the two parts, Newton is able to isolate the effect of the centripetal force and proceed with its analysis.

Let Δt be the time it takes for the point-mass to go from P to Q. Let F_P be the magnitude of the constant force acting on the point, and focus on the motion from R to Q. With m the mass of the point, its acceleration from R to Q is $a = \frac{F_P}{m}$ by Newton's second law. Since Δt is the duration of this motion, it follows from an observation in Section 6.5 that the displacement that the point-mass experiences is

$$RQ = \tfrac{1}{2}\frac{F_P}{m}(\Delta t)^2.$$

Because of the assumption made about the force F_P, this is in fact only an approximation. So Newton concludes that

$$F_P \approx \frac{2m\,QR}{(\Delta t)^2}.$$

Return to Figure 6.30. Newton drops the perpendicular QT from Q to SP. Since Q is close to P, he observes that the area of the pie-shaped sector PSQ is nearly equal to that of the triangle ΔPSQ. So

$$\text{Area sector } SPQ \approx \text{Area } \Delta SPQ = \tfrac{1}{2}(SP \times QT).$$

Since SP traces out the sector SPQ in the time interval Δt, it follows from the definition of Kepler's constant κ that

$$\frac{\frac{1}{2}(SP \times QT)}{\Delta t} \approx \frac{\text{Area sector } SPQ}{\Delta t} = \kappa.$$

So $\frac{1}{\Delta t} \approx \frac{2\kappa}{SP\times QT}$ and $\frac{1}{(\Delta t)^2} \approx \frac{4\kappa^2}{(SP\times QT)^2}$. By inserting this last approximation into $F_P \approx \frac{2m\,QR}{(\Delta t)^2}$, Newton concludes that

$$F_P \approx 8\kappa^2 m \frac{QR}{QT^2 \times SP^2}.$$

After rewriting the variable distance between S and P by r_P, this becomes

$$F_P \approx \frac{8\kappa^2 m}{r_P^2}\,\frac{QR}{QT^2}.$$

The two approximations that go into this result become tighter and tighter for smaller and smaller Δt. This is so for the first approximation, because the smaller the Δt, the closer Q is to P, and the less variability there is in the both the magnitude and direction of F_P. And it is so for the second approximation because the closer Q is to P, the better the approximation of the sector PSQ by the triangle ΔPSQ.

Therefore to obtain an exact formula for F_P, it remains to push Δt to zero, or equivalently Q to P and determine what happens to $\frac{8\kappa^2 m}{SP^2}\frac{QR}{QT^2}$ in the process. Since κ and m are constants and r_P is the distance from S to P, it is clear that the term $\frac{8\kappa^2 m}{r_P^2}$ does not depend on Q. Therefore

$$F_P = \frac{8\kappa^2 m}{r_P^2} \cdot \lim_{Q\to P} \frac{QR}{QT^2}.$$

Newton puts it this way

> the centripetal force will be reciprocally as the solid $\frac{SP^2 \times QT^2}{QR}$; if the solid is taken of that magnitude which it ultimately acquires, supposing the points P and Q continually to approach to each other.

The limit $\lim_{Q \to P} \frac{QR}{QT^2}$ is the remaining mystery. Go back to Newton's diagram in Figure 6.30, and notice that the answer to the question of what happens to the ratio $\frac{QR}{QT^2}$ when Q is pushed to P depends entirely on the shape of the orbit between P and Q. In order to compute this limit, Newton must assume something about the orbit of the point-mass. With the application to the planets of the solar system in mind, he takes the orbit to be an ellipse and places the center of force S at one of the focal points. Newton then extends Figure 6.30 to Figure 6.31. It is this diagram—one of the most significant in the history of science—on which Newton's discussion of the limit $\lim_{Q \to P} \frac{QR}{QT^2}$ is based. It marks one of the most important breakthroughs in our understanding of the workings of the solar system. The scientific relevance of Newton's analysis is underscored by the fact that this diagram was depicted on the British one-pound

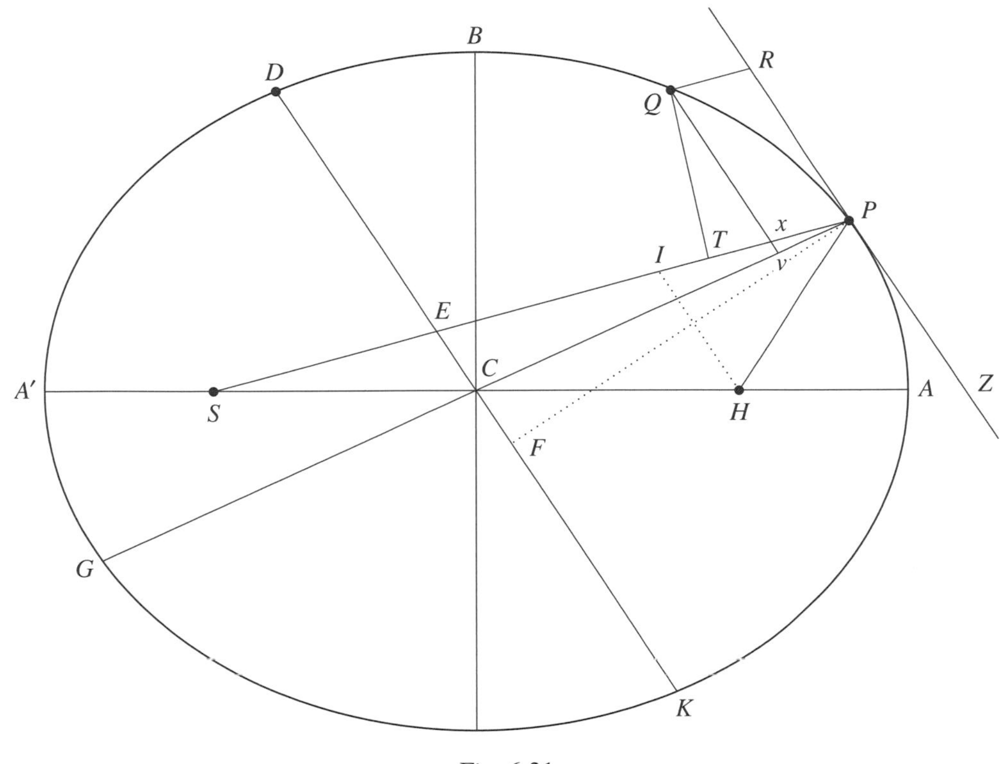

Fig. 6.31

note of the 1980s. See Figure 6.32. Curiously, the image on the banknote places the radiant Sun at the center C of the ellipse and not at its focus S. The banknote shows the great man with an opened *Principia* resting on his lap. Shown on the table behind him are the reflecting telescope that he invented as well as the prism that he used to study the properties of light.

Newton's analysis of $\lim_{Q \to P} \frac{QR}{QT^2}$ in the case of an elliptical orbit is ingenious, but technical. It makes use of several subtle facts about the ellipse that are already developed in the *Conics* of Apollonius, but are no longer emphasized in today's textbooks. While the analysis of the limit is complicated, it's conclusion is easily stated. Consider a segment that is perpendicular to the focal axis of the ellipse and goes through one of the focal points. The distance L between the two points of intersection of this segment and the ellipse is the *latus rectum* of the ellipse. See Figure 6.33. Newton shows that no matter what the

Fig. 6.32. The British one-pound note, in circulation from 1978 to 1988. A young Queen Elizabeth II appears on the other side. (The image was scanned from a banknote of the author.)

position P of the point-mass is in its orbit, the limit is always the same, namely,

$$\lim_{Q \to P} \frac{QR}{QT^2} = \frac{1}{L}.$$

Newton's verification of this fact would take us too far afield, but we will demonstrate it with a quick trigonometric argument in the special case of a circular orbit of radius r and center S. Refer to Figure 6.34.

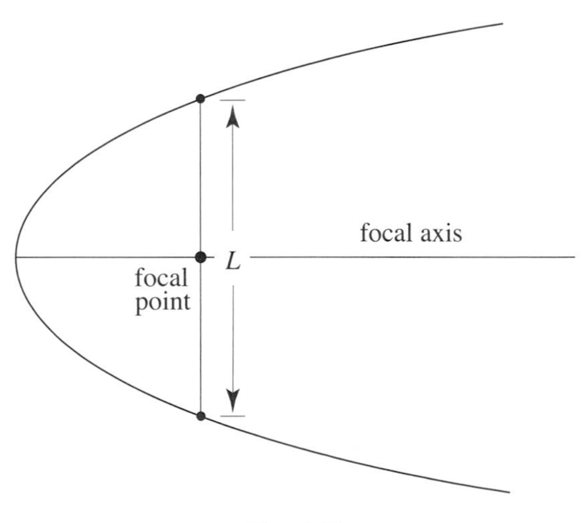

Fig. 6.33

Notice that $\cos\theta = \frac{ST}{r}$. So $ST = r\cos\theta$ and $QR = TP = r - r\cos\theta = r(1 - \cos\theta)$. Since $\sin\theta = \frac{QT}{r}$, $QT = r\sin\theta$. Therefore

$$\lim_{Q \to P} \frac{QR}{QT^2} = \lim_{\theta \to 0} \frac{r(1-\cos\theta)}{r^2\sin^2\theta} = \lim_{\theta \to 0} \frac{1-\cos\theta}{r(1-\cos^2\theta)} = \lim_{\theta \to 0} \frac{1-\cos\theta}{r(1+\cos\theta)(1-\cos\theta)} = \lim_{\theta \to 0} \frac{1}{r(1+\cos\theta)} = \frac{1}{2r}.$$

Observe that the latus rectum of a circle of radius r is $L = 2r$, so that the computation of the limit in the

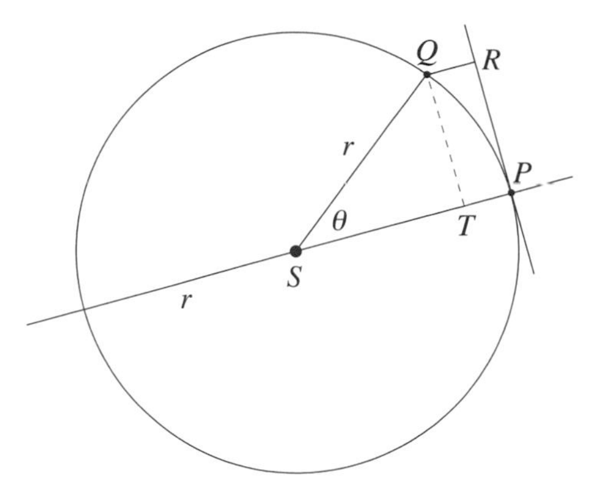

Fig. 6.34

case of a circle is complete.

Returning to the situation of the ellipse and inserting the result $\lim_{Q\to P} \frac{QR}{QT^2}$ into the earlier expression for F_P, Newton gets

$$F_P = \frac{8\kappa^2 m}{L}\frac{1}{r_P^2}.$$

This is Newton's famous *inverse square law*. With the point-mass at any position P in its elliptical orbit, it expresses the magnitude F_P of the centripetal force on the point-mass in terms of the Kepler constant κ of the orbit, the mass m of the point-mass, the latus rectum L of the ellipse, and the inverse of the square of the distance r_P from P to S. This inverse square gives the law its name. Note that both F_P and r_P vary with time, but that everything else is constant.

The inverse square law is more useful in rewritten form. Let a and b be the semimajor and semiminor axes of the elliptical orbit. We'll see in Problem 6.28 that the latus rectum of the ellipse is equal to $L = \frac{2b^2}{a}$. Since the point T in Figures 6.30 and 6.31 is no longer relevant, we'll now let T be the time it takes for the point-mass to complete one revolution around S. This time T is the *period* of the orbit. Using the fact that the area of the ellipse is $ab\pi$ (this was verified at the end of Section 5.7), it follows that Kepler's constant is equal to $\kappa = \frac{ab\pi}{T}$. By substituting $\frac{1}{L} = \frac{a}{2b^2}$ and $\kappa^2 = \frac{(ab\pi)^2}{T^2}$, Newton's inverse square law $F_P = \frac{8\kappa^2 m}{L}\frac{1}{r_P^2}$ takes the form

$$F_P = \frac{4\pi^2 a^3 m}{T^2}\frac{1}{r_P^2}.$$

If the ellipse is a circle, then $a = b = r$ and $\kappa = \frac{\pi r^2}{T}$, where r is the radius of the circle. Therefore in the case of the circle of radius r, the inverse square law reduces to

$$F_P = \frac{4\pi^2 rm}{T^2}.$$

Is the orbit of every body moving in the solar system an ellipse? No! When a comet passes appropriately close to a larger planet—Jupiter is the prime example—the gravitational pull of the planet will deflect the comet along a parabolic arc or hyperbolic arc around it until this pull becomes negligible with increased distance. Once deflected, the comet's orbit around the Sun can also be hyperbolic or parabolic. Newton is aware of this and also proves his inverse square law

$$F_P = \frac{8\kappa^2 m}{L}\frac{1}{r_P^2}$$

for point-masses describing parabolic and hyperbolic trajectories. The latus rectum for these curves is defined by Figure 6.33 in the same way as for the ellipse. Newton verifies that $\lim_{Q \to P} \frac{QR}{QT^2} = \frac{1}{L}$ for parabolic and hyperbolic trajectories in separate arguments.

Our look at some of the most important elements of Book I of the *Principia* is now complete. Let's summarize what Newton's famous treatise accomplishes: Suppose that a point-mass is propelled by a

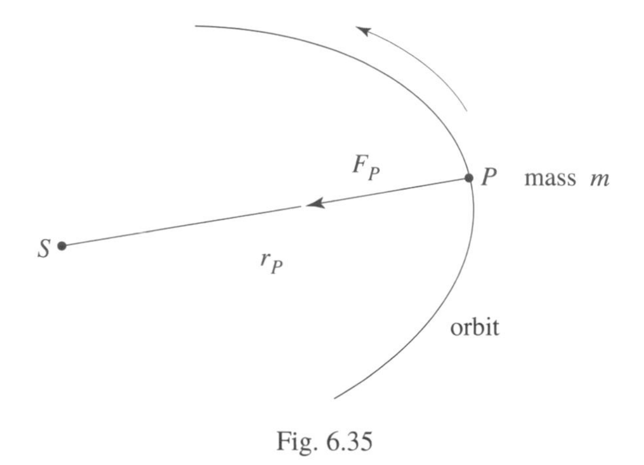

Fig. 6.35

centripetal force of magnitude F_P that has center of force S. See Figure 6.35.

Conclusion A. The segment SP sweeps out equal areas in equal times. In particular, if A_t is the area swept out by SP during some time t, then $\frac{A_t}{t}$ is the same constant κ, no matter what t is equal to and no matter where in the orbit this occurs.

Conclusion B. If the orbit is an ellipse, a parabola, or a hyperbola and the center of force S is at a focal point, then the centripetal force F_P satisfies the inverse square law

$$F_P = \frac{8\kappa^2 m}{L} \frac{1}{r_P^2},$$

where m is the mass of the point-mass, L is the latus rectum of the orbit, and r_P is the distance between P and S. If the orbit is an ellipse with semimajor axis a, semiminor axis b, and period T, then this reduces to

$$F_P = \frac{4\pi^2 a^3 m}{T^2} \frac{1}{r_P^2}.$$

Conclusion C. If the centripetal force F_P is given by an inverse square law—in other words, by an equation of the form

$$F_P = Cm\frac{1}{r_P^2},$$

where m is the mass of the point-mass and C is some constant—then the orbit is either an ellipse, a parabola, or a hyperbola, and the center of force S is at a focal point.

Conclusion C completes the picture. It tells us that Kepler's first law of elliptical orbits is a mathematical consequence of Newton's basic inverse square law. As already remarked, Newton's arguments are often terse and difficult to penetrate. In reference to Conclusion C, he outdoes even himself. The two sentences that Newton devotes to its verification do not succeed in providing clarity. We'll not attempt to provide clarity here, and move on instead.

Observe that Newton relies on the strategies of calculus both in the proof of Kepler's second law and also in the proof of the inverse square law. While Newton's approach is geometric, it does not make explicit use of functions and coordinate geometry. Hence it does not "look" like modern calculus. This is the reason why some historians of science make it a point to say (incorrectly in the view of this author) that Newton relies on geometry in the *Principia* but not on calculus.

6.10 THE LAW OF UNIVERSAL GRAVITATION

This final section turns to Newton's primary purpose: the application of his abstract mathematical theory of centripetal force to the explanation of the workings of the solar system. This is the aim of Book III of the *Principia*, the *System of the World*. The elliptical orbits that Kepler observed in combination with his Conclusions B and C (as well as the orbital data he had about the Moon) gave Newton strong evidence that centripetal force and his inverse square law

$$F_P = \frac{8\kappa^2 m}{L}\frac{1}{r_P^2}$$

is a valid quantitative description of the force with which astronomical bodies in the universe attract each other. He is convinced that his theory applies to Earth and the Moon, to the Sun and the planets, as well as to a planet and its satellites. He models any such body as a sphere that has its matter distributed in such a way that its density is the same at equal distances from the center. With this assumption, he proves that in terms of the dynamics of the motion and the forces involved, the body can be regarded as a point-mass with the entire mass at the center of the sphere.

Consider any two astronomical bodies S and P in the universe with P in orbit around S. Rewrite the expression for the magnitude of the force on P as

$$F_P = C_P m \frac{1}{r^2},$$

where m is the mass of P, r the distance between P and S, and C_P the constant $\frac{8\kappa^2}{L}$. If S exerts a pull on P, why should not also P exert a pull on S? Indeed, by Newton's third law, the force F_P with which S pulls on P has an equal and opposite reaction. In other words, P pulls on S with a force F_S of magnitude F_P. See Figure 6.36. For instance, a planet attracts the Sun with a force equal to that which the Sun exerts on

S F_S F_P P

Fig. 6.36

on the planet. The effects of the two forces are different, of course, because the Sun's mass is much greater than that of the planet. In any case, the situation is symmetric, and by this symmetry, F_S can be expressed as

$$F_S = C_S M \frac{1}{r^2},$$

where M is the mass of S and C_S is a constant. In the *System of the World*, Newton puts the matter this way:

> Since the action of the centripetal force upon bodies attracted is, at equal distances, proportional to the quantities of matter in those bodies, reason requires that it should be also proportional to the quantity of matter in the body attracting.

Since $F_S = F_P$, Newton knows that $C_S M = C_P m$ and hence that $\frac{C_P}{M} = \frac{C_S}{m}$. He lets $G = \frac{C_P}{M} = \frac{C_S}{m}$ and notices that $GM = C_P$ and $Gm = C_S$. With $F = F_P = F_S$ and a substitution, he gets

$$F = G\frac{mM}{r^2}.$$

In reference to this equation, Newton is convinced that the masses m and M of the two bodies and the distance r between them are the essential aspects that determine the magnitude F of the attractive gravitational force and that his equation should hold anywhere in the universe with the *same constant* G. This is Newton's law of universal gravitation.

Newton's Law of Universal Gravitation. Any two bodies in the universe with masses m and M and a distance r apart—whether one is in orbit around the other or not—attract each other with a force of

$$F = G\frac{mM}{r^2}.$$

We turn next to some of the incredible implications of Newton's insights. Consider any situation of a body P in the universe in elliptical orbit around a very massive one S (much more massive than P and other orbiting bodies), and let r be the (variable) distance between P and S. Let m be the mass of P, a the semimajor axis, and T the period of its orbit. Since S is assumed to be very massive relative to the bodies in orbit around it, the gravitational forces on S will have little effect on S. Thus S will move only little, so that we can assume that the attractive force of S on P is centripetal and directed to the center of S. By Newton's Conclusion B, the attractive force F between P and S satisfies

$$F = \frac{4\pi^2 a^3}{T^2} m \frac{1}{r^2},$$

and by the universal law of gravitation,

$$F = G\frac{mM}{r^2},$$

with M the mass of S. After a little algebra, we get

$$\frac{a^3}{T^2} = \frac{GM}{4\pi^2}.$$

Notice that the term $\frac{GM}{4\pi^2}$ on the right has nothing to do with the particulars of the body P and its orbit. In other words, it is the same for any P in orbit around S. It follows that the ratio $\frac{a^3}{T^2}$ of the cube of the semimajor axis a to the square of the period T of the orbit is the same for any body P in orbit around S. This is precisely what Kepler had asserted about the planets orbiting the Sun. (Note in this regard that the Sun contains about 99.85% of the mass of the solar system.) Therefore Kepler's third law is a consequence of Newton's theory of gravitation! Refer to Conclusions A and C of Section 6.9, and observe therefore that all three of Kepler's laws are consequences of Newton's theory.

The question of the value of the constant G still remained. As we will see in Segment 6J of the Problems and Projects section of this chapter, Newton did not think that the determination of G was possible. This time he was wrong. In the latter part of the 18th century, in 1798 to be exact, the English scientist Henry Cavendish succeeded in measuring G in the laboratory with an extremely delicate experiment.

Before we describe Cavendish's experiment, we'll take up the organization of the basic units of physics that occurred at around this time. At the time of French Revolution at the end of the 18th century, the meter, kilogram, and second are proposed as the scientific units of distance, mass, and time. The *meter* is defined to be equal to one ten-millionth of the distance between the North Pole and the Earth's equator. The *kilogram* is declared to be the mass of one liter (this is one-thousandth of a cubic meter) of water (at a temperature of a few degrees centigrade above freezing). Thereafter, the second was added as unit of time. It is defined in terms of the *solar day*, the time from the instant the Sun is highest in the sky to the instant this occurs the next day. Since the duration of the solar day varies, the average solar day is considered and the *second* is defined so that the average solar day is divided into precisely 86,400 seconds. With this definition and the understanding that 1 minute has 60 seconds and 1 hour has 60 minutes, the average solar day is exactly 24 hours long. In 1832, the famous German mathematician Carl Friedrich Gauss (we will encounter him in later discussions about the solar system) promotes this system as the appropriate set of units for the physical sciences. In 1875, the intergovernmental agency General Conference on Weights and Measures is organized (CGPM is its international acronym). Under its stewardship, the meter–kilogram–second system is extended coherently to the electrical realm with the addition of the *ampere* as a unit of current. In 1960, the CGPM launches the Système International d'Unités, or SI, with its seven coherent base units that include the meter, kilogram, second, and ampere, (as well as units that measure the amount of a substance in terms of elementary particles such as atoms or molecules, thermodynamic temperature, and light intensity). Meanwhile, the definitions of the units are refined. For instance, the second is defined to be the duration of 9,192,631,770 periods of the frequency at which atoms of the element cesium 133 change from one state to another. This atomic cesium clock is so precise that it looses/gains less than one second in a million years. The meter becomes the distance traveled by light in vacuum during a time interval of $\frac{1}{299{,}792{,}458}$ of a second. This provides light with a speed of 299,792,458 meters per second.

The system consisting of the meter, kilogram, and second is the international MKS *system of units*. We will continue to abbreviate the meter by m, the kilogram by kg, the second by s. An overview of the metric units that have been discussed as well as some of the units derived from them follows below. The American equivalents are included for readers whose intuitive sense is more aligned with them. Precise values are expressed in bold.

Length: 1 kilometer = **1000** meters, 1 centimeter = $\mathbf{\frac{1}{100}}$ meter, 1 millitimeter = $\mathbf{\frac{1}{1000}}$ meter, 1 inch = **2.54** centimeters, 1 foot = **0.3048** meters, 1 meter = 3.280840 feet, 1 mile = **1.609344** kilometers, 1 kilometer = 0.621371 miles, and 1 mile = **5280** feet. The units kilometer, centimeter, and millimeter are abbreviated by km, cm, and mm, respectively.

Mass: 1 tonne (or metric ton) = **1000** kilograms, 1 gram = $\mathbf{\frac{1}{1000}}$ kilogram, 1 slug = 14.593903 kilograms, 1 kilogram = 0.68522 slugs. The units tonne and gram are abbreviated by t and gm, respectively.

Time: 1 minute = **60** seconds, 1 hour = **60** minutes, a day = **24** hours = **86,400** seconds. We'll abbreviate the unit second by s and also sec.

Consider an object of mass 1 kg. Suppose that a constant force imparts an acceleration of $1\,\frac{\text{m}}{\text{s}}$ per second to the object. So during each second, the object's speed is increased by 1 meter per second. The equation $F = ma$ implies that the force on the object is $1\text{ kg}\cdot\frac{\text{m}}{\text{s}^2}$. The unit $\frac{\text{kg}\cdot\text{m}}{\text{s}^2}$ is the basic unit of force in MKS. In honor of Newton, it is called the newton and abbreviated by N.

Force: 1 pound force = 4.45359237 N, 1 newton = 0.22480894 pounds force. 1 kilonewton = **1000** newtons. The unit kilonewton is abbreviated by kN. The American ton is a unit of weight or force. It is the equivalent of 2000 pounds.

A big advantage of the MKS system is that its units and derived units (except for units of time) parallel our base 10 decimal number system. The American system with its inch, foot, yard, mile and its blob, slug, pound mass, and so on, is not. The metric MKS with its derivative units has been adopted internationally.

We can now turn to Cavendish's experiment. The essence was this. He suspended a fine wire from a fixed point A and attached a rigid "crossbar" BC. Refer to Figure 6.37. From BC, in turn, he suspended two heavy iron balls. They are shown in black. He now moved two more heavy iron balls (shown lightly shaded) in place near the two others. With this apparatus very delicately balanced and controlled, the gravitational force F between the two pairs of balls will bring about a rotation of the axis BC. This allows F to be measured. Since Cavendish knew the masses of the balls and the distances between them, the equation $F = G\frac{mM}{r^2}$ provided him with the estimate

$$G \approx 6.67 \times 10^{-11}\,\frac{\text{m}^3}{\text{kg}\cdot\text{s}^2}\,.$$

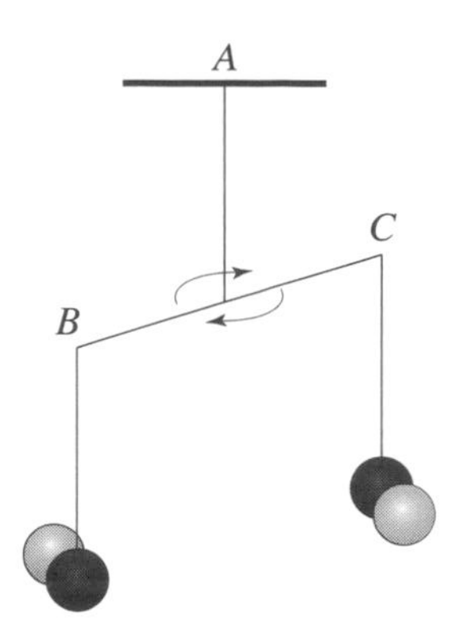

Fig. 6.37

Consider the rewritten version

$$M = \frac{4\pi^2 a^3}{GT^2}$$

of the equation that established Kepler's third law. Using the fact that the semimajor axis of the Earth's orbit about the Sun is 149.6×10^6 km, or 149.6×10^9 m, and that its period T is 365.25 days, Cavendish, after expressing T in seconds, could now compute the mass M of the Sun:

$$M = \frac{4\pi^2 a^3}{T^2 G} \approx \frac{4(3.14)^2(149.6 \times 10^9)^3}{[(365.25)(24)(60)(60)]^2(6.67 \times 10^{-11})} \approx \frac{1320 \times 10^{32}}{66 \times 10^3} \approx 2 \times 10^{30} \text{ kg}.$$

The same calculation (making use of information of the Moon's orbit about the Earth) shows that the mass of the Earth is about 6×10^{24} kg. Incredibly, Newton's theory of universal gravitation together with Cavendish's value of G has "served up" estimates for the masses of the Sun and Earth!

The universal law of gravitation can also be applied to an object of mass m on or near the surface of Earth. See Figure 6.18. The Earth is essentially spherical and its matter is distributed in such a way that its density is the same (at least approximately) at equal distances from its center. This means that with regard to gravity, all of its mass, let's set it equal to M, can be regarded to be concentrated at its center. Inserting the fact that Earth's average radius is $r = 6371$ km into the universal law of gravitation tells us

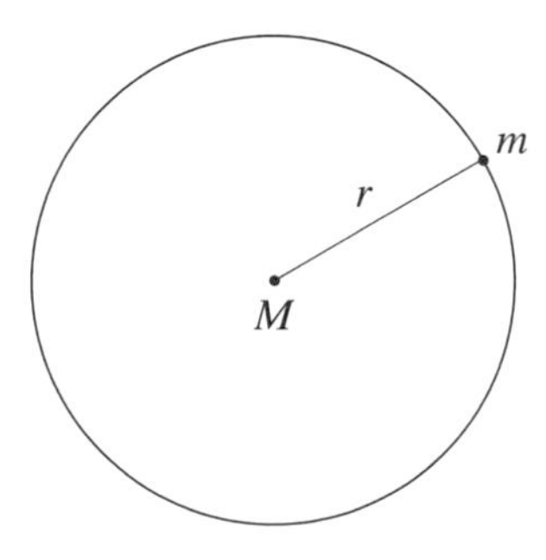

Fig. 6.38

that Earth's gravitational pull F on the object is approximately

$$F \approx G\frac{mM}{6{,}371{,}000^2} \text{ N}$$

in MKS. By Newton's second law, the magnitude of this force is also equal to the weight mg of the object, and it follows that

$$g \approx \frac{GM}{6{,}371{,}000^2} \approx \frac{6.67 \times 10^{-11} \cdot 6 \times 10^{24}}{(6.37 \times 10^6)^2} \approx 9.86 \text{ m/s}.$$

(The more accurate estimate 5.97×10^{24} kg for Earth's mass would have resulted in the more accurate estimate $g \approx 9.81$ m/s.)

By the early 17th century, Galileo had already observed that this acceleration g is the same no matter what the mass m of the falling object is. The approximate value of the acceleration generated by the force of gravity on an object falling near Earth's surface had been known long before Newton. But Newton's theory of universal gravitation, valid not only in outer space but also here at home, explains Galileo's observation. The value of g that it provides is further evidence for its validity. Incidentally, the fact that Earth is not a perfect sphere means that the distance r from the object to Earth's center depends on the location of the objects. It follows that g depends on the location on the Earth's surface as well. For instance, g is greater on the North Pole than on the equator.

Our look at some of the most important results of Book I of the *Principia* is now complete. Newton's arguments are at the same time both counterintuitive and ingenious. He assumes that the centripetal force acts intermittently (machine gun style) in bursts that are a small, fixed time interval apart. This assumption is surprising, given the continuous and smooth paths that the planets trace out in their motion around the Sun. But it has the practical advantage in that it provides an approximation of the path of P as a sequence of line segments and, consequently, of the area that P traces out as a sum of triangles. This simplified geometry provides Newton only with approximations in place of equalities. However, when he lets the time interval between the bursts shrink to zero, these approximations become the equalities he wants to verify. Historical commentary often asserts that even though Newton developed calculus, when it came to the mathematics of his magnum opus, *Principia Mathematica*, he uses geometry instead. The fact is that Newton derives his equations by letting things shrink to zero, and in so doing he uses a fundamental strategy of calculus. However, since Newton does not use the function concept (at least explicitly), the calculus in the *Principia* is a calculus that lives in geometric constructs and is not today's calculus of functions.

6.11 PROBLEMS AND PROJECTS

6A. Derivatives, Area Functions, and Antiderivatives. These problems use facts from Sections 6.1 and 6.2.

6.1. Use Newton's derivative formula to compute the derivatives of the following.

- **i.** $g(x) = 4x^5$
- **ii.** $h(x) = \frac{7}{x^{-\frac{2}{3}}}$
- **iii.** $f(x) = 5x^{\frac{1}{100}} - 4x^{\frac{1}{3}}$
- **iv.** $g(x) = -2x^{\frac{1}{3}} + 3x^5 - 6$
- **v.** $f(x) = 3(\sqrt{x})^7$
- **vi.** $y = x^{\frac{2}{7}} + 30x^4 - \frac{1}{4}x^{\frac{5}{3}}$

6.2. Use Newton's derivative formula to find antiderivatives for the following functions.

- **i.** $f(x) = 2x^3$
- **ii.** $f(x) = 5x^{\frac{1}{3}}$
- **iii.** $f(x) = 3x^5 + \frac{1}{4}x^{\frac{2}{7}}$
- **iv.** $f(x) = 6x^4 - \frac{3}{8}x^{\frac{5}{3}}$

6.3. Find the area function $A(x)$ for the following.

i. $f(x) = x^2$ with $x \geq 0$

ii. $f(x) = x^{\frac{1}{3}}$ with $x \geq 0$

iii. $f(x) = x^{\frac{5}{2}}$ with $x \geq 0$

6.4. Compute the areas under the graph of each of the following functions (over the indicated intervals).

i. $f(x) = 2x^2$ between $x = 4$ and $x = 8$

ii. $f(x) = 5x^3$ between $x = 1$ and $x = 4$

iii. $f(x) = 3\sqrt{x}$ between $x = 4$ and $x = 9$

iv. $f(x) = 4x^2 + 2x^{\frac{1}{3}}$ between $x = 1$ and $x = 8$

6.5. Study Newton's verification of the fact that the derivative of the function $y = f(x) = cx^r$ is $f'(x) = crx^{r-1}$. Reproduce his argument in the special cases $y = f(x) = x^{\frac{2}{3}}$ and $y = f(x) = x^{\frac{3}{4}}$.

6B. Newton's Use of Power Series. The next several problems call for the use of power series to solve definite integrals.

6.6. The power series

$$\frac{1}{1+x} = 1 - x + x^2 - x^3 + x^4 - x^5 + x^6 - x^7 + \ldots$$

converges for all x with $|x| < 1$. Use it to approximate

$$\int_0^{\frac{3}{4}} \frac{1}{1+x}\,dx \quad \text{and} \quad \int_0^{\frac{3}{4}} \frac{1}{1+x^2}\,dx$$

with an accuracy of three decimal places. Carry six decimal places in your computations and then use the rule of thumb to round off to three.

6.7. Multiplying the power series $\frac{1}{1+x} = 1 - x + x^2 - x^3 + x^4 - x^5 + x^6 - x^7 + \ldots$ by $x^{\frac{1}{2}}$ gives the series

$$\frac{x^{\frac{1}{2}}}{1+x} = x^{\frac{1}{2}} - x^{\frac{3}{2}} + x^{\frac{5}{2}} - x^{\frac{7}{2}} + x^{\frac{9}{2}} - \ldots.$$

Why does it converge for all x with $0 \leq x < 1$ but not at $x = 1$? Use this series to show that

$$\int_0^{\frac{1}{2}} \frac{x^{\frac{1}{2}}}{1+x}\,dx \approx 0.183.$$

Carry six decimal places in your computations and then use the rule of thumb to round off to three.

Newton studied special instances of the *binomial series*,

$$(1+x)^r = 1 + rx + \frac{r(r-1)}{2!}x^2 + \frac{r(r-1)(r-2)}{3!}x^3 + \cdots + \frac{r(r-1)(r-2)\cdots(r-(k-1))}{k!}x^k + \ldots,$$

where r is any positive rational number, and $k! = k \cdot (k-1) \cdot (k-2) \cdots 3 \cdot 2 \cdot 1$ for any positive integer k. For example, $3! = 3 \cdot 2 \cdot 1 = 6$; $5! = 5 \cdot 4 \cdot 3 \cdot 2 \cdot 1 = 120$, and so on. Expressed in sigma notation, the binomial series is

$$(1+x)^r = \sum_{k=0}^{\infty} \binom{r}{k} x^k,$$

where $\binom{r}{0} = 1$ and $\binom{r}{k} = \frac{r(r-1)(r-2)\cdots(r-(k-1))}{k!}$ for any integer $k > 0$. If r is a positive integer, then the series for $(1+x)^r$ is the multiplied-out version of the polynomial $(1+x)^r$. Since there are only a finite number of terms, it converges for all x. In all other cases, the binomial series converges for $|x| < 1$ and diverges for $|x| > 1$. In the cases of $x \pm 1$ that remain, the binomial series converges at $x = 1$ only for $r > -1$ and at $x = -1$ only for $r > 0$.

6.8. Write out the terms for the binomial series for the $(1+x)^3$ and $(1+x)^4$.

6.9. Suppose $r = -1$. Notice that for any k, k is also the number of terms in the numerator of

$$\binom{r}{k} = \frac{r(r-1)(r-2)\cdots(r-(k-1))}{k!}.$$

Check that this coefficient is equal to 1 if k is even and -1 if k is odd. So the binomial series in the case $r = -1$ is the same series

$$\frac{1}{1+x} = 1 - x + x^2 - x^3 + x^4 - x^5 + \ldots$$

as the one that Newton obtained with the division process.

6.10. Newton developed the case $r = \frac{1}{2}$ of the binomial series

$$\begin{aligned}(1+x)^{\frac{1}{2}} &= 1 + \binom{\frac{1}{2}}{1}x + \binom{\frac{1}{2}}{2}x^2 + \binom{\frac{1}{2}}{3}x^3 + \binom{\frac{1}{2}}{4}x^4 + \binom{\frac{1}{2}}{5}x^5 + \binom{\frac{1}{2}}{6}x^6 + \ldots \\ &= 1 + \tfrac{1}{2}x - \tfrac{1}{8}x^2 + \tfrac{1}{16}x^3 - \tfrac{5}{128}x^4 + \tfrac{7}{256}x^5 - \tfrac{21}{1024}x^6 + \ldots.\end{aligned}$$

by a repetitive squaring process. Check the coefficients included above and use this series to approximate $\int_0^{\frac{1}{2}} \sqrt{1+x}\,dx$ with an accuracy of three decimal places. Then try to do the same for $\int_0^5 \sqrt{1+x}\,dx$. Why is there a problem in the second situation?

Chapter 11 will study power series more extensively. Section 11.7 develops some general facts, and Section 11.8 discusses the contributions of the Englishman Brook Taylor (1685–1731) and the Scotsman Colin Maclaurin (1698–1746).

6C. Moving Points. The next group of problems considers points moving on a coordinate axis. In the first problem, the position function $p(t)$ and the starting time are given.

6.11. In each case, determine the particle's velocity $v(t)$ and acceleration $a(t)$ as functions of t, give a description of the motion of the particle, and sketch the motion as done in Figure 6.12.

i. $p(t) = 2t - 5$ starting at $t = 0$

ii. $p(t) = 2t^2 + 2t + 12$ starting at $t = -10$

iii. $p(t) = t^3 - 4t^2 - 21t$ starting at $t = -6$

iv. $p(t) = \frac{3}{t} = 3t^{-1}$ starting at $t = 1$

6.12. A point starts at the origin at time $t = 0$ with an initial velocity of zero. Its acceleration is given by $a(t) = 6t - 12$.

i. Determine its velocity function $v(t)$ and its position function $p(t)$.

ii. Draw a careful diagram of the motion over the time interval $[0, 7]$. Indicate the direction of the motion, and point out where and at what time(s) the point stops.

6.13. The acceleration $a(t)$ of a moving point is given by $a(t) = 2t - 6$ for $t \geq 0$. The initial velocity is $v(0) = 5$, and its initial position is $p(0) = 6$. Determine the velocity and position functions $v(t)$ and $p(t)$ for $t \geq 0$, and describe the motion of the point.

6.14. Each pair of functions below represents the x- and y-coordinates of a point moving in the xy-plane in terms of the elapsed time t. In each case, determine the equation of the curve along which the point moves. Sketch the curve. Compute the speed of the point at any time t, and describe how the point moves along its path.

i. $x(t) = 2,\ y(t) = 5$, and $t \geq 0$

ii. $x(t) = t,\ y(t) = t^2$, and $t \geq -2$

iii. $x(t) = t^{\frac{1}{3}},\ y(t) = t^{\frac{2}{3}}$, and $t \geq -8$

iv. $x(t) = t^3,\ y(t) = t^2$, and $t \geq -1$

v. Let $y = f(x)$ be a function defined for $x \geq b$, and let $x(t) = t$ and $y(t) = f(t)$ with $t \geq b$. What can you say about the times t for which the point's speed is a minimum?

6D. Forces. The next problems study forces and their vector representations.

6.15. Figure 6.39 depicts a force of 400 N and a direction at an angle of 35° with that of the force. Draw the component of this force in the indicated direction into the diagram and compute its magnitude.

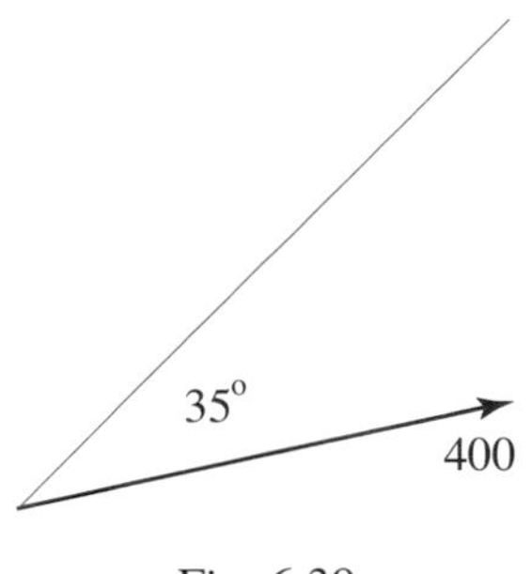

Fig. 6.39

6.16. Explain why Figures 6.40a and 6.40b show that the resultant of the two vectors in Figure 6.40a is the vector with endpoint $(-3, 1)$ of Figure 6.40b. Compute the magnitudes of the three vectors.

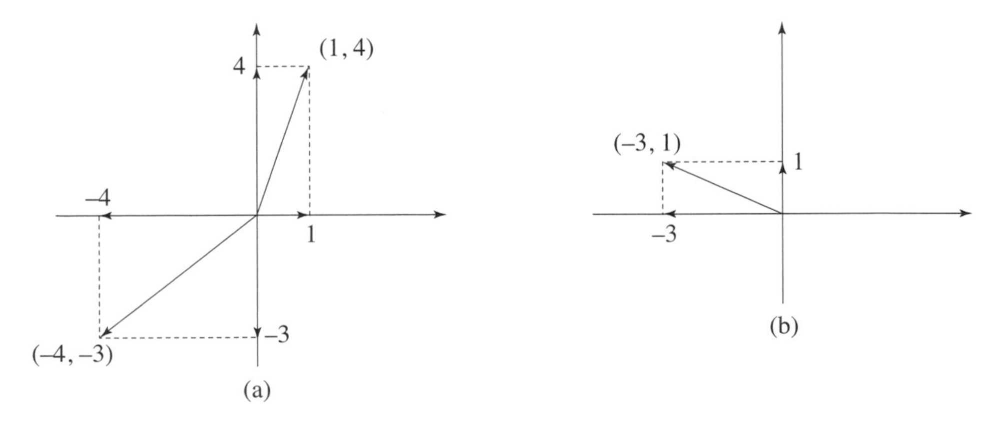

Fig. 6.40

6.17. Show that the resultant of the two vectors of Figure 6.41a is the vector in Figure 6.41b that the point $(1, 4)$ determines.

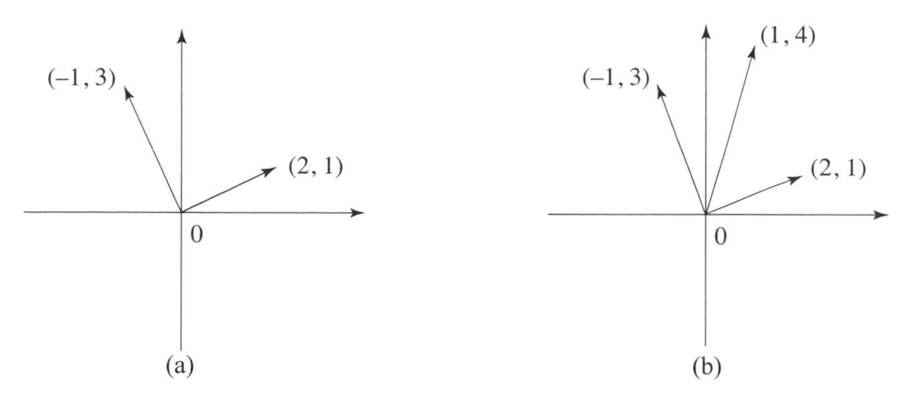

Fig. 6.41

6.18. The three forces in Figure 6.42 are in equilibrium. They all act on the point P. The dotted line is horizontal, and F acts vertically. Express the equilibrium of the horizontal components and then that of the vertical components with two equations.

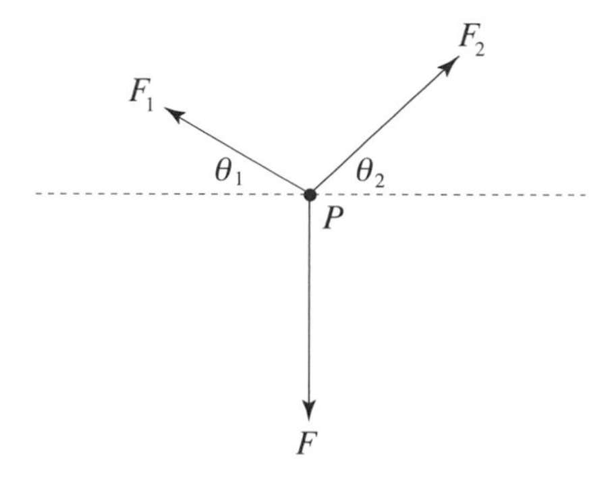

Fig. 6.42

i. If F_1 has a magnitude of 10 pounds and if the angles θ_1 and θ_2 are 30° and 60°, respectively, determine the magnitudes of F_2 and F.

ii. If the mass of the object attached at the point P is 2 kg with $\theta_1 = 30°$ and $\theta_2 = 45°$, determine the magnitudes of the forces F_1 and F_2.

6.19. i. Figure 6.43a represents two forces F_1 and F_2, their resultant of 115 pounds, as well as the angles between the forces and the resultant. Use the law of sines to determine the magnitudes of F_1 and F_2.

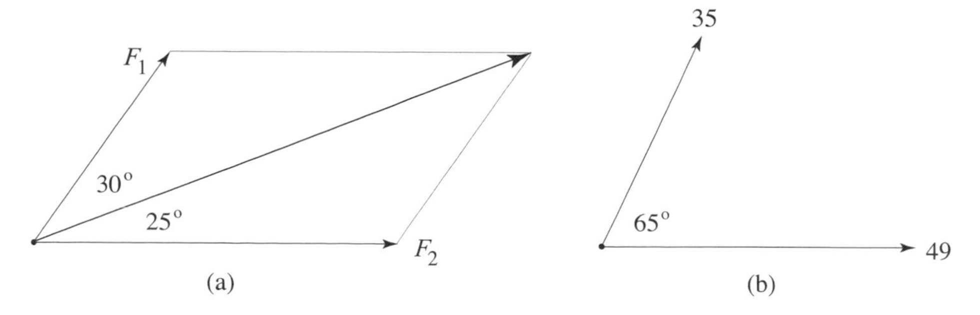

Fig. 6.43

ii. Complete Figure 6.43b to a parallelogram. Then use the law of cosines to determine the magnitude of the resultant of the two vectors.

6E. Projectiles. The next set of problems studies projectiles in situations where air resistance is assumed to be negligible.

6.20. Newton tosses an apple in the direction of his younger colleague, the astronomer Edmund Halley. When the apple leaves his hand, it is 3 feet above the ground, has an initial velocity of 25 feet per second, and has an angle of elevation of 45°. The ground is level, and Halley stands 18 feet away. Is he in a position to catch it? See Figure 6.44.

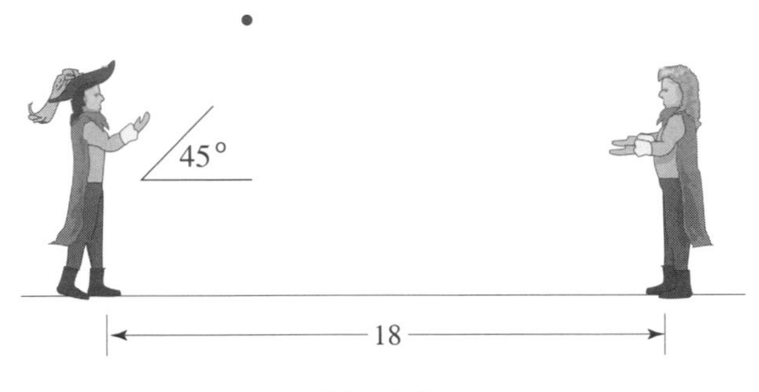

Fig. 6.44

i. Show that the apple will reach the the location of Halley in almost exactly 1 second and that it will be close to 4.4 feet above the ground at that time.

ii. So Halley is in position to catch it. Alas, Halley ducks! Show that the apple will hit the ground about 22 feet from where Newton is standing almost exactly $1\frac{1}{4}$ seconds after it leaves Newton's hand.

iii. Show that the speed of the apple at impact is about 28.5 feet/second.

6.21. Isaac Newton throws an apple in the direction of his nemesis, the scientist Robert Hooke, who stands on the same level ground 35 feet away. Newton throws the apple with an initial velocity of 40 feet per second and an angle of elevation of 20°. The apple is 5 feet above the ground at the moment of release. What is the maximal height above the ground reached by the apple? Will the apple hit Hooke (who was a small fellow)? If so, with what speed?

6.22. Suppose that a basketball player's jump shot is most accurate when he releases the ball from 8 feet above the floor at an angle of 45° and a velocity of 22 feet per second. See Figure 6.45. How far from the basket should he be taking his shots?

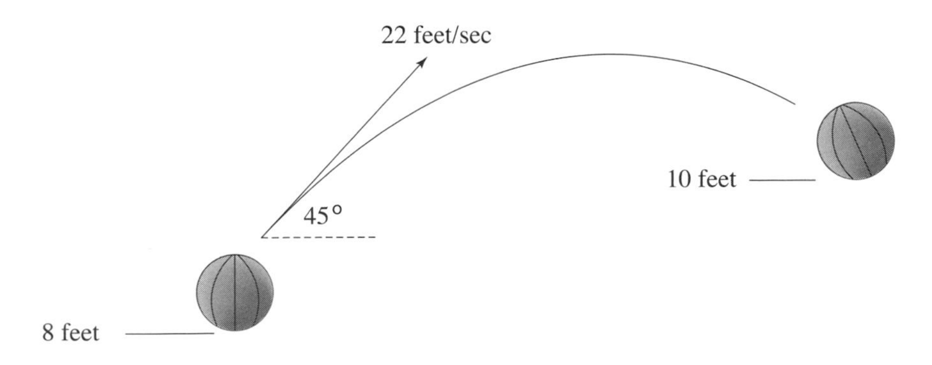

Fig. 6.45

6.23. Newton's pet parakeet escapes and flies over his garden straight toward his house. It has a speed of 6 m/s and flies upward at an angle of 30° with the horizontal when it suddenly releases, at a height of 15 m ... a dropping. How long after the release will the dropping reach its maximal height? What is the maximal height? Will it splatter against the white stucco wall of Newton's house, which is 6 m high and 9 m away (at the time of the release)? If so, with what velocity?

6.24. At the opening ceremonies of the 1992 Olympic games in Barcelona, an archer lit the Olympic flame by shooting a burning arrow into a circular pool of flammable liquid. Suppose the circular pool was affixed to a structure that put the center of the surface of the liquid 55 m downrange from the archer and 25 m above the ground. Now suppose the archer shot the arrow from 1.5 m above ground level at an angle of 70° with the horizontal, and that the arrow struck the center of the pool on its descent. With what initial speed did the arrow leave the bow? [Hint: Ignore air resistance and use Formula (6c) of Section 6.7.]

6.25. Suppose that the Olympic archer in Problem 6.24 shot the arrow from 6 feet above ground level and that the center of the circular pool was 240 feet downrange from the archer and 62 feet above the ground. If the initial speed of the arrow was 120 feet per second, determine the angle of elevation with which the marksman would have had to shoot the arrow in order to hit the center of the surface of the liquid. [Hint: Use the identity $\sec^2 \varphi = 1 + \tan^2 \varphi$.]

6F. Applications to Ballistics? *Ballistics* is the study of the motion of projectiles, primarily in the context of both light and heavy firearms. The modern science of ballistics has its origins in the middle of the 14th century when gunpowder came into use in Western Europe. At first, artillery was deployed mainly against fortifications. It was used extensively in the Thirty Years' War of 1618–1648, and it has played an increasingly important role in warfare ever since. In the 18th and 19th centuries, cannons were classified by the weight of the cannonball—usually made of iron—that they fired. For example, there were 6-, 9-, 12-, 24-, and 32-pounder cannons. Tables 6.1 and 6.2 contain data for 19th-century American 6-pounder and

Type of Ordinance	Amount of Powder (pounds)	Ammunition Type	Elevation (deg min)	Range (yards)	Time of Flight (seconds)
6-pounder field gun	1.25	shot	0 00	315	
		"	1 00	674	
		"	2 00	867	
		"	3 00	1138	
		"	4 00	1256	
		"	5 00	1523	
	1.00	spherical	2 00	650	2
		case shot	2 30	840	3
		"	3 00	1050	4
12-pounder field gun	2.50	shot	0 00	347	
		"	1 00	662	
		"	1 30	785	
		"	2 00	909	
		"	3 00	1269	
		"	4 00	1455	
		"	5 00	1663	
	1.50	spherical	1 00	670	2
		case shot	1 45	950	3
		"	2 30	1250	4

Table 6.1

12-pounder field guns. In Table 6.1, the word "ordinance" means "mounted gun" or "artillery piece," and the column "elevation" refers to the angle of elevation, namely the angle φ_0 of Figure 6.22. The information in

Type of Ordinance	Projectile		Amount of Powder	Initial Velocity
	Ammunition Type	Weight (pounds)	(pounds)	(feet/second)
6-pounder field gun	shot	6.15	1.25	1439
			1.50	1563
			2.00	1741
	spherical case shot	5.50	1.00	1357
	canister	6.80	1.00	1230
12-pounder field gun	shot	12.30	2.50	1486
			3.00	1597
			4.00	1826
	spherical case shot	11.00	2.00	1262
	canister	13.50	2.00	1392

Table 6.2

the two tables comes from the appendix of

> John Gibbon, *The Artillerist's Manual*, Greenwood Press, Westport CT, 1971, originally published by D. Van Nostrand Co., New York, and Trübner & Co., London in 1860.

This appendix also informs us that the muzzles of these guns are about $y_0 = 3.6$ feet above the ground. In the two problems that follow, use the value $g = 32$ feet/sec^2 for the gravitational constant.

6.26. Consider a 6-pounder field gun placed on horizontal ground. Let's put in 1.25 pounds of powder and load a cannonball. Table 6.2 tells us that the muzzle velocity is $v_0 = 1439$ feet/second.

i. Set the angle of elevation at $\varphi_0 = 0°$, and fire. Show that the predicted range of the cannonball is about 685 feet. Is this result consistent with the data in Table 6.1?

ii. Set the angle of elevation at $\varphi_0 = 1°$, and fire the cannon. Show that the range predicted by the range formula of Section 6.7 is approximately 2450 feet. The range of the cannonball listed in Table 6.1 is 674 yards, or 2022 feet.

iii. Set the angle of elevation at $\varphi_0 = 5°$, and fire the cannon. Show that the range predicted by the range formula of Section 6.7 is approximately 11300 feet. The range of the cannonball listed in Table 6.1 is 1523, yards or 4569 feet.

6.27. Fire a 12-pounder field gun with a cannonball using 2.50 pounds of powder. By Table 6.2, the muzzle velocity is 1486 feet/second. Fire three successive times with angles of departure of 0°, 1°, and 5°. Compare the ranges predicted by theory with those given in Table 6.1.

The solutions to Problems 6.26 and 6.27 tell us two things. First, that the data in the tables have inaccuracies. Secondly, that while air resistance has little effect when velocities are small, its effect at larger velocities can be substantial. In fact, in the context of artillery, the formulas of Section 6.7 are of little or no value. Indeed, at high velocities, the effect of air resistance is much greater than that of gravity. The shape or, more precisely, the *aerodynamics* of the projectile plays a crucial role. In general, the mathematical relationship between the aerodynamics and the velocity is a function of the velocity. It is hardly surprising that mathematical and computer models of trajectories that take air resistance into account are complicated.

6G. About Ellipses. The next set of problems may require a review of some basics from Section 4.4.

6.28. Consider an ellipse with equation $\frac{x^2}{a^2} + \frac{y^2}{b^2} = 1$, semimajor axis a, and the semiminor axis b. Determine the coordinates of the two focal points in terms of a and b. Show that the latus rectum is equal to $\frac{2b^2}{a}$.

6.29. Consider the ellipse the $\frac{x^2}{a^2} + \frac{y^2}{b^2} = 1$. Rewrite the equations as $b^2x^2 + a^2y^2 = a^2b^2$. Review Leibniz's tangent method from Section 5.2, and show that the slope of the tangent at any point (x_1, y_1) on the ellipse with $y_1 \neq 0$ (and hence $x_1 \neq \pm a$) is $-\frac{b^2}{a^2}\frac{x_1}{y_1}$.

6.30. Let P_1 and P_2 be any two points on the ellipse $\frac{x^2}{a^2} + \frac{y^2}{b^2} = 1$ such that the segment joining them goes through the center of the ellipse at the origin O. Show that if (x_1, y_1) are the coordinates of P_1, then $(-x_1, -y_1)$ are the coordinates of P_2. Use the conclusion of Problem 6.29 to conclude that the tangent lines to the ellipse at the points P_1 and P_2 are parallel. [Newton's proof of the fact that $\lim_{Q\to P} \frac{QR}{QT^2} = \frac{1}{L}$ in the elliptical case makes use of this fact.]

6H. Newton's Test Case: The Orbit of the Moon. To demonstrate that his conclusions apply in the real world, Newton tests them against available evidence. In particular, he verifies that basic observations about the Moon's orbit around the Earth are consistent with his theory.

Let's begin with some numerical data about the Moon's orbit. By Newton's time, these were much more accurate than the earlier estimates achieved by the Greeks. The French had measured the radius of the Earth at the equator to be R = 19,615,800 Paris feet. (Being equal to 0.9393 Paris feet, today's foot is a little smaller.) It was known that the Moon completes an orbit in 27 days, 7 hours, and 43 minutes, or 39,343 minutes. The average distance from the center of the Earth to the center of the Moon was known to be close to $60R$.

Newton assumes that the Moon is in a circular orbit of radius $60R$ around the center of the Earth. He takes the Moon in a typical position P and lets it be at Q exactly 1 minute later. In Figure 6.46, the motion of the Moon from P to Q is decomposed into the tangential component PQ' and the component $Q'Q$ in the direction of the Earth. Newton knows that the angle θ is equal to $\frac{360}{39343}$ degrees and is able to compute $1 - \cos\theta = 0.0000000127$ with remarkable accuracy. From the figure, $\frac{60R}{x+60R} = \cos\theta$, so that

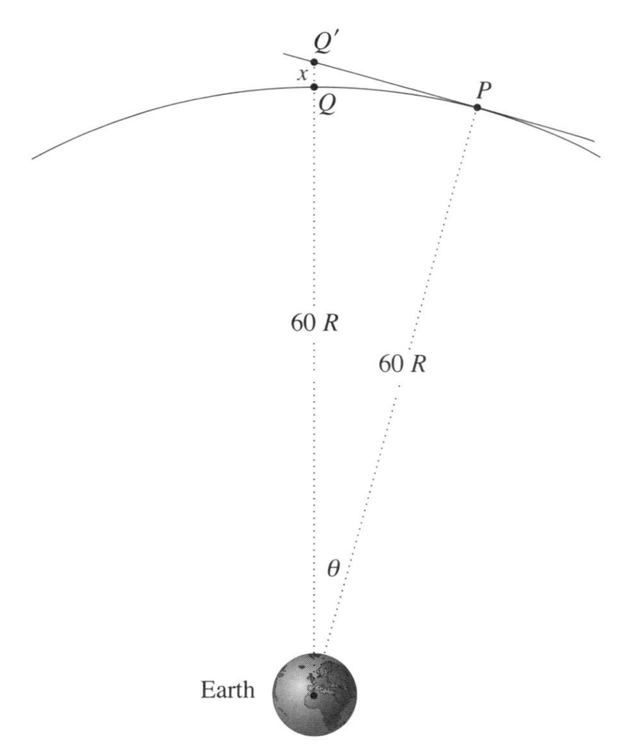

Fig. 6.46

$1 - \frac{60R}{x+60R} = 1 - \cos\theta = 0.0000000127$. Hence $\frac{x}{x+60R} = 0.0000000127$ and $x = (0.0000000127)(x + 60R)$. Solving for x and taking R = 19,615,800 Paris feet provides the value

$$x = 14.95 \text{ Paris feet.}$$

This estimate for the distance of the "fall" of the Moon toward the Earth in 1 minute is a consequence of observational data alone. Is the value provided by Newton's theory at least approximately the same?

An application of his second law of motion and his universal law of gravitation to an object of mass m on the Earth's surface tells Newton that the gravitational pull of the Earth on the object is $mg = \frac{GmM_E}{R^2}$, where g is the gravitational constant and M_E is the mass of the Earth. So $g = \frac{GM_E}{R^2}$. Applying the same two laws to the gravitational force of the Earth on the Moon informs him that this force is $ma = \frac{GmM_E}{(60R)^2}$, where m is the mass of the Moon and a the acceleration of the Moon's fall toward the Earth. Since $a = \frac{GM_E}{(60R)^2}$, Newton knows that

$$a = \frac{GM_E}{60^2R^2} = \frac{g}{60^2}.$$

At the latitude of Paris, the gravitational g was known to be equal to $g = 30.22$ Paris feet per second^2 (corresponding to 32.17 feet/sec^2) and hence to $g = (30.22)(60^2)$ Paris feet per minute^2. Therefore the acceleration of the Moon's fall is $a = 30.22$ Paris feet per minute^2. The initial velocity of the Moon's fall toward Earth from Q' to Q is zero. It follows that the velocity of the Moon along the line joining Q' to the center of the Earth is at and hence that the distance of this fall is $\frac{1}{2}at^2$. Taking $t = 1$ minute, the distance of fall x predicted by Newton's theory is

$$x = 15.11 \text{ Paris feet.}$$

Newton's theory has passed the test. The agreement between the observation of $x = 14.95$ Paris feet, and the result $x = 15.11$ Paris feet of Newton's theory is good. The discrepancy can be explained by the fact that simplifying assumptions were made. After all, the Moon's orbit was taken to be circular and not elliptical, and the gravitational effects of the Sun were ignored.

6I. The Moons of Jupiter and Saturn. In his study of the moons of Jupiter, Newton lists

$$5.578, \quad 8.876, \quad 14.159, \quad \text{and} \quad 24.903$$

for the maximal distances of the four largest moons of Jupiter from Jupiter's center. The unit of distance is the radius of Jupiter. The corresponding periods of the orbits of these moons in days, hours, minutes, and seconds are listed as

$$1^d\,18^h\,28'\,36'', \quad 3^d\,13^h\,17'\,54'', \quad 7^d\,3^h\,59'\,36'', \text{and } 16^d\,18^h\,5'\,13''.$$

This corresponds to

$$42.48, \quad 85.30, \quad 171.99, \quad \text{and } 402.09 \quad \text{hours, respectively.}$$

This information was provided to Newton by the astronomer Flamsteed. These are the four moons discovered by Galileo in 1610. Their current names Io, Europa, Ganymede, and Callisto were given to them in the 19th century.

6.31. Newton checked his theory of gravitation against this data. Which part of his theory did he test? How good was the fit?

The five largest moons of Saturn were discovered by Huygens and Cassini. Huygens discovered Titan in 1655, and Cassini discovered Iapetus in 1671, Rhea in 1672, and Tethys and Dione in 1684. They are discussed by Newton in later editions of the *Principia*. He credits Cassini with the data that he uses. The respective distances from Saturn's center are

$$1\tfrac{19}{2}, \; 2\tfrac{1}{2}, \; 3\tfrac{1}{2}, \; 8, \text{ and } 24,$$

where the unit is the radius of Saturn's outer ring. The periods of the orbits are listed in days, hours, minutes, and seconds as

$$1^d\,21^h\,18'\,27'',\;\; 2^d\,17^h\,41'\,22'',\;\; 4^d\,12^h\,25'\,12'',\;\; 15^d\,12^h\,41'\,14'', \text{ and } 79^d\,7^h\,48'\,0''.$$

This corresponds to

45.31, 65.69, 108.42, 372.69, and 1903.8 hours, respectively.

6.32. Which aspect of his theory of gravitation did Newton check using this data. How good was the fit?

6J. Project: A Speculation of Newton. Newton makes the following statement in the *System of the World*:

> For the attraction of homogeneous spheres near their surfaces are as their diameters. Whence a sphere of one foot in diameter, and of like nature to the Earth, would attract a small body placed near its surface with a force of 20,000,000 less than the Earth would do if placed near its surface. But so small a force could produce no sensible effect. If two such spheres were distant but by $\frac{1}{4}$ inch, they would not even in spaces void of resistance, come together by the force of their mutual attraction in less than a month's time.

Newton speculates about the possibility of estimating the constant G in some sort of experimental setting. He suggests that this would be an impossible task. Is what Newton is saying correct? In our discussion, we'll work in the units centimeters-grams-seconds (CGS). The unit of force in this system is the *dyne*, defined as $1\,\frac{\text{gr}\cdot\text{cm}}{\text{s}^2}$. In view of the assumption "of like nature to the Earth," we'll start by computing the Earth's density. (Recall that the average density of an object is the ratio of the object's mass over its volume.)

6.33. Use the values 6×10^{24} kg for the Earth's mass and 6371 km for its radius to estimate the Earth's average density in CGS.

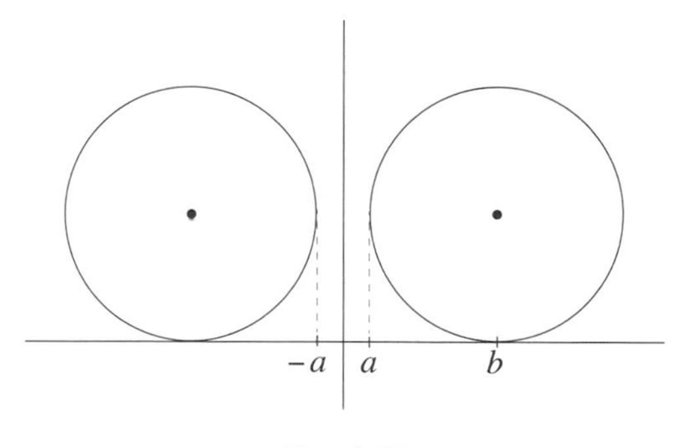

Fig. 6.47

Turn your attention to Newton's two spheres and to Figure 6.47. Suppose that they are of a uniform density equal to the average density of the Earth.

6.34. What are the dimensions in centimeters of the two spheres that Newton describes? What is the mass in grams of each of the two spheres?

6.35. What is the minimal distance between the centers of the two spheres? Convert the universal gravitational constant G into CGS, and derive an estimate for the maximal gravitational force of attraction between the two spheres.

6.36. Suppose that the two spheres start from rest in such a way that the distance $2a$ between them is the CGS equivalent of $\frac{1}{4}$ inch that Newton mentions. How long would the force of Problem 6.35 have to act so that the spheres move until they touch? How do your findings mesh with Newton's assertions?

6K. The Earth-Moon-Sun System. In discussing his theory of centripetal force, Newton offers the following insight.

> I have hitherto explained the motions of bodies attracted towards an immoveable centre, though perhaps no such motions exist in nature. For attractions are made towards bodies; and the actions of bodies attracting and attracted are always mutual and equal, by the third law of motion: so that, if there are two bodies, neither the attracting nor the attracted body can really be at rest; but both as it were by a mutual attraction, revolve about the common center of gravity.

To understand what Newton is saying, consider the Earth-Moon system. The center of mass (or gravity) of this system, the *barycenter* of the system, is about 4600 km (2900 miles) from the center of the Earth, or about 1700 km (1050 miles) below its surface. The centers of mass of both the Moon and the Earth travel along ellipses with the barycenter at a focus of each ellipse. To understand what is going on, think of the centers of the Earth and Moon as being connected with a horizontally placed lever with fulcrum at the barycenter **B**. See Figure 6.48 and note that the lever is balanced. Now let the lever revolve in the horizontal plane. This simulates the dynamics of the Earth-Moon system. The Moon is in a monthlong

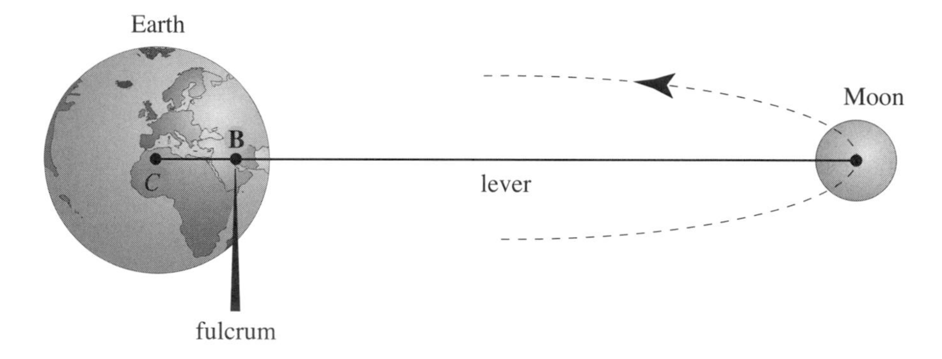

Fig. 6.48

orbit around **B**. This is a circular orbit in the simulation, but elliptical in fact. The center C of the Earth is also in "orbit" around **B**. In other words, the Moon's gravitational pull on the Earth causes it to wobble

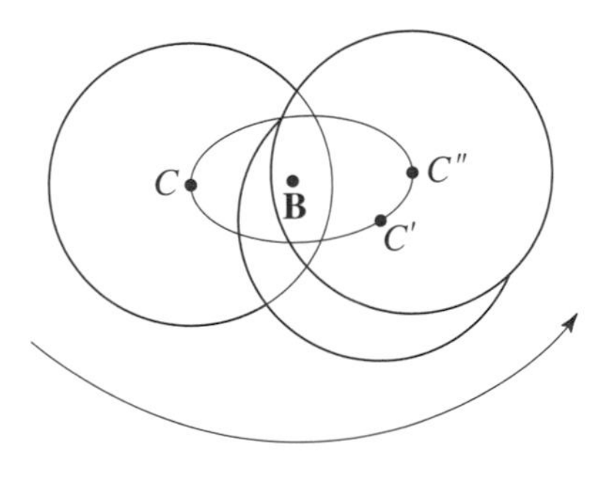

Fig. 6.49

about its center C in a monthly cycle. Figure 6.49 illustrates Earth's wobble in the context of Figure 6.48. The circle is Earth's horizontal cross section and C, C', and C'' is its moving center.

What the mass of the Moon is was a question that had interested astronomers for some time. What follows is an answer that the astronomer Sir George Airy (1801–1892) was able to provide by quantifying what goes on in Figure 6.48. Airy was director of the Royal Greenwich Observatory and, like Newton before him, professor of mathematics at Cambridge. Figure 6.50 abstracts what Figure 6.48 already observed. It shows the Moon in its orbit around the Moon-Earth barycenter **B**. Let M and m be the masses of the Earth and the Moon, respectively. Let d be the distance between the center of mass C of the Earth

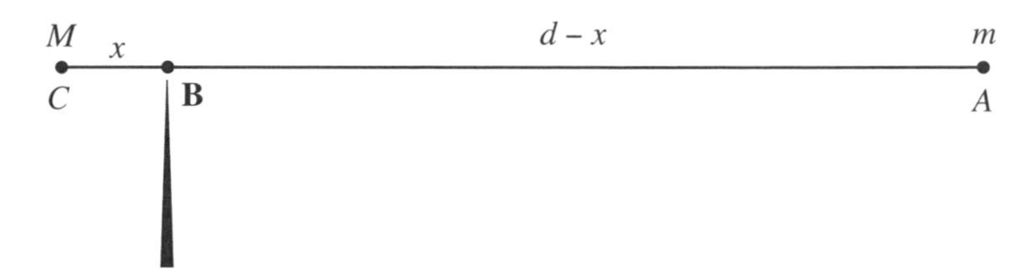

Fig. 6.50

and the center of mass A of the Moon, and let F be the gravitational force of attraction between them. By Newton's law of universal gravitation, $F = G\frac{mM}{d^2}$.

Let x be the distance between C and **B**. So $d - x$ is the distance between **B** and A. We'll assume that the Moon's orbit is a circle centered at **B** with radius $d - x$. Applying Conclusion B of Section 6.9 with $r_P = a = d - x$, we find that $F = \frac{4\pi^2 m}{T^2}(d - x)$, where T is the period of the Moon's orbit. By Archimedes's law of the lever, $Mx = m(d - x)$. So $md - (M + m)x = 0$. Therefore $Md + md - (M + m)x = Md$, so $(M + m)(d - x) = Md$, and it follows that $d - x = \frac{Md}{M+m}$. Therefore

$$G\frac{mM}{d^2} = F = \frac{4\pi^2 m}{T^2}(d - x) = \frac{4\pi^2 m}{T^2} \cdot \frac{Md}{M + m}.$$

After canceling and solving for $m + M$, we get $M + m = \frac{4\pi^2}{G} \cdot \frac{d^3}{T^2}$. Dividing by M, we get $1 + \frac{m}{M} = \frac{4\pi^2}{GM} \cdot \frac{d^3}{T^2}$.

Let g_e be the Earth's gravitational constant at the equator, and let r_e be the equatorial radius of the Earth. Since $m_e g_e$ and $G\frac{m_e M}{r_e^2}$ are both expressions of the Earth's gravitational force on any mass m_e on the equator, $GM = g_e r_e^2$. Therefore

$$\frac{m}{M} = \frac{4\pi^2}{g_e r_e^2} \cdot \frac{d^3}{T^2} - 1.$$

Inserting today's values (all in MKS), $g_e = 9.7803$ m/s^2, $r_e = 6.3781 \times 10^6$ m, $d = 3.844 \times 10^8$ m (the semimajor axis of the Moon's orbit), and $T = 2.3606 \times 10^6$ (the Moon's period in seconds), Airy's approach tells us that $\frac{m}{M} \approx 1.0114 - 1 = 0.0114$. Therefore the mass of the Moon is the fraction $m \approx \frac{1}{87.72}M$ of Earth's mass M.

Airy's estimate fell short of the mark. With $M = 6 \times 10^{24}$, it implies that $m = 6.84 \times 10^{22}$. As late as 1968, the value of the ratio was thought to be close to $\frac{1}{81.5}$. Only after the very accurate information about the orbits of the command and service modules of the Apollo Moon missions became available was it possible to measure the mass of the Moon and therefore the ratio $\frac{m}{M}$ accurately as $\frac{1}{81.300588}$. This value in turn implied that the mass of the Moon is 7.4×10^{22} kg.

Let's apply some of these considerations to the solar system as a whole. It is the barycenter **B** of the Earth-Moon system (rather than the center of the Earth) that describes an elliptical orbit about the Sun. The center of force is not the Sun, but rather the center of mass of the entire system of the Sun, planets, asteroids, and comets. This is the point that is at a focus of the ellipses of all the planets. Newton's explanations thus constitute a subtle refinement of Kepler's laws. Since the Sun comprises 99.85% of the mass of the solar system, the difference is only minor, however. These considerations explain other phenomena as well. The shape of the Earth is essentially spherical, but it is a sphere that is flatter at the poles and bulges out at the equator. Figure 6.51 provides an exaggerated look. This shape was brought about by the rotation of the

Earth about its axis. The Earth's axis is tilted by about 29° relative to the Moon's orbital plane and about 23.5° relative to Earth's orbital plane. This means that the gravitational pulls of both the Moon and the

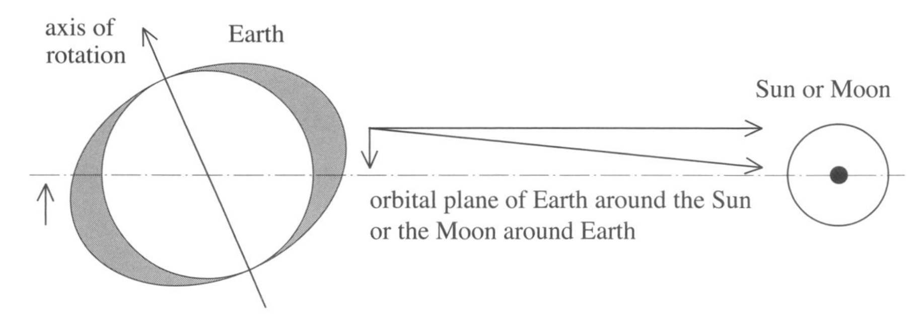

Fig. 6.51

Sun on Earth's bulge has a component that pulls Earth's axis of rotation in a direction perpendicular to its orbital plane. See Figure 6.51. This component is substantial enough to produce a very slow gyration of the Earth's axis of rotation. (A similar effect explains the tides of the oceans: they are produced by the tug of the Moon on Earth's surface.) The Earth's axis of rotation completes a full revolution every 26,000 years. This gyration of the axis is precisely what the Greeks observed as the "precession of the equinoxes." See Section 1.1.

6L. Maximal Speeds in the Solar System. A discussion about the speed of the planets, comets, and asteroids in the solar system follows. When an elliptical orbit of a planet, moon, comet, or asteroid is considered, a will be its semimajor axis, b its semiminor axis, $c = \sqrt{a^2 - b^2}$, $\varepsilon = \frac{c}{a}$ its eccentricity, T its period, and κ its Kepler constant.

We'll begin by illustrating how Newton would have developed the equation for the maximum speeds of the bodies in the solar system. His Figure 6.30 is the key.

6.37. Consider a planet (or comet or asteroid) in orbit around the Sun. Explain why Kepler's second law implies that the planet attains its maximum speed $v_{\max}$ at perihelion. Figure 6.52 shows the perihelion position P, a short stretch of the orbit, and the position Q of the planet a short time Δt

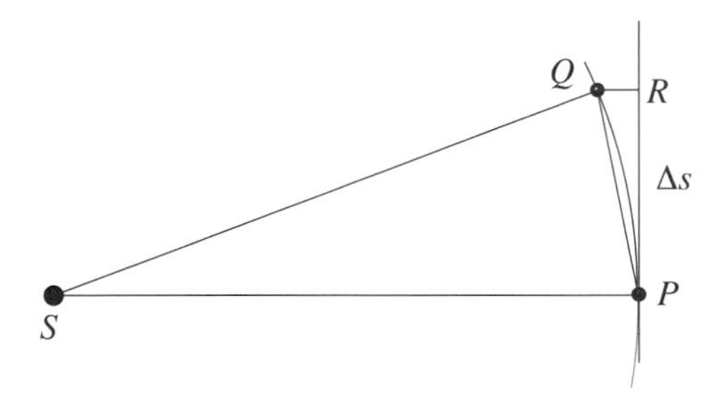

Fig. 6.52

later. The segment through P is tangent to the orbit at P, and R is chosen so that RQ is parallel to SP. The segment PR has length Δs.

i. Provide an expression for the precise value of the *average velocity* v_{av} of the motion of the planet from P to Q. Provide an approximation of v_{av} by using Δs.

ii. Use both the exact value $\kappa = \frac{ab\pi}{T}$ for Kepler's constant and an approximation for κ that arises from the diagram to verify that the average velocity satisfies $v_{\text{av}} \approx \frac{2ab\pi}{(a-c)T}$.

iii. What two things happen in (ii) when Δt is pushed to zero that result in the conclusion $v_{\max} = \frac{2ab\pi}{(a-c)T}$?

iv. Use the equalities $b = \sqrt{a^2 - c^2}$ and $c = \varepsilon a$ to conclude that $v_{\max} = \frac{2\pi a}{T}\sqrt{\frac{1+\varepsilon}{1-\varepsilon}}$.

6.38. Argue by using Kepler's second law that a planet (or comet or asteroid) attains its minimum speed $v_{\min}$ in its orbit at aphelion. Proceed as in Problem 6.37 to show that $v_{\min} = \frac{2ab\pi}{(a+c)T} = \frac{2\pi a}{T}\sqrt{\frac{1-\varepsilon}{1+\varepsilon}}$.

Recall that 1 au is very nearly the semimajor axis of Earth's orbit around the Sun. Using the approximations 1 au = 149,598,000 km (see the end of Section 3.7) and 1 year = 31,558,000 s (in current context, a year consists of 365.2564 days), we get

$$1\,\frac{\text{au}}{\text{year}} = \frac{149{,}597{,}892\text{ km}}{1\text{ year}} \times \frac{1\text{ year}}{31{,}558{,}000\text{ s}} \approx 4.74\,\frac{\text{km}}{\text{s}}.$$

6.39. Use the data in Table 3.3 to show that the maximum and minimum speeds of Earth in its orbit are approximately 30.29 km/s and 29.28 km/s, respectively, and that those of Mercury are 58.94 km/s and 38.81 km/s, respectively.

6.40. The elliptical orbit of the comet Halley has semimajor axis a = 17.83 au, eccentricity ε = 0.967, and period T = 75.32 years. Use this information to show that its maximum and minimum orbital speeds are approximately 54.43 km/s and 0.91 km/s.

The next set of problems establishes the fact that there is a "maximum speed limit" on the speed of all the planets, comets, or asteroids that are in elliptical orbits around the Sun.

6.41. Show that $\frac{a^3}{T^2} = \frac{GM}{4\pi^2}$ is equal to the same constant for all planets in the solar system and that this constant is equal to 1 in the unit $\frac{\text{au}^3}{\text{year}^2}$.

6.42. Consider any object in an elliptical orbit around the Sun. Use the results of Problems 6.37iv and 6.41 to show that

$$v_{\max} = 2\pi\sqrt{\frac{1+\varepsilon}{a(1-\varepsilon)}} < \frac{2\pi\cdot\sqrt{2}}{\sqrt{a(1-\varepsilon)}},$$

where $v_{\max}$ is expressed in au/year and $a(1-\varepsilon)$ is the perihelion distance expressed in au.

The orbits of most comets are very flat, so that their eccentricity ε is close to 1. This implies, in view of the previous problem, that their maximum speeds are primarily determined by their perihelion distances, and that the smaller its perihelion distance, the greater the maximum speed of the comet. In other words, the closer the comet approaches the Sun, the greater its maximum speed. Experts believe that larger comets can survive close encounters with the Sun provided their perihelion distances are at least 0.005 au or 750,000 km. Since the average radius of the Sun is 0.00465 au or 695,500 km, such comets approach the surface of the Sun to within about 50,000 km. Known as "sungrazing" comets, they are—when at perihelion—the fastest-moving objects in the solar system.

6.43. Show that the maximum speed of a comet in elliptical orbit around the Sun is bounded by 126 au/year or 598 km/s.

The first sungrazing comet that received a lot of attention was the Great Comet of 1680. Newton tracked it carefully. A painting by the Dutchman Lieve Verschuier (who was there to observe it himself), depicted in Figure 6.53, informs us that its tail swept across the sky in a spectacular arch. It shows Newton's Dutch contemporaries observing the comet with great interest. Some of them are seen pointing cross-staffs skyward. These were simple devices for measuring angles. The comet passed about 0.00134 au

Fig. 6.53. Lieve Verschuier, *The Great Comet of 1680 over Rotterdam*, oil on panel, 25.5 cm by 32.5 cm, Historisch Museum Rotterdam. Image courtesy of the Hesburgh Library, University of Notre Dame.

(or 200,000 km) above the Sun's surface, so that its perihelion distance was close to $0.00465 + 0.00134 \approx 0.006$ au. The fact that its perihelion distance was small meant that it reached a great maximal speed. Unlike some other sungrazers, Newton's comet survived its close encounter with the Sun.

6.44. Use the information developed above to show that the maximal speed that the Great Comet of 1680 attained was around 544 km/s.

Some of the most spectacular comets ever observed have been sungrazers. On the day the Great Comet of 1843 reached perihelion, it was widely seen in full daylight and described as "an elongated white cloud." Observations provided a tight perihelion distance of 0.005460 au (or 820,000 km) and an eccentricity of $\varepsilon = 0.999914$. After perihelion, the comet diminished in brightness but its tail grew enormously, eventually attaining a length of 320 million km or over 2 au. (Recall that the average Earth-Sun distance is 1 au.) The next super comet was the Great Comet of 1882. First spotted by a group of Italian sailors in the Southern Hemisphere, it brightened dramatically as it approached its rendezvous with the Sun and became visible in broad daylight. It had an orbital eccentricity of $\varepsilon = 0.999907$. During its perihelion passage at a distance of about 0.0076 au (or 1,150,000 km) from the center of the Sun, its nucleus broke into at least four separate parts. A day after perihelion, observers described the comet as a "blazing star" near the Sun. In the days and weeks that followed, its tail continued to shine brilliantly. The comet Ikeya-Seki was the brightest comet of the 20th century. It was discovered in the fall of the year 1965 by the two Japanese amateur astronomers Ikeya and Seki only a little over a month before its perihelion passage. The comet was glowing in the sky "ten times brighter than the Full Moon." Before perihelion, its nucleus was observed to break into two pieces. The two new nuclei emerged with slightly different orbits, both with eccentricity 0.999918 and perihelion distance 0.00778 au (or 1,200,000 km). After it passed the Sun, its tail of about 120 million km in length dominated the morning sky.

6.45. Estimate the maximal speeds of the great comets of 1843, 1882, and 1965 in km/s.

6M. Satellites (Mostly Artificial). The first group of problems focuses on artificial satellites of Earth. The Earth is a sphere that is slightly flattened at the poles. So the distance from the Earth's center to its surface varies slightly. The radius at the equator is 6378 km and that at the poles is 6357 km. The average radius is 6371 km.

6.46. The launching of the satellite Sputnik 1 by the Soviet Union on October 4, 1957, inaugurated the Space Age. In Sputnik's elliptical orbit, its distance from the Earth's surface ranged from 230 km to 942 km. It circled the Earth once every 96 minutes and had a mass of 83.6 kg. It remained in orbit until early in 1958, when it burned up in the Earth's atmosphere.

- **i.** Use the value of 6371 km for Earth's radius to compute the semimajor axis a and eccentricity ε of the orbit of Sputnik 1.
- **ii.** Use information about Sputnik's orbit to confirm that the mass of the Earth is approximately 6.00×10^{24} kg.

6.47. The satellite Explorer 1, launched from Cape Kennedy (then known as Cape Canaveral) in Florida on January 31, 1958, was America's response to Sputnik 1. Its orbit took it from a distance of 360 km to a distance of 2534 km above the Earth's surface. Its period was 114.9 minutes. It had a mass of 13.92 kg. It carried cosmic-ray and micrometeorite detection instruments and transmitted the data they collected back to Earth. Estimate the semimajor axis and eccentricity of Explorer's orbit. What estimate for the Earth's mass does its orbital data provide?

We'll next turn to the Moon and its orbit around Earth. Use the following information for the problems that follow. The eccentricity of the Moon's orbit is $\varepsilon = 0.0549$, its semimajor axis is $a \approx 384{,}400$ km $\approx 3.8 \times 10^8$ m, its period is $T = 27.32$ days $\approx 2{,}360{,}600$ s, and its mass is 7.4×10^{22} kg.

6.48. You are looking out at the rising Moon on a cloudless night. You observe it change position relative to the horizon. Does the Earth's rotation or the Moon's own motion have the greater effect on this change of position?

6.49. Make a calculation to show that the gravitational force of the Sun on the Moon is greater than the force of the Earth on the Moon. This being so, how is it that the Moon goes around the Earth and not around the Sun? Or does it? [Use the following data: the mass of the Sun is 2.0×10^{30} kg; the mass of the Earth is 6.0×10^{24} kg; the average distance from the Earth to the Sun is 1.5×10^{11} m.]

6.50. Use the formulas $v_{\max} = \frac{2\pi a}{T}\sqrt{\frac{1+\varepsilon}{1-\varepsilon}}$ and $v_{\min} = \frac{2\pi a}{T}\sqrt{\frac{1-\varepsilon}{1+\varepsilon}}$ to estimate the maximum and minimum orbital speeds of the Moon in km/s. Are your answers consistent with the fact that the average orbital speed of the Moon is 1.023 km/s?

6.51. The Soviet space craft Luna 10 was the first satellite to be placed into orbit around the Moon. It had a mass of about 1350 kg (including the fuel) attached to a 245-kg lunar orbiter module. It was launched in March of 1966, injected into an Earth orbit, and launched towards the Moon from its Earth-orbiting platform. Following a midcourse correction, Luna 10 turned around at a distance of 8000 km from the Moon and fired its rockets, slowing by 0.64 km/s. It entered lunar orbit on April 3, 1966. In its orbit it ranged from 350 km to 1017 km above the Moon's surface. The semimajor axis of its orbit was 2413 km with a period 178.05 minutes. Show that this information provides an estimate of 7.29×10^{22} kg for the mass of the Moon. [Hint: Be aware that you are provided with a lot of extraneous information.]

6.52. Gravity provides all bodies falling near the surface of the Moon with the same acceleration regardless of the mass of the body. Use the fact that the Moon's radius is 1740 km to show that this acceleration is about 1.63 m/s^2. (This is about 1/6 of the gravitational acceleration of Earth.)

The next two examples apply Newton's gravitational theory to determine the masses of the asteroids Eros and Eugenia. Asteroids have been observed to approach Earth rather closely (occasionally, closer than the orbiting Moon) and are regarded to be a potential threat. It is therefore relevant to have a sense of how massive they are.

6.53. On April 11, 2000, the *New York Times* (in the Science Times Section) reported that the craft NEAR Shoemaker, guided by gentle shoves from small thruster rockets, was taking up a circular orbit of 99.8 km from the center of the asteroid Eros in order to conduct a study of the asteroid. The 772 kg craft was named NEAR (Near Earth Asteroid Rendezvouz) Shoemaker in honor of Eugene Shoemaker, a pioneer in the study of asteroids and comets. Eros is one of the larger asteroids, measuring about 34 km long and 13 km thick. It is a potato-shaped rock covered with craters with a large gouge in the center. Instead of turning on its long axis, it rolls end over end. Eros's gravitational field is so slight that the space craft must keep its speed down to 4.8 km/hour in order to stay in orbit. Can you estimate the mass of Eros?

6.54. On October 12, 1999, the *New York Times* (in the Science Times Section) reported that a team of astronomers using a telescope on Mauna Kea, Hawaii, had discovered a satellite orbiting the asteroid Eugenia (in the main asteroid belt between Mars and Jupiter). The moon is about 13 km across, and orbits Eugenia at a distance of 1130 km once every 4.7 days. Use this information to estimate the mass of Eugenia.

We'll close our discussion about artificial Earth satellites with a look at some of the important information that they have provided.

6.55. An artificial satellite that has been spectacularly successful is the Hubble Space Telescope. It was put into orbit in 1990 by the Space Shuttle. The size of a large school bus, it has a mass of 11,110 kg. Its orbit is circular at a distance of 612 km above the Earth's surface, and the plane of the orbit makes an angle of 28.5° with the plane of the Earth's equator. Hubble orbits the Earth once every 97 minutes. It has taken an array of spectacular images of objects in the solar system and beyond. Explore the site http://www.nasa.gov/hubble/ for a look at the vast amount of important information that the Hubble has provided about the universe. Many of the striking images on the Astronomy Picture of the Day website (go to http://antwrp.gsfc.nasa.gov/apod/ and click on Archives) were taken by Hubble.

6.56. Suppose that a moving object P is propelled by a single force that is centripetal in the direction of a fixed point S. So by one of Newton's conclusions, the segment SP sweeps out equal areas in equal times. Suppose the orbit is an ellipse and that the segment SP also sweeps out equal angles in equal

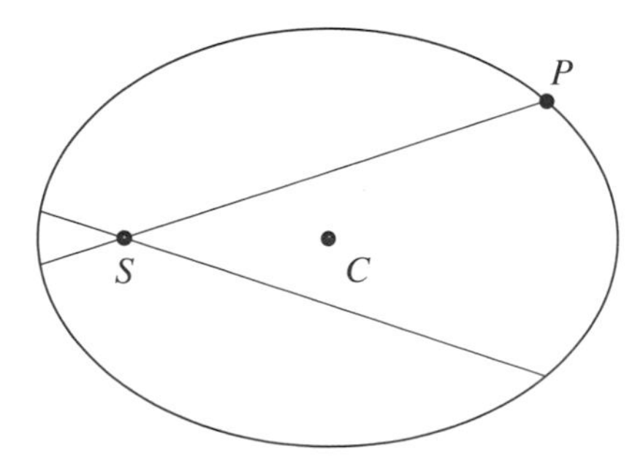

Fig. 6.54

times. Show that the orbit must be a circle with center S. [Hint: Make use of Figure 6.54. The point C in the figure is the center of the ellipse.]

6.57. A communications satellite is placed into an orbit that keeps it exactly over a designated fixed point on the Earth's equator. In such a "geosynchronous" orbit, a satellite travels around the Earth exactly once every 24 hours. Show that the orbit of such a satellite must be a circle and that its radius must be about 42,000 km. [Hint: Make use of the conclusion of Problem 6.56.]

6.58. The world's Global Positioning System (GPS) is an important application of satellite technology. The system consists of 24 satellites. The 24 satellites of the original system were launched between 1989 and 1994. (As older satellites are being replaced by newer ones, there are often more than 24 in orbit.) The 24 satellites are carefully spaced. The equator of the Earth is a circle with the Earth's center of mass C at its center. Take six diameters of this circle in such a way that the angle between consecutive diameters is 30°. Figure 6.55 shows that the six diameters are equally spaced over this

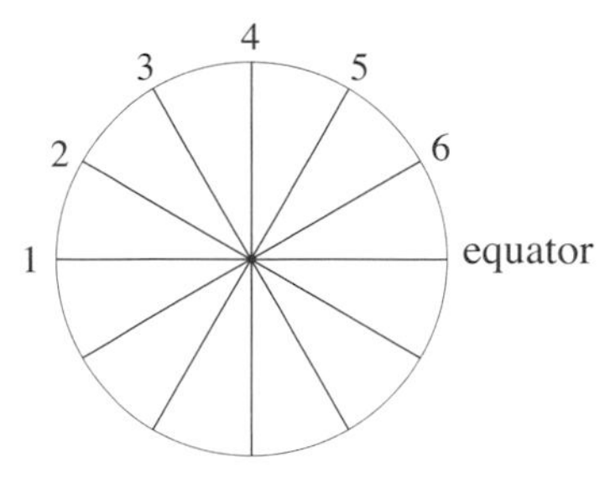

Fig. 6.55

equatorial circle. Let each of these diameters determine a plane that makes an angle of 55° with the plane of the Earth's equator. (For each diameter, there are two possibilities for the "lean" of the plane. Lean the planes of consecutive diameters in the same general direction.) Into each of these

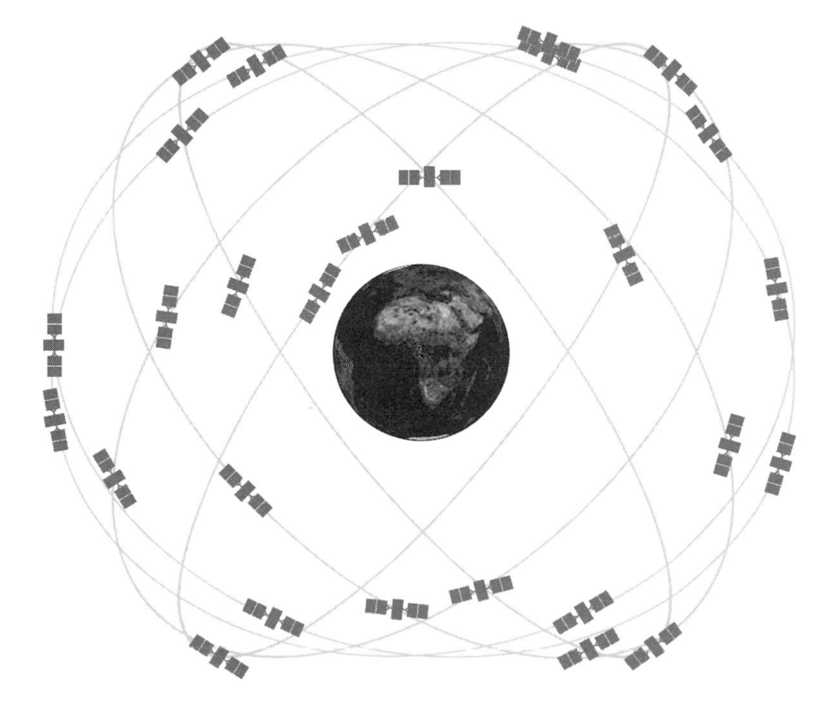

Fig. 6.56. The GPS. *Source*: U.S. Government, http://www.gps.gov/multimedia/images/.

six planes four satellites are placed into a circular orbit with C at the center. So there are 24 satellites in all. They are staggered relative to each other in such a way that they are spread out uniformly as they speed high over the surface of the Earth. Figure 6.56 gives an indication of the way they are configured. Each satellite is set to complete one revolution in 12 hours. Why does this mean that the radii of their circular orbits have to be the same? What is this common radius equal to? What is the speed of the satellites? The planes of the orbits are tilted at 55° relative to the equatorial plane so as to provide good coverage of the relevant areas of the Earth's surface. (Observe that if this angle were to be 90°, then all satellites would go over the two poles. It would seem that most of the residents there—penguins, walruses, polar bears—would have little use for the GPS.) The path of every satellite of the system is carefully monitored by ground stations. The system makes it possible for those carrying a GPS receiver—whether they are sailing somewhere in the Atlantic or exploring the jungles of the Amazon—to obtain precise coordinates of their positions. Here is how. From any point on Earth, there are at any time between five and eight (and almost always at least six) satellites within range of the receiver. The radio signals that each of the satellites emits allow the receiver to determine the position of the satellite and its distance from it. Suppose for instance that at a given time, four satellites are within range of the receiver R. Suppose they are in positions P_1, P_2, P_3, and P_4 in their orbits and that their distances from the receiver are d_1, d_2, d_3, and d_4. So the receiver lies simultaneously on four different spheres with centers P_1, P_2, P_3, and P_4 and radii d_1, d_2, d_3, and d_4. Since the intersection of two spheres is a circle (sit back and convince yourself of this), it follows that R lies at the intersection of two circles. So there are only two different possibilities for the position of R. This is the information that four satellites provide. The data from one or two more will rule out one of these possibilities, so that the location of R is determined.

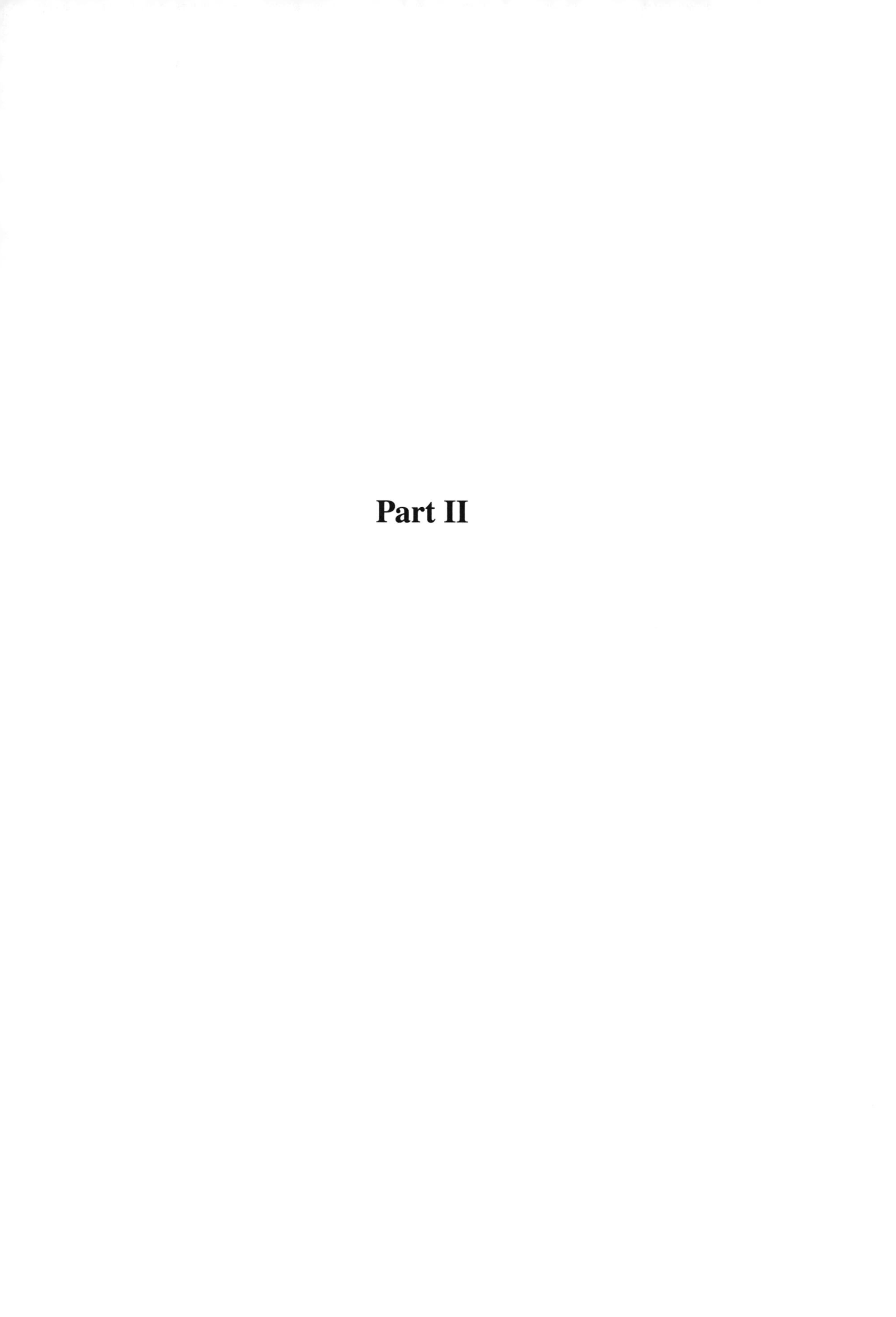

Part II

Chapter 7

Differential Calculus

Chapter 7 is an account of the essentials of differential calculus. It starts with functions and limits and considers basic properties such as continuity and differentiability and related theoretical concerns. It develops the rules for churning out derivatives, and it studies the information required for graphing functions, including critical points, their increase and decrease, maxima and minima, concavity, and inflection points. It attends to the derivatives and graphs of the trigonometric functions and introduces the exponential, logarithm, and hyperbolic functions and their graphs and derivatives.

Newton and Leibniz had a good intuitive sense of their calculus, and they developed it as a useful tool. However, they were not able to give it a rigorous foundational structure of the sort that Euclid's geometry has. The latter rises from a foundation of axioms and postulates about points and lines to the basic facts of plane geometry in a logically tight proof-driven way. The fluxions of Newton "by the ultimate ratio of evanescent quantities is to be understood the ratio of the quantities not before they vanish, not afterwards, but with which they vanish," and the derivative of Leibniz "provided that the difference is not assumed to be zero until the calculation is purged as far as possible by legitimate omissions, and reduced to ratios of non-evanescent quantities, and we finally ... apply our result to the ultimate case" were a big target for the critical mind and powerful pen of the Irish philosopher George Berkeley (1685–1753). Berkeley graduated from Trinity College, Cambridge. While there, he became familiar with the new science and mathematics that Newton and Leibniz (as well as their predecessors and collaborators) had developed. His critical reflections advanced a wide range of thoughts that included mathematics, physics, morals, economics, medicine, and the psychology of vision, and established him as one of the brilliant philosophers of the 18th century. (Berkeley was also ordained in the Anglican Church and would later become a bishop. His favorable views about the New World and his subsequent visit to America were the inspiration that provided the name Berkeley to the California university town in 1866.)

In the mathematical context, Berkeley is best known for his sharp attack on Newton's calculus that he published in *The Analyst* in 1734. Berkeley did not dispute the results of calculus and acknowledged its results were correct. But he contended that the practitioners of calculus introduced errors that canceled out to provide correct answers.

> And what are these Fluxions? The Velocities of evanescent Increments? And what are these same evanescent Increments? They are neither finite Quantities nor Quantities infinitely small, nor yet nothing. May we not call them the ghosts of departed quantities?

The Analyst stirred intense controversies in British mathematical circles. Numerous publications offered widely differing points of view. These ranged from the defense of the logical foundations of the calculus that Newton and Leibniz had developed to various suggestions for reformulating its fundamental concepts. The bottom line? In the words of a historian of mathematics of the 20th century, "Berkeley's criticisms of the rigor of the calculus were witty, unkind, and—with respect to the mathematical practices he was criticizing—essentially correct."

Calculus continued to develop in the 18th and 19th centuries in the hands of Johann and Jakob Bernoulli and Leonhard Euler, among others. In Euler's hands, calculus became what it is today—namely, the extensive study of mathematical functions—but his arguments still included nonrigorous, intuitive aspects. It was not until around 1830 (100 years after the publication of the *The Analyst* and the controversies that it touched off) when Augustin Cauchy, and later Bernhard Riemann and Karl Weierstrass, reformulated the fundamental concepts and provided the calculus of functions with the logical foundation that gave validity to its assertions. Within a definitive algebraic formulation of the notion of a limit, the important constructs such as continuity and differentiability became the foundation of a rigorous mathematical edifice. We will take a careful look at this rigorous formulation of the limit to get a sense of what is involved. However, in terms of the formulation of the key concepts and the development of the basics of the theory, we will still rely on the more easily understood intuitive approaches and the accompanying pictorial and geometric evidence. Lots of examples will illustrate the general facts and strategies that are laid out. They involve not only algebraic and trig functions, but also, given the applications in later chapters, exponential, log, and hyperbolic functions.

7.1 MATHEMATICAL FUNCTIONS

Mathematical functions arise explicitly and implicitly in all disciplines in which quantitative relationships are relevant. The mathematical function is a concept that is fundamental in many areas of mathematics, and in particular in calculus. The importance of this notion was already recognized by Gottfried Leibniz and Isaac Newton, but it was Leonhard Euler who understood its central relevance.

Let's begin with an example from economics. In the *New York Times* of January 19th, 2007, a bar graph supplied information about the changes in the Consumer Price Index (CPI) of the United States from January to December of the year 2006. This graph is reproduced in Figure 7.1. The CPI is an average of the prices of the items in a "basket" of goods and services of importance to consumers. (Economists of the U.S. government decide which goods and services go into the basket.) There is a bar, or block, for

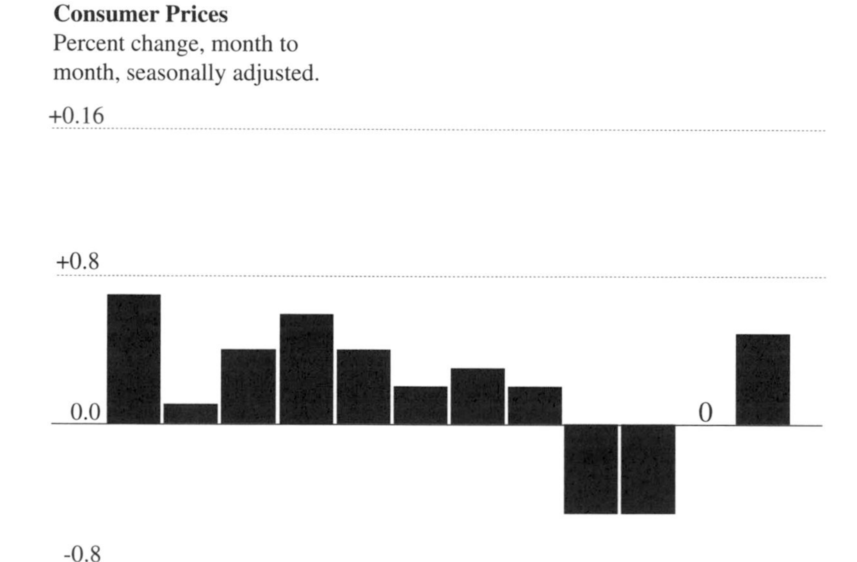

Fig. 7.1

each month. The height of the block (as given by the scale on the left in the figure) measures the increase in the CPI from the previous month. In the months of September and October, the blocks extend downward from the base line to indicate negative increase, or decrease, in the CPI. The bar graph tells us, for example, that the CPI increased by 0.7% during January 2006, by 0.4% in May, that it fell by 0.5% in both September and October, and that it did not change in November. The changes are seasonally adjusted (factors, such as the impact of cold winter weather, are corrected for).

Before we recognize the CPI as a function, let's recall the definition of this concept (the function concept was already introduced in Section 5.3 and used in Chapters 5 and 6). A *function* is a rule or correspondence that assigns *exactly one real number* to each number from a given set of real numbers. Such a rule is often denoted by a single letter, for example, by f, or g, or h. If x is a typical number from the given set, then the number that f assigns to x is written $y = f(x)$. For this reason, the equation $y = f(x)$ is also used as notation for the function (and in the same way for g and h). The *graph* of a function f is the graph of the equation $y = f(x)$. The *graph of an equation* in x and y is the set of all points in the coordinate plane with the property that their x- and y-coordinates satisfy the equation. The *domain* of a function is the set of all numbers for which its rule makes sense. Sometimes the domain is explicitly restricted to a smaller set.

Consider the function $g(x) = \sqrt[3]{x} = x^{\frac{1}{3}}$, for instance. The point $(-1, -1)$ is on its graph, and the domain is the set of all real numbers. Within a particular application, x may be specified to lie in the interval $-8 \le x \le 8$, for example. With regard to the function $y = f(x) = -\sqrt{12 - 3x^2}$, the point $(1, -3)$ is on the graph because $f(1) = -3$. For $f(x) = -\sqrt{12 - 3x^2}$ to make sense, we need to have $12 - 3x^2 \ge 0$, so that $3x^2 \le 12$ or $x^2 \le 4$. Therefore the domain of f is $-2 \le x \le 2$. If x denotes length, then $x \ge 0$, so that the domain would be restricted to $0 \le x \le 2$.

Now back to the consumer price index. By designating January, February, ..., December numerically by 1, 2, ..., 12, the information in Figure 7.1 can be represented as

$$1 \to 0.7,\ 2 \to 0.1,\ 3 \to 0.4,\ 4 \to 0.6, 5 \to 0.4, 6 \to 0.2,$$

$$7 \to 0.3,\ 8 \to 0.2,\ 9 \to -0.5,\ 10 \to -0.5,\ 11 \to 0,\ 12 \to 0.5.$$

Since there corresponds to each of the numbers 1, 2, 3, ..., 12 (representing the 12 months) only one number (expressing the percentage increase or decrease in the price index for that month), this rule defines a function. With the understanding that the x-axis represents time in months and the y-axis the monthly percentage change in the CPI, this function can be graphed in the xy-plane. See Figure 7.2. This price

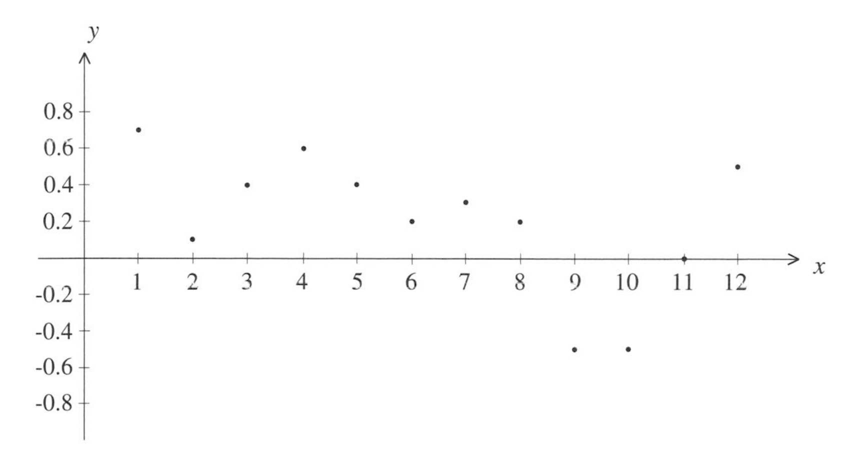

Fig. 7.2

index function, while of interest to an economist, provides only a few elements of data and is therefore of little interest in the context of calculus.

The functions that calculus studies are often given by algebraic equations. These include functions of the from $f(x) = a$ with a constant. They are the *constant functions*. Their graphs are horizontal lines.

Functions of the form $f(x) = ax + b$, where a and b are constants, are called *linear functions*. Their graphs are nonvertical lines. Linear functions and their graphs are studied in Section 5.1. A function given by an equation of the form $f(x) = ax^2 + bx + c$ with a, b, and c constants (and $a \neq 0$) is a *quadratic function*. Their graphs are parabolas. They are studied in Section 4.3. Section 5.3 provides a short catalogue of basic functions and their graphs, including $f(x) = x^2$, $f(x) = \sqrt{x} = x^{\frac{1}{2}}$, $f(x) = x^3$, $g(x) = \sqrt[3]{x} = x^{\frac{1}{3}}$, $g(x) = \frac{1}{x} = x^{-1}$, and $h(x) = \frac{1}{x^2} = x^{-2}$. Such algebraic functions, as well as trigonometric, exponential, and logarithmic functions and their inverses, are a focus of this chapter.

Consider the equation $x = y^2$. The pairs of numbers $x = 4$, $y = 2$ and $x = 4$, $y = -2$ are both solutions. So for a given x, this equation does not determine y uniquely. Therefore $x = y^2$ does not define a function of x. However, because $x \geq 0$, we can solve for y to get $y = \pm\sqrt{x}$. Both $f_+(x) = \sqrt{x}$ and $f_-(x) = -\sqrt{x}$ are functions of x. Their graphs are shown in Figure 7.3a. The graph of $f_+(x) = \sqrt{x}$ is above the x-axis, and that of $f_-(x) = -\sqrt{x}$ is below it. In each case, the domain is the set of all x with $x \geq 0$. The situation is similar with the equation $y^2 = x^2$. Its graph is not the graph of a function. Instead, it consists of the pair

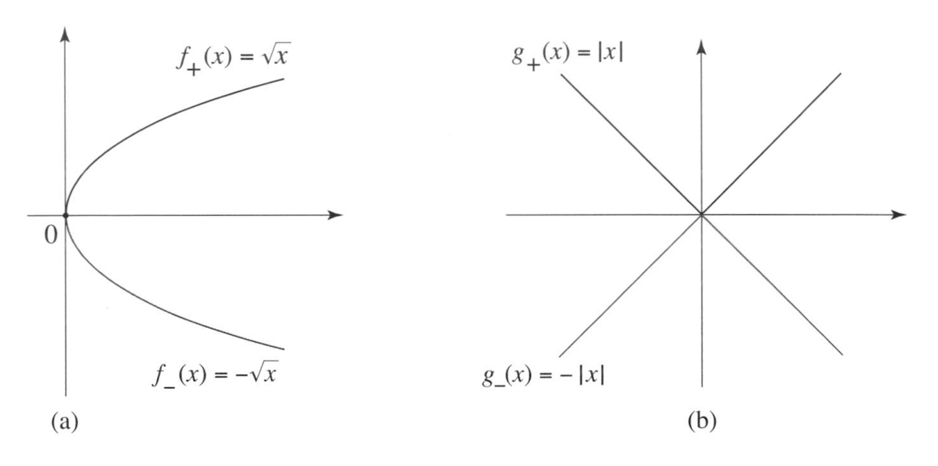

Fig. 7.3

of lines $y = x$ or $y = -x$, in other words, of the graphs of the two functions $f_+(x) = x$ and $f_-(x) = -x$ shown in Figure 7.3b. This combination of lines also gives rise to the functions $g_+(x) = |x|$ and $g_-(x) = -|x|$. Their graphs are the V and the inverted V above and below the x-axis.

7.2 A STUDY OF LIMITS

We begin with an initial look at limits. Figure 7.4 shows the graph of the function $f(x) = \frac{1}{2}x^2$. When the

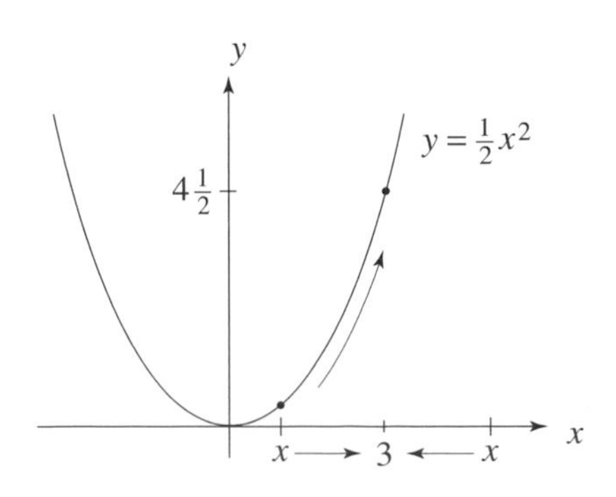

Fig. 7.4

point x on the horizontal axis is pushed to 3, the value $f(x) = \frac{1}{2}x^2$ goes to $\frac{1}{2}3^2 = \frac{9}{2} = 4\frac{1}{2}$. A look at the graph shows the same thing. This observation is abbreviated in mathematical shorthand by

$$\lim_{x \to 3} \tfrac{1}{2}x^2 = 4\tfrac{1}{2}\,.$$

In the same way, when x is pushed to 6, the expression $\sqrt{x-4}$ closes in on $\sqrt{2}$, so that

$$\lim_{x \to 6} \sqrt{x-4} = \sqrt{2}.$$

More generally, let $y = f(x)$ be a function and consider its graph. A typical situation is depicted in Figure 7.5. A point (c, L) is singled out. The fact that it is designated by a small circle (rather than a solid point) is meant to tell us that the point may, or may not, be on the graph of the function. Observe from the

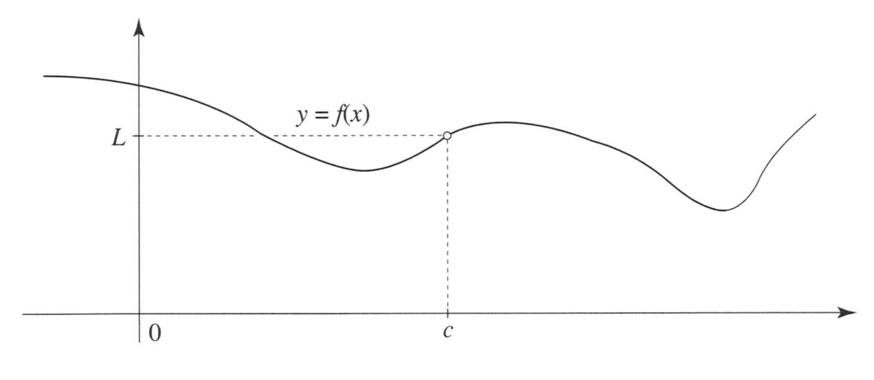

Fig. 7.5

graph that as x is pushed in on the number c, the values $f(x)$ close in on the number L. In limit notation this is expressed by writing

$$\lim_{x \to c} f(x) = L.$$

The observation that this expression represents is intuitive and geometric, but it does not have the precision that is normally required of a mathematical statement. We'll now describe how the notion of a limit is made precise and rigorous. Figure 7.6 shows a horizontal strip. It is chosen so that the point (c, L) lies on the centerline of the strip. The figure also shows a segment of the strip. It is chosen so that (c, L) is

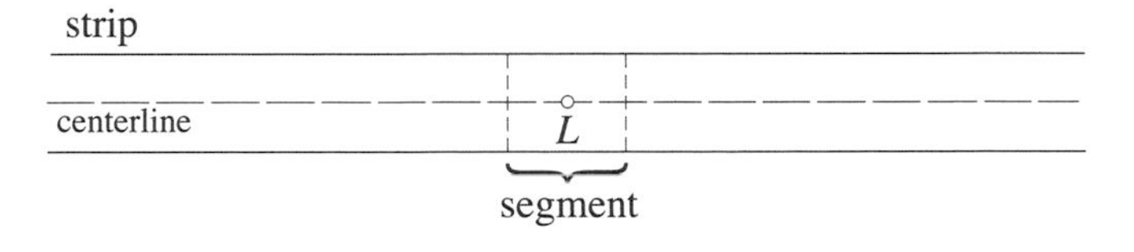

Fig. 7.6

at the center of the segment. The precise definition of $\lim\limits_{x \to c} f(x) = L$ is this: This limit exists if every horizontal strip that has (c, L) on its centerline has a segment with (c, L) at its center, so that the graph of $y = f(x)$ passes through the segment in the sense that it enters the segment through the dashed line on the left, leaves it through the dashed line on the right, and lies entirely within the segment between the two dashed lines (with the possible exception of the point on the graph that corresponds to $x = c$, because the concern is the values of $f(x)$ for x near c, not at $x = c$). For the limit $\lim\limits_{x \to c} f(x) = L$ to exist, this has to be so *for any given horizontal strip no matter how thin it is.* Think about this requirement and get your mental arms around the idea that this can only mean that when x is pushed to c, then the point $(x, f(x))$ has to head toward (c, L).

The rest of the definition of the limit is the translation of this geometric formulation into algebraic terms. Refer to Figure 7.7. Notice that the width of a horizontal strip is determined by the distance between its boundary and the centerline—designated by ε in the figure—and that the width of a segment is determined by the distance designated by δ in the figure. So the widths of the strip and the segment in the figure are 2ε and 2δ, respectively. Using the fact that $|a - b| = |b - a|$ is the distance between any two

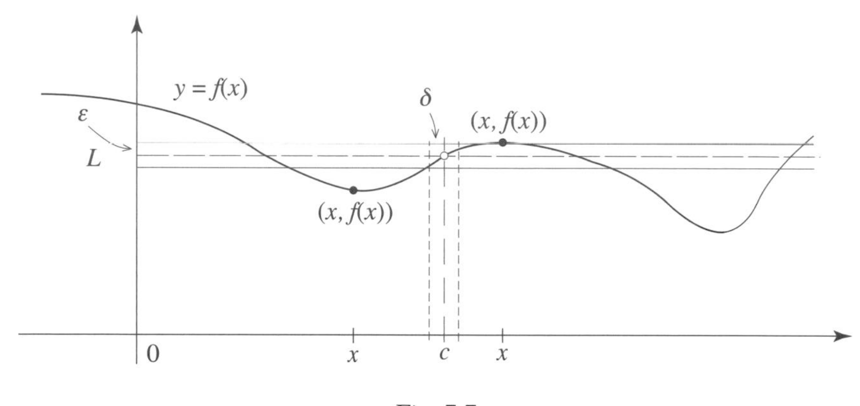

Fig. 7.7

numbers a and b on the number line, the strip/segment requirement can be translated as follows: For any $\varepsilon > 0$ (no matter how small), there is a corresponding $\delta > 0$ such that the inequality $|f(x) - L| < \varepsilon$ is achie-

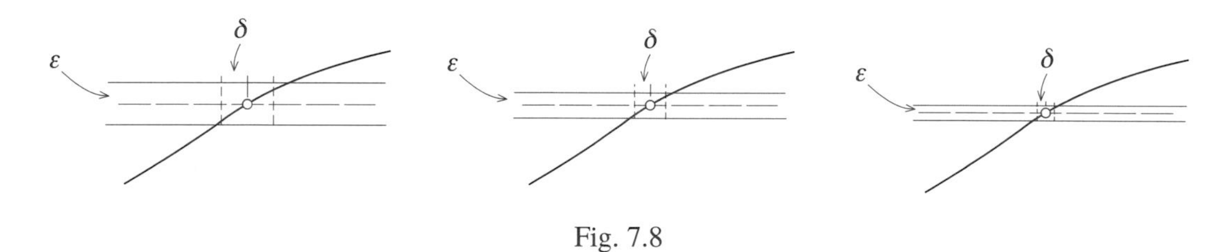

Fig. 7.8

ved for any x (with $x \neq c$) that satisfies $|x - c| < \delta$. Figure 7.8 illustrates how the δ is determined by the graph of $y = f(x)$ for a sequentially shrinking ε. We repeat what is required once more:

Limit Test: For the assertion $\lim_{x \to c} f(x) = L$ to be correct, it must pass the following test. For any accuracy requirement $\varepsilon > 0$, there must exist a $\delta > 0$ such that the accuracy $|f(x) - L| < \varepsilon$ is achieved with any $x \neq c$ that satisfies $|x - c| < \delta$.

Meditate over this statement—and the discussion that precedes it—again to convince yourself that if this test is satisfied, then $f(x)$ closes in on L when x is pushed to c.

Example 7.1. The assertion $\lim_{x \to 6} \sqrt{x-4} = \sqrt{2}$ is obvious. But let's put it to the test. By multiplying $\sqrt{x-4} - \sqrt{2}$ by its conjugate $\sqrt{x-4} + \sqrt{2}$ (often a clarifying maneuver when it comes to square roots),

$$(\sqrt{x-4} - \sqrt{2}) \cdot (\sqrt{x-4} + \sqrt{2}) = (\sqrt{x-4})^2 - (\sqrt{2})^2 = (x-4) - 2 = x - 6.$$

Take the absolute value of this expression and note that $\sqrt{2} \leq \sqrt{x-4} + \sqrt{2}$, to get $\sqrt{2} \cdot \left|\sqrt{x-4} - \sqrt{2}\right| \leq |x-6|$. Therefore $|\sqrt{x-4} - \sqrt{2}| \leq \frac{1}{\sqrt{2}}|x - 6|$. This inequality does what we need. Let $f(x) = \sqrt{x-4}$. For any $\varepsilon > 0$, let $\delta = \sqrt{2}\varepsilon$. If $|x - 6| < \delta$, then $\left|f(x) - \sqrt{2}\right| = \left|\sqrt{x-4} - \sqrt{2}\right| < \frac{1}{\sqrt{2}}\delta = \varepsilon$. So the limit test is passed.

This was relatively easy. But you are right if you suspect at this point that the application of the limit test can be a delicate matter, even in situations where—from the intuitive viewpoint—the limit is completely self-evident.

As was already pointed out in the introduction of this chapter, the standards of mathematical precision were raised in the 19th century. We have seen in Chapters 5 and 6 how pervasive limits are in differential as well as integral calculus. They are required in the formulations of the core ideas of both. (The discussions in the two chapters make this much more explicit than the original work of Newton and Leibniz.) In the recasting of the intuitive verbal "this goes to that" version of a limit into the precise algebraic concept described above, the cornerstone was laid for a new, rigorous differential calculus. Basic limit theorems formulated and proved with the new definition, concepts that were implicit or intuitive with Leibniz and Newton were made explicit and definitive, and the entire structure of calculus was rebuilt. (How, analogously, integral calculus was reformulated will be described in Chapter 9.) The development that was just described is both abstract and deep, and its pursuit is beyond the limits of any elementary text. However, today's basic calculus course concerns itself with aspects of these issues and therefore provides a glimpse of what is involved. The strategy of this chapter will be the following. We will work exclusively with the earlier intuitive notion of limits. In other words,

$$\lim_{x\to c} f(x) \text{ exists and is equal to } L \text{ or, equivalently, } \lim_{x\to c} f(x) = L$$

will continue to be the observation that if x is pushed to c from either the left or the right, then $f(x)$ closes in on the finite number L. If you combine the flow of the discussion of this chapter with the requirement that all limits that appear in it should be subjected to the limit test (directly or indirectly), then you will have a sense how the mathematicians of the 19th century rebuilt differential calculus.

We conclude this section with some additional examples of and comments about limits. The most difficult and important limit problems are those of $\frac{0}{0}$ type. These are the questions of the form

$$\lim_{x\to c} \frac{f(x)}{g(x)} = ?$$

where both $\lim\limits_{x\to c} f(x) = 0$ and $\lim\limits_{x\to c} g(x) = 0$. The answer to such limit questions invariably requires that $\frac{f(x)}{g(x)}$ be rewritten in such a way that a common term can be canceled from the numerator and denominator.

Example 7.2. Consider the problem $\lim\limits_{x\to -3} \frac{x^2-9}{x+3} = ?$ Observe that it is of $\frac{0}{0}$ type. The equality $\frac{x^2-9}{x+3} = \frac{(x-3)(x+3)}{(x+3)} = x - 3$ tells us that $\lim\limits_{x\to -3} \frac{x^2-9}{x+3} = \lim\limits_{x\to -3}(x-3) = -6$.

Notice that the equality $\frac{(x-3)(x+3)}{(x+3)} = x - 3$ requires that $x \neq -3$ (because cancellation by 0 is not allowed). But this is of no relevance here because the question is: "What happens to $\frac{x^2-9}{x+3}$ as x is pushed to -3?" and not "What is the value of $\frac{x^2-9}{x+3}$ at -3?" ("It closes in on -6" is the answer to the first question. "The value is not defined" answers the second.)

Example 7.3. The question $\lim\limits_{x\to 6} \frac{\sqrt{x-2}-2}{x-6} = ?$ is also of $\frac{0}{0}$ type. Before it can be solved, it must be rewritten. This involves the conjugate $\sqrt{x-2}+2$ of $\sqrt{x-2}-2$. Since

$$(\sqrt{x-2}-2)(\sqrt{x-2}+2) = (x-2) - 4 = x - 6,$$

we get $\frac{\sqrt{x-2}-2}{x-6} = \frac{\sqrt{x-2}-2}{x-6} \cdot \frac{\sqrt{x+2}+2}{\sqrt{x-2}+2} = \frac{x-6}{(x-6)(\sqrt{x-2}+2)} = \frac{1}{\sqrt{x-2}+2}$. Therefore $\lim\limits_{x\to 6} \frac{\sqrt{x-2}-2}{x-6} = \lim\limits_{x\to 6} \frac{1}{\sqrt{x-2}+2} = \frac{1}{4}$.

As in the previous example, the fact that the cancellation $\frac{x-6}{(x-6)(\sqrt{x-2}+2)} = \frac{1}{\sqrt{x-2}+2}$ requires $x \neq 6$ is of no consequence. The method of multiplying an expression involving a square root term by the conjugate of the term and thereby achieving a simplification is known as *rationalizing*.

Example 7.4. Consider the limit $\lim\limits_{x\to 0}\frac{\sin x}{x}$, where x is in radians. Notice that it is also of $\frac{0}{0}$ type. The last part of Section 1.6 considers this limit in an "experimental way" and provides evidence for the fact that $\lim\limits_{x\to 0}\frac{\sin x}{x}=1$. That this is indeed the case will be verified in Section 7.6.

Not all limit questions have answers. For example, when x is pushed to 0, $\frac{1}{x}$ becomes larger and larger, and does not close in on a finite number. In such a case, one says that the limit *does not exist*. In particular, the limits $\lim\limits_{x\to 0}\frac{1}{x}$ and $\lim\limits_{x\to 0}\frac{1}{\sin x}$ do not exist.

We conclude the discussion about limits with a notational matter. Observe that $\lim\limits_{x\to -4}\sqrt{x+4}=0$. But observe also that $\sqrt{x+4}$ is defined only for $x+4\geq 0$, and hence only for $x\geq -4$. Therefore x can be pushed to -4 from the right only, in other words, through values that are greater than -4. The exponent $^+$

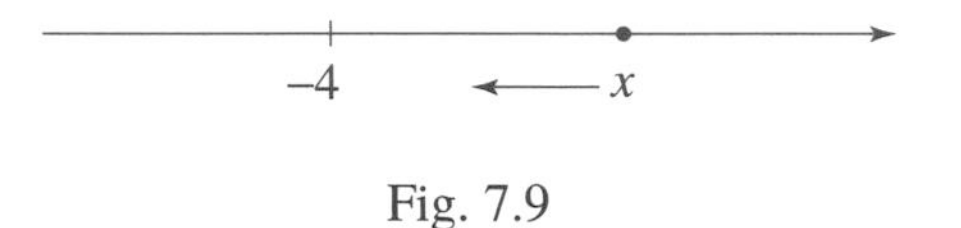

Fig. 7.9

in the notation $\lim\limits_{x\to -4^+}\sqrt{x+4}=0$ emphasizes that x approaches -4 from the positive side. In a similar way, $g(x)=\sqrt{5-x}$ is defined only for $5-x\geq 0$. So we must have $5\geq x$, or $x\leq 5$. Therefore $\lim\limits_{x\to 5}\sqrt{5-x}=0$

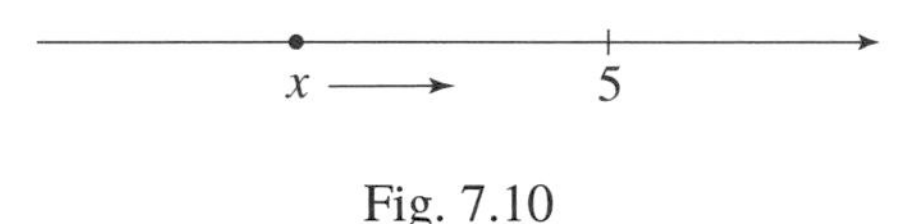

Fig. 7.10

can only make sense if x approaches 5 from the left. The notation $\lim\limits_{x\to 5^-}\sqrt{5-x}=0$ makes this explicit.

About Limit Notation. The assertion that $\lim\limits_{x\to c}f(x)=L$ means that x can be pushed to c from either the right or the left, unless the function f is not defined to the right or left of c. If the function is defined on both sides of c, then $\lim\limits_{x\to c}f(x)=L$ means the same thing as:

$$\lim_{x\to c^-}f(x) \text{ exists, } \lim_{x\to c^+}f(x) \text{ exists, and both are equal to } L.$$

7.3 CONTINUOUS FUNCTIONS

One of the assumptions that a number of basic and important assertions about a function require is the property of continuity. This section will explore this property.

Let $y=f(x)$ be a function. Taking the geometric point of view, we say that f is *continuous* at a number or point c on the x-axis if the graph of f has no gaps or breaks at the line $x=c$. This can be formulated algebraically as follows. Refer to Figure 7.11. Note first that if a gap over c is to be avoided, then c must be in the domain of f. For otherwise $f(c)$ does not make sense, and there is a gap in the graph of f at $x=c$. The assumption that c is in the domain of f provides the point $(c,f(c))$ on the graph of f. Suppose that c is not an endpoint of the domain of f, so that $f(x)$ is defined for x to the immediate left and right of c. Let x be a number in the domain of f to the left of c. Push x in on c, and consider the point $(x,f(x))$ on the graph in the process. Observe that if a gap in the graph is to be avoided, then $(x,f(x))$ must close in on $(c,f(c))$. But this is possible only if $f(x)$ closes in on $f(c)$. Put another way, it must be the case that $\lim\limits_{x\to c^-}f(x)=f(c)$. In the same way, if x is pushed to c from the right, then what is needed to avoid a gap in the graph is the condition $\lim\limits_{x\to c^+}f(x)=f(c)$. Therefore if $\lim\limits_{x\to c}f(x)=f(c)$, then both of these conditions are met and there is no gap in the graph of f at the line $x=c$. Suppose next that c is an endpoint of the

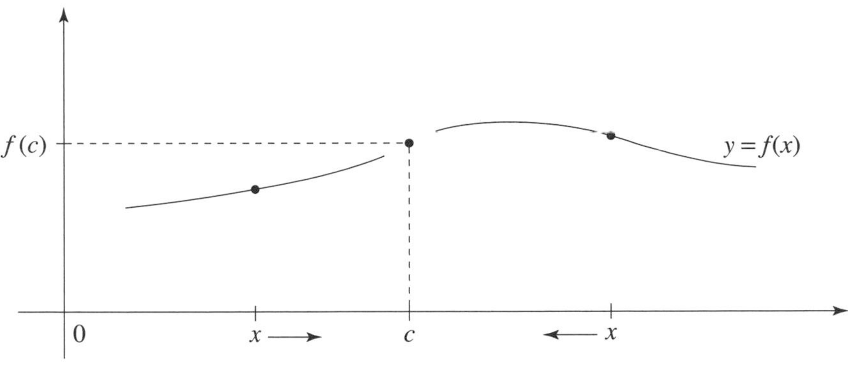

Fig. 7.11

domain of the function f. In this case, c can be approached from the left or the right, but not both. So $\lim_{x \to c} f(x) = f(c)$ is taken to mean either that $\lim_{x \to c^-} f(x) = f(c)$ or that $\lim_{x \to c^+} f(x) = f(c)$. It follows again that $\lim_{x \to c} f(x) = f(c)$ is the condition needed to avoid a gap in the graph at the line $x = c$.

The discussion above has established the continuity criterion.

Continuity Criterion. For a function $y = f(x)$ to be *continuous at a number or point* c on the x-axis, it must be the case that

i. c is in the domain of f, so that $f(c)$ is defined, and

ii. $\lim_{x \to c} f(x) = f(c)$.

Example 7.5. Consider the function $f(x) = 3x - 4$. Let's test for continuity at the number 5. Note first that $f(5) = 15 - 4 = 11$. So criterion (i) is satisfied. Since $\lim_{x \to 5} f(x) = \lim_{x \to 5}(3x - 4) = 3 \cdot 5 - 4 = 11$ and $11 = f(5)$, criterion (ii) is met as well. Therefore $f(x) = 3x - 4$ is continuous at 5. In the same way, for any number c, $f(c) = 3c - 4$ and

$$\lim_{x \to c} f(x) = \lim_{x \to c}(3x - 4) = 3c - 4 = f(c).$$

Therefore $f(x) = 3x - 4$ is continuous at any number c. Its graph, as we already know from our study in

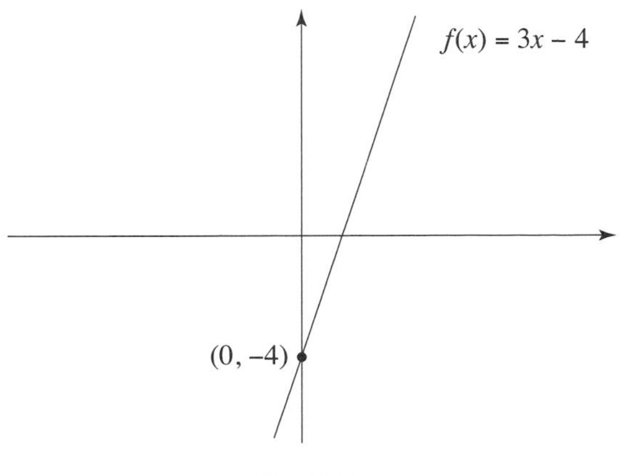

Fig. 7.12

Section 5.1, is the line with slope 3 and y-intercept -4. See Figure 7.12.

Example 7.6. Consider the function $f(x) = x^2$. Let's check for continuity at -3. Since $f(-3) = 9$, criterion (i) is satisfied. Since

$$\lim_{x \to -3} f(x) = \lim_{x \to -3} x^2 = (-3)(-3) = 9 = f(-3),$$

criterion (ii) holds as well. So $f(x) = x^2$ is continuous at -3. In exactly the same way, $f(c) = c^2$ and $\lim_{x \to c} f(x) = \lim_{x \to c} x^2 = c^2 = f(c)$. Again, both criteria are met, so that $f(x) = x^2$ is continuous at any real

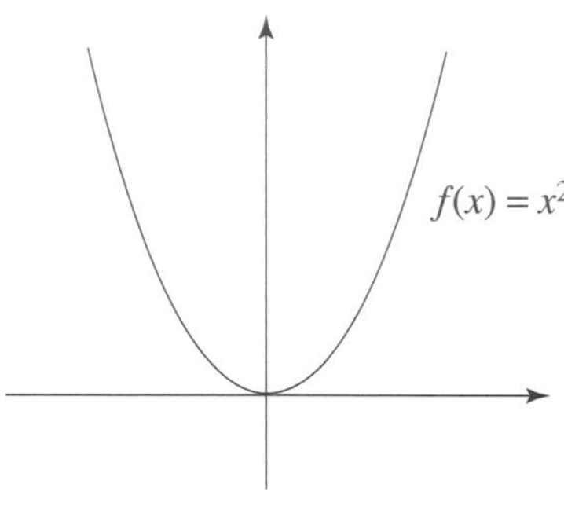

Fig. 7.13

number c. Its graph is the parabola shown in Figure 7.13.

Remark 7.1. The arguments in the two examples above apply more generally to show that any function $y = f(x)$ with $f(x)$ a polynomial in x is continuous at any real number c.

A similar argument shows that $g(x) = \frac{1}{x}$ is continuous at any $c \neq 0$. Observe, however, that since 0 is not in the domain, g is not continuous at 0. In the same way, the function $h(x) = \frac{1}{x^2}$ is continuous at any

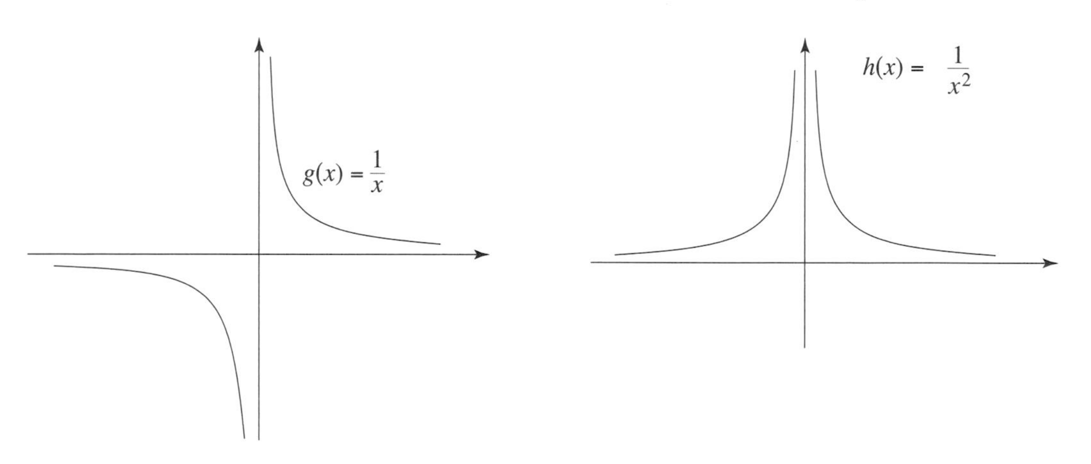

Fig. 7.14

real number c, except $c = 0$. The graphs of the two functions are sketched in Figure 7.14.

Example 7.7. Consider the function $f(x) = \frac{x^2-1}{x+1}$. Let $c \neq -1$. So $c + 1 \neq 0$, and hence c is in the domain of f. Notice that

$$\lim_{x \to c} f(x) = \lim_{x \to c} \frac{x^2 - 1}{x + 1} = \frac{c^2 - 1}{c + 1} = f(c).$$

So both parts of the continuity criterion are satisfied. It follows that f is continuous at c. Since f is not defined at -1, f cannot be continuous at -1. Notice that

$$\frac{x^2-1}{x+1} = \frac{(x-1)(x+1)}{x+1}$$

and hence that $f(x) = \frac{x^2-1}{x+1} = x - 1$ whenever $x \neq -1$. So observe that

$$\lim_{x \to -1} f(x) = \lim_{x \to -1} \frac{x^2-1}{x+1} = \lim_{x \to -1} (x-1) = -2.$$

So for $c = -1$, the limit exists and yet the function is not continuous. Consider the function $g(x) = x - 1$. Its graph is the line with slope 1 and y-intercept -1. Notice that $f(x) = \frac{x^2-1}{x+1} = x - 1 = g(x)$ for all $x \neq -1$.

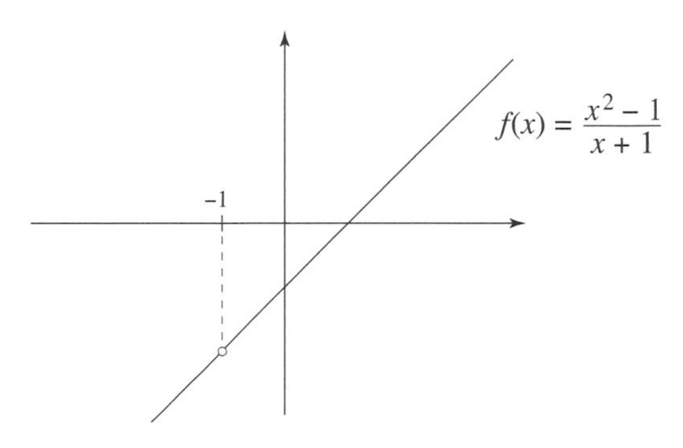

Fig. 7.15

It follows that the graph of f agrees with that of g, except when $x = -1$. At -1 they differ because the graph of f has a gap, but the graph of g does not. It follows that except for the gap at -1, the graph of $f(x) = \frac{x^2-1}{x+1}$ is the line with slope 1 and y-intercept -1. It is sketched in Figure 7.15.

Remark 7.2. In the same way as in the previous example, any function of the form $f(x) = \frac{g(x)}{h(x)}$, where $g(x)$ and $h(x)$ are polynomials in x, is continuous at any number c for which $h(c) \neq 0$.

Example 7.8. Consider the function $f(x) = \sqrt{x}$. Is f continuous at any $c \geq 0$? If $c > 0$, then $\sqrt{c}$ makes sense, so that c is in the domain of f. Therefore criterion (i) is satisfied. If x is pushed to c, it is clear that $\sqrt{x}$ gets pushed to $\sqrt{c}$. This is so whether x is pushed to c from the right with $x > c$, or whether x is pushed to c from the left with $0 < x < c$ (the first part of this inequality is needed to keep x in the domain of f). So $\lim_{x \to c} \sqrt{x}$ exists and is equal to $\sqrt{c}$. Therefore criterion (ii) is satisfied as well, and we can conclude that f is continuous at c. Finally, is $f(x) = \sqrt{x}$ continuous at $c = 0$? Since $\sqrt{0} = 0$, $c = 0$ is in the domain of f. Now take $x \neq 0$, and push x to 0. Since x must remain in the domain of f in order for this process to

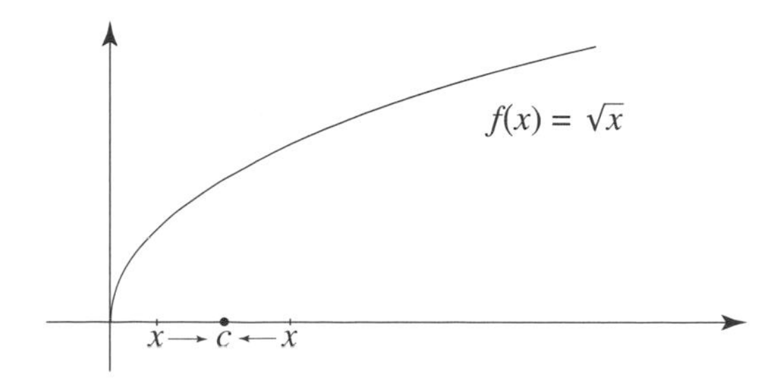

Fig. 7.16

take place, x can be pushed to 0 only from the right. Since $\lim\limits_{x\to 0}\sqrt{x} = \lim\limits_{x\to 0^+}\sqrt{x} = 0$, both criteria are again satisfied, and f is continuous at 0. We have established that $f(x) = \sqrt{x}$ is continuous at any point c with $c \geq 0$. The graph of f is sketched in Figure 7.16.

Suppose that c is an endpoint of the domain of a function $y = f(x)$. So f is not defined for x immediately to the left of c or for x immediately to the right of c. Example 7.8 tells us that so long as one of the limits $\lim\limits_{x\to c^-} f(x)$ or $\lim\limits_{x\to c^+} f(x)$ exists and is equal to $f(c)$, f is continuous at c.

Take two numbers $a \leq b$ on a number line, and consider the following sets. The set $[a, b]$ consists of all x on the line with $a \leq x \leq b$. The set (a, b) consists of all x that satisfy $a < x < b$. (Since (a, b) can also be a point in the coordinate plane, there is a notational ambiguity. The context will avoid possible

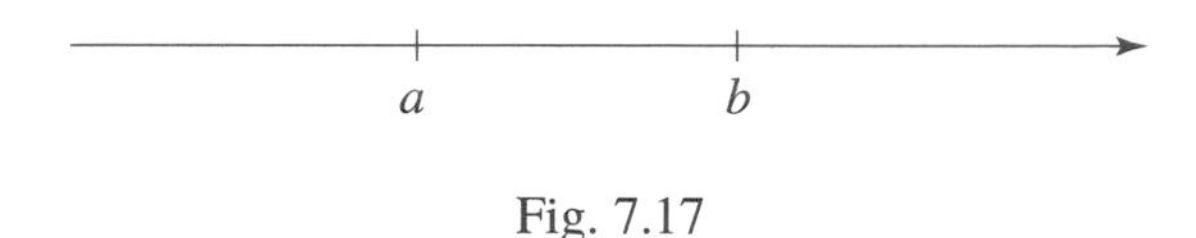

Fig. 7.17

confusions.) The set $(a, b]$ consists of all x with $a < x \leq b$, and the set $[a, b)$ is defined in a similar way. Each of these sets is an *interval*. For any number a, we will also consider the intervals $(-\infty, a]$ and $[a, \infty)$. The first consists of all x with $x \leq a$, and the second of all x with $a \leq x$. The intervals $(-\infty, a)$ and (a, ∞) are defined in a similar way. How do they differ from $(-\infty, a]$ and $[a, \infty)$? The interval $(-\infty, \infty)$ consists of all the points on the number line. The intervals $(a, b), (a, \infty), (-\infty, a)$, and $(-\infty, \infty)$ are all called *open*, and $[a, b]$ is said to be *closed*. The notation (a, b) excludes the possibility that $a = \pm\infty$ or $b = \pm\infty$.

It is not hard to construct functions that are not continuous at numbers in their domain. A standard example follows.

Example 7.9. The prices for some tickets (those for buses, trains, or movie theaters) depend on the age of the user of the ticket. Let $p(x)$ be the price of a certain ticket for someone x years old. Suppose that for $0 < x < a_1$, the price is $p(x) = p_0$; for $a_1 \leq x < a_2$, it is the highest amount $p(x) = p_1$; for $a_2 \leq x < a_3$, it is lower again at $p(x) = p_2$; and finally for $x \leq a_3$, the lowest price $p(x) = p_3$ is in effect. For example, with $a_1 = 5$, the price p_0 might be a child's fare, with $a_2 = 16$, p_1 might be a youth fare, and with $a_3 = 65$, the price p_2 might be a discounted fare for senior citizens. The graph of $p(x)$ is shown in Figure 7.18. Observe that it is not continuous at a_0, a_1, and a_2. The definition of the price function $p(x)$ can be restated

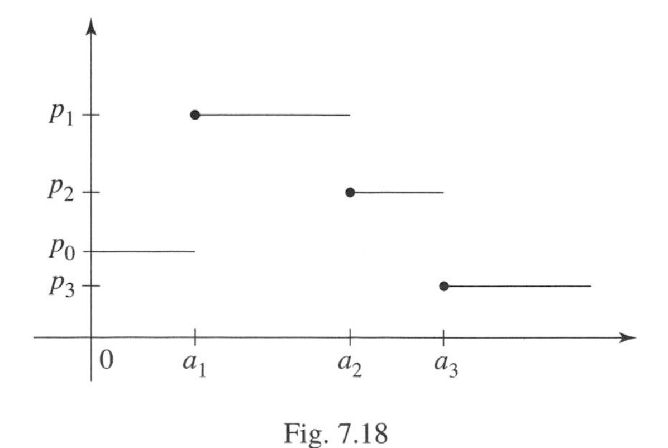

Fig. 7.18

as follows using interval notation:

$$p(x) = \begin{cases} p_0, & \text{for } x \text{ in } [0, a_1) \\ p_1, & \text{for } x \text{ in } [a_1, a_2) \\ p_2, & \text{for } x \text{ in } [a_2, a_3) \\ p_3, & \text{for } x \text{ in } [a_3, \infty) \end{cases}$$

What part of the continuity criterion is not satisfied by $p(x)$? For what numbers c?

Consider the function $f(x) = \sqrt{2x^3 - 6}$, and notice that it is a combination of two simpler functions. Apply $h(x) = 2x^3 - 6$ to any x first, and then apply the function $g(x) = \sqrt{x}$ to the result to get $f(x) = g(h(x)) = \sqrt{2x^3 - 6}$. This makes sense for any x for which $h(x) = 2x^3 - 6 \geq 0$ so that $h(x)$ is in the domain of $g(x) = \sqrt{x}$. In such a situation, we say that the function $y = f(x)$ is the *composite function* of $g(x)$ and $h(x)$. Notice that the order in which the functions are composed matters. The composite of $h(x)$ and $g(x)$ is the function $h(g(x)) = 2(\sqrt{x})^3 - 6$. The importance of this construction to our discussion is simple to state. Basic questions about the more complex function $f(x) = \sqrt{2x^3 - 6}$ can be reduced to the same basic questions about the two component functions. We will see in a moment that this applies to questions about the continuity of functions (and we will soon see that it also applies to questions about derivatives of functions).

Remark 7.3. Let f be the composite $f(x) = g(h(x))$ of the functions $g(x)$ and $h(x)$. If the function h is continuous at c and the function g is continuous at $h(c)$, then the composite function f is continuous at c. A proof of this fact (using the limit test) is sketched in Problem 7.15.

The definition of continuity at a single point or number c (this has been the focus so far) can be extended to sets of points. A function is *continuous on an interval I* if it is continuous at every point of I. A function is *continuous on its domain* if it is continuous at all points in its domain. It follows from Remarks 7.1 and 7.2 above that any function of the form $y = f(x)$ with $f(x)$ a polynomial or $f(x) = \frac{g(x)}{h(x)}$ with $g(x)$ and $h(x)$ polynomials is continuous on its domain. These (and other) facts about continuity can be proved by the repeated application of the continuity theorem.

Continuity Theorem. Let $y = f(x)$ and $y = g(x)$ be two functions that are continuous on their domains. Then their sum $y = f(x) + g(x)$, difference $y = f(x) - g(x)$, product $y = f(x) \cdot g(x)$, quotient $y = \frac{f(x)}{g(x)}$, and composite $y = f(g(x))$ are all continuous on their domains.

Example 7.10. Use the fact that constant functions and the functions $y = x$, $y = \sqrt{x} = x^{\frac{1}{2}}$, and $y = x^{\frac{1}{3}}$ are continuous on their domains together with the continuity theorem to explain why each of the functions

$$y = 3x^5 - 7x^2 + \sqrt{x},\ y = (3x^2 - 6x)^{\frac{2}{3}},\ y = \frac{3x^5 - 7x^2}{\sqrt{23x^4 - 9x^3}},\ \text{and } y = \frac{4(x-1)}{(3x(x-2))^{\frac{1}{3}}}$$

is continuous on its domain.

The verification, or proof, of the continuity theorem is carried out in the realm of limits. For instance, let c be any number in the domain of $y = \frac{f(x)}{g(x)}$. So $f(c)$ and $g(c)$ are defined, and $g(c) \neq 0$. If $f(x)$ and $g(x)$ are continuous at c, then $\lim_{x \to c} f(x) = f(c)$ and $\lim_{x \to c} g(x) = g(c)$. It follows that

$$\lim_{x \to c} \frac{f(x)}{g(x)} = \frac{\lim_{x \to c} f(x)}{\lim_{x \to c} g(x)} = \frac{f(c)}{g(c)}.$$

Hence the continuity criterion is satisfied, and $y = \frac{f(x)}{g(x)}$ is continuous at c. The facts used in the proof of the continuity theorem are known as limit theorems: the limit of a sum is equal to the sum of the limits, the limit of a product is the product of the limits, and the limit of a quotient is the quotient of the limits (this is the theorem that was just used). Rigorous proofs of these theorems need to deploy the limit test.

We close this discussion with some observations about continuous functions. Let $y = f(x)$ be continuous on a closed interval $[a, b]$. Then there are numbers c and d, both between a and b, such that

$$f(c) \leq f(x) \leq f(d) \text{ for all } x \text{ in } [a, b].$$

The values $f(d)$ and $f(c)$ are appropriately called the *maximum* and *minimum* values of the function $y = f(x)$ on $[a, b]$. The proof of this fact is more complicated than one might expect. We therefore appeal to our sense of geometry and Figure 7.19. In the situation of Figure 7.19a, both c and d occur in the interior of $[a, b]$. In the situation of Figure 7.19b, they occur at the endpoints. It is possible for the maximum to be provided by several different numbers c and the minimum by several different d. It is also possible for the

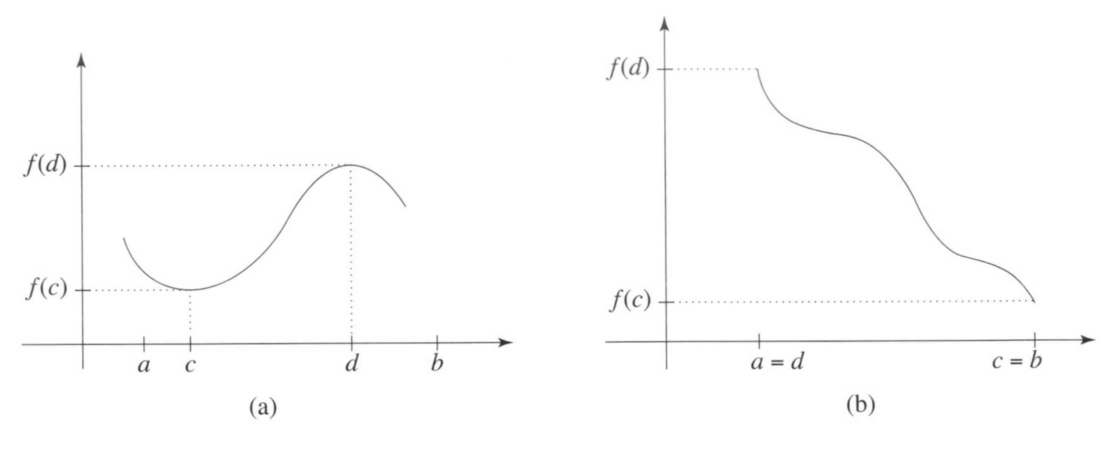

Fig. 7.19

maximum and the minimum value of a function to be the same. What can be concluded about the function in this case?

Intermediate Value Theorem. Let $y = f(x)$ be continuous on a closed interval $[a, b]$, and let M and m be the maximum and minimum values of f on $[a, b]$. Then for any number v with $m \leq v \leq M$, there is a number u with $a \leq u \leq b$ such that $f(u) = v$.

Figure 7.20 illustrates why this is so. Let $M = f(d)$ and $m = f(c)$ be the maximum and minimum values of $y = f(x)$ on $[a, b]$. For any number v between m and M, consider the horizontal segment through

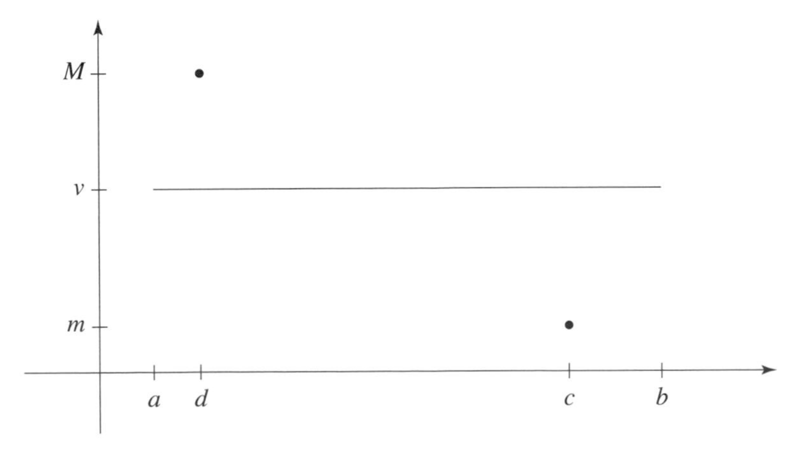

Fig. 7.20

v and over $[a, b]$. Because the graph of $y = f(x)$ is in one continuous, unbroken piece and the points (d, M) and (c, m) are both on the graph, this horizontal segment must intersect the graph at some point. Let (u, v) be this point. Since (u, v) is on the graph of f, it follows that $f(u) = v$. This is intuitively clear, but the details of the proof are subtle.

Example 7.11. Let $f(x) = x^2 + 3$. Why is there a u between 0 and 1 such that $f(u) = \pi$? Why is there a u between 2 and 3 such that $f(u) = \pi^2$?

The considerations that lead to the intermediate value theorem also give information about the zeros of a continuous function f. For example, let f be a function and let $a < b$ be numbers such that f is continuous on $[a, b]$. If $f(a) > 0$ and $f(b) < 0$, then there must be some number u, with $a < u < b$, such that $f(u) = 0$. Sketch a graph that shows why.

Example 7.12. Consider the polynomial

$$f(x) = x^5 + 10x^4 - 12x^3 + 14x^2 + 16x - 20.$$

Check that $f(1) = 9$ and $f(0) = -20$. Why is there a number u with $0 < u < 1$ such that $f(u) = 0$?

7.4 DIFFERENTIABLE FUNCTIONS

The approach to the derivative of a function as presented in Sections 5.4 and 6.2 was in essence this. After a nonvertical tangent line was assumed to exist at a point $P = (x, f(x))$ on the graph of a function $y = f(x)$, it was shown how the slope of this tangent line could be computed with a limit procedure, and the derivative $f'(x)$ was defined to be the slope of this tangent. See Figures 5.15 and 6.5. The tangent line to a curve at a point P is the line that "hugs" the curve most tightly near P. Intuitively, we can see when such a tangent should exist and which line it is if it does. But is it possible to formulate the matter of the existence of a tangent line in precise mathematical terms? Given the central relevance of the derivative, this is an important order of business. We will see that the limit $\lim\limits_{\Delta x \to 0} \frac{f(x+\Delta x)-f(x)}{\Delta x}$ deployed in Section 5.4 to compute the slope of a nonvertical tangent to the graph also informs us precisely when such a tangent exists.

Let f be a function and fix a point $(c, f(c))$ on its graph. It is our concern to formulate a condition that ensures that the graph has a nonvertical tangent at a point $(c, f(c))$. Refer to Figure 7.21. To start, take

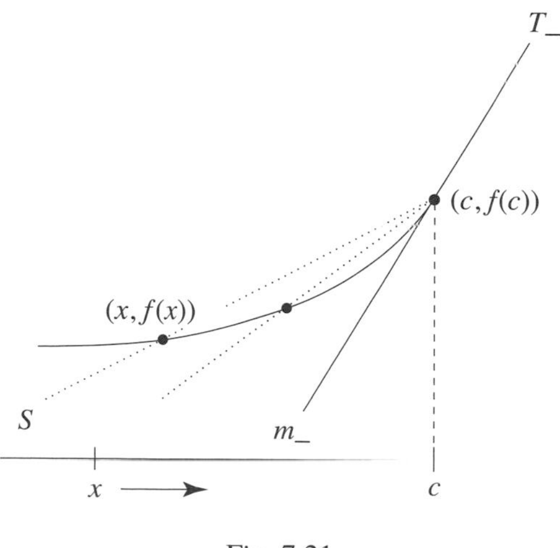

Fig. 7.21

$x < c$. Push x to c (from the left) and assume that

$$\lim_{x\to c^-} \frac{f(x)-f(c)}{x-c} \text{ exists and is equal to } m_-.$$

What does this mean? Fix the line T_- through $(c, f(c))$ with slope m_-. Take any point $(x, f(x))$ on the graph to the left of $(c, f(c))$, and consider the segment S from $(c, f(c))$ through $(x, f(x))$. Note that the slope of S is $\frac{f(x)-f(c)}{x-c}$. Now push x to c, and observe the segment S in the process. A look at Figure 7.21 shows that $\lim_{x\to c^-} \frac{f(x)-f(c)}{x-c} = m_-$ means that the slope of S closes in on the slope m_- of T_-, and that S rotates into the line T_-. Next take $x > c$. Push x to c (this time from the right) and assume that

$$\lim_{x\to c^+} \frac{f(x)-f(c)}{x-c} \text{ exists and is equal to } m_+.$$

This situation is analogous to the one just considered. Let T_+ be the line through $(c, f(c))$ with slope m_+. See Figure 7.22. The segment S connecting the points $(x, f(x))$ and $(c, f(c))$ has slope $\frac{f(x)-f(c)}{x-c}$. Since

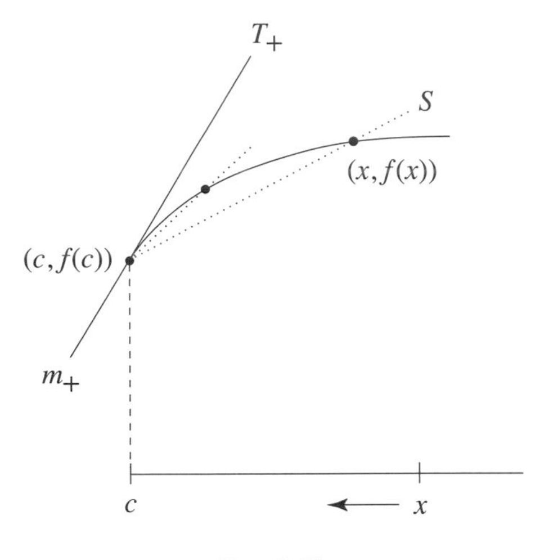

Fig. 7.22

$\lim_{x\to c^+} \frac{f(x)-f(c)}{x-c} = m_+$, it follows that S rotates into T_+ as x is pushed to c from the right. Suppose that $m_- = m_+$. This means that the lines T_- and T_+ have the same slope. Since they both go through $(c, f(c))$, they are the same line. Label this line T. Gluing together Figures 7.21 and 7.22 provides Figure 7.23, and

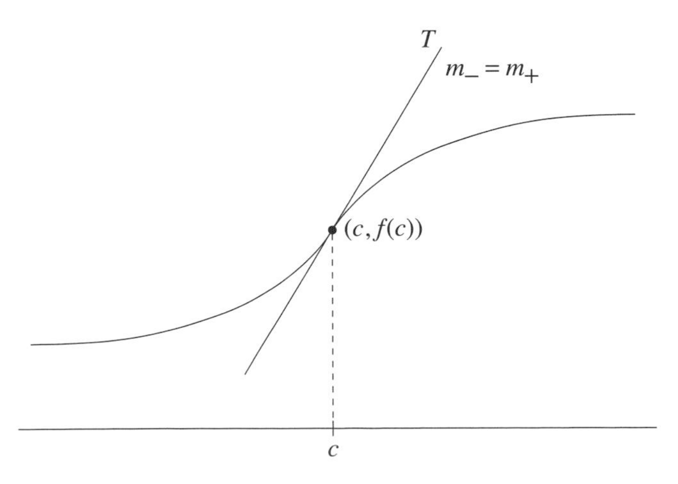

Fig. 7.23

as consequence, the nonvertical tangent line T to the graph at $(c, f(c))$. In summary, therefore, the condition for the existence of a nonvertical tangent line to the graph of $y = f(x)$ at the point $(c, f(c))$ is:

$$\text{The limits } \lim_{x\to c^-} \frac{f(x)-f(c)}{x-c} \text{ and } \lim_{x\to c^+} \frac{f(x)-f(c)}{x-c} \text{ both exist and are equal.}$$

The common value of these limits is the slope $m_- = m_+$ of the tangent.

Differentiability of a Function. The condition for the function $y = f(x)$ to have a nonvertical tangent at the point $(c, f(c))$ of its graph is for the limits

$$\lim_{x\to c^-} \frac{f(x)-f(c)}{x-c} \quad \text{and} \quad \lim_{x\to c^+} \frac{f(x)-f(c)}{x-c}$$

to both exist and be equal. The existence and equality of these two limits is equivalent to the existence of the single limit

$$\lim_{x\to c} \frac{f(x)-f(c)}{x-c}.$$

If the limit exists, then f is said to be *differentiable at* c, and the limit is equal to the slope of the tangent. Suppose that c is an endpoint of the domain of the function f, so that f is not defined for x immediately to the left c or for x immediately to the right of c. In this case, the function f is regarded to be *not differentiable* at c, even if one of the limits $\lim_{x\to c^-} \frac{f(x)-f(c)}{x-c}$ or $\lim_{x\to c^+} \frac{f(x)-f(c)}{x-c}$ exists. (The property of continuity is dealt with differently in this regard. Refer to Section 7.3.)

It is more common to express the limit $\lim_{x\to c} \frac{f(x)-f(c)}{x-c}$ in another way (in a way already done in Section 5.4). Set $x = c + \Delta x$. In reference to Figures 7.21 and 7.22, notice that if $\Delta x < 0$, then $x = c + \Delta x$ is to the left of c, and if $\Delta x > 0$, then x is to the right of c. Pushing x to c from the left is equivalent to pushing Δx to 0 through negative values, and pushing x to c from the right is equivalent to pushing Δx to 0 through positive values. Since $f(x) - f(c) = f(c + \Delta x) - f(c)$ and $\Delta x = x - c$, the preceding discussion translates as follows: The function $y = f(x)$ has a nonvertical tangent at the point $(c, f(c))$ of its graph if the limit

$$\lim_{\Delta x\to 0} \frac{f(c+\Delta x)-f(c)}{\Delta x}$$

exists. In this case, the function is *differentiable at* c, and the value of the limit is the slope of the tangent. To compute this limit, both $\Delta x > 0$ and $\Delta x < 0$ need to be considered. However, our later discussions will focus mostly on the case $\Delta x > 0$.

As in the case of continuity, the definition of differentiability can be extended from points to sets of real numbers. A function f is *differentiable on an interval* I if it is differentiable at all numbers in the interval I. This includes the open intervals (a, b), (a, ∞), $(-\infty, a)$, and $(-\infty, \infty)$, but not $[a, b]$, $[a, b)$, $[a, \infty)$, and so on (in view of the fact that a function is not differentiable at an endpoint of its domain).

Replacing c by x in the expression above provides the definition of the *derivative* of $f(x)$. This is the function f' defined by the rule

$$f'(x) = \lim_{\Delta x\to 0} \frac{f(x+\Delta x)-f(x)}{\Delta x}.$$

It is important to keep in mind that throughout this limit process, x is a fixed number in the domain of f and that only Δx varies (and is pushed to zero). If this limit exists, then x is in the domain of f', the set of all x such that the graph of f has a nonvertical tangent at $(x, f(x))$. The value $f'(x)$ is the slope of this nonvertical tangent. Notice that if $f(x) = C$ with C a constant, then $f'(x) = 0$ for all x.

Example 7.13. Consider the two functions $f(x) = x^2$ and $g(x) = x^2$ with x in $[-3, 5]$. Since

$$\frac{(x+\Delta x)^2 - x^2}{\Delta x} = \frac{x^2 + 2x\Delta x + (\Delta x)^2 - x^2}{\Delta x} = \frac{\Delta x(2x+\Delta x)}{\Delta x} = 2x + \Delta x$$

(the last equality requires $\Delta x \neq 0$), it follows that $f'(x) = 2x$ for all x. So $f(x) = x^2$ is differentiable on the interval $(-\infty, \infty)$. The function $g(x) = x^2$ is also differentiable, but on $(-3, 5)$ and not at -3 and 5.

The derivative $f'(x)$ of a function $y = f(x)$ is often designated by

$$\frac{dy}{dx}, \quad \frac{d}{dx}f(x), \quad \frac{df(x)}{dx}, \quad \text{or} \quad \frac{df}{dx}.$$

This notation originated with Leibniz. All these symbols have the same meaning. They all stand for the derivative of the function $y = f(x)$.

Suppose that the function f is differentiable at c. Geometrically this means that near the point $(c, f(c)$, its graph is tightly approximated by a straight nonvertical segment through $(c, f(c))$ with slope $f'(c)$. Numerically, $f(c + \Delta x) - f(c)$ is the change in the value of the function over the interval $[c, c + \Delta x]$ for Δx positive and over $[c + \Delta x, c]$ for Δx negative. The ratio $\frac{f(c+\Delta x)-f(c)}{\Delta x}$ is the average rate of this change. Taking the limit of this average for smaller and smaller Δx provides the *rate of change of f at c*. With the time variable t in place of x, this was the approach (in Sections 6.4 and 6.5) to the definitions of velocity and acceleration of a moving point at any instant.

This section has resolved the question as to when a graph has a nonvertical tangent at a point with a definitive criterion that is formulated algebraically in terms of limits. Going back to geometric intuition, observe that if the graph of a function is smooth near the point $(c, f(c))$, then a tangent line to the graph can be placed there. If this tangent is not vertical, then f is differentiable at c. This suggests that our intuitive sense of the differentiability of a function $y = f(x)$ at c should be:

the graph of f is smooth near $(c, f(c))$ with a nonvertical tangent.

This is in fact a good intuitive picture. It is valid for all the functions that we will encounter in this text. However, the example that follows injects a word of caution.

Example 7.14. Consider the *sawtooth function* $y = s(x)$ depicted in Figure 7.24. Its graph falls between the graph of $f(x) = x^2$ and the x-axis. It is defined as follows. Let m be a positive number. (Take $m = 3$ if you like.) We'll first proceed from right to left: for $x \geq 1$, the graph of $y = s(x)$ coincides with that of the line $y = 1$ until that line hits the parabola $f(x) = x^2$ at the point $(1, 1)$. From that point, the graph of $y = s(x)$ follows the line with slope m down to the x-axis. From there, the graph follows the line with slope $-m$ back up to the graph of $f(x) = x^2$. From that point, the graph of $y = s(x)$ goes down to the x-axis along the line with slope m. Then up again, and so forth, and so on. The points on the graph of $y = s(x)$

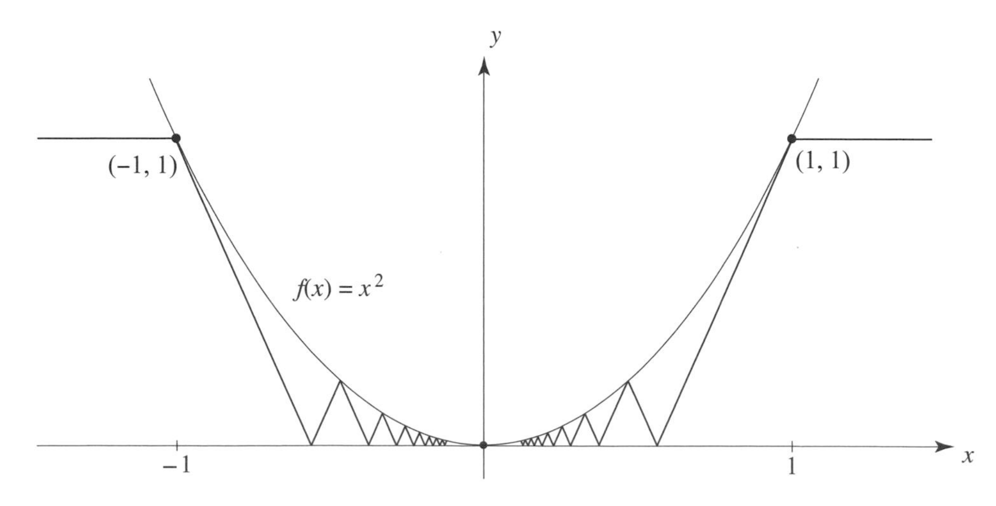

Fig. 7.24

that lie on the x-axis converge to the origin. (The figure makes this plausible, but we'll skip the challenging proof.) The function $y = s(x)$ is defined in the same way for negative x. Again, the points on the graph that lie on the negative x-axis converge to 0. To complete the definition of the sawtooth function, it remains to set $s(0) = 0$. Notice that

$$0 \le s(x) \le x^2$$

for all x. The function $y = s(x)$ is differentiable at all points, except those that lie either on the graph of $f(x) = x^2$ or the x-axis (because the graph turns sharply at all these points). But $y = s(x)$ *is differentiable* at $x = 0$. To see this, consider the limit

$$\lim_{\Delta x \to 0} \frac{s(0 + \Delta x) - s(0)}{\Delta x} = \lim_{\Delta x \to 0} \frac{s(\Delta x)}{\Delta x}.$$

Because $0 \le s(\Delta x) \le (\Delta x)^2$, it follows that

$$0 \le \frac{s(\Delta x)}{\Delta x} \le \frac{(\Delta x)^2}{\Delta x} = \Delta x$$

for all $\Delta x \neq 0$. When Δx is pushed to 0, the term $\frac{s(\Delta x)}{\Delta x}$, as it is caught between 0 and Δx (the proverbial rock and a hard place), must go to zero. So $\lim_{\Delta x \to 0} \frac{s(\Delta x)}{\Delta x} = 0$, and hence $y = s(x)$ *is differentiable* at $x = 0$. While the teeth of the graph of $y = s(x)$ become smaller as x approaches 0, they retain their sharp tips. You would therefore not be inclined to say that the graph is smooth near the point $(0, 0)$. But it does have a horizontal tangent at $(0, 0)$.

7.5 COMPUTING DERIVATIVES

This section will consider various strategies with which derivatives can be computed. As we have seen, the evaluation of the derivative of a function necessarily involves the computation of a limit. Such limit computations can be used to provide general rules that apply to many specific derivatives. One such rule tells us how to differentiate the function $f(x) = x^r$ for any rational constant r.

Power Rule: For any rational number r, the derivative of $f(x) = x^r$ is $f'(x) = rx^{r-1}$.

Newton's proof of the power rule for any positive rational r was presented in Section 6.2. We will see in a moment that for any negative rational number r, the power rule is a consequence of the rule for the derivative of the quotient of two functions. For $r = \frac{1}{3}$, the domain of f is the set of all real numbers x. In general, as $r = -\frac{1}{2}$ illustrates, the restriction $x > 0$ is required.

Example 7.15. The derivative of $g(x) = \sqrt[3]{x} = x^{\frac{1}{3}}$ is $g'(x) = \frac{1}{3}x^{-\frac{2}{3}} = \frac{1}{3}\frac{1}{x^{\frac{2}{3}}}$. Note that $g'(x)$ is not defined at $x = 0$. So $y = g(x)$ is not differentiable at 0. A look at Figure 7.25 tells us what the problem is. The

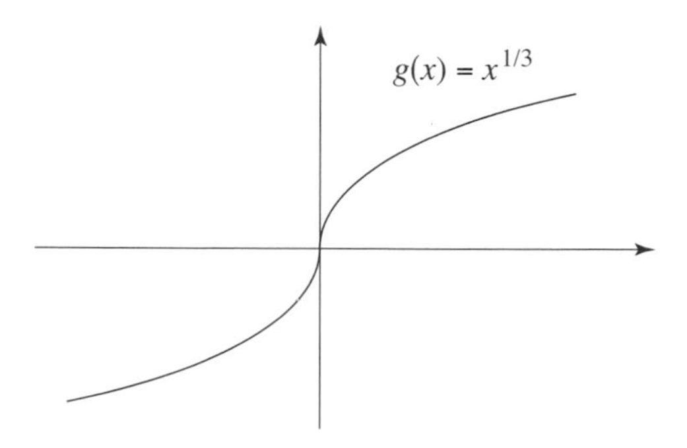

Fig. 7.25

graph of g is smooth at the origin and it has a tangent there, but this tangent is vertical.

Example 7.16. The derivative of $f(x) = x^{\frac{2}{3}}$ is $f'(x) = \frac{2}{3}x^{-\frac{1}{3}} = \frac{2}{3}\frac{1}{x^{\frac{1}{3}}}$. Observe that $f'(x)$ is not defined at $x = 0$, so that f is not differentiable for $x = 0$. Its graph (see Figure 7.26) shows why. It comes to a sharp point at $(0, 0)$, and the tangent there is vertical. As to the graph of $f(x) = x^{\frac{2}{3}}$, observe that $x^{\frac{2}{3}} = (x^{\frac{1}{3}})^2 \geq 0$, so that it lies above the x-axis. Since $0 \leq x^{\frac{1}{3}} \leq 1$ for $0 \leq x \leq 1$ and $1 \leq x^{\frac{1}{3}}$ for $1 \leq x$, the squaring operation decreases $y = x^{\frac{1}{3}}$ over $0 \leq x \leq 1$ and increases $y = x^{\frac{1}{3}}$ over $1 \leq x$. Accordingly, to get the graph of $y = x^{\frac{2}{3}}$ from the graph of $y = x^{\frac{1}{3}}$, the latter needs to be compressed over the first of these intervals and

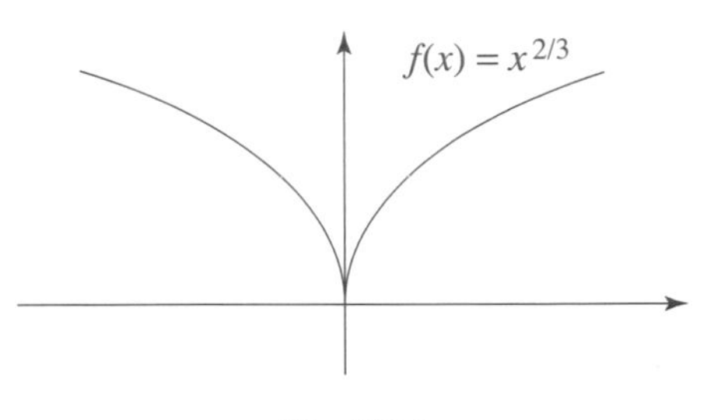

Fig. 7.26

stretched over the second (both vertically). The same thing works for $x \leq 0$.

In both of these examples, the derivative is defined for all x except $x = 0$. At all other points, the graphs of the two functions are smooth with a nonvertical tangents.

Many differentiations can be carried out by using the *sum, difference, constant, product,* and *quotient* rules, along with the power rule, and often in combination. We state these rules next and provide examples. Let two functions $y = f(x)$ and $y = g(x)$ be given.

Sum Rule: $\frac{d}{dx}(f(x) + g(x)) = \frac{d}{dx}f(x) + \frac{d}{dx}g(x)$

Difference Rule: $\frac{d}{dx}(f(x) - g(x)) = \frac{d}{dx}f(x) - \frac{d}{dx}g(x)$

Constant Rule: $\frac{d}{dx}(cf(x)) = c\frac{d}{dx}f(x)$, where c is any constant

Product Rule: $\frac{d}{dx}(f(x) \cdot g(x)) = \frac{d}{dx}f(x) \cdot g(x) + f(x) \cdot \frac{d}{dx}g(x)$

Quotient Rule: $\dfrac{d}{dx}\left(\dfrac{f(x)}{g(x)}\right) = \dfrac{\frac{d}{dx}f(x) \cdot g(x) - f(x) \cdot \frac{d}{dx}g(x)}{g(x)^2}$

The verification of each of these rules—as was already pointed out—involves limit computations. We will show how this is done for the product rule in Section 7.6.

Example 7.17. Let $f(x) = 3x^5 - 7x^2 + 2$. Using the constant, power, and difference rules, we get $f'(x) = 15x^4 - 14x$.

Example 7.18. Use of the product, constant, power, difference, and sum rules shows that the derivative of $y = (3x^{100} - 4x^{21})(50x^{10} + 9x^{\frac{1}{2}})$ is

$$\begin{aligned}&(\tfrac{d}{dx}(3x^{100} - 4x^{21})) \cdot (50x^{10} + 9x^{\frac{1}{2}}) + (3x^{100} - 4x^{21}) \cdot (\tfrac{d}{dx}(50x^{10} + 9x^{\frac{1}{2}}))\\ &\qquad = (300x^{99} - 84x^{20})(50x^{10} + 9x^{\frac{1}{2}}) + (3x^{100} - 4x^{21})(500x^9 + \tfrac{9}{2}x^{-\frac{1}{2}}).\end{aligned}$$

Example 7.19. Use of the quotient rule shows that the derivative of $y = \frac{3x^5-7x^2+2}{6x^3+9x}$ is

$$\frac{(\frac{d}{dx}(3x^5 - 7x^2 + 2)) \cdot (6x^3 + 9x) - (3x^5 - 7x^2 + 2) \cdot \frac{d}{dx}(6x^3 + 9x)}{(6x^3 + 9x)^2}$$

$$= \frac{(15x^4 - 14x)(6x^3 + 9x) - (3x^5 - 7x^2 + 2)(18x^2 + 9)}{(6x^3 + 9x)^2}.$$

The quotient rule can be applied to extend Newton's power rule $\frac{d}{dx}x^r = rx^{r-1}$ to negative r. Suppose that $r < 0$. Then $f(x) = x^r = \frac{1}{x^{-r}}$ with $-r > 0$ and $x \neq 0$. By using the quotient rule and the positive case of the power rule,

$$f'(x) = \frac{\frac{d}{dx}(1) \cdot x^{-r} - 1 \cdot \frac{d}{dx}(x^{-r})}{(x^{-r})^2} = \frac{0 - (-r)x^{-r-1}}{x^{-2r}} = rx^{-r-1+2r} = rx^{r-1}.$$

Only $f(x) = x^r$ with $r = 0$ remains to be considered. In this case, $f(x) = x^0 = 1$ for all $x \neq 0$ and $f(0) = 0^0$ is undefined. Since $y = 1$ is a horizontal line, $f'(x) = 0$ for $x \neq 0$. Therefore the power rule applies for $r = 0$ as well.

The following fact is a direct consequence of the rules: let $y = f(x)$ and $y = g(x)$ be two functions that are both differentiable on their domains. Then the sum $y = f(x) + g(x)$, the difference $y = f(x) - g(x)$, the product $y = f(x) \cdot g(x)$, and the quotient $y = \frac{f(x)}{g(x)}$ are all functions that are differentiable on their domains. For instance, let c be in the domain of $\frac{f(x)}{g(x)}$. This means that $f(c)$ makes sense, $g(c)$ makes sense, and $g(c) \neq 0$. Since f and g are both differentiable on their domains, both $f'(c)$ and $g'(c)$ make sense. It now follows that $(\frac{f(c)}{g(c)})' = \frac{f'(c)g(c) - f(c)g'(c)}{(g(c))^2}$ makes sense. So $\frac{f(x)}{g(x)}$ is differentiable at c. We have verified that $\frac{f(x)}{g(x)}$ is differentiable at every point in its domain.

The most important differentiation rule is the chain rule. We have already seen that more complicated functions can be "built up" from simpler functions by composition. Recall from Section 7.3 that the composite function of $f(x)$ and $g(x)$ is obtained by splicing the rule that defines g into the one that defines f, in other words, by forming $y = f(g(x))$. For instance, the composite of $f(x) = 3x^5 - 7x^2 + 2$ and $g(x) = \sqrt{x} = x^{\frac{1}{2}}$ is the function

$$\begin{aligned} y &= f(g(x)) = 3(g(x))^5 - 7(g(x))^2 + 2 = 3(x^{\frac{1}{2}})^5 - 7(x^{\frac{1}{2}})^2 + 2 \\ &= 3x^{\frac{5}{2}} - 7x + 2. \end{aligned}$$

The domain of the composite function is $x \geq 0$ in this case, since this restriction is needed for $g(x) = \sqrt{x} = x^{\frac{1}{2}}$. The composite $y = g(f(x))$ of $g(x)$ and $f(x)$ is the entirely different function

$$y = g(f(x)) = g(3x^5 - 7x^2 + 2) = \sqrt{3x^5 - 7x^2 + 2}.$$

For x to be in its domain, the condition $3x^5 - 7x^2 + 2 \geq 0$ needs to be met. It matters, therefore, in which order the composite is taken.

The chain rule tells us how the derivative of the composite function $y = f(g(x))$ is obtained from the derivatives of the two component functions.

Chain Rule: The derivative of the composite function $y = f(g(x))$ is

$$\boxed{\frac{d}{dx}(f(g(x))) = f'(g(x)) \cdot g'(x)}$$

Expressed more explicitly, the derivative of $f(g(x))$ is the composite of the functions $f'(x)$ and $g(x)$ multiplied by the derivative $g'(x)$.

Example 7.20. Let's begin with some evidence for the validity of the chain rule. Consider the functions

$$f(x) = 3x^5 - 7x^2 + 2 \text{ and } g(x) = \sqrt{x} = x^{\frac{1}{2}}.$$

We already know that

$$f(g(x)) = 3x^{\frac{5}{2}} - 7x + 2.$$

So by our previous rules, $\frac{d}{dx}(f(g(x)) = \frac{15}{2}x^{\frac{3}{2}} - 7$. On the other hand, since $f'(x) = 15x^4 - 14x$ and $g'(x) = \frac{1}{2}x^{-\frac{1}{2}}$, we get by applying the chain rule that

$$\begin{aligned}\frac{d}{dx}(f(g(x))) &= f'(g(x)) \cdot g'(x) = (15(g(x))^4 - 14g(x))(\tfrac{1}{2}x^{-\frac{1}{2}}) \\ &= (15(x^{\frac{1}{2}})^4 - 14x^{\frac{1}{2}})(\tfrac{1}{2}x^{-\frac{1}{2}}) = (15x^2 - 14x^{\frac{1}{2}})(\tfrac{1}{2}x^{-\frac{1}{2}}) \\ &= \tfrac{15}{2}x^{\frac{3}{2}} - 7.\end{aligned}$$

Observe that the results of the two computations are the same. Consider next the composite

$$y = g(f(x)) = \sqrt{3x^5 - 7x^2 + 2} = (3x^5 - 7x^2 + 2)^{\frac{1}{2}}.$$

In this case, the chain rule is the only method available. Using it, we get

$$\begin{aligned}\frac{d}{dx}(g(f(x))) &= g'(f(x)) \cdot f'(x) = \tfrac{1}{2}f(x)^{-\frac{1}{2}} \cdot (15x^4 - 14x) \\ &= \tfrac{1}{2}(3x^5 - 7x^2 + 2)^{-\frac{1}{2}} \cdot (15x^4 - 14x).\end{aligned}$$

A look back at the statement of the chain rule shows that the last function that is applied—this is the "outside" function f of the formula—is differentiated first, and the function that is applied first—this is the "inside" function g of the formula—is differentiated last. The examples that follow illustrate this explicitly and show how this "flow" expedites the computations.

Example 7.21. Consider $f(x) = (x^2 + 1)^{100}$. Here, the outside is $(\quad)^{100}$ and the inside is $x^2 + 1$. Differentiate the outside to get $100(\quad)^{99}$. Now reinsert the inside term and then differentiate it to get

$$f'(x) = 100(\quad)^{99} \cdot \tfrac{d}{dx}(\quad) = 100(x^2 + 1)^{99} \cdot 2x.$$

Example 7.22. Let's do this for $y = (3x^5 - 7x^2 + 2)^{\frac{1}{2}}$. The derivative of the outside is $\frac{1}{2}(\quad)^{-\frac{1}{2}}$. Since the derivative of the inside is $15x^4 - 14x$, we get

$$\frac{dy}{dx} = \tfrac{1}{2}(\quad)^{-\frac{1}{2}} \cdot \tfrac{d}{dx}(\quad) = \tfrac{1}{2}(3x^5 - 7x^2 + 2)^{-\frac{1}{2}} \cdot (15x^4 - 14x).$$

As expected, this agrees with the answer obtained earlier.

Generalized Power Rule: A combination of the chain rule and the power rule tells us that if $g(x)$ is a function and r a real number, then

$$\frac{d}{dx}((g(x)^r) = r(g(x))^{r-1} \cdot g'(x).$$

The chain rule can be extended to situations that involve more than two functions simply by applying it more than once. This is done by the "from the outside to the inside" strategy already explained.

Example 7.23. Let $y = ((x^2 + 6)^{100} + 2x^3)^{\frac{3}{2}}$. Differentiating the outside and knowing that the derivative of the inside follows next, we get

$$\frac{dy}{dx} = \tfrac{3}{2}((x^2 + 6)^{100} + 2x^3)^{\frac{1}{2}} \cdot \tfrac{d}{dx}((x^2 + 6)^{100} + 2x^3).$$

Turning to the remaining differentiation, we get

$$\begin{aligned}\tfrac{d}{dx}(x^2+6)^{100} &= 100(\quad\ )^{99}\tfrac{d}{dx}(\quad\ ) = 100(x^2+6)^{99}\tfrac{d}{dx}(x^2+6)\\ &= 100(x^2+6)^{99}2x,\end{aligned}$$

and $\frac{d}{dx}2x^3 = 6x^2$. Putting it all together,

$$\frac{dy}{dx} = \tfrac{3}{2}((x^2+6)^{100}+2x^3)^{\frac{1}{2}}(100(x^2+6)^{99}2x+6x^2).$$

To repeat, with the chain rule, derivatives of complicated functions can be computed efficiently by considering them to be composites of simpler functions that are more easily dealt with. Our next several examples illustrate how the derivatives of algebraic functions can be computed by combining the rules that have been discussed.

Example 7.24. Let $f(x) = (\sqrt{x+1})(x^7-1) = (x+1)^{\frac{1}{2}}(x^7-1)$. We get

$$\begin{aligned}f'(x) &= (\tfrac{d}{dx}(x+1)^{\frac{1}{2}})(x^7-1)+(x+1)^{\frac{1}{2}}\cdot\tfrac{d}{dx}(x^7-1) \text{ (product rule)}\\ &= (\tfrac{1}{2}(x+1)^{-\frac{1}{2}}(1))(x^7-1)+(x+1)^{\frac{1}{2}}\cdot 7x^6 \text{ (chain rule and power rule)}\\ &= \tfrac{1}{2}(x+1)^{-\frac{1}{2}}(x^7-1)+7x^6(x+1)^{\frac{1}{2}}.\end{aligned}$$

Example 7.25. Let $g(x) = \frac{x^3+7x^2+2}{(x^4-5x^2)^{\frac{2}{3}}}$. Use the quotient rule, and then the sum rule and the chain rule, to get

$$g'(x) = \frac{(\frac{d}{dx}(x^3+7x^2+2))\cdot(x^4-5x^2)^{\frac{2}{3}}-(x^3+7x^2+2)\cdot\frac{d}{dx}(x^4-5x^2)^{\frac{2}{3}}}{((x^4-5x^2)^{\frac{2}{3}})^2}.$$

After carrying out the remaining two differentiations, we see that this equals

$$\begin{aligned}&= \frac{(3x^2+14x)\cdot(x^4-5x^2)^{\frac{2}{3}}}{(x^4-5x^2)^{\frac{4}{3}}} - \frac{(x^3+7x^2+2)\cdot\frac{2}{3}(x^4-5x^2)^{-\frac{1}{3}}(4x^3-10x)}{(x^4-5x^2)^{\frac{4}{3}}}\\ &= \frac{(3x^2+14x)}{(x^4-5x^2)^{\frac{2}{3}}} - \frac{\frac{2}{3}(x^3+7x^2+2)(4x^3-10x)}{(x^4-5x^2)^{\frac{5}{3}}}.\end{aligned}$$

The meaning of the composite of two functions and the chain rule can be illustrated in concrete terms. Suppose that a balloon is being inflated. Assume that it retains the shape of a perfect sphere during the process. If r is its radius, then its volume is $V(r) = \frac{4}{3}\pi r^3$. Note that the radius in turn is a function $r = f(t)$ of time t. See Figure 7.27. The composite $V(f(t)) = \frac{4}{3}\pi(f(t))^3$ expresses the volume of the balloon as

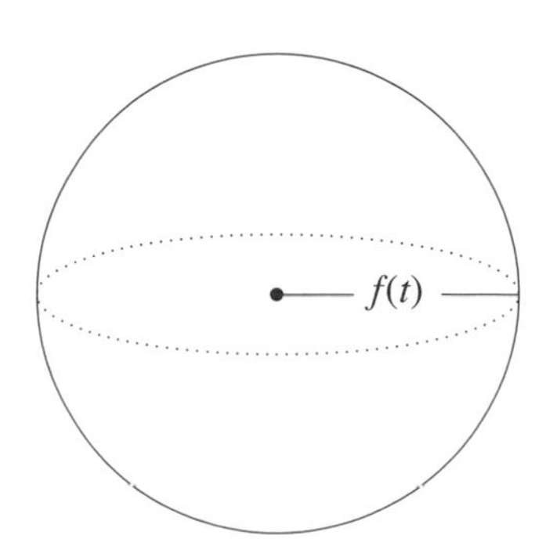

Fig. 7.27

a function of t. The derivative $\frac{dr}{dt} = f'(t)$ is the rate at which the radius is increasing at time t. (The connection between the derivative and rates of increase and decrease is dealt with in detail in Section 8.1.) The derivative $\frac{dV}{dr} = 4\pi r^2$ is the rate at which V is increasing relative to the radius r, and $\frac{dV}{dt} = \frac{d}{dt}V(f(t))$ is the rate at which V is changing relative to t. An application of the chain rule shows that

$$\frac{dV}{dt} = \frac{d}{dt}V(f(t)) = V'(f(t)) \cdot f'(t) = \tfrac{4}{3}\pi(3(f(t))^2) \cdot f'(t) = 4\pi r^2 \cdot f'(t) = \frac{dV}{dr} \cdot \frac{dr}{dt}.$$

Therefore $\frac{dV}{dt} = \frac{dV}{dr} \cdot \frac{dr}{dt}$. In this context, the chain rule is nothing but the assertion that the rate at which V changes relative to t is the product of the rate at which V changes relative to r times the rate at which r changes relative to t. Note the suggestive nature of the notation of Leibniz. By considering these quantities as differentials (see Section 5.8), the chain rule is the assertion that the differential dr can be canceled.

7.6 SOME THEORETICAL CONCERNS

Suppose that the function $y = f(x)$ is differentiable at a number c. Therefore the graph of the function has a nonvertical tangent at the point $(c, f(c))$, and the limit

$$\lim_{x \to c} \frac{f(x) - f(c)}{x - c}$$

is equal to the slope $f'(c)$ of this tangent. Since $f(c)$ arises in the expression of this limit, c must be in the domain of f. This, along with the fact that the graph is tightly approximated by the tangent near the point $(c, f(c)$, suggests at least intuitively that the graph should have no gaps or breaks at $(c, f(c))$, and that therefore f should be continuous at c. This is in fact the case: when a function is differentiable at a number, it is also continuous at that number. The verification is a quick "one-liner":

$$\lim_{x \to c}(f(x) - f(c)) = \lim_{x \to c}\left((x - c) \cdot \frac{f(x) - f(c)}{x - c}\right) = \lim_{x \to c}(x - c) \cdot \lim_{x \to c} \frac{f(x) - f(c)}{x - c} = 0 \cdot f'(c) = 0.$$

It follows that $\lim_{x \to c} f(x) = f(c)$. Therefore, by the continuity criterion of Section 7.3, f is continuous at c.

To repeat, if a function is differentiable at c, then it must be continuous at c. It follows that if a function f is differentiable on an interval I, then it is continuous on the interval I.

We turn next to the proof of the product rule for differentiation. Let f and g be two functions. From the definition of the derivative, we know that

$$\frac{d}{dx}(f(x) \cdot g(x)) = \lim_{\Delta x \to 0} \frac{f(x + \Delta x)g(x + \Delta x) - f(x)g(x)}{\Delta x}.$$

Adding $-f(x)g(x + \Delta x) + f(x)g(x + \Delta x) = 0$ to the numerator, and then rearranging terms, gives us

$$\begin{aligned}
&\lim_{\Delta x \to 0}\left(\frac{f(x + \Delta x)g(x + \Delta x) - f(x)g(x + \Delta x)}{\Delta x} + \frac{f(x)g(x + \Delta x) - f(x)g(x)}{\Delta x}\right)\\
&= \lim_{\Delta x \to 0} \frac{[f(x + \Delta x) - f(x)]g(x + \Delta x)}{\Delta x} + \lim_{\Delta x \to 0} \frac{f(x)[g(x + \Delta x) - g(x)]}{\Delta x}\\
&= \lim_{\Delta x \to 0} \frac{f(x + \Delta x) - f(x)}{\Delta x} g(x + \Delta x) + \lim_{\Delta x \to 0} f(x) \frac{g(x + \Delta x) - g(x)}{\Delta x}\\
&= \tfrac{d}{dx}f(x) \cdot g(x) + f(x) \cdot \tfrac{d}{dx}g(x).
\end{aligned}$$

This proof leads, step by step, to the conclusion that we're after. (Why is the continuity of the function g at x needed to complete the last step?) Be aware that a definitive version of the proof has to verify each of these steps by applying the limit test of Section 7.2. The proofs of the other rules of differentiation are similar. All rely on basic properties of limits.

Recall from Section 7.2 that limits of $\frac{0}{0}$ type are often tricky to compute. With L'Hospital's rule, many such limits can be handled effectively and efficiently.

L'Hospital's Rule. Suppose that f and g are both differentiable on an open interval containing c. If $\lim\limits_{x\to c}\frac{f(x)}{g(x)}$ is of $\frac{0}{0}$ type and $\lim\limits_{x\to c}\frac{f'(x)}{g'(x)}$ exists, then

$$\lim_{x\to c}\frac{f(x)}{g(x)} = \lim_{x\to c}\frac{f'(x)}{g'(x)}.$$

This result will play no role in our subsequent discussions, so we'll skip the proof. In terms of the logical interconnections, note that the rule depends on the derivative, which is itself a limit of $\frac{0}{0}$ type. Of historical interest is the fact that the rule is actually the discovery of Johann Bernoulli, who was hired by the Marquis de L'Hospital to teach him the fundamentals of calculus. (Johann and his brother Jakob were students of Leibniz and became eminent mathematicians themselves.) The marquis then chose to publish what he learned (and purchased) under his own name.

Example 7.26. Let $f(x) = x^5 + x^4 - x - 1$ and $g(x) = x^3 + x^2 - x - 1$. Consider $\lim\limits_{x\to 1}\frac{f(x)}{g(x)}$. Notice that it is of $\frac{0}{0}$ type. Since $f'(x) = 5x^4 + 4x^3 - 1$, $g'(x) = 3x^2 + 2x - 1$, and $\lim\limits_{x\to 1}\frac{f'(x)}{g'(x)} = \frac{5+4-1}{3+2-1} = \frac{8}{4} = 2$, it follows that $\lim\limits_{x\to 1}\frac{f(x)}{g(x)} = 2$. Notice that $\lim\limits_{x\to -1}\frac{f(x)}{g(x)}$ is also of $\frac{0}{0}$ type. Since $f'(-1) = 0$ and $g'(-1) = 0$, $\lim\limits_{x\to -1}\frac{f'(x)}{g'(x)}$ is again of $\frac{0}{0}$ type. Since we do not know whether this limit exists, L'Hospital's rule cannot be applied. But why not repeat the process once more? Since $f''(x) = 20x^3 + 12x^2$ and $g''(x) = 6x + 2$, $\lim\limits_{x\to -1}\frac{f''(x)}{g''(x)} = \frac{-8}{-4} = 2$. So by L'Hospital's rule, $\lim\limits_{x\to -1}\frac{f'(x)}{g'(x)} = 2$. By applying the rule again, $\lim\limits_{x\to -1}\frac{f(x)}{g(x)} = 2$.

We conclude this section with one more fundamental theoretical concern.

Mean Value Theorem. Let $y = f(x)$ be continuous on a closed interval $[a, b]$. If f is differentiable on the open interval (a, b), then there is a number c in (a, b) such that

$$f(b) - f(a) = f'(c)(a - b).$$

Let's consider this assertion by looking at the geometry involved. A look at Figure 7.28 shows that $\frac{f(b)-f(a)}{b-a}$ is the slope of the segment joining the points $(a, f(a))$ and $(b, f(b))$. The continuity and differentiability properties of the graph imply that at some point $(c, f(c))$, the slope of the tangent is equal to the slope of the segment. Since $f'(c)$ is the slope of this tangent, $f'(c) = \frac{f(b)-f(a)}{b-a}$ and the conclusion of

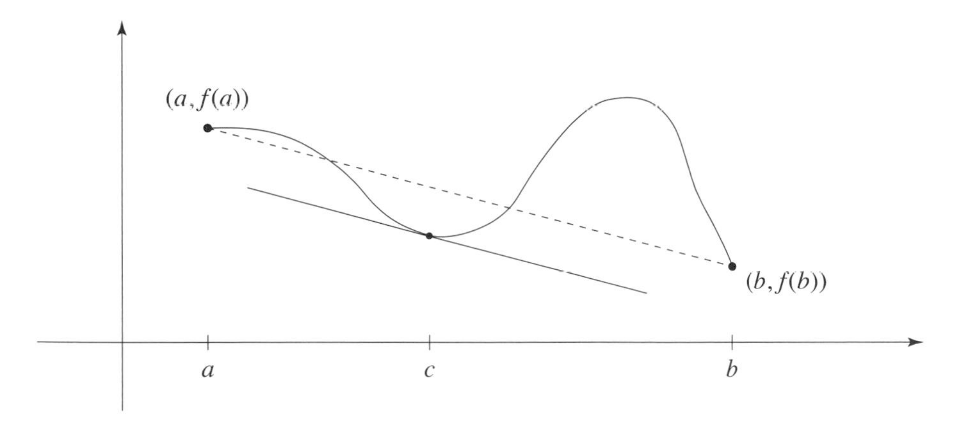

Fig. 7.28

the mean value theorem follows. Notice that there is a second point $(c, f(c))$ in the figure that satisfies the conclusion of the theorem. Incidentally, the word *mean* is used in the sense of average, as $\frac{f(b)-f(a)}{b-a}$ is the change in $f(x)$ averaged over the interval $a \leq x \leq b$.

Corollary i. If f is a function with the property that $f'(x) = 0$ for all x in an open interval I, then f is constant throughout I. [To see that this is so, let a and b be any two numbers in I with $a < b$. Since f is differentiable on I, f is continuous on I and hence on $[a, b]$. So by the mean value theorem there is a c with $a < c < b$ such that $f(b) - f(a) = f'(c)(b - a)$. Since $f'(c) = 0$, it follows that $f(b) = f(a)$.]
Corollary ii. Let $f(x)$ and $g(x)$ be two functions that are differentiable for all x in an open interval I. If $f'(x) = g'(x)$ for all x in I, then $g(x) = f(x) + C$ all x in I. [This follows quickly from Corollary i.]

As one might expect, the proof of the mean value theorem relies on an analysis of limits. It is beyond the aims of this text, except for the following point. In view of Corollary ii, an important fact that was verified "by picture" at the beginning of Section 5.8 is now a consequence of the mean value theorem. This provides an illustration of the transformation of calculus from the early approaches of Newton and Leibniz (that became the target of Bishop Berkeley) to the rigorous formulations of Cauchy and Weierstrass (that relied on theorems and proofs).

7.7 DERIVATIVES OF TRIGONOMETRIC FUNCTIONS

The trigonometric functions sine, cosine, tangent, and secant were studied in Section 4.6 as functions of the variable θ in radians. Their graphs—sketched in Figures 4.23, 4.24, 4.26, and 4.27—tell us that these trigonometric functions are continuous on their domains (but note that there are points where the tangent and secant are not defined). We will see shortly that each of these functions is not only continuous—but in fact differentiable—at every point in its domain. The limit

$$\lim_{\theta \to 0} \frac{\sin \theta}{\theta} = 1$$

is a cornerstone of the differential calculus of the trigonometric functions. It was already observed in Section 1.6 that $\sin \theta \approx \theta$ when θ is small, and that the small the θ is, the better the approximation gets. This makes the limit plausible. But now we will verify it.

We'll assume first that θ goes to 0 through positive θ. Since it is being pushed to zero, we can assume that $0 < \theta < \frac{\pi}{2}$. Use θ to construct Figure 7.29. The arc BC lies on the circle of radius 1 and center O. The segment AC is perpendicular to the radius OB, and the radius OC is extended to the point D with the

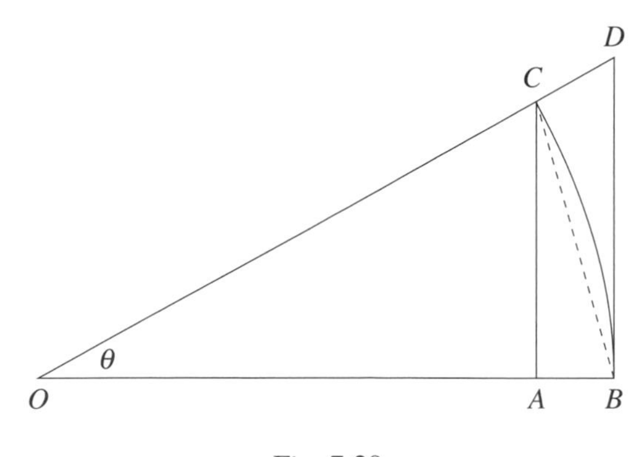

Fig. 7.29

property that BD is perpendicular to OB. Note that $\sin \theta = \frac{AC}{OC} = AC$ and that $\tan \theta = \frac{BD}{OB} = BD$. So the areas of the triangles ΔOBC and ΔOBD are $\frac{1}{2}OB \cdot AC = \frac{1}{2}\sin \theta$ and $\frac{1}{2}OB \cdot BD = \frac{1}{2}\tan \theta$, respectively.

Since the radius is 1, the area of the circular sector OBC is $\frac{1}{2}\theta$. (This was verified in Section 2.2.) From the figure,

$$\text{Area } \Delta OBD \geq \text{Area sector } OBC \geq \text{Area } \Delta OBC,$$

so that $\frac{1}{2}\tan\theta \geq \frac{1}{2}\theta \geq \frac{1}{2}\sin\theta$. Therefore $\frac{\sin\theta}{\cos\theta} = \tan\theta \geq \theta \geq \sin\theta$. Dividing through by $\sin\theta$ gives $\frac{1}{\cos\theta} \geq \frac{\theta}{\sin\theta} \geq 1$, and after inverting all terms, $\cos\theta \leq \frac{\sin\theta}{\theta} \leq 1$. Now push θ to 0. Since OA goes to $OB = 1$, we see that $\lim\limits_{\theta\to 0}\cos\theta = 1$. It follows that $\frac{\sin\theta}{\theta}$ is squeezed to 1, and therefore that

$$\lim_{\theta\to 0}\frac{\sin\theta}{\theta} = 1.$$

If θ is negative, $-\theta$ is positive. Since $\sin(-\theta) = -\sin\theta$, see Section 6.6, we get

$$\lim_{\theta\to 0}\frac{\sin\theta}{\theta} = \lim_{\theta\to 0}\frac{\sin(-\theta)}{-\theta} = 1.$$

Another fact needed to determine the derivatives of the trigonometric functions is the limit $\lim\limits_{\theta\to 0}\frac{\cos\theta - 1}{\theta} = 0$. The verification of this is based on the following sequence of equalities:

$$\frac{\cos\theta - 1}{\theta} = \frac{\cos\theta - 1}{\theta}\cdot\frac{\cos\theta + 1}{\cos\theta + 1} = \frac{\cos^2\theta - 1}{\theta}\cdot\frac{1}{\cos\theta + 1} = -\frac{\sin^2\theta}{\theta}\cdot\frac{1}{\cos\theta + 1} = -\frac{\sin\theta}{\theta}\cdot\frac{\sin\theta}{\cos\theta + 1}.$$

Since $\lim\limits_{\theta\to 0}\frac{\sin\theta}{\theta} = 1$ and $\lim\limits_{\theta\to 0}\sin\theta = 0$ (notice that AC goes to when θ does), it follows that

$$\lim_{\theta\to 0}\frac{\cos\theta - 1}{\theta} = \lim_{\theta\to 0}\left(-\frac{\sin\theta}{\theta}\cdot\frac{\sin\theta}{\cos\theta + 1}\right) = -1\cdot\frac{0}{1+1} = 0.$$

Since it relies on Figure 7.29, θ is positive in this argument. Why is $\lim\limits_{\theta\to 0}\frac{\cos\theta - 1}{\theta} = 0$ for negative θ as well?

We'll now change variables from θ to x, and turn to the computation of the derivative of $f(x) = \sin x$. By the addition formula for the sine (see Problem 1.24),

$$\begin{aligned}\sin(x + \Delta x) - \sin x &= (\sin x)(\cos\Delta x) + (\cos x)(\sin\Delta x) - \sin x\\ &= (\sin x)(\cos\Delta x - 1) + (\cos x)(\sin\Delta x).\end{aligned}$$

Therefore $\frac{\sin(x+\Delta x)-\sin x}{\Delta x} = \sin x\,\frac{(\cos\Delta x - 1)}{\Delta x} + \cos x\,\frac{\sin\Delta x}{\Delta x}$. By using $\lim\limits_{\Delta x\to 0}\frac{(\cos\Delta x-1)}{\Delta x} = 0$ and $\lim\limits_{\Delta x\to 0}\frac{\sin\Delta x}{\Delta x} = 1$,

$$\frac{d}{dx}\sin x = \lim_{\Delta x\to 0}\frac{\sin(x+\Delta x) - \sin x}{\Delta x} = \sin x\cdot\lim_{\Delta x\to 0}\frac{(\cos\Delta x - 1)}{\Delta x} + \cos x\cdot\lim_{\Delta x\to 0}\frac{\sin\Delta x}{\Delta x} = \sin x\cdot 0 + \cos x\cdot 1 = \cos x.$$

We have established that

$$\boxed{\frac{d}{dx}\sin x = \cos x}$$

The same argument (using the addition formula for the cosine) shows that

$$\boxed{\frac{d}{dx}\cos x = -\sin x}$$

A Matter of Notation: We will adopt a common notational convention and write $\sin^2 x$ to mean $(\sin x)^2$, $\cos^3 x$ to mean $(\cos x)^3$, $\tan^4 x$ to mean $(\tan x)^4$, and so on. This convention will be used for positive powers only. In particular, it will not be in force for -1, as $\sin^{-1}x$ will refer (later in Chapter 9) to the inverse sine function and not to $\frac{1}{\sin x}$. (The same is the case for the other trig functions.)

Example 7.27. Let $f(x) = \sin^2 x$. By the chain rule, $f'(x) = 2(\sin x)\frac{d}{dx}(\sin x) = 2(\sin x)(\cos x)$.

Example 7.28. Let $g(x) = \cos^2 x$. Again, by the chain rule, $g'(x) = 2(\cos x)(-\sin x) = -2(\cos x)(\sin x)$.

By a combination of Examples 7.27 and 7.28, $\frac{d}{dx}(\sin^2 x + \cos^2 x) = 0$. In view of Corollary i of the mean value theorem, this means that $\sin^2 x + \cos^2 x$ is a constant. (We already know that it is equal to 1.)

Example 7.29. Apply the quotient rule to $\tan x = \frac{\sin x}{\cos x}$ to show that

$$\boxed{\frac{d}{dx}\tan x = \sec^2 x}$$

Example 7.30. Check that $\frac{d}{dx}(\cos x)^{-1} = -(\cos x)^{-2}(-\sin x) = \frac{\sin x}{\cos^2 x} = \frac{\sin x}{\cos x} \cdot \frac{1}{\cos x}$, and conclude that

$$\boxed{\frac{d}{dx}\sec x = (\sec x)(\tan x)}$$

In the proofs of the formulas $\frac{d}{dx}\sin x = \cos x$ and $\frac{d}{dx}\cos x = -\sin x$, there is no restriction on x. It follows that both $y = \sin x$ and $y = \cos x$ are differentiable for all real numbers x. This means in turn that the formulas $\frac{d}{dx}\tan x = \sec^2 x$ and $\frac{d}{dx}\sec x = (\sec x)(\tan x)$ hold for all x in the domains of these two functions. Notice that in each case the domain is the set of all x for which $\cos \neq 0$, or all x except $x = \pm\frac{\pi}{2}, \pm\frac{3\pi}{2}, \pm\frac{5\pi}{2}, \ldots$.

7.8 UNDERSTANDING FUNCTIONS

Turning to a study of functions in general, we'll start with some elementary properties. Consider a function $y = f(x)$ defined for all x in some interval I. We will say that

f is *increasing on I* if for any numbers x_1 and x_2 in I with $x_1 < x_2$, $f(x_2) > f(x_1)$, and that

f is *decreasing on I* if for any numbers x_1 and x_2 in I with $x_1 < x_2$, $f(x_2) < f(x_1)$.

Observe (see Figure 7.30) that if f is increasing, then the graph rises from left to right over I, and that if

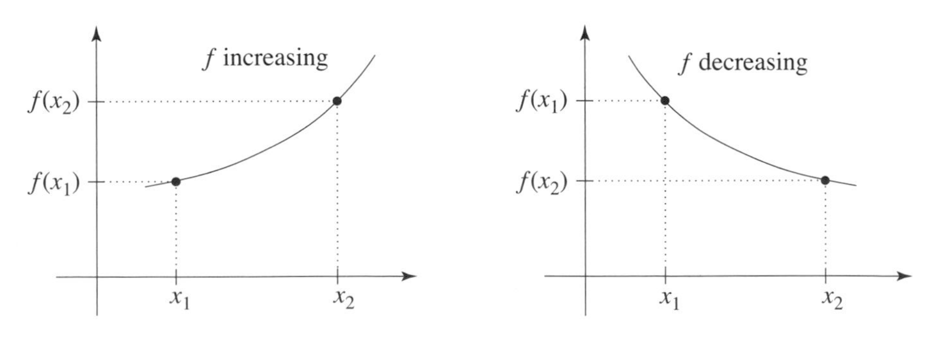

Fig. 7.30

f is decreasing, then the graph falls from left to right over I.

Suppose that the interval I is open and that f is differentiable on I. Let x_1 and x_2 be any numbers in I with $x_1 < x_2$. Since f is also continuous on I (see the beginning of Section 7.6), we can apply the mean value theorem to f and the interval $[x_1, x_2]$ to conclude that there is a number x satisfying $x_1 < x < x_2$ such that

$$f(x_2) - f(x_1) = f'(x)(x_2 - x_1).$$

Suppose that $f'(x) > 0$ for all x in I. Since $x_2 > x_1$, we know that $f'(x)(x_2 - x_1) > 0$, so that $f(x_2) > f(x_1)$. Since x_1 and x_2 were any numbers in I with $x_1 > x_2$, it follows that f is increasing on I. If, on the other hand, $f'(x) < 0$ for all x in I, then $f'(x)(x_2 - x_1) < 0$ and $f(x_2) < f(x_1)$. So f is decreasing on I. We have verified the first derivative test.

First Derivative Test. Consider a function $y = f(x)$ that is differentiable on an open interval I. If $f'(x) > 0$ for all x in I, then f is increasing on I. If $f'(x) < 0$ for all x in I, then f is decreasing on I.

Consider the function $y = f(x)$ that has the graph depicted in Figure 7.31. It is more complicated than anything we will ever encounter, but it allows for the illustration of the concepts that will be introduced next. Consider the point c_1. Observe that f is increasing immediately to the left of c_1 and decreasing immediately to the right of c_1. So there is an open interval I containing c_1 such that $f(c_1) \geq f(x)$ for all x in I. We will say that f has a *local maximum at* c_1 to describe this situation. Notice that f has local maxima at c_3, c_6, and c_9 as well. The values of f at c_1, c_3, c_6, and c_9 are maximal only "locally," meaning that for each of these numbers, there is only a segment of the x-axis (an open interval) around the number, such that $f(c_1), f(c_3), f(c_6)$, or $f(c_9)$ is the maximum value of f over that segment. The absolutely largest

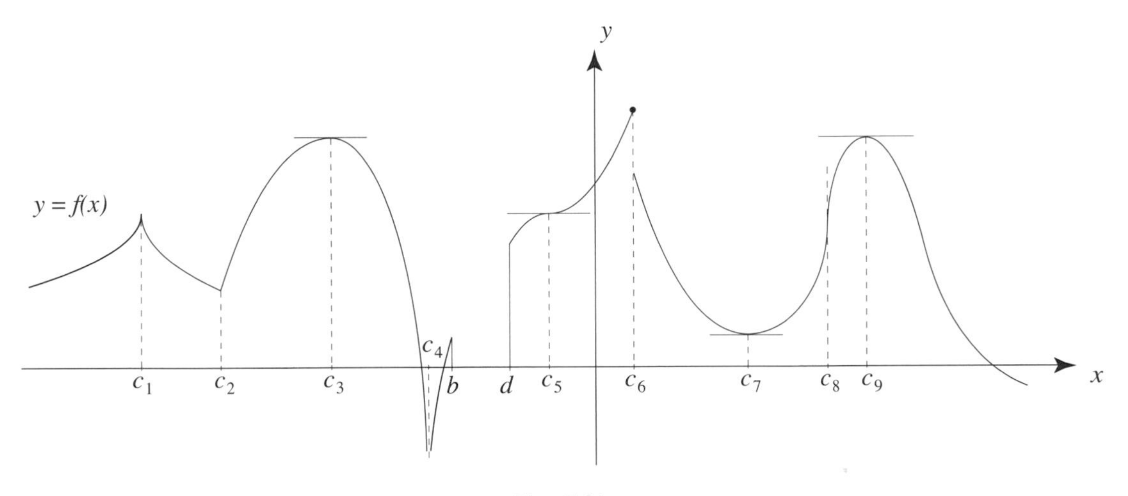

Fig. 7.31

value of the function f occurs at c_6 and is equal to $f(c_6)$. Consider the number c_2 next. Note that f is decreasing immediately to the left of c_2 and increasing immediately to the right of c_2. In this case, there is an open interval I containing c_2 such that $f(c_2) \leq f(x)$ for all x in I, and we say that f has a *local minimum at* c_2. There is also a local minimum at c_7. Since f is not defined at c_4, it has neither a local maximum nor a local minimum there. The function f has no absolutely smallest value because its graph has a "bottomless pit" at c_4. The points b and d are *endpoints* of the domain of the function. The function is defined at these points and immediately to their left or right, but not both. Notice, therefore, that they don't meet the open interval requirement for a local maximum or minimum. Therefore endpoints of a domain are *not regarded* to be local maxima or minima.

The numbers at which a given function f has a local maximum or a local minimum value are important for two reasons. They are transition points between the intervals of increase and decrease of f. They are also of direct interest if f has an absolutely largest or smallest value that needs to be determined. The key to the detection of the local maxima and minima of a function is the max-min theorem.

Max-Min Theorem. If f is differentiable at c and if f has either a local maximum or a local minimum at c, then $f'(c) = 0$.

Proof. Suppose that $f'(c)$ exists, and consider the case where f has a local maximum at c. So $f(c) \geq f(x)$ and hence $f(x) - f(c) \leq 0$ for all x in some open interval I of c. Consider any x in I with $x < c$. Since $x - c < 0$, we see that $\frac{f(x)-f(c)}{x-c} \geq 0$ for all such x. It follows that

$$\lim_{x\to c^-} \frac{f(x) - f(c)}{x - c} \geq 0.$$

Suppose x in I satisfies $x > c$. Now $x - c > 0$, and hence $\frac{f(x)-f(c)}{x-c} \leq 0$ for all such x. Therefore

$$\lim_{x\to c^+} \frac{f(x) - f(c)}{x - c} \leq 0.$$

Because f is differentiable at c, these two limits are equal. It follows that both must be equal to 0. Since $f'(c) = \lim\limits_{x\to c} \frac{f(x)-f(c)}{x-c}$ is their common value, $f'(c) = 0$. The same conclusion, for entirely similar reasons, also holds if f has a local minimum at c.

Consider a number c with the property that f is defined on some open interval containing c (so f is defined immediately to the right and left of c), but not necessarily at c itself. Then

c is a *critical number* for f, if either $f'(c) = 0$ or $f'(c)$ does not exist.

Note that an endpoint of the domain of f is not a critical number for f.

Assume that f has a local maximum or a local minimum at a number c. If f is not differentiable at c, then $f'(c)$ does not exist and c is a critical number. If f is differentiable at c, then by the max-min theorem $f'(c) = 0$, and c is a critical number also in this case. We have arrived at the following.

Critical Fact. If f has a local maximum or a local minimum at a number c, then c is a critical number for f. Therefore a strategy for finding all the numbers c at which f has either a local maximum or a local minimum is the following:

1. Find all critical numbers for f.
2. Check each critical number: Does it give a local maximum, a local minimum, or neither?

Example 7.31. What are the critical numbers for the functions $f(x) = x^2$ and $g(x) = x^3$? Since $f'(x) = 2x$ and $g'(x) = 3x^2$, both derivatives exist for all x and are equal to 0 only for $x = 0$. It follows that in each case, 0 is the only critical number. Refer to the graphs of the two functions in Figure 7.32 and observe that

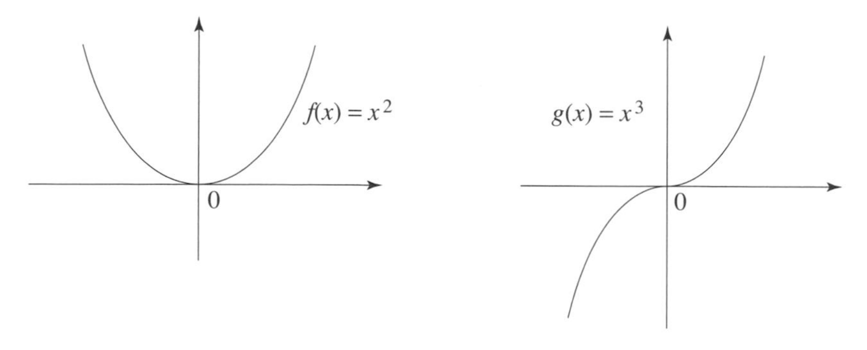

Fig. 7.32

$f(x) = x^2$ has a local minimum at 0, but that $g(x) = x^3$ has neither a local minimum nor a local maximum at $x = 0$.

Example 7.32. What are the critical numbers for the functions $f(x) = x^{\frac{1}{3}}$ and $g(x) = x^{\frac{2}{3}}$? Since $f'(x) = \frac{1}{3}x^{-\frac{2}{3}} = \frac{1}{3x^{\frac{2}{3}}}$ and $g'(x) = \frac{2}{3}x^{-\frac{1}{3}} = \frac{2}{3x^{\frac{1}{3}}}$, neither $f'(x)$ nor $g'(x)$ is ever equal to 0. However, both

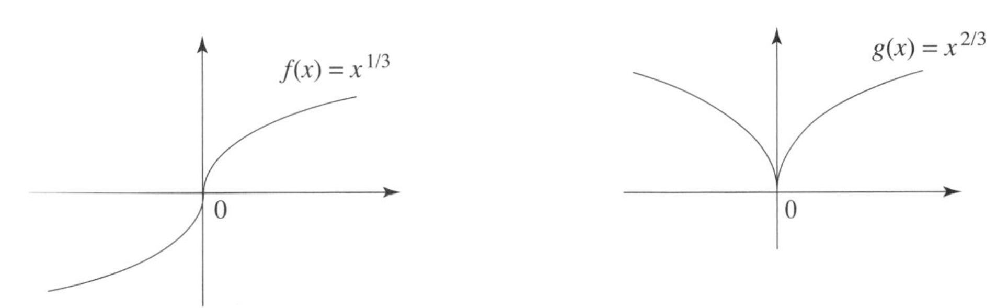

Fig. 7.33

are undefined at $x = 0$. Therefore in each case, 0 is the only critical number. Refer to the graphs of the two functions in Figure 7.33 and observe that $f(x) = x^{\frac{1}{3}}$ has neither a local maximum nor a local minimum at $x = 0$. On the other hand, $g(x) = x^{\frac{2}{3}}$ has a local minimum at $x = 0$.

Example 7.33. Consider the functions $g(x) = \frac{1}{x}$ and $h(x) = \frac{1}{x^2}$. Since $g'(x) = -x^{-2} = \frac{-1}{x^2}$, and $h'(x) = -2x^{-3} = \frac{-2}{x^3}$, both g and h have nonzero derivatives whenever $x \neq 0$. Since neither function is continuous at $x = 0$, neither is differentiable there. Since both functions are defined in an open interval of $x = 0$ with the exception of 0 itself, 0 is a critical number for both g and h. Their graphs are sketched in Figure 7.14.

Let's return to the function f of Figure 7.31. For each of the numbers c, f is defined on some open interval containing c, with the exception of c_4. In the case of c_4, the function is defined at all numbers in an open interval containing c_4 but not at c_4. The numbers c_1, c_2, c_4, c_6, and c_8 are all critical because the function f is not differentiable at these numbers. For c_1 and c_8, the graph has vertical tangents at the corresponding points $(c_1, f(c_1))$ and $(c_8, f(c_8))$. At $(c_2, f(c_2))$, there is both a left tangent and a right tangent, but their slopes are different. At c_4 and c_6, the function is not even continuous. The numbers c_3, c_5, c_7, and c_9 are also critical because at each of them the function f is differentiable with derivative equal to 0. Of the critical numbers on our list, c_1, c_3, c_6, and c_9 yield local maxima, and c_2 and c_7 yield local minima. At c_4, c_5, and c_8, there is neither a local maximum nor a local minimum. Let's suppose that Figure 7.31 captures all the drama of the graph of f (and that the graph continues its gentle downward flow both on the left and right). This means that all of its critical numbers have now been identified and that $c_1, c_2, \ldots, c_9$ is a complete list. (Since b and d are endpoints of the domain, they are not critical numbers). The critical numbers split the domain of f into the sequence

$$x < c_1,\ c_1 < x < c_2,\ c_2 < x < c_3,\ c_3 < x < c_4,\ c_4 < x < b \text{ and } d < x < c_5,\ c_5 < x < c_6,\ \ldots,\ c_9 < x$$

of open intervals. Why is it the case that the function f is differentiable over each of these open intervals with a derivative that is never equal to 0? Refer to the graph and notice that f is either increasing or decreasing over each of the intervals.

The discussion about the function of Figure 7.31 provides a general strategy for analyzing a function and its graph. Let f be any function and proceed as follows. Check where the function is not defined and determine its domain. Consider the derivative f', find all the critical numbers of f, and let $c_1 < c_2 < c_3, \ldots$ be a complete list of them. (In all cases that we will encounter, there are only finitely many such numbers. However, the function defined by Figure 7.24 has infinitely many critical numbers in the interval $[-1, 1]$.)

The critical numbers split the domain into a number of open intervals. Since the list of critical numbers is complete, the function f is differentiable on each interval with a derivative that is never 0.

Test for Increase/Decrease. Assume that the derivative f' is continuous on its domain, and let I be any of the open intervals that the critical numbers of f determine. If $f'(t) > 0$ for a single number t in I, then f is increasing on I, and if $f'(t) < 0$ for a single number t in I, then f is decreasing on I.

Let I be any of these open intervals, and suppose that $f'(t) > 0$. Let x be a number in I such that $f'(x) < 0$. Since f' is continuous on the closed interval determined by t and x, the intermediate value theorem tells us that there is a number u between t and x such that $f'(u) = 0$. So u is a critical number for f. But we know that there are no critical numbers for f in I. So $f'(x) > 0$ for all x in I, and hence f is increasing on I. The proof in the case $f'(t) < 0$ is similar.

Specific points t will be referred to as *test points*. By applying the increase/decrease test, all the intervals over which the function f is either increasing or decreasing can be identified. This information in turn tells us for each critical number whether the graph of f has a local maximum there, a local minimum there, or if neither of these two options occurs. Incidentally, the assumption that the derivative f' is continuous over each of the intervals I is met by all the functions that we will encounter. (But "pathological" examples can be constructed where this is not so.)

It is time for an example.

Example 7.34. Consider the function $f(x) = (3x^2 - 6x)^{\frac{2}{3}}$. By the chain rule,

$$f'(x) = \tfrac{2}{3}(3x^2 - 6x)^{-\frac{1}{3}}(6x - 6) = 4(x-1)(3x(x-2))^{-\frac{1}{3}} = \frac{4(x-1)}{(3x(x-2))^{\frac{1}{3}}}.$$

Notice that the critical numbers for f are 0, 1, and 2. They split the domain of f into the four subintervals

$$-\infty < x < 0,\ 0 < x < 1,\ 1 < x < 2, \text{ and } 2 < x < \infty.$$

By the continuity theorem and Example 7.10 of Section 7.3, the function $f'(x) = \frac{4(x-1)}{(3x(x-2))^{\frac{1}{3}}}$ is continuous on its domain. Pick the points $-1, \frac{1}{2}, \frac{3}{2}$, and 3 in these four intervals as test points. Since

$$f'(-1) = \frac{4(-2)}{(3(-1)(-3))^{\frac{1}{3}}} < 0,\ f'(\tfrac{1}{2}) = \frac{4(-\frac{1}{2})}{(\frac{3}{2}(-\frac{3}{2}))^{\frac{1}{3}}} > 0,\ f'(\tfrac{3}{2}) = \frac{4(\frac{1}{2})}{(\frac{9}{2}(-\frac{1}{2}))^{\frac{1}{3}}} < 0, \text{ and } f'(3) = \frac{4(2)}{(9(1))^{\frac{1}{3}}} > 0,$$

the increase/decrease test allows us to conclude that f is decreasing over $(-\infty, 0)$, increasing over $(0, 1)$, decreasing over $(1, 2)$, and increasing over $(2, \infty)$. It follows that f has local minima at 0 and 2, and a local maximum at 1. Since $f(x) = (3x^2 - 6x)^{\frac{2}{3}} = ((3x^2 - 6x)^{\frac{1}{3}})^2$, $f(x) \geq 0$ for all x. It follows that f attains its absolute minimum of 0 at both 0 and 2. It does not have an absolute maximum, since $f(x)$ increases without bound if x increases (or $-x$ decreases) without bound.

Continue to let $y = f(x)$ be any function. If there is a number c such that $f(c) \leq f(x)$ for *all* x in the domain of f, then $f(c)$ is called *the minimum value of* f. We then say that f attains its minimum value at c. Similarly, there may be a number d such that $f(d) \geq f(x)$ for *all* x in the domain of f. In this case, $f(d)$ is called *the maximum value of* f, and we say that f attains its maximum value at d. The maximum value of f is the largest of all the y-coordinates that a point on the graph of f can have, and the minimum value is the smallest. So the maximum value is the y-coordinate of the highest point on the graph, and the minimum value is the y-coordinate of the lowest. It is possible for the maximum vale of f to be attained at more than one number (the graph could have more than one highest point), and similarly for the minimum value (the graph could have more than one lowest point). It has already been observed that such maximum and minimum values may or may not exist.

It was pointed out in Section 7.3 that if $y = f(x)$ is a continuous function with domain a closed interval $[a, b]$, then there exist numbers c and d in $[a, b]$ such that

$$f(c) \le f(x) \le f(d), \text{ for all } x \text{ in } [a,b].$$

This tells us that in any such situation the existence of both a maximum and a minimum value is guaranteed.

Example 7.35. Return to the function $f(x) = (3x^2 - 6x)^{\frac{2}{3}}$ but restrict its domain to $[-1,3]$. It follows from the study of Example 7.34 that $f(0) = f(2) = 0$ is the minimum value of this function, and that its maximum value can only occur at the local maximum $x = 1$ or at the endpoints $x = -1, 3$ of the domain. Notice that $f(1) = (-3)^{\frac{2}{3}} = 9^{\frac{1}{3}}$, $f(-1) = (3+6)^{\frac{2}{3}} = 9^{\frac{2}{3}}$, and $f(3) = (27-18)^{\frac{2}{3}} = 9^{\frac{2}{3}}$. Since $9^{\frac{1}{3}} \approx 2.08$, the prize for the maximum goes to $f(-1) = f(3) = 9^{\frac{2}{3}} \approx 4.33$.

There is one property of a function that still needs to be explored before we can sketch its graph in definitive terms. Consider a function f that is differentiable on some open interval I. Suppose that the graph of f over I has one of the two shapes sketched in Figure 7.34. While f is increasing in both cases, it is clear that case (a) is fundamentally different from case (b). The difference is easy to see. In the first case, the increase is modest initially, gains speed, and then becomes dramatic. In the second case, the increase

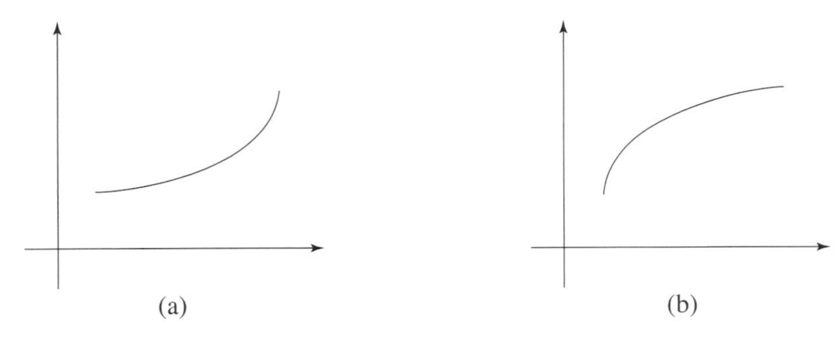

Fig. 7.34

is rapid at first, quickly levels off, and then proceeds slowly. Both the function f and and its graph are said to be *concave up* on I in the first situation and *concave down* on I in the second.

The key to the mathematical analysis of the two situations is the behavior of the slopes of the tangent lines to the graphs. Going from left to right, the slopes of the tangents are increasing in case (a) and decreasing in case (b). Equivalently, since the derivative measures slope, f' is an increasing function on I in case (a) and a decreasing function on I in case (b). For example, a look at the graph of Figure 7.31 tells us that over the interval (c_2, c_4) the slopes of the tangents are positive but decreasing as x varies from c_2 to $x = c_3$, and that (now negative) they continue to decrease thereafter. So the function f is concave down on (c_2, c_4). Over the interval (c_6, c_8), on the other hand, the slopes of the tangents increase (while negative) down to $x = c_7$, and then (now positive) continue to increase thereafter. So f is concave up on (c_6, c_8).

Suppose that f' is differentiable. The derivative $(f')'$ is *the second derivative* of f. The symbols

$$f''(x) = \frac{d}{dx}f'(x) = \frac{d}{dx}\Big(\frac{d}{dx}f\Big) = \frac{d^2}{dx^2}f$$

all refer to the second derivative of f. An application of the first derivative test to the derivative f' provides the concavity test.

Concavity Test. If $f''(x) > 0$ for all x in I, then f' is increasing on I and hence f is concave up on I. Similarly, if $f''(x) < 0$ for all x in I, then f' is decreasing on I and hence f is concave down on I.

Let's look at this more concretely for the functions $f(x) = x^2$ and $g(x) = x^3$ on the interval $(-4,4)$. Since $f''(x) = \frac{d}{dx}f'(x) = \frac{d}{dx}(2x) = 2$ is positive throughout $(-4,4)$, $f(x) = x^2$ is concave up on $(-4,4)$. Since $g''(x) = \frac{d}{dx}g'(x) = \frac{d}{dx}(3x^2) = 6x$, $g''(x)$ is negative over $(-4,0)$ and positive over $(0,4)$. It follows that $g(x) = x^3$ is concave down over $(-4, 0)$ and concave up over $(0, 4)$. A look at Figure 7.32 confirms these conclusions.

Any point $(c, f(c))$ on the graph of a function f with the property that f is concave in the one sense to the one side and in the opposite sense to the other side is called a *point of inflection*. So $(0, 0)$ is a point of inflection for the function $g(x) = x^3$. On the other hand, $f(x) = x^2$ has no point of inflection. Another look at Figure 7.31 tells us that $(c_2, f(c_2))$, $(c_5, f(c_5))$, and $(c_8, f(c_8))$ are all inflection points for f. Are these all of them? Or is there one more? Where are the inflection points for the two functions of Example 7.33?

Suppose that the graph of a function f has a horizontal tangent at the point $(c, f(c))$. A moment's thought shows that if the graph is concave down at $(c, f(c))$, then it must have a local maximum at c, and if it is concave up at $(c, f(c))$, then it must have a local minimum at c. This conclusion is known as the second derivative test.

Second Derivative Test. Let f be a function that is differentiable and with a differentiable derivative f' on an open interval containing a number c. Suppose that $f'(c) = 0$. If $f''(c) < 0$, then f has a local maximum at c. If $f''(c) > 0$, then f has a local minimum at c.

Example 7.36. Let $f(x) = x^4 - 4x^2 + 16$. Because $f'(x) = 4x^3 - 8x = 4x(x^2 - 2)$, observe that $f'(0) = 0$, $f'(\sqrt{2}) = 0$, and $f'(-\sqrt{2}) = 0$. Since $f''(x) = 12x^2 - 8$, we get $f''(0) = -8$, $f''(\sqrt{2}) = 24 - 8 = 16$, and $f''(-\sqrt{2}) = 24 - 8 = 16$. Therefore, by the second derivative test, f has a local minimum at both $\sqrt{2}$ and $-\sqrt{2}$, and a local maximum at 0.

Notice that the second derivative test says nothing when $f''(c) = 0$. To see why nothing can be said, analyze each of the three functions $y = x^3$, $y = x^4$, and $y = -x^4$ at $x = 0$.

The method for determining the intervals over which a function f is concave up or concave down uses the second derivative in the same way that the first derivative is used in determining the intervals of increase or decrease. In other words, the strategy is this: find the numbers for which f'' is either 0 or does not exist—these are the critical numbers for f'—and use these numbers to divide the domain of f into open intervals. Under the assumption that f'' is continuous, choose a test point in each interval. If f'' is positive at a test point, then f is concave up on the entire interval, and if f'' is negative at a test point, then f is concave down on the interval.

7.9 GRAPHING FUNCTIONS: SOME EXAMPLES

The examples that follow illustrate how the strategies developed in the previous section can be put to use in the examination of the graphs of specific functions.

Example 7.37. Consider the function $f(x) = x^3 - 3x$. Since $f'(x) = 3x^2 - 3 = 3(x^2 - 1) = 3(x + 1)(x - 1)$, $f(x)$ is differentiable for all x, and all critical numbers arise as zeros of $f'(x)$. So they are -1 and 1. They

critical numbers of $f(x) = x^3 - 3x$

x	-2	-1	0	1	2
sign of $f'(x)$:	$3(-3)(-1)$ +		$3(-1)(1)$ −		$3(1)(3)$ +
$f(x)$ is	increasing		decreasing		increasing

Table 7.1

divide the domain of f into the three intervals: $(-\infty, -1)$, $(-1, 1)$, and $(1, \infty)$. Take the test points -2 in $(-\infty, -1)$, 0 in $(-1, 1)$, and 2 in $(1, \infty)$. Since $f'(-2) = 3(-1)(-3) = 9$ is positive, f is increasing throughout the interval $(-\infty, -1)$. Since $f'(0) = -3$ is negative, f is decreasing throughout $(-1, 1)$. Finally, because $f'(2) = 9 > 0$, we know that f is increasing throughout $(1, \infty)$. Table 7.1 summarizes these findings. Notice that f has a local maximum at -1 and a local minimum at 1.

After plotting a few relevant points, we can sketch the graph of f. This is done in Figure 7.35. Since $f'(x) = 3x^2 - 3$, we see that $f''(x) = 6x$. It follows that $f''(x) < 0$ for all $x < 0$, and $f''(x) > 0$ for all

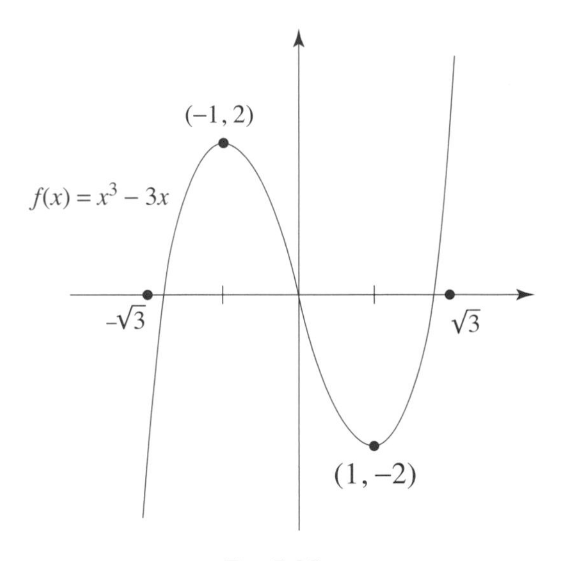

Fig. 7.35

$x > 0$. Therefore the graph is concave down to the left of the origin and concave up to the right of the origin. The origin $(0, f(0)) = (0, 0)$ is a point of inflection.

Example 7.38. Analyze the function $f(x) = 3x^4 - 4x^3 - 36x^2 + 3$. By routine computations,

$$\begin{aligned} f'(x) &= 12x^3 - 12x^2 - 72x = 12x(x^2 - x - 6) \\ &= 12x(x-3)(x+2). \end{aligned}$$

Therefore f is differentiable for all x. It follows that the critical numbers are those for which $f'(x) = 0$. So they are -2, 0, and 3. They divide the domain of f into the intervals $(-\infty, -2), (-2, 0), (0, 3)$, and $(3, \infty)$. Take $-3, -1, 1$, and 4 as test points. Since $f'(-3) = (-36)(-6)(-1) < 0$, f is decreasing throughout the interval $(-\infty, -2)$. The fact that $f'(-1) = (-12)(-4)(1) > 0$ tells us that f is increasing throughout $(-2, 0)$, and $f'(2) = 24(-1)(4) < 0$ means that f is decreasing throughout (0,3). Finally, since $f'(4) = 48(1)(6) > 0$,

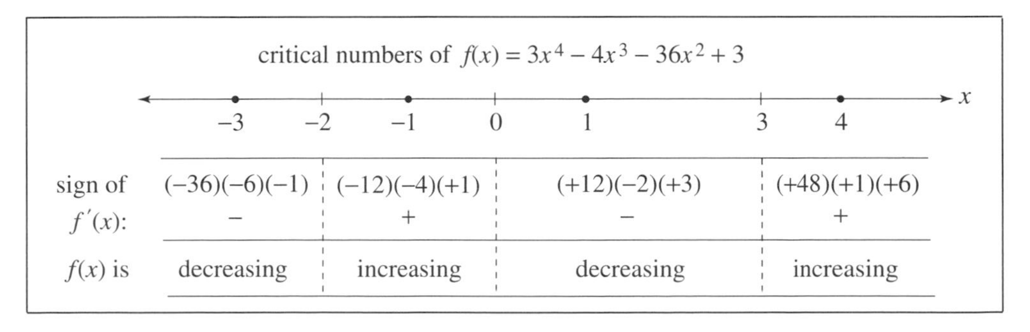
critical numbers of $f(x) = 3x^4 - 4x^3 - 36x^2 + 3$

x: -3, -2, -1, 0, 1, 3, 4

sign of $f'(x)$:	$(-36)(-6)(-1)$ $-$	$(-12)(-4)(+1)$ $+$	$(+12)(-2)(+3)$ $-$	$(+48)(+1)(+6)$ $+$
$f(x)$ is	decreasing	increasing	decreasing	increasing

Table 7.2

we see that f increases throughout $(3, \infty)$. Table 7.2 summarizes this information. The function has a local maximum at 0 and local minima at -2 and 3.

We turn to the matter of concavity. Because $f'(x) = 12x^3 - 12x^2 - 72x$, we see that

$$f''(x) = 36x^2 - 24x - 72 = 12(3x^2 - 2x - 6) = 36(x^2 - \tfrac{2}{3}x - 2).$$

To find the numbers for which $f''(x) = 0$, apply the quadratic formula to $x^2 - \frac{2}{3}x - 2$ to get

$$x = \frac{\frac{2}{3} \pm \sqrt{\frac{4}{9} + 8}}{2} = \frac{\frac{2}{3} \pm \sqrt{\frac{4+72}{9}}}{2} = \tfrac{1}{2}(\tfrac{2}{3} \pm \tfrac{\sqrt{76}}{3}) = \tfrac{1}{2}(\tfrac{2}{3} \pm \tfrac{2\sqrt{19}}{3}) = \tfrac{1 \pm \sqrt{19}}{3}.$$

Because $f''(x) = 36(x^2 - \frac{2}{3}x - 2)$ exists for all x, we know that $\frac{1-\sqrt{19}}{3}$ and $\frac{1+\sqrt{19}}{3}$ are the only critical numbers for f'. (Since $\sqrt{19} \approx 4.36$, these numbers are approximately -1.12 and 1.79, respectively.) The domain of f' is split into the intervals $(-\infty, \frac{1-\sqrt{19}}{3})$, $(\frac{1-\sqrt{19}}{3}, \frac{1+\sqrt{19}}{3})$, and $(\frac{1+\sqrt{19}}{3}, \infty)$. Take the test points -3, 0, and 3, and check that $f''(-3) = 36 \cdot 9 > 0$, $f''(0) = -72 < 0$, and $f''(3) = 36 \cdot 5 > 0$. It follows that f is concave up on the intervals $(-\infty, \frac{1-\sqrt{19}}{3})$ and $(\frac{1+\sqrt{19}}{3}, \infty)$ and concave down on $(\frac{1-\sqrt{19}}{3}, \frac{1+\sqrt{19}}{3})$. The

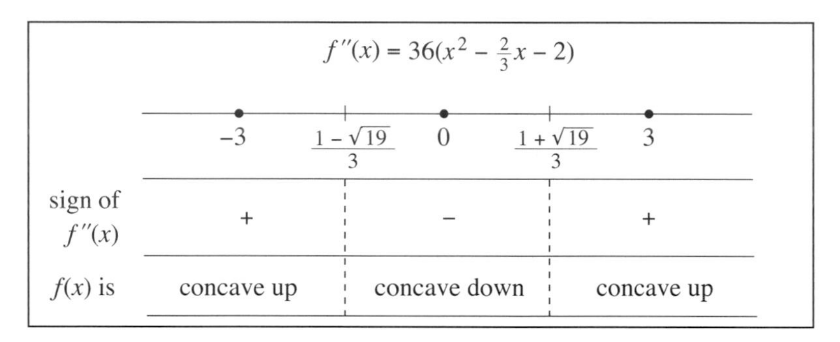

$f''(x) = 36(x^2 - \frac{2}{3}x - 2)$

	-3	$\frac{1-\sqrt{19}}{3}$	0	$\frac{1+\sqrt{19}}{3}$	3
sign of $f''(x)$	+		−		+
$f(x)$ is	concave up		concave down		concave up

Table 7.3

points on the graph with x-coordinates $\frac{1-\sqrt{19}}{3}$ and $\frac{1+\sqrt{19}}{3}$ are both points of inflection. Table 7.3 summarizes our conclusions about the concavity of the function, and the graph of Figure 7.36 incorporates all the infor-

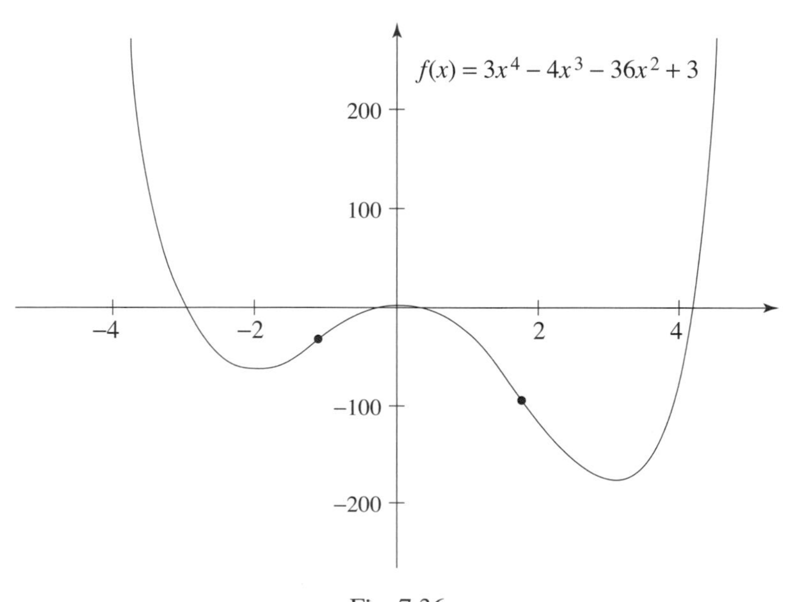

Fig. 7.36

mation that has been developed.

Example 7.39. Study $g(x) = x^{\frac{1}{3}}(x^2 - 14) = x^{\frac{7}{3}} - 14x^{\frac{1}{3}}$. Using the power rule and some algebra, we get

$$g'(x) = \tfrac{7}{3}x^{\frac{4}{3}} - \tfrac{14}{3}x^{-\frac{2}{3}} = \frac{7x^2}{3x^{\frac{2}{3}}} - \frac{14}{3x^{\frac{2}{3}}} = \frac{7x^2-14}{3x^{\frac{2}{3}}} = \frac{7(x^2-2)}{3x^{\frac{2}{3}}}.$$

Notice that both types of critical numbers occur. The derivative is zero at $-\sqrt{2}$ and $\sqrt{2}$, and the derivative is not defined at 0. Therefore the critical numbers are $-\sqrt{2}$, 0, and $\sqrt{2}$. They split the domain of g into the intervals

$$(-\infty, -\sqrt{2}), (-\sqrt{2}, 0), (0, \sqrt{2}), \text{ and } (\sqrt{2}, \infty).$$

Taking the test points -2, -1, 1, and 2 and proceeding as in the earlier examples, we get the information

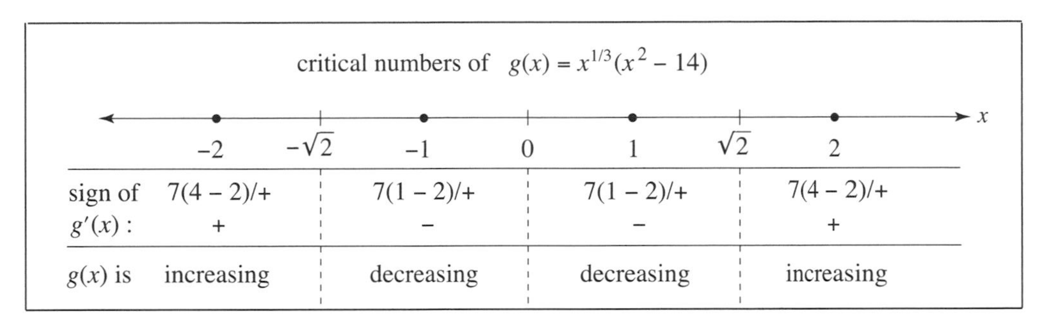

critical numbers of $g(x) = x^{1/3}(x^2 - 14)$

	-2	$-\sqrt{2}$	-1	0	1	$\sqrt{2}$	2
sign of $g'(x)$:	$7(4-2)/+$ $+$		$7(1-2)/+$ $-$		$7(1-2)/+$ $-$		$7(4-2)/+$ $+$
$g(x)$ is	increasing		decreasing		decreasing		increasing

Table 7.4

of Table 7.4. There is a local maximum at $-\sqrt{2}$ and a local minimum at $\sqrt{2}$.

Rewrite $g'(x)$ as $g'(x) = \frac{7}{3}x^{\frac{4}{3}} - \frac{14}{3}x^{-\frac{2}{3}}$. Combining the power rule with some algebra, we get

$$g''(x) = \tfrac{28}{9}x^{\frac{1}{3}} + \tfrac{28}{9}x^{-\frac{5}{3}} = \tfrac{28}{9}(x^{\frac{1}{3}} + x^{-\frac{5}{3}}) = \tfrac{28}{9}\Big(\frac{x^{\frac{6}{3}}+1}{x^{\frac{5}{3}}}\Big) = \tfrac{28}{9}\Big(\frac{x^2+1}{x^{\frac{5}{3}}}\Big).$$

Because $x^2 + 1$ is never zero and $x = 0$ is the only number for which $g''(x)$ is not defined, it follows that 0 is the only critical number for g'. Since $x^{\frac{5}{3}} = \sqrt[3]{x^5}$, observe that $g''(x) < 0$ when $x < 0$, and $g''(x) > 0$ when $x > 0$. Therefore the graph of $g(x)$ is concave down to the right of the origin $(0, 0)$ and concave up

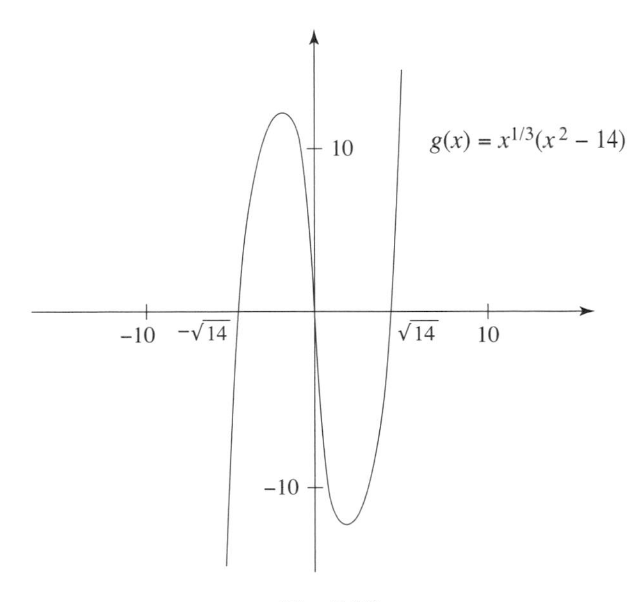

Fig. 7.37

to the left. The origin is a point of inflection. Plot a few points and use the information that was derived to see that the graph of g is as sketched in Figure 7.37.

The graph of Figure 7.36 departs from the practice of taking the same unit of length for both the x- and y-axes. Had this not been done, the y-coordinates $f(1) = -34$ and $f(-1) = -26$ would already have been off the charts, and those of the two points of inflection even more so.

It is useful to know that there are websites that feature sophisticated graphing calculators. The site

https://www.desmos.com/calculator

is a good example.

Up to now we have studied algebraic and trigonometric functions. Periodic and repetitive phenomena often require the trigonometric functions for their definitive explanation. The quantitative analysis of other phenomena however—for instance, those involving trends, and in particular processes of growth or decline—requires different functions. Of these, the exponential and logarithm functions are the most important. Leonhard Euler in his treatise *Introductio in Analysin Infinitorum* (1748) was first to treat them comprehensively. The next two sections introduce and study the exponential functions, and the closely related logarithm and hyperbolic functions.

7.10 EXPONENTIAL FUNCTIONS

The word "exponent" is the composite of the Latin *ex* (out of) and *ponere* (place). The literal meaning is to make something visible, obvious, or clear. This seems to be what happens when the index is raised "out of" the line. The English word "expound" has the same roots and means to make clear. As a mathematical term, exponent was first introduced in 1544 by the German mathematician and Lutheran cleric Michael Stifel (1487-1567) in his book *Arithmetica Integra.*

We already know that $3^1 = 3$, $3^2 = 3 \cdot 3 = 9$, and $3^0 = 1$. Recall also that the square root of 3, written $\sqrt{3} = 3^{\frac{1}{2}}$, is that number whose square is 3. Similarly, for any positive integer n, the nth root of three, denoted by $\sqrt[n]{3} = 3^{\frac{1}{n}}$, is that number which when raised to the nth power is 3. We know also that $3^{\frac{2}{3}} = (3^{\frac{1}{3}})^2 = (3^2)^{\frac{1}{3}}$. More generally, for any positive rational number $r = \frac{m}{n}$ (where m and n are both positive integers),

$$3^r = 3^{\frac{m}{n}} = (3^{\frac{1}{n}})^m = (3^m)^{\frac{1}{n}}.$$

If r is a negative rational number, then $-r$ is positive, and 3^r is defined to be equal to $\frac{1}{3^{-r}}$. For example,

$$3^{-\frac{5}{4}} = \frac{1}{3^{\frac{5}{4}}} = \frac{1}{\left(3^{\frac{1}{4}}\right)^5} = \frac{1}{\left(3^5\right)^{\frac{1}{4}}}.$$

Suppose now that x is a number that is not rational. For example, take $x = \sqrt{2}$ or $x = \pi$. What does 3^x mean then? In particular, what is the meaning of $3^{\sqrt{2}}$ or 3^{π}? Here is how we can give meaning to these expressions. Start with the decimal expansion

$$\sqrt{2} = 1.414213562\ldots.$$

Observe that this expansion gives rise to the sequence of rational numbers

$$1,\ 1.4 = \tfrac{14}{10},\ 1.41 = \tfrac{141}{100},\ 1.414 = \tfrac{1414}{1000},\ 1.4142 = \tfrac{14{,}142}{10{,}000},\ 1.41421 = \tfrac{141{,}421}{100{,}000},\ldots$$

which closes in on, or *converges to*, $\sqrt{2}$. Raising 3 to each of these terms makes sense and provides the sequence

$$3^1 = 3,\ 3^{1.4} = 4.655537\ldots,\ 3^{1.41} = 4.706965\ldots,\ 3^{1.414} = 4.727695\ldots,\ 3^{1.4142} = 4.728734\ldots,$$

$$3^{1.41421} = 4.728786\ldots,\ 3^{1.414213} = 4.728801\ldots,\ 3^{1.4142135} = 4.728804\ldots, \text{ and so on.}$$

It is apparent (at least intuitively) that this sequence closes in on some number that has a decimal expansion of the form 4.72880... . We will also say that the sequence *converges* to 4.72880... or *has the limit* 4.72880... to describe what is going on. This number is what $3^{\sqrt{2}}$ is defined to mean. Its decimal expansion can be computed as far out as a particular context might require. For example, up to 14 decimal place accuracy, $3^{\sqrt{2}} = 4.72880438783741\ldots$.

The same thing can be done with 3^{π}. Start with $\pi = 3.141592653\ldots$. Consider the sequence

$$3^3 = 27,\ 3^{3.1} = 30.135326\ldots,\ 3^{3.14} = 31.489136\ldots,\ 3^{3.141} = 31.523749\ldots,\ 3^{3.1415} = 31.541070\ldots,$$
$$3^{3.14159} = 31.544189\ldots,\ 3^{3.141592} = 31.544258\ldots, 3^{3.1415926} = 31.544278\ldots,$$

and define 3^{π} to be the limit of this sequence. With an accuracy of the first 14 decimal places, $3^{\pi} = 31.54428070019754\ldots$. In this way, the number 3^x can be defined for any real number x.

The rule $f(x) = 3^x$ determines a function. It is an increasing function. Its graph rises slowly through negative x and then rapidly through positive x. The general shape of the graph is shown in Figure 7.38.

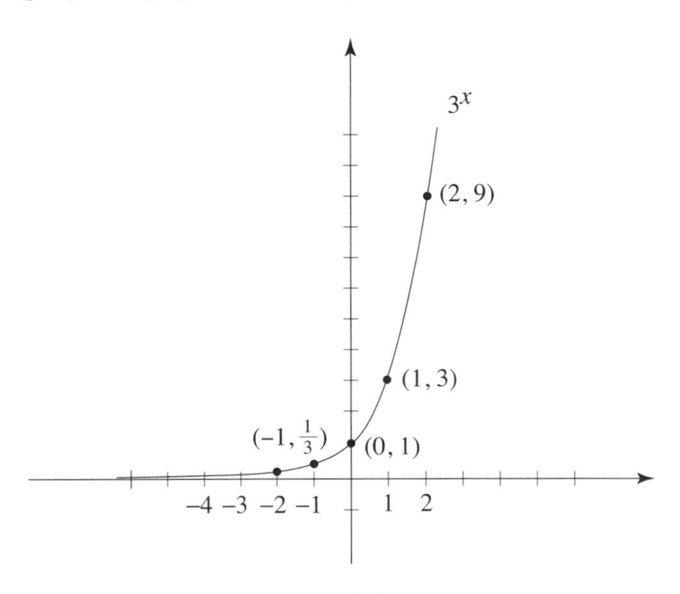

Fig. 7.38

The same strategy can be used to define a^x for any positive constant a. Namely, start with the decimal expansion of x to obtain a sequence of rational numbers $r_1, r_2, r_3, r_4, \ldots$ that converges to x. In mathematical shorthand, this is expressed by writing $x = \lim_{n\to\infty} r_n$. Then define

$$a^x = \lim_{n\to\infty} a^{r_n}.$$

Notice that this definition does not work for $a = 0$ or $a < 0$. For example, $0^{-1} = \frac{1}{0}$ and $(-3)^{\frac{1}{2}} = \sqrt{-3}$ are not defined. The rule

$$f(x) = a^x$$

defines a function for any constant $a > 0$. These are the *exponential functions*. All have the set of real numbers as their domain and all are continuous on their domains. For $a > 1$, $f(x) = a^x$ is an increasing function. The larger the a, the more rapid the increase. When $a = 1, f(x) = a^x = 1$ for all x. For $a < 1$, $f(x) = a^x$ is a decreasing function. The closer a is to 0, the more rapid the decrease. This follows by considering the equalities

$$\frac{1}{a^x} = \frac{1^x}{a^x} = \left(\frac{1}{a}\right)^x.$$

The cases $a = \frac{1}{2}, a = \frac{1}{3}, a = 1, a = 2$, and $a = 3$ are sketched in Figure 7.39. The fact is that for $a > 1$, the values $f(x) = a^x$ increase dramatically with increasing x. This is so even for relatively small a. For

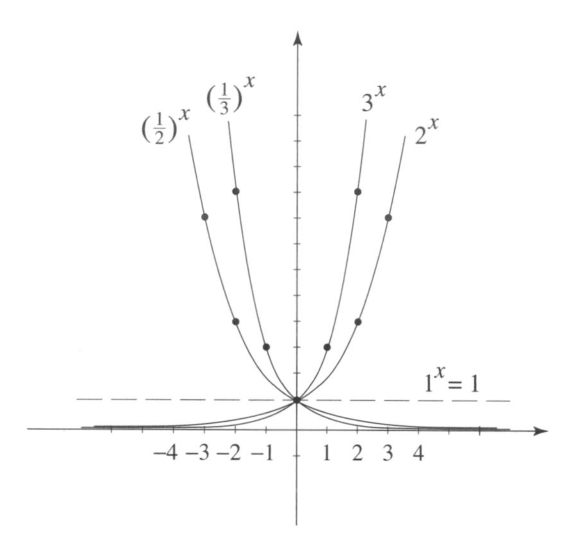

Fig. 7.39

example, for $x = 2, 4, 6, 8, 10, \ldots$, the values of 3^x (as a calculator provides them) are $9, 81, 729, 6561, 59049, .$ Even in everyday speech, we speak of an "exponential increase."

The process of raising positive numbers to powers satisfies a number of basic laws.

Rules for Exponents: For a and b any positive constants and x_1, x_2, and x any real numbers

i. $a^{x_1}a^{x_2} = a^{x_1+x_2}$ (for example, $a^2a^3 = (aa)(aaa) = a^5$),

ii. $a^{x_1}a^{-x_2} = a^{x_1-x_2} = \frac{a^{x_1}}{a^{x_2}}$ (for example, $a^2 \cdot a^{-7} = a^{-5} = a^{2-7} = \frac{a^2}{a^7}$),

iii. $(a^{x_1})^{x_2} = a^{x_1x_2}$, (for example, $(a^6)^3 = (a^6)(a^6)(a^6) = a^{18}$), and

iv. $(ab)^x = a^xb^x$ (for example, $(ab)^4 = (ab)(ab)(ab)(ab) = a^4b^4$).

We will now single out a particularly interesting exponential function. Begin with the sequence

$$\left(1 + \frac{1}{1}\right)^1, \left(1 + \frac{1}{2}\right)^2, \left(1 + \frac{1}{3}\right)^3, \left(1 + \frac{1}{4}\right)^4, \left(1 + \frac{1}{5}\right)^5, \ldots, \left(1 + \frac{1}{n}\right)^n, \ldots.$$

The first few terms are

$2^1 = 2$, $\left(\frac{3}{2}\right)^2 = 2.25000$, $\left(\frac{4}{3}\right)^3 = 2.3703...$, $\left(\frac{5}{4}\right)^4 = 2.44140...$, $\left(\frac{6}{5}\right)^5 = 2.48832...$,

$\left(\frac{101}{100}\right)^{100} = 2.70481..., \ldots, \left(\frac{1001}{1000}\right)^{1000} = 2.71692...$,

and shifting into much higher gears,

$$\left(\frac{1{,}000{,}001}{1{,}000{,}000}\right)^{1{,}000{,}000} = 2.718280469..., \ldots, \left(\frac{1{,}000{,}000{,}001}{1{,}000{,}000{,}000}\right)^{1{,}000{,}000{,}000} = 2.7182818271....$$

It should be evident that this sequence converges, but that it does so extremely slowly. The one-millionth term does not provide the correct value for the sixth decimal place, and the one-billionth term does not give the correct value for the eighth decimal place. The limiting value of the sequence is the number e, so designated in recognition of the work of the brilliant and prolific Leonhard Euler. The decimal expansion of e with an accuracy up to the first 16 decimal places, is

$$e = 2.7182818284590452\ldots.$$

Incidentally, e is not a rational number. Since $2 < e < 3$, the graph of the function e^x lies between those of 2^x and 3^x, as illustrated in Figure 7.40. Refer back to the definition of the number e and notice that it can

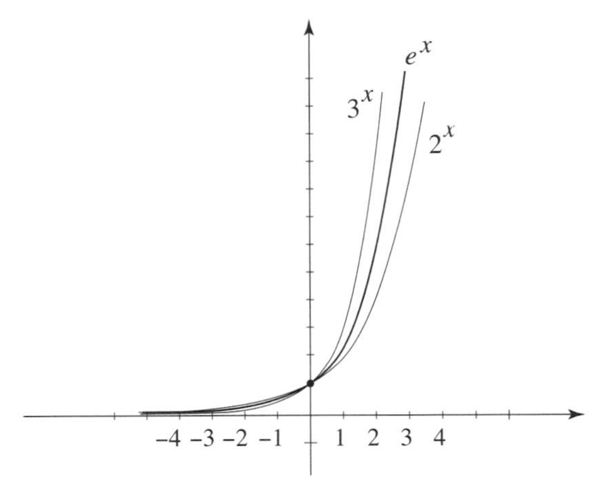

Fig. 7.40

be expressed as the limit

$$e = \lim_{n\to\infty}\left(1+\tfrac{1}{n}\right)^n = \lim_{n\to\infty}\left(\tfrac{n+1}{n}\right)^n.$$

It follows that $e \approx \left(1+\frac{1}{n}\right)^n$ for any very large positive integer n, and the larger the integer, the better this approximation. By raising both sides to the $\frac{1}{n}$ power and using exponential rule (iii), we get $e^{\frac{1}{n}} \approx 1 + \frac{1}{n}$. Again, the larger the n is, the better the approximation. Now let $\Delta x = \frac{1}{n}$ and assume that Δx is very small. So n is very large and therefore $e^{\Delta x} \approx 1 + \Delta x$. Hence $e^{\Delta x} - 1 \approx \Delta x$, and $\frac{e^{\Delta x}-1}{\Delta x} \approx 1$. Note that the smaller the Δx, the larger the n, and the better each of these approximations will be. We can conclude that if we let Δx slide to 0 through numbers of the form $\frac{1}{n}$, then

$$\lim_{\Delta x\to 0}\Big(\frac{e^{\Delta x}-1}{\Delta x}\Big) = 1.$$

But since we know how to raise positive real number to real number powers, this limit should also hold with Δx sliding to 0 through the real numbers. We'll assume this to be true and turn to the question: What is the derivative of the function $f(x) = e^x$?

Using exponential law (i) and algebra, we get

$$\begin{aligned} f'(x) &= \lim_{\Delta x\to 0}\frac{f(x+\Delta x)-f(x)}{\Delta x} = \lim_{\Delta x\to 0}\frac{e^{x+\Delta x}-e^x}{\Delta x} \\ &= \lim_{\Delta x\to 0}\frac{e^x e^{\Delta x}-e^x}{\Delta x} = \lim_{\Delta x\to 0} e^x\Big(\frac{e^{\Delta x}-1}{\Delta x}\Big) \\ &= e^x \lim_{\Delta x\to 0}\Big(\frac{e^{\Delta x}-1}{\Delta x}\Big) = e^x. \end{aligned}$$

Therefore the function $f(x) = e^x$ is its own derivative! In the notation of Leibniz,

$$\boxed{\frac{d}{dx}e^x = e^x}$$

Example 7.40. Use the chain rule to show that $\frac{d}{dx}e^{x^3} = e^{x^3}3x^2$ and $\frac{d}{dx}(e^{x^2}\cdot\sin x) = e^{x^2}2x\cdot\sin x + e^{x^2}\cos x$.

Let c be any constant, and consider the function $f(x) = e^{cx}$. By the chain rule once more, this function satisfies the equality $f'(x) = ce^{cx} = cf(x)$. We'll conclude our discussion about exponential functions by showing that if $g(x)$ is *any function* that satisfies the equation $g'(x) = cg(x)$, then $g(x) = Ae^{cx}$ for some constant A. This is easily done with one line. Take the quotient $\dfrac{g(x)}{e^{cx}}$ and differentiate it to get

$$\frac{d}{dx}\frac{g(x)}{e^{cx}} = \frac{g'(x)\cdot e^{cx} - g(x)\cdot ce^{cx}}{(e^{cx})^2} = \frac{cg(x)\cdot e^{cx} - g(x)\cdot ce^{cx}}{(e^{cx})^2} = 0.$$

By Corollary i at the end of Section 7.6, this implies that the function $\dfrac{g(x)}{e^{cx}}$ is equal to some constant A. Therefore $g(x) = Ae^{cx}$.

7.11 LOGARITHM FUNCTIONS

Closely related to the exponential function $y = e^x$ is the *logarithm* function $y = \ln x$. Start with the function $y = \frac{1}{x}$ and its graph. For any $x > 0$, define the *natural logarithm*

$$\ln x$$

as follows: for $x \geq 1$, let $\ln x$ be the number that equals the area under the graph of $y = \frac{1}{x}$, above the x axis, and between the vertical lines through 1 and x. Refer to Figure 7.41a. Notice that $\ln 1 = 0$. For x with $0 < x < 1$, define $\ln x$ to be the *negative* of the number that equals the area under the graph of $y = \frac{1}{x}$,

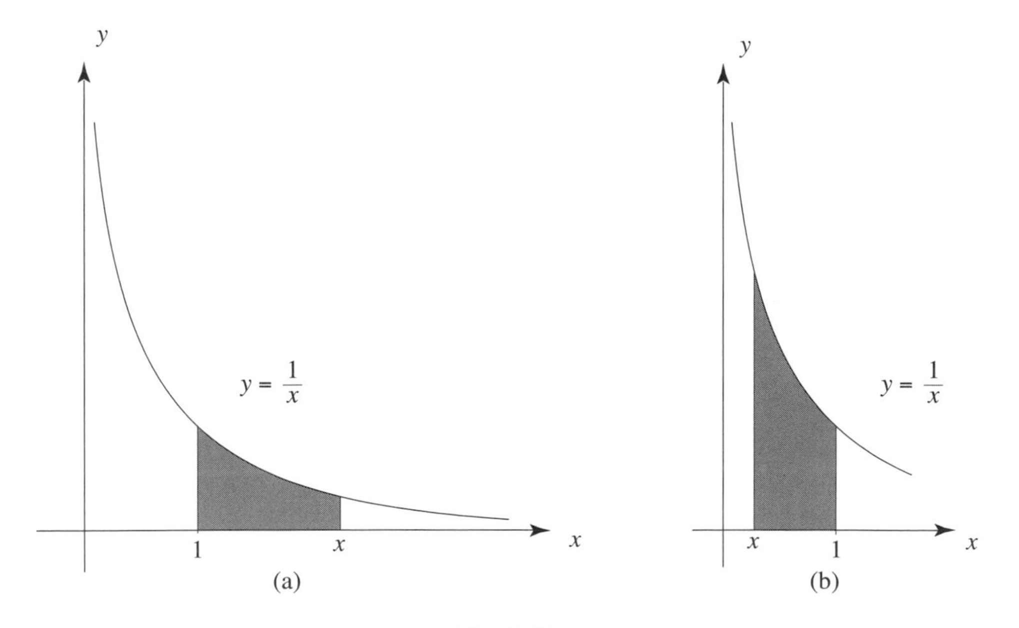

Fig. 7.41

above the x-axis, and between the vertical lines through x and 1. See Figure 7.41b.

This definition raises an important question. Assigning a number to the area of a rectangle or a triangle is simple, but how do we assign a number to the area of a region with curving boundaries? We will assume that we know how to do this. (This is a fundamental problem of integral calculus. In Section 5.6 we saw how Leibniz dealt with it, and we will return to it in Chapter 9.)

For a small positive x, the area highlighted in Figure 7.41b is relatively large, so that $\ln x$ is a relatively large negative number. The number $\ln x$ increases as x approaches 1, and $\ln 1 = 0$. For $x > 1$, $\ln x$ is positive. A look at Figure 7.41a shows that as x gets larger, the area under the curve $y = \frac{1}{x}$ increases, but

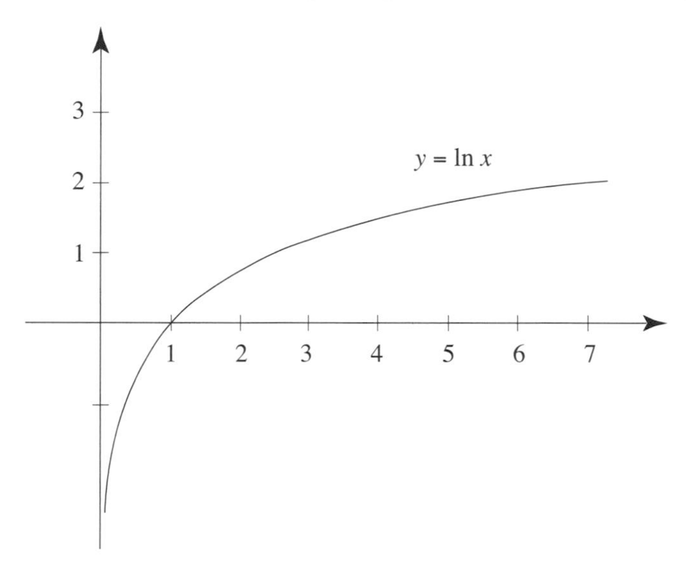

Fig. 7.42

slowly. So for $x \geq 1$, $\ln x$ is a slowly increasing function of x. The graph of $y = \ln x$ is sketched in Figure 7.42.

Example 7.41. Let $x > 1$, refer to Figure 7.41a, and use two different rectangles with base the interval $[1, x]$ to show that

$$1 - \tfrac{1}{x} < \ln x < x - 1.$$

It follows in particular that $\frac{1}{4} < \ln \frac{4}{3} < \frac{1}{3}$ and that $\frac{1}{2} < \ln 2 < 1$. The approximations $\ln \frac{4}{3} \approx 0.288$ and $\ln 2 \approx 0.693$ provide accurate values.

We will now focus our attention on the derivative of the function $g(x) = \ln x$. We'll look at the case $x > 1$ (but $x < 1$ is handled similarly). The difference

$$\ln(x + \Delta x) - \ln x$$

is the area depicted in light gray in Figure 7.43. Move the horizontal segment in the figure (up or down) over the interval from x to $x + \Delta x$ until it reaches a height v with the property that the rectangle that it determines is equal in area to $\ln(x + \Delta x) - \ln x$. Notice that $v = \frac{1}{u}$ for some u between x and Δx. (This u depends on Δx as well as x.) Since $\ln(x + \Delta x) - \ln x = v\Delta x = \frac{1}{u}\Delta x$, it follows that

$$\frac{\ln(x + \Delta x) - \ln x}{\Delta x} = \frac{1}{u}.$$

Figure 7.43 tells us that by pushing Δx to 0, u gets pushed to x. Expressed in terms of limits,

$$\frac{d}{dx}\ln x = \lim_{\Delta x \to 0} \frac{\ln(x + \Delta x) - \ln x}{\Delta x} = \frac{1}{x}.$$

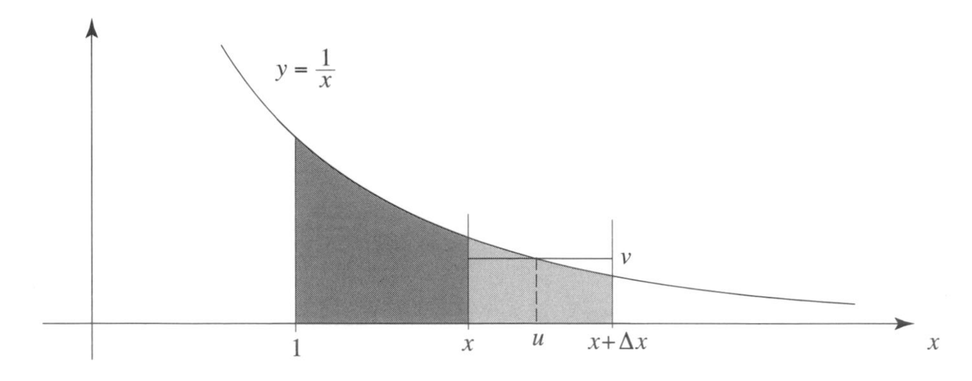

Fig. 7.43

A look at Section 6.1 tells us that this is precisely the argument that Newton devised to show that the derivative of the area function $A(x)$ of any "simple" function is that simple function.

The function $\ln x$ is defined only for $x > 0$, so that the formula

$$\frac{d}{dx}\ln x = \frac{1}{x}$$

holds only for $x > 0$. For a negative x, $-x > 0$, so that $\ln(-x)$ is defined. By the chain rule, $\frac{d}{dx}\ln(-x) = \frac{1}{-x}(-1) = \frac{1}{x}$. It follows that

$$\boxed{\frac{d}{dx}\ln|x| = \frac{1}{x}}$$

extends the earlier formula to all nonzero x.

We will next establish the special relationship that exists between the functions $f(x) = e^x$ and $g(x) = \ln x$. We first let $y = f(x)$ be any function that is defined and increasing on an interval I. So the graph of the function rises from left to right over I. See Figure 7.44 for a typical example. Suppose that c

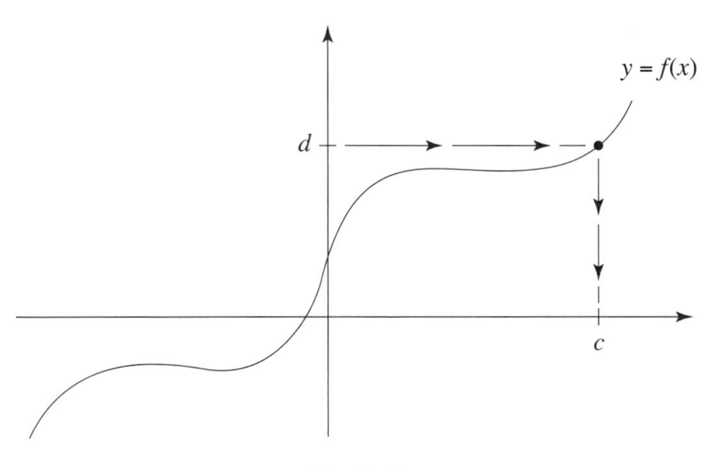

Fig. 7.44

is in I and that $f(c) = d$. The fact that f is increasing on I means that c is the only number in I such that $f(c) = d$. This means in turn that the rule $d \to c$ illustrated in the figure can be defined. It specifies a function, denoted f^{-1} and referred to as the inverse of f, that reverses the rule of f. In a similar way, the inverse f^{-1} can also be defined for a function that is decreasing over an interval I. In either situation, the *inverse function* f^{-1} of f is defined as follows:

$$f^{-1}(d) \text{ is precisely that number } c \text{ with the property that } f(c) = d.$$

So if $f(c) = d$, then $f^{-1}(d) = c$, and, conversely, if $f^{-1}(d) = c$, then $f(c) = d$. It follows that if f is defined at c and if f^{-1} is defined at d, then

$$f^{-1}(f(c)) = c \quad \text{and} \quad f(f^{-1}(d)) = d.$$

Be aware of the ambiguity with the notation $f^{-1}(x)$ for the inverse function. It is *not at all the same thing* as $\frac{1}{f(x)}$. (In mathematics it is not uncommon that the same notation has different meanings in different contexts.) Using x in place of c, and then in place of d, provides the two equations

$$f^{-1}(f(x)) = x \text{ and } f(f^{-1}(x)) = x.$$

This says that the composite of the functions f and f^{-1} as well as the composite of f^{-1} and f fix all x for which they are defined.

Let $y = f(x)$ be any differentiable function that has an inverse function $y = f^{-1}$. Applying the chain rule to the second equation above tells us that

$$\frac{d}{dx}(f(f^{-1}(x))) = f'(f^{-1}(x)) \cdot \frac{d}{dx}f^{-1}(x) = 1,$$

and therefore that

$$\boxed{\frac{d}{dx}f^{-1}(x) = \frac{1}{f'(f^{-1}(x))}}$$

It is implicit in this formula that if $f(c) = d$ and $f'(c) \neq 0$, then f^{-1} is differentiable at d and $(f^{-1})'(d) = \frac{1}{f'(c)}$.

We'll now return to the function $g(x) = \ln x$ with $x > 0$. Since this function is increasing, we know that it has an inverse $y = g^{-1}(x)$. Using the formula for the derivative of the inverse function and the fact that $g'(x) = \frac{1}{x}$, we get

$$\frac{d}{dx}g^{-1}(x) = \frac{1}{g'(g^{-1}(x))} = \frac{1}{\frac{1}{g^{-1}(x)}} = g^{-1}(x).$$

So the derivative of the function $y = g^{-1}(x)$ is the function itself! The concluding point of Section 7.10 (with $c = 1$) informs us that therefore $g^{-1}(x) = Ae^x$ for some constant A. We know that $g(1) = \ln(1) = 0$, so that $g^{-1}(0) = 1$. Hence $1 = Ae^0 = A$, and therefore

$$g^{-1}(x) = e^x.$$

This is a surprise! The functions $f(x) = e^x$ and $g(x) = \ln x$ have such very different starting points. One is defined by raising e to the variable power x, and the other is defined in terms of areas under the graph of $\frac{1}{x}$. The two contexts seem to be completely unrelated. Nonetheless, as we have just seen, the two functions are tightly connected: each is the inverse of the other. A comparison of their graphs illustrates this close relationship. The graph of one of the functions is obtained by revolving the graph of the other around the line $y = x$. Compare Figures 7.40 and 7.42.

Let's go back to the function $f(x) = a^x$ with $a > 0$. Figure 7.40 tells us that there is some constant c such that $a = e^c$. Therefore

$$f(x) = a^x = (e^c)^x = e^{cx}.$$

The fact that $a = e^c$ implies that $\ln a = c$ (since $y = e^x$ is the inverse of $y = \ln x$). Therefore

$$f(x) = a^x = e^{(\ln a)x}.$$

By applying the chain rule, we get $f'(x) = (\ln a)e^{(\ln a)x}$, and hence

$$\boxed{\frac{d}{dx}\, a^x = \ln a \cdot a^x}$$

Since a^x is always positive, it follows that $f'(x) = \ln a \cdot a^x$ is positive (for all x) when $a > 1$ and negative (for all x) when $a < 1$. Consequently, $f(x) = a^x$ is an increasing function for $a > 1$, and a decreasing function for $a < 1$. In either case, $f(x) = a^x$ has an inverse function. (If $a = 1$, then $a^x = 1^x = 1$ and there is no inverse.) The inverse function of $f(x) = a^x$, referred to as the *logarithm to the base a*, is denoted

$$f^{-1}(x) = \log_a x.$$

From the definition of inverse, we know that for any number $x > 0$, $\log_a x$ is that number y with the property that $a^y = x$. Taking $a = e$, we see that $\log_e x = \ln x$.

The tight connection between the functions $y = a^x$ and $y = \log_a x$ also means that the rules of exponentiation stated in Section 7.10 can be reformulated to provide the basic laws of logarithms.

Laws of Logarithms: Fix any two positive real numbers a and b with neither equal to 1. For any positive real numbers x_1, x_2, and x, and any real number x_3,

i. $\log_a(x_1x_2) = \log_a x_1 + \log_a x_2$,

ii. $\log_a \frac{x_1}{x_2} = \log_a x_1 - \log_a x_2$,

iii. $\log_a(x_1^{x_3}) = x_3 \log_a x_1$, and

iv. $\log_a x = \dfrac{\log_b x}{\log_b a}$.

The verification of these laws follows by "translation" from the rules for exponentiation. These verifications are all similar. They are also routine (if one is a little careful with the "bookkeeping"). We'll show that law (iv) holds. Set $\log_b x = u$ and $\log_b a = v$. It follows from the inverse relationship that $b^u = x$ and $b^v = a$. The second equality implies that $a^{\frac{1}{v}} = b$ and hence that $a^{\frac{u}{v}} = b^u = x$. From $a^{\frac{u}{v}} = x$, we get that

$$\log_a x = \frac{u}{v} = \frac{\log_b x}{\log_b a},$$

as we needed to show. Taking $b = e$ in law (iv) tells us that

$$\log_a x = \frac{\ln x}{\ln a} = \tfrac{1}{\ln a} \ln x,$$

and hence that every logarithm function is gotten by multiplying the natural log function $\ln x$ by a constant. The formula

$$\boxed{\frac{d}{dx}\log_a x = \frac{1}{\ln a} \cdot \frac{1}{x}}$$

follows directly.

We will see next that the natural logarithm $y = \ln x$ is of direct importance to differential calculus. Let g be any differentiable function. By an application of the formula $\frac{d}{dx}\ln|x| = \frac{1}{x}$ in combination with the chain rule, $\frac{d}{dx}\ln|g(x)| = \frac{1}{g(x)} \cdot g'(x) = \frac{g'(x)}{g(x)}$ for any x with $g(x) \neq 0$. So for any such x,

$$\boxed{g'(x) = \frac{d}{dx}\ln|g(x)| \cdot g(x)}$$

Example 7.42. Let $f(x) = \ln|\sec x|$. Since $f(x) = \ln|\cos x^{-1}| = -\ln|\cos x|^{-1} = -\ln|\cos x|$, we get $f'(x) = -\frac{1}{\cos x} \cdot (-\sin x) = \tan x$.

The formula derived above tells us that $g'(x)$ can be computed by first taking ln and then differentiating. This procedure, called *logarithmic differentiation*, facilitates differentiations that would otherwise be difficult and lengthy.

Example 7.43. Let $g(x) = \frac{(2x^5+6)^{100}(3x^4-7)^{90}}{(3x^3+8)^{80}(5x^2+9x)^{70}}$. After taking the natural logarithm of both sides, we get

$$\begin{aligned}\ln g(x) &= \ln(2x^5+6)^{100} + \ln(3x^4-7)^{90} - \ln(3x^3+8)^{80} - \ln(5x^2+9x)^{70} \\ &= 100\ln(2x^5+6) + 90\ln(3x^4-7) - 80\ln(3x^3+8) - 70\ln(5x^2+9x).\end{aligned}$$

By differentiating term by term,

$$\frac{d}{dx}\ln g(x) = 100\,\frac{10x^4}{2x^5+6} + 90\,\frac{12x^3}{3x^4-7} - 80\,\frac{9x^2}{3x^3+8} - 70\,\frac{10x+9}{5x^2+9x}.$$

Therefore $g'(x) = \left(100\,\frac{10x^4}{2x^5+6} + 90\,\frac{12x^3}{3x^4-7} - 80\,\frac{9x^2}{3x^3+8} - 70\,\frac{10x+9}{5x^2+9x}\right) \cdot \frac{(2x^5+6)^{100}(3x^4-7)^{90}}{(3x^3+8)^{80}(5x^2+9x)^{70}}$.

The conventional approach of differentiating $g(x)$ directly using the chain, product, and quotient rules would have been considerably more complicated. (Give it a try!)

The fact that the function $f(x) = x^r$ can be defined for any fixed real number r with domain $x > 0$, raises the question as to whether the power rule $f'(x) = rx^{r-1}$ holds in this generality. The next example tells us that the answer "yes" is another quick consequence of logarithmic differentiation.

Example 7.44. Differentiate both sides of the equality $\ln(x^r) = r\ln x$ that logarithm law (iii) provides to get $\frac{\frac{d}{dx}(x^r)}{x^r} = r\frac{1}{x}$. The power rule

$$\frac{d}{dx}(x^r) = rx^{r-1}$$

is an immediate consequence. What happens in the case of a variable exponent? Let $g(x) = x^x$ with $x > 0$. If you are tempted to write $g'(x) = x \cdot x^{x-1}$, don't. The power rule works only for constant exponents. Having no other recourse, let's try logarithmic differentiation again. Since $\ln x^x = x \cdot \ln x$, we get $\frac{d}{dx}\ln g(x) = 1 \cdot \ln x + x \cdot \frac{1}{x} = 1 + \ln x$. Therefore $g'(x) = (1 + \ln x)x^x$.

The logarithm concept was invented independently in 1614 by John Napier, a Scottsman; in 1620 by Jost Bürgi, a Swiss; and again in 1624 by Henry Briggs, an Englishman. It was clearly an idea whose time had come. The term *logarithm* was first used by Napier. What exactly Napier had in mind when he put together the two Greek words *logos* and *arithmos* does not appear to be known. The word *arithmos* means "number." The word *logos* has a broad range of meanings, extending from "word" or "speech" (as in the word "dialogue"), to "thought" and "reason" (as in "logic"), to "proportion" or "ratio." The fact is that before the invention of hand calculators (which came into wide use only in the early 1970s) and computers, logarithms were of enormous help to mathematicians and scientists as a computational tool. The French mathematician Laplace remarked in this regard that logarithms "by shortening the labors, doubled the life of the astronomer." What makes them a useful tool for computation? Observe that law (i) converts the multiplication of two numbers to an addition of two numbers, and law (ii) converts exponentiating into multiplying. So more complex operations are transformed into simpler ones. This in combination with log tables made it possible (for Kepler, for instance) to carry out laborious numerical calculations with good accuracy more simply and quickly. Incidentally, the slide rule (invented in the 17th century, used by engineers until the 1970s, and now obsolete) relied on logarithms.

Example 7.45. We will give an illustration how logarithms "shortened the labors" by computing (or at least estimating) the number $(384{,}937)^{23}$. Observe that

$$\begin{aligned}\log_{10}(384{,}937)^{23} &= 23\log_{10} 384{,}937 \;=\; 23\cdot\log_{10}(3.84937\times 10^5) = 23(\log_{10}(3.84937)+\log_{10}10^5)\\ &= 23(\log_{10}(3.84937)+5\cdot\log_{10}10) = 23(\log_{10}(3.84937)+5).\end{aligned}$$

Prepared "log" tables (readily available in the old days) contained approximations of $\log_{10}$ for numbers between 0 and 10. These could be used to determine that $\log_{10}(3.84937) \approx 0.5854$. Therefore

$$\log_{10}(384{,}937)^{23} \approx 23(5.5854) = 128.4642 = 128 + 0.4642 = \log_{10} 10^{128} + 0.4642.$$

Returning to the log tables, one would find that $0.4642 \approx \log_{10} 2.91$. Therefore

$$\log_{10}(384{,}937)^{23} \approx \log_{10} 2.91 + \log_{10} 10^{128} = \log_{10}(2.91\times 10^{128}).$$

Since $384{,}937^{23}$ and 2.91×10^{128} have approximately the same $\log_{10}$, they are approximately equal. So $384{,}937^{23} \approx 2.91\times 10^{128}$. (Try a calculator. If it is powerful enough, it will give you $2.910463\ldots \times 10^{128}$ for $384{,}937^{23}$.)

7.12 HYPERBOLIC FUNCTIONS

By combining the exponential functions e^x and e^{-x} in two simple ways, a pair of useful new functions can be defined. We will see that they behave very similarly to the trigonometric functions $\sin x$ and $\cos x$ in terms of their derivatives, the formulas that connect them, and the additional functions that they lead to.

Define

$$\sinh x = \frac{e^x - e^{-x}}{2} \quad\text{and}\quad \cosh x = \frac{e^x + e^{-x}}{2}$$

for any real number x. These functions are known as the *hyperbolic sine* and *hyperbolic cosine*, respectively. They get these names and the "suffix" h because they are related to the hyperbola $x^2 - y^2 = 1$ in the same way that $\sin x$ and $\cos x$ are related to the circle $x^2 + y^2 = 1$. (See Section 4.6 in this regard.) Their graphs are obtained by combining the graphs of $y = e^x$ and $y = e^{-x}$ of Figure 7.45. In this regard, recog-

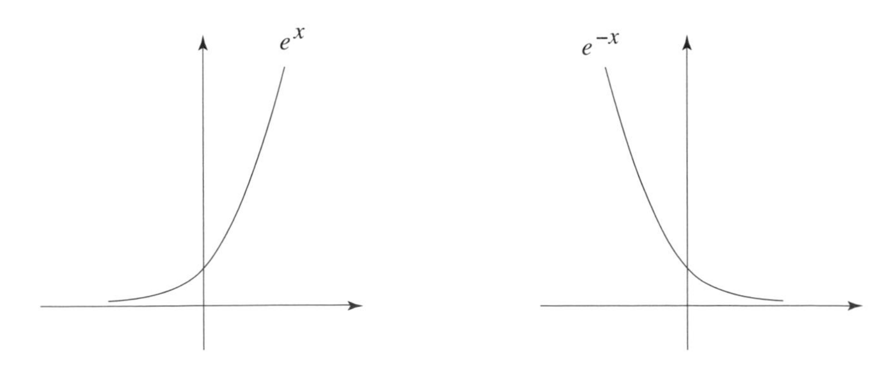

Fig. 7.45

nize that for x positive (especially large and positive), the term e^x dominates e^{-x}, and for x negative (especially large and negative), the term e^{-x} dominates e^x. This observation tells us that the graphs are as depicted in Figure 7.46 for small x. For larger x, they continue their rise (or fall) more and more rapidly.

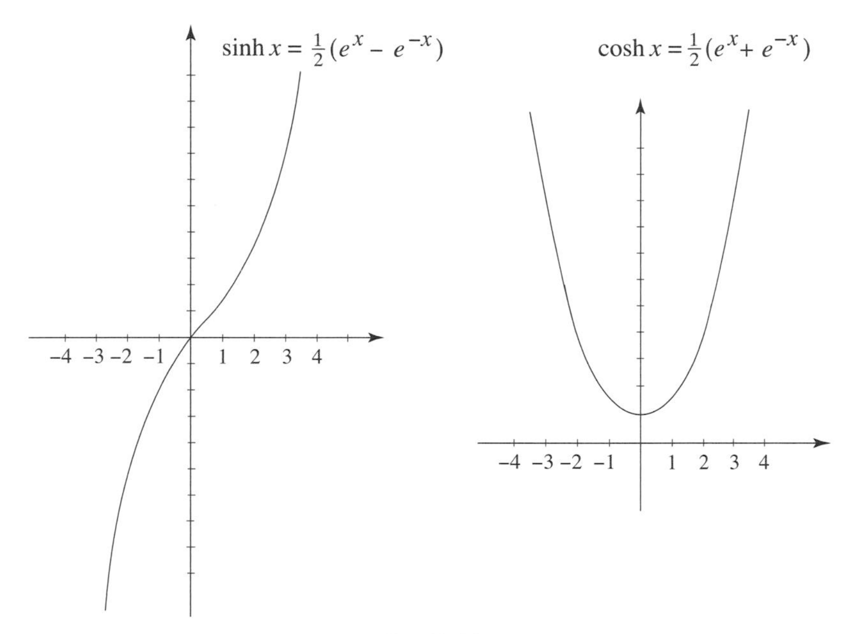

Fig. 7.46

We will see in Section 8.3 that the curve of the main cable of a suspension bridge is a parabola. The underlying reason is the fact (or approximative assumption) that the pull of the deck subjects the cable to a vertical load that is constant per unit length of deck. But what about a freely hanging flexible cable or chain under no load—a telephone wire or a power line—where this condition is not satisfied? The analysis of Section 11.12 will demonstrate that in this case, the shape of the cable is described by the graph of a function that is determined by the hyperbolic cosine.

The most basic relationship between $\sinh x$ and $\cosh x$ is

$$\cosh^2 x - \sinh^2 x = 1.$$

As in trigonometric situations, we write $\sinh^2 x$ in place of $(\sinh x)^2$, and similarly for the other hyperbolic functions to follow. The verification of this and other identities is easy, often much easier than the trigonometric versions. For instance,

$$\cosh^2 x - \sinh^2 x = \tfrac{1}{4}[(e^x)^2 + 2e^x e^{-x} + (e^{-x})^2)] - \tfrac{1}{4}[(e^x)^2 - 2e^x e^{-x} + (e^{-x})^2)] = e^x e^{-x} = 1.$$

Example 7.46. Review the basics about the hyperbola from Section 4.5, in particular the hyperbola $x^2 - y^2 = 1$. Show that the point $(\cosh u, \sinh u)$ lies on the right branch of this hyperbola for any real number u. Let u vary from $-\infty$ to ∞, and show that the point $(\cos u, \sin u)$ traces out the entire right branch of the hyperbola.

Example 7.47. Verify the following addition formulas.

i. $\sinh(x + y) = (\sinh x)(\cosh y) + (\cosh x)(\sinh y)$

ii. $\cosh(x + y) = (\cosh x)(\cosh y) + (\sinh x)(\sinh y)$

The striking similarities between the trigonometric sine and cosine and the hyperbolic sine and cosine suggest that additional *hyperbolic functions* analogous to their trigonometric counterparts, should be singled out and considered. These are the hyperbolic tangent, hyperbolic secant, cotangent, and cosecant functions, denoted by tanh, sech, coth, and csch, respectively. They are defined by

$$\tanh x = \frac{\sinh x}{\cosh x}, \quad \operatorname{sech} x = \frac{1}{\cosh x}, \quad \coth x = \frac{\cosh x}{\sinh x}, \text{ and } \operatorname{csch} x = \frac{1}{\sinh x}.$$

(The last two of these functions will not play a role in this text.) Dividing both sides of the equality $\cosh^2 x - \sinh^2 x = 1$ by $\cosh^2 x$ provides the identity

$$1 - \tanh^2 x = \operatorname{sech}^2 x.$$

Let's turn to the study of the graph of $y = \tanh x$. Observe that $\cosh x \geq 1$ for any x and that $\sinh x \geq 0$ for $x \geq 0$ and $\sinh x < 0$ for $x < 0$. Observe also that

$$\cosh x - \sinh x = e^{-x} > 0 \text{ and } \cosh x - (-\sinh x) = \cosh x + \sinh x = e^x > 0$$

for all x. In particular, $\cosh x > \sinh x$ and $\cosh x > -\sinh x$ for all x. Because $|\cosh x| = \cosh x$ and $|\sinh x| = \pm \sinh x$, it follows that $|\cosh x| > |\sinh x|$. Therefore

$$|\tanh x| < 1 \text{ for all } x.$$

So the graph of $y = \tanh x$ lies between the lines $y = 1$ and $y = -1$. Since $\lim_{x \to +\infty} e^{-x} = \lim_{x \to +\infty} \frac{1}{e^x} = 0$, it follows that $\lim_{x \to +\infty} \tanh x = \lim_{x \to +\infty} \frac{e^x - e^{-x}}{e^x + e^{-x}} = 1$. In a similar way, $\lim_{x \to -\infty} \tanh x = -1$. Therefore the lines $y = 1$

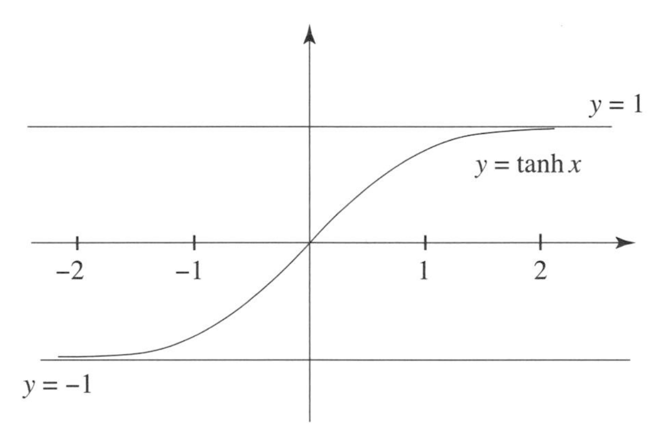

Fig. 7.47

and $y = -1$ are both horizontal asymptotes of the graph of $f(x) = \tanh x$. We will see in the next section that asymptotes are lines to which a graph converges. Figure 7.47 depicts the graph of $y = \tanh x$. Study the graph of $y = \cosh x$ and conclude that the graph of $y = \operatorname{sech} x$ has the shape shown in Figure 7.48.

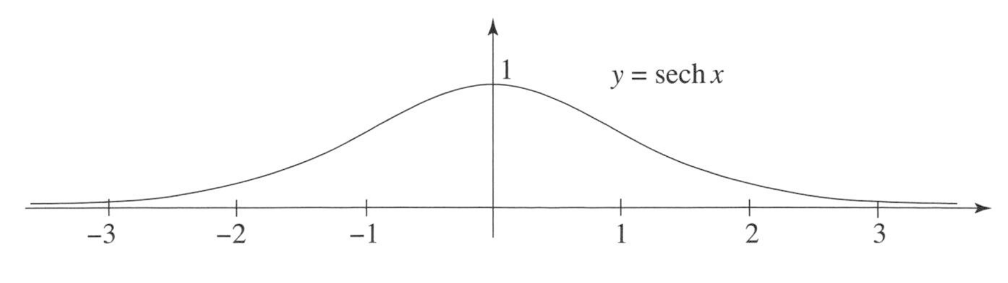

Fig. 7.48

The derivatives of the hyperbolic functions are easy to compute. Since $\frac{d}{dx} e^x = e^x$ and $\frac{d}{dx} e^{-x} = -e^{-x}$, we get $\frac{d}{dx} \sinh x = \frac{d}{dx}\left(\frac{e^x - e^{-x}}{2}\right) = \frac{e^x + e^{-x}}{2} = \cosh x$ and $\frac{d}{dx} \cosh x = \frac{d}{dx}\left(\frac{e^x + e^{-x}}{2}\right) = \frac{e^x - e^{-x}}{2} = \sinh x$. So

$$\frac{d}{dx}\sinh x = \cosh x \quad \text{and} \quad \frac{d}{dx}\cosh x = \sinh x$$

Therefore, by the quotient and chain rules, $\frac{d}{dx}\tanh x = \frac{d}{dx}\frac{\sinh x}{\cosh x} = \frac{\cosh^2 x - \sinh^2 x}{\cosh^2 x} = \frac{1}{\cosh^2 x} = \text{sech}^2 x$ and $\frac{d}{dx}\,\text{sech}\,x = \frac{d}{dx}(\cosh x)^{-1} = -(\cosh x)^{-2}\sinh x = -\text{sech}\,x \cdot \tanh x$. We have verified that

$$\frac{d}{dx}\tanh x = \text{sech}^2 x \quad \text{and} \quad \frac{d}{dx}\,\text{sech}\,x = -(\text{sech}\,x)(\tanh x)$$

Notice that $\frac{d}{dx}\tanh x$ is always positive, so that $y = \tanh x$ is an increasing function (as the graph of $y = \tanh x$ already shows). Check that $\frac{d^2}{dx^2}\tanh x = \frac{d}{dx}(1 - \tanh^2 x) = -2(\tanh x)(\text{sech}^2 x)$. Why does this confirm that the graph of $y = \tanh x$ is concave up for $x < 0$ and concave down for $x > 0$?

Example 7.48. Verify that $\frac{d^2}{dx^2}\,\text{sech}\,x = (\text{sech}\,x)(2\tanh^2 x - 1)$. Use a calculator to show that for $x = \pm 1$, the value of this expression is approximately equal to 0.1. Conclude that the x-coordinates of the points of inflection of the graph of $y = \text{sech}\,x$ are in the interval $[-1, 1]$.

7.13 FINAL COMMENTS ABOUT GRAPHS

We'll close this chapter by outlining a general strategy for the study of the graphs of functions. We'll include a few aspects that were not mentioned earlier. This discussion will be brief since it will play only a minor role in later chapters.

I. Determine the domain of the function. This set of numbers for which the function is defined, tells us over which intervals on the x-axis the function f has a graph.

II. Symmetry and shifting. Suppose that for any point (x, y) on the graph of f, the point $(-x, y)$ is also on the graph. This arises if $f(-x) = f(x)$ for all x in the domain of f. Observe that $f(x) = x^2$ is an example of such a function. Refer to Figure 7.50a. Notice that in such a situation the graph is *symmetric about the y-axis*. Another type of symmetry occurs if for any point (x, y) on the graph the point $(-x, -y)$ is also on the graph. This occurs if $f(-x) = -f(x)$ for all x in the domain of f. The functions $f(x) = x^3$ and $x^3 - 3x$ of Figure 7.49a and b, and $y = \frac{1}{x}$ of Figure 7.50b, satisfy this condition. In this case, we say that the graph of f is *symmetric about the origin*. Any line through the origin intersects the graph of such a function at two points that have the same distance from the origin. (Why does this brief discussion of symmetry not mention symmetry about the x-axis?)

The matter of shifting graphs is best illustrated by observing what happens with circles. Start with the circle $x^2 + y^2 = 4$ of radius 2 and center the origin $(0, 0)$. Replacing x by $x - 7$ and y by $y - 3$ in this equation gives the equation $(x - 7)^2 + (y - 3)^2 = 4$. This circle also has radius 2, but its center has been shifted to the point $(7, 3)$. In the same way, if we replace x and y by $x + 9$ and $y + 3$, respectively, then the center of the circle is shifted from $(0, 0)$ to $(-9, -3)$. See Figure 4.9 of Section 4.2. The same considerations apply to any graph. For example, a completion of the square procedure shows that the equation $y = x^2 + 3x + 4$ can be rearranged algebraically into the equation $y - \frac{7}{4} = (x + \frac{3}{2})^2$. It follows that the graph of $y = x^2 + 3x + 4$ is obtained by taking the graph of $y = x^2$ and shifting it without rotating it $\frac{3}{2}$ to the left and $\frac{7}{4}$ units up. Refer to Example 4.6 for the details. Let $y = f(x)$ be any function and $p > 0$ any constant. Replacing x by $x - p$ and leaving y as is shifts the graph of $y = f(x)$ by p units to the right. This is the graph of $y = f(x - p)$. In particular, if $f(x) = f(x - p)$, then these graphs are the same. We say in this case that the graph is *periodic*. The smallest $p > 0$ for which a periodic function satisfies the shift condition $f(x) = f(x - p)$ is called the *period* of f. The trigonometric functions exhibit periodic phenomena. See Section 4.6.

III. An exploration that often provides quick information about essential aspects of a graph involves the existence of *dominant terms*. Consider the function $f(x) = x^3 - 3x$ of Example 7.35. For large x (either positive or negative), x^3 is much larger than $3x$ in absolute value. So for large x, the term x^3 dominates the graph of f. For small x, on the other hand, x^3 is much smaller in absolute value than $3x$, so that $-3x$ dominates the graph. Figure 7.49 shows the graphs of x^3 and $-3x$ and highlights these dominant segments. Since the function is given by a polynomial, we know that it is continuous. So the

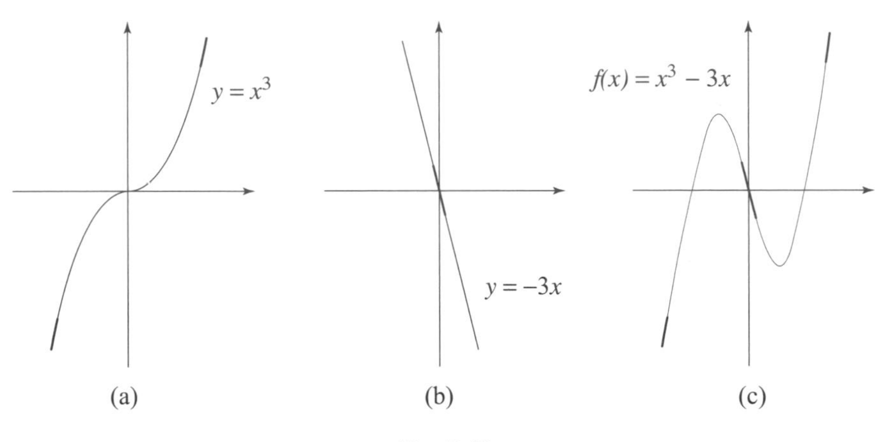

Fig. 7.49

graph needs to wind its way in one continuous piece from left to right through the three highlighted segments. The graph of $f(x) = x^3 - 3x$ is the simplest of such curves.

Consider the function $f(x) = x^2 + \frac{1}{x}$ next. Observe that when x is large in either a positive or negative sense, then the term x^2 will dominate and $\frac{1}{x}$ will contribute little. So for large x, the function f will behave like $y = x^2$. On the other hand, if x is small, then $\frac{1}{x}$ is large and x^2 is small. Now $\frac{1}{x}$ will control f, and x^2

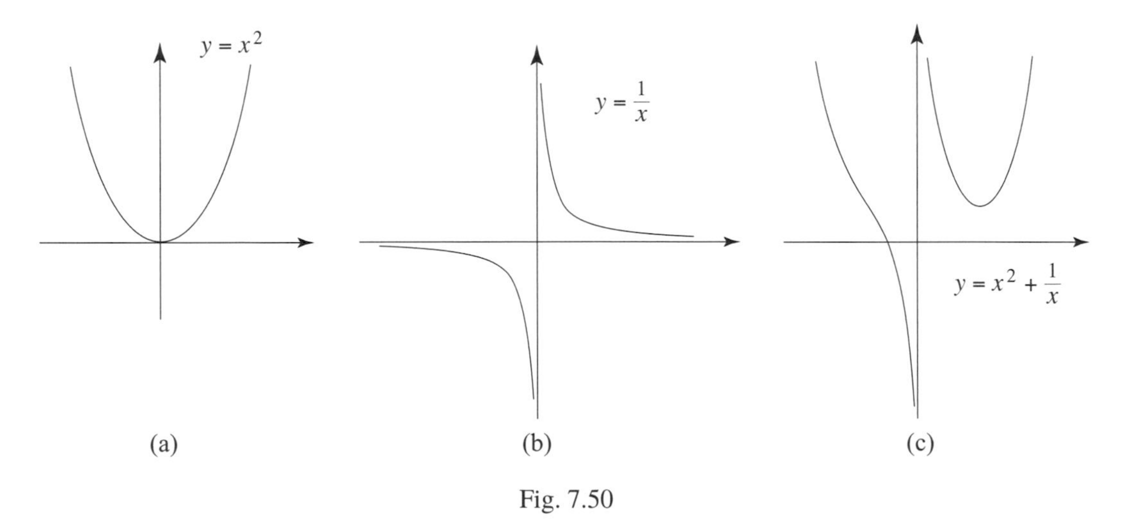

Fig. 7.50

will be negligible. It follows that the graph of f is approximated by that of $y = x^2$ when x is large and by $y = \frac{1}{x}$ when x is small. The graph of $f(x) = x^2 + \frac{1}{x}$ is sketched in Figure 7.50.

IV. Let's look at asymptotic behavior of graphs. An *asymptote* of a curve is a line to which the curve tends for large x or y. For example, Figure 7.50 tells us that the x-axis is a horizontal asymptote for the graph of $y = \frac{1}{x}$ and that the y-axis is a vertical asymptote for the graphs of both $y = \frac{1}{x}$ and $y = x^2 + \frac{1}{x}$.

How can one can test whether a function $y = f(x)$ has horizontal or vertical asymptotes, and where such lines are? In the horizontal case, the line $y = c$ is a horizontal asymptote for the graph of f if for large positive x or large negative x (or both) $f(x) \approx c$, or, more succinctly, if

$$\lim_{x \to +\infty} f(x) = c \quad \text{or} \quad \lim_{x \to -\infty} f(x) = c.$$

For example, the limits $\lim\limits_{x \to +\infty} \frac{1}{x} = 0$ and $\lim\limits_{x \to -\infty} \frac{1}{x} = 0$ confirm that the line $y = 0$ (the x-axis) is a horizontal asymptote of $y = \frac{1}{x}$.

In the vertical case, the first observation that has to be made is that if the line $x = d$ is a vertical asymptote of the graph of f, then f cannot be continuous at d. Suppose that it is. Then f is defined at d and $\lim\limits_{x \to d} f(x) = f(d)$. On the other hand, $y = f(x)$ must be large when x is near d. However, the vertical line test tells us that the graph of a function cannot satisfy both of these requirements. See Figure 7.51.

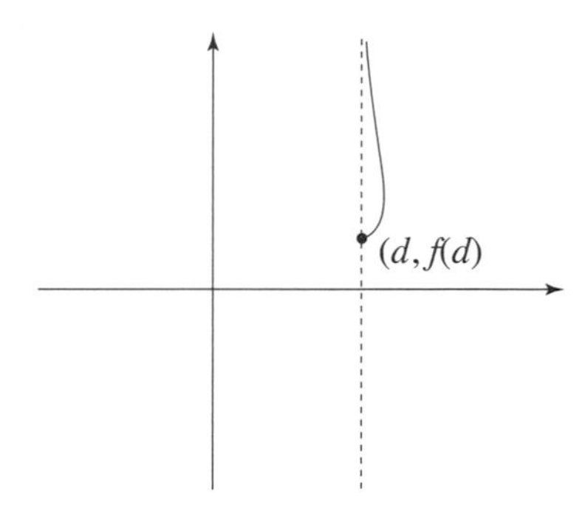

Fig. 7.51

Therefore, to find the vertical asymptotes of f, determine the numbers for which f is not continuous, and investigate the behavior of the function f near each of them. Consider a discontinuity at a number d. The line $x = d$ is a vertical asymptote of the graph of f if either

$$\lim_{x \to d^-} f(x) = \pm\infty \quad \text{or} \quad \lim_{x \to d^+} f(x) = \pm\infty$$

or both. For example, since $\lim\limits_{x \to 0^-} (x^2 + \frac{1}{x}) = -\infty$ and $\lim\limits_{x \to 0^+} (x^2 + \frac{1}{x}) = +\infty$, the line $x = 0$ (the y-axis) is a vertical asymptote of the graph of $y - x^2 + \frac{1}{x}$.

Example 7.49. Consider the function $f(x) = \frac{1}{x+2} + 3$. Notice that

$$\lim_{x \to +\infty} (\tfrac{1}{x+2} + 3) = 0 + 3 = 3 \quad \text{and} \quad \lim_{x \to -\infty} (\tfrac{1}{x+2} + 3) = 0 + 3 = 3.$$

Therefore the line $x = 3$ is a horizontal asymptote for $f(x) = \frac{1}{x+2} + 3$. Note that f is not defined and hence not continuous at -2. So the only possibility for a vertical asymptote is the line $x = -2$. To decide whether it is, we study the limits $\lim\limits_{x \to -2^-} (\frac{1}{x+2} + 3)$ and $\lim\limits_{x \to -2^+} (\frac{1}{x+2} + 3)$. Observe that as x closes in on -2, $\frac{1}{x+2} + 3$ becomes large. Does it become large and positive, or large and negative? When x closes in on -2 from the left, $x < -2$ and hence $x + 2 < 0$. Therefore $\frac{1}{x+2} + 3$ becomes large and negative. So

$$\lim_{x \to -2^-} (\tfrac{1}{x+2} + 3) = -\infty.$$

When x closes in on -2 from the right, $-2 < x$, so that $x + 2 > 0$. Therefore $\lim\limits_{x \to -2^+} (\frac{1}{x+2} + 3) = +\infty$. This confirms that the line $x = -2$ is a vertical asymptote of the graph of $f(x) = \frac{1}{x+2} + 3$ and describes the behavior of the graph of the function near it.

Example 7.50. Consider the function

$$f(x) = \frac{x-1}{x^2(x+2)}.$$

To find the horizontal asymptotes, we need to answer the questions

$$\lim_{x\to-\infty} \tfrac{x-1}{x^2(x+2)} = \ ? \quad \text{and} \quad \lim_{x\to+\infty} \tfrac{x-1}{x^2(x+2)} = \ ?$$

One common strategy to compute limits of this type is first to divide both the numerator and the denominator of the fraction by the highest power of the variable. Since $\frac{x-1}{x^2(x+2)} = \frac{x-1}{x^3+2x^2}$ in the current case, this power is x^3. Dividing both the numerator and denominator of $\frac{x-1}{x^2(x+2)} = \frac{x-1}{x^3+2x^2}$ by x^3, we get

$$\frac{x-1}{x^2(x+2)} = \frac{x-1}{x^3+2x^2} = \frac{\frac{x}{x^3} - \frac{1}{x^3}}{\frac{x^3}{x^3} + \frac{2x^2}{x^3}} = \frac{\frac{1}{x^2} - \frac{1}{x^3}}{1 + \frac{2}{x}}.$$

Pushing x to either $+\infty$ or $-\infty$ forces this quotient to $\frac{0-0}{1-0} = 0$ in either case. Therefore the line $y = 0$ is a horizontal asymptote of the graph of $f(x) = \frac{x-1}{x^2(x+2)}$. The vertical lines $x = 0$ and $x = -2$ are both possible vertical asymptotes. If x is pushed to 0, the denominator $x^2(x+2)$ goes to 0 through positive values, and the numerator $x - 1$ goes to -1. It follows that

$$\lim_{x\to0^-} \tfrac{x-1}{x^3+2x^2} = -\infty \quad \text{and} \quad \lim_{x\to0^+} \tfrac{x-1}{x^3+2x^2} = -\infty.$$

So the line $x = 0$ is a vertical asymptote for $f(x) = \frac{x-1}{x^2(x+2)}$. A similar analysis shows that

$$\lim_{x\to-2^-} \tfrac{x-1}{x^3+2x^2} = +\infty \quad \text{and} \quad \lim_{x\to-2^+} \tfrac{x-1}{x^3+2x^2} = -\infty.$$

Therefore $x = -2$ is also a vertical asymptote for $f(x) = \frac{x-1}{x^2(x+2)}$.

Example 7.51. What are the asymptotes of the graph of $f(x) = \frac{\sin x}{x}$? Since $-1 \le \sin x \le 1$, it follows that $\lim\limits_{x\to\pm\infty} \frac{\sin x}{x} = 0$. So the x-axis is a horizontal asymptote. Since the function is not defined at $x = 0$, it is not continuous there. Is the line $x = 0$ a vertical asymptote? No, because as we saw in Section 7.7, $\lim\limits_{x\to0} \frac{\sin x}{x} = 1$ (and not $\pm\infty$).

Example 7.52. Figure 7.39 suggests that the x-axis is a horizontal asymptote for the graph of $f(x) = a^x$ for any $a > 0$ except $a = 1$. If $a > 1$, write $a^x = \frac{1}{a^{-x}}$ and observe that the denominator is huge for large negative x. Therefore

$$\lim_{x\to-\infty} a^x = \lim_{x\to-\infty} \frac{1}{a^{-x}} = 0.$$

Hence the line $y = 0$ (the x-axis) is a horizontal asymptote for the graph. If $a < 1$, then $\lim\limits_{x\to+\infty} a^x = 0$, and the same is true.

Components I, II, III, and IV of the discussion above—when they apply—can provide important information about the graph of a function. This information can be supplemented and refined by the precise numerical data of the earlier analyses:

V. Intervals of increase or decrease and local maxima and minima

VI. Concavity and points of inflection

Our overview of the calculus of derivatives is complete. The next chapter uses this theory in a variety of different applied contexts.

7.14 PROBLEMS AND PROJECTS

7A. Domains of Functions. Determine the domains of the following functions.

7.1. $f(x) = \sqrt{7-x}$, $g(x) = \sqrt{x+5}$, and $h(x) = f(x)g(x)$

7.2. $f(x) = \frac{x^2}{x^2-3x+2}$ and $g(x) = \frac{x^3}{x^3+x^2-2x}$

7.3. $f(x) = \sqrt{3x-4}$, $g(x) = \sqrt{2x-3}$ and $k(x) = \frac{f(x)}{g(x)}$

7.4. $f(x) = \sqrt{3-\sqrt{x+6}}$ and $g(x) = \sqrt{\frac{x-5}{x+3}}$

7B. Evaluation of Limits. Evaluate the following limits. If a limit does not exist, say why.

7.5. Say what number the expression approaches as x closes in on the indicated number.

i. $\lim\limits_{x\to 2}(x^2+1)(x^2+4x)$

ii. $\lim\limits_{x\to 1}\frac{x-2}{x^2+4x-3}$

iii. $\lim\limits_{x\to 4}\sqrt{x+\sqrt{x}}$

iv. $\lim\limits_{x\to 3}\frac{x^2-x+12}{x+3}$

7.6. If a limit is of $\frac{0}{0}$ type, you may first need to carry out an algebraic step (possibly involving a multiplication by the conjugate, or a polynomial division) leading to a cancellation.

i. $\lim\limits_{x\to -3}\frac{x^2-x+12}{x+3}$

ii. $\lim\limits_{x\to -3}\frac{x^2-x-12}{x+3}$

iii. $\lim\limits_{t\to -1}\frac{t^3-t}{t^2-1}$

iv. $\lim\limits_{x\to 1}\frac{x^3-1}{x^2-1}$

v. $\lim\limits_{x\to 9}\frac{x^2-81}{\sqrt{x}-3}$

vi. $\lim\limits_{x\to 0}\frac{x}{\sqrt{1+3x}-1}$

vii. $\lim\limits_{s\to 16}\frac{4-\sqrt{s}}{s-16}$

viii. $\lim\limits_{x\to 2}\frac{|x-2|}{x-2}$ [Hint: Let x approach 2 from the left and right separately.]

ix. $\lim\limits_{h\to 0}\frac{(h-5)^2-25}{h}$

x. $\lim\limits_{h\to 0}\frac{\sin(\pi+h)-\sin\pi}{h}$

7.7. Given that $\lim\limits_{x\to a} f(x) = 5$ and $\lim\limits_{x\to a} g(x) = 3$, determine the following.

i. $\lim\limits_{x\to a}\frac{f(x)}{g(x)}$

ii. $\lim\limits_{x\to a}\frac{2f(x)}{g(x)-f(x)}$

iii. $\lim\limits_{x\to a}\frac{2f(x)}{5g(x)-3f(x)}$

7C. Continuity. The next set of problems takes on various concerns dealing with continuous functions and related issues.

7.8. Use the definition of continuity and the properties of limits to show that the given functions are continuous at the given number.

i. $f(x) = 1 + \sqrt{x^2 - 9}$, $c = 5$

i. $g(x) = \frac{x+1}{2x^2-1}$, $c = 4$

7.9. Determine the domains of the two functions and explain why they are continuous on their domains.

i. $G(x) = \frac{x^4+17}{6x^2+x-1}$

ii. $H(x) = \frac{1}{\sqrt{x+1}}$

7.10. For which value of the constant c is a function f defined below continuous for all real numbers x?

$$f(x) = \begin{cases} cx + 1, & \text{for } x \leq 3, \\ cx^2 - 11, & \text{for } 3 < x \end{cases}$$

7.11. Check whether the following function is continuous at all numbers on its domain.

$$f(x) = \begin{cases} \frac{x^2-2x-8}{x-4}, & \text{if } x \neq 4, \\ 6, & \text{if } x = 4 \end{cases}$$

7.12. You are given the following three functions with domains as prescribed: $g(x) = \frac{x^2+3x}{x+3}$ with $x < 0$ but $x \neq -3$, $h(x) = 4\sqrt{x}$ with $0 < x < 1$, and $j(x) = \frac{x^2-x-12}{x-4}$ with $1 \leq x$ but $x \neq 4$. Sketch the graphs of these functions on the same coordinate plane. Is there a function $f(x)$ that is continuous at all real numbers and has the same value as $g(x)$, $h(x)$, and $j(x)$ for any x in the domain of any one of these functions? Provide an explicit rule for the function $f(x)$.

7.13. Use the intermediate value theorem to show that $2x^3 + x^2 + 2 = 0$ for some x in the interval $(-2, -1)$.

7.14. Suppose that a function f is continuous on the interval $[-1, 1]$ and that $f(-1) = 2$ and $f(1) = 3$. Why is there a number x such that $|x| < 1$ and $f(x) = e$?

7.15. Project. Supply all the details for the proof of the limit theorem for a composite function: let $y = f(g(x))$ be the composite of two functions f and g. Assume that $\lim_{x \to c} g(x)$ exists and equals b and that f is continuous at b. Then

$$\lim_{x \to c} f(g(x)) = f(\lim_{x \to c} g(x)) = f(b).$$

We'll outline a proof of this theorem using the limit test of Section 7.2. We need to show that for any given $\varepsilon > 0$, there is a corresponding $\delta > 0$ such that whenever x satisfies $x \neq c$ and $|x - c| < \delta$, then $|f(g(x)) - f(b)| < \varepsilon$. Let $y = g(x)$. Since f is continuous at b, $\lim_{y \to b} f(y) = f(b)$. By the limit test, this means that for the given ε above there is a $\delta_1 > 0$, such that $|f(y) - f(b)| < \varepsilon$ for any y with $y \neq b$ and $|y - b| < \delta_1$. Since $\lim_{x \to c} g(x) = b$, the limit test provides for this δ_1 a corresponding $\delta_2 > 0$, such that $|g(x) - b| < \delta_1$ for any x with $x \neq c$ and $|x - c| < \delta_2$. For the $\delta > 0$ that we need for the given ε, we'll take $\delta = \delta_2$. We now need to check whether it is the case that when $x \neq c$ and $|x - c| < \delta$, then $|f(g(x)) - f(b)| < \varepsilon$. Suppose that $x \neq c$ and $|x - c| < \delta = \delta_2$. Then $|g(x) - b| < \delta_1$. But this means that $|f(g(x)) - f(b)| < \varepsilon$ (whether $g(x) = b$ or not).

7.16. The limit theorem of Problem 7.15 provides the missing details of the verification of the speed formula $v(t) = \sqrt{x'(t)^2 + y'(t)^2}$ at the end of Section 6.4. The concern was to move the $\lim\limits_{\Delta t \to 0}$ of

$$\lim_{\Delta t \to 0} \sqrt{\left[\frac{x(t+\Delta t) - x(t)}{\Delta t}\right]^2 + \left[\frac{y(t+\Delta t) - y(t)}{\Delta t}\right]^2}$$

past the $\sqrt{}$ and then again past the two squaring operations. In this limit, the time t is held fixed, only Δt moves (to 0). Now fix c and let $t = c - \Delta t$. This time, both t and Δt vary. Since $t + \Delta t = c$ and $\Delta t = c - t$, we get $\frac{x(t+\Delta t)-x(t)}{\Delta t} = \frac{x(c)-x(t)}{c-t} = \frac{x(t)-x(c)}{t-c}$. Similarly, $\frac{y(t+\Delta t)-y(t)}{\Delta t} = \frac{y(t)-y(c)}{t-c}$. Since $\lim\limits_{\Delta t \to 0}$ is equivalent to $\lim\limits_{t \to c}$, the limit above becomes

$$\lim_{t \to c} \sqrt{\left[\frac{x(t) - x(c)}{t - c}\right]^2 + \left[\frac{y(t) - y(c)}{t - c}\right]^2}.$$

Since $r(x) = \sqrt{x}$ is a continuous function, the Limit Theorem (and the additivity of limits) tells us that this limit is equal to

$$\sqrt{\lim_{t \to c} \left[\frac{x(t) - x(c)}{t - c}\right]^2 + \lim_{t \to c} \left[\frac{y(t) - y(c)}{t - c}\right]^2}.$$

Since $s(x) = x^2$ is continuous, this in turn (again by the limit theorem) is equal to

$$\sqrt{\left[\lim_{t \to c} \frac{x(t) - x(c)}{t - c}\right]^2 + \left[\lim_{t \to c} \frac{y(t) - y(c)}{t - c}\right]^2}.$$

7D. Computing Derivatives. This group of problems focuses on the definition of the derivative as well as the formulas that facilitate their computation.

7.17. Compute the derivative of the function by using the limit definition of the derivative.

i. $f(x) = \frac{1}{x} = x^{-1}$ [Hint: Take common denominators and cancel.]

ii. $f(x) = x^3$ [Hint: Use the formula $(a + b)^3 = a^3 + 3a^2b + 3ab^2 + b^3$.]

iii. $g(x) = \frac{1}{x^2}$ [Hint: Take common denominators.]

iv. $g(x) = x - \frac{2}{x}$

v. $h(x) = \sqrt{6 - x}$

7.18. Each of the limits below can be interpreted as the derivative of a function f at some number c. In each case, say which function and which number c.

i. $\displaystyle\lim_{\Delta x \to 0} \frac{(5 + \Delta x)^2 - 25}{\Delta x}$

ii. $\displaystyle\lim_{\Delta x \to 0} \frac{\sqrt{1 + \Delta x} - 1}{\Delta x}$

iii. $\displaystyle\lim_{x \to 1} \frac{x^9 - 1}{x - 1}$

iv. $\displaystyle\lim_{x \to 0} \frac{\sqrt{1 + x} - 1}{x}$

7.19. **i.** Sketch the graphs of $f(x) = x^2 - 9$ and $g(x) = |x^2 - 9|$. For what values of x does the function $g(x)$ fail to be differentiable? Sketch the graphs of both f' and g'.

ii. Sketch the graph of $f(x) = |x|$ and use it to sketch the graph of f'.

7.20. Use rules for differentiating to compute the slopes of the tangents to the graphs of the following functions at the given points.

i. $f(x) = x^3$ at $(-2, -8)$

ii. $g(x) = \sqrt[3]{x}$ at $(-3, \sqrt[3]{-3})$

iii. $f(x) = \frac{1}{x}$ at $(-\frac{1}{3}, -3)$

iv. $f(x) = \frac{1}{x^2}$ at $(-2, \frac{1}{4})$

7.21. By using rules for differentiating, compute the derivatives of each of the following functions.

i. $f(x) = -10$

iii. $f(x) = 7x^2 - 5x + 2$

iv. $y = 2\sqrt[3]{x} + \pi x^3$

v. $g(x) = \frac{3}{x} + 3x - 6$

vi. $f(x) = 2x^3 + 3x + 4 - \frac{1}{x^2}$

vii. $g(x) = 4x^{\frac{1}{2}} + 5x^{-1}$

viii. $h(x) = 6x^3 - 7x^{\frac{1}{3}}$

7.22. Use rules for differentiating to compute the following derivatives:

i. $G(x) = (x^2 + 1)(2x - 7)$, $G'(x) =$

ii. $f(x) = \frac{a-x^2}{1+x^2}$, $f'(x) =$

iii. $g(t) = \sqrt[3]{t}(t + 2)$, $\frac{ds}{dt} =$

iv. $h(x) = \frac{1}{x^4+x^2+1}$, $\frac{dy}{dx} =$

v. $y = (2x^3 + 4x^5)^6(7x^8 + 9x^{10})^{11}$, $\frac{dy}{dx} =$

vi. $f(x) = \sqrt{1 - x^2}$, $f'(x) =$

vii. $y = \frac{x}{\sqrt{9-4x}}$, $\frac{dy}{dx} =$

viii. $F(x) = \frac{(x^2+4x+6)^5}{\sqrt{x^3+4x^5}}$, $F'(x) =$

ix. $s(t) = \sqrt[4]{\frac{t^3+1}{t^3-1}}$, $s'(t) =$

x. $\frac{d}{dx}f(g(h(x))) =$

7E. About Tangent Lines. Problems 7.23–7.29 deal with tangent lines to the graphs of functions.

7.23. Find the equation of the tangent to the graph of the function at the indicated point.

i. $g(x) = 1 - x^3$ at $(0, 1)$

ii. $h(x) = \frac{1}{2x-1}$ at $(-1, -\frac{1}{3})$

iii. $y = \frac{x}{x-3}$ at $(6, 2)$

7.24. Find an equation of the tangent line to the graph of $y = x^2 - 1$ that is parallel to the line $x - 2y = 1$.

7.25. For what values of x does the graph of $f(x) = 2x^3 - 3x^2 - 6x + 87$ have a horizontal tangent?

7.26. Show that the curve $y = 6x^3 + 5x - 3$ has no tangent lines with slope 4.

7.27. What point P on the parabola $y = \frac{1}{10}x^2$ has the property that $(10, 5)$ lies on the tangent at P?

7.28. Draw a diagram showing that there are exactly two lines that are tangent to both of the parabolas $y = -1 - x^2$ and $y = 1 + x^2$. Find the coordinates of the points at which these tangents touch the parabolas.

7.29. Refer to Figure 7.52. It depicts the graphs of 8 functions. The first 4 graphs are labeled (i) to (iv). The second 4, labeled (a) through (d), are the graphs of the derivatives of the first 4. For each of the graphs (i) to (iv), insert the letter of the graph of the matching derivative into the blank:

(i) ____ (ii) ____ (iii) ____ (iv) ____.

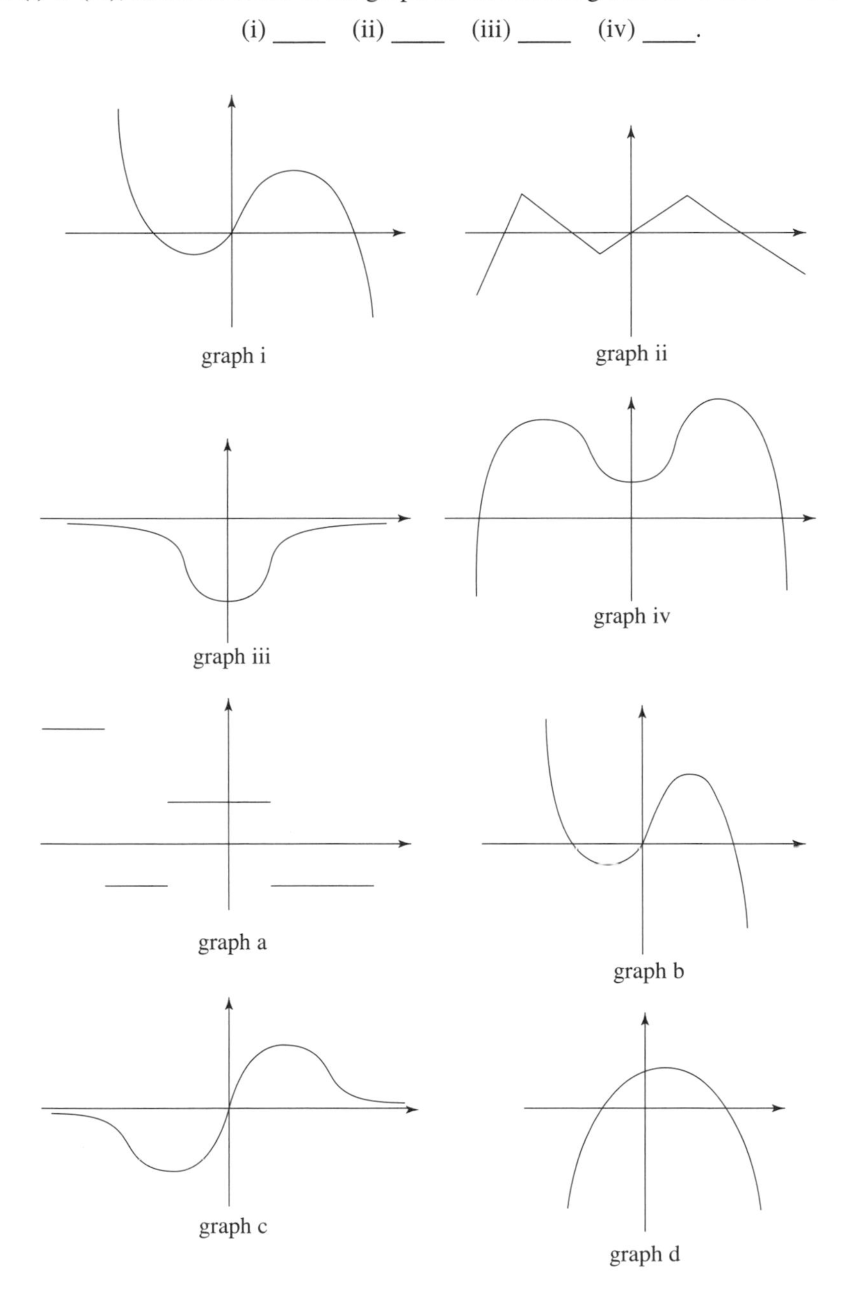

Fig. 7.52

7.30. Consider the circle $x^2 + y^2 = r^2$.

i. Let (x_0, y_0) be any point on the circle with $y_0 \neq 0$. Use the chain rule to show that the slope of the tangent line to the circle at any point (x_0, y_0) with $y_0 \neq 0$ is equal to $-\frac{x_0}{y_0}$.

ii. Suppose that $r = 1$. Consider a line $y = -\frac{1}{3}x + b$ with slope $-\frac{1}{3}$ and y-intercept b. For which constants b is the line tangent to the circle?

7F. Derivatives of Trigonometric Functions

7.31. Compute $\frac{dy}{dx}$ for the functions below.

i. $y = \sin \frac{1}{x}$

ii. $y = \sin^2(\cos 4x)$

iii. $y = \frac{\sin^2 x}{\cos x}$

iv. $y = x \sin \frac{1}{x}$

vi. $y = \tan 3x$

vii. $y = (\cos \sqrt{x^2 + 1})^{-5}$

viii. $y = (1 + \sec^3 x)^6$

ix. $y = \tan(x^2) + \tan^2 x$

x. $y = \sqrt{1 + 2 \tan x}$

7.32. Use the chain rule to differentiate the following.

i. $\sin(3x^2 + 1)$

ii. $\sin^2(\sqrt{t}) + \cos^2(\sqrt{t})$ [How could you have gotten your answer without any computation?]

iii. $\tan \sqrt{u^2 + 27u}$

iii. $\frac{d^2}{dx^2} \tan x = \frac{d}{dx} \sec^2 x$

7.33. Use the chain rule to differentiate the following functions of t.

i. $\sin \alpha(t)$, where $\alpha(t)$ is some unspecified differentiable function of t

ii. $\cos \beta(t)$, where $\beta(t)$ is some differentiable function of t

7.34. Express the limit $\lim\limits_{\theta \to \pi/3} \frac{\cos \theta - \frac{1}{2}}{\theta - \pi/3}$ as a derivative and evaluate it.

7.35. Find the equation of the tangent to the curve $y = \tan x$ at the point $(\frac{\pi}{3}, \sqrt{3})$.

7G. Derivatives and Velocity. Before turning to the next two problems, study Section 6.4 about the connection between the derivative and the velocity of a moving point.

7.36. Each pair of functions below represents the x- and y-coordinates of a point moving in the xy-plane in terms of the elapsed time t. In each case, determine the equation of the curve along which the point moves. Sketch the curve. Compute the speed of the point at any time t, and describe how the point moves along its path.

i. $x(t) = t$, $y(t) = t^2$, and $t \geq 0$

ii. $x(t) = t$, $y(t) = \sqrt{1 - t^2}$, and $0 \leq t \leq 1$

iii. $x(t) = t$, $y(t) = \cos t$, and $t \geq 0$

7.37. Each pair of functions below represents the x- and y-coordinates of a point moving in the xy-plane in terms of the elapsed time t. In each case, compute the distance $d(t)$ from the point to the origin at any time t and describe the curve in the xy-plane along which the point moves. Sketch the curve. Compute the velocity of the point at any time t, and describe how the point moves with a diagram. [Hint: Ask yourself if and how the motions of (ii) and (iii) are related to that of (i).]

i. $x(t) = \cos t$, $y(t) = \sin t$, and $t \geq 0$

ii. $x(t) = t\cos t$, $y(t) = t\sin t$, and $t \geq 0$

iii. $x(t) = \frac{1}{t}\cos t$, $y(t) = \frac{1}{t}\sin t$, and $t \geq \pi$

7H. More about Derivatives

7.38. Let $y = u^2$ and $u = x^2 + 2x + 3$. Use the chain rule in the form $\frac{dy}{dx} = \frac{dy}{du} \cdot \frac{du}{dx}$. Compute the value $\frac{dy}{dx}\big|_{x=1}$ of the derivative $\frac{dy}{dx}$ at $x = 1$.

7.39. Consider the function $f(x) = 5x^2 - 2x^3$, and let $P = (x, y)$ be some point on its graph. Determine an equation for the tangent line to the graph at P. [Suggestion: Pay attention to the notation.]

7.40. Consider the parabola $y = x^2$ and the line $y = 3x - 4$. Show that they do not intersect. Move the line toward the parabola without changing its slope. At what point will it first touch the parabola?

7.41. Use derivatives to compute the distance between the line $y = -\frac{1}{2}x + 5$ and the point $(-4, 3)$. [Hint: Let (x, y) be any point on the line. Express the distance between $(-4, 3)$ and (x, y) as a function of x. Then determine the smallest value of the square of this distance.]

7.42. Suppose that $f(3) = 4$, $g(3) = 2$, $f'(3) = -6$, $f'(2) = -3$, and $g'(3) = 5$. Find $\left(\frac{f}{g}\right)'(3)$. What is the value of the derivative of $f(g(x))$ at 3?

7.43. Sketch the graphs of $f(x) = x^2 - 9$ and $g(x) = |x^2 - 9|$. For what values of x does the function $g(x)$ fail to be differentiable? Sketch the graphs of f and g and also f' and g'.

7.44. Sketch the graphs of $f(x) = \sin x$ and $g(x) = |\sin x|$ over the interval $0 \leq x \leq 4\pi$. For what values of x does the function $g(x)$ fail to be differentiable? Sketch the graphs of f and g and also f' and g'.

7.45. For which values of the constants c and d is the function

$$f(x) = \begin{cases} cx^2 + 12, & \text{for } x \leq 1, \\ d\sqrt{x} + c, & \text{for } 1 < x \end{cases}$$

differentiable (and hence continuous) for all real numbers x?

7.46. Consider the function $y = x + 1$ for $x \leq -1$ and the function $y = x - 1$ for $1 \leq x$, and sketch both of their graphs. Next let $f(x)$ with $-1 \leq x \leq 1$ be a function that satisfies the following condition: the function $g(x)$ defined by $g(x) = x + 1$ for $x \leq -1$, $g(x) = f(x)$ for $-1 \leq x \leq 1$, and $g(x) = x - 1$ for $1 \leq x$ is differentiable. Add the graph of one such function $f(x)$ to your diagram.

i. How many functions $f(x)$ with the required property are there? Explain how you came to your conclusion.

ii. Determine an explicit function of the form $f(x) = ax^3 + bx^2 + cx + d$, where a, b, c, and d are constants, so that $f(x)$ with $-1 \leq x \leq 1$ satisfies the required condition.

iii. For $1 \leq x$, replace $y = x - 1$ by $y = 0$ and repeat (i) and (ii).

7.47. Consider the function $f(x) = x^2 + x$ with $x < 0$. For $x \geq 0$, let $g(x)$ be a function of the form $g(x) = ax^2 + bx + c$ where a, b, and c are constants.

i. For what values of a, b, and c do the functions $f(x)$ and $g(x)$ splice together to give a function that is continuous for all real numbers? Explain.

ii. For what values of a, b, and c do the functions $f(x)$ and $g(x)$ splice together to give a function that is differentiable for all real numbers? How many such functions $g(x)$ are there? Explain.

7.48. Let $y = f(x)$ be a function that is differentiable for all x. Draw a typical graph, and use it to give a geometric interpretation of the quotient $\frac{f(x+\Delta x)-f(x-\Delta x)}{2\Delta x}$ for a small positive Δx. What do you think this quotient closes in on when Δx is pushed to zero? Verify your thinking with a limit computation. [Hint: Show that $\lim\limits_{\Delta x \to 0} \frac{f(x-\Delta x)-f(x)}{-\Delta x} = f'(x)$. Make use of this and the definition of the derivative.]

7.49. Suppose that f is differentiable at a, where $a \geq 0$. Evaluate the limit $\lim\limits_{x \to a} \frac{f(x)-f(a)}{\sqrt{x}-\sqrt{a}}$ in terms of $f'(a)$.

7I. From L'Hospital's Rule to Implicit Differentiation

7.50. Use L'Hospital's rule to evaluate the following.

i. $\lim\limits_{x \to -3} \dfrac{x^2 - x - 12}{x + 3}$

ii. $\lim\limits_{x \to 1} \dfrac{x^3 - 1}{x^2 - 1}$

iii. $\lim\limits_{x \to 9} \dfrac{x^2 - 81}{\sqrt{x} - 3}$

iv. $\lim\limits_{s \to 4} \dfrac{s^3 - 7s^2 + 17s - 20}{s^2 - 5s + 4}$

7.51. Evaluate the limit of Example 7.26 after carrying out a polynomial division.

7.52. Let $f(x) = x + \sqrt{x}$. Use the mean value theorem to show that there is a number c between 0 and 9 such that $f'(c) = \frac{4}{3}$. Determine a c that works.

7.53. You are given a function $f(x)$ that is differentiable for all real numbers x and that has the points $(3, 17)$ and $(7, 9)$ on its graph.

i. Explain why there is some number d between 3 and 7 such that $f(d) = 4\pi$. (State the theorem that applies.)

ii. Explain why there is some number c between 3 and 7 such that $f'(c) = -2$. (State the theorem that applies.)

7.54. In each of the problems that follows, $y = f(x)$ is some differentiable function of x. Find the derivative of the given function in terms of x, y, and y'. This procedure is called *implicit differentiation*.

i. $g(x) = xy^3$

ii. $h(x) = 3\sqrt{y} + xy$

iii. $k(x) = \frac{4x}{y^2}$

iv. $g(x) = (4x + y^{-\frac{3}{2}})^4$

v. $g(x) = (2x^2 + 3y^{\frac{1}{2}})^2$

7J. Increase and Decrease of Functions

7.55. Find the critical numbers for the given function.

i. $f(x) = x^3 - 3x + 1$

ii. $F(x) = x^{\frac{4}{5}}(x-4)^2$

iii. $T(x) = x^2(2x-1)^{\frac{2}{3}}$

7.56. Find (a) the intervals on which f is increasing or decreasing and (b) the local maximum and minimum values of f.

i. $f(x) = x^3 - 2x^2 + x$

ii. $f(x) = x^4 - 4x^3 - 8x^2 + 3$

iii. $f(x) = x\sqrt{1-x^2}$

iv. $f(x) = x\sqrt{x-x^2}$

7.57. Verify that $a + \frac{1}{a} < b + \frac{1}{b}$ whenever $1 \le a < b$. [Hint: Show that the function $f(x) = x + \frac{1}{x}$ is increasing on $[1, \infty)$.]

7.58. Let n be any positive integer, and show that $(1+x)^n > 1 + nx$ for any $x > 0$.

7.59. Find (a) the intervals on which the given trigonometric function is increasing or decreasing and (b) the local maximum and minimum values of the function.

i. $f(x) = x - 2\sin x,\ 0 \le x \le \pi$

ii. $f(x) = x\sin x + \cos x,\ -\pi \le x \le \pi$

iii. $f(x) = 2\tan x - \tan^2 x,\ -\frac{\pi}{2} \le x \le \frac{\pi}{2}$

iv. $g(x) = \sin x + \cos x,\ -\frac{\pi}{2} \le x \le \frac{\pi}{2}$

7K. Finding Maximum and Minimum Values

7.60. Find the maximum and minimum values of f over the given interval.

i. $f(x) = 1 + (x+1)^2,\ -2 \le x \le 5$

ii. $f(x) = x^3 - 12x + 1,\ [-3, 5]$

iii. $f(x) = 3x^5 - 5x^3 - 1,\ [-2, 2]$

iv. $f(x) = \sqrt{9 - x^2},\ [-1, 2]$

7.61. Consider the function $f(x) = x^2 - x - 6$. Because its graph is a parabola that opens upward, this function has a minimum value (but not a maximum value). Find the minimum value of $f(x)$ as well as the number x at which it occurs. Do so by using calculus and also by completing the square.

7.62. Inscribe a rectangle in the ellipse $\frac{x^2}{a^2} + \frac{y^2}{b^2} = 1$, as shown in Figure 7.53. What are the dimensions in terms of a and b of the largest (in terms of area) such rectangle? What is its area?

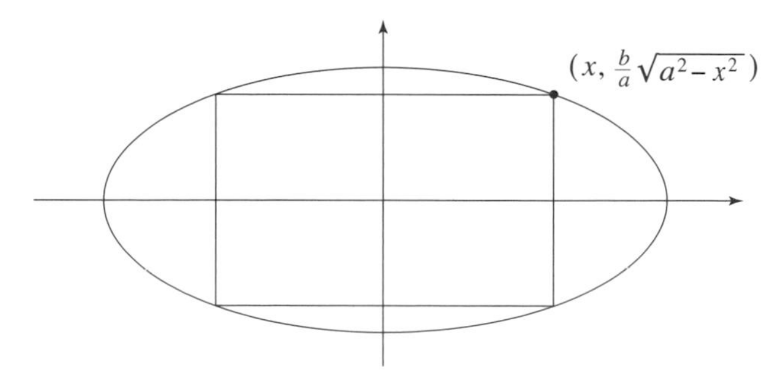

Fig. 7.53

7.63. Consider the circle with radius r centered at the origin. Rotate the semicircle and the rectangle shown in Figure 7.54 one complete revolution about the y-axis. This rotation generates a sphere of radius r and an inscribed cylinder. Notice that the base of the cylinder has area πx^2.

i. Show that the volume of this cylinder is $V(x) = 2\pi x^2\sqrt{r^2 - x^2}$.

ii. Find the value of x that makes $V(x)$ a maximum.

iii. What is the ratio of the volume of the sphere to that of the largest inscribed cylinder?

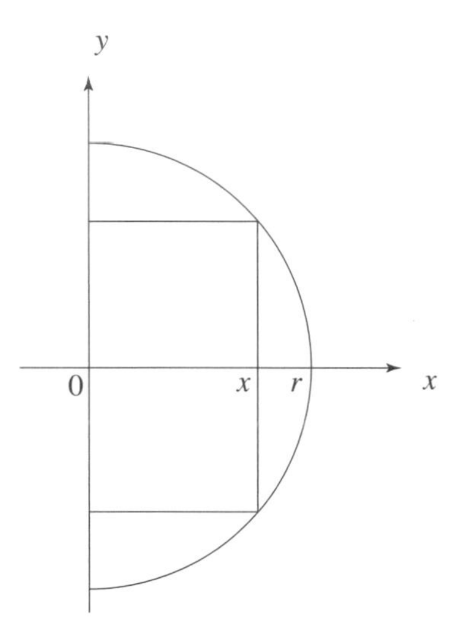

Fig. 7.54

7.64. The fixed point (a, b) in Figure 7.55 determines a rectangle. Consider a segment from one axis, through (a, b), to the other axis. What is the shortest length (in terms of a and b) that such a segment can have? [Here's one way to proceed: Let d be the length of such a segment, and let x be the base

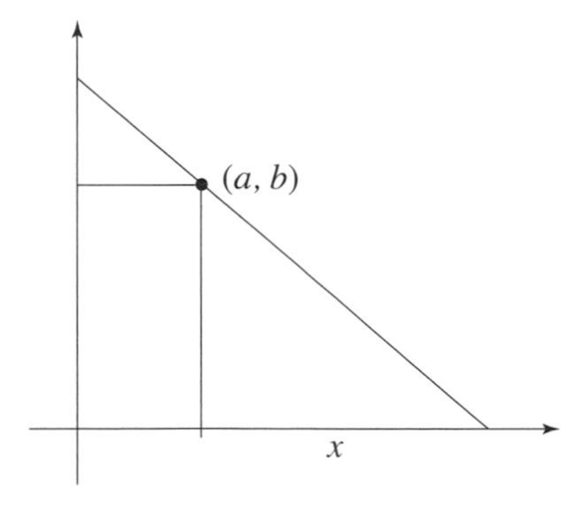

Fig. 7.55

of the right triangle that the point (a, b) and the segment determine. Use similar triangles to show that $d = d(x) = (x + a)^2\sqrt{1 + b^2x^{-2}}$. Now let $D(x) = d(x)^2$, and show that the relevant critical number of $D(x)$ is $x = a^{\frac{1}{3}}b^{\frac{2}{3}}$. Why does this value for x provide the smallest $D(x)$ and hence the smallest $d(x)$?]

7.65. Fix any two positive constants a and h. Of all the rectangular boxes with height h and volume a^3, find the dimensions of the one that has the smallest surface area. Have a go at this on your own first.

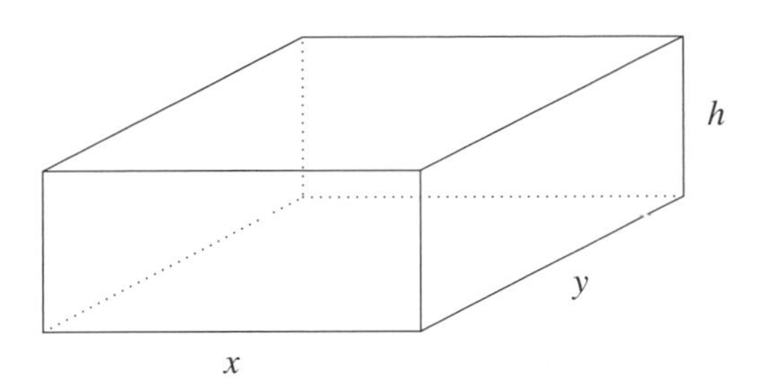

Fig. 7.56

Whether you succeed or not, complete the outline of the solution that follows. [Let x and y be the dimensions of the base, and note that $xyh = a^3$. So $y = \frac{a^3}{hx} = \frac{a^3}{h}x^{-1}$. Refer to Figure 7.56, and check that the surface area of the box is $2hx + 2hy + 2yx$. Substituting $y = \frac{a^3}{h}x^{-1}$ gives the surface area as the function

$$S(x) = 2hx + 2h\tfrac{a^3}{h}x^{-1} + 2\tfrac{a^3}{h}x^{-1}x = 2hx + 2a^3x^{-1} + 2\tfrac{a^3}{h}$$

of x. The question now is this: For which x does $S(x)$ attain its minimum value? Conclude that the base of the box with minimal surface area is a square.]

7L. Exponential and Logarithm Functions

7.66. Determine the following derivatives.

i. $f(x) = e^{\sqrt{x}}$

ii. $g(x) = e^{-5x}\cos 3x$

iii. $y = e^{x+e^x}$

iv. $f(x) = x^2e^x$

v. $y = xe^{x^2}$

vi. $y = e^{1/(1-x^2)}$

vii. $y = \tan(e^{3x-2})$

viii. $y = \frac{e^x+e^{-x}}{e^x-e^{-x}}$

7.67. **i.** Find the equation of the tangent line to the curve $y = x^2e^{-x}$ at the point $(1, \frac{1}{e})$.

ii. Show that the function $y = e^{2x} + e^{-3x}$ satisfies the equation $y'' + y' - 6y = 0$.

iii. Find the 100th derivative of $f(x) = xe^{-x}$. [Start with f', f'', and study the developing pattern.]

iv. Use the intermediate value theorem to show that there is a solution of the equation $e^x + x = 0$.

v. Use the graph of $y = e^x$ to draw the graphs of $y = e^{-x}$ and $y = 3 - e^x$.

vi. Find the absolute minimum value of the function $g(x) = \frac{e^x}{x}$, $x > 0$.

7.68. Sketch the graphs of the following functions in the same coordinate plane.

i. $y = \log_2 x$

ii. $y = \ln x$

iii. $y = \log_{10} x$

7.69. Express the given quantity as a single logarithm.

i. $\log_2 x + 3\log_2(x+1) + \frac{1}{4}\log_2(x-1)$

ii. $\frac{1}{3}\ln x - 4\ln(2x+3)$

7.70. Solve the given equation or inequality for x by making use of the inverse relationship between exponential and logarithm functions.

i. $\log_2 x = 3$

ii. $2^{x^2-5} = 3$

iii. $5^{x^2-1} = 2$

iv. $4^{x^2+1} = 3$

v. $\log_9(4x^2 - 11) = 7$

vi. $\log_5(\log_5 x) = 6$

vii. $\ln(x+6) + \ln(x-3) = \ln 5 + \ln 7$

viii. $\ln(\frac{x-2}{x+1}) = 1 + \ln(\frac{x-3}{x+1})$

ix. Solve the inequality $\ln(3x-2) \leq 0$.

x. Solve the equation $4^x - 2^{x+3} + 12 = 0$.

7.71. Find the domains of the following functions.

i. $f(x) = \log_{10}(1-x)$

ii. $F(t) = \sqrt{t}\ln(t^2-1)$

7.72. For each of the functions f below, find f' and the domains of f and f'.

i. $f(x) = \cos(\ln x)$

ii. $f(x) = \ln(2 - x - x^2)$

iii. $f(x) = \ln(\sqrt{x} - \sqrt{x-1})$

iv. $f(x) = \log_{11}(x^4 + 3x^2)$

v. $f(x) = \ln\sqrt{x + 3x^2}$

7.73. Find y' and y'' for each of the functions below.

i. $y = x\ln x$

ii. $y = \log_{10} x$

iii. $y = \ln(\sec x + \tan x)$

7.74. Differentiate the functions below.

i. $g(x) = \sqrt{\ln x}$

ii. $f(t) = \log_7(t^4 - t^2 + 1)$

iii. $f(x) = e^x \ln x$

iv. $h(t) = t^3 - 3^t$

v. $x^{\sin x}$ for $x > 0$

7.75. Find the absolute minimum value of the function $f(x) = x\ln x$.

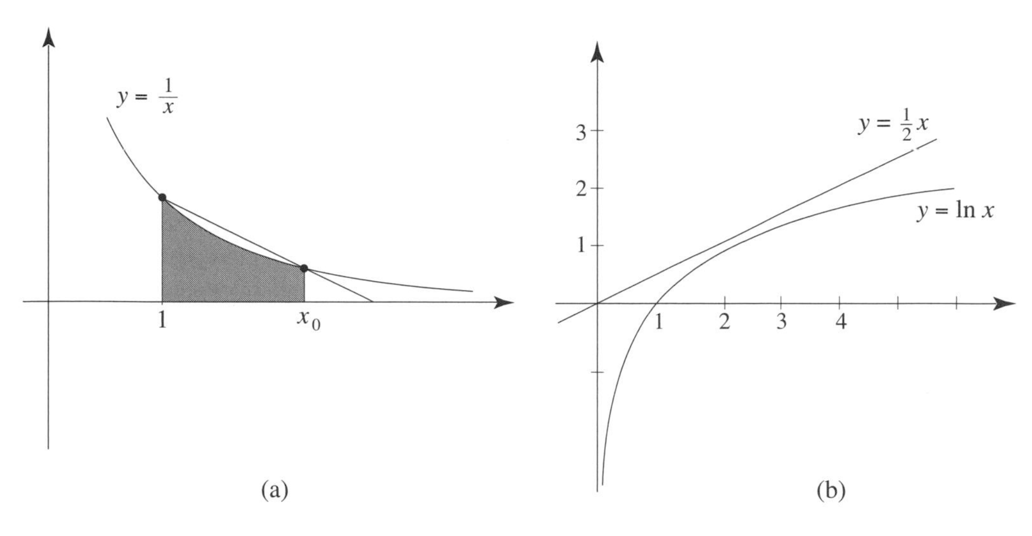

Fig. 7.57

7.76. Let $(x_0, \frac{1}{x_0})$ be a point on the graph of $y = \frac{1}{x}$ with $x_0 > 1$. Show that $y = -\frac{1}{x_0}x + (1 + \frac{1}{x_0})$ is an equation of the line determined by the points $(1, 1)$ and $(x_0, \frac{1}{x_0})$. Use Figure 7.57a and an area comparison to show that $\ln x_0 < \frac{1}{2}(x_0 - \frac{1}{x_0})$ and $\ln x_0 \approx \frac{1}{2}(x_0 - \frac{1}{x_0})$ for x_0 near 1. Use this to approximate $\ln 2$. What per a calculator is the actual value of $\ln 2$? Does the approximation apply when x_0 is much greater than 1? [Hint: Assess Figure 7.57b.]

7.77. Verify laws (i), (ii), and (iii) for logarithms by following the strategy of the verification of law (iv) carried out in the text.

7.78. Use one of the rules for exponents to show that if $\log_a x_1 = \log_b x_2$ for some x_1 and x_2, then $\log_{ab} x_1 x_2 = \log_a x_1 = \log_b x_2$.

7.79. Project. Supply all the details for the proof of the formula

$$e^x = \lim_{n\to\infty} (1 + \tfrac{x}{n})^n$$

for any real number x. It suffices to show that $\ln\big(\lim_{n\to\infty} (1 + \frac{x}{n})^n\big) = x$ for any x. Why? To verify this, we'll first set $h = \frac{x}{n}$ and rewrite

$$\ln\big(\lim_{n\to\infty} (1 + \tfrac{x}{n})^n\big) = \ln\big(\lim_{h\to 0} (1 + h)^{\frac{x}{h}}\big).$$

Because the natural log function is continuous at $x = 1$, this last term is equal to $\lim_{h\to 0} \ln(1 + h)^{\frac{x}{h}}$. This in turn is equal to $\lim_{h\to 0} \frac{x}{h} \cdot \ln(1 + h) = x \cdot \lim_{h\to 0} \frac{\ln(1+h)}{h}$. But what is $\lim_{h\to 0} \frac{\ln(1+h)}{h}$ equal to? The derivative of $\ln x$ evaluated at $x = 1$ is $\lim_{\Delta x\to 0} \frac{\ln(1+\Delta x)-\ln 1}{\Delta x}$. Use what you know about $\frac{d}{dx} \ln x$ to show that this limit is equal to 1. Since $\ln 1 = 0$, the verification of $\ln\big(\lim_{n\to\infty} (1 + \frac{x}{n})^n\big) = x$ is now complete.

7M. More about Hyperbolic Functions

7.80. Sketch the graphs of the functions $y = \operatorname{sech} x$, $y = \coth x$, and $y = \operatorname{csch} x$ by analyzing the flow and pattern of the graphs of $y = \cosh x$, $y = \tanh x$, and $y = \sinh x$.

7.81. Recall the trig identities $\sin^2\frac{\theta}{2} = \frac{1}{2}(1-\cos\theta)$, $\cos^2\frac{\theta}{2} = \frac{1}{2}(1+\cos\theta)$ and $\tan\theta = \frac{2\tan\frac{\theta}{2}}{1-\tan^2\frac{\theta}{2}}$ from Problem 1.22. Verify and in the process formulate the hyperbolic analogues of each of them.

7.82. Refer to Example 7.48. Use a calculator to evaluate $(\operatorname{sech} x)(2\tanh^2 x)$ and $x = 0.88$ and $x = 0.89$, and use this information to locate the inflection points of the graph of $y = \operatorname{sech} x$.

7N. Studying Some Graphs. This set of problems illustrates the strategies for graphing functions, including those discussed in Section 7.13.

7.83. Compare the functions $f(x) = x^2$ and $g(x) = 2^x$ by evaluating each of them for $x = 0, 1, 2, 3, 4, 5, 10, 15$, and 20. Then draw the graphs of f and g in the same coordinate plane for $-4 \leq x \leq 4$.

7.84. Consider the function $f(x) = 2x^2 - x^4 = x^2(2-x^2)$. Notice that the graph is symmetric about the x-axis.

i. Show that $f(x) \geq 0$ for $-\sqrt{2} \leq x \leq \sqrt{2}$, and that $f(x) < 0$ for all other x.

ii. Show that $f'(x) = 4x(1-x^2)$, and notice that $-1, 0$, and 1 are the critical numbers for $f(x)$. Determine the intervals of increase and decrease for $f(x)$.

iii. Start with a large negative x on the x-axis. Move toward $x = -1$. Is $f'(x)$ increasing or decreasing in the process? Now start at $x = 1$ and move to the right. Is $f'(x)$ increasing or decreasing? What do your answers tell you about the slopes of the tangent lines of the graph of f?

iv. Show that $f''(x) = -4(3x^2-1)$, and determine the intervals over which $f(x)$ is concave up and down.

v. Use all the information you have about f, f', and f'' to sketch the graph of f.

7.85. Consider the function $f(x) = 3x^{\frac{2}{3}} - x^2$. Figure 7.58 provides the relevant graphs. Notice that all three are symmetric about the y-axis. For large positive or negative x, the term $-x^2$ is dominant. For x close to 0, the term $3x^{\frac{2}{3}}$ is dominant. Since the function is continuous, this suggests that its

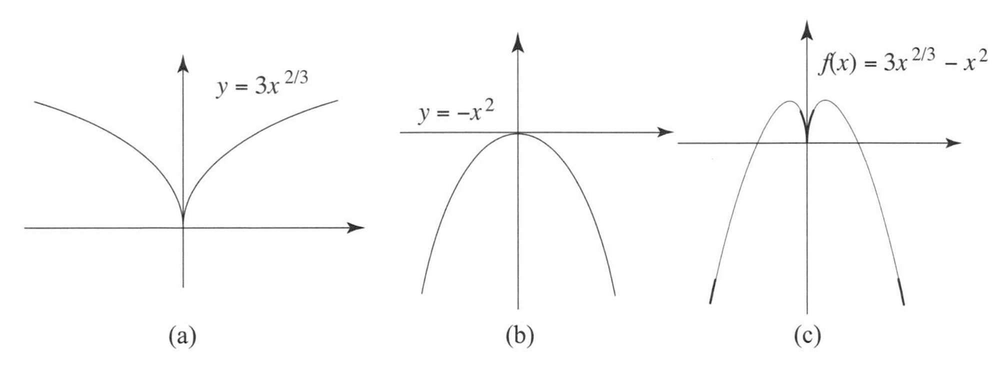

Fig. 7.58

graph winds its way as shown in Figure 7.58c. Analyze $f'(x) = 2x^{-\frac{1}{3}} - 2x$ and $f''(x) = -\frac{2}{3}x^{-\frac{4}{3}} - 2$, and use the information obtained to confirm that the graph of $f(x) = 3x^{\frac{2}{3}} - x^2$ is as shown in Figure 7.58c. Provide some explicit numerical details.

7.86. Match the graphs of the functions

(1) $g(x) = x^3 - 3x$, (2) $g(x) = x^3$, (3) $g(x) = x^3 - 1$, (4) $g(x) = x^3 + x$, (5) $g(x) = x^3 + 4x$, (6) $g(x) = x^3 + x^2$, (7) $g(x) = x^3 - x$, (8) $g(x) = x^3 - 2x$, (9) $g(x) = x^3 - x^2$.

to the graphs (a)–(i) displayed in Figure 7.59.

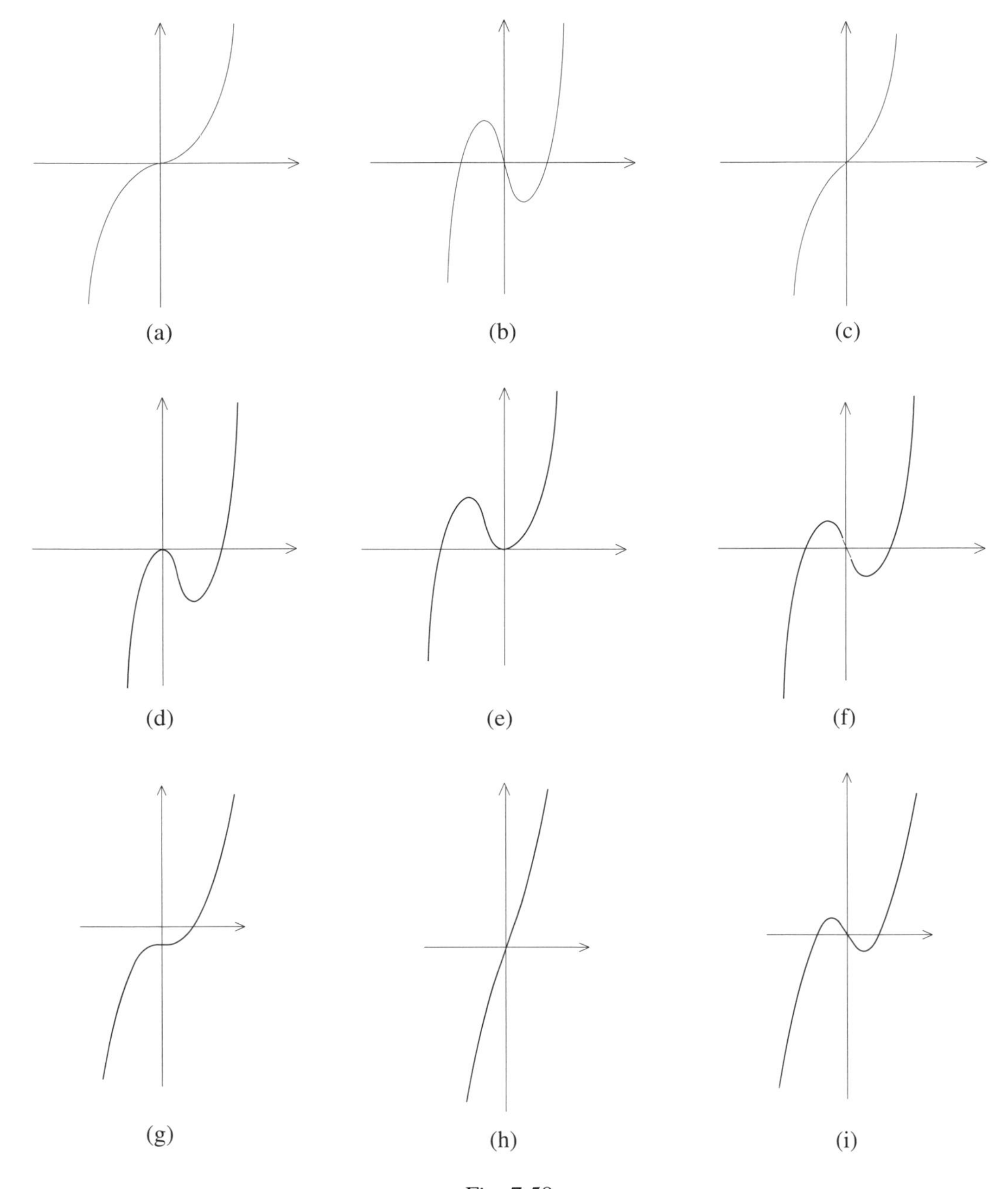

Fig. 7.59

7.87. Consider $f(x) = \frac{x^2-x-6}{x^2-9}$. Because $x^2 - x - 6 = (x-3)(x+2)$ and $x^2 - 9 = (x-3)(x+3)$,

$$f(x) = \frac{x+2}{x+3}$$

for all x except $x = 3$, the graphs of $f(x) = \frac{x^2-x-6}{x^2-9}$ and $g(x) = \frac{x+2}{x+3}$ are identical, except at the single point $(3, \frac{5}{6})$, where the graph of $f(x)$ has a hole and the graph of $g(x)$ does not.

- **i.** Explain why the line $x = 3$ is not a vertical asymptote of the graph of f.
- **ii.** Check that when x is pushed to -3 from the left, $\frac{x+2}{x+3}$ remains positive, and when the push is from the right, $\frac{x+2}{x+3}$ remains negative. Therefore

$$\lim_{x\to -3^-} \frac{x+2}{x+3} = +\infty \text{ and } \lim_{x\to -3^+} \frac{x+2}{x+3} = -\infty.$$

So $x = -3$ is a vertical asymptote of $f(x) = \frac{x^2-x-6}{x^2-9}$, and it is the only vertical asymptote. This analysis has also determined the behavior of the graph immediately to the left and right of -3. The fact that $\lim_{x\to\pm\infty} \frac{x+2}{x+3} = \lim_{x\to\pm\infty} \frac{1+\frac{2}{x}}{1+\frac{3}{x}} = 1$ tells us that $y = 1$ is a horizontal asymptote for the graph.

iii. Show that $g'(x) = \frac{1}{(x+3)^2}$ and $g''(x) = -\frac{-2}{(x+3)^3}$. Analyze these derivatives to obtain the increasing/decreasing and concavity information for the graph. Then sketch the graph.

70. Project. Newton's Method for Solving Equations. Consider the polynomial $f(x) = x^4+2x^3-12x-1$. Notice that $f(0) = -1$ and $f(2) = 2^4 + 2\cdot 2^3 - 12\cdot 2 - 1 = 16 + 16 - 24 - 1 = 7$. Since $f(0)$ is negative and $f(2)$ is positive, it follows that $f(c) = 0$ for some c between 0 and 2. (This is an application of the intermediate value theorem.) Newton developed a method of successive approximations that makes it

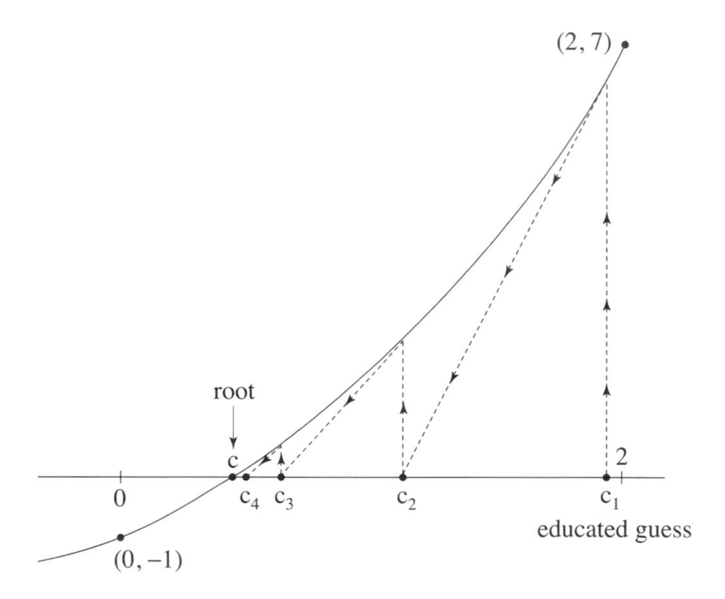

Fig. 7.60

possible to calculate the root c to any desired degree of accuracy. The essence of his method is illustrated in Figure 7.60.

i. Make an educated guess, say, c_1 for c.

ii. Use the following approximation step. From a point on the x-axis go up to the graph and down on the tangent to a new point on the x-axis.

iii. Begin by applying the approximation step to the guess c_1 to get c_2; apply the approximation step to c_2 to get c_3; apply it to c_3 to get $c_4, \ldots$. Continue in this way, and observe that the numbers c_1, c_2, c_3, $c_4, \ldots$ zero in on the root c.

We will now see how Newton makes this mathematically precise. Let's start with $c_1 = 2$. To determine c_2, the intersection of the tangent at $(c_1, f(c_1))$ with the x-axis needs to be found. Because $f'(x) = 4x^3 + 6x^2 - 12$, the slope of the tangent is $f'(c_1) = f'(2) = 4\cdot 2^3 + 6\cdot 2^2 - 12 = 44$. Since $f(c_1) = f(2) = 7$, the equation of the tangent is

$$y - f(c_1) = f'(c_1)(x - c_1) \quad \text{or} \quad y - 7 = 44(x-2).$$

The x-coordinate c_2 of the point of intersection is determined by setting $y = 0$ and solving for x. So $-f(c_1) = f'(c_1)(x - c_1)$, $x - c_1 = -\frac{f(c_1)}{f'(c_1)}$, and $x = c_1 - \frac{f(c_1)}{f'(c_1)}$. Therefore, working with an accuracy of four decimal places,

$$c_2 = c_1 - \frac{f(c_1)}{f'(c_1)} = 2 - \frac{7}{44} = \frac{81}{44} = 1.8410.$$

In exactly the same way, $c_3 = c_2 - \frac{f(c_2)}{f'(c_2)}$ and $c_4 = c_3 - \frac{f(c_3)}{f'(c_3)}$. In general, once the approximation c_i of the root has been found, then the next and better approximation is given by

$$c_{i+1} = c_i - \frac{f(c_i)}{f'(c_i)}.$$

In the specific case we are considering, we get

$$\begin{aligned}
c_3 &= c_2 - \frac{f(c_2)}{f'(c_2)} = 1.8410 - \frac{f(1.8410)}{f'(1.8410)} \\
&= 1.8410 - \frac{(1.8410)^4 + 2(1.8410)^3 - 12(1.8410) - 1}{4(1.8410)^3 + 6(1.8410)^2 - 12} \\
&= 1.8410 - \frac{0.8746}{33.2944} = 1.8147, \\
c_4 &= 1.8147 - \frac{(1.8147)^4 + 2(1.8147)^3 - 12(1.8147) - 1}{4(1.8147)^3 + 6(1.8147)^2 - 12} \\
&= 1.8147 - \frac{0.0205}{31.6630} = 1.8141, \\
c_5 &= 1.8141 - \frac{(1.8141)^4 + 2(1.8141)^3 - 12(1.8141) - 1}{4(1.8141)^3 + 6(1.8141)^2 - 12} \\
&= 1.8141 - \frac{0.0015}{31.6263} = 1.8141.
\end{aligned}$$

So 1.8141 ought to be a close approximation of a root of the polynomial $f(x) = x^4 + 2x^3 - 12x - 1$. Our computation has shown that up to four decimal places, $f(c_5) = 0.0015$. Carrying more decimal places would have provided greater accuracy. Since $f(1.8141) = 0.0015$ and $f(1.8140) = -0.0017$, note that the root is somewhere between 1.8141 and 1.8140.

7.88. Apply Newton's method to $f(x) = \frac{1}{2}x^2 - 1$. Start with $c_1 = 2$ and use four-decimal-place accuracy to approximate a root. Check what you got by observing that the roots are equal to $\pm\sqrt{2}$.

7.89. Apply Newton's method to $f(x) = x^3 + x^2 - 7x - 7$. Start with $c_1 = 3$, and use four-decimal-place accuracy to approximate a root. Compare what you got by solving for the roots. Start by noticing that $x = -1$ is a root, and carry out a polynomial division to find the others.

7.90. Apply Newton's method to $f(x) = x^3 + x^2 + 7x + 7$. As in the previous problem, start with $c_1 = 3$. What happens this time? As in the previous problem, $x = -1$ is again a root. What are the others?

7.91. Consider a function $y = f(x)$, and assume that it is differentiable on $(-\infty, \infty)$. Suppose $a < b$, and that c is the unique solution of the equation $f(x) = 0$ in $[a, b]$.

i. Suppose that the graph of f is either concave up and increasing on $[c, b]$, or concave down and decreasing on $[c, b]$. Draw a generic graph for each of these two situations that shows that for any guess c_1 with $c < c_1 < b$, Newton's method will converge to c.

ii. Suppose that the graph of f is either concave up and decreasing on $[a, c]$ or concave down and increasing on $[a, c]$. Draw a generic graph for each of these two cases that illustrates that Newton's method will converge to c for any guess c_1 with $a < c_1 < c$.

iii. Draw a graph that illustrates how Newton's method can fail for a guess c_1 satisfying $c < c_1 < b$. Draw another graph that illustrates the failure of Newton's method for a guess c_1 with $a < c_1 < c$.

Consider the polynomial function $f(x) = x^4 - 3x^2 + 2$. The factorization $x^4 - 3x^2 + 2 = (x^2 - 1)(x^2 - 2)$ shows that $-\sqrt{2}, -1, 1,$ and $\sqrt{2}$ are the roots of the polynomial.

7.92. Make a careful sketch of the graph of the function $f(x) = x^4 - 3x^2 + 2$, paying particular attention to the intervals over which the function is increasing, decreasing, concave up, and concave down. Check that the graph has horizontal tangents at $(-\sqrt{\frac{3}{2}}, -\frac{1}{4})$, $(0, 2)$, and $(\sqrt{\frac{3}{2}}, -\frac{1}{4})$. Include these points along with $(-\sqrt{2}, 0)$, $(-1, 0)$, $(1, 0)$, and $(\sqrt{2}, 0)$ in your graph. Check that $(-1, 0)$ and $(1, 0)$ are the points of inflection.

7.93. With regard to the application of Newton's method to $f(x) = x^4 - 3x^2 + 2 = 0$, consider the guesses $c_1 = 3, \sqrt{\frac{3}{2}}, 1.1, 0.9,$ and 0.1. By studying the graph of the function (and without doing any computations), predict for each of these five guesses which root of $x^4 - 3x^2 + 2 = 0$ Newton's method will converge to. Carry out the details of Newton's method for $c_1 = 0.1$ and $c_1 = 0.9$.

Chapter 8

Applications of Differential Calculus

Mathematics in general, and differential and integral calculus in particular, has been energized throughout its development by its relationship to basic science, engineering, and economics. In return, mathematics with its insistence on exactness has provided these fields with substantial information and defining clarity. It is one of the primary purposes of this text to illustrate this ongoing process of cross-fertilization. The topics of science and engineering that are involved are broad and complex, and this text can only provide a limited perspective. The role of the derivative in bacterial biology, nuclear physics, and economics is briefly illustrated. Broader and deeper is the narrative about the interaction between differential calculus and basic aspects of the disciplines of statics, dynamics, and optics. *Statics* is the branch of physics that studies the forces acting on a material object at rest under equilibrium conditions, *dynamics* investigates the motion of an object in relation to the physical factors such as force and mass that affect its motion, and *optics* is the study of light and the consequences of its basic properties. Each of these three concerns has developed into a large and important discipline. Statics underlies essential aspects of civil engineering and architecture. Dynamics is fundamental to mechanical and aerospace engineering. Optics in the context of instruments such as cameras, microscopes, and telescopes continues to have important impact, especially on astronomy. Mathematics has played a crucial role in the development of all three.

Our discussion of the subject of statics begins with one of the most interesting problems—the pulley problem—from the Marquis de L'Hospital's *Analyse des infiniment petits*, the world's first basic calculus text. We will discuss two solutions. One will illustrate the methods of calculus; the other will deploy basic properties of forces. A mathematical analysis of the suspension bridge follows. Early versions of such bridges using vines and ropes were built by natives of the Himalayas, Equatorial Africa, and South America. Suspension bridges supported by metal chains have spanned rivers in China since the 17th century. The initial mathematical analysis of the suspension bridge was undertaken late in the 18th century in response to a proposal to erect such a bridge in St. Petersburg, Russia. The basic fact is that a suspended cable subject to a uniform vertical load—this is the condition that the main cables of a suspension bridge satisfy—describes a parabola. We will describe this analysis, as well as some of the conclusions of the French mathematician and engineer Henri Navier, in the context of the George Washington Bridge. This bridge, constructed over the Hudson River in the period 1927–1931, was the first of a then-new generation of suspension bridges. To illustrate the connection between mathematics and dynamics, we will analyze an important experiment of Galileo. As late as the 1970s, historians of science could only speculate about the experiments that Galileo actually undertook. Indeed, some influential historians of science held that Galileo's experiments were primarily thought experiments, in other words, carefully conceived speculations. This assessment turned out to be incorrect. The evidence for this comes from Galileo's working papers, which were discovered to contain the results of a number of precise experiments with inclined planes. For instance, a page from the year 1608—certainly one of the most important pages in the history of science—records the numerical data of an experiment into the nature of trajectories. The goal of our discussion will be the analysis of this experiment with particular attention to the mathematics that underlies it. Perhaps most interesting is the application of calculus to optics. The development of the telescope energized astronomy and, in turn, investigations into the nature of light. It was realized that light was only one slice of the broader phenomenon of electromagnetic radiation ranging from the infrared, to the ultraviolet, to X-rays. The common fundamental feature is that such radiation consists of a combination of particles and waves. Larger and larger telescopes that rely on combinations of lenses, parabolic mirrors, and sophisticated technologies continue to be built. They gather light from galaxies that are so far away that it takes billions of years to reach its mirrors. These telescopes therefore provide information about the early stages of the development of the universe. Our study will take a look at the impact of calculus on basic optics.

When it comes to the application of mathematics in concrete situations, there are—almost always—inherent limitations. If mathematical principles and strategies are to be successfully applied, fundamental effects need to be incorporated, possibly in simplified form, but in order to avoid excessive complexity, secondary effects are often modified and sometimes ignored. The consequence is that the mathematical conclusions are almost always only approximations—often very accurate indeed—of what actually occurs. A second limitation has to do with the data that are used and the calculations that are performed. Unless the concern is the number of apples in a basket, goldfish in an aquarium, or dollars in a wallet, the measurements that an applied context provides as the starting point of a mathematical analysis are also only approximate. Think of something as simple as measuring the length of a room. Is it exactly 7 meters? Or 7 meters and $2\frac{1}{4}$ centimeters? So is it 7.0225 meters long? Let's say that the measured length is 7.0224, that the 7, 0, 2, and the next 2 are certain, but not the rest. Then these four figures are called *significant*. Additional measurements of the size of the room will also involve some number of significant figures. Suppose that n is the smallest number of the significant figures involved. When arithmetic operations are carried out with these numbers, the standard practice is to use the complete numerical data, but to round the final results off to n places. Suppose that the room in question is rectangular and that its other side is 5.3659 meters long with only the 5, 3, and 6 significant. If the concern is the area of the room, then $n = 3$, the area of the room is computed as $7.0224 \times 5.3659 = 37.68149616$ square meters, and this figure is rounded off to 37.7 square meters. (Since the central focus of this text is calculus and its applications, we will only loosely adhere to this roundoff strategy.)

8.1 DERIVATIVES AS RATES OF CHANGE

It is often of interest to study rates of increase and decrease of relevant quantities and to estimate their largest or smallest values. Such quantities can often be modeled with functions that vary smoothly. They therefore lend themselves to analysis by calculus. If a function $y = f(x)$ represents a specific physical quantity within a given context, then its derivative $f'(x)$ has a meaning derived from the context.

Before starting the discussion about statics, dynamics, and optics, we'll briefly consider examples that illustrate the reach of calculus. In the first two, time $t \geq 0$ is the basic variable. Think of t as elapsed time as given by a stopwatch that is started at time $t = 0$.

8.1.1 Growth of Organisms

Consider a culture of bacteria that has been introduced into a host organism or one that has been transferred to a test tube for study. Generally, there is an initial period, the *lag phase*, during which the bacteria adapt to the new medium and there is little if any growth. Eventually, the lag phase ends, and these one-celled organisms begin to divide. If nutriments and temperature are ideal and there are no agents (such as antibiotics) that might attack it, the culture will enter an *exponential growth phase*. At a regular interval called *doubling time*, one cell will split into two, two into four, four into eight, and so on. The doubling time depends on both the culture and the specifics of the medium (types of nutrients, acidity, and temperature, for instance). Under ideal conditions, most bacteria have a doubling time of 1 to 3 hours. For some microbes it is only a few minutes, and for others it can be many hours. For the bacterium *E. coli*, this maximal doubling time is about 16 minutes. After some time, this period of rapid increase in the number of bacteria will stop, the culture will enter a stationary phase, and, possibly later still, a death phase. Our focus will be on the exponential growth phase of the culture. It is during this phase that cellular reproduction and metabolic activity are at their peak. It is also the phase during which the cells are most susceptible to adverse conditions. Radiation or antibiotics, for example, exert their effect by interfering with some important step during this phase of the growth process.

Suppose that a culture is observed during its exponential growth phase. Designate the time at the beginning of the observation by $t = 0$. Let $y(t)$ be the number of cells in the culture at any time $t \geq 0$. So $y(0) = y_0$ is the number of cells when the culture is first observed. Let d be the doubling time. At times $t = d, 2d, 3d, \ldots, nd, \ldots$, the number of cells will have doubled, doubled twice, doubled three times, ..., doubled n times, ... Therefore

$$y(d) = 2y_0,\ y(2d) = 2(2y_0) = 2^2y_0,\ y(3d) = 2(2^2y_0) = 2^3y_0,\ \ldots,\ y(nd) = 2^ny_0, \ldots .$$

Note that this is the quantitative behavior of the culture as a whole and that it is not the case that all cells divide at the same instant. Take any time t and let $r = \frac{t}{d}$. So $t = rd$. Making use of the pattern just observed, we get that

$$y(t) = y(rd) = 2^ry_0 = 2^{\frac{t}{d}}y_0 = (2^{\frac{1}{d}})^ty_0.$$

This formula for $y(t)$ can be rewritten as follows. Start with the equality $2 = e^{\ln 2}$, and raise both sides to the power $\frac{1}{d}$ to get $2^{\frac{1}{d}} = e^{\frac{\ln 2}{d}}$. Now set $\mu = \frac{\ln 2}{d}$. So $2^{\frac{1}{d}} = e^{\mu}$. So by a substitution,

$$\boxed{y(t) = y_0e^{\mu t}}$$

The constant $\mu = \frac{\ln 2}{d} \approx \frac{0.693}{d}$ is called the *growth constant* of the culture. So during the exponential growth phase of a culture, the increase in the number of bacteria is given by an exponential function. See Fig-

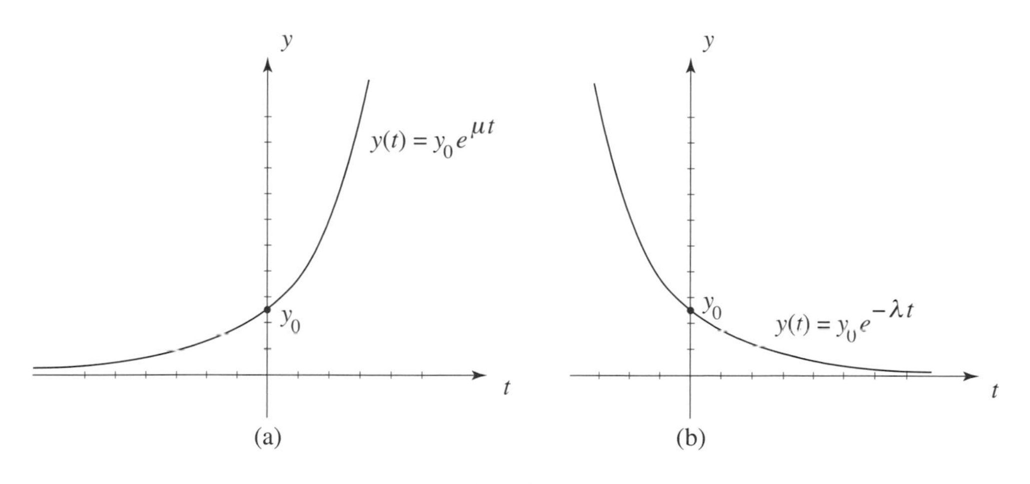

Fig. 8.1

ure 8.1a. If the exponential growth phase of a culture goes on for any length of time, an enormous number of cells can accumulate.

Let's continue the analysis of the number $y(t)$. Fix a time t. Now let an additional time Δt elapse. Then $y(t + \Delta t)$ is the number of bacteria at time $t + \Delta t$, and $y(t + \Delta t) - y(t)$ is the increase in this number during the time interval $[t, t + \Delta t]$. So

$$\frac{y(t + \Delta t) - y(t)}{\Delta t}$$

is the rate at which this increase unfolds. It is the *average rate of increase in the number of bacteria* per unit time during the time interval $[t, t + \Delta t]$. By taking this average rate of increase over smaller and smaller Δt, in other words by pushing Δt to 0, we get what is called the *rate of increase of the number of bacteria at the instant* t. Observe that this is nothing but the derivative

$$y'(t) = \lim_{\Delta t \to 0} \frac{y(t + \Delta t) - y(t)}{\Delta t}$$

of $y(t)$. By differentiating $y(t) = y_0 e^{\mu t}$ with the chain rule, we get $y'(t) = y_0 \mu e^{\mu t} = \mu y(t)$. The equation

$$y'(t) = \mu y(t),$$

for $t \geq 0$, is characteristic of the exponential phase of growth. It asserts that $y'(t)$, the rate of increase in the cell count at time t, is proportional to $y(t)$. The growth constant μ is the constant of proportionality. The proportionality $y'(t) = \mu y(t)$ implies by remarks that conclude Section 7.10 that $y(t) = Ae^{\mu t}$ for some constant A. By substituting $t = 0$, we get that $y_0 = y(0) = Ae^0 = A$. So we get what we already knew, namely, that $y(t) = y_0 e^{\mu t}$.

8.1.2 Radioactive Decay

The collection of chemical elements includes stable elements and radioactive elements. The former remain unchanged over time, while the radioactive elements become transformed. Consider a sample of a radioactive substance. Over time, the atoms of this substance will change or *decay* to another substance (and emit energy in the form of radiation in the process). For instance, carbon-14 decays to nitrogen-14, radium-226 to radon-222, and uranium-238 to thorium-234. The numbers refer to the numbers of protons and neutrons in the nucleus of one of the atoms. We begin to observe the sample at time $t = 0$, and let $y(t)$ be the amount of the radioactive substance—either in terms of the number of atoms or the mass in some units such as grams—at any time $t \geq 0$. So $y(0) = y_0$ is the initial amount. As the atoms of the sample decay, the number y_0 decreases. The decay of any radioactive substance satisfies the following basic law: there is a regular interval of time h, called the *half-life* of the substance, such that the amount of the substance at the beginning of this interval will have decayed to one-half this amount at the end of this time interval. So at times $h, 2h, 3h, \ldots, nh$ into the observation of the sample, the number of atoms will decay to one-half, one-half again, one-half once more, and so on, of the original amount y_0. Therefore

$$y(h) = \tfrac{1}{2}y_0,\ y(2h) = \tfrac{1}{2}(\tfrac{1}{2}y_0) = \tfrac{1}{2^2}y_0,\ y(3h) = \tfrac{1}{2}(\tfrac{1}{2^2}y_0) = \tfrac{1}{2^3}y_0,\ \ldots,\ y(nh) = \tfrac{1}{2}(\tfrac{1}{2^{n-1}}y_0) = \tfrac{1}{2^n}y_0, \ldots .$$

The half-lives of radioactive elements vary dramatically from the tiniest fraction of a second to billions of years. The half-life of radon-222 is 3.8 days, that of carbon-14 is about 5730 years, and that of uranium-238 is about 4.5 billion years. (The latter is therefore close to being completely stable.)

Take any time t and let $r = \frac{t}{h}$. So $t = rh$. Following the pattern already established, we get

$$y(t) = y(rh) = \frac{1}{2^r}y_0 = \frac{1}{2^{\frac{t}{h}}}y_0 = \Big(\frac{1}{2^{\frac{1}{h}}}\Big)^t y_0.$$

Because $2 = e^{\ln 2}$, we get $2^{\frac{1}{h}} = e^{\frac{\ln 2}{h}}$. With $\lambda = \frac{\ln 2}{h}$, $2^{\frac{1}{h}} = e^{\lambda}$. So $\frac{1}{2^{\frac{1}{h}}} = e^{-\lambda}$, and therefore

$$\boxed{y(t) = y_0 e^{-\lambda t}}$$

The constant $\lambda = \frac{\ln 2}{h} \approx \frac{0.693}{h}$ is called the *decay constant* of the radioactive substance. The decreasing number of atoms of the radioactive sample is given by a decreasing exponential function. See Figure 8.1b.

When the sample is first observed, the amount is $y(0) = y_0$. Fix a time t, and then let an additional Δt elapse. Since $y(t)$ is the amount of the substance at time t and $y(t + \Delta t)$ is the amount at time $t + \Delta t$, it follows that

$$y(t + \Delta t) - y(t)$$

is negative and equal to the decrease in the amount during the time interval $[t, t + \Delta t]$. The ratio

$$\frac{y(t + \Delta t) - y(t)}{\Delta t}$$

is the *average rate of decrease of the amount* per unit time over the time interval $[t, t + \Delta t]$. Since the quantity $y(t)$ is decreasing, this rate is negative. Pushing Δt to 0, we get the *rate of decrease of the amount at the instant* t. This is equal to the derivative

$$y'(t) = \lim_{\Delta t \to 0} \frac{y(t + \Delta t) - y(t)}{\Delta t}$$

of $y(t)$. It is negative for any t. Differentiating $y(t) = y_0 e^{-\lambda t}$, we get that $y'(t) = \lambda y_0 e^{-\lambda t}$, and hence that

$$y'(t) = -\lambda y(t).$$

So $y'(t)$ and $y(t)$ are proportional with $-\lambda$ the constant of proportionality.

The validity of the analysis of radioactive decay undertaken in this section requires the assumption that the initial amount consists of a large number of atoms. In particular, it says essentially nothing about an individual atom. Only assertions about probabilities are possible. The probability that a certain event occurs is some number between 0 and 1. A probability of 0 means that the event has no chance of occurring, and a probability of 1 means that it will definitely occur. For instance, if in a bag of 10 marbles, 3 are red and 7 are blue, then the probability of pulling a red marble out of the bag (blindly drawn) is 3 out of 10, or $\frac{3}{10} = 0.3$. What is the probability that a particular atom in some radioactive substance with y_0 atoms initially will decay during the time interval $[0, t]$? In other words, what is the probability that at time t the particular atom is no longer among the $y(t)$ atoms of the substance still remaining? Since there are y_0 atoms at the beginning and $y(t)$ at time t, it follows that $y_0 - y(t)$ have decayed, so that this probability is

$$\frac{y_0 - y(t)}{y_0} = 1 - \frac{y_0 e^{-\lambda t}}{y_0} = 1 - e^{-\frac{\ln 2}{h} t}.$$

The probability that a particular atom of a radioactive substance during the time from $t = 0$ to $t = h$ is $1 - e^{-\frac{\ln 2}{h} h} = 1 - e^{-\ln 2} = 1 - \frac{1}{e^{\ln 2}} = 1 - \frac{1}{2} = 0.5$, as one would have expected. This observation provides a glimpse at the fact that sophisticated statistical and probabilistic methods—known as quantum mechanics and quantum electrodynamics—lie at the heart of modern nuclear physics.

8.1.3 *Cost of Production*

Consider a manufacturing company, and suppose that the total cost of producing x units per year of a certain product is $C(x)$. Suppose that the company raises the output of the product by Δx units from x to $x + \Delta x$ units per year. The cost of producing the additional Δx units per year is $C(x + \Delta x) - C(x)$, and the average cost of these additional Δx units is

$$\frac{C(x + \Delta x) - C(x)}{\Delta x}.$$

Consider this ratio, and push Δx to 0. What you get is the derivative

$$C'(x) = \lim_{\Delta x \to 0} \frac{C(x + \Delta x) - C(x)}{\Delta x}$$

of $C(x)$. This derivative is called the *marginal cost* at x. By taking Δx to be a small number of units, we get an approximation

$$C'(x) \approx \frac{C(x + \Delta x) - C(x)}{\Delta x}$$

of the marginal cost. So the marginal cost $C'(x)$ should be thought of as the average cost per unit of a small increase in production from the base level of x units.

Many different cost situations can arise, but what often happens in the short term is this. Generally, when output is increased, the labor force needs to be added to. As more employees are hired, the existing infrastructure suffices initially, so that the average cost per each additional unit produced decreases and the marginal cost goes down. But at some point, inefficiencies and bottlenecks may appear. These may involve assembly lines, manufacturing machinery, and warehouse space for supplies and inventory. In such a situation, it is possible that the cost per each additional unit produced increases and that the marginal cost goes up. So in the short term, it may well be that the marginal cost falls initially, but that it levels off after a time, and then starts to increase. This pattern suggests that the marginal cost is given by a parabola or, more precisely, by a quadratic function of the form

$$C'(x) = ax^2 + bx + c,$$

with a, b, and c constants and $a > 0$. This means in turn that the cost function has the form $C(x) = \frac{a}{3}x^3 + \frac{b}{2}x^2 + cx + d$. Since $C(0) = d$ is the cost of maintaining and operating the firm before even a single unit of the item is produced, $C(0)$ is the *fixed cost*. The cost $C(x) - C(0) = \frac{a}{3}x^3 + \frac{b}{2}x^2 + cx$ is the *variable cost*. In the context just described, both the total and the variable costs are cubic functions. In the short-term situation that we have discussed, the fixed cost is often nearly constant. In the longer term, the investment that a firm might make in expanding its production facilities would add to the fixed cost (while at the same time reducing the marginal cost). The revenue from the sale of x units at a unit price of p is $R(x) = p \cdot x$, and the profit from these sales is $P(x) = R(x) - C(x)$. The study of revenue, profit, and also of supply and demand, proceeds in a way that is similar to the analysis of the cost as given above.

A word of clarification is in order for each of the three examples that were discussed above. The fact is that none of the functions that arise in the examples is continuous, and therefore none are differentiable. This is clear in the case of the functions that represent the numbers of bacteria in a culture, the number of atoms in a radioactive substance, and the cost of a number of units of a product. As t or x varies, the values of the functions $y(t)$ and $C(x)$ jump from one integer value to another. In order for the derivative of such function to make sense, it is assumed that there is a differentiable function that agrees with the data that the example presents. For instance, in the case of the bacteria culture, this was one in going from $y(d) = 2y_0,\ y(2d) = 2^2y_0,\ y(3d) = 2^3y_0,\ \ldots,$ to the function $y(t) = y_0e^{\mu t}$. That this is possible in general can best be seen geometrically: a set of points in the plane with different horizontal coordinates can be studied by fitting a smooth curve through the points and by analyzing the curve.

8.2 THE PULLEY PROBLEM OF L'HOSPITAL

Consider two distinct points B and C on the horizontal ceiling of a room. Suppose that the points are a distance c apart. At the point C, attach a cord of length a, which has a pulley affixed at the other end, the point F. At the point B, attach a cord of length l, pass it through the pulley at F, and connect a weight W at the other end, at D. Release the weight and allow the system to achieve its equilibrium position. This is shown in Figure 8.2a. The question that L'Hospital asks is this: What is the geometry of this equilibrium configuration? More precisely, what are the dimensions of the triangle BCF in terms of the constants c, a, l, and W? We will see that the solution of this seemingly elementary problem is surprisingly challenging. L'Hospital makes the simplifying assumption that the weight W is heavy enough so that the weights of the cords and the pulley are negligible and can be ignored.

The first remark that needs to be made is this. Consider the case $a \geq c$. Under this assumption, the cord BD will hang vertically and support the entire weight W. The reason is that the cord CF will not be under any tension, hang loosely, and have no ability with its pulley at F to deflect the cord BD. See Figure 8.2b. With this equilibrium situation out of the way, we will assume from now on that $a < c$.

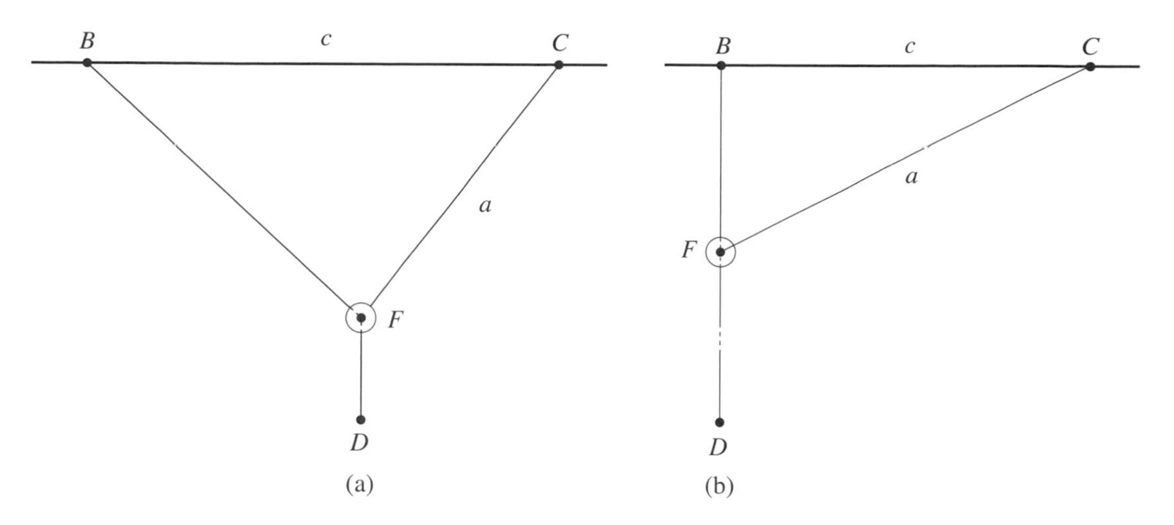

Fig. 8.2

8.2.1 The Solution Using Calculus

Let E be the intersection of the extension of the segment FD with BC, and set x equal to EC. See Figure 8.3. Applying the Pythagorean theorem to the right triangle CEF, we see that $EF = \sqrt{a^2 - x^2}$. Note

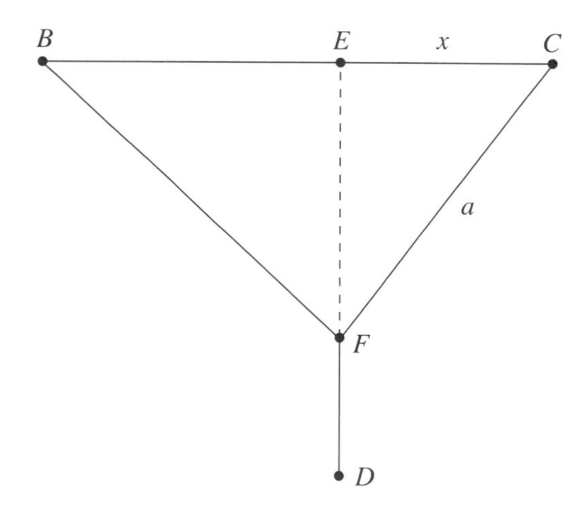

Fig. 8.3

therefore that $x \leq a$. Another application of the Pythagorean theorem shows that

$$BF = \sqrt{BE^2 + EF^2} = \sqrt{(c-x)^2 + (a^2 - x^2)}.$$

Because the length of the string from B to D is l, we see that $FD = l - BF = l - \sqrt{(c-x)^2 + (a^2 - x^2)}$. The fact that the system is in equilibrium means that the weight at D must be at the lowest point that the geometry of the situation will allow. Put another way, the distance ED must be a maximum. Observe that

$$\begin{aligned} ED &= EF + FD \\ &= \sqrt{a^2 - x^2} + l - \sqrt{(c-x)^2 + (a^2 - x^2)} \\ &= (a^2 - x^2)^{\frac{1}{2}} + l - \left((c-x)^2 + (a^2 - x^2)\right)^{\frac{1}{2}}. \end{aligned}$$

If L'Hospital can find the value of x for which ED is a maximum in terms of the constants c, a, and l, he will then have determined $EC = x$, $BE = c - x$, as well as EF, BF, EF, and FD in terms of these constants, and he will have solved his problem. It therefore remains to find the value (or, possibly, values) of x for which the function

$$f(x) = (a^2 - x^2)^{\frac{1}{2}} + l - \left((c-x)^2 + (a^2 - x^2)\right)^{\frac{1}{2}},$$

with $0 \le x \le a$, attains its maximum value. By the continuity theorem (and Example 7.10), we know that the function $y = f(x)$ is continuous of the interval $[0, a]$. So this function has a maximum value. Section 7.8 tells us how to find it. Compute $f'(x)$, and find all the numbers x for which f' is either zero or does not exist. These are the critical numbers of f as well as the endpoints 0 and a. Then evaluate f at the critical numbers as well as at the endpoints 0 and a. The largest of these values provides the required x.

By the chain rule,

$$\begin{aligned} f'(x) &= \tfrac{1}{2}(a^2 - x^2)^{-\frac{1}{2}}(-2x) - \tfrac{1}{2}\left((c-x)^2 + (a^2 - x^2)\right)^{-\frac{1}{2}}(-2(c-x) - 2x) \\ &= -x(a^2 - x^2)^{-\frac{1}{2}} - \tfrac{1}{2}\left((c-x)^2 + (a^2 - x^2)\right)^{-\frac{1}{2}}(-2c) \\ &= -x(a^2 - x^2)^{-\frac{1}{2}} + c\left((c-x)^2 + (a^2 - x^2)\right)^{-\frac{1}{2}} \\ &= \frac{c}{((c-x)^2 + (a^2 - x^2))^{\frac{1}{2}}} - \frac{x}{(a^2 - x^2)^{\frac{1}{2}}}. \end{aligned}$$

Since $x \le a < c$, we see that $x = a$ (and the other endpoint $x = 0$) are the only numbers at which $f'(x)$ does not exist. After setting the last expression for $f'(x)$ equal to zero and cross multiplying, we see that $f'(x)$ is equal to zero precisely when

$$c(a^2 - x^2)^{\frac{1}{2}} = x\left((c-x)^2 + (a^2 - x^2)\right)^{\frac{1}{2}}.$$

After squaring both sides, we get $c^2(a^2 - x^2) = x^2(c-x)^2 + x^2(a^2 - x^2)$. By rearranging things,

$$\begin{aligned} c^2a^2 - c^2x^2 &= x^2(c^2 - 2cx + x^2) + x^2a^2 - x^4 = x^2c^2 - 2cx^3 + x^4 + x^2a^2 - x^4 \\ &= c^2x^2 - 2cx^3 + a^2x^2. \end{aligned}$$

It follows that

$$(*) \qquad 2cx^3 - 2c^2x^2 - a^2x^2 + a^2c^2 = 0.$$

The solution of this equation for x begins by noticing that

$$2cc^3 - 2c^2c^2 - a^2c^2 + a^2c^2 = 0.$$

So $x = c$ is a root of the polynomial $(*)$, and hence the term $x - c$ divides it. (See segment 4E in the Problems and Projects section of Chapter 4.) By a polynomial division,

$$2cx^3 - 2c^2x^2 - a^2x^2 + a^2c^2 = (x - c)(2cx^2 - a^2x - a^2c).$$

(You can check this by multiplying the two factors out.) The quadratic formula tells us that the roots of $2cx^2 - a^2x - a^2c$ are

$$\begin{aligned} \frac{a^2 \pm \sqrt{a^4 - 4(2c)(-a^2c)}}{4c} &= \frac{a^2 \pm \sqrt{a^4 + 8a^2c^2}}{4c} = \frac{a^2 \pm a\sqrt{a^2 + 8c^2}}{4c} \\ &= \tfrac{a}{4c}(a \pm \sqrt{a^2 + 8c^2}). \end{aligned}$$

Since $x = c$ cannot occur, we have shown that $f'(x) = 0$ only for

$$x = \tfrac{a}{4c}(a + \sqrt{a^2 + 8c^2}) \quad \text{and} \quad x = \tfrac{a}{4c}(a - \sqrt{a^2 + 8c^2}).$$

Notice that $\sqrt{a^2 + 8c^2} > \sqrt{a^2} = a$. It follows that $a - \sqrt{a^2 + 8c^2} < 0$. This rules out the possibility that $x = \frac{a}{4c}(a - \sqrt{a^2 + 8c^2})$.

We summarize the results of our computations. The only critical number of the function

$$f(x) = \sqrt{a^2 - x^2} + l - \sqrt{(c - x)^2 + (a^2 - x^2)},\ 0 \le x \le a,$$

is $x = \frac{a}{4c}(a + \sqrt{a^2 + 8c^2})$.

By Section 7.8, the maximum value of the function f must occur either at one of the endpoints $x = 0$ and $x = a$, or at $x = \frac{a}{4c}(a + \sqrt{a^2 + 8c^2})$. Refer back to Figure 8.3. The geometry of the situation rules out both $x = 0$ and $x = a$. If $x = 0$, then $E = C$ and the weight at D hangs directly under C. But this is impossible, since the weight W pulls on the cord CF that holds the pulley. If $x = a$, then $E = F$. This too is impossible, because the weight W pulls down on the pulley at F. It follows that f attains its maximum value when

$$x = \tfrac{a}{4c}(a + \sqrt{a^2 + 8c^2}).$$

So this is the value of x that the configuration in Figure 8.3 has when it is in equilibrium. Note that W does not appear in this expression for x. Therefore the equilibrium configuration is the same regardless of the magnitude of the weight W. This was not obvious before our analysis began.

Example 8.1. Suppose specifically that $c = 4$ and $a = 3$. Then

$$\begin{aligned} x &= \tfrac{a}{4c}(a + \sqrt{a^2 + 8c^2}) = \tfrac{3}{16}(3 + \sqrt{9 + 128}) \\ &\approx \tfrac{3}{16}(3 + 11.70) = \tfrac{3}{16}(14.70) \approx 2.76. \end{aligned}$$

The pulley hangs $\sqrt{a^2 - x^2} \approx \sqrt{3^2 - 2.76^2} \approx 1.18$ below the point E.

8.2.2 *The Solution by Balancing Forces*

The *tension* in a cable or string is the magnitude of the force—called a *tensile force* —with which the cable or string pulls. Let T_1 be the tension in the cord BF, and let T_2 be the tension in the cord CF to which the pulley is affixed. See Figure 8.4. Let θ_1 and θ_2 be the indicated angles. Since the system is in equilibrium,

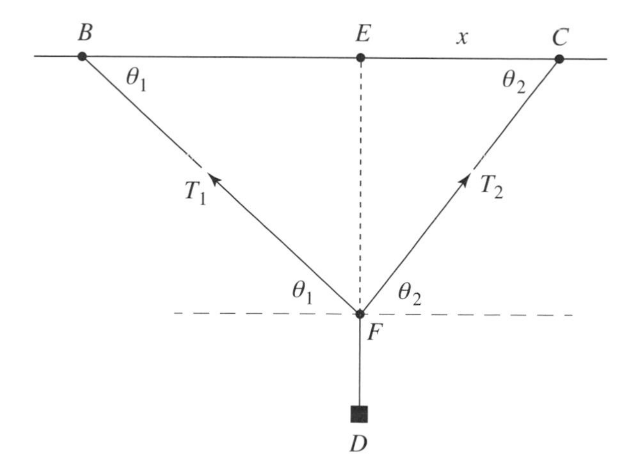

Fig. 8.4

the cord FD pulls with a force equal to the weight W of the suspended object. And because the pulley at F is free to rotate, the tension in the cord BFD must be constant. This implies that $T_1 = W$. (The discussion that follows relies on basic properties of forces. These are discussed in Section 6.6.)

Figure 8.5 provides a diagram of the forces acting at F. The forces T_1 and T_2 are decomposed into their horizontal and vertical components. From the definitions of the sine and cosine, we know that the two horizontal components have magnitudes $T_1 \cos\theta_1$ and $T_2 \cos\theta_2$, respectively, and the magnitudes of the two vertical components are $T_1 \sin\theta_1$ and $T_2 \sin\theta_2$. Since the forces are in balance, the two horizontal

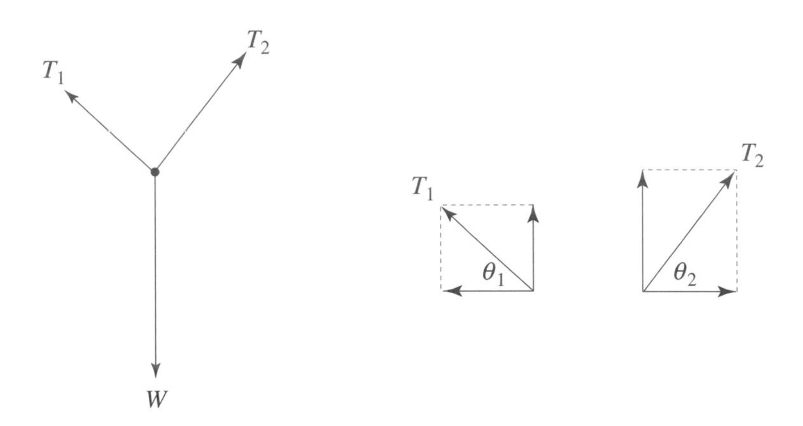

Fig. 8.5

components must be equal, and the two vertical components must add up to the weight W. Therefore,

$$T_1 \cos\theta_1 = T_2 \cos\theta_2 \quad \text{and} \quad T_1 \sin\theta_1 + T_2 \sin\theta_2 = W.$$

Using the second equation and the fact that $W = T_1$, we get

$$\frac{T_1 \sin\theta_1}{T_1 \cos\theta_1} + \frac{T_2 \sin\theta_2}{T_1 \cos\theta_1} = \frac{T_1 \sin\theta_1 + T_2 \sin\theta_2}{T_1 \cos\theta_1} = \frac{W}{T_1 \cos\theta_1} = \frac{T_1}{T_1 \cos\theta_1} = \frac{1}{\cos\theta_1}.$$

Because $T_1 \cos\theta_1 = T_2 \cos\theta_2$,

$$\frac{T_1 \sin\theta_1}{T_1 \cos\theta_1} + \frac{T_2 \sin\theta_2}{T_2 \cos\theta_2} = \frac{1}{\cos\theta_1}.$$

It follows that $\tan\theta_1 + \tan\theta_2 = \frac{1}{\cos\theta_1}$. Refer to Figure 8.4, and recall from the beginning of Section 8.2.1 that $BE = c - x$, $EF = \sqrt{a^2 - x^2}$, and $BF = \sqrt{(c-x)^2 + (a^2 - x^2)}$. Hence

$$\tan\theta_1 = \frac{EF}{BE} = \frac{\sqrt{a^2 - x^2}}{c - x}, \ \tan\theta_2 = \frac{EF}{EC} = \frac{\sqrt{a^2 - x^2}}{x}, \text{ and } \cos\theta_1 = \frac{BE}{BF} = \frac{c - x}{\sqrt{(c-x)^2 + (a^2 - x^2)}}.$$

After substituting, we get

$$\frac{\sqrt{a^2 - x^2}}{c - x} + \frac{\sqrt{a^2 - x^2}}{x} = \frac{\sqrt{(c-x)^2 + (a^2 - x^2)}}{c - x}.$$

Taking common denominators shows that

$$\frac{x\sqrt{a^2 - x^2} + (c - x)\sqrt{a^2 - x^2}}{x(c - x)} = \frac{\sqrt{(c-x)^2 + (a^2 - x^2)}}{c - x}.$$

Simplifying the numerator on the left, canceling $c - x$ from both sides, and multiplying through by x gives

$$c\sqrt{a^2 - x^2} = x\sqrt{(c-x)^2 + (a^2 - x^2)}.$$

By squaring both sides,

$$c^2(a^2 - x^2) = x^2\left((c-x)^2 + (a^2 - x^2)\right).$$

By routine algebra, we now see that

$$\begin{aligned} c^2a^2 - c^2x^2 &= x^2(c^2 - 2cx + x^2 + a^2 - x^2) = x^2(c^2 - 2cx + a^2) \\ &= x^2c^2 - 2cx^3 + a^2x^2. \end{aligned}$$

It follows, finally, that $2cx^3 - 2c^2x^2 - a^2x^2 + a^2c^2 = 0$. But this is equation $(*)$ that was derived in Section 8.2.1 with the methods of calculus, and we have already seen that the solution $x = \frac{a}{4c}(a + \sqrt{a^2 + 8c^2})$ of this equation solves L'Hospital's pulley problem.

Example 8.2. Suppose that $c = 4$ and $a = 3$, as in Example 8.1. Let $W = 100$ pounds, and determine the angles θ_1 and θ_2 as well as tensions T_1 and T_2. From Example 8.1,

$$x \approx 2.76 \quad \text{and} \quad \sqrt{a^2 - x^2} \approx 1.18.$$

So $\tan\theta_1 = \frac{\sqrt{a^2-x^2}}{c-x} \approx \frac{1.18}{4-2.76} \approx 0.95$ and $\tan\theta_2 = \frac{\sqrt{a^2-x^2}}{x} \approx \frac{1.18}{2.76} \approx 0.43$. By the inverse tan button of a calculator, $\theta_1 \approx 43.58°$ and $\theta_2 \approx 23.15°$. Since $T_1 = W$, we know that $T_1 = 100$ pounds. The tension of the cord holding the pulley is $T_2 = \frac{T_1\cos\theta_1}{\cos\theta_2} \approx \frac{(100)(0.72)}{0.92} \approx 72.26$ pounds.

8.3 THE SUSPENSION BRIDGE

During his second extended visit to Saint Petersburg, Russia, the great Swiss mathematician Leonhard Euler (1707–1783) invited his young countryman Nicolas Fuss (1755–1826) to join him. It was Fuss who in his exploratory studies for a suspension bridge for the Russian capitol discovered that the main supporting cables of such a bridge hung in the shape of a parabola. Later, the French mathematician-engineer Claude-Louis Navier (1785-1836) developed a mathematical theory of suspension bridges and earned fame as a builder of roads and bridges. He also put forward a design for a large suspension bridge over the Seine in Paris with a span of about 500 feet that incorporated his theoretical studies as well as what he learned from the pioneer bridge builders in Britain. However, this bridge was ultimately not successful. When problems with one of the major buttresses surfaced, the financial backing for the bridge was terminated, and the bridge was eventually dismantled.

The basic principle behind the suspension bridge is simple: a road bed is attached by numerous vertical cables to either two or four *main cables*. These are "draped" over two towers and they—together with the towers—support the entire load. See Figure 8.6. The great pioneer of the modern suspension

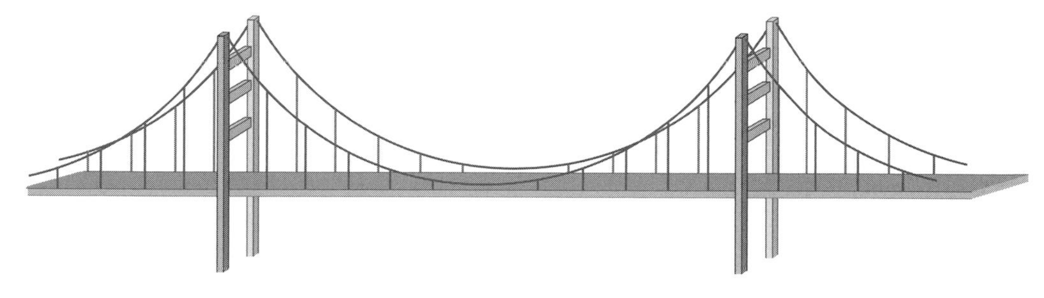

Fig. 8.6

bridge was John Roebling. He designed and supervised the construction of a number of impressive suspension bridges toward the latter part of the 19th century. The most famous of these is the Brooklyn

Bridge over the East River in New York. Opened to traffic in 1883, it has a *center span*—this is the section of the bridge between the two towers—of 1600 feet. The Brooklyn Bridge is generally recognized to be the ancestor of the modern suspension bridge.

A major advance came in 1931 with the George Washington Bridge over the Hudson River in New York. Its 3500-foot center span was almost twice as long as that of any previous suspension bridge. This bridge set a new standard for a series of great suspension bridges in America. These included the Golden Gate Bridge in San Francisco (completed in 1937) with its 4200-foot center span, and the Verrazano Narrows Bridge in New York City (completed in 1964) with a 4260-foot center span. Our discussion will focus on the George Washington Bridge from its inception to its completion. In the process, we will focus on essential elements of planning, engineering, and mathematics.

A need for additional bridges and tunnels connecting New Jersey and New York became apparent in the early 1920s. In 1923, the New York Port Authority was authorized to proceed with the planning, and by the mid-1920s, various preliminary plans had developed into a concrete project for a bridge spanning the Hudson River between New York City and Fort Lee, New Jersey. Given the requirements of shipping and the fact that a bridge of total length of about 5000 feet was required, a major design question was already answered: suspension bridges provide sufficient clearance for shipping and span such distances economically. The project began in September of 1927 with chief engineer Othmar Ammann in charge.

The geological properties of the soil and rock formations under the riverbed of the proposed sight were determined by borings, excavations, soil and compression tests, and so on. At this point, the location of the two main supporting towers was decided upon. This determined the record-setting length of 3500 feet for the *center span*. Refer to Figure 8.7 for the explanation of the terminology used in the discussion that follows. In addition to the center span of 3500 feet, the bridge was to consist of two *side spans* of 610

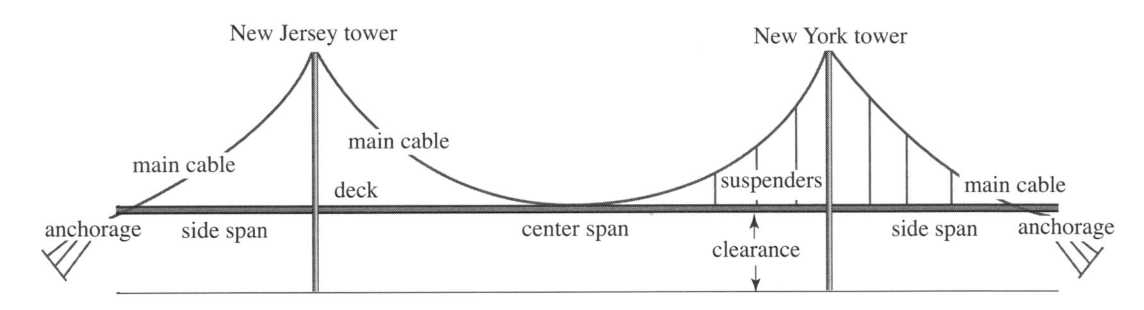

Fig. 8.7

feet and 650 feet each. So the total length of the bridge was to be 4760 feet. The design called for the *deck* to be about 200 feet over the water surface. This would provide enough clearance for the ships of the Hudson.

The deck was designed to accommodate a roadway of eight lanes for a vehicular traffic projected at 25,000,000 vehicles per year. (The plans included provisions for the possible later addition of a second deck. This was added in 1962.) These requirements called for the deck to have a *dead load* (this is the structural weight of the deck without traffic) of about 39,000 pounds per foot for the center span and 40,000 pounds for the two side spans. These figures included not only the weight of the deck, but also the estimated weights of the *main cables* and the *suspenders* (both of which the main cables have to bear). The design also called for a maximum *live load* capacity (the projected weight of the vehicular traffic at its maximum: cars, buses, and trucks, bumper to bumper, on all eight lanes) of 8000 pounds per foot for all spans.

It goes without saying that the execution of such a structure cannot proceed simply by sending out a crew of workers with shovels or suitable machinery, by taking some cables off a shelf, or by hoisting some beams into place. Quite obviously, an exacting and detailed mathematical analysis is required before any aspect of the construction can proceed. Automobiles can be crash tested, but an enormous structure such as the suspension bridge under discussion cannot! And catastrophic failure is not an option.

The mathematical analysis must answer a number of fundamental questions: What loads will the towers have to support? How high should the towers be? What tensions will the main cables be subjected to? How much sag should be allowed in the main cables over the center span? How should the side spans be configured? How much should the tension be in the main cables that support the side spans? What are the requirements on the structures—called *anchorages*—on either end of the bridge that affix the main cables to the ground? What about the suspender cables that attach the deck to the main cables? What effect do all these choices have on the stability of the structure as a whole? What materials should be used? What safety factors have to be built in?

Return to the essential elements of the George Washington Bridge. Estimates for the dead and live loads of the decks are already in place. The vertical suspenders connect the deck to the main cables and generate tensions in these cables. These tensions in turn produce a pull on the towers. It follows that the structural demands on the towers can be understood only if it is known what tensions the loads of the deck produce in the main cables. It is one of the purposes of the main cables over the side spans to counterbalance the forces on the towers that the main cables over the center span generate. Therefore only if the tensions in the main cables over the center span are known can the tension requirements of the main cables over the two side spans be determined. Once these are understood, then the specifications on the anchorages can be considered. It should now be apparent why the main cables over the center span will be the primary focus of the analysis to follow.

Note that the height h of the towers above the water level is largely determined by the *sag* s in the cable of the center span. See Figure 8.8. Concentrate on the New York tower of the George Washington Bridge. The deck near this tower is to clear the water level by about 195 feet. Letting r be the vertical distance from the bottom of the deck to the lowest point of the main cable over the center span, we get

$$h = 195 + r + s.$$

Since r is by design relatively small, the sag s in the main cable over the center span is the remaining factor in the height of the towers.

For the George Washington Bridge, it was decided to use a configuration of four parallel main cables, two on each side of the bridge. This was the strategy already deployed in the three large bridges in New York over the East River (the Brooklyn, Williamsburg, and Manhattan Bridges). Given the projected size of the George Washington Bridge, it was realized that the use of only two main cables would have required cables of a size far beyond that of any cables previously constructed.

To facilitate the mathematical analysis, we will make certain assumptions that are contrary to actual fact but do not have an appreciable effect on the quantitative considerations in question. These assumptions

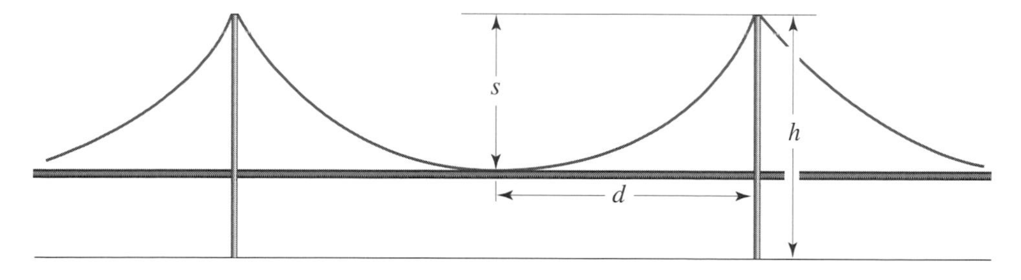

Fig. 8.8

are the following. The weight of the main cables and the suspenders are considered to be included in the weight of the deck. The main cables are assumed to be completely flexible (no resistance to bending) and inextensible (they do not stretch when under tension). The flexibility assumption is met by chains (to which this study also applies), but not by steel cables (for which our study leads to approximate solutions). Finally, we'll assume that there is a large number of vertical suspender cables, each close to the next.

We will see that the important constants—we will also call them *parameters*—are:

$w =$ the dead load of the deck (including the main cables and suspenders) plus the maximum live load capacity *per cable* in pounds per foot. In the case of the George Washington Bridge, this is obtained by adding $39{,}000 + 8000 = 47{,}000$ and distributing this over the four cables. So $w = \frac{47{,}000}{4} = 11{,}750$ pounds per foot.

$d =$ the horizontal distance from the low point of the cable over the center span to the center of a tower. In the case of the George Washington Bridge, this has already been determined as $d = 1750$ feet.

$s =$ the sag, i.e., the vertical distance from the top of the tower to the bottom of the cable at midspan. This is understood to be the sag under dead and live load. In the case of the George Washington Bridge, the sag still has to be determined.

Consider a single main cable over the center span of the bridge. Place an xy-coordinate system in such a way that the origin is at the lowest point of the cable. Let P be an arbitrary point on the cable and let x be its coordinate. Let $T(x)$ be the tension at x. This is the magnitude of the pull of the cable at P. The pull of the cable at P is tangent to the cable. The assumption that the system is in balance means that the cable above P pulls up at P with the same force $T(x)$ that the cable below P pulls down. See Figure 8.9. Let

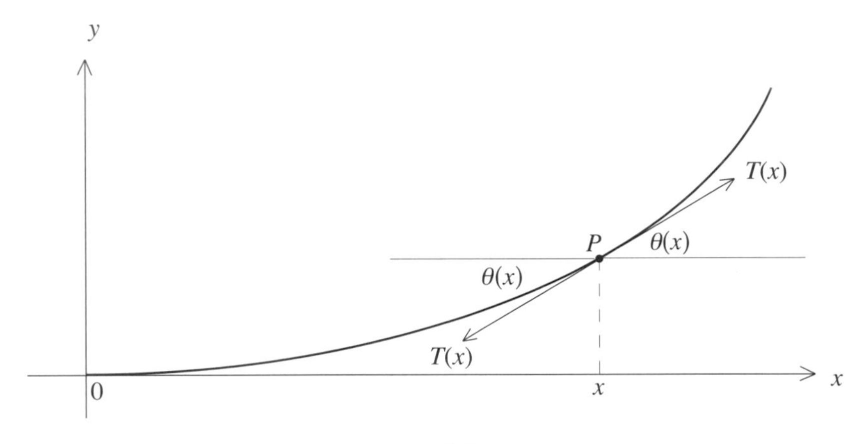

Fig. 8.9

$\theta(x)$ be the angle at P between the tangent and the horizontal.

Now let Δx be a short distance, and consider the deck from x to $x + \Delta x$. Refer to Figure 8.10. The difference between the vertical component of the downward tension at x and the vertical component of the upward tension at $x + \Delta x$ supports the weight of Δx units of the deck. To see this, it helps to regard the deck segment from x and $x + \Delta x$ to be cut off from the rest of the deck and supported by a single suspender cable (as the figure suggests). Since w is the weight of the deck per unit length,

$$T(x + \Delta x)\sin\theta(x + \Delta x) - T(x)\sin\theta(x) \approx w\Delta x.$$

The reason that this is only an approximation is that the three forces in question act at the same point only approximately. (That the forces act at the same point is required for the application of the strategy of balancing forces.) The smaller the Δx, the better the approximation. After dividing both sides by Δx, we get that

$$\lim_{\Delta x \to 0} \frac{T(x + \Delta x)\sin\theta(x + \Delta x) - T(x)\sin\theta(x)}{\Delta x} = w,$$

and therefore that $\frac{d}{dx}(T(x)\sin\theta(x)) = w$. Therefore the functions $y = T(x)\sin\theta(x)$ and $y = wx$ have the same derivative, and it follows from Corollary ii that concludes Section 7.6 that $T(x)\sin\theta(x) = wx + C$, with C a constant. Since $\sin\theta(0) = \sin 0 = 0$, it follows that

$$T(x)\sin\theta(x) = wx.$$

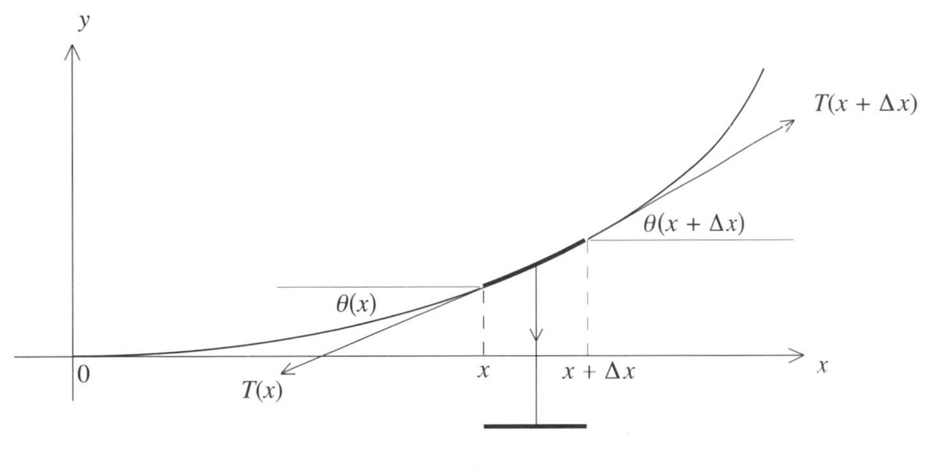

Fig. 8.10

Having dealt with the vertical components of the tensions in the main cable, we turn to the horizontal components. Let's have a look at the difference

$$T(x + \Delta x)\cos\theta(x + \Delta x) - T(x)\cos\,\theta(x)$$

between the two horizontal forces that act on the segment of the deck between x and $x + \Delta x$. If m is the mass of the deck per unit length (including the dead plus live loads, the cables), then by Newton's second law,

$$T(x + \Delta x)\cos\,\theta(x + \Delta x) - T(x)\cos\,\theta(x) \approx (m\Delta x)\cdot a,$$

where a is the acceleration (depending on x and Δx) that the segment undergoes. As in the vertical situation, this is an approximation that becomes more and more accurate for smaller and smaller Δx. Consider the limit

$$\lim_{\Delta x\to 0}\frac{T(x + \Delta x)\cos\theta(x + \Delta x) - T(x)\cos\theta(x)}{\Delta x} = m\lim_{\Delta x\to 0} a.$$

If this limit is not zero, then for some small Δx, the deck segment will be accelerated horizontally. As before, think of the deck segment from x and $x+\Delta x$ to be cut off from the rest of the deck and supported by a single suspender cable. The assumption that the structure is stable tells us that there can be no horizontal acceleration, and it follows that

$$\frac{d}{dx}T(x)\cos\theta(x) = \lim_{\Delta x\to 0}\frac{T(x + \Delta x)\cos\theta(x + \Delta x) - T(x)\cos\theta(x)}{\Delta x} = 0.$$

Calling again on Corollary ii at the end of Section 7.6, we now know that $T(x)\cos\theta(x)$ is constant. Since $\theta(0) = 0$, the substitution $x = 0$ tells us that

$$T(x)\cos\theta(x) = T_0,$$

where T_0 is the tension at the low point of the cable at midspan.

Now let $y = f(x)$ be a function whose graph represents the curve of the cable. A look at Figure 8.11 tells us that $f'(x)$ is equal to the slope of the tangent to the curve at $P = (x, f(x))$. It follows that

$$f'(x) = \tan\theta(x) = \frac{T(x)\sin\theta(x)}{T(x)\cos\theta(x)} = \frac{wx}{T_0} = \frac{w}{T_0}x.$$

Since the functions $y = f(x)$ and $y = \frac{w}{2T_0}x^2$ have the same derivative, one more application of Corollary ii

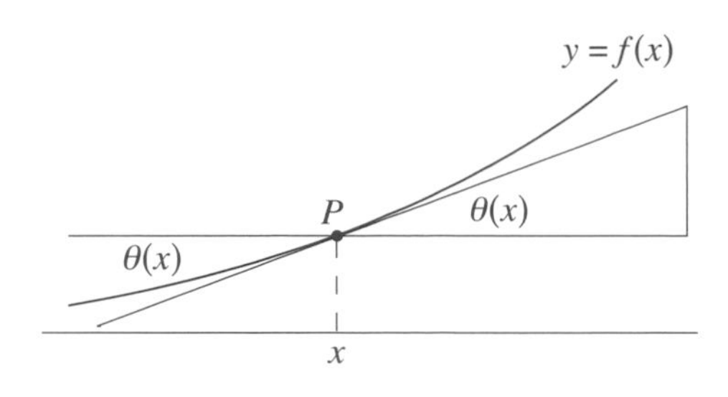

Fig. 8.11

at the end of Section 7.6, tells us that $y = f(x)$ must have the form $f(x) = \frac{w}{2T_0}x^2 + C$ for some constant C. Since $f(0) = 0$, we see that $C = 0$. Therefore the curve of the cable over the center span is the graph of the function

$$f(x) = \frac{w}{2T_0}x^2.$$

We know from Section 4.3 that the graph of such a function is a parabola. Refer to Figure 8.8, and observe that the point (d, s) is on the graph of $f(x) = \frac{w}{2T_0}x^2$. Therefore $s = \frac{w}{2T_0}d^2$. Hence $\frac{s}{d^2} = \frac{w}{2T_0}$ and $T_0 = \frac{1}{2}\frac{wd^2}{s}$. So by a substitution,

$$f(x) = \frac{s}{d^2}x^2 \quad \text{and} \quad f'(x) = \frac{2s}{d^2}x.$$

Having determined the geometry of the cable, we now turn to to the question of the tension. By squaring both sides of the equations $T(x)\cos\theta(x) = T_0$ and $T(x)\sin\theta(x) = wx$, and then adding the results, we get

$$T_0^2 + (wx)^2 = T(x)^2\cos^2\theta(x) + T(x)^2\sin^2\theta(x) = T(x)^2(\sin^2\theta(x) + \cos^2\theta(x)) = T(x)^2.$$

So $T(x) = \sqrt{T_0^2 + (wx)^2}$. After substituting $T_0 = \frac{1}{2}\frac{wd^2}{s}$,

$$T(x) = \sqrt{\frac{1}{4}\frac{w^2d^4}{s^2} + w^2x^2} = w\sqrt{\frac{1}{4}\frac{d^4}{s^2} + x^2}.$$

Because we are considering the cable only from its lowest point to the tower, x is restricted to the interval $0 \le x \le d$. A look at the expression for $T(x)$ just derived shows that $T(x)$ has its minimum value when $x = 0$ and its maximum value when $x = d$. So the minimal tension in the cable is $T_0 = \frac{1}{2}\frac{wd^2}{s}$. It occurs at its lowest point. The maximal tension that the main cable over the center span is subject to is

$$T_d = w\sqrt{\frac{1}{4}\frac{d^4}{s^2} + d^2} = wd\sqrt{\frac{1}{4}\frac{d^2}{s^2} + 1} = wd\sqrt{\left(\frac{d}{2s}\right)^2 + 1},$$

and this occurs at the point at which the cable meets the tower.

Let α be the angle that the cable makes at the tower with the horizontal. Observe that

$$\tan\alpha = f'(d) = \frac{2s}{d^2}d = \frac{2s}{d}.$$

We have now determined—in terms of the parameters w, d, and s—both the magnitude and the direction of the force with which the main cable pulls on the tower. See Figure 8.12. The downward vertical component

$$T_d \sin\alpha$$

of the pull of a main cable over the center span subjects the tower to a downward compressive force. The towers can resist considerable forces of *compression*, but they have much lesser resistance to horizontal forces. To ensure the stability of the towers, the main cables of over the side spans must be calibrated so as to counterbalance the horizontal components

$$T_d \cos\alpha$$

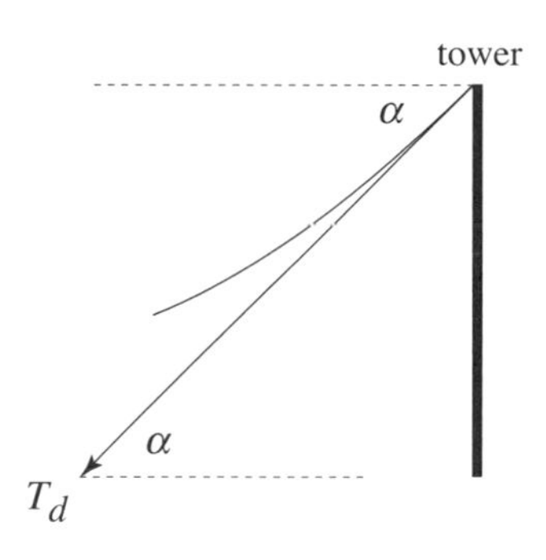

Fig. 8.12

of the pull of the corresponding main cable over the side span. The towers must also be able to withstand the changes in the tensions in the cable brought about by the variations in live loads and temperatures that the bridge is subjected to. They must be designed to be flexible enough to allow them to bend slightly as the bridge adjusts to such variations.

It should now be apparent that our analysis has provided crucial information about essential components of a suspension bridge: the main cables over the center span, the towers, the cables over the side span, and hence also about the demands on the anchorages.

Let's return to the specifics of the George Washington Bridge. We already know that $d = 1750$ feet and $w = 11{,}750$ pounds. So by substituting,

$$T_d = wd\sqrt{(\tfrac{d}{2s})^2 + 1} = (11{,}750)(1750)\sqrt{(\tfrac{1750}{2s})^2 + 1}.$$

The important remaining question is: What allowance should be made for the sag s in the main cable (under dead plus live load)? Taking s larger will make the term $(\frac{1750}{2s})^2$ smaller. This would decrease the tension T_d in the main cable and the stresses on the towers. On the other hand, a larger sag requires higher towers and longer cables. After carefully weighing all the critical factors, including the impact on the stability of the structure and its cost, the engineers stipulated a sag of $s = 327$ feet. This would give the George Washington Bridge a *sag-to-center-span ratio* of $\frac{s}{2d} = \frac{327}{3500} = 0.093$ (less than the more conventional $\frac{1}{8} = 0.125$ of the time). Since the height h of the towers above the water level is determined by the equation $h = 195 + r + s$, the choice of $s = 327$ feet meant that the height of the towers would exceed 500 feet. The choice $s = 327$ feet would result in a tension of

$$T_0 = \frac{1}{2}\frac{wd^2}{s} = \frac{1}{2}\frac{(11{,}750)(1750)^2}{327} \approx 55{,}000{,}000 \text{ pounds}$$

at the bottom of the center span, and a maximal tension in the main cable of

$$\begin{aligned} T_d &= (11,750)(1,750)\sqrt{(\tfrac{1750}{2s})^2 + 1} = (11{,}750)(1750)\sqrt{(\tfrac{1750}{654})^2 + 1} \\ &= (20{,}562{,}500)(2.85659) \approx 58{,}700{,}000 \text{ pounds.} \end{aligned}$$

(These projections for the maximum and minimum tensions in the main cable match well with the values $T_d = 58{,}500{,}000$ pounds and $T_0 = 55{,}000{,}000$ pounds that the New York Port Authority supplied for the completed bridge.) The four main cables over the center span would subject each tower to a force of

$$4T_d \approx 235{,}000{,}000 \text{ pounds.}$$

Because $\tan\alpha = \frac{2s}{d} = \frac{654}{1750} = 0.374$, we get $\alpha = 20.5°$. Since $\sin\alpha = \sin 20.5° = 0.350$, these four cables would produce a compression in each tower of

$$4T_d \sin\alpha \approx (235{,}000{,}000)(0.350) \approx 82{,}000{,}000 \text{ pounds.}$$

The four main cables over the side span would roughly double this compression.

What has our mathematical study achieved? An outline of a design for the bridge has emerged! In addition, estimates have been derived for the tensions and stresses that a bridge built in accordance with such a design would be subjected to. The engineers determined that such tensions and stresses—as well as the implications that they would have for the suspender cables, the main cables over the side span, the anchorages, and the foundations of the towers—could be dealt with. This analysis and the study of small-scale models formed the basis of a detailed design for the George Washington Bridge. Thereafter, the plans for its construction were drawn up.

The construction of the bridge began on September 21, 1927. It was a huge undertaking. The large concrete foundational structures in which the towers were to be lodged were put in place under the riverbed. The two towers, 591 feet high and consisting of a combination of carbon and silicone steels, were erected one section at a time. The main cables were made from thousands of parallel galvanized steel wires. Much too massive and long to be hauled to the bridge, they were manufactured on site by a process called spinning. Long strands were spun from the wires, and these in turn were grouped to form the cables. (We'll discuss the particulars of the production of the cable in Section 10.2.) The strands at the end of each cable were locked into the anchorages with the appropriate tensions. Then the vertical suspender cables were put into place. Each consists of a single wire rope of about 3 inches in diameter. They are placed every 60 feet in sets of sixteen, four for each main cable. Then the floor steel for the deck of the center span was hoisted into position by cranes. This was done in sections, starting from the towers and proceeding toward the center of the bridge. The floor sections for the side spans followed. Stiffening girders were built into the deck. Running parallel to the deck, they serve to distribute the live load more evenly over the main cables. This controls the amount of deformation in the cables at heavy live loads. The construction of the bridge was completed on October 24, 1931, four years after it began. The famous architect Le Corbusier referred to the George Washington Bridge as "the most beautiful bridge in the world. It is blessed. It is the only seat of grace in the disorderly city."

Our mathematical study of the suspension bridge made simplifying assumptions and represents only a start. For instance, there was the assumption that the weight of the cable and the suspenders should be folded in with that of the deck. In practice, the curve is not quite a parabola, but the deviations are small. This is an issue that we will return to in Section 11.12. The complete mathematical analysis of the suspension bridge is complicated. It includes the investigation of the stresses, strains, and torques in the girders and towers. (The concept of *torque*—the capacity of a force to rotate an object or a structure—will be discussed in the next section.) It also includes the analysis of the deformations of the structure under various load and temperature conditions. These must be clarified to ensure a safe and stable structure. A so-called *deformation theory* has been developed in response to these concerns. Such mathematical theories are supplemented with computer analyses. Because wind conditions can destabilize a bridge, wind tunnel tests are undertaken on scale models. A number of failures testify to the complexity of both design and construction. The most dramatic of these occurred in 1940 when the Tacoma Narrows Bridge collapsed just four months after its completion. Winds of 40 miles per hour caused the deck of the 2800-foot center span to sway and roll. Later, the deck began to rotate back and forth so violently that a section of the deck dropped into the water some 208 feet below. The wind had subjected the structure to motions that its steady force progressively amplified, until failure occurred.

8.4 AN EXPERIMENT OF GALILEO

In 1608, Galileo took an inclined plane with a groove running along it and placed it on top of a table. He took a bronze ball (of about 2 centimeters = 0.02 meters in diameter), set it in the groove, and let it roll repeatedly down the plane. The ball always started from rest, but Galileo let it descend from various heights. Each time it would roll down the plane, then briefly along the horizontal table, before falling to the ground along a parabolic arc. Galileo carefully measured his observations and recorded his findings in his notebook. His unit of length was the *punto* (about 0.94 millimeters), and his unit of time was the *tempo* (about 0.011 seconds). Figure 8.13 depicts one of his diagrams. Galileo's data have been converted to meters, and the point B designating the foot of the table has been added. We see that the edge of the table was 0.778 meters above the ground. The starting heights of the ball (above the surface of the table) appear in the column on the right; the corresponding distances of the point of impact from the foot

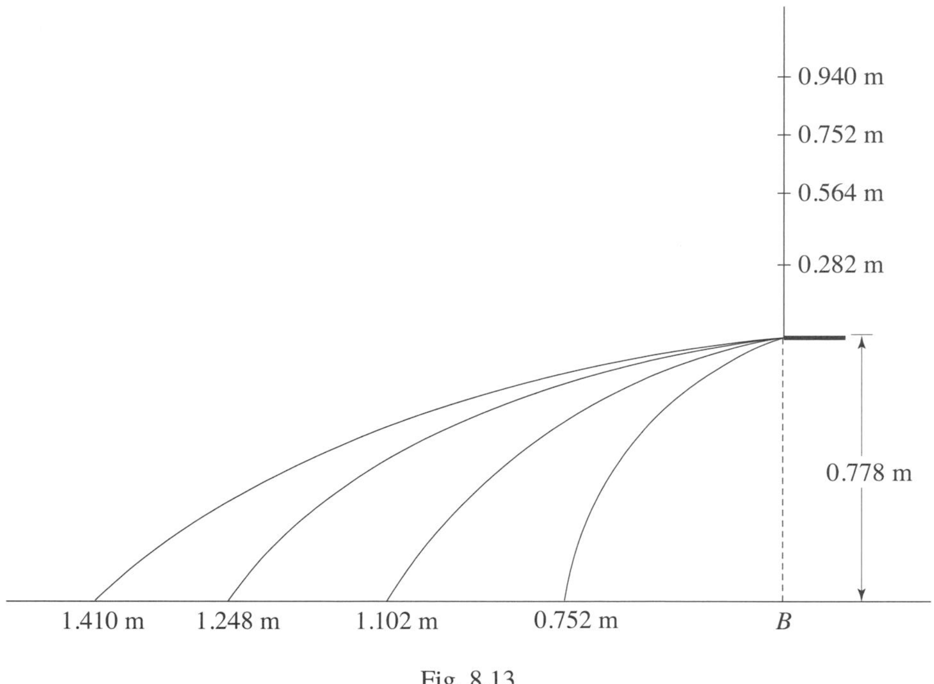

Fig. 8.13

of the table are listed at the bottom of the diagram. This experiment was briefly described in segment 3E and Figure 3.48 of the Problems and Projects section of Chapter 3. It is the goal of this section to develop in detail the mathematical theory that underlies this experiment and to compare Galileo's observations with the predictions of the theory.

8.4.1 Sliding Ice Cubes and Spinning Wheels

Consider Galileo's inclined plane, and let its angle of inclination be β. Let's start by letting an ice cube of mass m slide down the plane. We'll assume that the bottom of the ice cube is melting and that it slides in a completely frictionless way. See Figure 8.14. Note that gravity acts on the ice cube. If the ice cube were in free fall, it would experience a downward acceleration of g. If m is the mass of the ice cube, then the corresponding downward force is its weight mg. This force also acts on the sliding ice cube. Consider this force split into two components: one in the direction of the inclination of the plane and the other perpen-

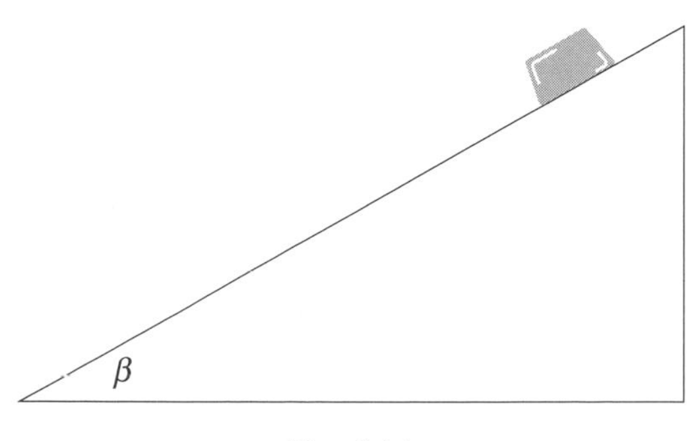

Fig. 8.14

dicular to it. The first propels the ice cube down the plane; the second is evenly balanced by the plane pushing up on the ice cube. Refer to Figure 8.15a. (For a review of some of the basic facts that are now relevant, refer to Section 6.6.) That the angle labeled by β in Figure 8.15a is equal to the angle β that determines the elevation of the inclined plane follows by comparing the two right triangles in Figure 8.15b. They are both right triangles that have the angle γ in common. So the remaining two angles must be equal

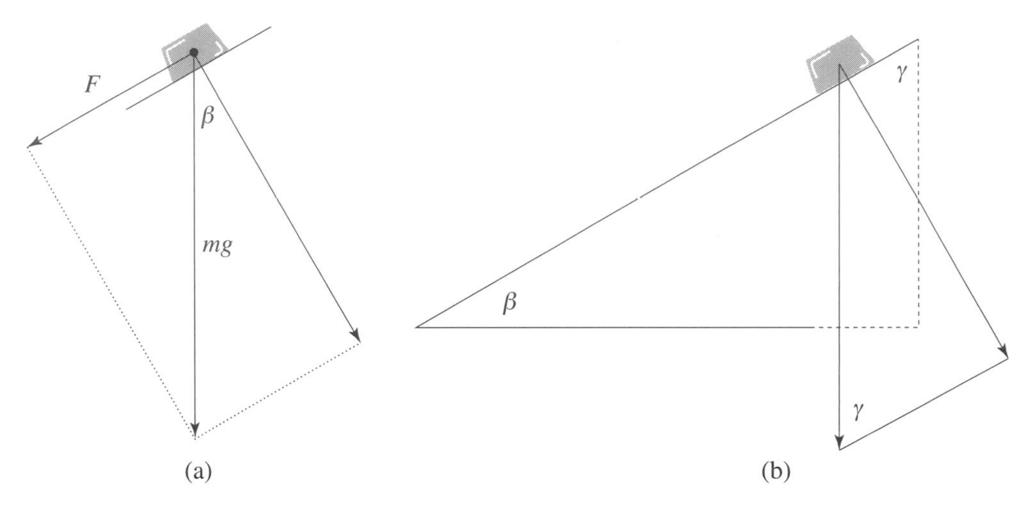

Fig. 8.15

as well. So these remaining angles are both equal to β. Let F be the magnitude of the force in the direction of the inclination of the plane. From Figure 8.15a, $\sin\beta = \frac{F}{mg}$, and therefore

$$F = mg\sin\beta.$$

So this is the magnitude of the force that accelerates the ice cube down the plane.

We know that Galileo did not experiment with ice cubes and that he did not have frictionless inclined planes. He used metal balls and real planes. But one key element remains the same. If m is the mass of the ball, then the component of the gravitational force that pushes the ball down the plane is the same $F = mg\sin\beta$ already observed for the ice cube. But there is an important difference. The ice cube on the frictionless surface slides down the plane. On a frictionless plane, Galileo's bronze ball would also slide. But his plane is not frictionless. It is friction that causes the ball to roll.

We will now begin to analyze the rotation of the ball. Suppose that a point P is moving on a circle with center O and radius r. Think of P on the perimeter of a wheel that is spinning counterclockwise around its axis at O. (The assumption that the spin is counterclockwise is not necessary, but it streamlines our discussion.) Let A be a fixed point lying just outside the wheel's perimeter. Start observing the point P at time $t = 0$. Consider the rotating segment determined by O and the position of P. See Figure 8.16. For any time $t \geq 0$, let $\theta(t)$ be the angle between the segments OP and OA measured counterclockwise in radians. This angle is the *angular position* of the point P at time t. Let $s(t)$ be the distance along the circle from the point A to the point P at time t. This distance is also measured counterclockwise from A. It provides the *linear position* of P. Each time the point P goes past A, an angle of 2π is added to $\theta(t)$, and a distance $2\pi r$ is added to $s(t)$. The definition of the radian measure of an angle shows that the linear and angular positions $s(t)$ and $\theta(t)$ are related by the equation $\theta(t) = \frac{1}{r}s(t)$ or

$$s(t) = r\theta(t)$$

for any $t \geq 0$. Fix a time t, and let an additional Δt elapse. Let the point be in position P at time t and in position P' at time $t + \Delta t$. Refer to Figure 8.16 once more. Observe that the angle that the point traces out

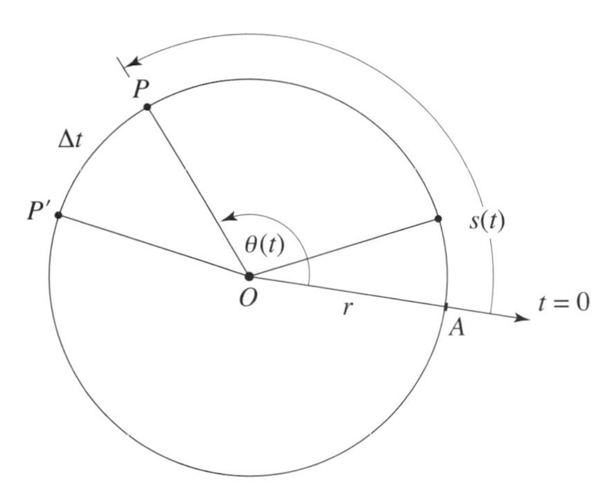

Fig. 8.16

during the time interval $[t, t + \Delta t]$ is $\theta(t + \Delta t) - \theta(t)$. The ratio

$$\frac{\theta(t + \Delta t) - \theta(t)}{\Delta t}$$

is the *average rate of change of the angular position* per unit time over the time interval $[t, t + \Delta t]$. The limit

$$\lim_{\Delta t \to 0} \frac{\theta(t + \Delta t) - \theta(t)}{\Delta t}$$

is *rate of change of the angular position* at the instant t. This is the derivative $\theta'(t)$ of the angular position $\theta(t)$. It is the *angular velocity* of the point P at time t and is typically denoted by $\omega(t)$. So

$$\omega(t) = \theta'(t).$$

Taking derivatives of both sides of the equation, $s(t) = r\theta(t)$, we get $s'(t) = r\theta'(t) = r\omega(t)$. Therefore the *linear velocity* $v(t) = s'(t)$ of the point along the circle is related to its angular velocity $\omega(t)$ by

$$v(t) = r\omega(t)$$

for any $t \geq 0$. Repeating the derivation of the derivation of $\omega(t)$ provides the derivative $\omega'(t)$ of the angular velocity. This is the *angular acceleration*. Denoting it by $\alpha(t)$, we get

$$\alpha(t) = \omega'(t).$$

Differentiating $v(t) = r\omega(t)$ shows that the *linear acceleration* $a(t) = v'(t)$ of the point P along the circle is equal to $a(t) = v'(t) = r\omega'(t) = r\alpha(t)$ at any time $t \geq 0$. It follows that the linear and angular acceleration of P are related by the equation

$$\boxed{a(t) = r\alpha(t)}$$

8.4.2 Torque and Rotational Inertia

A force can have a rotational effect on an object. Think, for example, of the rotational effect of the weight of an Olympic diver on the springboard at the time of his jump. The fulcrum and the lever tell us how to quantify such rotational effects. A *lever* is a thin, rigid rod. We'll assume that the weight of the lever is negligible compared to the loads that it is subjected to. The lever does not bend or break when forces are applied to it. The *fulcrum* is a fixed point around which the lever is free to rotate without slippage, friction, and obstruction.

Suppose a force of magnitude f acts perpendicularly to the lever at a distance d from the fulcrum. The capacity of the force to produce a rotation about the fulcrum is measured by the product $f \cdot d$. Let a second

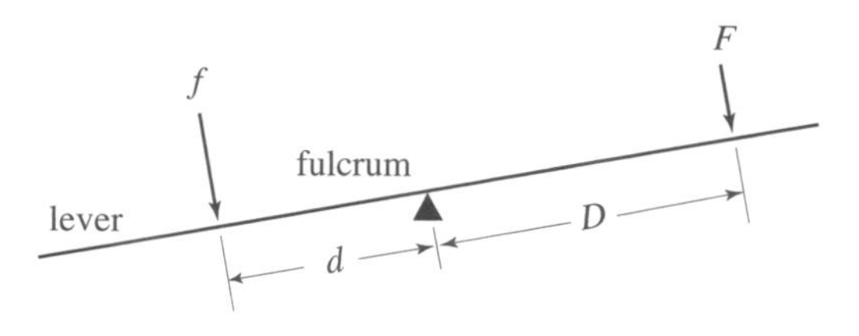

Fig. 8.17

force with magnitude F act perpendicularly to the lever on the other side of the fulcrum at a distance D from the fulcrum. See Figure 8.17. Notice that the two forces are in competition. The first is trying to rotate the lever in a counterclockwise direction and the second in a clockwise direction. Archimedes's law of the lever informs us that if fd and FD are equal, then the lever will not rotate, so that the system is in balance. (We saw in Section 2.5 that Archimedes made ingenious use of the law of the lever in his computations of areas and volumes.)

Consider two forces f_1 and f_2 that act—*one at a time*—perpendicularly to the lever on the same side of the fulcrum as f at the respective distances d_1 and d_2 from the fulcrum. Refer to Figure 8.18. If $f_1 d_1 = FD$ and $f_2 d_2 = FD$, then by Archimedes's law, each of the two forces, either f_1 or f_2, will balance the system.

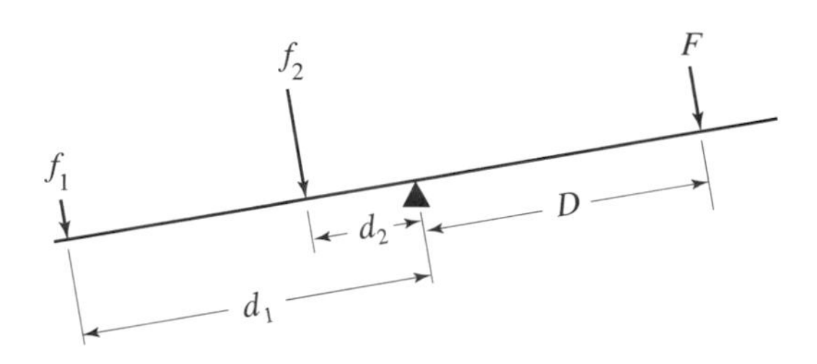

Fig. 8.18

It follows that the two forces f_1 and f_2 have the *same capacity* to effect a rotation of the lever about the fulcrum, if

$$f_1 d_1 = f_2 d_2.$$

Pushing against a door gives a sense of what we have discussed. For instance, if you push perpendicularly against the plane of a door with a force of 5 pounds at a distance of 3 feet from the vertical line determined by the hinges, then the rotational effect is equal to $5 \times 3 = 15$ pound-feet. If you push with a force of 90 pounds at a distance of 2 inches, or $\frac{1}{6}$ of a foot, from that line, then the rotational effect $90 \times \frac{1}{6} = 15$ pound-feet is the same as before. This is why in order to close a door, a much stronger push is required near the hinges than near the door handle.

So far we have considered only forces that act perpendicularly to a lever. We will now consider a force of magnitude f pushing at an angle θ against a rigid beam fixed at the point O but free to rotate there. As depicted in Figure 8.19, the force acts on the beam a distance of l units from O. The components of the force perpendicular to the beam and along the beam have magnitudes $f \sin\theta$ and $f \cos\theta$, respectively. The perpendicular component produces a rotational effect of $(f \sin\theta)l$ around O. The component $f \cos\theta$ puts the beam under compression or tension (compression in the case illustrated in the figure) but does not have any capacity to rotate the beam. Continue the line of force past the beam, and let d be the distance from O to this line. A look at Figure 8.19 shows that $d = l\sin\theta$. Therefore $(f \sin\theta)l = fd$. So if a force

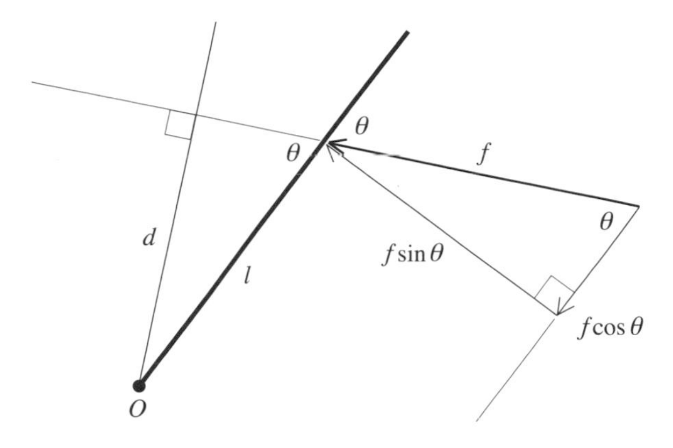

Fig. 8.19

of magnitude f is acting on beam fixed at O, then its capacity to effect a rotation of the beam around O is equal to

$$f \cdot d,$$

where d is the perpendicular distance from O to the line of action of the force. The product fd is called the *torque* produced by the force. It is also referred to as *moment of force*.

Consider a particle of mass m attached to a rigid rod of length r. Suppose that the mass of the rod is negligible, that its other end is fixed at O, and that it can rotate freely about O. Suppose that at any time $t \geq 0$, a force of variable magnitude $f(t)$ is acting on the particle perpendicularly to the rod. This force will cause the particle to move along a circle of radius r and center O. See Figure 8.20. If $a(t)$ is the linear acceleration of the particle produced by $f(t)$, then by Newton's second law, $f(t) = ma(t)$. Because $a(t) = r\alpha(t)$, where $\alpha(t)$ is the angular acceleration of the particle, it follows that $f(t) = (mr) \cdot \alpha(t)$. After

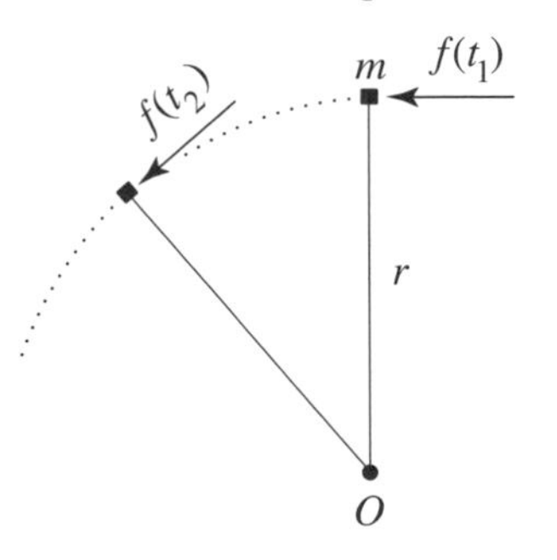

Fig. 8.20

multiplying through by r, we get

$$f(t) \cdot r = (mr^2) \cdot \alpha(t).$$

The equation $f(t) \cdot r = (mr^2) \cdot \alpha(t)$ that was just established has the form

$$\text{torque} = \text{constant} \times \text{angular acceleration}.$$

Such an equality holds in any situation where a force moves an object along a circle with a fixed center and radius. The constant of proportionality that relates the torque to the angular acceleration is called the *index of inertia* of the object. In the situation of Figure 8.20, the index of inertia is mr^2. By denoting the index of inertia by I, we can write the equation above in the form

$$\boxed{f(t) \cdot r = I \cdot \alpha(t)}$$

where r is the distance from the point of action of the force to the axis of rotation of the circular motion.

This equation is the rotational equivalent of Newton's force = mass × acceleration. Torque is rotational force, and index of inertia is rotational mass. Notice that for a fixed torque, the larger the index of inertia I, the less the angular acceleration $\alpha(t)$, so the greater the resistance of the object to being rotated.

Take a circle in a plane with radius r and center O. Regard the circle as a physical object with mass m. Suppose that the mass is evenly distributed as a thin circular strip around the circumference. Suppose also that the circle can rotate freely about O. Think of a bicycle wheel whose spokes have negligible mass. Let $f(t)$ be the magnitude of a force acting on the circle tangentially at any time $t \geq 0$. This force will cause the circle to rotate. Consider the circle to be subdivided into many small, equal segments, say, n of them, with each segment attached to the center O by a "spoke" as shown in Figure 8.21a. Instead of having the

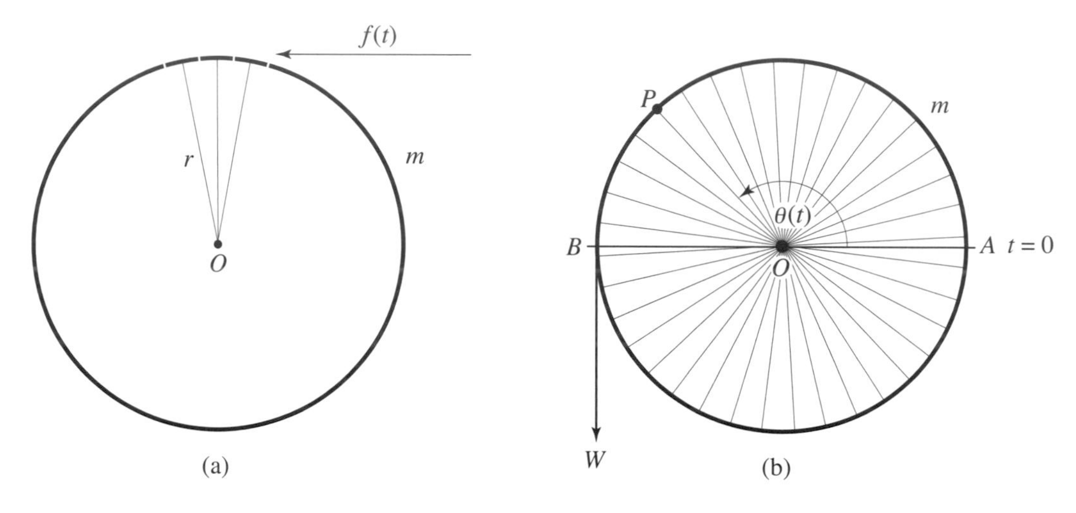

Fig. 8.21

force act on the entire circle, consider it as acting tangentially on each of the n segments with a magnitude of $\frac{f(t)}{n}$. Figure 8.20 illustrates this for a single segment. Since the mass of each particle is $\frac{m}{n}$, we see by the equality already established for a single rotating mass that $r \cdot \frac{f(t)}{n} = (\frac{m}{n} r^2) \cdot \alpha(t)$. After multiplying through by n, we get

$$f(t) \cdot r = (mr^2) \cdot \alpha(t).$$

So the index of inertia of the rotating circular mass is given by the same expression $I = mr^2$ as the index of inertia of the rotating point-mass.

Example 8.3. Consider the wheel of radius r and mass m of Figure 8.21a in vertical position. Take one end of a string (regarded to be of negligible weight), attach it to the rim of the wheel, wrap it around counterclockwise on the rim a number of times, and attach a weight W at the other end. See Figure 8.21b. The point O is the fixed center of the wheel, BA is a fixed horizontal segment through O, and P is a typical point on the rim. Initially, the wheel is held fixed with the point P at A. At time $t = 0$, the wheel is released, and—pulled by the tension W in the string—begins to rotate counterclockwise (without friction). The figure shows the point P in position at time $t > 0$ and the angle $\theta(t) = AOP$ in radians. As shown in Section 8.4.1, the angular acceleration of the wheel is $\alpha(t) = \omega'(t) = \theta''(t)$. Since the torque produced by the tension is Wr and the index of inertia is mr^2, it follows that $Wr = mr^2\alpha(t)$. Therefore

$$\alpha(t) = \frac{W}{mr}.$$

Since the wheel is fixed at time $t = 0$, $\omega(0) = 0$. Letting $\theta(0) = 0$, and applying Corollary ii of Section 7.6 twice, we get that the angular speed and the angular positions of the point P are

$$\omega(t) = \frac{W}{mr}t \quad \text{and} \quad \theta(t) = \frac{W}{2mr}t^2,$$

respectively. Let's become specific, and take $r = 0.30$ m, $m = 1.40$ kg, $W = 5.00$ N, and t in seconds. (Because weight = mass × g and $9.80 \text{ N} = 1 \text{ kg} \cdot 9.80 \text{ m/s}^2$, this corresponds to a mass of about 0.50 kg.) With these values, $\frac{W}{mr} \approx 11.90 \text{ m/s}^2$. Therefore $\omega(t) \approx 11.90t$ radians/s and $\theta(t) \approx \frac{11.90}{2}t^2 \approx 5.95t^2$ radians. Since 2π radians corresponds to 1 revolution, $\omega(t) \approx 11.90t$ radians/s corresponds to a rotational speed of about $1.89t$ revolutions/s. For $t = 3$ for instance, we get a rotational speed of the wheel of 5.68 revolutions/s. Let L be the length that has been added to the hanging string during these 3 seconds. Since $\theta(3) = \frac{L}{r}$, we get that $L = \theta(3)r \approx (5.95)(9)(0.30) \approx 16.07$ m.

There is one fundamental difference between Newton's famous equation force = mass × acceleration and its rotational counterpart. In Newton's equation, it can usually be assumed that the mass of the object is located at the center of mass, and that beyond this assumption it does not matter how the mass of the object is distributed. In the rotational analogue, it matters how the mass of the rotating object is distributed. In particular, the indices of inertia of two objects are usually different even if they have the same mass and are rotating about their centers of mass.

Consider a circular disc that has its mass evenly distributed and can rotate freely about its center. Compared to the case of the circle, more of the mass is now closer to the center of rotation. The formula $I = mr^2$ for the index of inertia of the circular mass suggests that there should be less resistance to rotation and hence that the index of inertia of the disc should be smaller. This is indeed so. It turns out that the index of inertia of the homogeneous disc is

$$I = \tfrac{1}{2}m \cdot r^2.$$

We'll turn next to a solid sphere of radius r with a mass of m that is uniformly distributed. We'll rotate this sphere about an axis through its center. See Figure 8.22. Compared to the uniform disc, even more of

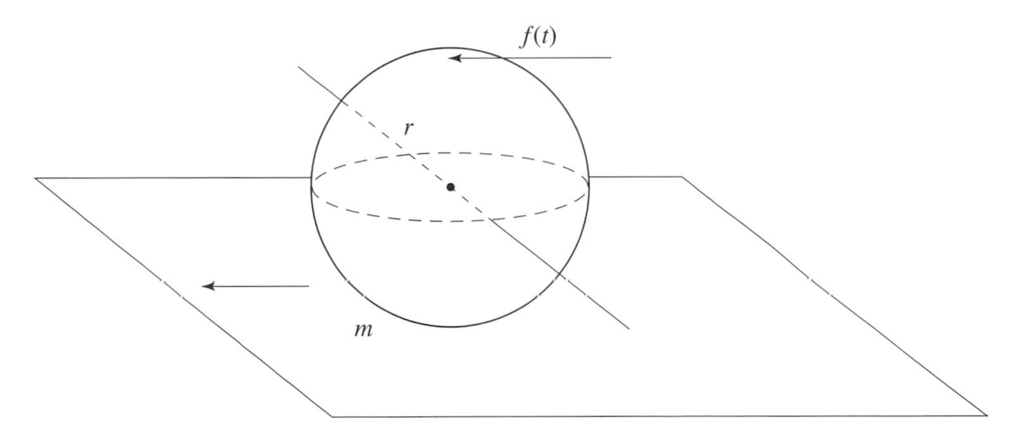

Fig. 8.22

the mass is, proportionally, closer to the axis of rotation. So the expectation is that the index of inertia of the sphere is less than than of the disc. This is indeed so. The index of inertia of the homogeneous sphere is

$$I = \tfrac{2}{5}mr^2.$$

The computations of the indices of inertia of the homogeneous disc and sphere rely on summation processes that are studied in integral calculus. They will be taken up in segment 10K of the Problems and Projects section of Chapter 10.

The fact that the index of inertia of a solid homogeneous sphere of radius r and mass m is $\frac{2}{5}mr^2$ tells us that the connection between a force $f(t)$ acting tangentially on the sphere and the angular acceleration $\alpha(t)$ that this force produces is given by the equation $f(t) \cdot r = (\frac{2}{5}mr^2) \cdot \alpha(t)$, so that

$$f(t) = \tfrac{2}{5}mr \cdot \alpha(t).$$

We are now in a position to undertake the mathematical analysis of Galileo's experiment.

8.4.3 The Mathematics behind Galileo's Experiment

Let's return to Galileo and the bronze ball rolling down the inclined plane. We'll assume that the bronze ball is homogeneous, that it has mass m, and that its radius is r. Place a coordinatized axis along the incline with the positive part pointing downward as shown in Figure 8.23. Suppose that the origin 0 is at a height h above the base of the plane. Position the ball at the origin, and release it from a state of rest at time $t = 0$. For any time $t \geq 0$, let $p(t)$ be the position of the ball on the inclined axis. So $p(t)$ is the distance that the ball has moved during the time interval $[0, t]$. As already discussed in Section 8.4.1, the ball is propelled down the plane by the component of the force of gravity in the direction of the plane, and the magnitude of this force is $F = mg \sin\beta$. The vector in the figure labeled $f(t)$ represents friction.

The first concern is the explicit determination of the function $p(t)$. The frictional force $f(t)$ is the result of the contact of the bronze ball with the plane. Therefore this is the force that rotates the ball. Begin by thinking of the counterclockwise rotation of the bronze ball as occurring with its center fixed in place

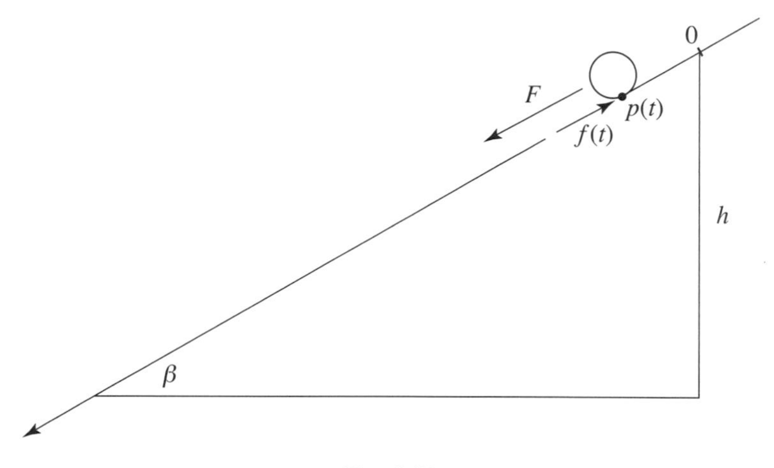

Fig. 8.23

and disregard (for the moment) its motion down the plane. So the point singled out in Figure 8.23 moves around the circular perimeter of the ball. It follows from the discussion in Section 8.4.2 that the relationship between $f(t)$ and the angular acceleration $\alpha(t)$ of the ball is $f(t) = \frac{2}{5}mr \cdot \alpha(t)$. Because Section 8.4.1 tells us that $a(t) = r\alpha(t)$, we get the connection

$$f(t) = \tfrac{2}{5}ma(t)$$

between the frictional force $f(t)$ and the linear acceleration $a(t)$ of the ball.

Now let the ball roll down the plane. The same linear acceleration $a(t)$ is now the linear acceleration of the point of the figure down the plane. Another look at Figure 8.23 tells us that the total force on the ball is $F - f(t)$, so that by Newton's second law, $F - f(t) = ma(t)$. After substituting $F = mg \sin\beta$ and $f(t) = \frac{2}{5}ma(t)$, we get $ma(t) = F - f(t) = mg \sin\beta - \frac{2}{5}ma(t)$. So $a(t) = g \sin\beta - \frac{2}{5}a(t)$, and hence $\frac{7}{5}a(t) = g \sin\beta$. Therefore, the acceleration $a(t)$ of the ball down the inclined plane is

$$a(t) = \tfrac{5g}{7} \sin\beta.$$

The rest is routine! Let $v(t)$ be the velocity of the ball down the plane. Since $v(t)$ and $(\frac{5g}{7}\sin\beta)t$ have the same derivative $a(t) = \frac{5g}{7}\sin\beta$, we know from Corollary ii at the end of Section 7.6 that $v(t) = (\frac{5g}{7}\sin\beta)t + v(0)$. Because the ball started from rest, $v(0) = 0$, and therefore

$$v(t) = (\tfrac{5g}{7}\sin\beta)t.$$

Since $p'(t) = v(t)$, we get by a repetition of this argument that $p(t) = \frac{1}{2}\left(\frac{5g}{7}\sin\beta\right)t^2 + p(0)$. The ball started from the origin, so that $p(0) = 0$, and therefore

$$p(t) = \left(\tfrac{5g}{14}\sin\beta\right)t^2.$$

Refer back to Galileo's diagram in Figure 8.13 and recall that his observations relate the starting height of the ball with the distance from the ball's point of impact to the foot of the table. We will continue by establishing a connection between these two quantities.

To compute the distance between the point of impact of the ball and the foot of the table, we need to know the velocity that the ball has when it reaches the bottom of the inclined plane. Suppose that this occurs at time t_b. Notice that t_b is the moment at which $p(t_b)$ is equal to the length of the hypotenuse of the right triangle of Figure 8.23. It follows that $\sin\beta = \frac{h}{p(t_b)}$. Combining $p(t_b) = \frac{h}{\sin\beta}$ with the preceding formula $p(t_b) = \left(\frac{5g}{14}\sin\beta\right)t_b^2$, we get $\left(\frac{5g}{14}\sin\beta\right)t_b^2 = \frac{h}{\sin\beta}$. So $t_b^2 = \frac{14h}{5g}\frac{1}{\sin^2\beta}$, and hence

$$t_b = \sqrt{\frac{14h}{5g}\frac{1}{\sin\beta}}.$$

Substituting t_b for t in the velocity formula $v(t) = \left(\frac{5g}{7}\sin\beta\right)t$ shows us that the velocity of the ball at the instant that it reaches the bottom of the inclined plane is

$$v(t_b) = \left(\tfrac{5g}{7}\sin\beta\right)t_b = \left(\tfrac{5g}{7}\sin\beta\right)\left(\sqrt{\frac{14h}{5g}\frac{1}{\sin\beta}}\right) = \frac{5g}{7}\cdot\sqrt{\frac{14h}{5g}} = \sqrt{\frac{5^2g^2 14h}{7^2\cdot 5g}} = \sqrt{\tfrac{10}{7}gh}\ .$$

Observe that—amazingly—the velocity $v(t_b)$ depends only on the starting height h of the ball. It is independent of the mass of the ball, its radius, and even the angle β of the inclined plane!

The situation that we have arrived at is this. When Galileo releases his bronze ball from rest at a starting height h above the table, it arrives at the bottom of the plane with a velocity of $\sqrt{\frac{10}{7}gh}$. See Figure 8.24. At this point, its motion is subject to the theory that was developed in Section 6.7. In particular, the ball will fly to the ground in a parabolic arc. The distance R from the its point of impact and the base B of the

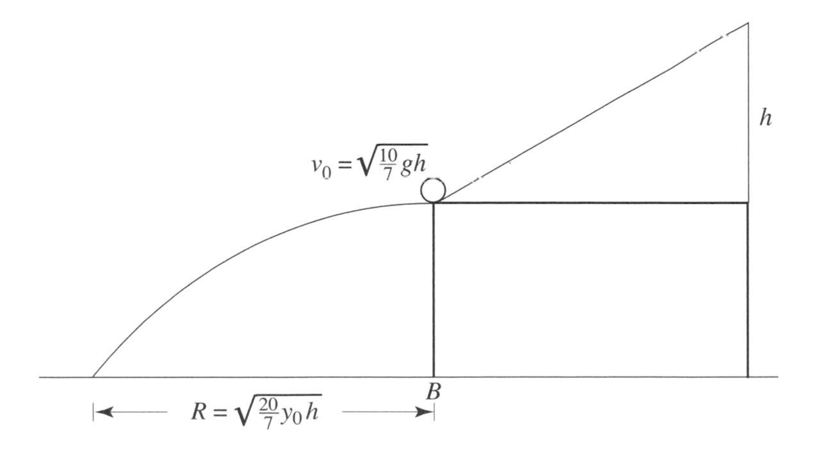

Fig. 8.24

table is given by the range formula

$$R = \tfrac{v_0}{g}\cos\varphi_0[v_0\sin\varphi_0 + \sqrt{v_0^2\sin^2\varphi_0 + 2gy_0}\,].$$

What constants have to be inserted? The v_0 is the velocity with which the ball flies off the edge of the table. So $v_0 = v(t_b) = \sqrt{\frac{10}{7}gh}$. Because the tabletop is horizontal, the angle of elevation is $\varphi_0 = 0$. Since $\sin\varphi_0 = \sin 0 = 0$ and $\cos\varphi_0 = \cos 0 = 1$, the distance from the point of impact of the ball to the point B is

$$R = \tfrac{v_0}{g}\sqrt{2gy_0} = \sqrt{\tfrac{2y_0}{g}}\,v_0 = \sqrt{\tfrac{2y_0}{g}}\cdot\sqrt{\tfrac{10}{7}gh} = \sqrt{\tfrac{20}{7}y_0h}.$$

The initial height y_0 is the distance $y_0 = 0.778$ m from the top of the table to the ground. By substitution into the formula above, we get—with an accuracy of three decimal places—that

$$R = 1.491\sqrt{h}\text{ m}.$$

Our analysis of Galileo's experiment has arrived at the key question: Are the data that Galileo displays in the diagram of Figure 8.13 "in tune" with this formula? For the starting heights h equal to 0.282 m, 0.564 m, 0.752 m, and 0.940 m, the formula provides the distances

$$R = 0.792\text{ m}, 1.120\text{ m}, 1.293\text{ m}, \text{ and } 1.446\text{ m}.$$

Galileo's measurements for these distances are

0.752, 1.102, 1.248, and 1.410 m, respectively.

The differences between theory and experiment are 4 cm, 1.8 cm, 4.5 cm, and 3.6 cm. All of Galileo's measurements fall a little short of the distances predicted by the theory. But this is to be expected! The ball that Galileo used was certainly not perfectly spherical, the groove in the inclined plane was not perfectly smooth, so that the actual velocity of the ball would have been less than that given by the equation. Therefore one would expect the ball's actual points of impact to have been closer to the foot of the table than those predicted by the mathematical study.

It was pointed out earlier that some prominent historians of science had expressed the view that Galileo's experiments were largely based on speculations. The experiment that we have discussed is presented on folio 116v of Galileo's working papers. See Figure 8.25. The calculations on the folio are the details of Galileo's successful effort to show that the experiment confirms his theory of motion (as it is described in Figure 3.46 and Problem 3.24). The folio, together with the results of the mathematical analysis carried out above, suggests quite clearly that Galileo did experiment and that he did so with precision. Historians have not been able to determine whether Galileo was aware of the difference between rolling and sliding descent. In any case, it is relevant to note that the physics and mathematics of torque and rotational acceleration were not developed until Leonhard Euler and others did so more than 100 years after Galileo died.

8.5 FROM FERMAT'S PRINCIPLE TO THE REFLECTING TELESCOPE

The final section of this chapter is devoted to optics. Optics is one of the oldest sciences to which mathematical analysis was applied. Studies undertaken by Euclid around 300 B.C. and later by Archimedes revealed that when a ray of light strikes a mirror, the angle that the approaching ray makes with the perpendicular to the mirror is the same as the angle that the reflected ray makes with the perpendicular. More briefly put, the angle of incidence is equal to the angle of reflection. Claudius Ptolemy realized that a light ray coming from a star or planet near the horizon is refracted—in other words, it is bent—as it travels through the atmosphere. Attempts to reach a quantitative understanding of this phenomenon failed. Galileo wrestled with the question as to whether the speed of light is finite or whether it is propagated instantaneously, and he tried to find the answer experimentally. The dominant view in the 17th century,

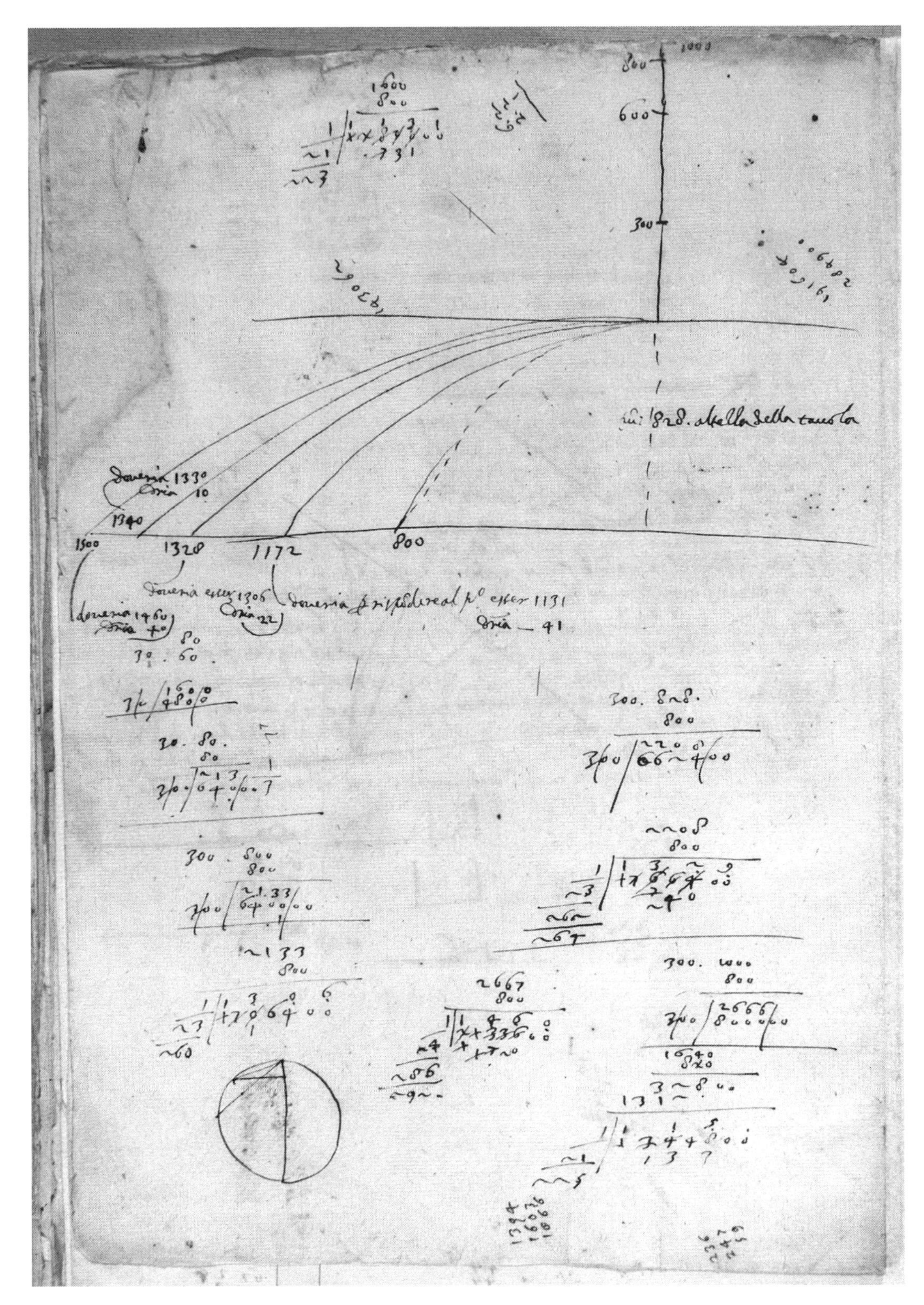

Fig. 8.25. Folio 116v of Galileo's working papers. Image courtesy of the Ministero dei beni e delle attività culturali e del turismo/Biblioteca Nazionale Centrale di Firenze. Any further reproduction or duplication by any means is prohibited. See http://www.imss.fi.it/ms72/INDEX.HTM for the entire collection of Galileo's working papers.

advocated strongly by René Descartes, favored an infinite speed. His French contemporary Pierre de Fermat (1601–1665) assumed that the speed of light is finite in proposing his *principle of least time*: if a ray of light travels from a point A to another point B, then of all neighboring paths from A to B, it chooses the path that requires the least time. From his principle, Fermat deduced the *law of refraction*, which describes how light is bent by transparent media and lenses in particular. The Dutch mathematician-scientist Willebrord Snell (1580–1626) had discovered the law earlier experimentally.

8.5.1 Fermat's Principle and the Reflection of Light

We'll begin with the speed of light. The Danish astronomer Ole Rømer (1644–1710), working at the Paris Observatory, was compiling extensive observations of the orbit of Io, the innermost of the four big moons of Jupiter discovered by Galileo in 1610. By timing the eclipses of Io by Jupiter, Rømer hoped to determine a more accurate value of Io's orbital period. Such observations had a practical importance in the 17th century. The fact that the location of stars (especially the North Star) could be used to determine the latitude (north-south location) of a ship at sea was already known to ancient navigators. However, the determination of longitude (east-west position) had been an intractable problem. Galileo had suggested that tables of the orbital motion of Jupiter's satellites would provide a kind of "clock" in the sky, and that such a clock would provide the local time on board of the ship, and hence the time difference with a place of known longitude (like the observatories in Paris or Greenwich). With this information, the longitude of the ship's position could be estimated. This approach to longitude turned out to lack the required precision. It was not until the 18th century, after clocks were invented that could run accurately at sea, that the problem of longitude was finally solved. However, the eclipse data that Rømer collected unexpectedly solved another important scientific problem—that of the speed of light.

In the 1670s, the astronomers Giovanni Domenico Cassini and John Flamsteed made observations that provided the estimate of 140 million km for the average distance between Earth and the Sun. (See Section 3.7 for the details.) This had been by far the best estimate of this elusive distance. (Today's accurate value is close to 150 million km.) The satellite Io is eclipsed by Jupiter once for each orbit and hence about every 1.8 Earth days. Figure 8.26 shows Io emerging from an eclipse with Earth in position A. By timing these eclipses for 10 years, Rømer noticed that the time interval between successive eclipses became steadily shorter as Earth in its orbit moves toward Jupiter and steadily longer as Earth moved away from Jupiter. He measured the difference between successive eclipses of Io with Earth in position A to be about 22

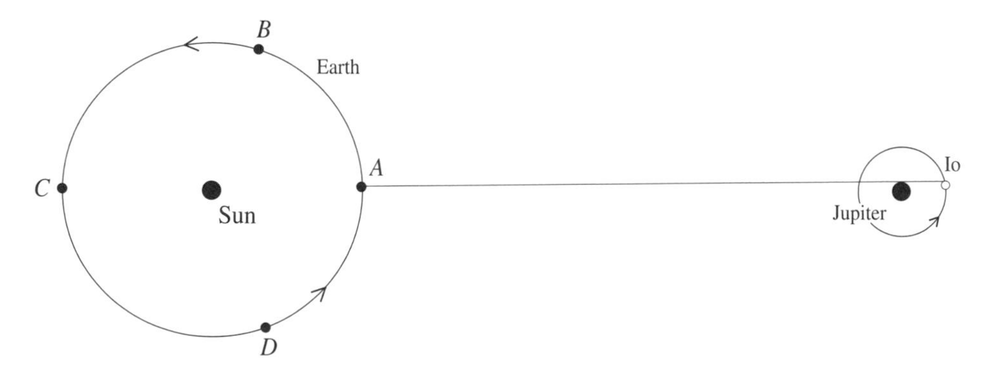

Fig. 8.26

minutes shorter than the time between successive eclipses with Earth in position C (opposite A in Earth's orbit). Rømer knew that the orbital period of Io around Jupiter could have nothing to do with the relative positions of Earth and Jupiter. The inescapable conclusion was that the time difference of 22 minutes that he observed must be the time it takes for light from Io to travel from A to C. The conversion of Rømer's measurement to $22 \cdot 60$ seconds in combination with the Cassini-Flamsteed estimate of the Earth-Sun distance, provided the approximate value of

$$\frac{AC}{22 \cdot 60} \approx \frac{2 \cdot 140 \times 10^6}{22 \cdot 60} \approx 212{,}000 \text{ km/s}$$

for the speed of light in a vacuum. (With the exception of the planets, satellites, asteroids, and comets, the space of the solar system is vacuous.) While Rømer's procedure was correct in principle, his measurement of 22 minutes was off. Use of the accurate value of about 16.7 minutes and the correct 150 million km for the average Earth-Sun distance would have provided the correct value of 300,000 km/s (or 186,000 miles per second) for the speed of light in a vacuum. But Rømer did provide a figure for the speed of light that was in the right ballpark. Today, the speed of light is known with precision as 299,792.458 km/s. As one might expect, light travels slightly slower in air than in a vacuum. In air at sea level, its speed is 299,705.543 km/s.

It seems consistent with our intuition that if a light ray is reflected from the surface of a flat mirror, then the *angle of incidence* α is equal to the *angle of reflection* β, as illustrated in Figure 8.27a. We will begin our study of the geometric properties of light rays by showing that this conclusion is a mathematical consequence of Fermat's principle of least time. Suppose that a light ray proceeds from some source, strikes a mirror, and is reflected. Let A be a point on the ray before it strikes the mirror, and let B be a point on the ray after the reflection. Suppose that the surrounding medium is air or a vacuum, and let v be the speed of light in this medium. The light ray determines a plane that is perpendicular to the mirror.

Place a coordinate system in this plane in such a way that the x-axis runs along the mirror's surface and the y-axis goes through A. Let $A = (0, a)$, $B = (b, d)$, and suppose that the ray reflects off the mirror at x. See Figure 8.27b. Let D_1 be the distance from A to x, and let t_1 be the time it takes for the ray to travel

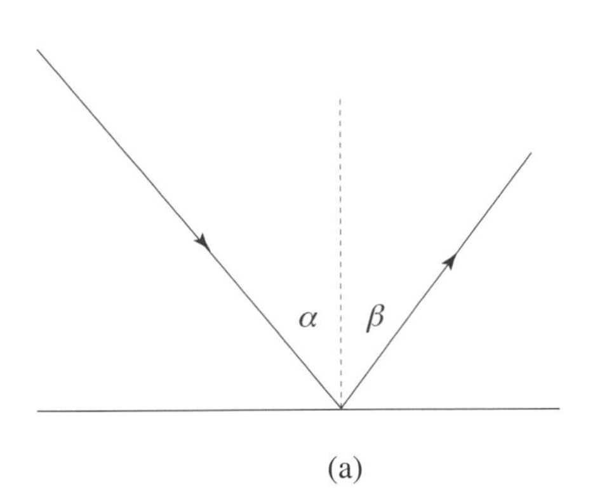

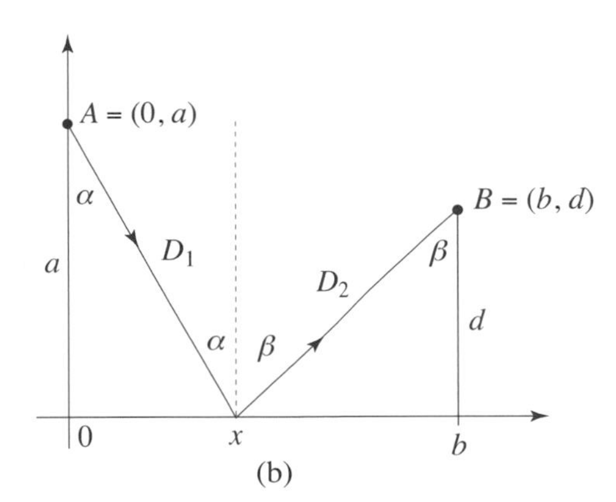

Fig. 8.27

this distance. Similarly, let D_2 be the distance from x to B, and let t_2 be the time of travel for this distance. Observe that $D_1 = vt_1$, and hence $t_1 = \frac{D_1}{v}$. In the same way, $t_2 = \frac{D_2}{v}$. By applying the Pythagorean theorem twice, we obtain

$$D_1 = \sqrt{x^2 + a^2} = (x^2 + a^2)^{\frac{1}{2}} \quad \text{and} \quad D_2 = \sqrt{d^2 + (b - x)^2} = (d^2 + (b - x)^2)^{\frac{1}{2}}.$$

Therefore the time t that it takes for the ray to travel from A to B is

$$\begin{aligned} t &= t_1 + t_2 = \frac{D_1}{v} + \frac{D_2}{v} \\ &= \tfrac{1}{v}(x^2 + a^2)^{\frac{1}{2}} + \tfrac{1}{v}(d^2 + (b - x)^2)^{\frac{1}{2}}. \end{aligned}$$

So the time of travel is determined by the function

$$t(x) = \tfrac{1}{v}(x^2 + a^2)^{\frac{1}{2}} + \tfrac{1}{v}(d^2 + (b - x)^2)^{\frac{1}{2}},$$

where x is the point of incidence. With Fermat's principle of least time in mind, the question is: For which x is the travel time $t(x)$ a minimum? To solve this problem, we will analyze the derivative $t'(x)$. By the chain rule,

$$\begin{aligned} t'(x) &= \tfrac{1}{2v}(x^2+a^2)^{-\frac{1}{2}}(2x) + \tfrac{1}{2v}(d^2+(b-x)^2)^{-\frac{1}{2}}2(b-x)(-1) \\ &= \frac{x}{v(x^2+a^2)^{\frac{1}{2}}} - \frac{b-x}{v(d^2+(b-x)^2)^{\frac{1}{2}}}. \end{aligned}$$

What are the critical numbers for $t(x)$? Note that $t'(x)$ is not defined when

$$(x^2+a^2)^{\frac{1}{2}} = 0 \quad \text{or} \quad (d^2+(b-x)^2)^{\frac{1}{2}} = 0.$$

In the first case, a must be zero. This means that A is on the x-axis and hence on the mirror. But this is not the case. If $(d^2+(b-x)^2)^{\frac{1}{2}}$ is zero, d must be zero, and B is on the mirror. Again, this is not the case. It follows that the only critical numbers are those x for which $t'(x) = 0$. So this has to be the value of x at which $t(x)$ has its minimum value. Setting $t'(x) = 0$, we get

$$\frac{x}{v(x^2+a^2)^{\frac{1}{2}}} = \frac{b-x}{v(d^2+(b-x)^2)^{\frac{1}{2}}}.$$

From a look at Figure 8.27b, we see that

$$\sin\alpha = \frac{x}{D_1} = \frac{x}{(x^2+a^2)^{\frac{1}{2}}} \quad \text{and} \quad \sin\beta = \frac{b-x}{D_2} = \frac{b-x}{(d^2+(b-x)^2)^{\frac{1}{2}}}.$$

It follows directly that $\frac{\sin\alpha}{v} = \frac{\sin\beta}{v}$, and hence that $\sin\alpha = \sin\beta$. Because both α and β are between 0 and 90°, we can conclude that

$$\alpha = \beta.$$

(A look at the graph of the sine function in Figure 4.23 of Section 4.6 confirms that different angles in the range $0 \le \theta \le \frac{\pi}{2}$ have different sines.) The above argument has demonstrated that the property "the angle of incidence is equal to the angle of reflection" is a consequence of Fermat's principle of least time.

Fermat's principle of least time requires a comment. The physics of quantum mechanics describes a light ray as a combination of waves and particles called *photons*. Some experiments that study light reveal its particle nature and others its wave nature. The fact that light is a wave-particle duality leads to surprising conclusions. For example, if light shines on an imperfectly transparent sheet of glass, it may happen that 95% of the light transmits through the glass while 5% is reflected back. This makes good sense if light is a wave (the wave splits and a smaller wave is reflected back). But if light is a stream of identical particles, then all that can be said is that each photon arriving at the surface of the glass has a 95% chance of passing through and a 5% chance of being reflected. In particular, it is not possible to predict a single definite outcome for a photon, but two different outcomes, each of which has some nonzero probability of occurring. The situation is similar to what happens to a radioactive substance. We saw in Section 8.1.2 that it is possible to predict when one of its atoms will decay only in terms of probabilities. Another consequence of the probabilistic nature of light is that it follows not just a single path of shortest time from a point A to a point B, but that every path from A to B has a nonzero probability of being followed. The reason we are not aware of these effects is that an observed beam of light consists of a huge number of photons. For example, a typical laser beam is a flow of about 10^{15} (this is one million billions) photons per second. Although the position of each individual photon is highly uncertain, the fact that there are so many means that the combined probabilities add up to a predictable aggregate. So in the context of Fermat's principle of least time, we will regard a ray of light to be a thin beam of many photons.

Lasers make it possible to measure the speed of light with great accuracy. Light emitted by a flashlight or a bulb is *incoherent*, meaning that photons of many wave frequencies are oscillating in different directions. Such light scatters and spreads. Light emitted by a laser (an abbreviation of "light amplification by stimulated emission of radiation"), on the other hand, is *coherent*, meaning that it consists of waves of photons that have the same frequency and move in sync. A coherent beam of light with a known frequency can be split to follow two paths and then recombined. By adjusting the path length and observing how the two waves interfere with each other, data can be extracted that allow for the computation of the speed of light with near perfect precision. In fact, since the 1980s, the meter has been defined by international agreement as the distance traveled by light in vacuum during a time interval of one 299,792,458th of a second. So the meter is defined in such a way that the speed of light is exactly 299,792.458 km/s. (This value is equivalent to 186,282.397 miles per second.)

When light travels in a single medium, its speed is constant, so that *least time* is equivalent to *shortest distance*. This was apparent in the solution of the reflection problem above. When different speeds are involved, this equivalence fails.

8.5.2 The Refraction of Light

Consider two homogeneous transparent mediums of different densities. For example, let one of them be air and the other a type of glass. Suppose that the boundary between them is a plane. Let A be a point in the lighter medium, and let B be a point in the denser medium. Neither A nor B lies on the boundary. The path of a light ray traveling from A to B lies in a plane perpendicular to the plane of the boundary. At the boundary, the ray bends as shown in Figure 8.28. Such a change in direction is known as *refraction*. The

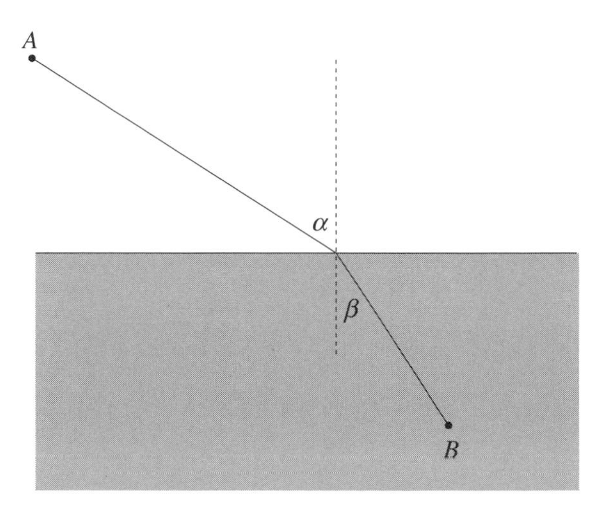

Fig. 8.28

angles α and β in the figure are known as the *angle of incidence* and the angle *angle of refraction*, respectively. We saw that for a reflected ray, the angles of incidence and reflection are equal. The relationship between the angles α and β in the case of a refracted ray is different, but it too is brought "to light" by Fermat's principle of least time. Of all the possible paths from A to B, a light ray will pick the path of shortest time.

Let v_A be the speed of light in the lighter medium, and let v_B be the speed of light in the heavier medium. Consider the plane determined by the refracted light ray, and place a coordinate system so that the x-axis is on the boundary and the y-axis goes through A. Refer to Figure 8.29. Let $A = (0, a)$ and $B = (b, d)$, and suppose that the light ray crosses the boundary at x. Let D_1 be the distance from A to x, and let t_1 be the time it takes for the ray to travel this distance. Similarly, let D_2 be the distance from x to B, and let t_2 be the time of travel through this distance. Note that $D_1 = v_A t_1$, so $t_1 = \frac{D_1}{v_A}$. In the same way, $t_2 = \frac{D_2}{v_B}$. By the Pythagorean theorem,

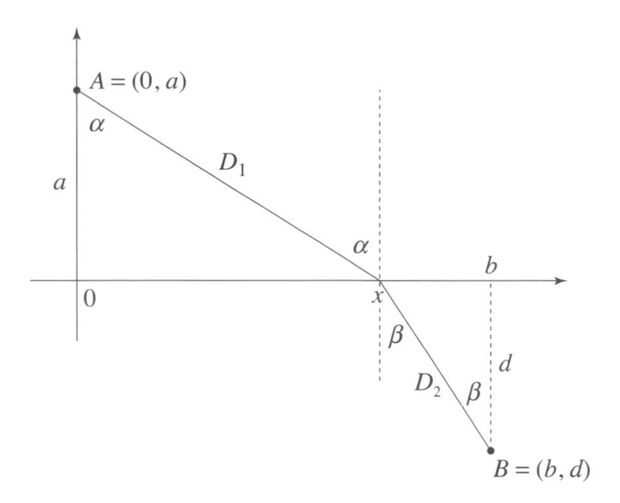

Fig. 8.29

$$D_1 = \sqrt{x^2 + a^2} = (x^2 + a^2)^{\frac{1}{2}} \text{ and } D_2 = \sqrt{d^2 + (b - x)^2} = (d^2 + (b - x)^2)^{\frac{1}{2}}.$$

So the time it takes for the ray to travel from A to B is

$$\begin{aligned} t &= t_1 + t_2 = \tfrac{D_1}{v_A} + \tfrac{D_2}{v_B} \\ &= \tfrac{1}{v_A}(x^2 + a^2)^{\frac{1}{2}} + \tfrac{1}{v_B}(d^2 + (b - x)^2)^{\frac{1}{2}}. \end{aligned}$$

For which x is t a minimum? The calculations are exactly the same as in the earlier situation of reflection:

$$\begin{aligned} t'(x) &= \tfrac{1}{2v_A}(x^2 + a^2)^{-\frac{1}{2}}(2x) + \tfrac{1}{2v_B}(d^2 + (b - x)^2)^{-\frac{1}{2}}2(b - x)(-1) \\ &= \frac{x}{v_A(x^2 + a^2)^{\frac{1}{2}}} - \frac{b - x}{v_B(d^2 + (b - x)^2)^{\frac{1}{2}}}. \end{aligned}$$

As before, the only critical points are those where $t'(x) = 0$. Setting $t'(x) = 0$ gives us

$$\frac{x}{v_A(x^2 + a^2)^{\frac{1}{2}}} = \frac{b - x}{v_B(d^2 + (b - x)^2)^{\frac{1}{2}}}.$$

It is this is the equality that the x-coordinate of the point on the boundary must satisfy so that the ray's travel time from A to B is least. A look at Figure 8.29 shows that

$$\sin\alpha = \frac{x}{D_1} = \frac{x}{(x^2 + a^2)^{\frac{1}{2}}} \text{ and } \sin\beta = \frac{b - x}{D_2} = \frac{b - x}{(d^2 + (b - x)^2)^{\frac{1}{2}}}.$$

It follows that the connection between the angle of incidence α and the angle of refraction β is

$$\boxed{\frac{\sin\alpha}{v_A} = \frac{\sin\beta}{v_B}}$$

This is *Snell's law of refraction.* The same relationship holds for the light ray at the point of exit from the transparent material. See Figure 8.30. Since $\beta' = \beta$ and $\alpha' = \alpha$, Snell's law of refraction applies to

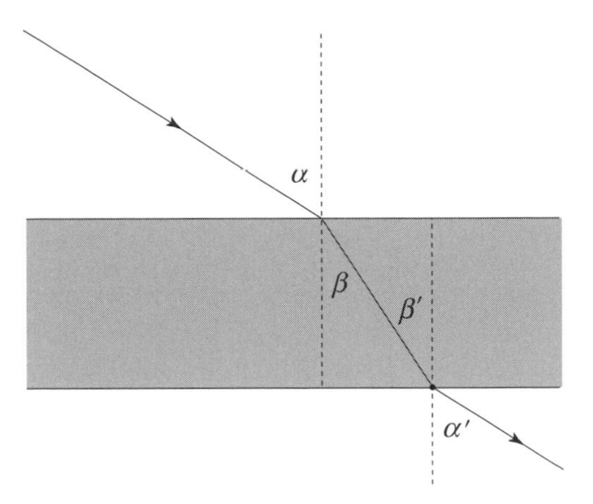

Fig. 8.30

inform us that here too

$$\frac{\sin\alpha}{v_A} = \frac{\sin\beta}{v_B}.$$

In air, water, glass, and in fact in any transparent medium, light travels more slowly than the $c =$ 299,792.458 km/s with which it travels in a vacuum. The *index!of refraction n* of any medium is defined to be

$$n = \frac{c}{v},$$

where v is the speed of light in that medium. The index of refraction of a vacuum is $n = \frac{c}{c} = 1$. There is no medium in which light propagates faster than it does in a vacuum. So the index of refraction n of any medium satisfies $n \geq 1$. Table 8.1 lists some examples. Think of the index of refraction as a measure of the density of the medium: the denser the medium, the less the speed v at which light will travel through it, and hence the higher its index of refraction n. The index of refraction of a medium can be determined by

Medium	Index of Refraction
vacuum	1.00000
air at sea level at 0° C	1.00029
ice	1.31
water	1.33
acrylic plastic	1.50
crown glass	1.52
heavy flint glass	1.66
diamond	2.42

Table 8.1

a careful analysis of the path of light through a prism made of the material in question. By rewriting the formula $n = \frac{c}{v}$ as $v = \frac{c}{n}$ and referring to Table 8.1, we see, for example, that light travels with a speed of $\frac{299{,}792.458}{1.00029} = 299{,}705.543$ km/s in through air and a speed of $\frac{299{,}792.458}{2.42} = 123{,}881.18$ km/s through a diamond.

Let the indices of refraction of the two mediums in Figure 8.28 be n_A and n_B, respectively. The speeds of light in the two mediums are $v_A = \frac{c}{n_A}$ and $v_B = \frac{c}{n_B}$. By substituting into Snell's law, we get $\frac{n_A \sin\alpha}{c} = \frac{n_B \sin\beta}{c}$. So Snell's law can be reformulated as

$$n_A \sin\alpha = n_B \sin\beta.$$

Suppose that a light ray travels from a vacuum or air into a denser transparent medium and strikes this medium at angle of incidence α. Let β be the angle of refraction. If n_A and n_B are the respective indices of

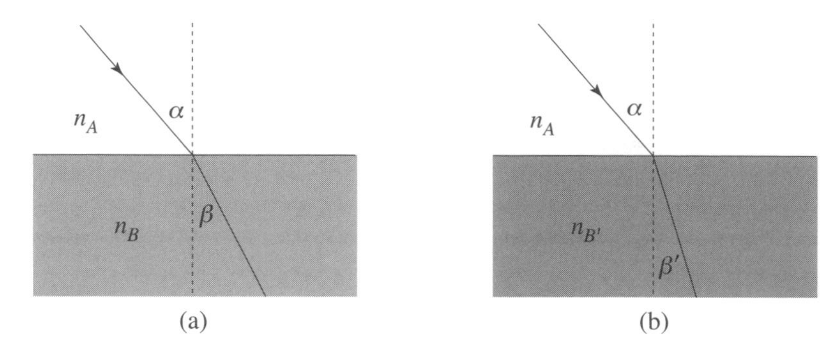

Fig. 8.31

refraction, then $n_A < n_B$, so that

$$\sin\beta = \tfrac{n_A}{n_B}\sin\alpha < \sin\alpha.$$

Since the sine is an increasing function over the interval from 0 to $\frac{\pi}{2}$ (look at Figure 4.23 in Section 4.6), it follows that $\beta < \alpha$. Suppose next that the light ray enters a denser medium than before with the same angle of incidence α and angle of refraction β'. Let the index of refraction of the denser medium be $n_{B'}$, and observe that $n_{B'} > n_B$. As before, $\sin\beta' = \frac{n_A}{n'_B}\sin\alpha < \sin\alpha$. Since $\frac{n_A}{n'_B} < \frac{n_A}{n_B}$, it follows that $\sin\beta' < \sin\beta$ and hence that $\beta' < \beta$. Our conclusions are summarized in Figure 8.31. In both cases, the angle of incidence α is greater than the angle of refraction. The denser medium, shown in Figure 8.31b, gives rise to a smaller angle of refraction, so that it bends the ray more.

Example 8.4. Suppose that a light ray passing through air strikes a slab of plate glass at an angle of incidence $\alpha = 20°$. If the glass is crown glass, then from Table 8.1, $n_A = 1.00029$ and $n_B = 1.52$, so that the angle of refraction β satisfies

$$\sin\beta = \tfrac{n_A}{n_B}\sin\alpha = \tfrac{1.00029}{1.52}\sin 20° = (0.658)(0.342) = 0.225.$$

A calculator shows that $\beta = 13.00°$. If the glass is the denser flint glass instead, then $n_B = 1.66$, and

$$\sin\beta = \tfrac{n_A}{n_B}\sin\alpha = \tfrac{1.00029}{1.66}\sin 20° = (0.603)(0.342) = 0.206.$$

This time, $\beta = 11.89°$. As expected, the denser glass bends the light ray more.

Given the initial angle of incidence, it is possible to determine the path of a light ray through any sequence of transparent materials, provided that we know their indices of refraction and their shapes. For example, suppose that a light ray travels through air, then through a lens, and then through air again, as shown in Figure 8.32. Suppose that the initial angle of incidence is α and the index of refraction of the glass is n_B. Consider the entry of the ray into the glass. As we have already observed,

$$\sin\beta = \tfrac{n_A}{n_B}\sin\alpha = \tfrac{1.00029}{n_B}\sin\alpha.$$

So $\sin\beta$ is determined by α and n_B, and hence β is determined by α and n_B. The dimensions and curvatures of the lens determine the point of exit from the lens and the angle β'. By an application of Snell's law to the exiting ray, $\sin\alpha' = \frac{n_B}{1.00029}\sin\beta'$. This equality provides the angle of exit α'. In this way, it possible to predict the path of a light ray through a glass lens and more generally through a combination of lenses.

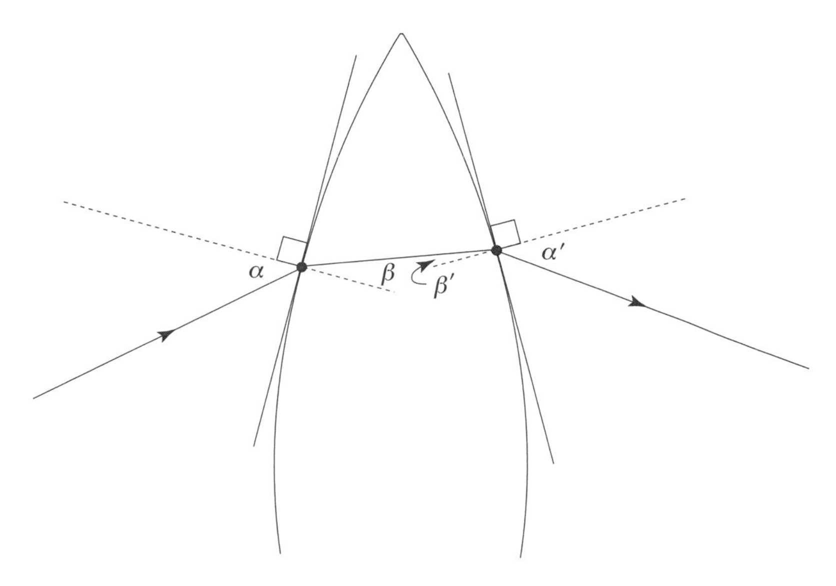

Fig. 8.32

8.5.3 About Lenses

Consider a glass lens, and assume that it is rotationally symmetric. So its outer edge is a cylinder (possibly very thin), and the central axis of the cylinder is an *axis of symmetry* of the lens. This means that after any rotation of the lens about this axis, the lens looks exactly as it did before. It follows that we can capture the full geometry of the lens with a diagram of its cross section. All lenses that our discussion considers have such an axis of symmetry. Suppose that the lens is thickest at its axis of symmetry and that it gets thinner away from axis. Such a lens is called *convex*. As a consequence of Snell's law, a collection of incoming light rays that are parallel to the axis of symmetry will be bent inward, as shown in Figure 8.33a. If a lens is to produce a sharp image, these incoming rays need to be bent in such a way that they meet at a point

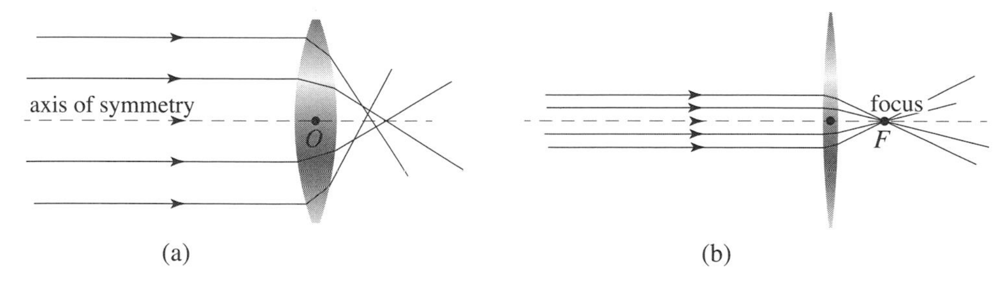

Fig. 8.33

on the other side. Such a point is a *focus* or *focal point* of the lens. In Figure 8.33b, the focal point is denoted by F. The distance between F and the center O of the lens is its *focal length* f.

Traditionally, the manufacture of optical instruments has deployed *spherical* lenses. This means that the two lens surfaces are sections of either a plane or a sphere. Such lenses are relatively easy to manufacture. It turns out that *thin* convex spherical lenses—thin meaning that the thickness of the lens (at its axis of symmetry) is negligible compared to the radii of the spherical surfaces (or the one radius if the other surface is planar)—have the property that *paraxial* rays converge to a focus. A paraxial ray is one that approaches the lens *close to and parallel to* the axis of symmetry. Reversing the direction of the paraxial rays coming into the lens provides another focus of the lens. The distances from the two focal points to the center of the lens—in other words, the two focal lengths—are equal. For a thin convex spherical lens, it is also true that a light ray that comes to the lens after having passed through a focus will exit on the other side parallel to the axis (provided that its angle with the axis of symmetry at the focus is flat enough).

With an application of Fermat's principle of least time, a simple property of the circle, and some algebra, one can derive the *lens maker's equation* for a thin spherical lens. This asserts that the focal length f of the lens is determined by the index of refraction n of the glass and the radii R_1 and R_2 of the two spherical surfaces by the formula

$$\frac{1}{f} = (n-1)\left(\frac{1}{R_1} + \frac{1}{R_2}\right).$$

For a planar lens surface, the radius is regarded to be infinite, so that one over the radius is 0.

Figure 8.34 illustrates the properties of a thin convex spherical lens. (Note that as the lens in the figure is depicted, it is not thin.) Consider an object—the arrow AB—positioned as shown. The arrow is lit (naturally or artificially), and light emanates from it in all directions. The two rays depicted in dashed lines are both parallel to the axis of symmetry (one on its approach to the lens, the other after it exist). Both rays are assumed to be paraxial, so that each goes through a focal point. All light rays that emerge from a given point on the arrow AB and travel through the lens between the two paraxial rays converge at some point on the other side. This point is the *image* of the given point. For example, the light rays from point A in the figure meet at the image point A'. All the image points are aligned, and when the various light rays

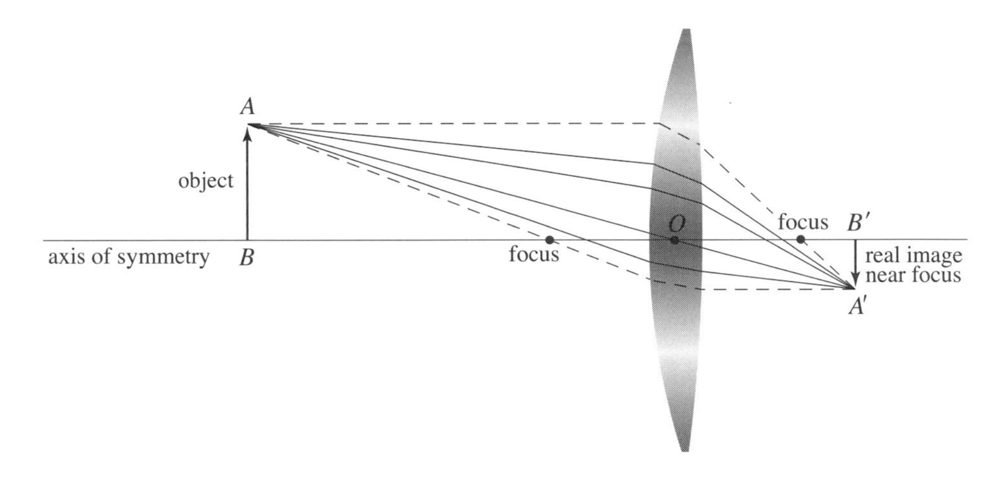

Fig. 8.34

carry the "pictorial" properties of the arrow AB (such as the colors, textures, and shadows) to the other side of the lens, an image of the arrow is produced. This image is the arrow $A'B'$ of the figure. Since the image point of A is A' and that of B is B', observe that the image is inverted. If a screen is positioned at the place where the image is formed, then the image will be visible on the screen. Since the image can be realized in this way, it is called *real*. If the screen is a light-sensitive film, then the image will be recorded. What we have described are the essential elements of a simple camera. If the lens is the lens of an eye and the screen is the retina, then what has been described is the first step of the process of seeing.

A lens of the type described can be used as a magnifying glass. Place the arrow AB within the focus of the lens. See Figure 8.35. A light ray emanating from the object at A and striking the lens parallel to the axis is bent as shown. The eye perceives this ray to be coming from some point A'. Since this happens for all the light rays emanating from AB, it follows that the eye sees a noninverted enlarged image $A'B'$. A moment's thought tells us that a film placed at $A'B'$ will not record anything. This noninverted image is therefore not a real image, but only a perceived, or *virtual*, image.

The manufacture of thin spherical lenses was developed in the 19th century. Molten glass is poured

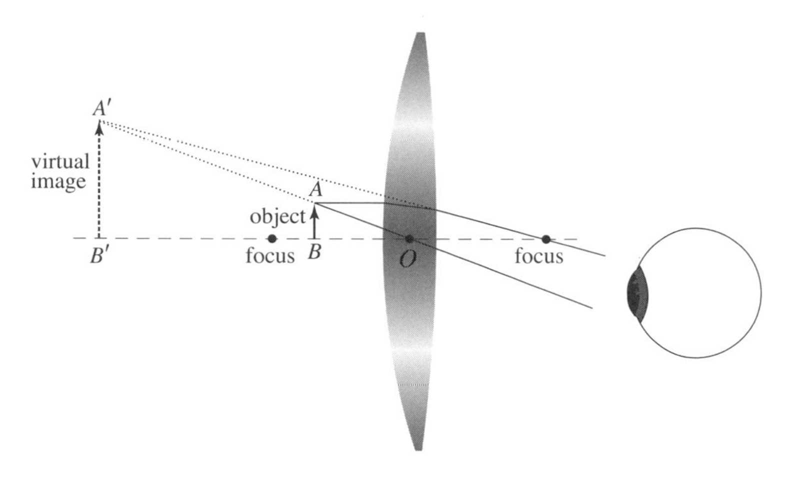

Fig. 8.35

into a lens-shaped mold. After chemicals are added to alter the properties of the glass (for instance, the addition of lead raises the index of refraction), the glass is allowed to cool and harden under carefully controlled conditions designed to keep imperfections to a minimum. The resulting piece of glass is then ground and polished into the desired spherical shape.

A thin convex spherical lens focuses only paraxial rays with any precision. Other rays, especially those far from the axis of symmetry, are bent too much and blur the image. This effect is called *spherical aberration*. Let's return to Figure 8.33b for a moment. Suppose that the parallel light rays come from a point source S that is relatively far from the lens. By Fermat's principle, it must be the case that all the rays from S take the same time to arrive at the focus F. (The idea that nature does not select a light ray from S to F that takes longer is incorporated into the principle.) The ray that travels from S to F along the axis of symmetry has the shortest path of travel, but it goes through the thickest part of the lens. Therefore the slowing effect of the glass is greater there than elsewhere. A ray that comes in away from the axis will travel a longer distance to F, but if the lens is crafted so that the path of the ray through the glass is shorter in just the right way, it can arrive at F at the same time as the one along the axis. The questions then are these: Is there such a thing as a "just so" lens that avoids spherical aberration and brings *all rays* parallel to the axis of symmetry to converge to a point? If so what is the geometry of its surfaces? We already know that these surfaces cannot be spherical. It turns out that "just so" lenses exist. They are called *aspheric* lenses. We'll illustrate what is involved by applying Fermat's principle of least time to study a simpler, but related, question.

Consider a rotation-symmetric, solid block of glass that has the property that it refracts *all* incoming light rays parallel to the axis of symmetry to meet at a point. What is the geometry of the surface of such a block of glass? Figure 8.36 depicts a cross section through the axis of symmetry of such a block as well as the point F on the axis at which the refracted rays converge. The curve C depicts the cross section of the curving surface of the block. Place an xy-coordinate system into the figure as shown, and let $F = (f, 0)$. We'll suppose that all the parallel rays coming to the y-axis emerge from a "point at infinity" S far from the block, that they travel the same long distance d from S to get to the y-axis, and that they travel from S to the block in a vacuum. By Fermat's principle, their times of travel from S to F are all the same. Designate this common time by T. With c the speed of light in a vacuum and n the index of refraction of the glass, the speed of light in the glass is $\frac{c}{n}$. The ray into $P = (x, y)$ is the typical ray from S. It follows that the time it takes for this ray to travel to P and then to F is

$$\frac{d + x}{c} + \frac{\sqrt{(x - f)^2 + y^2}}{\frac{c}{n}}.$$

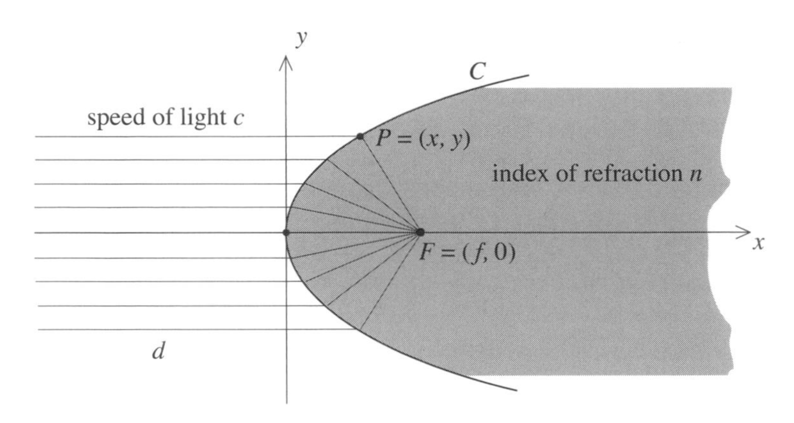

Fig. 8.36

Since this time is equal to T, we get $(d+x)+n\sqrt{(x-f)^2+y^2} = Tc$ and hence $\sqrt{(x-f)^2+y^2} = \frac{Tc-d}{n} - \frac{x}{n}$. Since $(0,0)$ is on the curve C, $f = \frac{Tc-d}{n}$. Therefore $\sqrt{(x-f)^2+y^2} = f - \frac{x}{n}$. After several algebraic moves that include the completion of a square,

$$\begin{aligned}
x^2 - 2fx + f^2 + y^2 &= f^2 - \tfrac{2f}{n}x + \tfrac{x^2}{n^2} \\
(1 - \tfrac{1}{n^2})x^2 - 2f(1 - \tfrac{1}{n})x + y^2 &= 0 \\
\tfrac{n^2-1}{n^2}x^2 - 2\tfrac{n-1}{n}fx + y^2 &= 0 \\
x^2 - 2\tfrac{fn}{n+1}x + \tfrac{y^2}{\frac{n^2-1}{n^2}} &= 0 \\
x^2 - 2\tfrac{fn}{n+1}x + (\tfrac{fn}{n+1})^2 + \tfrac{y^2}{\frac{n^2-1}{n^2}} &= (\tfrac{fn}{n+1})^2 \\
(x - \tfrac{fn}{n+1})^2 + \tfrac{y^2}{\frac{n^2-1}{n^2}} &= (\tfrac{fn}{n+1})^2,
\end{aligned}$$

so that finally, with $a = \frac{fn}{n+1}$, $(x-a)^2 + \frac{y^2}{\frac{n^2-1}{n^2}} = a^2$. Therefore $\frac{(x-a)^2}{a^2} + \frac{y^2}{a^2 \cdot \frac{n^2-1}{n^2}} = 1$. After setting $b^2 = a^2 \cdot \frac{n^2-1}{n^2}$, and hence $b = a\,\frac{\sqrt{n^2-1}}{n}$, we have

$$\frac{(x-a)^2}{a^2} + \frac{y^2}{b^2} = 1.$$

Because $n^2 - 1 < n^2$, $\sqrt{n^2-1} < n$, and therefore $b < a$. Since $a^2 = f^2\frac{n^2}{(n+1)^2}$,

$$b^2 = f^2\frac{n^2}{(n+1)^2} \cdot \frac{n^2-1}{n^2} = f^2\frac{n-1}{n+1},$$

so that $b = f\sqrt{\frac{n-1}{n+1}}$. It follows that the curve C is the left half of an ellipse with semimajor axis $a = \frac{fn}{n+1}$ and semiminor axis $b = f\sqrt{\frac{n-1}{n+1}}$. It is the left half of the ellipse $\frac{x^2}{a^2} + \frac{y^2}{b^2} = 1$ shifted a units to the right. (To see why, review Section 7.13.) So the center of the ellipse of Figure 8.36 is $(a, 0)$. Since $a = f\frac{n}{n+1}$, it lies to the left of $F = (f, 0)$.

Since there were no restrictions on $f > 0$, it follows that for every choice of f there is such an elliptical geometry for the surface of the block, namely, a geometry such that all incoming rays parallel to the axis of symmetry are pulled to the point $F = (f, 0)$. This geometry is determined by $a = f\frac{n}{n+1}$ and $b = f\sqrt{\frac{n-1}{n+1}}$, so that—as expected—it depends on f (and the index of refraction n). Given the meaning of the semiminor axis b, observe that incoming rays that are more than b units away from the axis will fail to hit the lens.

The study above does not provide the geometry of an aspheric lens because it does not consider the other surface. The fact is that the surfaces of aspheric lenses are complex. The geometry of the surfaces of aspheric lenses is often designed by relying on the *aspheric lens formula*

$$z(r) = \boxed{\frac{r^2}{R\left[1+\sqrt{1-(1-k)\frac{r^2}{R^2}}\right]}} + \boxed{A_1r^2 + A_2r^4 + A_3r^6 + \cdots}$$

The term R is the radius of a spherical surface within the lens that determines the geometry of the lens near the axis of symmetry. The variable r measures the distance from the axis of symmetry and the function $z(r)$ the deviation of the lens surface from the vertical, as illustrated in Figure 8.37. The term in the first box above represents a *base surface*. Its cross section is a conic section that depends on the parameter k in the way described in Table 8.2. The term $A_1r^2 + A_2r^4 + A_3r^6 + \cdots$ in the second box is a *correction polynomial*. The $A_1, A_2, A_3, \ldots$ are constants that allow for aspheric lenses of high order of precision to be defined. The more terms that are added to the base equation, the greater the precision that is achieved.

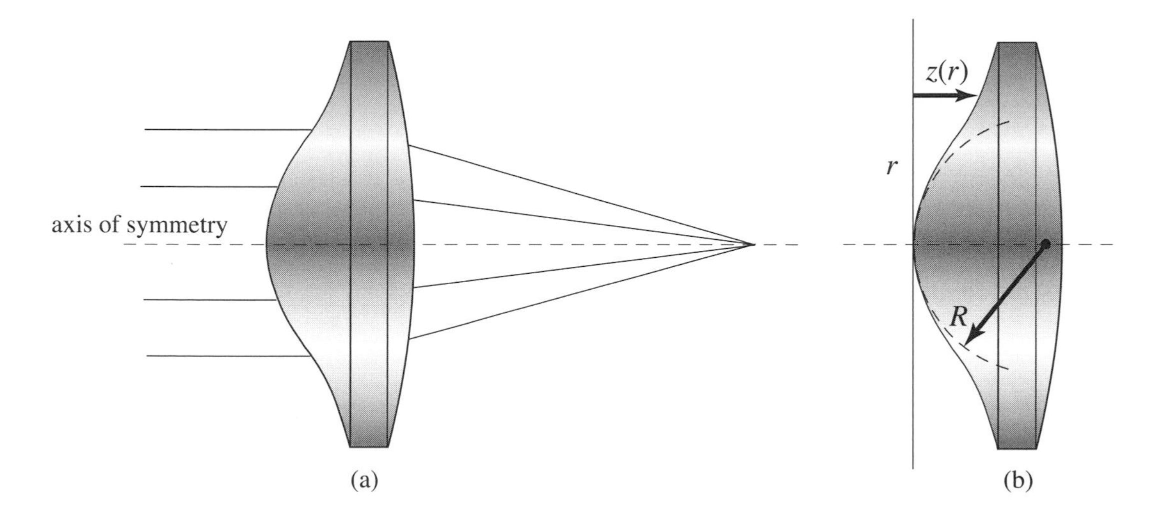

Fig. 8.37. Aspheric lens, adapted from Wikipedia, https://en.wikipedia.org/wiki/Aspheric_lens. Image Credit to ArtMechanic. Use under the Creative Commons Attribution-ShareAlike License.

Turn to Figure 8.37b, and consider the left boundary of the lens above the horizontal axis. The inflection point that this curve has must involve a contribution from the correction polynomial since (refer back to Chapter 4) conic sections have no inflection points. The information that the table specifies

Parameter k	Conic Section
$k = 0$	the circle of radius R and center $(R, 0)$
$0 < k < 1$	an ellipse with major axis on the axis of symmetry
$k = 1$	a parabola with focal axis on the axis of symmetry
$k > 1$	a hyperbola with focal axis on the axis of symmetry
$k < 0$	an ellipse with minor axis on the axis of symmetry

Table 8.2

is analyzed in Problems 8.42 and 8.43 of the Problems and Projects section of this chapter.

Aspheric lenses are used in optical systems for all kinds of applications. They have made it possible for the optics industry to produce lens systems with reduced numbers of elements (and hence with reduced size and weight) and with enhanced performance. Modern high-end optics depends more and more on the use of aspheric surfaces. Their manufacture has seen an overwhelming increase, in part due to the development of new fabrication technologies, including computer-controlled polishing and finishing. These technologies have speeded up the production process and made possible the cost-effective manufacture of precise lens surfaces.

8.5.4 Refracting and Reflecting Telescopes

The simple telescope combines the two properties of lenses illustrated in Figures 8.34 and 8.35. Light coming from an object is collected by a thin convex spherical lens called the *objective lens*, and a real inverted image is formed. If we assume that the distance of the object from the lens is very large compared to the size of the object (think of a planet as viewed from Earth), then this real image will form near the focus of the objective lens. When a second lens—the *eyepiece*—is placed in an appropriate position, a magnified virtual image of this real image can be seen by an observer. See Figure 8.38. Such a combination of lenses placed in a suitable housing is the *refracting telescope* in its most simple form. This is the kind of telescope that Kepler pointed to the sky in the early 17th century. The telescope that Galileo had pointed skyward before Kepler was similar but used a *concave* spherical lens for an eyepiece. A concave lens has an axis of symmetry and lens surfaces that curve inward from its cylindrical rim. It is narrowest at the

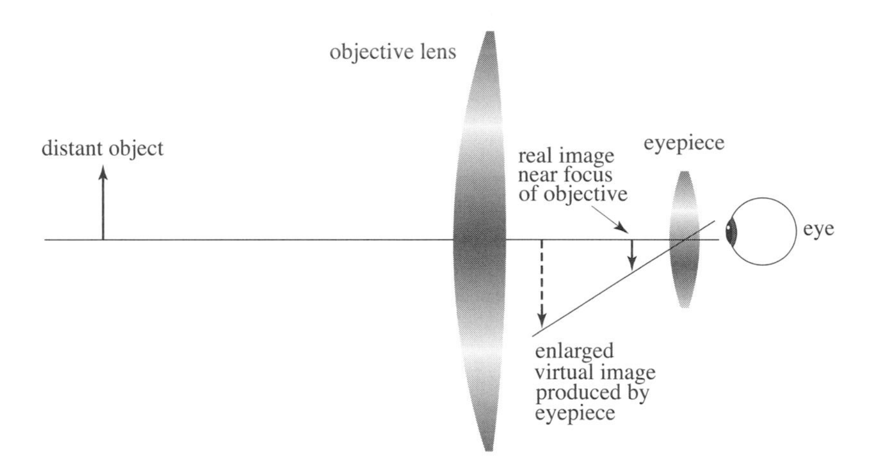

Fig. 8.38

axis. Such lenses bend incoming paraxial rays outward.

Newton experimented to discover that ordinary light is composed of light of different colors ranging from violet, to blue, to green, to yellow, to pink, to red, namely, the colors of the rainbow. When a beam of ordinary light passes through a glass prism, these colors are separated out. This is so because glass (or water droplets in the case of a rainbow) refracts the component colors red, green, blue, and so on progressively more. Conventional lenses, including the objective lens of a simple telescope, also exhibit such distortions, called *chromatic aberration*, primarily at the edges of the lenses. Figure 8.39a illustrates

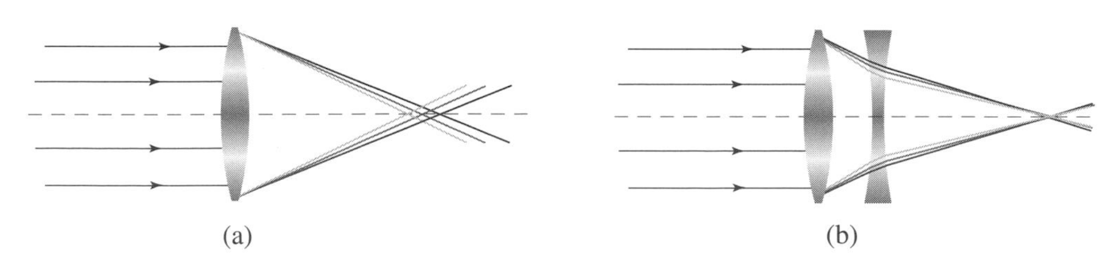

Fig. 8.39

this phenomenon. It can be corrected by placing a concave lens between the convex lens and its focus. See Figure 8.39b. Such lens combinations are called *compound lenses*.

A look at the images of Saturn of Figure 8.40 tells us that the quality of the image that a telescope produces depends on several factors: its brightness, sharpness, resolution, and magnification. The brightness of the image depends on the amount of light that the objective lens pulls into the instrument, and this depends directly on the size of the objective lens. The sharpness, or ability of the instrument to resolve details, depends on the size of the objective lens, as well as the precision of both the objective lens and the eyepiece. The magnifying power of the telescope depends on both the objective lens and the

Fig. 8.40. Image reproduced with permission from Joe Roberts, Amateur Astronomer's Notebook, http://www.rocketroberts.com/astro/first.htm

eyepiece. It is given by the ratio $\frac{f_o}{f_e}$, where f_o and f_e are the focal lengths of the objective lens and the eyepiece.

The world's largest refracting telescope is housed at the Yerkes Observatory of the University of Chicago. It was installed late in the 19th century and still functions. It has a compound objective lens of over 1 m in diameter and provides a magnification of about 1000. But it also illustrates the limitations of refracting telescopes. The fact that the focal length of the objective lens is a lengthy 19 m means that the tube containing the optics needs to be about as long. In addition, the objective lens of the Yerkes telescope is so heavy—over 225 kg—that it sags under its own weight and has to be rotated periodically to preserve its symmetry.

In the second half of the 17th century, Isaac Newton presented a telescope to the scientists of the Royal Society of London based on a different idea. Instead of collecting and bending light by using a lens, why not devise a telescope that does so with a curved mirror? This idea had been proposed before, but it was Newton who first built such an instrument. The general schematic for such a reflecting telescope is shown in Figure 8.41. Light enters in essentially parallel rays from a distant object, strikes a curving primary mirror, and is deflected inward. A secondary mirror deflects the light again, brings it to a focal point, and produces an image that is seen, magnified by an eyepiece, by the observing astronomer. Newton's telescope was small. It was only 16 cm long and had a concave primary mirror 5 cm in diameter in the shape of a section of a sphere of radius 32 cm. It magnified by a factor of about 35.

We saw in Section 8.5.3 that the geometry of a lens, designed to bend all light rays (and not just paraxial rays) that come in parallel to its axis of symmetry so that they come to a focus, has a complex geometry. The geometry that a primary mirror has to have so that it reflects all light rays parallel to its axis of symmetry to converge at a point is much simpler. We will see that the geometry is parabolic.

Consider any concave mirror that has an axis of symmetry. Its surface is obtained by revolving a curve C, such as the one depicted in Figure 8.42, one complete revolution around some axis (its axis of symmetry). Suppose that the mirror has the following property: all light rays from some distant object S coming in parallel to the axis of symmetry that strike the mirror are reflected inward and converge at

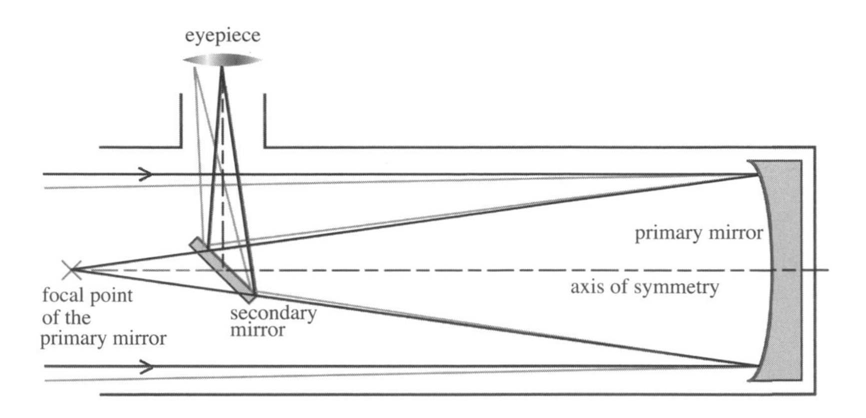

Fig. 8.41. Newtonian telescope. From the website https://en.wikipedia.org/wiki/Reflecting_telescope.

some point P. Suppose that they all leave the object at the same time $t = 0$. By Fermat's principle, all the rays in this "front" of light rays arrive at the point P at the same time. Since Fermat's principle selects the one that takes the least time, any latecomer would not be selected, as it would follow a path that is not allowed. Suppose that all the light rays reach P at time t_1. Suppose also that all the rays reach the line line 0 at some earlier time t_0. Let line 1 be the line that the front of light rays would reach at time t_1 if the mirror were not there. (Both line 0 and line 1 are perpendicular to the axis of symmetry.) This means that each light ray takes the same time $t_1 - t_0$ to go from line 0 to P or to go from line 0 to line 1 (if the mirror were removed). Let Q be any point on the mirror. Since the object can be seen by an observer at Q, there is a light ray from the object to Q (that goes on to P). Given the fact that the travel times of the light

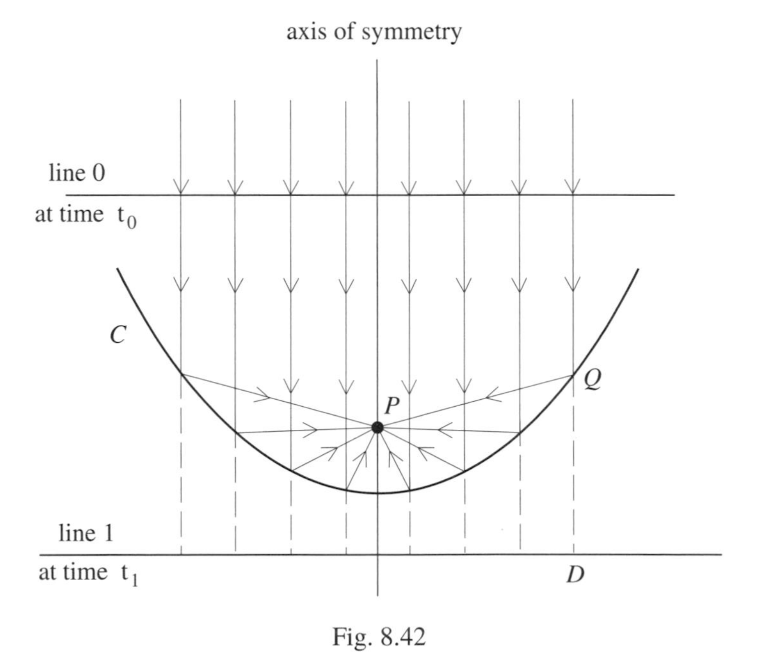

Fig. 8.42

ray from Q to P and from Q to the point D on line 1 are the same, it follows that the distance from Q to P is the same as the distance from Q to D. So all the points Q on the curve C of the mirror satisfy the condition that the distance from Q to P is the same as the distance from Q to line 1. A look at the definition of a parabola tells us, therefore, that the curve C is a part of the parabola with focus the point P and directrix the line 1. We already knew that parabolas have the property that we are discussing. By Proposition P1 of Section 2.1, the angle of "incidence" at any point on the parabola is the same as the angle of "reflection" to the focus. Therefore all light rays coming into a parabolic mirror parallel to the axis of symmetry are reflected to the focal point of the parabola. Therefore, for a parabolic mirror, the focal point of the parabola and the focal point of the mirror are the same point.

The Scotsman James Gregory, a mathematician and astronomer working in the 17th century, proposed a different design for a reflecting telescope a few years before Newton built his. But a working instrument of this design was not constructed until several years later. The essential scheme of the Gregorian telescope is shown in Figure 8.43. The primary mirror collects the light and brings it to a focus before the second

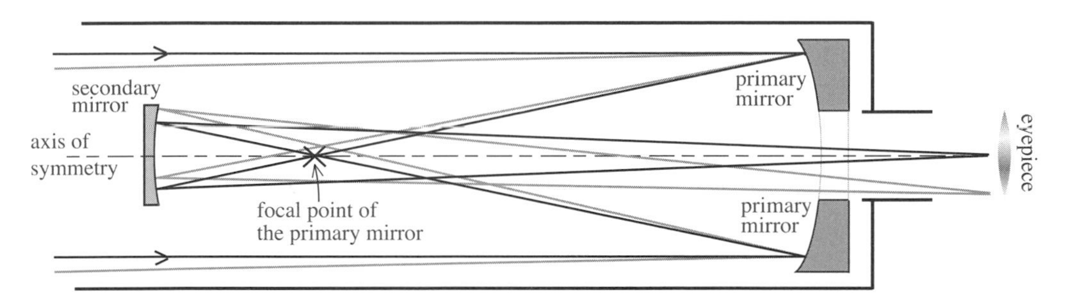

Fig. 8.43. Gregorian telescope. From the website https://en.wikipedia.org/wiki/Reflecting_telescope.

inward curving mirror reflects it back through a hole in the center of the primary mirror to form an image that is enlarged by the eyepiece.

Reflecting telescopes have several important advantages over refracting telescopes. In a lens, the volume of glass or plastic has to have a precise geometry, be homogeneous, and be free of imperfections throughout. In an optical mirror, only the surface has to be perfectly shaped and polished. A large objective lens is heavy and can be distorted by gravity. In contrast, a large mirror can be supported by a lighter frame on the other side of its reflecting face with no gravitational sag. In other words, in terms of the critical light-collecting ability of a telescope, it is not feasible to build huge lenses, but it is feasible to build huge light-collecting mirrors.

All large telescopes built in the 20th and 21st centuries, as well as those still under construction, are reflecting telescopes, and most of these are Gregorian telescopes (or modified versions of the Gregorian design) with primary mirrors that have central holes. The largest are three super telescopes: the Giant Magellan Telescope, the Thirty Meter Telescope, and the European Extremely Large Telescope. All three are scheduled to come on line in the 2020s.

The primary mirror of the Giant Magellan Telescope (GMT) will consist of a configuration of seven circular mirrors all of the same diameter. One mirror at the center (with the hole that the Gregorian design requires) is surrounded by six more, aligned in such a way that their combined surface will lie on a parabola of revolution. All of these mirrors are to be manufactured by the Mirror Lab at the University of Arizona. We will use the mirror at the center to describe the manufacturing process. A cylindrical form of about 8.4 m in diameter is placed in horizontal position so that its central axis is vertical. A tight, tiled arrangement of hundreds of hexagonal boxes made of heat-resistant material forms the supporting base of the mirror during its manufacture. Chunks of the purest glass, about 20,000 kg in all, are placed into the cylinder on top of this base. See Figure 8.44. The cylinder sits inside a tub that forms the bottom half of a furnace. After the lid is put in place, the furnace encloses the cylinder. Powerful heating elements melt the glass in the cylinder at a temperature of 1160 degrees centigrade. The furnace assembly is rotated in the horizontal plane around the central axis of the cylinder at a constant rate of 5 revolutions per minute. This rotation pushes the liquid glass outward toward the rim of the cylinder. The molten glass reaches steady state with its upper surface curving from the center of the cylinder up to its edge. After the correct mirror geometry has been achieved, the cooling process begins. It takes about three months to cool the glass to room temperature. The slowness of the process ensures that the glass will not develop cracks. To preserve the geometry, the furnace assembly continues to rotate during this time. The process that has been described is called *spin casting*.

The remarkable fact is that the spin-casting process provides the upper surface of the glass and hence the eventual mirror with the parabolic geometry that it needs to have! Let's verify this. Suppose that the

Fig. 8.44. Casting the central mirror GMT4 for the Giant Magellan Telescope. Image credit Ray Bertram, Richard F. Caris Mirror Lab, University of Arizona. See http://www.gmto.org/gallery/.

glass surface has reached steady state. Take a plane through the vertical axis of rotation. Think of this vertical plane to be fixed, and consider the curve obtained by intersecting the upper surface of the molten glass with this vertical plane. Figure 8.45 shows this plane along with an xy-coordinate system. The coordinates 4.2 and 1.15 represent the radii of the mirror and its central hole (both in meters). The y-axis

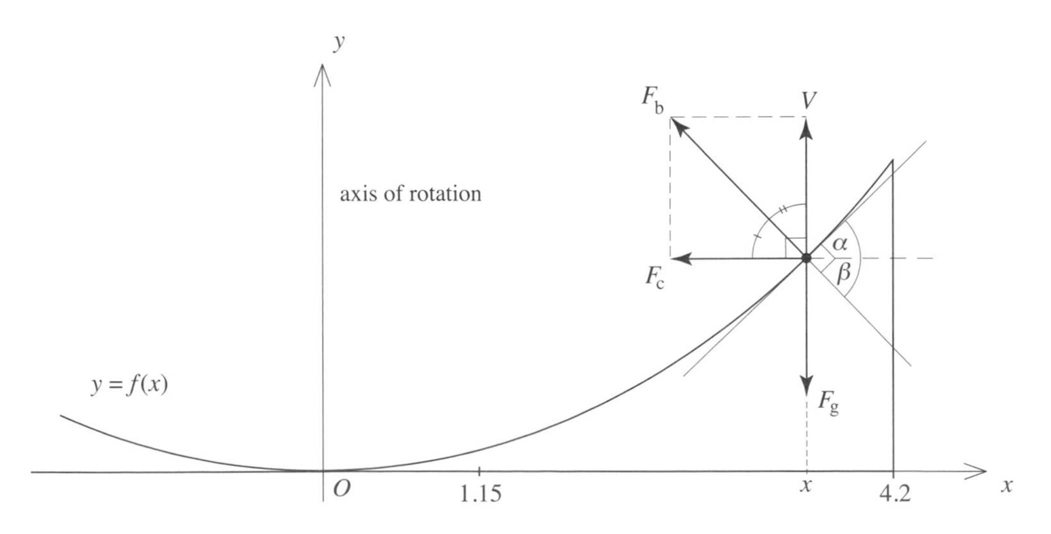

Fig. 8.45

is the axis of rotation of the furnace assembly, and the curve that is depicted as the graph of a function $y = f(x)$ represents the surface of the molten glass. Consider a small particle of molten glass of mass m riding on this surface. Let α be the angle between the tangent line to the graph at the glass particle and the horizontal. Let F_b be the buoyant force (refer to segment 2G of the Problem and Projects section for Chapter 2) with which the molten glass pushes against the particle. It acts perpendicularly to the tangent line. Because steady state has been reached, the vertical component V of the buoyant force is equal in magnitude to the force of gravity F_g acting on the glass particle. Let F_c be the horizontal component of the buoyant force. Extend the line of force of F_b, and draw in the angle β. Notice that $\alpha + \beta = 90°$. The angle marked with the single stroke is equal to β so that the angle marked with the double stroke is equal to α. Since the tangent of this angle is equal to $\frac{F_c}{V} = \frac{F_c}{F_g}$, it follows that

$$\tan\alpha = \frac{F_c}{F_g}.$$

Suppose that the cylindrical form turns at a constant rate of τ revolutions per unit time. Since steady state has been reached, the particle moves at this rate on a fixed horizontal plane in a circle of constant radius x. The particle is kept in "orbit" on this horizontal plane by the horizontal component of the buoyant force of magnitude F_c, which acts centripetally in the direction of the axis of rotation. Refer to the discussion of centripetal force in Sections 6.8 and 6.9. Since the mass m moves in a circle of radius x, we know by the circular case of the inverse square law in Section 6.9 that

$$F_c = \frac{4\pi^2 xm}{T^2},$$

where T is the time of one complete revolution. If the glass particle takes the time T for 1 revolution, it completes $\frac{1}{T}$ revolutions in 1 unit of time. Therefore $\frac{1}{T} = \tau$, and it follows that

$$F_c = 4\pi^2\tau^2 mx.$$

Because $f'(x)$ is the slope of the tangent,

$$f'(x) = \tan\alpha = \frac{F_c}{F_g} = \frac{4\pi^2\tau^2 mx}{mg} = \frac{4\pi^2}{g}\tau^2 x.$$

Since the derivatives of $y = f(x)$ and $y = \frac{2\pi^2}{g}\tau^2 x^2$ are the same and $f(0) = 0$, it follows by Corollary ii at the end of Section 7.6 that the upper surface of the molten glass is obtained by rotating the parabola

$$f(x) = \frac{2\pi^2}{g}\tau^2 x^2$$

one complete revolution around the y-axis. The focal point of the parabola has coordinates $(0, \frac{g}{8\pi^2\tau^2})$. This follows from the conclusion of Problem 2.1 (or from the analysis of Section 4.3). Notice that the shape of the parabola depends (in addition to π and g) only on the speed of the rotation τ and not on the density of the molten glass. So the shape is the same regardless of the liquid that is being rotated.

Notice that the constant τ controls the geometry of the mirror. For the central mirror of the GMT, the rotational speed was set at $\tau = \frac{1}{12}$ revolutions per second (the equivalent of 5 revolutions per minute). Since $g \approx 9.80$ m/s^2 (at the location of the University of Arizona in Tucson, Arizona), it follows that in meters,

$$f(x) \approx 0.014x^2.$$

The parabolic surface that has now been described only approximates the final shape of the mirror. The precision requirements on the mirror are extraordinary. After the glass mass has cooled, its parabolic surface is made smooth with diamond grinding wheels. This brings the accuracy of the surface to within about $\frac{1}{10}$ of 1 millimeter of what is needed. Finally, after polishing the glass, the necessary tolerance of less that $\frac{1}{10{,}000}$ of 1 millimeter is achieved. The final step—undertaken on location of the GMT high in the Chilean Andes—is the application of a thin, fragile, reflective aluminum coating to the glass surface. Only now is the manufacture of the mirror complete.

Example 8.5. Consider the shape of the central mirror of the Giant Magellan Telescope as it is described by Figure 8.45 and the function $f(x) \approx 0.014x^2$.

i. The depth of the mirror—namely, the vertical distance between the horizontal plane at the mirror's rim and its central hole—is

$$f(4.2) - f(1.15) \approx 0.014[(4.2)^2 - (1.15)^2] \approx 0.23 \text{ m}.$$

ii. The focus of the central mirror is

$$\frac{g}{8\pi^2\tau^2} \approx \frac{(9.80)(12^2)}{8(9.87)} \approx 17.87 \text{ m}$$

above the vertex of the parabola.

We have described the manufacture of the central primary mirror of the GMT. The six mirrors surrounding the central mirror are constructed in the same way. But there is one important difference. The parabola of the central mirror determines the parabola of the entire configuration, and its central axis coincides with the focal axis of this parabola. The surfaces of the other six mirrors lie higher on the parabola. So they are off axis and not rotationally symmetric. The spin-cast glass form from the furnace needs to be ground and polished more carefully to achieve the delicate geometry that it needs to have. When complete, the seven-mirror configuration of the GMT will have a diameter of 25 m.

An even larger telescope, the Thirty Meter Telescope (TMT), is currently being built on Mauna Kea in Hawaii with a primary mirror of 30 m in diameter. The strategy for building this mirror is completely different. It will be a composite of 492 individual hexagonal mirror segments that measure 1.44 m from corner to opposite corner. The hexagons, slightly different in shape, are carefully aligned to form the hyperbolic primary mirror of the design. (The fact that the primary and secondary mirrors are both curved and work in tandem means that the parabolic geometry is not the only option for the primary mirror.) The advantage of this approach is that the smaller mirrors are more quickly manufactured and more easily shipped to the construction site than the huge mirrors of the GMT. On the other hand, it will be much easier to control the few moving parts of the GMT (seven primary mirror segments and seven secondary mirror segments) with the necessary accuracy. The European Extremely Large Telescope (E-ELT), to be built in Chile, will have the same mirror design as the TMT. With its 798 hexagonal mirrors and a diameter of 39 m, it will be the largest of the three new super telescopes. Within another 10 years, these telescopes will look deeply into space to unravel the mysteries surrounding the evolution and current state of the universe (its age, galaxy formation, dark matter and dark energy, black holes, and planets of distant stars).

Earth-based telescopes need to compensate for atmospheric interference, such as air currents and turbulence, as well as refraction. These distort the information carried by the light from the object being observed. The powerful computers that are integrated within the optics of high-tech telescopes can measure the distortions, and the many actuators can continuously adjust the shape of the mirrors in response. Such systems are referred to as *adaptive optics*. They transform twinkling stars into clear, steady points of light. In this way, all three super telescopes will produce images that are 10 times sharper than those of the Hubble Space Telescope. (The amazingly successful Hubble with its 2.4 m mirror orbits Earth and does not have to deal with atmospheric conditions.)

8.6 PROBLEMS AND PROJECTS

8A. Rates of Change. This first set of problems deals with rates of change in the context of bacterial growth, radioactive decay, and the cost of a firm's production. We'll start by examining bacteria cultures during their exponential phase of growth.

8.1. A medical laboratory investigates a culture of bacteria. There are 4000 cells after 4 hours into the experiment, 62,000 cells after 7 hours, and 154,000 cells after 8 hours. Compute the average rates of change in the number of bacteria during the time intervals $[4, 7]$ and $[7, 8]$ per hour. Find a formula that that expresses the number of bacteria in the exponential growth phase of the culture as a function $y(t)$ of time for $t \geq 0$ in hours.

8.2. A culture of *E. coli* bacteria is growing in its exponential phase under ideal conditions. It has a doubling time of 15 minutes. When it is first observed at time $t = 0$, it has 20,000 cells.

- **i.** Find an expression for the number of cells after t hours.
- **ii.** When will the population reach one billion?
- **iii.** Find the number of cells after 6 hours.

8.3. A bacteria culture is in its exponential growth phase. It starts with 10,500 bacteria at time $t = 0$. Two hours later, there are 23,000 bacteria.

- **i.** Express the number $y(t)$ of bacteria at any time $t \geq 0$ in terms of an exponential function.
- **ii.** What is the size of the population after 3 hours?
- **iii.** At what time will the population reach 130,000?

8.4. A culture of bacteria is being studied in a lab in a situation of exponential growth. Initially, there were 5000 cells in the culture. Two hours later, the cells were increasing in number at a rate of 10,000 cells per hour. Estimate the growth constant of the culture. [Hint: Let x_0 be the x-coordinate of the point of intersection of the graphs of the $y = e^{2x}$ and $y = \frac{2}{x}$. Show that $\frac{1}{2} < x_0 < 1$. Find an estimate by exploration.]

8.5. A culture of bacteria is being studied during its phase of exponential growth. When first measured, there were $75,000$ cells in the culture, and exactly 24 hours later, the cells were increasing in number at a rate of 150,000 cells per hour. Estimate the growth constant of the culture. [Hint: Let the "day" be the working unit of time, and consider the hint for the previous problem.]

One of the numbers that is often attached to the label of a chemical element is the number of protons and neutrons in its nucleus. This number is known as the element's atomic mass. The designation polonium-210 signifies that this radioactive element has an atomic mass of 210. *Avogadro's number* relates the amount of a sample of a pure element in grams to the number of atoms in the sample. The relationship is that for every gram of the sample, there are $\frac{1}{m}(6.02 \times 10^{23})$ atoms, where m is the atomic mass of the element.

8.6. The radioactive element polonium-210 with a half-life of 138 days decays to a stable form of lead.

- **i.** Determine the decay constant of polonium-210.
- **ii.** Compute the number of atoms in a 30 milligram sample of polonium-210.
- **iii.** Express the number of polonium-210 atoms that remain in this sample after t days in terms of an exponential function.
- **iv.** How many polonium-210 atoms will remain after 6 weeks?

8.7. Assume that the decay equation for the radioactive element radon-222 is $y(t) = y_0e^{-0.18t}$, with t given in days.

i. What value for the half-life is this equation based upon?

ii. About how long will it take for a sample of pure radon-222 to decay to 90% of the original amount?

iii. How long will it take for the sample to decay to $\frac{1}{3}$ of the original amount?

8.8. One series of measurements gave a value of 3.7×10^{10} disintegrations per second as the activity of 1 gram of pure radium-226. What half-life for radium-226 can be deduced from this measurement?

8.9. An unknown radioactive substance is tested in a lab with a radiation counter. The counter measures 3200 disintegrations per minute at 8:00 a.m. and 900 disintegrations per minute at 5:00 p.m. of the same day. Determine the half-life of the radioactive substance.

8.10. Let $C(x)$ be a company's cost for manufacturing x units of a certain product per year. Show that if the average cost per unit is constant, then this average cost is equal to the marginal cost.

8.11. A firm sells a certain product for \$6652 per unit, and it sells all the units that it produces. Let x be the number of units produced per quarter. It is known that the firm's marginal cost for the product is given by the quadratic function $C'(x) = 0.000012x^2 - 0.002x + 2800$. It is known that the total cost for producing 10,000 units is \$39,476.

i. Determine the cost function $C(x)$.

ii. Express the revenue R as well as the profit P as a function of the number x of units sold per quarter.

iii. Determine the production level at which the firm's quarterly profit is a maximum. What is the average cost per unit at that production level?

iv. What is the maximal profit per quarter that the firm earns?

8B. Stanley Tests Pulleys. The piece of fiction that follows is adapted from Jerry Van Amerongen's cartoon *Ballard Street* from December 2005 and January 2006. The cartoons are reproduced here with the artist's permission.

Stanley is intrigued by L'Hospital's pulley problem. Is the analysis of the hanging contraption undertaken in Section 8.2 correct? The solution there seems too complicated for what is surely a simple problem. Why not experiment? To this end, Stanley engages his wife Bertha to join him in an effort to test ropes and

Stanley gears up and thinks deeply.

"How did he do that?" wonders Bertha.

"Stanley, watch it. This is some sort of trick!"

"There he goes!"

pulleys. Initially they run into some difficulties, but soon they begin to sort things out and make progress.

Convinced that he has a handle on things, Stanley sets up L'Hospital's contraption in a tree in his back yard. He grabs the end of the rope, pulls himself up, and suspends himself. Alas, Stanley has made a

"Whoa!"

Stanley gets TP'd.

conceptual error. The system does not respond as he had intended. The neighborhood kids add insult to injury and further bruise his ego.

8C. About L'Hospital's Pulley

8.12. Consider the pulley system of L'Hospital with $c = 10$ feet and $a = 8$ feet and turn to Figure 8.4. Observe that $CB = 10$ and $CF = 8$, and determine the distances EC, EF, and BF and the angles θ_1 and θ_2.

8.13. In the situation of Problem 8.12 and Figure 8.4, suppose that the weight at D is 150 pounds, and determine the tensions T_1 and T_2.

8.14. Refer to Figure 8.3, and suppose that ΔBCF is an isosceles triangle. Show that the length a of the string to which the pulley is attached is equal to $\frac{c}{\sqrt{3}}$. [Hint: First rule out the possibility that $BF = c$. So only the case where the two slanting sides are equal remains. Continue by showing that the two component triangles are similar.]

8.15. Is it possible for $\angle BFC$ in Figure 8.3 to be a right angle? [Hint: Assume this to be so. Apply the Pythagorean theorem and explore the consequences.]

8D. About Tension in Strings

8.16. A weight W is suspended from a cable as shown in Figure 8.46. The segment AB is horizontal. The angles that the cable makes with the horizontal at A and B are α and β, respectively. The system is

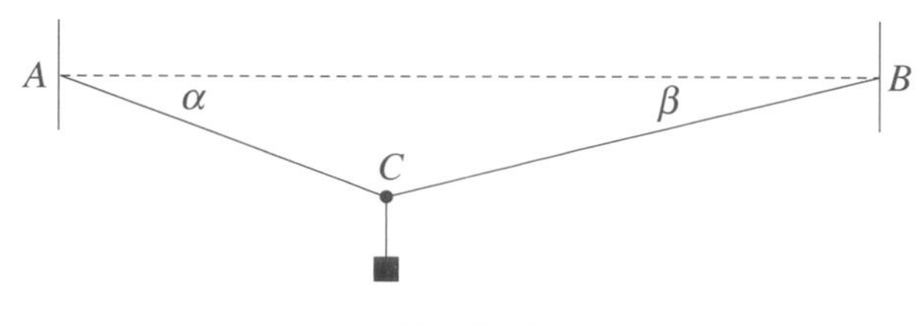

Fig. 8.46

assumed to be in equilibrium. Compute the tensions T_1 and T_2 in the respective cable segments AC and CB in terms of the weight W and the angles α and β.

8.17. Look up the addition formula for the sine (see Problem 1.25), and use it to simplify the expressions for T_1 and T_2 that you derived in Problem 8.16 to

$$T_1 = \frac{W\cos\beta}{\sin(\alpha+\beta)} \quad \text{and} \quad T_2 = \frac{W\cos\alpha}{\sin(\alpha+\beta)}.$$

Under the assumption that the weight W is attached at C with a pulley wheel that rotates freely, show that $\alpha = \beta$, and hence that the point C is the midpoint of the cable.

8.18. Assume that $W = 160$ pounds, $\alpha = 10°$, and $\beta = 5°$ and use the formulas of Problem 8.17 to compute the tensions T_1 and T_2. Repeat your computation of T_1 and T_2 with $W = 200$ pounds, $\alpha = 4°$, and $\beta = 2°$. [What is the message behind these numbers?]

8.19. A jolly circus clown walks on a high wire from platform A to platform B. See Figure 8.47. The edge of the platforms are at the same height, and they are separated by a distance of 20 m. When he stops exactly one-quarter of the way to B, the cable is deflected 15 cm from the horizontal AB. Let T be the tension in the cable segment AC. By what factor is T greater than the weight W of the jolly guy? [Hint: use the formula from Problem 8.17 in combination with the geometry of the wire.]

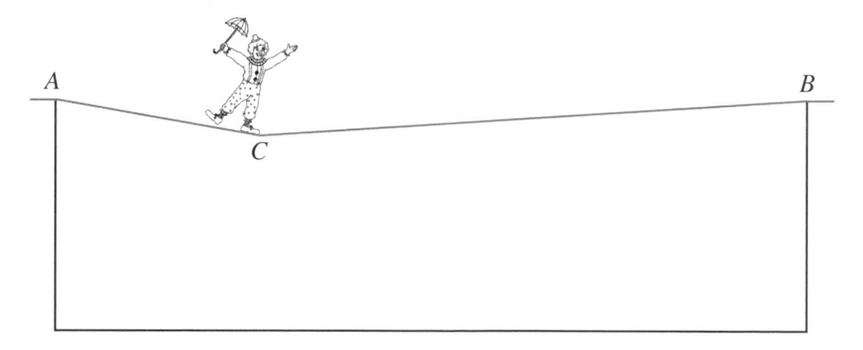

Fig. 8.47. Thanks to Greg Young, School of Architecture, University of Notre Dame, for creating the clown.

8.20. A sphere 20 cm in diameter is made of homogeneous material and has a mass of 12 kg. It is held by a cord against a frictionless vertical wall. See Figure 8.48.

i. Express the tension in the string in terms of α.

ii. Express the force with which the sphere pushes against the wall in terms of α.

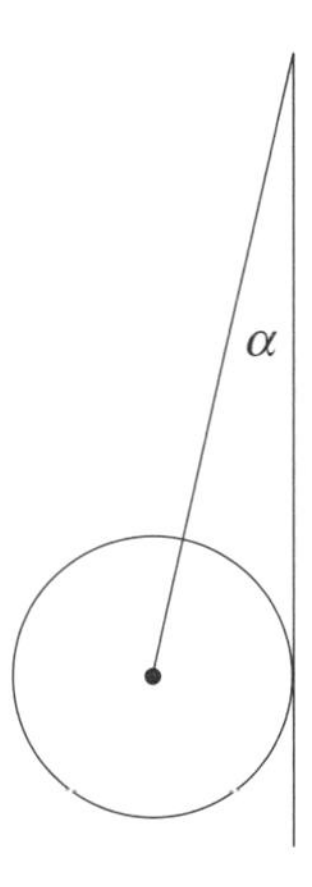

Fig. 8.48

iii. Compute the tension and the force if $\alpha = 15°$.

8E. Archimedes's Law of Hydrostatics. This basic fact about buoyant force was introduced in segment 2H of the Problems and Projects section of Chapter 2.

8.21. In Marilyn vos Savant's column in *Parade Magazine* of August 3, 2003, a reader asks the following question: "Say that I row my bowling ball to the middle of a lake. I drop the ball over the side and watch as it sinks to the bottom. What happens to the water level of the lake? Does it rise, fall, or remain the same?" Marilyn begins her answer as follows: "Bowling balls are heavier than water, so the level of the lake will fall." Is she right or wrong? Provide a "water tight" argument for the right answer. [Hint: The two cases to consider are (1) the bowling ball submersed, and (2) the bowling ball in a floating, weightless (except for the bowling ball) "boat." Which situation displaces more water?]

8.22. A basketball weighing 1.3 pounds with a radius of $r = 0.39$ feet floats in a pool of calm water. How much of the ball is submersed? Start by making a sketch of what has been described.

i. Let h be the distance in feet from the bottom of the ball to the surface of the water. Use results of Section 5.9 to show that volume of the part of the ball that is under water is equal to $\pi rh^2 - \frac{\pi}{3}h^3$. Use this to show that $1.047h^3 - 1.225h^2 + 0.021 = 0$.

ii. Consider the function $f(x) = 1.047x^3 - 1.225x^2 + 0.021$. Note that the h above is a zero of $f(x)$. To find h, turn to Newton's method of segment 7O of the Problems and Projects section of Chapter 7. A preliminary guess for this zero is $c_1 = 0.39$ (the radius of the ball). Carry out three steps in Newton's method with accuracy to six decimal places to get $c_4 = 0.139533$. Show that c_4 is extremely close to being a zero of $f(x)$. Is $h \approx 0.14$ feet a reasonable answer to the question? Take a basketball to a tub of water and experiment.

8F. Suspension Bridges. The theory of the suspension bridge developed in Section 8.3 relies on the assumption that the vertical load on the bridge is uniformly distributed over the length of the bridge. The constant w is taken to be the maximum weight (dead load plus maximum live load) per foot that the bridge needs to support, distributed over the number of cables. This w is relevant for determining the greatest loads, tensions, and compressions that the structure is subject to. However, the analysis applies more generally. So w could be the weight per foot (distributed over the number of cables) in the situation of zero live load, maximum live load, or anything in between (with slightly varying sag s). Indeed, the theory applies to any situation of a cable under uniform vertical load (as long as the cable is completely flexible and does not lengthen when stretched).

8.23. A clothesline is supported by two posts of the same height 30 feet apart. Fifty pigeons have made it a temporary home. They sit roughly evenly spaced from post to post. A typical bird weighs 14 ounces, or $\frac{13}{16} \approx 0.875$ pounds. They cause the string to sag by 1/2 foot at its center. Determine the maximal and minimal tension in the string.

8.24. You have joined the Peace Corps and are working with natives in Peru. In order to improve their ability to move through the mountainous terrain, they wish to build a primitive suspension bridge across a deep 90-foot-wide gorge. The plan for the bridge is simple. Solid wooden boards are to be cut, each about 5 feet long, $1\frac{1}{2}$ feet wide, 2 inches thick, and weighing about 10 pounds. Two heavy ropes tied to large trees on each side are to span the gorge in a precisely parallel way. The boards are to be attached by two hemp cords at each of their ends to the two arching ropes in such a way that they form a continuous path across. Additional cords are to be used to attach the boards to each other. The structure is designed to support the crossing of 30 natives in single file equally spaced. A native typically weighs about 130 pounds. The natives begin the construction by shooting individual cords across the gorge with bows and arrows. They will weave the individual cords into the two heavy ropes. Since you are the sophisticated American (with some knowledge about suspension bridge design), they had asked you to advise them. You know that the heavy ropes are the key to the structure. What computations did you undertake, and what counsel did you give them?

8.25. A new Tacoma Narrows Bridge was built in the years 1948–1950. The data for the new bridge are as follows. It has a total length of 5000 feet and a center span of 2800 feet. Its two main cables support a single deck that carries 4 lanes of automobile traffic. The dead load is 8680 pounds per foot. Assume that it is designed for a live load capacity of 4000 pounds per foot and that the sag in the main cable over the center span is 280 feet. Compute T_d and α. Compute the compression that one of the main cables over the center span produces in a tower. (The information is based on data supplied by the Washington State Department of Transportation.)

The Golden Gate Bridge is a suspension bridge that spans the Golden Gate, the opening of the San Francisco Bay to the Pacific Ocean. When it was completed in 1937, it was the world's longest suspension bridge. It has a center span of 4200 feet and two side spans of 1125 feet each, for a total length of 6450 feet. It has one deck and two main cables supporting it. A suspension bridge is designed to move in response to changes in loads, temperatures, and wind conditions. For example, its towers lean and its deck moves. For example, under a sustained transverse wind, the deck of the Golden Gate can move laterally by as much as 28 feet. In part in response to such challenges, the Golden Gate has undergone major structural modifications since the time it was built. In 1953 a new bottom lateral bracing system was added, and in 1985 the original concrete bridge deck was replaced with a steel deck. This reduced the dead load to 20,170 pounds per foot while retaining the original live load capacity of 4000 pounds per foot. (This information was supplied in 2003 by Jerry Kao, supervising civil engineer at the Golden Gate Bridge District.)

8.26. The towers of the Golden Gate have a height of 746 feet, and the sag in the cables over the center span is 470 feet. Compute T_d under dead load only and then under dead plus live load. (The Golden Gate Bridge, Highway and Transportation District, San Francisco, provides the information that the tension in the cable over the center span at the towers is 54,000,000 pounds under dead load only and 64,100,000 pounds under dead plus live load.)

8.27. The Verrazano Narrows Bridge, New York City, was completed in 1964. It has two decks, a center span of 4260 feet, a dead load of 37,000 pounds per foot, and a live load capacity of 4800 pounds per foot. It has four main cables, and the sag in each of them is 385 feet at midspan. Compute T_d and α. Compute the compression that the four main cables over the center span produce in one of the towers. (Data supplied by the Triborough Bridge and Tunnel Authority.)

The Humber Bridge in England, completed in 1981 with a center span of 4625 feet, held the record for the center span of a suspension bridge until it was eclipsed by two "super" suspension bridges, both completed in 1998. One is the "missing link" of a highway system that connects Copenhagen with mainland Denmark. It has a center span of 5300 feet. The second, with its enormous center span of 6500 feet, spans the Akashi Straits in Japan and is an integral part of a highway system connecting the main island of Honshu to some smaller islands.

8.28. The Akashi Straits Bridge is has an overall length of 3910 m (about 2.5 miles) and a center span of 1990 m (about 6530 feet). Each of its two towers rises to a height of 297 m (about 975 feet) above sea level. Its two main cables support a deck that carries 6 lanes of traffic. The sag in the main cables is about 201 m. It is estimated that the two main cables together (over the main and side spans) put each tower under a total compression of 980 million newtons. Show that the angle α is about 22°. Then show that $T_d \approx 650$ million newtons (under the assumption that the compressions generated by a main cable are the same for a side span and the center span) and approximate the dead load (plus the much smaller live load) of the deck in newtons per meter. (Data based on a publication of the Tarumi Construction Office, First Construction Bureau, Honshu-Shikoku Bridge Authority.)

8G. Rotating Things

8.29. The counterclockwise motion of a particle along a circle starting from rest at time $t = 0$ is given by $\theta(t) = \frac{t^3}{125} + \frac{t}{5}$, with t in seconds and θ in radians. See Figure 8.49. Calculate the angle θ at $t = 10$, the average angular velocity between $t = 0$ to $t = 10$, the angular velocity ω at the instant $t = 10$, the average angular acceleration from $t = 0$ to $t = 10$, and the angular acceleration α at $t = 10$.

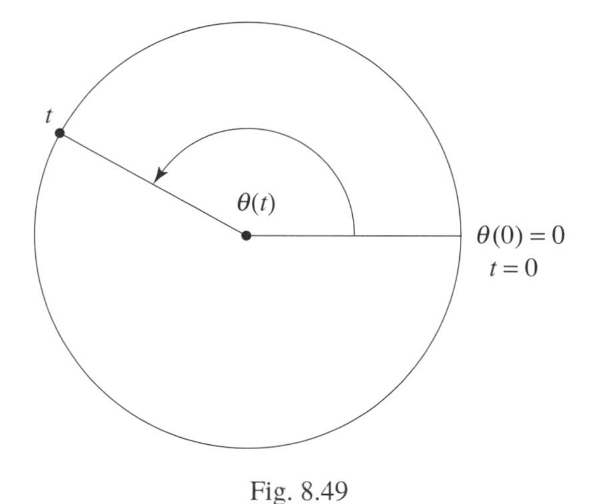

Fig. 8.49

8.30. Consider an equilateral triangle of side r and vertices A_1, A_2, and A_3. Draw in the three circular arcs C_1, C_2, and C_3 each of radius r centered at A_1, A_2, and A_3, respectively, shown in Figure 8.50. The focus will be on the curve consisting of the three circular arcs. We'll call the points A_1, A_2, and A_3 the vertices of this curve.

i. Turn to Figure 8.50. Determine the functions $y = f_1(x), y = f_2(x)$, and $y = f_3(x)$ that have the three circular arcs C_1, C_2, and C_3, respectively, as graphs. Notice that their domains are $[0, \frac{r}{2}], [-\frac{r}{2}, 0]$, and $[-\frac{r}{2}, \frac{r}{2}]$, respectively. Compute the derivatives of the three functions and show that

$$-\infty < f_1'(x) < \tfrac{-1}{\sqrt{3}} \text{ for all } x \text{ in } (0, \tfrac{r}{2}),$$
$$\tfrac{-1}{\sqrt{3}} < f_2'(x) < \tfrac{1}{\sqrt{3}} \text{ for all } x \text{ in } (-\tfrac{r}{2}, \tfrac{r}{2}), \text{ and}$$
$$\tfrac{1}{\sqrt{3}} < f_3'(x) < \infty \text{ for all } x \text{ in } (-\tfrac{r}{2}, 0).$$

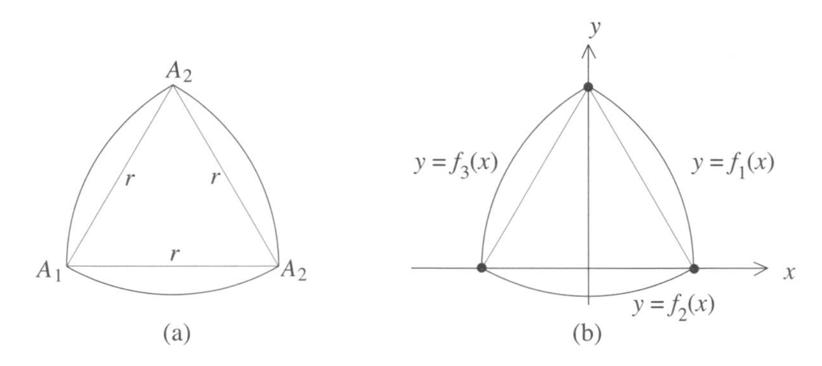

Fig. 8.50

ii. Let P be any point on the curve that is not a vertex, and let L be the tangent to the curve at P. Now let L' be any line parallel to L that touches the curve at a single point, say, P'. Use the conclusion of part (i) to verify that P' must be a vertex of the curve. Show that P' cannot be

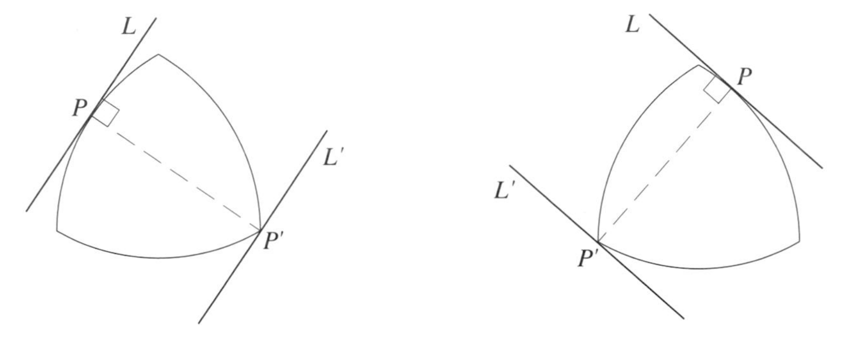

Fig. 8.51

either one of the two vertices of the circular arc on which P lies. So P' is the vertex that is the center of the circular arc on which P lies. Figure 8.51 provides two examples.

iii. Let P and P' be any pair of points that satisfy the assumptions and hence the conclusion of part (ii). Show that the segment $P'P$ is a radius of the circular arc involved. Conclude that r is the distance between P and P'.

Parts (ii) and (iii) together imply the following. Suppose that two cylinders with cross sections given by Figure 8.51 are placed in parallel on a smooth horizontal plane. Suppose that a flat board is placed on top of the two cylinders. If the two cylinders are rolled forward, then the board that they carry will move parallel to the plane and remain completely level at a fixed distance r above the plane. See Figure 8.52.

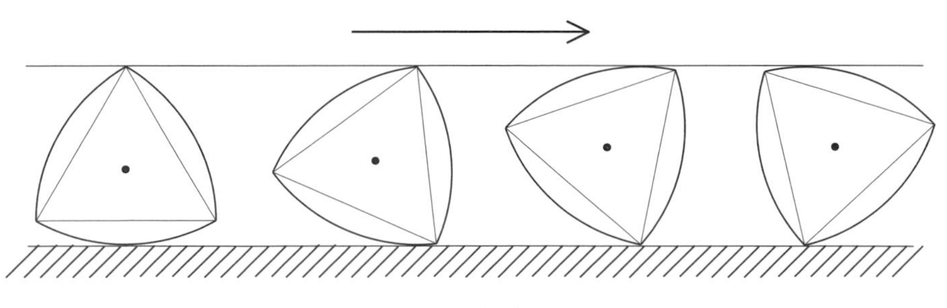

Fig. 8.52

It may come as a surprise that a circular cylinder with diameter r is not the only cylinder with this property.

8.31. Consider an equilateral triangle with vertices A, B, and C. Let A' and B' be the midpoints of the sides BC and AC, respectively. Refer to Section 2.5 for the fact that the point of intersection M of AA' and BB' is the centroid (or center of mass) of the equilateral triangle and that $A'M = \frac{1}{3}AA'$.

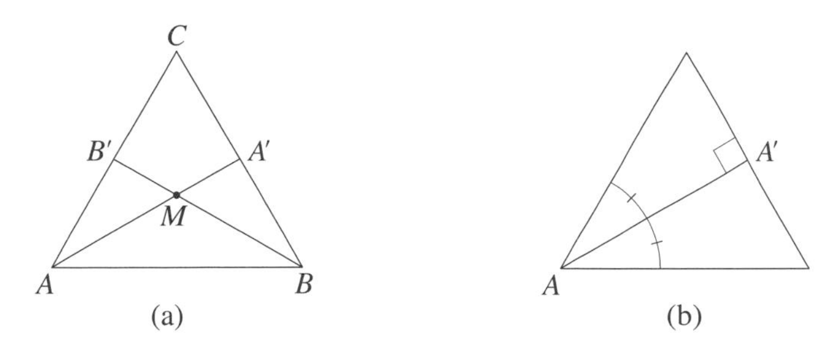

Fig. 8.53

See Figure 8.53a. Show that the segment AA' is perpendicular to BC and that it bisects $\angle BAC$. See Figure 8.53b.

8.32. Return to Figure 8.52 and assume that $r = 2$. Now go to Figure 8.53. The point at the center of the moving figure is the centroid of the triangle. Use the result of Problem 8.31 to show that the height of the center of the figure above the horizontal surface varies from a minimum of $2 - \frac{2}{3}\sqrt{3} \approx 0.8453$ to a maximum of $\frac{2}{3}\sqrt{3} \approx 1.1547$.

A curve with shape given by Figure 8.52 also forms the cross-section of the piston of the Wankel engine, named after Felix Wankel, its German inventor. In this internal combustion engine, the piston

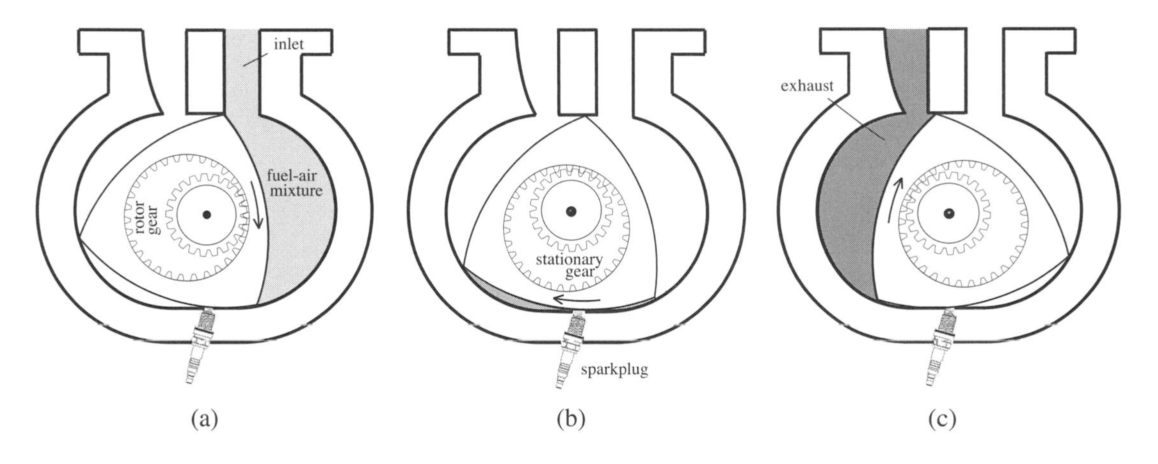

Fig. 8.54

revolves in the cylinder. Figure 8.54 illustrates how the engine functions. In Figure 8.54a, the fuel air mixture has been drawn into the cylinder. In Figure 8.54b, the mixture is compressed and ignited. The explosion drives the rotation forward. Figure 8.54c shows the burned fuel being expelled. An off-centered drive shaft transfers the power. The Wankel engine is being phased out as an engine for automobiles, but improved versions are still in development. Fewer moving parts and general simplicity are its advantages over standard combustion engines. Disadvantages include the difficulty with the seals at the vertices, as well as the relatively high fuel consumption and relatively dirty exhaust. It is used to power motorcycles, snowmobiles, and jet skis.

8H. Galileo's Inclined Planes

8.33. The slanting length of an inclined plane is 5 m and its angle with the horizontal is 15°. Suppose that an ice cube and a ball both of mass 0.25 kg descend down the inclined plane side by side on separate tracks from the top of the plane with zero initial speed. The track of the ice cube is wet and heated so that the ice cube glides frictionlessly. The track that the ball rolls down is perfectly smooth and flat, so that the only force slowing the motion is the frictional force that rotates the ball. Take $g = 9.81$ m/s^2.

i. What are the magnitudes of the forces that drive the ice cube and the ball down the plane?

ii. What are the respective accelerations of the ice cube and the ball?

iii. What is the rotational velocity of the ball at the bottom of the plane? (Assume that the diameters of the ball is 6 cm.)

iv. How much sooner does the ice cube arrive at the bottom of the plane?

v. By what distance does the ice cube win the race?

8.34. Consider a ball rolling down from the top of an inclined plane having started from rest. Assume that it rolls without slippage and retardation (other than the frictional force that rotates the ball).

i. Show that the velocity of the ball is proportional to the square root of the distance it has rolled. What is the constant of proportionality?

ii. Show that the average velocity of the ball during its roll to any point on the plane is $\frac{1}{2}$ the velocity that it has the instant it reaches that point.

iii. Show that the velocity of the ball at any point depends only on the (vertical) drop h to that point.

iv. Show that two balls rolling with the same initial speed from the top to the bottom of two inclined planes of the same height have the same velocity when they arrive at the bottom of their planes.

8.35. Consider two balls rolling down two different ramps of the same heights. They both start from rest. One ramp is straight, the other curving. See Figure 8.55. Draw a third ramp of the same height between the two in the figure that illustrates that the two balls have the same velocity whenever they

Fig. 8.55

they are the same vertical distance from the ground. [Hint: Use one of the conclusions of Problem 8.34, and draw the third ramp as a composite of short straight segments.]

8.36. In his *Discorsi*, Galileo describes the following method for sketching parabolas. "I use an exquisitely round bronze ball, no larger than a nut; this is rolled on a metal mirror held not vertically but somewhat tilted, so that the ball in motion runs over it and presses it lightly. ... To describe parabolas in this way, the ball must be somewhat warmed and moistened ... so that the traces it will leave shall be more apparent on the mirror." Figure 8.56 depicts what Galileo describes. Show that Galileo's procedure does generate parabolas. [Hint: Place a coordinate system as shown, and make use of a combination of the discussions in Sections 6.7 and 8.4.]

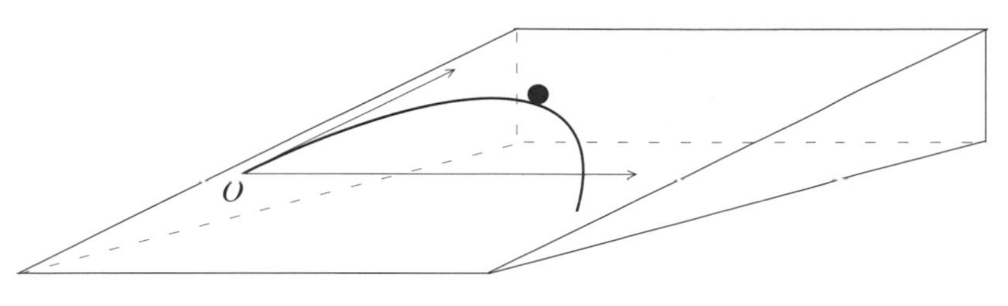

Fig. 8.56

8.37. Figure 8.57a shows a circle of radius r and center $(0, -r)$, and a line through the origin O with equation $y = mx$. The origin and another point P are the points of intersection of the circle and the line. Show that $P = (\frac{-2mr}{1+m^2}, \frac{-2m^2r}{1+m^2})$ and that the distance between P and O is $\frac{2mr}{\sqrt{1+m^2}}$.

8.38. Consider an inclined plane given by the cord of a vertical circle of radius r. The highest point of the inclined plane is the point O at the top of the circle. See Figure 8.57b. Suppose that a ball is

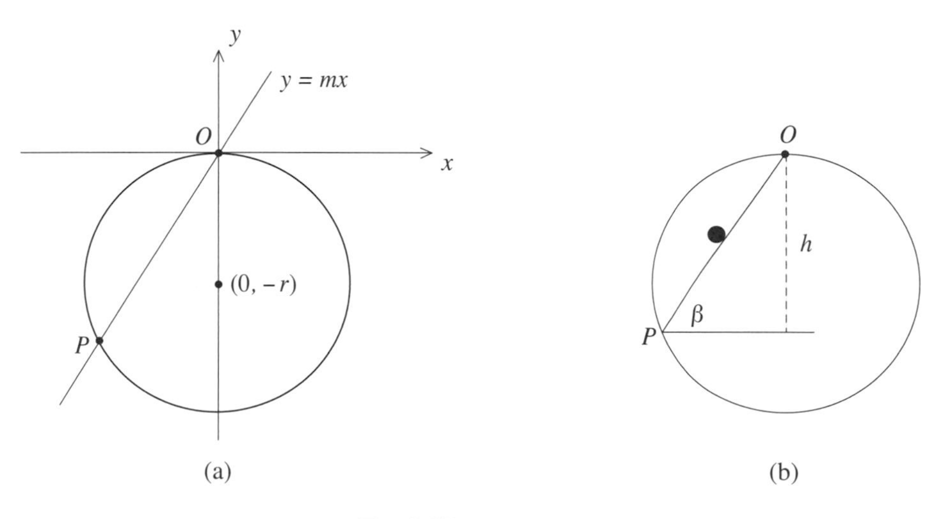

Fig. 8.57

released from rest from the point O at time $t = 0$. Use the results of Section 8.4.3 and Problem 8.37 to show that it arrives at the point P at time $t = 2\sqrt{\frac{7r}{5g}}$.

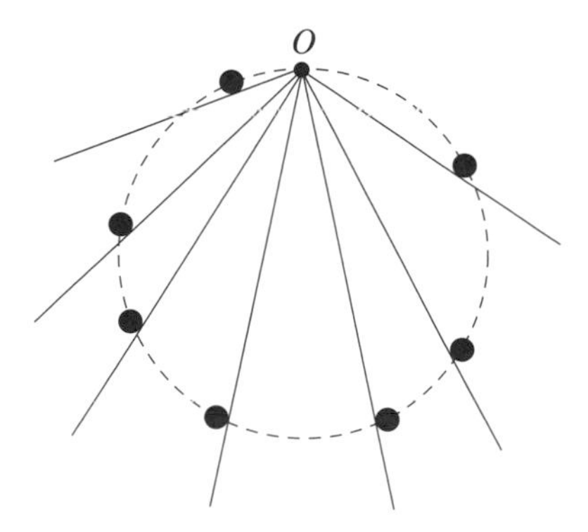

Fig. 8.58

8.39. Consider any number of inclined planes all with the same highest point O and all lying in the same vertical plane. Simultaneously release a ball from rest at point O on all of the planes at time $t = 0$. Let a time t elapse, and consider the positions of the balls. See Figure 8.58. Galileo concluded correctly that O and all the points of contact of the balls with their planes lie on a circle. Verify this using the conclusion of Problem 8.38. Determine the radius of this circle in terms of the elapsed time t.

8I. More About Optics

8.40. Suppose that a light ray passes through the air and strikes a sheet of crown glass at an angle of incidence $\alpha = 30°$. Determine the angle of refraction.

8.41. Conventional mirrors consist of a piece of plate glass supplied with a silver coating as shown in Figure 8.59. Verify that the angle of incidence α is equal to the angle of reflection β.

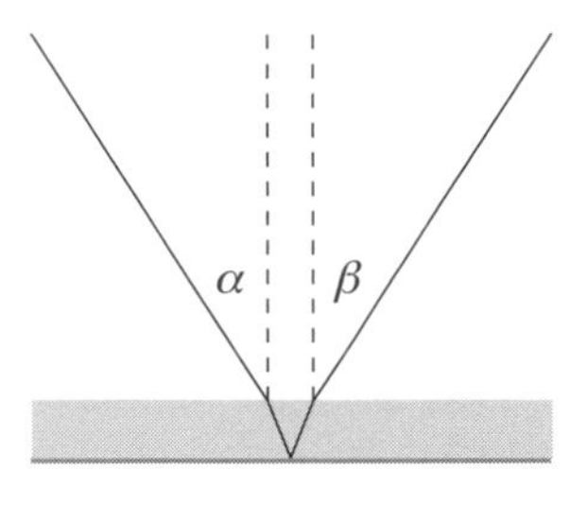

Fig. 8.59

The aspheric lens formula of Section 8.5.3 consists of a complicated first term and a correction polynomial. Figure 8.60 depicts the aspheric lens of Figure 8.39b, but adds an xy-coordinate system. In

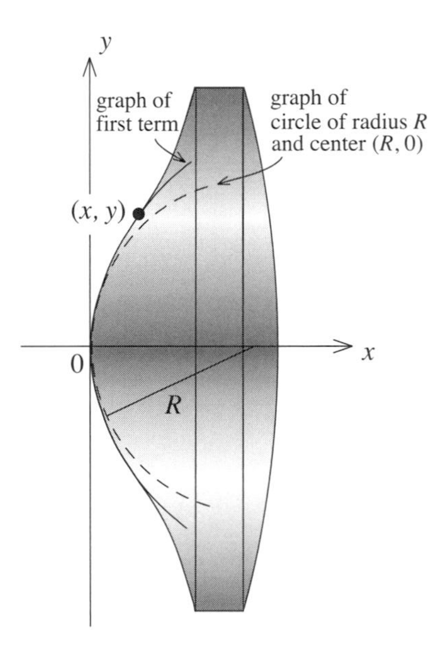

Fig. 8.60

terms of x and y, the aspheric lens formula becomes

$$x = \frac{y^2}{R[1 + \sqrt{1 - (1-k)\frac{y^2}{R^2}}]} + (A_1 y^2 + A_2 y^4 + A_3 y^6 + \cdots).$$

We now turn to the first term of the aspheric and show that its graph is a conic section.

8.42. Show that the equation $x = \frac{y^2}{R\left[1+\sqrt{1-(1-k)\frac{y^2}{R^2}}\right]}$ can be put into the form $y^2 = 2Rx - (1-k)x^2$. [Hint: Move the denominator $R[1+\sqrt{1-(1-k)\frac{y^2}{R^2}}]$ to the left side of the equation, multiply, and move xR to the right side. Then square both sides and simplify.] Verify the following.

i. If $k = 0$, complete the square to show that $y^2 = 2Rx - (1-k)x^2$ is the circle $(x-R)^2 + y^2 = R^2$.

ii. If $k \neq 0$, then the graph of $y^2 = 2Rx - (1-k)x^2$ intersects the circle $(x-R)^2 + y^2 = R^2$ only at the point $(0,0)$.

iii. If $k > 0$, then the graph of $y^2 = 2RX - (1-k)x^2$ lies completely outside the circle $(x-R)^2 + y^2 = R^2$ (except for the point $(0,0)$). This is the situation sketched in Figure 8.60.

iv. If $k < 0$, then the graph of $y^2 = 2Rx - (1-k)x^2$ lies completely inside the circle $(x-R)^2 + y^2 = R^2$ (except for $(0,0)$).

8.43. Verify that the graph of $y^2 = 2Rx - (1-k)x^2$ is a parabola, an ellipse, or a hyperbola. Proceed as follows.

i. If $k = 1$, the graph of $y^2 = 2Rx$ is the parabola with focal point $(\frac{R}{2}, 0)$ and directrix $x = -\frac{R}{2}$. [Hint: Derive the equation of the parabola with focus $(\frac{R}{2}, 0)$ and directrix $x = -\frac{R}{2}$.]

ii. For $k \neq 1$, show (after some algebra that includes the completion of a square) that

$$\frac{(x - \frac{R}{1-k})^2}{\frac{R^2}{(1-k)^2}} + \frac{y^2}{\frac{R^2}{1-k}} = 1.$$

iii. For $k < 0$, the graph is the ellipse with focal points $(\frac{R}{1-k}, \pm\frac{\sqrt{-k}R}{1-k})$ and eccentricity $\varepsilon = \sqrt{\frac{-k}{1-k}}$.

iv. For $k = 0$, the graph is the circle of radius R and center $(R, 0)$ (as already verified in Problem 8.42i.)

v. For $0 < k < 1$, the graph is the ellipse with focal points $(\frac{R}{1\pm\sqrt{k}}, 0)$ and eccentricity $\varepsilon = \sqrt{k}$.

vi. For $k > 1$, the graph is the right branch of the hyperbola with focal points $(\frac{R}{1\pm\sqrt{k}}, 0)$ and eccentricity $\varepsilon = \sqrt{k}$.

With modern measurement and manufacturing technology, it has become possible to produce aspheric lenses in a cost-effective way. They are now used in optical products from cameras, to eyeglasses, to handheld magnifiers. "End-to-end" component design and manufacture are highly automated and rely on computer-aided design (CAD) and computer-aided manufacturing (CAM). This relies on computer files that contain the commands for the operation of the machines. These often combine multiple tools such as drills, saws, and so on into a single "cell." Different machines are used with an external controller and human or robotic operators that move the manufacturing process from machine to machine to its completion.

8J. Atmospheric Refraction. Any assessment of the location of an object in the solar system or beyond it with a telescope, a pair of binoculars, or the naked eye has to compensate for the fact that light rays are refracted by the Earth's atmosphere. There is a simple way of measuring this effect that makes use of the discussion of Section 8.5.2.

Consider a light ray that travels through a vacuum, passes through a transparent slab of a denser medium with parallel sides, and reemerges on the other side into a vacuum. Section 8.5.2 (refer to Figure 8.30 and the discussion that explains it) informs us that the angle that the ray makes with the perpendicular to the slab at the point of entry is the same as the angle it makes with the perpendicular at the point of exit. Suppose that a light ray passes through two successive parallel slabs of transparent mediums that are separated by a slight amount of vacuum. They can have different indices of refraction. Applying the fact just mentioned twice, tells us that the angle the ray makes with the perpendicular at the point of entry into the first slab is equal to the angle it makes with the perpendicular at the point of exit from the second slab.

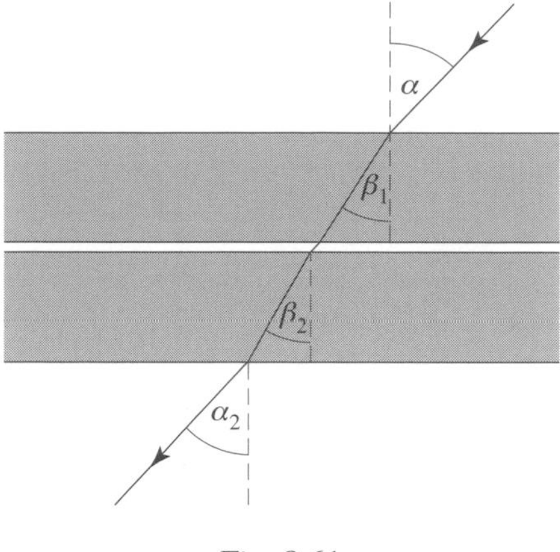

Fig. 8.61

In reference to Figure 8.61, the assertion is that the angles α and α_2 are the same. Suppose that a light ray passes through four successive parallel slabs of transparent mediums, each separated from the other by a slight amount of vacuum. Let the initial angle of incidence be α. Applying the above argument to Fig-

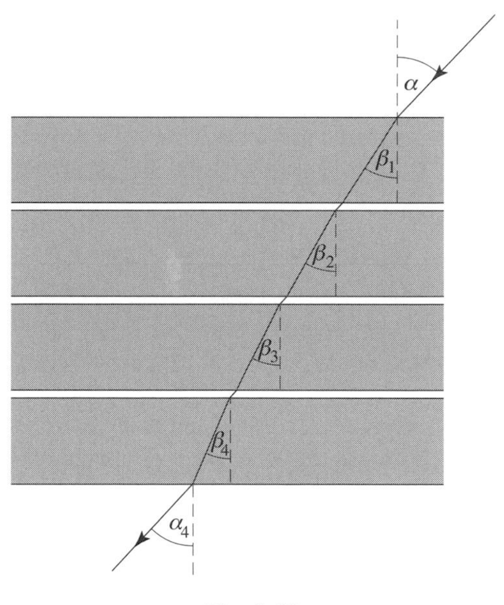

Fig. 8.62

ure 8.62 tells us that the angles α and α_4 are equal. Let n_4 be the index of refraction of the slab at the bottom. Since the index of refraction of a vacuum is 1, Snell's law of refraction (applied to the ray exiting the bottom slab) tells us that $\sin \alpha = n_4 \sin \beta_4$.

The fact that the ray of both Figures 8.61 and 8.62 become steeper as they descend through the slabs tells us that the indices of refraction increase from a given slab to the lower one. Consider a light ray that passes through k successive parallel slabs of transparent mediums, each separated from the other by a slight amount of vacuum. Suppose the index of refraction of the last medium is n_k. It should now be clear that if the initial angle of incidence is α and β_k is the last angle of refraction, then $\sin \alpha = n_k \sin \beta_k$.

The density of Earth's atmosphere diminishes as the height above the Earth increases. Studies show that there is some atmosphere at a height of 800 km above the Earth. The density is still appreciable enough at 150 km for air resistance to produce changes in the orbit of a satellite. Refraction of light rays begins to be a factor at about 100 km. We will model these 100 km of the Earth's atmosphere as follows. Above the atmosphere is the vacuum of space. We will neglect the curvature of the Earth. This is reasonable since the radius of the Earth is about 6400 km as compared to the relevant layers of atmosphere of 100 km. Then we will regard the atmosphere as being made up of many—say, k—layers of different densities, the density being greatest in the lowest layer and constant within each layer. The lower the layer, the greater its index of refraction. We will think of consecutive layers as being separated by a tiny strip of vacuum. A ray of light from a star S strikes the top layer at the point T with an angle of incidence of α. It is then refracted by the successive layers of the atmosphere until it reaches the observer at O. Refer to Figure 8.63. The axis OZ points straight up (along a plumb line) from the observer. The notation Z refers to *zenith*, a word that comes from the Arabic and means "direction of the head, or path above the head." Draw the line OS parallel to the light ray coming in to T. This is the line from the observer in the

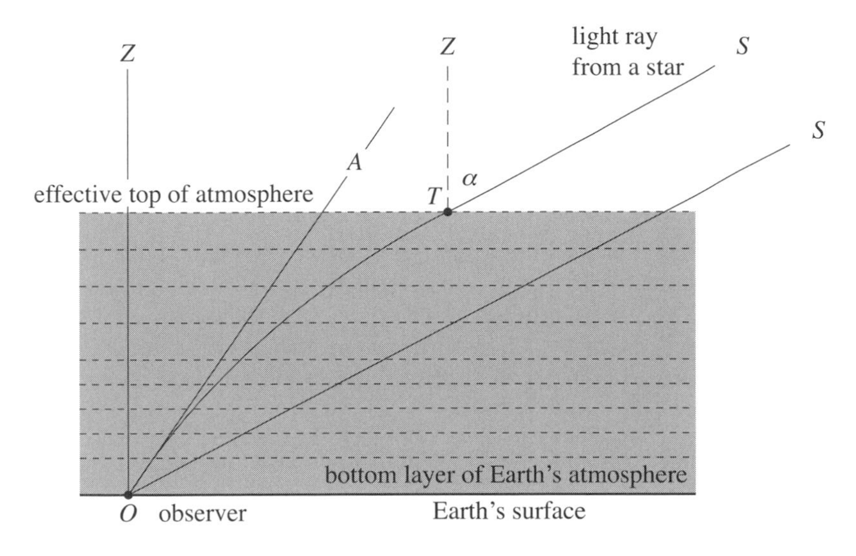

Fig. 8.63

direction of the actual position of the star. The line OA points in the observed direction of the star. Notice that the star will appear to be higher in the sky than it actually is. The angle $z_{true} = \angle ZOS$ is called the *true zenith distance* of the star. Observe that $z_{\text{true}} = \alpha$. The angle $z_{\text{app}} = \angle ZOA$ is called the *apparent zenith distance* of the star. (For now, both z_{true} and z_{app} are given in radians.) The angle

$$\rho = z_{\text{true}} - z_{\text{app}}$$

measures the difference between the true position and the observed position of the star. It measures the effect of the refraction of the atmosphere.

Consider the enlargement of the bottom layer in Figure 8.64. Recall that α is the angle of incidence of the first layer and that β_k is equal to the angle of refraction of the ray in the last layer. Observe also that $z_{\text{app}} = \beta_k$. By applying what we already know,

$$\sin z_{\text{true}} = \sin\alpha = n_k \sin\beta_k = n_k \sin z_{\text{app}}.$$

Since $z_{\text{true}} = \rho + z_{\text{app}}$, we get by the addition formula for the sine (refer to Problem 1.24) that

$$\sin z_{\text{true}} = (\sin\rho)(\cos z_{\text{app}}) + (\cos\rho)(\sin z_{\text{app}}).$$

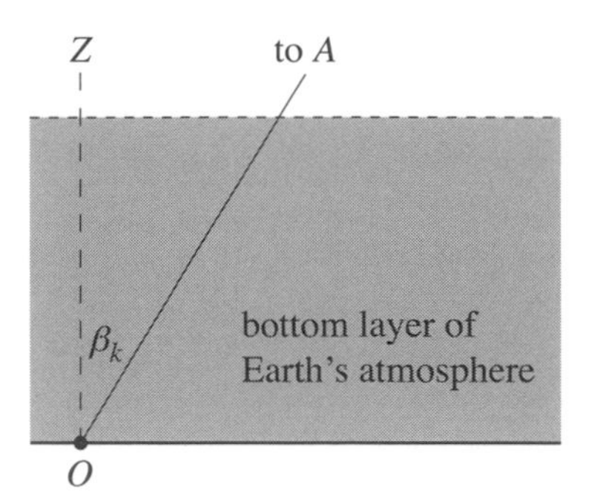

Fig. 8.64

So $(\sin\rho)(\cos z_{\text{app}}) + (\cos\rho)(\sin z_{\text{app}}) = n_k \sin z_{\text{app}}$. Because ρ is usually small and often very small, we will assume that $\sin\rho \approx \rho$ and $\cos\rho \approx 1$ to obtain the approximation $\rho\cos z_{\text{app}} + \sin z_{\text{app}} \approx n_k \sin z_{\text{app}}$. So $\rho \approx (n_k - 1)\frac{\sin z_{\text{app}}}{\cos z_{\text{app}}} = (n_k - 1)\tan z_{\text{app}}$. The index of refraction of the air at sea level at 10° centigrade is 1.000283. Inserting this for n_k gives the approximation $\rho \approx 0.000283\tan z_{\text{app}}$. Converting from radians tells us that $\rho \approx 0.016214\tan z_{\text{app}}$ in degrees, and that

$$\rho \approx 58.372941\tan z_{\text{app}} \approx 58.37\tan z_{\text{app}}$$

in seconds.

Let's summarize. To find the true zenith distance z_{true} of a heavenly body in degrees, measure z_{app} in degrees, take the corresponding ρ, and (being careful with units) compute $z_{\text{true}} = z_{\text{app}} + \rho$. The approximation $\rho \approx 58.37\tan z_{\text{app}}$ for the difference between the true position z_{true} of a star and its observed position z_{app} is accurate for z_{app} up to 45° but becomes progressively less accurate for larger angles z_{app}. Table 8.3 provides—for selected angles z_{app} in degrees (in the columns on the left)—more

z_{app} in °	ρ in ″	z_{app} in °	ρ in ″	z_{app} in °	ρ in ″	z_{app} in °	ρ in ″
0	0	50	70	80	319	86	706
10	10	55	84	81	353	87	863
20	21	60	101	82	394	88	1103
30	34	65	125	83	444	89	1481
40	49	70	159	84	509	$89\frac{1}{2}$	1760
45	59	75	215	85	593	90	2123

Table 8.3

accurate values for ρ in seconds (in the columns on the right).

8.44. Compare the values for ρ provided by the approximation $\rho \approx 58.37\tan z_{\text{app}}$ with those of Table 8.3 for $z_{\text{app}} = 20°, 40°, 60°$, and $80°$. Suppose that stars are observed at these apparent zenith distances. What are the true zenith distances?

Observe the Sun at sunset, precisely when the apparent zenith position of the bottom of the Sun's disc is 90°. Consider the angle determined at the eye of an observer by a vertical diameter of the Sun's disc. This apparent "vertical angular diameter" of the Sun is equal to about 29 minutes of arc. Use the data in Table 8.3 to show that the true zenith position of the top of the Sun's disc is greater than 90°. This means that the Sun is in fact already below the horizon when it is observed at sunset! Draw a diagram of what you are observing. Use Table 8.3 again to show that the true vertical angular diameter of the Sun is about

35 minutes. Since the Sun is circular, this is also the true horizontal angular diameter of the Sun. But the horizontal angular diameter will not be affected appreciably by refraction. So the apparent horizontal diameter is also about 35 minutes. Therefore, when the Sun is observed at sunset, its apparent vertical and horizontal diameters will be 29 and 35 minutes, respectively. This is why the Sun at sunset is seen as a slightly flattened disc, in other words, as an ellipse.

8K. More about the Speed of Light. Recall from Section 8.5.1 that in the latter part of the 17th century, Ole Rømer's measurements provided the first "ballpark" estimate of the speed of light. About 200 years later, in the 1870s, the physicists Albert Michelson and Edward Morley measured the speed of light with great accuracy using an apparatus consisting of a lens and two mirrors. The lens and its two focal points S'

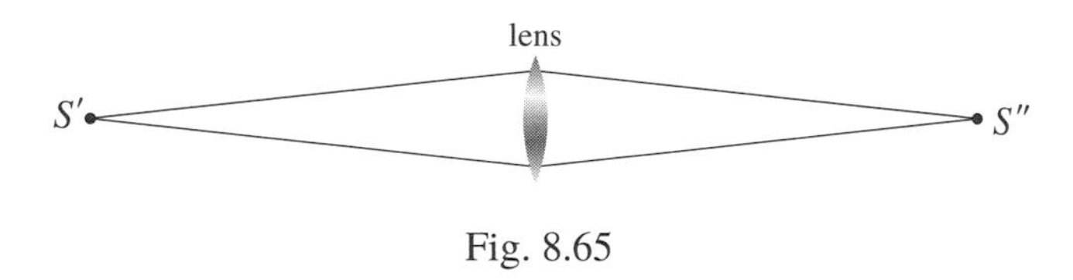

Fig. 8.65

and S'' are shown in Figure 8.65. The two mirrors are placed as shown in Figure 8.66. Mirror 1 is free to rotate about an axis perpendicular to its plane. The point S is the reflection of the focal point S'. Now suppose that a point light source is placed at S. The light rays that emanate from it strike mirror 1, go through the lens, refocus at S'', are reflected by mirror 2, go through the lens again, are reflected again by mirror 1, and finally reconverge at the point S. Let c be the speed of light, and let t be the time it takes for

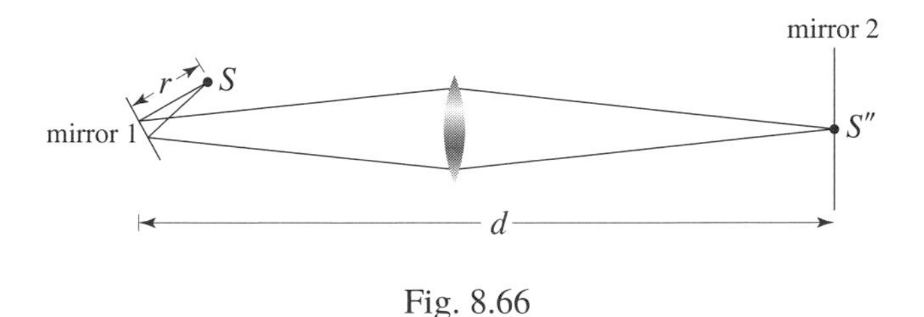

Fig. 8.66

the rays to travel from mirror 1 back to mirror 1. Since the rays cover the distance $2d$ during this time, it follows that $2d = ct$. Hence $c = \frac{2d}{t}$. Now suppose that mirror 1 is allowed to rotate. When the axis of rotation is placed appropriately and the speed of rotation is high enough, the light rays will hit a different position on mirror 1 on their return. Thus they no longer reconverge at the point S, but at a point that is a small, but measurable, distance δ away. The larger the time t, the larger the deviation δ. For his apparatus, Michelson was able to show that $\delta = 4\pi rnt$, where r is the distance indicated in Figure 8.66 and n is the number of revolutions per second of the mirror. Measuring δ allowed him to compute t and hence c. He was able to conclude that light travels at 299,895 ± 30 km/s. This corresponds tightly to today's value of 299,792.458 km/s. For this and other achievements, Michelson was awarded the Nobel Prize for physics in 1907.

8L. Unravelling the Mysteries of the Universe. For about 150 years, from 1543 when Copernicus's *De Revolutionibus* introduced the idea to the time of the publication of Newton's *Principia* of 1687, the key question of astronomy was: Is a fixed Earth the center of the system of the Sun and planets, or is the Sun the fixed point around which the planets including the Earth move? If it is the latter, would not the picture of the canopy of stars change when observed from Earth at opposite points of its orbit? The astronomer Tycho Brahe had observed no such change. The firmament of stars looked exactly the same from month to month. While this had been a credible argument against the motion of the Earth, the fact is that the distances to the stars are so vast that any change in their observed positions—known as *stellar parallax*—is almost imperceptible. It was 150 years after the appearance of the *Principia* when the German astronomer

Friedrich Bessel was able to observe and measure the parallax of the star 61 Cygni with a small but precise refracting telescope equipped with a split image viewfinder. It turned out that compared to the angular width of the Moon of about $\frac{1}{2}^\circ$, the largest angles of parallax observed for a star were a minuscule $\frac{1}{5000}^\circ$. (See Section 3.4 and Problem 3.26.) It is no wonder that Tycho was not able to detect such small angles. This measurement, and similar ones, confirmed that our planet Earth does indeed move relative to the Sun. But it also confirmed that the universe is huge.

We saw in Section 3.7 that the astronomical unit, 1 au = 149,597,892 km, is a convenient unit for measuring the already considerable distances in the solar system. But the stars and galaxies of the universe are so far away that a much larger unit of distance than the au was called for. This unit is the *light-year*, the distance that light travels in one year in the vacuum of space. As defined by the International Astronomical Union, the light-year, abbreviated by ly, is the product of the year of 365.25 days (the measure of the year of the old Julian calendar) expressed in seconds times the speed of light of 299,792,458 m/s. A little arithmetic establishes the connection,

$$1 \text{ ly} \approx 63{,}241 \text{ au}.$$

The nearest star, Proxima Centauri, is about 4.22 ly from our Sun. So it takes light moving at its enormous speed 4.22 years to travel from Proxima Centauri to Earth. By way of comparison, it takes light about 8 minutes to reach us from the Sun. The reflecting telescopes of the early part of the 20th century began to provide astonishing information. The Sun is part of a huge cluster of stars, the spiraling Milky Way galaxy, consisting of hundreds of millions of stars. The Milky Way measures 300,000 ly across. In the 1920s, Edwin Hubble pointed his 2.54 m reflecting telescope skyward and observed stars far beyond the Milky Way, one million ly away. In 1929, he discovered that the universe is not fixed in place and time, but that it is expanding. Galaxies are moving away from each other and the farther they away they are away, the faster they are receding. If a movie of the expansion of the universe were to be played backward into the past, the universe would be seen to contract until finally, its beginning state of infinite density and energy would be revealed. Restart the movie from there, and you see the initial explosion of the universe at the beginning of time. This is what astronomers call the *Big Bang*. It has since been established that the galaxies at the outer reaches of the universe are an incomprehensible 13.8 billion ly away. Since light takes 13.8 billion years to reach us from such galaxies, this tells us that 13.8 billion years have passed since the universe sprang into being with the Big Bang.

In the 1970s, it was confirmed that galaxies like the Milky Way are rotating with such speed that the gravitational forces between their observed matter are not strong enough to hold them together. Astronomers have come to the conclusion that there is additional *dark matter* that provides these galaxies with the extra mass and extra gravity they need to stay intact. Unlike normal matter, dark matter does not absorb, reflect, or emit light, making it extremely hard to detect. The evidence suggests that dark matter is more massive than visible matter by a factor of more than 5. Scientists think that dark matter comprises 27% of the universe, and that all the stars and galaxies account for only 5%.

Astronomers had reasoned that the mutual gravitational attraction between all the matter in the universe must be slowing down the expansion that Hubble had observed. But what would the ultimate fate of the universe be? Would these gravitational effects be comparatively slight, weaken with the expansion, and unable to keep the universe from expanding aggressively? Or would it balance the expansion in such a way that the universe would slow down and come to a virtual standstill? Or, finally, would the collective gravitational attraction be so strong as to stop the expansion and reverse it, so that all the matter in the universe will eventually collapse in a *Big Crunch*? Throughout the 1990s, two teams of astrophysicists pursued this problem. They analyzed a number of exploding stars, or supernovas, and used these unusually bright, short-lived distant objects to gauge the growth of the universe. They observed and studied supernovae with the Hubble Space Telescope that were six or seven billion light-years away, or halfway across the universe, and found that these pulses of light were dimmer and therefore more distant than expected. Both teams concluded that the expansion of the universe must be speeding up! This discovery meant that the dominant force in the evolving universe isn't gravity, it's something else. This something else was given the name *dark energy*. Dark energy is thought to make up approximately 68% of the universe, appears to be distributed evenly throughout the universe, and its effect is not diluted as the universe expands. With dark energy comprising 68% of the universe and dark matter another 27%, only 5% of the universe consists of the galaxies that can be directly observed.

One approach to understanding dark energy involves a method called gravitational lensing. According to Albert Einstein's theory of general relativity, a beam of light traveling through space is bent by the gravitational pull of matter. More precisely, it is space itself that bends, and light just goes along for the ride. This means that if two clusters of galaxies lie along a single line of sight, then the cluster in the foreground will act like a lens on the light coming from the cluster in the background in much the same way as Figure 8.65 illustrates for ordinary lenses. The images of the background galaxies that gravitational lenses produce are faint and distorted, but the analysis of these images provides information about the lenses and hence masses of the clusters in the foreground. Many millions of images of such galaxies will need to be identified and captured. It will be an upcoming agenda to use the wide and deep views of the universe that the next generation of super telescopes—the GMT, TMT, and E-ELT discussed in Section 8.5.4—will provide to observe the gravitational lens systems in all parts of the universe. The goal will be to measure the distortion of the background galaxies and the distances of the foreground galaxies and to construct a three-dimensional mass map of the universe. Since masses seen at great distance are masses seen at a much earlier time, this will provide a chart of the evolution of the dark matter structure over cosmic time. This chart should enable astronomers to estimate the rate at which galaxies have clumped into clusters and how fast the universe expanded at different points in its history. This in turn will be a measure of the presence and impact of dark matter and dark energy.

Chapter 9

The Basics of Integral Calculus

Chapter 2 told us how Archimedes computed the area of a parabolic section. He did it by covering the region with an infinite array of triangles that get smaller and smaller near the curving boundary and adding up the areas of all the triangles. His study *The Method* also computed the area of a parabolic section, this time by slicing it into thin parallel strips. These two early examples illustrate the basic strategy of integral calculus. In the 16th and 17th centuries, mathematicians used similar ideas to solve various problems. They understood that the relationship between the process for finding tangents—the derivative—is inverse to the process for finding areas—the definite integral. In Chapters 5 and 6, we saw how Leibniz and Newton approached definite integrals, areas, volumes, and the fundamental theorem of calculus. The reason that history has given Newton and Leibniz the primary credit for inventing calculus is that these two geniuses were first to recognize differentiation and integration as general methods, not just as techniques that could solve particular problems. In their hands, calculus was no longer an offshoot of classical Greek geometry. Significantly broader in scope, it had developed into a useful and systematic mathematical tool.

Having comprehensively dealt with the calculus of differentiation in Chapter 7, we now turn to give a similar account of the calculus of integration. The criticism that Bishop Berkeley aimed at the work of Leibniz and Newton—namely, that its foundational notions were vague—was valid in the context of integral calculus as well. The definitive, rigorous conceptual framework that was required was provided in the 19th century, when the German Bernhard Riemann (1826–1866) among others defined the summation process—the core idea of the definite integral—in terms of a formal limit. Chapter 9 examines this summation process beginning with its visualization as area, but it also highlights the role of Riemann's formal limit. The chapter includes the geometric applications to volumes, lengths of curves, and surface areas, and it establishes the relationship between the summation process and antiderivatives that the fundamental theorem of calculus expresses. The questions that this fundamental theorem raises deserve attention. Does any function have an antiderivative? If so, can it be found explicitly? In this regard, our library of basic functions with algebraic, trig, exponential, log, and hyperbolic functions already on its shelves is expanded to include the inverses of these functions. The standard techniques for finding antiderivatives—including integration by substitution, by partial fractions, by parts, and by trigonometric and hyperbolic substitution—are analyzed. The trapezoidal and Simpson rules—methods with which any definite integral can be approximated as accurately as any application might be required—are presented. They are of special relevance in situations where explicit antiderivatives are impossible to come by.

It was one of the central aims of the previous chapter to make the point that calculus infuses several basic areas of science and engineering with important and also concept-defining information. This chapter begins the extension of this discussion to integral calculus. We will see that the definite integral of a function—in essence a sum consisting of a huge number of very small terms—has many interpretations in different contexts, both geometric (volumes, lengths of curves, and areas of surfaces) as well as applied (in science and engineering). The fundamental theorem provides one of the methods with which such sums can be computed or estimated. The Simpson and trapezoidal rules provide two more. This chapter and the next will focus on both the meaning of the definite integral as well as the methods for evaluating it.

The difficulty that a student faces in trying to understand what this chapter involves has both a conceptual and a notational aspect. The latter is perhaps even more challenging than the former, because the array of terminology and symbols required to write down things down is formidable and often intimidating. The attempt to internalize things is difficult. It requires a concentrated and time-consuming engagement with pencil and paper in hand: study, reflection, examples, and back again, and again. The initial sections of this chapter amount to an illustration of this strategy. They provide a repetitive and expanding look at the central ideas of the chapter: the summation process of the definite integral.

9.1 THE DEFINITE INTEGRAL OF A FUNCTION

The definite integral is a number that is determined in a certain way for a function defined over a closed interval. The number is closely related to the area bounded by the function's graph. A brief version of what follows was already discussed in Section 5.6 in the context of the work of Leibniz.

Let $y = f(x)$ be a function defined over an interval $[a, b]$. Recall that $[a, b]$ designates the set of all x with $a \le x \le b$. We will assume that the function f is continuous on $[a, b]$. So its graph is a connected curve, a curve with no breaks or gaps in it. Let n be a positive integer and select $n-1$ points

$$a < x_1 < x_2 < \cdots < x_i < x_{i+1} < \cdots < x_{n-2} < x_{n-1} < b$$

between a and b. These points are numbers on the x-axis. For notational purposes, let $a = x_0$ and $b = x_n$. Since they subdivide $[a, b]$, such a set of points or numbers is called a *partition* of $[a, b]$. Figure 9.1 depicts

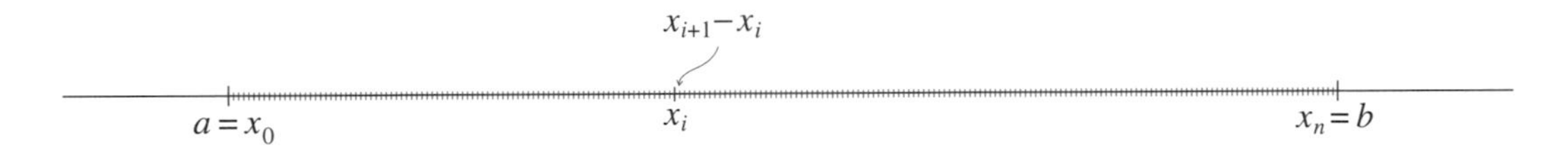

Fig. 9.1

such a partition. Consider the lengths

$$\Delta x_0 = x_1 - x_0, \Delta x_1 = x_2 - x_1, \ldots, \Delta x_i = x_{i+1} - x_i, \ldots, \Delta x_{n-2} = x_{n-1} - x_{n-2}, \text{ and } \Delta x_{n-1} = x_n - x_{n-1}$$

of the n subintervals that the partition determines. The largest of the Δx_i is the *norm* of the partition. Let $\mathcal{P}$ denote the partition, and let $||\mathcal{P}||$ denote its norm. Observe that $\Delta x_i \le ||\mathcal{P}||$ for all i, with $0 \le i \le n-1$.

For each x_i with $0 \le i \le n-1$, multiply $f(x_i)$ by the distance $\Delta x_i = x_{i+1} - x_i$ from x_i to the next point x_{i+1}. Form the sum

$$\sum_{i=0}^{n-1} f(x_i)\Delta x_i.$$

The symbol Σ in the notation $\sum_{i=0}^{n-1} f(x_i)\Delta x_i$ is the Greek version of a capital S. The idea is that Σ suggests "sum." The letter i is *the summation index*. This Σ notation works as follows. Start at the bottom of the Σ, put $i = 0, 1, 2, \ldots$ into the term $f(x_i)\Delta x_i$, keep adding, and stop at the top with $i = n-1$. In this way,

$$\sum_{i=0}^{n-1} f(x_i)\Delta x_i = f(x_0)\Delta x_0 + f(x_1)\Delta x_1 + \cdots + f(x_i)\Delta x_i + \cdots + f(x_{n-2})\Delta x_{n-2} + f(x_{n-1})\Delta x_{n-1}.$$

Suppose that $||\mathcal{P}||$ is small relative to the length $b-a$ of the interval $[a, b]$, so that all the Δx_i are also small. If $f(x_i) \ge 0$, then $f(x_i)\Delta x_i$ is the area of a thin rectangle that lies above the x-axis. If $f(x_i) < 0$, then the thin rectangle falls below the x-axis and its area is $-f(x_i)\Delta x_i$. (Notice that Δx_i need not be the same for all values of i. In other words, the width of the rectangles can differ.) It follows that the sum

$$\sum_{i=0}^{n-1} f(x_i)\Delta x_i$$

is an approximation of the area that falls under the graph of f and above the x-axis minus the area that falls above the graph of f and below that x-axis. See Figure 9.2. (The alternating black and gray colors

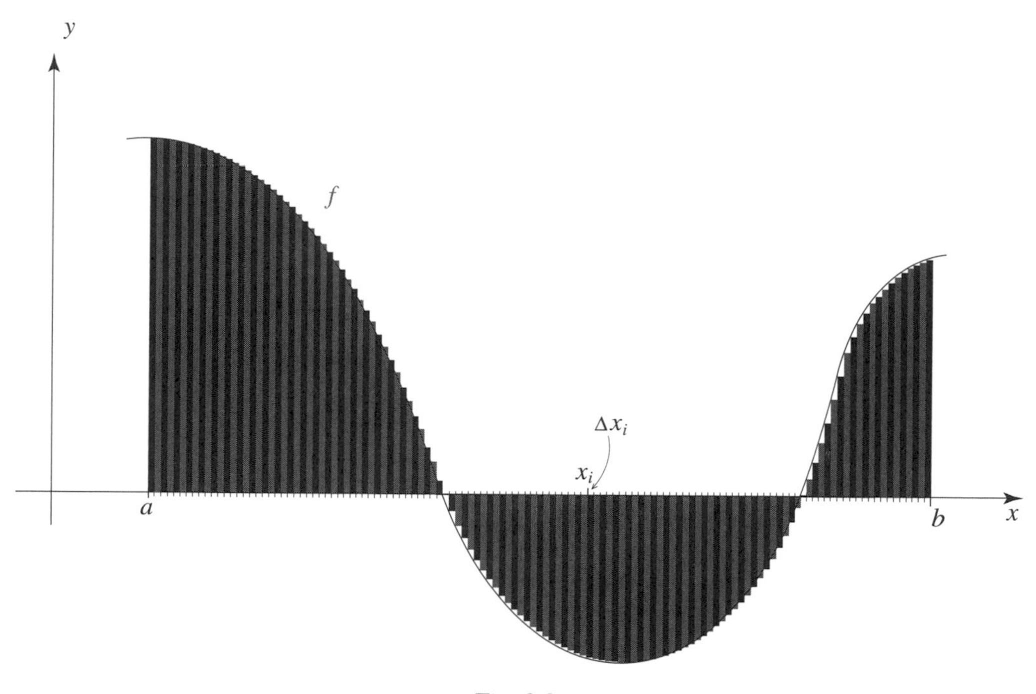

Fig. 9.2

are intended to distinguish the rectangles.)

We'll continue by changing to a more succinct notation. We'll write x for the typical point x_i from the list $a = x_0 < x_1 < x_2 < \cdots < x_{n-1} < x_n = b$, and we'll write dx for the distance $\Delta x_i = x_{i+1} - x_i$ to the very next point. So $f(x_i)\Delta x_i$ becomes $f(x)dx$, and the sum

$$\sum_{i=0}^{n-1} f(x_i)\Delta x_i \quad \text{is now more informally written as} \quad \sum_a^b f(x)\,dx.$$

Repeat the construction just described by taking partitions $\mathcal{P}$ of increasingly smaller norm, so that distances $dx = \Delta x_i = x_{i+1} - x_i$ between consecutive points get smaller each time, and form the sum

$$\sum_a^b f(x)\,dx$$

for each partition. Each time, the rectangles are thinner, and the sum is a tighter approximation of the difference between the area under the graph and above the x-axis minus the area above the graph and below the x-axis. By repeating this construction with partitions $\mathcal{P}$ of norms that shrink to 0, the result is a sequence of numbers of the form $\sum_a^b f(x)\,dx$ that close in on, or *converges to*, a number that is denoted

$$\int_a^b f(x)\,dx.$$

This number, called *the definite integral* of the function $f(x)$ from a to b, represents the difference between the area under the graph of f and above the x-axis minus the area above the graph of f and below the x-axis (between a and b). The function $f(x)$ is the *integrand*, and the numbers a and b are the *limits*—the *lower* and *upper* limits—of integration, respectively. In mathematical shorthand, we can express what we have described by

$$\lim_{\|\mathcal{P}\|\to 0} \sum_a^b f(x)\,dx = \int_a^b f(x)\,dx.$$

We have established that the number $\int_a^b f(x)\,dx$ is closely related to area, but we will soon see that it has many other interpretations as well.

An important fact—so important that it is known as *the fundamental theorem of calculus*—tells us that the definite integral from a to b of the function $y = f(x)$ can be computed as follows: let $y = F(x)$ be any function such that its derivative $F'(x)$ is equal to $f(x)$—so F is *an antiderivative* of f—then

$$\boxed{\int_a^b f(x)\,dx = F(b) - F(a)}$$

The difference $F(b) - F(a)$ does not depend on the particular antiderivative F of f. If F_1 is any other antiderivative of f, then $F_1'(x) = f(x) = F'(x)$. So by Corollary ii at the end of Section 7.6, $F_1(x) - F(x) = C$, with C a constant. Therefore $F_1(x) = F(x) + C$. It follows that $F_1(b) - F_1(a) = (F(b) + C) - (F(b) + C) = F(b) - F(a)$. The difference $F(b) - F(a)$ is often written as $F(x)\big|_a^b$.

We will prove the fundamental theorem later in Section 9.5. For now, we'll point out some basic consequences. Let c be any number in $[a, b]$. Since $\int_a^c f(x)\,dx = F(c) - F(a)$ and $\int_c^b f(x)\,dx = F(c) - F(b)$, it follows that

$$\int_a^c f(x)\,dx + \int_c^b f(x)\,dx = \int_a^b f(x)\,dx.$$

Let C be any constant. Since $\frac{d}{dx}CF(x) = C\frac{d}{dx}F(x) = Cf(x)$, the function $y = CF(x)$ is an antiderivative of $y = Cf(x)$. Now let a second function $y = g(x)$ be given, and assume that it is continuous on the interval $[a, b]$. Let G be an antiderivative of g, and note that $F(x) + G(x)$ is an antiderivative of $f(x) + g(x)$. Two applications of the fundamental theorem inform us that

$$\int_a^b Cf(x)\,dx = C\int_a^b f(x)\,dx \quad\text{and}\quad \int_a^b (f(x) \pm g(x))\,dx = \int_a^b f(x)\,dx \pm \int_a^b g(x)\,dx.$$

We'll illustrate this initial look at the definite integral with a number of examples.

Example 9.1. Since $F(x) = \frac{1}{3}x^3$ is an antiderivative of $f(x) = x^2$, $\int_0^4 x^2\,dx = F(4) - F(0) = \frac{64}{3} - 0 = 21\frac{1}{3}$. Make a sketch of the area that this number represents.

Example 9.2. Since the function $F(x) = \frac{2}{3}x^{\frac{3}{2}}$ is an antiderivative of $f(x) = \sqrt{x} = x^{\frac{1}{2}}$, it follows that $\int_4^9 \sqrt{x}\,dx = \frac{2}{3}9^{\frac{3}{2}} - \frac{2}{3}4^{\frac{3}{2}} = \frac{2}{3}(27 - 8) = \frac{38}{3}$. Sketch the area that this number represents.

Example 9.3. Make use of the two previous examples to show that

$$\int_1^4 (2x^2 - 5x^{\frac{1}{2}})\,dx = \int_1^4 2x^2\,dx - \int_1^4 5x^{\frac{1}{2}}\,dx = \frac{2}{3}x^3\Big|_1^4 - \frac{10}{3}x^{\frac{3}{2}}\Big|_1^4 = \left(\frac{2\cdot 4^3}{3} - \frac{2}{3}\right) - \left(\frac{10\cdot\sqrt{4}^3}{3} - \frac{10}{3}\right) = \frac{56}{3}.$$

Example 9.4. Show that $\int_{-\frac{\pi}{2}}^{0} \cos x\, dx = 1$. Refer to the graph of $y = \cos x$ in Figure 4.24 and say what $\int_{0}^{\frac{\pi}{2}} \cos x\, dx$, $\int_{0}^{\pi} \cos x\, dx$, and $\int_{0}^{\frac{3\pi}{2}} \cos x\, dx$ are equal to without undertaking any further calculations.

Example 9.5. Let's consider $\int_{-r}^{r} \sqrt{r^2 - x^2}\, dx$. An antiderivative for $y = \sqrt{r^2 - x^2}$ is is not (as yet) on our list of available functions. (We will determine an antiderivative of the function $f(x) = \sqrt{r^2 - x^2}$ in Section 9.10.) Note, however, that the graph of $y = \sqrt{r^2 - x^2}$ is the upper half of the circle $x^2 + y^2 = r^2$ sketched in Figure 9.3a. The connection between the definite integral and area tells us that $\int_{-r}^{r} \sqrt{r^2 - x^2}\, dx = \frac{1}{2}\pi r^2$.

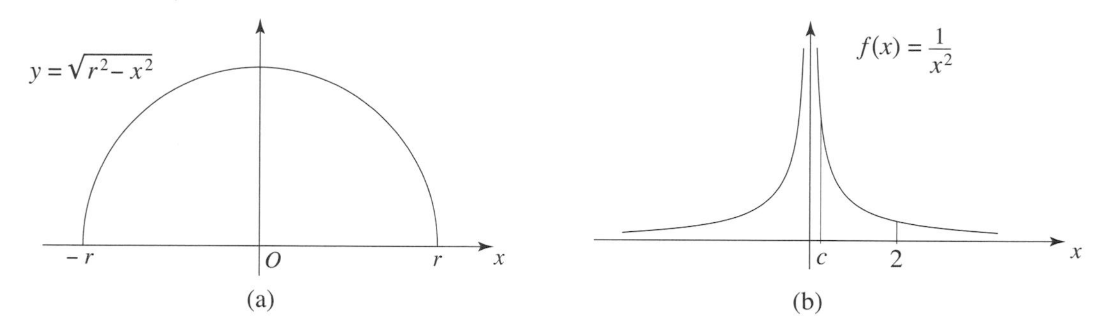

Fig. 9.3

Example 9.6. The function $F(x) = -\frac{1}{x} = -x^{-1}$ is an antiderivative of $f(x) = \frac{1}{x^2} = x^{-2}$. Therefore, by the fundamental theorem, $\int_{-1}^{2} \frac{1}{x^2}\, dx = F(2) - F(-1) = (-\frac{1}{2}) - (-(-1)) = -1\frac{1}{2}$. But Figure 9.3b shows us that the graph of $f(x) = \frac{1}{x^2}$ lies above the x-axis. So how can it be that the definite integral is negative? What has gone wrong? Notice that the requirement that $f(x) = \frac{1}{x^2}$ has to be continuous on the interval $[-1, 2]$ is not met! The correct evaluation of the integral tells us that the area under the graph is in fact infinite (as one might have expected). To see this, let c be a number satisfying $0 < c < 2$. Then

$$\int_{c}^{2} \frac{1}{x^2}\, dx = F(2) - F(c) = -\tfrac{1}{2} - (-\tfrac{1}{c}) = \tfrac{1}{c} - \tfrac{1}{2}$$

is equal to the area under the graph from c to 2, since there is no problem with discontinuities now. Taking c to be small means that the area $\frac{1}{c} - \frac{1}{2}$ is large. Pushing c to 0 pushes this area to infinity. A similar argument shows that the area under the graph between the line $x = -1$ and the y-axis is infinite as well. Can you compute the area under the graph of $f(x) = \frac{1}{x^2}$ from $x = 2$ to $x = c > 2$? What happens when c is pushed to infinity?

Example 9.7. The sum $(1)^2 \cdot \frac{1}{1000} + (1 + \frac{1}{1000})^2 \frac{1}{1000} + (1 + \frac{2}{1000})^2 \frac{1}{1000} + \cdots + (1 + \frac{999}{1000})^2 \frac{1}{1000}$ follows the pattern that the first three terms and the last term establish. For what a, b, n, and $x_1, \ldots, x_{n-1}$, and what $y = f(x)$ is this sum of the form $\sum_{i=0}^{n-1} f(x_i)\Delta x_i$? Show that the sum is closely approximated by $\frac{7}{3}$.

It has been our approach to form the sums leading to the definition of the definite integral by adding the terms $f(x_i)\Delta x_i$ with x_i the *left endpoints* of the intervals $[x_0, x_1], [x_1, x_2], \ldots, [x_i, x_{i+1}], \ldots$. This choice was motivated by the convention of our culture to give preference to the direction " left to right." We could just as well have taken the right endpoints x_{i+1} of these intervals and added the terms $f(x_{i+1})\Delta_i$ instead. However, we will see later in Section 9.5 that it is advantageous to allow flexibility in this regard by taking the sum of all $f(c_i)\Delta x_i$, where c_i in $[x_i, x_{i+1}]$ can be completely arbitrary.

9.2 VOLUME AND THE DEFINITE INTEGRAL

We'll turn next to some applications of the discussion of the previous section. We begin with the fact that the volume of a cylinder is equal to the area of the base times the height. So if its height is h and the circular base has radius r, then the volume of the cylinder is $\pi r^2 h$.

Let f be a continuous function that satisfies $f(x) \geq 0$ for all x in an interval $[a, b]$. Let $\mathcal{P}$ be a partition of $[a, b]$ with a norm $\|\mathcal{P}\|$ that is small compared to the distance $b - a$. Let x be a typical point of $\mathcal{P}$, and let dx be the distance to the very next point of $\mathcal{P}$ to its right. Since $dx \leq \|\mathcal{P}\|$, we know that dx is small. As we saw in Section 9.1, the points of the partition and the graph of f determine thin rectangles. The rectangle that has its left edge at x, thickness dx, and height $f(x)$ is shown in Figure 9.4a. Now revolve the region bounded by the graph, the x-axis, and the lines $x = a$ and $x = b$ once around the x-axis. Observe that the volume V of the resulting pear-shaped solid is tightly approximated by adding the volumes of the

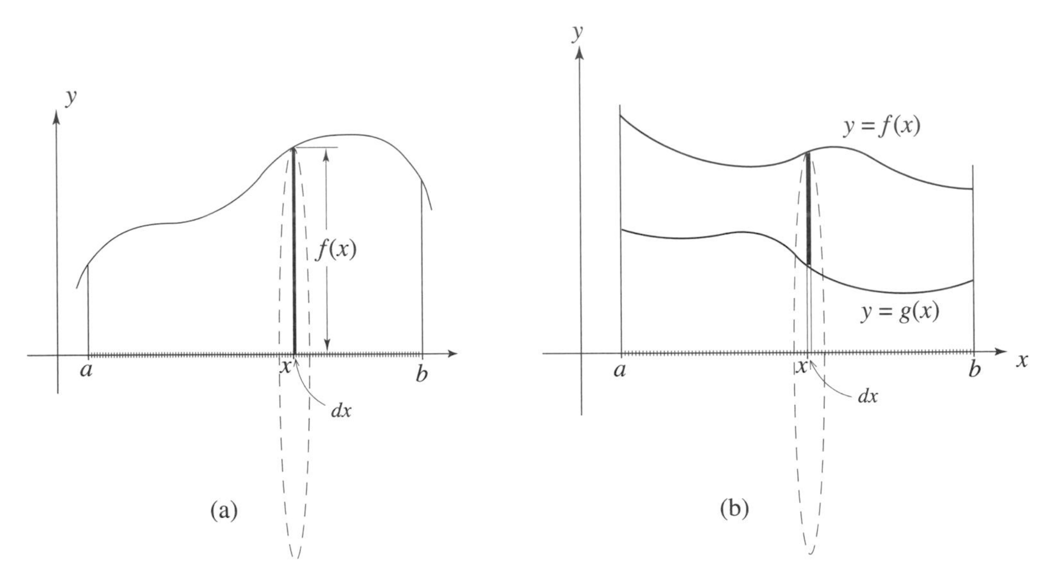

Fig. 9.4

thin cylindrical discs obtained by revolving all the rectangles that the points of the partition determine. The typical disc has circular area $\pi f(x)^2$, horizontal height dx, and volume $\pi f(x)^2\,dx$. So V is approximated by the sum $\sum_a^b \pi f(x)^2\,dx$. Now take a new partition $\mathcal{P}$ with many more points and even smaller norm, and repeat the process just described. The typical distance dx is smaller now, the rectangles are thinner, and there are many more of them. It follows that the sum $\sum_a^b \pi f(x)^2 dx$ is an even better approximation of the volume V of revolution that we are considering. By repeating this process with partitions of smaller and smaller norm and going to the limit described in Section 9.1, we see that the volume V of the solid obtained by revolving the graph of $y = f(x)$ once around the x-axis is given by

$$(V_1) \qquad \boxed{V = \pi \int_a^b f(x)^2\,dx}$$

Since the key to Formula (V_1) is a disc, the method just described is the *disc method*. Refer back to the connection between the definite integral and area, and notice that this definite integral is also equal to the area under the graph of the function $\pi(f(x))^2$ from a to b.

Turn to Figure 9.4b next. Let $y = f(x)$ and $y = g(x)$ be two continuous functions on the interval $[a, b]$ and assume that $f(x) \geq g(x) \geq 0$ for all x in the interval. Applying the disc method twice, once to the volume of revolution generated by the graph of $f(x)$ and then again to the volume of revolution generated by the graph of $g(x)$, tells us that the volume of revolution V determined by the region under the graph of $f(x)$ over the graph of $g(x)$ and between the lines $x = a$ and $x = b$ is equal to

$$(V_2) \qquad \boxed{V = \pi \int_a^b [f(x)^2 - g(x)^2]\, dx}$$

Example 9.8. Consider the ellipse $\frac{x^2}{a^2} + \frac{y^2}{b^2} = 1$, where either a or b can be the semimajor axis. Solving for y^2, we get $y^2 = b^2(1 - \frac{x^2}{a^2}) = \frac{b^2}{a^2}(a^2 - x^2)$. It follows that the graph of $f(x) = \frac{b}{a}\sqrt{a^2 - x^2}$ is the upper half of this ellipse. Therefore, by Formula (V_1), the volume obtained by revolving the upper part of the ellipse around the x-axis is

$$V = \pi \int_{-a}^{a} \tfrac{b^2}{a^2}(a^2 - x^2)\, dx = \tfrac{b^2}{a^2}\pi \int_{-a}^{a} (a^2 - x^2)\, dx.$$

By applying the fundamental theorem, we get that

$$V = \tfrac{b^2}{a^2}\pi(a^2x - \tfrac{x^3}{3})\Big|_{-a}^{a} = \tfrac{b^2}{a^2}\pi[(a^3 - \tfrac{a^3}{3}) - (-a^3 + \tfrac{a^3}{3})] = \tfrac{b^2}{a^2}\pi(\tfrac{4}{3}a^3) = \tfrac{4}{3}\pi ab^2.$$

For the particular ellipse drawn in Figure 9.5a, this solid has the shape of a rugby ball (a football with rounded ends). In the special case $a = b$ of a circle, this is the formula $V = \frac{4}{3}\pi a^3$ for the volume of a sphere of radius a.

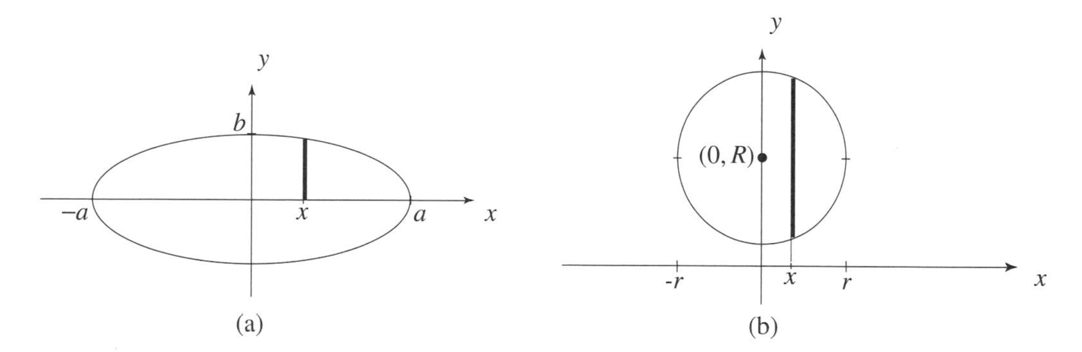

Fig. 9.5

Example 9.9. Figure 9.5b depicts a circle with center $(0, R)$ and radius r. As shown in the figure, we have assumed that $r \leq R$. The equation of the circle is $x^2 + (y - R)^2 = r^2$. Solving for y, we get $y = R \pm \sqrt{r^2 - x^2}$. Notice that the graph of $f(x) = R + \sqrt{r^2 - x^2}$ is the upper part of the circle and the graph of $g(x) = R - \sqrt{r^2 - x^2}$ is the lower part of the circle. It follows from Formula (V_2) that the volume V obtained by revolving the circle on complete revolution around the x-axis is

$$V = \pi \int_{-r}^{r} [(R + \sqrt{r^2 - x^2})^2 - (R - \sqrt{r^2 - x^2})^2]\, dx.$$

The integrand simplifies to $4R\sqrt{r^2 - x^2}$. Therefore $V = 4\pi R \int_{-r}^{r} \sqrt{r^2 - x^2}\, dx$. The integral $\int_{-r}^{r} \sqrt{r^2 - x^2}\, dx$ was shown to be equal to $\frac{1}{2}\pi r^2$ in Example 9.5. Therefore $V = 2\pi^2 r^2 R$. The surface of this donut-shaped volume of revolution is called a torus by mathematicians.

A region in the xy-plane is bounded by two vertical lines $x = a$ and $x = b$ where $0 \leq a \leq b$, and the graphs of two functions $y = f(x)$ and $y = g(x)$ that are both continuous over the interval $[a, b]$ and satisfy $g(x) \leq f(x)$ for all x in $[a, b]$. Consider the volume V obtained by revolving this region once around the y-axis. To determine V, we'll again turn to the strategy that gave rise to the definite integral. Let $\mathcal{P}$ be a partition of $[a, b]$ with a norm $||\mathcal{P}||$ that is small relative to the distance $b - a$. Let x be a typical point of $\mathcal{P}$, and let dx be the distance to the very next point of $\mathcal{P}$ on its right. In a way similar to previous situations, the points of the partition and the graphs of f and g determine thin rectangles. The rectangle that has its left edge at x, thickness dx, and height $f(x) - g(x)$ is shown in Figure 9.6. Let's study the circular band that is formed by the revolution of this rectangle around the y-axis. When this band is cut along the vertical

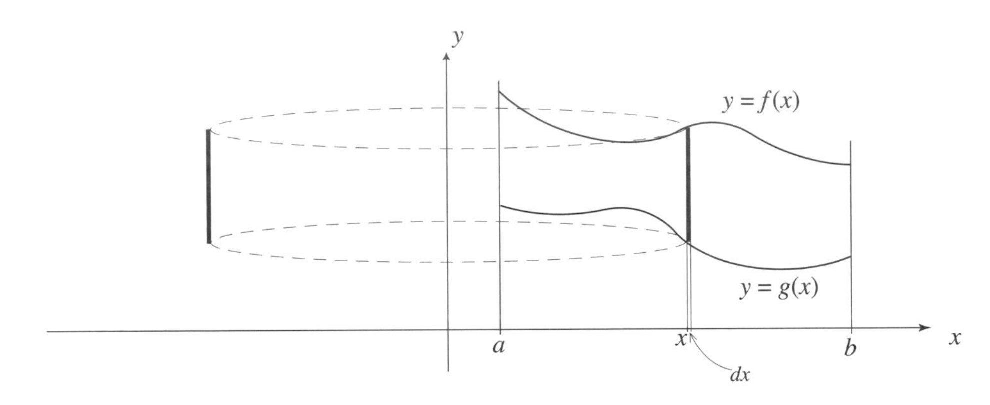

Fig. 9.6

line through x and flattened, a rectangular region with base $2\pi x$, height $f(x) - g(x)$, and thickness dx is obtained. Its volume is $2\pi x(f(x) - g(x))dx$. The addition of the volumes of all the rectangular regions that the points of the partition $\mathcal{P}$ determine in this way provides the approximation

$$V \approx \sum_{a}^{b} 2\pi x(f(x) - g(x))\, dx$$

of the volume that we are looking for. By repeating this construction with partitions $\mathcal{P}$ of smaller and smaller norm, and going to the limit, we get that

$$(V_3) \qquad \boxed{V = 2\pi \int_a^b x[f(x) - g(x)]\, dx}$$

The circular band involved in the derivation of the volume Formula (V_3) is often referred to as a *shell* and the method that gave rise to the formula as the *shell method*.

Example 9.10. Return to Figure 9.5a. Revolve the upper right quarter of the ellipse one revolution around the y-axis. From Example 9.8, we know that the upper half of the ellipse is the graph of the function $f(x) = \frac{b}{a}\sqrt{a^2 - x^2}$. By applying volume Formula (V_3) to this $f(x)$ and $g(x) = 0$, we get that this volume of revolution is

$$V = 2\pi \int_0^a x(\tfrac{b}{a}\sqrt{a^2 - x^2})\, dx = 2\pi \tfrac{b}{a} \int_0^a x(\sqrt{a^2 - x^2})\, dx.$$

It follows by the chain rule that $F(x) = -\frac{1}{3}(a^2 - x^2)^{\frac{3}{2}}$ is an antiderivative of the integrand. Therefore

$$V = 2\pi \tfrac{b}{a}\Big[-\tfrac{1}{3}(a^2 - x^2)^{\frac{3}{2}}\Big|_0^a\Big] = 2\pi \tfrac{b}{a}[0 + \tfrac{1}{3}(a^2)^{\frac{3}{2}}] = \tfrac{2}{3}\pi a^2 b.$$

So the solid obtained by revolving the entire right half of the ellipse of Figure 9.5a around the y-axis is $\frac{4}{3}\pi a^2 b$. Our planet Earth has the shape of such a flattened sphere (with a only slightly larger than b).

Example 9.11. Go back to Figure 9.5b. Move the circle so that its center is the point $(R, 0)$ on the x-axis. The equation of the circle in this new position is $(x - R)^2 + y^2 = r^2$. So $y^2 = r^2 - (x - R)^2$, and hence $y = \pm\sqrt{r^2 - (x - R)^2}$. The functions that have the upper and lower halves of this circle as their graphs are $f(x) = \sqrt{r^2 - (x - R)^2}$ and $g(x) = -\sqrt{r^2 - (x - R)^2}$, respectively. An application of the volume Formula (V_3) tells us that the volume V obtained by revolving this circle once around the y-axis is

$$V = 2\pi \int_{R-r}^{R+r} x(2\sqrt{r^2 - (x - R)^2})\, dx.$$

We will develop methods that will solve this integral later in the chapter. Thanks to Example 9.10, we already know that its value is $2\pi^2 r^2 R$.

9.3 LENGTHS OF CURVES AND THE DEFINITE INTEGRAL

The definite integral also provides a way to compute the length of the graph of a function between any two points.

Let f be a function, and let $A = (a, c)$ and $B = (b, d)$ be two points on its graph. This time we'll need to assume not only that f is continuous on $[a, b]$, but also that f has a derivative that is continuous on $[a, b]$. Figure 9.7 shows a typical situation. The length L of the curve between the points A and B can be analyzed as follows. Once more, take a partition $\mathcal{P}$ with a norm that is small relative to the length of the

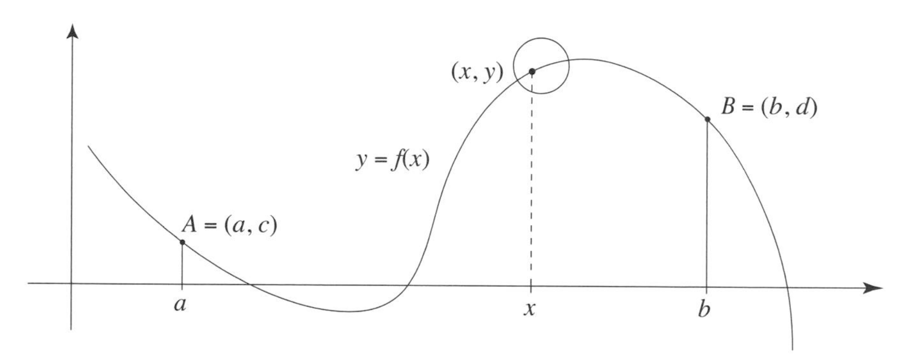

Fig. 9.7

interval $[a, b]$. For a typical point x of the partition, let dx be the distance to the next point of the partition on its right. The fact that the norm of the partition is small means that dx is small. Let (x, y) be the point on the graph corresponding to x, and use a segment of length dx and the tangent line at (x, y) to construct a right triangle at (x, y). Denote its height by dy. Figure 9.8 shows this triangle "under a microscope." Notice that the slope of the tangent at (x, y) is the ratio $\frac{dy}{dx}$ of the lengths of the two segments. Therefore $f'(x) = \frac{dy}{dx}$. By the Pythagorean theorem, the length of the hypotenuse of the triangle is $\sqrt{(dx)^2 + (dy)^2}$. A factoring maneuver tells us that $(dx)^2 + (dy)^2 = [1 + (\frac{dy}{dx})^2](dx)^2$, so that

$$\sqrt{(dx)^2 + (dy)^2} = \sqrt{[1 + (\tfrac{dy}{dx})^2](dx)^2} = \sqrt{1 + (\tfrac{dy}{dx})^2}\, dx = \sqrt{1 + f'(x)^2}\, dx.$$

Since dx is small, the length of the arc of the graph of the function from $(x, f(x))$ to $(x + dx, f(x + dx))$ is closely approximated by the hypotenuse of the triangle (as Figure 9.8 illustrates). Since the length L of the curve from A to B is equal to the sum of the lengths of all these tiny arcs from the point A to the point B, it follows that L is tightly approximated by the sum $\sum_b^a \sqrt{1 + f'(x)^2}\, dx$ of the lengths of the hypotenuses

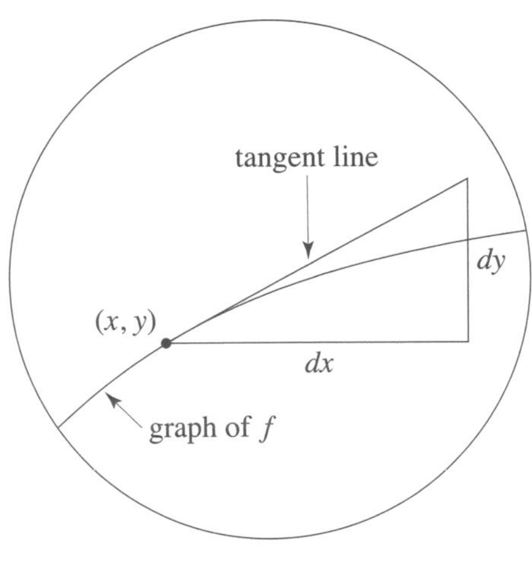

Fig. 9.8

of all the right triangles that the points x of the partition determine. Figure 9.9 shows what is involved. So that we can see what is going on, the figure illustrates a partition that has only a handful of points. Therefore all the dx are relatively large. Even in this situation, the figure tells us that the sum of the lengths of the tangential segments add up to approximate the length L of the graph. The figure and your imagination

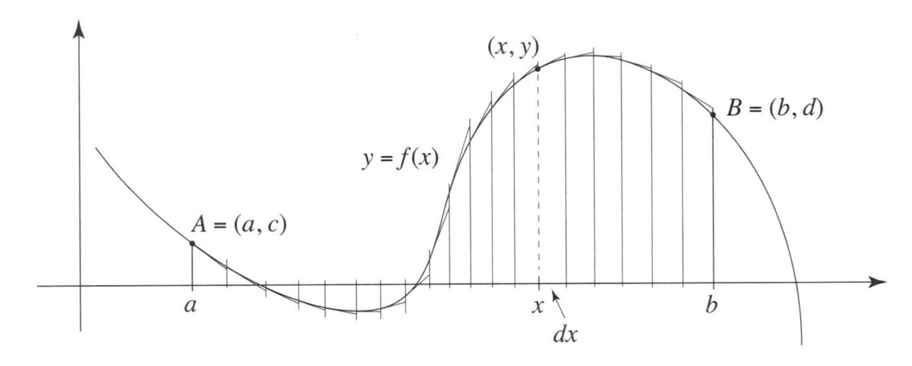

Fig. 9.9

should convince you that if the number of points in the partition $\mathcal{P}$ is huge with all the dx tiny, then the sum $\sum_{b}^{a} \sqrt{1 + f'(x)^2}\, dx$ of all the little slanting segments will be a tight approximation of L. By repeating this analysis with partitions of smaller and smaller norm and going to the limit, we can conclude that

$$\boxed{L = \int_a^b \sqrt{1 + f'(x)^2}\, dx}$$

Example 9.12. Let's check this formula in a simple case. Consider the points $A = (-1, -5)$ and $B = (5, 7)$ in the plane. The distance between A and B is

$$\sqrt{(-1-5)^2 + (-5-7)^2} = \sqrt{36 + 144} = \sqrt{180} = 6\sqrt{5}.$$

Since the points $A = (-1, -5)$ and $B = (5, 7)$ are both on the graph of the function $f(x) = 2x - 3$, it follows that the distance between A and B is also given by

$$\int_a^b \sqrt{1+f'(x)^2}\,dx = \int_{-1}^5 \sqrt{1+2^2}\,dx = \int_{-1}^5 \sqrt{5}\,dx.$$

Because $G(x) = \sqrt{5}x$ is an antiderivative of $\sqrt{5}$,

$$\int_{-1}^5 \sqrt{5}\,dx = G(5) - G(-1) = 5\sqrt{5} - (-1)\sqrt{5} = 6\sqrt{5},$$

as expected.

Example 9.13. Consider the circle $x^2 + y^2 = 6^2$ of radius 6 with center the origin. Since $y = \pm\sqrt{6^2 - x^2}$, it follows that the graph of the function $f(x) = (6^2 - x^2)^{\frac{1}{2}}$ is the upper half of the circle. Take the points A and B on the circle with x-coordinates 3 and -3, respectively. What is the length L of the circular arc from A to B? By the length formula,

$$L = \int_a^b \sqrt{1+f'(x)^2}\,dx = \int_{-3}^3 \sqrt{1+f'(x)^2}\,dx.$$

By an application of the chain rule, $f'(x) = \frac{1}{2}(6^2 - x^2)^{-\frac{1}{2}}(-2x) = -\frac{x}{\sqrt{6^2-x^2}}$. So $f'(x)^2 = \frac{x^2}{6^2-x^2}$, and hence $1 + f'(x)^2 = \frac{6^2-x^2+x^2}{6^2-x^2} = \frac{6^2}{6^2-x^2}$. Therefore

$$L = \int_{-3}^3 \sqrt{\frac{6^2}{6^2-x^2}}\,dx = \int_{-3}^3 \frac{6}{\sqrt{6^2-x^2}}\,dx.$$

We will develop methods for finding an antiderivative of $\frac{6}{\sqrt{6^2-x^2}}$ later in this chapter. This will then allow us to evaluate the integral by using the fundamental theorem of calculus. However, the length of arc AB can be computed directly. Look at Figure 9.10 and observe that the angle determined by the segment OB and the positive x-axis has cosine equal to $\frac{1}{2}$. This tells us (refer to Section 1.6) that this angle is 60°. For

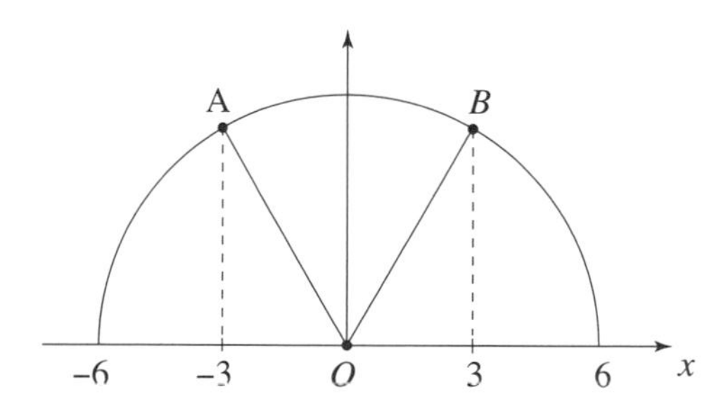

Fig. 9.10

the same reason, the angle that OA makes with the negative x-axis is also 60°. Therefore $\angle AOB$ is 60°. This means that the length of the arc AB is $\frac{1}{6}$ of the circumference of a circle of radius 6. It follows that

$$L = \int_{-3}^3 \frac{6}{\sqrt{6^2-x^2}}\,dx = \tfrac{1}{6}(2\pi\cdot 6) = 2\pi.$$

Since the upper half of the circumference of this circle is $\frac{1}{2}(2\pi\cdot 6) = 6\pi$, it should also be the case that

$$\int_{-6}^6 \frac{6}{\sqrt{6^2-x^2}}\,dx = 6\pi.$$

And it is! But there is a fly in the ointment: the fact that $\frac{6}{\sqrt{6^2-x^2}}$ is not defined at $x = -6$ and $x = 6$ violates the continuity condition that the integrand of a definite integral needs to satisfy. We'll have a look at this difficulty in Section 9.9.

9.4 SURFACE AREA AND THE DEFINITE INTEGRAL

This section turns to the computation of the area of a surface obtained by revolving the graph of a function $y = f(x)$ around the x-axis.

As the first order of business, we will compute the surface area of a cone. Consider a circle placed horizontally, and let C be a point over its center. Consider a segment in a vertical plane from C to the circle. While keeping one endpoint fixed at C, rotate the other end of the segment around the perimeter of the circle. The surface that the segment generates is a *right circular cone*. Such a cone can be constructed as follows. Let r be the radius of the circle that generates the cone, and let s be the slanting height of the cone. Refer to Figure 9.11a. Now take a circular sector of radius s and angle $\theta = \frac{2\pi r}{s}$. The definition of the

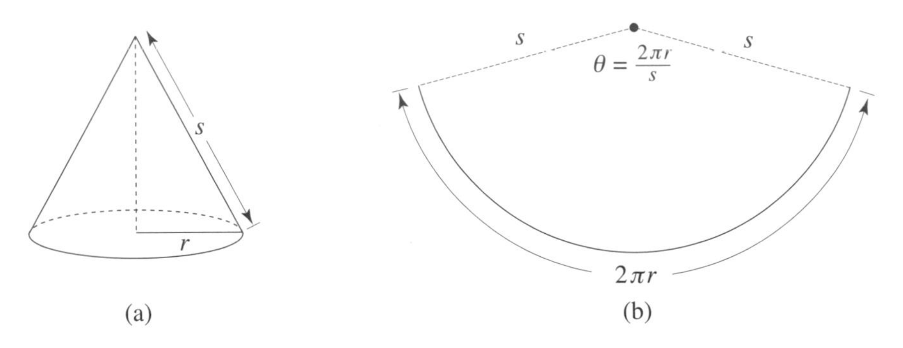

Fig. 9.11

radian measure of an angle tells us that the circular arc that θ determines has length $\theta s = 2\pi r$. Refer to Figure 9.11b and shape the circular sector by joining and aligning the two sides of lengths s. This results in the cone depicted in Figure 9.11a. By applying the formula in Section 2.2 for the area of a circular sector, we get that the surface area of the slanting side of the cone of Figure 9.11a is equal to

$$\tfrac{1}{2}\theta s^2 = \tfrac{1}{2}\tfrac{2\pi r}{s}s^2 = \pi r s.$$

The shape that will soon be relevant is the *truncated cone*. It is obtained by slicing off the top of a right circular cone with a cut that is parallel to its circular base. This is shown in Figure 9.12a. Comparing Figures 9.11a and 9.12a, we see that $r = r_2$ and $s = z_1 + z$. Remove the smaller upper cone. The piece that is left is called a *truncated cone*. We'll compute the surface area T (slanting side only, not including the

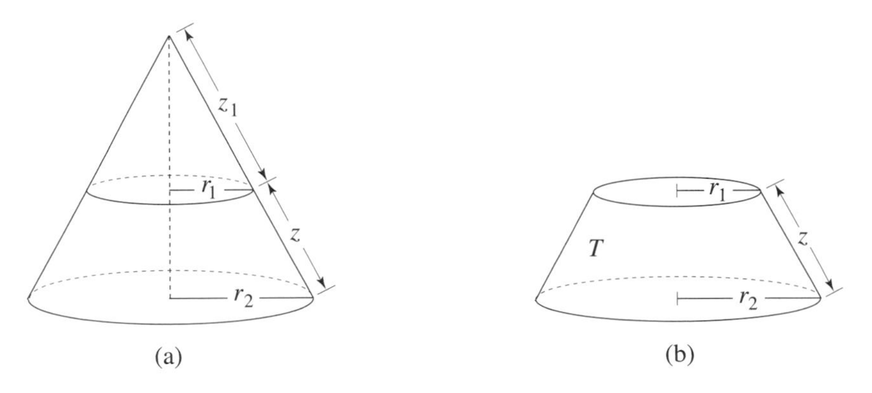

Fig. 9.12

two circular areas at the top and bottom) of this truncated cone. Because it is the difference between two cones, it follows that

$$T = \pi r_2 s - \pi r_1 z_1 = \pi r_2(z_1 + z) - \pi r_1 z_1 = \pi(r_2 z_1 + r_2 z - r_1 z_1).$$

By similar triangles, $\frac{z_1}{r_1} = \frac{z_1+z}{r_2}$, and therefore $r_2z_1 = r_1(z_1+z) = r_1z_1+r_1z$. Popping this into the expression for T gives us

$$T = \pi(r_1z_1 + r_1z + r_2z - r_1z_1) = \pi(r_1z + r_2z) = \pi(r_1 + r_2)z.$$

Now turn to the graph of a function $y = f(x)$ that is continuous on a closed interval $[a, b]$. Suppose that $f(x) \geq 0$ for all x in $[a, b]$. Assume also that f has a derivative that is continuous on $[a, b]$. As we have done before (but less explicitly), let $\mathcal{P}$ be a partition

$$a = x_0 < x_1 < x_2 < x_3 < \cdots < x_i < x_{i+1} < \cdots < x_{n-2} < x_{n-1} = b$$

of $[a, b]$ with a norm $\|\mathcal{P}\|$ that is very small relative to the length $b - a$ of the interval $[a, b]$. Note that $\left|x_{i+1} - x_i\right| \leq \|\mathcal{P}\|$ for all i. Figure 9.13 depicts the graph of a typical function $y = f(x)$ as well as the vertical lines that the points of the partition determine. (In order to let us "see" what is going on, the figure

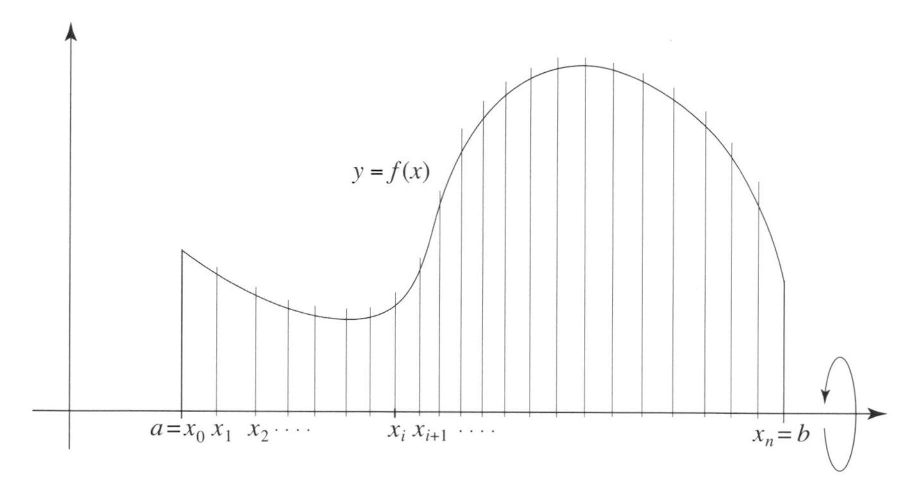

Fig. 9.13

depicts a partition with only a few points.)

Now revolve the graph of the function once around the x-axis to get the "vase" shown in Figure 9.14. We will focus on its curving surface between the two vertical circular sections. The $n + 1$ rotated vertical segments from $a = x_0$ to $x_n = b$ divide the area S of this surface into n thin sections. Labeling the surface areas of these thin sections by $S_0, S_1, \ldots, S_i, \ldots, S_{n-1}$, respectively, we get that

$$S - S_0 + \cdots + S_i + \cdots + S_{n-1}.$$

In Figure 9.14, the typical point x_i of the partition is written as x, and dx is the distance to the next point x_{i+1} of the partition. So $x_{i+1} = x + dx$. The corresponding slice with surface area S_i is also depicted in the figure. This slice is shown enlarged and in horizontal position in Figure 9.15a.

Also inserted into Figure 9.14—one for each of the intervals—are the segments that are tangent to the graph at the points that the left endpoints of these intervals determine. These tangential segments are revolved around the x-axis along with the graph. Included in Figure 9.14 is the slice obtained by revolving the tangential segment over the typical interval $[x, x + dx]$. The analysis of Section 9.3 tells us that the length of this tangential segment is $\sqrt{1 + f'(x)^2}\,dx$. This slice is enlarged and placed horizontally in Figure 9.15b. Notice that it is a truncated cone. The radius of its upper circular cross section is $f(x)$, and a study of Figure 9.8 tells us that the radius of its lower circular cross section is $f(x) + dy = f(x) + f'(x)\,dx$. Taking $z = \sqrt{1 + f'(x)^2}\,dx, r_1 = f(x)$ and $r_2 = f(x) + f'(x)\,dx$ in the formula for the surface area of a truncated cone, we get that the surface area of the truncated cone of Figure 9.15b is equal to

$$\pi(2f(x) + f'(x)\,dx)\sqrt{1 + f'(x)^2}\,dx.$$

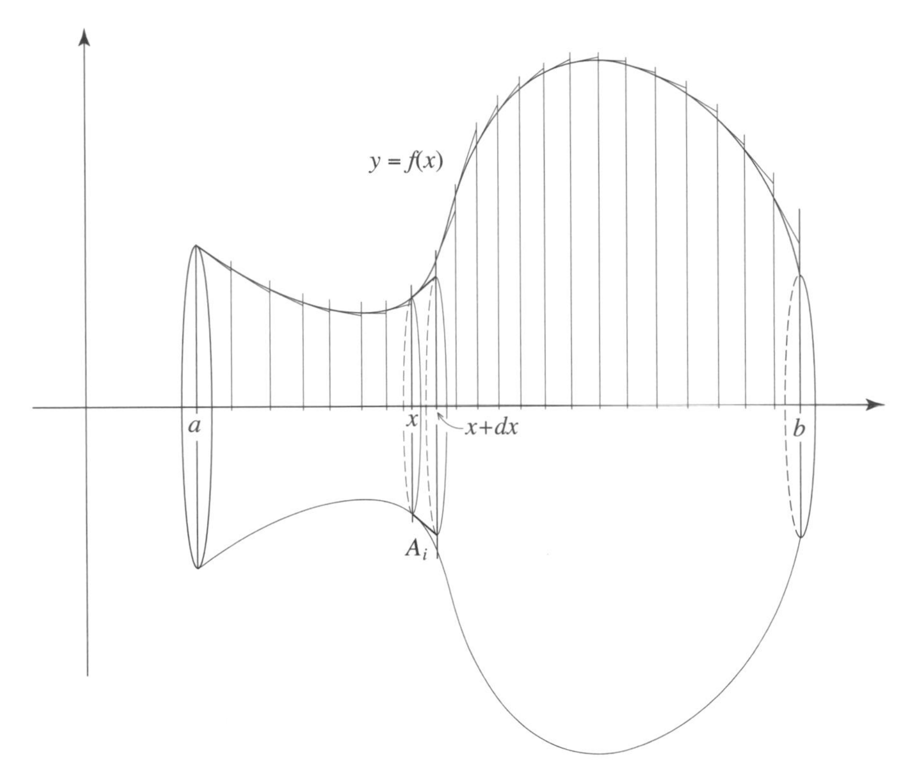

Fig. 9.14

A comparison of the slices depicted in Figures 9.15a and 9.15b tells us that the area S_i is closely approxi-

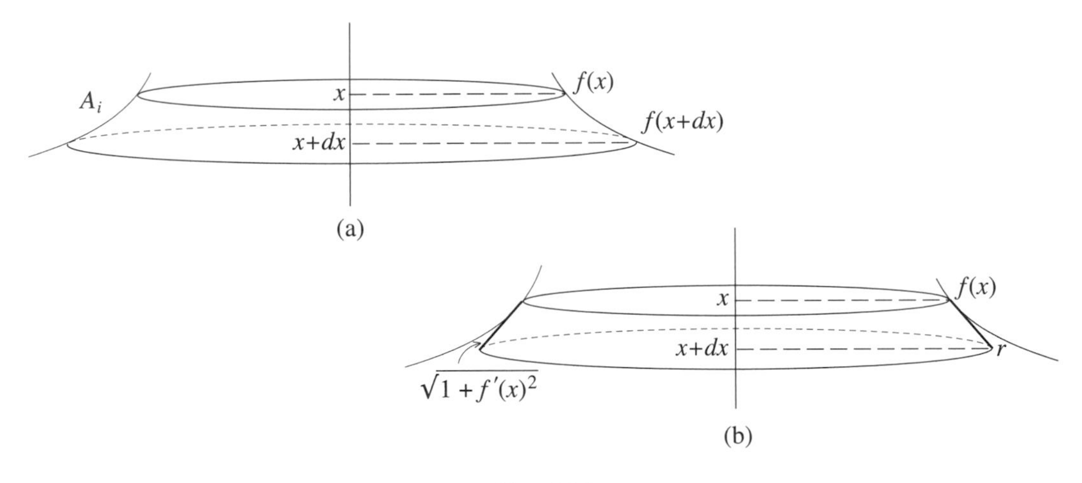

Fig. 9.15

mated by the slanted area of the truncated cone. It follows that

$$S_i \approx \pi(2f(x) + f'(x)\,dx)\sqrt{1 + f'(x)^2}\,dx = 2\pi f(x)\sqrt{1 + f'(x)^2}\,dx + \pi f'(x)\,dx\sqrt{1 + f'(x)^2}\,dx.$$

(We'll see that the first term of this sum is the important part, and we'll regard the second part to be an error term.) Doing this for each of the n slices and making use of the Σ notation of Section 9.1, we get that

$$S = S_0 + \cdots + S_i + \cdots + S_{n-1} \approx 2\pi \sum_a^b f(x)\sqrt{1 + f'(x)^2}\, dx + \pi \sum_a^b f'(x)\, dx \sqrt{1 + f'(x)^2}\, dx.$$

As in previous similar situations, the smaller the norm $\|\mathcal{P}\|$ of the partition $\mathcal{P}$, the thinner the slices, and the better this approximation. It will be important to know how the error terms add up and accumulate. Since

$$\Big|\sum_a^b f'(x)\, dx \sqrt{1 + f'(x)^2}\, dx\Big| = dx \Big|\sum_a^b f'(x)\sqrt{1 + f'(x)^2}\, dx\Big| \le \|\mathcal{P}\| \Big|\sum_a^b f'(x)\sqrt{1 + f'(x)^2}\, dx\Big|,$$

we see that

$$\lim_{\|\mathcal{P}\|\to 0} \Big|\sum_a^b f'(x)\, dx \sqrt{1 + f'(x)^2}\, dx\Big| \le \lim_{\|\mathcal{P}\|\to 0} \|\mathcal{P}\| \Big|\sum_a^b f'(x)\sqrt{1 + f'(x)^2}\, dx\Big| = 0 \cdot \Big|\int_a^b f'(x)\sqrt{1 + f'(x)^2}\, dx\Big|$$

and therefore that $\lim\limits_{\|\mathcal{P}\|\to 0} \sum\limits_a^b f'(x)\, dx \sqrt{1 + f'(x)^2}\, dx = 0$. So in the limit, the sum of the error terms is zero. It follows that

$$S = \lim_{\|\mathcal{P}\|\to 0} 2\pi \sum_a^b f(x)\sqrt{1 + f'(x)^2}\, dx,$$

and therefore that the area of the surface of our "vase" is equal to

$$\boxed{S = 2\pi \int_a^b f(x)\sqrt{1 + f'(x)^2}\, dx}$$

Example 9.14. Consider the circle of radius r centered at the origin. Its equation is $x^2 + y^2 = r^2$, and the upper semicircle shown in Figure 9.16 is the graph of the function $y = f(x) = \sqrt{r^2 - x^2} = (r^2 - x^2)^{\frac{1}{2}}$ with domain $[-r, r]$. The surface obtained by revolving this semicircle around the x-axis is a sphere of radius r.

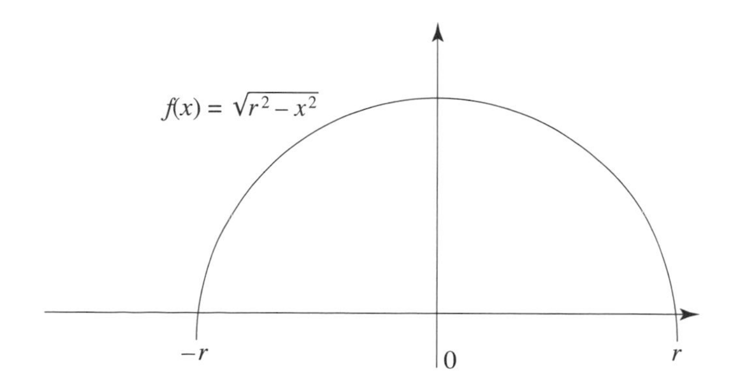

Fig. 9.16

It follows that the surface area of the sphere of radius r can be expressed as the definite integral

$$2\pi \int_{-r}^{r} f(x)\sqrt{1 + f'(x)^2}\, dx,$$

where $f(x) = (r^2 - x^2)^{\frac{1}{2}}$. Let's see if we can compute it. By the chain rule,

$$f'(x) = \tfrac{1}{2}(r^2 - x^2)^{-\frac{1}{2}}(-2x) = \tfrac{-x}{(r^2-x^2)^{\frac{1}{2}}}.$$

So $f'(x)^2 = \frac{x^2}{r^2-x^2}$, and hence

$$\sqrt{1 + f'(x)^2} = \sqrt{1 + \tfrac{x^2}{r^2-x^2}} = \sqrt{r^2 - x^2 + \tfrac{x^2}{r^2-x^2}} = \tfrac{r}{(r^2-x^2)^{\frac{1}{2}}}.$$

So the surface area of a sphere of radius r is given by

$$2\pi \int_{-r}^{r} f(x)\sqrt{1 + f'(x)^2}\,dx = 2\pi \int_{-r}^{r} (r^2 - x^2)^{\frac{1}{2}} \tfrac{r}{(r^2-x^2)^{\frac{1}{2}}}\,dx = 2\pi \int_{-r}^{r} r\,dx.$$

Because $2\pi rx$ is an antiderivative of $2\pi r$, we get

$$2\pi \int_{-r}^{r} r\,dx = 2\pi rx \Big|_{-r}^{r} = 2\pi r^2 - (-2\pi r^2) = 4\pi r^2.$$

We have shown that the surface area of a sphere of radius r is $4\pi r^2$.

Example 9.15. Instead of rotating the full semicircle of radius r, let's rotate only the part between a and b. See Figure 9.17. By the argument of Example 9.14, the area of the section of the surface of the sphere

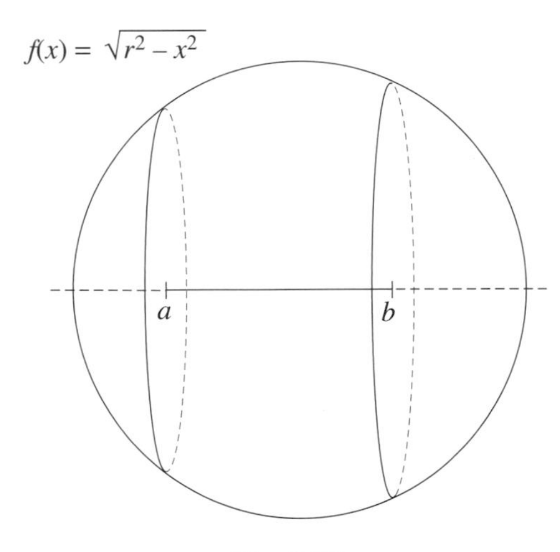

Fig. 9.17

shown in the figure is equal to

$$2\pi \int_{a}^{b} r\,dx = 2\pi rx \Big|_{a}^{b} = 2\pi rb - 2\pi ra = 2\pi r(b - a).$$

Notice that for a sphere of given radius r, the area of this surface depends only on the width $b - a$ of the section and not on its location on the sphere. If you have a taste for crusty bread, you might like the following consequence of this fact. Suppose you have a loaf of bread in the shape of a sphere. Cut any two slices from the loaf (making sure that the two cuts that make a slice are parallel to each other). If the slices have the same thickness, then they have the same amount of crust, no matter which part of the loaf they are taken from!

The problematic fly that we discovered in Example 9.13 is also buzzing around in Examples 9.14 and 9.15 when either $a = -r$ or $b = r$. Since $f'(x) = -\frac{x}{(r^2-x^2)^{\frac{1}{2}}}$ is not defined and hence not continuous at

$x = -r$ and $x = r$, the integrand $2\pi f(x)\sqrt{1+f'(x)^2}$ does not meet the continuity requirement in these situations. Let's illustrate the remedy in the special case $2\pi\int_0^r f(x)\sqrt{1+f'(x)^2}\,dx$ of Example 9.15. Let c be any number satisfying $0 \le c < r$. Over the interval $0 \le x \le c$, the problem that we have identified does not arise, so that

$$2\pi\int_0^c f(x)\sqrt{1+f'(x)^2}\,dx = 2\pi\int_0^c (r^2-x^2)^{\frac{1}{2}}\frac{r}{(r^2-x^2)^{\frac{1}{2}}}\,dx = 2\pi\int_0^c r\,dx = 2\pi r x\Big|_0^c = 2\pi rc.$$

(For $x = r$, this cancellation would not have been valid.) By defining

$$2\pi\int_0^r f(x)\sqrt{1+f'(x)^2}\,dx = \lim_{c\to r} 2\pi\int_0^c f(x)\sqrt{1+f'(x)^2}\,dx,$$

we can conclude that

$$2\pi\int_0^r f(x)\sqrt{1+f'(x)^2}\,dx = \lim_{c\to r} 2\pi rc = 2\pi r^2$$

is the surface area of one-half of the sphere of radius r. We will see in Section 9.9 that a similar strategy removes the fly from Example 9.13.

An integral that is defined by a limit procedure of the sort just illustrated is called an *improper integral.* We have already computed an improper integral in Example 9.6, and we will encounter more of them in Chapter 10.

9.5 THE DEFINITE INTEGRAL AND THE FUNDAMENTAL THEOREM

The concept of the limit of a function $f(x)$ as x approaches some fixed value is the key idea underlying differential calculus. In Section 7.2, we saw how $\lim_{x\to c} f(x) = L$ makes sense intuitively and geometrically, but also how this notion can be formulated abstractly in terms of "epsilons and deltas" in such a way that it breaks free of words and pictures and becomes grounded solely in the realm of rigorous logic.

The discussion in Section 9.1 that led to the definition of the definite integral presents a similar challenge. It tells us that

> the repetition of this construction again and again with partitions $\mathcal{P}$ of norms $\|\mathcal{P}\|$ that shrink to zero, produces a sequence of numbers $\sum_a^b f(x)\,dx$ that close in on, or converge to, a number that is denoted by
>
> $$\int_a^b f(x)\,dx.$$
>
> It is called *the definite integral* of the function $f(x)$ from a to b. In mathematical shorthand, we can express what has been described as
>
> $$\lim_{\|\mathcal{P}\|\to 0}\sum_a^b f(x)\,dx = \int_a^b f(x)\,dx\,.$$

This description of the limit process that underlies integral calculus makes sense in a verbal or geometric way, but it would not be able to stand when measured against the standards of rigorous modern mathematics. Such a rigorous approach also involves an epsilon-delta mechanism that allows the convergence in question to be tested. It was the achievement of mathematicians of the 19th century. The work of the Frenchman Augustin Louis Cauchy in the 1820s and that of the Germans Lejeune Dirichlet and Georg Bernhard Riemann later in the century deserve to be singled out. What follows describes what is involved without going into the difficult details. While we will state the epsilon-delta criterion of convergence, we will not make use of it.

Let $y = f(x)$ be a function that is continuous on a closed interval $[a, b]$.

1. Take a partition $\mathcal{P}$

$$a = x_0 < x_1 < x_2 < x_3 < \cdots < x_i < x_{i+1} < \cdots < x_{n-1} < x_n = b$$

of the interval $[a, b]$. As before, let

$$\Delta x_0 = x_1 - x_0, \Delta x_1 = x_2 - x_1, \Delta x_2 = x_3 - x_2, \ldots, \Delta x_i = x_{i+1} - x_i, \ldots, \Delta x_{n-1} = x_n - x_{n-1},$$

and let the *norm* $\|\mathcal{P}\|$ of the partition be the largest of all the Δx_i. So $\Delta x_i \leq \|\mathcal{P}\|$ for all i.

2. Next pick a point, any point, c_0 in the first subinterval, any c_1 in the second, any c_2 in the third, ..., any c_i in the $(i+1)$st interval, and so on. See Figure 9.18. So $x_i \leq c_i \leq x_{i+1}$ for any i with

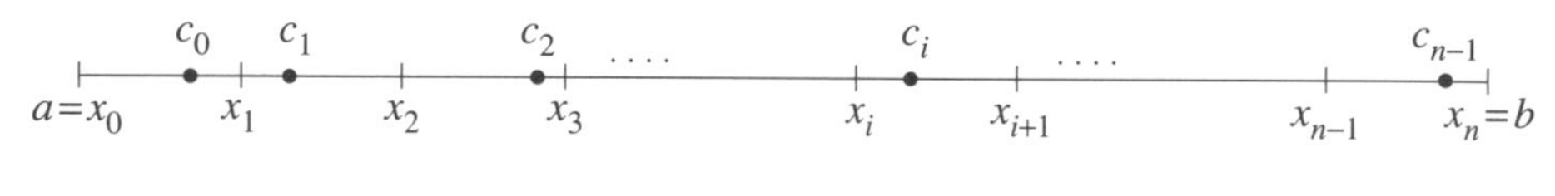

Fig. 9.18

$0 \leq i \leq n-1$. Form the sum

$$\sum_{i=0}^{n-1} f(c_i)\Delta x_i = f(c_0)\Delta x_0 + f(c_1)\Delta x_1 + \cdots + f(x_i)\Delta x_i + \cdots + f(c_{n-2})\Delta x_{n-2} + f(c_{n-1})\Delta x_{n-1}.$$

3. This construction can be repeated again and again with partitions of smaller and smaller norm. Each step produces a sum $\sum_{i=0}^{n-1} f(c_i)\Delta x_i$, and the fact is that the numbers $\sum_{i=0}^{n-1} f(c_i)\Delta x_i$ converge to a limit. Formulated rigorously, this means that there is a number S such that for any number $\varepsilon > 0$, there is a corresponding number $\delta > 0$ with the property that for every partition $\mathcal{P} = \{x_0, x_1, \ldots, x_n\}$ of $[a, b]$ with norm $\|\mathcal{P}\| < \delta$ and any choice of c_i in $[x_i, x_{i+1}]$, it is the case that

$$\left| \sum_{i=0}^{n-1} f(c_i)\Delta x_i - S \right| < \varepsilon.$$

This number S is the *definite integral of the function* f from a to b. It is denoted by

$$\int_a^b f(x)\,dx.$$

The equality $\lim_{\|\mathcal{P}\| \to 0} \sum_{i=0}^{n-1} f(c_i)\Delta x_i = \int_a^b f(x)\,dx$ summarizes step 3.

The construction outlined in steps 1 to 3 can be visualized by considering areas (in much the same way that the discussion in Section 9.1 was illustrated by the connection with area). Figure 9.19 tells us that if $f(c_i) \geq 0$, then $f(c_i)\Delta x_i$ is the area of a rectangle that lies above the x-axis, and if $f(c_i) < 0$, then $f(c_i)\Delta x_i$ is equal to the negative of the area of a rectangle that lies below the x-axis. The figure tells us that the sum $\sum_{i=0}^{n-1} f(c_i)\Delta x_i$ is a number that approximates the sum of the areas below the graph and above

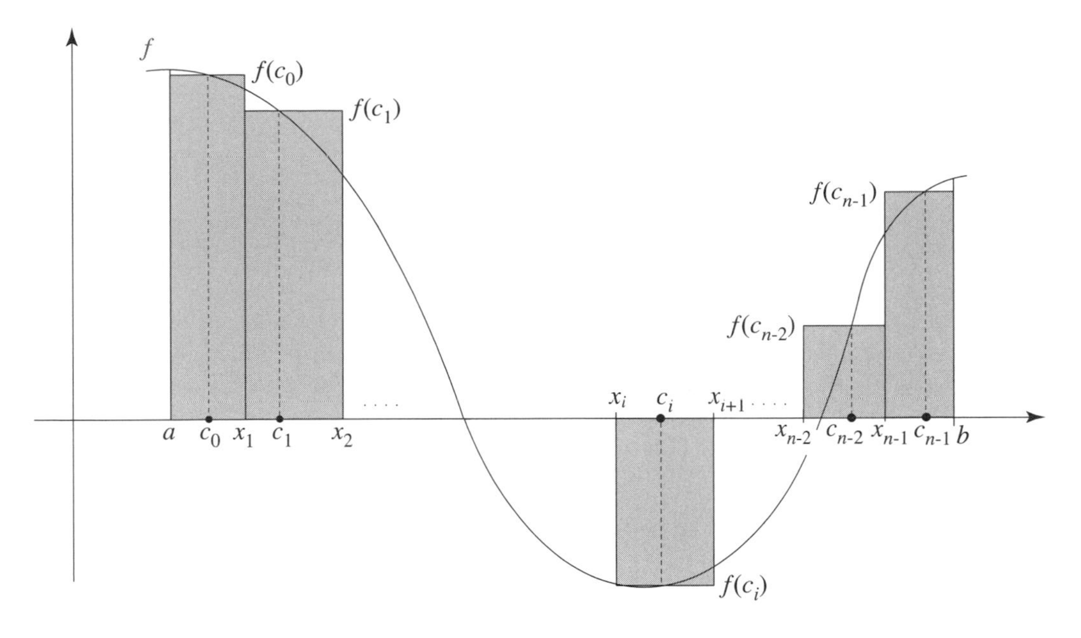

Fig. 9.19

the x-axis minus the sum of the areas above the graph and below the x-axis. Figure 9.20 illustrates how the approximation is tighter if the norm of the partition, and hence the distances Δx_i, are smaller. Going

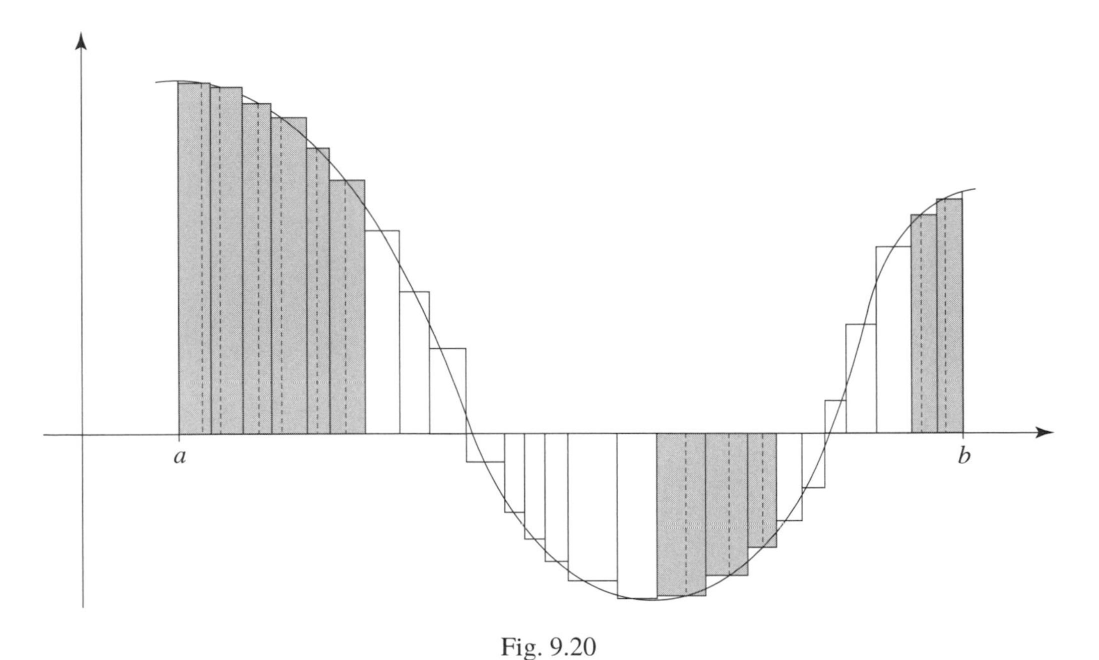

Fig. 9.20

to the limit, we can see that

$$\lim_{\|\mathcal{P}\|\to 0} \sum_{i=0}^{n-1} f(c_i)\Delta x_i = \int_a^b f(x)\,dx$$

is equal to

the sum of areas above the axis (and below the graph) *minus*

the sum of the areas below the axis (and above the graph),

with x varying from $x = a$ to $x = b$.

This is the same conclusion reached in Section 9.1 with the specific choice

$$c_0 = x_0, c_1 = x_1, c_2 = x_2, \ldots, c_i = x_i, \ldots c_{n-2} = x_{n-2}, \text{ and } c_{n-1} = x_{n-1}$$

of the left endpoints of the intervals. The right endpoints could have been picked as well. However, again, the limit exists no matter how the numbers $c_0, c_1, \ldots, c_{n-2}, c_{n-1}$ are chosen in the intervals. But why keep these numbers general, when a particular choice does the same thing? The importance of the flexibility in the choice of the points c_i in the intervals $[x_i, x_{i+1}]$ will now become apparent. The quick proof of the fundamental theorem of calculus that follows relies on the careful selection of numbers c_i that the mean value theorem provides.

Let $y = F(x)$ be an antiderivative of the function $y = f(x)$. Consider any partition

$$a = x_0 < x_1 < x_2 < x_3 < \cdots < x_i < x_{i+1} < \cdots < x_{n-1} < x_n = b$$

of the interval $[a, b]$. By an application of the mean value theorem of Section 7.6 to the function $y = F(x)$ over the subinterval $[x_i, x_{i+1}]$, there is a number c_i with $x_i \leq c_i \leq x_{i+1}$ such that

$$F'(c_i) = \frac{F(x_{i+1}) - F(x_i)}{x_{i+1} - x_i} = \frac{F(x_{i+1}) - F(x_i)}{\Delta x_i}.$$

Figure 9.21 illustrates this equality. The slope of the line determined by $(x_i, F(x_i))$ and $(x_{i+1}, F(x_{i+1}))$ is

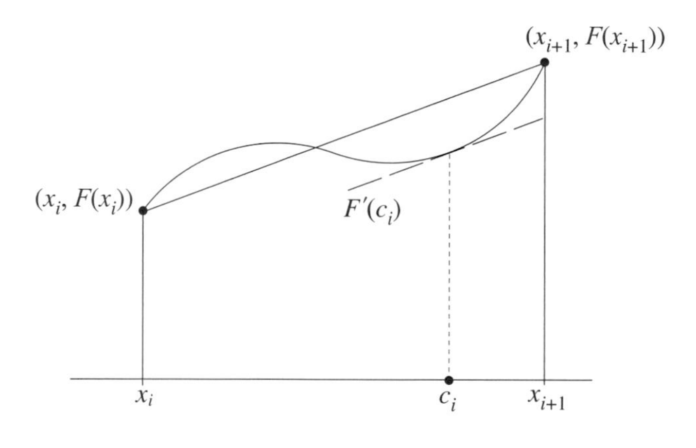

Fig. 9.21

equal to the slope $F'(c_i)$ of the tangent. Because $F'(c_i) = f(c_i)$, we get that

$$f(c_i)\Delta x_i = F'(c_i)(x_{i+1} - x_i) = F(x_{i+1}) - F(x_i).$$

Using these c_i, we get

$$\begin{aligned}\sum_{i=0}^{n-1} f(c_i)\Delta x_i &= f(c_0)\Delta x_0 + f(c_1)\Delta x_1 + f(c_2)\Delta x_2 + \ldots \\ &\quad + f(c_{i-1})\Delta x_{i-1} + f(c_i)\Delta x_i + \cdots + f(c_{n-2})\Delta x_{n-2} + f(c_{n-1})\Delta x_{n-1} \\ &= [F(x_1) - F(x_0)] + [F(x_2) - F(x_1)] + [F(x_3) - F(x_2)] + \ldots\end{aligned}$$

$$+ [F(x_i - F(x_{i-1})] + [F(x_{i+1} - F(x_i)] + \dots + [F(x_{n-1} - F(x_{n-2})] + [F(x_n - F(x_{n-1}]$$
$$= -F(x_0) + F(x_n) = F(b) - F(a).$$

It has been demonstrated for any partition

$$a = x_0 < x_1 < x_2 < x_3 < \cdots < x_i < x_{i+1} < \cdots < x_{n-1} < x_n = b$$

of $[a, b]$ that if $c_0, c_1, c_2, \dots c_{n-2}, c_{n-1}$ are the numbers that the mean value theorem provides, then

$$\sum_{i=0}^{n-1} f(c_i)\Delta x_i = F(b) - F(a).$$

Since we are assuming that the limit $\lim_{\|\mathcal{P}\|\to 0} \sum_{i=0}^{n-1} f(c_i)\Delta x_i$ exists, there is no place for it to be except at $F(b) - F(a)$. Therefore

$$\int_a^b f(x)\,dx = F(b) - F(a).$$

This is the fundamental theorem of calculus.

9.6 AREA AS ANTIDERIVATIVE

The fundamental theorem of calculus invites an important question: Does the antiderivative $y = F(x)$ that it requires for the given continuous function $y = f(x)$ always exist?

We saw in Section 6.1 that Isaac Newton answered this question in the affirmative for the simple functions that he considered. The area determined by the graph of the function played an essential role. The same approach was used in Section 7.11 with $y = \frac{1}{x}$ to define the natural log. We will now see that Newton's strategy applies to show that an antiderivative exists for any continuous function. The argument makes important use of the fact—a consequence of the discussion of Section 9.1—that an area with a boundary given by the graph of a function can be expressed in terms of a number.

Let f be a function that is continuous on an interval I. Fix a number a in this interval, and define the function A for any x in I as follows. If $x \geq a$, set

$A(x)$ = the sum of the areas *above* the x-axis (and *below* the graph of $y = f(x)$) *minus* the sum of the areas *below* the x-axis (and *above* the graph of $y = f(x)$) over the interval from a to x.

Notice that $A(a) = 0$. For $x < a$, set

$A(x)$ = the sum of the areas *below* the x-axis (and *above* the graph of $y = f(x)$) *minus* the sum of the areas *above* the x-axis (and *below* the graph of $y = f(x)$) over the interval from x to a.

Figure 9.22 illustrates the definition of the function $y = A(x)$. In the figure, $A_1, A_2, \dots, A_x$ are the areas of the indicated regions to the right of $x = a$, and $B_1, B_2, \dots, B_x$ are the areas of the indicated regions to the left of $x = a$. In the situation of the figure, the value of $A(x)$ is given by

$$A(x) = -A_1 + A_2 - \cdots + A_x, \text{ for } x \geq a \text{ and}$$

$$A(x) = B_1 - B_2 + \cdots - B_x, \text{ for } x < a.$$

We will show that the function $A(x)$ is differentiable on the interval I (except at endpoints) and that

$$A'(x) = f(x)$$

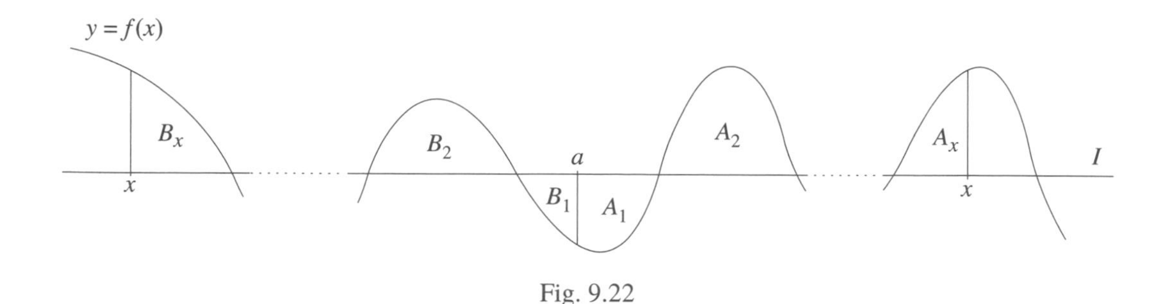

Fig. 9.22

for all x in I (except at endpoints). In other words, the function $A(x)$ is an answer to the question that was asked. Given the definition of the derivative, the proof of this assertion requires that we verify

$$\lim_{\Delta x \to 0} \frac{A(x + \Delta x) - A(x)}{\Delta x} = f(x)$$

for any x in I. Take any such x and consider it fixed (throughout the argument that follows). We'll take $x \geq a$, but a similar argument works for $x < a$. Pick any Δx, but notice that Δx must be appropriately small (for instance, if $I = [c, d]$, then $c \leq x + \Delta x \leq d$), but this is no problem since the focus is on what happens when $\Delta x \to 0$. Let's also assume that Δx is positive, so that $x + \Delta x > x$. (This assumption is not essential to the argument.) Consider the definitions of $A(x)$ and $A(x + \Delta x)$, study Figures 9.23a and 9.23b, and ob-

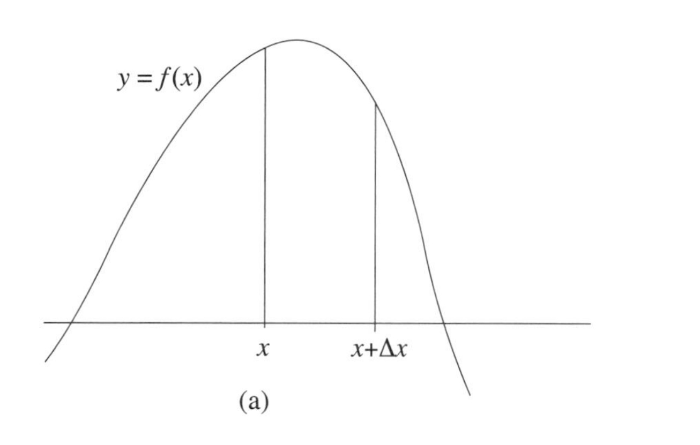

(a)

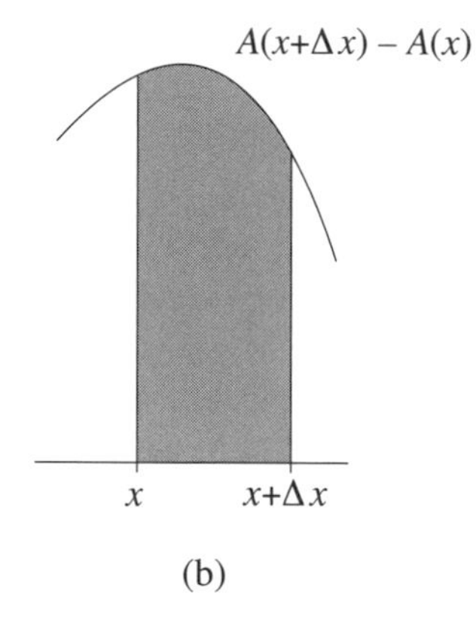

(b)

Fig. 9.23

serve that $A(x + \Delta x) - A(x)$ is equal to the difference $A_{x+\Delta x} - A_x$. So

$$A(x + \Delta x) - A(x)$$

is the shaded area in Figure 9.23b. Since the function f is continuous on the interval $[x, x + \Delta x]$, it achieves a minimum value m and a maximum value M on this closed interval. Refer to Section 7.8. It follows, as illustrated in Figures 9.24a and 9.24b, that

$$m \cdot \Delta x \leq A(x + \Delta x) - A(x) \leq M \cdot \Delta x.$$

Therefore

$$m \leq \frac{A(x + \Delta x) - A(x)}{\Delta x} \leq M.$$

Let $v = \frac{A(x+\Delta x)-A(x)}{\Delta x}$. The continuity of f over $[x, x + \Delta x]$ and the intermediate value theorem of Section 7.3

tell us that there is a number u with $x \le u \le x + \Delta x$ such that $f(u) = v = \frac{A(x+\Delta x)-A(x)}{\Delta x}$. So $A(x+\Delta x)-A(x) = f(u)\Delta x$. Refer to Figure 9.24c. Now push $\Delta x \to 0$. Since $u \to x$ in the process and f is continuous at x,

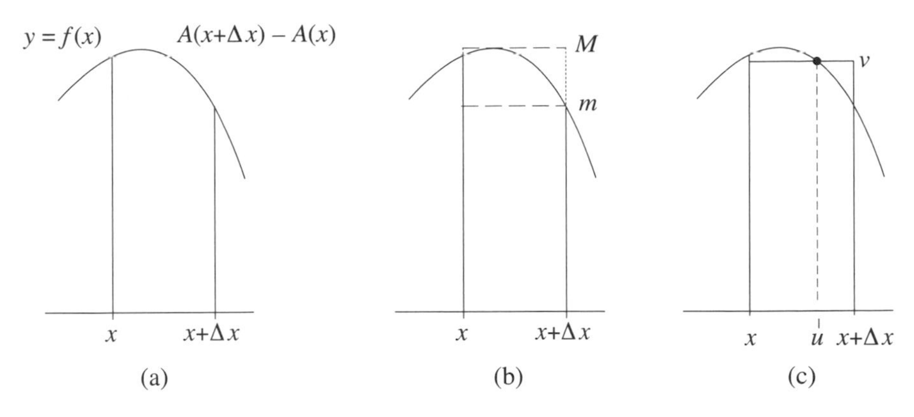

Fig. 9.24

it follows that

$$A'(x) = \lim_{\Delta x \to 0} \frac{A(x + \Delta x) - A(x)}{\Delta x} = \lim_{\Delta x \to 0} \frac{f(u)\Delta x}{\Delta x} = \lim_{\Delta x \to 0} f(u) = f(x).$$

Therefore, as asserted,

$$A'(x) = f(x).$$

A similar argument works in case the region with area A_x lies below the x-axis. We now know that as long as $y = f(x)$ is continuous, then the antiderivative of f that the fundamental theorem of calculus requires exists!

What if a different choice for a is made? Let b be some other constant in I, and let the corresponding area function be $B(x)$. By applying what was just verified, we know that $B'(x) = f(x)$ for all x in I. It follows by Corollary ii at the end of Section 7.6 that

$$B(x) = A(x) + C$$

for some constant C. Since $B(c) = 0$, $C = -A(c)$ and hence $B(x) = A(x) - A(c)$.

Example 9.16. Consider the function $f(x) = x^2$ over the interval $I = (-\infty, \infty)$. Define the area function $A(x)$ by taking $a = 3$. Since $A(x)$ and $F(x) = \frac{1}{3}x^3$ are both antiderivatives of $f(x) = x^2$, we know that $A(x) = \frac{1}{3}x^3 + C$ with C a constant. The fact that $A(3) = 0$ tells us that $C = -9$. So $A(x) = \frac{1}{3}x^3 - 9$.

Example 9.17. Consider the function $f(x) = \sqrt{x}$ over the interval $I = [0, \infty)$. Use the constant $a = 4$ to define the area function $A(x)$. The fact that both $A(x)$ and $F(x) = \frac{2}{3}x^{\frac{3}{2}}$ are both antiderivatives of $f(x) = \sqrt{x}$ means that $A(x) = \frac{2}{3}x^{\frac{3}{2}} + C$, for some constant C. Since $A(4) = 0$, $C = -\frac{2}{3}2^3 = -\frac{16}{3}$. Hence $A(x) = \frac{2}{3}x^{\frac{3}{2}} - \frac{16}{3}$.

The next point that needs to be made is that the definite integral does not depend on the notation for the variable of the integrand. This should not be a surprise because the way a mathematical expression is written down should not affect any conclusions.

Example 9.18. Using the fundamental theorem of calculus, tells us that

$$\int_1^3 x^2\,dx = \left[\tfrac{1}{3}\,x^3 \Big|_1^3\right] = \tfrac{1}{3}(3^3 - 1^3) = \tfrac{28}{3} \quad \text{and} \quad \int_1^3 \theta^2\,d\theta = \left[\tfrac{1}{3}\,\theta^3 \Big|_1^3\right] = \tfrac{1}{3}(3^3 - 1^3) = \tfrac{28}{3}.$$

So the variable that is used for the integrand has no impact on the answer.

Suppose that we are dealing with a function that is continuous on an interval I. For a fixed a and a variable x in I with $a \leq x$, consider the definite integral

$$\int_a^x f(x)\,dx.$$

Observe that in this expression x plays a dual role. It is both the upper limit of the integral and the variable of the function. Mathematicians understandably stay away from a situation where a symbol has multiple meanings in the same discussion. Since—as we have just seen—the value of the integral is the same no matter how the variable of the function is labeled, we'll use t for the variable of the function and use x only for the upper limit of limit of the integral. We therefore write

$$\int_a^x f(t)\,dt \quad \text{in place of} \quad \int_a^x f(x)\,dx.$$

By combining the definition of the area function $A(x)$ with the connection between area and the definite integral established in Section 9.1, we see that for any $x \geq a$,

$$A(x) = \int_a^x f(t)\,dt.$$

But what about $x < a$? Our definition of the definite integral required that the upper limit is greater than or equal to the lower limit. So for $x < a$, it makes sense to set

$$\int_a^x f(x)\,dx = -\int_x^a f(x)\,dx.$$

Another look at the definition of the area function $A(x)$ confirms that the equality

$$A(x) = \int_a^x f(t)\,dt$$

now holds for $x < a$ as well. This equality is therefore valid for all x in the interval I. Since $A'(x) = f(x)$, we have derived the basic fact

$$\boxed{\frac{d}{dx}\left(\int_a^x f(t)\,dt\right) = f(x)}$$

that illustrates the inverse relationship between integration and differentiation explicitly: the one operation undoes the effect of the other.

We will use the notation

$$\int f(x)\,dx$$

to represent the typical or generic antiderivative of a function $f(x)$ and refer to it as the *indefinite integral* of $f(x)$. We saw earlier that if $F(x)$ is any antiderivative of $f(x)$, then any other antiderivative is equal to $F(x) + C$ for some constant C. This fact provides the equality

$$\int f(x)\,dx = F(x) + C,$$

where C is a constant, the *constant of integration*.

In the context of the fundamental theorem of calculus, the fact that any continuous function has an antiderivative is reassuring. However, when it comes to the value of a specific definite integral, it is quite another matter to determine an explicit antiderivative of the integrand to which the fundamental theorem can then be applied. The remainder of this chapter will focus on the evaluation of definite integrals, and in particular on methods for finding antiderivatives of functions.

9.7 FINDING ANTIDERIVATIVES

In view of the discussion that concludes the previous section, finding an antiderivative is the same thing as solving an indefinite integral. In particular, the problem of finding an antiderivative of a function is often referred to as the problem of integrating the function. It needs to be said that the task of integrating a function is frequently difficult and often impossible. The tools that are available consist of a combination of a few general approaches and a bag of tricks. The most basic general strategies are the methods of integration by substitution and integration by parts. Both are described in this section. Our discussion will make use of differentials, meaning, for example, that for $y = f(x)$ and $\frac{dy}{dx} = f'(x)$, both dx and $dy = f'(x)\,dx$ are regarded as separate entities. We'll skip the theoretical discussion that legitimizes this strategy, but point out that this was already done in a different context in Section 9.3.

9.7.1 Integration by Substitution

Consider the problem of finding an antiderivative of the function $f(x) = (x+2)^{11}$, in other words, the question

$$\int (x+2)^{11} dx = ?$$

This problem can be solved by multiplying out the polynomial $(x+2)^{11}$ and then integrating term by term. But this would be a rather tedious thing to do. A much better way is to let $u = x + 2$, so that $\frac{du}{dx} = 1$, and hence $dx = du$. Rewrite the integral in the variable u, and observe that $\int u^{11}\, du = \frac{1}{12}u^{12} + C$. To return to the problem we started with, substitute $u = x + 2$ to get back to the variable x and to the solution,

$$\int (x+2)^{11} dx = \tfrac{1}{12}(x+2)^{12} + C.$$

The chain rule confirms that this answer is correct.

Let's see why this substitution procedure works. Consider two functions $f(x)$ and $g(x)$, and assume that $f(x)$ is continuous and $g(x)$ is differentiable on some interval I. Let $F(x)$ be an antiderivative $f(x)$, and form the composite function $F(g(x))$. By the chain rule,

$$\frac{d}{dx}F(g(x)) = F'(g(x)) \cdot g'(x) = f(g(x)) \cdot g'(x).$$

So $F(g(x))$ is an antiderivative of the function $f(g(x)) \cdot g'(x)$, and therefore

$$\int f(g(x)) \cdot g'(x)\, dx = F(g(x)) + C.$$

Let $u = g(x)$. Since $\frac{du}{dx} = g'(x)$, and hence $du = g'(x)dx$, this integral can be rewritten as

$$\int f(u)\, du = F(u) + C$$

in the variable u. This strategy can be adapted to definite integrals by using the previous integral formula and noticing that for two constants c and d,

$$\int_c^d f(g(x)) \cdot g'(x)\,dx \;=\; F(g(d)) - F(g(c)) \;=\; \int_{g(c)}^{g(d)} f(u)\,du.$$

A carefully chosen substitution $u = g(x)$—and this is the point of the substitution method —can transform a complicated integral into one that is readily solved. The idea is to have a careful look at the integral and to try to find two functions $f(x)$ and $g(x)$ such that the integrand has the form $f(g(x)) \cdot g'(x)$ (usually, times some constant). Start by choosing a $u = g(x)$ such that $\frac{du}{dx} = g'(x)$ also appears in the integrand. To complete the solution, locate $f(x)$ and try to find an antiderivative of $f(x)$. The examples below provide several instances in which this can be carried out successfully. Unfortunately, success is not guaranteed. It may well be that an integral cannot be solved by this method (or any other).

Example 9.19. We'll solve $\int 3(x^2-5)^8 x\,dx$. With

$$u = x^2 - 5 \quad \text{and} \quad \tfrac{du}{dx} = 2x,$$

we see that $du = 2x\,dx$ and hence that $3x\,dx = \frac{3}{2}du$. By substituting, $\int 3(x^2-5)^8 x\,dx = \int \frac{3}{2}u^8\,du$. But $\int \frac{3}{2}u^8\,du = \frac{3}{18}u^9 + C$. Since $u = x^2 - 5$, we get $\int 3(x^2-5)^8 x\,dx = \frac{1}{6}(x^2-5)^9 + C$.

Example 9.20. Let's solve $\int x\sqrt{x+3}\,dx$. Taking

$$u = x + 3 \quad \text{and hence} \quad \tfrac{du}{dx} = 1,$$

we get $du = dx$ and, since $x = u - 3$, the equality $\int x\sqrt{x+3}\,dx = \int (u-3)u^{\frac{1}{2}}\,du$. Has anything meaningful been accomplished? Yes! Noting that $(u-3)u^{\frac{1}{2}} = u^{\frac{3}{2}} - 3u^{\frac{1}{2}}$, we get that

$$\int x\sqrt{x+3}\,dx = \int (u^{\frac{3}{2}} - 3u^{\frac{1}{2}})\,du = \tfrac{2}{5}u^{\frac{5}{2}} - 2u^{\frac{3}{2}} + C.$$

Because $u = x + 3$, it follows that

$$\int x\sqrt{x+3}\,dx = \tfrac{2}{5}(x+3)^{\frac{5}{2}} - 2(x+3)^{\frac{3}{2}} + C.$$

Consider the definite integral $\int_{-2}^{5} x\sqrt{x+3}\,dx$. Since $\frac{2}{5}(x+3)^{\frac{5}{2}} - 2(x+3)^{\frac{3}{2}}$ is an antiderivative of $x\sqrt{x+3}$,

$$\int_{-2}^{5} x\sqrt{x+3}\,dx = \left(\tfrac{2}{5}(x+3)^{\frac{5}{2}} - 2(x+3)^{\frac{3}{2}}\right)\Big|_{-2}^{5} = \left(\tfrac{2}{5}8^{\frac{5}{2}} - 2\cdot 8^{\frac{3}{2}}\right) - \left(\tfrac{2}{5} - 2\right) \approx 28.75.$$

Using the substitution to transform the limits 5 and -2 is more efficient. Substituting $x = 5$ and $x = -2$ into $u = x + 3$ gives us $u = 8$ and $u = 1$, and

$$\int_{-2}^{5} x\sqrt{x+3}\,dx = \int_{1}^{8} (u-3)u^{\frac{1}{2}}\,du = \left(\tfrac{2}{5}u^{\frac{5}{2}} - 2u^{\frac{3}{2}}\right)\Big|_{1}^{8}.$$

Notice that the answer is the same as before.

Example 9.21. Let's look at the integral $\int (x+6)^2(2x-3)^{50}\,dx$. Let

$$u = 2x - 3 \quad \text{and} \quad \tfrac{du}{dx} = 2.$$

Since $x = \frac{1}{2}(u+3)$ and $dx = \frac{1}{2}du$, the integral is transformed into

$$\begin{aligned}\int \tfrac{1}{2}(\tfrac{1}{2}(u+3)+6)^2\, u^{50}\, du &= \int \tfrac{1}{2}(\tfrac{1}{2}u + \tfrac{15}{2})^2\, u^{50}\, du = \int \tfrac{1}{2}(\tfrac{1}{4}u^2 + \tfrac{15}{2}u + \tfrac{15^2}{4})u^{50}\, du \\ &= \int (\tfrac{1}{8}u^{52} + \tfrac{15}{4}u^{51} + \tfrac{15^2}{8}u^{50})\, du = \tfrac{1}{8\cdot 53}u^{53} + \tfrac{15}{4\cdot 52}u^{52} + \tfrac{15^2}{8\cdot 51}u^{51} + C \\ &= \tfrac{1}{424}(2x-3)^{53} + \tfrac{15}{208}(2x-3)^{52} + \tfrac{15^2}{408}(2x-3)^{51} + C.\end{aligned}$$

Example 9.22. Consider the integral $\int 3x\sin(2x^2)\,dx$. Take

$$u = 2x^2 \quad \text{and} \quad \tfrac{du}{dx} = 4x.$$

Since $x\,dx = \frac{1}{4}du$, the integral is transformed to $\int \frac{3}{4}\sin u\,du$. From Section 7.7, $\frac{d}{dx}\cos x = -\sin x$, so that $\int \frac{3}{4}\sin u\,du = -\frac{3}{4}\cos u + C$. Hence $\int 3x\sin(2x^2)\,dx = -\frac{3}{4}\cos(2x^2) + C$.

Example 9.23. Let's solve $\int \frac{\ln x}{x}\,dx$. Turn to Section 7.11. Notice that the definition of $\ln x$ requires that $x > 0$. The formula $\frac{d}{dx}\ln x = \frac{1}{x}$ tells us that the equalities

$$u = \ln x \quad \text{and} \quad \tfrac{du}{dx} = \tfrac{1}{x}$$

along with $\int u\,du = \frac{1}{2}u^2 + C$ provide the solution $\int \frac{\ln x}{x}\,dx = \frac{1}{2}(\ln x)^2 + C$.

Efforts to use the substitution method on integrals obtained by modifying those of the two last examples give a sense of the limitations of this method. For instance, if in Example 9.22 the term $3x$ is replaced by $3x^{\frac{1}{2}}, 3x^2, 3x^3$, and so on, then the method runs into a wall because the substitution $u = 2x^2$ results in integrals in the variable u that are—with some exceptions (Example 9.27, for instance)—more complicated than the original integral. The same thing is the case if the denominator x in Example 9.23 is replaced by x^2, x^3, and so forth.

9.7.2 *Integration by Parts*

In the same way that the chain rule gives rise to the method of integration by substitution, the product rule gives rise to the method of integration by parts.

Consider an integral of the form $\int f(x)\cdot g(x)\,dx$, where $f(x)$ is differentiable and $g(x)$ continuous on some interval I. Suppose that $G(x)$ is any antiderivative of $g(x)$. By the product rule,

$$\frac{d}{dx}f(x)G(x) = f'(x)G(x) + f(x)g(x) \quad \text{so that} \quad \int (f'(x)G(x) + f(x)g(x))\,dx = f(x)G(x) + C.$$

Therefore

$$\int f(x)\cdot g(x)\,dx = f(x)G(x) - \int f'(x)G(x)\,dx.$$

(The constant of integration is implicit on both sides.) Make the two substitutions $u = f(x)$ and $v = G(x)$. Then $\frac{du}{dx} = f'(x)$ and $\frac{dv}{dx} = G'(x) = g(x)$. Therefore $du = f'(x)\,dx$, $dv = g(x)\,dx$, and the equality above simplifies to the integration by parts formula

$$\boxed{\int u\,dv = uv - \int v\,du}$$

If things work out as intended, then this formula reduces the solution of a more complicated integral—namely, $\int u\,dv$—to one that is simpler namely, $\int v\,du$. It is often helpful if $u = f(x)$ is a function such that $\frac{du}{dx}$ is simpler than u. Let's illustrate this method with a few examples.

Example 9.24. Let's solve $\int x \sin x\,dx$. Let

$$u = x \quad \text{and} \quad dv = \sin x\,dx.$$

Since $\frac{dv}{dx} = \sin x$, we can take $v = -\cos x$ to be the simplest antiderivative of $\sin x$ (on occasion, it is convenient to add a constant). Because $\frac{du}{dx} = 1, du = dx$. By the integration by parts formula,

$$\int x \sin x\,dx = -x\cos x - \int (-\cos x)dx = -x\cos x + \int \cos x\,dx = \sin x - x\cos x + C.$$

Example 9.25. Consider $\int e^x \sin x\,dx$. Let

$$u = e^x \quad \text{and} \quad dv = \sin x\,dx.$$

As in the previous example, we take $v = -\cos x$. Because $du = e^x dx$, the integration by parts formula tells us that

$$\int e^x \sin x\,dx = -e^x\cos x - \int (-\cos x)e^x\,dx = -e^x\cos x + \int e^x \cos x\,dx.$$

It seems that this did not get us anywhere, because $\int e^x \cos x\,dx$ is no easier than $\int e^x \sin x\,dx$. But let's plod ahead and try integration by parts on $\int e^x \cos x\,dx$. With

$$u = e^x \quad \text{and} \quad dv = \cos x\,dx,$$

$v = \sin x$ and $du = e^x dx$, so that

$$\int e^x \cos x\,dx = e^x \sin x - \int e^x \sin x\,dx.$$

Putting together the formulas for the integrals of $e^x \sin x$ and $e^x \cos x$, we get

$$\int e^x \sin x\,dx = -e^x \cos x + \int e^x \cos x\,dx = -e^x\cos x + e^x \sin x - \int e^x \sin x\,dx.$$

It follows that $2\int e^x \sin x\,dx = e^x \sin x - e^x \cos x$ plus a constant, and therefore that

$$\int e^x \sin x\,dx = \tfrac{1}{2}[e^x \sin x - e^x \cos x] + C.$$

Example 9.26. Let's try the method of integration by parts on $\int \ln x\,dx$. The only possible choice is

$$u = \ln x \quad \text{and} \quad dv = dx.$$

So $du = \frac{1}{x}\,dx$ and $v = x$. By the integration by parts formula, $\int \ln x\,dx = x\ln x - \int x\frac{1}{x}\,dx = x\ln x - x + C$.

The method of integration by parts does not guarantee success. The method may fail either because appropriate u and dv may not exist, or because the integral on the other side is too complicated. One final example shows that some integrals can be solved by using integration by substitution and integration by parts in sequence.

Example 9.27. Consider $\int 3x^3\sin(2x^2)\,dx$. Starting with the substitution

$$z = 2x^2 \quad \text{and} \quad dz = 4x\,dx,$$

we get $x^2 = \frac{1}{2}z$ and $x\,dx = \frac{1}{4}dz$. This transforms the integral to $\int \frac{3}{2}z\sin z\,\frac{1}{4}dz = \int \frac{3}{8}z\sin z\,dz$. By Example 9.24,

$$\int 3x^3\sin(2x^2)\,dx = \int \tfrac{3}{8}z\sin z\,dz = \tfrac{3}{8}(\sin z - z\cos z) + C = \tfrac{3}{8}\sin(2x^2) - \tfrac{3}{4}x^2\cos(2x^2) + C.$$

9.7.3 *Some Algebraic Moves*

We'll now consider the indefinite integral of a function that is given as a fraction of polynomials where the numerator is a constant or a linear polynomial and the denominator is quadratic. An algebraic move makes it possible to split such a fraction into two parts that are both easily integrated by using the natural log. With later applications in mind, we'll consider the two examples

$$\int \frac{x-1}{x^2+x}\,dx \quad \text{and} \quad \int \frac{dx}{s^2-x^2}, \text{ where } s \text{ is a constant.}$$

The method illustrated below works in any situation where the quadratic polynomial in the denominator can be factored. A similar (but more complicated) approach applies to integrals with denominators consisting of quadratic polynomials that cannot be factored. Since we will not encounter such integrals, we won't elaborate. We'll turn instead to the solution of the two integrals above.

Notice that $x^2 + x = x(x+1)$. Let A and B be constants and set

$$\frac{x-1}{x(x+1)} = \frac{A}{x} + \frac{B}{x+1}.$$

Of course, this might not go anywhere, but let's just go with it. At the end we will see that it does. Take common denominators to get

$$\frac{x-1}{x(x+1)} = \frac{A}{x} + \frac{B}{x+1} = \frac{A(x+1)+Bx}{x(x+1)} = \frac{(A+B)x+A}{x(x+1)}.$$

A comparison of the first numerator with the last suggests that we set $A + B = 1$ and $A = -1$. So $B = 2$, and we get $\frac{x-1}{x(x+1)} = \frac{-1}{x} + \frac{2}{x+1}$. (In reference to the validity of the initial move to split the fraction, it is easy to check that this last equality is in fact correct.) The method just illustrated has split a more complicated fraction into two simpler ones. It is known as the *method of partial fractions*. Using basic facts from Section 7.11, we now get that

$$\int \frac{x-1}{x(x+1)}\,dx = \int \frac{-1}{x}\,dx + \int \frac{2}{x+1}\,dx = -\ln|x| + 2\ln|x+1| + C,$$

and that this answer can be rewritten as $\ln|x|^{-1} + \ln(x+1)^2 + C = \ln\left(|x|^{-1}(x+1)^2\right) + C$.

The integral $\int \frac{dx}{x^2-s^2}$ with s a constant is handled in the same way. Let A and B be constants, and consider

$$\frac{1}{x^2-s^2} = \frac{1}{(x-s)(x+s)} = \frac{A}{x-s} + \frac{B}{x+s} = \frac{A(x+s)+B(x-s)}{(x-s)(x+s)} = \frac{(A+B)x+(A-B)s}{x^2-s^2}\,.$$

Compare the numerators 1 and $(A+B)x + (A-B)s$ of the first and last terms. Since there is no x term, $A+B$ must be 0. So $B = -A$, and the remaining term $(A-B)s = 2As = 1$. It follows that $A = \frac{1}{2s}$ and $B = -\frac{1}{2s}$. Therefore, by applying basic facts from Section 7.11,

$$\int \frac{dx}{x^2-s^2} = \tfrac{1}{2s}\int \Big(\frac{1}{x-s} - \frac{1}{x+s}\Big)dx = \tfrac{1}{2s}\big(\ln|x-s| - \ln|x+s| + C\big) = \tfrac{1}{2s}\ln\left|\tfrac{x-s}{x+s}\right| + C.$$

Example 9.28. We'll use a substitution to evaluate $\int \frac{e^{at}-1}{e^{at}+1}\,dt$, where a is a constant. Let $u = e^{at}$. So $\frac{du}{dt} = ae^{at}$ and $\frac{1}{a}du = e^{at}dt$. Notice that

$$\int \frac{e^{at}-1}{e^{at}+1}\,dt = \int \frac{e^{at}-1}{e^{at}+1}\cdot\frac{e^{at}}{e^{at}}\,dt = \int \frac{e^{at}-1}{e^{at}(e^{at}+1)}\cdot e^{at}\,dt = \tfrac{1}{a}\int \frac{u-1}{u(u+1)}\,du\,.$$

From an earlier conclusion in this section, we see that $\frac{1}{a}\int \frac{u-1}{u(u+1)}\,du = -\frac{1}{a}\ln u + \frac{2}{a}\ln(u+1) + C$. After substituting $u = e^{at}$ back in,

$$\int \frac{e^{at}-1}{e^{at}+1}\,dt = -\tfrac{1}{a}\ln e^{at} + \tfrac{2}{a}\ln(e^{at}+1) + C = \tfrac{2}{a}\ln(e^{at}+1) - t + C.$$

Chapter 7 added exponential functions, logarithm functions, and hyperbolic functions to the holdings of algebraic functions (polynomials and roots, for instance) and trigonometric functions that our library of functions already contained. The fact that products, quotients, and composites can be formed with all of them makes for a substantial collection. However, the problem of having explicit functions available with which basic integrals can be solved calls for another expansion of this library.

9.8 INVERSE FUNCTIONS

Let $y = f(x)$ be a function that is defined over an interval I. Suppose that the function is either increasing over I or decreasing over I. The graph of an increasing function rises from left to right over I, and the graph of a decreasing function falls from left to right over I. Refer to Section 7.8 for more of the basics. Our initial focus will be on increasing functions, but the essence of our discussion applies to decreasing functions as well. Figure 9.25a depicts the graph of a typical increasing function.

For any c in I, let (c, d) be the corresponding point on the graph of f. The definition of the graph of a function tells us that the rule of the function is given by $c \to d$. The assumption that f is increasing means that for the value d there is only one c such that $f(c) = d$. This tells us in turn that the rule of f can be reversed. This reversed rule sends $d \to c$, as illustrated in Figure 9.25b. It defines a function that is denoted by f^{-1} and referred to as the *inverse function* of f. Observe that f^{-1} is defined for any number d for which there is some c in I such that $f(c) = d$, and if there is such a c, then $f^{-1}(d) = c$. To repeat, the inverse function f^{-1} is defined by the following rule:

$$f^{-1}(d) \text{ is precisely that number } c \text{ with the property that } f(c) = d.$$

So if $f(c) = d$, then $f^{-1}(d) = c$ and, conversely, if $f^{-1}(d) = c$, then $f(c) = d$. It follows that

$$f^{-1}(f(c)) = c \quad\text{and}\quad f(f^{-1}(d)) = d \text{ whenever } f^{-1} \text{ is defined at } d.$$

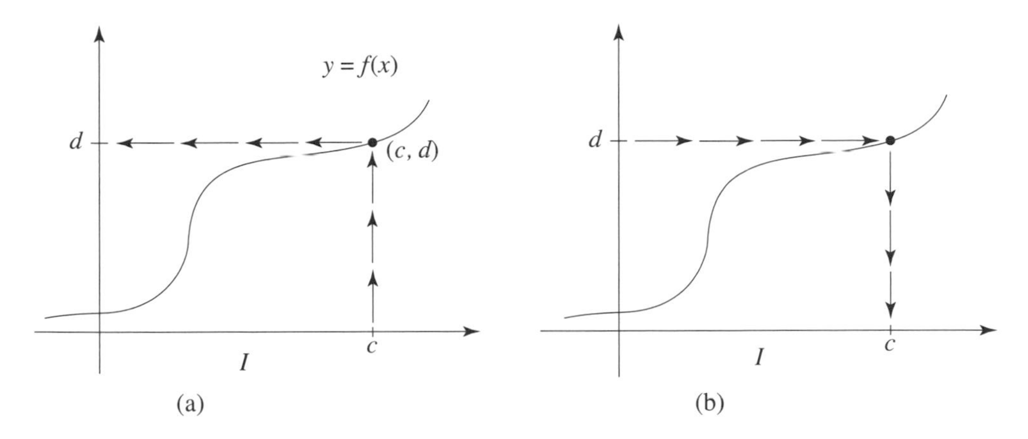

Fig. 9.25

Be aware of the ambiguity with the notation $f^{-1}(x)$ for the inverse function. It is *not at all the same thing* as $\frac{1}{f(x)} = f(x)^{-1}$. (In mathematics it is not uncommon that the same notation has different meanings in different contexts.) The inverse function is defined in the same way for any function that is decreasing on an interval I.

Let f be any function that is increasing or decreasing on an interval I, and let $y = f^{-1}$ be its inverse. Using x in place of the c and then d above provides the two equations

$$f^{-1}(f(x)) = x \quad \text{and} \quad f(f^{-1}(x)) = x.$$

Therefore both of these composite functions fix all x for which they are defined. Suppose that f is differentiable on the interval I. Applying the chain rule to the second of these equations tells us that

$$\frac{d}{dx}(f(f^{-1}(x))) = f'(f^{-1}(x)) \cdot \frac{d}{dx}f^{-1}(x) = 1,$$

and therefore that

$$\boxed{\frac{d}{dx}f^{-1}(x) = \frac{1}{f'(f^{-1}(x))}}$$

Let x be in the domain of f^{-1}. Notice that as long as $f'(f^{-1}(x)) \neq 0$, then f^{-1} is differentiable at x.

We have already encountered an example of the function–inverse function relationship. It was verified in Section 7.12 that the natural log function $y = \ln x$ and the exponential function $y = e^x$ are inverses of each other. The formula for $\frac{d}{dx}f^{-1}(x)$ played an important role in this analysis.

Let's pause to look at another example. Consider the function $f(x) = x^2$. Observe from Figure 9.26a that the graph is decreasing for $x \leq 0$ and increasing for $x \geq 0$. The fact that for a given $d > 0$ there are two points on the graph with y-coordinate d means that the strategy for reversing the rule as depicted in Figure 9.25b runs into an ambiguity. So it is not possible to define an inverse. However, the restricted function $f_+(x) = x^2$ for any $x \geq 0$ is increasing so that it has an inverse, as Figures 9.26b illustrates. Let $d \geq 0$ be any number. Note that c satisfies $c \geq 0$ and $c^2 = d$. So $c = \sqrt{d}$ and $f_+^{-1}(d) = \sqrt{d}$. Therefore $f_+^{-1}(x) = \sqrt{x} = x^{\frac{1}{2}}$ is the inverse function of $y = f_+(x)$. Since $f_+'(x) = 2x$, the formula $\frac{d}{dx}f_+^{-1}(x) = \frac{1}{f_+'(f_+^{-1}(x))}$ confirms what we already know, namely, that

$$\frac{d}{dx}x^{\frac{1}{2}} = \frac{1}{f_+'(x^{\frac{1}{2}})} = \frac{1}{2\,(x^{\frac{1}{2}})} = \tfrac{1}{2}x^{-\frac{1}{2}}.$$

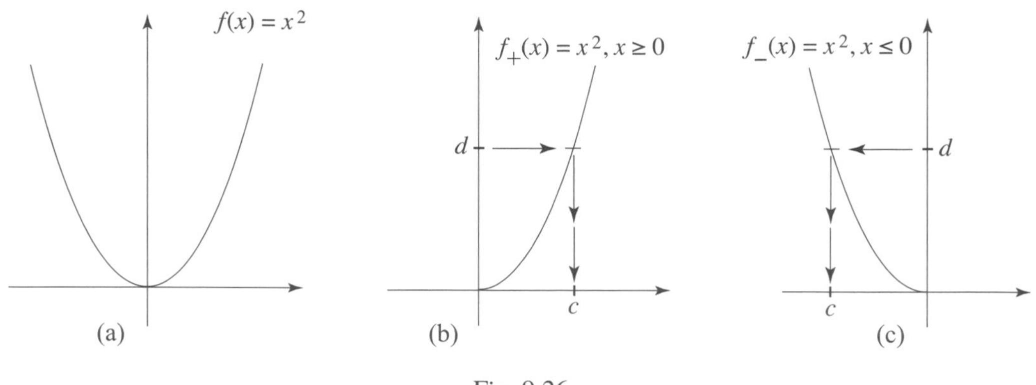

Fig. 9.26

The inverse is also defined for the decreasing function $f_-(x) = x^2$ for $x \leq 0$. A look at Figure 9.26c tells us that in this case, $f_-^{-1}(d) = -\sqrt{d}$ for any $d \geq 0$. So $f_-^{-1}(x) = -\sqrt{x} = -x^{\frac{1}{2}}$ is the inverse of $y = f_-(x)$.

We'll return to any function $y = f(x)$ that is either increasing or decreasing on an interval I, and we'll let $y = f^{-1}(x)$ be its inverse. Let (c, d) be any point on the graph of f. So $f(c) = d$. Therefore $f^{-1}(d) = c$ and (d, c) is on the graph of f^{-1}. So whenever (c, d) is a point on the graph of f, then (d, c) is a point on the graph of f^{-1}. Plot both of the points (c, d) and (d, c) and also (c, c) and (d, d). Notice that these four points determine a square and that the segment from (c, d) to (d, c) is a diagonal of the square. As a

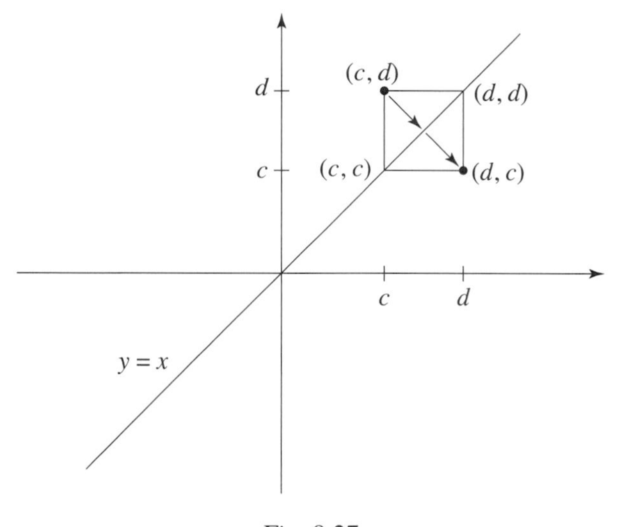

Fig. 9.27

consequence, the graph of f^{-1} is obtained by the following steps. See Figure 9.27.

i. Start with any point on the graph of f.

ii. Move from this point perpendicularly toward the line $y = x$. Stop at this line and record the distance moved.

iii. Now continue in the same direction away from the line $y = x$. Stop after you have moved the distance recorded in step (ii). The point you have reached is on the graph of f^{-1}.

This "reflection" procedure for sketching the graph of f^{-1} is illustrated in Figure 9.28. Note that the graph of f^{-1} is the mirror image (with the mirror placed on the line $y = x$) of the graph of f. In particular,

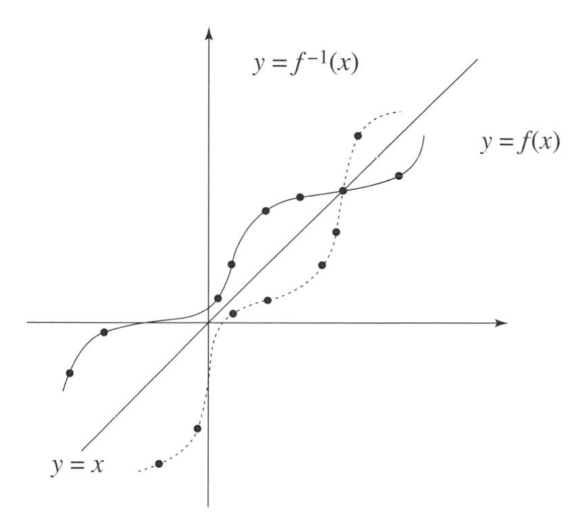

Fig. 9.28

if the graph of f is continuous, then the graph of f^{-1} is also continuous. If we do this for the graph of the function $f_+(x) = x^2$, $x \geq 0$, we get the graph of the inverse $f_+^{-1}(x) = x^{\frac{1}{2}} = \sqrt{x}$. Refer to Figure 9.29.

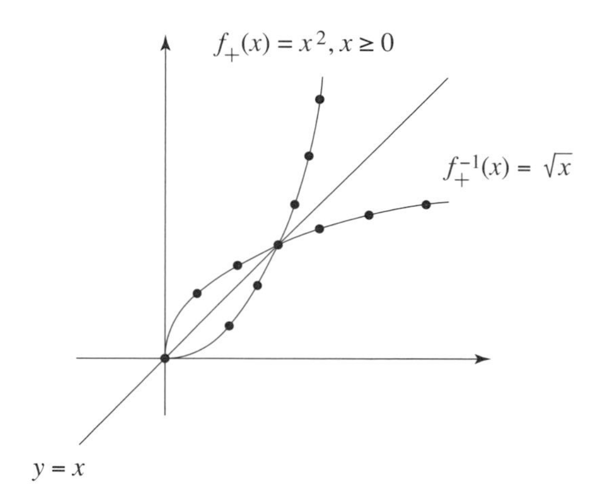

Fig. 9.29

9.9 INVERSE TRIGONOMETRIC AND HYPERBOLIC FUNCTIONS

Among the most relevant and interesting inverse functions are those of the trigonometric and hyperbolic functions. We will develop these inverses next and see that they are a powerful tool in the solution of integrals that involve the terms $x^2 + a^2, x^2 - a^2$, and $a^2 - x^2$ raised to powers, including fractional powers such as $\frac{1}{2}$.

9.9.1 Trigonometric Inverses

Refer to Sections 4.6 and 7.7 for the basics about the trigonometric functions and their graphs. We'll start our pursuit of their inverses with $f(x) = \sin x$. The fact that the graph of $f(x) = \sin x$ rises and falls and is neither increasing nor decreasing tells us that we'll need to restrict its domain. Note that $f(x) = \sin x$ is increasing for $-\frac{\pi}{2} \leq x \leq \frac{\pi}{2}$ (see Figure 9.30) and it is the conventional approach to restrict x to the interval

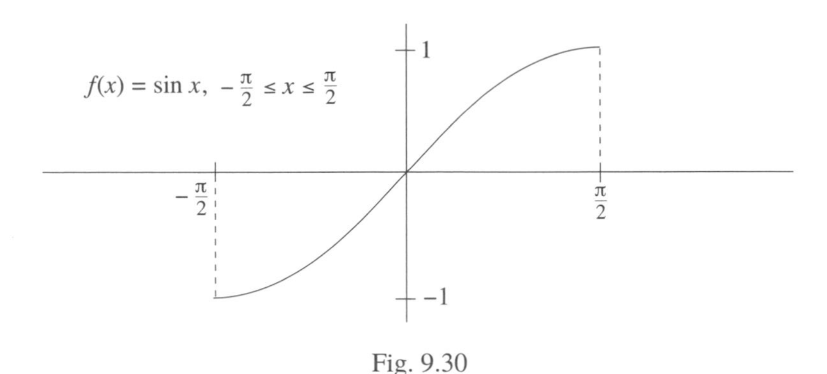

Fig. 9.30

$-\frac{\pi}{2} \le x \le \frac{\pi}{2}$. This restricted sine function has the inverse

$$f^{-1}(x) = \sin^{-1} x.$$

It is called the *inverse sine*. The reflection procedure of Section 9.8 provides its graph. It is sketched in Figure 9.31. Since $-1 \le \sin x \le 1$, we know that the domain of $f^{-1}(x) = \sin^{-1} x$ is the interval $[-1, 1]$.

For a number d with $-1 \le d \le 1$, $\sin^{-1} d$ is precisely that number c between $-\frac{\pi}{2}$ and $\frac{\pi}{2}$ such that $\sin c = d$. The question "$\sin^{-1} d = ?$" translated into words is "what angle between $-\frac{\pi}{2}$ and $\frac{\pi}{2}$ has sine

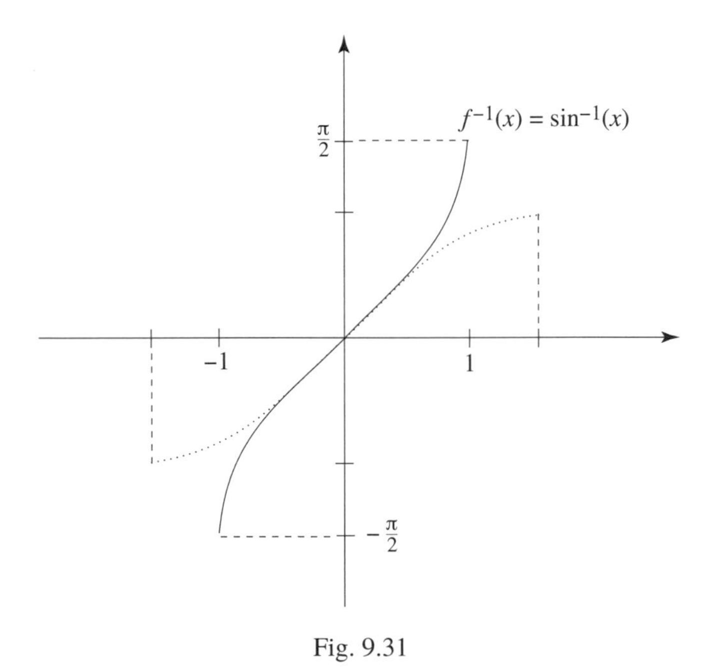

Fig. 9.31

equal to d?" For example, $\sin \frac{\pi}{4} = \frac{\sqrt{2}}{2}$, so the answer to $\sin^{-1} \frac{\sqrt{2}}{2} = ?$ is $\frac{\pi}{4}$. The function $\sin^{-1} x$ is the function behind the $\sin^{-1}$ button (or "arcsine" button) on your calculator. Use it to show that $\sin^{-1}(0.750) \approx 0.848$, and reconfirm this by checking that $\sin(0.848) \approx 0.750$. (Be sure that your calculator is in "radian" mode when you do this.)

Combining the fact that $f'(x) = \cos x$ with the formula

$$\frac{d}{dx} f^{-1}(x) = \frac{1}{f'(f^{-1}(x))}$$

shows that $\frac{d}{dx}\sin^{-1}x = \frac{1}{\cos(\sin^{-1}x)}$. This expression can be made transparent. From $\sin^2 x + \cos^2 x = 1$, we get $\cos^2 x = 1 - \sin^2 x$. But $\cos x \geq 0$ for all x with $-\frac{\pi}{2} \leq x \leq \frac{\pi}{2}$, so $\cos x = \sqrt{1-(\sin x)^2}$. Therefore $\cos(\sin^{-1}x) = \sqrt{1-(\sin(\sin^{-1}x))^2} = \sqrt{1-x^2}$, and hence

$$\boxed{\frac{d}{dx}\sin^{-1}x = \frac{1}{\sqrt{1-x^2}}}$$

Example 9.29. Find an antiderivative of $f(x) = \frac{1}{\sqrt{1-(3x)^2}}$. By the chain rule,

$$\frac{d}{dx}\sin^{-1}(3x) = \frac{1}{\sqrt{1-(3x)^2}} \cdot \frac{d}{dx}(3x) = \frac{3}{\sqrt{1-(3x)^2}}.$$

So we get three times what we need. It follows that $F(x) = \frac{1}{3}\sin^{-1}(3x)$ is an antiderivative of $f(x) = \frac{1}{\sqrt{1-(3x)^2}}$.

Example 9.30. Let's solve $\int \frac{6}{\sqrt{6^2-x^2}}\,dx$. By factoring, we get $6^2 - x^2 = 6^2(1-(\frac{x}{6})^2)$, hence $\sqrt{6^2-x^2} = 6\sqrt{1-(\frac{x}{6})^2}$, and therefore $\frac{6}{\sqrt{6^2-x^2}} = \frac{1}{\sqrt{1-(\frac{x}{6})^2}}$. The integral falls to the method of substitution. With $u = \frac{x}{6}$, $\frac{du}{dx} = \frac{1}{6}$, so that

$$\int \frac{6}{\sqrt{6^2-x^2}}\,dx = \int \frac{1}{\sqrt{1-(\frac{x}{6})^2}}\,dx = \int 6\frac{1}{\sqrt{1-u^2}}\,du = 6\sin^{-1}u + C = 6\sin^{-1}\tfrac{x}{6} + C.$$

With this result, we can clean up the "fly in the ointment" that was encountered in Example 9.13 in the verification of the fact that $\int_{-6}^{6} \frac{6}{\sqrt{6^2-x^2}}\,dx = 6\pi$. The problem was that the integrand is not defined and hence not continuous at either limit of integration. As in Example 9.15, we'll use the strategy of improper integrals. Choose c with $0 \leq c < 6$. Since there are no discontinuities over the interval $[0, c]$, we see that

$$\int_0^c \frac{6}{\sqrt{6^2-x^2}}\,dx = 6\sin^{-1}\tfrac{x}{6}\Big|_0^c = 6\sin^{-1}\tfrac{c}{6} - 0 = 6\sin^{-1}\tfrac{c}{6}.$$

Pushing c to 6 and considering Figure 9.31, we get

$$\int_0^6 \frac{6}{\sqrt{6^2-x^2}}\,dx = \lim_{c\to 6}\int_0^c \frac{6}{\sqrt{6^2-x^2}}\,dx = \lim_{c\to 6} 6\sin^{-1}\tfrac{c}{6} = 6\cdot\tfrac{\pi}{2} = 3\pi.$$

Doing the same thing for $\int_{-6}^{0} \frac{6}{\sqrt{6^2-x^2}}\,dx$ provides the answer of 6π that we had gotten before (by not quite legitimate means).

The inverse cosine is defined to be the inverse of the restriction of $f(x) = \cos x$ to $0 \leq x \leq \pi$. Because its calculus is the same as that of the inverse sine, we'll turn to $f(x) = \tan x$ next.

A look at its graph tells us that $f(x) = \tan x$ has to be restricted as well. It is customary to take the restriction $f(x) = \tan x$, $-\frac{\pi}{2} < x < \frac{\pi}{2}$. See Figure 9.32. The inverse of this increasing function is denoted by

$$f^{-1}(x) = \tan^{-1}x$$

and called the *inverse tangent.* It follows from Figure 9.32 that the domain of $f^{-1}(x) = \tan^{-1}x$ consists of all real numbers. For any number d, $\tan^{-1}d$ is the number c between $-\frac{\pi}{2}$ and $\frac{\pi}{2}$ such that $\tan c = d$. So "$\tan^{-1}d = ?$" is the same question as "what angle between $-\frac{\pi}{2}$ and $\frac{\pi}{2}$ has tangent equal to d?" For example, the fact that $\tan\frac{\pi}{4} = 1$ means that the answer to $\tan^{-1}1 = ?$ is $\frac{\pi}{4}$.

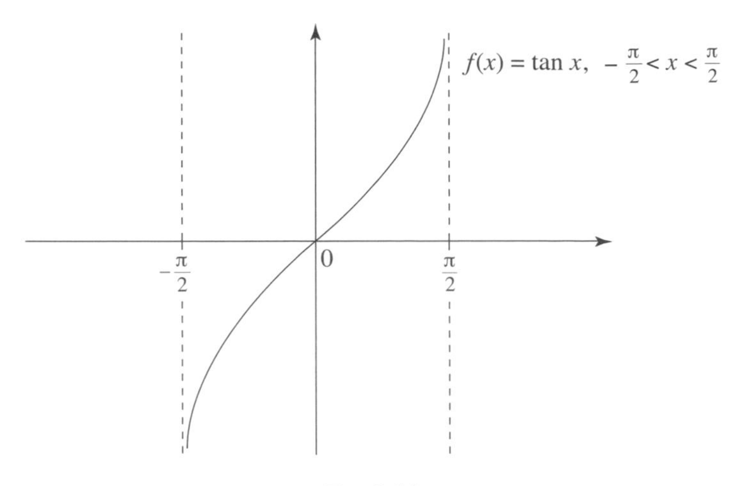

Fig. 9.32

The graph of inverse tan is obtained by the reflection procedure and is sketched in Figure 9.33. Use the $\tan^{-1}$ button (or "arctan" button) on a calculator to verify that $\tan^{-1}(0.5) \approx 0.46$, $\tan^{-1}(2) \approx 1.11$,

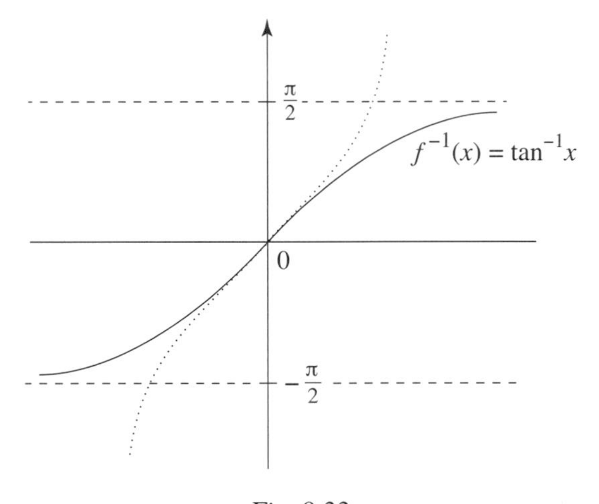

Fig. 9.33

$\tan^{-1}(10) \approx 1.47$, and $\tan^{-1}(1000) \approx 1.57$.

It was shown in Example 7.29 that the derivative of $f(x) = \tan x$ is $f'(x) = \sec^2 x$. Therefore by an application of the formula $\frac{d}{dx}f^{-1}(x) = \frac{1}{f'(f^{-1}(x))}$,

$$\frac{d}{dx}\tan^{-1}x = \frac{1}{\sec^2(\tan^{-1}(x))}.$$

Recall that $\sec^2 x = \tan^2 x + 1$. So $\sec^2(\tan^{-1}x) = (\tan(\tan^{-1}x))^2 + 1 = x^2 + 1$. It follows that

$$\boxed{\frac{d}{dx}\tan^{-1}x = \frac{1}{x^2+1}}$$

We turn next to $f(x) = \sec x$. Its graph is depicted in Figure 4.27. In order to define an inverse, we'll restrict the domain of $f(x) = \sec x$ to the intervals $0 \le x < \frac{\pi}{2}$ and $\frac{\pi}{2} < x \le \pi$. See Figure 9.34. This restricted function is not increasing throughout its domain, but it is increasing on the intervals $[0, \frac{\pi}{2})$ and $(\frac{\pi}{2}, \pi]$. More importantly, a look at its graph shows that for any value d with $|d| \geq 1$, there is but one

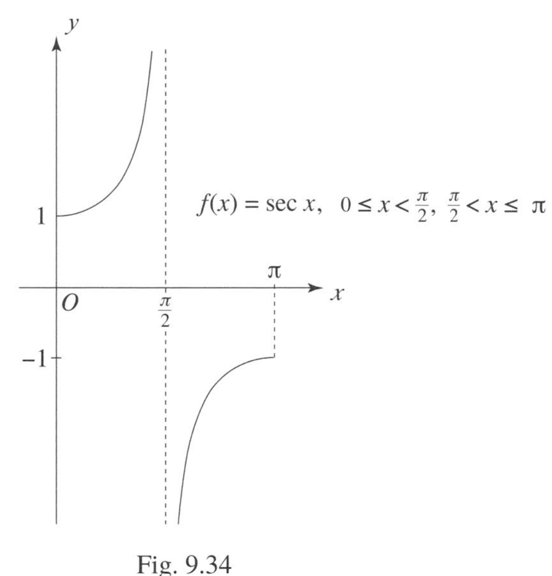

Fig. 9.34

number c such that $\sec c = d$. This means that the discussion in Section 9.8 can be applied and that this restricted secant has an inverse. It is denoted

$$f^{-1}(x) = \sec^{-1} x$$

and referred to as the *inverse secant*.

For any number d with $|d| \geq 1$, $\sec^{-1} d$ is the number c between 0 and π such that $\sec c = d$. So "$\sec^{-1} d = ?$" is the same question as "what angle between 0 and π has secant equal to d?" Because $\sec \frac{\pi}{6} = \frac{1}{\cos \frac{\pi}{6}} = \frac{2}{\sqrt{3}}$, we know that $\sec^{-1} \frac{2}{\sqrt{3}} = \frac{\pi}{6}$. By applying the reflection principle, we see that the graph

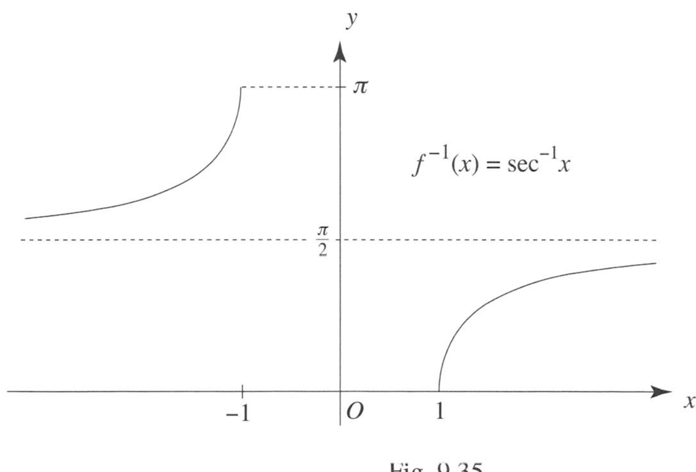

Fig. 9.35

of $f^{-1}(x) = \sec^{-1} x$ has the form depicted in Figure 9.35.

To determine the derivative of $f^{-1}(x) = \sec^{-1} x$, we'll proceed slightly differently than we did for the inverse sine and tangent. Recall from Example 7.30 that $\frac{d}{dx} \sec x = (\sec x)(\tan x)$. Now let $y = \sec^{-1} x$.

So $\sec y = x$. Differentiate both sides of the equation

$$\sec y = x$$

using the chain rule for the composite function $\sec y$ (this is *implicit differentiation*) to get $(\sec y \cdot \tan y)\frac{dy}{dx} = 1$. Therefore $\sec y \cdot \tan y \neq 0$, and

$$\frac{dy}{dx} = \frac{1}{\sec y \cdot \tan y}.$$

The identity $\tan^2 y + 1 = \sec^2 y$ tells us that $\tan y = \pm\sqrt{\sec^2 y - 1} = \pm\sqrt{(\sec(\sec^{-1} x))^2 - 1} = \pm\sqrt{x^2 - 1}$. It follows that

$$\frac{dy}{dx} = \pm\frac{1}{x\sqrt{x^2 - 1}}.$$

So $y = \sec^{-1} x$ is differentiable for all x with $|x| \geq 1$, except for $x = \pm 1$, where the graph has vertical tangents. If $x > 1$, then $0 < y = \sec^{-1} x < \frac{\pi}{2}$, so that $\tan y > 0$ by Figure 9.32. If $x < -1$, then $\frac{\pi}{2} < y = \sec^{-1} x < \pi$, so that $\tan y < 0$, again by Figure 9.32. So $\tan y = \sqrt{x^2 - 1}$ for $x > 1$ and $\tan y = -\sqrt{x^2 - 1}$ for $x < -1$. We have therefore established the formula

$$\boxed{\frac{d}{dx}\sec^{-1} x = \frac{1}{|x|\sqrt{x^2 - 1}}}$$

The inverses of the cotangent and cosecant functions can be dealt with in similar fashion. Since we will not need them in this text, we'll omit their study. Instead, we now turn to the inverses of the hyperbolic functions.

9.9.2 *Hyperbolic Inverses*

We'll concentrate on the hyperbolic sine, hyperbolic cosine, hyperbolic tangent, and hyperbolic secant functions and refer to Section 7.12 for their basic properties and their graphs.

Let's start with $f(x) = \sinh x$. Since this function is increasing (refer back to Figure 7.46), it has an inverse

$$f^{-1}(x) = \sinh^{-1} x$$

called the *inverse hyperbolic sine*. As in the earlier cases, for any real number d, $\sinh^{-1} d$ is that number c such that $\sinh c = d$. The graph of $f^{-1}(x) = \sinh^{-1} x$, as provided by the reflection procedure, is sketched

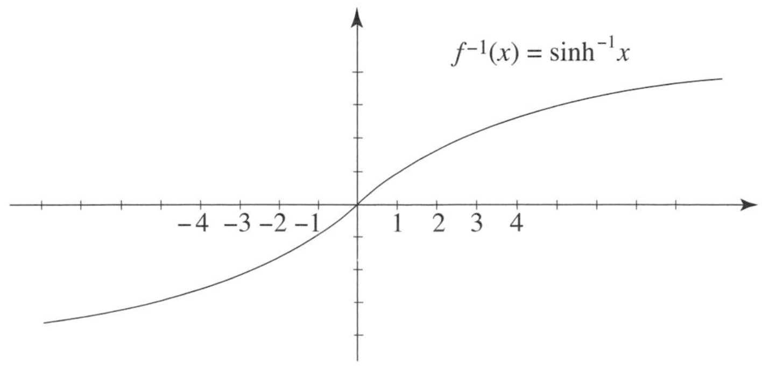

Fig. 9.36

in Figure 9.36.

We turn to the computation of the derivative of $f^{-1}(x) = \sinh^{-1} x$. The general formula for the derivative of the inverse is

$$\frac{d}{dx} f^{-1}(x) = \frac{1}{f'(f^{-1}(x))}.$$

Recall that $f'(x) = \cosh x$. Combining $\cosh^2 x - \sinh^2 x = 1$ with the fact that $\cosh x > 0$, we get $\cosh x = \sqrt{\sinh^2 x + 1}$. It follows that

$$f'(f^{-1}(x)) = \cosh(\sinh^{-1} x) = \sqrt{(\sinh(\sinh^{-1} x))^2 + 1} = \sqrt{x^2 + 1},$$

and therefore that

$$\boxed{\frac{d}{dx} \sinh^{-1} x = \frac{1}{\sqrt{x^2 + 1}}}$$

We'll turn to the function $f(x) = \cosh x$ next. Figure 7.46 tells us that its graph is increasing for $x \geq 0$. In view of the graph, we can take any $x \geq 1$ and define

$$f^{-1}(x) = \cosh^{-1} x$$

to be the number y with $y \geq 0$ such that $\cosh y = x$. Applying the reflection principle once more provides the graph of $f^{-1}(x) = \cosh^{-1} x$. It is sketched in Figure 9.37. The computation of the derivative of $\cosh^{-1} x$

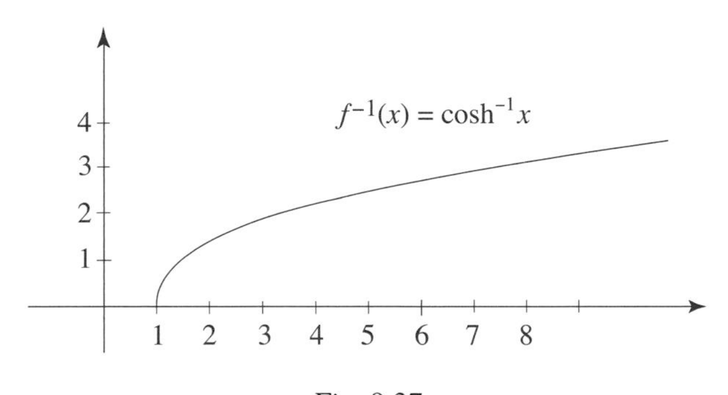

Fig. 9.37

is similar to that of $\sinh^{-1} x$. Recall that $\cosh^2 y - \sinh^2 y = 1$ for any y. So $\sinh^2 y = \cosh^2 y - 1$. For $y \geq 0$, $\sinh y \geq 0$, so that $\sinh y = \sqrt{\cosh^2 y - 1}$. Because $\frac{d}{dx} f(x) = \frac{d}{dx} \cosh x = \sinh x$, it follows that

$$f'(f^{-1}(x) = \sinh(\cosh^{-1} x) = \sqrt{(\cosh(\cosh^{-1} x))^2 - 1} = \sqrt{x^2 - 1}.$$

Therefore, by the formula for $\frac{d}{dx} f^{-1}(x)$,

$$\boxed{\frac{d}{dx} \cosh^{-1} x = \frac{1}{\sqrt{x^2 - 1}}}$$

Now to the inverse hyperbolic tangent. From Figure 7.47 we know that $f(x) = \tanh x$ is increasing for all x. Given its graph, we can define

$$f^{-1}(x) = \tanh^{-1} x$$

for any real number x with $-1 < x < 1$ to be that number y such that $\tanh y = x$. By an application of the reflection process, the graph of $f^{-1}(x) = \tanh^{-1} x$ has the form depicted in Figure 9.38. To obtain the

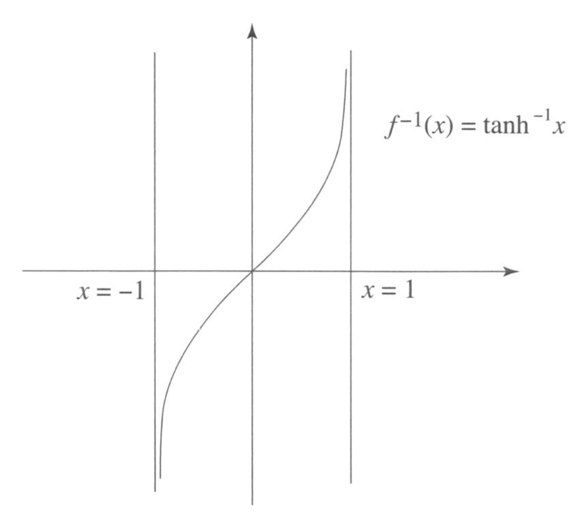

Fig. 9.38

derivative of $f^{-1}(x) = \tanh^{-1} x$, we note first that

$$f'(x) = \frac{d}{dx} \tanh x = \frac{d}{dx} \frac{\sinh x}{\cosh x} = \frac{\cosh^2 x - \sinh^2 x}{\cosh^2 x} = \frac{1}{\cosh^2 x} = \operatorname{sech}^2 x.$$

Because $1 - \tanh^2 x = \operatorname{sech}^2 x$ (to get this, simply divide $\cosh^2 x - \sinh^2 x = 1$ through by $\cosh^2 x$), it follows that

$$f'(f^{-1}(x)) = (\operatorname{sech}(\tanh^{-1}(x)))^2 = 1 - (\tanh(\tanh^{-1}(x))^2 = 1 - x^2.$$

Therefore by the formula $\frac{d}{dx} f^{-1}(x) = \frac{1}{f'(f^{-1}(x))}$,

$$\boxed{\frac{d}{dx} \tanh^{-1} x = \frac{1}{1 - x^2}}$$

The last inverse that we'll consider is that of the hyperbolic function $f(x) = \operatorname{sech} x$. Study Figure 7.46 to get a sense of the graph of $\operatorname{sech} x = \frac{1}{\cosh x}$. Observe that it is a flat bell-shaped curve, that $0 < \operatorname{sech} x \leq 1$, and that $f(x) = \operatorname{sech} x$ is decreasing for $x \geq 0$. For any x with $0 < x \leq 1$, define

$$f^{-1}(x) = \operatorname{sech}^{-1} x$$

to be that value $y \geq 0$ such that $\operatorname{sech} x = y$. The reflection principle provides the graph. See Figure 9.39.

We'll compute the derivative of $f^{-1}(x)$ by relying on the formula $\frac{d}{dx} f^{-1}(x) = \frac{1}{f'(f^{-1}(x)}$ once more. The derivative of $f(x) = \operatorname{sech} x$ is easily computed to be

$$\frac{d}{dx} \operatorname{sech} x = \frac{d}{dx}(\cosh x)^{-1} = -(\cosh x)^{-2}(\sinh x) = -(\cosh x)^{-1} \cdot \tanh x = -(\operatorname{sech} x)(\tanh x).$$

Let $y = f^{-1}(x) = \operatorname{sech}^{-1} x$. So $f'(f^{-1}(x)) = f'(y) = -\operatorname{sech} y \cdot \tanh y$. Recall the formula $\tanh^2 y = 1 - \operatorname{sech}^2 y$. Since $y \geq 0$, $\tanh y \geq 0$ from Figure 9.38, and it follows that $\tanh y = \sqrt{1 - \operatorname{sech}^2 y}$. Since $y = \operatorname{sech}^{-1} x$, we get $f'(f^{-1}(x)) = -\operatorname{sech}(\operatorname{sech}^{-1} x) \cdot \tanh(\operatorname{sech}^{-1} x) = -x \cdot \sqrt{1 - (\operatorname{sech}(\operatorname{sech}^{-1} x))^2} = -x\sqrt{1 - x^2}$.

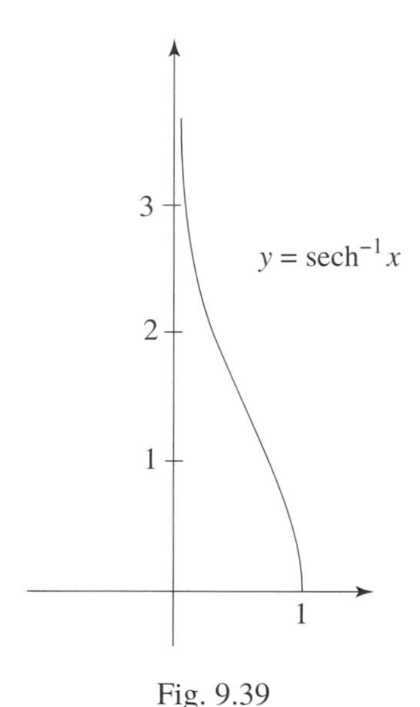

Fig. 9.39

Therefore

$$\frac{d}{dx}\operatorname{sech}^{-1}x = \frac{-1}{x\sqrt{1-x^2}}$$

Inverses can also be defined for the hyperbolic cotangent and cosecant functions, and similar analyses can be carried out for them. Since they will not be relevant later in this text, we'll omit this. One final word of caution about our notation. For example, $\cosh^{-1}x$ refers to the inverse function of the hyperbolic cosine *and not* to $(\cosh x)^{-1} = \frac{1}{\cosh x} = \operatorname{sech} x$.

All the inverse hyperbolic functions can be expressed in terms of the natural logarithm $\ln x$. This should not come as a surprise, since all the hyperbolic functions are defined by various combinations of the exponential function e^x. The facts are as follows.

1. $\sinh^{-1}x = \ln(x + \sqrt{x^2+1}) = \ln|x + \sqrt{x^2+1}|$ for all real numbers x

2. $\cosh^{-1}x = \ln(x + \sqrt{x^2-1})$ for all $x \geq 1$

3. $\tanh^{-1}x = \frac{1}{2}\ln\left(\frac{1+x}{1-x}\right)$ for $-1 < x < 1$

4. $\operatorname{sech}^{-1}x = \ln\left(\frac{1+\sqrt{1-x^2}}{x}\right)$ for $0 < x \leq 1$

The proofs of these formulas are all similar. They involve nothing more than the properties of the exponential and logarithm functions in combination with some algebra. We will prove facts 2 and 4. With regard to fact 1, notice that $x + \sqrt{x^2+1} > 0$. If $x \geq 0$ this is clear, and if $x < 0$, then $\sqrt{1+x^2} > \sqrt{x^2} = -x$, so that again $x + \sqrt{1+x^2} > 0$.

As for fact 2, let $y = \cosh^{-1}x$ to get $\cosh y = x$. Since $\cosh y = \frac{e^y+e^{-y}}{2} = x$, we obtain $e^y - 2x + e^{-y} = 0$. Multiply through by e^y to get $(e^y)^2 - 2xe^y + 1 = 0$. So e^y is a root of the polynomial

$$X^2 - 2xX + 1.$$

Therefore, by the quadratic formula, $e^y = \frac{1}{2}(2x \pm \sqrt{4x^2-4}) = x \pm \sqrt{x^2-1}$. Suppose that the $-$ option were possible. Since $y \geq 0$, this would imply that $x - \sqrt{x^2-1} \geq 1$, and hence that $x - 1 \geq \sqrt{x^2-1}$. So

$(x-1)^2 \geq x^2 - 1$, and therefore $x - 1 \geq x + 1$. Since this cannot be so, we get the + option $e^y = x + \sqrt{x^2 - 1}$. After taking ln of both sides, we get $y = \ln(x + \sqrt{x^2 - 1})$.

Fact 4 remains. Proceeding as in the case of fact 2, we'll set $y = \operatorname{sech}^{-1} x$. So $\operatorname{sech} y = x$ and hence $\frac{1}{\frac{1}{2}(e^y + e^{-y})} = x$. So $x(e^y + e^{-y}) = 2$, and we get by multiplying by e^y that $x(e^y)^2 + x = 2e^y$. Therefore $x(e^y)^2 - 2e^y + x = 0$, so that e^y is a root of

$$xX^2 - 2X + x.$$

By the quadratic formula, $e^y = \frac{2 \pm \sqrt{4 - 4x^2}}{2x} = \frac{1 \pm \sqrt{1 - x^2}}{x}$. Since $\operatorname{sech}^{-1} x = y > 0$, we know that $e^y > 1$. The possibility that $e^y = \frac{1 - \sqrt{1 - x^2}}{x}$ would imply that $\frac{1 - \sqrt{1 - x^2}}{x} > 1$, and since $x > 0$, that $1 - \sqrt{1 - x^2} > x$. So $1 - x > \sqrt{1 - x^2}$, and therefore $(1 - x)^2 > 1 - x^2 = (1 - x)(1 + x)$. But this would mean that $x \neq 1$ and $1 - x > 1 + x$. But this is not possible since $x > 0$. It follows that $e^y = \frac{1 + \sqrt{1 - x^2}}{x}$. Talking ln of both sides tells us that $y = \ln\left(\frac{1 + \sqrt{1 - x^2}}{x}\right)$.

9.10 TRIGONOMETRIC AND HYPERBOLIC SUBSTITUTIONS

We'll see next that the trigonometric and hyperbolic functions are the key to a powerful substitution strategy with which a broad class of integrals—those involving terms of the form $x^2 + a^2$, $x^2 - a^2$, and $a^2 - x^2$ among others—can be solved. We'll start with a look at the role that the trigonometric functions play.

With $x = a \sin \theta$,

$$a^2 - x^2 = a^2 - a^2 \sin^2 \theta = a^2(1 - \sin^2 \theta) = a^2 \cos^2 \theta,$$

with $x = a \tan \theta$,

$$a^2 + x^2 = a^2 + a^2 \tan^2 \theta = a^2(1 + \tan^2 \theta) = a^2 \sec^2 \theta, \text{ and}$$

with $x = a \sec \theta$,

$$x^2 - a^2 = a^2 \sec^2 \theta - a^2 = a^2(\sec^2 \theta - 1) = a^2(\tan^2 \theta).$$

The idea in each case is to use one of the formulas above to turn the term in question into a square. After such a substitution is made for x, all the other terms within the integrand need to be expressed in the variable θ. The hope is (going into a particular problem there is little else but hope) that this transformed integral can be solved. Briefly, this is the method of trigonometric substitutions. Diagrams of the type depicted in Figure 9.40 can facilitate the process.

Let's apply trig substitution to the integral

$$\int \sqrt{a^2 - x^2}\, dx.$$

As we already saw in Example 9.5, for instance, this integral is closely related to the area of the circle. The formula for the area of a circle of radius r was easily verified in Section 2.2 by filling the circle with triangles. Given that integral calculus is based on the rectangle, it should not come as a surprise that the determination of the area of a circle with integral calculus is more complicated.

To solve the integral, we'll take $a \geq 0$ and try the substitution $x = a \sin \theta$ with $-\frac{\pi}{2} \leq \theta \leq \frac{\pi}{2}$ that Figure 9.40 suggests. Since $\sin^2 \theta + \cos^2 \theta = 1$ and $\cos \theta \geq 0$ for this range of θ,

$$\sqrt{a^2 - x^2} = \sqrt{a^2 - a^2 \sin^2 \theta} = a\sqrt{1 - \sin^2 \theta} = a\sqrt{\cos^2 \theta} = a \cos \theta.$$

Because $\frac{dx}{d\theta} = a \cos \theta$,

$$\int \sqrt{a^2 - x^2}\, dx = \int a^2 \cos^2 \theta\, d\theta.$$

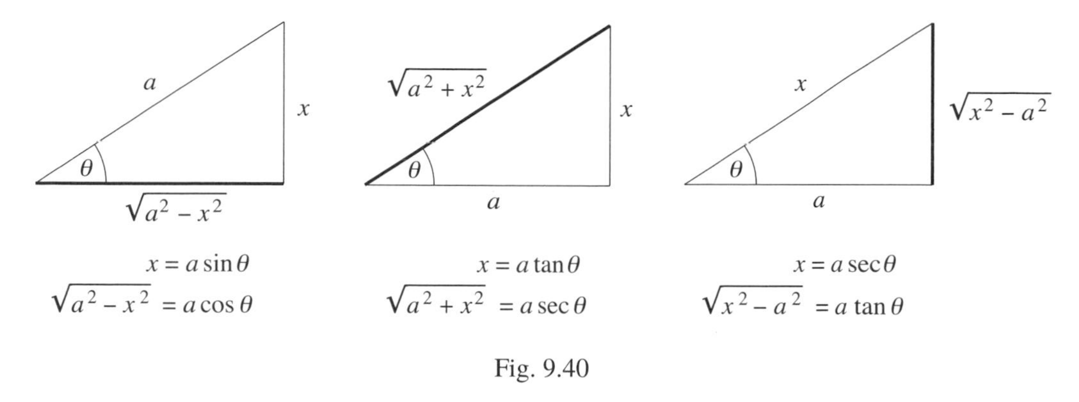

$x = a\sin\theta$ $\quad$ $x = a\tan\theta$ $\quad$ $x = a\sec\theta$

$\sqrt{a^2-x^2} = a\cos\theta$ $\quad$ $\sqrt{a^2+x^2} = a\sec\theta$ $\quad$ $\sqrt{x^2-a^2} = a\tan\theta$

Fig. 9.40

By using the trig formula $\cos^2\theta = \frac{1}{2}(1+\cos 2\theta)$ of Problem 1.22i,

$$\int a^2\cos^2\theta\,d\theta = \int \tfrac{a^2}{2}(1+\cos 2\theta)\,d\theta = \tfrac{a^2}{2}[\theta + \tfrac{1}{2}\sin 2\theta] + C.$$

We'll rewrite this solution in a more transparent form. Since $\sin 2\theta = 2\sin\theta\cos\theta$ (by Problem 1.26i), we get

$$\tfrac{a^2}{2}[\theta + \tfrac{1}{2}\sin 2\theta] = \tfrac{a^2}{2}[\theta + \sin\theta\cos\theta].$$

Since $\sin\theta = \frac{x}{a}$ with $-\frac{\pi}{2} \leq \theta \leq \frac{\pi}{2}$, we know that $\theta = \sin^{-1}\frac{x}{a}$. In combination with $\cos\theta = \frac{\sqrt{a^2-x^2}}{a}$, this implies that

$$\tfrac{a^2}{2}[\theta + \sin\theta\cos\theta] = \tfrac{a^2}{2}[\sin^{-1}\tfrac{x}{a} + \tfrac{x}{a}\tfrac{\sqrt{a^2-x^2}}{a}] = \tfrac{x}{2}\sqrt{a^2-x^2} + \tfrac{a^2}{2}\sin^{-1}\tfrac{x}{a}.$$

Putting the last several equalities together, we get

$$\int \sqrt{a^2-x^2}\,dx = \tfrac{1}{2}[x\sqrt{a^2-x^2} + a^2\sin^{-1}\tfrac{x}{a}] + C.$$

Example 9.31. Let's return to the circle $x^2+y^2=r^2$ of Example 9.5. The computation

$$\int_{-r}^{r}\sqrt{r^2-x^2}\,dx = [\tfrac{x}{2}\sqrt{r^2-x^2} + \tfrac{r^2}{2}\sin^{-1}\tfrac{x}{r}]\Big|_{-r}^{r} = \tfrac{r^2}{2}(\sin^{-1}1 - \sin^{-1}(-1)) = \tfrac{r^2}{2}(\tfrac{\pi}{2} - (-\tfrac{\pi}{2})) = \tfrac{1}{2}\pi r^2$$

confirms that the area of the upper half of this circle is $\frac{1}{2}\pi$.

In reference to the formula for $\int\sqrt{a^2-x^2}\,dx$ just derived, it would have been sufficient to deal with the case $a = 1$. This is the consequence of the method of integration by substitution discussed earlier. Given the equality

$$\boxed{\int \sqrt{1-u^2}\,du = \tfrac{1}{2}[u\sqrt{1-u^2} + \sin^{-1}u] + C}$$

we can deal with $\int\sqrt{a^2-x^2}\,dx$ as follows. Write $\sqrt{a^2-x^2} = a\sqrt{1-(\frac{x}{a})^2}$ and let $u = \frac{x}{a}$. Then $du = \frac{1}{a}\,dx$, and therefore

$$\int\sqrt{a^2-x^2}\,dx = \int a\sqrt{1-(\tfrac{x}{a})^2}\,dx = \int a^2\sqrt{1-u^2}\,du = a^2\tfrac{x}{2a}\sqrt{1-(\tfrac{x}{a})^2} + \tfrac{a^2}{2}\sin^{-1}\tfrac{x}{a} + C.$$

After a little algebra with the last term, we get the earlier formula for $\int\sqrt{a^2-x^2}\,dx$.

Let's use another trig substitution to solve the indefinite integral $\int \sqrt{1+x^2}\,dx$. It will be relevant in connection with the computation of the lengths of curves. Refer to Figure 9.40, and let $x = \tan\theta$ with θ satisfying $-\frac{\pi}{2} < \theta < \frac{\pi}{2}$. We get $1 + x^2 = 1 + \tan^2\theta = \sec^2\theta$, and since $\sec\theta > 0$ for $-\frac{\pi}{2} \le \theta < \frac{\pi}{2}$, that $\sqrt{1+x^2} = \sec\theta$. Differentiating $x = \tan\theta$ gives us $\frac{dx}{d\theta} = \sec^2\theta$. After substituting,

$$\int \sqrt{1+x^2}\,dx = \int \sec\theta \cdot \sec^2\theta\,d\theta = \int \sec^3\theta\,d\theta.$$

To solve $\int \sec^3\theta\,d\theta$, we'll start with integration by parts. Let $u = \sec\theta$ and $dv = \sec^2\theta\,d\theta$. Since $\frac{du}{d\theta} = \sec\theta \cdot \tan\theta$ and $v = \tan\theta$ is an antiderivative of $\frac{dv}{dx} = \sec^2\theta$, we get

$$\int \sec^3\theta\,d\theta = \sec\theta \cdot \tan\theta - \int \sec\theta \cdot \tan^2\theta\,d\theta$$

by the integration by parts formula. Because $\tan^2\theta = \sec^2\theta - 1$, we now see that

$$\begin{aligned}\int \sec^3\theta\,d\theta &= \sec\theta \cdot \tan\theta - \int \sec\theta \cdot \sec^2\theta\,d\theta + \int \sec\theta\,d\theta \\ &= \sec\theta \cdot \tan\theta - \int \sec^3\theta\,d\theta + \int \sec\theta\,d\theta,\end{aligned}$$

and therefore

$$2\int \sec^3\theta\,d\theta = \sec\theta \cdot \tan\theta + \int \sec\theta\,d\theta.$$

The integral $\int \sec x\,dx$ is solved by a trick. Begin by observing that

$$\sec x = \sec x\,\frac{\sec x + \tan x}{\sec x + \tan x} = \frac{\sec^2 x + \sec x \tan x}{\sec x + \tan x}.$$

By Examples 7.29 and 7.30, $\frac{d}{dx}\sec x = (\sec x)(\tan x)$ and $\frac{d}{dx}\tan x = \sec^2 x$. Let $g(x) = \sec x + \tan x$, and notice that $\sec x = \frac{g'(x)}{g(x)}$. By applying the formula $\frac{d}{dx}\ln|g(x)| = \frac{g'(x)}{g(x)}$ from Section 7.11,

$$\frac{d}{dx}\ln|\sec x + \tan x| = \frac{d}{dx}\ln|g(x)| = \frac{g'(x)}{g(x)} = \sec x.$$

Therefore

$$\int \sec x\,dx = \ln|\sec x + \tan x| + C.$$

It has now been verified that

$$\int \sec^3\theta\,d\theta = \tfrac{1}{2}[\,\sec\theta \cdot \tan\theta + \ln|\sec\theta + \tan\theta|] + C.$$

Since $\tan\theta = x$ and $\sec\theta = \sqrt{1+x^2}$, we finally get

$$\int \sqrt{1+x^2}\,dx = \tfrac{1}{2}[x\sqrt{1+x^2} + \ln|x + \sqrt{1+x^2}|] + C.$$

The fact that $\sqrt{1+x^2} > \sqrt{x^2} = \pm x$ tells us that $x + \sqrt{1+x^2} > 0$. By changing the variable to u and using fact 1 from Section 9.9.2, we get

$$\boxed{\int \sqrt{1+u^2}\,du = \tfrac{1}{2}[u\sqrt{1+u^2} + \sinh^{-1}u] + C}$$

Given the fact that the hyperbolic functions behave completely analogously to the trig functions, it should not be surprising that hyperbolic substitutions can also be used to solve integrals of the type that we are considering. The equality $1 + \sinh^2 z = \cosh^2 z$ for any real number z suggests that the substitution $x = \sinh z$ might also be used to establish that

$$\int \sqrt{1+x^2}\,dx = \tfrac{1}{2}[x\sqrt{1+x^2} + \sinh^{-1}x] + C.$$

With $x = \sinh z$, we get $\cosh^2 z = 1+\sinh^2 z = 1+x^2$. Since $\cosh z \geq 0$, it follows that $\cosh z = \sqrt{1+x^2}$. Since $\frac{dx}{dz} = \cosh z$, $dx = \cosh z\,dz$, and therefore $\sqrt{1+x^2}\,dx = \cosh^2 z\,dz$. Therefore

$$\int \sqrt{1+x^2}\,dx = \int \cosh^2 z\,dz.$$

Because $\frac{d}{dz}(\sinh z \cdot \cosh z + z) = (\cosh^2 z + \sinh^2 z) + 1 = 2\cosh^2 z$, the function $\frac{1}{2}(\sinh z \cdot \cosh z + z)$ is an antiderivative of $\cosh^2 z$. So

$$\int \cosh^2 z\,dz = \tfrac{1}{2}(\sinh z \cdot \cosh z + z) + C.$$

After substituting $\sinh z = x$, $\cosh z = \sqrt{1+x^2}$, and $z = \sinh^{-1}x$, we can conclude that

$$\int \sqrt{1+x^2}\,dx = \tfrac{1}{2}[x\sqrt{1+x^2} + \sinh^{-1}x] + C.$$

We have seen that the hyperbolic substitution $x = \sinh z$ also solves

$$\int \sqrt{1+x^2}\,dx$$

and that—in this particular case—it does so much more efficiently than the trig substitution $x = \tan\theta$.

9.11 SOME INTEGRAL FORMULAS

This section lists some integral formulas that were already verified and develops a few more. In order to "set things up" for integration by substitution, we'll write them in the variable u.

In Section 6.2, we saw how Newton verified that $\frac{d}{dx}x^r = rx^{r-1}$ for a positive rational number r (with $x \geq 0$ when needed in the definition of x^r). Example 7.44 used logarithmic differentiation to extend this formula to any real number r (under the assumption that $x > 0$). This, along with the formula $\frac{d}{dx}\ln|x| = \frac{1}{x}$, implies after a change of variables to u, that

$$(1) \qquad \int u^r\,du = \tfrac{1}{r+1}u^{r+1} + C \text{ for } r \neq -1 \text{ and } \int u^{-1}\,du = \int \tfrac{1}{u}\,du = \ln|u| + C.$$

Since u^r makes sense—for a general constant r—only for $u > 0$, there is the implied restriction $u > 0$ on the variable u in Formula (1). Similarly in Formula (9), $\frac{1}{\sqrt{1-u^2}}$ is defined only for $-1 < u < 1$. Since they are determined by the functions involved, we'll no longer mention such restrictions explicitly.

We'll turn to formulas that follow from our earlier studies of the trig, exponential, log, and hyperbolic functions. From results in Section 7.7, we know that

$$(2) \qquad \int \sin u\,du = -\cos u + C \quad\text{and}\quad \int \cos u\,du = \sin u + C.$$

From Section 7.10 and Example 9.26, we know that

$$\int e^u\,du = e^u + C \quad \text{and} \quad \int \ln u\,du = u\ln u - u + C, \tag{3}$$

and from Section 7.12,

$$\int \sinh u\,du = \cosh u + C \quad \text{and} \quad \int \cosh u\,du = \sinh u + C. \tag{4}$$

Since the indefinite integrals for the sine and cosine follow the formulas for their derivatives, we turn to the tangent next. Consider $f(u) = \ln|\sec u|$. By properties of the natural log, $f(u) = \ln|\cos u|^{-1} = -\ln|\cos u|$. By applying the chain rule, $f'(u) = -\frac{1}{\cos u}\cdot(-\sin u) = \tan u$. Therefore

$$\int \tan u\,du = \ln|\sec u| + C. \tag{5}$$

In Section 9.10, it was shown that

$$\int \sec u\,du = \ln|\sec u + \tan u| + C. \tag{6}$$

The indefinite integrals for the hyperbolic sine and cosine follow directly from their derivatives. The formula

$$\int \tanh u\,du = \ln\cosh u + C \tag{7}$$

does as well. For the indefinite integral of the hyperbolic secant, use the same trick that was used for the secant (but take $\operatorname{sech} u - \tanh u$ in place of $\sec u + \tan u$) to get

$$\int \operatorname{sech} u\,du = \ln|\operatorname{sech} u - \tanh u| + C. \tag{8}$$

The integral formulas

$$\int \frac{1}{\sqrt{1-u^2}}\,du = \sin^{-1}u + C, \tag{9}$$

$$\int \frac{1}{u^2+1}\,du = \tan^{-1}u + C, \tag{10}$$

as well as

$$\int \frac{1}{|u|\sqrt{u^2-1}}\,du = \sec^{-1}u + C \tag{11}$$

follow directly from the derivative results of Section 9.9.1.

The next three are consequences of the formulas for the derivatives of the inverse hyperbolic functions in Section 9.9.2.

$$\int \frac{1}{\sqrt{u^2+1}}\,du = \sinh^{-1}u + C \tag{12}$$

$$\int \frac{1}{\sqrt{u^2-1}}\,du = \cosh^{-1}u + C, \text{ for } u > 1 \tag{13}$$

(14) $$\int \frac{1}{1-u^2}\,dx = \tanh^{-1}u + C$$

We saw in Section 9.9.2 that $\frac{d}{dx}\,\text{sech}^{-1}x = \frac{-1}{x\sqrt{1-x^2}}$ for any x with $0 < x < 1$. If $-1 < x < 0$, then by the chain rule, $\frac{d}{dx}\,\text{sech}^{-1}(-x) = \frac{(-1)(-1)}{-x\sqrt{1-x^2}} = \frac{-1}{x\sqrt{1-x^2}}$. So $\frac{d}{dx}\,\text{sech}^{-1}|x| = \frac{-1}{x\sqrt{1-x^2}}$ for any nonzero x with $-1 < x < 1$. Therefore

(15) $$\int \frac{1}{u\sqrt{1-u^2}}\,du = -\text{sech}^{-1}|u| + C.$$

The two formulas that follow were established in Section 9.10.

(16) $$\int \sqrt{1-u^2}\,du = \tfrac{u}{2}\sqrt{1-u^2} + \tfrac{1}{2}\sin^{-1}u + C$$

(17) $$\int \sqrt{1+u^2}\,du = \tfrac{1}{2}[u\sqrt{1+u^2} + \ln(u+\sqrt{1+u^2})] + C$$

Three routine applications of integration by parts provide the indefinite integrals

(18) $$\int \sin^{-1}u\,du = u\sin^{-1}u + \sqrt{1-u^2} + C,$$

(19) $$\int \tan^{-1}u\,du = u\tan^{-1}u - \tfrac{1}{2}\ln(u^2+1) + C,$$

and

(20) $$\int \sec^{-1}u\,du = u\sec^{-1}u - \cosh^{-1}u + C \text{ for } u > 1.$$

Three more applications of integration by parts result in the formulas

(21) $$\int \sinh^{-1}u\,du = u\sinh^{-1}u - \sqrt{u^2+1} + C,$$

(22) $$\int \tanh^{-1}u\,du = u\tanh^{-1}u + \tfrac{1}{2}\ln(1-u^2) + C,$$

and

(23) $$\int \text{sech}^{-1}u\,du = u\,\text{sech}^{-1}u + \sin^{-1}u + C.$$

We have seen that finding antiderivatives of basic functions can be challenging, but the list of formulas above and their derivation suggest that such integrals can be determined if there is enough ingenuity and persistence. This raises a question. Let $y = f(x)$ be any *elementary function*, meaning that it is a function put together from a finite number of algebraic functions, trigonometric functions, exponential functions (including hyperbolic functions), logarithms, and all their inverses by a combination of the operations of addition, subtraction, multiplication, division, taking roots, and composition of functions. Is it the case that the integral of any elementary function is also an elementary function? Consider the functions

$$y = \sqrt{1-x^4},\ y = \sin x^2,\ y = \sqrt{1-\sin^2 x},\ y = \frac{\sin x}{x},\ y = \frac{1}{\ln x},\ y = \ln(\ln x),\ y = \frac{e^x}{x}, \text{ and } y = e^{-x^2},$$

for example. Each is an elementary function, and each has an antiderivative that the corresponding area function provides (as decribed in Section 9.6). However, in *none* of these eight cases is the integral elementary.

There are useful websites that compute integrals of functions explicitly. The site

http://www.integral-calculator.com/#

is a good example. What answers does it provide for the integrals of the eight functions just listed? The answer for the first asserts that the "antiderivative or integral could not be found," and adds "many functions don't have an elementary antiderivative." In the other seven cases, the answers involve "error" or "gamma" functions that are not elementary. These cases are analogous to the status of $\int \frac{1}{x}\,dx$ before the logarithm function $\ln x$ was introduced to provide an antiderivative of $\frac{1}{x}$.

Given the fundamental theorem of calculus, the conclusion that might be drawn from what was just said is that it is impossible to evaluate a definite integral in general. That this is not so is primarily due to the fact that there are approximation methods with which any definite integral can be computed to any required degree of accuracy. The trapezoidal rule and Simpson's rule are two such methods. Both make use of the connection between the definite integral and area.

9.12 THE TRAPEZOIDAL AND SIMPSON RULES

Again let $y = f(x)$ be a function that is continuous on a closed interval $[a, b]$. Let's recall the discussion of Section 9.5 and some of its consequences. Let $\mathcal{P}$ be any partition

$$a = x_0 < x_1 < x_2 < x_3 < \cdots < x_i < x_{i+1} < \cdots < x_{n-1} < x_n = b$$

of the interval $[a, b]$, and let

$$\Delta x_0 = x_1 - x_0, \Delta x_1 = x_2 - x_1, \Delta x_2 = x_3 - x_2, \ldots, \Delta x_i = x_{i+1} - x_i, \ldots, \Delta x_{n-1} = x_n - x_{n-1}$$

If the norm $||\mathcal{P}||$ of the partition (this is the largest of the Δx_i) is small, then for any set of numbers $c_0, c_1, \ldots, c_{n-1}$ with c_i in the interval $[x_i, x_{i+1}]$ for $i = 0, 1, \ldots n-1$, the sum

$$\sum_{i=0}^{n-1} f(c_i)\Delta x_i = f(c_0)\Delta x_0 + f(c_1)\Delta x_1 + \cdots + f(x_i)\Delta x_i + \cdots + f(c_{n-2})\Delta x_{n-2} + f(c_{n-1})\Delta x_{n-1}$$

provides an approximation for $\int_a^b f(x)\,dx$. Different choices for the points c_i lead to different approximations.

In Section 9.1, we took the left endpoint $c_i = x_i$ of $[x_i, x_{i+1}]$ for $i = 0, 1, \ldots n-1$ and worked with the approximation

$$\int_a^b f(x)\,dx \approx \sum_{i=0}^{n-1} f(x_i)\Delta x_i.$$

This approximation is known as the *left endpoint approximation*. Taking the right endpoints x_{i+1} of the intervals $[x_i, x_{i+1}]$ instead gives the *right endpoint approximation*

$$\int_a^b f(x)\,dx \approx \sum_{i=0}^{n-1} f(x_{i+1})\Delta x_i.$$

With c_i the midpoint $c_i = \frac{x_i + x_{i+1}}{2}$ of $[x_i, x_{i+1}]$, we get the *midpoint approximation*.

We will now restrict our attention to the case where all the lengths Δx_i are equal. We'll let this equal length be Δx. The average

$$\tfrac{1}{2}\Big[\sum_{i=0}^{n-1} f(x_i)\Delta x + \sum_{i=0}^{n-1} f(x_{i+1})\Delta x\Big] = \sum_{i=0}^{n-1} \tfrac{1}{2}[f(x_i) + f(x_{i+1}]\Delta x$$

of the left and right endpoint approximations is the nth *trapezoidal approximation*. It is denoted by T_n. Why trapezoidal? Consider Figure 9.41 and a typical region shaded in gray. Notice that its shape is a trapezoid

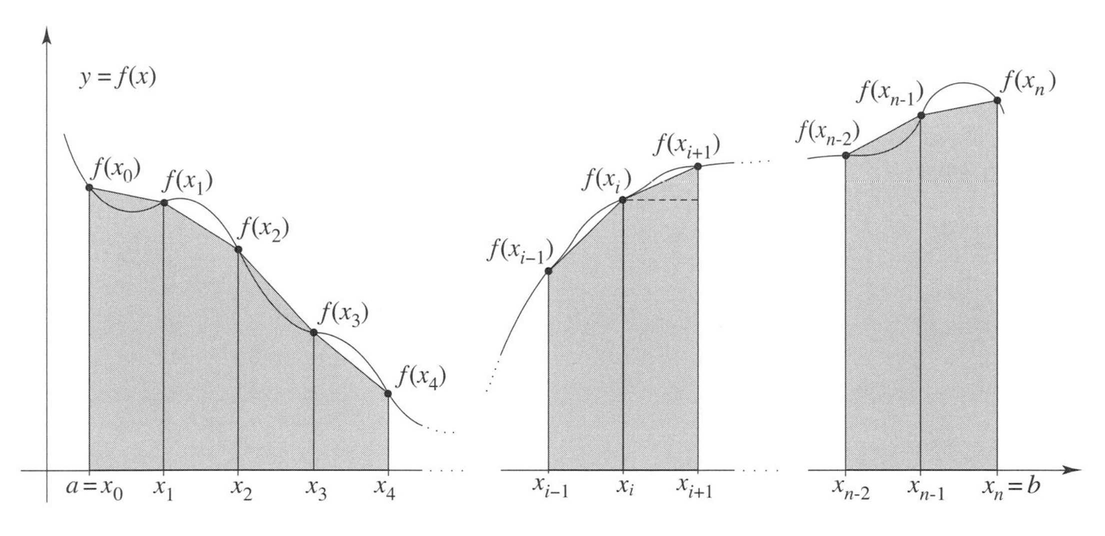

Fig. 9.41

determined by the graph. The figure tells us that each trapezoid consists of a rectangle and a triangle. The trapezoid over the interval $[x_i, x_{i+1}]$ has area $f(x_i)\cdot\Delta x$ and a triangle of area $\frac{1}{2}(f(x_{i+1}) - f(x_i))\cdot\Delta x$. It follows that the area of the trapezoid is

$$f(x_i)\Delta x + \tfrac{1}{2}(f(x_{i+1}) - f(x_i))\Delta x = \tfrac{1}{2}(f(x_i) + f(x_{i+1}))\Delta x.$$

By explicitly writing out the sum for T_n, we get the *trapezoidal rule*

$$\int_a^b f(x)\,dx \approx T_n = \tfrac{\Delta x}{2}[f(x_0) + 2f(x_1) + 2f(x_2) + \cdots + 2f(x_{n-2}) + 2f(x_{n-1}) + f(x_n)]$$

where $\Delta x = \frac{b-a}{n}$ and $x_i = a + i\Delta x$ for $i = 0, 1, \ldots, n$. Figure 9.41 illustrates the trapezoidal rule for a function with graph above the x-axis by showing that the sum of the areas of the trapezoidal regions highlighted in gray approximates the area under the graph. Note, however, that the trapezoidal rule is valid for any continuous function.

The error E_n^T of the nth trapezoidal approximation is defined to be

$$E_n^T = \Big|\int_a^b f(x)\,dx - T_n\Big|.$$

Advanced texts establish a bound on this error. Suppose that the second derivative $f''(x)$ of $y = f(x)$ is continuous on $[a, b]$. Then

$$E_n^T \le \tfrac{n}{12}(\Delta x)^3[\max|f''(x)|,\ a \le x \le b].$$

Another approximation of $\int_a^b f(x)\,dx$ relies on the fact that any three distinct points in the plane lie on the graph of a function of the form $y = Ax^2 + Bx + C$, where A, B, and C are constants. If the points do not all lie on a line, then $A \neq 0$ and the graph is a parabola. Suppose that the number n of subintervals that the partition $\mathcal{P}$ divides $[a, b]$ into is even. Group the n subintervals into $\frac{n}{2}$ consecutive pairs. Each pair determines three points on the graph of the function, and these determine a parabolic arc or a line segment. The curve obtained by connecting all of these arcs (or line segments) approximates the graph of the function. In turn, this graph determines an approximation of $\int_a^b f(x)\,dx$. For example, Figure 9.42 illustrates how the areas of the parabolic regions together approximate the area under the graph of the

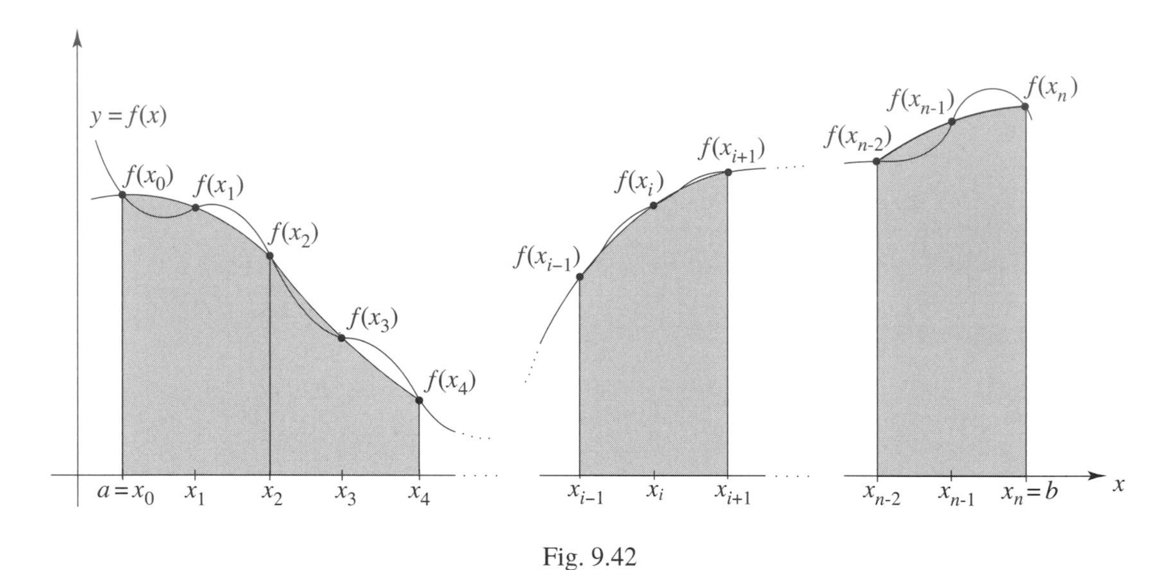

Fig. 9.42

function. What was just described can be quantified, and the result is Simpson's rule. We'll carry out the details by referring to Figure 9.43. To start, let $\Delta x = h$, move the typical parabolic section (the one over the interval $[x_{i-1}, x_{i+1}]$ in Figure 9.43a), and position it along the y-axis so that $y_{i-1} = f(x_{i-1})$, $y_i = f(x_i)$, and $y_{i+1} = f(x_{i+1})$ as shown in Figure 9.43b. Let's take the parabolic arc of Figure 9.43b to be the graph of the function $y = Ax^2 + Bx + C$, and find the area $\int_{-h}^{h} (Ax^2 + Bx + C)\,dx$ under it. This is the easy

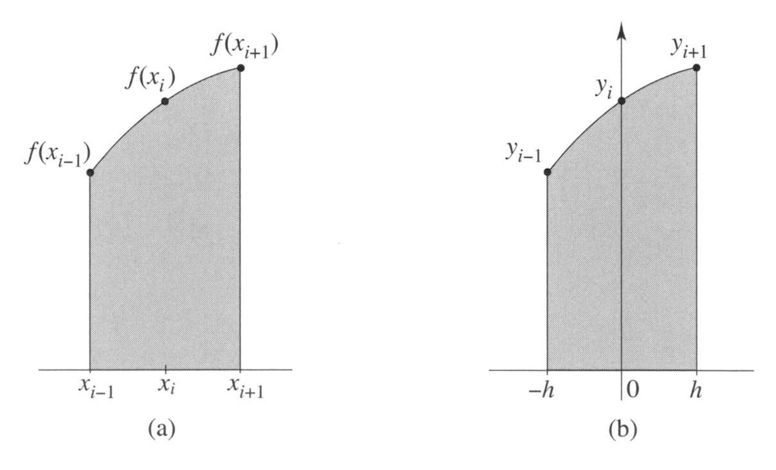

Fig. 9.43

computation

$$\int_{-h}^{h}(Ax^2+Bx+C)\,dx = \tfrac{A}{3}x^3+\tfrac{B}{2}x^2+Cx\Big|_{-h}^{h} = \tfrac{A}{3}h^3+\tfrac{B}{2}h^2+Ch-\left(\tfrac{A}{3}(-h)^3+\tfrac{B}{2}(-h)^2+C(-h)\right)$$
$$= \tfrac{2A}{3}h^3+2Ch.$$

Since the parabola (or line) passes through $(-h, y_{i-1})$, $(0, y_i)$, and (h, y_{i+1}),

$$\begin{aligned} y_{i-1} &= A(-h)^2+B(-h)+C = Ah^2-Bh+C \\ y_i &= C \\ y_{i+1} &= Ah^2+Bh+C. \end{aligned}$$

So $y_{i-1}+4y_i+y_{i+1} = (Ah^2-Bh+C)+4C+(Ah^2+Bh+C) = 2Ah^2+6C$, and therefore

$$\int_{-h}^{h}(Ax^2+Bx+C)\,dx = \tfrac{2A}{3}h^3+2Ch = \tfrac{h}{3}(2Ah^2+6C) = \tfrac{\Delta x}{3}(y_{i-1}+4y_i+y_{i+1}).$$

Since this is the area of Figure 9.43b, it is also the area of Figure 9.43a. Doing this for the odd numbers $i = 1, 3, \ldots, n-1$ and adding areas provides the Simpson approximation S_n. This is *Simpson's rule*

$$\int_a^b f(x)\,dx \approx S_n = \tfrac{\Delta x}{3}[f(x_0)+4f(x_1)+2f(x_2)+4f(x_3)+\cdots+2f(x_{n-2})+4f(x_{n-1})+f(x_n)]$$

Simpson's rule is not only valid in the case of Figure 9.42, where $f(x) \geq 0$, but extends to any continuous function. The error E_n^S of the nth Simpson approximation is defined to be

$$E_n^S = \left|\int_a^b f(x)\,dx - S_n\right|.$$

In advanced texts we find that if the fourth derivative $f^{(4)}(x)$ of $y = f(x)$ is continuous on $[a, b]$, then

$$E_n^S \leq \tfrac{n}{180}(\Delta x)^5[\max|f^{(4)}(x)|,\ a \leq x \leq b].$$

9.13 ONE LOOP OF THE SINE CURVE

We'll illustrate the trapezoidal and Simpson rules by studying various aspects of one loop of the sine curve, namely, the graph of the function $f(x) = \sin x$ over the interval $[0, \pi]$.

Let's start with the area under this graph (and above the x-axis). This area is equal to

$$\int_0^{\pi}\sin x\,dx = -\cos x\Big|_0^{\pi} = 1-(-1) = 2.$$

Since this integral has been evaluated "on the nose," there is no need to use approximation methods. But we'll do so regardless to gain a sense of what is involved. We'll start by exploring the trapezoidal rule in the context of this integral using the calculator that the website

http://www.emathhelp.net/calculators/calculus-2/trapezoidal-rule-calculator/

provides. (Before you start the computation, visit the Show Instructions page of the site. If this link is not operative, an Internet search for "trapezoidal rule calculator" will lead to viable alternatives.)

Example 9.32. Consider $f(x) = \sin x$ with $0 \le x \le \pi$. Let's take the trapezoidal approximation with $n = 10$ to estimate the value of $\int_0^\pi \sin x\,dx$. The partition in question is

$$0 = x_0 < x_1 = \tfrac{\pi}{10} < x_2 = \tfrac{2\pi}{10} < \cdots < x_9 = \tfrac{9\pi}{10} < x_{10} = \pi,$$

so that

$$\begin{aligned}
f(x_0) &= \sin x_0 &&= \sin 0 = 0\\
2f(x_1) &= 2\sin x_1 &&= 2\sin(\tfrac{\pi}{10}) \approx 0.618033988749895\\
2f(x_2) &= 2\sin x_2 &&= 2\sin(\tfrac{\pi}{5}) \approx 1.17557050458495\\
2f(x_3) &= 2\sin x_3 &&= 2\sin(\tfrac{3\pi}{10}) \approx 1.61803398874989\\
2f(x_4) &= 2\sin x_4 &&= 2\sin(\tfrac{2\pi}{5}) \approx 1.90211303259031\\
2f(x_5) &= 2\sin x_5 &&= 2\sin(\tfrac{\pi}{2}) = 2\\
2f(x_6) &= 2\sin x_6 &&= 2\sin(\tfrac{3\pi}{5}) \approx 1.90211303259031\\
2f(x_7) &= 2\sin x_7 &&= 2\sin(\tfrac{7\pi}{10}) \approx 1.61803398874989\\
2f(x_8) &= 2\sin x_8 &&= 2\sin(\tfrac{4\pi}{5}) \approx 1.17557050458495\\
2f(x_9) &= 2\sin x_9 &&= 2\sin(\tfrac{9\pi}{10}) \approx 0.618033988749895\\
f(x_{10}) &= \sin x_{10} &&= \sin \pi = 0.
\end{aligned}$$

Adding these numbers and multiplying the sum by the required $\frac{\Delta x}{2} = \frac{1}{2} \cdot \frac{\pi}{10} = \frac{\pi}{20}$ gives the result

$$\int_0^\pi \sin x\,dx \approx T_{10} \approx 1.98352353750946.$$

The actual error E_{10}^T is approximately equal to $|2 - 1.98352353750946|$ but less than 0.01648. Since $f(x) = \sin x$, $f''(x) = -\sin x$ and $|\max f''(x)| = 1$. Therefore the bound on the error term E_{10}^T is

$$E_{10}^T \le \tfrac{10}{12}\left(\tfrac{\pi}{10}\right)^3 \cdot 1 \approx 0.02584.$$

We'll now apply Simpson's rule to the integral $\int_0^\pi \sin x\,dx$ by using the calculator on the website

http://www.emathhelp.net/calculators/calculus-2/simpsons-rule-calculator/.

(Before you launch the computation, visit the Show Instructions page of the site. If the link is not accessible, an Internet search for "Simpson rule calculator" will provide a viable alternative.)

Example 9.33. Let's return to $f(x) = \sin x$ over $0 \le x \le \pi$ and use the Simpson approximation S_{10} to estimate $\int_0^\pi \sin x\,dx$. The partition is

$$0 = x_0 < x_1 = \tfrac{\pi}{10} < x_2 = \tfrac{2\pi}{10} < \cdots < x_9 = \tfrac{9\pi}{10} < x_{10} = \pi.$$

Since

$$\begin{aligned}
f(x_0) &= \sin x_0 &&= \sin 0 = 0\\
4f(x_1) &= 4\sin x_1 &&= 4\sin(\tfrac{\pi}{10}) \approx 1.23606797749979\\
2f(x_2) &= 2\sin x_2 &&= 2\sin(\tfrac{\pi}{5}) \approx 1.17557050458495\\
4f(x_3) &= 4\sin x_3 &&= 4\sin(\tfrac{3\pi}{10}) \approx 3.23606797749979\\
2f(x_4) &= 2\sin x_4 &&= 2\sin(\tfrac{2\pi}{5}) \approx 1.90211303259031
\end{aligned}$$

$$\begin{aligned}
4f(x_5) &= 4\sin x_5 = 4\sin(\tfrac{\pi}{2}) = 4 \\
2f(x_6) &= 2\sin x_6 = 2\sin(\tfrac{3\pi}{5}) \approx 1.90211303259031 \\
4f(x_7) &= 4\sin x_7 = 4\sin(\tfrac{7\pi}{10}) \approx 3.23606797749979 \\
2f(x_8) &= 2\sin x_8 = 2\sin(\tfrac{4\pi}{5}) \approx 1.17557050458495 \\
4f(x_9) &= 4\sin x_9 = 4\sin(\tfrac{9\pi}{0}) \approx 1.23606797749979 \\
f(x_{10}) &= \sin x_{10} = \sin\pi = 0,
\end{aligned}$$

we get by adding these numbers and multiplying the sum by the coefficient $\frac{\Delta x}{3} = \frac{1}{3} \cdot \frac{\pi}{10} = \frac{\pi}{20}$ that

$$\int_0^\pi \sin x\, dx \approx S_{10} \approx 2.000109517315.$$

The actual error E_{10}^S is approximately equal to $|2 - 2.000109517315|$ but less than 0.00011. Since $f(x) = \sin x$, $f^{(4)}(x) = \sin x$ and $|\max f''(x)| = 1$. Therefore the bound on the error E_{10}^S is

$$E_{10}^S \leq \tfrac{10}{180}\left(\tfrac{\pi}{10}\right)^5 \cdot 1 \approx 0.00017.$$

We turn next to the volume determined by revolving the region under the graph of $f(x) = \sin x$ and over the interval $0 \leq x \leq \pi$ once about the x-axis. By Formula (V_1) of Section 9.2, this volume is equal to

$$\pi \int_0^\pi \sin^2 x\, dx.$$

The trig substitution $\sin^2 x = \frac{1}{2}(1 - \cos 2x)$ that Problem 1.23i provides solves this integral as

$$\pi \int_0^\pi \sin^2 x\, dx = \pi \int_0^\pi \tfrac{1}{2}(1 - \cos 2x)\, dx = \tfrac{\pi}{2}\left[x - \tfrac{1}{2}\sin 2x\right]\Big|_0^\pi = \tfrac{\pi^2}{2} \approx 4.9348.$$

As in the case of the computation of the area under one loop, we do not need the approximations that the trapezoidal and Simpson rules provide.

The matter is quite different with regard to the analysis of the length of one loop of the sine curve. We know from Section 9.3 that the length of the graph of $f(x) = \sin x$ over $0 \leq x \leq \pi$ is equal to

$$\int_0^\pi \sqrt{1 + f'(x)^2}\, dx = \int_0^\pi \sqrt{1 + \cos^2 x}\, dx.$$

No effort to solve this definite integral on the nose via the fundamental theorem of calculus can bear fruit, because the integrand $\sqrt{1 + \cos^2 x}$ of this "elliptic integral" is not an elementary function. Problems 9.82 and 9.83 tell us why the integral is called "elliptic." The calculator available at the website

http://www.integral-calculator.com/#

responds to the problem of finding an antiderivative of $\sqrt{1 + \cos^2 x}$ with "antiderivative or integral could not be found."

Example 9.34. Let's estimate the integral

$$\int_0^\pi \sqrt{1 + \cos^2 x}\, dx$$

that represents the length of one loop of the sine curve. Using the trapezoidal and Simpson rules calculators already listed, we get with $n = 10, 20$, and 100 that

$$T_{10} = 3.82019778749287,\ T_{20} = 3.82019778902771, \text{ and } T_{100} = 3.82019778902771,$$

as well as

$$S_{10} = 3.82018762368766,\ S_{20} = 3.82019778953933,\ \text{and}\ S_{100} = 3.82019778902771.$$

It follows that 3.820197789 is a close estimate of the length of one loop of the sine curve.

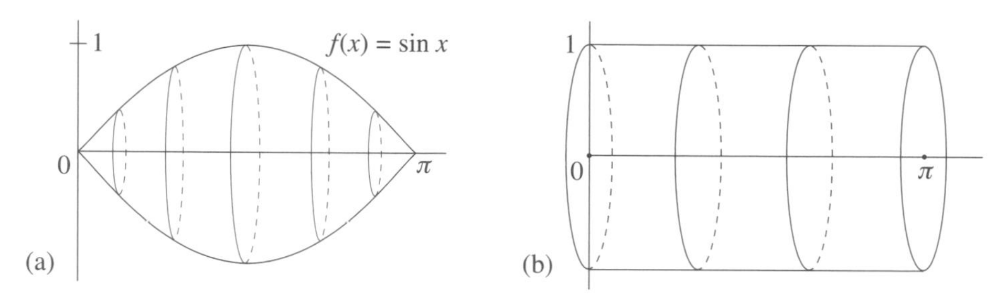

Fig. 9.44

We'll close this discussion by computing the area of the surface obtained by revolving one loop of the sine curve around the x-axis. See Figure 9.44a. Since $f(x) = \sin x$ with $0 \le x \le \pi$ and $f'(x) = \cos x$, we know from Section 9.4 that this surface area is given by the integral

$$2\pi \int_0^{\pi} f(x)\sqrt{1 + f'(x)^2}\, dx = 2\pi \int_0^{\pi} \sin x\sqrt{1 + \cos^2 x}\, dx.$$

Let $u = \cos x$. So $\frac{du}{dx} = -\sin x$. Since the limits $x = 0$ and $x = \pi$ become $u = 1$ and $u = -1$, respectively, under the transformation from x to u, we get that

$$\int_0^{\pi} \sin x\sqrt{1 + \cos^2 x}\, dx = -\int_1^{-1} \sqrt{1 + u^2}\, du = \int_{-1}^{1} \sqrt{1 + u^2}\, du.$$

Formula (17) of Section 9.11 informs us that this integral is equal to $\frac{1}{2}[u\sqrt{1 + u^2} + \ln(u + \sqrt{1 + u^2})]\Big|_{-1}^{1}$. So

$$\begin{aligned} 2\pi \int_0^{\pi} \sin x\sqrt{1 + \cos^2 x}\, dx &= 2\pi \cdot \tfrac{1}{2}[\sqrt{2} + \ln(1 + \sqrt{2}) - (-\sqrt{2} + \ln(-1 + \sqrt{2}))] \\ &= \pi[2\sqrt{2} + \ln(\tfrac{\sqrt{2}+1}{\sqrt{2}-1})] \approx 14.42360. \end{aligned}$$

By way of comparison, the tenth trapezoidal and the tenth Simpson approximations are

$$T_{10} = 14.2768263796265 \ \text{and}\ S_{10} = 14.4260450433179,$$

respectively.

Example 9.35. By way of comparison, show that a cylinder with base radius 1 and and height π has surface area $2\pi \cdot \pi \approx 19.73880$ (the two circular pieces at the ends are excluded). Refer to Figure 9.47b.

In the examples that have been considered, Simpson's rule often gives a better approximation than the trapezoidal rule for a given (even) n. Example 9.34 is an exception. There, the trapezoidal rule closes in on the correct value more quickly.

9.14 PROBLEMS AND PROJECTS

9A. Integrals and Areas

9.1. Determine antiderivatives of the following functions.

i. $f(x) = 2x^3$

ii. $f(x) = 5x^{\frac{1}{3}}$

iii. $f(x) = 3x^5 + \frac{1}{4}x^{\frac{2}{7}}$

iv. $f(x) = 6x^4 - \frac{3}{8}x^{\frac{5}{3}}$

9.2. Integrate the functions $f(x) = 1 - 3x^3 + 2x^{-\frac{1}{2}}$, $g(x) = -\frac{1}{3}x^{-2} + 8x^{\frac{1}{3}}$, and $h(x) = -4 + 3x^{-2} + 7x^{\frac{1}{2}}$.

9.3. Use the fundamental theorem of calculus to evaluate the definite integrals

$$\int_0^3 x^2\,dx, \quad \int_{-8}^{-2} \frac{1}{x^2}\,dx, \quad \text{and} \quad \int_3^{12} \sqrt{x}\,dx.$$

Sketch the areas that these integrals represent.

9.4. Sketch the graph of the function $f(x) = \sqrt{x} + 2$. Compute the area under the graph and over the x-axis from $x = 0$ to $x = 9$.

9.5. Consider the function $f(x) = x^3 + 1$ for $0 \le x \le 4$. Sketch its graph, and compute the area under the graph and over the x-axis.

9.6. Express the area between the parabola $y = -3x^2 + 2x + 1$ and the x-axis as a definite integral. Then compute the area.

9.7. Compute the area between the parabola $y = -x^2 + 9x - 6$ and the line $y = 2$.

9.8. Consider the parabola $y = x^2$, and cut it with the line $y - 5x - 8 = 0$ as shown in Figure 9.45.

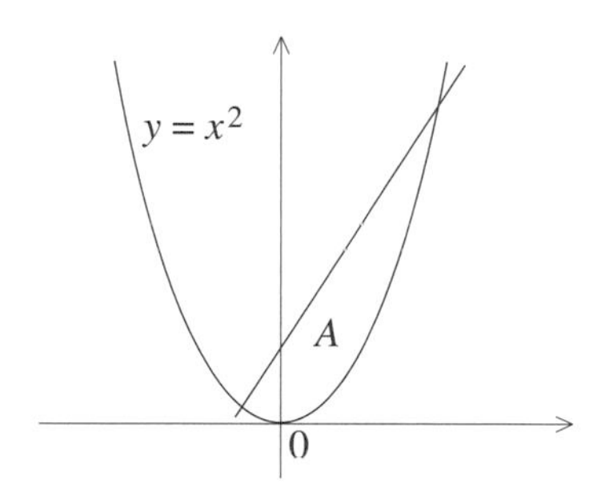

Fig. 9.45

Compute the resulting area A.

9.9. The equation of the circle with radius 2 and centered at the origin is $x^2 + y^2 = 4$. Use the geometry of the situation to evaluate the definite integral

$$\int_0^2 \sqrt{4 - x^2}\,dx$$

after drawing a picture of the area that it represents.

9.10. Evaluate the definite integral $\int_0^5 \frac{5}{2}\sqrt{5^2 - x^2}\,dx$ by making use of the area of an appropriate circle. What area does the integral represent?

9.11. Find the area under one loop of the cosine curve by considering $f(x) = \cos x, -\frac{\pi}{2} \leq x \leq \frac{\pi}{2}$.

9.12. Compute the area under the graphs of the following.

i. $f(x) = \frac{1}{x}$ over the interval $[2, 8]$

ii. $f(x) = e^x$ over the interval $[-1, 4]$

9.13. Show that $\int_{\ln 2}^{\ln 5} e^x\,dx = 3$.

9.14. Show that $\int_3^7 \frac{1}{x}\,dx = \ln 7 - \ln 3 = \ln \frac{7}{3} \approx 0.847$.

9B. About Sums and Definite Integrals. After studying Section 9.1, have a go at the following problems.

9.15. Consider the function $g(x) = 4 - x^2$ with $0 \leq x \leq 1$.

i. Take the partition

$$0 = x_0 < x_1 = 0.3 < x_2 = 0.5 < x_3 = 0.7 < x_4 = 1$$

and compute $\sum_{i=0}^{3} g(x_i)\Delta x_i$ with two-decimal-place accuracy. Draw the graph of the function as well as the rectangles that the terms of the sum determine.

ii. Take the finer partition

$$0 = x_0 < x_1 = 0.2 < x_2 = 0.4 < x_3 = 0.5 < x_4 = 0.7 < x_5 = 0.9 < x_6 = 1$$

and compute of the sum $\sum_{i=0}^{5} g(x_i)\Delta x_i$ with accuracy to two decimal places. Again draw the graph of the function as well as the rectangles that the terms of the sum determine.

iii. Why are the sums of the areas only rough approximations of $\int_0^1 (4 - x^2)\,dx$? Which of the two approximations is better, and why? Compute the integral to find the precise value of the area under the graph and over $0 \leq x \leq 1$.

9.16. Consider the function $f(x) = \frac{1}{x}$ with $\frac{1}{3} \leq x \leq 3$.

i. Take the partition

$$\tfrac{1}{3} = x_0 < x_1 = \tfrac{1}{2} < x_2 = 1 < x_3 = 2 < x_4 = 2 < x_5 = 3,$$

and compute $\sum_{i=0}^{4} f(x_i)\Delta x_i$ with accuracy to two decimal places.

ii. Add the points $\frac{2}{3}$ and $\frac{3}{2}$ to the partition, and compute the sum in this case (again with accuracy to two decimal places).

iii. Compute $\int_{\frac{1}{3}}^{3} \frac{1}{x}\,dx$. Notice that the two sums computed above are only rough estimates of this integral.

9.17. Take the function $f(x) = \frac{1}{x}$ over the interval from 2 to 4. Select the points

$$2 \leq 2.3 \leq 2.5 \leq 2.9 \leq 3.4 \leq 3.6 \leq 4,$$

and compute the sum of the products $f(x) \cdot dx$ that the selection of points determines. This sum is an approximation of $\int_2^4 \frac{1}{x}\, dx$. Compute the precise value of this integral.

9.18. Consider the function $g(x) = \sqrt{x}$ over the interval from 0 to 2. Select the points

$$0 \leq \tfrac{1}{9} \leq \tfrac{2}{9} \leq \tfrac{4}{9} \leq \tfrac{7}{9} \leq \tfrac{11}{9} \leq \tfrac{16}{9} \leq 2,$$

and compute the sum of all of the products $g(x) \cdot dx$ that this selection determines. This sum is a rough estimate of $\int_0^2 \sqrt{x}\, dx$. What is the precise value of this integral?

9.19. Let $f(x) = \sqrt{4 - x^2}$. Use a formula from Section 9.10 to evaluate the integrals

$$\int_0^2 \sqrt{4 - x^2}\, dx \quad \text{and} \quad \int_{-\sqrt{2}}^{\sqrt{2}} \sqrt{4 - x^2}\, dx.$$

Evaluate them again by interpreting them as parts of the area of a circle.

9.20. Study the sum $4(0)^3 \cdot \frac{1}{10{,}000} + 4(\frac{1}{10{,}000})^3 \frac{1}{10{,}000} + 4(\frac{2}{10{,}000})^3 \frac{1}{10{,}000} + \cdots + 4(5 + \frac{9{,}999}{10{,}000})^3 \frac{1}{10{,}000}$ and the pattern that the first three terms along with the last set. What are the fourth and next to last terms of this sum? For what n, $a = x_0, x_1, \ldots, x_{n-1}, x_n = b$, and $y = f(x)$ is the sum of the form $\sum_{i=0}^{n-1} f(x_i)\Delta x_i$? Use the fundamental theorem of calculus to provide a tight estimate for the sum.

9.21. The sum $\sqrt{5} \cdot \frac{1}{10{,}000} + \sqrt{5 + \frac{1}{10{,}000}} \cdot \frac{1}{10{,}000} + \sqrt{5 + \frac{2}{10{,}000}} \cdot \frac{1}{10{,}000} + \cdots + \sqrt{5 + \frac{9{,}999}{10{,}000}} \cdot \frac{1}{10{,}000}$ exhibits a pattern that is determined by the first three and last terms. For what n, $a = x_0, x_1, \ldots, x_{n-1}, x_n = b$, and $y = f(x)$ is this sum of the form $\sum_{i=0}^{n-1} f(x_i)\Delta x_i$? Explain why the sum is closely approximated by $\frac{2}{3}(\sqrt{10}^{\,3} - \sqrt{5}^{\,3}) \approx 13.63$.

9C. A Project with Sums. Study the principle of mathematical induction and then turn to the next two problems.

9.22. Restrict the function $f(x) = x^2$ to an interval $[0, b]$ with $b \geq 0$. Let n be a positive integer, and consider the partition

$$0 = x_0 < x_1 = \tfrac{b}{n} < x_2 = \tfrac{2b}{n} < x_3 = \tfrac{3b}{n} < \cdots < x_i = \tfrac{ib}{n} < \cdots < x_{n-1} = \tfrac{(n-1)b}{n} < x_n = \tfrac{nb}{n} = b.$$

i. Show that $\sum_{i=0}^{n-1} f(x_i)\Delta x_i = \sum_{i=0}^{n-1} (\frac{ib}{n})^2 \frac{b}{n} = (\frac{b^3}{n^3}) \sum_{i=0}^{n-1} i^2$.

ii. Use mathematical induction to prove the formula $1 + 2^2 + 3^2 + \ldots + (n-1)^2 = \frac{(n-1)n(2n-1)}{6}$.

iii. Use the formula to show that

$$\sum_{i=0}^{n-1} f(x_i)\Delta x_i = \tfrac{b^3}{3}(1 - \tfrac{3}{2n} + \tfrac{1}{2n^2}).$$

iv. Push $n \to \infty$, and conclude that $\int_0^b x^2\,dx = \frac{b^3}{3}$.

9.23. Restrict the function $f(x) = x^3$ to an interval $[0, b]$ with $b \geq 0$. Let n be a positive integer, and consider the partition

$$0 = x_0 < x_1 = \tfrac{b}{n} < x_2 = \tfrac{2b}{n} < x_3 = \tfrac{3b}{n} < \cdots < x_i = \tfrac{ib}{n} < \cdots < x_{n-1} = \tfrac{(n-1)b}{n} < x_n = \tfrac{nb}{n} = b.$$

i. Show that $\sum_{i=0}^{n-1} f(x_i)\Delta x_i = \sum_{i=0}^{n-1} (\frac{ib}{n})^3 \frac{b}{n} = (\frac{b^4}{n^4}) \sum_{i=0}^{n-1} i^3$.

ii. Use mathematical induction to prove the formula $1^3+2^3+3^3+\cdots+(n-2)^3+(n-1)^3 = \frac{1}{4}(n-1)^2n^2$.

iii. Use the formula to show that

$$\sum_{i=0}^{n-1} f(x_i)\Delta x_i = \tfrac{b^4}{4}(1 - \tfrac{2}{n} + \tfrac{1}{n^2}).$$

iv. Push $n \to \infty$, and conclude that $\int_0^b x^3\,dx = \frac{b^4}{4}$.

9D. Definite Integrals as Areas, Volumes, and Lengths of Curves

9.24. Sketch the graph of the function $f(x) = \sqrt{x}$. Let $Q = (c, \sqrt{c})$ be any point on the graph, and let $P = (c, 0)$. Let A be the area of the region under the graph of f and over the segment from the origin O to P. Let B be the area of the triangle determined by the tangent to the graph at Q, the segment PQ, and the x-axis. Show that $A = \frac{2}{3}B$ no matter where Q is taken.

9.25. Compute the volume of the solid obtained by rotating the graph of $f(x) = e^x$, $-2 \leq x \leq 1$, one revolution about the x-axis.

9.26. Evaluate $\int_0^{\frac{\pi}{4}} \tan x\,dx$. Sketch the area that this integral represents.

9.27. The definite integral $\int_1^5 \sqrt{1+4x^2}\,dx$ is both

i. the area under the graph of the function $f(x) =$ ____________ over the interval ____ $\leq x \leq$ ____, and

ii. the length of the graph of the function $f(x) =$ __________ from the point (,) to the point (,).

9.28. The definite integral $\int_0^3 \sqrt{1+x}\,dx$ is

i. the area under the graph of $f(x) =$ ____________ from $x =$ ____ to $x =$ ____, as well as

ii. the volume obtained by rotating a region under the graph of $g(x) =$ ____________ one revolution about the x-axis, and also

iii. the length of a piece of the graph of $h(x) =$ ____________ .

9.29. Consider the upper half of the circle of radius r and center $(r, 0)$. Use it to show that the volume of a sphere of radius r is $\frac{4}{3}\pi r^3$.

9.30. Use Figure 9.46 to show that the volume of a cone of height h and base of radius r is $\frac{\pi}{3}r^2h$.

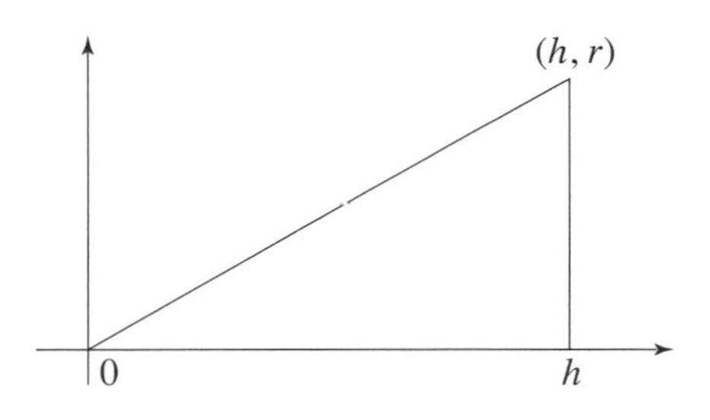

Fig. 9.46

9.31. Show that the volume obtained by revolving the graph of $f(x) = \sqrt{x}$, $0 \le x \le 4$, once about the x-axis, is equal to 8π.

9.32. Express as a definite integral the volume obtained by rotating the graph of the function $y = \sin x$, $0 \le x \le \frac{\pi}{2}$, one revolution around the x-axis. Then do the same thing for the function $y = \cos x$, $0 \le x \le \frac{\pi}{2}$. Show that the sum of the two integrals is equal to $\frac{\pi^2}{2}$. Use this conclusion to show that the two volumes are each equal to $\frac{\pi^2}{4}$.

9.33. Show that the length of the arc of the parabola $y = x^2$ from the point $(2, 4)$ to the point $(5, 25)$ is equal to the area under the graph of the function $f(x) = \sqrt{1 + 4x^2}$ from $x = 2$ to $x = 5$.

9.34. Consider the circle $x^2 + y^2 = 4$. Refer to Figure 9.47, and show that the x-coordinate of the

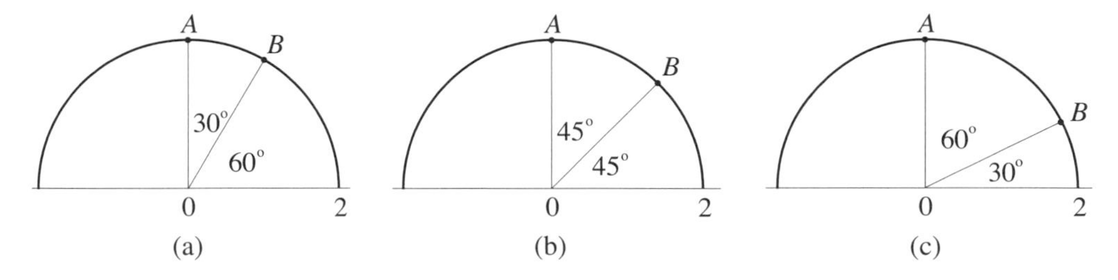

Fig. 9.47

point B is 1 in (a), $\sqrt{2}$ in (b), and $\sqrt{3}$ in (c). Show that the derivative of $f(x) = \sqrt{4 - x^2}$ is $f'(x) = \frac{-x}{\sqrt{4-x^2}}$. Deduce from this and the diagrams that

$$\int_0^1 \frac{1}{\sqrt{4 - x^2}}\, dx = \frac{\pi}{6}, \quad \int_0^{\sqrt{2}} \frac{1}{\sqrt{4 - x^2}}\, dx = \frac{\pi}{4}, \text{ and } \int_0^{\sqrt{3}} \frac{1}{\sqrt{4 - x^2}}\, dx = \frac{\pi}{3}.$$

9E. More Volumes and Surface Areas

9.35. Figure 9.48 depicts a solid that has two of its sides lie on parallel planes. The y-axis of the figure is perpendicular to the two planes, and the planes cut the axis at $y = c$ and $y = d$, as shown. For any y between c and d, we let $A(y)$ be the cross-sectional area of the solid that a plane perpendicular to the axis at y determines. We'll assume that $A(y)$ is a continuous function on the interval $c \le y \le d$. Let $\mathcal{P}$ be a partition $c = y_0 < y_1 < \cdots y_{n-1} < y_n = d$ of the interval $[c, d]$, and let $\Delta y_i = y_{i+1} - y_i$. At each y_i, cut the solid perpendicularly to the y-axis, and let V_i be the volume of the slice that the cuts at y_i and y_{i+1} determine. This slice is Δy_i thick, and Figure 9.48 informs us that it is approximated by a cylinder with base area $A(y_i)$ and height Δy_i. Therefore $V_i \approx A(y_i)\Delta y_i$, and the sum $\sum_{i=0}^{n-1} A(y_i)\Delta y_i$

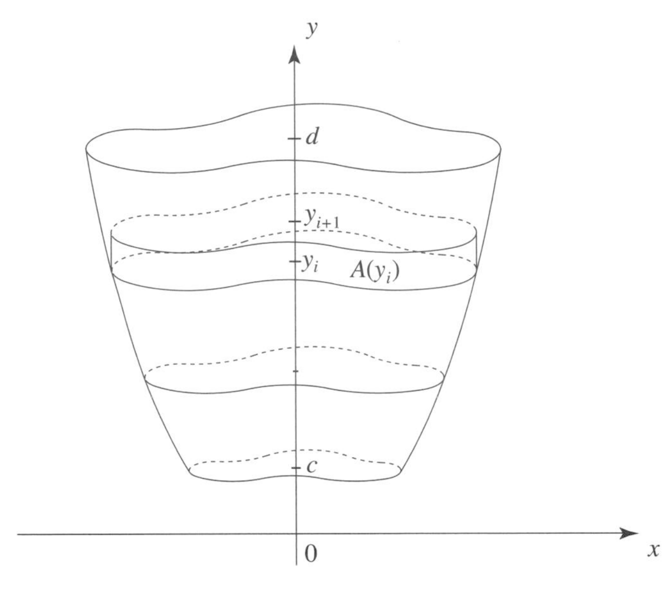

Fig. 9.48

approximates the volume V of the solid. Repeating this construction with partitions $\mathcal{P}$ of smaller and smaller norm improves this approximation for V and (as described in Section 9.1) in the limit $||\mathcal{P}|| \to 0$,

$$V = \int_c^d A(y)\,dy.$$

9.36. Consider a sphere of radius r. Suppose that its center is at the origin of an xy-coordinate system. The intersection of the sphere with the xy-plane is the circle $x^2 + y^2 = r^2$. Check that for any y between $-r$ and r, the cross section of the sphere at y perpendicular to the y-axis is a circle of radius $x = \sqrt{r^2 - y^2}$. Use the formula of Problem 9.35 to show that the volume of the sphere is $V = \frac{4}{3}\pi r^3$.

9.37. Consider the area S of the surface obtained by revolving the graph of $f(x) = \sqrt{x}$, $a \le x \le b$, one

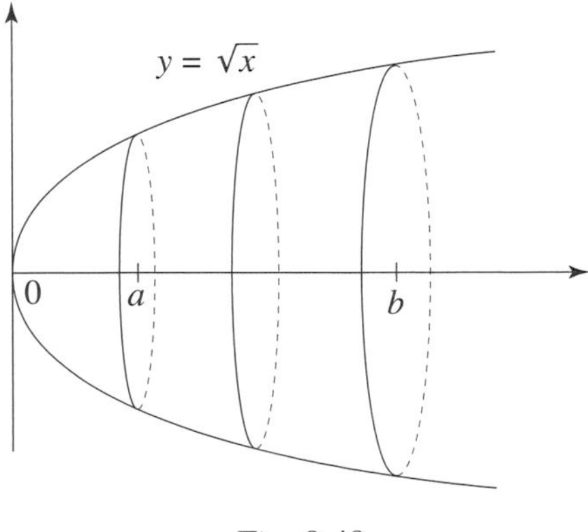

Fig. 9.49

revolution around the x-axis. See Figure 9.49. Show that S is given by the integral

$$S = 2\pi \int_a^b \sqrt{x + \tfrac{1}{4}}\,dx.$$

By evaluating it, show that $S = \frac{4}{3}\pi[(b+\frac{1}{4})^{\frac{3}{2}} - (a+\frac{1}{4})^{\frac{3}{2}}]$.

9.38. Consider a sphere of fixed radius R. Place an xy-coordinate system so that the center of the sphere is at the point $(R, 0)$. Figure 9.50 shows the intersection of the sphere with the xy-plane. This intersection is the circle $(x-R)^2 + y^2 = R^2$. Let (x, y) be a point on the upper half of this circle, and use it to inscribe a cone into the sphere as illustrated in the figure.

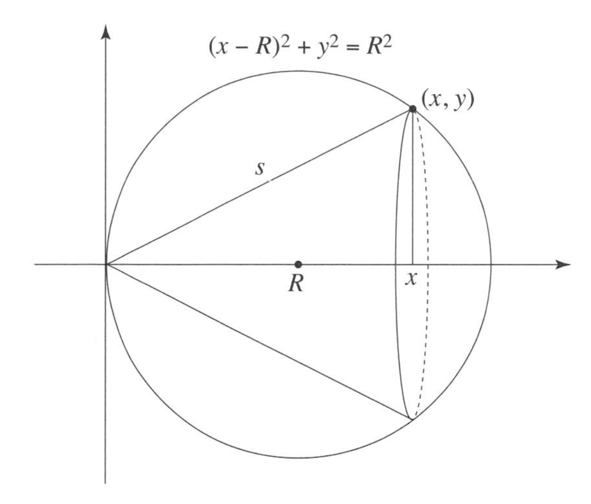

Fig. 9.50

i. Use the conclusion of Problem 9.30 to show that the volume of the cone is $V(x) = \frac{\pi}{3}x(2Rx - x^2)$.

ii. Use a formula from Section 9.4 to show that the surface area of the cone (without its circular base) is $S(x) = \pi\sqrt{2Rx(2Rx - x^2)}$.

iii. Show that both $V(x)$ and $S(x)$ have their maximum values when $x = \frac{4}{3}R$. Show that the maximum volume V and surface area S that such a cone can have are $V = \frac{32\pi}{81}R^3$ and $S = \frac{8\pi}{3\sqrt{3}}R^2$.

9F. Area Functions

9.39. The function $f(x) = 1 + x + x^2$ is defined over the interval $(-\infty, \infty)$. Determine its area function $A(x)$ for $a = 3$.

9.40. Find the area function $A(x)$ of $\cos x, (-\infty, \infty)$ for $a = \frac{\pi}{2}$.

9.41. Find the area function $A(x)$ of $\sinh x, (-\infty, \infty)$ for $a = 1$.

9.42. Consider the functions $f(x) = x^2 + 3x$, $F(y) = y^2 + 3y$, and $\phi(z) = z^2 + 3z$. Each is defined for all real numbers. Are the three functions identical? Why or why not?

9.43. Show that $\int_0^4 \sqrt{x}\,dx = \frac{16}{3}$. What are $\int_0^4 \sqrt{t}\,dt$, $\int_0^4 \sqrt{u}\,du$, and $\int_0^4 \sqrt{z}\,dz$ equal to?

9.44. The function $f(x) = \sqrt{x}$ is given. The definite integral $\int_0^t \sqrt{x}\,dx$ is equal to a number that depends only on _____ . So the rule _____ $\longrightarrow \int_0^t \sqrt{x}\,dx$ defines a function. The values that this function assigns to 1, 4, and 100 are _____ , _____ , and _____ .

9.45. Define the functions F, G, and H by $F(x) = \int_2^x \frac{1}{t^2}\,dt$, $G(x) = \int_2^x \frac{1}{z^2}\,dz$, and $H(t) = \int_2^t \frac{1}{x^2}\,dx$. Evaluate each of these functions at 4. Evaluate each of them at any number $c \geq 2$. Explain why these functions are the same by interpreting them in terms of areas. Define the function $K(x) = \int_1^x \frac{1}{t^2}\,dt$. Show that F and K differ by a constant C and interpret C as an area.

9.46. Consider the function defined by the rule $x \longrightarrow \int_3^x (t^2+5)\,dt$. Evaluate it at $x = 5$. What is the problem with defining this function by $x \longrightarrow \int_3^x (x^2+5)\,dx$?

9.47. Consider the functions $F(x) = \int_0^x (t^2+3t)\,dt$, $G(x) = \int_2^x \frac{1}{t^2}\,dt$, and $K(x) = \int_5^x \sqrt{t^3+5}\,dt$. Determine the derivatives $F'(x)$, $G'(x)$, and $K'(x)$.

9.48. Use a definite integral to define a function that has $f(x) = \frac{1}{x}$ as its derivative.

9.49. Use a definite integral to define an antiderivative of $g(x) = \sqrt{2x^2+4}$.

9.50. Consider the upper half of the circle of radius 1 with center at the origin. For any x with $-1 \le x \le 1$, let $G(x)$ be the area under this curve (and above the x-axis) from -1 to x. Express $G(x)$ in terms of a definite integral. What is the derivative of $G(x)$ equal to?

9G. Integration by Substitution

9.51. Solve $\int (4x-5)^{\frac{1}{2}}\,dx$.

9.52. With regard to the solution of the indefinite integral $\int 10x(1-5x^2)^{\frac{2}{3}}\,dx$, which of the choices $u = 10x$, $u = 5x^2$, or $u = 1-5x^2$ is most likely to succeed? Experiment and solve the integral.

9.53. Solve $\int x\cos x^2\,dx$ and $\int \sin^3 t\cos t\,dt$.

9.54. Which of the choices $u = x-1$ or $u = x+1$ is more likely to solve $\int (x-1)(x+1)^{\frac{1}{2}}\,dx$. Try both and check.

9.55. Solve $\int x^2(x+3)^{\frac{1}{2}}\,dx$ and $\int \frac{x^2}{(x-2)^3}\,dx$.

9.56. Use a trig identity to solve $\int \frac{\sec^2\varphi}{\tan^2\varphi+1}\,d\varphi$.

9.57. Try $u = x^7+9$ and a fact about logarithms to solve $\int \frac{5x^6}{x^7+9}dx$.

9.58. Solve $\int (1+4x)(1+2x+4x^2)^{\frac{1}{2}}dx$.

9.59. Try the substitution $u = \cos x$ to solve $\int \tan x\,dx$.

9.60. Use $u = e^z+1$ to solve both $\int (e^z+1)^{\frac{1}{2}}e^z dz$ and $\int (e^z+1)^{\frac{1}{2}}e^{2z}dz$.

9.61. Evaluate the definite integrals $\int_{-1}^{1} \sqrt{1-x^2}\,x\,dx$ and $\int_{-1}^{3} (3+5x^3)^{\frac{2}{3}}x^2\,dx$.

9.62. The most promising choice for evaluating $\int_1^8 x^{-\frac{2}{3}}\sqrt{1+4x^{\frac{1}{3}}}\,dx$ is $u =$ __________. Experiment and show that the integral is equal to $\frac{1}{2}(27-5\sqrt{5})$.

9.63. Evaluate $\int_3^1 \frac{(\ln x)^2}{x}\,dx = -\int_1^3 \frac{(\ln x)^2}{x}dx$.

9H. Integrating by Parts

9.64. Try integration by parts with $u = x$ and $dv = \cos x\,dx$, and hence $du = dx$ and $v = \sin x$, to solve $\int x\cos x\,dx$.

9.65. Show that $\int xe^{5x}dx = \frac{x}{5}e^{5x} - \frac{1}{25}e^{5x} + C$ by starting with $u = x$ and $dv = e^{5x}\,dx$. Using a similar strategy, go on to show that $\int x^2e^{5x}dx = \frac{x^2}{5}e^{5x} - \frac{2x}{25}e^{5x} + \frac{2}{125}e^{5x} + C$.

9.66. Solve both $\int x\ln x\,dx$ and $\int x^2\ln x\,dx$ by integration by parts starting with $u = \ln x$.

9.67. Consider $\int \ln(x^2+1)dx$. One way to proceed to a solution is outlined below. Check the details.

- **i.** An integration by parts with $u = \ln(x^2+1)$ and $dv = dx$ transforms this integral to $x\ln(x^2+1) - 2\int \frac{x^2}{x^2+1}\,dx$.
- **ii.** By a polynomial division (or by taking common denominators), $\frac{x^2}{x^2+1} = 1 - \frac{1}{x^2+1}$. Therefore the integral of (i) becomes $x\ln(x^2+1) - 2x + 2\int \frac{1}{x^2+1}\,dx$.
- **iii.** Since $\frac{d}{dx}\tan^{-1}x = \frac{1}{x^2+1}$ we now get that $\int \ln(x^2+1)dx = x\ln(x^2+1) - 2x + 2\tan^{-1}x + C$.
- **iv.** Check the correctness of this result by differentiating.

9.68. Solve $\int \cos t^{\frac{1}{2}}\,dt$ by following the substitution $z = t^{\frac{1}{2}}$ by the integration by parts procedure with $u = z$ and $dv = \cos z\,dz$.

9I. Algebraic Strategies

9.69. Use integration by partial fractions to show that

$$\int \frac{1}{(x-2)(x-3)}\,dx = -\ln|x-2| + \ln|x-3| + C\,.$$

9.70. Show that

$$\int \frac{x+1}{(x+2)(x-3)}\,dx = \tfrac{1}{5}\ln|x+2| + \tfrac{4}{5}\ln|x-3| + C\,.$$

9.71. Use the conclusion $\int \frac{1}{x^2-s^2}\,dx = \frac{1}{2s}\ln\left|\frac{x-s}{x+s}\right| + C$ of Section 9.7.3 and fact 3 of Section 9.9.2 to verify Formula (14)

$$\int \frac{1}{1-x^2}\,dx = \tanh^{-1}x + C$$

of Section 9.11.

9J. About Inverse Functions

9.72. The graph of the function $f(x) = \sqrt{r^2 - x^2}, 0 \le x \le r$ is the upper right quarter of the circle $x^2 + y^2 = r^2$. See Figure 9.51. Since f is decreasing, it has an inverse f^{-1}. Why does the graph of f suggest that $f^{-1}(x) = f(x)$ for all x with $0 \le x \le r$? Verify this by discussing the connection

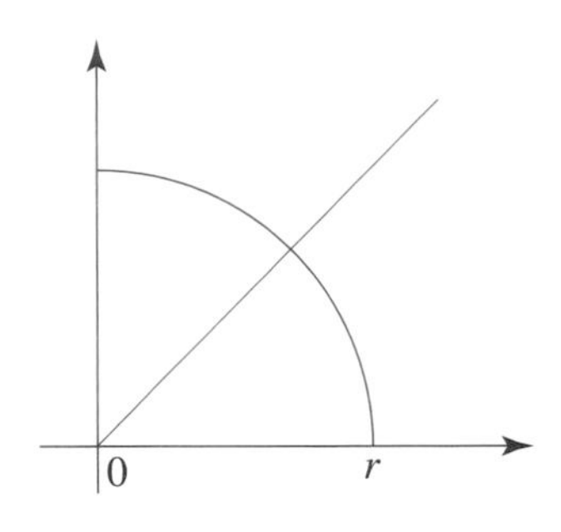

Fig. 9.51

between f and f^{-1}.

9.73. The function $f(x) = \cos x, 0 \le x \le \pi$ is decreasing. Let $f^{-1}(x) = \cos^{-1}(x)$ for x with $-1 \le x \le 1$ be its inverse function. The graphs of f and f^{-1} are depicted in Figure 9.52. Show that $\frac{d}{dx} \cos^{-1} x =$

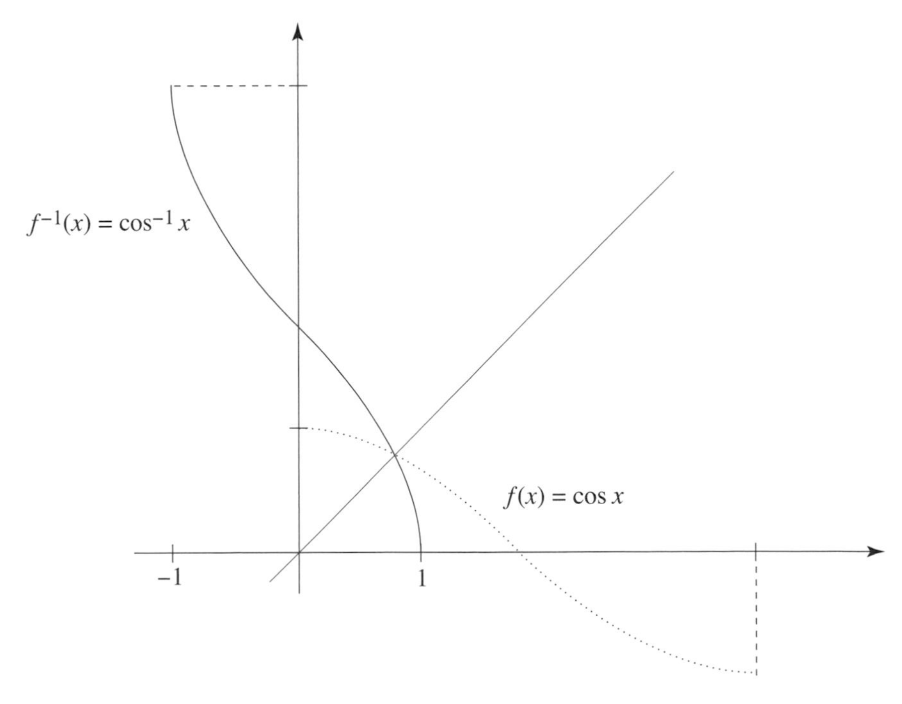

Fig. 9.52

$\frac{1}{\sqrt{1-x^2}}$. Conclude from this that $\cos^{-1} x = \sin^{-1} x + C$, and in particular that $\cos^{-1} x = \sin^{-1} x + \frac{\pi}{2}$.

9.74. Consider the function $f(x) = x^2, 0 \leq x$ and its inverse $f^{-1}(x) = \sqrt{x}$, and let (c, d) and (d, c) be

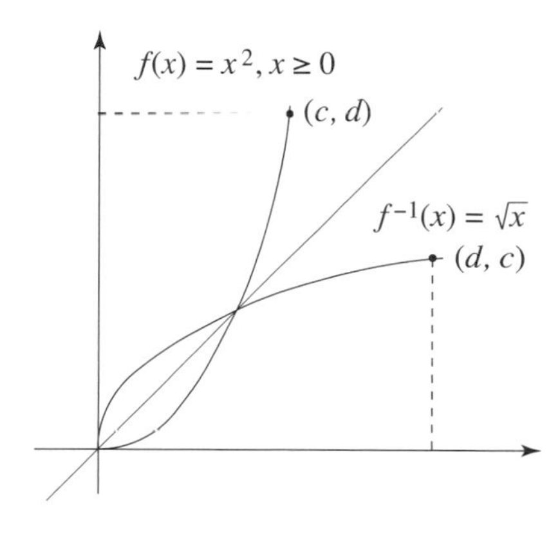

Fig. 9.53

points on their respective graphs. See Figure 9.53. Explain without computing anything why

$$\int_0^d \sqrt{x}\,dx = cd - \int_0^c x^2\,dx \quad \text{and} \quad \int_0^d \sqrt{1 + \tfrac{1}{4x}}\,dx = \int_0^c \sqrt{1 + 4x^2}\,dx.$$

Then verify the the two equalities by doing the computations.

9.75. Suppose that $y = f(x)$ is a continuous function that is increasing or decreasing over an interval $[a, b]$ with $0 \leq a$. If f is increasing, let $f(a) = c$ and $f(b) = d$, and if f is decreasing, let $f(a) = d$ and $f(b) = c$. Use the conclusion of Problem 9.35 to show that the volume obtained by revolving the region bounded by the graph of $y = f(x)$ and the lines $y = c$ and $y = d$ around the y-axis is equal to

$$V = \int_c^d \pi(f^{-1}(y))^2\,dy.$$

One of our basic properties asserts that if two functions have the same derivative, then they differ by a constant. The next problem tells us that this fact must be used with caution.

9.76. Consider the function $f(x) = \tan^{-1}\left(\frac{x-1}{x+1}\right)$.

i. Show that $f'(x) = \frac{1}{x^2+1}$.

ii. Because $f(x)$ and $\tan^{-1} x$ have the same derivative, we conclude that $\tan^{-1}\left(\frac{x-1}{x+1}\right) = \tan^{-1} x + C$ for some constant C.

iii. Plugging $x = 0$ into the equality in (ii) and using $\tan^{-1}(-1) = -\frac{\pi}{4}$, we get $\tan^{-1}\left(\frac{x-1}{x+1}\right) = \tan^{-1} x - \frac{\pi}{4}$.

iv. From Figure 9.33 we know that $\lim\limits_{x\to\infty} \tan^{-1} x = \frac{\pi}{2}$ and $\lim\limits_{x\to-\infty} \tan^{-1} x = -\frac{\pi}{2}$. Verify that $\lim\limits_{x\to\pm\infty} \frac{x-1}{x+1} = 1$, and use this fact to compute $\lim\limits_{x\to\infty} \tan^{-1}\left(\frac{x-1}{x+1}\right)$ and $\lim\limits_{x\to-\infty} \tan^{-1}\left(\frac{x-1}{x+1}\right)$.

v. Are the conclusions of (iii) and (iv) consistent? Why is there a problem with (ii)?

9K. Using and Deriving Integral Formulas

9.77. Consider the parabola $f(x) = x^2$. Let (c, c^2) with $c \geq 0$ be a point on the parabola, and let $L(c)$ be the length of the parabola from the origin $(0, 0)$ to (c, c^2). See Figure 9.54. Use one of the formulas developed in Section 9.10 to show that $L(c) = \frac{1}{4}(2c\sqrt{1 + (2c)^2} + \ln(2c + \sqrt{1 + (2c)^2})$.

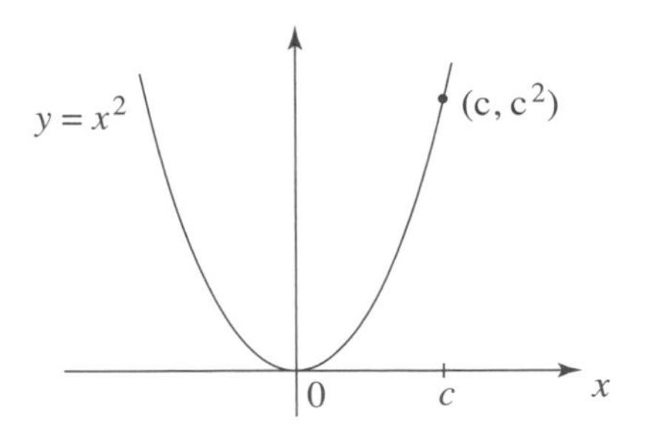

Fig. 9.54

9.78. Project. Carefully study Kepler's analysis of wine barrels in Sections 5.5 and 5.9. Figure 5.39a depicts his model of the standard Austrian barrel. Kepler's model was reasonable for his analysis of the way the local wine merchants of his day estimated the volume of barrels, his model is not exactly barrel shaped. This project will study a barrel shape that is obtained from Kepler's by replacing the two slanting segments by a circular arc. Figure 9.55 adds this circular arc to Kepler's diagrams in Section 5.9 and summarizes much of the relevant information about them.

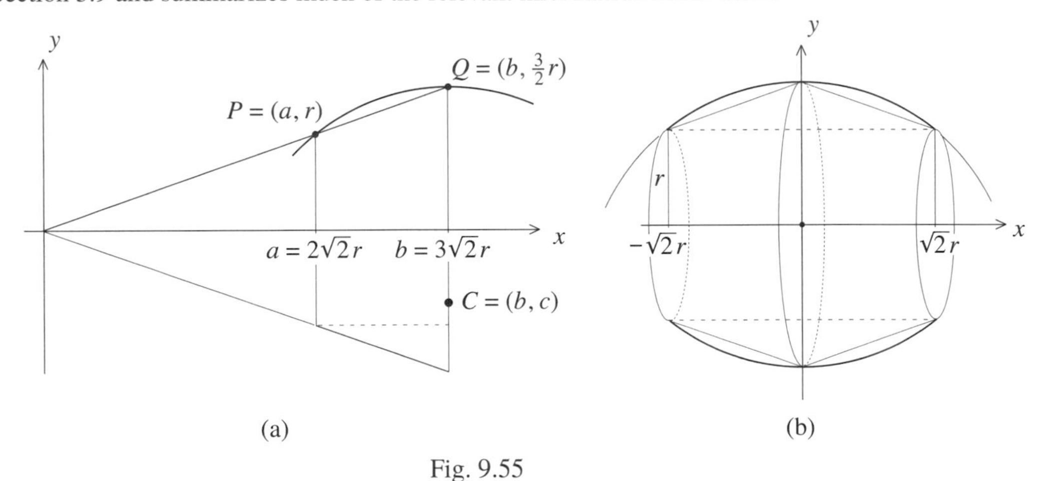

Fig. 9.55

i. Figure 9.55 shows Kepler's diagrams with the circular arc inserted. In Figure 9.55a, P and Q are points on the arc and C is the center of the circle on which the arc lies. Determine the equation of this circle in terms of the constant r (and the variables x and y).

ii. Shift the circle horizontally to the position it has in Figure 9.55b. Its center now lies on the y-axis. Show that the upper half of this shifted circle is the graph of $f(x) = \sqrt{(\frac{9}{4}r)^2 - x^2} - \frac{3}{4}r$.

iii. Consider the barrel-shaped solid obtained by revolving the region bounded by the graph of $f(x) = \sqrt{(\frac{9}{4}r)^2 - x^2} - \frac{3}{4}r$, the x-axis, and the lines $x = -\sqrt{2}r$ and $x = \sqrt{2}r$ around the x-axis. Use one of the integral formulas of Section 9.10 to show that its volume V is given by

$$V = 2\pi \int_0^{\sqrt{2}r} \left(\sqrt{(\tfrac{9}{4}r)^2 - x^2} - \tfrac{3}{4}r\right)^2 dx \approx 16.18r^3.$$

iv. Let s be the length of the slanting diagonal on which the Austrian wine merchants based their measurement of the volumes of barrels. See Figure 5.40 and notice that $r = \frac{2}{\sqrt{33}}s$ also holds for the barrel shape described in (iii). Show that $V \approx 0.68s^3$. Discuss the connection between this approximation and the assessment $V_{\text{rule}} = 0.6s^3$ used by the merchants, as well as the implications for the cost of the wine in a full barrel of this shape.

9.79. Solve $\displaystyle\int \frac{1}{\sqrt{x^2+1}}\,dx.$

i. Use the substitution $x = \tan\theta$ with $-\frac{\pi}{2} < \theta < \frac{\pi}{2}$ to show that

$$\int \frac{1}{\sqrt{x^2+1}}\,dx = \int \sec\theta\,d\theta = \ln|\sec\theta + \tan\theta| + C.$$

ii. Use the identity $\sec^2\theta = 1 + \tan^2\theta$ and the fact that $\theta = \tan^{-1}x$ to get

$$\int \frac{1}{\sqrt{x^2+1}}\,dx = \ln|x + \sqrt{1+x^2}| + C.$$

9.80. Solve $\displaystyle\int \frac{x^2}{\sqrt{x^2-1}}\,dx$ for $x > 1$.

i. Let $x = \sec\theta$ with $0 < \theta < \frac{\pi}{2}$. So $\sqrt{x^2-1} = \sqrt{\sec^2\theta - 1} = \tan\theta$ and $\frac{dx}{d\theta} = \sec\theta\cdot\tan\theta$. Hence

$$\int \frac{x^2}{\sqrt{x^2-1}}\,dx = \int \sec^3\theta\,d\theta = \tfrac{1}{2}[\sec\theta\cdot\tan\theta + \ln|\sec\theta + \tan\theta|] + C.$$

ii. Conclude that $\displaystyle\int \frac{x^2}{\sqrt{x^2-1}}\,dx = \tfrac{1}{2}[x\sqrt{x^2-1} + \text{ł}(x + \sqrt{x^2-1})] + C.$

9.81. Solve $\displaystyle\int \frac{\sqrt{1-x^2}}{x}\,dx.$

i. Use the substitution $x = \sin\theta$, $\frac{\pi}{2} \le \theta \le \frac{\pi}{2}$, to show that

$$\int \frac{\sqrt{1-x^2}}{x}\,dx = \int \frac{1}{\sin\theta}\,d\theta \;+\; \cos\theta + C.$$

The solution of the integral $\int \frac{1}{\sin\theta}\,d\theta$ is complicated. It is solved in a way analogous to the solution

$$\int \frac{1}{\cos\theta}\,d\theta = \int \sec\theta\,d\theta = \ln|\sec\theta + \tan\theta| + C,$$

with $\csc\theta$ and $\cot\theta$ in place of $\sec\theta$ and $\tan\theta$.

ii. This complication can be avoided by the substitution $x = \cos\theta, 0 \le \theta \le \pi$. Use it and the identity $\sin\theta = \sqrt{1-\cos^2\theta}$ to show that

$$\int \frac{\sqrt{1-x^2}}{x}\,dx = \int \cos\theta\,d\theta - \int \sec\theta d\theta = \sqrt{1-x^2} - \ln\left|\tfrac{1+\sqrt{1-x^2}}{x}\right| + C.$$

9L. Some Elliptic Integrals. Consider the ellipse $\frac{x^2}{a^2} + \frac{y^2}{b^2} = 1$. Assume that a and b are the semimajor and semiminor axes, respectively. So $a \ge b$. Let $c = \sqrt{a^2 - b^2}$. Recall from Section 4.4 that $(-c, 0)$ and $(c, 0)$ are the two focal points of the ellipse and that $\varepsilon = \frac{c}{a}$ is its eccentricity.

9.82. Show that the graph of $f(x) = \frac{b}{a}(a^2 - x^2)^{\frac{1}{2}}$ is the upper half of the ellipse and that the length of the upper half of the ellipse is given by

$$\int_{-a}^{a} \sqrt{1 + \tfrac{b^2}{a^2}\tfrac{x^2}{a^2-x^2}}\,dx.$$

Show that the trig substitution $x = a\sin\theta$ transforms this integral to

$$a\int_{-\frac{\pi}{2}}^{\frac{\pi}{2}} \sqrt{1 - \varepsilon^2\sin^2\theta}\,d\theta.$$

9.83. In Problem 9.82, let $a = 1, b = \frac{1}{\sqrt{2}}$, and check that $\varepsilon = \frac{1}{\sqrt{2}}$. Show that

$$\sqrt{2}\int_{-\frac{\pi}{2}}^{\frac{\pi}{2}} \sqrt{1-\varepsilon^2 \sin^2\theta}\, d\theta = \int_{-\frac{\pi}{2}}^{\frac{\pi}{2}} \sqrt{1+\cos^2\theta}\, d\theta,$$

and that $\int_{-\frac{\pi}{2}}^{\frac{\pi}{2}} \sqrt{1+\cos^2\theta}\, d\theta$ is the length of one loop of the sine curve. Refer to Section 9.13.

The integrals that arise in Problems 9.82 and 9.83 are known as elliptic integrals. Their integrands do not have antiderivatives that are elementary functions.

9M. Using the Trapezoidal and Simpson Rules

Use the calculators available at the websites

http://www.emathhelp.net/calculators/calculus-2/trapezoidal-rule-calculator/ and
http://www.emathhelp.net/calculators/calculus-2/simpsons-rule-calculator/

to approximate the integrals below by finding the indicated trapezoidal and Simpson approximations.

9.84. For $\int_0^3 \cosh x^2\, dx$, show that

i. $T_{10} \approx 894.6303$, $T_{20} \approx 767.5961$, $T_{50} \approx 729.9901$, $T_{100} \approx 724.5377$, and $T_{200} \approx 723.1714$.

ii. $S_{10} \approx 752.6349$, $S_{20} \approx 725.2513$, $S_{50} \approx 722.7877$, $S_{100} \approx 722.7202$, and $S_{200} \approx 722.7160$.

The close agreement between S_{50}, S_{100}, and S_{200} suggests that $S_{200} \approx 722.7160$ is a close approximation of the value of the integral. The convergence of the trapezoidal approximations, on the other hand, is a bit sluggish. (The trapezoidal and Simpson rules calculators referred to above have a limit of 200 on the number of intervals allowed.)

9.85. For $\int_1^{100} \frac{1}{x}\, dx$, show that

i. $T_{100} \approx 4.6809$ and $T_{200} \approx 4.6251$.

ii. $S_{100} \approx 4.6176$ and $S_{200} \approx 4.6025$.

iii. We know that $\int_1^x \frac{1}{t}\, dt$ defines the function $\ln x$ for $x \geq 1$. Therefore the actual value of the integral is $\ln 100$, and this is approximated by 4.6052 with accuracy to four decimal places. Once again, the Simpson rule wins the accuracy race.

9N. Trying the Integral Calculator

Use the calculator at the website

http://www.integral-calculator.com/#

to evaluate

$$\int \frac{1}{9+x^3}\, dx \quad \text{and} \quad \int \sin\sqrt{x}\, dx.$$

Then try the calculator on

$$\int e^{-x^2}\, dx, \int \cosh x^2\, dx, \text{ and } \int \sqrt{1-\tan^2 x}\, dx.$$

Explain the answers to these three integrals.

Chapter 10

Applications of Integral Calculus

Chapter 10 applies calculus to a diverse set of topics ranging from the structure of masonry domes, to the cables of a suspension bridge, to some concerns of Euclidean geometry, to the motion of the planets in their elliptical orbits, to the dynamics of a bullet in a gun barrel.

Until the 19th century, humans built their public buildings, churches, and temples out of stone, brick, and concrete. The arches, vaults, and domes that were used to cover their interior spaces presented significant problems. The sloping shapes of these heavy masonry elements generated significant outward forces that the surrounding architectural formations had to counteract in order to ensure the stability of the structure as a whole. The Gothic cathedral's flying buttress balances the lateral thrust of the Gothic vault and dramatically illustrates what is involved. We'll start this chapter with a look at the domes of two of the most famous existing structures of classical architecture: the Pantheon in Rome from the 2nd century A.D. and the Hagia Sophia in Constantinople (now Istanbul) from the 6th century. Integral calculus informs us about the volumes and hence the weights of these domes, and these in turn shed light on the lateral thrusts.

In Section 8.3, the considerable tension that the massive loads of a suspension bridge put on a main cable was analyzed. The second segment of the current chapter studies the manufacture of such cables and their strength, and it applies integral calculus to estimate their lengths. This is followed by another problem that involves a cord and weight. It was presented to Leibniz during the time he was a diplomat in Paris. On a visit to a learned colleague, the latter took out his pocket watch, put it and its chain on a horizontal table, and stretched the chain to a line. He then pulled the free end of the chain along a straight line perpendicular to the initial line of the chain. The watch followed and traced a curve that converged to this straight line. His colleague challenged Leibniz to determine this curve. We will see that the curve is a combination of an algebraic and a logarithmic term. This "pocket watch" curve also clarifies a historic problem of Euclidean geometry having to do with parallel lines.

The next topic that this chapter tackles is the analysis of the planetary orbits. We learned in Chapter 6 that Newton deduced Kepler's three observations—now known as Kepler's laws—from a combination of fundamental laws of force and motion and the methods of calculus. Kepler's first laws tells us that a planetary orbit is elliptical, the second informs us that the segment from the planet to the Sun sweeps out equal areas in equal times, and the third links the orbital data of all the planets. The collection of planets is larger now than it was at the time of Galileo, Kepler, Cassini, and Flamsteed. The planets Uranus, Neptune, and Pluto—in order of their increasing distance from the Sun, as well as the time of their discovery—have since been added to the list. Pluto, discovered in 1930, was the last. In the 1950s, astronomers predicted that other "icy rocks" similar to Pluto would be found beyond the orbit of Neptune. The discovery of several such objects led to the expulsion of Pluto from the family of planets in 2007. The fact that we understand the geometry of each planet's orbit and the way it is traced out, suggests that it ought to be possible to compute a planet's position relative to the Sun in terms of the elapsed time from some fixed position. It turns out that such an explicit description of the motion of a planet is possible, but not without a challenging mathematical analysis.

The last topic continues the study of the physics of motion. It explores the relationship between two fundamental paired concepts, that of work and kinetic energy, and that of impulse and momentum. The definition of the definite integral is central to the understanding of both of these relationships. In the context of ballistics, the equations that are developed provide information about the force with which the pressure wave generated by the detonation of the explosive powder in a cartridge propels the bullet down the barrel of a gun.

After everything in this chapter—including the extensive Problems and Projects section—is said and done, there is one point that will need no further elaboration: the basic concepts of calculus give us a way of thinking about fundamental physical phenomena and structures. We will see in addition that when these connections are combined with the computational methods that calculus supplies, then a wealth of quantitative information about our world emerges. In this way, integral calculus is basic to the physical sciences, the disciplines of engineering, and it is central to the understanding of our physical environment.

10.1 ESTIMATING THE WEIGHT OF DOMES

A critical concern for the stability of a masonry dome is its weight. The weight of a dome and its curving, downward-sloping geometry generate considerable outward forces. These put parts of the *shell*—the word "shell" refers specifically to the brick, stone, or concrete material structure of the dome—under tension that can lead to cracks that propagate downward to its base. Finding ways to contain these outward forces was a formidable challenge for architects. Until the industrial revolution of the latter part of the 18th century, domes were often fitted (and retrofitted) with surrounding iron chains. Thereafter, iron was produced in large quantities and became a common building material. Nowadays, even stronger materials, such as steel, are used in the construction of domes. This section will study two of the most celebrated domes in the history of architecture: the concrete dome of the Pantheon in Rome, built during the rule of the emperor Hadrian in the 2nd century A.D., and the brick and mortar dome of the Hagia Sophia in Constantinople, built during the reign of the Byzantine emperor Justinian in the 6th century A.D. While there are differences in the stated dimensions of these structures in the historical record, the discussion below relies on values commonly found.

10.1.1 The Hagia Sophia

The masterpiece of Byzantine architecture, the Hagia Sophia (in Greek, *hagia* = "holy," *sophia* = "wisdom") is one of the world's historic buildings. It was constructed in an incredibly short time between 532 and 537. Its two architects were mathematicians skilled in the engineering of the day. All of their talents were needed for the execution of the unprecedented design of this monumental church. It would be the largest and most magnificent Christian church for about 900 years. The Hagia Sophia is topped by a dome in the shape of a section of a hemisphere. About 30 m in diameter at its base, it rises 55 m above the floor at its highest point. Two of the large arches supporting the dome open into half domes and in turn into recesses that together form a continuous clear space of 76 m in length. Figure 10.1 shows the architectural features of this space in cross section through the dome, half domes, and recesses. The circular arcade of 40 windows around the base of the dome gives the impression that the dome floats above the soaring space that it tops. The elaborate Greek Orthodox religious ceremonies that the church hosted must have been impressive. At around the year 1000, Prince Vladimir of Kiev sent emissaries to the centers of the four great religions of this region of the world: Islam, Judaism, and Latin and Byzantine Christianity. The glowing descriptions of their visit to the Hagia Sophia surely influenced the prince when he decided that his Russian state would embrace Byzantine Christianity.

Let's have a look at the basic structural aspects of the dome. The shell of the dome is made of brick and mortar. Its outer and inner surfaces are sections of spheres that have the same center and radii of $R = 16.0$ m and $r = 15.2$ m, respectively. This leaves 0.8 m for the thickness of the shell. The combination of masonry and mortar used in the construction of the shell has a mass of 1760 kg/m^3. (For a reader more at home with the American system of units, R and r are 52.5 and 50 feet respectively, and the masonry of the shell weighs 110 pounds per cubic foot.) The segments from the common center of the inner and outer circles of the cross

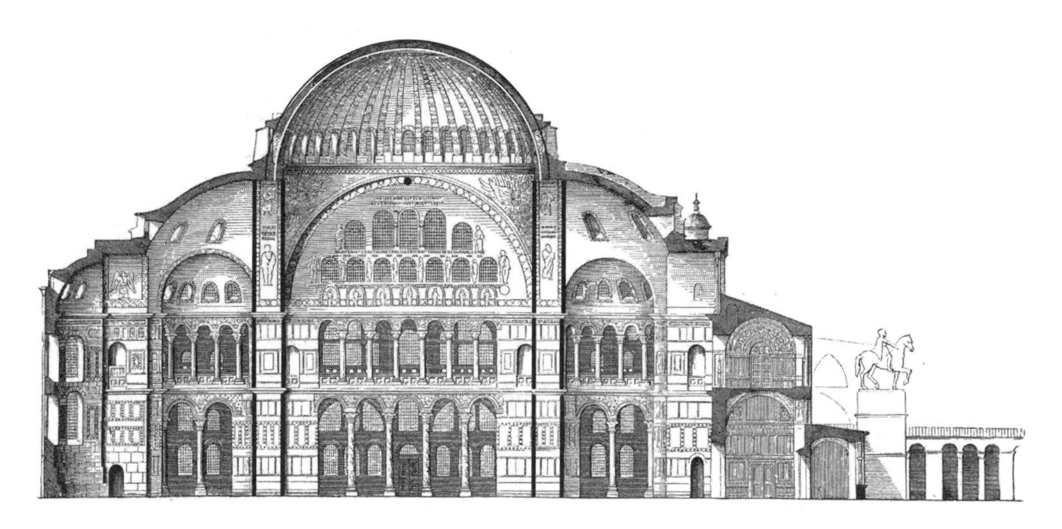

Fig. 10.1. Section of the Hagia Sophia. From Wilhelm Luebke and Max Semrau, *Grundriß der Kunstgeschichte.* Paul Neff Verlag, Esslingen, 1908.

sections emanate at about 20° from the horizontal to determine the boundaries of the dome above the row of 40 windows. Forty ribs radiate down from the top of the dome not unlike the ribs of an umbrella. As shown in Figure 10.1, they descend between the 40 windows to support the dome and to anchor it to its circular base. The basic structural challenges facing the builders of the Hagia Sophia were the same as those faced by the Roman architects of the Pantheon four centuries earlier. To achieve a stable dome, the outward push of the ribs of the shell needed to be counteracted. This was in part achieved by the buttresses between each pair of windows. Figure 10.2 provides a detail of the cross section of the dome that is abstracted from Figure 10.1. The two circular arcs are the cross sections of the inner and outer

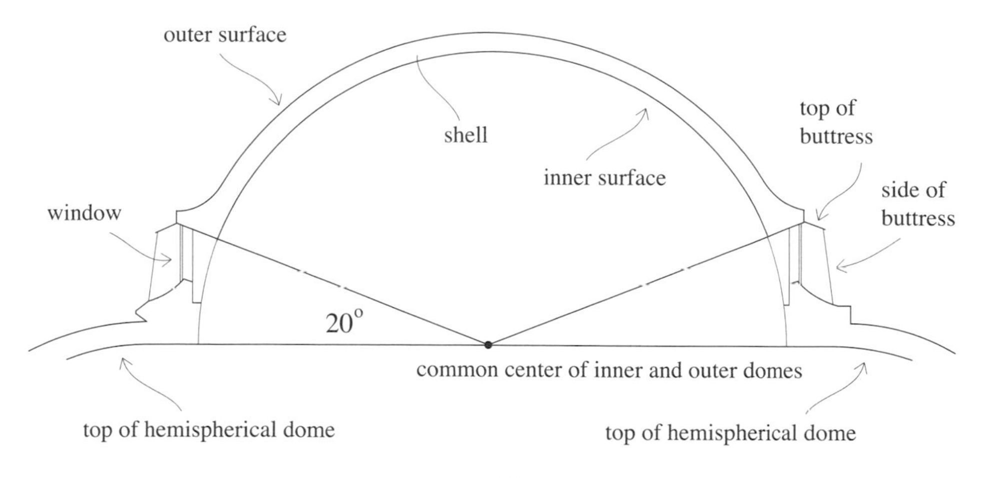

Fig. 10.2

surfaces of the shell. The location of the windows, the ribs between them, and the supporting buttresses are also shown.

The goal of our mathematical study of the dome of the Hagia Sophia will be twofold. We'll first calculate the volume of the shell above the gallery of 40 windows. This will then provide us with estimates of the weight of the dome and the force with which each of the 40 ribs pushes against its base.

The volume V of the shell of the dome can be obtained as follows. Reposition the cross section of the dome as indicated in Figure 10.3, and add an xy-coordinate system. Compute the volume V_1 obtained by revolving about the x-axis the region below the upper semicircle and above the segment $[a, R]$. Then compute the volume V_2 obtained by revolving about the x-axis the region below the lower semicircle and above the segment $[a, r]$. Notice that the difference $V = V_1 - V_2$ is an estimate of the volume of the shell.

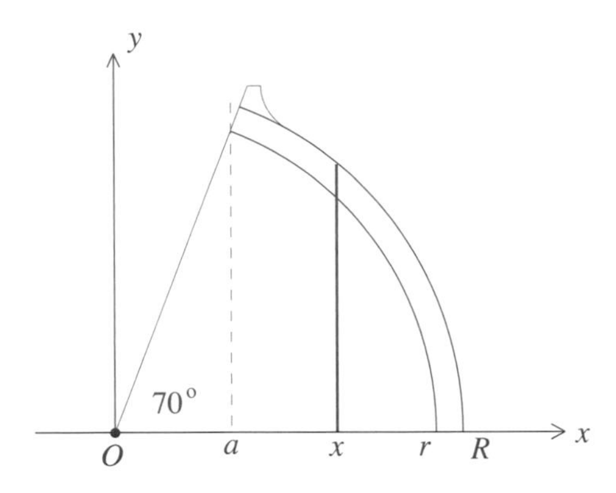

Fig. 10.3

The upper half of the outer circle is the graph of the function $f(x) = \sqrt{R^2 - x^2}$, and the upper half of the inner circle is the graph of the function $g(x) = \sqrt{r^2 - x^2}$. Since $\cos 70° = \frac{a}{r}$, $a = r\cos 70° \approx (15.2)(0.342) \approx 5.2$ m. By combining what we learned in Section 9.2 with the fundamental theorem of calculus, we get

$$\begin{aligned} V_1 &= \pi\int_a^R (\sqrt{R^2 - x^2})^2\,dx = \pi\int_a^R (R^2 - x^2)\,dx = \pi[R^2x - \tfrac{1}{3}x^3\big|_a^R] \\ &= \pi[(R^3 - \tfrac{1}{3}R^3) - (R^2a - \tfrac{1}{3}a^3)] = \pi[\tfrac{2}{3}R^3 - R^2a + \tfrac{1}{3}a^3]. \end{aligned}$$

In the same way,

$$\begin{aligned} V_2 &= \pi\int_a^r (\sqrt{r^2 - x^2})^2\,dx = \pi\int_a^r (r^2 - x^2)\,dx = \pi[r^2x - \tfrac{1}{3}x^3\big|_a^r] \\ &= \pi[(r^3 - \tfrac{1}{3}r^3) - (r^2a - \tfrac{1}{3}a^3)] = \pi[\tfrac{2}{3}r^3 - r^2a + \tfrac{1}{3}a^3]. \end{aligned}$$

Therefore

$$V = V_1 - V_2 = \pi[\tfrac{2}{3}R^3 - R^2a + \tfrac{1}{3}a^3] - \pi[\tfrac{2}{3}r^3 - r^2a + \tfrac{1}{3}a^3] = \pi[\tfrac{2}{3}(R^3 - r^3) - (R^2 - r^2)a].$$

Putting in the values $R = 16$, $r = 15.2$, and $a = 5.2$ all in meters, we get

$$V \approx \pi[\tfrac{2}{3}(16^3 - 15.2^3) - (16^2 - 15.2^2)5.2] \approx \pi(389.46 - 129.79) \approx 815.78 \text{ m}^3.$$

We'll assume that the data has an accuracy of three significant figures—so that $R = 16.0$ m, $r = 15.2$ m, $a = 5.20$ m, and 1760 kg/m^3 for the mass of the shell. This means that the derived numerical information needs to be rounded to three significant figures. So our estimate for the volume of the shell of the dome above the row of 40 windows is $V \approx 816$ m^3, and the estimate for its mass m is

$$m = 815.78 \times 1760 \approx 1{,}435{,}773 \approx 1{,}440{,}000 \text{ kg}.$$

The weight of the shell is $W = mg \approx (1{,}435{,}778)(9.81) \approx 14{,}084{,}933 \approx 141{,}000{,}000$ N. Assuming the even distribution of this weight over the 40 supporting ribs, we get a weight of

$$\frac{14{,}084{,}933}{40} \approx 352{,}123 \approx 352{,}000 \text{ N}$$

for each rib. Review the discussion of forces in Section 6.6. Let P be the magnitude of the push of one rib against the base of the dome, and turn to Figure 10.4. The vertical component of P is equal to $P\sin 70°$, and the horizontal component H of P is equal to $P\cos 70°$. This is the lateral thrust of one rib against the

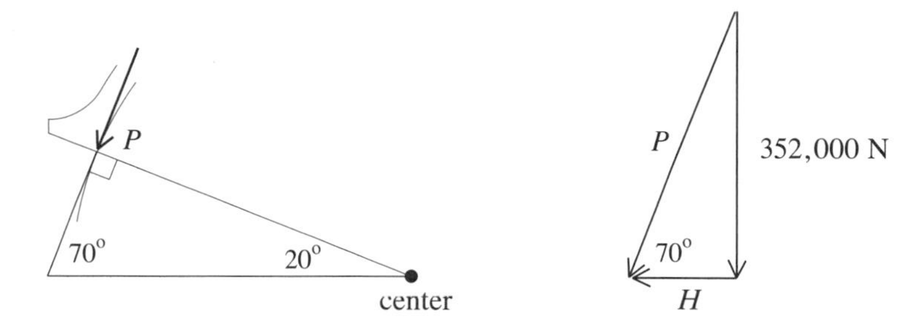

Fig. 10.4

base of the dome. Figure 10.4 tells us that $\sin 70° \approx \frac{352{,}123}{P}$ and $\tan 70° \approx \frac{352{,}123}{H}$. Therefore

$$P \approx \frac{352{,}123}{\sin 70°} \approx 375{,}000 \text{ N} \quad \text{and} \quad H \approx \frac{352{,}123}{\tan 70°} \approx 128{,}000 \text{ N}.$$

These are estimates of the force P with which each rib pushes against the base of the dome as well as the horizontal component H of this force. (In the American system of units, these forces are about 75,000 pounds and 27,000 pounds, respectively.) The two architects were aware of the fact that in order for its dome to be structurally sound, the Hagia Sophia had to be constructed in such a way that the building as a whole, and its buttresses in particular, could resist the forces generated by the ribs. (The force estimates undertaken above are based on the simplifying assumption that a pair of opposite ribs and the loads that they support are modeled by the *simple truss*. See Problem 10.1.)

10.1.2 The Roman Pantheon

One of the most impressive structures of antiquity is the Pantheon in Rome. Its supervising architect was Emperor Hadrian. It was built between 118 and 128 A.D. and dedicated to all Roman gods (in Greek, *pan* = "all," *theon* = "of the gods"). Its *elevation*, this is a representation of the facade of a building, is shown in Figure 10.5. The entrance hall is a large portico in the Greek Corinthian style. The portico fronts a large cylindrical structure that is capped by a hemispherical dome. The circular floor of the Pantheon has a diameter of 43.2 m and its dome rises to a height of 43.2 m above the floor at its highest point. To let in light and air, there is a circular opening, or *oculus* (in Latin, *oculus* = "eye"), with a diameter of about 8.2 m at the top. Other than the entrance, the oculus is the sole source of light for the interior. The oculus also reduces the dome's weight.

The Romans were aware of the challenges that the large, heavy dome that they envisioned for the Pantheon would present. In order to support the dome and to counter the outward forces with which it pushes against its base, they built it on top of a massive, symmetric, essentially closed cylinder with walls 6 m thick. This means that the dome of the Pantheon is supported uniformly around its base. The Hagia Sophia is structurally more complex in this regard. Its dome rests on four large arches and the curving triangular structures between them. Two of these arches open into half domes to form the long interior space of the church that Figure 10.1 depicts. The other two arches (as Figure 10.1 also indicates) are closed off by walls that are perforated by rows of windows and arcades. Thus, unlike the dome of the Hagia Sophia, the dome of the Pantheon receives uniform support from the powerful cylinder on which it rests.

The Romans built the Pantheon out of concrete interspersed by courses of bricks and with brick facing all around the exterior. The use of concrete in much of their construction was a Roman innovation. They discovered that when a certain volcanic powder is mixed with lime, sand, fragments of stone and masonry,

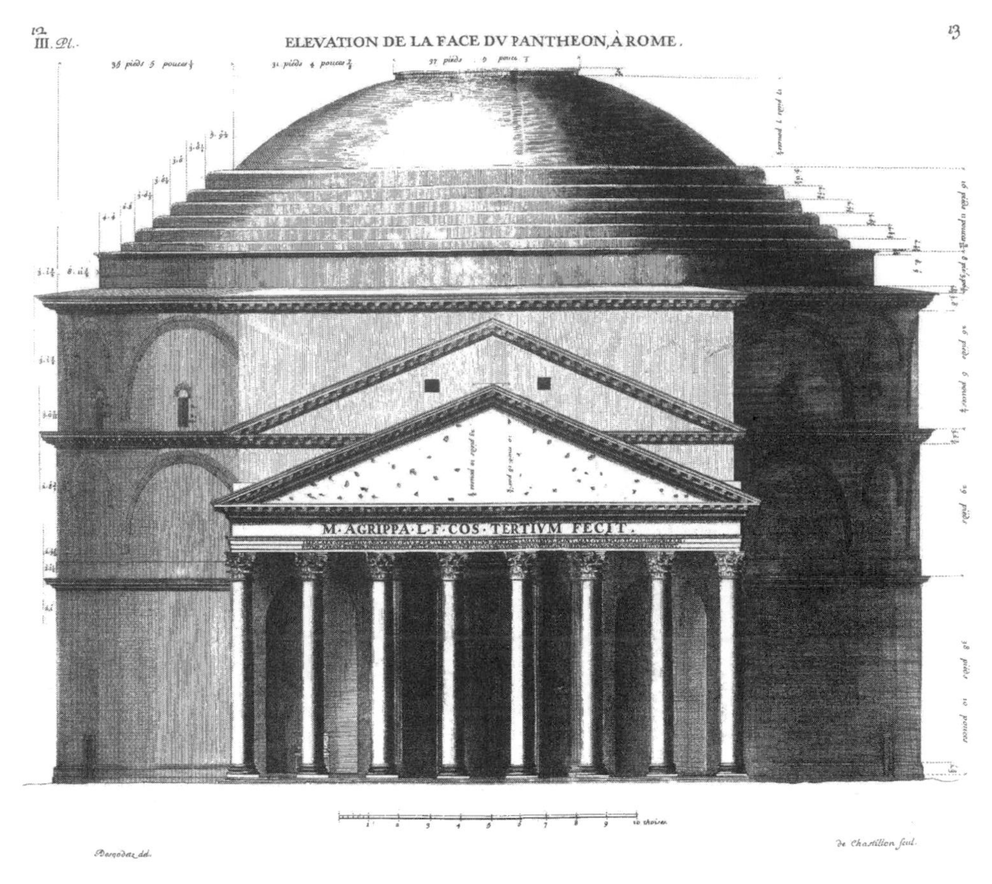

Fig. 10.5. Antoine Desgodetz, engraving of the elevation of the Pantheon, from *Les edifices antiques de Rome*, Claude-Antoine Jambert, Paris, 1779.

and subsequently with water, the mix hardens to a substance of stonelike consistency. Roman concrete had a thick consistency before it set, but it could be shaped and molded and was easy to build with. Because concrete was not attractive enough for some applications, the Romans became expert at the use of brick, stucco, marble, and mosaic finishes. (Modern concrete differs from Roman concrete in several respects. Modern concrete is fluid and homogeneous, can be poured into forms, and can be given significant additional strength by embedding steel rods and wire mesh into it.) One important advantage that concrete offered in the context of the dome of the Pantheon was that it could be made lighter or heavier simply by adding lighter or heavier *aggregate*—stone, gravel, or masonry fragments—to the mix. By using light volcanic rock such as *pumice* for the upper sections of the shell, they could reduce its weight and the resulting lateral forces. By adding heavier rock such as basalt to the aggregate for the concrete of the supporting cylinder and the substantial foundation on which it rests, they could make both stronger. Figure 10.6 tells us how the density of the concrete varies from light at the top of the dome to heavy at the bottom of the Pantheon's cylindrical wall. Forming an aggregate of volcanic rock of increasing density and successively mixing in brick, tile, and marble fragments, the mass of the concrete decreases from 1300 kg/m^3 in region 1, to 1500 kg/m^3 in region 2, to 1600 kg/m^3 in region 3, 1700 kg/m^3 in region 4, and finally to 1800 kg/m^3 in region 5.

The Romans often placed masonry and concrete masses on top of the lower, outer sections of arches and vaults. These masses were intended to increase the stability of such structures. The Romans may have intended for the step rings they built into the lower exterior of the dome of the Pantheon—see Figure 10.5—to serve such a function. However, studies have indicated that the step rings seem to play no significant role in this regard. The dome of the Pantheon was constructed with the use of a wooden *centering structure*, namely, an elaborate forest of timbers that reached upward from the floor to support

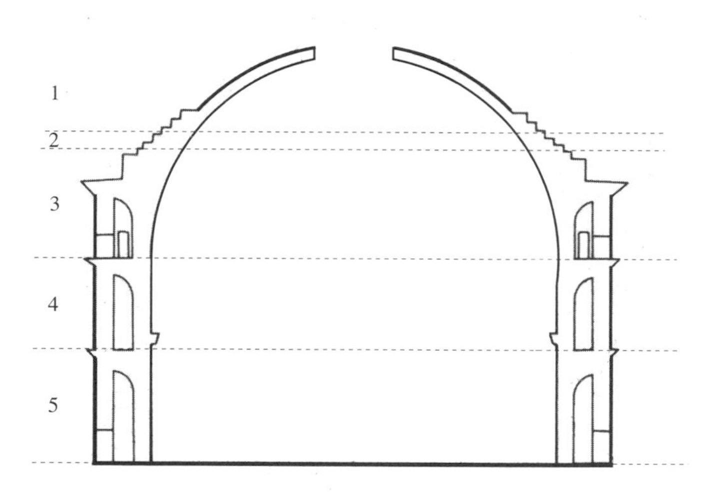

Fig. 10.6. A section of the Pantheon indicating the distribution of concrete of different densities

the growing shell until the construction was completed and the concrete had set.

Figure 10.7 depicts half of a vertical section of the dome of the Pantheon. It shows the structure of the shell, part of the cylindrical supporting wall, the oculus, and the step rings. It also shows the rising array of indentations on the inside of the dome. This *coffering* is decorative and serves no structural purpose. The vectors flowing down the shell represent the downward transmission of the weight of the dome. The vectors pointing upward represent the support of the shell from its cylindrical base. The two circular arcs are parts of the cross sections of the dome's inner and outer spherical surfaces. The radius of

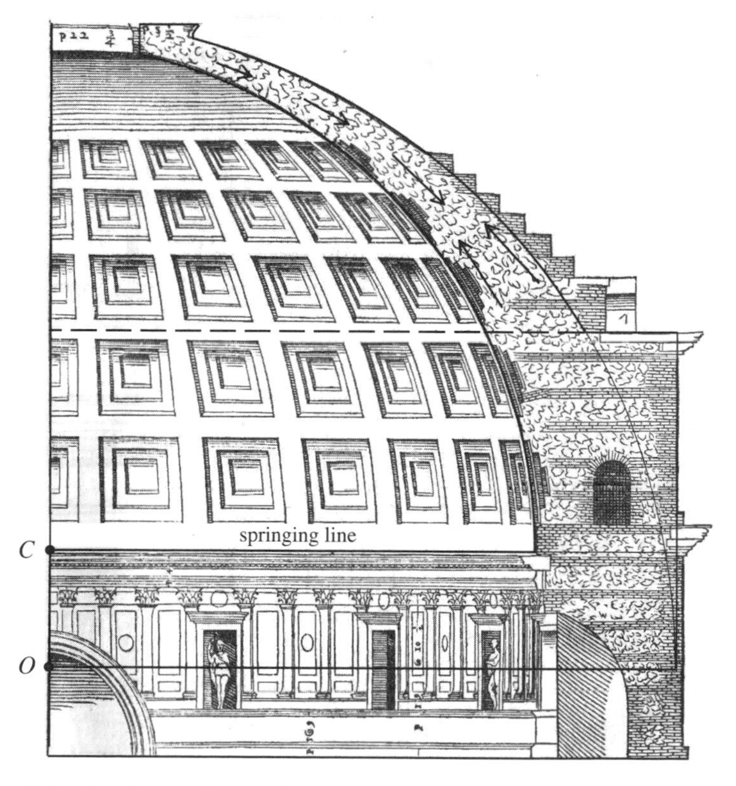

Fig. 10.7. Section of the Pantheon adapted from the *I Quattro Libri dell' Architettura* (*The Four Books of Architecture*, Venice, 1570) of the famous Renaissance architect Andrea Palladio.

the inner sphere is $\frac{1}{2} \cdot 43.2 = 21.6$ m. Its center C is the point on the centering structure from which the builders of the Pantheon stretched ropes in all upward directions to guide the spherical shape of the shell during construction.

In spite of all the efforts, the dome of the Pantheon did experience extensive cracking. Nonetheless, the fact that the Pantheon still stands after almost 1900 years tells us how well the Romans succeeded. The Pantheon is one of the most influential buildings in the history of architecture.

We'll now turn to the quantitative study of the shell of the dome. Figure 10.8 is an abstraction of Figure 10.7. It depicts parts of the circular cross sections of the inner and outer spherical surfaces of the shell along with an xy-coordinate system. The coordinate system has the center of the outer sphere at the

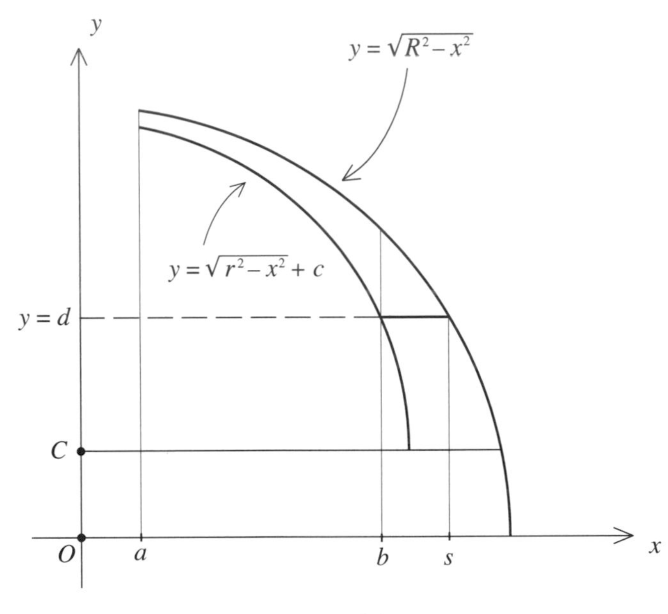

Fig. 10.8

origin O. The meaning of the notation in the figure is as follows:

r = the radius of the circular cross section the inner surface of the shell.

C = $(0, c)$ is the center of the sphere that determines the inner surface of the shell.

d = the y-coordinate of the lower horizontal boundary of the shell. The supporting cylinder begins below the line $y = d$.

a = the x-coordinate of the boundary of the oculus.

b = the x-coordinate of the intersection of the lower boundary of the shell with the circular cross section of its inner spherical surface.

s = the x-coordinate of the intersection of the lower boundary $y = d$ of the shell with the circular cross section of its outer spherical surface.

R = the radius of the outer spherical surface of the shell. The center of this sphere is the origin O.

The equation of the outer circle is $x^2 + y^2 = R^2$. Solving for y gives us $y = \pm\sqrt{R^2 - x^2}$. Because only the upper half of the circle is being considered, the relevant equation is $y = \sqrt{R^2 - x^2}$. The equation of the inner circle is $x^2 + (y - c)^2 = r^2$. So $y - c = \pm\sqrt{r^2 - x^2}$. We will consider only that part of the inner circle that lies above the line $y = d$. Therefore $y \geq d$, hence $y - c = \sqrt{r^2 - x^2}$, and the relevant equation is $y = \sqrt{r^2 - x^2} + c$.

Now that we have an understanding of the geometry of the dome of the Pantheon, let's take on the problem of estimating both the volume and the weight of its shell. We will focus on the section of the shell above the line $y = d$. Both the step rings on the exterior of the shell and the coffering on its interior will be ignored.

We'll use the shell method of Section 9.2, in particular, volume Formula V_3. In doing so, refer to Figure 10.9. We'll first apply the shell method to the functions $f(x) = \sqrt{R^2 - x^2}$ and $g(x) = \sqrt{r^2 - x^2} + c$

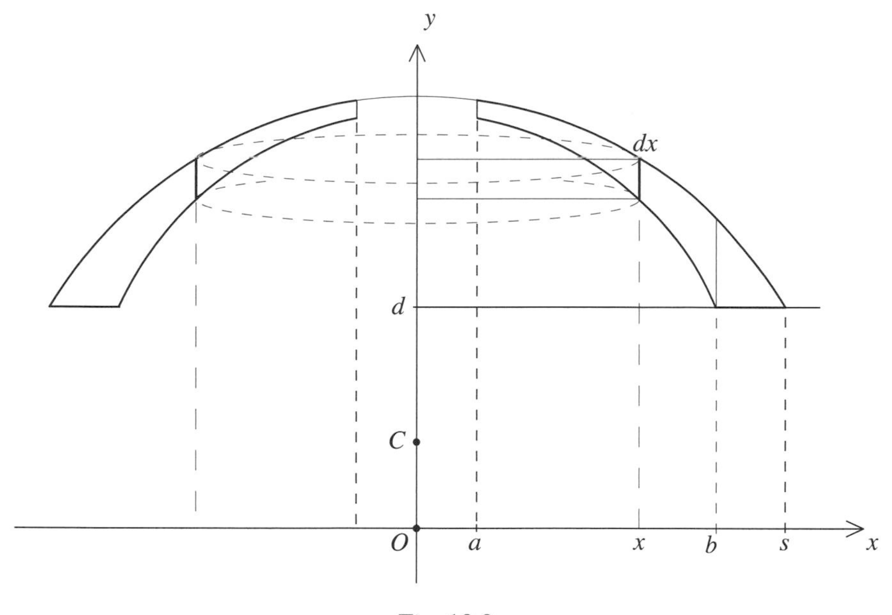

Fig. 10.9

over the interval $a \le x \le b$ to get the volume

$$V_1 = 2\pi \int_a^b x(\sqrt{R^2 - x^2} - (\sqrt{r^2 - x^2} + c))\,dx$$

of that part of the shell that falls inside the cylinder that the vertical line $x = b$ generates. To determine the rest of the volume of the shell, it remains to apply the shell method with $f(x) = \sqrt{R^2 - x^2}$ and $g(x) = d$ over the interval $b \le x \le s$ to obtain the volume

$$V_2 = 2\pi \int_b^s x(\sqrt{R^2 - x^2} - d)\,dx$$

and to observe that the volume of the shell of the Pantheon is the sum $V = V_1 + V_2$ of the the volumes V_1 and V_2. The integrals involved are not difficult to compute. Let's start with V_1. Notice that

$$V_1 = \pi \int_a^b 2x\sqrt{R^2 - x^2}\,dx - \pi \int_a^b 2x(\sqrt{r^2 - x^2} + c)\,dx\,.$$

To evaluate the first of these integrals, use the substitution $u = R^2 - x^2$. So $\frac{du}{dx} = -2x$ and hence $du = -2xdx$. When $x = a$, $u = R^2 - a^2$, and when $x = b$, $u = R^2 - b^2$. It follows that

$$\begin{aligned}
\pi \int_a^b 2x\sqrt{R^2 - x^2}\,dx &= \pi \int_{R^2-a^2}^{R^2-b^2} -u^{\frac{1}{2}}\,du = \pi\Big[-\tfrac{2}{3}u^{\frac{3}{2}}\Big|_{R^2-a^2}^{R^2-b^2}\Big] \\
&= -\tfrac{2}{3}\pi[(R^2 - b^2)^{\frac{3}{2}} - (R^2 - a^2)^{\frac{3}{2}}] \\
&= \tfrac{2}{3}\pi[(R^2 - a^2)^{\frac{3}{2}} - (R^2 - b^2)^{\frac{3}{2}}].
\end{aligned}$$

The second integral is solved in a similar way. Let $u = r^2 - x^2$. So $\frac{du}{dx} = -2x$ and $du = -2xdx$. When $x = a$, $u = r^2 - a^2$, and when $x = b$, $u = r^2 - b^2$. By substituting, we get

$$\begin{aligned}\pi\int_a^b 2x(\sqrt{r^2-x^2}+c)\,dx &= \pi\int_{r^2-a^2}^{r^2-b^2} -(u^{\frac{1}{2}}+c)\,du = -\pi\Big[\big(\tfrac{2}{3}u^{\frac{3}{2}}+cu\big)\Big|_{r^2-a^2}^{r^2-b^2}\Big]\\ &= -\pi\big[\tfrac{2}{3}(r^2-b^2)^{\frac{3}{2}} - \tfrac{2}{3}(r^2-a^2)^{\frac{3}{2}} + c(-b^2+a^2)\big]\\ &= \pi\big[\tfrac{2}{3}(r^2-a^2)^{\frac{3}{2}} - \tfrac{2}{3}(r^2-b^2)^{\frac{3}{2}} + c(b^2-a^2)\big].\end{aligned}$$

Therefore

$$V_1 = \tfrac{2}{3}\pi[(R^2-a^2)^{\frac{3}{2}} - (R^2-b^2)^{\frac{3}{2}}] - \tfrac{2}{3}\pi[(r^2-a^2)^{\frac{3}{2}} - (r^2-b^2)^{\frac{3}{2}}] - \pi c(b^2-a^2).$$

Now on to the integral for V_2. Again, let $u = R^2 - x^2$. So $\frac{du}{dx} = -2x$ and hence $du = -2xdx$. When $x = b$, $u = R^2 - b^2$, and when $x = c$, $u = R^2 - c^2$. After substituting as before, we get

$$V_2 = \pi\int_b^s 2x(\sqrt{R^2-x^2}-d)\,dx = \tfrac{2}{3}\pi[(R^2-b^2)^{\frac{3}{2}} - (R^2-s^2)^{\frac{3}{2}}] - \pi d(s^2-b^2).$$

It remains to insert the data specific to the Pantheon. Recall from above that the radius of the inner dome is $r = 21.6$ m and that the oculus is 8.2 m in diameter. These data and the diagram of Figure 10.7 can be used to extract the estimates

$$a = 4.10 \text{ m},\ b = 19.4 \text{ m},\ s = 23.8 \text{ m},\ c = 5.20 \text{ m},\ d = 14.9 \text{ m, and } R = 28.1 \text{ m}.$$

Plugging $R = 28.1$, $a = 4.10$, $b = 19.4$, $r = 21.6$, and $c = 5.2$ into the expression for V_1 and rounding off tells us that

$$V_1 \approx 27{,}400 - 18{,}200 - 5870 \approx 3300 \text{ m}^3.$$

Plugging $R = 28.1$, $b = 19.4$, $s = 23.8$, and $d = 14.9$ into the expression for V_2 and rounding off, we get

$$V_2 \approx 10{,}600 - 8900 = 1700 \text{ m}^3.$$

Therefore

$$V \approx 3300 + 1700 \approx 5000 \text{ m}^3$$

is an estimate for the volume of the shell of the Pantheon.

The problem of estimating the mass M of the shell of the Pantheon runs into a difficulty. In the situation of Hagia Sophia, the masonry materials of the shell have uniform density, which makes the estimate of its mass simple. It is just a matter of multiplying the volume of the shell by the constant density of the masonry. We have seen that the concrete of the shell of the Pantheon varies in density from 1300 kg/m^3 at the top to 1600 kg/m^3 at the base. See Figure 10.6. Given the estimate of 5000 m^3 for the volume of the shell, this provides the range

$$1300 \times 5000 \leq 6{,}500{,}000 \text{ kg} \leq M \leq 1600 \times 5000 \leq 8{,}000{,}000 \text{ kg}$$

for the mass M of the shell.

Several questions remain. Could the estimate for the mass of the shell of the Pantheon be improved if the information in Figure 10.6 were to be used more fully? Recall that the mass of the shell of the dome of the Hagia Sophia is approximately 1,440,000 kg. Is it reasonable that the mass of the shell of the dome of the Pantheon is greater by a factor of about five? Does a comparison of the cross sections provide a clue? The problem of estimating the forces with which the shell of the Pantheon pushes against its base is also more subtle. Unlike the situation of the Hagia Sophia, the weight of the dome of the Pantheon is not carried by ribs. Can the forces pushing against the base be estimated if this is assumed? Some of these questions are considered in the Problems and Projects section at the end of this chapter.

10.2 THE CABLES OF A SUSPENSION BRIDGE

We'll now return to our study of the suspension bridge of Section 8.3 with a particular focus on the main cables. The four main cables of the George Washington Bridge (as well as those of all other large suspension bridges before and since) were manufactured on-site. Galvanized steel wires with a diameter of 0.196 inches were laid in big loops from one anchorage, over the two towers, to the other anchorage, and back again. This process is called *spinning*. A *spinning wheel* riding on its own set of wires pulled the wires from one end of the bridge to the other. Special tongs then compressed 217 loops at a time into *strands* of about 4.5 inches in diameter. So each strand consisted of 434 wires. For each of the main cables of the George Washington Bridge, 61 strands were spun. Squeezers, each consisting of 12 hydraulic jacks mounted on a circular frame, pressed the 61 strands together into the main cable. Thus the cable has 61 strands of 434 wires each, for a total of $61 \times 434 = 26{,}474$ individual wires! After a coat of lead paste was applied, each cable was wrapped with additional wires. The cables were now complete. Each had a smooth, round surface, a diameter of about 3 feet, and a weight of 2780 pounds per foot. The strands at the end of each cable were locked into the massive concrete anchorages built to keep the main cables firmly and securely in place.

The placement of the vertical suspender cables followed. Each consists of a single wire rope of 2.875 inches in diameter. The are set every 60 feet in groups of 16, four for each main cable. As already described in Section 8.3, the floor steel for the deck of the center span was hoisted into position by cranes and attached to the suspender cables. This proceeded in sections from the towers toward the center of the bridge. The deck sections for the side spans followed. Stiffening girders were built into the deck to distribute the loads more evenly over the main cables. They control the amount of deformation in the cables at heavy live loads. Once the deck was in place, the George Washington Bridge was finished. It goes without saying that the cables of the bridge are essential, in terms of both the structural and the aesthetic aspect.

The same strategy had been used earlier to build the Brooklyn Bridge, as well as the Williamsburg and Manhattan Bridges over the East River in New York. It was also used later to build the Golden Gate Bridge in San Francisco and the Verrazano in New York. The number of strands in the main cables varied from 19 for the Brooklyn, to 37 for the Williamsburg, to the 61—already spun earlier for the George Washington—for the Golden Gate and the Verrazano. Why the numbers 19, 37, and 61? Figure 10.10 provides the answer. Strands in these numbers can be set into an hexagonal arrangements. These configurations can be wrapped and completed to cables with a circular cross section. Notice that $1 + 6 = 7$, that $7 + 2 \cdot 6 = 19$, that $19 + 3 \cdot 6 = 37$, and that $37 + 4 \cdot 6 = 61$. The progressing pattern is clear. If more than 61 strands are

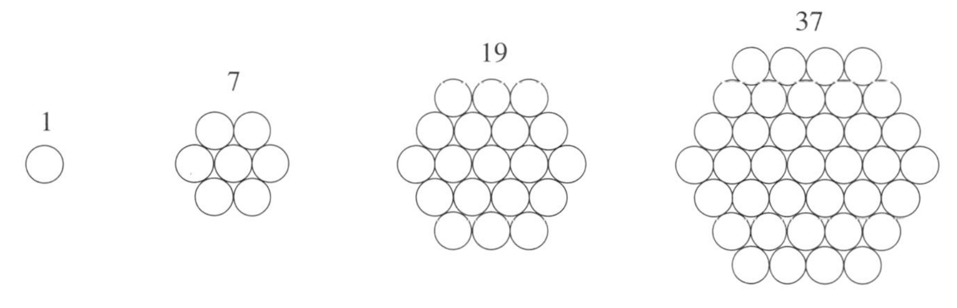

Fig. 10.10

called for, their number could be $61 + 5 \cdot 6 = 91$ or $91 + 6 \cdot 6 = 127$. Notice that the horizontal central diameter of each configuration has an odd number of strands. In the Problems and Projects section of this chapter, it is verified that if the number of strands in the horizontal diameter is $2k + 1$ for some k, then the number of strands in the configuration is $3k^2 + 3k + 1$. Notice that for $k = 0, 1, 2, 3, 4$, and 5, the numbers that arise are $1, 7, 19, 37, 61$, and 91.

The galvanized steel wire that has been used to fashion the strands of American suspension bridges has an *ultimate strength*—this is the tension supported just before the point of failure—of 220,000 pounds per square inch. Because it has cross-sectional area $\pi\left(\frac{0.196}{2}\right)^2 \approx 0.03$ square inches, a single wire can bear an ultimate load of $0.03(220{,}000) \approx 6600$ pounds. Therefore each main cable of the George Washington Bridge with its 26,474 wires has an *ultimate strength* of

$$6600 \times 26{,}474 \approx 175{,}000{,}000 \text{ pounds.}$$

A comparison of this number with the estimate of $T_d = 58{,}700{,}000$ pounds, derived in Section 8.3 for the maximum tension that a main cable is actually subjected to, shows that a *safety factor* of about 3 is built into the bridge's main cables.

How long should the main cable over the center span of a suspension bridge be so that it achieves the sag that the design calls for? Recall from Section 8.3 that the shape of the main cable over the center span is that of the parabola $f(x) = \frac{s}{d^2}x^2$, where d is one-half of the length of the center span and s is the sag. See Figure 10.11. A lower sag-to-span ratio means that the towers are shorter and that the bridge is vertically more stable, but it implies a higher tension in the cables and requires that the anchorages need

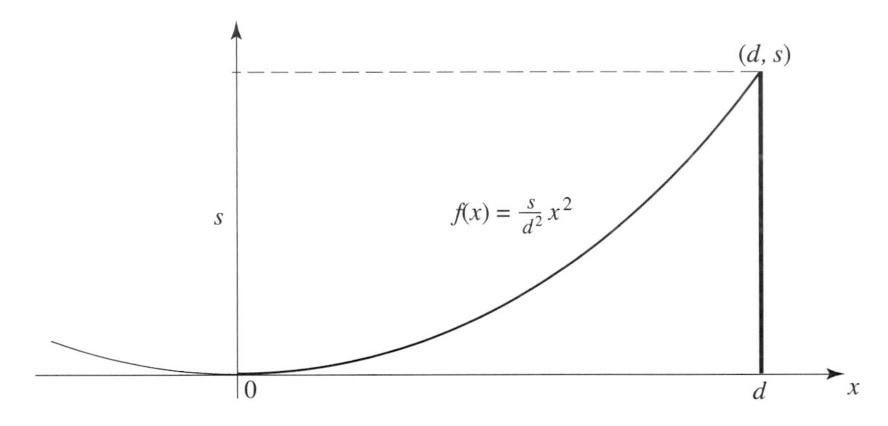

Fig. 10.11

to be stronger. A high ratio means that the towers are higher, but that there is less stress on the cables and the anchorages. Since $f'(x) = \frac{2s}{d^2}x$, the length formula of Section 9.3 informs us that the length L of the cable from its low point at midspan to the top of the tower is

$$L = \int_0^d \sqrt{1 + f'(x)^2}\, dx = \int_0^d \sqrt{1 + (\tfrac{2s}{d^2}x)^2}\, dx\,.$$

The entire length of the main cable over the center span is $2L$. To compute L, we'll turn to the method of substitution of Section 9.7.1 and let $u = \frac{2s}{d^2}x$. So $du = \frac{2s}{d^2}\, dx$ and hence $dx = \frac{d^2}{2s}\, du$. By Formula (17) of Section 9.11,

$$\int \sqrt{1 + u^2}\, du = \tfrac{1}{2}[u\sqrt{1 + u^2} + \ln(u + \sqrt{1 + u^2})] + C.$$

The limits of integration in the variable u are determined by the observation that when $x = 0, u = 0$, and when $x = d, u = \frac{2s}{d}$. Using the fact that $\ln 1 = 0$, we now get

$$\int_0^d \sqrt{1 + (\tfrac{2s}{d^2}x)^2}\, dx = \frac{d^2}{2s}\int_0^{\frac{2s}{d}} \sqrt{1 + u^2}\, du = \frac{d^2}{4s}\Big[\tfrac{2s}{d}\sqrt{1 + (\tfrac{2s}{d})^2} + \ln\big(\tfrac{2s}{d} + \sqrt{1 + (\tfrac{2s}{d})^2}\big)\Big].$$

Therefore

$$\boxed{L = \tfrac{d}{2}\sqrt{1 + (\tfrac{2s}{d})^2} + \tfrac{d^2}{4s}\ln\Big(\tfrac{2s}{d} + \sqrt{1 + (\tfrac{2s}{d})^2}\Big)}$$

For the George Washington Bridge, $d = 1750$ feet and $s = 327$ feet. Inserting these values into the formula gives

$$L = \tfrac{1750}{2}\sqrt{1+\left(\tfrac{2\cdot 327}{1750}\right)^2} + \tfrac{1750^2}{4\cdot 327}\ln\left(\tfrac{2\cdot 327}{1750} + \sqrt{1+\left(\tfrac{2\cdot 327}{1750}\right)^2}\right) \approx 1789.92 \text{ feet.}$$

For the Brooklyn Bridge, $d = 798$ feet and $s = 128$ feet. Put these values into the formula to get

$$L = \tfrac{798}{2}\sqrt{1+\left(\tfrac{2\cdot 128}{798}\right)^2} + \tfrac{798^2}{4\cdot 128}\ln\left(\tfrac{2\cdot 128}{798} + \sqrt{1+\left(\tfrac{2\cdot 128}{798}\right)^2}\right) \approx 811.48 \text{ feet.}$$

Finally, for the Golden Gate $d = 2100$ feet and $s = 470$ feet. Therefore

$$L = \tfrac{2100}{2}\sqrt{1+\left(\tfrac{2\cdot 470}{2100}\right)^2} + \tfrac{2100^2}{4\cdot 470}\ln\left(\tfrac{2\cdot 470}{2100} + \sqrt{1+\left(\tfrac{2\cdot 470}{2100}\right)^2}\right) \approx 2168.16 \text{ feet.}$$

There is a formula for the length L that is much simpler than the one developed above. It provides approximations only, but these are tight. The formula is derived from Newton's binomial series

$$(1+x)^r = 1 + rx + \tfrac{r(r-1)}{2!}x^2 + \tfrac{r(r-1)(r-2)}{3!}x^3 + \tfrac{r(r-1)(r-2)(r-3)}{4!}x^4 + \ldots.$$

Newton knew that the infinite sum on the right converges to the number on the left whenever $|x| < 1$. (The binomial series is studied in detail in segment 11F of the Problems and Projects section of Chapter 11.) In the current context, $r = \frac{1}{2}$ is relevant, and it is easy to check that in this case

$$\sqrt{1+x} = (1+x)^{\frac{1}{2}} = 1 + \tfrac{1}{2}x - \tfrac{1}{8}x^2 + \tfrac{1}{16}x^3 - \tfrac{5}{128}x^4 + \ldots$$

Now consider the term $\left(\frac{2s}{d^2}x\right)^2$ of the definite integral for the length L. Since the ratio $\frac{s}{2d}$ of sag to center span satisfies $\frac{s}{2d} \le \frac{1}{6}$ for any standard suspension bridge, we can assume that $\frac{2s}{d} \le \frac{2}{3}$. Since $0 \le x \le d$, this tells us that

$$\left(\tfrac{2s}{d^2}x\right)^2 = \left(\tfrac{2s}{d}\right)^2\left(\tfrac{x}{d}\right)^2 \le \left(\tfrac{2s}{d}\right)^2 \le \left(\tfrac{2}{3}\right)^2 = \tfrac{4}{9}.$$

This allows us to replace x by $\left(\frac{2s}{d^2}x\right)^2$ in the binomial series for $\sqrt{1+x}$ to get

$$\sqrt{1+\left(\tfrac{2s}{d^2}x\right)^2} = 1 + \tfrac{1}{2}\left(\tfrac{2s}{d^2}x\right)^2 - \tfrac{1}{8}\left(\tfrac{2s}{d^2}x\right)^4 + \tfrac{1}{16}\left(\tfrac{2s}{d^2}x\right)^6 - \tfrac{5}{128}\left(\tfrac{2s}{d^2}x\right)^8 + \ldots$$

for any x in the interval $0 \le x \le d$. Since $\left(\frac{2s}{d}\right)^2 \le \frac{4}{9}$, the terms

$$\left(\tfrac{2s}{d^2}x\right)^{2k} = \left(\tfrac{2s}{d^2}\right)^{2k}x^{2k} = \left(\tfrac{2s}{d}\right)^{2k}\left(\tfrac{x}{d}\right)^{2k} \le \left(\tfrac{2s}{d}\right)^{2k} \le \left(\tfrac{4}{9}\right)^k$$

get smaller quickly with increasing k. The result is the approximation

$$\sqrt{1+\left(\tfrac{2s}{d^2}x\right)^2} \approx 1 + \tfrac{1}{2}\left(\tfrac{2s}{d^2}x\right)^2 - \tfrac{1}{8}\left(\tfrac{2s}{d^2}x\right)^4 + \tfrac{1}{16}\left(\tfrac{2s}{d^2}x\right)^6$$

for any x in the interval $0 \le x \le d$. By applying the fundamental theorem of calculus term by term, we get

$$\begin{aligned} L &= \int_0^d \sqrt{1+\left(\tfrac{2s}{d^2}x\right)^2}\,dx \approx \left[x + \tfrac{1}{2}\cdot\tfrac{1}{3}\left(\tfrac{2s}{d^2}\right)^2 x^3 - \tfrac{1}{8}\cdot\tfrac{1}{5}\left(\tfrac{2s}{d^2}\right)^4 x^5 + \tfrac{1}{16}\cdot\tfrac{1}{7}\left(\tfrac{2s}{d^2}\right)^6 x^7\right]\Big|_0^d \\ &\approx d + \tfrac{1}{6}\tfrac{4s^2}{d} - \tfrac{1}{40}\tfrac{16s^4}{d^3} + \tfrac{1}{112}\tfrac{64s^6}{d^5} = d + \tfrac{2}{3}\tfrac{s}{d}s - \tfrac{2}{5}\left(\tfrac{s}{d}\right)^3 s + \tfrac{4}{7}\left(\tfrac{s}{d}\right)^5 s. \end{aligned}$$

We'll now compute this approximate value of L for each of the three bridges that we are studying. For the George Washington Bridge, $s = 327$ feet and $\frac{s}{d} = \frac{327}{1750} \approx 0.187$, so that

$$\begin{aligned} L &\approx 1750 + \tfrac{2}{3}(0.187)(327) - \tfrac{2}{5}(0.187^3)(327) + \tfrac{4}{7}(0.187^5)(327) \\ &\approx 1750 + 40.77 - 0.86 + 0.04 \\ &\approx 1789.95 \text{ feet.} \end{aligned}$$

For the Brooklyn Bridge, $s = 128$ feet and $\frac{s}{d} = \frac{128}{798} \approx 0.160$. So

$$\begin{aligned} L &\approx 798 + \tfrac{2}{3}(0.160)(128) - \tfrac{2}{5}(0.160^3)(128) + \tfrac{4}{7}(0.160^5)(128) \\ &\approx 798 + 13.65 - 0.21 + 0.01 \\ &\approx 811.45 \text{ feet.} \end{aligned}$$

For the Golden Gate Bridge, $s = 470$ feet, $\frac{s}{d} = \frac{470}{2100} \approx 0.224$, and

$$\begin{aligned} L &\approx 2100 + \tfrac{2}{3}(0.224)(470) - \tfrac{2}{5}(0.224^3)(470) + \tfrac{4}{7}(0.224^5)(470) \\ &\approx 2100 + 70.19 - 2.11 + 0.15 \\ &\approx 2168.23 \text{ feet.} \end{aligned}$$

These computations tell us that that the term $\frac{4}{7}(\frac{s}{d})^5 s$ contributes little to the length L and that in the context of suspension bridges, the approximation

$$\boxed{L \approx d + \tfrac{2}{3}\tfrac{s}{d}s - \tfrac{2}{5}(\tfrac{s}{d})^3 s}$$

is very accurate.

The binomial series that informed the analysis above is a special case of a *power series*. Power series lead to convenient approximation formulas such as the one developed above for the length of the cable of a suspension bridge. We'll see that they also provide solutions for indefinite integrals and differential equations in situations where solutions with elementary functions are not possible. Refer to the discussion that concludes Section 9.11. We will study power series in some detail in Section 11.7.

10.3 FROM POCKET WATCH TO PSEUDOSPHERE

This story begins in Paris toward the end of the 17th century with Leibniz and his French colleague Claude Perrault. Perrault, himself a distinguished scholar, presented Leibniz with the following problem. Refer to Figure 10.12. A weight B is placed on a horizontal plane, and a string is attached to the weight. When the other end A of the string is moved along the fixed straight line $AA'A''$ in the plane, the weight is pulled along and describes a curve $BB'B''$. Can a precise description of this curve be given? Perrault illustrated the curve by pulling his pocket watch by the free end of its chain along a straight line across a table. Leibniz observed the path of the pocket watch with some care and noticed that the chain was tangent to the

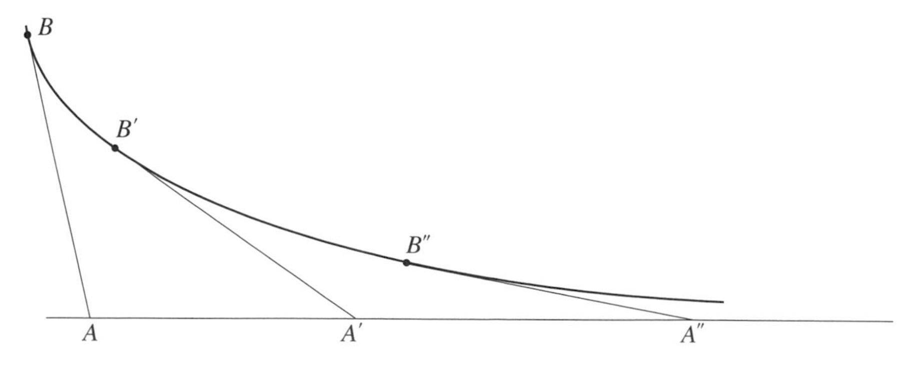

Fig. 10.12

curve throughout the motion. He saw that the problem was to find a curve satisfying the condition that the length of the part of the tangent AB between the axis AA' and the curve BB' is equal to a fixed constant.

To analyze the problem, we'll assume that the weight B moves in a horizontal xy-plane. Start by placing the weight on the positive x-axis and the stretched string along the x-axis so that the free end is at the origin O. Let a be the length of the string and note that $(a, 0)$ is the initial position of the weight B. Now move the free end of the string from the origin up the positive y-axis. See Figure 10.13. The curve that B traces out in the process is the object of our study. This curve was later called *tractrix* (from the

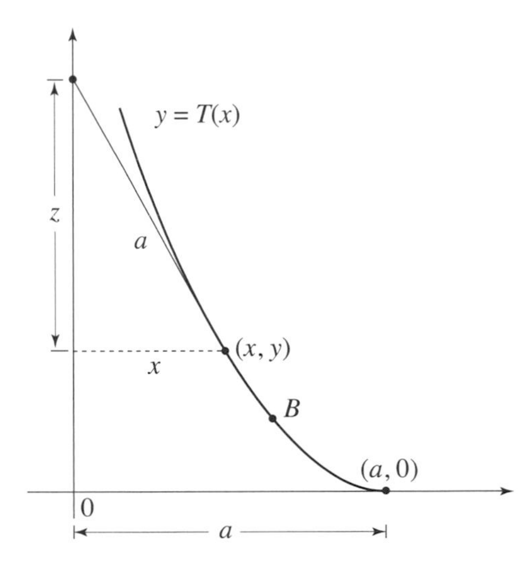

Fig. 10.13

Latin verb *trahere* = "to pull" or "to drag") by Christian Huygens, the famous Dutch astronomer, physicist, and mathematician. Huygens worked in Paris at the time, met Leibniz there, and also studied this curve.

Let (x, y) represent a typical position of the weight B, and let the segment from (x, y) to the y-axis be the string. Suppose that $y = T(x)$ is the function that has the tractrix as its graph. Observe, as Leibniz did, that the string is always tangent to the curve. Figure 10.13 informs us therefore that $\frac{dy}{dx} = T'(x) = -\frac{z}{x}$. By the Pythagorean theorem, $z = \sqrt{a^2 - x^2}$, and therefore

$$\frac{dy}{dx} = T'(x) = -\frac{\sqrt{a^2 - x^2}}{x}.$$

So T is an antiderivative of $y = -\frac{\sqrt{a^2-x^2}}{x}$, and the task is to solve the integral $\int -\frac{\sqrt{a^2-x^2}}{x}\,dx$. Following the hyperbolic substitution strategy illustrated in Section 9.10, we'll let $x = a \operatorname{sech} u$. (The trig substitution $x = a \sin\theta$ could also have been used.) Refer to Figure 7.48. Since $x > 0$ decreases throughout the motion of B, we know that $u \geq 0$. Dividing the identity $\cosh^2 u - \sinh^2 u = 1$ by $\cosh u$, we get $1 - \tanh^2 u = \operatorname{sech}^2 u$, and therefore that $a^2 - x^2 = a^2 - a^2\operatorname{sech}^2 u = a^2\tanh^2 u$. Because $u \geq 0$, we know from Figure 7.47 that $\tanh u \geq 0$, so that $\sqrt{a^2 - x^2} = a\tanh u$. Taking the derivative of $x = a\operatorname{sech} u$, we get

$$\frac{dx}{du} = \frac{d}{du}a(\cosh u)^{-1} = -a(\cosh u)^{-2}\sinh u = -a\frac{\sinh u}{\cosh u}\cdot\frac{1}{\cosh u} = -a\tanh u\cdot\operatorname{sech} u.$$

Transforming the integral into an integral in the variable u, we get

$$\int -\frac{\sqrt{a^2 - x^2}}{x}\,dx = \int -\frac{a\tanh u}{a\operatorname{sech} u}(-\,a\tanh u\cdot\operatorname{sech} u)\,du = \int a\tanh^2 u\,du.$$

It was shown at the end of Section 7.12 that $\frac{d}{du}\tanh u = \operatorname{sech}^2 u = 1 - \tanh^2 u$. It follows therefore that $\frac{d}{du}(u - \tanh u) = 1 - (1 - \tanh^2 u) = \tanh^2 u$ and hence that

$$\int a\tanh^2 u\,du = au - a\tanh u + C.$$

Since $x = a\,\text{sech}\,u$, we know that $u = \text{sech}^{-1}\frac{x}{a}$. So $au - a\tanh u = a\,\text{sech}^{-1}\frac{x}{a} - \sqrt{a^2 - x^2}$. Next, use the formula $\text{sech}^{-1}x = \ln\left(\frac{1+\sqrt{1-x^2}}{x}\right)$ from Section 9.9 to get

$$\int -\frac{\sqrt{a^2 - x^2}}{x}\,dx = a\ln\left(\tfrac{a}{x} + \tfrac{1}{x}\sqrt{a^2 - x^2}\right) - \sqrt{a^2 - x^2} + C.$$

Since $y = T(x)$ is an antiderivative of $-\frac{\sqrt{a^2-x^2}}{x}$, it follows that there is a constant C such that

$$T(x) = a\ln\left(\tfrac{a}{x} + \tfrac{1}{x}\sqrt{a^2 - x^2}\right) - \sqrt{a^2 - x^2} + C.$$

From Figure 10.13, $T(a) = 0$. So $0 = T(a) = a\ln\left(\frac{a}{a} + 0\right) - 0 + C = 0 + 0 + C$. Therefore $C = 0$ and hence

$$T(x) = a\,\text{sech}^{-1}\tfrac{x}{a} - \sqrt{a^2 - x^2} = a\ln\left(\tfrac{a}{x} + \tfrac{1}{x}\sqrt{a^2 - x^2}\right) - \sqrt{a^2 - x^2}$$

This is the explicit description of the tractrix or "pocket watch" curve that Perrault, Leibniz, and Huygens had been looking for.

Refer back to Figure 10.13. Let time $t \geq 0$ elapse, and pull the weight B in such a way that its x-coordinate is given by $x(t) = a\,\text{sech}\,t$ at any time t. Plug $x(t)$ into $T(x) = a\,\text{sech}^{-1}\frac{x}{a} - \sqrt{a^2 - x^2}$ for x. Using the identity $\tanh^2 t = 1 - \text{sech}^2 t$, we get that $T(x(t)) = a(t - \tanh t)$. This informs us that as t advances, the point with coordinates

$$x(t) = a\,\text{sech}\,t \quad \text{and} \quad y(t) = a(t - \tanh t)$$

lies on the graph of $y = T(x)$ of Figure 10.13. Figures 7.47 and 7.48 tell us that $\tanh(0) = 0$ and $\text{sech}(0) = 1$, so that the point is at its initial position $(a, 0)$ at time $t = 0$, and moves up along the curve for increasing $t > 0$. Since the points determined by the equations $x(t) = a\,\text{sech}\,t$ and $y(t) = a(t - \tanh t)$ trace out the tractrix, they are referred to as *parametric equations* of the tractrix. The variable t is the *parameter*. It follows directly from the speed formula $v(t) = \sqrt{x'(t)^2 + y'(t)^2}$ of Section 6.4, the derivative formulas of Section 7.12, and the identity $\tanh^2 t + \text{sech}^2 t = 1$ that

$$v(t) = a\,\tanh t.$$

Another look at Figure 7.47 informs us that the speed $v(t)$ of the point steadily increases from 0 but that it is bounded by a.

The upcoming discussion of the tractrix requires information about the limit

$$\lim_{x\to 0}(x\,\text{sech}^{-1}x) = \lim_{x\to 0}\left(\frac{\text{sech}^{-1}x}{\frac{1}{x}}\right).$$

By Figure 9.39, this is a limit of "$\frac{\infty}{\infty}$" type. An application of L'Hospital's rule together with the differentiation formula for $\text{sech}^{-1}x$ from Section 9.9, and the fact that $\lim\limits_{x\to 0}\frac{\frac{-1}{x\sqrt{1-x^2}}}{-\frac{1}{x^2}} = \lim\limits_{x\to 0}\frac{x^2}{x\sqrt{1-x^2}} = \lim\limits_{x\to 0}\frac{x}{\sqrt{1-x^2}} = 0$, tells us that $\lim\limits_{x\to 0}(x\,\text{sech}^{-1}x) = 0$.

10.3.1 Volume and Surface Area of Revolution of the Tractrix

The shape obtained by revolving the tractrix once about its asymptote—this is the straight line along which the string is pulled—is of historic significance in the development of geometry. So we'll revolve the tractrix of Figure 10.13 once around the y-axis and study its volume and its surface area.

Let's start with the volume. We'll use the strategy of improper integrals (as discussed at the end of Section 9.4) to get around the fact that $y = T(x)$ has a discontinuity at $x = 0$. Let's fix a number c with $0 < c \leq a$ and compute the volume of revolution that the region from $x = c$ to $x = a$ below the curve and

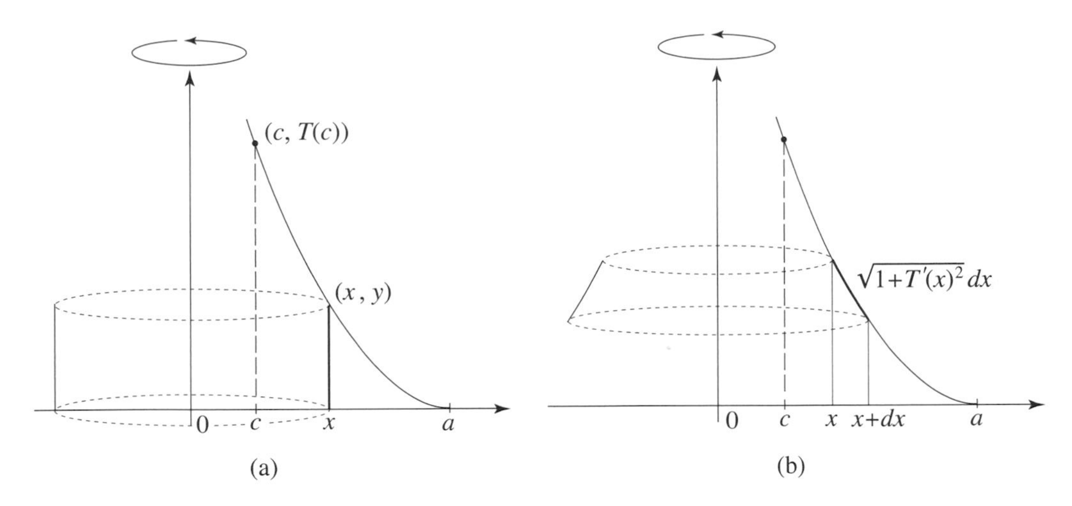

Fig. 10.14

above the x-axis determines. By Formula (V_3) of Section 9.2 with $f(x) = T(x)$ and $g(x) = 0$, the volume of revolution in question is equal to

$$V(c) = 2\pi \int_c^a xT(x)\,dx = 2\pi \int_c^a x(a\,\text{sech}^{-1}\tfrac{x}{a} - \sqrt{a^2 - x^2})\,dx.$$

We'll start the analysis of this integral with $2\pi \int ax\,\text{sech}^{-1}\frac{x}{a}\,dx = 2\pi a \int x\,\text{sech}^{-1}\frac{x}{a}\,dx$. This integral falls to integration by parts. For $u = \text{sech}^{-1}\frac{x}{a}$ and $dv = x\,dx$, we find by making use of the results in Section 9.9.2 that

$$du = \frac{-1}{\frac{x}{a}\sqrt{1 - \frac{x^2}{a^2}}} \cdot \tfrac{1}{a}\,dx = \frac{-1}{\frac{x}{a}\sqrt{a^2 - x^2}}\,dx = \frac{-a}{x\sqrt{a^2 - x^2}}\,dx \quad \text{and} \quad v = \tfrac{1}{2}x^2.$$

By the integration by parts formula of Section 9.7.2,

$$\int x\,\text{sech}^{-1}\tfrac{x}{a}\,dx = \tfrac{1}{2}x^2\text{sech}^{-1}\tfrac{x}{a} + \tfrac{1}{2}a\int \frac{x}{\sqrt{a^2 - x^2}}\,dx.$$

To solve this last integral use the substitution $u = a^2 - x^2$ and $du = -2x\,dx$, to get

$$\tfrac{1}{2}a\int \frac{x}{\sqrt{a^2 - x^2}}\,dx = \tfrac{1}{2}a\int \frac{-\frac{1}{2}}{u^{\frac{1}{2}}}\,du = -\tfrac{1}{4}a\int u^{-\frac{1}{2}}\,du = -\tfrac{1}{4}a \cdot 2u^{\frac{1}{2}} + C = -\tfrac{1}{2}a\sqrt{a^2 - x^2} + C.$$

Therefore

$$\int ax\,\text{sech}^{-1}\tfrac{x}{a}\,dx = \tfrac{1}{2}ax^2\text{sech}^{-1}\tfrac{x}{a} - \tfrac{1}{2}a^2\sqrt{a^2 - x^2} + C.$$

The integral $\int x\sqrt{a^2-x^2}\,dx$ remains. The substitution $u = a^2 - x^2$ and $du = -2x\,dx$ works here too and transforms the integral to $\int -\frac{1}{2}\,u^{\frac{1}{2}}\,du = -\frac{1}{3}u^{\frac{3}{2}} + C$. So

$$\int x\sqrt{a^2-x^2}\,dx = -\tfrac{1}{3}(a^2-x^2)^{\frac{3}{2}} + C.$$

After assembling all the information above, we get

$$\begin{aligned} V(c) &= 2\pi\int_c^a x(a\,\mathrm{sech}^{-1}\tfrac{x}{a} - \sqrt{a^2-x^2})\,dx \\ &= 2\pi(\tfrac{1}{2}ax^2\mathrm{sech}^{-1}\tfrac{x}{a} - \tfrac{1}{2}a^2\sqrt{a^2-x^2} + \tfrac{1}{3}(a^2-x^2)^{\frac{3}{2}})\Big|_c^a \\ &= 2\pi(0 - \tfrac{1}{2}ac^2\mathrm{sech}^{-1}\tfrac{c}{a} + \tfrac{1}{2}a^2\sqrt{a^2-c^2} - \tfrac{1}{3}(a^2-c^2)^{\frac{3}{2}}) \\ &= \pi(a^2\sqrt{a^2-c^2} - \tfrac{2}{3}(a^2-c^2)^{\frac{3}{2}} - ac^2\mathrm{sech}^{-1}\tfrac{c}{a}). \end{aligned}$$

To get the full volume V obtained by revolving the region bounded by the graph of $y = T(x)$ and the x-axis around the y-axis, we push $c \to 0$ and study what happens to $V(c)$. By the observation that concludes the previous section, $\lim_{c\to 0}(ac\,\mathrm{sech}^{-1}\frac{c}{a}) = a\lim_{c\to 0}(c\,\mathrm{sech}^{-1}\frac{c}{a}) = 0$. Therefore $\lim_{c\to 0}(ac^2\,\mathrm{sech}^{-1}\frac{c}{a}) = 0$ as well, and it follows that $\lim_{c\to 0} V(c) = \pi(a^3 - \frac{2}{3}a^3 - 0) = \frac{1}{3}\pi a^3$. We have verified that V is the solution of the improper integral

$$V = 2\pi\int_0^a x(a\,\mathrm{sech}^{-1}\tfrac{x}{a} - \sqrt{a^2-x^2})\,dx = \tfrac{1}{3}\pi a^3.$$

We'll next compute the surface area S of the shape that we have been discussing. Again, the computation relies on the strategy of improper integrals.

Again, fix a number c with $0 < c \le a$, and consider first the surface area $S(c)$ of the solid of revolution in question x to restricted to $c \le x \le a$. In Figure 10.14b, x is a typical coordinate and dx is a small distance. Notice that the surface area formula of Section 9.4 does not apply because we are revolving about the y-axis rather than the x-axis. But the ideas that lead to this formula apply. Turn to Section 9.3 for the fact that the length of the segment of the tractrix depicted in Figure 10.14b is very nearly equal to $\sqrt{1+T'(x)^2}\,dx$. Apply the formula $\pi(r_1 + r_2)z$ for the surface area of the truncated cone (of Figure 9.12b), and conclude that the surface area of the region obtained by revolving this segment of the tractrix once around the y-axis is approximately equal to

$$\pi(x + (x+dx))\sqrt{1+T'(x)^2}\,dx \approx \pi(2x)\sqrt{1+T'(x)^2}\,dx \approx \pi(2x)\sqrt{1+(\tfrac{-\sqrt{a^2-x^2}}{x})^2}\,dx = 2\pi x\sqrt{\tfrac{a^2}{x^2}}\,dx.$$

Combining this with the summation process of the definite integral (as this was desribed several times in Sections 9.2, 9.3, and 9.4) and the fundamental theorem of calculus, we get that

$$S(c) = 2\pi\int_c^a x\sqrt{1+T'(x)^2}\,dx = 2\pi\int_c^a x\sqrt{\tfrac{a^2}{x^2}}\,dx = \int_c^a 2\pi a\,dx = 2\pi ax\Big|_c^a = 2\pi a(a-c).$$

By pushing c to 0, we get $\lim_{c\to 0} S(c) = 2\pi a^2$, and therefore that the area S of the surface obtained by revolving the tractrix $y = T(x)$ around the y-axis is

$$S = 2\pi\int_0^a x\sqrt{1+T'(x)^2}\,dx = 2\pi a^2.$$

10.3.2 *The Pseudosphere*

We saw in the previous section that—even though the solid obtained by revolving the region between the graph of the tractrix $y = T(x)$ and the x-axis one revolution around the y-axis has infinite extent—both its volume V and its surface area S are finite.

Recall the definition and properties of the inverse of a function from Section 9.8. Figure 10.15 shows the graph of the inverse $y = T^{-1}(x)$ of the tractrix on the right of the y-axis. The rule $y = T^{-1}(|x|)$ extends

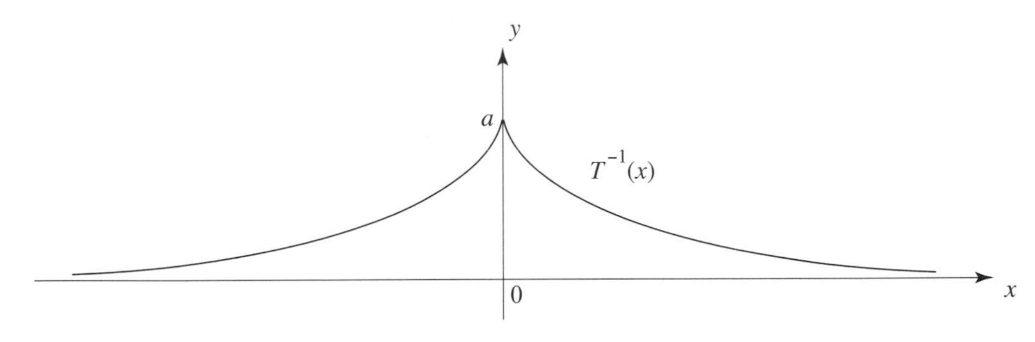

Fig. 10.15

this inverse to a function that is defined for all x. The graph of this extended function is depicted in the figure. Figure 10.16 shows the surface obtained by revolving this extended graph one revolution about the

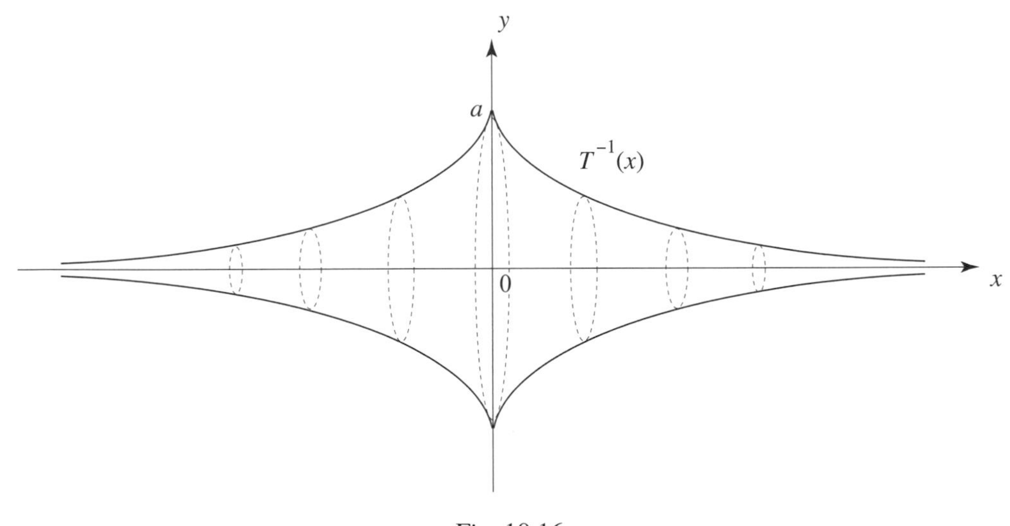

Fig. 10.16

x-axis. This surface is the *pseudosphere* of *radius* a. It follows from the results of Section 10.3.1 that the volume V enclosed by the pseudosphere and its surface area S the pseudosphere are

$$V = \tfrac{2}{3}\pi a^3 \quad \text{and} \quad S = 4\pi a^2$$

respectively. A comparison of these values with the expressions for the volume and surface area of a sphere of radius a—these are $\frac{4}{3}\pi a^2$ and $4\pi a^2$, respectively—suggests that the behavior of these two surfaces is analogous. For this and similar reasons, history has given the name "pseudosphere" to the surface depicted in Figure 10.16.

The pseudosphere has played an important role in the development of geometry. Euclid's *Elements* had provided plane geometry with the structure of an axiomatic system (the word *axiom* is derived from the Greek word *axioma*, meaning "that which is worthy of thought" or "that which commends itself as evident") dominated by five important statements called *Postulates*. Current versions of the five postulates that Euclidsingled out are:

1. Any two points can be joined by a line segment.
2. Any line segment can be extended indefinitely to a straight line.
3. Given any line segment, one can draw a circle with the segment as radius and one of its endpoints as center.
4. All right angles are congruent (any one can be superimposed on the other).
5. Given a straight line and a point not on the line, there is exactly one straight line through the given point parallel to the given line.

In the first three postulates, "line segment" refers to a path of shortest distance connecting two points. In the fifth postulate, "parallel" lines are understood to be lines that do not intersect no matter how far they are extended.

The fifth postulate—it would become known as the *parallel postulate*—was the object of much study by geometers soon after the *Elements* was composed around 300 B.C. Geometers generally thought that only the first four postulates were necessary, and that it should be possible to deduce the fifth postulate from the first four. Over the centuries, many so-called proofs of the parallel postulate were presented. However, none of them was correct. Finally in the early 19th century, over 2000 years after the matter first became an issue, a trio of mathematicians, the German Carl Friedrich Gauss, the Hungarian Janos Bolyai, and the Russian Nicolai Lobachevsky, came to the conclusion that such a proof is impossible. How can such an impossibility be established? In the 1820s, Bolyai and Lobachevsky each published a geometric system as consistent as Euclid's own, in which the first four postulates hold, but the fifth is replaced by its negation: in the plane formed by a line and a point (not on the line), there is more than one line through the point that is parallel to the given line. Gauss was reluctant to discuss even in mathematical circles a geometry that contradicted what seemed so obvious. It took another 40 years before the Italian Eugenio Beltrami discovered what would convince the skeptics: the pseudosphere is a concrete realization of the geometries that Bolyai and Lobachevsky had devised. It is an example of a geometry for which the first four postulates hold but the fifth does not. For a given line and a given point not on the line, there are infinitely many lines on the pseudosphere through the point and parallel to the given line.

The condition "for any line and a point not on the line there are no lines through the point parallel to the given line" is another way to negate the parallel postulate. In this regard, consider the surface of a sphere. A circle on the sphere is a *great circle* if the plane that the circle determines goes through the center of the sphere. A line segment on the surface of the sphere is a path that lies on some great circle. The fact that any two points on the sphere together with the sphere's center determine a plane, tells us that the first postulate holds on the sphere. The next three postulates hold as well. But any two great circles intersect, so that the parallel postulate fails in this spherical geometry.

Consider any three distinct points on either the Euclidean plane, the pseudosphere, or the sphere. Since Postulate 1 provides a line segment between any two of these points, the three points form a triangle. Let α, β, and γ be its three angles. Let's start with a look at the Euclidean case of Figure 10.17a. By the parallel postulate, there is a unique straight line through C that is parallel to the segment AB. Since the extension of the line through C does not meet the extension of the line determined by AB, it follows that the three angles at C are α, γ, and β (from left to right). Therefore $\alpha + \beta + \gamma = 180°$. So in the Euclidean situation, the parallel postulate implies that the interior angles of a triangle add to $180°$. It is conversely the case, that if the interior angles of any triangle add to $180°$, then the parallel postulate holds and the geometry is Euclidean . On the other hand, as Figures 10.17b and 10.17c illustrate, in the case of the pseudosphere, $\alpha + \beta + \gamma < \pi$, and in the case of the sphere, $\alpha + \beta + \gamma > \pi$. What has just been observed is related to the concept of curvature as defined by Gauss. It quantifies how much a surface curves. This

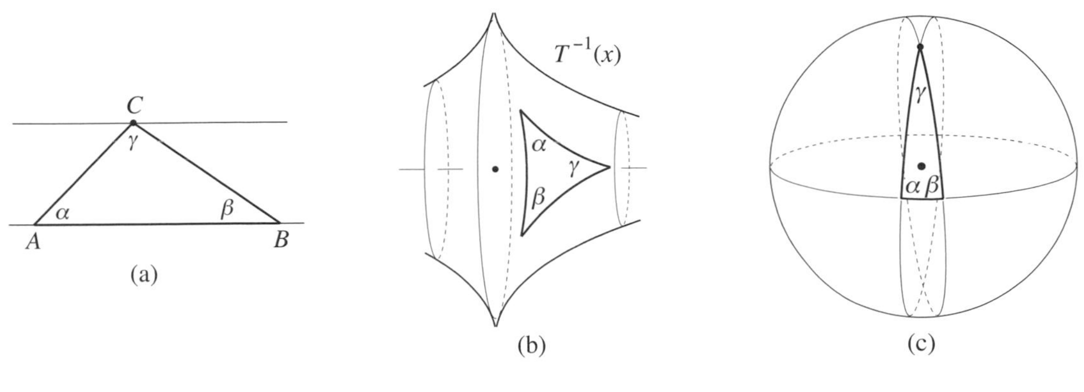

Fig. 10.17

Gaussian curvature is constant in all three cases: Euclidean plane, pseudosphere, and sphere. It is equal to 0 for the plane, $\frac{1}{a^2}$ for the sphere of radius a, and $-\frac{1}{a^2}$ for the pseudosphere of radius a. In the situation of the sphere, this definition of curvature is consistent with what we already knew. The larger the radius of the sphere, the flatter it is.

10.4 CALCULATING THE MOTION OF A PLANET

We learned in Section 3.5 that Kepler's painstaking analysis of the jigsaw puzzle of data that Tycho Brahe had amassed allowed him to conclude that the planets move in accordance with what would later be called Kepler's three laws of planetary motion. We saw in Sections 6.8, 6.9, and 6.10 how Newton provided the explanations for Kepler's observations by combining the underlying laws of physics with the mathematics of calculus. The starting point of this section is Kepler's first law, which informs us that planets and other bodies of the solar system orbit in ellipses that have the Sun at a focal point. After a brief review of some basics, our discussion turns to the study of the mathematical consequences of Kepler's second law, which says that the segment from any planet to the Sun sweeps out equal areas in equal times. We will see how this law provides numerically precise and predictive information about how the orbits are traced out. Our discussion pursues these historic insights from today's point of view.

Let P be a planet, comet, or asteroid in an elliptical orbit around the Sun S. Our discussion also applies to the situation of a satellite (natural or artificial) P in an elliptical orbit around a planet or asteroid S. By considering the mass of P to be concentrated at its center, we regard P to be a point-mass (an assumption also made in Sections 6.8 and 6.9), and we'll often refer to P as a "point." The body S, also regarded as a point-mass, is located at one of the focal points of the ellipse. The place in an elliptical orbit where P is closest to S is called *periapsis* in general, and *perihelion* if S is the Sun. The place in the orbit where P is farthest from S is called *apoapsis*, and *aphelion* if S is the Sun. (This terminology comes from the Greek via Latin: *apsis* = arch or vault, *peri* = "near," *apo* = "away from," and *helio* = "Sun.")

We begin with some basic properties of the ellipse described in Section 4.4. The focal axis of the ellipse is the line that the two focal points determine, and the center of the ellipse is the point on the focal axis that lies halfway between the two focal points. Let a and b be the semimajor and semiminor axes of the ellipse, respectively, and let c be the distance between the center of the ellipse and either focal point. Recall that $a^2 = b^2 + c^2$. Note also that a is the distance between the center of the ellipse and the periapsis position or, equivalently, the apopasis position. The ratio $\varepsilon = \frac{c}{a}$ is the eccentricity of the ellipse. The eccentricity satisfies $0 \leq \varepsilon < 1$. If $\varepsilon = 0$, then the ellipse is a circle and there are no periapsis and apoapsis positions to single out. The closer ε is to 0, the closer the ellipse is to a circle, and the closer ε is to 1, the flatter the ellipse. Since the eccentricity is $\varepsilon = \frac{c}{a}$, it follows that $a - c = a - a\varepsilon = a(1 - \varepsilon)$ and $a + c = a + a\varepsilon = a(1 + \varepsilon)$ are the respective distances of the periapsis and apoapsis positions from S. Since

$$\frac{(a - c) + (a + c)}{2} = \frac{2a}{2} = a,$$

the semimajor axis of the orbit is the average of the distances from P to S at periapsis and apoapsis.

We'll now begin to quantify the motion of P around S. When P is at periapsis, click your stopwatch. This is time $t = 0$. We'll let the time $t \geq 0$ flow forward from this instant. Consider P in typical position at elapsed time t. Let $A(t)$ be the area the segment from S to P has swept out during time t. We saw in Section 6.8 that for a given orbit, $\frac{A(t)}{t}$ is a constant. We called it Kepler's constant for the orbit and denoted it by κ. So

$$A(t) = \kappa t, \text{ for any } t \geq 0.$$

Different orbits, different Kepler's constants κ. The *period* T of the orbit is the time it takes for the ellipse to be traced out exactly once. It was shown at the end of Section 5.7 that the area of an ellipse with semimajor axis a and semiminor axis b is $ab\pi$. Therefore

$$\kappa = \frac{ab\pi}{T}.$$

At any time $t \geq 0$, let $r(t)$ be the distance between P and S and $\alpha(t)$ the angle in radians that the segment PS makes with the focal axis (in the direction of periapsis). Figure 10.18 captures what has been

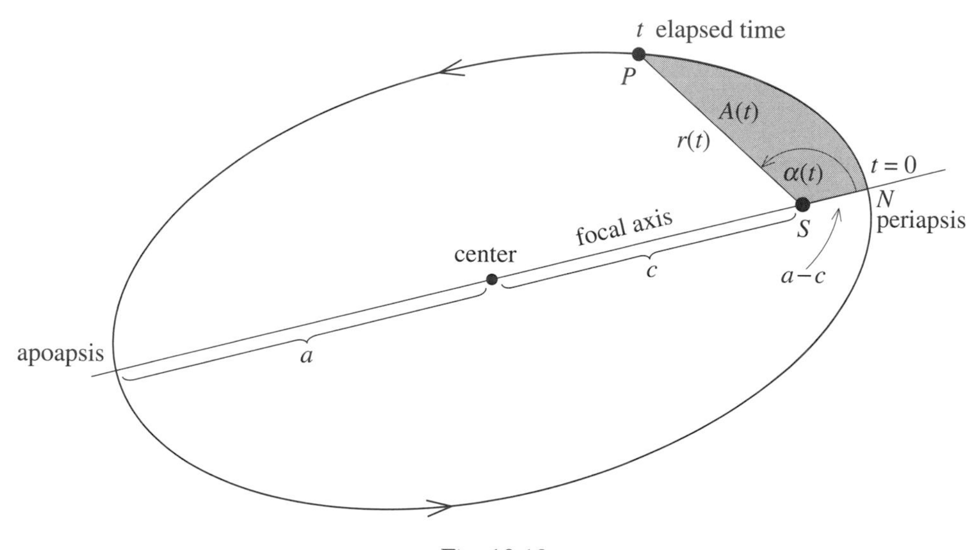

Fig. 10.18

described. The question that will be pursued is this: How exactly does P trace the ellipse out? Can the position of P in its orbit be determined by knowing the elapsed time t as well as the constants a, ε, and T of the orbit? Put another way, the question is this: since the functions

$$r(t) \quad \text{and} \quad \alpha(t)$$

provide the precise position of P in its orbit relative to S, can $r(t)$ and $\alpha(t)$ be determined explicitly in terms of the elapsed time t (and the orbital constants)? We'll ask similar questions about the speed and the direction of P at any time in its orbit. The only fact that will be needed from Newton's description of the connection between the geometry of the orbit of P and the magnitude of the gravitational force that keeps it there is his formulation

$$\frac{a^3}{T^2} = \frac{GM}{4\pi^2}$$

of Kepler's third law. Notice that it tells us how the orbit of P is related to the product of the universal gravitational constant G and the mass M of the body S that P orbits. The units that we will use are MKS—refer to Section 6.10—but the year as unit of time and the kilometer as unit of distance will also be relevant. Recall (again from from Section 6.10) that $G \approx 6.67384 \times 10^{-11} \frac{\text{m}^3}{\text{kg}\cdot\text{s}^2}$ in MKS.

10.4.1 Determining Position in Terms of Time

The first thing that we'll do is place an xy-coordinate system in such a way that the origin is at the center of the ellipse and the focal axis coincides with the x-axis. In this xy-coordinate system, the equation of the ellipse is $\frac{x^2}{a^2} + \frac{y^2}{b^2} = 1$. Figure 10.19 shows the orbiting body P in typical position at elapsed time t. We have chosen a vantage point from which the point P moves counterclockwise around the ellipse of the figure. The coordinates x and y of P vary with time. They are therefore functions $x = x(t)$ and $y = y(t)$

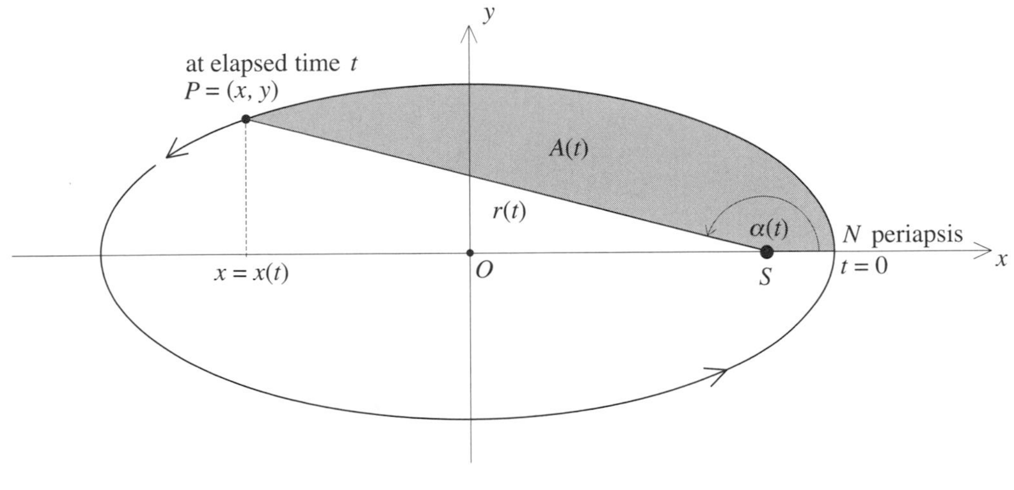

Fig. 10.19

of t. The point N (N for near) designates the periapsis position. The shaded sector of the ellipse is defined by the segments SP and SN. We know from our earlier discussion that it has area $A(t) = \kappa t$.

As important facilitating construction, let's surround the ellipse of Figure 10.19 with a circle of radius a, center the origin, and hence equation $x^2 + y^2 = a^2$. See Figure 10.20. It is of historical interest to note that when Kepler first studied these matters, he also surrounded the ellipse with such a circle. The point P_0 is obtained by projecting P vertically onto the circle. Its coordinates are $x = x(t)$ and $y_0 = y_0(t)$. Let $\beta = \beta(t)$ be the angle in radians determined by P_0, O, and the positive x-axis. The area function $B(t)$ is defined by the shaded region in the figure. The formula for the area of a circular sector established in Section 2.2 tells us that $B(t) = \frac{1}{2}a^2\beta(t)$.

Observe that each completed orbit adds 2π to the angles α and β, and $ab\pi$ and πa^2, respectively, to the areas A and B. In some situations, it may be of interest to know where P is before it reaches periapsis at time $t = 0$. Such a position of P is assigned a negative time t (where $|t|$ is the time for P to reach periapsis). For $t \geq 0$, the angles $\alpha(t)$ and $\beta(t)$, are measured counterclockwise. So both are positive. For $t < 0$, the angles $\alpha(t)$ and $\beta(t)$ are measured clockwise, and are negative. (Refer to Section 4.6 for the details in this regard.) If $t < 0$, then $A(t)$ and $B(t)$ are the negatives of the areas that the segments SP and OP_0 trace out during the motion of P from time t to $t = 0$.

We will assume—for any planet, moon, comet, asteroid, or satellite in elliptical orbit—that P moves smoothly along its path. There are no fits, stops, and starts. So $r = r(t)$, $\alpha = \alpha(t)$, $x = x(t)$, $y = y(t)$, and $\beta = \beta(t)$ are smooth—that is to say, differentiable—functions of t.

The solution of the problem to determine $r(t)$ and $\alpha(t)$ and hence the position of P proceeds as follows. Step 1 expresses r and α in terms of β. Step 2 finds an equation that links β and t. Step 3 solves this equation for β in terms of t. After this β is inserted into the equations for r and α derived in Step 1, both r and α are determined as functions of t. Notice that the angle β is the key to the solution.

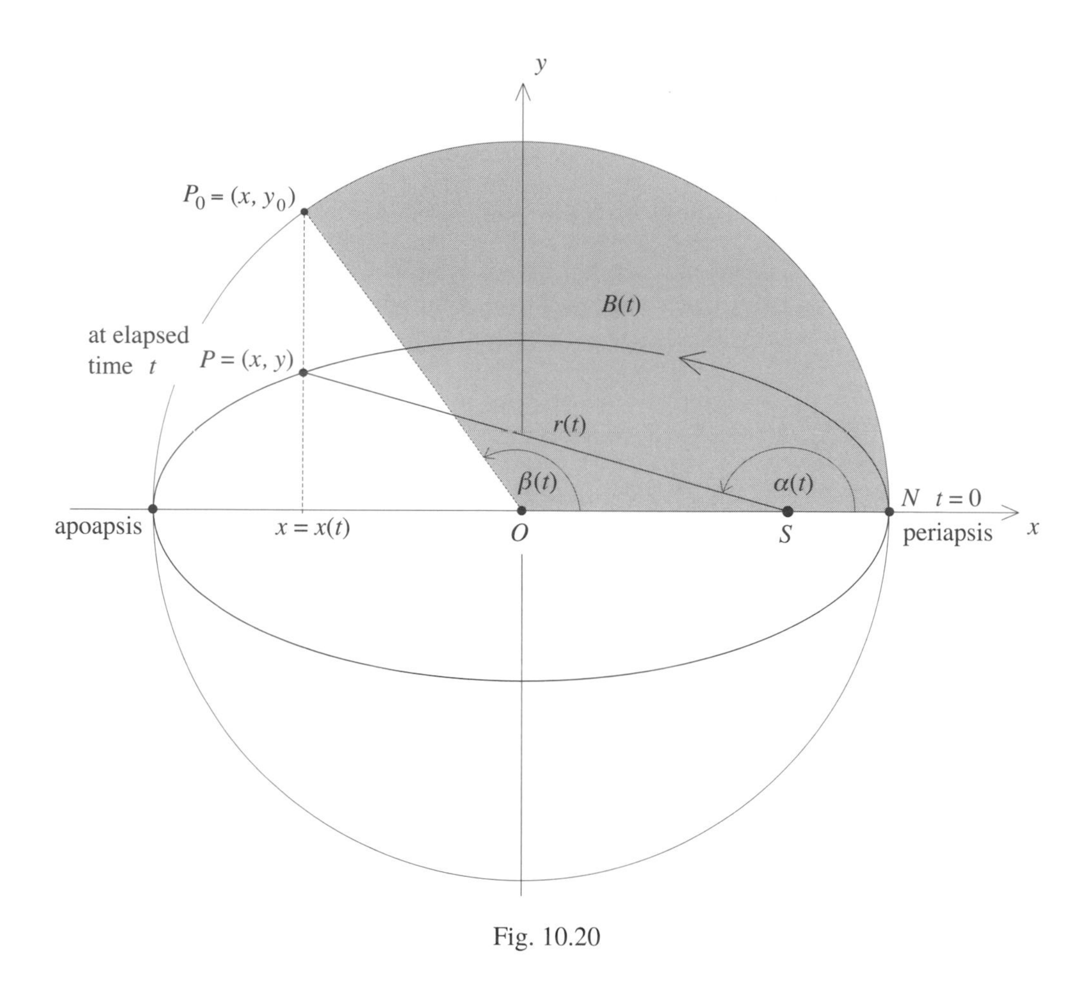

Fig. 10.20

Steps 1 and 2 rely on an analysis of Figures 10.19 and 10.20. The argument includes the possibility that P might have already completed many orbits since $t = 0$. But to enhance the clarity of the illustrations and the presentation, the primary focus is on the situation where P is in its first orbit. This is implies in particular that t is positive. However, the discussion that follows also applies (with some modifications) to the situation of a negative t.

Step 1. We first relate the ellipse to the circle. Solving $\frac{x^2}{a^2} + \frac{y^2}{b^2} = 1$ for y, we get $\frac{y^2}{b^2} = 1 - \frac{x^2}{a^2} = \frac{a^2-x^2}{a^2}$. So $y^2 = \frac{b^2}{a^2}(a^2 - x^2)$ and hence $y = \pm\frac{b}{a}\sqrt{a^2 - x^2}$. Since $x^2 + y_0^2 = a^2$, we get $y_0 = \pm\sqrt{a^2 - x^2}$. A look at Figure 10.20 tells us that P and P_0 are always on the same side of the x-axis. In other words, y and the corresponding y_0 always have the same sign. Therefore

$$y = \frac{b}{a}\, y_0.$$

It follows from the discussion about the sine and cosine in Section 4.6 that the x- and y-coordinates of the point P_0 are $x(t) = a\cos\beta(t)$ and $y_0(t) = a\sin\beta(t)$, respectively. Thus

$$\boxed{x(t) = a\cos\beta(t) \quad \text{and} \quad y(t) = b\sin\beta(t)}$$

We'll turn to Figure 10.20 and apply the Pythagorean theorem. Since the distance OS between the center and the focus of the ellipse is equal to c with $c = a\varepsilon$ and $a^2 = b^2 + c^2$, we get

$$\begin{aligned} r(t)^2 &= (c-x(t))^2+y(t)^2 = c^2-2cx(t)+x(t)^2+y(t)^2 = c^2-2cx(t)+x(t)^2+\tfrac{b^2}{a^2}y_0(t)^2 \\ &= c^2-2cx(t)+x(t)^2+\tfrac{b^2}{a^2}(a^2-x(t)^2) = a^2-2cx(t)+x(t)^2-\tfrac{b^2}{a^2}x(t)^2 \\ &= a^2-2cx(t)+x(t)^2-\tfrac{a^2-c^2}{a^2}x(t)^2 = a^2-2cx(t)+x(t)^2-x(t)^2+\tfrac{c^2}{a^2}x(t)^2 \\ &= a^2-2a\varepsilon x(t)+\varepsilon^2x(t)^2 = (a-\varepsilon x(t))^2. \end{aligned}$$

Because $a \geq x(t) \geq \varepsilon x(t)$, $a - \varepsilon x(t) \geq 0$. Since $r(t) \geq 0$ and $r(t)^2 = (a-\varepsilon x(t))^2$, it follows that $r(t) = a - \varepsilon x(t)$. The substitution $x(t) = a\cos\beta(t)$ provides the equality

$$\boxed{r(t) = a(1-\varepsilon\cos\beta(t))}$$

The link between $\alpha(t)$ and $\beta(t)$ relies on basic properties of the cosine. Note first that

$$\cos\alpha(t) = -\cos(\pi-\alpha(t)) = -\frac{c-x(t)}{r(t)} = \frac{x(t)-c}{r(t)} = \frac{a\cos\beta(t)-a\varepsilon}{a(1-\varepsilon\cos\beta(t))} = \frac{\cos\beta(t)-\varepsilon}{1-\varepsilon\cos\beta(t)}.$$

By a double application of the trig identity

$$\tan^2\tfrac{\theta}{2} = \frac{1-\cos\theta}{1+\cos\theta}$$

established in Problem 1.23i,

$$\begin{aligned} \tan^2\tfrac{\alpha(t)}{2} &= \frac{1-\cos\alpha(t)}{1+\cos\alpha(t)} = \frac{1-\frac{\cos\beta(t)-\varepsilon}{1-\varepsilon\cos\beta(t)}}{1+\frac{\cos\beta(t)-\varepsilon}{1-\varepsilon\cos\beta(t)}} = \frac{\frac{1-\varepsilon\cos\beta(t)-\cos\beta(t)+\varepsilon}{1-\varepsilon\cos\beta(t)}}{\frac{1-\varepsilon\cos\beta(t)+\cos\beta(t)-\varepsilon}{1-\varepsilon\cos\beta(t)}} \\ &= \frac{(1+\varepsilon)-(1+\varepsilon)\cos\beta(t)}{(1-\varepsilon)+(1-\varepsilon)\cos\beta(t)} = \frac{(1+\varepsilon)(1-\cos\beta(t))}{(1-\varepsilon)(1+\cos\beta(t))} \\ &= \left(\tfrac{1+\varepsilon}{1-\varepsilon}\right)\tan^2\tfrac{\beta(t)}{2}. \end{aligned}$$

Starting at $t = 0$, follow the motion of P around the ellipse. Note that when $\alpha(t) = 0, \pi, 2\pi, 3\pi, 4\pi, \ldots$, then $\beta(t) = 0, \pi, 2\pi, 3\pi, 4\pi, \ldots$, and when $\alpha(t)$ lies in one of the intervals $(0,\pi), (\pi, 2\pi), (2\pi, 3\pi), \ldots$, then $\beta(t)$ lies in the same interval. Therefore if $\frac{\alpha(t)}{2}$ lies in one of the intervals $(0, \frac{\pi}{2}), (\frac{\pi}{2}, \frac{3\pi}{2}), (\frac{3\pi}{2}, 2\pi), (2\pi, \frac{5\pi}{2}), \ldots$, then $\frac{\beta(t)}{2}$ lies in that interval as well. It follows from the graph of the tangent—refer to Figure 4.26—that $\tan\frac{\alpha(t)}{2}$ and $\tan\frac{\beta(t)}{2}$ are either both positive or both negative. Since $\tan^2\frac{\alpha(t)}{2} = \left(\frac{1+\varepsilon}{1-\varepsilon}\right)\tan^2\frac{\beta(t)}{2}$, we can conclude that

$$\boxed{\tan\tfrac{\alpha(t)}{2} = \sqrt{\tfrac{1+\varepsilon}{1-\varepsilon}}\,\tan\tfrac{\beta(t)}{2}}$$

When $\alpha(t)$ and $\beta(t)$ are multiples of π, then neither $\tan\frac{\alpha(t)}{2}$ nor $\tan\frac{\beta(t)}{2}$ is defined. But in this case, $\alpha(t)$ and $\beta(t)$ are the same multiple of π. So $\alpha(t) = \beta(t)$. The equation derived above is *Gauss's equation*. It is named after its discoverer, the German mathematician-astronomer whom we met in a completely different context in Section 10.3. Carl Friedrich Gauss (1777–1855) had an exceptional influence on many fields of mathematics and science and is often ranked along with Archimedes and Newton as one of history's three most brilliant mathematicians.

Recall that T is the period of the orbit. Since the elapsed times from $t = 0$ to $t = \frac{T}{2}$ and from $t = \frac{T}{2}$ to $t = T$ are the same, it follows that $A(\frac{T}{2}) + A(\frac{T}{2}) = A(T) = ab\pi$, so that $A(\frac{T}{2}) = \frac{ab\pi}{2}$. This means that P is at its apoapsis position when $t = \frac{T}{2}$ (and at times $t = \frac{T}{2} + T, \frac{T}{2} + 2T, \ldots$ as well). Let k_t be the number of complete orbits that P has traced out during the elapsed time t, and observe that $t = t_1 + k_tT$, where $0 \leq t_1 < T$. Suppose that $0 < t_1 < \frac{T}{2}$. Observe that in this case, $\alpha(t) = \varphi(t) + 2\pi k_t$ for an angle $\varphi(t)$ with $0 < \varphi(t) < \pi$. Suppose that $\frac{T}{2} < t_1 < T$. In this case, $\pi < \alpha(t) - 2\pi k_t < 2\pi$. So $-\pi < \alpha(t) - 2\pi k_t - 2\pi < 0$, and hence $-\pi < \alpha(t) - 2\pi(k_t + 1) < 0$. With $\varphi(t) = \alpha(t) - 2\pi(k_t + 1)$, we get that $\alpha(t) = \varphi(t) + 2\pi(k_t + 1)$, where $-\pi < \varphi(t) < 0$. Now set $n_t = k_t$ when P is in motion from periapsis to apoapsis, and $n_t = k_t + 1$ when P moves from apoapsis to periapsis. Notice that $\frac{\alpha(t)}{2} = \frac{\varphi(t)}{2} + \pi n_t$ in either situation. Since $-\frac{\pi}{2} < \frac{\varphi(t)}{2} < \frac{\pi}{2}$, it follows from the definition of the inverse tangent function in Section 9.9.1 that $\tan^{-1}(\tan \frac{\alpha(t)}{2}) = \frac{\varphi(t)}{2}$. So $\alpha(t) = 2\tan^{-1}(\tan \frac{\alpha(t)}{2}) + 2\pi n_t$. Together with Gauss's equation, this implies that

$$\boxed{\alpha(t) = 2\tan^{-1}\left(\sqrt{\tfrac{1+\varepsilon}{1-\varepsilon}}\,\tan \tfrac{\beta(t)}{2}\right) + 2n_t\pi}$$

As noted before, when $\beta(t)$ is a multiple of π, then the term $\tan \frac{\beta(t)}{2}$ is not defined. But as we have already seen, in this case $\alpha(t) = \beta(t)$.

Step 2. The connection between t and $\beta(t)$ that needs to be established comes from an analysis that relates the areas $A(t)$ and $B(t)$ for any elapsed time t.

Assume that P is in the first half of its initial orbit, and refer back to Figure 10.20. Study the circular section determined by the segments $x(t)N$ and $x(t)P_0$. Notice that it consists of the circular sector NOP_0 and the triangle $\Delta Ox(t)P_0$. Since the area of $B(t)$ is $\frac{1}{2}\beta(t)a^2$, we see that the area of this circular section is equal to

$$\int_{x(t)}^{a} \sqrt{a^2 - x^2}\,dx = \tfrac{1}{2}\beta(t)a^2 - \tfrac{1}{2}x(t)y_0(t)\,.$$

Notice that this holds for $x(t)$ negative (as in Figure 10.20) or positive. By multiplying through by $\frac{b}{a}$, we get that

$$\int_{x(t)}^{a} \tfrac{b}{a}\sqrt{a^2 - x^2}\,dx = \tfrac{1}{2}\beta(t)ab - \tfrac{1}{2}x(t)y(t)$$

is the area of the elliptical section of Figure 10.20 determined by the segments $x(t)N$ and $x(t)P$. After subtracting the area $\frac{1}{2}(c - x(t))y(t)$ of the triangle $\Delta Sx(t)P$ from this elliptical section, we get that $A(t) = \frac{1}{2}\beta(t)ab - \frac{1}{2}x(t)y(t) - \frac{1}{2}(c - x(t))y(t) = \frac{1}{2}\beta(t)ab - \frac{1}{2}cy(t)$. Because $c = \varepsilon a$ and $y(t) = b\sin\beta(t)$, we can conclude that

$$A(t) = \tfrac{1}{2}ab\beta(t) - \tfrac{1}{2}\varepsilon ab\sin\beta(t).$$

(Note that if $x(t) \geq c$, then $\frac{1}{2}(c - x(t))y(t)$ is negative, so that the subtraction adds the area of $\Delta Sx(t)P$ to the elliptical section. But this is exactly what needs to be done in this case.)

We'll next verify this equality without the restriction on the position of P. Let's suppose that P is in the second half of its first orbit at time t. Let t' be the previous moment in the orbit for which $x(t') = x(t)$. By considering the positions of P at the two instants t and t' in Figure 10.19, notice that $A(t) + A(t') = ab\pi$. Doing the same thing in Figure 10.20 shows that $\beta(t) + \beta(t') = 2\pi$. By applying the result already verified,

$$A(t') = \tfrac{1}{2}ab\beta(t') - \tfrac{1}{2}\varepsilon ab\sin\beta(t').$$

Since $A(t') = ab\pi - A(t)$ and $\beta(t') = 2\pi - \beta(t)$, we get by using basic properties of the sine that

$$ab\pi - A(t) = \tfrac{1}{2}ab(2\pi - \beta(t)) - \tfrac{1}{2}\varepsilon ab\sin(2\pi - \beta(t)) = ab\pi - \tfrac{1}{2}ab\beta(t) + \tfrac{1}{2}\varepsilon ab\sin\beta(t).$$

Therefore

$$A(t) = \tfrac{1}{2}ab\beta(t) - \tfrac{1}{2}\varepsilon ab\sin\beta(t).$$

Suppose that P is in its second orbit. So the elapsed time is $t = t_1 + T$ with $0 \le t_1 \le T$, the area swept out is $A(t) = A(t_1 + T) = A(t_1) + ab\pi$, and $\beta(t) = \beta(t_1 + T) = \beta(t_1) + 2\pi$. Since P is in its first orbit at time t_1,

$$\begin{aligned} A(t) &= ab\pi + A(t_1) = ab\pi + \tfrac{1}{2}ab\,\beta(t_1) - \tfrac{1}{2}\varepsilon ab\sin\beta(t_1) \\ &= \tfrac{1}{2}ab(\beta(t_1) + 2\pi) - \tfrac{1}{2}\varepsilon ab\sin(\beta(t_1) + 2\pi) \\ &= \tfrac{1}{2}ab(\beta(t) - \tfrac{1}{2}\varepsilon ab\sin(\beta(t)). \end{aligned}$$

Doing this for every additional orbit tells us that

$$A(t) = \tfrac{1}{2}ab\beta(t) - \tfrac{1}{2}\varepsilon ab\sin\beta(t)$$

holds for any $t \ge 0$. From the definition of Kepler's constant, $\frac{A(t)}{t} = \kappa = \frac{ab\pi}{T}$. Therefore $A(t) = \frac{tab\pi}{T}$, and hence $\frac{1}{2}ab\beta(t) - \frac{1}{2}\varepsilon ab\sin\beta(t) = \frac{tab\pi}{T}$. So $\beta(t) - \varepsilon\sin\beta(t) = \frac{2\pi t}{T}$. Because $\frac{4\pi^2}{T^2} = \frac{GM}{a^3}$, this last term is also equal to $\frac{1}{a}\sqrt{\frac{GM}{a}}\,t$. We have arrived at the equation

$$\boxed{\beta(t) - \varepsilon\sin\beta(t) = \frac{2\pi t}{T} = \frac{1}{a}\sqrt{\frac{GM}{a}}\,t}$$

It is known as *Kepler's equation* to celebrate its discoverer. It provides the connection between $\beta(t)$ and t that Step 2 called for. (As a historical aside, we note that Kepler referred to the quantities $\frac{2\pi t}{T}$, $\beta(t)$, and $\alpha(t)$ as the *mean anomaly, eccentric anomaly,* and *true anomaly,* respectively. These terms are still in use today.)

Example 10.1. We'll illustrate our discussion with a look at Earth's orbit. We know, as a consequence of Kepler's equal area law, that Earth moves faster in its orbit when it is near perihelion than when it is farther away. Figure 10.21 considers Earth in position E at the "top" of its orbit, exactly halfway between its perihelion and aphelion positions. This corresponds to precisely one-quarter of Earth's orbit. So will it take Earth exactly one-quarter of a year after perihelion to arrive there? Or will it be less than that, since Earth moves faster near perihelion? If so, how much less? Kepler's equation provides the answer. Let t be

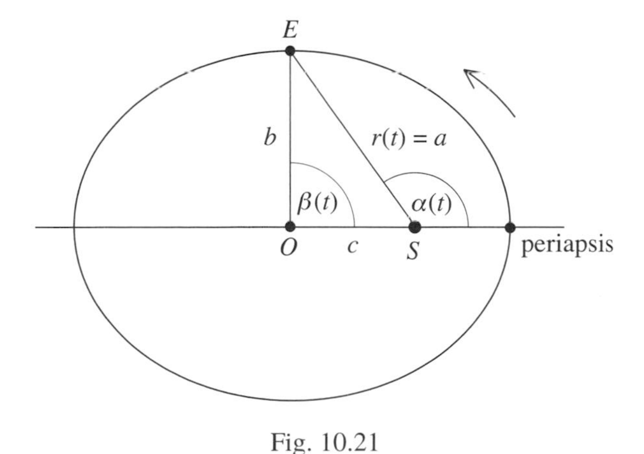

Fig. 10.21

the time its takes for Earth to move from perihelion to position E. Refer to Figure 10.20 and observe that at this position, $\beta(t) = \frac{\pi}{2}$. By inserting this value and Earth's orbital eccentricity $\varepsilon \approx 0.0167$ into Kepler's equation, we get

$$\tfrac{\pi}{2} - (0.0167) \cdot 1 \approx \tfrac{2\pi t}{T}.$$

We know that Earth's period T is one year. Taking T to be equal to 365.25 days, and solving Kepler's equation for t, we get

$$t \approx \tfrac{365.25}{2\pi}(\tfrac{\pi}{2} - 0.0167) = 365.25(\tfrac{1}{4} - \tfrac{0.0167}{2\pi}) \approx 90.34 \text{ days.}$$

Since $\frac{365.25}{4} \approx 91.31$ days, this is close to a day less than a quarter of a year. We now also know that Earth takes $\frac{365.25}{2} - 90.34 \approx 92.29$ days to traverse the second quarter of its orbit.

Let's continue on to Step 3. Consider the function $f(x) = x - \varepsilon \sin x$. This function is continuous because both x and $\varepsilon \sin x$ are. Think about the values of $f(x)$ for large positive and negative x and observe—because $|\varepsilon \sin x| \le \varepsilon < 1$—that for any real number y there is an x such that $f(x) = y$. Since, as we will see in a moment, $f(x) = x - \varepsilon \sin x$ is an increasing function, it follows that for any given t, Kepler's equation has a unique solution $\beta(t)$. The trick is to find it, or at least to provide an estimate for it.

This is the goal of Step 3. It applies a method of successive approximations to tell us for any given elapsed time t what the corresponding $\beta(t)$ is that satisfies Kepler's equation. The method of successive approximations in question will rely critically on the inequality $|\sin x_1 - \sin x_2| \le |x_1 - x_2|$ for any real numbers x_1 and x_2. We will verify it by showing that

$$|\sin x_1 - \sin x_2| < |x_1 - x_2| \quad \text{whenever} \quad x_1 \neq x_2.$$

Consider the functions $f(x) = x - \sin x$ and $g(x) = x + \sin x$. Their derivatives are $f'(x) = 1 - \cos x$ and $g'(x) = 1 + \cos x$. Because $1 > \cos x$, except when $x = 0, \pm 2\pi, \pm 4\pi, \ldots$ for which $\cos x = 1$, it follows that $f'(x) > 0$, except for the isolated points for which $f'(0) = 0$. So $y = f(x)$ is an increasing function of x and hence $f(x_1) > f(x_2)$ whenever $x_1 > x_2$. A similar argument shows that $g(x_1) > g(x_2)$ whenever $x_1 > x_2$. To verify that

$$|\sin x_1 - \sin x_2| < |x_1 - x_2|$$

for any x_1 and x_2 with $x_1 \neq x_2$, we may take $x_1 > x_2$ (or else we can work with $x_2 > x_1$). It follows from what was established about $f(x)$ and $g(x)$ that

$$x_1 - \sin x_1 > x_2 - \sin x_2 \quad \text{and} \quad x_1 + \sin x_1 > x_2 + \sin x_2.$$

So

$$x_1 - x_2 > \sin x_1 - \sin x_2 \quad \text{and} \quad x_1 - x_2 > \sin x_2 - \sin x_1.$$

Since $|\sin x_1 - \sin x_2|$ is equal to either $\sin x_1 - \sin x_2$ or $\sin x_2 - \sin x_1$, the verification is complete.

Step 3. We now solve $\beta(t) - \varepsilon \sin \beta(t) = \frac{2\pi t}{T}$ for $\beta(t)$ by successive approximations. The application of any such method to the solution of an equation starts with an educated guess, or an informed initial stab, at a solution. This initial educated guess is then refined step by step to any desired or required degree of accuracy.

1. The first stab at $\beta(t)$ is $\beta_1 = \frac{2\pi t}{T}$. By Kepler's equation, $|\beta(t) - \beta_1| = \left|\beta(t) - \frac{2\pi t}{T}\right| = |\varepsilon \sin \beta(t)| \le \varepsilon$. Because $\varepsilon < 1$ and usually much smaller (see Table 10.1), β_1 approximates $\beta(t)$. So $\beta_1 = \frac{2\pi t}{T}$ is a good first guess.

2. The approximation step: after the angle β_i has been determined (in radians), the next angle β_{i+1} is given by $\beta_{i+1} = \frac{2\pi t}{T} + \varepsilon \sin \beta_i = \beta_1 + \varepsilon \sin \beta_i$ (in radians).

Applying the approximation step (2) to $\beta_1 = \frac{2\pi t}{T}$ gives the new angle $\beta_2 = \frac{2\pi t}{T} + \varepsilon \sin \beta_1$. Repeating this with β_2, we get $\beta_3 = \frac{2\pi t}{T} + \varepsilon \sin \beta_2$. Doing this again and again, we get $\beta_4 = \frac{2\pi t}{T} + \varepsilon \sin \beta_3$, $\beta_5 = \frac{2\pi t}{T} + \varepsilon \sin \beta_4, \ldots$, and $\beta_i = \frac{2\pi t}{T} + \varepsilon \sin \beta_{i-1}, \ldots$. The big question is this: does the sequence of numbers

$$\beta_1, \beta_2, \beta_3, \ldots, \beta_i, \ldots$$

close in on the solution $\beta(t)$ of $\beta(t) - \varepsilon \sin\beta(t) = \frac{2\pi t}{T}$? This is indeed the case. Since $\beta(t) = \frac{2\pi t}{T} + \varepsilon \sin\beta(t)$ by Kepler's equation, it follows that

$$\beta(t) - \beta_2 = (\tfrac{2\pi t}{T} + \varepsilon \sin\beta(t)) - (\tfrac{2\pi t}{T} + \varepsilon \sin\beta_1) = \varepsilon(\sin\beta(t) - \sin\beta_1).$$

Therefore, using inequalities established earlier,

$$|\beta(t) - \beta_2| = \varepsilon|\sin\beta(t) - \sin\beta_1| \leq \varepsilon|\beta(t) - \beta_1| \leq \varepsilon^2.$$

In the same way,

$$\beta(t) - \beta_3 = (\tfrac{2\pi t}{T} + \varepsilon \sin\beta(t)) - (\tfrac{2\pi t}{T} + \varepsilon \sin\beta_2) = \varepsilon(\sin\beta(t) - \sin\beta_2),$$

so that

$$|\beta(t) - \beta_3| = \varepsilon|\sin\beta(t) - \sin\beta_2| \leq \varepsilon|\beta(t) - \beta_2| \leq \varepsilon^3.$$

Repeating this computation again and again shows that

$$|\beta(t) - \beta_4| \leq \varepsilon^4,\ |\beta(t) - \beta_5| \leq \varepsilon^5, \ldots,\ |\beta(t) - \beta_i| \leq \varepsilon^i, \ldots .$$

Since $\varepsilon < 1$, the powers $\varepsilon^2, \varepsilon^3, \varepsilon^4, \ldots$ become progressively smaller. Therefore the distances $|\beta(t) - \beta_1|$, $|\beta(t) - \beta_2|, |\beta(t) - \beta_3|, \ldots$ between $\beta_1, \beta_2, \beta_3, \ldots$ and $\beta(t)$ get progressively smaller. So the numbers β_1, β_2, $\beta_3, \ldots$ close in on $\beta(t)$ as required. Since $0 \leq \varepsilon < 1$ for any ellipse, this successive approximation process will always converge to the solution $\beta(t)$.

In situations where the eccentricity ε is close to 0, the convergence will be rapid. Table 10.1 tells us that this is so for all planets except Mercury. In situations with ε close to 1—the comet Halley, for instance—thousands of iterations may be necessary (so that the computations need to rely on computers).

Example 10.2. Of the planets, Mercury's eccentricity ε is the largest. But even in this case, the process described converges quickly. Squaring $\varepsilon \approx 0.2057$ four consecutive times, we get

$$\varepsilon^2,\ \varepsilon^4,\ \varepsilon^8,\ \varepsilon^{16} \approx 1 \times 10^{-11}.$$

Since $|\beta(t) - \beta_{16}| < \varepsilon^{16}$, the sixteenth iteration of the process provides an approximation of $\beta(t)$ that is accurate to within a tiny fraction of a radian.

Example 10.3. For a relatively large $\varepsilon < 1$, achieving good accuracy for a particular t will usually require many steps. For Halley's comet, $\varepsilon \approx 0.967$. By repeatedly squaring 0.97,

$$\varepsilon^{128} \approx 0.0203, \ldots,\ \varepsilon^{256} \approx 0.0004, \ldots,\ \varepsilon^{512} \approx 1.68 \times 10^{-7},\ \varepsilon^{1024} \approx 2.85 \times 10^{-14}.$$

So for tight accuracy, a 1000 iterations might be necessary. But this is hardly a problem for a computer.

The solution of the problem of determining the position of P is complete: take the given elapsed time t, and solve Kepler's equation for $\beta(t)$ by the method just described. Then substitute $\beta(t)$ into the equations derived in Step 1 to get the corresponding $r(t)$ and $\alpha(t)$. In this way, given the elapsed time of the orbiting object from periapsis, it is possible to determine its position in the orbit in terms of the orbital constants. Rather than an approximation procedure that closes in on $\beta(t)$, it would be of interest to have an explicit function $\beta = f(t)$ that provides the angle β for a given time t. A function that solves this problem was defined in terms of power series by Friedrich Bessel (the astronomer who was first able to detect stellar parallax for some near stars).

Orbiting Body	Semimajor Axis (km)	Eccentricity	Orbit Period (years)	Angle of Orbit Plane to Earth's	Average Speed (km/sec)
Mercury	57,909,227	0.20563593	0.2408467	7.00°	47.362
Venus	108,209,475	0.00677672	0.6151973	3.39°	35.021
Earth	149,598,262	0.01671123	1.0000174	0.00°	29.783
Mars	227,943,824	0.0933941	1.8808476	1.85°	24.077
Jupiter	778,340,821	0.04838624	11.862651	1.31°	13.056
Saturn	1,426,666,422	0.05386179	29.447498	2.49°	9.639
Uranus	2,870,658,186	0.04725744	84.016846	0.77°	6.873
Neptune	4,498,396,441	0.00859048	164.79132	1.77°	5.435
Pluto	5,906,440,628	0.2488273	247.92065	17.14°	4.669
Halley	2,667,950,000	0.9671429	75.32	162.26°	

Table 10.1. Orbital information provided by NASA and the Jet Propulsion Laboratory as of the year 2015.

It should not come as a surprise that the mathematics in Steps 1 to 3 can be used to construct a computer model of the solar system. This is done in *Computer Model of Elliptical Orbits Generated by Kepler's Equations*. Go to the website

http://learning.nd.edu/orbital/orbital-info.html

and experiment with the simulations for the inner and outer planets.

10.4.2 Determining Speed and Direction

Let $v(t)$ denote the speed of the planet, comet, asteroid, or satellite depicted in Figure 10.18 at time t in its orbit. We will use the formula $v(t) = \sqrt{x'(t)^2 + y'(t)^2}$ that was derived at the end of Section 6.4 in combination with facts from the previous section to establish that

$$v(t) = \tfrac{2\pi a}{T}\sqrt{\tfrac{2a}{r(t)} - 1}.$$

Recall the equations $x(t) = a\cos\beta(t)$, $y(t) = b\sin\beta(t)$, $r(t) = a(1 - \varepsilon\cos\beta(t))$ as well as Kepler's equation $\beta(t) - \varepsilon\sin\beta(t) = \frac{2\pi t}{T}$ from Section 10.4.1. By the chain rule, $x'(t) = -(a\sin\beta(t))\beta'(t)$ and $y'(t) = (b\cos\beta(t))\beta'(t)$. So

$$\begin{aligned} v(t)^2 &= x'(t)^2 + y'(t)^2 = [a^2\sin^2\beta(t) + b^2\cos^2\beta(t)]\beta'(t)^2 \\ &= [a^2\sin^2\beta(t) + (a^2 - c^2)\cos^2\beta(t)]\beta'(t)^2 \\ &= [a^2 - c^2\cos^2\beta(t)]\beta'(t)^2 = [a^2 - a^2\varepsilon^2\cos^2\beta(t)]\beta'(t)^2 \\ &= a^2[1 - \varepsilon^2\cos^2\beta(t)]\beta'(t)^2 = a^2[(1 + \varepsilon\cos\beta(t))(1 - \varepsilon\cos\beta(t))]\beta'(t)^2. \end{aligned}$$

Differentiate Kepler's equation $\beta(t) - \varepsilon\sin\beta(t) = \frac{2\pi t}{T}$ to get $\beta'(t) - \varepsilon\cos\beta(t)\beta'(t) = \frac{2\pi}{T}$. So $\beta'(t) = \frac{2\pi}{T(1-\varepsilon\cos\beta(t))}$ and hence $\beta'(t)^2 = \frac{4\pi^2}{T^2(1-\varepsilon\cos\beta(t))^2)}$. Substitute this into the last term of the expression for $v(t)^2$ just derived, and cancel the term $1 - \varepsilon\cos\beta(t)$ to obtain

$$v(t)^2 = \tfrac{4\pi^2a^2}{T^2} \cdot \tfrac{1+\varepsilon\cos\beta(t)}{1-\varepsilon\cos\beta(t)}\,.$$

Because $r(t) = a(1 - \varepsilon\cos\beta(t))$, we get $1 - \varepsilon\cos\beta(t) = \frac{r(t)}{a}$ and hence that $\varepsilon\cos\beta(t) = 1 - \frac{r(t)}{a}$. After two more substitutions,

$$v(t)^2 = \tfrac{4\pi^2a^2}{T^2} \cdot \tfrac{1+(1-\frac{r(t)}{a})}{\frac{r(t)}{a}} = \tfrac{4\pi^2a^2}{T^2}\Big(2 - \tfrac{r(t)}{a}\Big)\tfrac{a}{r(t)} = \tfrac{4\pi^2a^2}{T^2}\Big(\tfrac{2a}{r(t)} - 1\Big).$$

Taking square roots finishes the verification of the formula $v(t) = \frac{2\pi a}{T}\sqrt{\frac{2a}{r(t)} - 1}$. Because $\frac{4\pi^2}{T^2} = \frac{GM}{a^3}$ and hence $\frac{4\pi^2 a^2}{T^2} = \frac{GM}{a}$, we get that $v(t) = \sqrt{\frac{GM}{a}}\sqrt{\frac{2a}{r(t)} - 1} = \sqrt{GM}\sqrt{\frac{2}{r(t)} - \frac{1}{a}}$. Therefore the speed $v(t)$ of P at any time t in its orbit is equal to

$$v(t) = \frac{2\pi a}{T}\sqrt{\frac{2a}{r(t)} - 1} = \sqrt{GM}\sqrt{\frac{2}{r(t)} - \frac{1}{a}}$$

Example 10.4. Use the speed formula to confirm what Kepler's second law already told us: the maximum speed achieved by P is $v_{\text{max}} = \sqrt{\frac{GM}{q}}\sqrt{1+\varepsilon}$, where $q = a(1-\varepsilon)$ is the orbit's periapsis distance, and that its minimum speed is $v_{\text{min}} = \sqrt{\frac{GM}{d}}\sqrt{1-\varepsilon}$, where $d = a(1+\varepsilon)$ is its apoapsis distance. Notice that if $\varepsilon = 0$, then the ellipse is a circle with radius a, and the speed $v(t) = \sqrt{\frac{GM}{a}}$ is constant.

It remains to study the direction of the motion of the planet, satellite, asteroid, or comet P in its elliptical orbit. We'll continue to let P be represented by a point-mass, and we'll refer to it as point. Figure 10.22 shows it in two typical positions at elapsed times t and t' from the perihelion at $t = 0$. Let $\gamma(t)$ be the angle between the tangent at P in the direction of the motion and the segment from P to S. If $\varepsilon = 0$, then the ellipse is a circle with center S. In this case, $\gamma(t) = \frac{\pi}{2}$ at any point in the orbit. We will therefore take $\varepsilon > 0$ in the discussion that follows. When P is moving toward S in its approach to perihelion, we'll measure $\gamma(t)$ counterclockwise from the tangent to the segment, and when P is moving away from S on its way to aphelion, we'll measure it clockwise from the tangent to the segment. After sketching several instances of what has been described, you will notice that $0 < \gamma(t) \leq \frac{\pi}{2}$ when P is moving toward S, and $-\frac{\pi}{2} \leq \gamma(t) < 0$ when P is moving away from S. Let's focus our analysis of the angle $\gamma(t)$ when the point is

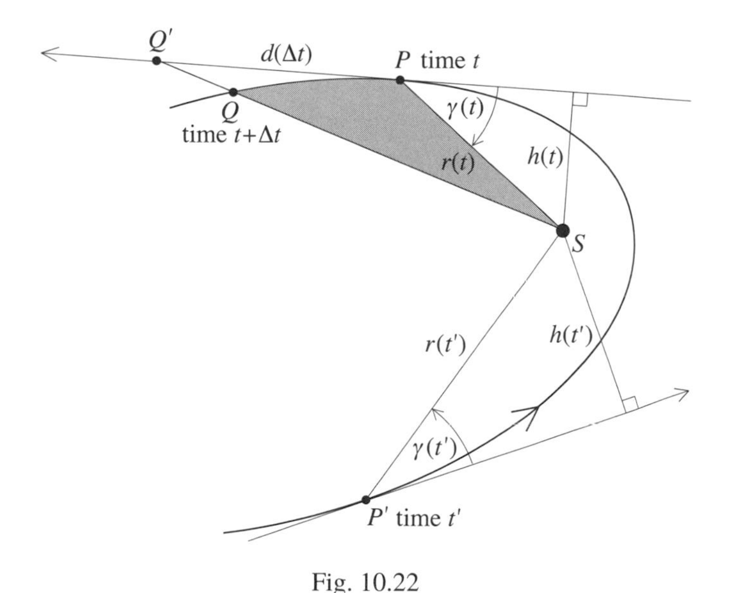

Fig. 10.22

in position P at time t as shown in the figure. We'll let $h(t)$ be the length of the perpendicular segment from S to the tangent.

Regard P and t to be fixed and let an additional short time Δt elapse. Let the point be in position Q at time $t + \Delta t$. The area swept out by SP during the motion from P to Q is $A(t + \Delta t) - A(t)$. This area is shaded in the figure. The point Q' is the intersection of the continuation of the segment SQ and the tangent at P. Let d be the distance from P to Q'. With t fixed, d is a function $d = d(\Delta t)$ of Δt. Since Δt is small, Q is close to P, so that

$$d(\Delta t) = PQ' \approx \text{arc } PQ \quad \text{and} \quad A(t + \Delta t) - A(t) \approx \text{area } \Delta PQ'S = \tfrac{1}{2} d(\Delta t) \cdot h(t).$$

It is apparent from the figure that the smaller the Δt is, the closer the distances PQ and PQ' are to each other, and the tighter the above approximations are. The average speed of the point during its motion from P to Q is $\frac{\text{arc } PQ}{\Delta t}$, so that the speed $v(t)$ at P is

$$v(t) = \lim_{\Delta t \to 0} \frac{\text{arc } PQ}{\Delta t} = \lim_{\Delta t \to 0} \frac{d(\Delta t)}{\Delta t}.$$

Since $\kappa = \frac{A(t+\Delta t)-A(t)}{\Delta t}$ for any Δt, we now get

$$\kappa = \lim_{\Delta t \to 0} \frac{A(t + \Delta t) - A(t)}{\Delta t} \approx \lim_{\Delta t \to 0} \tfrac{1}{2} \frac{d(\Delta t)}{\Delta t} \cdot h(t) = \tfrac{1}{2} v(t)\, h(t).$$

From another look at the figure and the definition of $\gamma(t)$, we see that $\sin \gamma(t) = \pm \frac{h(t)}{r(t)}$ with the $+$ in place when P is on approach to S and perihelion, and the $-$ in place when P is moving away from S and toward aphelion. Therefore $\kappa = \pm \frac{1}{2} v(t) r(t) \sin \gamma(t)$, and hence $\sin \gamma(t) = \frac{\pm 2\kappa}{r(t)v(t)}$. Since $\kappa = \frac{ab\pi}{T}$ and $v(t) = \frac{2\pi a}{T} \sqrt{\frac{2a}{r(t)} - 1}$, and because $a^2 = b^2 + c^2 = b^2 + (a\varepsilon)^2$ implies that $b = a\sqrt{1 - \varepsilon^2}$, it follows that

$$\boxed{\sin \gamma(t) = \frac{\pm a\sqrt{1-\varepsilon^2}}{\sqrt{r(t)(2a-r(t))}}}$$

After applying $\sin^{-1}$ to both sides of this equation and recalling that $-\frac{\pi}{2} \le \gamma(t) \le \frac{\pi}{2}$, we get the formulas

$$\boxed{\gamma(t) = \sin^{-1}\Big(\frac{a\sqrt{1-\varepsilon^2}}{\sqrt{r(t)(2a-r(t))}}\Big) \quad \text{or} \quad \gamma(t) = -\sin^{-1}\Big(\frac{a\sqrt{1-\varepsilon^2}}{\sqrt{r(t)(2a-r(t))}}\Big)}$$

for the orbital angle $\gamma(t)$. The first case applies during the motion of P from apoapsis to periapsis, and the second applies as P moves from periapsis to apoapsis.

Example 10.5. Consider the definition of the angle $\gamma(t)$ at the periapsis and apoapsis positions. At either position, both $\frac{-\pi}{2}$ and $\frac{\pi}{2}$ are possible, depending on whether the position is regarded to be a point of arrival or a point of departure. Suppose $\sin \gamma(t) = \pm 1$. Use the formula for $\sin \gamma(t)$ as well as the quadratic formula to show that $r(t) = a \pm a\varepsilon$. Conclude that $\gamma(t) = \pm \frac{\pi}{2}$ occurs only at periapsis and apoapsis.

Example 10.6. Consider $\gamma(t)$ as P moves from apoapsis to periapsis on its approach to S. We already know that the maximum value of $\gamma(t) = \sin^{-1}\Big(\frac{a\sqrt{1-\varepsilon^2}}{\sqrt{r(t)(2a-r(t))}}\Big)$ is $\frac{\pi}{2}$. What is the minimum value of $\gamma(t)$ over this stretch? Figure 9.31 tells us that the inverse sine is an increasing function, so that $\sin^{-1}\Big(\frac{a\sqrt{1-\varepsilon^2}}{\sqrt{r(t)(2a-r(t))}}\Big)$ reaches its minimum value when $\frac{a\sqrt{1-\varepsilon^2}}{\sqrt{r(t)(2a-r(t))}}$ is at its minimum, and hence when $\sqrt{r(t)(2a - r(t))}$ is at its maximum. Since the parabola $y = x(2a - x) = -x^2 + 2ax$ has its maximum value at $x = a$, it follows that $\sin^{-1}\Big(\frac{a\sqrt{1-\varepsilon^2}}{\sqrt{r(t)(2a-r(t))}}\Big)$ reaches its minimum value when $r(t) = a$. Conclude that this minimum value is $\gamma_{\min} = \sin^{-1} \sqrt{1 - \varepsilon^2}$. What position in the orbit corresponds to this value? What is the situation in this regard for P in motion from periapsis to apoapsis?

10.4.3 Earth, Jupiter, and Halley

We will now turn to several concrete examples to illustrate the information that Sections 10.4.1 and 10.4.2 provide. The examples involve Earth, Jupiter, and Halley's comet so that S is the Sun in each case. We'll use the data in Table 10.1 as well as the value $GM = 1.327124 \times 10^{11}$ km^3/s^2 for the Sun provided by the National Aeronautic and Space Administration (NASA).

We'll begin by studying our orbiting Earth. Relying on Table 10.1, we'll take $a = 1.495983 \times 10^8$ km (close to 150 million kilometers) and $\varepsilon = 0.016711$. Before evaluating the distance $r(t)$, the velocity $v(t)$, and the orbital angle $\gamma(t)$ for specific times t, let's gain a sense of these values by examining their range from minimum to maximum in each case. From the expressions for the periapsis and apoapsis distances, we know that

$$(1.495982 \times 10^8)(1 - 0.016712) < a(1-\varepsilon) \le r(t) \le a(1+\varepsilon) < (1.495984 \times 10^8)(1.016712)$$

and therefore that

$$1.470981 \times 10^8 < r(t) < 1.520985 \times 10^8$$

in kilometers. By substituting the data we have into the formulas of Example 10.4, we get the bounds

$$29.29099 < \sqrt{\tfrac{GM}{a(1+\varepsilon)}}\sqrt{1-\varepsilon} = v_{\min} \le v(t) \le v_{\max} = \sqrt{\tfrac{GM}{a(1-\varepsilon)}}\sqrt{1+\varepsilon} < 30.28661$$

on the speed $v(t)$ of Earth in its orbit in kilometers per second. Finally, turning to the orbital angle $\gamma(t)$, we find from Example 10.6 that

$$1.554084 < \sin^{-1}\sqrt{1-\varepsilon^2} = \gamma_{\min} \le \gamma(t) \le \gamma_{\max} = \tfrac{\pi}{2} < 1.570797$$

in radians for Earth in its orbit from apoapsis to periapsis. The angle 1.554084 expressed in degrees is 89.04°. The Earth's near-circular orbit is the reason why $\gamma(t)$ is close to 90° throughout.

We know that the period of Earth's orbit is $T = 365.259636$ days. This is the time it takes for Earth to move from one perihelion position to the next. This is a little longer than the "year of the seasons" that two consecutive summer solstices or two consecutive spring equinoxes determine. The positions of the solstices and equinoxes in Earth's orbit are described in Section 3.1.

In 2013, it took Earth $t = 77.497916$ days to travel from perihelion to spring equinox. For this t, we will determine $r(t)$, $v(t)$, and $\gamma(t)$. For the angle $\beta(t)$, we'll apply the approximation machine of Step 3 of Section 10.4.1 using six-decimal-place accuracy at each step to get

$$\begin{aligned}
\beta_1 &= \tfrac{2\pi t}{T} \approx \tfrac{2\pi(77.497916)}{365.259636} \approx 1.333117 \text{ radians},\\
\beta_2 &= \beta_1 + \varepsilon\sin\beta_1 \approx 1.333117 + 0.016711(0.971887) \approx 1.349358 \text{ radians},\\
\beta_3 &= \beta_1 + \varepsilon\sin\beta_2 \approx 1.333117 + 0.016711(0.975583) \approx 1.349420 \text{ radians, and}\\
\beta_4 &= \beta_1 + \varepsilon\sin\beta_3 \approx 1.333117 + 0.016711(0.975596) \approx 1.349420 \text{ radians.}
\end{aligned}$$

Since the process has terminated, $\beta(t) \approx \beta_4 \approx 1.349420$ radians with the required six-decimal-place accuracy. By substitution into the equations $r(t) = a(1 - \varepsilon\cos\beta(t))$ and $\alpha(t) = 2\tan^{-1}\!\left(\sqrt{\tfrac{1+\varepsilon}{1-\varepsilon}}\,\tan\tfrac{\beta(t)}{2}\right)$ of Section 10.4.1 (we have taken $n_t = 0$ by assuming that Earth is on its way from perihelion to aphelion in its initial orbit), we find that

$$\begin{aligned}
r(t) &\approx (1.495983 \times 10^8)(1 - 0.016711(\cos 1.349420)) \approx 1.490494 \times 10^8 \text{ km and}\\
\alpha(t) &= 2\tan^{-1}(1.016853\tan(0.674710)) \approx 2(0.682877) \approx 1.365754 \text{ radians.}
\end{aligned}$$

So at spring equinox in 2013, $r(t)$ and $\alpha(t)$ were close to 149 million kilometers and 78.25°, respectively. Substituting into the formula $v(t) = \sqrt{\tfrac{GM}{a}}\sqrt{\tfrac{2a}{r(t)} - 1}$, we get

$$v(t) \approx \sqrt{8.871137 \times 10^2}\,\sqrt{2.007365 - 1} \approx 29.8939 \text{ km/s}$$

for the Earth's orbital speed. Finally, the formula $\gamma(t) = -\sin^{-1}\big(\frac{a\sqrt{1-\varepsilon^2}}{\sqrt{r(t)(2a-r(t))}}\big)$ tells us that Earth's orbital angle $\gamma(t)$ was

$$\gamma(t) \approx -\sin^{-1}\Big(\frac{(1.495983\times10^8)\sqrt{0.999721}}{\sqrt{(1.490494\times10^8)(2(1.495983\times10^8)-(1.490494\times10^8))}}\Big) \approx -\sin^{-1}(0.999867) = -1.554501 \text{ radians},$$

close to $-89.07°$. The angle is negative because the Earth moved away from perihelion in the direction of aphelion at the time.

In 2013, it took the Earth 170.000046 days to move from perihelion to summer solstice. Applying Step 3 to compute $\beta(t)$ for $t = 170.000046$, we get

$$\begin{aligned}
\beta_1 &= \tfrac{2\pi t}{T} \approx \tfrac{2\pi(170.000046)}{365.259636} \approx 2.924336 \text{ radians},\\
\beta_2 &= \beta_1 + \varepsilon \sin\beta_1 \approx 2.924336 + 0.016711(0.215552) \approx 2.927938 \text{ radians},\\
\beta_3 &= \beta_1 + \varepsilon \sin\beta_2 \approx 2.924336 + 0.016711(0.212033) \approx 2.927879 \text{ radians},\\
\beta_4 &= \beta_1 + \varepsilon \sin\beta_3 \approx 2.924336 + 0.016711(0.212090) \approx 2.927880 \text{ radians, and}\\
\beta_5 &= \beta_1 + \varepsilon \sin\beta_4 \approx 2.924336 + 0.016711(0.212089) \approx 2.927880 \text{ radians}.
\end{aligned}$$

So $\beta(t) \approx \beta_5 \approx 2.927880$ radians on summer solstice in 2013. Check that the corresponding distance $r(t)$ from the Sun and angle $\alpha(t)$ were $r(t) \approx 1.520414 \times 10^8$ km and $\alpha(t) \approx 2.931396$ radians, or 167.96°. Check also that the Earth's velocity was $v(t) \approx 29.3019$ km/s on that day. Why is this distance close to the maximum and the velocity close to a minimum?

Example 10.7. Use the data in Table 10.1 and follow the above computations for Earth to check that Jupiter's distance from the Sun varies from about 7.41×10^8 km to about 8.16×10^8 km, that its smallest and greatest orbital speeds are $v_{\min} = 12.44$ km/s and $v_{\max} = 13.71$ km/s, and that the smallest angle of Jupiter's approach to the Sun is $\gamma_{\min} = 87.23°$. Verify that when Jupiter is 7.5×10^8 km from the Sun, it has a speed of about 13.5 km/s. How far is Jupiter from the Sun exactly 2 years after it passes periapsis? How fast is it moving at that time?

Example 10.8. Halley's comet passed its periapsis position at $t = 0$. At the precise time t years later, it completed the first quarter of its orbit. Refer to Example 10.1 and use the data of Table 10.1 to show that $t \approx 7.24$ years. Then show that Halley completed the second quarter of its orbit in about 30.42 years.

10.5 INTEGRAL CALCULUS AND THE ACTION OF FORCES

This last section of this chapter is devoted to an analysis of the action of forces. It begins with the mathematics of work and energy as well as impulse and momentum. The summation process that defines the definite integral is precisely what is needed to define these basic concepts of physics. The primary example will be the analysis of the propulsion of a bullet in the barrel of a gun with a focus on the famous Springfield rifle. This five-shot repeating rifle (five bullets can be fired in rapid succession) was closely modeled after a rifle used by the German military. The Springfield rifle was first issued in 1904 and remained in the U.S. arsenal until 1938. It was one of the most reliable and accurate firearms in military history. It was used by Ernest Hemingway on his big-game safaris in Africa. A version of it is still popular with drill teams and color guards.

Refer to Section 6.6 for the basics about forces. We'll use the metric MKS system of units as described in Section 6.10, with occasional references to American equivalents, for example, the foot and the pound. In MKS, distance is in meters, mass in kilograms, and force in newtons. The weight of an object is the magnitude of the force of gravity acting on it. By Newton's second law, this is equal to $m \cdot g$, where m is the object's mass and g is the gravitational constant. As was already pointed out in Section 6.10, g depends on the distance of the object from Earth's center of mass and is slightly smaller at the equator than at the poles. At a latitude of 40°—the cities San Francisco, New York, Madrid, Rome, Athens, Ankara, Beijing, and Tokyo all have approximately this latitude—g is close to 9.81 m/s^2 (or 32.17 feet/s^2). This is the value that we will use.

10.5.1 Work and Energy, Impulse and Momentum

Suppose that a force of magnitude f acts on an object and propels it along a straight line through a distance d. Take the line to be a coordinate axis, and let the force act from a to b as illustrated in Figure 10.23. So

Fig. 10.23

$d = b - a$. If f is constant, then the *work* W done by the force is defined to be the product

$$W = f \cdot d = f \cdot (b - a).$$

The newton is defined as the unit $\frac{\text{kg}\cdot\text{m}}{\text{s}^2}$. It is often abbreviated by N (see Section 6.10). Because work = force × distance, the corresponding unit of work is the newton-meter, abbreviated N-m. The newton-meter is called *joule*. The connection with the foot-pound of the American system is

$$1 \text{ joule} = 1 \text{ newton-meter} = (0.2248)(3.2808) \text{ foot-pounds} = 0.7375 \text{ foot-pounds}.$$

The work done by a constant force can be negative if the force acts in the direction opposite to the direction of the motion. Consider friction, for example. When an object is pushed or pulled along a rough surface, the frictional force acts in the direction opposite to the displacement. Thus the work done by the frictional force will be negative. The same is the case in the situation of an object in free fall. Gravity will be the dominant force, but air resistance will provide an upward push against the downward direction of the fall. The work done by the upward push of the resisting air will therefore be negative.

Forces generally vary in magnitude, and the question that will concern us now is this. If an object moves along a straight line and a variable force acts on it along the line of its motion, then what is the work done by the force on the object in this situation? A "classical" application of the summation strategy of the definite integral—see Chapter 9—provides the answer.

Suppose that the object moves from a to b on a coordinatized line. Let the force be given by a function $f(x)$ for $a \leq x \leq b$. See Figure 10.24. If $f(x) > 0$, then the force acts from left to the right at x with magnitude $f(x)$; if $f(x) < 0$, then it acts from right to left with a magnitude of $-f(x)$. If $f(x) = 0$, then the force has zero magnitude at x. How can the work done on the object be computed? The strategy is to break up the segment from a to b through which the object moves into small intervals and to assume that

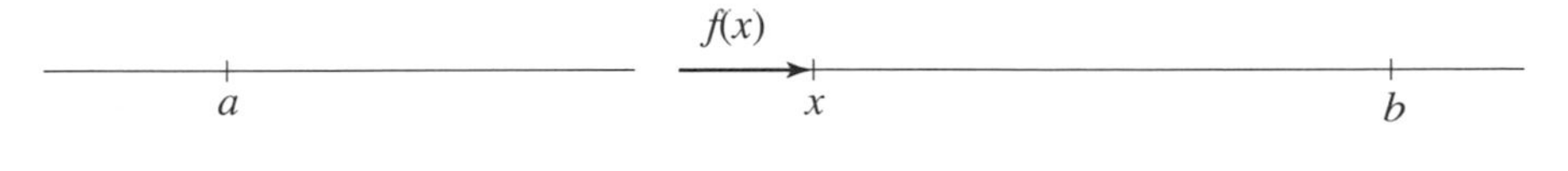

Fig. 10.24

the force is constant over each of them. Under this assumption, the work done on the object over each of these small intervals is computed, and finally all these little "bits" of work are added up. Taking this to the limit, we get the work done by the variable force on the object. The details follow.

With a partition $\mathcal{P}$

$$a = x_0 < x_1 < x_2 < \cdots < x_i < x_{i+1} < \cdots < x_{n-2} < x_{n-1} < x_n = b$$

break the interval of $[a, b]$ into n subintervals. Let ΔW_0 be the work done on the object over the first subinterval, ΔW_1 the work done over the second, ΔW_2 that over the third, and so on, and finally let ΔW_{n-1}

ΔW_0 ΔW_1 ΔW_2 ... ΔW_i ... ΔW_{n-1}

a b

Fig. 10.25

be the work done over the last subinterval. See Figure 10.25. So the total work W done on the object as it moves from a to b is

$$W = \Delta W_0 + \Delta W_1 + \Delta W_2 + \cdots + \Delta W_{n-1} = \sum_{i=0}^{n-1} \Delta W_i.$$

Let $\Delta x_0 = x_1 - x_0$ be the length of the first subinterval, $\Delta x_1 = x_2 - x_1$ the length of the second, $\Delta x_2 = x_3 - x_2$ that of the third, and so on, and $\Delta x_{n-1} = x_n - x_{n-1}$ that of the last. As already done in the definition of the integral in Section 9.5 and as illustrated in Figure 10.26, pick a point c_0 in the first subinterval, c_1 in the

c_0 c_1 c_2 ... c_i ... c_{n-1}

a Δx_0 Δx_1 Δx_2 Δx_i Δx_{n-1} b

Fig. 10.26

second, ..., and c_{n-1} in the last. Since all the distances Δx_i are small, we'll assume—as an approximation—that the force is constant over each of the subintervals. So the force is $f(c_0)$ over the first interval, $f(c_1)$ over the second, $f(c_2)$ over the third, and so on. It follows that

$$\Delta W_0 \approx f(c_0)\Delta x_0,\ \Delta W_1 \approx f(c_1)\Delta x_1, \ldots,\ \Delta W_i \approx f(c_i)\Delta x_i, \ldots,\ \Delta W_n \approx f(c_{n-1})\Delta x_{n-1}.$$

Therefore

$$W \approx f(c_0)\Delta x_0 + f(c_1)\Delta x_1 + \cdots + f(c_{n-1})\Delta x_{n-1} = \sum_{i=0}^{n-1} f(c_i)\Delta x_i.$$

The smaller a subinterval is, the less variation there will be in the force as it acts through it, so the more accurately the force is approximated by a constant. It follows that if what was just described is repeated with finer and finer partitions, the resulting approximation of W become tighter and tighter. Therefore, as in Section 9.5,

$$W = \lim_{\|\mathcal{P}\| \to 0} \sum_{i=0}^{n-1} f(c_i)\Delta x_i.$$

This last step requires that the force function $y = f(x)$ be continuous over the interval $[a, b]$. Given this assumption and the definition of the definite integral of Section 9.5, it follows that the work W done by the force on the object as it moves from a to b is equal to

$$\boxed{W = \int_a^b f(x)\,dx}$$

What happens if we apply the formula $W = \int_a^b f(x)\,dx$ in a situation where the force $f(x)$ is constant, say, equal to C? Because

$$W = \int_a^b f(x)\,dx = Cx\Big|_a^b = C(b-a),$$

we get the expected result: the work W done on the object is the product of the force C times the distance $b - a$ that the object has moved.

The *kinetic energy* of a moving object with mass m and velocity v is defined to be $\frac{1}{2}mv^2$. A look at the units involved tells us that

$$1 \text{ kilogram-(meters/second)}^2 = 1 \text{ kilogram-(meters/second}^2\text{)-meter} = 1 \text{ newton-meter}.$$

Observe therefore that kinetic energy and work are expressed in the same units! This suggests the possibility of a connection. Indeed, the action of a force on an object will bring about a change in its speed and hence a change in its kinetic energy. But what is the exact mathematical relationship?

Let the object have mass m, and assume that the only force that acts on the object along the line of its motion is given by the function $f(x)$ for $a \leq x \leq b$. Take a stopwatch and time what is happening. Suppose that the object is at a at time t_0 and at b at time t_1. Let its position at any time t with $t_0 \leq t \leq t_1$ be given by $x(t)$. See Figure 10.27. Recall from Sections 6.4 and 6.5 that the velocity $v(t)$ of the object at any time t is the derivative $v(t) = x'(t)$, and its acceleration is $a(t) = v'(t) = x''(t)$. The force can also be

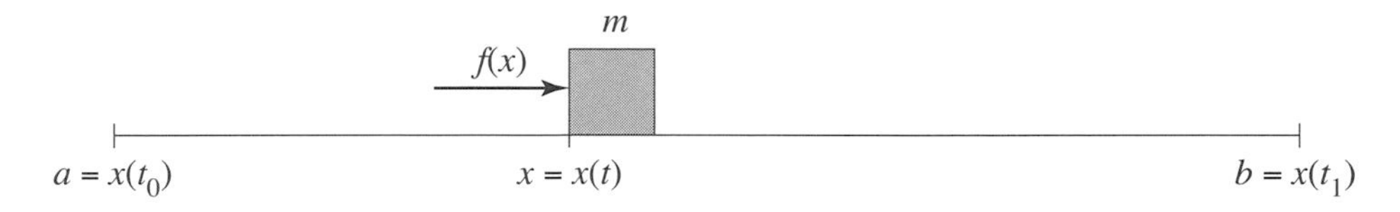

Fig. 10.27

expressed as a function of time as the composite $f(x(t))$ of the two functions $f(x)$ and $x(t)$. Note that by Newton's second law, $f(x(t)) = ma(t)$ at any time t with $t_0 \leq t \leq t_1$. Consider the work

$$W = \int_a^b f(x)\,dx$$

done on the object. Refer back to Section 9.7.1, and observe that the substitution $x = x(t)$ converts this definite integral to

$$W = \int_a^b f(x)\,dx = \int_{t_0}^{t_1} f(x(t))x'(t)\,dt.$$

Differentiate the kinetic energy function $\frac{1}{2}mv(t)^2$ using the chain rule to get

$$\tfrac{d}{dt}(\tfrac{1}{2}mv(t)^2) = mv(t) \cdot v'(t) = ma(t)v(t) = f(x(t))x'(t).$$

So $\frac{1}{2}mv(t)^2$ is an antiderivative of $f(x(t))x'(t)$. Therefore, by the fundamental theorem of calculus,

$$\boxed{W = \int_a^b f(x)\,dx = \int_{t_0}^{t_1} f(x(t))x'(t)\,dt = \tfrac{1}{2}mv(t_1)^2 - \tfrac{1}{2}mv(t_0)^2}$$

This equality provides the precise connection between the kinetic energy of a moving object and the acting force that we were looking for. If an object is moving along a coordinatized line and a variable force acts on it along the line of its motion, then

> the *work* done by the force in moving the object from a to b is equal to
> the *kinetic energy* of the object at b *minus* the kinetic energy of the object at a.

In other words, the work done by the force on the object is equal to the change that it affects in the kinetic energy of the object. If the object is at rest when it starts from a, then $v(t_0) = 0$ and $\frac{1}{2}mv(t_0)^2 = 0$. In this case, $W = \frac{1}{2}mv(t_1)^2$, and the work done on the object is equal to the kinetic energy that the object has at b.

Momentum is another concept of fundamental importance in the analysis of motion. The *momentum* of an object of mass m moving with velocity v is defined to be the product mv. In the sense that it measures the "quantity" of motion, the concept of momentum is similar to that of kinetic energy. When a force acts on an object, it will affect a change in its velocity and therefore also in its momentum. Can this change be computed? More fundamentally, what is the precise connection between force and momentum? The key to the answers to these questions lies in the consideration of force as a function of time.

Again, let an object move along a coordinatized line, and suppose that a variable force acts on it along the line of its motion. Suppose that the object moves from a to b on the line. Let the object have mass m, and let the force be given by the function $f(x)$ for any x with $a \leq x \leq b$. Suppose that the object is at a at time t_0 and at b at time t_1. Let its position at any time t with $t_0 \leq t \leq t_1$ be $x = x(t)$. See Figure 10.28. As

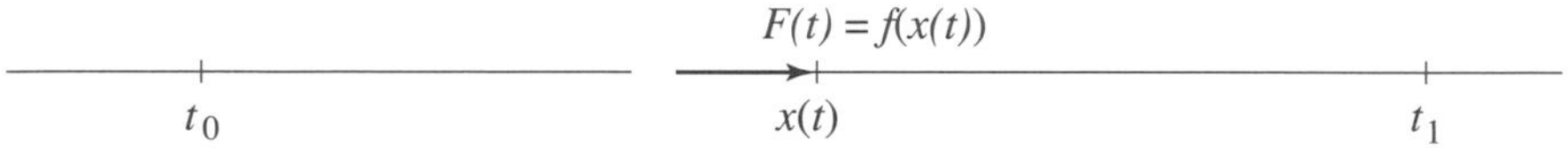

Fig. 10.28

before, the velocity of the object at any time t is $v(t) = x'(t)$, and its acceleration is $a(t) = v'(t) = x''(t)$. The force as a function of time is, again, the composite $F(t) = f(x(t))$. If $F(t) = C$ is a constant over the time interval $[t_0, t_1]$, then the *impulse* of the force is the product $C(t_1 - t_0)$. If the force varies, then by proceeding as in the definition of work (with a partition of the time interval $[t_0, t_1]$ replacing that of the space interval $[a, b]$), the *impulse* of the force over the time interval $[t_0, t_1]$ is the integral

$$\boxed{J = \int_{t_0}^{t_1} F(t)\,dt}$$

By Newton's second law,

$$F(t) = ma(t) = mv'(t) = \tfrac{d}{dt}mv(t),$$

so that the momentum $mv(t)$ is an antiderivative of the force function $F(t)$. By the fundamental theorem of calculus,

$$\boxed{J = \int_{t_0}^{t_1} F(t)\,dt = mv(t_1) - mv(t_0)}$$

The equality that we have just derived says that

> the *impulse* of the force acting on the object from time t_0 to time t_1 is equal to the *momentum* of the object at t_1 *minus* the momentum of the object at t_0.

In other words, the impulse of the force is equal to the change that it affects in the momentum of the object. If the object is at rest when it starts from $a = x(t_0)$, then $v(t_0) = 0$ and $mv(t_0) = 0$. In this case, $J = mv(t_1)$, and the impulse is equal to the momentum that the object has at $b = x(t_1)$. In particular, the relationship between momentum and impulse is analogous to the relationship between energy and work.

It is worth emphasizing that the fundamental theorem of calculus has just come to life within fundamental concepts of physics: it is the link between work and kinetic energy and also between impulse and momentum. The remarkable fact is that the definite integral, arising as it does as an abstract mathematical construction related to area, plays a concrete and conceptual role not only in physics, but also in various disciplines of science, engineering, and economics.

10.5.2 Analysis of Springs

Consider a coil spring made of a resilient metal such as steel. The length of the spring, when it is not subject to an external force, is its *natural length*. Put the spring on a horizontal plane, and anchor one end into place. Put an x-axis along the length of the spring in such a way that when the spring is at its natural length, its free end falls on the origin. What has been described is illustrated in Figure 10.29a. Now stretch the spring as illustrated in Figure 10.29b by applying a force at the free end. *Hooke's law* for springs asserts that the force $f(x)$ on the spring necessary to keep it stretched x units beyond its natural

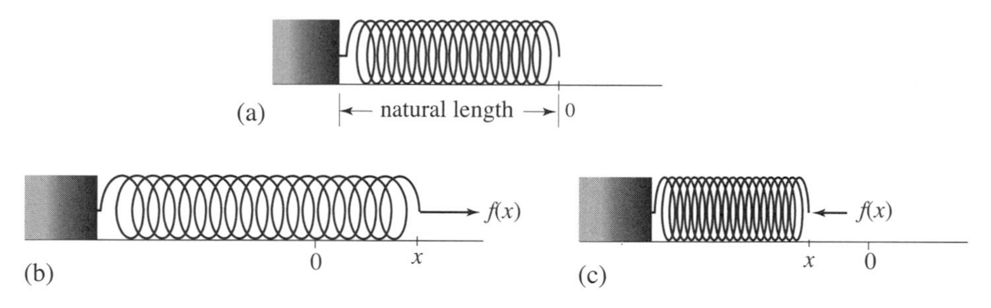

Fig. 10.29

length is proportional to x. More precisely, it says that there is a positive constant k—it is called the *spring constant* of the spring—such that

$$f(x) = kx$$

for any displacement x of the spring beyond its natural length. This equation also holds in the situation of Figure 10.29c, where the spring is compressed. In this case, both x and the force $f(x)$ are negative. In MKS, the unit for a spring constant is newtons/meter. The spring constant k depends on the qualities of the metal and the nature of the coils. Observe that a rigid or stiff spring has a large spring constant and a more flexible or soft spring a smaller one. When stretched or compressed too far, a spring may deform and no longer perform as just described. We will assume that the springs we are considering are not stretched (or compressed) beyond such a critical length.

Example 10.9. Consider a spring that has a natural length of 30 cm. Determine its spring constant if it takes a force of 300 N to keep it stretched at a length of 34 cm. The fact that we're in MKS means that we need to convert centimeters to meters. Since 4 cm = 0.04 m, we find by Hooke's law, that $300 = k(0.04)$. So $k = 7500$ N/m.

Example 10.10. Let's compute the work W done in stretching the spring of the previous example from its natural length of 30 cm to 34 cm. By applying the analysis above,

$$W = \int_0^{0.04} f(x)\,dx = \int_0^{0.04} kx\,dx = \int_0^{0.04} 7500x\,dx = \tfrac{7500}{2}x^2\Big|_0^{0.04} = (3750)(0.04)^2$$
$$\approx 6 \text{ N-m}.$$

We will use this spring to illustrate the connection between work and kinetic energy. Compress the spring by 6 cm, and hold it in this compressed state. Place a 3-kg block of ice in front of the spring. The block and the horizontal track on which it is about to slide have been warmed. The flat, bottom surface of the block is smooth and has started to melt. When the spring is released to push the block, it glides frictionlessly. Propelled by the force of the spring, the block accelerates and picks up speed. The block is pushed until the spring reaches its natural length. At that point, the block loses contact with the spring and travels at its maximum velocity. Since the essential aspects of what is happening are known, we should be able to determine this final velocity of the block. But how?

Let's start by computing the work that the spring does on the block. Place a coordinate axis as in Figure 10.30. In particular, the right end of the spring is at the origin when the spring is at its natural length. Note that when the spring is compressed by 0.06 m, its right end has coordinate −0.06. Figure 10.30 shows the block at its starting point. By Figure 10.29, the force with which the spring pushes on the block is the

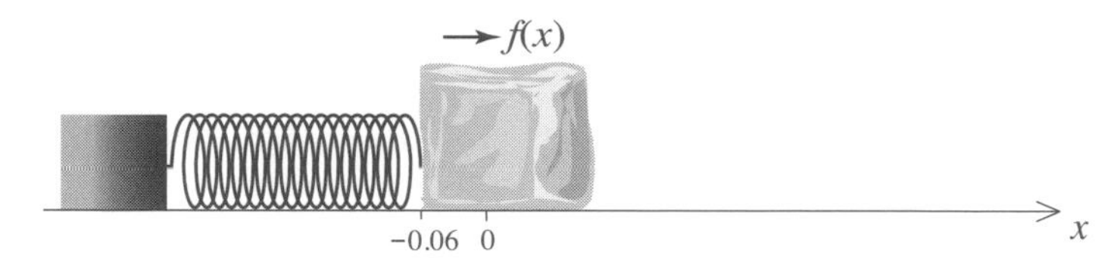

Fig. 10.30

positive force $-f(x) = -kx$. Therefore the work done on the block by the spring is

$$W = \int_{-0.06}^{0} -7500x\,dx = -3750\,x^2\Big|_{-0.06}^{0} = 0 - (-(3750)(-0.06)^2) = (3750)(0.06)^2$$
$$= 13.5 \text{ newton-meters.}$$

Take a stopwatch and time this "event." Let $t = 0$ be the instant the compressed spring is released. The velocity of the ice cube at this time is $v(0) = 0$. Let t_1 be the instant that the spring and the ice cube part company, and let $v(t_1)$ be the velocity of the ice cube at that time. The connection between work and kinetic energy developed earlier tells us that

$$13.5 \approx W = \tfrac{1}{2}mv(t_1)^2,$$

where $m = 3$ kg is the mass of the block of ice. It follows that $\frac{3}{2}v(t_1)^2 = 13.5$. Therefore $v(t_1)^2 = 9$, and hence the final velocity of the ice cube is

$$v(t_1) = 3 \text{ m/s.}$$

An analysis of the impulse of the force also provides important information about the moving block of ice. Let time t with $0 \le t \le t_1$ be arbitrary. As was already noted, since $x(t) \le 0$ and $F(t) \ge 0$, $F(t) = -kx(t)$. By Newton's second law, $-kx(t) = mx''(t)$. So the function $x(t)$ satisfies the equation

$$(*) \qquad x''(t) + \frac{k}{m}x(t) = 0.$$

Such equations are called *differential equations*. Consider the functions $f_1(t) = \sin\left(\sqrt{\frac{k}{m}}\,t\right)$ and $f_2(t) = \cos\left(\sqrt{\frac{k}{m}}\,t\right)$. Differentiate them each twice to see that

$$\tfrac{d^2}{dt^2}\cos\left(\sqrt{\tfrac{k}{m}}\,t\right) = -\tfrac{d}{dt}\sin\left(\sqrt{\tfrac{k}{m}}\,t\right)\cdot\sqrt{\tfrac{k}{m}} = -\tfrac{k}{m}\cos\left(\sqrt{\tfrac{k}{m}}\,t\right) \text{ and}$$
$$\tfrac{d^2}{dt^2}\sin\left(\sqrt{\tfrac{k}{m}}\,t\right) = \tfrac{d}{dt}\cos\left(\sqrt{\tfrac{k}{m}}\,t\right)\cdot\sqrt{\tfrac{k}{m}} = -\tfrac{k}{m}\sin\left(\sqrt{\tfrac{k}{m}}\,t\right).$$

So the functions $f_1(t) = \sin\left(\sqrt{\frac{k}{m}}\,t\right)$ and $f_2(t) = \cos\left(\sqrt{\frac{k}{m}}\,t\right)$ both satisfy equation $(*)$. Our study of differential equations in Section 11.6, Theorem 2 in particular, tells us that any solution of $(*)$ and hence the function $x(t)$ has the form

$$x(t) = D_1 \sin\left(\sqrt{\tfrac{k}{m}}\, t\right) + D_2 \cos\left(\sqrt{\tfrac{k}{m}}\, t\right)$$

for some constants D_1 and D_2. After one more differentiation,

$$v(t) = x'(t) = D_1\sqrt{\tfrac{k}{m}} \cos\left(\sqrt{\tfrac{k}{m}}\, t\right) - D_2\sqrt{\tfrac{k}{m}} \sin\left(\sqrt{\tfrac{k}{m}}\, t\right).$$

With $t = 0$, we get $x(0) = D_2$ and $v(0) = D_1\sqrt{\frac{k}{m}}$. Since we know that $x(0) = -0.06$ and $v(0) = 0$, we get $D_1 = 0$ and $D_2 = -0.06$. Because $k = 7500$ and $m = 3$, $\sqrt{\frac{k}{m}} = 50$. Therefore

$$x(t) = -(0.06 \cos 50\,t) \text{ m} \quad \text{and} \quad v(t) = (3 \sin 50\,t) \text{ m/s}.$$

Now we know exactly how the block of ice of Figure 10.30 moves. Take $t = t_1$ to be the moment that the spring has returned to its natural length and its push on the block of ice is over. Since $x(t_1) = 0$, it follows that $50t_1 = \frac{\pi}{2}$, and hence that

$$t_1 = \tfrac{\pi}{100} \approx 0.03 \text{ s}.$$

The push of the spring is over in a flash because the spring is stiff and pushes the block through only 6 cm. The fact that $v(t_1) = v(\frac{\pi}{100}) = 3\sin\frac{\pi}{2} = 3$ m/s confirms what we already learned from the study of the kinetic energy of the block.

10.5.3 The Force in a Gun Barrel

This section applies what we have learned to the study of the force inside the barrel of a gun. Exactly how is a bullet or shell propelled forward in the barrel? Can this force be analyzed quantitatively? Can the velocity with which the bullet exits the barrel—this is the *muzzle velocity*—be computed? These are difficult questions. Such questions are the concern of *interior ballistics*, the discipline that studies what happens inside the barrel of a gun. A basic law of interior ballistics together with the analysis of Section 10.5.1 will provide some of the answers to these questions. Once the shell leaves the muzzle and takes flight, it becomes the focus of *exterior ballistics*.

Suppose that an explosion occurs inside a closed chamber. The forces that are released will push against all surfaces of the chamber. If one of the walls of the chamber is free to move, it will be "blown away." This, in principle, is the situation inside the cartridge of the bullet of a gun, where it is the shell or bullet that is free to move. The force of the explosion will drive the shell or bullet down and through the barrel. See Figures 10.31a and 10.31b. The situation is entirely analogous to that of the block of ice

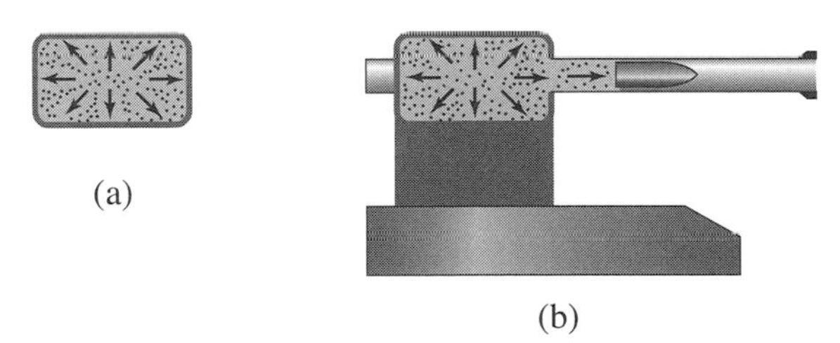

Fig. 10.31

considered earlier.

Place an x-axis inside the barrel, and let the shell be in typical position x in the barrel as shown in Figure 10.32. A formula of historical and practical interest in interior ballistics is the formula of Le Duc. It says that the position x and the velocity $v(x)$ of the shell at x are related by the formula

$$v(x) = \frac{ax}{x+b},$$

where a and b are positive constants. The constants depend on the specifics of the situation: properties of the barrel, type of bullet or shell, burn rate of the gun powder, the temperature of the gun, and so on. Le Duc was a captain in the French artillery of the 19th century when he discovered his formula empirically.

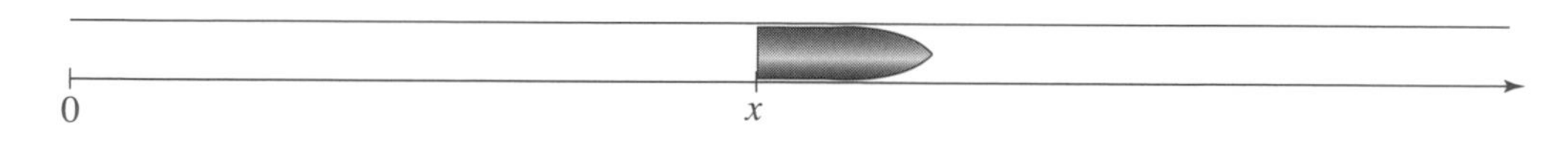

Fig. 10.32

His formula is grounded in numerical observations from test firings rather than the physics and chemistry of the explosion of the powder and the resulting propulsion of a bullet through the barrel. This is too complex to be captured by such a simple formula. Nonetheless, we will see that Le Duc's formula does capture the essence (but not all the specifics) of the dynamics of what goes on inside the barrel.

Let's take out our stopwatch again, and let $t = 0$ be the instant that the explosion of the gun powder first begins. As in the case of the block of ice, the position x of the shell in the barrel is a function of time $t \geq 0$ so that we can set $x = x(t)$. The velocity of the shell as a function of time is the composite function $v(t) = v(x(t))$. But we also know that $v(t) = x'(t)$. It follows that Le Duc's formula can be rewritten as the differential equation

$$x'(t) = v(x(t)) = \frac{ax(t)}{x(t)+b}$$

in the variable t. The acceleration of the shell at any time $t \geq 0$ is $a(t) = x''(t)$. Differentiating $x'(t)$, we get

$$x''(t) = \tfrac{d}{dt}x'(t) = \tfrac{d}{dt}\frac{ax(t)}{x(t)+b} = \frac{ax'(t)(x(t)+b) - ax(t)x'(t)}{(x(t)+b)^2} = \frac{abx'(t)}{(x(t)+b)^2}.$$

Inserting the formula of Le Duc, we get

$$x''(t) = \frac{ab(\frac{ax(t)}{x(t)+b})}{(x(t)+b)^2},$$

so that after a simplification,

$$x''(t) = \frac{a^2bx(t)}{(x(t)+b)^3}.$$

This is the acceleration produced by the net force on the shell, as generated by the explosion of the powder minus the frictional forces (that are negligible in comparison). This differential equation plays a role that is analogous to the role played by differential equation $(*)$ in Section 10.5.2 that relates the force of the spring to the motion of the block of ice.

Let m be the mass of the shell. By Newton's second law, the net force on the shell is equal to

$$F(t) = ma(t) = mx''(t)$$

at any time $t \geq 0$. Inserting the expression for $x''(t)$ just derived, we get that

$$F(t) = \frac{ma^2bx(t)}{(x(t)+b)^3}.$$

Notice that as a function of its position x in the barrel, the magnitude of the force on the shell is given by

$$f(x) = \frac{ma^2bx}{(x+b)^3}.$$

Let's analyze the function $f(x)$. By the quotient rule,

$$f'(x) = \frac{ma^2b \cdot (x+b)^3 - ma^2bx \cdot 3(x+b)^2}{(x+b)^6} = \frac{ma^2b(x+b)^2(x-3x+b)}{(x+b)^6} = \frac{ma^2b(b-2x)}{(x+b)^4}.$$

So $f'(x) = 0$ when $x = \frac{b}{2}$. Observe that if $x < \frac{b}{2}$, that is, if $2x < b$, then $f'(x) > 0$; and if $x > \frac{b}{2}$, then $f'(x) < 0$. It follows that f increases over the interval $[0, \frac{b}{2}]$, has an absolute maximum at $\frac{b}{2}$, and decreases thereafter. Using the quotient rule once more shows us that the second derivative of f is

$$\begin{aligned} f''(x) &= \frac{-2ma^2b \cdot (x+b)^4 - ma^2b(b-2x) \cdot 4(x+b)^3}{(x+b)^8} = \frac{-2ma^2b(x+b) + 4ma^2b(2x-b)}{(x+b)^5} \\ &= \frac{6ma^2bx - 6ma^2b^2}{(x+b)^5} = \frac{6ma^2b(x-b)}{(x+b)^5}. \end{aligned}$$

So $f''(x) = 0$ when $x = b$. Observe that $f''(x) < 0$ when $x < b$, and $f''(x) > 0$ when $x > b$. It follows that the graph of f is concave down over $[0, b]$, has a point of inflection at b, and is concave up thereafter.

The general shape of the graph of $y = f(x)$ is shown in Figure 10.33. The point of inflection is highlighted. The domain of $y = f(x)$ is the interval $[0, h]$, where h is the length of the barrel of the gun. Suppose that the shell of Figure 10.32 exits the barrel at time $t = t_1$. This t_1 is the "barrel time." Since

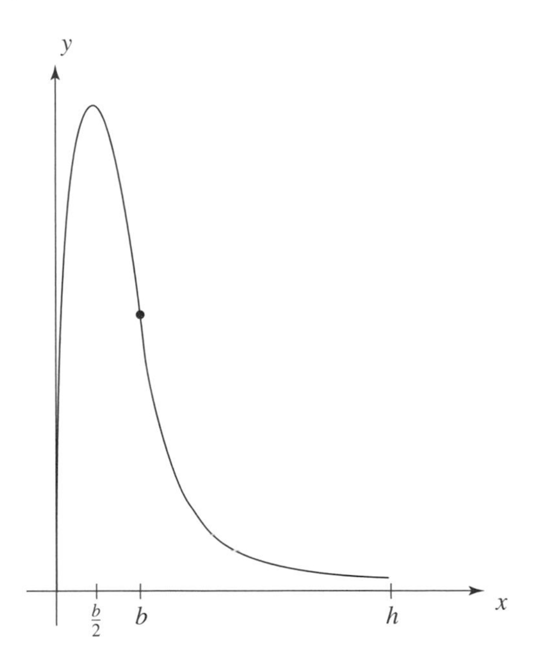

Fig. 10.33

$x(0) = 0$, $x(t_1) = h$, and $x'(t) = \frac{ax(t)}{x(t)+b}$, the velocity of the shell is $x'(0) = 0$ at $t = 0$ and $x'(t_1) = \frac{ah}{(h+b)}$ at $t = t_1$. It follows that

$$\tfrac{1}{2}mv(h)^2 - \tfrac{1}{2}mv(0)^2 = \tfrac{1}{2}m\frac{a^2h^2}{(h+b)^2} = \frac{ma^2h^2}{2(h+b)^2}$$

and that the work-energy and impulse-momentum equations of Section 10.5.1 are

$$W = \int_0^h \frac{ma^2b\,x}{(x+b)^3}\,dx = \frac{ma^2h^2}{2(h+b)^2} \quad \text{and} \quad J = \int_0^{t_1} \frac{ma^2bx(t)}{(x(t)+b)^3}\,dt = \frac{mah}{(h+b)},$$

respectively.

Let's conclude by analyzing the pressure that is generated inside the barrel. This is a concern of obvious importance in the design of the gun. The inside of the barrel is a cylinder. Let c be the diameter of its circular cross section. This diameter is known as the *caliber* of the gun. The area of the cross section is $\pi(\frac{c}{2})^2 = \frac{\pi c^2}{4}$. Now let $p(x)$ be the pressure in the barrel at x, where (as in Figure 10.32) x marks the location of the shell. Because force = pressure × area, we see that $f(x) = \frac{\pi c^2}{4}p(x)$. Therefore

$$p(x) = \frac{4}{\pi c^2} f(x).$$

So the pressure curve has the same general shape as the force curve. In particular, in reference to the location of the shell, the pressure in the barrel rises rapidly and reaches a maximum at $\frac{b}{2}$. It then drops off, rapidly at first, and then more gradually.

10.5.4 The Springfield Rifle

Let's load a rifle by placing a cartridge into what is called the firing chamber. A typical cartridge is shown in Figure 10.34. It consists not only of a bullet (shown in black) but also of a cylinder that contains an

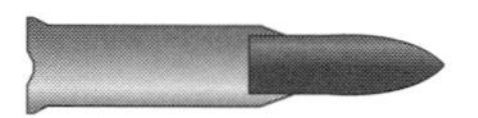

Fig. 10.34

explosive powder. The cartridge is made to fit precisely into the chamber of the rifle. The action of pulling the trigger pushes a spring-driven pin into the cartridge. This detonates a small charge, which starts the explosion of the powder in the cylinder. The explosion separates the bullet from the rest of the cartridge

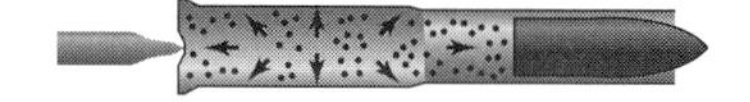

Fig. 10.35

and sends it hurtling down the barrel. This process is illustrated in Figure 10.35.

A rifle derives its name from the fact that it is *rifled*. This means that the inside of the barrel has grooves that spiral like the threading of a screw. These grooves cause the bullet to rotate rapidly as it rushes through the barrel. This rotation gives the bullet greater stability in flight and the weapon greater accuracy.

We will now turn to the famous Springfield rifle. It has a barrel length h of 60.96 cm and a caliber of $c = 0.76$ cm. The barrel has four grooves that make one complete turn every 25.4 cm. One bullet type that it fires has a mass of 10.70 grams. Carefully devised experiments have verified the explicit form

$$v(x) = \frac{1133x}{x + 0.20}$$

of the formula of Le Duc for the Springfield rifle. The position x in Figure 10.32 is given in meters and the velocity $v(x)$ in meters/second. After substituting $h = 0.61$ m, we get

$$v(0.61) = 853 \text{ m/s}.$$

for the muzzle velocity. This is equivalent to 2800 feet/second and corresponds precisely to the official 2800 feet/second for this bullet type. The kinetic energy of the bullet when it leaves the barrel at $h = 0.61$ m is

$$W = \int_0^h f(x)\,dx = \tfrac{1}{2}mv(h)^2 \approx \tfrac{1}{2} \cdot 0.0107 \cdot v(0.61)^2 \approx \tfrac{1}{2} \cdot 0.0107 \cdot 853^2 \approx 3893 \text{ N-m},$$

where $y = f(x)$ is the force on the shell in the barrel. This value for the kinetic energy corresponds closely to the listed kinetic energy of 3894 newton-meters.

The force on the bullet in the barrel reaches its maximum when $x = \frac{b}{2} = \frac{0.20}{2} = 0.10$ m. This maximum force is

$$f(0.10) = \frac{ma^2b(0.10)}{(0.10+b)^3} = \frac{(0.0107)(1133^2)(0.20)(0.10)}{(0.10+0.20)^3} = 10{,}174 \text{ N}.$$

The resulting graph of the force function of the Springfield rifle is depicted in Figure 10.36. The corresponding maximum pressure in the barrel is given by the formula $p(x) = \frac{4}{\pi c^2} f(x)$ and is equal to

$$\frac{4}{\pi c^2} \cdot 10{,}174 \approx \frac{4}{(3.14)(0.0076^2)}(10{,}174) \approx 224 \times 10^6 \text{ N/m}^2.$$

This value is less than the official 415×10^6 N/m^2 listed in the literature. While the predictions of the

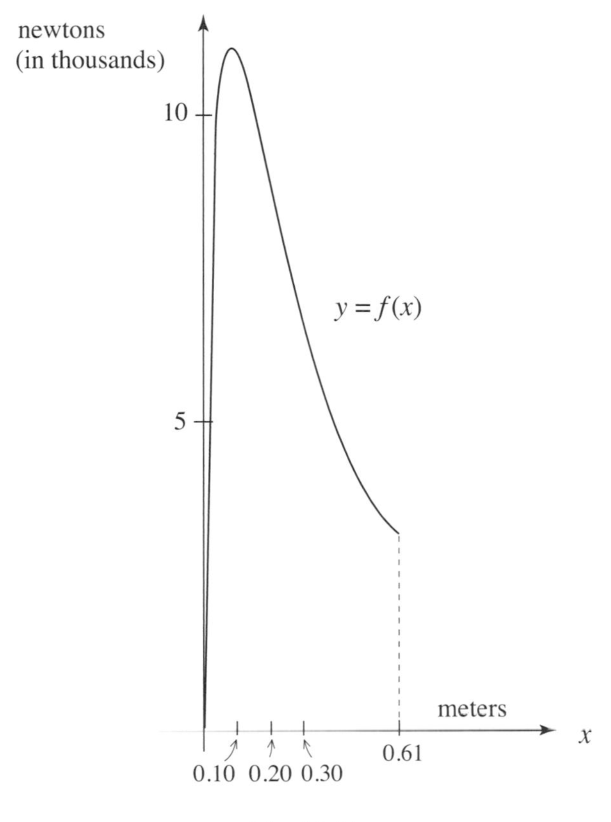

Fig. 10.36

model are in the general "ballpark," this discrepancy confirms the earlier comment that Le Duc's formula and the mathematical model that it gives rise to do not capture the full complexity of the dynamics of the bullet in the barrel.

10.6 PROBLEMS AND PROJECTS

10A. About the Hagia Sophia and the Pantheon. A *simple truss* consists of a triangle of beams each assumed to be rigid and attached to each other by pins. See Figure 10.37. The entire gravitational load L on the truss (including the weight of the beams) is assumed to act vertically at C. The triangular structure is supported at the points A and B. A load of $\frac{L}{2}$ is transmitted downward along each of two slanting beams.

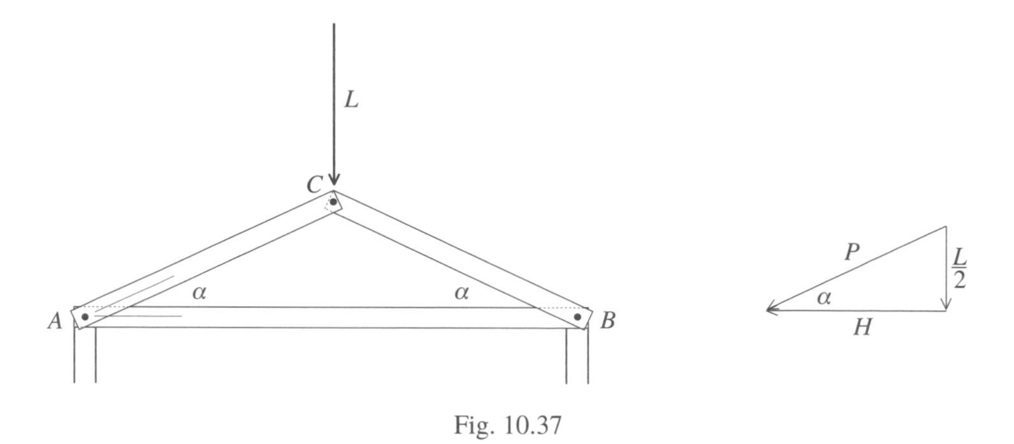

Fig. 10.37

Their angles at A and B with the horizontal are both α. We'll let P be the magnitude of the slanting push by each of the two beams. The force diagram in the figure separates the push P at A into its horizontal and vertical components. The horizontal beam of the truss, called a *tie-beam*, pulls inward to counteract the horizontal component of P.

10.1. Use the force diagram of the figure to show that $P = \frac{L}{2\sin\alpha}$ and $H = \frac{L}{2\tan\alpha}$.

Return to the shell of the dome of the Hagia Sophia and consider the combination of a typical rib and its opposite to be modeled by a simple truss. The inward pull of the tie-beam represents the push of the base of the dome against the two ribs that counteracts their outward thrust. The thrust computation of Section 10.1.1 assumed that a typical pair of opposite ribs is modeled by a simple truss in this way.

The Hagia Sophia has had a difficult history. The stresses that the shell of the dome is under makes it particularly vulnerable to the earthquakes that are common in both Greece and Turkey. An earthquake led to a partial collapse of the dome only 20 years after its completion. By 563 A.D., the dome had been completely rebuilt. This rebuilt dome is the current dome discussed in Section 10.1.1. The 40 buttresses that brace the 40 ribs between the windows of the dome were added at the time of the reconstruction. Additional earthquakes in the 10th and 14th centuries did major damage to the dome, and extensive repairs were required each time.

10.2. Figure 10.38 depicts a model of the shell of today's dome of the Hagia Sophia above the gallery of 40 windows. The radius of the inner surface of the shell is denoted by r, and the angle between the indicated radii by θ. The radius of the horizontal circular cross section of the dome (just above the windows) is b, and the distance from the center of this circle to the top of the inner surface of the shell is a. Section 10.1.1 informs us that for the current dome, $r = 15.2$ m and $\theta = 140°$. Conclude from this that $b \approx 14.3$ and $a \approx 10.0$ m.

Not much is known about the original dome of the Hagia Sophia other than that it was about 3.1 m lower than the rebuilt current dome. It is likely that the inner and outer surfaces of the original dome were also given by concentric hemispheres, that this dome also had a circular arcade of 40 windows, and that the configuration of its rib and support structures were basically the same as those of the rebuilt current dome.

10.3. Suppose that the original dome had the general shape depicted in Figure 10.38, with $b = 14.3$ m and $a = 10.0 - 3.1 = 6.9$ m.

i. Derive the estimates $r \approx 18.3$ m, $\sin\frac{\theta}{2} \approx 0.78$, and hence $\theta \approx 103°$ for the original dome.

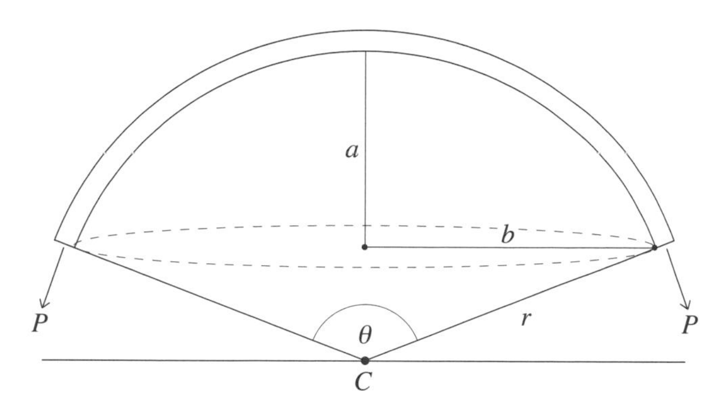

Fig. 10.38

ii. Assuming that the shell of the original dome was—like the shell of today's dome—0.8 m thick, so that the outer radius of the original shell was $18.3 + 0.8 \approx 19.1$ m, derive an estimate for the volume of the original shell above the row of 40 windows. Assume that the masonry of the original shell had the same density 1760 kg/m^3 as the current shell and provide an estimate for the mass of the original shell.

iii. Use the study of Section 10.1.1 to derive estimates for the magnitude P of the force with which a single rib of the original dome pushed against its base and the magnitude of the horizontal component H of this force. Compare these values with the estimates for the P and H of the current dome and discuss the differences.

10.4. Use Figure 10.8 and the data for the shell of the Pantheon provided in Section 10.1.2 to determine the minimal and maximal thickness of the shell.

10.5. Figures 10.39a and 10.39b are both extracted from Figure 10.8. Describe (without carrying out the details) how the strategy for computing the volume of the shell of the Pantheon that these figures

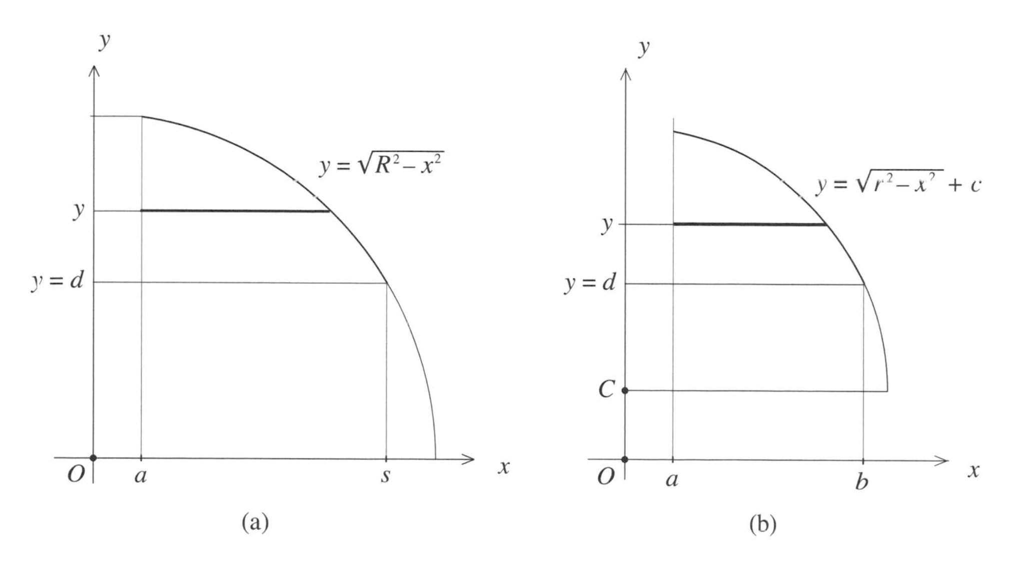

Fig. 10.39

suggest (see the formula of Problem 9.35 in this regard) might be applied to derive a better estimate for the mass of the shell of the Pantheon than the estimate developed in Section 10.1.2.

10B. About Clotheslines and Cables. A clothesline of length 20.5 feet is attached to two poles at points A and B. The poles are 20 feet apart. The straight line that connects the points A and B is horizontal. See

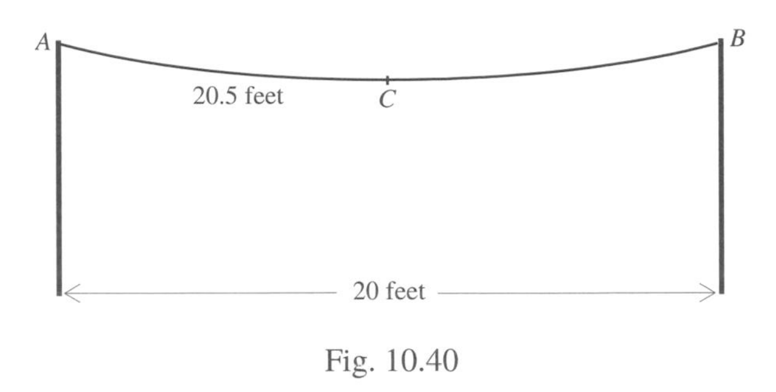

Fig. 10.40

Figure 10.40. The clothesline is completely flexible and inelastic.

10.6. A laundry bag containing 120 wet socks (60 pairs) and weighing 26 pounds has been attached to the clothesline at its midpoint C. Compared to the weight of the socks, the weights of both the clothesline and the laundry bag are negligible.

i. Show that in the situation just described, the point C is 2.25 feet below the horizontal AB.

ii. Let T be the tension in the clothesline, and draw a diagram of the forces acting at C. Combine the force diagram and the geometry to show that $T \approx 59.22$ pounds.

The 120 socks are taken out of the bag and suspended individually on the clothesline at approximately 6 socks per horizontal foot. Since 120 socks weigh 26 pounds, 6 socks weigh 1.3 pounds. Therefore the clothesline supports $w = 1.3$ pounds per foot. Let $d = 10$ be one-half the distance between the two posts. The point C has shifted from its previous situation. It is now s feet below the segment AB.

10.7. Place a coordinate system into the figure in such a way that the origin is at C and the x-axis is parallel to the segment AB.

i. Let L be the length of the part of the clothesline from C to B. Explain why

$$L = \int_0^d \sqrt{1 + \left(\tfrac{2s}{d^2}x\right)^2}\, dx\,.$$

ii. Recall the approximation

$$\int_0^d \sqrt{1 + \left(\tfrac{2s}{d^2}x\right)^2}\, dx \approx d + \tfrac{2}{3}\left(\tfrac{s}{d}\right)^2 d - \tfrac{2}{5}\left(\tfrac{s}{d}\right)^4 d$$

from Section 10.2. Use it and the quadratic formula to derive an estimate for s^2. From this, deduce the approximation $s \approx 1.96$ feet.

iii. Use the conclusion of (ii) to show that the maximum tension $T_{\max}$ and the minimum tension $T_{\min}$ in the clothesline are 35.62 pounds and 33.16 pounds, respectively. Compare this with the tension computed in Problem 10.6.

10.8. Construct a cable with a hexagonal arrangement of strands as follows. Start with any odd number $2k + 1$ of circles to form the horizontal diameter. Put $2k$ circles in the row above the diameter, put $2k - 1$ circles in the row above that, and continue the pattern in this way. Do a similar thing below the horizontal diameter. Why will the ith row above or below the diameter consist of $2k - (i - 1)$ circles? How many rows i above the diameter are needed so that the cluster of circles on and above the diameter form the upper part of an hexagonal array? Count the total number of circles above

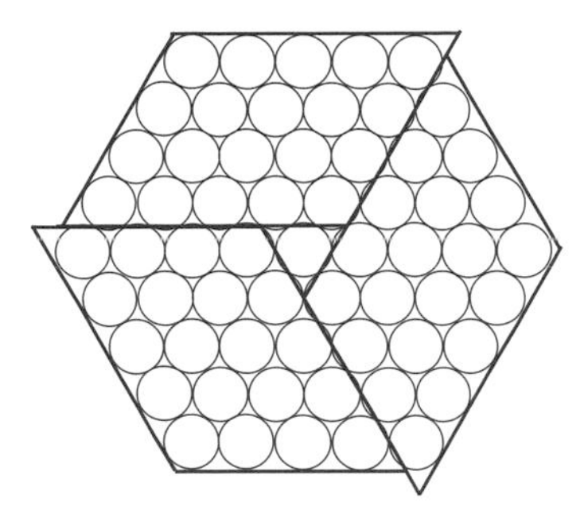

Fig. 10.41

and below the horizontal diameter by matching up the first row above the diagonal with the last row below the diagonal, the second row above the diagonal with the next to last row below the diagonal, and so on. Show that the total number of circles in the hexagonal array is $3k^2 + 3k + 1$. Does this provide the correct number of circles for the patterns in Figure 10.10? Alternatively, study Figure 10.41. It depicts the case $k = 4$ and confirms the expression $3k(k + 1) + 1 = 3k^2 + 3k + 1$ for the number of circles in the arrangement.

10.9. A new Tacoma Narrows Bridge was built in the years 1948–1950. The new bridge has a total length of 5000 feet and a center span of 2800 feet. Its two main cables support a single deck that carries four lanes of automobile traffic. The dead load of the deck is 8680 pounds per foot, and it is designed for a live load capacity of 4000 pounds per foot. The corresponding sag in the main cable over the center span is about 280 feet. Estimate the length of a main cable over the center span by applying the length formula of Section 10.2 and also by using its approximation. (The data for the bridge were supplied by the Washington State Department of Transportation.)

10.10. The Verrazano Narrows Bridge, New York City, was completed in 1964. It has two decks, a center span of 4260 feet, a dead load of 37,000 pounds per foot, and a live load capacity of 4800 pounds per foot. It has four main cables, and the sag in each of them is 385 feet at mid-span. Estimate the length of a main cable over the center span by using the length formula as well as its approximation. (Data supplied by the Triborough Bridge and Tunnel Authority of New York.)

10.11. Each of the two main cables of the Golden Gate Bridge has a diameter of 36 inches. Each cable consists of 61 strands with 452 steel wires of diameter 0.196 inches, for a total of 27,572 wires per cable. The ultimate strength of each wire is 220,000 pounds per square inch. Show that the ultimate strength of a main cable is 183,000,000 pounds. What is the safety factor built into the main cable? Use the fact (supplied by the Golden Gate Bridge, Highway and Transportation District, San Francisco, California) that the tension in the cable over the center span at the towers is 54,000,000 pounds under dead load only and 64,100,000 pounds under dead plus live load.

By far the largest suspension bridge in the world is the Akashi Straits Bridge in Japan. Since its completion in 1998, the bridge has held the world record of 1991 m (or 6532 feet) for the longest center span of a suspension bridge. (Note that Japan uses the metric system.) Each of the two side spans is 960 m long. The two main supporting towers rise 282.8 m above sea level. The expansion due to heat can stretch the bridge by up to 2 m over the course of a day. It took two million workers 10 years to construct the bridge. They used 181,000 tons (about 400 million pounds) of steel and poured 1.4 million cubic meters of concrete. The main cables made use of the hexagonal geometry already described. But rather than a hexagonal configuration of strands to form the main cable, it is each strand that is a hexagonal configuration of steel wires. Each steel wire has a diameter of 5.23 millimeters and an ultimate strength of 1800 N per square millimeter. Each strand is a hexagonal arrangement of 127 steel wires, and each main cable of the Akashi Straits Bridge consists of 290 such strands grouped into a circular arrangement. So each main cable consists of $290 \times 127 = 36{,}830$ steel wires and has an overall diameter of 1.122 m.

10.12. Consider one of the main cables of the Akashi Straits Bridge.

i. How many wires does the central diameter of a hexagonal strand of the cable consist of?

ii. Compute the cross-sectional area of one wire. Use your result to show that a single wire has an ultimate strength of about 38,669 N.

iii. Show that the ultimate strength of a main cable of the Akashi Straits Bridge is approximately

$$36{,}830 \times 38{,}669 \approx 1{,}424{,}000{,}000 \text{ N}.$$

iv. Show that a main cable of the Akashi Straits Bridge is almost twice as strong as a main cable of the George Washington Bridge.

10C. Studying the Tractrix and the Trumpet. This segment explores the tractrix and another classic curve with similar properties.

10.13. Turn to Figure 10.13, and let c be any number with $0 < c \leq a$. Use a formula from Section 9.3 to show that the length of the tractrix from $(c, T(c))$ to $(a, 0)$ is equal to $a \ln \frac{a}{c}$.

Since $\ln \frac{a}{c}$ goes to ∞ when c is pushed to zero, it follows—certainly as expected—that the graph of the tractrix has infinite length. We will see next, however, that the area of the region under the graph is finite, even though the extent of this region is infinite.

Continue to let c be any number with $0 < c \leq a$. We'll start by computing the part of the area $A(c)$ under the graph of $y = T(x)$ over the interval $[c, a]$. This requires that we compute

$$A(c) = \int_c^a T(x)\, dx = \int_c^a \left(a \operatorname{sech}^{-1} \tfrac{x}{a} - \sqrt{a^2 - x^2}\right) dx.$$

10.14. i. Use Formulas (23) and (16) of Section 9.11 to show that

$$A(c) = \tfrac{\pi a^2}{4} - ac \operatorname{sech}^{-1} \tfrac{c}{a} - \tfrac{a^2}{2} \sin^{-1} \tfrac{c}{a} + \tfrac{c}{2}\sqrt{a^2 - c^2}.$$

ii. Use a limit formula from Section 10.3.1 and a property of $y = \sin^{-1} x$ to show that

$$\lim_{c \to 0} A(c) = \tfrac{1}{4}\pi a^2,$$

and hence that the area under the full tractrix of Figure 10.15 is $\frac{1}{2}\pi a^2$.

The area under the full tractrix is therefore finite. We have already observed quantitative similarities between the sphere and the pseudosphere. These two surfaces are generated by revolving the half-circle and the full tractrix. The expression $\frac{1}{2}\pi a^2$ for both the area under the tractrix and the area of the half-circle (of radius a) is another example of such a similarity.

There is another curve with properties similar to those of the tractrix. It was studied by Evangelista Torricelli (1608–1647), the important Italian mathematician and physicist (and student of Galileo), before Perrault, Leibniz, and Huygens investigated the tractrix. Consider the graph of the function

$$y = f(x) = \tfrac{1}{x} \text{ for } 1 \leq x.$$

See Figure 5.12e. Let $c \geq 1$, and focus on the region under the graph and above the x-axis from $x = 1$ to $x = c$. The area of this region is

$$\int_1^c \tfrac{1}{x}\, dx = \ln x \Big|_1^c = \ln c - \ln 1 = \ln c.$$

Pushing c to infinity tells us that the entire area under the graph to the right of $x = 1$ is infinite. This is a point of difference between this curve and the tractrix.

Let's consider the solid of infinite extent obtained by rotating the graph of $f(x) = \frac{1}{x}$ for $1 \leq x$ one revolution around the x-axis. This solid derives its name *Torricelli's trumpet* from its shape. Again, let $c \geq 1$ and consider the finite solid obtained by rotating the graph of $f(x) = \frac{1}{x}$ with $1 \leq x \leq c$ around the x-axis. By a volume formula in Section 9.2, the volume $V(c)$ of this solid is

$$V(c) = \int_1^c \pi\left(\tfrac{1}{x}\right)^2 dx = -\pi x^{-1} \Big|_1^c = -\tfrac{\pi}{c} + \pi.$$

Since $\lim_{c \to \infty} V(c) = \pi$, the volume π of Torricelli's trumpet is finite even though it has infinite extent.

Let's turn next to the surface area of the trumpet. Again let $c \geq 1$. The fact that the surface area of a solid obtained by revolving the graph of a function $y = f(x)$ with $a \leq x \leq b$ one revolution around the x-axis is given by

$$\int_a^b 2\pi f(x)\sqrt{1 + f'(x)^2}\, dx$$

was established in Section 9.4.

10.15. Consider the function $f(x) = \frac{1}{x} = x^{-1}$ with $1 \leq x \leq c$.

i. Show that the area of the surface obtained by revolving its graph once around the x-axis is

$$S(c) = \int_1^c 2\pi \tfrac{1}{x}\sqrt{1 + f'(x)^2}\, dx = \int_1^c 2\pi \tfrac{1}{x^3}\sqrt{x^4 + 1}\, dx.$$

ii. Use the fact that any definite integral always has an interpretation as area under a curve together with the inequality $\frac{1}{x}\sqrt{1 + f'(x)^2} > \frac{1}{x}$ to show that $\lim_{c \to \infty} S(c) = \infty$ and hence that this surface area is infinite.

We saw in Section 10.3 that the pseudosphere has finite volume and finite surface area. The situation of the trumpet is different. It has, amazingly, finite volume but infinite surface area. Torricelli's trumpet is also known as Gabriel's horn. According to some religious traditions, it is Gabriel the Archangel who blows the horn to announce Judgment Day, the day on which humans, by nature limited and finite, are given the opportunity to ascend to God's infinite realm.

10.16. Try to solve the integral $\int_1^c \frac{1}{x^3}\sqrt{x^4 + 1}\, dx$ that arose in the study above. If you can't, turn to

http://www.integral-calculator.com/#

(or an equivalent website) for the answer. Then check that the response

$$\tfrac{1}{4}\Big[\ln\Big(\tfrac{\sqrt{x^4+1}}{x^2} + 1\Big) - \ln\Big(\tfrac{\sqrt{x^4+1}}{x^2} - 1\Big) - \tfrac{2\sqrt{x^4+1}}{x^2}\Big] + C$$

for the antiderivative of $\frac{1}{x^3}\sqrt{x^4 + 1}$ is correct. (This is a slog.)

10D. Moving Points. A review of Sections 6.4 and 6.5 will facilitate efforts to solve the next set of problems.

10.17. A point starts moving on the x-axis at time $t = -5$. Its position at any time $t \geq -5$ is given by $x(t) = 3t^5 - 65t^3 + 540t + 3950$. Determine the velocity function and the times at which the point comes to a stop. Over what time intervals is the point moving to the left and to the right? Suppose that the point has mass, and that its motion is the consequence of a force. Discuss the connection between the velocity and the action of the force on the particle over each of these time intervals. [You should get $x'(t) = 0$ for $t = -3, -2, 2$, and 3. So what are the time intervals? Use force = $ma(t)$ and $a(t) = 60t^3 - 390t = 30t(2t^2 - 13)$.]

10.18. Verify the following assertion of Galileo. Consider a body that starts from rest at time $t = 0$ and moves along a straight line with constant acceleration. Show that the velocity at any time $t \geq 0$ is equal to twice the average velocity over the time interval from 0 to t.

The next several problems study the motion of a point in an xy-plane. The x- and y-coordinates of its position are functions $x = x(t)$ and $y = y(t)$ of time t. The speed of the point is given by the formula $v(t) = \sqrt{x'(t)^2 + y'(t)^2}$ of Section 6.4.

10.19. A point's position is given by $x(t) = t$ and $y(t) = 1 - t$ at any time $t \geq 0$. Along what curve does it move? What is its speed at any time t? In what direction does it move along its path?

10.20. A point's position is given by $x(t) = t$ and $y(t) = \frac{1}{5}t^2$ during $-10 \leq t \leq 10$. Describe the path along which the point travels, and compute the speed of the point along its path. Suppose that the point's mass is 1 unit. Explain the motion by describing the horizontal and vertical components of the force acting on the point.

10.21. A point's position is given by $x(t) = \sqrt{t}$ and $y(t) = \sqrt{1 - t}$. The motion starts at time $t = 0$. At what time must it end? Show that the point travels along the circle of radius 1 with center the origin. Discuss the motion of the point around the circle. What curve do the parametric equations $x(t) = \sqrt{t}$ and $y(t) = \sqrt{1 - t}$ describe?

10.22. A point moves from time $t = -1000$ to time $t = 1000$ with position given by $x(t) = t^{\frac{1}{3}}$ and $y(t) = t$. Along what path does it move? Compute its accelerations $x''(t)$ and $y''(t)$. Suppose that the point has mass is 1 unit. Explain its motion by analyzing the force that acts on it.

10.23. A point starts at $(-4, 3)$ at time $t = 0$. It is known that at any time $t \geq 0$ its velocities in the x and y directions are $2t$ and $t^3 + 4t$, respectively.

i. Determine the functions $x(t)$ and $y(t)$ in precise terms. Where is the point at time $t = 2$?

ii. Determine an equation for the path of the point, and discuss the shape of the path.

10.24. A point's position is given by $x(t) = t$ and $y(t) = \sin t$ for $t \geq 0$. So the point travels along the graph of $y = \sin x$ for $x \geq 0$. Suppose that the point has mass 1. Explain the motion of the point by analyzing the forces $x''(t)$ and $y''(t)$.

10.25. The position of a point P is given by $x(t) = \cos t$ and $y(t) = \sin t$ for all $t \geq 0$. Along what path does the point move? Suppose that the point has a mass of 1 unit, so that force and acceleration are equal.

i. Compute $x'(t)$ and $y'(t)$, and show that the speed $v(t)$ of the point is equal to 1. So the acceleration of the point along its path is zero.

ii. Show that $x''(t) = -x(t)$ and $y''(t) = -y(t)$. Place P in a typical position. Regard $x''(t)$ as force, and represent it with an arrow that starts at P. Repeat this for $y''(t)$.

iii. Use the parallelogram law to draw an arrow that represents the composite of the two forces $x''(t)$ and $y''(t)$. Show that the magnitude of this composite is equal to 1. Check the slope of this force vector to show that it points from P in the direction of the origin O.

With regard to the last problem, how do you reconcile the fact that there is a force acting on the point but that the point moves with constant velocity along its path? What role does the force play? If the force were to stop acting at time t_0, what would the motion of the point be for $t \geq t_0$? In general, a force on a point-mass will determine both changes in direction of its path as well as changes in the velocity along its path.

10E. From the Popular Literature. Marilyn vos Savant in her column in the November 19, 2000, issue of *Parade Magazine* posed the following problem.

10.26. "Say that two motorboats on opposite shores of a river start moving toward each other, but at different speeds. When they pass each other the first time, they are 700 yards from the shore line. They continue to the opposite shore, then turn around and start moving toward each other again. When they pass a second time, they are 300 yards from the other shore line. How wide is the river?"

This question, already asked in an earlier column, had generated controversy because readers—including "plenty of mathematicians and other professionals"—pointed out that the solution leads to a system of four equations in five unknowns, and that such systems could not be solved. But Marilyn persisted in her claim that the width of this river can be determined from what is given. In fact, she claimed that this is a mighty river of 1800 yards across. Is she right? Have a go at this problem (even though it is not a calculus problem). Assume that the two shorelines are parallel and that the boats move perpendicularly to them. Assume that the boats move at constant speeds, and in particular that they bounce off the shore like billiard balls without loss of speed. Ignore other factors such as the current.

Here's a strategy you might try. Let the width of the river be w. Let the speed of boat A be a, and let the speed of boat B be b. Suppose that the boats start moving at time $t = 0$. Let the first and second encounters occur at times t_1 and t_2, respectively. Consider Figure 10.42 and write four equations involving

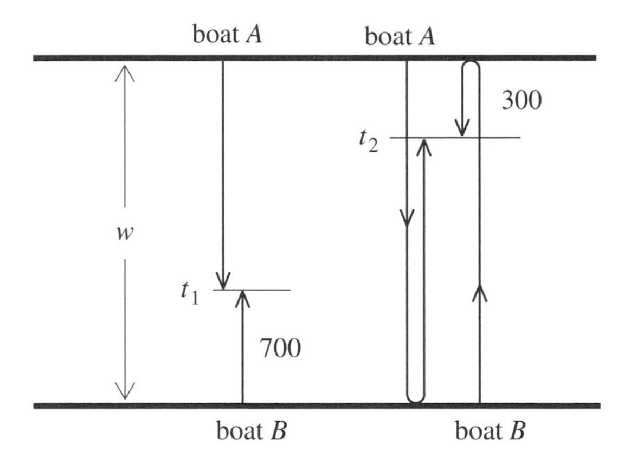

Fig. 10.42

at_1, bt_1, at_2, and bt_2 and w. So we have four equations in five unknowns, just as the experts had pointed out. (However, they reduce to two equations in $\frac{a}{b}$ and w.)

10.27. On occasion, there is a piece of fiction that gains sense, meaning, and humor from basic concerns of mathematics and physics. The author David Foster Wallace wove "The Bricklayer's Accident Report," a story from the last years of the 19th century, into his award-winning novel *Infinite Jest*. (There is nothing to do here but to read and be amused.)

> Dear Sir,
>
> I am writing in response to your request for additional information. In block #3 of the accident reporting form, I put "trying to do the job alone" as the cause of my accident. You said in your letter that I should explain more fully and I trust that the following details will be sufficient.

> I am a bricklayer by trade. On the day of the accident, March 27, I was working alone on the roof of a new six story building. When I completed my work, I discovered that I had about 900 kg of brick left over. Rather than laboriously carry the bricks down by hand, I decided to lower them in a barrel by using a pulley which fortunately was attached to the side of the building at the sixth floor. Securing the rope at ground level, I went up to the roof, swung the barrel out and loaded the brick into it. Then I went back to the ground and untied the rope, holding it tightly to insure a slow descent of the 900 kg of bricks. You will note in block number #11 of the accident reporting form that I weigh 75 kg.
>
> Due to my surprise at being jerked off the ground so suddenly, I lost my presence of mind and forgot to let go of the rope. Needless to say, I proceed up at a rather rapid rate up the side of the building. In the vicinity of the third floor, I met the barrel coming down. This explains the fractured skull and broken collarbone.
>
> Slowed only slightly, I continued my rapid ascent, not stopping until the fingers of my right hand were two knuckles deep into the pulley. Fortunately, by this time I regained my presence of mind and was able to hold tightly to the rope in spite of considerable pain. At approximately the same time, however, the barrel of bricks hit the ground and the bottom fell out of the barrel from the force of hitting the ground.
>
> Devoid of the weight of the bricks, the barrel now weighed approximately 30 kg. I refer you again to my weight of 75 kg in block #11. As you could imagine, still holding the rope, I began a rapid descent from the pulley down the side of the building. In the vicinity of the third floor, I met the barrel coming up. This accounts for the two fractured ankles and the laceration of my legs and lower body.
>
> The encounter with the barrel slowed me enough to lessen my impact with the brick-strewn ground below. I am sorry to report, however, that as I lay there on bricks in considerable pain, unable to stand or move and watching the empty barrel six stories above me, I again lost my presence of mind and unfortunately let go of the rope, causing the barrel to begin a ...

10F. On the Orbital Mathematics of Kepler and Newton. Let a body P be in an elliptical orbit around a much more massive one S located a focus of the orbit. Suppose that the semimajor and semiminor axes of the orbit are a and b, that ε is the eccentricity, $c = a\varepsilon$ is the distance between S and the center of the ellipse, and T is the period of the orbit. The problems below rely on facts from Section 10.4, in particular on Table 10.1, and Newton's version of Kepler's third law $\frac{a^3}{T^2} = \frac{GM}{4\pi^2}$, where M is the mass of S.

10.28. How does Table 10.1 inform us that Mercury's ellipse is flatter than the ellipses of the other planets? (Recall that Pluto lost its status of planet and is now classified as a dwarf planet.) Use the data from the table and a scale of 10,000,000 km = 3 cm to carefully draw the $2a \times 2b$ box that contains Mercury's ellipse. Then draw the circle of radius a of Figure 10.20 that surrounds the ellipse. Can you visually distinguish the rectangle from a square and the circle from the ellipse?

The *astronomical unit* au is a common unit for measuring distances in the solar system. Section 3.7 told us that it is based on the average distance from Earth to the Sun and that 1 au = 149,597,892 km.

10.29. On August 19, 2003, the *New York Times* reported that on August 27, at 5:51 a.m. Eastern time, the distance between Mars and Earth would be 55,758,006 km. This would be the closest that Mars had been to Earth in 59,619 years, but that in the year 2287, Mars and Earth would be even closer. The closest Mars approaches to Earth for the next several hundred years will be:

au	kilometers	date
0.37200418	55,651,033	Sept. 8, 2729
0.37200785	55,651,582	Sept. 3, 2650
0.37217270	55,676,243	Sept. 5, 2934
0.37225400	55,688,405	Aug. 28, 2287
0.37230224	55,695,623	Sept. 11, 2808

Use the information in Table 10.1 to show that the geometrically shortest possible distance between Earth and Mars is about 54.6 million kilometers.

10.30. A 2010 article in the journal *Icarus* discusses the orbits of the asteroid Eugenia and its two satellites. Both satellites have nearly circular orbits of radii 1165 km and 611 km, respectively. The more distant satellite was discovered in 1999, the second in 2004. The outer satellite has a diameter of about 7 km, and the diameter of the inner satellite is about 5 km. They both have about same mass of 2.5×10^{14} kg. It had been known earlier that the outer satellite completes its orbit once every 4.7 days. Select from the data provided to estimate both the mass of Eugenia and the period of the inner satellite.

10.31. With its eccentricity of $\varepsilon \approx 0.0068$, Venus's orbit is closer to being a circle than the orbit of any other planet. Turn to Figure 10.19, and focus on the angular position α of Venus. How many days does it take for α to rotate from 0° to 60°? From 60° to 120°? And finally from 120° to 180°? [Hint: Proceed by combining the equations of Gauss and Kepler. Answers will depend on the number of decimal places carried and the roundoff procedures that are used, but they should be close to 37.03 days, 74.48 − 37.03 = 37.45 days, and 112.35 − 74.48 = 37.87 days, respectively.]

Since the Moon orbits Earth and Earth orbits the Sun, you might expect the orbit of the Moon about the Sun to occur in loops. But there are no loops. Why not? Recall from Section 10.4.3 that Earth's orbital

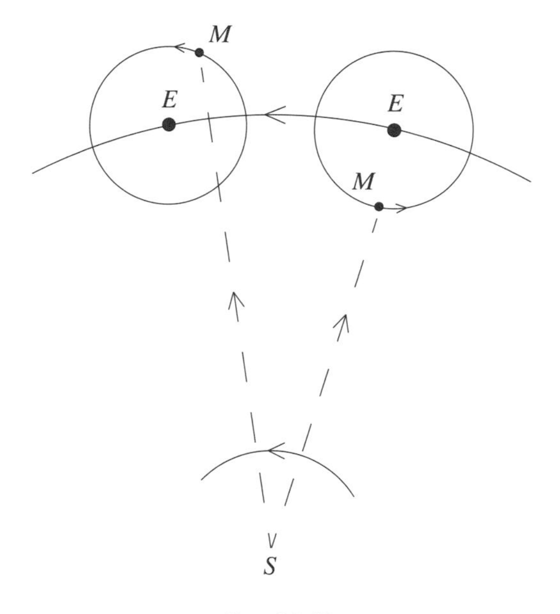

Fig. 10.43

velocity satisfies $v_{\max} \approx 30.2866$ km/s and $v_{\min} \approx 29.2910$ km/s. The website

http://nssdc.gsfc.nasa.gov/planetary/factsheet/moonfact.html

provides the following information about our Moon's orbit about the Earth. Its semimajor axis is 384,400 km, its eccentricity is 0.05490, and its period is $T = 27.3217$ days.

10.32. Show that the formulas in Example 10.4 for the maximum and minimum orbital speeds can be expressed as $v_{\max} = \frac{2\pi a}{T}\sqrt{\frac{1+\varepsilon}{1-\varepsilon}}$ and $v_{\min} = \frac{2\pi a}{T}\sqrt{\frac{1-\varepsilon}{1+\varepsilon}}$. Use these formulas to show that the maximum and minimum speeds of the Moon in its orbit around the Earth $v_{\max} = 1.0810$ km/s and $v_{\min} = 0.9684$ km/s.

It follows that the orbital speed of the Earth around the Sun is about 30 times greater than the orbital speed of the Moon around Earth. This means, as Figure 10.43 illustrates, that the segment from the Sun S to the Moon M always rotates in the same direction (with close to the same rotational speed as the segment from S to Earth E). This in turn implies that the Moon's orbit about the Sun has no loops.

10.33. Table 10.1 provides the orbital data $a = 17.83$ au, $\varepsilon = 0.967$, and T = 75.32 years for the comet Halley. It follows that the semiminor axis and perihelion distance are $b = a\sqrt{1-\varepsilon^2} = 4.543$ au and $q = a(1-\varepsilon) = 0.588$ au, respectively. Figure 10.44 shows an xy-coordinate system with the center of Halley's ellipse $\frac{x^2}{a^2} + \frac{y^2}{b^2} = 1$ at the origin O. The Sun S is at $(c, 0)$, where $c = a - q = 17.24$ au.

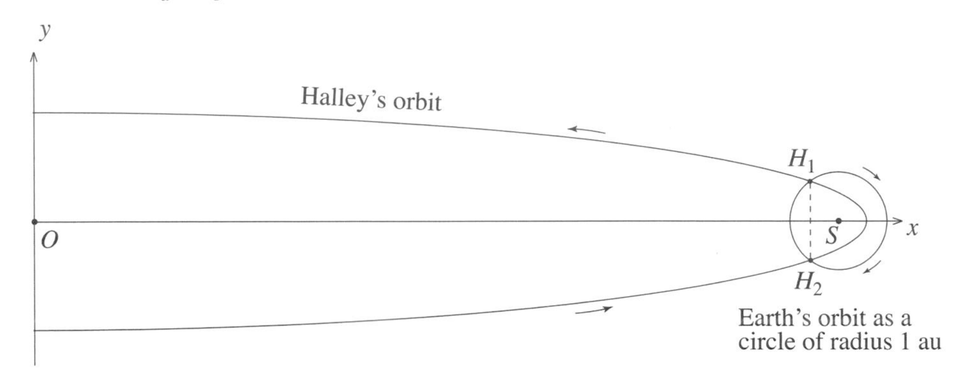

Fig. 10.44

Figure 10.44 shows Earth's orbit as a circle with center S and radius 1 au. (In terms of the distances involved, the figure is to scale.) Table 10.1 also tells us that Halley's orbital plane is separated by 17.7° from Earth's orbital plane, and that the two bodies orbit in opposite directions. However, in what follows we'll assume that the orbital planes of Halley and Earth are the same.

i. Determine an equation of Earth's circular orbit. Show that the x-coordinate of the points of intersection H_1 and H_2 of the two orbits is $x = \frac{a^2-a}{c}$. Determine the numerical values of the x- and y-coordinates of H_1 and H_2. [Hints: In reference to the equation of the ellipse, it is better to work with the parameters a, b, c, etc., rather than their numerical values. Use the identity $c^2 = a^2 - b^2$. Since $\frac{a^2+a}{c} = \frac{a^2}{c} + \frac{a}{c} > a$, note that $x = \frac{a^2+a}{c}$ is not possible.]

ii. Consider Sections 10.4.1 and 10.4.2 in the context of Halley's orbit. Let t be the time Halley requires to travel from perihelion to $P = H_1$. Use the x- and y-coordinates computed in (i) to compute $r(t)$ and $\alpha(t)$. Then use Gauss's equation to compute $\beta(t)$ and Kepler's equation to determine t. For how many days will Halley be inside the Earth's orbit?

iii. Consider the successive approximation method of Step 3 for computing the angle $\beta(t)$ for Halley's orbit. Give an estimate for the number of iterations necessary to do so with an accuracy of 0.0002. [Hint: Refer to Example 10.3.]

10.34. Look up the dates and times for Earth's perihelion and aphelion for the current year on the website

http://aa.usno.navy.mil/data/docs/EarthSeasons.php

(or another website that provides such data). In the table of the website, the entry "Jan 2 04 38" (for example) refers to January 2nd, 38 minutes after 4 a.m. Similarly, "July 6 19 40" refers to July 6th, 40 minutes after 7:00 p.m., and so on. Use the data and strategy of Section 10.4.3 to do the following.

i. Take today's date, and count the number of days and hours t that have elapsed since Earth's last perihelion.

ii. Use the strategy of Section 10.4.1 to estimate the current distance $r(t)$ of Earth from the Sun and the corresponding angle $\alpha(t)$.

iii. Turn to Section 10.4.2 to compute the current speed $v(t)$ of Earth in its orbit around the Sun.

Return to a body P in elliptical orbit around a much more massive one S at a focus of the orbit. Suppose that the semimajor and semiminor axes of the orbit are a and b, that ε is the eccentricity, $c = a\varepsilon$ is the distance between S and the center of the ellipse, and T is the period. Let $\kappa = \frac{ab\pi}{T}$ be Kepler's constant of the orbit, and $L = \frac{2b^2}{a}$ its latus rectum. (See Section 6.9.) Recall Newton's version of Kepler's third law, namely that $K = \frac{a^3}{T^2}$ is equal to $\frac{GM}{4\pi^2}$, where M is the mass of S and G is the universal gravitational constant, for any body P in orbit around S.

10.35. Verify the equalities $\kappa^2 = \frac{\pi^2}{2}LK = \frac{GLM}{8}$ and $\kappa = \pi\sqrt{\frac{LK}{2}} = \frac{1}{2}\sqrt{\frac{GLM}{2}}$ relating the orbital parameters κ, L, and K.

10.36. Use Kepler's equation $\beta(t) - \varepsilon\sin\beta(t) = \frac{2\pi t}{T}$ with $\beta(t) = \pi$ to show that the body P takes exactly as long—namely, $\frac{T}{2}$—to go from periapsis to apoapsis as from apoapsis to periapsis.

10.37. A look at Figure 10.20 tells us that, starting from perihelion, P completes the first fourth of its orbit at the moment t for which $\beta(t) = \frac{\pi}{2}$. Show that $t = \frac{T}{4} - \frac{T\varepsilon}{2\pi}$. Check that if the orbit is a very flat ellipse—one for which ε is close to 1—then $t \approx \frac{1}{11}T$.

Kepler's original strategy for dealing with his equation was indirect. For a tight sequence of angles β between 0° and 180°, he used his equation to compute the corresponding times t, and he tabulated this information in his *Rudolphine Tables*. See the website

http://www.hps.cam.ac.uk/starry/keplertables.html.

With this table, he could then go in the other direction and read off—for a given time t—angles β that provide approximate solutions of his equation. He would then fine-tune and sharpen these solutions by extrapolation.

10.38. Review Newton's method for finding the zeros of a differentiable function from the last segment of the Problems and Projects section of Chapter 7. Consider it for the function $f(x) = x - \varepsilon\sin x - \frac{2\pi t}{T}$, where t is a given elapsed time. Then show that for an approximation β_i of the solution $\beta(t)$ of Kepler's equation $x - \varepsilon\sin x - \frac{2\pi t}{T} = 0$, the approximation $\beta_{i+1} = \beta_i + \frac{\frac{2\pi t}{T} - (\beta_i - \varepsilon\sin\beta_i)}{1-\varepsilon\cos\beta_i}$ is a better approximation of $\beta(t)$.

10.39. Turn to Section 10.4.3 and the computation of $\beta(t)$ for Earth's summer solstice position in 2013. Check that Newton approximation with $\beta_1 = \frac{2\pi t}{T} = 2.924336$ radians provides the correct result

$$\beta_2 = \beta_1 + \frac{\frac{2\pi t}{T} - (\beta_1 - \varepsilon\sin\beta_1)}{1-\varepsilon\cos\beta_1} = 2.927880$$

after a single step. It is apparent that Newton's method converges more quickly in general than the successive approximation method of Step 3. (But the latter is simpler to apply.)

10G. The Meaning of Average. The definite integral provides a definition for the average value of a function. Start by considering a list of numbers, for example, 5, 3, 6, 4, 2, and 8. Their average is

$$\frac{5+3+6+4+2+8}{6} = \frac{28}{6} = \frac{14}{3} = 4\tfrac{2}{3}.$$

A graphical interpretation of this average is given in Figure 10.45. The area under the graph of the function that the horizontal segments of length 1 with respective y-coordinates 5, 3, 6, 4, 2, and 8 determine is 28. So the average $4\frac{2}{3}$ is this area divided by its extent—namely, 6—along the x-axis.

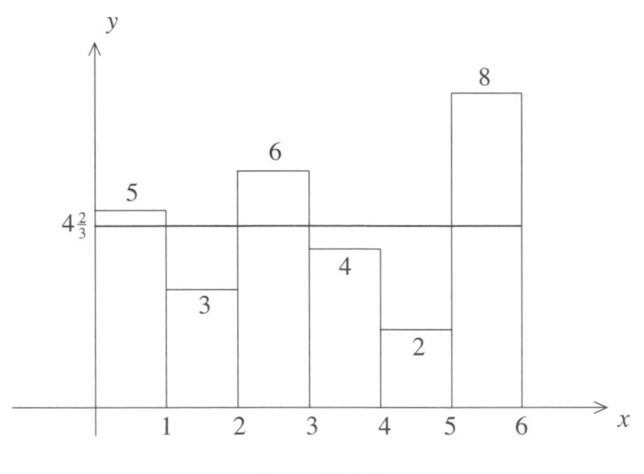

Fig. 10.45

Consider next any continuous function $y = f(x)$ defined over a closed interval $[a, b]$. Turn to Figure 10.46. Let c be precisely that number such that the area of the rectangle with base $b - a$ and height c

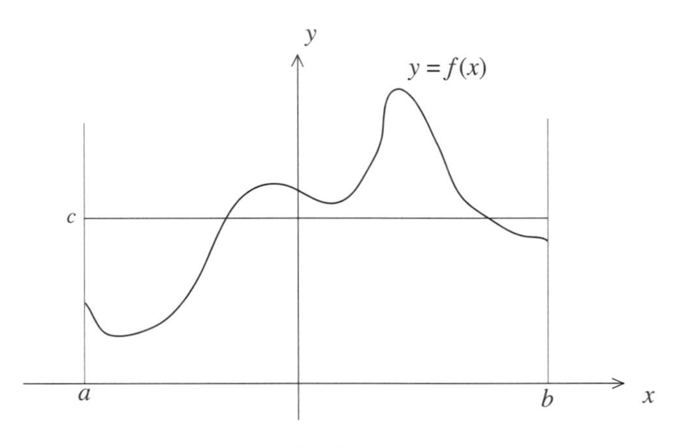

Fig. 10.46

is equal to the area under the graph of $y = f(x)$ over $[a, b]$. Since the area under this graph is equal to $\int_a^b f(x)\,dx$, it follows that

$$c = \tfrac{1}{b-a} \int_a^b f(x)\,dx\,.$$

The *average value* of $f(x)$ over $[a, b]$ is defined to be this value.

10.40. Compute the average values of the following functions. In each case, sketch the graph and put in the relevant rectangle.

i. $f(x) = x^2$ over $[-1, 1]$
ii. $f(x) = x^3$ over $[-2, 2]$
iii. $f(x) = x^{\frac{1}{3}}$ over $[0, 8]$

10.41. Find the average value of the function $f(x) = \frac{1}{x}$ from $x = 1$ to $x = 10$.

10.42. Compute the average values of the following functions. Sketch the graphs, and assess the reasonableness of your answers. If an answer is "out of whack," explain what went wrong.

i. $f(x) = x^2 - 4$ over $[-5, 5]$

ii. $f(x) = \frac{1}{x^2}$ over $[-1, 1]$

iii. $f(x) = \sqrt{1 - x^2}$ over $[-1, 1]$

10.43. Compute the average value of the function $f(x) = \sin x$ over $[0, \pi]$. Sketch the graph and put in the relevant rectangle.

Consider a point moving along a coordinate axis starting at time $t = 0$. Let $p(t)$ be its position at any time $t \geq 0$. We know from Section 6.4 that the velocity $v(t)$ and acceleration $a(t)$ of the point at any time $t \geq 0$ are given by

$$v(t) = p'(t) \quad \text{and} \quad a(t) = v'(t).$$

We also know, that for two instants of time t_1 and t_2 with $0 \leq t_1 < t_2$, the average velocity and the average acceleration over the time period $[t_1, t_2]$ are

$$\frac{p(t_2) - p(t_1)}{t_2 - t_1} \quad \text{and} \quad \frac{v(t_2) - v(t_1)}{t_2 - t_1},$$

respectively. How do these averages compare to the average values

$$\tfrac{1}{t_2 - t_1} \int_{t_1}^{t_2} v(t)\,dt \quad \text{and} \quad \tfrac{1}{t_2 - t_1} \int_{t_1}^{t_2} a(t)\,dt$$

of the velocity and acceleration functions $v(t)$ and $a(t)$ over $[t_1, t_2]$ as these were just defined? Since $p(t)$ and $v(t)$ are antiderivatives of $v(t)$ and $a(t)$, respectively, the fundamental theorem of calculus tells us that these average values are

$$\tfrac{1}{t_2 - t_1} \cdot (p(t_2) - p(t_1)) \quad \text{and} \quad \tfrac{1}{t_2 - t_1} \cdot (v(t_2) - v(t_1)).$$

Therefore the two definitions of average velocity and average acceleration are the same.

Let P be an object that moves in an elliptical orbit with the attracting body S at a focus. It is often asserted that the semimajor axis a of the ellipse is equal to the average distance from P to S. This is justified by the observation that the minimum and maximum distances from P to S are $a - c$ and $a + c$, respectively, and that

$$\tfrac{1}{2}((a + c) + (a - c)) = a$$

is the average of these two distances. Let's think about this for a moment. Suppose you're driving on a highway from point A to point B. You drive at a steady 40 miles per hour until you're almost at B, but for the last part of your trip to B, you increase your speed to 70 miles per hour. Surely, your average speed is not $\frac{1}{2}(70 + 40) = 55$. But this is the thinking behind the point just made about the semimajor axis of the orbit. The fact is that the distance r from P to S varies all along the orbit, so that it is the *average value of a function* that needs to be determined.

The challenge that this problem faces is that r can be expressed as a function of several different variables. One of the formulas of Step 1 of Section 10.4.1 expresses the distance r between P and S as the function

$$r(t) = a(1 - \varepsilon \cos \beta(t))$$

of time t. By ignoring the dependence on t, we also get r as the function

$$r(\beta) = a(1 - \varepsilon \cos \beta)$$

of β. The identity $\cos\alpha(t) = \frac{\cos\beta(t)-\varepsilon}{1-\varepsilon\cos\beta(t)}$ (also established in Section 10.4.1) can be solved for $\cos\beta(t)$ with the result that $\cos\beta(t) = \frac{\varepsilon+\cos\alpha(t)}{1+\varepsilon\cos\alpha(t)}$. Inserting this into $r(t) = a(1-\varepsilon\cos\beta(t))$, we get $r(t) = \frac{a(1-\varepsilon^2)}{1+\varepsilon\cos\alpha(t)}$, and hence r as the function

$$r(\alpha) = \frac{a(1-\varepsilon^2)}{1+\varepsilon\cos\alpha}$$

of α. We'll compute of the average value of r with respect to each of the variables β, α, and t.

Figure 10.20 tells us that as β varies over $[0, 2\pi]$, the object P traces out the entire ellipse, and that the same is true for α. By the symmetry of the situation, the average values of $r(\beta)$ and $r(\alpha)$ over $[0, 2\pi]$ are the same as those over $[0, \pi]$.

10.44. Show that the average value of $r(\beta)$ over $[0, \pi]$ is equal to $\frac{1}{\pi}\int_0^\pi r(\beta)\,d\beta = a$.

While this confirms the earlier average of a, the angle β—historically known as the "eccentric anomaly"—is not the most meaningful variable over which to average r. Figure 10.19 suggests that the average value of r as a function of α is a better choice.

10.45. Show that the average value of the function $r(\alpha)$ over $[0, \pi]$ is

$$\frac{1}{\pi}\int_0^\pi r(\alpha)\,d\alpha = \frac{a(1-\varepsilon^2)}{\pi}\int_0^\pi \frac{1}{1+\varepsilon\cos\alpha}\,d\alpha = b.$$

The verification is tricky and proceeds in several steps.

i. Use the substitution $u = \tan\frac{\alpha}{2}$. Consider the identities $\tan^2\frac{\alpha}{2} = \frac{1-\cos\alpha}{1+\cos\alpha}$ and $\tan\frac{\alpha}{2} = \frac{\sin\alpha}{1+\cos\alpha}$ (both provided by Problem 1.23), and solve them for $\cos\alpha$ and $\sin\alpha$, respectively, to get $\cos\alpha = \frac{1-u^2}{1+u^2}$ and $\sin\alpha = \frac{2u}{1+u^2}$. Use these equations to show that

$$\int \frac{1}{1+\varepsilon\cos\alpha}\,d\alpha = \int \frac{2}{u^2+1+\varepsilon(1-u^2)}\,du = \frac{1}{1+\varepsilon}\int \frac{2}{1+\left(\sqrt{\frac{1-\varepsilon}{1+\varepsilon}}\,u\right)^2}\,du.$$

ii. Substitute $z = \sqrt{\frac{1-\varepsilon}{1+\varepsilon}}\,u$ to show that the integral of (i) is equal to

$$\frac{2}{\sqrt{1-\varepsilon^2}}\int \frac{dz}{1+z^2} = \frac{2}{\sqrt{1-\varepsilon^2}}\tan^{-1}z + C = \frac{2}{\sqrt{1-\varepsilon^2}}\tan^{-1}\left(\sqrt{\frac{1-\varepsilon}{1+\varepsilon}}\tan\frac{\alpha}{2}\right) + C.$$

iii. Note that the antiderivative of (ii) is not defined for $\alpha = \pi$. But use the strategy of improper integrals to conclude that with $0 \le \varphi < \pi$,

$$\int_0^\pi \frac{1}{1+\varepsilon\cos\alpha}\,d\alpha = \lim_{\varphi\to\pi}\int_0^\varphi \frac{1}{1+\varepsilon\cos\alpha}\,d\alpha = \lim_{\varphi\to\pi}\frac{2}{\sqrt{1-\varepsilon^2}}\tan^{-1}\left(\sqrt{\frac{1-\varepsilon}{1+\varepsilon}}\tan\frac{\varphi}{2}\right) = \frac{2}{\sqrt{1-\varepsilon^2}}\cdot\frac{\pi}{2} = \frac{\pi}{\sqrt{1-\varepsilon^2}}.$$

iv. Finish the computation of the average value of $r(\alpha)$ over $[0, \pi]$.

The substitution $u = \tan\frac{x}{2}$ used above is widely applicable as a method that transforms integrals involving trig functions into integrals involving polynomials.

We turn finally to the average of r as function of time t. The relevant interval of time is $[0, T]$, where T is the period of the orbit. This computation also involves some trickery and confirms what has been said before. Finding the antiderivative of a function is often a matter of pulling rabbits out of a hat.

10.46. **i.** Use the derivative of the equation $\beta(t) - \varepsilon\sin\beta(t) = \frac{2\pi t}{T}$ as well as the equality $r(t) = a(1-\varepsilon\cos\beta(t))$ to show that $\frac{T}{2a\pi}\beta'(t)r(t) = 1$.

ii. Show that the average value of $r(t)$ over $[0, T]$ is

$$\frac{1}{T}\int_0^T r(t)\,dt = \frac{1}{2a\pi}\int_0^T r(t)^2\beta'(t)\,dt = \frac{a}{2\pi}\int_0^T (1-\varepsilon\cos\beta(t))^2\beta'(t)\,dt.$$

iii. Use the substitution $u = \beta(t)$ and the formula $\cos^2 u = \frac{1}{2}(1+\cos 2u)$ to check that this last integral is equal to

$$\frac{a}{2\pi}\left[\int_0^{2\pi} du - 2\varepsilon\int_0^{2\pi}\cos u\,du + \varepsilon^2\int_0^{2\pi}\cos^2 u\,du\right] = a(1+\tfrac{1}{2}\varepsilon^2).$$

The average value $a(1+\frac{1}{2}\varepsilon^2)$ of $r(t)$ as function of t over $0 \le t \le T$ reflects the fact that the orbiting object P moves more quickly at periapsis and more slowly at apoapsis. So the larger value of $r(t)$ near apoapsis has a greater impact on the average than the smaller value of $r(t)$ near periapsis. This accounts for the fact that this average value is larger than the previously computed averages a and b.

10.47. Reflect about the conclusions of the analysis above, and discuss the question: What is the average distance r from P to S over one complete orbit of P around S?

10H. Work and Impulse. The next two problems deal with work and impulse in the context of a point-mass P of mass $m = 1$ moving in a straight line. The point of Problem 10.48 moves along the slanting line of Figure 10.47a, and the point of Problem 10.49 moves along the z-axis of Figure 10.47b.

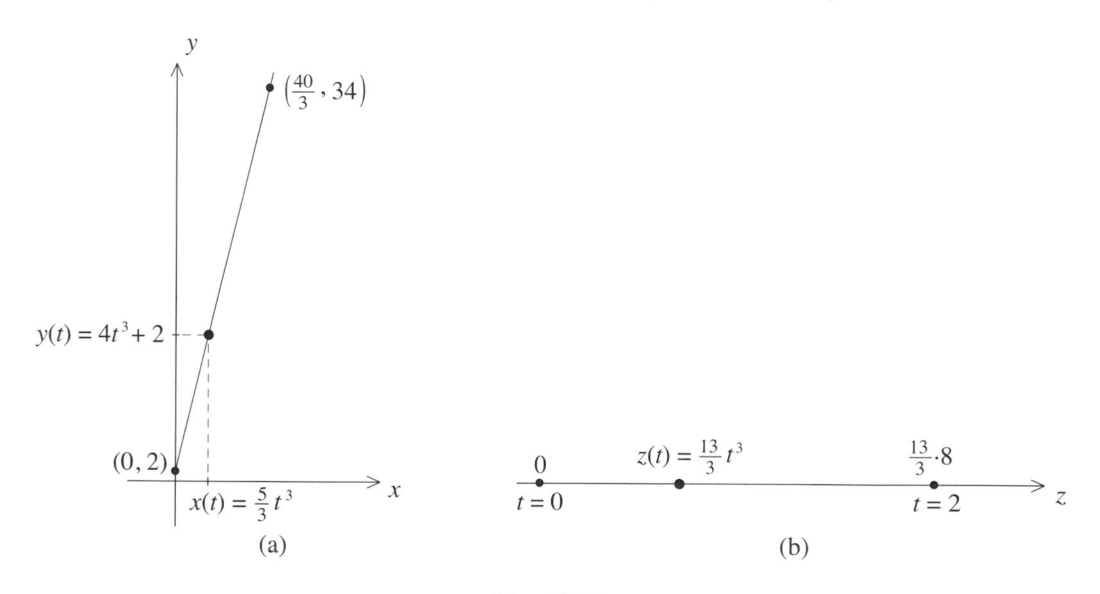

Fig. 10.47

10.48. Let the motion of P in the xy-plane be given by the equations $x(t) = \frac{5}{3}t^3$ and $y(t) = 4t^3 + 2$ with $t \ge 0$.

i. Check that from time $t = 0$ to time $t = 2$, the point moves from $(0, 2)$ to $(\frac{40}{3}, 34)$. Show that the point moves on the line $y = \frac{12}{5}x + 2$.

ii. Show that the point moves with a speed of $v(t) = 13t^2$.

iii. Show that $x''(t) = 10t$ and $y''(t) = 24t$, and determine the magnitude of the force that pushes the point along its path.

iv. Compute the work done by the force in moving the point-mass from $(0, 2)$ to $(\frac{40}{3}, 34)$ by making use of the connection between work and kinetic energy.

v. Compute the impulse of the force between time $t = 0$ and $t = 2$ by using the connection between impulse and momentum.

10.49. Let P move on the z-axis with position given by $z(t) = \frac{13}{3}t^3$ at any time $t \geq 0$.

i. Show that over the time interval $[0, 2]$, the point moves from $z = 0$ to $z = \frac{104}{3}$.

ii. Show that the speed of P at any time t is $v(t) = 13t^2$.

iii. Show that the force acting on P is equal to $F(t) = 26t$ as a function of t and $f(z) = 2(3\cdot 13^2)^{\frac{1}{3}}z^{\frac{1}{3}}$ as a function of z.

iv. Compute the work done by the force in moving the point from $z = 0$ to $z = \frac{104}{3}$ by evaluating the integral $\displaystyle\int_0^{\frac{104}{3}} f(z)\,dz$.

v. Compute the impulse of the force from $t = 0$ to $t = 2$ by evaluating the integral $\displaystyle\int_0^2 F(t)\,dt$.

10.50. Explain some of the numerical coincidences that your answers to Problems 10.48 and 10.49 have brought to light, and more generally how the motion of the point-mass of Figure 10.47a is related to the motion of the point-mass of Figure 10.47b.

10I. Centroids and the Theorems of Pappus of Alexandria. When the actions of forces on an object are considered, it can often be assumed that the mass of the object is concentrated at a single point and that all the forces act on this point. This point is the object's *center of mass*. This important concept does more than simplify such studies, it makes them possible. We'll define the center of mass of an object "experimentally" as follows. For a one- or two-dimensional object, the center of mass is that point C such that the object is in balance when it is supported or suspended at C. Figures 10.48a and 10.48b illustrate what is involved. To pinpoint the center of mass of a three-dimensional object, suspend the object on two

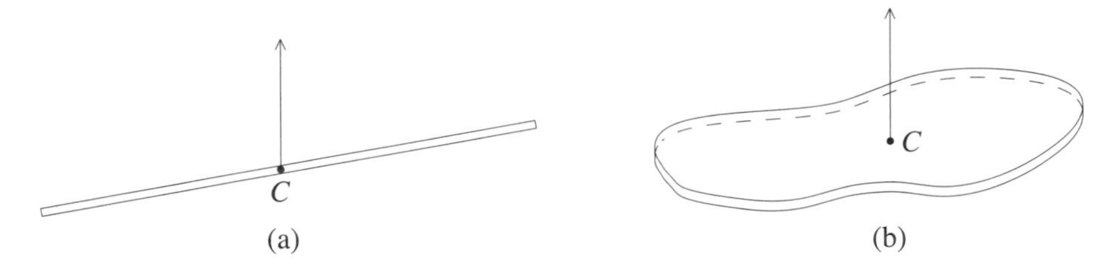

Fig. 10.48

strings that are attached at two different points on the surface. The intersection of the object with the vertical plane that the two strings determine is shown in Figure 10.49a. Now pick a point P on the surface

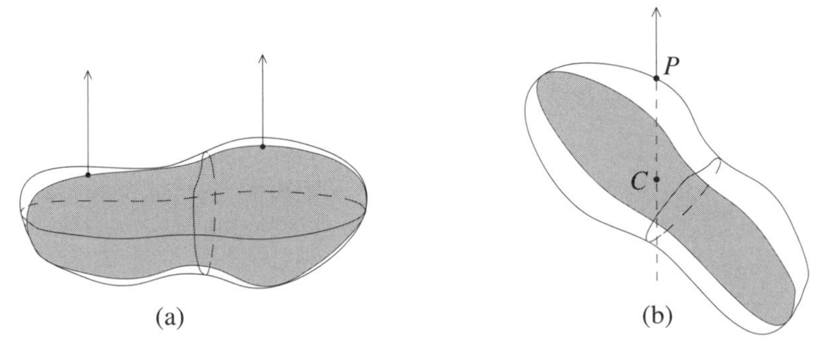

Fig. 10.49

of the object that is not in this plane, and suspend the object on a string attached at P. The center of mass of the object is the point of intersection C of the extension of the line of this string with the plane identified earlier. Refer to Figure 10.49b. If the object is made of a completely homogeneous material—material of constant density—then the center of mass is the geometric center, or *centroid*, of the object.

We saw in Section 2.5 how Archimedes determined the centroid of a triangle. He analyzed the centroids of other shapes as well. The last of the great Greek mathematicians, Pappus of Alexandria (around 300 A.D.), also studied the centroids of general shapes and their properties. His important insights in this regard were rediscovered much later by the Jesuit priest Paul Guldin (1577–1643) in a more definitive form. (Neither Guldin nor his contemporaries, including Galileo and Kepler, seemed to be aware of this work of Pappus). The Theorems A and B that follow are often named after both Pappus and Guldin.

Theorem A. If an arc lying in a plane is revolved one complete revolution about an axis that lies in the plane but does not cross the arc, then the area of the surface that is formed is equal to the product of the length of the arc times the length of the circle traced out by the centroid of the arc.

10.51. Consider the semicircular arc of radius r of Figure 10.50a, and let the axis lie on the diameter of the circle. If the semicircular arc is rotated one complete revolution about this axis, the surface that is formed is a sphere. The centroid C of the arc will trace out a circle. Use Pappus's Theorem A to determine the location of the centroid of the semicircular arc in terms of r. [Hint: Use the fact that the surface area of a sphere of radius r is $4\pi r^2$.]

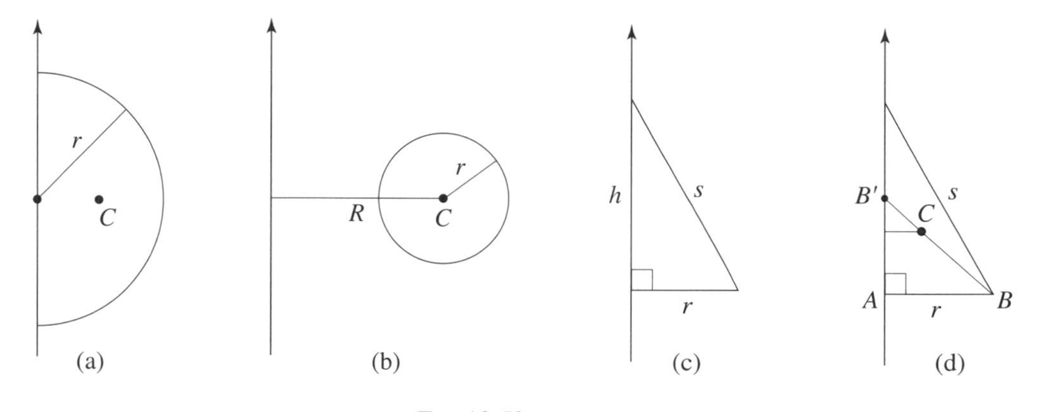

Fig. 10.50

10.52. Consider a circle of radius r, and choose an axis such that the distance from its center C to to the axis is R with $R \geq r$. See Figure 10.50b. Use Pappus's Theorem A to determine the area of the surface of the perfect geometric donut that is generated by revolving the circle once about the axis.

10.53. Consider a triangle with base r, height h, and slanting side s. See Figure 10.50c. When it is revolved around the indicated axis, a cone of base radius r, height h, and slanting side s is generated. Consider Figure 10.50c, and use Pappus's Theorem A to show that the surface area of the cone (without the area of its circular base) is πrs.

Theorem B. If a region lying in a plane is revolved one complete revolution about an axis that lies in the plane but does not intersect the region, then the volume of the solid that is formed is equal to the product of the area of the region times the length of the path traced out by its centroid.

10.54. Consider the semicircular region of Figure 10.50a, and let C be the centroid of the region. Let the axis lie on the diameter of the circle. Use Pappus's Theorem B to determine the location of the centroid of the semicircular region in terms of its radius r. [Hint: Use the fact that the volume of a sphere of radius r is $\frac{4}{3}\pi r^3$.]

10.55. Use Pappus's Theorem B to determine the volume of a perfect geometric donut generated by Figure 10.50b. Compare your answer with the result of Example 9.9 in Section 9.2.

10.56. In Figure 10.50d, C is the centroid of the triangle, AB is its base, and B' is the midpoint of its height. Use a conclusion from Archimedes's study in Section 2.5, a property of similar triangles, and Pappus's Theorem B to show that the volume of the cone that one revolution of the triangle generates is $\frac{1}{3}\pi r^2 h$.

The proofs of the theorems of Pappus and Guldin require mathematical versions of the definitions of centers of mass and centroids of shapes in the plane or in space, rather than the "experimental" definitions supplied above. These definitions combine the approaches of Archimedes (as illustrated by Figure 2.28) with those of integral calculus. The next discussion illustrates what is involved.

10J. Torque and Center of Mass. Consider a rectangular beam of the sort depicted in Figure 10.51. Its cross-sectional area is fixed and equal to A. An x-axis is placed along its edge. Suppose that for any x, the material of the beam is homogeneous along the cut through x. We'll define the density $\rho(x)$ of the beam at x as follows. Let a be some positive constant, and consider the rectangular segment of length $2a$ shown in

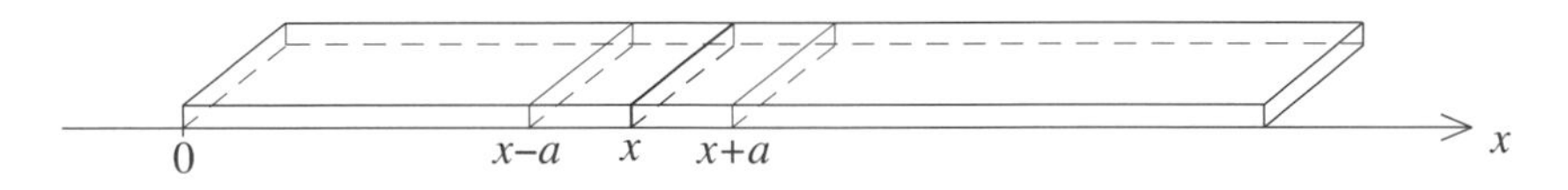

Fig. 10.51

the figure. Let $v_a = 2aA$ and m_a be the volume and mass of this segment, respectively, and consider the ratio $\frac{m_a}{v_a}$. Now let a shrink to 0, and define $\rho(x)$ by

$$\rho(x) = \lim_{a \to 0} \frac{m_a}{v_a}.$$

We'll assume that the density function $\rho(x)$ is continuous along the beam. Figure 10.52a shows the beam in place in the xy-plane. The beam has length L. One end is fixed at the origin O, but the beam can rotate

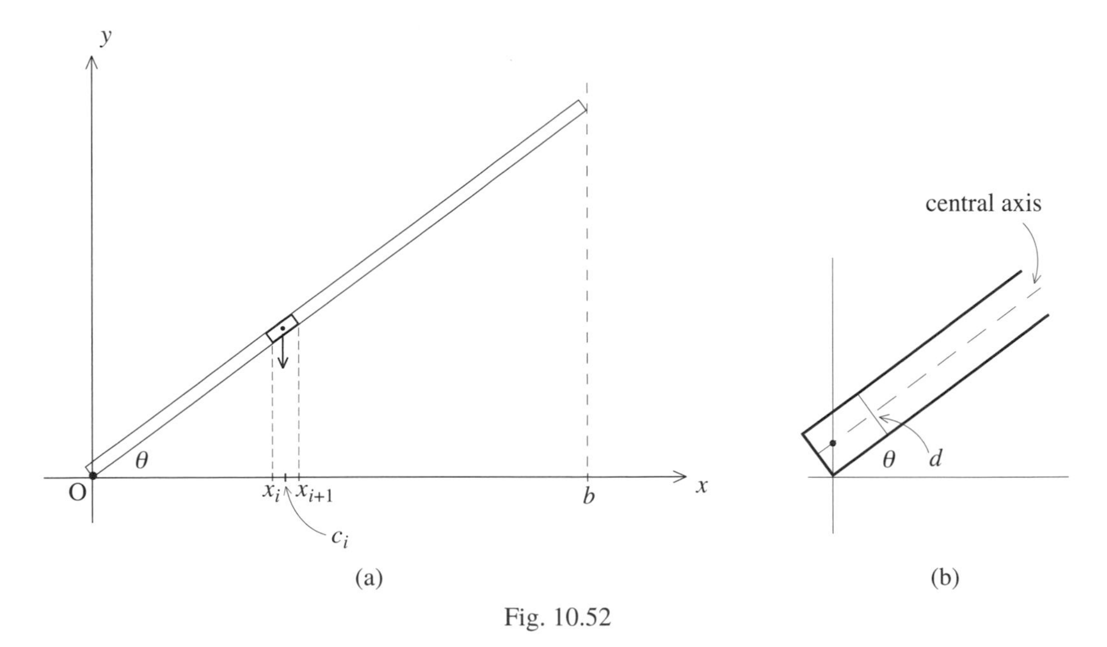

Fig. 10.52

freely there. The coordinate b is highlighted and the angle θ satisfies $\cos\theta = \frac{b}{L}$.

We'll compute the torque with respect to the origin O generated by the weight of the beam. The methods of integral calculus as developed in Sections 9.1 and 9.5 are central. Let $\mathcal{P}$ be a partition of the interval $[0, b]$. Let the partition divide the interval into n subintervals with subdivision points

$$0 = x_0 < x_1 < x_2 < \cdots < x_i < x_{i+1} < \cdots < x_{n-1} < x_n = b.$$

As on earlier occasions, we'll let $\Delta x_i = x_{i+1} - x_i$. We'll assume that the norm $||\mathcal{P}||$ of the partition is small and hence that all the Δx_i are small. As Figure 10.52a illustrates, the interval $[x_i, x_{i+1}]$ determines a segment of the beam. If l is the length of this segment, then $\cos\theta = \frac{\Delta x_i}{l}$, so that $l = \frac{L}{b}\,\Delta x_i$. It follows that the volume of the segment is $A \cdot l = \frac{AL}{b}\,\Delta x_i$. Select a point c_i in each interval $[x_i, x_{i+1}]$. Multiplying the volume by the density $\rho(c_i)$ provides the approximation $\rho(c_i)\frac{AL}{b}\,\Delta x_i = \frac{AL}{b} \cdot \rho(c_i)\Delta x_i$ of the mass of the segment. So the sum

$$\frac{AL}{b}\sum_{i=0}^{n-1} \rho(c_i)\,\Delta x_i$$

is an approximation of the entire mass of the beam. The smaller the norm $||\mathcal{P}||$, the better this approximation, so that the mass M of the beam is equal to

$$M = \lim_{||\mathcal{P}||\to 0} \frac{AL}{b}\sum_{i=0}^{n-1}\rho(c_i)\,\Delta x_i = \frac{AL}{b} \cdot \lim_{||\mathcal{P}||\to 0}\sum_{i=0}^{n-1}\rho(c_i)\,\Delta x_i = \frac{AL}{b}\int_0^b \rho(x)\,dx\,.$$

Since weight = mass $\times\, g$, where g is the gravitational constant, $g(\frac{AL}{b} \cdot \rho(c_i)\Delta x_i)$ is an approximation of the weight of the segment, and

$$W = \frac{gAL}{b}\int_0^b \rho(x)\,dx$$

is the entire weight of the beam. Notice that c_i approximates the distance between the line of action of the gravitational force on the segment and the origin O. So by Figure 8.19 (and the discussion that explains it), $c_i(\frac{gAL}{b} \cdot \rho(c_i)\Delta x_i) = \frac{gAL}{b} \cdot c_i\rho(c_i)\Delta x_i$ approximates the torque with respect to O that the weight of the segment determines. Therefore

$$T_0 = \frac{gAL}{b}\int_0^b x\rho(x)\,dx$$

is the entire torque with respect to O that the weight of the beam generates. Consider the ratio

$$\bar{x} = \frac{T_0}{W} = \frac{\displaystyle\int_0^b x\rho(x)\,dx}{\displaystyle\int_0^b \rho(x)\,dx}\,.$$

The diagram of Figure 10.52b implies that the central axis of the beam has equation $y = (\tan\theta)x + \frac{dL}{2b}$. Let $\bar{y} = (\tan\theta)\bar{x} + \frac{dL}{2b}$. The point

$$C = (\bar{x}, \bar{y})$$

is defined to be the *center of mass* of the beam. Suppose that the entire mass M of the beam is concentrated at the point $C = (\bar{x}, \bar{y})$. Under this assumption, the principle that Figure 8.19 illustrates implies that the torque of the beam about O is

$$gM \cdot \bar{x} = W\bar{x} = T_0.$$

Therefore the torque of the beam about the point O can be computed by assuming that its mass is concentrated at its center of mass C.

Suppose that the material of the beam is homogeneous. In this case, the density of the beam is constant. So $\rho(x) = c$ and C is the centroid of the beam. By the fundamental theorem of calculus,

$$\int_0^b x\rho(x)\,dx = c\frac{x^2}{2}\Big|_0^b = c\frac{b^2}{2} \quad\text{and}\quad \int_0^b \rho(x)\,dx = cx\Big|_0^b = cb.$$

It follows that $\bar{x} = \frac{b}{2}$. A look at Figure 10.52a tells us that in this case C is the geometric center of the beam, and hence that the torque of the beam around the origin O is $W\frac{b}{2}$. If the homogeneous beam is in horizontal position, then $b = L$, and the torque it generates is equal to $W\frac{L}{2}$.

By way of an example, let's assume that a homogeneous rigid beam is supported horizontally at two points B and D, as shown in Figure 10.53. It is not bolted down at B or D. Only the weight of the beam keeps it in place. The beam reaches 3 m past B on the left and 1 m past D on the right. It has an overall length of 9 m and a mass of 4 kg/m. A clown with a mass of 80 kg is walking on the beam between B and D

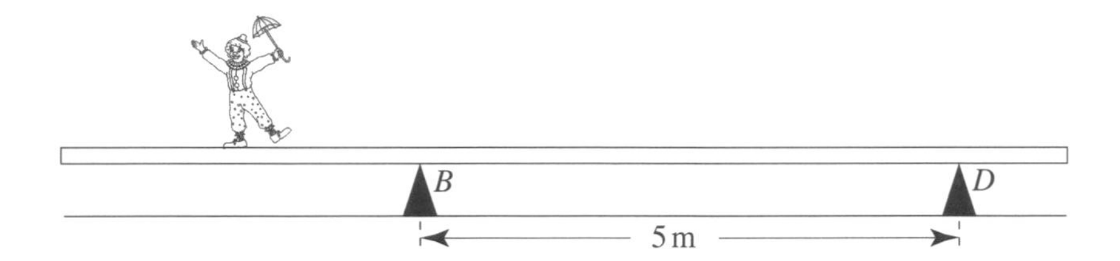

Fig. 10.53. Thanks to Greg Young, School of Architecture, University of Notre Dame, for the clown.

toward B. How far past B will he be able to walk before the beam tilts and he falls off?

The mass of the part of the beam to the right of B is $m = 4 \cdot 6 = 24$ kg and the mass to the left of B is $m = 4 \cdot 3 = 12$ kg. In terms of the weights involved, this corresponds to $mg = (9.81)(24) = 235.44$ N and $mg = (9.81)(12) = 117.72$ N, respectively. By the conclusion about torques and centroids reached above, the torque of the beam to the right of the point B is $(235.44) \cdot (\frac{6}{2}) = 706.32$ newton-meters (clockwise), and the torque of the beam to the left of B is $(117.72) \cdot (\frac{3}{2}) = 176.58$ newton-meters (counterclockwise). The weight of the clown is $9.81 \cdot 80 = 784.80$ N. If the clown is x meters to the left of B, then the torque generated by his weight is $784.8x$ newton-meters (counterclockwise). The clown reaches a tipping point when the counterclockwise torques are equal to the clockwise torques, so when

$$784.80x + 176.58 = 706.32.$$

This gives $x = 0.675$ meters, or about $\frac{2}{3}$ of a meter.

10.57. What smallest mass placed on the beam at the point D will allow the clown to walk all the way to the end of the beam?

10.58. Suppose that the beam is 3 m longer and that it extends 4 m to the right of D (instead of 1 m). Suppose also that a lighter clown of 65 kg is walking past the point B toward the left end of the beam. Will there be a tipping point before the clown reaches the end of the beam?

10K. The Indexes of Inertia of the Disc and Sphere. We'll now turn to some unfinished business from the discussion of Galileo's experiment with inclined planes in Chapter 8. Consider a circle of radius r made of a thin, homogeneous material, and let m be its mass. Suppose that the circle can rotate freely around the axis perpendicular to the plane of the circle through its center. Think of a bicycle wheel with spokes of negligible mass. We saw in Section 8.4.2 that when a force of magnitude $f(t)$ (regarded to be a function of time t) is applied tangentially to the circle, the rotation around the central axis that the force affects satisfies

$$f(t) \cdot r = (mr^2) \cdot \alpha(t),$$

where $\alpha(t)$ is the angular acceleration of the circle. The quantity mr^2 is the circle's index of inertia. This equality was established by subdividing the circle into many small segments and by adding their individual indexes of inertia.

The principle that the index of inertia of a rotating object is equal to the sum of the indexes of inertia of all the particles that constitute it is a generally valid fact. We will now use it in combination with the methods of integral calculus to demonstrate that the index of inertia of a homogeneous sphere of radius r and mass is m is

$$\tfrac{2}{5}mr^2.$$

This fact was an important ingredient in the mathematical analysis of Galileo's experiment in Section 8.4.3, but was used there without having been verified. The demonstration proceeds in two steps: from the circle to the disc and from the disc to the sphere.

A disc is given by a circle and the region inside it. Consider a thin, homogeneous disc of radius r and mass m. Since the area of the disc is πr^2, the density of the disc is $\frac{m}{\pi r^2}$ per unit area.

10.59. Place the center of the disc at the origin of a coordinate axis that lies in the plane of the disc. Take a partition $\mathcal{P}$

$$0 = x_0 < x_1 < \cdots < x_i < x_{i+1} < \cdots < x_{n-1} < x_n = r$$

of the interval $[0, r]$, and suppose that its norm $\|\mathcal{P}\|$ is small. This divides the disc into lots of thin circular pieces each with center 0. See Figure 10.54a. The circular piece with inner radius x_i and thickness $\Delta x_i = x_{i+1} - x_i$ is shown in the figure.

i. Check that the area of this circular piece is given by

$$\pi x_{i+1}^2 - \pi x_i^2 = \pi(x_i + \Delta x_i)^2 - \pi x_i^2 \approx 2\pi x_i \Delta x_i,$$

that its mass is approximately $\frac{m}{\pi r^2}(2\pi x_i \Delta x_i) = \frac{2m}{r^2} x_i \Delta x_i$, and that its index of inertia is approximately $(\frac{2m}{r^2} x_i \Delta x_i) x_i^2 = \frac{2m}{r^2} x_i^3 \Delta x_i$.

ii. Conclude that the index of inertia of a homogeneous disc of radius r and mass m is $\int_0^r \frac{2m}{r^2} x^3\, dx = (\frac{2m}{r^2} \frac{x^4}{4})\Big|_0^r = \frac{1}{2}mr^2.$

Finally, there is the step from the disc to the sphere. The sphere in this context includes the solid that it encloses. Suppose that the sphere is homogeneous with radius r and mass m. Since the volume of the

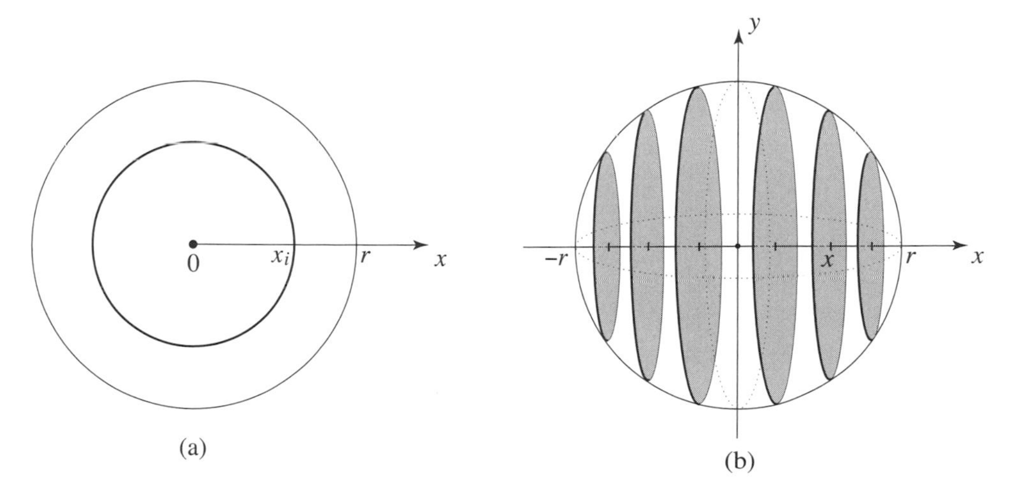

Fig. 10.54

sphere is $\frac{4}{3}\pi r^3$, the sphere has a density of $\frac{m}{\frac{4}{3}\pi r^3} = \frac{3}{4}\frac{m}{\pi r^3}$ per unit volume. Problem 10.60 presents an outline of the argument that its index of inertia is $\frac{2}{5}mr^2$.

10.60. Provide the plane already described with an xy-coordinate system with origin the center of the sphere. This plane cuts the sphere in a circle with equation $x^2 + y^2 = r^2$. Consider the interval $[-r, r]$ along the x-axis, and take a partition of it with small norm. The partition divides the interval into small subintervals. Let x be a typical division point, and let dx be the distance to the next one. Cuts through x and $x + dx$ perpendicular to the x-axis determine a thin slice of the sphere. It is approximated by a disc of radius $y = \sqrt{r^2 - x^2}$ and thickness dx. In this way, the sphere is approximated by a large number of thin parallel discs. Figure 10.54b shows what is involved and highlights several of these discs.

i. Show that the volume of the typical disc is $\pi y^2 dx = \pi(r^2 - x^2)dx$, that its mass is equal to $\frac{3}{4}\frac{m}{\pi r^3}\pi(r^2 - x^2)dx = \frac{3}{4}\frac{m}{r^3}(r^2 - x^2)dx$, and that its index of inertia is

$$\tfrac{1}{2}\left(\tfrac{3}{4}\tfrac{m}{r^3}(r^2 - x^2)dx\right)y^2 = \left(\tfrac{3}{8}\tfrac{m}{r^3}(r^2 - x^2)dx\right)(r^2 - x^2) = \tfrac{3}{8}\tfrac{m}{r^3}(r^2 - x^2)^2 dx.$$

ii. Conclude that the index of inertia of the homogeneous sphere of radius r and mass m is

$$\int_{-r}^{r} \tfrac{3}{8}\tfrac{m}{r^3}(r^2 - x^2)^2 dx = \tfrac{3}{8}\tfrac{m}{r^3}\left(r^4 x - \tfrac{2}{3}r^2 x^3 + \tfrac{1}{5}x^5\right)\Big|_{-r}^{r} = \tfrac{2}{5}mr^2.$$

Chapter 11

Basics of Differential Equations

Chapter 11 presents an introduction to differential equations. Because of its wide range of applications, this is one of the most important pursuits in all of mathematics. Falling within its scope is most any problem that calls for the analysis of a physical process that unfolds smoothly or, more generally, any construct that can be modeled with smoothly varying functions. At the core of such constructs and processes are relationships between key magnitudes, such as forces, accelerations, positions, speeds, and others. Once the relationships are understood and appropriate simplifying assumptions are made, it is often possible to express these simplified relationships in terms of one or more equations involving relevant functions and their rates of change. The realization of such "differential equations" belongs to the context of the discipline involved, be it physics, chemistry, biology, economics, finance, or one of the fields of engineering. The solution of the equations calls for the determination of the functions that arise, and this is where mathematics moves into the spotlight. Such solutions may be explicit, implicit, and they may rely on numerical approximations. The differential equations as well as their solutions are as complicated as the problem that is being considered. Once—or, better, if—they are successfully analyzed, they can provide a quantitative understanding of what is going on.

Let f be a function that can be differentiated again and again. The successive derivatives $f', f'' = (f')'$, $f''' = ((f')')', \ldots$ are called the first, second, third, ... derivatives of f. The word "order" is used in this context with f'' a derivative of order two, f''' of order three, and so on. A *differential equation* is any equation involving constants, variables (often x and y), derivatives (often $y' = \frac{dy}{dx}$), second derivatives (often $y'' = \frac{d^2y}{dx^2}$), as well as higher derivatives. The *order* of a differential equation is the order of the highest derivative that occurs in the equation. For instance, $\frac{dy}{dx}(4 - y^2) = 2x$ is a first-order differential equation, $5\frac{d^2y}{dx^2} - 7\frac{dy}{dx} + 6 = 0$ is of the second order, and $4y''' = 5y + 3x$ has order three. A *solution* of a differential equation is a function that satisfies the equation. A solution of a differential equation *on an interval* I is a function that satisfies the equation for all real numbers in I. Given a differential equation, it is usually a routine matter to check whether a function is a solution of the equation or not. Consider $\frac{dy}{dx} - y^2 - 1 = 0$, for instance. Let $y = \tan x$. Because $\frac{dy}{dx} = \sec^2 x$ and $\sec^2 x = \tan^2 x + 1$, it follows that $y = \tan x$ is a solution of this equation. A look at its graph tells us that $y = \tan x$ is a solution on the interval $I = (-\frac{\pi}{2}, \frac{\pi}{2})$. For another example, let $y = f(t) = e^t + t$. Since $f'(t) = e^t + 1$ and $f''(t) = e^t$,

$$(1 - t)f''(t) + tf'(t) - f(t) = (1 - t)e^t + t(e^t + 1) - e^t - t = 0.$$

Because the derivatives $f'(t) = e^t + 1$ and $f''(t) = e^t$ both exist for all t, it follows that $y = f(t) = e^t + t$ is a solution of $(1 - t)y'' + ty' - y = 0$ on the interval $I = (-\infty, \infty)$. Given a function, it is often not difficult to find a differential equation that it satisfies. The hard part is the other direction: to find—for a given differential equation—all the functions that satisfy it. This task is usually far from routine! The theory of differential equations is a response to this challenge. The *general solution* of a differential equation is a solution that involves one or more constants and satisfies the following requirement: every solution of the differential equation has this general form. *Particular solutions* are solutions that satisfy additional conditions, known as *initial conditions*, such as $f(1) = 2, f'(3) = 4$, and $f''(5) = 6$ that pinpoint these constants. The problem of finding a solution to a differential equation that satisfies initial conditions is known as an *initial value problem*.

We have already encountered a number of examples of first- and second-order differential equations in previous chapters. The equation $\frac{dy}{dx} = g(x)$, where $g(x)$ is any function, is an example. Its solutions are the functions $y = f(x)$ such that $\frac{dy}{dx} = g(x)$. They are, therefore, the antiderivatives of the function $y = g(x)$. So methods of integration—strategies for finding antiderivatives of functions—belong to the theory of differential equations. The differential equation $\frac{dy}{dx} = Cy$ arose in several contexts that were described in Section 8.1. Its general solution, $y = y_0e^{Ct}$, was derived at the end of Section 7.10. The differential equation $x''(t) + \frac{k}{m}x(t) = 0$ appeared in Section 10.5.2 and captured the dynamics of a spring pushing a mass. The differential equation $\frac{dx}{dt} = \frac{ax}{b+x}$, with a and b constants, was important in the development of interior ballistics in Section 10.5.3.

This chapter first examines some aspects of first-order differential equations. These include the study of separable equations, integrating factors, slope fields, and Euler's method. The chapter goes on to construct the complex numbers and relies on them in the solutions of certain second-order equations. Power series and the important role they play are also examined. The applications focus on an analysis of free fall with air resistance, the suspension system of a car, and the geometry of cables that includes cables under loads and feely hanging cables as special cases. The Problems and Projects section explores additional applications.

11.1 FIRST-ORDER SEPARABLE DIFFERENTIAL EQUATIONS

There is a general class of first-order differential equations that can be solved, at least in the sense that they can be reduced to the problem of finding antiderivatives of functions. They are the equations of the form

$$\frac{dy}{dx} = g(x) \cdot h(y)$$

for some functions $g(x)$ of x and $h(y)$ of y. Since the variables x and y can be separated, such equations are called *separable differential equations*. For example, the equation $\frac{dy}{dx} = ky$, where k is a constant, is of this form with $g(x) = k$ and $f(y) = y$. So is the equation $\frac{dy}{dx} - y^2 - 1 = 0$. Take $g(x) = 1$ and $h(y) = y^2 + 1$.

Let $\frac{dy}{dx} = g(x) \cdot h(y)$ be any separable differential equation. Assume that $y = f(x)$ is a solution. So $f'(x) = g(x) \cdot h(y)$. Taking $k(y) = \frac{1}{h(y)}$, we have $k(y)f'(x) = g(x)$. Now let $K(y)$ be an antiderivative of $k(y)$. The function $K(y) = K(f(x))$ is also a function of x. Differentiating it with the chain rule gives us $(K(f(x))' = K'(y) \cdot f'(x)$. Therefore

$$(K(f(x))' = K'(y) \cdot f'(x) = k(y) \cdot f'(x) = g(x).$$

So $K(f(x))$ is an antiderivative of $g(x)$. If $G(x)$ is another antiderivative of $g(x)$, then $K(f(x))$ and $G(x)$ differ by a constant, and therefore

$$(*) \qquad K(y) = K(f(x)) = G(x) + \text{constant}.$$

We have shown that any solution $y = f(x)$ of the equation $\frac{dy}{dx} = g(x) \cdot h(y)$ satisfies equation $(*)$. The point is this: find antiderivatives $K(y)$ of the function $k(y) = \frac{1}{h(y)}$ and $G(x)$ of the function $g(x)$, and then look for a function $y = f(x)$ that satisfies equation $(*)$. Differentiating equation $(*)$ with respect to x (using the chain rule) shows that $k(y) \cdot \frac{dy}{dx} = k(y) \cdot f'(x) = g(x)$. So $f'(x) = \frac{g(x)}{k(y)} = g(x) \cdot h(y)$, and therefore $y = f(x)$ is a solution of $\frac{dy}{dx} = g(x) \cdot h(y)$.

More briefly formulated, the strategy is this: separate the variables of the differential equation $\frac{dy}{dx} = g(x) \cdot h(y)$ by rewriting it as $\frac{1}{h(y)}dy = g(x)\,dx$, and consider the equality

$$\int \frac{1}{h(y)}\,dy = \int g(x)\,dx.$$

If both of these indefinite integrals can be solved, then an equation in the variables y and x is obtained. If this equation can be solved for y in terms of x, then an explicit solution $y = f(x)$ of the differential equation $\frac{dy}{dx} = g(x) \cdot h(y)$ has been found. If it is not possible to explicitly solve the equation for y in terms of x, then the relationship between x and y is implicit, and we speak of an implicit solution.

The differential equation $\frac{dy}{dt} = cy$ that arose in Section 8.1 in the context of bacterial growth and nuclear decay is a separable equation with $g(t) = c$ and $h(y) = y$. According to the method just described,

$$\int \frac{dy}{y} = \int c\,dt,$$

and after taking antiderivatives of both sides, $\ln y = ct + C$. Can this equation be solved for y? Yes! The inverse relationship between the exponential and log functions (established in Section 7.11) tells us that $y = e^{\ln y} = e^{ct+C} = e^C e^{ct}$. By substituting $t = 0$, we see that $e^C = y(0)$. So $y = y(0)e^{ct}$ is the general solution of the equation. (It was obtained in Section 7.10 with a different strategy.)

Consider the equation $\frac{dy}{dx} - x^2y^2 = 0$. By separating variables, $y^{-2}dy = x^2dx$. So

$$\int y^{-2}\,dy = \int x^2\,dx.$$

Taking antiderivatives of both sides, we get $-y^{-1} = \frac{1}{3}x^3 + C$. So $y = -(\frac{1}{3}x^3 + C)^{-1}$. In this case of a first-order equation, the general solution $y = -(\frac{1}{3}x^3 + C)^{-1}$ contains only a single constant. For the particular solution $y = f(x)$ with initial condition $f(0) = -4$, we need to have $-4 = -(\frac{1}{3}0^3 + C)^{-1}$, and hence $C = \frac{1}{4}$. Therefore this particular solution is $f(x) = -(\frac{1}{3}x^3 + \frac{1}{4})^{-1}$. Notice that it is explicit.

Example 11.1. Let's solve the equation $\frac{dy}{dx} - y^2 - 1 = 0$ considered earlier. Because $\frac{dy}{dx} = y^2 + 1$, we get $\frac{dy}{y^2+1} = dx$. By a result from Section 9.9, $\frac{d}{dy}\tan^{-1}y = \frac{1}{y^2+1}$, so that

$$\int \frac{dy}{y^2+1}\,dy = \tan^{-1}y + C.$$

We can therefore conclude that $\tan^{-1}y = x + C$ for some constant C. It follows from Figure 9.33 that $-\frac{\pi}{2} < x + C < \frac{\pi}{2}$. Taking the tangent of both sides, we find that $y = \tan(\tan^{-1}y) = \tan(x + C)$. Therefore $y = \tan(x + C)$ is the general solution of the equation $\frac{dy}{dx} - y^2 - 1 = 0$. Taking $C = 0, 1$, and 2, we get the particular solutions $y = \tan x$, $y = \tan(x + 1)$, and $y = \tan(x + 2)$.

Example 11.2. Let's try to solve $y' = xy + x$. This is a separable equation because $y' = x(y + 1)$. In rewritten form, it is equal to $\frac{dy}{y+1} = x\,dx$. After two antidifferentiations, we get $\ln(y + 1) = \frac{1}{2}x^2 + C$. Since $y + 1 = e^{\ln(y+1)} = e^{\frac{1}{2}x^2+C} = e^C \cdot e^{\frac{1}{2}x^2}$, it follows that $y + 1 = De^{\frac{1}{2}x^2}$, where $D = e^C$. Note that $D = e^C$ is positive, but since $y = -1$ is also a solution, the general solution of $y' = x(y + 1)$ is $y = De^{\frac{1}{2}x^2} - 1$ with $D \geq 0$. Taking $D = 1$ and $D = 2$, we get the particular solutions $y = e^{\frac{1}{2}x^2} - 1$ and $y = 2e^{\frac{1}{2}x^2} - 1$.

Example 11.3. Let's take the differential equation $y' = \frac{x^2}{1+\cos y}$ next. After rewriting it, we get

$$\int (1 + \cos y)\,dy = \int x^2\,dx.$$

Taking antiderivatives of each side, we get $y + \sin y = \frac{1}{3}x^3 + C$. In this case, there is no explicit solution for y in terms of x, so that this solution is only implicit.

Suppose that the surface of a concave mirror is obtained by rotating a curve in the plane around an axis. This axis is an axis of symmetry for the mirror. Suppose that such a mirror has the following property: there is a point such that all light rays coming into the mirror parallel to the axis of symmetry are reflected through this point. We saw in Section 8.5.4 that a curve that generates the surface of a mirror with this property is a parabola and that the point is its focus. The verification was based on Fermat's principle of least time in combination with Figure 8.42. We will now consider this problem again, this time in the context of differential equations.

Example 11.4. Consider a curve in the plane that has the property described above. Let the axis of rotation be the x-axis of an xy-coordinate system in this plane, and let the point in question be the origin. Suppose that the curve is the graph of a function $y = f(x)$ with $x > 0$ and $y \geq 0$, and that the graph has—at all points (except where it touches the x-axis)—tangent lines that are not parallel to the y-axis. This tangent line condition means that f is differentiable. Can the functions $y = f(x)$ that have a graph that meets these requirements be explicitly determined?

Let $y = f(x)$ be such a function, and let $P = (x_0, y_0)$ be any point on its graph with $y_0 > 0$. See Figure 11.1. The tangent line to the graph at P has slope $f'(x_0)$ and equation $y - y_0 = f'(x_0)(x - x_0)$. The point $Q = (x_1, 0)$ is the intersection of the tangent and the x-axis. It was verified in Section 8.5.1 that when light is reflected by a planar surface, then the angle of incidence of a ray is equal to the angle reflection. Therefore

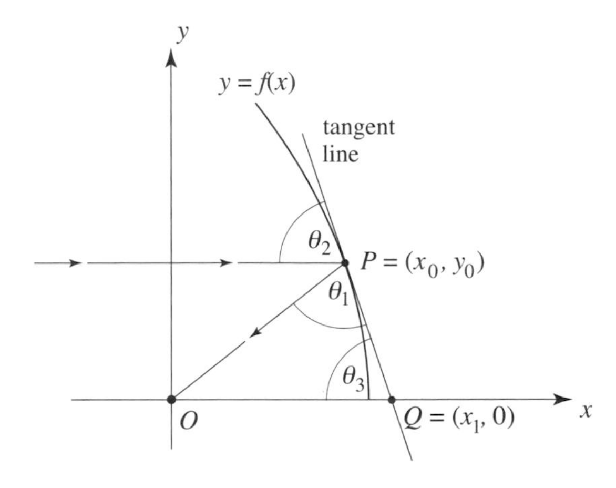

Fig. 11.1

the angles θ_1 and θ_2 are equal. Since $\theta_3 = \theta_2$, it follows that $\theta_1 = \theta_3$ and hence that the triangle ΔOPQ is isosceles. Therefore $x_1 = \sqrt{x_0^2 + y_0^2}$. Since $Q = (x_1, 0)$ lies on the tangent, x_1 satisfies $-y_0 = f'(x_0)(x_1 - x_0)$. So $x_1 = x_0 - \frac{y_0}{f'(x_0)}$, and hence $\sqrt{x_0^2 + y_0^2} = x_0 - \frac{y_0}{f'(x_0)}$. Therefore $y_0 = f'(x_0)(x_0 - \sqrt{x_0^2 + y_0^2})$. Since this equality is satisfied by any point P of the graph, the function $y = f(x)$ satisfies the differential equation $y = \frac{dy}{dx}(x - \sqrt{x^2 + y^2})$ or

$$\frac{dy}{dx} = \frac{y}{x - \sqrt{x^2 + y^2}}.$$

The determination of the function $y = f(x)$ calls for the solution of this differential equation. As a first step, we'll rewrite it—with an algebraic maneuver involving the conjugate $x + \sqrt{x^2 + y^2}$ of $x - \sqrt{x^2 + y^2}$—as

$$\frac{dy}{dx} = \frac{y}{x - \sqrt{x^2 + y^2}} \cdot \frac{x + \sqrt{x^2 + y^2}}{x + \sqrt{x^2 + y^2}} = \frac{y(x + \sqrt{x^2 + y^2}}{x^2 - (x^2 + y^2)} = \frac{x + \sqrt{x^2 + y^2}}{-y}.$$

So $y\frac{dy}{dx} = -x - \sqrt{x^2 + y^2}$. The next step is the substitution $u = x^2 + f(x)^2 = x^2 + y^2$. Since $\frac{du}{dx} = 2x + 2y\frac{dy}{dx}$

and hence $y\frac{dy}{dx} = \frac{1}{2}\frac{du}{dx} - x$, this substitution transforms the differential equation to $\frac{1}{2}\frac{du}{dx} - x = -x - u^{\frac{1}{2}}$ and hence to $\frac{1}{2}\frac{du}{dx} = -u^{\frac{1}{2}}$. After a separation of variables, we get

$$\int \tfrac{1}{2}u^{-\frac{1}{2}}du = -\int dx.$$

Therefore $u^{\frac{1}{2}} = -x + C$. Since $u^{\frac{1}{2}} = \sqrt{x^2 + y^2}$ and x are positive, the constant C is positive. Since $u = (-x + C)^2$, we get $x^2 + y^2 = x^2 - 2Cx + C^2$, and therefore

$$y^2 = -2Cx + C^2.$$

Consider a parabola in the xy-plane with focus the origin $O = (0, 0)$ and directrix the vertical line $x = C$. We know from the definition of such a parabola that for any point (x, y) on it, $\sqrt{x^2 + y^2} = C - x$. After squaring both sides, $x^2 + y^2 = C^2 - 2Cx + x^2$, so that $y^2 = -2Cx + C^2$. Notice that this is the same equation as the one just derived. We have therefore verified that $y = f(x) = \sqrt{-2Cx + C^2}$ with $C > 0$ and that its graph is the upper half of a parabola with focus the origin O and directrix the line $x = C$.

11.2 THE METHOD OF INTEGRATING FACTORS

The *method of integrating factors* is a strategy that can be applied (sometimes successfully, sometimes not) to first-order differential equations of the form

$$y' + p(x)y = q(x),$$

where, as indicated, $p(x)$ and $q(x)$ are functions of x. Like separation of variables, this approach reduces the solution of such an equation to the problem of finding antiderivatives. The rewritten form $y' = q(x) - p(x)y$ tells us that if $q(x) = cp(x)$ for a constant c, then such an equation is separable. The equations $y' - 2y = 5$, $y' = xy + x$, and $xy' = x^2 - 3y$ can be quickly rearranged (as we will see in a moment) so as to bring them into the form required for the method of integrating factors. The strategy proceeds as follows:

1. Rearrange the equation so that it is in the form $y' + p(x)y = q(x)$, and note that the coefficient of the derivative term y' needs to be 1.
2. Assume that $P(x)$ is an antiderivative of $p(x)$. The term $e^{P(x)}$ is the corresponding *integrating factor*.
3. Multiply both sides of the differential equation by the integrating factor to get
$$e^{P(x)}y' + e^{P(x)}p(x)y = e^{P(x)}q(x).$$
4. Make the important observation that the left side is equal to $\frac{d}{dx}(y \cdot e^{P(x)})$. It follows that $y \cdot e^{P(x)}$ is an antiderivative of the right side $e^{P(x)}q(x)$ of the equation. So $\displaystyle\int e^{P(x)}q(x)\,dx = y \cdot e^{P(x)} + C$.
5. If this integral can be solved, then $ye^{P(x)}$, and therefore y can be determined as a function of x, and the explicit general solution of $y' + p(x)y = q(x)$ is at hand.

Example 11.5. Let's illustrate this approach by solving $y' - 2y = 5$. Here $p(x) = -2$ and $q(x) = 5$. Because $P(x) = -2x$ is an antiderivative of $p(x)$, we can take e^{-2x} as integrating factor. So $y'e^{-2x} - 2ye^{-2x} = 5e^{-2x}$ and hence

$$\frac{d}{dx}(y \cdot e^{-2x}) = 5e^{-2x}.$$

Since $-\frac{5}{2}e^{-2x}$ is an antiderivative of the right side, we get

$$y \cdot e^{-2x} = -\tfrac{5}{2}e^{-2x} + C.$$

Multiplying through by e^{2x} gives us the general solution $y = Ce^{2x} - \frac{5}{2}$.

Example 11.6. Let's solve $y' = xy + x$. Putting the equation into the required form, we get $y' - xy = x$. So $p(x) = -x$ and $q(x) = x$. Because $P(x) = -\frac{x^2}{2}$ is an antiderivative of $p(x) = -x$, we take $e^{-\frac{x^2}{2}}$ as integrating factor. So $y'e^{-\frac{x^2}{2}} - xye^{-\frac{x^2}{2}} = xe^{-\frac{x^2}{2}}$, and it follows that

$$\frac{d}{dx}(y \cdot e^{-\frac{x^2}{2}}) = xe^{-\frac{x^2}{2}}.$$

Since $-e^{-\frac{x^2}{2}}$ is an antiderivative of the right side, we get

$$y \cdot e^{-\frac{x^2}{2}} = -e^{-\frac{x^2}{2}} + C.$$

Multiplying through by $e^{\frac{x^2}{2}}$ provides the general solution $y = Ce^{\frac{x^2}{2}} - 1$.

These two examples illustrate the method of integrating factors, but they provide nothing new in the sense that both equations can be solved by separating the variables. Check this for the first example, and review Example 11.2 for the second. However, the two examples that follow solve equations that are beyond the reach of separation of variables.

Example 11.7. Let's have a go at $xy' = x^2 - 3y$. Since this equation transforms to $y' + \frac{3}{x}y = x$, it is in the form $y' + p(x)y = q(x)$, with $p(x) = \frac{3}{x}$ and $q(x) = x$. Taking $P(x) = 3\ln x = \ln x^3$, we get $e^{P(x)} = e^{\ln x^3} = x^3$, and therefore,

$$y'e^{P(x)} + p(x)ye^{P(x)} = q(x)e^{P(x)} \quad \text{or} \quad y' \cdot x^3 + \frac{3}{x}y \cdot x^3 = x \cdot x^3 = x^4.$$

So $\frac{d}{dx}(y \cdot x^3) = x^4$ and hence $yx^3 = \frac{1}{5}x^5 + C$. By dividing through by x^3, we get $y = \frac{1}{5}x^2 + Cx^{-3}$.

Example 11.8. Consider the first-order differential equation $\cos\theta \cdot \frac{dr}{d\theta} - \sin\theta \cdot r = \cos^3\theta$. Here, the function r that the solution calls for is a function of θ. After dividing through by $\cos\theta$, the equation is in the required form

$$\frac{dr}{d\theta} + \frac{-\sin\theta}{\cos\theta} \cdot r = \cos^2\theta.$$

The first thing to do is to find an antiderivative of $p(\theta) = \frac{-\sin\theta}{\cos\theta} = -\tan\theta$. Note that the substitution $u = \cos\theta$ transforms the integral $\int \frac{-\sin\theta}{\cos\theta}\,d\theta$ into $\int \frac{1}{u}\,du$. It follows that $\int \frac{-\sin\theta}{\cos\theta}\,d\theta = \ln u + C = \ln(\cos\theta) + C$. Taking the simplest antiderivative, we set $P(\theta) = \ln(\cos\theta)$. So the integrating factor is $e^{P(\theta)} = e^{\ln(\cos\theta)} = \cos\theta$. Multiplying the equation through by $e^{P(\theta)} = \cos\theta$, we get

$$\frac{dr}{d\theta} \cdot \cos\theta - \sin\theta \cdot r = \cos^3\theta.$$

So $\frac{dr}{d\theta}(r \cdot \cos\theta) = \cos^3\theta$. It remains to find an antiderivative of $\cos^3\theta$. To do this, write $\cos^3\theta = \cos\theta \cdot \cos^2\theta = \cos\theta(1 - \sin^2\theta)$. The substitution $u = \sin\theta$ and $\frac{du}{d\theta} = \cos\theta$ tells us that

$$\int \cos^3\theta\,d\theta = \int \cos\theta(1 - \sin^2\theta)\,d\theta = \int (1 - u^2)\,du = u - \tfrac{1}{3}u^3 + C = \sin\theta - \tfrac{1}{3}\sin^3\theta + C,$$

and it follows that $r\cos\theta = \sin\theta - \frac{1}{3}\sin^3\theta + C$. So $r = \tan\theta - \frac{1}{3}(\tan\theta)(\sin^2\theta) + C\sec\theta$ is the general solution of the given differential equation.

The mathematics of differential equations includes not only various methods of solution, but also theoretical results. One of these is a theorem that is relevant to the method of integrating factors.

Theorem 1. If the functions $p(x)$ and $q(x)$ are continuous for all x in an open interval (a, b), then the initial value problem

$$y' + p(x)y = q(x) \quad \text{and} \quad y(x_0) = y_0, \text{ where } x_0 \text{ is in } (a, b)$$

has a unique particular solution $y = f(x)$ for all x in the interval (a, b).

The theorem is the consequence of the primary result of Section 9.6. Namely, the assumption that $p(x)$ and $q(x)$ are continuous guarantees the existence of the antiderivatives of $p(x)$ and $q(x)e^{P(x)}$ that the method of integrating factors relies on. Finding these functions in specific situations is quite another matter. See the discussion that concludes Section 9.11. Power series and the functions they define are central in this regard. This topic is taken up in a subsequent section of this chapter.

11.3 DIRECTION FIELDS AND EULER'S METHOD

We will consider any first-order differential equation of the form $\frac{dy}{dx} = F(x, y)$, where F is a function of the two variables x and y. Since separable differential equations satisfy $\frac{dy}{dx} = g(x) \cdot h(y)$ for some functions $g(x)$ and $h(y)$, they have this form. Equations to which the method of integrating factors applies can be rewritten as $\frac{dy}{dx} = -p(x)y + q(x)$ for some functions $p(x)$ and $q(x)$. So they too are of this form.

There is an approach that gives pictorial insight into the solutions of any equation of the form $\frac{dy}{dx} = F(x, y)$. Start with a rectangular grid of points in the xy-plane. The center of the grid can be any point, but it is often the origin. See Figure 11.2. At each point (x, y) of the grid, place a short, straight line segment with center the point and slope $\frac{dy}{dx} = F(x, y)$. Some of these short lines are shown in the

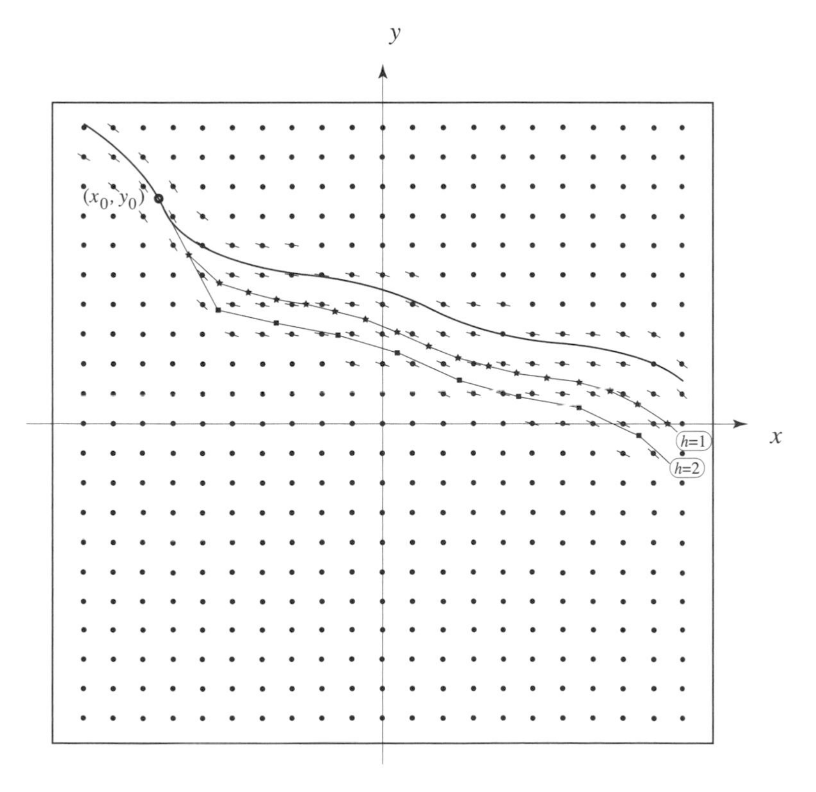

Fig. 11.2

figure. A grid with all of them in place is called the *slope field* or the *direction field* of the differential equation $\frac{dy}{dx} = F(x, y)$.

The general solution of any first-order differential equation—whether explicit or implicit—involves a constant C. Different values for C determine different particular solutions of the differential equations. Now let (x_0, y_0) be any point in the rectangle of the grid, and regard it to represent the initial condition $x = x_0$ and $y = y_0$. Substituting $x = x_0$ and $y = y_0$ into the general solution of $\frac{dy}{dx} = F(x, y)$ determines a specific constant C and hence a particular solution of the equation. The curving graph of such a particular solution along with the point (x_0, y_0) is shown in Figure 11.2. Since the flow of this graph is determined by the slopes of the points on it, its direction is generally aligned with the flow of the short lines at the points on the grid in its vicinity. This observation means that the particular solution that the point (x_0, y_0) determines can be approximated in numerical terms. This is *Euler's method.* It is named after the prolific and powerful Swiss mathematician Leonhard Euler (1707–1783). We turn to the details next.

Let $h > 0$ represent a distance. For example, h could be 1 (the distance between consecutive points in a horizontal or vertical row of the grid), or 2, or 3, or—and this is more common—$\frac{1}{2}$, or $\frac{1}{5}$, or $\frac{1}{10}$. With h fixed, consider the x-coordinates

$$x_0, x_1 = x_0 + h, x_2 = x_0 + 2h, x_3 = x_0 + 3h, \ldots, x_i = x_0 + i \cdot h, \ldots.$$

Start with the segment that begins at (x_0, y_0) with slope $F(x_0, y_0)$ and stops at the point with x-coordinate x_1. The point-slope form of the equation of the line on which this segment lies is

$$y - y_0 = F(x_0, y_0)(x - x_0).$$

With $x = x_1$, we get the point (x_1, y_1), where

$$y_1 = y_0 + F(x_0, y_0)(x_1 - x_0) = y_0 + hF(x_0, y_0).$$

The segment from (x_0, y_0) to (x_1, y_1) is a tangential approximation of the particular solution near the point (x_0, y_0). (See Figure 11.2 in the situations $h = 1$ and $h = 2$.) It has the same starting point and starting slope as the particular solution. This is the first step of Euler's method. The second step is a segment that corrects the direction of the first. It starts at the endpoint (x_1, y_1) of the first segment, has slope $F(x_1, y_1)$, and ends at the point with x-coordinate $x_2 = x_1 + h$. Since this second segment lies on the line

$$y - y_1 = F(x_1, y_1)(x - x_1),$$

its right endpoint is (x_2, y_2), where

$$y_2 = y_1 + hF(x_1, y_1).$$

The third segment begins at (x_2, y_2) with slope $F(x_2, y_2)$ and lies on the line

$$y - y_2 = F(x_2, y_2)(x - x_2).$$

Its right endpoint is (x_3, y_3), where $x_3 = x_2 + h$ and

$$y_3 = y_2 + hF(x_2, y_2).$$

This pattern can be continued until the grid is traversed. Each segment is determined by the preceding one. The ith segment starts at the point (x_{i-1}, y_{i-1}) with slope $F(x_{i-1}, y_{i-1})$ and has endpoint (x_i, y_i), where

$$y_i = y_0 + hF(x_{i-1}, y_{i-1}).$$

This scheme of segments can be extended to the left of the point (x_0, y_0). Select the points $x'_1 = x_0 - h, x'_2 = x_0 - 2h, \ldots$. Start with the tangent line $y - y_0 = F(x_0, y_0)(x - x_0)$ and the segment from (x_0, y_0) to (x'_1, y'_1), where $y'_1 = y_0 + F(x_0, y_0)(x'_1 - x_0) = y_0 - hF(x_0, y_0)$, and continue the pattern.

What we have arrived at is a chain of line segments that approximates (over the region of the grid) the graph of the particular solution of the equation $\frac{dy}{dx} = F(x, y)$ that the point (x_0, y_0) determines. Suppose that in Figure 11.2, the distance between consecutive points of each horizontal and vertical row of the grid is 1. The figure depicts the chains that the step-lengths $h = 2$ and $h = 1$ determine. The endpoints of the segments for $h = 2$ are denoted by black squares and those for $h = 1$ by black stars. Given any equation $\frac{dy}{dx} = F(x, y)$ and an initial condition (x_0, y_0), only a *step-length h* is needed to determine an approximation of the particular solution. In general, the shorter the h, the more corrections in the directions and the better the approximation.

Let's have a look at the specific differential equation $\frac{dy}{dx} = F(x, y) = 2x - 3y$. Figure 11.3 depicts the slope field determined by the grid of points (m, n), where m and n are integers with $-6 \leq m \leq 6$ and $-6 \leq n \leq 6$. The slope field shows that the solutions of the equations have a strong vertical aspect in each quadrant. However, there is a path of points through the first and fourth quadrants of the diagram where this verticality is broken. This is the path roughly described by connecting the points $(-6, -4), (-3, -2), (0, 0), (3, 2)$, and $(6, 4)$. At each of these points, the slope segment is horizontal. How

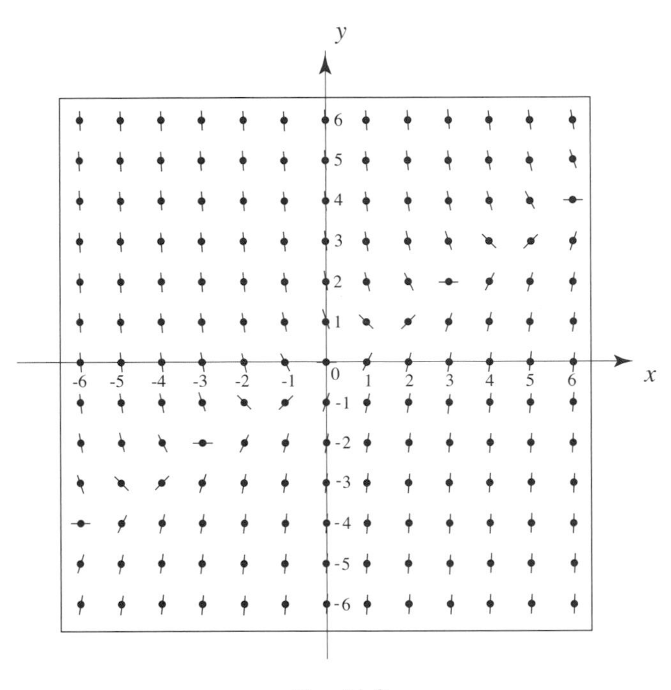

Fig. 11.3

exactly do the solutions respond to this apparent disruption?

To find out what happens, we'll solve the equation. With $p(x) = 3$ and $q(x) = 2x$, the differential equation has the form

$$\frac{dy}{dx} + p(x)y = q(x)$$

that the method of integrating factors requires. So take $P(x) = 3x$, and multiply the equation through by $e^{P(x)} = e^{3x}$ to get

$$\frac{dy}{dx} \cdot e^{3x} + 3ye^{3x} = 2xe^{3x}.$$

After taking antiderivatives of both sides,

$$ye^{3x} = \int 2xe^{3x}\, dx.$$

The integral $\int 2xe^{3x}$ can be dispatched with an integration by parts. (See Section 9.7.2.) Let $u = 2x$ and

$dv = e^{3x}$. So $du = 2dx$ and $v = \frac{1}{3}e^{3x}$. Since $\displaystyle\int u\,dv = uv - \int v\,du$,

$$\int 2xe^{3x}dx = \tfrac{2}{3}xe^{3x} - \int \tfrac{2}{3}e^{3x}dx = \tfrac{2}{3}xe^{3x} - \tfrac{2}{9}e^{3x} + C.$$

It follows that $ye^{3x} = \frac{2}{3}xe^{3x} - \frac{2}{9}e^{3x} + C$. Multiplying through by e^{-3x}, we get the general solution

$$y = \tfrac{2}{3}x - \tfrac{2}{9} + Ce^{-3x}$$

of the differential equation $\frac{dy}{dx} = 2x - 3y$.

We'll now consider each of the four sets of initial conditions:

$$x_0 = -6, y_0 = 6;\ x_0 = -3, y_0 = -2;\ x_0 = -2, y_0 = -2;\ \text{and, finally, } x_0 = 4, y_0 = 1.$$

1. Substituting $x = -6, y = 6$ into the general solution, we get $6 = \frac{2}{3}(-6) - \frac{2}{9} + Ce^{18}$, and hence that $C = \frac{92}{9}e^{-18}$. The corresponding particular solution is $y = \frac{2}{3}x - \frac{2}{9} + \frac{92}{9}e^{-3(x+6)}$.

2. The values $x = -3, y = -2$ give us $-2 = \frac{2}{3}(-3) - \frac{2}{9} + Ce^{9}$. So $C = \frac{2}{9}e^{-9}$. The particular solution in this case is $y = \frac{2}{3}x - \frac{2}{9} + \frac{2}{9}e^{-3(x+3)}$.

3. For $x = -2, y = -2$, we get $-2 = \frac{2}{3}(-2) - \frac{2}{9} + Ce^{6}$ and therefore that $C = -\frac{4}{9}e^{-6}$. For this C, the particular solution is $y = \frac{2}{3}x - \frac{2}{9} - \frac{4}{9}e^{-3(x+2)}$.

4. After substituting $x = 4, y = 1$, we see that $1 = \frac{2}{3}(4) - \frac{2}{9} + Ce^{-12}$ and therefore that $C = -\frac{13}{9}e^{12}$. The corresponding particular solution is $y = \frac{2}{3}x - \frac{2}{9} - \frac{13}{9}e^{-3(x-4)}$.

A graphing calculator, as supplied by the website https://www.desmos.com/calculator, for instance, provides the four graphs depicted in Figure 11.4. The fact that $\lim_{x\to\infty} Ce^{-3x} = \lim_{x\to\infty} \frac{C}{e^{3x}} = 0$ for any C tells

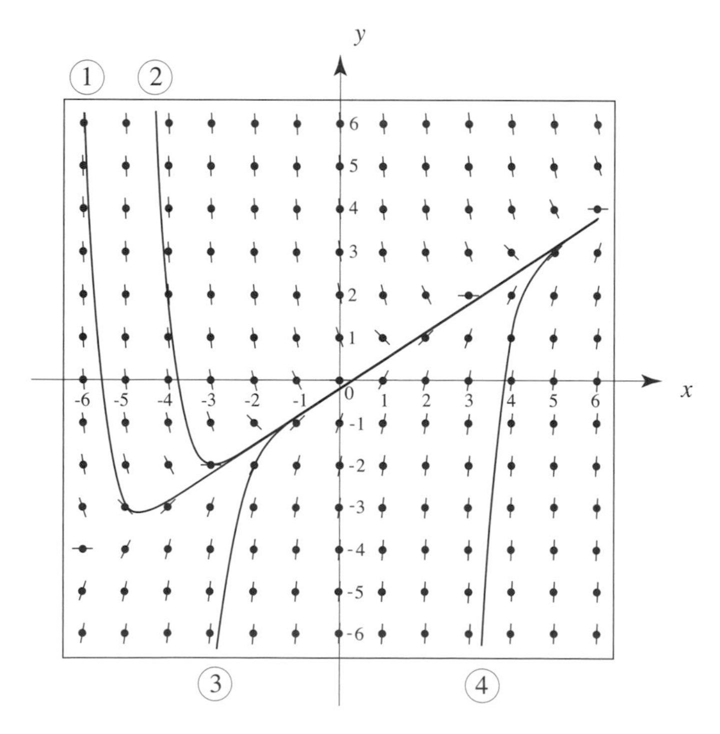

Fig. 11.4

us that the linear part $y = \frac{2}{3}x - \frac{2}{9}$ of the general solution $y = \frac{2}{3}x - \frac{2}{9} + Ce^{-3x}$ is an asymptote of the graph of every particular solution. This means that the graphs of particular solutions coming in steeply from the top through positive y (see curves 1 and 2) turn sharply in response to this asymptote. Solutions coming in through negative y (see curves 3 and 4) also turn, but less sharply. Figure 11.4 tells us that the grid of Figure 11.3 and the resulting direction field are not tight enough to fully capture the behavior of the solutions near this asymptotic line.

Let's turn to Euler's method to see what happens in the situation of case 1, the particular solution $y = \frac{2}{3}x - \frac{2}{9} + \frac{92}{9}e^{-3(x+6)}$ of the equation $\frac{dy}{dx} = F(x,y) = 2x - 3y$ that the initial condition $(-6,6)$ determines. What sort of approximation do we get by applying Euler's method with $h = 1$ starting at $(-6,6)$? The relevant x-coordinates are $x_0 = -6, x_1 = -5, x_2 = -4, x_3 = -3, x_4 = -2, x_5 = -1, x_6 = 0, x_7 = 1, x_8 = 2, x_9 = 3, x_{10} = 4, x_{11} = 5$, and $x_{12} = 6$. The slope at $(-6,6)$ that $F(x,y) = 2x - 3y$ provides is $F(-6,6) = -30$. So y_1 is given by $y_1 = y_0 - 30(x_1 - x_0) = 6 - 30 = -24$. Since $F(-5,-24) = -10 + 72 = 62$, we get that $y_2 = y_1 + 62(x_2 - x_1) = -24 + 62 = 38$. The fact that $F(-4,38) = -8 - 114 = -122$ tells us that $y_3 = y_2 - 122(x_3 - x_2) = 38 - 122 = -84$. One more step gives $y_4 = 162$. Therefore the first four segments of the approximation are given by the points $(-6,6), (-5,-24), (-4,38), (-3,-84)$, and $(-2,162)$. Notice that all these segments are far off the grid and that the Euler approximation is spinning out of control.

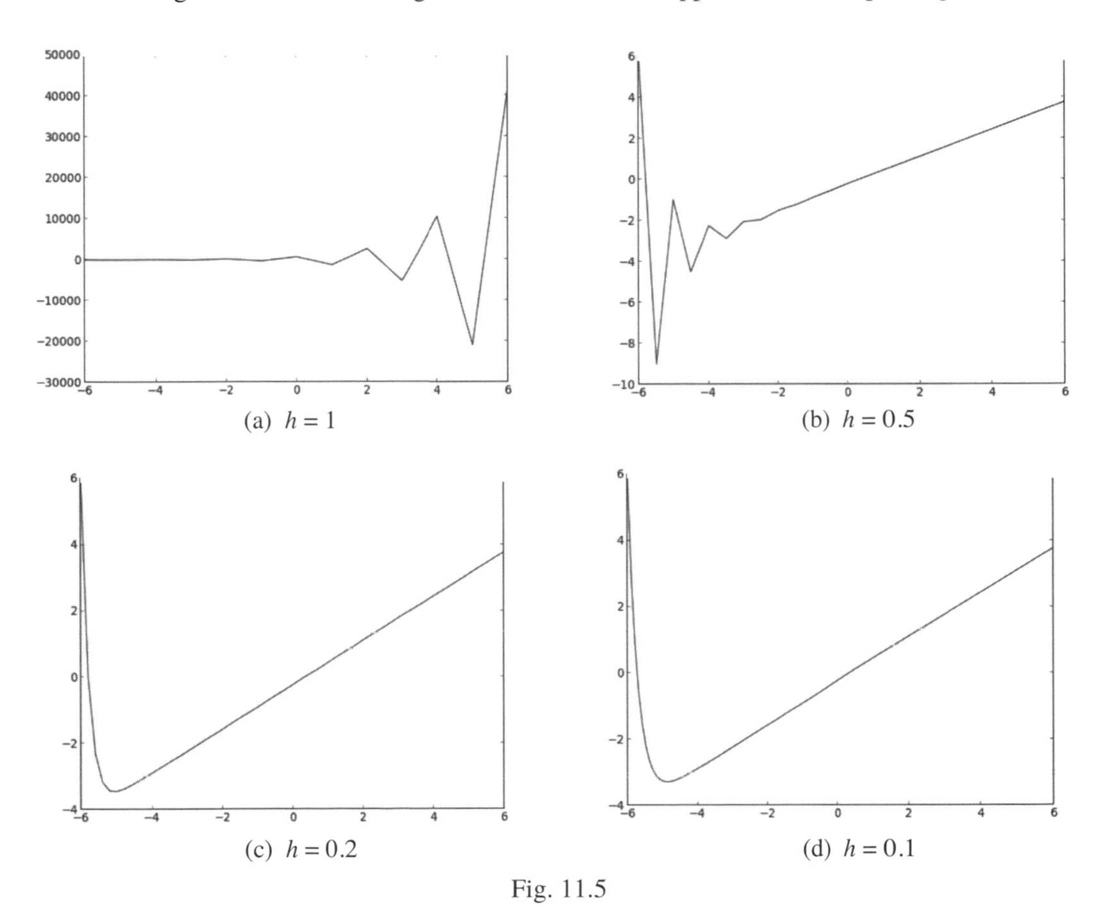

Fig. 11.5

How far out of control is confirmed by Figure 11.5a. The vertical scale needs to be huge to realize the graph of the Euler approximation. So huge, that the part of the approximation from $x = -6$ to $x = -2$ that was just established appears as a straight line. The last segment shoots out of sight! We can conclude that the step-length $h = 1$ is much too coarse to provide anything of value.

We'll consider $h = 0.5, h = 0.2$, and $h = 0.1$ and see what happens. If $h = 0.5$, it takes two steps to get to $x = -5$, two more to get to $x = -4$, and so on, and a total of 24 steps to get to $x = 6$. For $h = 0.2$, 60 computations are needed, and for $h = 0.1$, it takes 120. The computations for carrying out Euler's method are becoming progressively extensive. But there are programs that carry out these computations at an instant. The website http://www.math-cs.gordon.edu/~senning/desolver/ / supplies one of them (and an Internet search provides others). After the equation $\frac{dy}{dx} = F(x, y)$, the point (x_0, y_0), and the step-length h are fed into the program (in the variables t and y rather than x and y), the required y-coordinates and the graph of the resulting Euler approximation appear at the touch of a button. Doing this in the current situation for $h = 0.5, 0.2$, and 0.1 provides the graphs of Figures 11.5b, 11.5c, and 11.5d. Comparing the graph of case 1 of Figure 11.4 with Figure 11.5d suggests that the last approximation gives good agreement. But the example below shows that discrepancies remain.

Example 11.9. Consider the function $y = \frac{2}{3}x - \frac{2}{9} + \frac{92}{9}e^{-3(x+6)}$. Show that $\frac{dy}{dx} = \frac{2}{3} - \frac{92}{3}e^{-3(x+6)}$. It follows that $\frac{dy}{dx} = 0$ for $x = -\frac{1}{3}(18 - \ln 46) \approx -4.72$. The corresponding y-coordinate is $y \approx -3.15$. This tight approximation $(-4.72, -3.15)$ of the minimum of the graph of $y = \frac{2}{3}x - \frac{2}{9} + \frac{92}{9}e^{-3(x+6)}$ tells us that it is depicted a bit too low in the approximation of Figure 11.5d. The step-length $h = 0.1$ is still not small enough to grasp the graph accurately.

The bottom line with Euler's method is this. It is simple in principle, but gives accurate results only for comparatively small h. While the approximation for a small h requires lots of computations, these are easily and quickly carried out by computer programs such as the one mentioned.

Our discussion of differential equations will now turn to second-order equations of the form

$$A\frac{d^2y}{dx^2} + B\frac{dy}{dx} + Cy = 0,$$

where A, B, and C are constants. Such equations arise in the study of planetary motion (we will see how in Section 12.9), in the study of motions that are driven by springs (as we have already seen in Section 10.5.2), and in the study of electric circuits (an example that we will not take up). The analysis of second-order equations of this form relies on the *complex number system*. This system will be developed with the aid of *polar coordinates*. This coordinate system is also central to the concerns of Chapter 12.

11.4 THE POLAR COORDINATE SYSTEM

The connection that the rectangular, or Cartesian, coordinate system makes between points in the plane and their numerical coordinates has been crucial to much that has been done in this text so far. However, this system is not the only one that makes this connection. We will now discuss a coordinate system—the polar coordinate system—that provides a better framework for certain mathematical investigations. It turns out, in particular, that this system is tailor-made for the mathematical description of the orbits of the planets about the Sun. We will also use it to define the complex numbers. This section will introduce the polar coordinate system and present its basic properties.

Start with a plane, and fix a point in it. Call it the *pole* or *polar origin*, and designate it by O. Next, fix a ray—this is a "half line" with a direction—that emanates from the pole. Fix a unit of length and, starting with 0 at the pole, mark off points at distances 1, 2, 3, ... on the ray, and label them accordingly. Complete this ray to a positive real number system, and call it the *polar axis*. The *polar coordinate system*

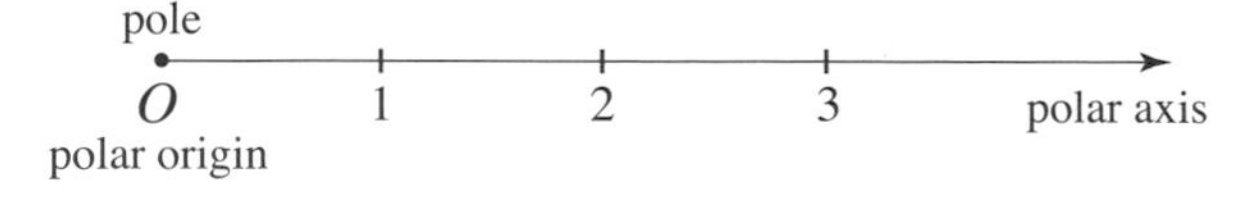

Fig. 11.6

is now complete. The polar axis is customarily drawn horizontally and directed from the origin O to the right. See Figure 11.6.

Let P be any point in the plane, and draw the segment OP. Let r be the length of this segment, and let θ be the angle in radians that it makes with the polar axis. See Figure 11.7. As in Section 4.6, θ is positive if it is measured in the counterclockwise direction and negative if it is measured in the clockwise direction.

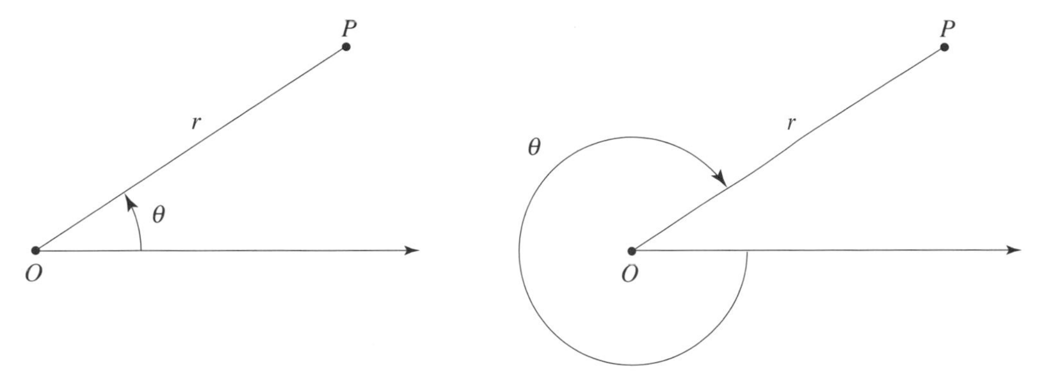

Fig. 11.7

In this way, the point P determines a pair (r, θ) of real numbers. The numbers r and θ are called *polar coordinates* of P. This procedure can be reversed. Namely, to any pair of real numbers (r, θ) there corresponds a point P in the plane. This is illustrated in Figure 11.8. For instance, the point corresponding to the pair $(2, \frac{3}{4}\pi)$ is obtained by selecting the ray corresponding to the positive angle $\frac{3}{4}\pi$ and choosing the point on this ray that is a distance 2 from the polar origin O. See Figure 11.8a. The understanding is that the point that corresponds to a pair (r, θ) with a negative r is obtained by taking the ray that θ determines

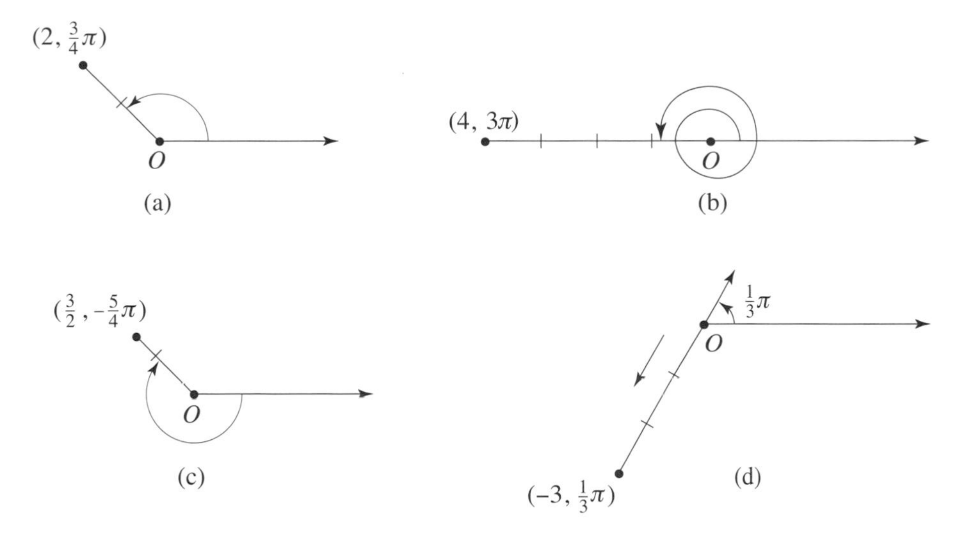

Fig. 11.8

and marking off the distance $|r|$ in the direction opposite to that of the ray, in other words, along the ray $\theta + \pi$. Figure 11.8d provides an example. One difference between the Cartesian coordinate system and the polar coordinate system is the fact that a point P can be represented in many ways (in fact in infinitely many ways) by a pair (r, θ). Figure 11.9 shows how four different pairs of polar coordinates all determine the same point.

Let's continue to focus on the way a point P in the polar plane is assigned to a given pair of real numbers (r, θ). We saw in Section 4.6 how any real number θ can be interpreted as an angle in radian measure by using the unit circle and the point 1 on the polar axis (and the unit circle). If $\theta \geq 0$, start at 1 and go counterclockwise around the circumference for a distance θ. If $\theta < 0$, start at 1 and go clockwise

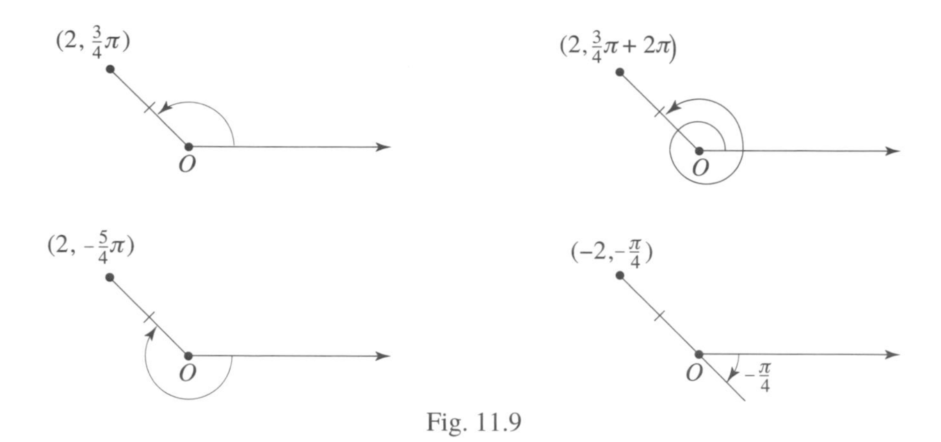

Fig. 11.9

around the circumference for a distance $-\theta = |\theta|$. In either case, θ determines a point P_θ on the unit circle, and hence a ray from the polar origin O through P_θ. If $r \geq 0$, measure off a distance r from O along this ray to locate the point P. If $r < 0$, take the ray in the opposite direction of the one that θ determines, and mark off a distance $-r = |r|$ from O in that direction to locate P_θ. Take the pair of numbers $(-7.4, 32.6)$, for example. To locate the point P_θ for $\theta = 32.6$, notice that $10\pi \approx 31.416$. Start at 1 on the polar axis and go counterclockwise around the circumference of the unit circle for 5 complete revolutions to get to $P_{10\pi}$, then an additional distance of $\frac{\pi}{3} \approx 1.047$ to get close to $P_{32.463}$, and, finally, a little farther to $P_\theta = P_{32.6}$. Consider the ray from O through $P_{32.6}$. Now look at the first coordinate -7.4. Since -7.4 is negative start at O, go in the direction opposite to this ray, measure off 7.4 units, and mark the point. This is the point P with polar coordinates $(-7.4, 32.6)$.

In the polar scheme of things, the second coordinate—or angle coordinate—of a point is more fundamental. It provides the ray from O on which the point lies. The first coordinate—or distance coordinate—serves to locate the point on this ray.

When analyzing a problem in the polar setting, it is often of advantage to make use of the information that a superimposed Cartesian coordinate system provides. Place an xy-coordinate system on top of the polar coordinate system in such a way that the two origins coincide, the polar axis coincides with the

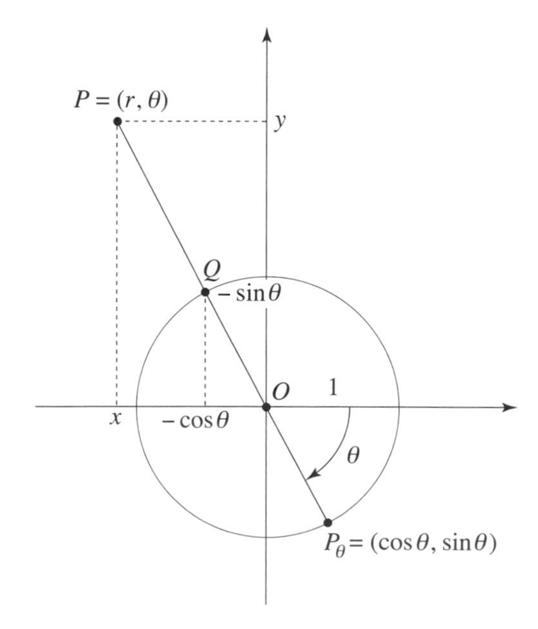

Fig. 11.10

positive x-axis, the negative x-axis is obtained by extending this axis to the other side of the origin O, and the y-axis is perpendicular to the x-axis at O. Suppose that this has been done as depicted in Figure 11.10. Now let P be any point in the plane, and let (r, θ) be any pair of polar coordinates for P. For purposes of illustration, we'll assume that both θ and r are negative. How can the x- and y-coordinates of P be found? The figure shows the location of the point P, the polar coordinate θ, and the point P_θ on the unit circle. The definitions of $\cos\theta$ and $\sin\theta$ in Section 4.6 tell us that in terms of its xy-coordinates $P_\theta = (\cos\theta, \sin\theta)$. This means that the point Q on the unit circle opposite to P_θ has x- and y-coordinates $-\cos\theta$ and $-\sin\theta$. Since r is negative, the point P lies on the ray from O through Q, a distance $-r$ from O. Because the two triangles determined by the segments PO and QO, the vertical dashed lines, and the negative x-axis are similar, it follows that $\frac{-r}{1} = \frac{-x}{\cos\theta}$. In the same way, that $\frac{-r}{1} = \frac{y}{-\sin\theta}$. Therefore

$$x = r\cos\theta \quad \text{and} \quad y = r\sin\theta.$$

Notice also that

$$r^2 = x^2 + y^2 \quad \text{and} \quad \tan\theta = \tfrac{y}{x}.$$

These relationships hold regardless of the location of the point P and regardless of the choice of its polar coordinates. For instance, the fact that the polar coordinates $(-r, \theta + \pi)$ represent the same point as (r, θ) means that the corresponding Cartesian coordinates of the point are the same. So it must be the case that

$$x = r\cos\theta = -r\cos(\theta + \pi) \quad \text{and} \quad y = r\sin\theta = -r\sin(\theta + \pi).$$

But this follows directly from the identities $\sin(\theta + \pi) = -\sin\theta$ and $\cos(\theta + \pi) = -\cos\theta$. These, in turn,

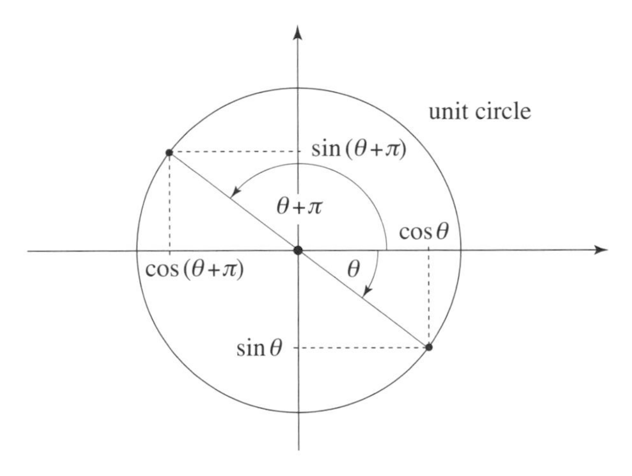

Fig. 11.11

follow from the definitions of the sine and cosine functions and Figure 11.11.

In summary, if (r, θ) are *any* polar coordinates of P, then the *one and only* set of Cartesian coordinates of P is given by $x = r\cos\theta$ and $y = r\sin\theta$.

Example 11.9. Consider the point P with polar coordinates $(-7.4, 32.6)$. To determine its Cartesian coordinates (x, y), simply take $x = -7.4\cos 32.6$ and $y = -7.4\sin 32.6$. With the aid of a calculator, $x \approx -2.79$ and $y \approx -6.85$.

Going in the other direction, we'll now let (x, y) be the Cartesian coordinates of P, and determine a pair of polar coordinates (r, θ) for P. This involves the equations $r^2 = x^2 + y^2$ and $\tan\theta = \frac{y}{x}$. Suppose first that P lies on the y-axis. In this case, check that (r, θ) with $r = y$ and $\theta = \frac{\pi}{2}$ are polar coordinates of P. If P is not on the y-axis, then it is not hard to see that θ can be chosen to satisfy $-\frac{\pi}{2} < \theta < \frac{\pi}{2}$. Figure 11.9

provides an example. Because $\tan\theta = \frac{y}{x}$, the definition of the inverse tangent function (in Section 9.9.1) implies that θ is determined by

$$\theta = \tan^{-1}\tfrac{y}{x}.$$

The r we need satisfies $r^2 = x^2 + y^2$. So $r = \pm\sqrt{x^2+y^2}$, and either

$$(\sqrt{x^2+y^2},\ \theta) \quad \text{or} \quad (-\sqrt{x^2+y^2},\ \theta),$$

are polar coordinates for P. Notice that the Cartesian point $P' = (-x, -y)$ provides the same values for $x^2 + y^2$ and $\tan^{-1}\frac{y}{x}$ as P does. Therefore, the other pair of polar coordinates is a pair of polar coordinates for P'. The fact that P and P' lie in opposite quadrants of the Cartesian plane tells us which of the two possibilities $(\sqrt{x^2+y^2},\ \theta)$ or $(-\sqrt{x^2+y^2},\ \theta)$ are polar coordinates of P (and which belong to P').

Example 11.10. Let's find polar coordinates for the Cartesian point $P = (-4, -5)$. Following the strategy just described, start by considering $\tan^{-1}(\frac{-5}{-4}) = \tan^{-1}1.25$. A calculator (be sure that it is in radian mode and *not* in degree mode) provides the angle $\theta = \tan^{-1}1.25 \approx 0.90$ radians. (Given that 1 radian $= \frac{180°}{\pi} \approx 57.3°$, this angle corresponds to about $51.5°$.) Since this is a positive angle and $(-4, -5)$ lies in the third quadrant of the Cartesian plane, we get r by taking the negative option $r = -\sqrt{(-4)^2 + (-5)^2} = -\sqrt{41} \approx -6.40$. So $(-\sqrt{41}, \tan^{-1}1.25) \approx (-6.40, 0.90)$ are polar coordinates of P.

Example 11.11. Consider the point P with Cartesian coordinates $(11.7, -8.3)$. As before, consider $\frac{-8.3}{11.7}$, and push the $\tan^{-1}$ button of a calculator to get $\theta \approx -0.95$ radians. This corresponds to approximately $-54.5°$. The polar coordinate r satisfies $r^2 = 11.7^2 + 8.3^2$ and hence $r \approx \pm 14.3$. Because P is in the fourth quadrant, it is reached by going in the direction of the ray that $\theta \approx -0.95$ radians determines. It follows that P has polar coordinates (r, θ), where $r = \sqrt{11.7^2 + 8.3^2} \approx 14.3$ and $\theta = \tan^{-1}(\frac{-8.3}{11.7}) \approx -0.95$ radians.

There were earlier formulations of the concept of a polar coordinate system, but Isaac Newton, starting with a paper written in 1671 (not published until 1736), was first to make extensive use of it. The terminology "pole" and "polar axis" and the designation of coordinates by specifying the distance from the pole and the angle from the polar axis were contributions of Jacob Bernoulli (a Swiss student of Leibniz) in the early 1690s. Even though the polar coordinate system is tailor-made for the study of the orbits of the planets (as we will see in Chapter 12), Newton does not make explicit use of it in his studies of the planetary orbits. See Sections 6.8 and 6.9.

11.5 THE COMPLEX PLANE

Consider a number line. Sufficient for its construction is a point, a ray emanating from it, and a unit of length. The point is the origin labeled with the number 0, the ray designates the positive direction, and the unit of length allows for the placement of the positive real numbers, including $1, 2, 3$, and so on, on the ray. Extending the ray in the opposite, or negative, direction provides the negative real numbers including -1, -2, and so forth, and completes the construction. In this way, points correspond to real numbers, and the decimal number system provides a way for any two points on the line to be added and multiplied. In short, the points on the line have become a number system. Take any two points P_1 and P_2 on the line, and let x_1 and x_2 be their coordinates, namely, the real numbers that they correspond to. The sum $x_1 + x_2$ and the

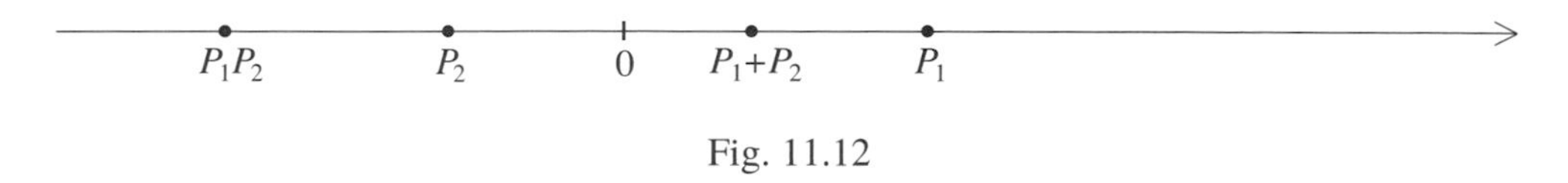

Fig. 11.12

product x_1x_2 are numbers on the line. They determine the sum $P_1 + P_2$ and product P_1P_2 of the two points P_1 and P_2. See Figure 11.12. In this way, any two points on the line can be added and multiplied to get—in each case—a third point on the line.

The complex number system is an extension of this idea to the plane. Again, start with the ray consisting of the origin O and the positive real numbers. Notice that such a ray is nothing but a polar coordinate system for the plane. We will now use it to turn the set of *all points in the plane* into a number system.

Let P_1 and P_2 be any two points in the plane. The point that represents the sum $P_1 + P_2$ of P_1 and P_2 is determined as follows. If the three points O, P_1, and P_2 do not lie on the same line as in Figure 11.13a, they determine a parallelogram, and the fourth point of the parallelogram is the sum $P_1 + P_2$ of P_1 and P_2. Alternatively, shift the arrow OP_2 without changing its direction, so that its initial point O ends up iat P_1. Then the sum $P_1 + P_2$ is the endpoint of this shifted arrow. This is also the strategy that is used to define the sum $P_1 + P_2$ in the situation where O, P_1, and P_2 lie on the same line. they do lie on the same line.

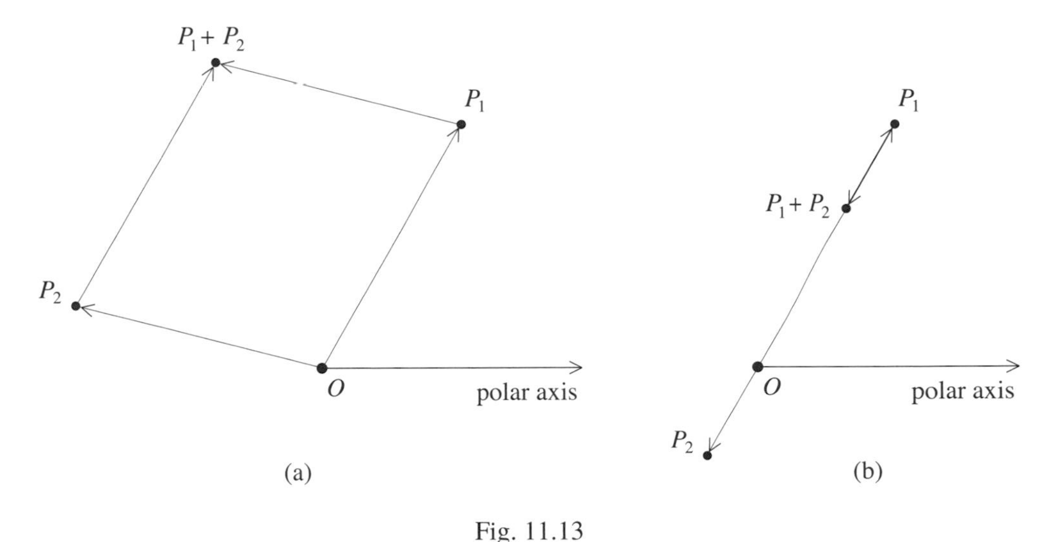

Fig. 11.13

See Figure 11.13b. Notice that this addition is nothing but the vector addition of Section 6.6. This recipe of adding points in the plane has only used the designated point O, but not the polar coordinate system.

To describe the product P_1P_2 of the points P_1 and P_2, the polar coordinate system comes into play. So let $P_1 = (r_1, \theta_1)$ and $P_2 = (r_2, \theta_2)$ be polar coordinates of P_1 and P_2. The product of P_1 and P_2 is the point P_1P_2 determined by the polar coordinates $(r_1r_2, \theta_1 + \theta_2)$, as shown in Figure 11.14. This definition does not depend on the choice of polar coordinates. For example, with $P_1 = (-r_1, \theta_1 - \pi)$ and $P_2 = (-r_2, \theta_2 - \pi)$, we get the product $P_1P_2 = (r_1r_2, \theta_1 + \theta_2 - 2\pi)$. This is a different set of polar coordinates, but the point P_1P_2 in the polar plane is the same as before.

With this addition and multiplication, the set of points in the plane becomes a number system that has all the basic properties that the real number system has. In particular, the commutative laws of addition and multiplication, the associative laws of addition and multiplication, and the distributive law of multiplication over addition all hold. Also, any number has an additive inverse, and any nonzero number has a multiplicative inverse. This set of numbers with addition and multiplication as described is the system of *complex numbers*. The plane of complex numbers is called the *complex plane*.

Extend the polar axis of Figure 11.14 to the left to obtain a horizontal number line in the plane of complex numbers. Putting in a vertical number line through O provides the complex plane with a Cartesian coordinate system. See Figure 11.15. Consider any two points on the horizontal axis, and let a and b be their coordinates. Observe that $(a, 0)$ and $(b, 0)$ are polar coordinates of the two points for any values of a and b. (Note that they are also the Cartesian coordinates of these two points.) The fact that the complex sum and product of $(a, 0)$ and $(b, 0)$ are $(a + b, 0)$ and $(ab, 0 + 0) = (ab, 0)$ respectively, tells us that the restriction of the addition and multiplication of complex numbers to the horizontal number line agrees with the ordinary addition and multiplication of real numbers. Check that the numbers $(0, 0)$ and $(1, 0)$ are the additive and multiplicative identity of the complex numbers. The horizontal axis of the complex plane is called the *real axis*. A typical number $(a, 0)$ on the real axis is often denoted simply by a.

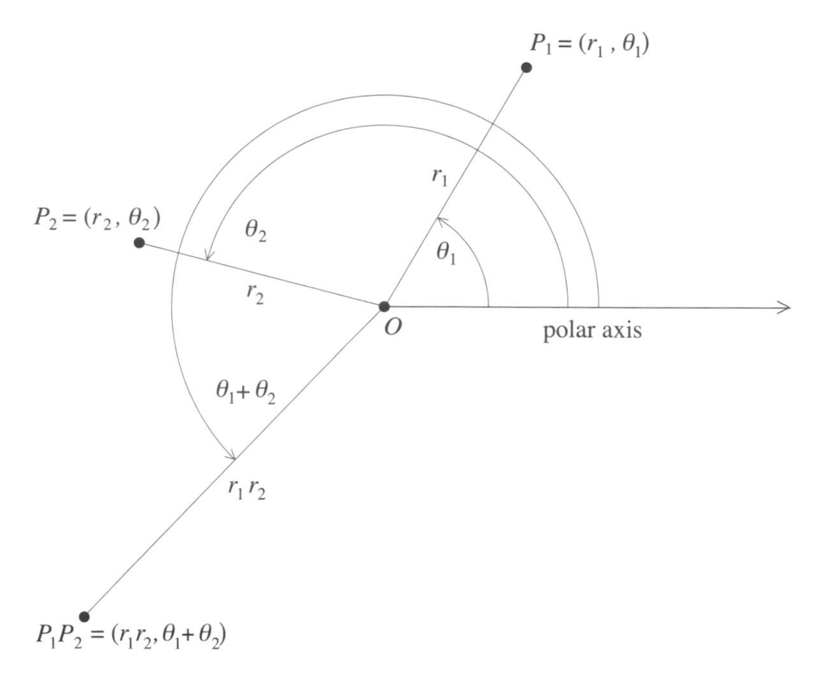

Fig. 11.14

Consider the point i with polar coordinates $(1, \frac{\pi}{2})$. Notice that i lies on the vertical axis of Figure 11.15, one unit above the origin. Apply the rule of complex multiplication to the product $i \cdot i$ to get

$$i^2 = i \cdot i = (1, \tfrac{\pi}{2})^2 = (1 \cdot 1, \tfrac{\pi}{2} + \tfrac{\pi}{2}) = (1, \pi) = (-1, 0) = -1.$$

Since $i^2 = -1$, it is a square root of -1. Accordingly, we'll also use the notation $i = \sqrt{-1}$. The number $i = \sqrt{-1}$ is commonly referred to as an *imaginary number* (even though it is concretely placed in the complex plane and there is nothing imaginary about it). The vertical axis through 0 and i is the *imaginary axis*. The product of any real number $a = (a, 0)$ and i is the imaginary number $ai = (a, 0)(1, \frac{\pi}{2}) = (a, \frac{\pi}{2})$.

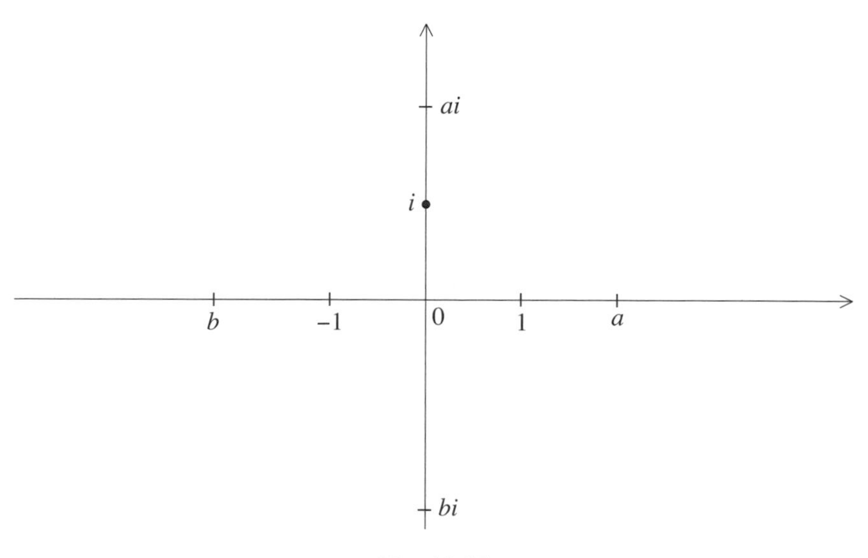

Fig. 11.15

Refer to Figure 11.15 once more. For a positive, ai is on the positive imaginary axis, and for a negative, ai is on the negative side of the imaginary axis. The observation that a and ai are at the same distance from the origin allows us to conclude that any number on the imaginary axis has the form ai for some real number a.

Let c be any point in the complex plane. Project the point c onto the real and imaginary axes as shown in Figure 11.16, and let a and bi be the respective real and imaginary numbers that arise. These numbers

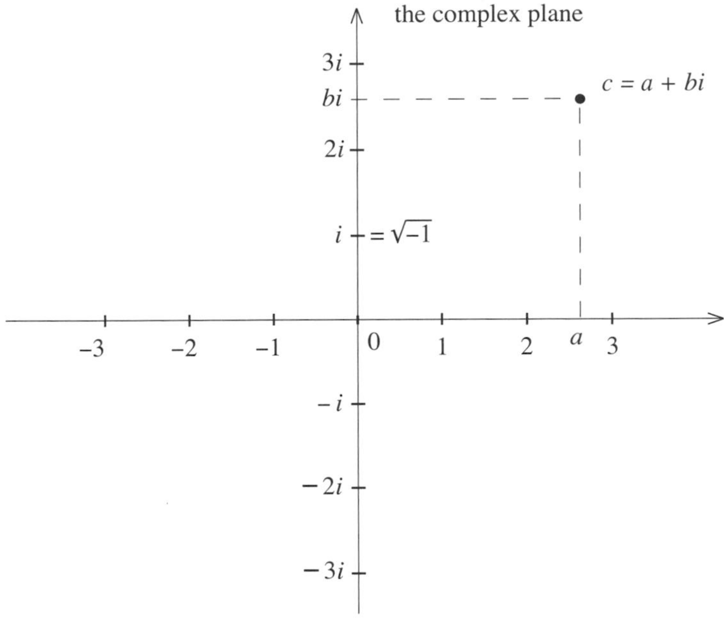

Fig. 11.16

are called the *real* and *imaginary parts* of c. The parallelogram definition of complex addition provides the fact that

$$c = a + bi = a + b\sqrt{-1}$$

is the sum of its real and imaginary parts.

Consider two complex numbers $c_1 = a_1 + b_1 i$ and $c_2 = a_2 + b_2 i$. Figure 11.16 tells us that c_1 and c_2 can only be equal if $a_1 = a_2$ and $b_1 = b_2$. The representation of a complex number as the sum of its real and imaginary parts makes it possible to express the addition and multiplication of complex numbers as follows. With

$$c_1 = a_1 + b_1 i \quad \text{and} \quad c_2 = a_2 + b_2 i,$$

the sum $c_1 + c_2$ and the product $c_1 c_2$ are, respectively, equal to

$$\begin{aligned} c_1 + c_2 &= (a_1 + b_1 i) + (a_2 + b_2 i) = (a_1 + a_2) + (b_1 + b_2)i, \text{ and} \\ c_1 c_2 &= (a_1 + b_1 i)(a_2 + b_2 i) = a_1 a_2 + a_1 b_2 i + a_2 b_1 i + (b_1 i)(b_2 i) \\ &= (a_1 a_2 - b_1 b_2) + (a_1 b_2 + a_2 b_1)i. \end{aligned}$$

The addition formula follows from the properties of vector addition. Refer to Section 6.6, and add the horizontal and vertical components of c_1 and c_2 to get the horizontal and vertical components of $c_1 + c_2$. The verification of the multiplication formula makes use of the connection between the expressions of a complex number in polar coordinates on the one hand, and its real and imaginary parts on the other.

Let the polar coordinates of a complex number c be (r, θ). It is a consequence of observations in Section 11.4 that the real and imaginary parts of c are $r\cos\theta$ and $r\sin\theta$, respectively. It follows that $c = r\cos\theta + (r\sin\theta)i$. See Figure 11.17. Let $c_1 = (r_1, \theta_1)$ and $c_2 = (r_2, \theta_2)$. Expressed in terms of their

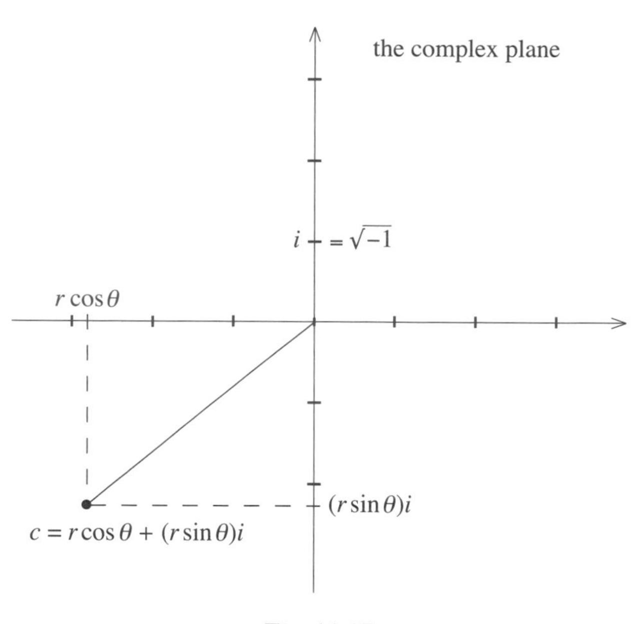

Fig. 11.17

real and imaginary parts, $c_1 = r_1\cos\theta_1 + (r_1\sin\theta_1)i$ and $c_2 = r_2\cos\theta_2 + (r_2\sin\theta_2)i$. From the definition of complex multiplication, $c_1c_2 = (r_1r_2, \theta_1 + \theta_2)$, so that

$$c_1c_2 = r_1r_2\cos(\theta_1 + \theta_2) + (r_1r_2\sin(\theta_1 + \theta_2))i.$$

The product formula of complex numbers in terms of their real and imaginary parts becomes

$$\begin{aligned}&r_1r_2\cos(\theta_1 + \theta_2) + (r_1r_2\sin(\theta_1 + \theta_2))i\\&\quad = [(r_1\cos\theta_1)(r_2\cos\theta_2) - (r_1\sin\theta_1)(r_2\sin\theta_2)] + [(r_1\cos\theta_1)(r_2\sin\theta_2) + (r_2\cos\theta_2)(r_1\sin\theta_1)]i.\end{aligned}$$

The fact that this equality holds follows from the addition formulas for the sine and cosine of Problem 1.25.

Consider the polynomial $Ax^2 + Bx + C$, where A, B, and C are real numbers. The quadratic formula tells us that it has the two roots

$$x = \tfrac{-B\pm\sqrt{B^2-4AC}}{2A} = \tfrac{-B}{2A} \pm \tfrac{\sqrt{B^2-4AC}}{2A}.$$

If $B^2 - 4AC \geq 0$, then both roots are real numbers. If $B^2 - 4AC = 0$, then $\frac{-B}{2A}$ is the only root. If $B^2 - 4AC < 0$, then $B^2 - 4AC = (4AC - B^2)(-1)$ with $4AC - B^2 > 0$, and the two roots are the complex numbers

$$x = \tfrac{-B}{2A} \pm \tfrac{\sqrt{4AC-B^2}\cdot\sqrt{-1}}{2A} = \tfrac{-B}{2A} \pm \tfrac{\sqrt{4AC-B^2}}{2A}\,i.$$

It follows that every quadratic polynomial $Ax^2 + Bx + C$ with A, B, and C real numbers has roots in the set of complex numbers. The question that arises is this: are there polynomials with real coefficients or, more generally, with complex coefficients that do not have roots in the complex numbers, and does the set of complex numbers need to be expanded to provide them? The answer is no. The *fundamental theorem of algebra* informs us that any polynomial $p(z) = c_nx^n + c_{n-1}x^{n-1} + \cdots + c_2x^2 + c_1x + c_0$, where $c_n, c_{n-1}, \ldots c_2, c_1$, and c_0 are complex numbers, can be factored $p(z) = c_n(x-z_1)(x-z_2)\cdots(x-z_{n-1})(x-z_n)$, where $z_1, z_2, \ldots, z_n$ are n complex roots of $p(z)$. In other words, $p(z) = c_nx^n+c_{n-1}x^{n-1}+\cdots+c_2x^2+c_1x+c_0$ has a complete set of roots in the set of complex numbers.

11.6 SECOND-ORDER DIFFERENTIAL EQUATIONS

This section describes the solutions of any second-order differential equation of the form

$$(*) \qquad Ay'' + By' + Cy = 0,$$

where A, B, and C are real constants. So that the equation is of the second order, we assume that $A \neq 0$.

Notice that the "zero" function $y = f(x) = 0$ for all x is a solution of equation $(*)$, but it is of little interest. Our discussion of equation $(*)$ will rely on the following theorem. We will use it without proof.

Theorem 2. Suppose that $y = f_1(x)$ and $y = f_2(x)$ are two nonzero solutions of $(*)$ and that it is not the case that there is a real constant d such that $f_2(x) = df_1(x)$ for all x. Then the general solution of $(*)$ has the form $f(x) = D_1 f_1(x) + D_2 f_2(x)$ for some real constants D_1 and D_2.

The polynomial $Ax^2 + Bx + C$ is called the *characteristic polynomial*, and $Ax^2 + Bx + C = 0$ is known as the *characteristic equation* of the differential equation $(*)$. We saw at the end of the previous section that there are three different possibilities for the roots of the characteristic polynomial:

1. If $B^2 - 4AC > 0$, then the characteristic polynomial has the two distinct real roots $r_1 = \frac{-B}{2A} + \frac{\sqrt{B^2-4AC}}{2A}$ and $r_2 = \frac{-B}{2A} - \frac{\sqrt{B^2-4AC}}{2A}$.
2. If $B^2 - 4AC = 0$, the characteristic polynomial has only one root, the real root $r = \frac{-B}{2A}$.
3. If $B^2 - 4AC < 0$, then $4AC - B^2 > 0$, and the characteristic polynomial has the two distinct complex roots $a \pm bi$, where $a = \frac{-B}{2A}$ and $b = \frac{\sqrt{4AC-B^2}}{2A}$.

Case 1. If the characteristic polynomial $Ax^2 + Bx + C$ has two distinct real roots r_1 and r_2, then the general solution of the differential equation $Ay'' + By' + Cy = 0$ is

$$y = D_1 e^{r_1 x} + D_2 e^{r_2 x},$$

where D_1 and D_2 are real constants.

In view of Theorem 2, it needs to be verified that $f(x) = e^{rx}$ is a solution of equation $(*)$ when r is a root of the characteristic polynomial. Because $f'(x) = re^{rx}$ and $f''(x) = r^2 e^{rx}$, we get

$$Af''(x) + Bf'(x) + Cf(x) = Ar^2 e^{rx} + Bre^{rx} + Ce^{rx} = (Ar^2 + Br + C)e^{rx} = 0,$$

as required. It also needs to be shown that if r_1 and r_2 are the two distinct roots of the characteristic equation, then there is no fixed constant d such that $e^{r_2 x} = de^{r_1 x}$. Such a constant would have to equal $d = e^{(r_2 - r_1)x}$, but the function $y = e^{(r_2 - r_1)x}$ is constant only if $r_2 - r_1 = 0$, or only if $r_1 = r_2$.

Example 11.12. Consider the differential equation $y'' + y' - 6y = 0$. The characteristic equation is $x^2 + x - 6 = 0$. After factoring, we get $(x-2)(x+3) = 0$, so that the roots of the characteristic polynomial are $r_1 = 2$ and $r_2 = -3$. By Case 1, the general solution of equation $(*)$ is

$$y = D_1 e^{2x} + D_2 e^{-3x}.$$

Example 11.13. Consider the differential equation $3y'' + y' - y = 0$. The characteristic polynomial is $3x^2 + x - 1$. By the quadratic formula the roots of the characteristic polynomial are $\frac{-1 \pm \sqrt{13}}{6}$. So the roots are the real numbers $r_1 = \frac{-1+\sqrt{13}}{6}$ and $r_2 = \frac{-1-\sqrt{13}}{6}$. By Case 1, the general solution of equation $(*)$ is

$$y = D_1 e^{\frac{-1+\sqrt{13}}{6}x} + D_2 e^{\frac{-1-\sqrt{13}}{6}x}.$$

Case 2. If the characteristic polynomial $Ax^2 + Bx + C$ has one real root r, then the general solution of $Ay'' + By' + Cy = 0$ is

$$y = D_1e^{rx} + D_2xe^{rx},$$

where D_1 and D_2 are real constants.

In view of Theorem 2 and the analysis of Case 1, it needs to be checked that $f(x) = xe^{rx}$ is a solution of equation $(*)$ and that an equality of the form $e^{rx} = dxe^{rx}$ for all x and some constant d is impossible. Notice that $f'(x) = e^{rx} + x \cdot re^{rx} = (1 + rx)e^{rx}$ and that $f''(x) = re^{rx} + (1 + rx)re^{rx} = (2r + r^2x)e^{rx}$. Therefore

$$Af''(x) + Bf'(x) + Cf(x) = A(2r + r^2x)e^{rx} + B(1 + rx)e^{rx} + Cxe^{rx} = (2rA + B)e^{rx} + (Ar^2 + Br + C)xe^{rx}.$$

Because $r = \frac{-B}{2A}$ is a root of $Ax^2 + Bx + C$, this last term is equal to 0. So $f(x) = xe^{rx}$ is a solution of $(*)$. If e^{rx} were to be equal to dxe^{rx} for a constant d and all x, then $dx = 1$. But then $x = d^{-1}$ for all x.

Example 11.14. Consider the second-order equation $y'' - 4y' + 4y = 0$. The characteristic equation is $x^2 - 4x + 4 = 0$. The quadratic formula informs us that the roots are $\frac{4\pm\sqrt{4^2-4\cdot1\cdot4}}{2} = 2$ and, in particular, that there is only one real root. By Case 2, the general solution of the equation $(*)$ is

$$y = D_1e^{2x} + D_2xe^{2x}.$$

Suppose that we are looking for the particular solution $y = f(x)$ that satisfies $f(0) = 3$ and $f'(0) = -2$. From $f(x) = (D_1+D_2x)e^{2x}$, we get $f'(x) = D_2e^{2x}+(D_1+D_2x)2e^{2x}$. Since $f(0) = 3$ and $f'(0) = -2$, we get $D_1 = 3$ and $D_2+2D_1 = -2$. So $D_2 = -2-6 = -8$. Therefore the particular solution is $f(x) = 3e^{2x} - 8xe^{2x}$.

Example 11.15. Let's study the second-order equation $4y'' + 12y' + 9y = 0$. The characteristic polynomial is $4x^2 + 12x + 9$. The quadratic formula informs us that its roots are $\frac{-12\pm\sqrt{12^2-4\cdot4\cdot9}}{8} = \frac{-12\pm\sqrt{144-144}}{8} = -\frac{3}{2}$ and, in particular, that there is one real root. By Case 2, the general solution of the equation is

$$y = D_1e^{-\frac{3}{2}x} + D_2xe^{-\frac{3}{2}x}.$$

The case where the roots of the characteristic polynomial $Ax^2 + Bx + C = 0$ are complex remains. In this case, $B^2 - 4AC < 0$, and the roots are the two complex numbers $a + bi$ and $a - bi$, where $a = \frac{-B}{2A}$ and $b = \frac{\sqrt{4AC-B^2}}{2A}$.

Case 3. If the characteristic polynomial $Ax^2 + Bx + C$ has the complex roots $r_1 = a + bi$ and $r_2 = a - bi$, where a and b are real numbers, then the general solution of $Ay'' + By' + Cy = 0$ is

$$y = e^{ax}(D_1 \cos bx + D_2 \sin bx),$$

where D_1 and D_2 are real constants.

To verify this, start with the equality $A(a + bi)^2 + B(a + bi) + C = 0$ and multiply it out to get $A(a^2 + 2abi + b^2i^2) + B(a + bi) + C = 0$. So $A(a^2 - b^2 + 2abi) + B(a + bi) + C = 0$, and therefore $(A(a^2 - b^2) + Ba + C) + (2Aab + Bb)i = 0$. It follows that

$$A(a^2 - b^2) + Ba + C = 0 \text{ and } 2Aab + Bb = 0.$$

Consider the function $y = e^{ax}\cos bx$. By the product and chain rules,

$$y' = ae^{ax}\cos bx + e^{ax}(-\sin bx)b = ae^{ax}\cos bx - be^{ax}\sin bx.$$

In a similar way,

$$\begin{aligned} y'' &= (a^2e^{ax}\cos bx - ae^{ax}(\sin bx)b) - (abe^{ax}\sin bx + be^{ax}(\cos bx)b) \\ &= (a^2e^{ax}\cos bx - abe^{ax}\sin bx) - (abe^{ax}\sin bx + b^2e^{ax}\cos bx) \\ &= (a^2 - b^2)e^{ax}\cos bx - 2abe^{ax}\sin bx. \end{aligned}$$

Therefore

$$\begin{aligned} Ay'' + By' + Cy &= A(a^2 - b^2)e^{ax}\cos bx - 2Aabe^{ax}\sin bx + Bae^{ax}\cos bx - Bbe^{ax}\sin bx + Ce^{ax}\cos bx \\ &= (A(a^2 - b^2) + Ba + C)e^{ax}\cos bx - (2Aab + Bb)e^{ax}\sin bx. \end{aligned}$$

The fact that $A(a^2 - b^2) + Ba + C = 0$ and $2Aab + Bb = 0$ therefore implies that $y = f_1(x) = e^{ax}\cos bx$ is a solution of the differential equation $Ay'' + By' + Cy$.

A similar computation shows that the function $y = f_2(x) = e^{ax}\sin bx$ is also a solution of $Ay'' + By' + Cy$. Now turn to Theorem 2. If there were to be a constant d such that $f_2(x) = df_1(x)$, then $\sin bx = d\cos bx$ for all x. With $x = 0$, this implies that $d = 0$. But $\sin bx = 0$ for all x implies that $b = 0$. Since $b = \frac{\sqrt{4AC-B^2}}{2A}$, this is not the case. Therefore Theorem 2 applies to tell us that the general solution of the second-order equation $Ay'' + By' + Cy = 0$ is

$$f(x) = D_1e^{ax}\cos bx + D_2e^{ax}\sin bx = e^{ax}(D_1\cos bx + D_2\sin bx),$$

where both D_1 and D_2 are real constants.

Notice in the context of Case 3 that we started with a problem, namely, the solution of the differential equation $Ay'' + By' + Cy = 0$ that involved real numbers only, then dove into the pool of complex numbers, to reemerge with a solution that involved only the realm of real numbers.

Example 11.16. Consider the second-order equation $y'' - 6y' + 13y = 0$. The characteristic polynomial is $x^2 - 6x + 13$. The quadratic formula informs us that the roots are $\frac{6\pm\sqrt{6^2-4\cdot1\cdot13}}{2} = \frac{6\pm\sqrt{36-52}}{2} = \frac{6\pm\sqrt{-16}}{2} = \frac{6\pm4\sqrt{-1}}{2} = 3 \pm 2\sqrt{-1}$. So the roots are the complex numbers $r_1 = 3 + 2i$ and $r_2 = 3 - 2i$. By Case 3, the general solution of the equation is

$$f(x) = e^{3x}(D_1\cos 2x + D_2\sin 2x).$$

We'll close our study of second-order differential equations by considering the equation

$$y'' + y = 0.$$

The characteristic polynomial is $x^2 + 1$. Since its roots are i and $-i$, Case 3 applies with $a = 0$ and $b = 1$. Therefore the general solution of $y'' + y = 0$ is

$$e^0(D_1\cos x + D_2\sin x) = D_1\cos x + D_2\sin x.$$

Because of its importance in the theory of planetary motion (we will see its relevance in this regard in Chapter 12), we'll present an approach to the general solution of $y'' + y = 0$ that is complete, brief, and *relies neither* on Case 3 *nor* on Theorem 2. Start with the observation that

$$\frac{d}{dx}\sin x = \cos x \quad \text{and} \quad \frac{d}{dx}\cos x = -\sin x$$

and the consequence that all functions of the form $y = D_1\cos x + D_2\sin x$, where D_1 and D_2 can be any real constants, are examples of functions that satisfy $y'' + y = 0$. We will now let $f(x)$ be *any solution* of $y'' + y = 0$ and show that

$$f(x) = D_1\cos x + D_2\sin x$$

for some real constants D_1 and D_2. This is the consequence of some trickery that is rather like pulling a rabbit out of a hat. Start by defining the two functions $D_1(x)$ and $D_2(x)$ as follows:

$$D_1(x) = (\cos x)f(x) - (\sin x)f'(x) \quad \text{and} \quad D_2(x) = (\sin x)f(x) + (\cos x)f'(x).$$

Check that $f(x) = D_1(x)(\cos x) + D_2(x)(\sin x)$. If we can show that the functions $D_1(x)$ and $D_2(x)$ are both constant, then we will have accomplished what we set out to do. Because $f''(x) + f(x) = 0$,

$$D_2'(x) = [(\cos x)f(x) + (\sin x)f'(x)] + [(-\sin x)f'(x) + (\cos x)f''(x)] = 0.$$

Therefore $D_2(x)$ is a constant. Let's put $D_2(x) = D_2$. Check with a similar computation that $D_1'(x) = 0$ as well, so that $D_1(x) = D_1$ is also a constant. We now have verified that the arbitrary solution of $y'' + y = 0$ has the form $f(x) = D_1 \cos x + D_2 \sin x$, where D_1 and D_2 are real constants.

11.7 THE BASICS OF POWER SERIES

We saw in Section 6.3 (and in the Problems and Projects section of Chapter 6) that Newton relied on power series to study functions and to evaluate definite integrals. This approach was applied in Section 10.2 to derive an estimate for the length of the main cable of a suspension bridge. The discussions just referred to are a good introduction to what follows below: a look at the theory of power series that concentrates on the basic aspects. A subsequent section will illustrate the relevance of power series to the solution of differential equations.

A *power series* is an expression of the form

$$\sum_{k=0}^{\infty} a_k(x - x_0)^k = a_0 + a_1(x - x_0) + a_2(x - x_0)^2 + a_3(x - x_0)^3 + \dots,$$

where $a_0, a_1, a_2, \dots$ as well as x_0 are real constants and x is a real variable. The constants $a_0, a_1, a_2, \dots$ are the *coefficients*, and x_0 is the *center* of the power series. We can also write this power series as

$$a_0 + \sum_{k=1}^{\infty} a_k(x - x_0)^k \quad \text{or} \quad a_0 + a_1(x - x_0) + \sum_{k=2}^{\infty} a_k(x - x_0)^k.$$

The parameter k is called the *summation index*. Notice that relabeling the index by j or n, for example, does not change the power series. A power series is said to *converge* at the number c if the sequence of numbers

$$a_0,\ a_0 + a_1(c-x_0),\ a_0 + a_1(c-x_0) + a_2(c-x_0)^2,\ a_0 + a_1(c-x_0) + a_2(c-x_0)^2 + a_3(c-x_0)^3, \dots$$

obtained by starting with a_0 and adding in more and more of these terms closes in on some finite number. This definition tells us that if $\sum_{k=0}^{\infty} a_k(x - x_0)^k$ converges at c, then the sum of the infinitely many terms of the power series obtained by taking $x = c$ adds up to a finite number in the same way that the numbers $1, \frac{1}{4}, \frac{1}{4^2}, \frac{1}{4^3}, \frac{1}{4^4}, \frac{1}{4^5}, \dots$ add up to $\frac{4}{3}$. (See Section 2.3 in this regard.) If the continuing sum does not close in on some (finite) number, then the power series *diverges* at $x = c$.

The special case

$$\sum_{k=0}^{\infty} ax^k = a + ax + ax^2 + ax^3 + \dots,$$

where $x_0 = 0$ and the coefficients $a_0 = a_1 = a_2, \dots$ are all equal and nonzero, is called a *geometric series*. We saw in Section 6.3 that such series are easily studied. Let a be a constant, and consider

$$S_n = a + ax + ax^2 + \cdots + ax^{n-2} + ax^{n-1}.$$

Since $xS_n = ax + ax^2 + \cdots + ax^{n-1} + ax^n$, we get $S_n(1-x) = S_n - xS_n = a - ax^n$. Therefore

$$S_n = \frac{a - ax^n}{1-x}.$$

Suppose that $|x| < 1$ and push n to infinity. Since x^n goes to zero, we have established that the geometric series

$$\sum_{k=0}^{\infty} ax^k = a + ax + ax^2 + \ldots = \frac{a}{1-x}$$

converges for all x with $|x| < 1$, and that for any such x, this infinite sum is equal to the number $\frac{a}{1-x}$. The geometric series diverges for all x with $|x| \geq 1$. For instance, for $x = 1$, the sum $a + a + a + \cdots = a(1 + 2 + 3 + \cdots)$ keeps getting larger and larger. For $x = -1$, the sum $a - a + a - a + \ldots$ jumps back and forth between a and 0 and hence does not add up to a (single) finite number.

The essence of what was just observed for a geometric series pertains to any power series.

Theorem 3 (Radius of Convergence). For any power series $\sum_{k=0}^{\infty} a_k(x - x_0)^k$, there is a number $R \geq 0$ such that

(1) the power series converges for all x such that $|x - x_0| < R$, and

(2) the power series diverges for all x such that $|x - x_0| > R$.

This number R is the *radius of convergence* of the power series.

The power series $\sum_{k=0}^{\infty} a_k(x - x_0)^k$ always converges for $x = x_0$ (it does so to a_0). If this is the only point for which it converges, its radius of convergence is $R = 0$. At the other extreme, if the power series converges for all real numbers x, then we say that its radius of convergence R is infinite and write $R = \infty$.

Let $\sum_{k=0}^{\infty} a_k(x - x_0)^k$ be any power series with a positive radius of convergence R. The *interval of convergence* of the power series is the set of all numbers for which it converges. If $R = \infty$, then this is $(-\infty, \infty)$. If R is finite, then x satisfies the inequality $|x - x_0| < R$, precisely if x lies in the open interval $(x_0 - R, x_0 + R)$. Given the assertion of Theorem 3, there are only the four possibilities

$$(x_0 - R, x_0 + R),\ [x_0 - R, x_0 + R),\ (x_0 - R, x_0 + R],\ [x_0 - R, x_0 + R],$$

for the interval of convergence. Which is it? The answer depends on the specific series, and in particular whether it converges or diverges at the two endpoints $x_0 - R$ and $x_0 + R$. We saw, for example, that the radius of convergence of a geometric series is $R = 1$ and that the series diverges at 1 and -1. It follows that its interval of convergence is the open interval $(-1, 1)$.

An important criterion for determining the radius of convergence of a power series is the *ratio test*. Let $\sum_{k=0}^{\infty} a_k(x - x_0)^k$ be any power series with the property that all coefficients a_k are nonzero. Consider the sequence of absolute values of the ratios

$$\left|\frac{a_1}{a_0}\right|, \left|\frac{a_2}{a_1}\right|, \left|\frac{a_3}{a_2}\right|, \ldots, \left|\frac{a_{k+1}}{a_k}\right|, \ldots,$$

and suppose that this sequence converges to a limit L. Include the case $\lim_{k\to\infty}\left|\frac{a_{k+1}}{a_k}\right| = +\infty$, where the terms $\left|\frac{a_{k+1}}{a_k}\right|$ grow without bound. For $x \neq x_0$, put in the variable terms to get

$$\left|\frac{a_1(x-x_0)^1}{a_0(x-x_0)^0}\right|, \left|\frac{a_2(x-x_0)^2}{a_1(x-x_0)^1}\right|, \left|\frac{a_3(x-x_0)^3}{a_2(x-x_0)^2}\right|, \dots, \left|\frac{a_{k+1}(x-x_0)^{k+1}}{a_k(x-x_0)^k}\right|, \dots,$$

and consider the limit

$$\lim_{k\to\infty}\left|\frac{a_{k+1}(x-x_0)^{k+1}}{a_k(x-x_0)^k}\right| = \lim_{k\to\infty}\left|\frac{a_{k+1}}{a_k}(x-x_0)\right| = L|x-x_0|.$$

The basic fact supplied by the ratio test is that $\sum_{k=0}^{\infty} a_k(x-x_0)^k$ converges at x if $L|x-x_0| < 1$, and it diverges if $L|x-x_0| > 1$. By bringing L to the other side of these inequalities, we have our conclusion:

$$R = \tfrac{1}{L} \text{ is the radius of convergence of the power series.}$$

Our discussion implies that if $L = 0$, then $R = \infty$, and if $L = +\infty$, then $R = 0$. Since the decisive aspect is the limit, observe that when it comes to the radius of convergence, it does not matter whether the power series is started at $k = 0$, or $k = 1$, or $k = 100$. With regard to the condition that the coefficients a_k be nonzero, it therefore suffices that $a_k \neq 0$ for all k greater than some positive integer N.

Example 11.17. Consider the power series $\frac{1}{3}x + \frac{2}{9}x^2 + \frac{3}{27}x^3 + \frac{4}{81}x^4 + \dots$. Expressed in sigma notation, this is the power series $\sum_{k=1}^{\infty}\frac{k}{3^k}x^k$ with center $x_0 = 0$. Let's turn to the ratio test. The nonzero coefficient assumption is met for $N = 1$. The relevant sequence of ratios is

$$\frac{\frac{2}{(3^2)}}{\frac{1}{3}} = \frac{2}{3^2}\cdot\frac{3}{1} = \frac{2}{3}, \frac{\frac{3}{(3^3)}}{\frac{2}{(3^2)}} = \frac{3}{3^3}\cdot\frac{3^2}{2} = \frac{1}{2}, \frac{\frac{4}{(3^4)}}{\frac{3}{(3^3)}} = \frac{4}{3^4}\cdot\frac{3^3}{3} = \frac{4}{9}, \dots, \frac{\frac{k+1}{(3^{k+1})}}{\frac{k}{(3^k)}} = \frac{k+1}{3^{k+1}}\cdot\frac{3^k}{k} = \frac{1}{3}(1+\frac{1}{k}), \dots.$$

Since $L = \lim_{k\to\infty}\frac{a_{k+1}}{a_k} = \lim_{k\to\infty}\frac{1}{3}(1+\frac{1}{k}) = \frac{1}{3}$, it follows that the radius of convergence of the power series in $R = \frac{1}{L} = 3$. Therefore $\sum_{k=1}^{\infty}\frac{k}{3^k}x^k$ converges for any x with $-3 < x < 3$.

Example 11.18. Turn next to the power series $\frac{1}{8}x^3 + \frac{2}{8^2}x^6 + \frac{3}{8^3}x^9 + \frac{4}{8^4}x^{12} + \dots$. In reference to the notation $\sum_{k=0}^{\infty} a_k(x-x_0)^k$, notice that $x_0 = 0$. But notice also that $a_0 = a_1 = a_2 = 0, a_4 = a_5 = 0, a_7 = a_8 = 0$, and so on. It follows that the ratio test does not apply. However, the power series can be rewritten as

$$\sum_{j=1}^{\infty}\frac{j}{8^j}x^{3j}.$$

Now all the coefficients are nonzero and the absolute values of the ratios of the coefficients are

$$\frac{\frac{2}{(8^2)}}{\frac{1}{8}} = \frac{1}{8}\cdot 2, \frac{\frac{3}{(8^3)}}{\frac{2}{(8^2)}} = \frac{1}{8}\cdot\frac{3}{2}, \frac{\frac{4}{(8^4)}}{\frac{3}{(8^3)}} = \frac{1}{8}\cdot\frac{4}{3}, \dots, \frac{\frac{j+1}{(8^{j+1})}}{\frac{j}{(8^j)}} = \frac{1}{8}(1+\frac{1}{j}), \dots.$$

Putting in the variable terms, we get

$$\frac{\frac{2}{(8^2)}x^6}{\frac{1}{8}x^3} = \frac{1}{8}\cdot 2x^3, \frac{\frac{3}{(8^3)}x^9}{\frac{2}{(8^2)}x^6} = \frac{1}{8}\cdot\frac{3}{2}x^3, \frac{\frac{4}{(8^4)}x^{12}}{\frac{3}{(8^3)}x^9} = \frac{1}{8}\cdot\frac{4}{3}x^3, \dots, \frac{\frac{j+1}{(8^{j+1})}x^{3(j+1)}}{\frac{j}{(8^j)}x^{3j}} = \frac{1}{8}(1+\frac{1}{j})x^3, \dots.$$

The limit of this sequence is $\frac{1}{8}x^3$. As in the earlier discussion about the ratio test, the power series $\sum_{i=j}^{\infty}\frac{j}{8^j}x^{3j}$ converges when $|\frac{1}{8}x^3| < 1$ and diverges when $|\frac{1}{8}x^3| > 1$. After multiplying through by 8 and then taking cube roots, we get that this power series converges for $|x| < 2$ and diverges for $|x| > 2$. Therefore its radius of convergence is 2.

Let $\sum_{k=0}^{\infty} a_k(x-x_0)^k$ be a power series with a positive radius of convergence R. In view of the fact that the series converges for any number x with $|x-x_0| < R$, we can *define a function* $y = f(x)$ by the rule

$$f(x) = \sum_{k=0}^{\infty} a_k(x - x_0)^k$$

for any such x.

Example 11.19. Consider the geometric series $\sum_{k=0}^{\infty} x^k$. We saw earlier in this section that its radius of convergence is 1 and that it converges to $\frac{1}{1-x}$ for all x in its interval of convergence $(-1, 1)$. Therefore the function that this power series defines for all such x is $f(x) = \frac{1}{1-x}$. Notice that x is in $(-1, 1)$ precisely when $-x$ is in $(-1, 1)$. It follows that the power series $\sum_{k=0}^{\infty}(-x)^k = \sum_{k=0}^{\infty}(-1)^k x^k$ also has interval of convergence $(-1, 1)$. The function that it defines is $\frac{1}{1+x}$ for all x in $(-1, 1)$.

Theorem 4. Suppose that the power series $\sum_{k=0}^{\infty} a_k(x - x_0)^k$ has a positive radius of convergence R, and consider the function defined by

$$f(x) = \sum_{k=0}^{\infty} a_k(x - x_0)^k$$

for all x with $|x - x_0| < R$. Then the two power series $\sum_{k=1}^{\infty} ka_k(x - x_0)^{k-1}$ and $\sum_{k=0}^{\infty} \frac{a_k}{k+1}(x - x_0)^{k+1}$ obtained by taking derivatives and antiderivatives term by term both have radius of convergence R, and for all x with $|x - x_0| < R$,

$$f'(x) = \sum_{k=1}^{\infty} ka_k(x - x_0)^{k-1} \quad \text{and} \quad \int f(x)\,dx = \sum_{k=0}^{\infty} \frac{a_k}{k+1}(x - x_0)^{k+1} + C.$$

Theorem 4 starts with a power series, uses it to define a function, and then asserts that the derivative and the antiderivative of the function are represented, respectively, by the power series obtained by differentiating and antidifferentiating the original power series term by term.

To illustrate Theorem 4, consider the fact that the power series series $\sum_{k=0}^{\infty} x^k$ and $\sum_{k=0}^{\infty}(-1)^k x^k$ define the functions $f(x) = \frac{1}{1-x}$ and $g(x) = \frac{1}{1+x}$, respectively, both for all x in the interval of convergence $(-1, 1)$ that they share. Since $f'(x) = \frac{1}{(1-x)^2}$, Theorem 4 applied to the power series for $f(x) = \frac{1}{1-x}$, tells us that

$$\frac{1}{(1-x)^2} = \sum_{k=1}^{\infty} kx^{k-1} = 1 + 2x + 3x^2 + 4x^3 + \dots$$

for all x in $(-1, 1)$. By a fact from Section 7.11, $\frac{d}{dx}\ln x = \frac{1}{x}$, so that by the chain rule, $y = \ln(1 + x)$ is an antiderivative of $g(x) = \frac{1}{1+x}$. Therefore by an application of Theorem 4 to $g(x) = \frac{1}{1+x}$,

$$\ln(1 + x) = \sum_{k=0}^{\infty} \frac{(-1)^k}{k+1}\, x^{k+1} = x - \tfrac{1}{2}x^2 + \tfrac{1}{3}x^3 - \tfrac{1}{4}x^4 + \dots$$

for all x in $(-1, 1)$.

The fact that the derivatives $f^{(k)}(x)$ of the function $f(x) = 0$ are equal to zero for all $k \geq 1$ and the repeated application of the "derivative part" of Theorem 4, provide the following corollary.

Corollary. If $\sum_{k=0}^{\infty} a_k(x - x_0)^k$ converges to 0, for all x in some open interval containing x_0, then all the coefficients a_k are equal to zero.

11.8 TAYLOR AND MACLAURIN SERIES

Power series are extended polynomials, and polynomials are relatively easy to study. The expression of a complicated function in terms of a converging power series is a powerful tool with which the analysis of the complicated function can often proceed. Initial segments of the power series of such an expression are polynomials that approximate the function. The larger the degree of the polynomial of such an initial segment, the better the approximation. The study of the main cable of a suspension bridge in Section 10.2 provided a concrete example. In the previous section, we saw that a given converging power series defines a function. What is needed in the current context is a mechanism that goes the other way. If power series are to be generally useful for the pursuits of calculus—for solving indefinite integrals and differential equations for instance—then there needs to be a method for determining converging power series for a broad range of functions. Such a method is the focus of the upcoming discussion.

Let $y = f(x)$ be a function defined on an open interval I, and assume that the function can be differentiated again and again—infinitely many times—for all x in I. For a fixed x_0 in I, let $f^{(0)}(x_0) = f(x_0), f^{(1)}(x_0) = f'(x_0)$, and for $k = 2, \ldots,$ let $f^{(k)}(x_0)$ be the kth derivative of f evaluated at x_0. The *Taylor polynomial* of degree n of the function $y = f(x)$ *centered at* x_0 is defined by

$$\begin{aligned} T_n(x) &= \sum_{k=0}^{n} \frac{f^{(k)}(x_0)}{k!}(x - x_0)^k \\ &= f(x_0) + f'(x_0)(x - x_0) + \tfrac{f^{(2)}(x_0)}{2!}(x - x_0)^2 + \cdots + \tfrac{f^{(n)}(x_0)}{n!}(x - x_0)^n. \end{aligned}$$

The *Taylor series* of the function $y = f(x)$ centered at x_0 is the power series defined by

$$\begin{aligned} T_\infty(x) &= \sum_{k=0}^{\infty} \frac{f^{(k)}(x_0)}{k!}(x - x_0)^k \\ &= f(x_0) + f'(x_0)(x - x_0) + \tfrac{f^{(2)}(x_0)}{2!}(x - x_0)^2 + \cdots + \tfrac{f^{(k)}(x_0)}{k!}(x - x_0)^k + \cdots. \end{aligned}$$

This important construction is named after Brook Taylor (1685–1731), an English mathematician, who published it along with other insights (including the method of integration by parts) in his *Methodus Incrementorum Directa et Inversa* (London, 1715). The full recognition that Taylor's construction was a fundamental aspect of differential and integral calculus came only much later in the 18th century. In the special case $x_0 = 0$, Taylor polynomials and the Taylor series are known as Maclaurin polynomials and Maclaurin series. Colin Maclaurin (1698–1746), a Scotsman, included their development in his two-volume *Treatise of Fluxions* of 1742. This was a systematic exposition of Newton's methods written in response to Bishop Berkeley's charge (see the introduction to Chapter 7) that calculus lacked a rigorous foundation. In his work, Maclaurin acknowledged the earlier and more general theory of Taylor.

Let's look at the simple example of the polynomial function $f(x) = 2x^3 - 5x^2 + 4x + 7$. Since

$$f'(x) = 6x^2 - 10x + 4,\ f^{(2)}(x) = 12x - 10,\ f^{(3)}(x) = 12, \text{ and } f^{(k)}(x) = 0 \text{ for all } k \geq 4,$$

we see that $f(0) = 7, f'(0) = 4, f^{(2)}(0) = -10, f^{(3)}(0) = 12$, and $f^{(k)}(0) = 0$ for all $k \geq 4$. Therefore the Taylor series of $f(x) = 2x^3 - 5x^2 + 4x + 7$ centered at $x_0 = 0$,

$$7 + 4(x - 0) - \tfrac{10}{2!}(x - 0)^2 + \tfrac{12}{3!}(x - 0)^3 = 7 + 4x - 5x^2 + 2x^3,$$

is the polynomial we started with. It's not hard to see that it is always the case that the Taylor series centered at $x_0 = 0$ of any polynomial is that polynomial. Since $f(-1) = -4, f'(-1) = 20, f^{(2)}(-1) = -22, f^{(3)}(-1) = 12$, and $f^{(k)}(0) = 0$ for all $k \geq 4$, the Taylor series of $f(x) = 2x^3 - 5x^2 + 4x + 7$ centered at $x_0 = -1$ is

$$-4 + 20(x + 1) - 11(x + 1)^2 + 2(x + 1)^3.$$

Simplifying this expression shows that this is merely a rewritten version of $f(x) = 2x^3 - 5x^2 + 4x + 7$.

Example 11.20. Let $f(x) = e^x$. Take $I = (-\infty, \infty)$ and $x_0 = 0$. Since $f^{(k)} = e^x$ for all $k \geq 1$, we know that $f(0) = f^{(k)}(0) = e^0 = 1$ for all $k \geq 1$. So the Taylor polynomial of degree n of $f(x) = e^x$ centered at $x_0 = 0$ is

$$T_n(x) = \sum_{k=0}^{n} \tfrac{1}{k!}\, x^k = 1 + x + \tfrac{1}{2!}x^2 + \cdots + \tfrac{1}{n!}x^n,$$

and the Taylor series of $f(x) = e^x$ centered at $x_0 = 0$ is

$$T_\infty(x) = \sum_{k=0}^{\infty} \tfrac{1}{k!}\, x^k = 1 + x + \tfrac{1}{2!}x^2 + \cdots + \tfrac{1}{k!}x^k + \ldots.$$

Example 11.21. Let r be a constant, and consider the function $f(x) = (1 + x)^r$. Keep taking derivatives to get

$$f^{(1)}(x) = f'(x) = r(1+x)^{r-1}, f^{(2)}(x) = r(r-1)(1+x)^{r-2}, f^{(3)}(x) = r(r-1)(r-2)(1+x)^{r-3},$$
$$f^{(4)}(x) = r(r-1)(r-2)(r-3)(1+x)^{r-4}, \ldots, f^{(k)}(x) = r(r-1)(r-2)\ldots(r-(k-1))x^{r-k} + \ldots.$$

By evaluating $f(x)$ and these derivatives at $x = 0$,

$$f(0) = 1,\ f^{(1)}(0) = r,\ f^{(2)}(0) = r(r-1),\ f^{(3)}(0) = r(r-1)(r-2), \ldots,\ f^{(k)}(0) = r(r-1)\ldots(r-(k-1)) + \ldots,$$

and it follows that the Taylor series of the function $f(x) = (1 + x)^r$ centered at $x_0 = 0$ is nothing but the binomial series

$$1 + rx + \tfrac{r(r-1)}{2!}x^2 + \tfrac{r(r-1)(r-2)}{3!}x^3 + \cdots + \tfrac{r(r-1)(r-2)\cdots(r-(k-1))}{k!}x^k + \ldots$$

that was already considered in the Problems and Projects section for Chapter 6. The case $r = \frac{1}{2}$ was applied in Section 10.2 in the study of the cable of a suspension bridge. Expressed in sigma notation, the binomial series is

$$\sum_{k=0}^{\infty} \binom{r}{k} x^k = \binom{r}{0} + \binom{r}{1} x + \cdots + \binom{r}{k} x^k + \cdots,$$

where $\binom{r}{0} = 1$ and $\binom{r}{k} = \frac{r(r-1)\ldots(r-(k-1))}{k!}$ for any $k \geq 1$. We will have a careful look at the binomial series in the Problems and Projects section of this chapter.

Example 11.22. Consider the function $f(x) = \sin x$. Show that its Taylor series centered at $x_0 = 2$ is

$$T_\infty(x) = \sin 2 + (\cos 2)(x-2) - \tfrac{\sin 2}{2!}(x-2)^2 - \tfrac{\cos 2}{3!}(x-2)^3 + \tfrac{\sin 2}{4!}(x-2)^4 + \tfrac{\cos 2}{5!}(x-2)^5 - \ldots.$$

Study the pattern, and conclude that the term with exponent k is given by $(-1)^j \cdot \frac{\sin 2}{k!}(x-2)^k$ if $k = 2j$ is even and $(-1)^j \cdot \frac{\cos 2}{k!}(x-2)^k$ if $k = 2j+1$ is odd. Consider the Taylor polynomials $T_5(x)$ (this is the sum of terms from $k = 0$ to $k = 5$ already listed above), as well as $T_8(x), T_{11}(x), T_{14}(x), T_{17}(x)$, and $T_{20}(x)$. Write down the polynomial $T_8(x)$ by following the pattern and without doing any further computations. Figure 11.18 shows the graphs of all six of these Taylor polynomials. The point $(2, \sin 2)$ determined by the center $x_0 = 2$ of the Taylor expansion is highlighted in each case. Notice that their graphs hug the graph of $f(x) = \sin x$ tightly over wider and wider intervals before they zoom off the charts. There is every reason to believe that the pattern of the closer and closer approximation of the graph of $f(x) = \sin x$ will continue and, in particular, that the Taylor series $T_\infty(x)$ is equal to the function $f(x) = \sin x$ over the entire $-\infty < x < \infty$. We'll turn to this issue next.

The collection of the library of functions that a calculus course builds and uses includes algebraic functions (as given by polynomials and roots), trigonometric functions, exponential functions, hyperbolic functions, and the inverses of these functions (including log functions), as well as the sums, differences, products, quotients, and the composites of all these functions. But can it happen that a function required for the solution of an indefinite integral or a differential equation is not on the shelves of this collection? Is

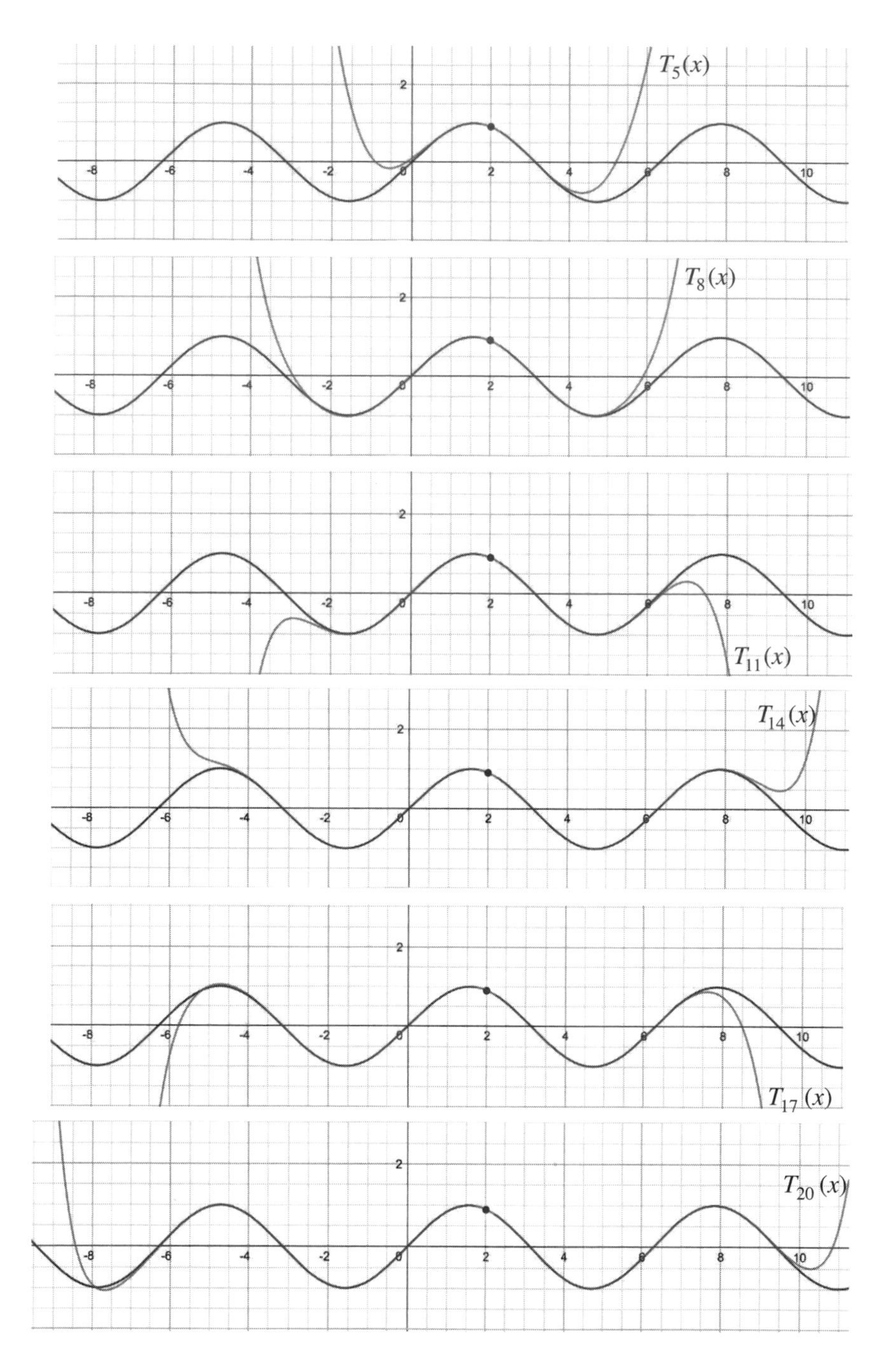

Fig. 11.18. These graphs of Taylor polynomials centered at $x_0 = 2$ for the function $f(x) = \sin x$ were produced by the graphing calculator of the website https://www.desmos.com/calculator/hmmutc9ija.

it then the case that the integral or differential equation cannot be solved? The answers are "yes" to the first question (as already discussed at the end of Section 9.11), and "power series provide a possible approach" to the second. Theorem 4 informed us that functions defined by power series are as easy to differentiate and integrate as polynomials. The discussion above told us that any function that can be differentiated infinitely many times has a Taylor series. In order for the promise of power series to be realized, it will be important to close the loop and to understand when a Taylor series converges to the function that gives rise to it. We know that the ratio test provides effective information about the range of x for which a power series converges. But again: will a Taylor series converge to the function that generates it?

The initial answer to this question is no! Consider the function $y = f(x)$ defined by

$$f(x) = \begin{cases} e^{-1/x^2}, \text{ if } x \neq 0, \text{ and} \\ 0, \text{ if } x = 0 \end{cases}$$

It turns out that this function is differentiable for all x. It is also the case that $0 \leq f(x) < 1$ for all x. The fact that $\lim_{x\to+\infty} \frac{1}{x^2} = 0$ and $\lim_{x\to-\infty} \frac{1}{x^2} = 0$ tells us that $y = 1$ is a horizontal asymptote of the graph both on the right and the left of the y-axis. What is important about this function in the current context is that it is so flat near the origin (see Figure 11.19) that all of its derivatives $f^{(1)}(0), f^{(2)}(0), f^{(3)}(0), \ldots$ at 0 are equal

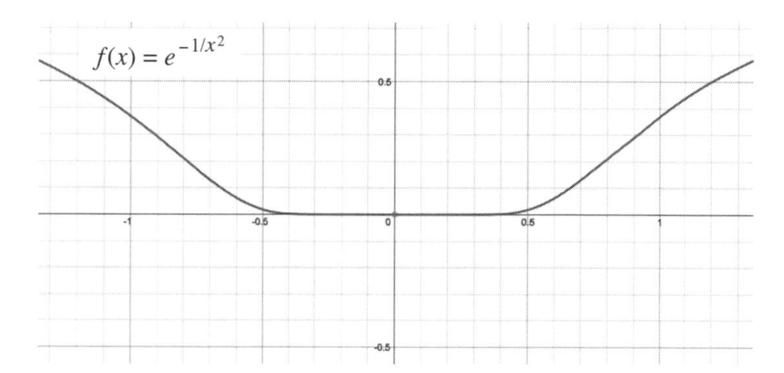

Fig. 11.19. The graph of the function $f(x) = e^{-1/x^2}$ for $x \neq 0$ and $f(0) = 0$ was produced with the Desmos graphing calculator of the website https://www.desmos.com/calculator.

to 0. (The proof of this is technical, and we'll forgo it.) This means that all the coefficients of the Taylor series of this function are 0. Therefore this function and the function defined by its Taylor series agree only at the point $x = 0$. This appears to be a serious setback to the goal that has been set out. So now what?

Let $y = f(x)$ be a function that is infinitely differentiable in an open interval I, and let

$$T_n(x) = \sum_{k=0}^{n} \tfrac{f^{(k)}(x_0)}{k!}(x - x_0)^k = f(x_0) + f'(x_0)(x - x_0) + \tfrac{f^{(2)}(x_0)}{2!}(x - x_0)^2 + \cdots + \tfrac{f^{(n)}(x_0)}{n!}(x - x_0)^n$$

be its nth Taylor polynomial centered at x_0 in I. The function determined by the difference $R_n(x) = f(x) - T_n(x)$ is the nth remainder. Observe that if $\lim_{n\to\infty} R_n(x) = 0$, then $f(x) = T_\infty(x)$, so that the value of the Taylor series at x is $f(x)$.

Theorem 5 (Taylor's Remainder Theorem). Suppose that the function $y = f(x)$ and its first $n + 1$ derivatives exist for all x in an open interval I containing x_0 and that $f^{(n+1)}(x)$ is continuous for all x in I. Then for any x in I there is a value z (depending on n) between x and x_0, such that

$$\boxed{R_n(x) = \frac{f^{(n+1)}(z)}{(n+1)!}(x - x_0)^{n+1}}$$

As Brook Taylor himself would recall later, his remainder theorem was motivated by coffeehouse conversations about works of Newton on planetary motion and works of Halley on roots of polynomials. (This is the Halley of comet fame, whom we met in the introduction to Chapter 6.) Instead of proving Taylor's theorem, we'll apply it to show that the Taylor series of each of the functions e^x, $\sin x$, and $\cos x$ (centered at $x_0 = 0$, and in one case at $x_0 = 2$) converge to the corresponding function for all x.

We start by verifying that for any fixed real number x, $\lim_{n\to\infty}\frac{x^n}{n!} = 0$. This can be established as follows. Take the smallest integer M such that $M \geq |x|$. Now let $n \geq M+2$ be any positive integer. Since $n-1, \ldots, M+1$ are all greater than $|x|$, the product $\frac{|x|}{n-1}\cdots\frac{|x|}{M+1}$ is less than 1. Therefore

$$0 \leq \frac{|x|^n}{n!} = \frac{|x|}{n}\cdot\left(\frac{|x|}{n-1}\cdots\frac{|x|}{M+1}\right)\cdot\frac{|x|}{M}\cdot\frac{|x|}{M-1}\cdots\frac{|x|}{2}\cdot\frac{|x|}{1} < \frac{|x|}{n}\cdot\frac{|x|^M}{M!}.$$

Since x is fixed, $\frac{|x|^M}{M!}$ is constant. Since $\lim_{n\to\infty}\frac{|x|}{n} = 0$, it follows that $\lim_{n\to\infty}\frac{|x|^n}{n!} = 0$. So $\lim_{n\to\infty}\frac{x^n}{n!} = 0$ as well.

Example 11.23. Let $f(x) = e^x$. The Taylor series of e^x centered at $x_0 = 0$ satisfies

$$1 + x + \frac{1}{2!}x^2 + \cdots + \frac{1}{k!}x^k + \ldots = e^x$$

for all x in $I = (-\infty, \infty)$. This follows quickly from Taylor's remainder theorem. Let x be in I. By the theorem, $R_n(x) = \frac{e^z}{(n+1)!}x^{n+1}$, where z lies between 0 and x (and depends on n). If $x \geq 0$, then $0 \leq z \leq x$, so that $e^z \leq e^x$. Therefore $0 \leq R_n(x) = \frac{e^z}{(n+1)!}x^{n+1} \leq \frac{e^x}{(n+1)!}x^{n+1}$. Using the fact that $\lim_{n\to\infty}\frac{x^n}{n!} = 0$, we can conclude that $\lim_{n\to\infty} R_n(x) = 0$. If $x < 0$, then $x \leq z \leq 0$, so $e^z \leq e^0 = 1$ and $0 \leq |R_n(x)| \leq \frac{|x|^{n+1}}{(n+1)!}$. Again, $\lim_{n\to\infty} R_n(x) = 0$. Therefore, in either case, $\lim_{n\to\infty} R_n(x) = 0$, and we are done.

Example 11.24. Let $f(x) = \sin x$. The Taylor series of $\sin x$ centered at $x_0 = 2$ satisfies

$$\sin 2 + (\cos 2)(x-2) - \frac{\sin 2}{2!}(x-2)^2 - \frac{\cos 2}{3!}(x-2)^3 + \frac{\sin 2}{4!}(x-2)^4 + \ldots = \sin x$$

for all x in $(-\infty, \infty)$. It is easy to see that $f^{(k)}(x) = (-1)^j \sin x$ if $k = 2j$ is even, and $f^{(k)}(x) = (-1)^j \cos x$ if $k = 2j+1$ is odd. Therefore $R_n(x) = (-1)^j\frac{\sin z}{(n+1)!}(x-2)^{n+1}$ if $n+1 = 2j$ is even and $R_n(x) = (-1)^j\frac{\cos z}{(n+1)!}(x-2)^{n+1}$ if $n+1 = 2j+1$ is odd. In each case, z is between the given x and 2. Since both $|\sin z| \leq 1$ and $|\cos z| \leq 1$ for any z, it follows that in either case

$$|R_n(x)| \leq \left|\frac{(x-2)^{n+1}}{(n+1)!}\right|.$$

Since $\lim_{n\to\infty}\frac{x^n}{n!} = 0$ for any x, we see that $\lim_{n\to\infty} R_n(x) = 0$ for any x. Therefore the Taylor series of $f(x) = \sin x$ centered at $x_0 = 2$ converges to $\sin x$ for any real number x. The sequence of graphs of Figure 11.18 already provided pictorial evidence for this conclusion.

Example 11.25. Show that the Taylor series of $f(x) = \cos x$ centered at $x_0 = 0$ is

$$1 - \frac{x^2}{2!} + \frac{x^4}{4!} - \frac{x^6}{6!} + \frac{x^8}{8!} - \ldots.$$

Check that the general term of the series is $(-1)^k\frac{x^{2k}}{(2k)!}$. The remainder term $R_n(x)$ for any n is equal to $R_n(x) = \frac{f^{(n+1)}(z)}{(n+1)!}x^{n+1}$. Since $f^{(n+1)}(z) = \pm\sin z$ or $f^{(n+1)}(z) = \pm\cos z$, it follows that $|f^{(n+1)}(z)| \leq 1$. Hence

$$|R_n(x)| \leq \left|\frac{x^{n+1}}{(n+1)!}\right|.$$

Therefore, as in the two previous examples, $\lim_{n\to\infty} R_n(x) = 0$. So the Taylor series of $f(x) = \cos x$ centered at $x_0 = 0$ converges to $\cos x$ for all x in the interval $(-\infty, \infty)$.

Suppose that a function $y = f(x)$ is defined by a power series with center $x = x_0$ and a positive radius of convergence R. We will close our introduction to power series by showing that this power series must be the Taylor series with center $x = x_0$. So let $y = f(x)$ be defined by

$$f(x) = \sum_{k=0}^{\infty} a_k(x-x_0)^k = a_0 + a_1(x-x_0) + a_2(x-x_0)^2 + a_3(x-x_0)^3 + \ldots,$$

where the power series has a positive radius of convergence R. Notice that $f(x_0) = a_0$. Repeatedly using

Theorem 4, we get

$$f'(x) = \sum_{k=1}^{\infty} ka_k(x-x_0)^{k-1},\ f^{(2)}(x) = \sum_{k=2}^{\infty}(k-1)ka_k(x-x_0)^{k-2},\ldots,$$
$$f^{(n)}(x) = \sum_{k=n}^{\infty}(k-(n-1))\cdots(k-1)ka_k(x-x_0)^{k-n},\ldots,$$

with each equality valid for all x satisfying $x_0 - R < x < x_0 + R$. Since the first coefficient of each of these power series is obtained by evaluating it at $x = x_0$, it follows that

$$f'(x_0) = a_1, f^{(2)}(x_0) = (1\cdot 2)a_2, f^{(3)}(x_0) = (1\cdot 2\cdot 3)a_3,\ldots, f^{(n)}(x_0) = 1\cdot 2\cdot 3\cdots(n-1)na_n,\ldots.$$

The fact that $a_0 = f(x_0)$, $a_1 = f'(x_0)$, $a_2 = \frac{f^{(2)}(x_0)}{2}$, $a_3 = \frac{f^{(3)}(x_0)}{3!}, \ldots, a_n = \frac{f^{(n)}(x_0)}{n!}, \ldots$ tells us that the power series we started with is nothing but the Taylor series centered at $x = x_0$ of the function that the power series defines.

11.9 SOLVING A SECOND-ORDER DIFFERENTIAL EQUATION

We'll now illustrate how power series can be used to solve differential equations that may be beyond the reach of other approaches. Let's analyze the second-order equation

$$y'' + x^2y' + 4xy = 0.$$

We will be looking for a solution $y = f(x)$ that satisfies $f(0) = 0$. We'll start with the assumption that there is a solution $y = f(x)$ defined by a power series

$$y = f(x) = \sum_{k=0}^{\infty} a_k x^k$$

with center $x_0 = 0$ and positive radius of convergence R. Then we'll use the constraints that the differential equation provides to explicitly determine the power series. The assumption $f(0) = 0$ tells us that $a_0 = 0$. By Theorem 4,

$$y' = f'(x) = \sum_{k=1}^{\infty} ka_k x^{k-1} \quad\text{and}\quad y'' = f''(x) = \sum_{k=2}^{\infty}(k-1)ka_k x^{k-2},$$

where the radius of convergence of both of these power series is R. Notice that $f'(0) = a_1$. After substituting the power series for y, y', and y'' into the differential equation, we get

$$\sum_{k=2}^{\infty}(k-1)ka_k x^{k-2} + \sum_{k=1}^{\infty} ka_k x^{k+1} + \sum_{k=0}^{\infty} 4a_k x^{k+1} = 0.$$

The only constant term in the expression above arises in the first series with $k = 2$. Since this term is equal to $1\cdot 2\cdot a_2 = 2a_2$, it follows (by taking $x = 0$ in the expression) that $a_2 = 0$. We'll now rewrite the first series

$$\sum_{k=2}^{\infty}(k-1)ka_k x^{k-2} = \sum_{k=3}^{\infty}(k-1)ka_k x^{k-2}$$

with a different summation index in order to align it with the other two series. With $k = i+3$, this rewritten version of the series is $\sum_{i=0}^{\infty}(i+2)(i+3)a_{i+3}x^{i+1}$. Now change the notation for the summation index of this series back from i to k, and write the sum of the three series above as

$$\sum_{k=0}^{\infty}(k+2)(k+3)a_{k+3}x^{k+1} + \sum_{k=1}^{\infty} ka_k x^{k+1} + \sum_{k=0}^{\infty} 4a_k x^{k+1} = 0.$$

Rearranging the first series and using the fact that in the third series $a_0 = 0$, we get

$$6a_3x + \sum_{k=1}^{\infty}(k+2)(k+3)a_{k+3}x^{k+1} + \sum_{k=1}^{\infty} ka_kx^{k+1} + \sum_{k=1}^{\infty} 4a_kx^{k+1} = 0,$$

and therefore

$$6a_3x + \sum_{k=1}^{\infty}[(k+2)(k+3)a_{k+3} + (k+4)a_k]x^{k+1} = 0.$$

Since this equality holds for all x with $|x| < R$, this power series has a positive radius of convergence, so that by an application of the corollary of Theorem 4 of Section 11.7, a_3 as well as the coefficients $(k+2)(k+3)a_{k+3} + (k+4)a_k$ for $k \geq 1$ are all equal to 0. So $a_3 = 0$ and

$$a_{k+3} = \frac{-(k+4)}{(k+2)(k+3)}\, a_k \text{ for all } k \geq 1.$$

Since $a_2 = 0$ and $a_3 = 0$, it follows from this formula that

$$a_5 = 0, a_8 = 0, a_{11} = 0, \ldots \text{ and } a_6 = 0, a_9 = 0, a_{12} = 0, \ldots.$$

So all coefficients of the power series $\sum_{k=0}^{\infty} a_kx^k$ are 0 except possibly $a_1, a_4, a_7, a_{10}, \ldots$ and these are of the form a_{3j+1} for $j = 0, 1, 2, \ldots$. So the power series for $y = f(x)$ that we started with has the form

$$f(x) = \sum_{j=0}^{\infty} a_{3j+1}x^{3j+1}.$$

By repeatedly using the link between a_{k+3} and a_k, we get

$$a_4 = \tfrac{-5}{3\cdot 4}\, a_1,\ a_7 = \tfrac{-8}{6\cdot 7}\, a_4,\ a_{10} = \tfrac{-11}{9\cdot 10}\, a_7, \ldots,$$

and (by observing the unfolding pattern) that for any $j \geq 0$,

$$a_{3(j+1)+1} = \frac{-(3(j+1)+2)}{3(j+1)(3(j+1)+1)}\, a_{3(j+1)-2} = \frac{-(3j+5)}{(3j+3)(3j+4)}\, a_{3j+1}.$$

Because $\left|\frac{a_{3(j+1)+1}x^{3(j+1)+1}}{a_{3j+1}x^{3j+1}}\right| = \left|\frac{-(3j+5)}{(3j+3)(3j+4)}\, x^3\right| = \left|\frac{3j+5}{9j^2+21j+12}\, x^3\right| = \left|\frac{3+\frac{5}{j}}{9j+21+\frac{12}{j}}\, x^3\right|$ (for the last step, take $j > 0$ and divide both the numerator and denominator by j) and $\lim_{j\to\infty}\left|\frac{3+\frac{5}{j}}{9j+21+\frac{12}{j}}\, x^3\right| = 0$, the radius of convergence of $\sum_{j=0}^{\infty} a_{3j+1}x^{3j+1}$ is infinite by the ratio test of Section 11.7. Therefore $y = f(x)$ is defined for all x in $(-\infty, \infty)$.

We have all the information we need to determine the coefficient a_{3j+1} for any j. From our discussion,

$$a_4 = \tfrac{-5}{3\cdot 4}\, a_1 = \tfrac{-2\cdot 5\cdot 5}{2\cdot 3\cdot 4\cdot 5} a_1 = \tfrac{(-1)2\cdot 5^2}{5!} a_1,\ a_7 = \tfrac{(-1)8}{6\cdot 7} a_4 = \tfrac{(-1)2\cdot 5^2}{5!}\tfrac{(-1)8}{6\cdot 7} a_1 = \tfrac{2\cdot 5^2 8^2}{8!} a_1,$$

$$a_{10} = \tfrac{(-1)11}{9\cdot 10} a_7 = \tfrac{(-1)2\cdot 5^2 8^2 11^2}{11!} a_1, a_{13} = \tfrac{(-1)14}{12\cdot 13} a_{10} = \tfrac{2\cdot 5^2 8^2 11^2 14^2}{14!} a_1,\ a_{16} = \tfrac{(-1)17}{15\cdot 16}\, a_{13} = \tfrac{(-1)2\cdot 5^2 8^2 11^2 14^2 17^2}{17!} a_1,$$

and so on. Look at the pattern, and notice that for $j = 1, 2, 3, 4$ and 5, the term a_{3j+1} is given by the formula

$$a_{3j+1} = \frac{(-1)^j 2\cdot 5^2 \cdots (3j+2)^2}{(3j+2)!}\, a_1.$$

Assuming that this formula holds for a_{3j+1}, we now get that

$$a_{3(j+1)+1} = \frac{(-1)(3j+5)}{(3j+3)(3j+4)}\, a_{3j+1} = \frac{(-1)^j 2\cdot 5^2 \cdots (3j+2)^2}{(3j+2)!}\frac{(-1)(3j+5)}{(3j+3)(3j+4)}\, a_1 = \frac{(-1)^{(j+1)} 2\cdot 5^2 \cdots (3j+2)^2)(3j+5)^2}{(3j+5)!}\, a_1,$$

and therefore by the principle of induction that the formula holds for all a_{3j+1}. Therefore

$$f(x) = \sum_{j=0}^{\infty} a_{3j+1}x^{3j+1} = \sum_{j=0}^{\infty} \frac{(-1)^j 2\cdot 5^2 \cdots (3j+2)^2}{(3j+2)!} a_1 x^{3j+1},$$

where $a_1 = f'(0)$. By the discussion that concludes Section 11.8, we know that this is the Taylor series $T_\infty(x)$ centered at $x_0 = 0$ of the function $f(x)$.

To the initial condition $f(0) = 0$ already in place, we'll now add $a_1 = f'(0) = 1$. Insert $a_1 = 1$ into the

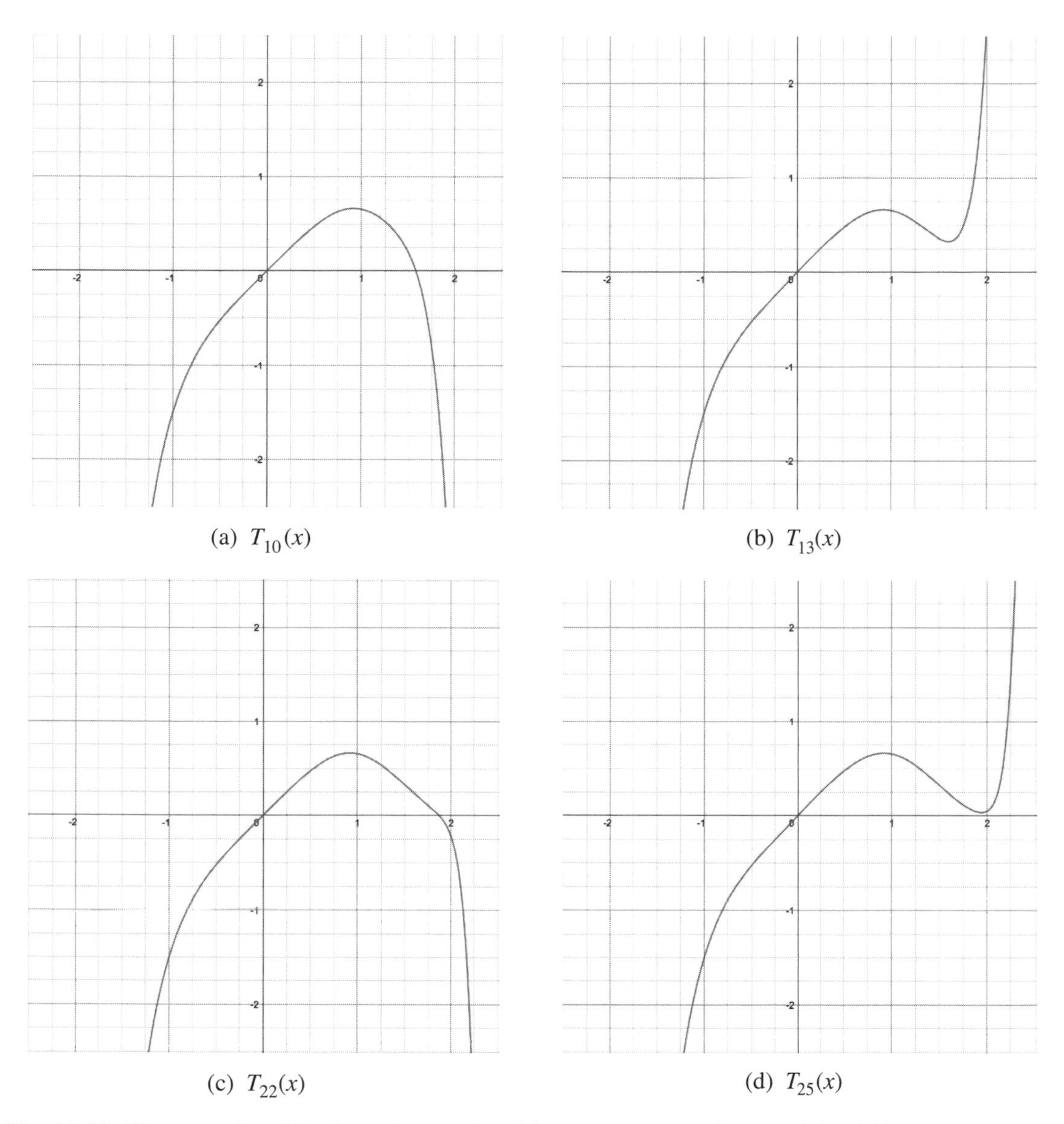

(a) $T_{10}(x)$ (b) $T_{13}(x)$ (c) $T_{22}(x)$ (d) $T_{25}(x)$

Fig. 11.20. These graphs of Taylor polynomials of the power series solution of the differential equation $y'' + x^2y' + 4xy = 0$ with initial conditions $y(0) = 0$ and $y'(0) = 1$ were produced with the graphing calculator available on the website https://www.desmos.com/calculator.

Taylor series above, and check that the Taylor polynomials $T_{10}(x), T_{13}(x), T_{22}(x)$, and $T_{25}(x)$ are

$$T_{10}(x) = x - \tfrac{5}{12}x^4 + \tfrac{5}{63}x^7 - \tfrac{11}{1134}x^{10}$$

$$T_{13}(x) = x - \tfrac{5}{12}x^4 + \tfrac{5}{63}x^7 - \tfrac{11}{1134}x^{10} + \tfrac{11}{12{,}636}x^{13}$$

$$T_{22}(x) = x - \tfrac{5}{12}x^4 + \tfrac{5}{63}x^7 - \tfrac{11}{1134}x^{10} + \tfrac{11}{12{,}636}x^{13} - \tfrac{11}{1{,}783{,}920}x^{16} + \tfrac{187}{51{,}858{,}144}x^{19} - \tfrac{391}{2{,}178{,}042{,}048}x^{22}, \text{ and}$$

$$T_{25}(x) = x - \tfrac{5}{12}x^4 + \tfrac{5}{63}x^7 - \tfrac{11}{1134}x^{10} + \tfrac{11}{12{,}636}x^{13} - \tfrac{11}{1{,}783{,}920}x^{16} + \tfrac{187}{51{,}858{,}144}x^{19} - \tfrac{391}{2{,}178{,}042{,}048}x^{22} + \tfrac{391}{50{,}262{,}508{,}800}x^{25}.$$

The evolving graphs of Figure 11.20 tell us that these last two Taylor polynomials approximate the solution $y = f(x)$ of the differential equation $y'' + x^2y' + 4xy = 0$ with initial conditions $f(0) = 0$ and $f'(0) = 1$ closely over the interval $-1 \le x \le 2$. Taylor polynomials of much higher degree would have to be considered before reliable information over larger intervals appears. See Figure 11.18 (in the context of the sine function) in this regard. The website

https://www.wolframalpha.com/examples/DifferentialEquations.html

can be used to provide more information about the equation $y'' + x^2y' + 4xy = 0$.

We now turn to some applications of the analysis of first- and second-order differential equations that has been carried out in this chapter.

11.10 FREE FALL WITH AIR RESISTANCE

This section undertakes a study of an object moving in a resisting medium. This is an extremely difficult mathematical problem that becomes tractable only after a number of simplifying assumptions are made. The first two of these are that the object is not self-propelled and that the medium is air. In particular, gravity and air resistance are the only forces on the object. This restricted problem still includes the motion of the artillery shell, the curving trajectory of a soccer ball propelled with heavy spin, the reentry of a space capsule into the atmosphere (after the thrusters are turned off), and the descent of a parachutist. Next we'll assume that spin has no influence on the motion and that the object has the shape of a smooth sphere (in recognition of the fact that air resistance affects a pointed artillery shell differently than a parachute). Finally, we'll assume that the object moves vertically (either upward or downward) in the direction of the force of gravity. Having stripped away enough of the complexity, we have arrived at a problem that is solvable with the methods that have been developed. Galileo already told us how gravity behaves so that the focus is on the resistance of the air. We know from experience (ever stretched your arm out of the window of a moving car?) how powerful the air's resisting force on an object can be. Our discussion makes use of the MKS system of units. Recall that its basic units include m = meters, kg = kilograms, s = seconds, and N = newtons.

Let's begin with a look at the factors that determine the impact of air resistance. There is the *density* ρ of the air that we'll take as $\rho = 1.22$ kg/m^3 (at standard atmospheric pressure at sea level and at 18°C or 65°F). Then there is its *viscosity* μ. This is a measure of the air's resistance to being displaced by an applied force. We'll take $\mu = 1.83 \times 10^{-5}$ kg/(m·s) (again at standard atmospheric pressure at sea level at 18°C). The *Reynolds number* Re combines the impact of ρ and μ, the radius r of the spherical object, and its speed v in the formula

$$\text{Re} = 2\tfrac{\rho}{\mu}rv \approx \left(\tfrac{2.44}{1.83} \times 10^5\right)rv \approx (1.33 \times 10^5)\,rv.$$

A study of the units involved tells us that they cancel out, so that the Reynolds number has no units. It is therefore a *dimensionless* number. Finally, there is the *drag coefficient* C_d, a dimensionless number that depends on Re. The important facts in the current context are that over the range $10^3 < \text{Re} < 2 \times 10^5$, the drag coefficient is close to being constant with $C_d \approx 0.4$, and that the resisting force or *drag* exerted by the air on the sphere is given by

$$D = \tfrac{1}{2}C_d\,\rho(\pi r^2)v^2 \approx 0.767\,r^2v^2 \text{ N}.$$

That the cross-sectional area πr^2 of the sphere should play a role is not surprising. (In the much more complicated situation of a falling, tumbling odd-shaped object, the cross-sectional area could be a complicated function of time.) An important aspect of the formula for the drag D is the assertion that it is proportional to v^2.

The Reynolds number is a measure of the way air streams around the moving sphere. For Re ≤ 2100, the flow is *laminar*, with the air opening in smooth layers around the sphere. When Re ≥ 4000, the flow is *turbulent*, moving chaotically and forming whirling eddies. In the gap $2100 \leq$ Re ≤ 4000, both laminar and turbulent flows occur. It is surprising—given the variable nature of the air flow—that the drag coefficient C_d remains essentially constant until Re reaches 2×10^5. In view of the relationship Re $\approx (1.33 \times 10^5)\,rv$, the range $10^3 <$ Re $< 2 \times 10^5$ of the Reynolds number at which the drag coefficient C_d is nearly constant, corresponds to the range $1.33 \times 10^{-2} < rv < 1.5 \times 10^2$ in m^2/s.

We know that the force of gravity acts downward on the sphere—we will now call it a ball—with magnitude mg, where m is the mass of the ball and g is the gravitational acceleration near the Earth's surface. At many locations in the Northern Hemisphere, $g \approx 9.81$ m/s^2 (equivalent to $g \approx 32$ feet/s^2). Since the gravitational force on the ball acts downward, it is equal to $-W = -mg$, where W is the ball's weight.

We now turn to the mathematics of the moving ball. Let y_0 be the height of the bottom of the ball above the ground at the beginning of its vertical motion at time $t = 0$. Since D acts against the motion, it acts downward when the ball is moving up and upward when the ball is on its way down. These two parts of the vertical motion of the ball split the study into two segments. Accordingly, for any time $t \geq 0$, we'll let the distance of the bottom of the ball above the ground be given by $y_\uparrow(t)$ when it is on its way up, and $y_\downarrow(t)$ when it is on its way down. Figure 11.21 captures the essence of the matter. It depicts the motion on the left and a diagram of forces on the right. The starting height of the ball is $y_\uparrow(0) = y_0$, and its initial velocity is v_0. Windstill conditions are assumed throughout. We'll set $s_\infty = \sqrt{\frac{2mg}{C_d \rho \pi r^2}}$. The parameter s_∞ simplifies the formulas that govern both the upward and downward motion of the ball (as well as their

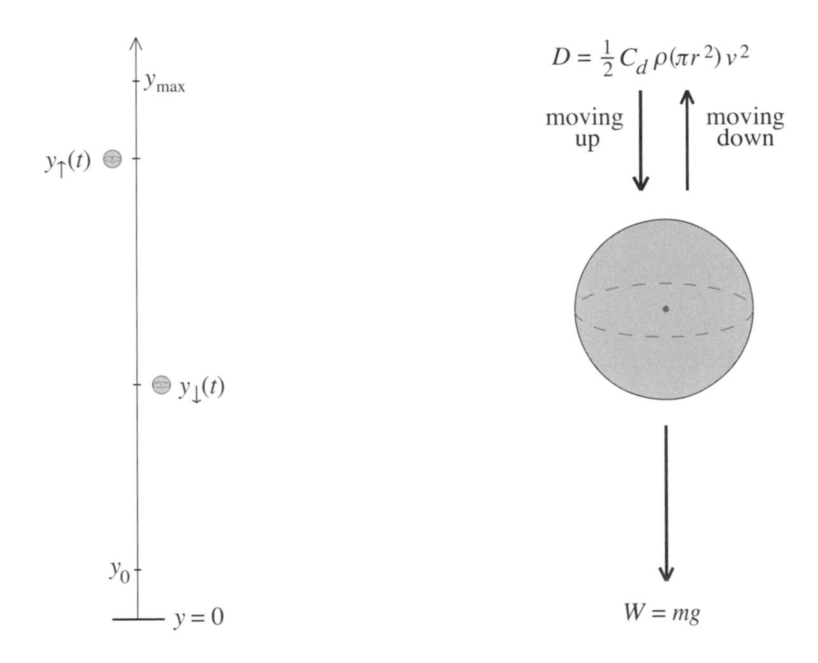

Fig. 11.21

derivations). It turns out that s_∞ is the speed limit of the ball on its way down. It is the speed that the ball would attain if it were to continue its fall forever.

It will be the goal of the next two sections to capture the specifics of the motion of the ball by determining the two functions $y_\uparrow(t)$ and $y_\downarrow(t)$, as well as the corresponding velocity functions $v_\uparrow(t)$ and $v_\downarrow(t)$, explicitly in terms of the basic parameters, y_0, v_0, g, and s_∞. Our analysis relies heavily on the trigonometric and hyperbolic functions, especially their inverses. See Sections 7.12 and 9.9.

11.10.1 Going Up

We will assume that the ball starts its upward flight at time $t = 0$, that it reaches its highest point at time t_{top}, and that it is on its descent thereafter. So the upward motion of the ball takes place during the time interval $0 \leq t \leq t_{\text{top}}$. The velocity of the ball during this time is denoted by $v_\uparrow(t)$. Since $y_\uparrow(t)$ is an increasing function, $v_\uparrow(t) = y'_\uparrow(t) \geq 0$. The ball's initial velocity is $v_0 = v_\uparrow(0) \geq 0$ and $v_\uparrow(t_{\text{top}}) = 0$. The total force on the ball for $0 \leq t \leq t_{\text{top}}$ is

$$F_{\text{up}} = -mg - \tfrac{1}{2}C_d\,\rho(\pi r^2)v_\uparrow^2 \text{ N}$$

(in newtons, as the MKS system is in use). By Newton's second law, $m\,\frac{dv_\uparrow}{dt} = -(mg + \frac{1}{2}C_d\,\rho(\pi r^2)v_\uparrow^2)$. Multiply this equation through by $\frac{2}{C_d\rho\pi r^2}$ to get $\frac{2m}{C_d\rho\pi r^2}\frac{dv_\uparrow}{dt} = -(\frac{2mg}{C_d\rho\pi r^2} + v_\uparrow^2)$. Since $\frac{2mg}{C_d\rho\pi r^2} = s_\infty^2$, we get

$$\tfrac{s_\infty^2}{g}\frac{dv_\uparrow}{dt} = -(s_\infty^2 + v_\uparrow^2).$$

This is a separable first-order differential equation that is not difficult to solve (given the groundwork that was already laid in Section 9.11). After separating variables,

$$-dt = \tfrac{s_\infty^2}{g}\frac{dv_\uparrow}{s_\infty^2 + v_\uparrow^2} = \tfrac{s_\infty^2}{g}\frac{dv_\uparrow}{s_\infty^2\left(1 + \left(\frac{1}{s_\infty}\right)^2 v_\uparrow^2\right)} = \tfrac{1}{g}\frac{dv_\uparrow}{1 + \frac{1}{s_\infty^2}v_\uparrow^2},$$

and therefore $\frac{dv_\uparrow}{1+\frac{1}{s_\infty^2}v_\uparrow^2} = -g\,dt$. The integral

$$\int \frac{dv_\uparrow}{1 + \frac{1}{s_\infty^2}v_\uparrow^2}$$

falls to a substitution. Recall Formula (10) $\displaystyle\int \frac{1}{1+u^2}\,du = \tan^{-1}u + C$ from Section 9.11, and let $u = \frac{1}{s_\infty}v_\uparrow$. Since $du = \frac{1}{s_\infty}\,dv_\uparrow$, we get $dv_\uparrow = s_\infty\,du$, and therefore

$$\int \frac{dv_\uparrow}{1 + \frac{1}{s_\infty^2}v_\uparrow^2} = s_\infty \int \frac{du}{1+u^2} = s_\infty \tan^{-1}u + C_1 = s_\infty \tan^{-1}(\tfrac{1}{s_\infty}v_\uparrow) + C_1.$$

It follows that $s_\infty \tan^{-1}(\frac{1}{s_\infty}v_\uparrow(t)) = -gt + C_2$. With $t = 0$ and $v_\uparrow(0) = v_0$, we get $C_2 = s_\infty \tan^{-1}\frac{v_0}{s_\infty}$ and $\tan^{-1}(\frac{1}{s_\infty}v_\uparrow(t)) = -\frac{g}{s_\infty}t + \tan^{-1}\frac{v_0}{s_\infty}$. By applying tan to both sides, $\frac{1}{s_\infty}v_\uparrow(t) = \tan(-\frac{g}{s_\infty}t + \tan^{-1}\frac{v_0}{s_\infty})$, and therefore

$$\boxed{v_\uparrow(t) = s_\infty \tan(-\tfrac{g}{s_\infty}t + \tan^{-1}\tfrac{v_0}{s_\infty})}$$

At the instant t_{top}, the ball is the top of its flight and $v_\uparrow(t_{\text{top}}) = 0$. Since $\tan^{-1}(\frac{1}{s_\infty}v_\uparrow(t)) = -\frac{g}{s_\infty}t + \tan^{-1}\frac{v_0}{s_\infty}$,

$$\boxed{t_{\text{top}} = \tfrac{s_\infty}{g}\tan^{-1}\tfrac{v_0}{s_\infty}}$$

Since the derivative of the height function $y_\uparrow(t)$ satisfies $\frac{dy_\uparrow}{dt} = v_\uparrow(t)$ for any time t during $0 \leq t \leq t_{\text{top}}$, we get

$$s_\infty \int \tan(-\tfrac{g}{s_\infty}t + \tan^{-1}\tfrac{v_0}{s_\infty})\,dt = \int v_\uparrow(t)\,dt = y_\uparrow(t) + C_3.$$

To solve this integral, use Formula (5) $\int \tan u\, du = \ln|\sec u| + C = \ln|\cos u|^{-1} + C = -\ln|\cos u| + C$ of Section 9.11 with $u = -\frac{g}{s_\infty}t + \tan^{-1}\frac{v_0}{s_\infty}$. The fact that $du = -\frac{g}{s_\infty}dt$ and hence $dt = -\frac{s_\infty}{g}du$ tells us that

$$\int \tan(-\tfrac{g}{s_\infty}t + \tan^{-1}\tfrac{v_0}{s_\infty})\, dt = \tfrac{s_\infty}{g}\ln\left|\cos(-\tfrac{g}{s_\infty}t + \tan^{-1}\tfrac{v_0}{s_\infty})\right| + C_4.$$

Since $\tan^{-1}(\frac{1}{s_\infty}v_\uparrow(t)) = -\frac{g}{s_\infty}t + \tan^{-1}\frac{v_0}{s_\infty}$ lies in the interval $(\frac{-\pi}{2}, \frac{\pi}{2})$ for all t (have a look at the graph of $\tan^{-1}x$ in Figure 9.33), we can conclude that $0 < \cos(-\frac{g}{s_\infty}t + \tan^{-1}\frac{v_0}{s_\infty}) \le 1$, and therefore that

$$y_\uparrow(t) = \tfrac{s_\infty^2}{g}\left(\ln\cos(-\tfrac{g}{s_\infty}t + \tan^{-1}\tfrac{v_0}{s_\infty})\right) + C_5.$$

With $t = t_{top} = \frac{s_\infty}{g}\tan^{-1}\frac{v_0}{s_\infty}$, we get $\ln\cos(-\frac{g}{s_\infty}t_{top} + \tan^{-1}\frac{v_0}{s_\infty}) = \ln\cos 0 = \ln 1 = 0$, so that $C_5 = y_\uparrow(t_{top})$. Because $y(t_{top}) = y_{max}$ is the maximal height that the ball reaches,

$$y_\uparrow(t) = \tfrac{s_\infty^2}{g}\ln\cos(-\tfrac{g}{s_\infty}t + \tan^{-1}\tfrac{v_0}{s_\infty}) + y_{max}.$$

Taking $t = 0$ in the equation above, we get $y_0 = y_\uparrow(0) = \frac{s_\infty^2}{g}\ln\cos(\tan^{-1}\frac{v_0}{s_\infty}) + y_{max}$, and therefore that $y_{max} - y_0 = -\frac{s_\infty^2}{g}\ln\cos(\tan^{-1}\frac{v_0}{s_\infty})$. Because $\cos x = \frac{1}{\sec x}$ and $\tan^2 x + 1 = \sec^2 x$, it follows that $\sec x = \sqrt{\tan^2 x + 1}$ for any x with $\frac{-\pi}{2} < x < \frac{\pi}{2}$, and hence that

$$\cos(\tan^{-1}\tfrac{v_0}{s_\infty}) = \frac{1}{\sec\left(\tan^{-1}\frac{v_0}{s_\infty}\right)} = \frac{1}{\sqrt{\tan\left(\tan^{-1}\frac{v_0}{s_\infty}\right)^2+1}} = \frac{1}{\sqrt{\left(\frac{v_0}{s_\infty}\right)^2+1}}.$$

So $y_{max} - y_0 = \frac{s_\infty^2}{g}\ln\left(\cos(\tan^{-1}\frac{v_0}{s_\infty})\right)^{-1} = \frac{s_\infty^2}{g}\ln\sqrt{\left(\frac{v_0}{s_\infty}\right)^2 + 1}$, and therefore

$$\boxed{y_{max} = y_0 + \tfrac{s_\infty^2}{g}\ln\sqrt{\left(\tfrac{v_0}{s_\infty}\right)^2 + 1}}$$

It follows from a combination of this equality, the earlier expression for $y_\uparrow(t)$, and a basic property of the natural log that

$$\boxed{y_\uparrow(t) = \tfrac{s_\infty^2}{g}\ln\left[\sqrt{\left(\tfrac{v_0}{s_\infty}\right)^2 + 1}\cdot\cos(-\tfrac{g}{s_\infty}t + \tan^{-1}\tfrac{v_0}{s_\infty})\right] + y_0}$$

Having determined explicit formulas for both $v_\uparrow(t)$ and $y_\uparrow(t)$ in terms of y_0, v_0, g, and s_∞, we have achieved what we set out to do.

11.10.2 Coming Down

After the ball reaches the top of its flight at the instant t_{top}, it begins its descent. For any $t \ge t_{top}$ during the ball's fall, the height of the ball above the ground and its velocity are given by the functions $y_\downarrow(t)$ and $v_\downarrow(t)$, respectively. Observe that $y_\downarrow(t_{top}) = y_{max}$, $v_\downarrow(t_{top}) = 0$, and since $y_\downarrow(t)$ is decreasing, that $v_\downarrow(t) = y'_\downarrow(t) < 0$ for $t > t_{top}$. From Figure 11.21, we know that the force on the ball during its descent is

$$F_{down} = -mg + \tfrac{1}{2}C_d\rho(\pi r^2)v_\downarrow^2 \text{ N}.$$

The fact that the term $\frac{1}{2}C_d\,\rho(\pi r^2)v^2$ has changed signs from the minus in the expression for F_{up} to the plus in the expression for F_{down} means—as we will see in a moment—that the hyperbolic functions take the place of the trigonometric functions that were decisive in the study of "going up." Other than that, the analysis of the ball "coming down" is virtually identical.

By Newton's second law, $m\frac{dv_\downarrow}{dt} = -mg + \frac{1}{2}C_d\,\rho(\pi r^2)v_\downarrow^2$. Multiply this equation through by $\frac{2}{C_d\rho\pi r^2}$ to get $\frac{2m}{C_d\rho\pi r^2}\frac{dv_\downarrow}{dt} = v_\downarrow^2 - \frac{2mg}{C_d\rho\pi r^2}$. Since $\frac{2mg}{C_d\rho\pi r^2} = s_\infty^2$, we see that

$$\frac{s_\infty^2}{g}\frac{dv_\downarrow}{dt} = v_\downarrow^2 - s_\infty^2.$$

By separating variables, we get

$$dt = \frac{s_\infty^2}{g}\frac{dv_\downarrow}{v_\downarrow^2 - s_\infty^2} = \frac{s_\infty^2}{g}\frac{dv_\downarrow}{s_\infty^2\left(\left(\frac{1}{s_\infty}\right)^2 v_\downarrow^2 - 1\right)} = \frac{1}{g}\frac{dv_\downarrow}{\frac{1}{s_\infty^2}v_\downarrow^2 - 1},$$

and therefore $\dfrac{dv_\downarrow}{1 - \frac{1}{s_\infty^2}v_\downarrow^2} = -g\,dt$. The integral

$$\int \frac{dv_\downarrow}{1 - \frac{1}{s_\infty^2}v_\downarrow^2}$$

is solved by substitution. Recall Formula (14) $\displaystyle\int \frac{1}{1-u^2}\,du = \tanh^{-1}u + C$ from Section 9.11, and let $u = \frac{1}{s_\infty}v_\downarrow$. Since $du = \frac{1}{s_\infty}dv_\downarrow$, we get $dv_\downarrow = s_\infty\,du$, and therefore

$$\int \frac{dv_\downarrow}{1 - \frac{1}{s_\infty^2}v_\downarrow^2} = s_\infty \int \frac{du}{1-u^2} = s_\infty \tanh^{-1}u + C_1 = s_\infty \tanh^{-1}\left(\tfrac{1}{s_\infty}v_\downarrow\right) + C_1.$$

So $s_\infty \tanh^{-1}(\frac{1}{s_\infty}v_\downarrow(t)) = -gt + C_2$. Since $v_\downarrow(t_{\text{top}}) = 0$, we get $C_2 = gt_{\text{top}}$. Therefore $\tanh^{-1}(\frac{1}{s_\infty}\,v_\downarrow(t)) = -\frac{g}{s_\infty}(t - t_{\text{top}})$, and after applying tanh to both sides, $v_\downarrow(t) = s_\infty \tanh(-\frac{g}{s_\infty}(t - t_{\text{top}}))$. From the definitions of the hyperbolic functions in Section 7.12, $\tanh(-x) = \frac{\sinh(-x)}{\cosh(-x)} = \frac{-\sinh x}{\cosh x} = -\tanh x$, so that

$$\boxed{v_\downarrow(t) = -s_\infty \tanh\left(\tfrac{g}{s_\infty}(t - t_{\text{top}})\right)}$$

Recall that in terms of the initial velocity v_0 of the ball on its way up, $t_{\text{top}} = \frac{s_\infty}{g}\tan^{-1}\frac{v_0}{s_\infty}$. From Figure 7.47, we know that $y = \tanh x \le 1$, so that $v_\downarrow(t) \le -s_\infty$ for all $t \ge t_{\text{top}}$. Suppose that the ball keeps falling, and consider $\lim\limits_{t\to\infty} v_\downarrow(t)$. By Figure 7.47, $\lim\limits_{x\to\infty}\tanh x = 1$, and hence

$$\boxed{\lim_{t\to\infty} v_\downarrow(t) = -s_\infty}$$

Using earlier data, it follows that this terminal speed of the ball is equal to $s_\infty = \sqrt{\frac{2mg}{C_d\rho\pi r^2}} \approx 3.577\frac{\sqrt{m}}{r}$ m/s. The longer the ball's flight, the closer it approaches this speed.

The determination of the function $y_\downarrow(t)$ follows. Since $\frac{dy_\downarrow}{dt} = v_\downarrow(t)$,

$$-s_\infty \int \tanh\left(\tfrac{g}{s_\infty}(t - t_{\text{top}})\right)dt = \int v_\downarrow(t)\,dt = y_\downarrow(t) + C_3.$$

To solve this integral, turn to Formula (7) $\int \tanh u\, du = \ln\cosh u + C$ of Section 9.11, and let $u = \frac{g}{s_\infty}(t - t_{\text{top}})$. The fact that $du = \frac{g}{s_\infty}\, dt$ and hence $dt = \frac{s_\infty}{g}\, du$ tells us that

$$\int \tanh\left(\tfrac{g}{s_\infty}(t - t_{\text{top}})\right) dt = \tfrac{s_\infty}{g} \ln\cosh\left(\tfrac{g}{s_\infty}(t - t_{\text{top}})\right) + C_4.$$

It follows that $y_\downarrow(t) = -\frac{s_\infty^2}{g}\ln\cosh\left(\frac{g}{s_\infty}(t - t_{\text{top}})\right) + C_5$. Since $y_\downarrow(t_{\text{top}}) = y_{\max}$ and $\ln\cosh 0 = \ln 1 = 0$, we get $C_5 = y_{\max}$. Therefore

$$\boxed{y_\downarrow(t) = y_{\max} - \tfrac{s_\infty^2}{g}\ln\cosh\left(\tfrac{g}{s_\infty}(t - t_{\text{top}})\right)}$$

After inserting the formula $y_{\max} = y_0 + \frac{s_\infty^2}{g}\ln\sqrt{(\frac{v_0}{s_\infty})^2 + 1}$ from "going up,"

$$y_\downarrow(t) = y_0 + \tfrac{s_\infty^2}{g}\ln\sqrt{(\tfrac{v_0}{s_\infty})^2 + 1} - \tfrac{s_\infty^2}{g}\ln\cosh(\tfrac{g}{s_\infty}(t - t_{\text{top}})),$$

and by a basic property of the natural log that

$$\boxed{y_\downarrow(t) = \tfrac{s_\infty^2}{g}\ln\left[\frac{\sqrt{\left(\frac{v_0}{s_\infty}\right)^2+1}}{\cosh\left(\frac{g}{s_\infty}(t-t_{\text{top}})\right)}\right] + y_0}$$

Unless it strikes something on its way up or down, the ball will eventually hit the ground at $y = 0$. The instant this occurs is the time of impact, designated by t_{imp}. The use of the earlier formula for $y_\downarrow(t)$ with $y_\downarrow(t_{\text{imp}}) = 0$ tells us that $\frac{s_\infty^2}{g}\ln\cosh\left(\frac{g}{s_\infty}(t_{\text{imp}} - t_{\text{top}})\right) = y_{\max}$. So $\ln\cosh\left(\frac{g}{s_\infty}(t_{\text{imp}} - t_{\text{top}})\right) = \frac{g}{s_\infty^2}y_{\max}$, hence $\cosh\left(\frac{g}{s_\infty}(t_{\text{imp}} - t_{\text{top}})\right) = e^{\frac{g}{s_\infty^2}y_{\max}}$, and therefore $\frac{g}{s_\infty}(t_{\text{imp}} - t_{\text{top}}) = \cosh^{-1} e^{\frac{g}{s_\infty^2}y_{\max}}$. It follows finally that

$$\boxed{t_{\text{imp}} = t_{\text{top}} + \tfrac{s_\infty}{g}\cosh^{-1} e^{\frac{g}{s_\infty^2}y_{\max}}}$$

Our list of formulas describing the motion of the ball from its start to the top of its flight to its impact on the ground is now complete. An important point that needs to be made is that these formulas provide only approximations. One difficulty in this regard deserves further comment. We learned earlier that in order for the drag D on the moving ball to be proportional to v^2 and in particular for D to be given by the equation

$$D = \tfrac{1}{2}C_d\,\rho(\pi r^2)v^2$$

with $C_d \approx 0.4$, it is necessary for the Reynolds number Re to be restricted to the range $10^3 < \text{Re} < 2\times 10^5$. We saw that this restriction on the Reynolds number of the moving ball translates to the restriction $1.33\times 10^{-2} < rv < 1.5\times 10^2$ on the product of its radius r and speed v in MKS, and hence to the restriction

$$\frac{1.33\times 10^{-2}}{r} < v < \frac{1.5\times 10^2}{r}$$

on its speed in m/s. This bound on the speed of the ball cannot be met at the top of its flight where $v = 0$ nor by its slow motion near it. How serious are the consequences of this for the formulas that we have derived?

We'll explore this question in the situations of (i) a bullet fired by the Springfield rifle, (ii) a ping-pong ball dropped from a great height, and (later in the Problems and Projects section) (iii) a cannonball dropped from a moderate height. In the first case, $r = 0.0038$ m (or $\frac{1}{2}$ the caliber of a bullet); in the second, $r = 0.02$ m (or 2 cm); and in the third, we'll take $r = 0.08$ m (or 8 cm). It follows that

$$\text{(i)}\quad 3.50 = \tfrac{1.33\times10^{-2}}{0.0038} < v < \tfrac{1.5\times10^{2}}{0.0038} \approx 39{,}474$$

$$\text{(ii)}\ 0.665 = \tfrac{1.33\times10^{-2}}{0.02} < v < \tfrac{1.5\times10^{2}}{0.02} = 7500$$

$$\text{(iii)}\ 0.166 \approx \tfrac{1.33\times10^{-2}}{0.08} < v < \tfrac{1.5\times10^{2}}{0.08} = 1875$$

all in m/s. The numbers of the column on the left convert to 12.6 km/hr, 2.39 km/hr, and 0.60 km/hr, respectively. To gain a sense of these speeds, note that 12.6 km/hr is the speed reached by a fast walker (the average speed of the record-setting 50-km Olympic walk was over 14 km/hr), and think of 2.39 km/hr as the speed of an easy stroll. As to the column on the right, the speeds of meteorites and returning space craft are higher (much higher) than 1875 m/s, and there are experimental guns developed in laboratories that fire bullets at speeds greater than 1875 m/s. But the speeds encountered in our daily experience fall far short of this figure. In the context of our study, the conclusion to be taken away is that the speeds of the three objects are outside the required range only on the low end and for a relatively short period of time. As a consequence, the impact on our analysis is minor.

11.10.3 Bullets and Ping-Pong Balls

Let's fire the Springfield rifle straight up into the air. We know from Section 10.5.4 that its bullets (now assumed to be spherical) have a radius of 0.38 centimeters, a mass $m = 10.7$ grams, and that they emerge from the barrel with an initial velocity of $v_0 = 853$ m/s. Let's assume that the top of the barrel is $y_0 = 2.5$ m above the ground. By applying our formulas, we know that the bullet's terminal speed is

$$s_\infty = \sqrt{\frac{2mg}{C_d\,\rho\pi r^2}} = \sqrt{\frac{2(0.0107)(9.81)}{(0.4)(1.22)\pi(0.0038^2)}} \approx 97.38 \text{ m/s},$$

and that the maximal height that the bullet attains is

$$y_{\max} = y_0 + \tfrac{s_\infty^2}{g}\ln\sqrt{(\tfrac{v_0}{s_\infty})^2 + 1} \approx 2.5 + \tfrac{97.38^2}{9.81}\ln\sqrt{(\tfrac{853}{97.38})^2 + 1} \approx 2106.53 \text{ m}.$$

The time it takes for the bullet to reach this height is $t_{\text{top}} = \frac{s_\infty}{g}\tan^{-1}\frac{v_0}{s_\infty} \approx \frac{97.38}{9.81}\tan^{-1}\frac{853}{97.38} \approx 14.46$ s. This tells us how considerable the air resistance on the bullet is. If it could sustain a speed close to its initial speed, the bullet would reach this height in less than 3 seconds. (The assumption that the shell is spherical, rather than bullet shaped with a smaller drag coefficient, is relevant here.) The time of flight of the bullet from $y_{\max}$ down to the ground $y = 0$ is

$$t_{\text{imp}} - t_{\text{top}} = \tfrac{s_\infty}{g}\cosh^{-1}e^{\frac{g}{s_\infty^2}y_{\max}} \approx \tfrac{97.38}{9.81}\cosh^{-1}e^{\frac{9.81}{97.38^2}2106.53} \approx 28.48 \text{ s}.$$

Therefore the instant of impact is $t_{\text{imp}} \approx 28.48 + 14.46 \approx 42.94$ s. This is the full duration of the bullet's flight. The velocity of the bullet at any time t during its descent is $v_\downarrow(t) = -s_\infty \tanh\left(\frac{g}{s_\infty}(t - t_{\text{top}})\right)$, so that the velocity of impact is

$$v_\downarrow(t_{\text{imp}}) = -s_\infty \tanh\tfrac{g}{s_\infty}(t_{\text{imp}} - t_{\text{top}}) = -97.38\tanh\tfrac{9.81}{97.38}(28.48) \approx -96.75 \text{ m/s}.$$

Notice that the bullet's speed at impact is close to its terminal speed s_∞. This is not surprising since its descent starts at a height of 2106.53 m (more than 2 km).

Suppose it were possible to lean over the railing and drop a ping-pong ball from the observation deck on the 102th floor of the Empire State Building. (Since this deck is undoubtedly fully enclosed, this is most likely not possible. But it is interesting to consider as a mathematical exercise.) The observation deck is 381 m above street level. How long will it take for the ping-pong to fall to the ground, and what will be its speed of impact? The equations derived in Section 6.5 (for fall in a vacuum) tell us that if there were no air resistance, the ball would reach the street in approximately 8.81 seconds and would hit the ground at 86.46 m/s. We will now see that the impact of air resistance on the ball's fall is considerable. We'll again assume that it is perfectly windstill. Who knows where the ball would end up if there were a breeze or a gale. Hoboken? The Hamptons?

The radius of a ping-pong ball is $r = 0.02$ m (or 2 cm) and it has a mass of $m = 0.0027$ kg (the equivalent of about 0.006 pounds). In the current situation, the formulas from the "coming down" segment apply, with $t_{\text{top}} = 0$, $v_{\downarrow}(0) = 0$, and $y_0 = y(0) = y_{\max} = 381$ m. The ball's terminal speed is

$$s_\infty = \sqrt{\frac{2mg}{C_d\rho\pi r^2}} \approx \sqrt{\frac{2(0.0027)(9.81)}{(0.4)(1.22)\pi(0.02^2)}} \approx 9.29 \text{ m/s}.$$

The ball reaches this speed quickly. Its velocity at any time $t \geq 0$ is $v_{\downarrow}(t) = -s_\infty \tanh \frac{g}{s_\infty}t$. For $t = 1$ second,

$$v_{\downarrow}(1) = -s_\infty \tanh(\tfrac{g}{s_\infty} \cdot 1) \approx -9.29 \tanh \tfrac{9.81}{9.29} \approx -7.28 \text{ m/s}.$$

One second later, the velocity is $v_{\downarrow}(2) \approx -9.02$ m/s, and one second after that it is $v_{\downarrow}(3) \approx -9.26$ m/s, already very close to the terminal velocity.

Using the formula $y_{\downarrow}(t) = y_{\max} - \frac{s_\infty^2}{g} \ln \cosh(\frac{g}{s_\infty}t)$, we find that the ping-pong ball is still

$$y_{\downarrow}(8.81) = 381 - \tfrac{9.29^2}{9.81} \ln(\cosh \tfrac{9.81}{9.29} \cdot 8.81) \approx 305.25 \text{ m}$$

above street level after having fallen $t = 8.81$ seconds (the full time for the ball's fall if there were no air resistance). The formula

$$t_{\text{imp}} = t_{\text{top}} + \tfrac{s_\infty}{g} \cosh^{-1} e^{\frac{g}{s_\infty^2} y_{\max}} = 0 + \tfrac{9.29}{9.81} \cosh^{-1} e^{\frac{9.81}{9.29^2} 381} \approx 41.67 \text{ s}$$

provides the time of impact. The velocity at impact? The information that we already have tells us that it is indistinguishable from the terminal velocity. This can be confirmed by plugging $t = 41.67$ s into the expression $v_{\downarrow}(t) = -s_\infty \tanh \frac{g}{s_\infty}t$.

In the Problems and Projects section at the end of this chapter, we'll apply the discussion above in an analysis of a problem related to the legendary experiment of Galileo (which in all likelihood never took place). Suppose two identical cannonballs are dropped side by side from the Leaning Tower of Pisa, one under conditions of vacuum and the other in air. How much sooner will the ball falling in a vacuum arrive? By how much, in terms of distance, will it win the race?

11.11 SYSTEMS WITH SPRINGS AND DAMPING ELEMENTS

The roads, streets, and highways on which our cars travel are not perfectly flat. Even when freshly paved, they have imperfections. A bump or a depression in the road causes the wheel of a car to move up and down perpendicularly to the road surface. In extreme cases, this could be severe, and the wheel could lose contact with the road entirely. In order to ensure a smooth and safe ride for their passengers, cars have systems that absorb the energy of the vertically moving wheel as it follows the bumps in the road so that the frame and body of the car can move with as little disturbance as possible. This is the role of a car's suspension system. It is designed to isolate the vehicle's body from road bumps and vibrations, while keeping the wheels in contact with the road. Such a system consists of two basic components: *spring* and *shock absorber*. Heavy springs, most often coiled and made of steel, allow a car to absorb the energy of a bump or pothole without jarring the occupants of the car. The shock absorber, often a hydraulic cylinder with a piston, subsequently works to dampen a spring's tendency to keep oscillating. There is one such system for each wheel. The spring and shock absorber are attached to both the wheel and the body of the car. Suppose, for example, that a wheel of the car encounters a pothole. The spring, partially compressed by the weight of the car, expands immediately to push the wheel into the hole, so that the wheel retains contact with

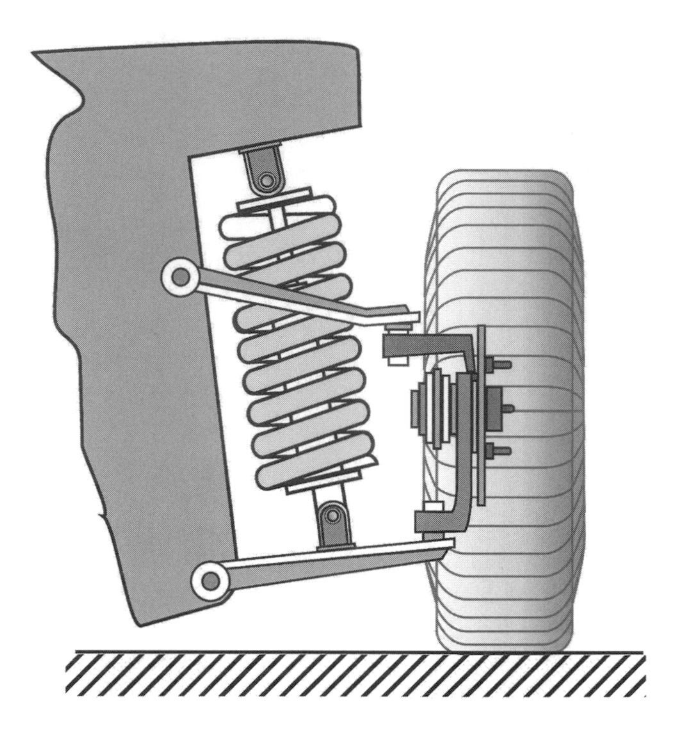

Fig. 11.22. Image from *Suspension System Design*, Powering the Future With Zero Emission and Human Powered Vehicles, Terrassa 2011, Erasmus LLP Intensive Programme, used with permission from the author Paul J. Aisopoulos, Alexander Technological Educational Institute, Thessaloniki, Greece.

the road surface. As the wheel emerges from the pothole, it is pushed up and the spring is compressed. This action of the spring absorbs the forces unleashed by the pothole and softens their transmission to the frame (and hence the body) of the car. So far so good. But suppose that the oscillating compressions and expansions of the spring that are generated are not quickly lessened and stopped. With as many as four wheels involved, the car's body will rock up and down as it moves forward, making it difficult to control. This is where the shock absorber comes in. It puts the breaks on the up-and-down vertical motion. Briefly put, the shock absorber works in tandem with the spring. The spring transforms the energy from the

Fig. 11.23. Image from an August, 2010 article in the online magazine Motorward.com. Many thanks to Mr. Bruno Silva, CEO of Motorward.com (http://www.motorward.com/), for permitting its use.

bumps in the road into kinetic energy, and this energy is dissipated by the shock absorber. A *strut* is a unit that combines spring and shock absorber into one compact assembly. Figure 11.22 shows a strut installed, and Figure 11.23 depicts it in detail.

The diagram of Figure 11.24 models the spring and shock absorber system just described. The strut is assumed to be in vertical position and to maintain its vertical position. (Figure 11.22 tells us that a strut is normally installed at a slant.) The block represents the part of the car that the wheel's suspension system supports. Its mass is m, its weight is mg, so that $-mg$ is the force of gravity acting on it. We'll now recall a basic fact about springs from Section 10.5.2. The length of a spring when no forces act on it is its natural length. Hooke's law asserts that there is a constant $k > 0$—the spring constant of the spring—such that if a force parallel to the direction of the spring has stretched or compressed the spring x units beyond its natural length, then the magnitude the force is $F(x) = kx$. We'll suppose that the spring of our system has spring constant k. It pushes up on the block if it is compressed and pulls down on it if it is stretched. The block, in turn, is rigidly attached to a piston that is free to move up and down in a cylinder containing a fluid. The fluid works to resist the piston's motion. This is the shock absorber. The damping force of the piston is quantified by the assumption that it is proportional to the velocity of the piston. We'll let $d \geq 0$ be the constant of proportionality. The piston's cylindrical housing is fixed in place.

A vertical y-axis runs parallel to this assembly. The top of the spring at its natural length determines

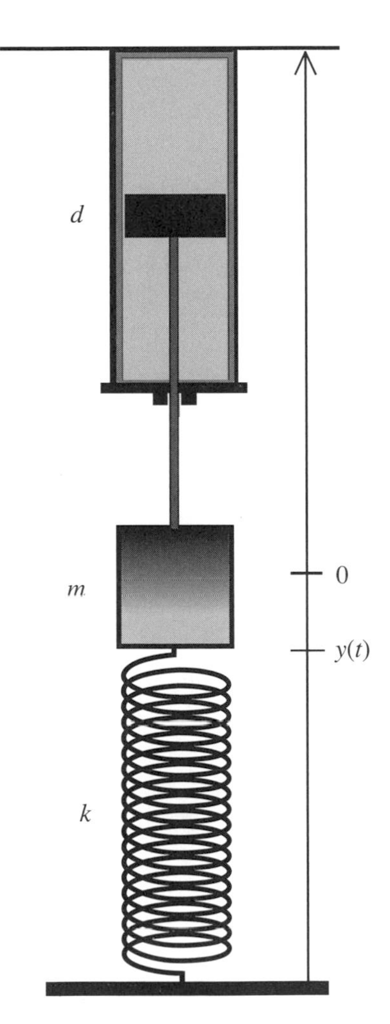

Fig. 11.24. A spring with spring constant k, a block of mass m, and a damping cylinder with damping constant d.

its origin 0. At any time t, the bottom of the block has coordinate $y(t)$. When $y(t)$ is negative, the spring is compressed and pushes upward. When $y(t)$ is positive, the spring is stretched and pulls downward. So the force of the spring on the block is $-ky(t)$ in either case. The velocity of the block is $y'(t)$. Since the fluid pushes down on the piston if $y'(t)$ is positive and up on the piston if $y'(t)$ is negative, the force of the connecting rod of the piston on the block is equal to $-dy'(t)$.

Suppose that our car travels on a stretch of road that is perfectly smooth, flat, and horizontal. The wheel that our study will now focus on moves without any vertical displacement. In reference to the wheel's suspension system, the spring is compressed, and pushes up with a force equal to the weight mg that it supports. So $y(t) < 0$, $-ky(t) = mg$, and hence $y(t) = -\frac{mg}{k}$. Since $y'(t) = 0$, the hydraulic shock absorber plays no role. We'll now suppose that the car's wheel encounters a bump. The part of the car's body that it supports is driven upward. At the top of this motion—let this occur at time $t = 0$—the upward force stops and the wheel's suspension system begins its response.

Turning to the model of Figure 11.24, we know that at any time $t \geq 0$, the forces acting on the block are gravity $-mg$, the force $-ky(t)$ of the spring, and the resistance $-dy'(t)$ of the piston. So the total force on the block is $F(t) = -dy'(t) - ky(t) - mg$. By Newton's second law, $F(t) = my''(t)$, and therefore

$$my''(t) + dy'(t) + ky(t) + mg = 0.$$

This differential equation governs the response of the suspension. To simplify it, write $ky(t) + mg = k(y(t) + \frac{mg}{k})$ and put $z(t) = y(t) + \frac{mg}{k}$. Notice that $z = 0$ represents the wheel's equilibrium position before it runs into the bump. Since $z'(t) = y'(t)$ and $z''(t) = y''(t)$, the differential equation becomes

$$mz''(t) + dz'(t) + kz(t) = 0.$$

This second-order differential equation is solved in Section 11.6. The particulars depend on the characteristic polynomial $mx^2 + dx + k$ and its roots $\frac{-d\pm\sqrt{d^2-4mk}}{2m}$. If $d < 2\sqrt{mk}$, the roots are complex. If $d \geq 2\sqrt{mk}$, they are real. Consider the complex situation first (with $d = 0$ separately), then turn to the real situation (with the single root $\frac{-d}{2m}$ first), to arrive at the following four possibilities:

1. If $d = 0$, the roots of the characteristic polynomial are $\pm\frac{2\sqrt{mk}}{2m}i = \pm\sqrt{\frac{k}{m}}i$ and they are both imaginary. The piston does not retard the motion, and the system is said to be *undamped*.
2. If $0 < d < 2\sqrt{mk}$, then the two roots of the characteristic polynomial are $a \pm bi$ with $a = -\frac{d}{2m}$ and $b = \frac{\sqrt{4mk-d^2}}{2m}$. The two roots are complex, and the system is *underdamped*.
3. If $d = 2\sqrt{mk}$, then the characteristic polynomial has the single real root $-\frac{d}{2m}$. In this situation, the system is *critically damped*.
4. If $d > 2\sqrt{mk}$, then the characteristic polynomial has the distinct real roots $\frac{-d\pm\sqrt{d^2-4mk}}{2m}$. In this situation, the system system is *overdamped*.

These four possibilities for the solution of the differential equation $z(t)$ are illustrated in Figure 11.25 in terms of the graphs of $z(t)$. In case 1, $d = 0$, the shock absorber provides no damping force, and the car's body over the wheel bounces uncontrollably up and down. In cases 3 and 4, d is too large and the suspension is too stiff. The complex case 2, with d less than but relatively close to $2\sqrt{mk}$, turns out to be the "Goldilocks" situation. The constant d can be chosen so that suspension is neither too stiff nor too soft.

11.11.1 The Family Sedan and the Stock Car

We'll now get into specifics and compare in quantitative terms the responses of the suspension systems of a front wheel of a standard family sedan with that of a NASCAR-type stock car. These responses depend on m, k, and d, and we begin by discussing the values of these parameters in each case. As in most of the studies in these text, our units are the meter, kilogram, second, and the newton as unit of force. The standard abbreviations are m, kg, s, and N, respectively.

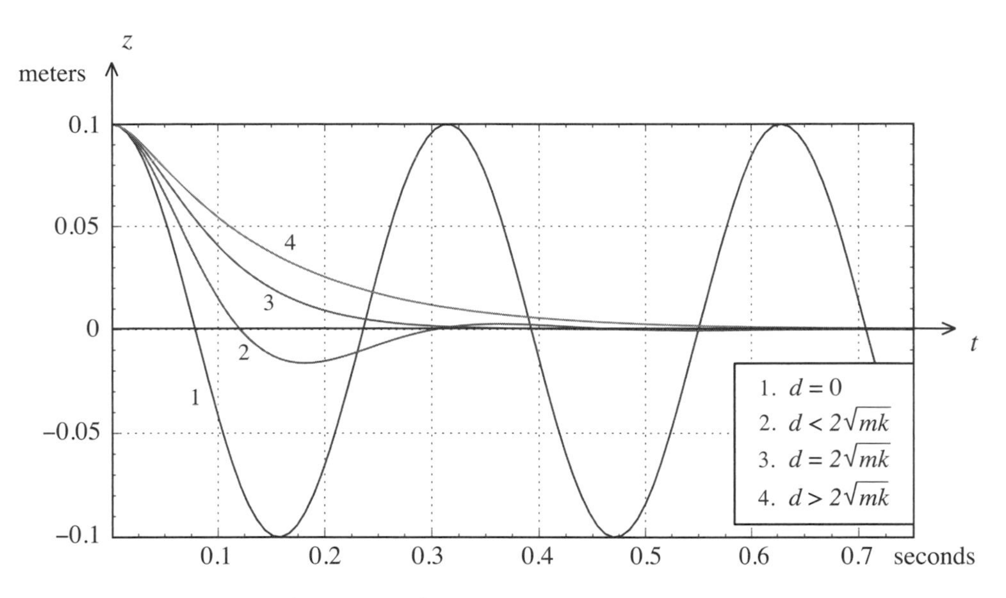

Fig. 11.25. A modified version of a diagram from the website https://en.wikipedia.org/wiki/Damping. Its reuse falls under the Creative Commons Attribution-ShareAlike License.

It is a NASCAR regulation that a competing vehicle weigh a minimum of 3400 pounds, including a full tank of fuel and the driver. This minimum weight requirement corresponds to $\frac{3400}{2.20} \approx 1545$ kg. So we'll take a stock car of mass 1600 kg and assume that our family sedan has the same mass. About 57%, or 912 kg, of this mass is carried by the front wheels. So each of the two front wheels carries about 456 kg. The *sprung mass* is the mass of the part of a vehicle that is supported by the springs. This includes the body, the frame, the engine, and related components. (The wheels, brake assemblies, and some components of the steering and suspension are not supported by the springs.) The sprung mass of a vehicle is about 85% to 87% of the total mass. For the two types of vehicles that are we are considering, we'll take

$$m = 390 \text{ kg}$$

as the mass that the suspension of each of the two front wheels supports. The spring constant k is usually called the *spring rate* in the context of a suspension system. The spring rate for most front-wheel-drive family sedans falls into the range

$$30{,}000 \leq k \leq 60{,}000 \text{ N/m}.$$

(To get a sense of the stiffness of such a spring in American units, use the approximations $10 \text{ N} \approx 2\frac{1}{4}$ pounds and $1 \text{ m} \approx 40$ inches, to see that that a force in the range of 170 to 340 pounds stretches or compresses such a spring by about 1 inch.) For our sedan, we'll take $k = 40{,}000$ N/m. For a stock car, k lies in the range

$$80{,}000 \leq k \leq 120{,}000 \text{ N/m}.$$

For our stock car, we'll take $k = 90{,}000$ N/m. As expected, the spring of the front suspension of a stock car is much stiffer than that of a typical sedan. The *damping ratio* ζ of a shock absorber is defined to be

$$\zeta = \frac{d}{2\sqrt{mk}},$$

where d is its damping constant (and—see option 3 of the previous section—$2\sqrt{mk}$ is the damping constant of a critically damped system for the given mass m and spring constant k). As a comparison of the options presented in Figure 11.25 suggests, the damping constant d of any shock absorber is less than $2\sqrt{mk}$, so that $0 < \zeta < 1$. With a smaller ζ, d is smaller, and the shock absorber is softer; with a ζ closer to 1, d is larger, and the shock absorber is stiffer. The damping ratio for the front suspension of a typical front-wheel-drive sedan lies in the range

$$0.2 \le \zeta \le 0.4.$$

This provides a soft and comfortable ride. The damping for the front suspension of the stock car needs to be tighter. Its damping ratio ζ lies in the range

$$0.5 \le \zeta \le 0.7.$$

We will take the values $\zeta = 0.3$ for our sedan and $\zeta = 0.6$ for the stock car. To summarize,

(1) for the family sedan, $m = 390$ kg, $k = 40{,}000$ N/m, $\zeta = 0.3$, so that its damping constant is

$$d = \zeta \cdot 2\sqrt{mk} \approx 0.6\sqrt{390 \cdot 40{,}000} \approx 2370\,\tfrac{\text{N·s}}{\text{m}}, \text{ and}$$

(2) for the stock car, $m = 390$ kg, $k = 90{,}000$ N/m, $\zeta = 0.6$, and

$$d = \zeta \cdot 2\sqrt{mk} \approx 1.2\sqrt{390 \cdot 90{,}000} \approx 7110\,\tfrac{\text{N·s}}{\text{m}}.$$

Since $0 < d < 2\sqrt{mk}$ for the front suspensions of the family sedan and the stock car, both are underdamped in the sense of option 2 of the previous section. Therefore Case 3 of Section 11.6 applies to inform us that in both situations, the solution of the equation $mz''(t) + dz'(t) + kz(t) = 0$ has the form

$$z(t) = e^{at}(D_1 \cos bt + D_2 \sin bt),$$

where $a = -\frac{d}{2m}$ and $b = \frac{\sqrt{4mk-d^2}}{2m}$. What about D_1 and D_2? The product rule tells us that

$$\begin{aligned} z'(t) &= ae^{at}(D_1 \cos bt + D_2 \sin bt) + e^{at}(-bD_1 \sin bt + bD_2 \cos bt) \\ &= e^{at}((aD_1 + bD_2)(\cos bt) + (aD_2 - bD_1)(\sin bt). \end{aligned}$$

So $z(0) = e^0(D_1 \cos 0 + D_2 \sin 0) = D_1$ and $z'(0) = e^0(aD_1 + bD_2) = aD_1 + bD_2$. Therefore

$$D_1 = z(0) \quad \text{and} \quad D_2 = \tfrac{1}{b}(z'(0) - az(0)).$$

The specifics for the two cars are as follows. For the family sedan, $a = -\frac{d}{2m} \approx -\frac{2370}{780} \approx -3.04$ and $b = \frac{\sqrt{4mk-d^2}}{2m} \approx \frac{\sqrt{4 \cdot 390 \cdot 40{,}000 - 2370^2}}{780} \approx 9.66$. Therefore the general solution for the family sedan is

$$\textbf{(1)} \qquad z(t) = e^{-3.04t}(D_1 \cos 9.66t + D_2 \sin 9.66t).$$

For the stock car, $a = -\frac{d}{2m} = -\frac{7110}{780} \approx -9.12$ and $b = \frac{\sqrt{4mk-d^2}}{2m} = \frac{\sqrt{4 \cdot 390 \cdot 90{,}000 - 7110^2}}{780} \approx 12.18$. So for the stock car, the general solution is

$$\textbf{(2)} \qquad z(t) = e^{-9.12t}(D_1 \cos 12.15t + D_2 \sin 12.15t).$$

We are ready to analyze the reaction of the two suspension systems to the encounter of the front wheel with a bump in the road. In each case, we'll suppose that the mass the wheel supports is driven up by 0.15 m (or about 6 inches) from its earlier equilibrium state. Such a large displacement means that the wheel meets a substantial bump at a good speed. Let the suspension systems begin their responses at time $t = 0$ and $z(0) = 0.15$. So $D_1 = 0.15$ in both cases. We'll assume that at this instant $z(t)$ is at a maximum, and take $z'(0) = 0$. Inserting this information into the expression for D_2 for the sedan, we get

$$D_2 = \tfrac{1}{9.66}(0 + (3.04)(0.15)) \approx 0.047,$$

and therefore (at least approximately)

$$z(t) = e^{-3.04t}(0.15 \cos(9.66t) + 0.047 \sin(9.66t)).$$

For the stock car,

$$D_2 = \tfrac{1}{12.15}(0 + (9.12)(0.15)) \approx 0.113,$$

and (again approximately),

$$z(t) = e^{-9.12t}(0.15\cos(12.15t) + 0.113\sin(12.15t)).$$

The graph at the top of Figure 11.26 is the graph of $z(t)$ for the sedan. The graph at the bottom of the figure is the graph of the $z(t)$ for the stock car. A comparison of the two graphs tells us that the restoration of the car's smooth and level motion occurs much more quickly for the stock car than for the family sedan. While the level ride is restored in 0.5 s for the stock car, it has not been completely restored for the sedan

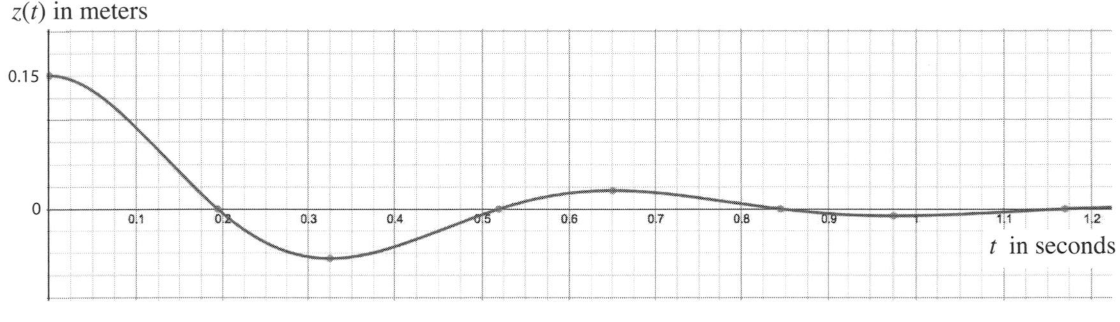

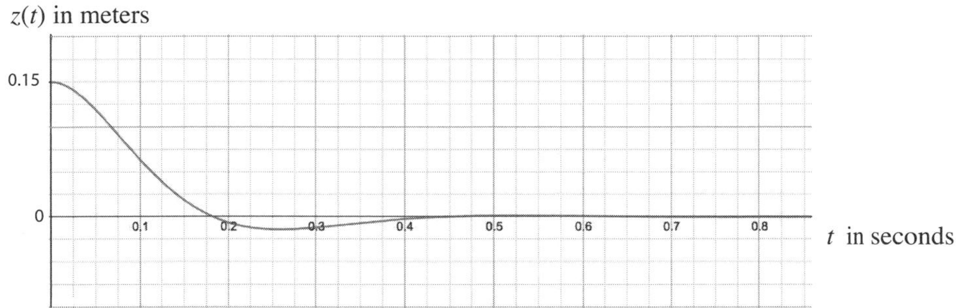

Fig. 11.26. The graph of $z(t)$ for the front suspension of the sedan is sketched on top and that of $z(t)$ for the front suspension of the stock car sketched below it. Both were graphed with the graphing calculator available on the website https://www.desmos.com/calculator.

even after 1.2 s. The stiffer suspension system of the stock car is the reason. Of course, the family sedan's softer suspension results in a more comfortable ride.

11.12 MORE ABOUT HANGING CABLES

Section 8.3 studied the main cable of a suspension bridge, computed the tension it is under, and verified that the curve it describes is a parabola. The simplifying assumption that made these conclusions possible is that at for all points on the cable, the magnitude of the downward pull of the horizontal deck is constant over the entire center span of the bridge. That the weight of the deck should be constant per unit length approximates a traffic situation in steady state. The related additional assumption that was made is that the weight of the deck also include the weight of the powerful cables. It is necessary to take the weight of the cable into account because this contributes to the tension that the cable is under. So there is the implicit assumption that the downward pull of the weight of the cable is also constant. The fact that the main cable of a suspension bridge curves upward from its lowest point in the direction of the tower means that this assumption is problematic. Since the curvature of the cable increases as the cable nears the towers of the bridge, the weight of the cable per horizontal unit of length will also increase in the direction of the towers. The fact that the cables of a large suspension bridge are massive means that this increase is not negligible.

Recall, for instance, that in the case of the George Washington Bridge (before a second deck was added), the collective weight of the four main cables was about 11,120 pounds per foot and the dead load plus the live load capacity of the deck was 47,000 pounds per foot. So the weight of the cables was around one-fourth of the dead plus live load of the deck. In the discussion that follows, the gravitational pull by the cable and the deck are separated, and it is no longer assumed that the downward gravitational pull of the cable is constant. This separation complicates the analysis, but it is possible to describe the shape of the cable under these assumptions. When all is said and done, however, this more refined shape deviates only little from the parabola of Section 8.3. As before, it will be assumed that the cable is completely flexible and that it does not change its length when placed under tension. (The assumption of complete flexibility, or lack of resistance to bending, is met by chains and light ropes, but leads only to approximate solutions for steel cables.)

The argument in Section 8.3 considered a small horizontal stretch Δx of deck and compared the vertical pull of the weight of this small piece of deck to the tangential pulls of the cable. This resulted in approximations for both the vertical and horizontal components of the tension of the cables. The push of Δx to zero snapped these approximations into equalities that were critical to the analysis. This procedure illustrates a point that this text has made again and again. Derivatives and integrals involve approximations of quantities that "snap" to equalities only after limits are taken. The approach that we'll use now is an abbreviated version of this strategy. Instead of taking a quantity Δx, getting approximations, and letting Δx shrink to zero, we'll take a tiny dx and regard the approximations that result to be equalities. Such an abbreviated argument, carefully attended to, leads to the same conclusion as the more formal version. (Such arguments were already used in Section 9.2.)

To separate the weight of the cable (or also a cord, chain, or rope) from the weight of the load that it supports, we'll assume that the cable weighs c units per unit length and that the suspended horizontal load weighs w units per unit length. (Our discussion does not depend on the choice of units, so these can be either MKS or American.) Figure 11.27 shows the cable in an xy-coordinate plane from its lowest point at the origin upward. The curve of the cable is the graph of a function $y = f(x)$. The horizontal x-axis is tangent to the cable at the origin. By the symmetry of the situation, we'll restrict x to the interval $0 \leq x \leq d$ from the low point of the cable at $x = 0$ to $x = d$, the x-coordinate of the point where the main cable meets the supporting tower. The point x is a typical point under the cable, dx is very small, and $x + dx$ is the point that x and dx determine. The corresponding segments of the cable and deck are highlighted. We'll let $\theta(x)$ be the angle that the tangent to cable at $(x, f(x))$ makes with the horizontal, and we'll let $T(x)$ be the tension in the cable at $(x, f(x))$. A suspender cable supports the short piece of

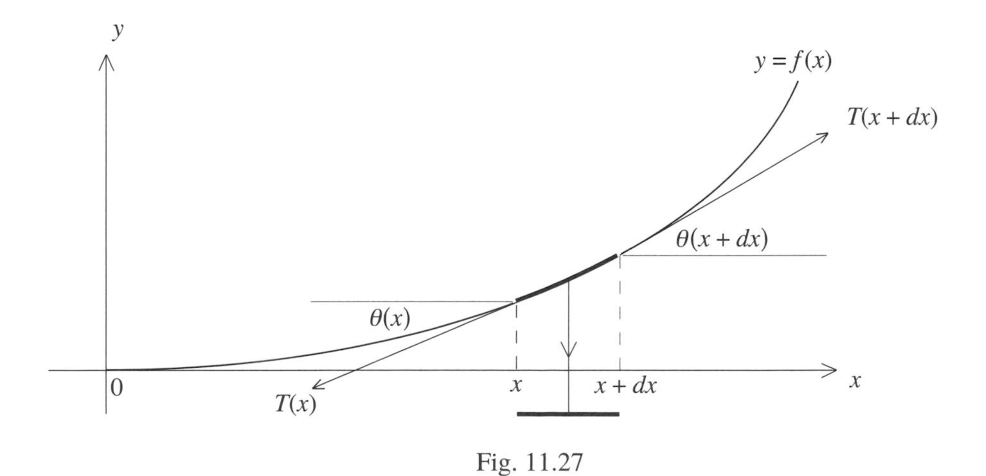

Fig. 11.27

the deck of length dx. Figure 11.28 takes the segment of length dx and completes it to a right triangle with the hypotenuse on the tangent line. We'll let dy be the height of the triangle. In the spirit of our abbreviated approach, we'll take the length of the cable over the interval $[x, x + dx]$ to be equal to the length $\sqrt{(dx)^2+(dy)^2}$ of the hypotenuse of the triangle. Factor out dx to see that this length is equal to $\sqrt{1+(\frac{dy}{dx})^2}\,dx = \sqrt{1+f'(x)^2}\,dx$. The cable and the weight it supports are assumed to be stable—nothing

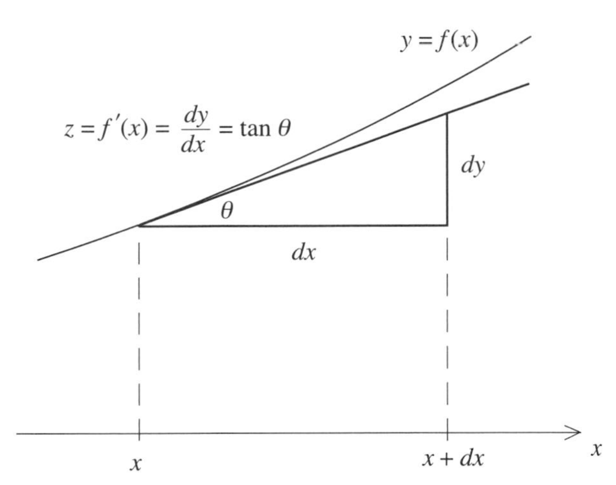

Fig. 11.28

moves—so that the difference $T(x+dx)\sin\theta(x+dx) - T(x)\sin\theta(x)$ between the upward pull of the cable at the point $(x+dx, f(x+dx))$ and the downward pull of the cable at $(x, f(x))$ is equal to the weight $w\cdot dx$ of the short piece of deck plus the weight $c\sqrt{(dx)^2+(dy)^2}$ of the short piece of cable. Therefore

$$T(x+dx)\sin\theta(x+dx) - T(x)\sin\theta(x) = wdx + c\sqrt{1+f'(x)^2}\,dx$$

and hence

$$\frac{d}{dx}T(x)\sin\theta(x) = \frac{T(x+dx)\sin\theta(x+dx) - T(x)\sin\theta(x)}{dx} = w + c\sqrt{1+f'(x)^2}.$$

(Here again, our abbreviation strategy is at work. The derivative is equal to $\lim\limits_{\Delta x\to 0}\frac{T(x+\Delta x)\sin\theta(x+\Delta x)-T(x)\sin\theta(x)}{\Delta x}$, so that the term in the middle is only an approximation.) In any case, $T(x)\sin\theta(x)$ is an antiderivative of $w+c\sqrt{1+f'(x)^2}$. By the discussion that concludes Section 9.6, $\int_0^x\left(w+c\sqrt{1+f'(t)^2}\right)dt$ is also an antiderivative of $w+c\sqrt{1+f'(x)^2}$, so that

$$T(x)\sin\theta(x) = \int_0^x\left(w+c\sqrt{1+f'(t)^2}\right)dt + C_1.$$

Substituting $x=0$ and noticing that $\theta(0)=0$, we get that $C_1=0$. Since no part of the load moves in a horizontal direction, the horizontal component $T(x)\cos\theta(x)$ of the tension $T(x)$ pulling to the right (see Figure 8.11) is equal to the horizontal tension $T(0)$ pulling at the origin $x=0$ to the left. So

$$T(x)\cos\theta(x) = T(0).$$

Setting the horizontal tension at the bottom of the cable at $x=0$ equal to $T(0)=T_0$, we now get

$$\tan\theta(x) = \frac{T(x)\sin\theta(x)}{T(x)\cos\theta(x)} = \frac{1}{T_0}\int_0^x\left(w+c\sqrt{1+f'(t)^2}\right)dt.$$

Since $\tan\theta(x)$ is the slope $\frac{dy}{dx}=f'(x)$ of the cable at x, it follows that $\frac{dy}{dx} = \frac{1}{T_0}\int_0^x\left(w+c\sqrt{1+f'(t)^2}\right)dt$. After differentiating this equation with respect to x, we finally arrive at the differential equation

$$\frac{d^2y}{dx^2} = \tfrac{1}{T_0}\Big(w + c\sqrt{1 + (\tfrac{dy}{dx})^2}\Big)$$

This second-order differential equation is solved by letting $z = f'(x) = \frac{dy}{dx}$ and reducing it to the first-order equation

$$\tfrac{dz}{dx} = \tfrac{1}{T_0}(w + c\sqrt{1 + z^2}).$$

Let $a = \frac{w}{T_0}$ and $b = \frac{c}{T_0}$ and separate variables to get

$$\frac{dz}{a + b\sqrt{1 + z^2}} = dx.$$

The substitutions $z = \tan\theta$ and (see Section 7.7) $dz = \sec^2\theta\, d\theta$, in combination with the identity $\tan^2\theta + 1 = \sec^2\theta$ and the fact that $\sec\theta > 0$ for $-\frac{\pi}{2} < \theta < \frac{\pi}{2}$, transform this equation to

$$\frac{\sec^2\theta\, d\theta}{a + b\sec\theta} = dx \text{ and hence to } \frac{d\theta}{a\cos^2\theta + b\cos\theta} = dx.$$

Now use the method of partial fractions as illustrated in Section 9.7.3 to show that

$$\frac{b}{a\cos^2\theta + b\cos\theta} = \frac{b}{\cos\theta(b + a\cos\theta)} = \frac{1}{\cos\theta} - \frac{a}{b + a\cos\theta}$$

(alternatively, take common denominators to check the second equality). This implies that

$$\sec\theta\, d\theta - \frac{a\, d\theta}{b + a\cos\theta} = b\, dx.$$

By Formula (6) of Section 9.11, $\int \sec\theta\, d\theta = \ln|\sec\theta + \tan\theta| + C_2$. Since $\sec\theta > 0$ for $-\frac{\pi}{2} < \theta < \frac{\pi}{2}$, $\sec\theta = \sqrt{\tan^2\theta + 1}$ and hence $\sec\theta + \tan\theta > 0$. So $\ln|\sec\theta + \tan\theta| + C_2 = \ln(\sec\theta + \tan\theta) + C_2$. We therefore focus on the solution of

$$\int \frac{a\, d\theta}{b + a\cos\theta}\, d\theta = a\int \frac{d\theta}{b + a\cos\theta}\, d\theta.$$

Here, the substitution $u = \tan\frac{\theta}{2}$ is the key. (This substitution was critical to the solution of the trigonometric integral that arose in Problem 10.45.) Use the identities $\tan^2\frac{\theta}{2} = \frac{1-\cos\theta}{1+\cos\theta}$ and $\tan\frac{\theta}{2} = \frac{\sin\theta}{1+\cos\theta}$ (see Problem 1.23), and solve them for $\cos\theta$ and $\sin\theta$, respectively, to verify that $\cos\theta = \frac{1-u^2}{1+u^2}$ and $\sin\theta = \frac{2u}{1+u^2}$. (Start by showing that $1 + u^2 = \frac{2}{1+\cos\theta}$.) Since $\frac{du}{d\theta} = \frac{1}{2}\sec^2\frac{\theta}{2} = \frac{1}{2}(\tan^2\frac{\theta}{2} + 1) = \frac{1}{2}(u^2 + 1)$, it is quickly checked that

$$a\int \frac{d\theta}{b + a\cos\theta} = a\int \frac{2du}{(u^2 + 1)\big(b + a(\frac{1-u^2}{1+u^2})\big)} = 2a\int \frac{du}{(a + b) - (a - b)u^2} = \tfrac{2a}{a+b}\int \frac{du}{1 - \frac{a-b}{a+b}u^2}.$$

1. We'll now assume that $w > c$ or that the weight per unit length of the deck is greater than the weight per unit length of the cable. This is an assumption easily met by any suspension bridge. Since $a = \frac{w}{T_0} > \frac{c}{T_0} = b$, the last integral above can be rewritten as

$$\tfrac{2a}{a+b}\int \frac{1}{1 - \Big(\sqrt{\frac{a-b}{a+b}}\, u\Big)^2}\, du\,.$$

By Formula (14) of Section 9.11 and the substitution $v = \sqrt{\frac{a-b}{a+b}}\, u$ and $dv = \sqrt{\frac{a-b}{a+b}}\, du$, this integral equals

$$\frac{2a}{\sqrt{a^2-b^2}} \int \frac{1}{1-v^2}\, dv = \frac{2a}{\sqrt{a^2-b^2}} \tanh^{-1} v + C_3.$$

By combining the various computations above, and using the equality $\tanh^{-1} x = \frac{1}{2}\ln(\frac{1+x}{1-x})$ from Section 9.9.2, we obtain that

$$bx = \ln(\sec\theta + \tan\theta) - \frac{a}{\sqrt{a^2-b^2}} \ln\Big(\frac{1+\sqrt{\frac{a-b}{a+b}}\tan\frac{\theta}{2}}{1-\sqrt{\frac{a-b}{a+b}}\tan\frac{\theta}{2}}\Big) + C_4.$$

Since $c = 0$ means that the cable is weightless, $b = \frac{c}{T_0} \neq 0$. Plug in $x = 0$ into the equation and recall that $\theta = 0$ at $x = 0$, to get $0 = \ln 1 - \frac{a}{\sqrt{a^2-b^2}} \ln 1 + C_4$ and hence that $C_4 = 0$. The rest is algebra. After multiplying the top and bottom of the expression inside the second natural log term by $1 + \sqrt{\frac{a-b}{a+b}}\tan\frac{\theta}{2}$, then inserting the trig identities $\tan\frac{\theta}{2} = \frac{\sin\theta}{1+\cos\theta}$ and $\tan^2\frac{\theta}{2} = \frac{1-\cos\theta}{1+\cos\theta}$, simplifying the algebraic terms that arise, and inserting $a = \frac{w}{T_0}$ and $b = \frac{c}{T_0}$, we have finally arrived at the equation

$$x = \frac{T_0}{c}\ln(\sec\theta + \tan\theta) - \frac{wT_0}{c\sqrt{w^2-c^2}} \ln\left(\frac{w + c\cos\theta + \sqrt{w^2-c^2}\sin\theta}{c + w\cos\theta}\right)$$

in the case where $w > c$. This equation expresses the x-coordinate of any point (x, y) on the cable in terms of the angle $\theta = \theta(x)$ that the tangent of the cable makes with the horizontal at that point. See Figure 11.27. To complete this equation for x to a pair of parametric equations for the curve of the cable, we'll now derive a similar equation for y. In this derivation, the assumption $w > c$ is not needed.

Let's begin by returning to

$$\frac{dz}{a + b\sqrt{1+z^2}} = dx.$$

Since $z = \frac{dy}{dx} = f'(x)$, we know that $\frac{dz}{dx} = f''(x)$, so that $dz = f''(x)\,dx$. From the fact that $\frac{dz}{dx} \cdot \frac{dx}{dy} = (a + b\sqrt{1+z^2}) \cdot \frac{1}{z}$, we get $\frac{dz}{dy} = \frac{a+b\sqrt{1+z^2}}{z}$. After separating variables,

$$\int \frac{z\,dz}{a + b\sqrt{1+z^2}} = y + C_1.$$

To solve the integral and to establish the connection between y and θ that we are after, we'll let $z = \tan\theta$ as before. Since $dz = \sec^2\theta\, d\theta$, $\tan^2\theta + 1 = \sec^2\theta$, and $\sec\theta \geq 0$ (because $-\frac{\pi}{2} < \theta < \frac{\pi}{2}$), we get that

$$\int \frac{z\,dz}{a + b\sqrt{1+z^2}} = \int \frac{(\tan\theta)(\sec^2\theta)\,d\theta}{a + b\sqrt{1+\tan^2\theta}} = \int \frac{(\tan\theta)(\sec^2\theta)\,d\theta}{a + b\sec\theta}.$$

Now let $u = a + b\sec\theta$. So $\sec\theta = \frac{1}{b}(u-a)$. Since $du = b\sec\theta \cdot \tan\theta\, d\theta$ and hence $\sec\theta \cdot \tan\theta\, d\theta = \frac{1}{b}\, du$, the last integral above becomes

$$\int \frac{1}{b^2} \frac{(u-a)\,du}{u} = \int \frac{1}{b^2}(1 - \frac{a}{u})\,du = \frac{1}{b^2}(u - a\ln u) + C_2.$$

Therefore

$$y + C_1 = \int \frac{1}{b^2} \frac{(u-a)\,du}{u} = \frac{1}{b^2}(u - a\ln u) + C_2 = \frac{1}{b^2}(a + b\sec\theta - a\ln(a + b\sec\theta)) + C_2.$$

So $y = \frac{1}{b^2}(a + b\sec\theta - a\ln(a + b\sec\theta)) - C_3$. Since $\theta = 0$ at the origin $(0, 0)$, it follows that $C_3 = \frac{1}{b^2}(a + b - a\ln(a+b))$. Therefore

$$y = \tfrac{1}{b^2}(a + b\sec\theta - a\ln(a + b\sec\theta)) - \tfrac{1}{b^2}(a + b - a\ln(a + b)).$$

After simplifying, using basic properties of the natural log, and substituting $a = \frac{w}{T_0}$ and $b = \frac{c}{T_0}$, we have arrived at the conclusion

$$\boxed{y = \tfrac{T_0}{c}(\sec\theta - 1) - \tfrac{wT_0}{c^2}\ln\left(\tfrac{w+c\sec\theta}{w+c}\right)}$$

In the case $w > c$—an inequality that holds for any suspension bridge—we have derived two parametric equations, one for x and one for y, that specify each point (x, y) on the curve of the cable in terms of the angle θ that the tangent to the curve of the cable at that point makes with the horizontal. Refer to Figure 11.27 one final time, and notice that $0 \leq \theta \leq \theta_1$, where $\theta_1 < \frac{\pi}{2}$ is the angle of the cable at the tower.

We'll now return to the original design of the George Washington Bridge and compare the curve of the main cable as given by these two parametric equations against the parabola derived by the much simpler approach of Section 8.3. This parabola is the graph of the function $y = f(x) = \frac{s}{d^2}$, where $s = 327$ is the sag in the main cable and $d = 1750$ the horizontal distance from the lowest point of the cable over the center span to the center of a tower. Given these values for s and d, we'll let this function be $f(x) = 0.0001068\,x^2$.

We know from Section 8.3 that each of the four main cables of the George Washington Bridge had a weight of $c = \frac{11{,}120}{4} = 2780$ pounds per linear foot. The weight of the dead load of the deck plus maximal live load capacity of the deck *per cable* was 11,750 pounds per foot. Since this included the weight of the cables, $w = 11{,}750 - 2780 = 8970$ pounds per foot. For the George Washington Bridge, $T_0 = 55{,}000{,}000$ pounds (as provided by Section 8.3 and the New York Port Authority). For the sake of accuracy, the computations below make use of seven significant figures (far more than the data allows) and round off the results (all in feet) to two decimal places.

The terms needed in the equations for x and y are $\frac{T_0}{c} = 19{,}784.17$, $\sqrt{w^2 - c^2} = 8528.335$, $\frac{wT_0}{c\sqrt{w^2-c^2}} = 20{,}808.75$, and $\frac{wT_0}{c^2} = 63{,}835.98$. Inserting them, we get

$$x = 19784.17\ln(\sec\theta + \tan\theta) - 20808.75\ln\left(\tfrac{8970+2780\cos\theta+8528.335\sin\theta}{2780+8970\cos\theta}\right) \text{ and}$$

$$y = 19784.17(\sec\theta - 1) - 63835.98\ln\left(\tfrac{8970+2780\sec\theta}{8970+2780}\right).$$

The focus is on the angles θ, $\frac{\pi}{72}$ (or 2.5°), $\frac{\pi}{36}$ (or 5°), $\frac{\pi}{24}$ (or 7.5°), $\frac{\pi}{18}$ (or 10°), $\frac{5\pi}{72}$ (or 12.5°), $\frac{\pi}{12}$ (or 15°), $\frac{7\pi}{72}$ (or 17.5°), $\frac{\pi}{9}$ (or 20°), and, finally, $\theta_1 = \frac{20.5\pi}{180}$ (or 20.5°) (see Section 8.3). The values that the parametric equations for x and y provide for each of these angles are listed in Table 11.1. The two coordinates determine the point on the curve of the cable (as described by the two parametric equations) at which the

θ (radians)	x (feet)	y (feet)	$f(x)$ (feet)
$\frac{\pi}{72}$ (or 2.5°)	204.36	4.46	4.46
$\frac{\pi}{36}$ (or 5°)	409.40	17.91	17.90
$\frac{\pi}{24}$ (or 7.5°)	615.83	40.52	40.50
$\frac{\pi}{18}$ (or 10°)	824.36	72.63	72.58
$\frac{5\pi}{72}$ (or 12.5°)	1035.73	114.70	114.57
$\frac{\pi}{12}$ (or 15°)	1250.73	167.33	167.07
$\frac{7\pi}{72}$ (or 17.5°)	1470.20	231.33	230.85
$\frac{\pi}{9}$ (or 20°)	1695.04	307.69	306.85
$\frac{20.5\pi}{180}$ (or 20.5°)	1740.73	324.55	323.62

Table 11.1

angle between the tangent at the point and the horizontal is the given θ. The final column of the table lists the value of the function $f(x) = 0.0001068\,x^2$ for each of the x-coordinates of the second column.

The information that these nine data points offer seems surprising. Since it merely supplies the x-coordinate that corresponds to the given θ, there is nothing surprising about the column of x-coordinates. However, the agreement for each of these x-coordinates between the y-coordinate of the curve supplied for the cable by the parametric equations on the one hand, and the corresponding y-coordinate of the parabolic model of the curve of the cable on the other, is closer than one might have thought. For all except the three largest angles, the difference falls within expected round-off errors. For the largest three angles, the differences are a bit larger, but all are less than one foot. The reason for this close agreement is the relative flatness or "horizontality" of the main cable. Table 11.1 tells us that the angle between the tangent to the cable and the horizontal is less than 2.5° for the first 200 feet, less than 5° for the first 400 feet, and less than 10° for the first 800 feet (all measured from the cable's lowest point). The close agreement just referred to validates the assumption made in Section 8.3 of folding in the weight of the cable with that of the deck. It also needs to be said that the earlier, simpler approach to the study of the suspension bridge determined the tension in the main cable. The more complicated parametric approach does not.

2. Having attended to the case $w > c$, we'll now assume that $c \geq w$. So $b = \frac{c}{T_0} \geq \frac{w}{T_0} = a$. Now the cable, cord, rope, or chain weighs as least as much per unit length as the load that it supports. The expression for y in terms of the angle θ that was already derived is valid under the assumption $b \geq a$, but the determination of the x-coordinate in terms of θ needs to be revisited. The conclusions

$$\sec\theta\, d\theta - \frac{a\, d\theta}{b + a\cos\theta} = b\, dx,$$

$\int \sec\theta\, d\theta = \ln(\sec\theta + \tan\theta) + C_1$, and

$$\int \frac{a d\theta}{b + a\cos\theta} = 2a\int \frac{du}{(a+b) - (a-b)u^2} = \frac{2a}{a+b}\int \frac{du}{1 - \frac{a-b}{a+b}u^2},$$

where $u = \tan\frac{\theta}{2}$, did not require $a > b$, and are all valid in the current situation. Rewriting this integral, we get

$$\frac{2a}{a+b}\int \frac{du}{1 - \frac{a-b}{a+b}u^2} = \frac{2a}{a+b}\int \frac{du}{1 + \frac{b-a}{a+b}u^2} = \frac{2a}{a+b}\int \frac{1}{1 + \left(\sqrt{\frac{b-a}{a+b}}\,u\right)^2}\, du.$$

The case $b = a$, is easy. The last integral above is equal to $\frac{2a}{a+b}u + C_2 = \frac{2a}{a+b}\tan\frac{\theta}{2} + C_2$, so that $bx = \ln(\sec\theta + \tan\theta) - \frac{2a}{a+b}\tan\frac{\theta}{2} + C_3$. Since $\theta = 0$ when $x = 0$, $0 = \ln 1 - 0 + C_3$ and $C_3 = 0$. Therefore

$$x = \frac{T_0}{c}\ln(\sec\theta + \tan\theta) - \frac{2wT_0}{c(w+c)}\tan\frac{\theta}{2}.$$

Now to $b > a$. By substituting $v = \sqrt{\frac{b-a}{a+b}}\,u$ and $dv = \sqrt{\frac{b-a}{a+b}}\,du$ and using Formula (10) of Section 9.11, the last integral becomes

$$\frac{2a}{a+b}\cdot\frac{\sqrt{a+b}}{\sqrt{b-a}}\int \frac{dv}{1+v^2} = \frac{2a}{(\sqrt{a+b})^2}\cdot\frac{\sqrt{a+b}}{\sqrt{b-a}}\int \frac{dv}{1+v^2} = \frac{2a}{\sqrt{b^2-a^2}}\int \frac{dv}{1+v^2} = \frac{2a}{\sqrt{b^2-a^2}}\tan^{-1}v + C_4.$$

We therefore get $bx = \ln(\sec\theta + \tan\theta) - \frac{2a}{\sqrt{b^2-a^2}}\tan^{-1}\left(\sqrt{\frac{b-a}{a+b}}\tan\frac{\theta}{2}\right) + C_5$. With the substitution $x = 0$, $0 = \ln(1+0) - \tan 0 + C_5$, so that $C_5 = 0$. Since $a = \frac{w}{T_0}, b = \frac{c}{T_0}$, we finally get

$$\boxed{x = \frac{T_0}{c}\ln(\sec\theta + \tan\theta) - \frac{2wT_0}{c\sqrt{c^2-w^2}}\tan^{-1}\left(\sqrt{\frac{c-w}{w+c}}\tan\frac{\theta}{2}\right)}$$

We'll now consider at the special case $w = 0$. This means that the cable, cord, rope, or chain supports no load other than itself. Telephone wires or electric power lines are examples. The parametric equations for x and y simplify to

$$x = \tfrac{T_0}{c}\ln(\sec\theta + \tan\theta) \quad \text{and} \quad y = \tfrac{T_0}{c}(\sec\theta - 1).$$

It follows that $e^{\frac{c}{T_0}x} = \sec\theta + \tan\theta$ and hence that $e^{-\frac{c}{T_0}x} = \frac{1}{\sec\theta+\tan\theta}$. So

$$e^{\frac{c}{T_0}x} + e^{-\frac{c}{T_0}x} = (\sec\theta + \tan\theta) + \frac{1}{\sec\theta+\tan\theta} = \frac{\sec^2\theta+2\sec\theta\tan\theta+\sec^2\theta}{\sec\theta+\tan\theta} = \frac{2\sec\theta(\sec\theta+\tan\theta)}{\sec\theta+\tan\theta} = 2\sec\theta.$$

Therefore $y = \frac{T_0}{c}(\sec\theta - 1) = \frac{T_0}{c}\Big(\frac{e^{\frac{c}{T_0}x}+e^{-\frac{c}{T_0}x}}{2} - 1\Big)$. With a look at the definition of $\cosh x$ in Section 7.12, we can conclude that

$$\boxed{y = \tfrac{T_0}{c}\big(\cosh(\tfrac{c}{T_0}x) - 1\big)}$$

where T_0 is the tension in the cable at its bottom point and c is its weight per unit length. Such a curve is called a *catenary*. The word comes from the Latin *catena*, meaning "chain." This label is appropriate because it is the curve of a freely hanging chain.

The analysis of the special case $w = 0$ has determined the general shape of a cable, cord, rope, or chain that hangs freely and supports no load (except itself) in terms of its weight c per unit length and the tension T_0 at the bottom of the cable. To find c for a particular situation is not a problem. Just weigh a segment. But what about T_0? Without insight into this constant, the understanding of the curve is not complete. This is a piece of the puzzle that is still missing. In the case of the suspension bridge of Section 8.3, the tension T_0 is easily determined with the formula $T_0 = \frac{1}{2}\frac{wd^2}{s}$, where w is weight per unit length and (d, s) is the point on the cable given by one-half the center span d and the sag s. Something like this ought to hold in the case of the catenary as well. "Something like" turns out to be correct. But the conclusions are not explicit and the details much more complicated. They are discussed at the end of the Problems and Projects section of this chapter.

11.13 PROBLEMS AND PROJECTS

11A. This first section of problems deals with first-order differential equations. It starts with the relatively simple matter of checking whether a proposed solution is a solution or not. It then turns to the problem of finding solutions. For each of the problems, check whether separation of variables and the method of integrating factors both apply. If both apply, carry out both.

11.1. Show that $y = \tan x + \sec x$ is a solution of the equation $2\frac{dy}{dx} - y^2 = 1$ on the interval $-\frac{\pi}{2} < x < \frac{\pi}{2}$.

11.2. Show that $y = \sin x$ and $y = \cos x$ are both solution of $(\frac{dy}{dx})^2 = 1 - y^2$.

11.3. Find the particular solution $y = f(x)$ of $\frac{dy}{dx} = xy$ that satisfies $f(0) = 4$.

11.4. Find the particular solution $y = f(x)$ of $\frac{dy}{dx} = \frac{y}{x}$ that satisfies $f(1) = 2$. Is there a particular solution $y = f(x)$ with $f(0) = 1$?

11.5. Find the general solution of the equation $y' = y\sin x - \sin x$. Then find the particular solution $y = f(x)$ that satisfies the initial condition $f(\pi) = 3$.

11.6. Find the general solution of the equation $\frac{dy}{dx} - x^2y^2 = 0$. Then find the particular solution $y = f(x)$ that satisfies the initial condition $f(0) = 8$.

11.7. Show that the general solution of $x\frac{dy}{dx} = 3y + x^2$ with $x > 0$ is $y = -x^2 + Cx^3$.

11.8. Show that the particular solution $y = f(x)$ of $3xy' - y = \ln\ x + 1$ with initial condition $f(1) = 5$, is $y = f(x) = 9x^{\frac{1}{3}} - \ln x - 4$. [Hint: One of the steps involves integration by parts.]

11.9. Show that the general solution of $(t^2 + 1)y' - (1 - t)^2 y = te^t$ is $y(t) = \frac{\frac{1}{2}t^2 + C}{(1+t^2)e^{-t}}$. [Hint: Use the equality $\frac{1-2t+t^2}{1+t^2} = 1 - \frac{2t}{1+t^2}$.]

11.10. Solve $(y - 3)\frac{dy}{dt} = 1$. Find an implicit solution first, and then use the quadratic formula to find an explicit solution $y = f(t)$ that satisfies $f(0) = 4$.

11.11. Find the particular solution $y = f(x)$ of $\frac{dy}{dx} = \frac{\sin x - \cos x}{1+y}$ that satisfies $f(\pi) = 0$. [Hint: Use the quadratic formula.]

11.12. Find the general solution of $(\ln y)^2 \frac{dy}{dx} = x^2 y$.

Use the calculator of the website

https://www.symbolab.com/solver/ordinary-differential-equation-calculator/

(or an equivalent website) for the last problem of this segment.

11.13. Show that the general solution of $2y' - y = 4 \sin 3t$ is $y = e^{\frac{t}{2}}[C - \frac{4}{37}e^{-\frac{t}{2}}(\sin(3t) + 6\cos(3t))]$. Then try to obtain this solution on your own. Use the calculator again to find the particular solution $y = f(x)$ of $y' + \frac{4}{x}y = x^2y^3$ that satisfies $f(1) = 2$.

11B. The next problems examine slope fields and Euler's method.

11.14. Consider the differential equation $y' = x - \frac{1}{2}y$.

i. For a few of the points, check the slopes that Table 11.2 lists. Then use the table and the grid

point	(-4, 4)	(-3, 4)	(-2, 4)	(-1, 4)	(0, 4)	(1, 4)	(2, 4)	(3, 4)	(4, 4)
slope	-6	-5	-4	-3	-2	- 1	0	1	2
point	(-4, 3)	(-3, 3)	(-2, 3)	(-1, 3)	(0, 3)	(1, 3)	(2, 3)	(3, 3)	(4, 3)
slope	-5.5	-4.5	-3.5	-2.5	-1.5	- 0.5	0.5	1.5	2.5
point	(-4, 2)	(-3, 2)	(-2, 2)	(-1, 2)	(0, 2)	(1, 2)	(2, 2)	(3, 2)	(4, 2)
slope	-5	-4	-3	-2	-1	0	1	2	3
point	(-4, 1)	(-3, 1)	(-2, 1)	(-1, 1)	(0, 1)	(1, 1)	(2, 1)	(3, 1)	(4, 1)
slope	-4.5	-3.5	-2.5	-1.5	-0.5	0.5	1.5	2.5	3.5
point	(-4, 0)	(-3, 0)	(-2, 0)	(-1, 0)	(0, 0)	(1, 0)	(2, 0)	(3, 0)	(4, 0)
slope	-4	-3	-2	-1	0	1	2	3	4
point	(-4, -1)	(-3, -1)	(-2, -1)	(-1, -1)	(0, -1)	(1, -1)	(2, -1)	(3, -1)	(4, -1)
slope	-3.5	-2.5	-1.5	-0.5	0.5	1.5	2.5	3.5	4.5
point	(-4, -2)	(-3, -2)	(-2, -2)	(-1, -2)	(0, -2)	(1, -2)	(2, -2)	(3, -2)	(4, -2)
slope	- 3	-2	-1	0	1	2	3	4	5
point	(-4, -3)	(-3, -3)	(-2, -3)	(-1, -3)	(0, -3)	(1, -3)	(2, -3)	(3, -3)	(4, -3)
slope	-2.5	-1.5	-0.5	0.5	1.5	2.5	3.5	4.5	5.5
point	(-4, -4)	(-3, -3)	(-2, -3)	(-1, -3)	(0, -3)	(1, -3)	(2, -3)	(3, -3)	(4, -4)
slope	-2	-1	0	1	2	3	4	5	6

Table 11.2

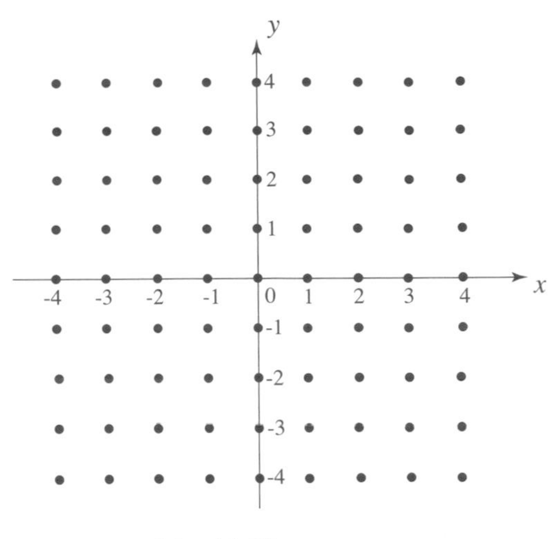

Fig. 11.29

of Figure 11.29 to carefully construct the slope field of the differential equation $y' = x - \frac{1}{2}y$.

ii. Use Euler's method, first with $h = 1$ and then with $h = 0.5$, to sketch two rough approximations of the graph of the solution $y = f(x)$ of the equation with initial condition $f(0) = -1$. Place both into the slope field of Figure 11.29.

iii. Use http://www.math-cs.gordon.edu/~senning/desolver/ to obtain the approximations of the graph of the solution $y = f(x)$ of the equation with initial condition $f(0) = -1$ that Euler's method gives for $h = 0.2$ and $h = 0.1$.

iv. Show that $f(x) = 2x - 4 + 3e^{-\frac{1}{2}x}$ is the particular solution of $y' = x - \frac{1}{2}y$ satisfying $f(0) = -1$.

v. Use https://www.desmos.com/calculator to graph $f(x) = 2x - 4 + 3e^{-\frac{1}{2}x}$ for $-4 \le x \le 4$ and compare the graph against the results of (ii) and (iii).

11.15. Consider the differential equation $y' = xy + 4x$ and show that $f(x) = e^{\frac{1}{2}(x^2-9)} - 4$ is the particular satisfying $f(-3) = -4$. Carry out parts (i), (ii), (iii), and (v) of Problem 11.14 in this situation.

11C. The next set of problems explores the polar plane and its relationship with the Cartesian plane.

11.16. The points below are all given in polar coordinates. Plot the points. Then find their Cartesian coordinates.

i. $(-3, \frac{\pi}{4})$

ii. $(-2, -\frac{\pi}{6})$

iii. $(3, \frac{7\pi}{3})$

iv. $(5, 0)$

v. $(-4, \frac{7\pi}{2})$

vi. $(5, -\frac{9\pi}{2})$

vii. $(0, \frac{6\pi}{7})$

viii. $(3, 8)$

ix. $(-1, 23)$

x. $(3, -32)$

11.17. Find sets of polar coordinates for each of the following points, all given in Cartesian coordinates. If the point is not on the y-axis, find coordinates (r, θ) with $-\frac{\pi}{2} < \theta < -\frac{\pi}{2}$. Use the equation $\tan\theta = \frac{y}{x}$ to find θ, and then use $r = \pm\sqrt{x^2 + y^2}$ to find r.

i. $(0, 5)$

ii. $(-4, 0)$

iii. $(3, -3)$

iv. $(4, -5)$

v. $(-3, 7)$

vi. $(7, -13)$

vii. $(-5, 9)$

viii. $(-6, -11)$

ix. $(8, 23)$

x. $(9, -36)$

11.18. Suppose a point P (other than the origin O) has polar coordinates (r, θ) with $-\frac{\pi}{2} < \theta \leq \frac{\pi}{2}$. Describe how all the polar coordinates for the point P can be obtained from the given one.

11D. The next set of problems deals with the complex plane.

11.19. Make a copy of Figure 11.30. Use a ruler to carefully place the points $P_1 + P_2$, $P_1 + P_3$, $P_1 + P_4$, $P_2 + P_3$, and $P_3 + P_4$ into your copy, and label them accordingly. Then place $2P_3 = P_3 + P_3$ and $2P_3 + P_4$. (Again, use a ruler and label the points.)

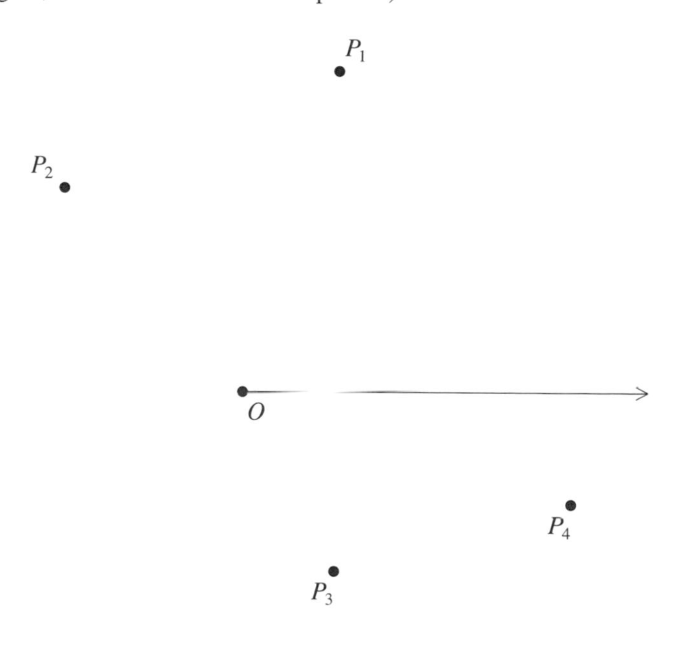

Fig. 11.30

11.20. Continue with another copy of Figure 11.30. Carefully locate a point Q_1 in the plane with the property that $P_1 + Q_1 = O$. This point is the additive inverse of P_1 and is denoted by $-P_1$. Locate the point $-P_2$ for P_2. Then locate $P_1 - P_2$. (Again, use a ruler and label the points.)

11.21. Consider the points P_1, P_2, and P_3 in the complex number plane of Figure 11.31. Make a copy of the figure and carefully place the points P_1P_2, P_1P_3, P_2P_3. (Use a ruler and a protractor from the Internet, and label the points.)

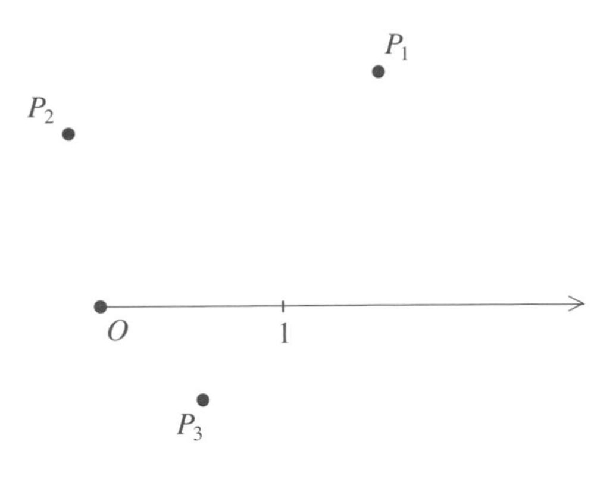

Fig. 11.31

11.22. Carefully locate the points P_1^2, P_2^2, and $P_1^2P_3$ in a copy of Figure 11.31. (Use a ruler and a protractor from the Internet, and label the points.)

11.23. Consider the complex plane of Figure 11.31 once more. What point in the plane serves as the multiplicative identity 1? Verify that your choice is correct. Locate a point Q_1 with the property that $P_1Q_1 = 1$. This point Q_1 is the multiplicative inverse of P_1 and is denoted by P_1^{-1}. Locate the points P_2^{-1} and P_3^{-1} and then $P_1P_3^{-1}$. (Use a ruler and a protractor from the Internet, and label the points.)

11.24. Refer to Figure 11.15, and consider the product $bi = (b, 0) \cdot (1, \frac{\pi}{2}) = (b, \frac{\pi}{2})$ on the imaginary axis. Check that b is the imaginary coordinate of this point regardless of whether b is positive, negative, or zero.

11.25. Refer to Figure 11.16, and place the points $-1 + i, 1 - 2i, -2 + 2i$, and $1 + \frac{1}{2}i$ into the complex plane of the figure.

11.26. Consider the complex numbers $c_1 = -1 + i$ and $c_2 = 1 - 2i$, and compute $c_1 + c_2$ and c_1c_2 in terms of their real and imaginary coordinates. Then place these points into a copy of Figure 11.15. Check that the addition and multiplication defined in terms of polar coordinates provide the same points $c_1 + c_2$ and c_1c_2. Repeat this with the complex numbers $c_1 = -2 + 2i$ and $c_2 = 1 + \frac{1}{2}i$.

11.27. Consider the complex numbers $c_1 = 1 + i$ and $c_2 = a_2 + b_2i$. Compute c_1c_2. Determine the complex number c_2 with the property that $c_1c_2 = 1$. This c_2 is the multiplicative inverse c_1^{-1} of c_1. Repeat these calculations with $c_1 = 2 - i$.

Let $c = a + bi$ be any complex number. Define the *conjugate* of c by $\bar{c} = a - bi$ and the *norm* of c by $N(c) = c\bar{c}$. Notice that $N(c) = a^2 + b^2$ is a real number that is not equal to zero if $c \neq 0$.

11.28. Compute the conjugate and the norm for the numbers $3, -i, 3 + i$, and $5 - i$.

11.29. Let $c = a + bi$ be any nonzero complex number. Use the definition of the conjugate and the norm to derive an expression for the multiplicative inverse c^{-1} of c in terms of a and b.

11.30. Consider the set of positive integers along with 0. Equations such as $x+7=0, 2x-3=0, 3x+5=0$, as well as $x^2-3=0$ and $6x^2+5=0$, and, finally, $x^2+1=0$ and $x^2+5=0$, can be solved only after the set of positive integers is successively enlarged. Describe the successive enlargements that are required.

The complex numbers are obtained by turning the set of points of the plane into a number system with an addition and multiplication that satisfies the same laws as the real numbers. It is not possible to turn the set of points in three-dimensional space into such a system. However, it is possible to do this in four dimensions (if the commutative law of multiplication is scrapped). Such a system, called *quaternions*, was invented (or discovered) by the Irishman William Rowan Hamilton (1805–1865). It is a four-dimensional version of the complex numbers. It has three elements i, j, and k that each play the role of the imaginary number i, and its elements have the form $a+bi+cj+dk$, where a, b, c, and d are real numbers. The addition is given by

$$(a+bi+cj+dk)+(a'+b'i+c'j+d'k)=(a+a')+(b+b')i+(c+c')j+(d+d')k,$$

and the multiplication is determined by the rules $i^2=-1, j^2=-1, k^2=-1$, along with $ij=k, jk=i, ki=j$ and, finally, $ji=-k, kj=-i$, and $ik=-j$.

11.31. Compute the product $(1+2i+3j+4k)(-3-2i-3k)$.

11.32. Determine a formula for the product of $a+bi+cj+dk$ and $a'+b'i+c'j+d'k$ that expresses this product in the form $a''+b''i+c''j+d''k$.

11.33. Let $q=a+bi+cj+dk$ be any quaternion. Define the *conjugate* and the *norm* of q by $\bar{q}=q=a-bi-cj-dk$ and $N(q)=q\bar{q}$, respectively. For any nonzero q (nonzero means that at least one of the real numbers a, b, c, or d is not zero), use these definitions to derive an expression for q^{-1} in terms of a, b, c, and d.

11E. The next several problems involve second-order differential equations of the form $Ay''+By'+Cy=0$ where A, B, and C are real, constant coefficients.

11.34. Show that $y=f(x)=\sin 2x-\cos 2x$ is a solution of $\frac{d^2y}{dx^2}+4y=0$ on $(-\infty,\infty)$.

11.35. Verify that $y=xe^{-2x}$ is a solution of $\frac{d^2y}{dx^2}+4\frac{dy}{dx}+4y=0$ on $(-\infty,\infty)$.

11.36. Suppose $a+bi$ is a root of the polynomial Ax^2+Bx+C. Verify that $y=f(x)=e^{ax}\sin bx$ is a solution of the differential equation $Ay''+By'+Cy$.

11.37. Refer to Cases 1, 2, and 3 of Section 11.6 to find the general solution of the second-order differential equations below.

i. $y''-6y'+8y=0$

ii. $y''+2y'+y=0$

iii. $y''-4y'+3y=0$

iv. $9y''+y=0$

v. $y''+y=0$

vi. $2y''+4y'+7y=0$

11.38. Consider the equation $Ay''+By'=0$ (with $A\neq 0$). Show that it reduces to a first-order differential equation. By solving it first, find the general solution of $Ay''+By'=0$ and check that your results agree with those of Section 11.6.

Study the discussion "Because of its importance in the theory of planetary motion (a topic that will be taken up in Chapter 12), we'll take an approach to the general solution of $y'' + y = 0 \ldots$" at the end of Section 11.6. After you have done so, turn to the next four problems.

11.39. Consider the second-order equation $y'' + 4y = 0$. Show that $y = \sin 2x$ and $y = \cos 2x$ are solutions. Now let $f(x)$ be any solution of the equation. Set $D_1(x) = (\sin 2x)f(x) + (\frac{1}{2}\cos 2x)f'(x)$ and $D_2(x) = (\cos 2x)f(x) - (\frac{1}{2}\sin 2x)f'(x)$. Proceed as in the solution of the equation $y'' + y = 0$ described above to show that $f(x) = D_1 \sin 2x + D_2 \cos 2x$, where D_1 and D_2 are real constants.

11.40. Consider the equation $y'' + 9y = 0$, and modify the argument outlined in Problem 11.39 to show that every solution $f(x)$ of this equation has the form $f(x) = D_1 \sin 3x + D_2 \cos 3x$ for some real constants D_1 and D_2.

11.41. Consider the equation $y'' + Cy = 0$ with $C > 0$. In this case, the characteristic polynomial is $x^2 + C$. Since its roots are $r_1 = \sqrt{C}i$ and $r_2 = -\sqrt{C}i$, Case 3 applies with $a = 0$ and $b = \sqrt{C}$ to tell us that general solution of $y'' + Cy = 0$ is $y = D_1 \cos \sqrt{C}x + D_2 \sin \sqrt{C}x$ for some real constants D_1 and D_2.

i. Show that $y = \sin \sqrt{C}x$ and $y = \cos \sqrt{C}x$ are both solutions.

ii. Study the outline of the solution of the special case $y'' + 4y = 0$ of Problem 11.39. Modify it to confirm that every solution $f(x)$ of the equation has the form $f(x) = D_1 \sin \sqrt{C}x + D_2 \cos \sqrt{C}x$.

We close this segment with a brief discussion of second-order differential equations of the form $Ay'' + By' + Cy = g(x)$. Note first that it suffices to find a single solution. Assume that $y = f_1(x)$ is a solution, and let $y = f_2(x)$ be any other. Then

$$A(f_2''(x) - f_2''(x)) + B(f_2'(x) - f_1'(x)) + C(f_2(x) - f_1(x)) = 0.$$

But this means $f_2(x) - f_1(x)$ can be determined by the method of Section 11.6. So in turn, $f_2(x)$ is determined by $f_1(x)$. Let's consider the example

$$(*) \qquad y'' + 3y' + 2y = 2\cos(3x).$$

The solution of this equation can be found by a *method of undetermined coefficients*. It tells us to start with $g(x) = 2\cos(3x)$, and suggests that the more general form $E_1 \sin(3x) + E_2 \cos(3x)$, where E_1 and E_2 are constants, should be tried as a solution of $(*)$. So let $y = f(x) = E_1 \sin(3x) + E_2 \cos(3x)$. By direct computation:

$$y' = f'(x) = 3E_1 \cos(3x) - 3E_2 \sin(3x) \text{ and } y'' = f''(x) = -9E_1 \sin(3x) - 9E_2 \cos(3x), \text{ so that}$$

$$\begin{aligned}
y'' + 3y' + 2y &= -9E_1 \sin(3x) - 9E_2 \cos(3x) + 3(3E_1 \cos(3x) - 3E_2 \sin(3x)) + 2(E_1 \sin(3x) + E_2 \cos(3x)) \\
&= (-9E_1 - 9E_2 + 2E_1)\sin(3x) + (-9E_2 + 9E_1 + 2E_2)\cos(3x) \\
&= (-7E_1 - 9E_2)\sin(3x) + (-7E_2 + 9E_1)\cos(3x).
\end{aligned}$$

Setting this equal to $2\cos(3x)$, we get $-7E_1 - 9E_2 = 0$ and $9E_1 - 7E_2 = 2$. By adding the two equations, $2E_1 - 16E_2 = 2$, so that $E_2 = \frac{1}{8}E_1 - \frac{1}{8}$. Therefore $9E_1 - (\frac{7}{8}E_1 - \frac{7}{8}) = 2$ and hence $\frac{65}{8}E_1 = \frac{23}{8}$. So $E_1 = \frac{23}{65}$ and $E_2 = \frac{23}{520} - \frac{65}{520} = -\frac{21}{260}$. With the coefficients E_1 and E_2 now determined, it follows by "going backward" that

$$f(x) = \tfrac{23}{65}\sin(3x) - \tfrac{21}{260}\cos(3x)$$

is a solution of equation $(*)$.

11.42. Use the method of undetermined coefficients to find one solution of $2y'' - 3y' + 5y = 4\sin(3t)$. Then use https://www.symbolab.com/solver/ordinary-differential-equation-calculator/ to see that the general solution is $y = e^{\frac{3}{4}t}[c_1 \cos(\frac{\sqrt{31}}{4}t) + c_2 \sin(\frac{\sqrt{31}}{4}t)] + \frac{18\cos(3t)}{125} - \frac{26\sin(3t)}{125}$, where c_1 and c_2 are constants. Similar approaches work if $g(x)$ is an algebraic or exponential term, or a combination of such terms.

11F. More about Power Series. This section pursues power series with a focus on questions about the radius and interval of convergence and the representation binomial series and its application to the solution of a difficult integral.

11.43. Use the ratio test to compute radii of convergence R for the power series below.

i. $\sum_{k=1}^{\infty} \frac{(x-4)^k}{k^2}$

ii. $\sum_{k=1}^{\infty} \frac{(x+2)^k}{k}$

iii. $\sum_{k=0}^{\infty} \frac{(-1)^k k}{3^k}(x+5)^k$

iv. $\sum_{k=1}^{\infty} \frac{2^k}{k}(2x-4)^k$

v. $\sum_{k=0}^{\infty} k!(2x+3)^k$

vi. $\sum_{k=2}^{\infty} \frac{(x-7)^k}{k^k}$

vii. $\sum_{k=3}^{\infty} \frac{k^k}{k!}(x-1)^k$

11.44. Show first that the radius of convergence of the power series $S = \sum_{k=1}^{\infty} \frac{(-1)^k(k+1)}{5^k}(x-2)^k$ is $R = 5$.

i. Write the power series that consists of the first, third, fifth, ... terms of the power series S in $\sum$ notation, and compute its radius of convergence. [Hint: Think about $k = 2j + 1$.]

ii. Write the power series that consists of the first, fourth, seventh, ... terms of the power series S in $\sum$ notation, and compute its radius of convergence.

11.45. Recall from Problem 6.7 that the series

$$\frac{x^{\frac{1}{2}}}{1+x} = x^{\frac{1}{2}} - x^{\frac{3}{2}} + x^{\frac{5}{2}} - x^{\frac{7}{2}} + x^{\frac{9}{2}} - \dots$$

converges for all x with $0 \le x < 1$, but not at $x = 1$. Why does this series not have a radius and interval of convergence as described in Section 11.7? Why is there a conflict with this description?

In the context of Maclaurin and Taylor series, it is not only a matter of determining the radius and interval of convergence, but also whether the series converges to the function that gives rise to it. Taylor's remainder theorem of Section 11.8 is an important tool in this regard.

11.46. Consider the Taylor series centered at $x_0 = 0$ for $f(x) = \sin x$, and show that its radius of convergence is infinite. Then use Taylor's remainder theorem to show that the series converges to $\sin x$ for all x.

We know from Example 11.19 that the power series $\sum_{k=0}^{\infty}(-1)^k x^k$ has radius of convergence 1 and that it converges to $\frac{1}{1+x}$ for all x in the interval $(-1,1)$. It follows from the discussion at the end of Section 11.8 that this series is the Taylor series centered at $x_0 = 0$ of the function $y = \frac{1}{1+x}$. So in this case,

$$\frac{1}{1+x} = \sum_{k=0}^{\infty}(-1)^k x^k$$

for all x satisfying $-1 < x < 1$, and the Taylor series converges to the function that defines it. At the two endpoints, this equality does not hold because at $x = -1$, $y = \frac{1}{1+x}$ is not defined, and at $x = 1$, the power series diverges. By an application of Theorem 4 of Section 11.7 the antiderivative $y = \ln(1+x)$ of $y = \frac{1}{1+x}$ is given by

$$\ln(1+x) = \sum_{k=0}^{\infty} \frac{(-1)^k}{k+1} x^{k+1}$$

for all x with $-1 < x < 1$. The suggestion from above is that this equality holds for neither $x = -1$ nor $x = 1$. For $x = -1$, this is the case because $y = \ln x$ is not defined. So what about $x = 1$? The fact is that

$$\ln 2 = 1 - \tfrac{1}{2} + \tfrac{1}{3} - \tfrac{1}{4} + \tfrac{1}{5} - \tfrac{1}{6} + \cdots$$

and that the equality does hold for $x = 1$. This infinite series was discovered by Nicholas Mercator (1620–1687), who published this result in his 1668 paper *Logarithmotechnia*. We will not go into the proof, and only point out that the convergence of the series to $\ln 2 \approx 0.693$ is slow:

$$1 - \tfrac{1}{2} + \tfrac{1}{3} - \tfrac{1}{4} \approx 0.583,\ 1 - \tfrac{1}{2} + \tfrac{1}{3} - \tfrac{1}{4} + \tfrac{1}{5} \approx 0.783,$$
$$1 - \tfrac{1}{2} + \tfrac{1}{3} - \tfrac{1}{4} + \cdots - \tfrac{1}{50} \approx 0.683,\ 1 - \tfrac{1}{2} + \tfrac{1}{3} - \tfrac{1}{4} + \cdots - \tfrac{1}{50} + \tfrac{1}{51} \approx 0.703,$$
$$1 - \tfrac{1}{2} + \tfrac{1}{3} - \tfrac{1}{4} + \cdots - \tfrac{1}{100} \approx 0.688,\ 1 - \tfrac{1}{2} + \tfrac{1}{3} - \tfrac{1}{4} + \cdots - \tfrac{1}{100} + \tfrac{1}{101} \approx 0.698.$$

Let r be a real constant and x a real variable. For what r and x does x^r make sense? If $x > 0$, then x^r can be defined as a limit or by noticing that $x^r = e^{r\ln x}$. Refer to Section 7.10. If $x = 0$, then $0^r = 0$ for any $r \neq 0$, but 0^0 is not defined. So let's assume that $x < 0$. If r is a positive integer then x^r is defined as the product $x \cdot x \cdots x$ with r factors. If r is 0, then, $x^r = 1$ (as in the case of a positive x). For a negative integer r, we can set $x^r = \frac{1}{x^{-r}}$. What about the case where $r = \frac{n}{m}$ is a rational number? For instance, let's consider $r = \frac{1}{m}$ for a positive integer m. If m is odd, the continuity of the function $y = f(x) = x^m$ together with the fact that $\lim_{x\to\pm\infty} f(x) = \pm\infty$ tells us that for any real number y there is some x such that $x^m = y$. Therefore $y^{\frac{1}{m}} = x$ makes sense. So (after changing notation) $x^{\frac{1}{m}}$ makes sense. But if $m = 2i$ is even, then if $x^{\frac{1}{m}} = z$ were to make sense as a real number z, then $(z^2)^i = z^m = x$. But this cannot be, since $z^2 \geq 0$ and $x < 0$. So $x^{\frac{1}{m}}$ makes sense only when $x \geq 0$.

We will study the Taylor series centered at $x_0 = 0$ of the function $f(x) = (1+x)^r$ for any real constant r. Given the discussion above, we'll assume that $r \neq 0$ and that $1 + x \geq 0$. So $x \geq -1$. We saw in Example 11.21 that the Taylor series of $y = f(x) = (1+x)^r$ centered at $x_0 = 0$ is given by

$$T_\infty(x) = \sum_{k=0}^{\infty}\binom{r}{k}x^k = 1 + rx + \frac{r(r-1)}{2}x^2 + \cdots + \frac{r(r-1)\cdots(r-(k-1))}{k!}x^k + \cdots$$

and that this is the *binomial series* for the function $f(x) = (1+x)^r$. The terms $\binom{r}{k}$ are the *binomial coefficients*. If $r > 0$ is an integer, then $f(x) = (1+x)^r = 0$ is a polynomial of degree r, so that $f^{(k)}(x) = 0$ for any $k > r$. In this case, the binomial series only has finitely many terms, so that it converges for all x. A look at its definition tells us that if r is not a positive integer, then all the binomial coefficients are nonzero.

11.47. Let $f(x) = (1+x)^r$, and assume that r is not a positive integer. Show that the radius of convergence of the binomial series $T_\infty(x) = \sum\limits_{k=0}^{\infty} \binom{r}{k} x^k$ is $R = 1$.

It follows that the binomial series $T_\infty(x)$ converges for all x with $|x| < 1$ and diverges for all x with $|x| > 1$. By using convergence tests that we have not considered, it can be shown that the interval of convergence of the binomial series is

$$(-1,1) \text{ for } r \le -1,\ (-1,1] \text{ for } -1 < r < 0, \text{ and } [-1,1] \text{ for } r > 0.$$

11.48. Suppose $r = -1$. Show that $\binom{-1}{k} = (-1)^k$, so that the binomial series in this case is

$$T_\infty(x) = \sum_{k=0}^{\infty} (-1)^k x^k = 1 - x + x^2 - x^3 + x^4 - x^5 + \dots .$$

We have already seen that in this case the Taylor series centered at $x_0 = 0$ converges to the function $y = \frac{1}{1+x}$ that defines it.

Is it always true that the Taylor series centered at $x_0 = 0$ converges to the function $(1+x)^r$ that gives rise to it for all x in its interval of convergence? This is the question we turn to next. Previous such explorations have made use of Taylor's remainder theorem. But in this case a direct approach is not only simpler, but also an instructive illustration of how to maneuver with power series. The maneuvers in this case consists of some binomial trickery.

Define the function $g(x)$ by $g(x) = \sum\limits_{k=0}^{\infty} \binom{r}{k} x^k$ for x in the interval of convergence. By Theorem 4 of Section 11.7,

$$g'(x) = r + r(r-1)x + \tfrac{r(r-1)(r-2)}{2!}x^2 + \tfrac{r(r-1)(r-2)}{3!}x^3 + \dots + \tfrac{r(r-1)\cdots(r-(k-1))}{(k-1)!}x^{k-1} + \dots = r\sum_{k=0}^{\infty}\binom{r-1}{k}x^k.$$

Consider the equality

$$(1+x)\sum_{k=0}^{\infty}\binom{r-1}{k}x^k = \sum_{k=0}^{\infty}\binom{r-1}{k}x^k + \sum_{k=0}^{\infty}\binom{r-1}{k}x^{k+1}.$$

Let $j = k+1$, and write $\sum\limits_{k=0}^{\infty}\binom{r-1}{k}x^{k+1}$ as $\sum\limits_{j=1}^{\infty}\binom{r-1}{j-1}x^j$. Changing the index of summation of this power series back to k, we get that

$$(1+x)\sum_{k=0}^{\infty}\binom{r-1}{k}x^k = \sum_{k=0}^{\infty}\binom{r-1}{k}x^k + \sum_{k=1}^{\infty}\binom{r-1}{k-1}x^k = 1 + \sum_{k=1}^{\infty}\left[\binom{r-1}{k} + \binom{r-1}{k-1}\right]x^k.$$

Using the definition of the binomial coefficient, we get that

$$\begin{aligned}\binom{r-1}{k} + \binom{r-1}{k-1} &= \frac{(r-1)(r-2)\cdots(r-1-(k-1))}{k!} + \frac{(r-1)(r-2)\cdots(r-1-(k-2))}{(k-1)!} = \frac{(r-1)(r-2)\cdots(r-1-(k-2))}{(k-1)!}\left[\frac{r-1-(k-1)}{k} + 1\right]\\ &= \frac{(r-1)(r-2)\cdots(r-1-(k-2))}{(k-1)!}\cdot\frac{r}{k} = \frac{r(r-1)(r-2)\cdots(r-(k-1))}{k!} = \binom{r}{k},\end{aligned}$$

and therefore that

$$(1+x)\sum_{k=0}^{\infty}\binom{r-1}{k}x^k = 1 + \sum_{k=1}^{\infty}\binom{r}{k}x^k = \sum_{k=0}^{\infty}\binom{r}{k}x^k.$$

Since

$$\frac{g'(x)}{r} = \sum_{k=0}^{\infty}\binom{r-1}{k}x^k \quad\text{and}\quad g(x) = \sum_{k=0}^{\infty}\binom{r}{k}x^k,$$

we get $(1+x)\frac{g'(x)}{r} = g(x)$. So $(1+x)g'(x) = rg(x)$. Now let $h(x) = (1+x)^{-r}g(x)$. Since

$$h'(x) = -r(1+x)^{-r-1}g(x) + (1+x)^{-r}g'(x) = -(1+x)^{-r-1}(1+x)g'(x) + (1+x)^{-r}g'(x) = 0,$$

we know that $h(x) = C$, a constant. Since $h(0) = (1+0)^{-r}g(0) = 1\cdot 1 = 1$, it follows that $h(x) = 1$.

11.49. Explain why it has now been demonstrated that for any $r \neq 0$, the binomial series $\sum_{k=0}^{\infty}\binom{r}{k}x^k$ converges to $(1+x)^r$ for all x in the interval of convergence of the series.

Turn to Section 11.9 as a guide for the solution of the next two problems.

11.50. Use a power series centered at $x_0 = 0$ to solve the differential equation $y' = y$.

11.51. Use a power series centered at $x_0 = 0$ to solve the differential equation $y'' - 2xy' + y = 0$.

11G. The Circumference of the Ellipse. Section 10.4 undertook a study that provided much essential information about the elliptical orbits of the planets. However, this discussion was silent about a basic question: What is the distance that a planet travels in one complete trip around the Sun? In other words, what is the length of an elliptical orbit? The computation of the area of an ellipse was a relatively simple problem. Not so, the computation of its circumference. This computation is the pursuit of this segment.

11.52. Consider the ellipse with equation $\frac{x^2}{a^2} + \frac{y^2}{b^2} = 1$, where a is the major axis, b is the minor axis. With $c = \sqrt{a^2-b^2}$, $\frac{a}{c} = \varepsilon$ is the eccentricity of the ellipse. Use the length of a curve formula from Section 9.3 and the trig substitution $x = a\sin\theta$ to show that the circumference of the ellipse is given by the integral

$$4a\int_0^{\frac{\pi}{2}} \sqrt{1-\varepsilon^2\sin^2\theta}\,d\theta.$$

The integral above is an elliptic integral that cannot be solved in closed form. This means that $\sqrt{1-\varepsilon^2\sin^2\theta}$ does not have an antiderivative that is given by an elementary function (a function that can be expressed as a combination of algebraic, trig, and exponential functions and their inverses). Refer to the discussion that concludes Section 9.11. It is not unusual for the solutions of indefinite integrals and differential equations to involve functions that are not elementary. Power series open the door to the exploration such functions.

To solve the integral above, the relevant series is the binomial series with $r = \frac{1}{2}$. For $k \geq 1$, the binomial coefficient $\binom{r}{k} = \frac{r(r-1)(r-2)\cdots(r-(k-1))}{k!}$ is now equal to

$$\binom{\frac{1}{2}}{k} = \frac{\frac{1}{2}(\frac{1}{2}-1)(\frac{1}{2}-2)\cdots(\frac{1}{2}-(k-1))}{k!} = \frac{\frac{1}{2}(-\frac{1}{2})(-\frac{3}{2})\cdots(-\frac{2k-3}{2})}{k!}.$$

Notice that the numerator has k factors and that $k-1$ of them are negative. So

$$\binom{\frac{1}{2}}{k} = \frac{(-1)^{k-1}\cdot 1\cdot 3\cdots(2k-3)}{2^k\cdot k!}.$$

It follows from the previous segment, in particular Problem 11.49, that

$$\sqrt{1+x} = (1+x)^{\frac{1}{2}} = 1 + \sum_{k=1}^{\infty}\frac{(-1)^{k-1}\cdot 1\cdot 3\cdots(2k-3)}{2^k\cdot k!}x^k$$

for all x with $-1 < x \leq 1$. Since $\varepsilon < 1$, we know that $\varepsilon^2\sin^2\theta < 1$ and $-1 < -\varepsilon^2\sin^2\theta < 1$. Therefore

$$\sqrt{1-\varepsilon^2\sin^2\theta} = 1 + \sum_{k=1}^{\infty}\frac{(-1)^{k-1}1\cdot 3\cdot 5\cdots(2k-3)}{2^k k!}(-1\cdot\varepsilon^2\sin^2\theta)^k = 1 - \sum_{k=1}^{\infty}\frac{1\cdot 3\cdot 5\cdots(2k-3)\varepsilon^{2k}}{2^k k!}\sin^{2k}\theta.$$

By an application of Theorem 4 of Section 11.7,

$$\int_0^{\frac{\pi}{2}} \sqrt{1-\varepsilon^2\sin^2\theta}\,d\theta = \tfrac{\pi}{2} - \sum_{k=1}^{\infty} \tfrac{1\cdot 3\cdot 5\cdots(2k-3)\varepsilon^{2k}}{2^k k!} \int_0^{\frac{\pi}{2}} \sin^{2k}\theta\,d\theta.$$

11.53. Show that the integration by parts substitution $u = \sin^{n-1}\theta$ and $dv = \sin\theta\,d\theta$ results in the formula

$$\int \sin^n\theta\,d\theta = -\tfrac{1}{n}\cos\theta\cdot\sin^{n-1}\theta + \tfrac{n-1}{n}\int \sin^{n-2}\theta\,d\theta,$$

and that by the repeated use of this formula, $\int_0^{\frac{\pi}{2}} \sin^{2k}\theta\,d\theta = \frac{1}{2}\cdot\frac{3}{4}\cdot\frac{5}{6}\cdots\frac{2k-1}{2k}\cdot\frac{\pi}{2}$ for any $k \ge 1$.

11.54. Show that the circumference of the ellipse with semimajor axis a and eccentricity ε is given by

$$\begin{aligned} 4a\int_0^{\frac{\pi}{2}} \sqrt{1-\varepsilon^2\sin^2\theta}\,d\theta &= 2\pi a\Big[1 - \sum_{k=1}^{\infty} \tfrac{[1\cdot 3\cdots(2k-1)]^2}{[2^k(k!)]^2}\,\tfrac{\varepsilon^{2k}}{2k-1}\Big] \\ &= 2\pi a\Big[1 - \big(\tfrac{1}{2}\big)^2\tfrac{\varepsilon^2}{1} - \big(\tfrac{1\cdot 3}{2\cdot 4}\big)^2\tfrac{\varepsilon^4}{3} - \big(\tfrac{1\cdot 3\cdot 5}{2\cdot 4\cdot 6}\big)^2\tfrac{\varepsilon^6}{5} - \big(\tfrac{1\cdot 3\cdot 5\cdot 7}{2\cdot 4\cdot 6\cdot 8}\big)^2\tfrac{\varepsilon^8}{7} - \cdots\Big]. \end{aligned}$$

A look at Figure 10.20 tells us that the circumference of the surrounding circle $x^2 + y^2 = a^2$ is greater than that of the ellipse $\frac{x^2}{a^2} + \frac{y^2}{b^2} = 1$ inside it. The circumference of the circle is $2a\pi$, and the formula above tells us how much has to be subtracted to get the smaller circumference of the ellipse. If the ellipse is a circle, then $\varepsilon = 0$, and the formula—as expected—tells us that the circumference is $2a\pi$. The next two problems require data from Table 10.1 and a calculator.

11.55. Let's compute the circumference of Earth's orbit around the Sun. This ellipse has major axis $a =$ 149,598,000 km and eccentricity $\varepsilon = 0.016711$. Show that the length of the orbit is

$$2\pi(149{,}598{,}000)(1 - \tfrac{1}{4}\varepsilon^2 - \tfrac{3}{64}\varepsilon^4 - \tfrac{45}{2304}\varepsilon^6 - \tfrac{1575}{147456}\varepsilon^8 - \cdots)$$

$$\approx 939{,}952{,}000(1 - 0.000070 - 0.000000004 - 0.000000000000000065) \approx 939{,}886{,}000 \text{ km}.$$

Since the eccentricity ε is close to zero, the infinite series for the circumference of the Earth's ellipse converges rapidly. Comets usually have eccentricities that are close to one. In such situations, the convergence is slower. Halley's comet is an example. Since 1 au (an astronomical unit) is equal to 149,597,892 km (refer to Section 3.7), it follows that Halley's semimajor axis is $a \approx 17.83$ au.

11.56. With $a \approx 17.83$ and $\varepsilon = 0.967$, we get that Halley's orbit has a circumference of

$$2\pi(17.83)(1 - \tfrac{1}{4}\varepsilon^2 - \tfrac{3}{64}\varepsilon^4 - \tfrac{45}{2304}\varepsilon^6 - \tfrac{1575}{147456}\varepsilon^8 - \tfrac{99225}{14745600}\varepsilon^{10} - \tfrac{9823275}{2123366400}\varepsilon^{12} - \cdots)$$

$$\approx 112.03(1 - 0.23377 - 0.04099 - 0.01597 - 0.00817 - 0.00481 - 0.00309 - 0.00211 - 0.00150)$$

$$\approx 77.25 \text{ au}.$$

Given that one year has 31,558,000 seconds, we get $\frac{939{,}886{,}000}{31{,}558{,}000}$ km/s ≈ 29.78 km/s for the average speed of Earth in its orbit. Since the period of Halley's orbit is 75.32 years, its average orbital speed is $\frac{77.25}{75.32} \approx 1.03$ au/year. This translates to

$$1.03\,\tfrac{\text{au}}{\text{year}} \approx \tfrac{149{,}598{,}000\,\text{km}}{1\,\text{au}} \cdot \tfrac{1\,\text{yr}}{31{,}558{,}000\,\text{s}} \approx 4.88\,\tfrac{\text{km}}{\text{s}}.$$

A look at the last column of Table 10.1 confirms the average speed of 29.78 km/s for the orbiting Earth, and the value 4.88 km/s for Halley fills in the blank of the lower right corner of the table.

11.57. Show that for any $\varepsilon < 1$,

i. $\sqrt{1-\varepsilon^2\cos^2\theta} = 1 - \sum_{k=1}^{\infty} \frac{1\cdot 3\cdot 5\cdots(2k-3)\varepsilon^{2k}}{2^k k!}\cos^{2k}\theta$ and

ii. $\int_0^{\frac{\pi}{2}} \sqrt{1-\varepsilon^2\cos^2\theta}\,d\theta = \frac{\pi}{2} - \sum_{k=1}^{\infty} \frac{1\cdot 3\cdot 5\cdots(2k-3)\varepsilon^{2k}}{2^k k!}\int_0^{\frac{\pi}{2}}\cos^{2k}\theta\,d\theta.$

11.58. Show by integration by parts that for any $n \geq 2$,

$$\int \cos^n\theta\,d\theta = \tfrac{1}{n}\sin\theta\cos^{n-1}\theta + \tfrac{n-1}{n}\int\cos^{n-2}\theta\,d\theta,$$

and that by the repeated use of this formula, $\int_0^{\frac{\pi}{2}}\cos^{2k}\theta\,d\theta = \frac{1}{2}\cdot\frac{3}{4}\cdot\frac{5}{6}\cdots\frac{2k-1}{2k}\cdot\frac{\pi}{2}$ for any $k \geq 1$. Is it surprising that this is the same conclusion as that of Problem 11.53 for $\sin^{2k}\theta$?

11.59. Use the above calculations to conclude that for $\varepsilon < 1$,

$$\int_0^{\frac{\pi}{2}}\sqrt{1-\varepsilon^2\cos^2\theta}\,d\theta = \int_0^{\frac{\pi}{2}}\sqrt{1-\varepsilon^2\sin^2\theta}\,d\theta.$$

11H. Galileo Drops Cannonballs. According to a biography written by his student Vincenzo Viviani, Galileo dropped two balls of different weights from the Leaning Tower of Pisa late in the 16th century to demonstrate that their time of descent does not depend on their weight. Today's historians have largely dismissed this account and think that Galileo most likely never dropped balls from the Leaning Tower. But let's suppose that Galileo did carry out this experiment. Let's assume that in this hypothetical experiment the ball of lesser weight was also smaller, so that the resisting drag of the air on it would have been less. This raises the question as to whether air resistance would have interfered with Galileo's experiment in any significant way. This is the question that we'll now consider.

An English "gunner's rule" from about 1590 (this is a compass-like instrument made of brass, used for measuring gun elevations and the sizes of cannonballs, and for land surveying) lists spherical cannonballs made of cast iron weighing from $3\frac{1}{4}$ to 36 pounds with diameters ranging from $2\frac{7}{8}$ inches to $6\frac{3}{8}$ inches. It is known that the density of one type of cast iron used at the time was 0.2682 pounds/in^3. It seems reasonable that cannonballs of such sizes and made of cast iron would have existed elsewhere in Europe. Using the conversions 1 pound $\approx$ 0.45 kg and 1 inch $\approx$ 0.025 m, we can therefore assume that Galileo could have gotten his hands on cast iron cannonballs of mass about 16 kg, diameter of about 0.16 m, and density of about 7700 kg/m^3. We'll take the gravitational constant is g to be $g = 9.81$ m/s^2.

We'll let two such cannonballs fall side by side from a height of 56 meters (the approximate height of the Leaning Tower of Pisa), one in a vacuum and the other in air, and both with zero initial velocity. See Figure 11.32. We already know that the ball falling in a vacuum will win the race. But by how much, both in terms of time and distance? The one on the left will have the greater speed at impact. Is this difference substantial?

We'll start with the simpler analysis of the cannonball falling in a vacuum. It will do so from rest at time $t = 0$ from the top of the tower. We saw in Section 6.5 that the cannonball's velocity and height above the ground at any time $t \geq 0$ (in seconds) into its fall are

$$v_{\text{vac}}(t) = -gt + v(0) = -9.81t \quad\text{and}\quad y_{\text{vac}}(t) = -\tfrac{1}{2}gt^2 + y_0 = -4.905t^2 + 56$$

in meters per second and meters, respectively. At the time t_1 of impact, $y_{\text{vac}}(t_1) = -4.905t_1^2 + 56 = 0$, so that $t_1 = \sqrt{\frac{56}{4.905}} \approx 3.3789$ s. The velocity of impact is $v(t_1) = (-9.81)(3.3789) \approx 33.15$ m/s.

11.60. Check the figures in the second and fourth columns of Table 11.3. (Round off to two decimal places.)

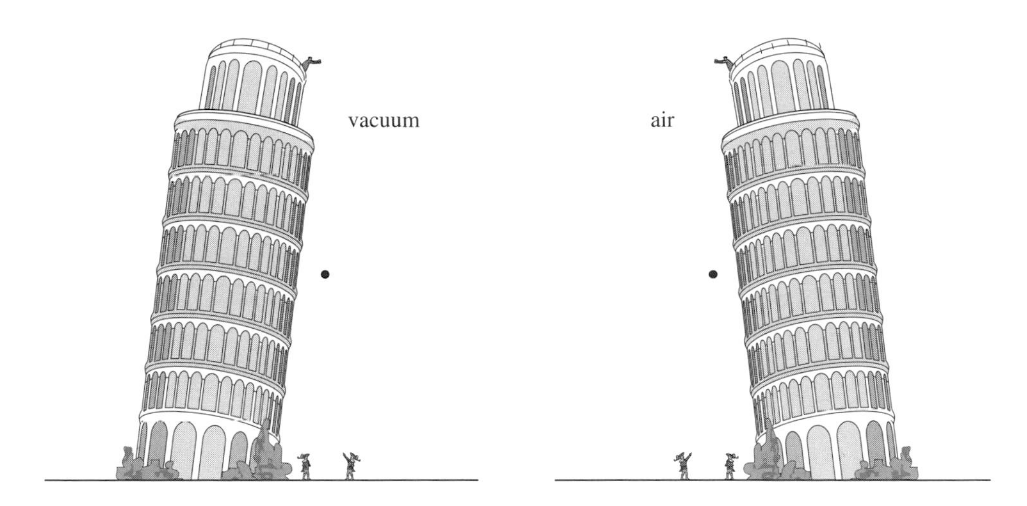

Fig. 11.32

Our discussion next turns to the cannonball in Figure 11.32 on the right, the one falling in air. We'll make use of the results developed in Section 11.10, in particular, the "coming down" segment already illustrated by the ping-pong ball of Section 11.10.3. In the situation of the falling cannonball, $t_{\text{top}} = 0$ and $y_{\max} = y_0 = 56$. Therefore the velocity $v_{\downarrow}(t) = v_{\text{air}}(t)$ of the cannonball falling in air and its height $y_{\downarrow}(t) = y_{\text{air}}(t)$ above ground at any time $t \geq 0$ are given by

$$v_{\text{air}}(t) = -s_\infty \tanh \tfrac{g}{s_\infty} t \quad \text{and} \quad y_{\text{air}}(t) = y_0 - \tfrac{s_\infty^2}{g} \ln \cosh\left(\tfrac{g}{s_\infty} t\right).$$

The terminal speed s_∞ of the cannonball (this is the speed it would reach if it could fall forever) is

$$s_\infty = \sqrt{\tfrac{2mg}{C_d \rho \pi r^2}} = \sqrt{\tfrac{2(16)(9.81)}{(0.4)(1.22)\pi(0.08^2)}} \approx 178.868827 \text{ m/s}.$$

It follows that $\frac{g}{s_\infty} \approx 0.054845 \, \frac{1}{\text{s}}$ and $\frac{s_\infty^2}{g} \approx 3261.371791$ m. Therefore, at any time $t \geq 0$,

$$v_{\text{air}}(t) \approx -(178.868827)\tanh(0.054845\, t) \quad \text{and} \quad y_{\text{air}}(t) \approx 56 - (3261.371791)\ln\cosh(0.054845\, t).$$

The formula for the time t_{imp} of impact tells us that $t_{\text{imp}} = 0 + \frac{s_\infty}{g} \cosh^{-1} e^{\frac{g}{s_\infty^2} y_{\max}} = 3.388550$ s. Table 11.3 collects the information for the flights of the two cannonballs, for $t = 1, 2,$ and 3 seconds as well as the

Time (seconds)	Height $y_{\text{vac}}(t)$ in Vacuum (m)	Height $y_{\text{air}}(t)$ in Air (m)	Speed $v_{\text{vac}}(t)$ in Vacuum (m/sec)	Speed $v_{\text{air}}(t)$ in Air (m/sec)
$t = 0$	56.00	56.00	0	0
$t = 1$	51.10	51.10	9.81	9.80
$t = 2$	36.38	36.42	19.62	19.54
$t = 3$	11.86	12.05	29.43	29.17
$t = 3.3789$	0	0.32	33.15	32.73
$t = 3.3886$	—	0	—	32.86

Table 11.3

two moments of impact. This information is numerically delicate. Even small changes in the constants in combination with round-off errors can produce very different results. In the calculations above it is assumed that the given data has an accuracy of four significant figures (for example, the radius of the cannonball is understood to be 0.08 = 0.08000 m), that the computations were carried out with six decimal places, and then rounded off at two decimal places.

11.61. Check the figures of the third and fifth columns of Table 11.3. (Work with six decimal places and round off.)

As expected, the cannonball falling in a vacuum reaches the ground first. But not by much! At the instant the cannonball traveling in a vacuum hits the ground, the one falling in air is only about 0.32 m (a little over 1 foot) above the ground. The cannonball falling in a vacuum smacks into the ground about $\frac{1}{100}$th of a second before the one falling in air. At this instant, the difference in the speeds of the two balls is about 33.15 − 32.73 m/s (less than 1.5 feet per second). If these differences seem surprisingly small, they shouldn't. The distance of fall of 56 m is not enough to bring them out. The time of fall would have to be greater for the impact of air resistance to become apparent. In any case, had Galileo carried this experiment out as described, the impact of air resistance would have been minimal. Galileo's estimate for the value of g was not accurate. Had he carried this experiment out and been able to measure the time of fall to be 3 s (or its equivalent in the units he used), what value for g would he have achieved?

11.62. **i.** Suppose the two cannonballs are dropped from rest from the 381 m high deck of the Empire State Building. Show that the ball falling in a vacuum takes about 8.81 s to fall to the street and the one falling in air about 8.99 s. Show that they make impact at the speeds of about 86.46 and 81.65 m/s, respectively. (The differences are still relatively small.)

ii. Suppose the two cannonballs are dropped from a height of 2000 m from a hot air balloon. Show that the ball falling in a vacuum takes about 20.19 s to fall to the ground and the one falling in air about 22.31 s. Show that they hit the ground at speeds of about 198.09 and 150.36 m/s, respectively. (Now the differences are more pronounced.)

11I. Going Up and Coming Down. Return to the discussion of the ball of Section 11.10. Recall that it is propelled upward at time $t = 0$ from a starting height y_0 off the ground with an initial velocity of v_0, that it reaches its maximal height y_{max} at time $t_{top} = \frac{s_\infty}{g}\tan^{-1}\frac{v_0}{s_\infty}$, and that it then begins its descent. The formula

$$y_\downarrow(t) = y_0 + \frac{s_\infty^2}{g}\ln\left[\frac{\sqrt{\left(\frac{v_0}{s_\infty}\right)^2+1}}{\cosh\left(\frac{g}{s_\infty}(t-t_{top})\right)}\right]$$

tells us how high off the ground it is at any time t after t_{top}. Let t_{ret} be the instant at which the ball returns to its starting point y_0.

11.63. Use a property of the natural logarithm to show that $t_{ret} - t_{top} = \frac{s_\infty}{g}\cosh^{-1}\sqrt{(\frac{v_0}{s_\infty})^2 + 1}$.

The fact that the ball starts its ascent at time $t = 0$ and reaches y_{max} at the instant $t_{top} = \frac{s_\infty}{g}\tan^{-1}\frac{v_0}{s_\infty}$ tells us that the time it takes for the ball to ascend from y_0 to its highest point is $t_{up} = \frac{s_\infty}{g}\tan^{-1}\frac{v_0}{s_\infty}$. The formula for $t_{ret} - t_{top}$ above provides the time t_{down} of travel of the ball from its highest point y_{max} to its starting point y_0. Which is greater, t_{down} or t_{up}? Think about the two forces involved. Going up, gravity and drag act in the same direction against the motion of the object. Going down, they act in opposite directions with gravity the stronger. This suggests that the ball will take longer to go down that to go up. A comparison of the functions

$$\tan^{-1}x \quad\text{and}\quad \cosh^{-1}\sqrt{x^2+1}$$

for $x \geq 0$ confirms this.

11.64. **i.** Make use of facts from Section 9.9 to show that $\frac{d}{dx}\tan^{-1}x = \frac{1}{x^2+1}$ and that for $x > 0$,

$$\frac{d}{dx}\cosh^{-1}\sqrt{x^2+1} = \frac{1}{\sqrt{\left(\sqrt{x^2+1}\right)^2-1}}\cdot\tfrac{1}{2}(x^2+1)^{-\frac{1}{2}}2x = \frac{1}{\sqrt{x^2+1}}.$$

ii. Explain why $\frac{d}{dx}\cosh^{-1}\sqrt{x^2+1} > \frac{d}{dx}\tan^{-1}x$ for $x > 0$. Since $\cosh^{-1}\sqrt{0+1} = 0 = \tan^{-1}(0)$, conclude that $\cosh^{-1}\sqrt{x^2+1} > \tan^{-1}x$ for all $x > 0$.

iii. It follows that $t_{\text{down}} = \frac{s_\infty}{g}\cosh^{-1}\sqrt{(\frac{v_0}{s_\infty})^2+1} > \frac{s_\infty}{g}\tan^{-1}\frac{v_0}{s_\infty} = t_{\text{up}}$.

So as we had suspected, $t_{\text{down}} > t_{\text{up}}$. The ratio $(\frac{v_0}{s_\infty})^2$ has a concrete interpretation in terms of the forces involved. Refer to Section 11.10 for the equality $s_\infty^2 = \frac{2mg}{C_d\rho\pi r^2}$ and the fact that $(\frac{v_0}{s_\infty})^2 = \frac{\frac{1}{2}C_d\rho\pi r^2 v_0^2}{mg}$ is the ratio of the initial drag on the ball over the pull of gravity on it. Our final concern is the velocity of the ball at time t_{ret}.

11.65. **i.** Show that the ball's velocity at time t_{ret} is $v_\downarrow(t_{\text{ret}}) = -s_\infty \tanh\left(\cosh^{-1}\sqrt{(\frac{v_0}{s_\infty})^2+1}\right)$.

ii. Use the identity $\cosh^2 x - \sinh^2 x = 1$ to show that for any $x \geq 0$, $\tanh x = \sqrt{1 - \frac{1}{(\cosh x)^2}}$ and $\tanh(\cosh^{-1}x) = \sqrt{1-\frac{1}{x^2}} = \sqrt{\frac{x^2-1}{x^2}}$.

iii. Conclude that $v_\downarrow(t_{\text{ret}})$ is more concretely expressed as $v_\downarrow(t_{\text{ret}}) = \frac{-v_0}{\sqrt{\left(\frac{v_0}{s_\infty}\right)^2+1}}$.

11J. More about Springs and Suspension Systems. Let a spring with spring constant k be suspended vertically with its upper end fixed in place. Attach a block of mass m to the free end of the spring, and let the system move into a state of balance. The spring is now stretched h units beyond its natural length, and the block at the end of the spring hangs motionlessly. Explain why $kh = mg$, where g is the gravitational constant. Figure 11.33 shows the suspended spring and the block in this position. A vertical y-axis has

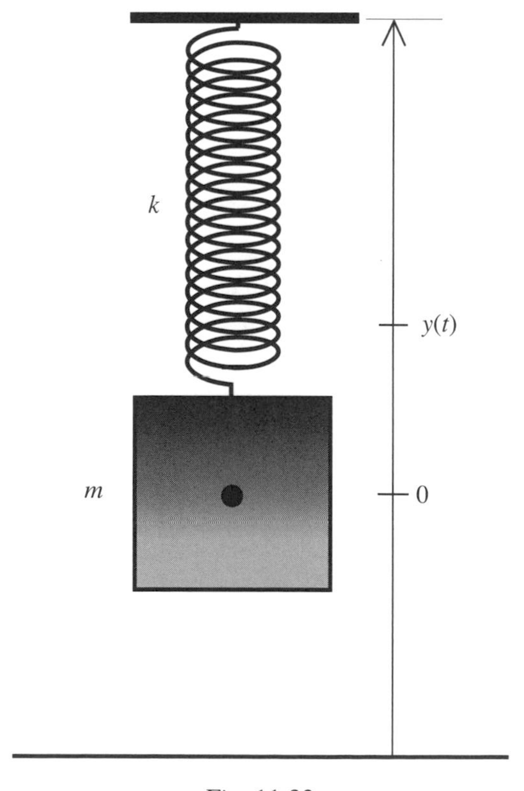

Fig. 11.33

been placed in such a way that the center of mass of the block is at $y = 0$.

Push or pull the block into position $y = d_0$ and release it at time $t = 0$ with initial velocity v_0. Let $y(t)$ be the position of the center of mass of the block at any time $t \geq 0$. Observe that $y(0) = d_0$ and $y'(0) = v_0$.

11.66. **i.** After carefully studying the figure, find an expression for the net force $F(t)$ on the block at any time $t \geq 0$ (analyzing both $y(t) \geq 0$ and $y(t) < 0$), and show that $y(t)$ satisfies the differential equation

$$y''(t) = \tfrac{-k}{m}\, y(t).$$

ii. Use results from Section 11.6 to show that $y(t) = d_0 \cos\left(\sqrt{\tfrac{k}{m}}t\right) + v_0\sqrt{\tfrac{m}{v}} \sin\left(\sqrt{\tfrac{k}{m}}t\right)$.

iii. If $v_0 = 0$, how long will it take for the block to move through one complete up-and-down cycle? [Hint: Refer to the graph of the cosine in Section 4.6 and observe that for one complete cycle, we need to have $0 \leq \sqrt{\tfrac{k}{m}}\, t \leq 2\pi$.]

We'll now make specific assumptions about Figure 11.33, including the spring, the block suspended from it, and the initial state of the system. Solve Problem 11.67 by applying the conclusion of Problem 11.66.

11.67. Suppose that the mass of the block is $m = 0.5$ kg, and that a force of 40 N is necessary to keep the spring extended 0.05 m beyond its natural length. Suppose that the block is pulled down 0.1 m from the position that it has at equilibrium and released at time $t = 0$ with a downward speed of 2 m/s. Find the position function $y(t)$ explicitly.

A high-performance race car has a total mass of about 726 kg, including the fuel and driver. Its distributed sprung mass is $m = 118.0$ kg and $m = 186.0$ kg per front and rear wheel, respectively. The suspension of each front wheel has a spring rate of $k = 525{,}000$ N/m and a damping ratio of $\zeta = 0.70$ and that of each rear wheel has a spring rate a damping ratio of 612,000 N/m and $\zeta = 0.90$, respectively. The race car encounters bumps on the track. One of them drives the chassis over a front wheel up by 5 cm = 0.05 m and another does the same thing to the chassis over one of the rear wheels.

11.68. Use Section 11.11.1 as a guide to study the responses of the two suspensions. In each case, assume that the chassis reaches the top of its displacement at time $t = 0$ and that the position function $z(t)$ satisfies $z(0) = 0.05$ m and $z'(0) = 0$ m/s.

i. Determine the coefficients of the differential equation $mz''(t) + dz'(t) + kz(t) = 0$ in each case.

ii. Find the particular solution of each equation.

iii. Use the graphing calculator https://www.desmos.com/calculator (or an equivalent version) to graph the two solutions.

iv. Is a comparison of the two graphs consistent with the data that were supplied?

11K. About the Tension in the Catenary. In Part 2 of the discussion in Section 11.12 of the cable (rope, cord, or chain) that hangs freely and supports no load except itself, the tension T_0 at the bottom of the cable was an unspecified parameter. This includes the determination of the function

$$y = f(x) = \tfrac{T_0}{c}\left(\cosh(\tfrac{c}{T_0}x) - 1\right),$$

where c the weight c of the cable per unit length, that pinpoints the curve of the cable. Without knowing T_0, this conclusion is incomplete. In the case of the parabolic cable of a suspension bridge, this tension is determined by $T_0 = \frac{1}{2}\frac{wd^2}{s}$, where w is the weight per unit length of the deck, and the location (d, s) of the end of the cable (where d is one-half of the length of the center the span, and s is the sag.) Refer to Section 8.3 for the details. This raises the question as to whether there is a similar expression for the tension T_0 of the cable that hangs freely. Is it the case that T_0 is determined by c and its endpoint? We will see that this is so, but that it is much less explicit and more complex than in the earlier case of the suspension bridge.

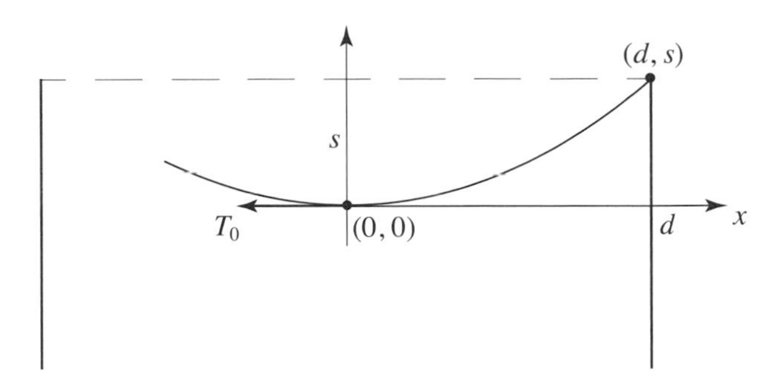

Fig. 11.34

Figure 11.34 represents the freely hanging cable in the coordinate plane as the graph of the function $y = f(x) = \frac{T_0}{c}(\cosh(\frac{c}{T_0}x) - 1)$. The point (d, s) is the point where the cable is supported by the post.

Let $b = \frac{c}{T_0}$. Since (d, s) is on the graph, $s = \frac{1}{b}(\cosh bd - 1)$. It follows from the definition $\cosh x = \frac{e^x + e^{-x}}{2}$ of the hyperbolic cosine, that $e^{bd} + e^{-bd} = 2(1 + bs)$.

11.69. Show that $(e^{bd})^2 - 2(1 + bs)e^{bd} + 1 = 0$ and that therefore

$$e^{bd} = \frac{2(1+bs) \pm \sqrt{4(1+bs)^2 - 4 \cdot 1 \cdot 1}}{2} = 1 + bs \pm \sqrt{(bs)^2 + 2bs}.$$

i. Show that $e^{bd} > 1$, and hence that $e^{bd} = 1 + bs + \sqrt{(bs)^2 + 2bs}$. With $x = bs$, we get $e^{\frac{d}{s}x} = 1 + x + \sqrt{x^2 + 2x}$.

ii. Therefore $x = bs = \frac{cs}{T_0}$ is the x-coordinate of a point of intersection of the curves $y = e^{\frac{d}{s}x}$ and $y = 1 + x + \sqrt{x^2 + 2x}$.

11.70. Consider the function $g(x) = 1 + x + \sqrt{x^2 + 2x}$ with domain restricted to $x \geq 0$.

i. Show that $g'(x) = 1 + \frac{x+1}{(x^2+2x)^{\frac{1}{2}}} = 1 + \sqrt{\frac{x^2+2x+1}{x^2+2x}} > 2$. Conclude that the graph of $y = g(x)$ increases over $[0, \infty)$ and that it has a vertical tangent at the point $(0, 1)$.

ii. Show that $g''(x) = \frac{-1}{(x^2+2x)^{\frac{3}{2}}}$, and conclude that the graph of $y = g(x)$ is concave down.

iii. Check that $2x + 2 - g(x) = 1 + x - \sqrt{x^2 + 2x} = \frac{1}{1+x+\sqrt{x^2+2x}}$, and use this result to verify that the line $y = 2x + 2$ is an asymptote of the graph of $y = g(x)$. Show that $2x + 2 > g(x)$ for all $x \geq 0$.

iv. Sketch the graph of $y = g(x)$.

11.71. Check that the function $g(x) = 1 + x + \sqrt{x^2 + 2x}$ is also defined for $x \leq -2$ (but not for $-2 < x < 0$). Use the strategies of Problem 11.70, to show that its graph is decreasing over $(-\infty, -2]$, that it is concave down, that it lies below the x-axis, and that the x-axis is a horizontal asymptote. Draw the graph of the function $g(x)$ for $x \leq -2$.

Since the interest is in points of intersection of $y = e^{\frac{d}{s}x}$ and $y = g(x) = 1 + x + \sqrt{x^2 + 2x}$, we need to consider $x \geq 0$ only (in view of Problem 11.71). Figure 11.35 shows the graphs of a typical $y = e^{\frac{d}{s}x}$ and $g(x) = 1 + x + \sqrt{x^2 + 2x}$ together and illustrates how they intersect. Our discussion has shown that $x = \frac{cs}{T_0}$ is the x-coordinate of the unique point of intersection of the curves $y = e^{\frac{d}{s}x}$ and $y = 1 + x + \sqrt{x^2 + 2x}$ that satisfies $x > 0$. This point depends only on $y = e^{\frac{d}{s}x}$ and hence only on the ratio $\frac{d}{s}$. It follows that $\frac{d}{s}$ determines $\frac{cs}{T_0}$. As a consequence, the tension T_0 is determined by the parameters c, d, and s. For any $\frac{d}{s}$, the graphing calculator https://www.desmos.com/calculator provides the corresponding x, namely, the

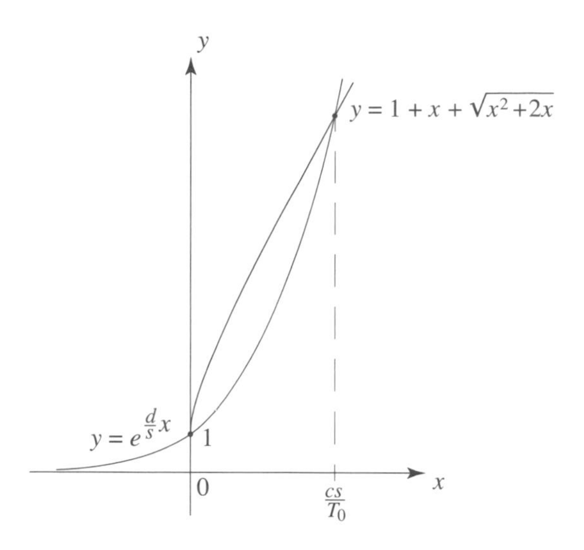

Fig. 11.35

solution of the equation $e^{\frac{d}{s}x} - 1 - x - \sqrt{x^2 + 2x} = 0$.

Recognizing that the distance d is usually much larger than s (see Figure 11.34), we'll consider $\frac{d}{s} = \frac{1}{10}, \frac{1}{5}, 1, 5, 10, 20, \ldots, 100$. Table 11.4 collects approximations for the values $\frac{cs}{T_0}$ that correspond to each of these values of $\frac{d}{s}$. Notice that the larger the ratio $\frac{d}{s}$, the smaller the term $\frac{cs}{T_0}$ and hence the larger

$\frac{d}{s}$	$\frac{1}{10}$	$\frac{1}{5}$	1	5	10	20	40	60	80	100
$\frac{cs}{T_0}$	45.28	18.25	1.616	0.0790	0.0199	0.0050	0.00125	0.00056	0.00031	0.00020

Table 11.4. T_0 for the cable.

the tension T_0.

Now that we have a way of determining T_0, we'll continue our earlier analysis to obtain more information about the freely hanging cable. By results of Sections 9.3 and 9.6, the length $L(x)$ of the cable over the interval $[0, x]$ is given by

$$L(x) = \int_0^x \sqrt{1 + f'(t)^2}\, dt.$$

The solution of this integral turns out to be an easy consequence of the properties of the hyperbolic functions in Section 7.12. Since $f(x) = \frac{T_0}{c}(\cosh(\frac{c}{T_0}x) - 1)$, we find that $f'(x) = \frac{T_0}{c}\sinh(\frac{c}{T_0}x) \cdot \frac{c}{T_0} = \sinh(\frac{c}{T_0}x)$, hence

$$\sqrt{1 + f'(t)^2} = \sqrt{1 + \sinh^2(\tfrac{c}{T_0}t)} = \sqrt{\cosh^2(\tfrac{c}{T_0}t)} = \cosh \tfrac{c}{T_0}t,$$

and therefore

$$L(x) = \int_0^x \sqrt{1 + f'(t)^2}\, dt = \tfrac{T_0}{c}\sinh \tfrac{c}{T_0}t\,\Big|_0^x = \tfrac{T_0}{c}\sinh \tfrac{cx}{T_0}.$$

We will let $L = L(d) = \frac{T_0}{c}\sinh\frac{cd}{T_0}$ be the length of the cable (rope, cord, or chain) over $[0, d]$. See Figure 11.34.

Refer to the discussion at the beginning of Section 11.12. Let $T(x)$ be the tension in the cable for any x with $0 \le x \le d$. Using the fact that $w = 0$ in the current situation of a cable that supports only itself and no other load, we get

$$T(x)\sin\theta(x) = \int_0^x c\sqrt{1 + f'(t)^2)}\, dt \text{ and } T(x)\cos\theta(x) = T_0.$$

It follows that $T(x)\sin\theta(x) = cL(x) = T_0 \sinh\frac{cx}{T_0}$, and therefore that

$$T(x)^2 = T(x)^2\sin^2\theta(x) + T(x)^2\cos^2\theta(x) = T_0^2\sinh^2\tfrac{cx}{T_0} + T_0^2 = T_0^2\cosh^2\tfrac{cx}{T_0}.$$

Therefore

$$T(x) = T_0\cosh\tfrac{cx}{T_0}.$$

Using the formula $L(x) = \frac{T_0}{c}\sinh\frac{cx}{T_0}$, we get $\sinh\frac{cx}{T_0} = \frac{c}{T_0}L(x)$, and therefore

$$T(x) = T_0\cosh\tfrac{cx}{T_0} = T_0\sqrt{1+\sinh^2\tfrac{cx}{T_0}} = T_0\sqrt{1+\tfrac{c^2}{T_0^2}L(x)^2} = \sqrt{T_0^2 + c^2L(x)^2}.$$

The substitutions $x = 0$ and $x = d$ tell us the minimal and maximal tensions are T_0 and $T(d) = \sqrt{T_0^2 + c^2L^2}$, respectively.

Consider a chain that has a mass m of 0.15 kg/m. Since its weight in newtons per meter is $mg = (0.15)(9.81) \approx 1.47$, we'll take $c = 1.47$ N/m. The two supporting posts of the chain have the same height and are a distance $2d = 20$ m apart. The next three problems study this chain under different sag conditions (and lengths).

11.72. **i.** Suppose that the sag in the chain is $s = 10$ m. Since $\frac{d}{s} = 1$, Table 11.4 tells us that $\frac{cs}{T_0} \approx 1.616$. Show that $T_0 \approx \frac{1}{1.616}cs \approx \frac{(1.47)(10)}{1.616} = 9.10$ N.

ii. Show that the length of the chain is $2L \approx 29.91$ m. So the weight of the chain is approximately 43.96 N, and the maximal tension that the chain is under is 23.79 N.

11.73. **i.** Let's shorten the chain so that its sag is $s = 1$ m. Now $\frac{d}{s} = 10$, and the table tells us that $\frac{cs}{T_0} \approx 0.0199$. Show that $T_0 \approx 73.87$ N.

ii Verify that the length of the chain is $2L \approx 20.13$ m. Hence the weight of the chain is approximately 29.59 N, and the maximal tension that the chain is under is 75.35 N.

11.74. **i.** Let's shorten the chain some more so that its sag is $s = 0.1$ m. Since $\frac{d}{s} = 100$, the table tells us that $\frac{cs}{T_0} \approx 0.0002$. Show that $T_0 \approx 735$ N.

ii. Show that the length of the chain is $2L \approx 20.00$ m. So the weight of the chain is approximately 29.40 N, and the maximal tension that the chain is under is 735.15 N.

Notice that the tension in the chain of Problem 11.72 varies considerably. The tensions of the chains in Problems 11.73 and 11.74 are close to being constant. The tension in the chain increases dramatically each time the ratio $\frac{d}{s}$ is increased. Can you provide explanations for these observations?

Chapter 12

Polar Calculus and Newton's Planetary Orbits

This chapter flows as follows. It starts with a study of equations and functions in polar coordinates and their graphs. The trigonometric functions get special attention and the uniform approach they provide to parabolas, ellipses, and hyperbolas is developed. The chapter then turns to the differential and integral calculus of polar functions. The meaning of the polar derivative is examined. The definite integral is explored in the context of the lengths of graphs of polar functions and the areas that the graphs encircle. Equiangular spirals are introduced and seen to play a role in the shell structures of the nautilus and the shapes of whirling low pressure weather systems. The chapter then comes full circle by returning to astronomical themes already explored in Chapters 1, 3, and 6. It applies polar functions to the study of centripetal force and revisits the connection between the magnitude of a centripetal force and the resulting trajectory of the object that it propels. We will see—now in the polar context—that if the force satisfies an inverse square law, then the trajectory is an ellipse, a parabola, or a hyperbola, with the force acting in the direction of a focal point, and, conversely, that if the trajectory has these properties, then the force satisfies an inverse square law. The chapter concludes with a study of the connection between gravitational force and equiangular spirals in the context of the structure of galaxies.

Newton approached the planetary orbits by combining the three fundamental laws of motion that he formulated with the mathematics of calculus that he developed. His success in doing so is a remarkable achievement. It is brilliant both in terms of its grand vision and the details of its execution. The universal law of gravitation is an example of the former, and the assumption that gravitational force acts intermittently "machine-gun-style" is an example of the latter. His law of gravitation shows that the physics developed here on Earth penetrates to the far reaches of the universe. His assumption about the way gravitational force acts leads to a triangulation of the curving orbit that enabled Newton to overcome the complexities of the analysis. (See Sections 6.8 and 6.9.) It seems miraculous that this assumption—at odds with the continuous and smooth pull of the Sun's gravitational force—allowed Newton to extract the essence of the matter and to provide the correct explanations. The basic situation of a fixed center of force and an object curving around it, suggests that polar coordinates (introduced in Section 11.4) should be a more appropriate construct for this study than the rectangular Cartesian framework. And they are! Polar coordinates turn out to be tailor-made for the mathematical investigation of a centripetal force and the geometry of the trajectories that it generates. This chapter revisits the subject of centripetal force, the inverse square law, and the geometry of orbits, but this time under the assumption of the continuity and smoothness of the action of the force. Unlike the discussion of Newton's work in Chapter 6, which restricted itself to elliptical orbits and left some important points unattended to—the verification that $\lim_{Q\to P} \frac{QR}{QT^2} = \frac{1}{L}$ for example—the analysis in this chapter is complete and includes the parabolic and hyperbolic cases within the uniform approach that polar calculus makes possible. The chapter also considers an issue that had vexed Newton for some time (and apparently led to delays in the publication of the *Principia*). This is the question as to whether it is possible to assume in the context of gravitational attraction that the Sun as well as the planets have their masses concentrated at their centers.

12.1 GRAPHING POLAR EQUATIONS

We are given a plane equipped with a polar coordinate system. See Section 11.4 for the details. The *graph* of an equation involving the variables r and θ is the set of all points in the plane for which at least one pair of polar coordinates satisfies the equation.

For example, the graph of the equation $\theta = \frac{5\pi}{4}$ is the set of all (r, θ) with $\theta = \frac{5\pi}{4}$. This set of points consists of the line through the ray $\theta = \frac{5\pi}{4}$, in other words, the combination of the two rays $\theta = \frac{\pi}{4}$ and $\theta = \frac{5\pi}{4}$. Refer to Figure 12.1a. Notice that all points with coordinates $(r, -\frac{3\pi}{4})$ are on this line (even though none of these coordinates satisfies the given equation). The graph of the equation $r = 5$ is the set of all

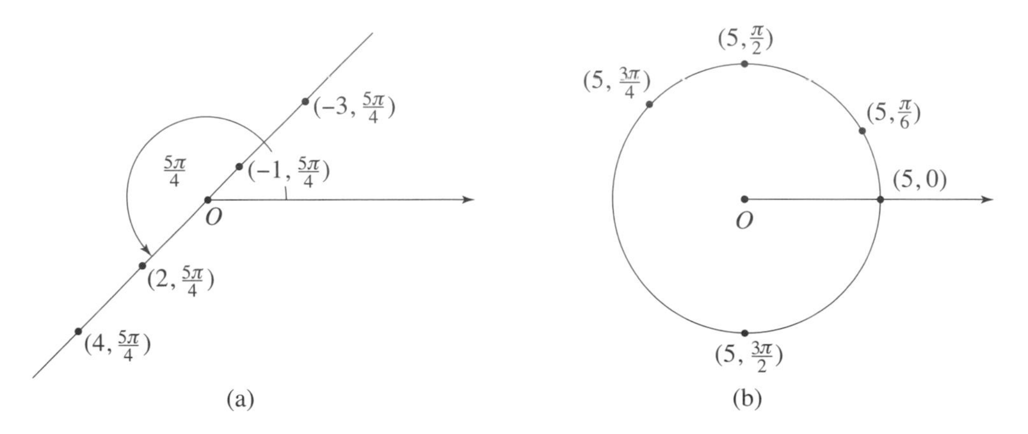

Fig. 12.1

(r, θ) with $r = 5$. Its graph is the circle of radius 5 with center the origin. See Figure 12.1b.

Let's look at the graph of the equation $r = \theta$ next. We'll consider $\theta \geq 0$ first and then $\theta \leq 0$. Plotting points tells us that in each case the graph is a spiral. With $\theta \geq 0$ and increasing, the ray that θ determines rotates counterclockwise, and the corresponding r moves out along this ray. So for $\theta \geq 0$, the spiral opens in a counterclockwise way, as shown in Figure 12.2a. When $\theta \leq 0$, the spiral expands in the same way, but in a clockwise direction. See Figure 12.2b. This spiral was studied by the Greek genius Archimedes,

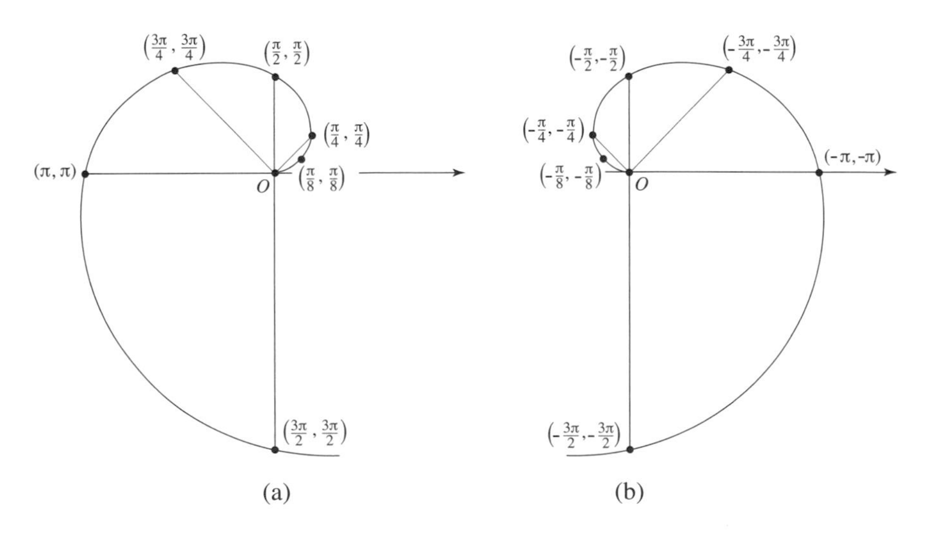

Fig. 12.2

whom we saw in action in Chapter 2. It is called *the spiral of Archimedes.*

When considering the polar graphs of more complicated functions, it will often be of advantage to make use of Cartesian information. To do this, place an xy-coordinate system on top of the polar coordinate system in such a way that the two origins coincide, the polar axis coincides with the positive x-axis, the negative x-axis is obtained by extending this axis to the other side of the origin O, and the y-axis is perpendicular to the x-axis at O. Suppose that this has been done. Let P be any point in the plane, and let (r, θ) be any pair of polar coordinates for P. We saw in Section 11.4 that the Cartesian coordinates of P are

$$x = r\cos\theta \text{ and } y = r\sin\theta\,.$$

Notice therefore that

$$r^2 = x^2 + y^2 \text{ and } \tan\theta = \frac{y}{x}.$$

These relationships hold regardless of the location of the point P and regardless of the way its polar coordinates are chosen.

So if (r, θ) is any set of polar coordinates of P, then the corresponding Cartesian coordinates are directly computed by $x = r\cos\theta$ and $y = r\sin\theta$. Suppose instead that we are given Cartesian coordinates (x, y) for P and need to determine a set of polar coordinates. If P is on the y-axis, then $P = (0, y)$ so that $(y, \frac{\pi}{2})$ are polar coordinates for P. If P is not on the y-axis, then θ can be chosen to satisfy $-\frac{\pi}{2} < \theta < \frac{\pi}{2}$. Because $\tan\theta = \frac{y}{x}$, this θ is determined by

$$\theta = \tan^{-1}\frac{y}{x}.$$

(Refer to Section 9.9.1.) The corresponding r needs to satisfy $r^2 = x^2 + y^2$. So either $r = \sqrt{x^2 + y^2}$ or $r = -\sqrt{x^2 + y^2}$. The location of P in the plane determines which of these two possibilities applies. Refer to the discussion and the examples of Section 11.4.

Let's apply what we have learned by analyzing the polar function $r = f(\theta) = \sin\theta$. We'll begin by recalling the Cartesian graph of $y = \sin x$ from Figure 4.23 of Section 4.6. Replacing the variable x by θ and y by r provides Figure 12.3 and hence the basic relationship between θ and r. The graph of the sine function is increasing and concave down over the interval $0 \le x \le \frac{\pi}{2}$, decreasing and concave down over

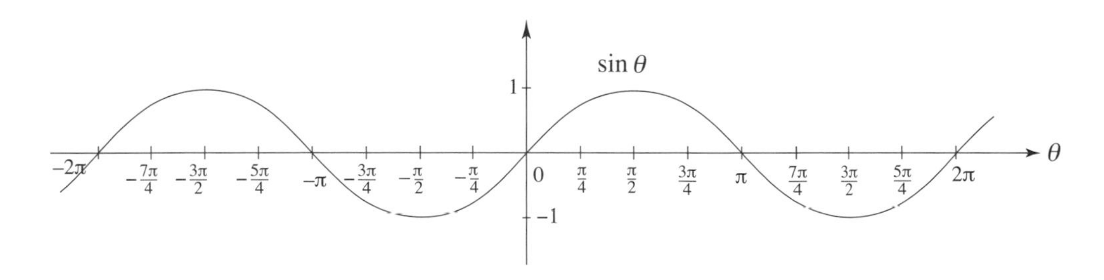

Fig. 12.3

$\frac{\pi}{2} \le x \le \pi$, decreasing and concave up over $\pi \le x \le \frac{3\pi}{2}$, and so on. So the essential properties of the sine function have a uniform description over each of these intervals, as well as the analogous ones in the negative direction. In order to understand the graph of the polar function $r = \sin\theta$, we will therefore consider θ over the interval $0 \le \theta \le \frac{\pi}{2}$, then over $\frac{\pi}{2} \le \theta \le \pi$, then over $\pi \le \theta \le \frac{3\pi}{2}$, and so on. The infor-

1	2	3	4	5	6	7	8
$0 \le \theta \le \frac{\pi}{2}$	$\frac{\pi}{2} \le \theta \le \pi$	$\pi \le \theta \le \frac{3\pi}{2}$	$\frac{3\pi}{2} \le \theta \le 2\pi$	$0 \ge \theta \ge -\frac{\pi}{2}$	$-\frac{\pi}{2} \ge \theta \ge -\pi$	$-\pi \ge \theta \ge -\frac{3\pi}{2}$	$-\frac{3\pi}{2} \ge \theta \ge -2\pi$
$0 \xrightarrow{\sin\theta} 1$	$1 \xrightarrow{\sin\theta} 0$	$0 \xrightarrow{\sin\theta} -1$	$-1 \xrightarrow{\sin\theta} 0$	$0 \xrightarrow{\sin\theta} -1$	$-1 \xrightarrow{\sin\theta} 0$	$0 \xrightarrow{\sin\theta} 1$	$1 \xrightarrow{\sin\theta} 0$

Table 12.1

mation in Table 12.1 comes directly from the graph of Figure 12.3.

Now to the polar graph of $f(\theta) = \sin\theta$. Notice that as the ray determined by the polar axis and θ rotates from $\theta = 0$ to $\theta = \frac{\pi}{2}$, r stretches from $r = 0$ to $r = 1$. So the polar graph for this range of θ is an arc that goes from the point $(0, 0)$ to $(1, \frac{\pi}{2})$. As the ray continues its swing from $\theta = \frac{\pi}{2}$ to $\theta = \pi$, r goes from $r = 1$ to $r = 0$. The corresponding polar graph is an arc from $(1, \frac{\pi}{2})$ back down to $(0, 0)$. It follows that as θ varies from 0 to π, the polar graph is a loop of the form shown in Figure 12.4. As it is sketched, it looks as though this loop is a circle. Is it a circle? How can we be sure? To answer, place an xy-coordinate sys-

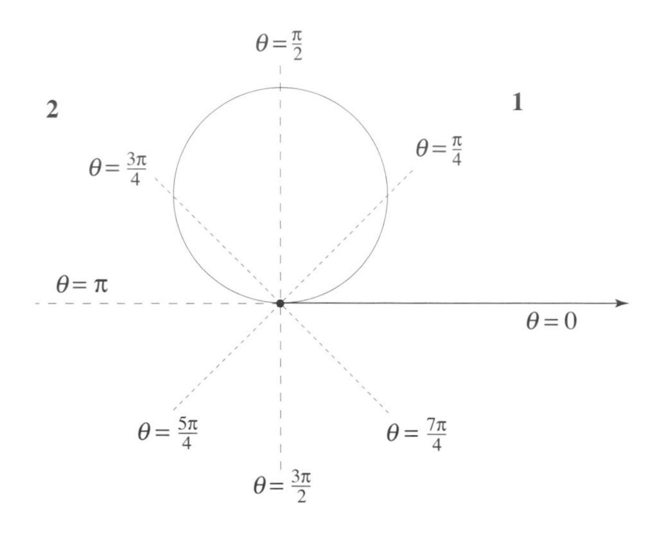

Fig. 12.4

tem over the polar coordinate system as already described. Now let P be a point in the plane. Let (r, θ) be any set of polar coordinates of P and let (x, y) be the Cartesian coordinates of P. In view of the relationships $r^2 = x^2 + y^2$ and $y = r\sin\theta$, we'll multiply $r = \sin\theta$ by r to get $r^2 = r\sin\theta$, so that $x^2 + y^2 = y$. So $x^2 + y^2 - y = 0$ is a Cartesian version of the polar equation $r = \sin\theta$. After completing squares in y, we get $x^2 + y^2 - y + (\frac{1}{2})^2 - (\frac{1}{2})^2 = 0$ or $x^2 + (y - \frac{1}{2})^2 = (\frac{1}{2})^2$. This confirms that the loop in Figure 12.4 is a circle with radius $\frac{1}{2}$ and center $(0, \frac{1}{2})$ in our xy-coordinates (or $(\frac{1}{2}, \frac{\pi}{2})$ in polar coordinates).

Did we arrive at the complete graph of $f(\theta) = \sin\theta$ by considering only θ with $0 \leq \theta \leq \pi$? Or is there more? Is the discussion that identifies the graph as the circle $x^2 + (y - \frac{1}{2})^2 = (\frac{1}{2})^2$ relevant in this regard? Think about the location of the graph of the function $f(\theta) = \sin\theta$ for θ in the intervals of columns 3 and 4 of Table 12.1, then again for the columns 5 and 6, and finally for 7 and 8.

Example 12.1. Consider the polar function $r = f(\theta) = \tan\theta$ for $-\frac{\pi}{2} < \theta < \frac{\pi}{2}$. Refer to Figure 4.26 of Section 4.6. Plot the points corresponding to $\theta = 0, \frac{\pi}{6}, \frac{\pi}{4}, \frac{\pi}{3}, \frac{5\pi}{12}, \frac{11\pi}{24}$, and finally $\frac{23\pi}{48}$. Let $0 \leq \theta < \frac{\pi}{2}$ and convert $r = \tan\theta$ into an equation in the coordinates x and y. Show that $\sqrt{x^2 + y^2} = \frac{y}{x}$ and then that $x = \frac{\frac{y}{x}}{\sqrt{1+(\frac{y}{x})^2}}$. Conclude that the graph of $r = \tan\theta$ for $0 < \theta < \frac{\pi}{2}$ lies between the vertical lines $x = 0$ and $x = 1$. Show that as $\theta \to \frac{\pi}{2}$, the graph of $r = \tan\theta$ approaches the line $x = 1$. Do a similar analysis for $-\frac{\pi}{2} < \theta \leq 0$. Then sketch the polar graph of the equation $r = \tan\theta$ for $-\frac{\pi}{2} < \theta < \frac{\pi}{2}$.

This paragraph has illustrated a strategy that will pay dividends again shortly: to study a polar function, place a Cartesian xy-coordinate system on top of the polar coordinate system, convert the equation in r and θ that the function provides into an equation in the variables x and y, and see what information this equation provides.

12.2 THE CONIC SECTIONS IN POLAR COORDINATES

A plane is provided with both a polar and a Cartesian xy-coordinate system. The origins coincide, and the positive x-axis is the polar axis.

Let $\varepsilon \geq 0$ and $d > 0$ be constants. We'll examine the graph of the polar equation

$$(*) \qquad r = \frac{d}{1+\varepsilon\cos\theta}.$$

Note that if $\varepsilon = 0$, then the graph is the circle $r = d$ with center the origin O. (This case will also be included in part ii below.)

Let's start with some general properties of this equation. Suppose that $\varepsilon \leq 1$. Because $-1 \leq \cos\theta$, we see that $-1 \leq -\varepsilon \leq \varepsilon\cos\theta$, and therefore that $1+\varepsilon\cos\theta \geq 0$. So if $\varepsilon \leq 1$, then $r > 0$ whenever r is defined (whenever $1+\varepsilon\cos\theta > 0$). If $\varepsilon > 1$, then both $r > 0$ and $r < 0$ are possible. Take $\theta = 0$ and then $\theta = \pi$, for instance. Now return to equation $(*)$ for any $\varepsilon \geq 0$, but consider it in the form $r + \varepsilon r\cos\theta = d$. Because $x = r\cos\theta$, we get $r + \varepsilon x = d$. Suppose that $r > 0$. Then $r = \sqrt{x^2+y^2} \geq x$, and hence $d = r + \varepsilon x \geq x + \varepsilon x$. So $x \leq \frac{d}{1+\varepsilon}$. If $r < 0$ (in this case $\varepsilon > 1$), then $r = -\sqrt{x^2+y^2} \leq -x$. So $d = r + \varepsilon x \leq -x + \varepsilon x$, and hence $\frac{d}{\varepsilon-1} \leq x$. Putting all this information together, we have arrived at the following: if $\varepsilon \leq 1$, then $r > 0$ and the entire graph lies to the left of the line $x = \frac{d}{1+\varepsilon}$. If $\varepsilon > 1$, then the graph has two separate components determined by the vertical line $x = \frac{d}{\varepsilon+1}$ and the vertical line $x = \frac{d}{\varepsilon-1}$ to its right. All points (r,θ) with $r > 0$ lie to the left of the line $x = \frac{d}{\varepsilon+1}$, and all points (r,θ) with $r < 0$ lie to the right of the line $x = \frac{d}{\varepsilon-1}$.

Let's turn to the specifics of the graph of equation $(*)$ or, equivalently, the equation $r + \varepsilon r\cos\theta = d$.

i. Start with the case $\varepsilon = 1$. We begin by plotting some of the points of the graph of

$$r = \frac{d}{1+\cos\theta}.$$

The values of $1+\cos\theta$ for $\theta = 0, \frac{\pi}{4}, \frac{\pi}{2}$, and $\frac{3\pi}{4}$ are $1+1 = 2, 1+\frac{\sqrt{2}}{2}, 1+0 = 1$, and $1-\frac{\sqrt{2}}{2}$, respectively. This implies, for the same sequence of angles, that $r = \frac{d}{2}, r \approx 0.6d, r = d$, and $r \approx 3.4d$. The fact that $\cos(-\theta) = \cos\theta$ provides all the points on the graph of $r = \frac{d}{1+\cos\theta}$ that are plotted in Figure 12.5a. This cluster of points and the fact that $r = f(\theta)$ is not defined for $\theta = \pi$ suggests that we are dealing with a

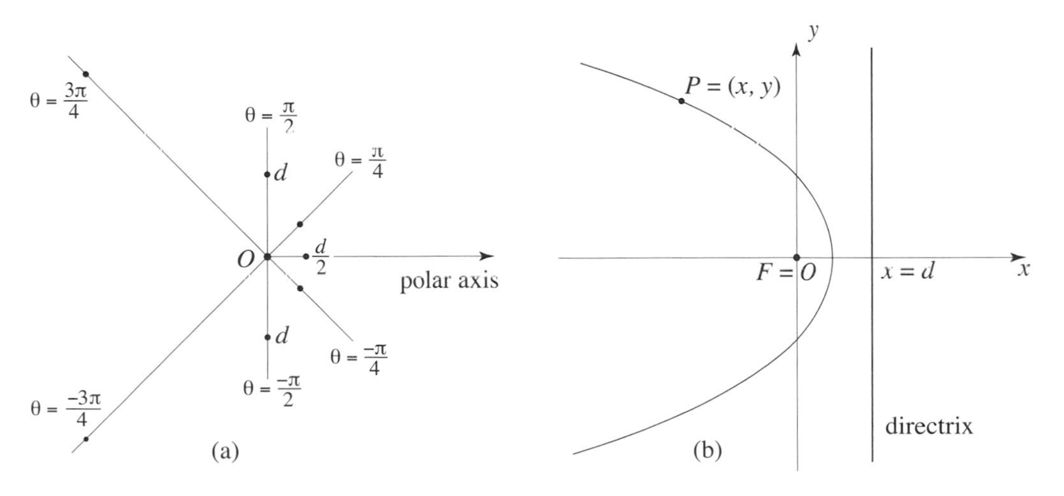

Fig. 12.5

parabola that has a vertical directrix to the right of the point $x = \frac{d}{2}$. In fact, it looks as though the parabola has the origin O as focal point and the line $x = d$ as directrix. Could this actually be so? Since

$r + r\cos\theta = d$ and since $r > 0$, we get $\sqrt{x^2+y^2} + x = d$. Now consider the parabola with focal point the origin O and directrix the vertical line $x = d$, and refer to Figure 12.5b. A point $P = (x, y)$ is on this parabola precisely when its distance to the origin O is equal to its distance to the line $x = d$. Notice that this is the case precisely when $\sqrt{(x-0)^2 + (y-0)^2} = \sqrt{x^2+y^2}$ is equal to $-x + d$. So the Cartesian equation of the parabola is $\sqrt{x^2+y^2} + x = d$. Since this is the Cartesian version of the equation $r + r\cos\theta = d$, it follows that the graph of equation $(*)$ with $\varepsilon = 1$ is the parabola with focal point O and directrix $x = d$.

We'll consider the remaining cases $\varepsilon < 1$ and $\varepsilon > 1$ together (at least initially). Starting with $\pm\sqrt{x^2+y^2} + \varepsilon x = d$, we get by the following sequence of steps (one is a completion of a square):

$$x^2 + y^2 = d^2 - 2\varepsilon dx + \varepsilon^2 x^2$$

$$(1-\varepsilon^2)x^2 + 2\varepsilon dx + y^2 = d^2$$

$$x^2 + \frac{2\varepsilon d}{1-\varepsilon^2}x + \frac{y^2}{1-\varepsilon^2} = \frac{d^2}{1-\varepsilon^2}$$

$$x^2 + \frac{2\varepsilon d}{1-\varepsilon^2}x + \frac{\varepsilon^2 d^2}{(1-\varepsilon^2)^2} + \frac{y^2}{1-\varepsilon^2} = \frac{d^2}{1-\varepsilon^2} + \frac{\varepsilon^2 d^2}{(1-\varepsilon^2)^2}$$

$$\left(x + \frac{\varepsilon d}{1-\varepsilon^2}\right)^2 + \frac{y^2}{1-\varepsilon^2} = \frac{(1-\varepsilon^2)d^2 + \varepsilon^2 d^2}{(1-\varepsilon^2)^2}$$

$$\left(x + \frac{\varepsilon d}{1-\varepsilon^2}\right)^2 + \frac{y^2}{1-\varepsilon^2} = \left(\frac{d}{1-\varepsilon^2}\right)^2, \text{ and finally}$$

$$\frac{(x + \frac{\varepsilon d}{1-\varepsilon^2})^2}{(\frac{d}{1-\varepsilon^2})^2} + \frac{y^2}{\frac{d^2}{1-\varepsilon^2}} = 1.$$

ii. Suppose $\varepsilon < 1$. So $1 - \varepsilon^2 > 0$. Put $a = \frac{d}{1-\varepsilon^2}$, $b = \frac{d}{\sqrt{1-\varepsilon^2}}$, and let $c = \sqrt{a^2 - b^2}$. Since $1 - \varepsilon^2 \le 1$, we see that $\sqrt{1-\varepsilon^2} \ge 1 - \varepsilon^2$, and hence that $a \ge b$. Since $a^2 - b^2 = \frac{d^2}{(1-\varepsilon^2)^2} - \frac{d^2}{1-\varepsilon^2} = \frac{d^2-(1-\varepsilon^2)d^2}{(1-\varepsilon^2)^2} = \frac{\varepsilon^2 d^2}{(1-\varepsilon^2)^2}$, it follows that $c = \frac{\varepsilon d}{1-\varepsilon^2} = \varepsilon a$. After a substitution, we get the equation $\frac{(x+c)^2}{a^2} + \frac{y^2}{b^2} = 1$. Now turn to Section 4.4. The equation $\frac{x^2}{a^2} + \frac{y^2}{b^2} = 1$ represents an ellipse centered at O with semimajor axis $a = \frac{d}{1-\varepsilon^2}$ and semiminor axis $b = \frac{d}{\sqrt{1-\varepsilon^2}}$ (or a circle with center O and radius a, if $a = b$). By Section 4.2, the graph of $\frac{(x+c)^2}{a^2} + \frac{y^2}{b^2} = 1$ is the same ellipse (or circle) shifted $c = \sqrt{a^2 - b^2}$ units to the left. Since c is the distance

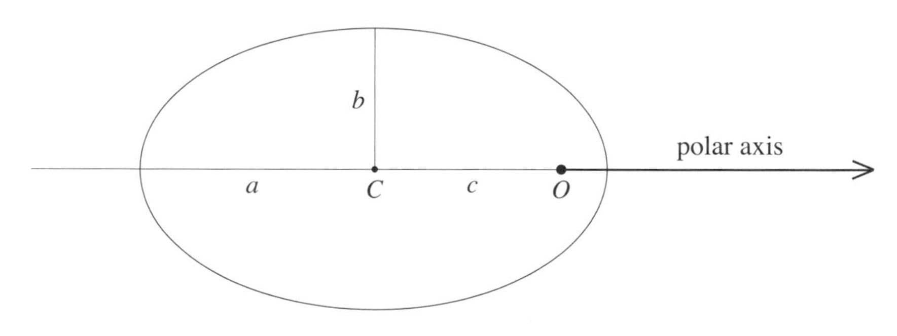

Fig. 12.6

between the center C of the ellipse and its right focal point, the ellipse has been shifted so that its right focal point is at polar origin O. See Figure 12.6. We have described the graph of equation $(*)$ in the situation $\varepsilon < 1$. Because $c = \varepsilon a$, the ε of equation $(*)$ is the eccentricity $\varepsilon = \frac{a}{c}$ of the ellipse. The focal points are the origin O and (in either Cartesian or polar coordinates) the point $(-2c, 0)$.

iii. Suppose $\varepsilon > 1$. So $\varepsilon^2 - 1 > 0$. Let $a = \frac{d}{\varepsilon^2-1}$, $b = \frac{d}{\sqrt{\varepsilon^2-1}}$, and set $c = \sqrt{a^2 + b^2}$. Because $a^2 + b^2 = \frac{d^2}{(\varepsilon^2-1)^2} + \frac{d^2}{\varepsilon^2-1} = \frac{d^2+(\varepsilon^2-1)d^2}{(\varepsilon^2-1)^2} = \frac{\varepsilon^2 d^2}{(\varepsilon^2-1)^2}$, we get $c = \frac{\varepsilon d}{\varepsilon^2-1}$. This time, after substituting carefully, the earlier equation becomes $\frac{(x-c)^2}{a^2} - \frac{y^2}{b^2} = 1$. Turn to Section 4.5. Consider the hyperbola $\frac{x^2}{a^2} - \frac{y^2}{b^2} = 1$ with semimajor axis a and semiminor axis b. We know that its asymptotes intersect at the origin O and that its focal points are both c units from O. The graph of $\frac{(x-c)^2}{a^2} - \frac{y^2}{b^2} = 1$, and hence that of equation $(*)$, is obtained by shifting this hyperbola c units to the righft. It follows that the left focal point of this shifted hyperbola is at the origin O. See Section 4.2. Its graph is sketched in Figure 12.7. It follows from the

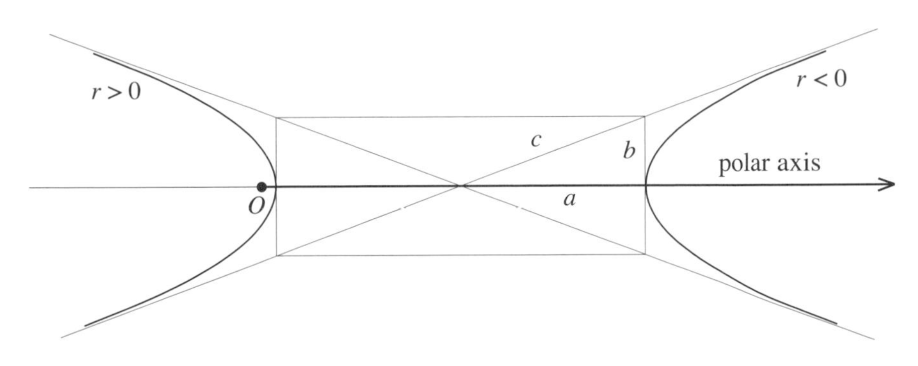

Fig. 12.7

equalities defining a, b, and c that the ε of equation $(*)$ is the eccentricity $\varepsilon = \frac{c}{a}$ of the hyperbola.

Return to equation $(*)$ for a general $\varepsilon \geq 0$. We have found that its graph is an ellipse if $\varepsilon < 1$, a parabola if $\varepsilon = 1$, and a hyperbola if $\varepsilon > 1$. If we define the eccentricity of a parabola to be equal to 1, then ε is the *eccentricity* of the conic section in all cases. We saw that if $\varepsilon \leq 1$, then $r > 0$ automatically. If $\varepsilon > 1$, then both $r > 0$ and $r < 0$ are possible. This is the situation of the hyperbola of Figure 12.7. Using information we already have, we know that the left branch of the hyperbola is the set of all (r, θ) with $r > 0$, and the right branch is the set of all (r, θ) with $r < 0$.

The distance from the origin and focal point O to the graph is a minimum when r is positive and as small as possible. This is so when the term $\cos\theta$ in equation $(*)$ is as large as possible, namely, for $\cos\theta = 1$. So this minimal distance is $\frac{d}{1+\varepsilon}$, and it is attained when $\theta = 0, \pm 2\pi, \ldots$. For the ellipse, $d = a(1 - \varepsilon^2)$, so that the minimal distance is $a(1 - \varepsilon)$. For the hyperbola, $d = a(\varepsilon^2 - 1)$, so that the minimal distance is $a(\varepsilon - 1)$. When $\theta = \frac{\pi}{2}$, then $\cos\theta = \cos\frac{\pi}{2} = 0$. The corresponding point on the graph of $(*)$ is $(d, \frac{\pi}{2})$. The length of the segment through O, perpendicular to the polar axis and bounded on both sides by the graph, is the latus rectum (refer to Section 6.9). Notice that the latus rectum is equal to $2d$.

Example 12.2. Consider the equation $r = \frac{4}{1+\frac{2}{5}\cos\theta}$. Why is its graph an ellipse? What can you say about the shape of the ellipse or, more precisely, about its semimajor and semiminor axes, and its eccentricity?

Example 12.3. Consider the hyperbola determined by the data $a = 2$ and $b = 7$. For which values of d and ε does the graph of $(*)$ have the shape of this hyperbola? [Hint: Set $\frac{d}{\varepsilon^2-1} = 2$ and $\frac{d}{\sqrt{\varepsilon^2-1}} = 7$, and solve for d and ε.]

Example 12.4. Make use of the Cartesian versions of each of the three equations $r = \frac{4}{1+\cos\theta}$, $r = \frac{5}{1+\frac{1}{2}\cos\theta}$, and $r = \frac{3}{1+3\cos\theta}$ to sketch their graphs.

Example 12.5. For each of the three cases,

(a) a parabola with distance between focal point and directrix equal to 7,

(b) an ellipse with semimajor axis $a = 6$ and semiminor axis $b = 4$, and

(c) a hyperbola with semimajor axis $a = 6$ and semiminor axis $b = 4$

determine the equation $(*)$ that has a graph of the given shape.

The equations $r = \frac{d}{1-\varepsilon\cos\theta}$, $r = \frac{d}{1+\varepsilon\sin\theta}$, and $r = \frac{d}{1-\varepsilon\sin\theta}$ also represent all the conic sections that have the origin as a focal point. Their study is analogous to the study above. Except for rotations and shifts, the conclusions are the same as before.

12.3 THE DERIVATIVE OF A POLAR FUNCTION

We know that the derivative of a function in Cartesian coordinates measures the slope of the tangent line to the graph. We will now see that the derivative of a function in polar coordinates is also related to the tangent of the graph. But the connection is more complicated.

Let $r = f(\theta)$ be a function in polar coordinates, and let $f'(\theta)$ be its derivative. A graph of a typical function is sketched in Figure 12.8. On occasion we will restrict the domain to an interval $\theta_1 \le \theta \le \theta_2$ as indicated in the figure. Suppose that $f'(\theta) > 0$ over such an interval. We know that this means that the function $r = f(\theta)$ increases with increasing θ over $\theta_1 \le \theta \le \theta_2$. So $r = f(\theta)$ grows as θ rotates from θ_1 to

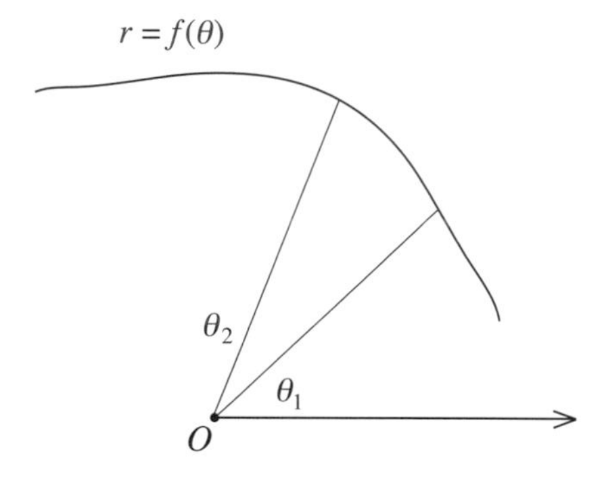

Fig. 12.8

θ_2. In a similar way, if $f'(\theta) < 0$ over the interval, then $r = f(\theta)$ decreases as θ rotates from θ_1 to θ_2.

Let $P = (f(\theta), \theta)$ be any point on the graph of $f(\theta)$. Assume that P is not the origin O, so that $f(\theta) \neq 0$. Let γ be the angle measured in the counterclockwise direction from the tangent at $P = (f(\theta), \theta)$ to the seg-

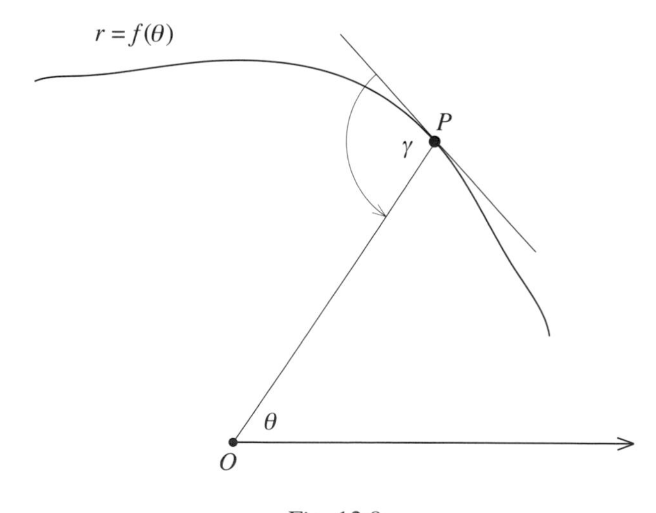

Fig. 12.9

ment from O to $P = (f(\theta), \theta)$. See Figure 12.9. Observe that $0 \le \gamma < \pi$ and that $\gamma = \gamma(\theta)$ is a function of θ.

Consider the constant function $r = f(\theta) = 5$. Here $f'(\theta) = 0$ for all θ and, because the graph of this function is a circle with center the origin, $\gamma(\theta) = \frac{\pi}{2}$ for all θ. For any function $r = f(\theta)$, notice that $f'(\theta) > 0$ and $\gamma(\theta) > \frac{\pi}{2}$ both tell us that $r = f(\theta)$ is increasing at θ and that $f'(\theta) < 0$ and $\gamma(\theta) < \frac{\pi}{2}$ both tell us that $r = f(\theta)$ is decreasing at θ. If $f'(\theta) = 0$ or $\gamma(\theta) = \frac{\pi}{2}$, then there is no growth in $r = f(\theta)$ at θ. Given what was just observed, there ought to be an explicit connection between $f'(\theta)$ and $\gamma(\theta)$. This is the connection that we will now explore.

Let's begin with an analysis of the limit

$$f'(\theta) = \lim_{\Delta\theta \to 0} \frac{f(\theta + \Delta\theta) - f(\theta)}{\Delta\theta}$$

that defines the derivative of the function $f(\theta)$. Consider the point $(f(\theta + \Delta\theta), \theta + \Delta\theta)$ for a small $\Delta\theta$. Draw the segment from O to this point into Figure 12.9, and put in the circular arc with center O and radius $f(\theta)$ between the rays determined by θ and $\theta + \Delta\theta$. This circular arc and the segment of length $f(\theta + \Delta\theta) - f(\theta)$ from this arc to the graph are both shown in Figure 12.10a. We'll call the curving triangle that is formed the *beak* at P. Let the length of the circular arc be Δs, and observe that the radian measure of $\Delta\theta$ is equal to $\Delta\theta = \frac{\Delta s}{f(\theta)}$. So $\frac{1}{\Delta\theta} = \frac{1}{\Delta s} \cdot f(\theta)$. After a substitution,

$$f'(\theta) = \lim_{\Delta\theta \to 0} \frac{f(\theta + \Delta\theta) - f(\theta)}{\Delta s} \cdot f(\theta).$$

In order to understand $f'(\theta)$, we need to come to grips with $\lim\limits_{\Delta\theta \to 0} \frac{f(\theta+\Delta\theta)-f(\theta)}{\Delta s}$. Put in the tangent line *to the graph* of $r = f(\theta)$ at P, and let A be the point of intersection of the tangent with the ray determined by $\theta + \Delta\theta$. Also put in the tangent line *to the circle* at P, and let B be the point of intersection of this tangent and the same ray. The two tangent lines and the ray form the triangle ΔAPB that we call the *triangle* at P.

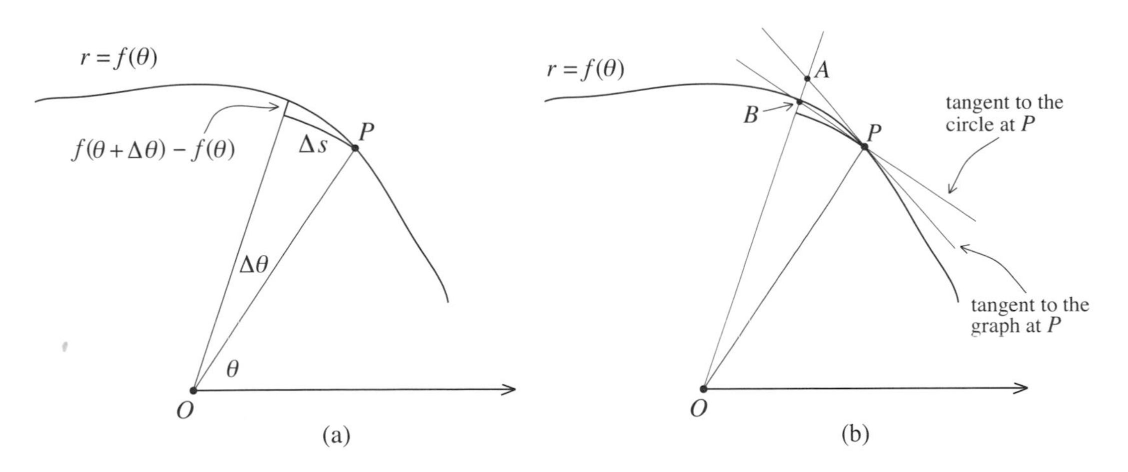

Fig. 12.10

It is shown in Figure 12.10b. The diagram of Figure 12.11 is an enlarged composite of the two diagrams of Figure 12.10. It shows both the beak at P and the triangle ΔAPB at P.

We'll now push $\Delta\theta$ to 0 and investigate $\lim\limits_{\Delta\theta \to 0} \dfrac{f(\theta + \Delta\theta) - f(\theta)}{\Delta s}$. What happens as $\Delta\theta$ shrinks to 0 is illustrated in Figure 12.12. It is a blowup of the central part of Figure 12.11. As $\Delta\theta$ is pushed to 0, the segment OBA rotates toward the segment OP. Both the beak at P and the triangle at P shrink in the direction of their tips at P. As Figure 12.12 illustrates, the shrinking triangle approximates the shrinking beak better and better as the gap between OBA and OP closes. In the process, Δs gets closer to BP and

$f(\theta + \Delta\theta) - f(\theta)$ to AB. Therefore, as $\Delta\theta$ is pushed to 0,

$$\frac{f(\theta + \Delta\theta) - f(\theta)}{\Delta s} \text{ closes in on the ratio } \frac{AB}{BP}.$$

Because the tangent line to a circle at a point is perpendicular to its radius to the point, we know that the angle at P between PO and PB is $\frac{\pi}{2}$. So as $\Delta\theta$ shrinks to 0, the angle $\angle PBA$ approaches $\frac{\pi}{2}$, and the triangle ΔAPB approaches a right triangle with right angle at B. Again refer to Figure 12.12. It follows that the ratio $\frac{AB}{BP}$ closes in on the tangent of the angle $\angle APB$. Because $\angle APO = \gamma$ and $\angle BPO = \frac{\pi}{2}$, the

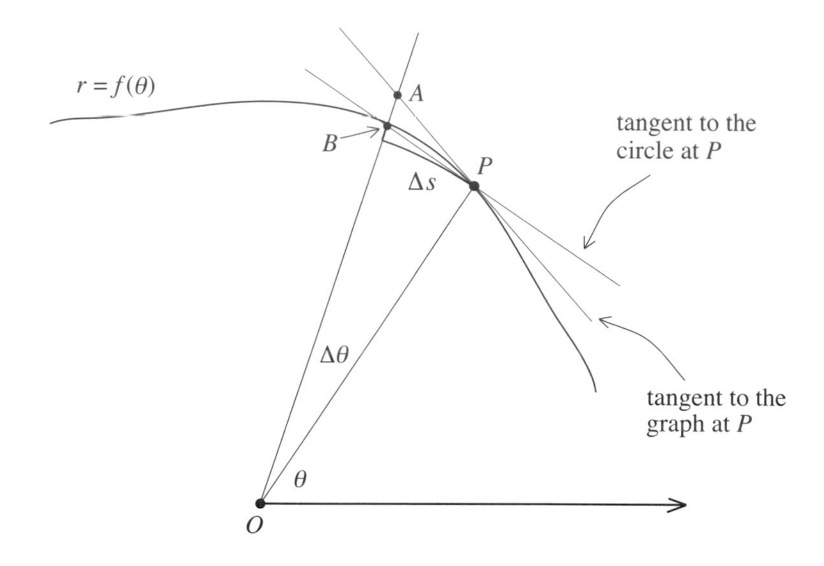

Fig. 12.11

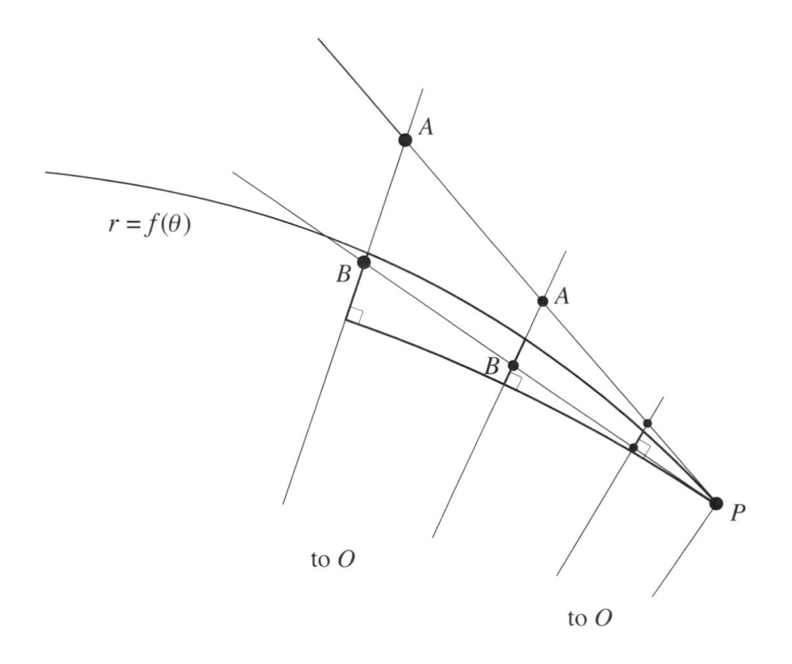

Fig. 12.12

angle $\angle APB = \gamma - \frac{\pi}{2}$. By putting it all together, we have demonstrated that as $\Delta\theta$ shrinks to 0

$\dfrac{f(\theta+\Delta\theta)-f(\theta)}{\Delta s}$ closes in on the ratio $\dfrac{AB}{BP}$ and this in turn on $\tan(\gamma-\frac{\pi}{2})$.

We have verified that

$$\lim_{\Delta\theta\to 0}\frac{f(\theta+\Delta\theta)-f(\theta)}{\Delta s} = \lim_{\Delta\theta\to 0}\frac{AB}{PB} = \tan(\gamma-\tfrac{\pi}{2})$$

and have arrived at the geometric interpretation

$$\boxed{f'(\theta)=f(\theta)\cdot\tan(\gamma(\theta)-\tfrac{\pi}{2})}$$

of the derivative $f'(\theta)$ of a function $r=f(\theta)$ for any θ with $f(\theta)\neq 0$.

The fact that $0\le\gamma(\theta)<\pi$ implies that $-\frac{\pi}{2}\le\gamma(\theta)-\frac{\pi}{2}<\frac{\pi}{2}$. A look at the graph of $\tan\theta$ in Figure 4.26 informs us that $f'(\theta)$ does not exist when $\gamma(\theta)=0$. A look back at the definition of $\gamma(\theta)$ tells us that this happens precisely when the line from O through the point $(f(\theta),\theta)$ is tangent to the graph. If $f'(\theta)=0$, then $\tan(\gamma(\theta)-\frac{\pi}{2})=0$. From Figure 4.26, we know that this means that $\gamma(\theta)=\frac{\pi}{2}$.

Example 12.6. Let $c>0$ be a constant, and investigate the equation $f'(\theta)=f(\theta)\cdot\tan(\gamma(\theta)-\frac{\pi}{2})$ for the circle $r=f(\theta)=c$ of radius c.

Example 12.7. Consider the function $r=f(\theta)=\frac{1}{\cos\theta}$. Show that the corresponding Cartesian equation is $x=1$. For any point $P=(f(\theta),\theta)$ on the graph, check that $\gamma(\theta)-\frac{\pi}{2}=\theta$. Use this to confirm the identity $f'(\theta)=f(\theta)\cdot\tan(\gamma(\theta)-\frac{\pi}{2})$.

Example 12.8. Consider the function $r=f(\theta)=\sin\theta$. Its graph is the circle of Figure 12.4. Let $P=(f(\theta),\theta)$ be any point on the circle, and draw a radius from the center to P. Show that $\gamma(\theta)-\frac{\pi}{2}=\frac{\pi}{2}-\theta$ by using a property of isosceles triangles, and that $\sin(\frac{\pi}{2}-\theta)=\cos\theta$ and $\cos(\frac{\pi}{2}-\theta)=\sin\theta$ by using the conclusion of Problem 4.80. Confirm the identity $f'(\theta)=f(\theta)\cdot\tan(\gamma(\theta)-\frac{\pi}{2})$.

Example 12.9. Let $r=f(\theta)=\theta$ be the spiral of Archimedes of Figure 12.2a. With each counterclockwise revolution, a point adds 2π to its distance from the origin O. Since $f'(\theta)=1$, the formula for $f'(\theta)$ tells us that $\tan(\gamma(\theta)-\frac{\pi}{2})=\frac{1}{\theta}$. By Section 9.9.1, $\gamma(\theta)=\frac{\pi}{2}+\tan^{-1}(\frac{1}{\theta})$, and $\gamma(\theta)$ approaches $\frac{\pi}{2}$ for increasing θ. To illustrate this numerically, suppose that the spiral has completed 10 revolutions from $\theta=0$ through $\theta=20\pi$. The corresponding point on the spiral is now a distance of $r=20\pi$ from O. With the eleventh revolution, the point adds 2π to this distance. Check that during this revolution the angle $\gamma(\theta)$ decreases from about 1.5867 radians, or about 90.016°, to about 1.5853 radians, or about 90.014°.

12.4 THE LENGTHS OF POLAR CURVES

Let's extract some more information from the analysis that verified the equality $f'(\theta)=f(\theta)\cdot\tan(\gamma-\frac{\pi}{2})$. Let a and b be constants with $a<b$. Assume that the function $r=f(\theta)$ is differentiable for all θ with $a<\theta<b$ and that $f(\theta)$ and its derivative $f'(\theta)$ are both continuous for all $a\le\theta\le b$. The focus will be on the length L of the graph of $r=f(\theta)$ between the points $(f(a),a)$ and $(f(b),b)$.

We'll apply the strategy of integral calculus of Sections 9.1 and 9.2. Let n be a large number, and consider a partition

$$a=\theta_0<\theta_1<\theta_2<\cdots<\theta_i<\theta_{i+1}<\cdots<\theta_{n-1}<\theta_n=b$$

of the angle $b-a$. Suppose that the norm of the partition is small relative to the angle $b-a$. Each θ_i is an angle in radian measure, and we'll let $d\theta=\theta_{i+1}-\theta_i$ be the typical difference between consecutive angles.

The rays determined by $\theta_0, \ldots, \theta_{n-1}, \theta_n$ divide the graph of $r = f(\theta)$ between the points $(f(a), a)$ and $(f(b), b)$ into n pieces. Letting the lengths of these pieces be $L_0, \ldots L_{n-1}$, we see that $L = L_0 + \cdots + L_{n-1}$. Now turn to Figure 12.13. Let $\theta = \theta_i$ be any of the angles selected, and note that $\theta_{i+1} = \theta + d\theta$. The point $P = (f(\theta), \theta)$, the segment of the graph of length L_i, and the arc of a circle of radius $f(\theta)$ are highlighted. We will now derive an approximation of L_i. Return to Figure 12.11 and its explanation. Add the tangent lines at P to both the graph of $r = f(\theta)$ and the circular arc to Figure 12.13. The figure shows both the *beak* at P and the triangle ΔAPB at P. The earlier $\Delta\theta$ and Δs are now written as $d\theta$ and ds. Note that $d\theta$ is small. A look at Figures 12.12 and 12.13 tells us that L_i is essentially equal to the length of the segment AP on the tangent to the graph. It follows from the earlier discussion that ΔAPB is essentially a right

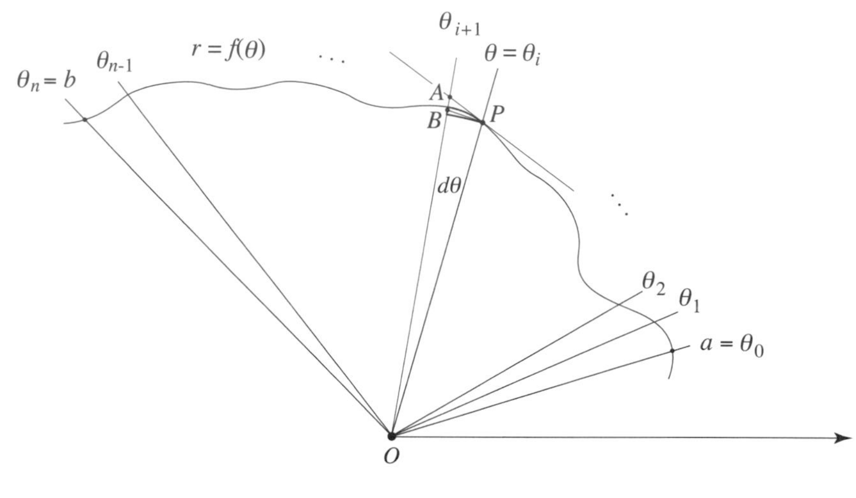

Fig. 12.13

triangle with hypotenuse AP and hence that $AP \approx \sqrt{AB^2 + BP^2}$. Return to the earlier discussion once more, and observe that $BP \approx ds$ and $AB \approx f(\theta + d\theta) - f(\theta)$. Because $d\theta = \frac{ds}{f(\theta)}$, it follows that

$$BP \approx ds = f(\theta)d\theta\,.$$

Because $d\theta$ is small, $\frac{f(\theta+d\theta)-f(\theta)}{d\theta} \approx f'(\theta)$ and hence

$$AB \approx f(\theta + d\theta) - f(\theta) \approx f'(\theta)\,d\theta.$$

Therefore

$$L_i \approx AP \approx \sqrt{BP^2 + AB^2} \approx \sqrt{f(\theta)^2(d\theta)^2 + f'(\theta)^2(d\theta)^2} = \sqrt{f(\theta)^2 + f'(\theta)^2}\,d\theta\,.$$

Because $L = L_1 + \cdots + L_n$, it is a consequence of the definition of the definite integral—refer to Section 9.1—that the length L of the graph of $r = f(\theta)$ between the points $(f(a), a)$ and $(f(b), b)$ is

$$L = \int_a^b \sqrt{f(\theta)^2 + f'(\theta)^2}\,d\theta$$

Example 12.10. Use the length formula with the function $f(\theta) = 5$, $0 \leq \theta \leq \pi$, to show that the length of half of a circle of radius 5 is 5π.

Example 12.11. Apply the length formula to determine the lengths of the graph of $r = f(\theta) = \sin\theta$ from $\theta = \frac{\pi}{4}$ to $\theta = \frac{3\pi}{4}$, then from $\theta = 0$ to $\theta = \pi$, and finally from $\theta = 0$ to $\theta = 2\pi$. Check your answers by studying the graph of $f(\theta) = \sin\theta$ as depicted in Figure 12.4.

Example 12.12. Use the fact that the Cartesian equation of the polar equation $r = \frac{1}{\sin\theta}$ is the line $y = 1$ to show that $\int_{\frac{\pi}{4}}^{\frac{3\pi}{4}} \frac{1}{\sin^2\theta}\, d\theta = 2$.

Example 12.13. Apply integral Formula (17) $\int \sqrt{1+x^2}\, dx = \frac{1}{2}[x\sqrt{1+x^2} + \ln|x + \sqrt{1+x^2}|] + C$ of Section 9.11 to show that the length of the spiral of Archimedes of Figure 12.2a from $\theta = 0$ to $\theta = \frac{3\pi}{2}$ is approximately 12.48.

12.5 AREAS IN POLAR COORDINATES

This section studies the area of a region determined by the graph of a function in polar coordinates. The conclusions will play an important role in the analysis of the dynamics of planetary motion in Section 12.8.

Let $r = f(\theta)$ be a continuous function defined on a closed interval $a \le \theta \le b$. The graph of a typical situation is shown in Figure 12.14. Our concern is the computation of the area A of the highlighted region.

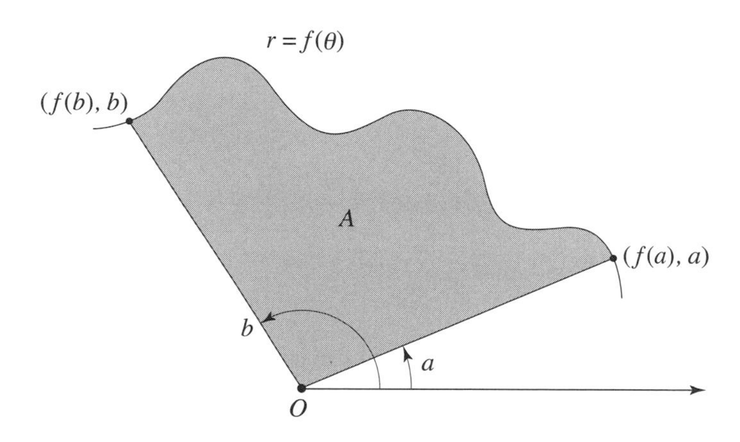

Fig. 12.14

It is the area that the segment from the origin O to the point $(f(\theta), \theta)$ sweeps out as it rotates from the ray $\theta = a$ to $\theta = b$. In general, this means that areas overlap, possibly multiple times, and that each of these overlaps is counted. If it is assumed that $b - a \le 2\pi$, then overlap is avoided. To facilitate the argument that follows, we will assume that $f(\theta) \ge 0$ (but the conclusions hold without this assumption).

As in Section 12.4, let n be a large number (relative to the difference $b - a$), and partition the angle $b - a$ into n very small pieces. As before,

$$a = \theta_0 < \theta_1 < \cdots < \theta_{i-1} < \theta_i < \cdots < \theta_{n-1} < \theta_n = b$$

with $d\theta = \theta_{i+1} - \theta_i$ for $0 \le i \le n-1$, the typical difference between consecutive angles. The rays determined by $\theta_0, \ldots, \theta_n$ divide the region into n pie-shaped wedges. We'll let $A_0, \ldots, A_i, \ldots, A_{n-1}$ denote their respective areas. This is illustrated in Figure 12.15. Let $\theta = \theta_i$ be any of the angles selected, and draw a circular arc with center the origin O and radius $f(\theta)$ from the ray determined by θ to the ray given by $\theta_{i+1} = \theta + d\theta$. Consider the circular sector determined by this arc and the two rays. Figure 12.16 shows this circular sector and the relevant part of the graph of the function f. Since $f(\theta)$ is the radius of the circular arc, the area of the circular sector is $\frac{1}{2}f(\theta)^2 d\theta$. Because $d\theta$ is very small, A_i is essentially equal

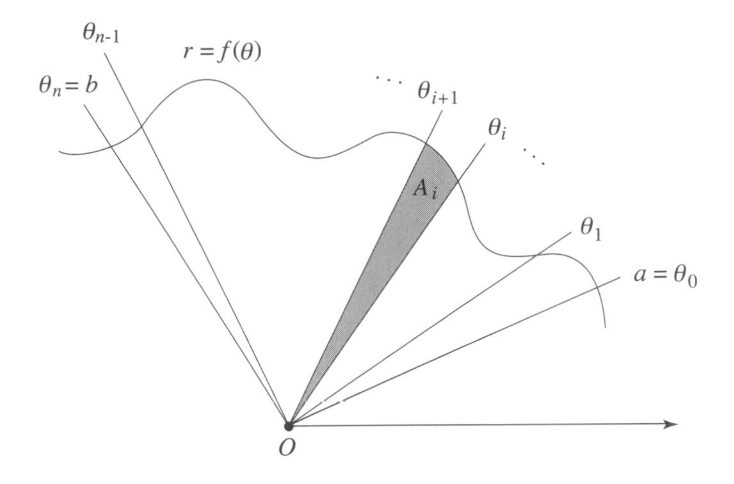

Fig. 12.15

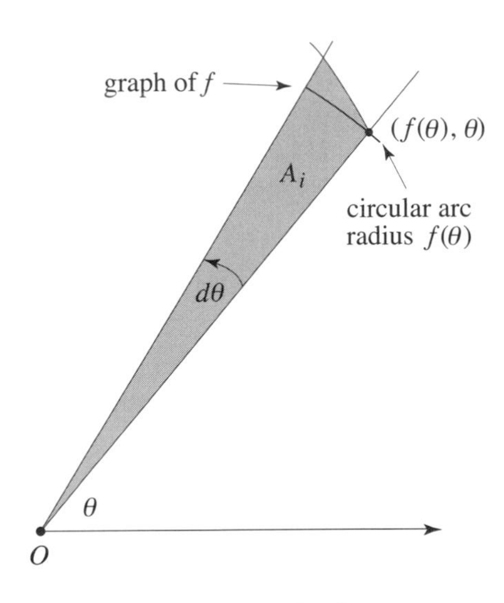

Fig. 12.16

to the area of the sector. The fact that $A = A_1 + \cdots + A_n$, in combination with the definition of the definite integral, informs us that

$$A = \int_a^b \tfrac{1}{2} f(\theta)^2 \, d\theta$$

Example 12.14. Use the area formula with the function $r = f(\theta) = 5$ to show that the area of a quarter circle of radius 5 is $\frac{25}{4}\pi$.

Example 12.15. Check that the area enclosed by the spiral of Archimedes $r = f(\theta) = \theta$ of Figure 12.2a over the interval $0 \leq \theta \leq 2\pi$ is equal to $\displaystyle\int_0^{2\pi} \tfrac{1}{2}\theta^2 \, d\theta = \tfrac{1}{6}\theta^3 \Big|_0^{2\pi} = \tfrac{1}{6}(2\pi)^3 = \tfrac{4}{3}\pi^3 \approx 41.34.$

The two examples that follow test the area formula in situations where its conclusion can be determined by other means. Consider the function $r = f(\theta) = c$ for $0 \le \theta \le 2\pi$ and $c > 0$ a constant. The graph is a circle of radius c. So it must be the case that

$$\int_0^{2\pi} \tfrac{1}{2} f(\theta)^2 d\theta = \int_0^{2\pi} \tfrac{1}{2} c^2 d\theta = \pi c^2.$$

Because $G(\theta) = \frac{1}{2}c^2\theta$ is an antiderivative of $g(\theta) = \frac{1}{2}c^2$, the fundamental theorem of calculus confirms that

$$\int_0^{2\pi} \tfrac{1}{2} f(\theta)^2 d\theta = \tfrac{1}{2}c^2\theta \Big|_0^{2\pi} = \pi c^2.$$

Consider the area of the region bounded by the graph of the function $f(\theta) = \sin\theta$ and the rays $\theta = \frac{\pi}{4}$ and $\theta = \frac{3\pi}{4}$. Refer to Figure 12.4, and notice that the area in question consists of a half circle of radius $\frac{1}{2}$ plus a triangle of height $\frac{1}{2}$ and base 1. This area is therefore equal to $A = \frac{1}{2}\pi(\frac{1}{2})^2 + \frac{1}{4} = \frac{\pi}{8} + \frac{1}{4}$. The area formula applied to this situation tells us that

$$A = \int_{\frac{\pi}{4}}^{\frac{3\pi}{4}} \tfrac{1}{2}\sin^2\theta \, d\theta\,.$$

Does this provide the same result? By the half-angle formula $\sin^2\theta = \frac{1-\cos 2\theta}{2}$ of Problem 1.22i,

$$\begin{aligned} A &= \int_{\frac{\pi}{4}}^{\frac{3\pi}{4}} \tfrac{1}{2}\sin^2\theta \, d\theta = \int_{\frac{\pi}{4}}^{\frac{3\pi}{4}} \tfrac{1}{4}(1 - \cos 2\theta)\, d\theta = \left(\tfrac{1}{4}\theta - \tfrac{1}{8}\sin 2\theta\right)\Big|_{\frac{\pi}{4}}^{\frac{3\pi}{4}} \\ &= \left(\tfrac{3\pi}{16} + \tfrac{1}{8}\right) - \left(\tfrac{\pi}{16} - \tfrac{1}{8}\right) = \tfrac{\pi}{8} + \tfrac{1}{4}. \end{aligned}$$

Again, the formula agrees with what the geometry provided. In order to illustrate the matter of overlap, consider the area that the segment from the origin O to the point $(f(\theta), \theta)$ traces out as θ varies from $\frac{\pi}{4}$ to $\frac{7\pi}{4}$. How much area do you think has been added to the area $\frac{\pi}{8} + \frac{1}{4}$ just computed? Confirm the correctness of your answer by computing $\int_{\frac{\pi}{4}}^{\frac{7\pi}{4}} \frac{1}{2}\sin^2\theta \, d\theta$. Notice also that the fact that $f(\theta) = \sin\theta$ is negative for $\pi \le \theta \le \frac{7\pi}{4}$ has no impact.

The final two examples find areas involving conic sections in polar form by converting them into Cartesian integrals.

Example 12.16. Study the parabola $r = f(\theta) = \frac{2}{1+\cos\theta}$ by referring to part i of Section 12.2. Verify that the upper half of this parabola is the graph of the function $y = \sqrt{4-4x}$ in Cartesian coordinates. Use this fact and integration by substitution to show that $\displaystyle\int_0^{\frac{\pi}{2}} \frac{2}{(1+\cos\theta)^2}\, d\theta = \frac{4}{3}$.

Example 12.17. Study the graph of the function $r = f(\theta) = \frac{4}{1+\frac{1}{5}\cos\theta}$ by referring to part ii of Section 12.2. Then use the area formula for the ellipse (see Section 5.7) to show that $\displaystyle\int_0^{\pi} \frac{8}{(1+\frac{1}{5}\cos\theta)^2}\, d\theta = \left(\frac{5}{\sqrt{6}}\right)^3 \pi$.

Return to Figure 12.14, and assume that the graph of $r = f(\theta)$ is traced out by a point P moving counterclockwise around O. So the angle θ that determines the position $(f(\theta), \theta)$ of the point is an increasing function $\theta = \theta(t)$ of time t. Suppose that P is at $(f(a), a)$ at time t_1 and that it is at $(f(b), b)$ at time t_2. So $\theta(t_1) = a$ and $\theta(t_2) = b$. Figure 12.17 captures the added information. It follows by integration by substitution (see Section 9.7.1), in combination with the polar area formula, that the area that the

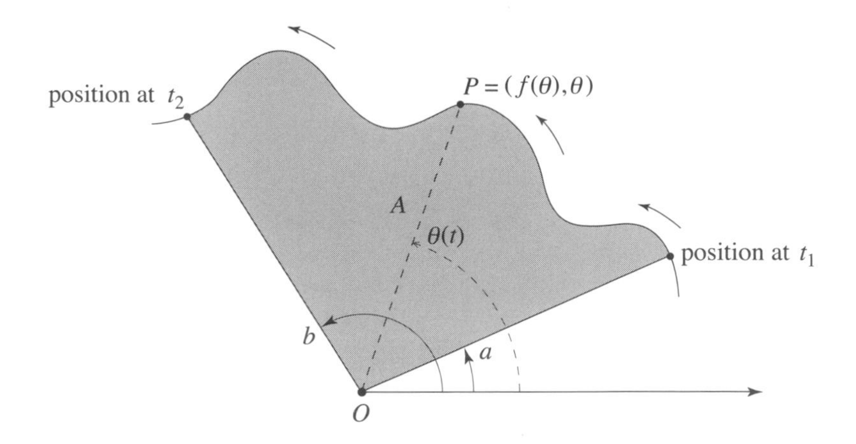

Fig. 12.17

segment OP traces out over the time interval $[t_1, t_2]$ is equal to

$$\int_{t_1}^{t_2} \tfrac{1}{2} f(\theta(t))^2\, \theta'(t)\, dt \;=\; \int_{\theta(t_1)}^{\theta(t_2)} \tfrac{1}{2} f(\theta)^2\, d\theta \;=\; \int_a^b \tfrac{1}{2} f(\theta)^2\, d\theta \;=\; A.$$

Suppose that P moves along the circle $f(\theta) = \sin\theta$ of Figure 12.4. Recall that the entire circle of area $\pi(\frac{1}{2})^2 = \frac{1}{4}\pi \approx 0.7854$ is traced out over $0 \le \theta \le \pi$. Assume that the point starts at O and moves counterclockwise at a constant angular speed of $\theta'(t) = \frac{1}{2}$ radians per second. So $\theta(0) = 0$ and $\theta(t) = \frac{1}{2}t$ radians for any $t \ge 0$. During the first 4 seconds of the motion, θ changes from $\theta(0) = 0$ radians to $\theta(4) = 2$ radians. By the formula above, the area A traced out by the segment OP during the time from $t = 0$ to $t = 4$ is

$$A = \int_0^4 \tfrac{1}{2} \sin^2\theta(t)\, \theta'(t)\, dt = \int_0^2 \tfrac{1}{2} \sin^2\theta\, d\theta.$$

Using the half-angle formula $\sin^2\theta = \frac{1-\cos 2\theta}{2}$ once more, we get that $A = \frac{1}{4}(\theta - \frac{1}{2}\sin 2\theta)\Big|_0^2 = \frac{1}{2} - \frac{1}{8}\sin 4 \approx 0.5000 + 0.0946 = 0.5946$. Since $\frac{0.5946}{0.7854} \approx 0.7571$, the segment sweeps out about 75% of the area of the circle during the 4 seconds.

12.6 EQUIANGULAR SPIRALS

Section 12.1 introduced the smoothly curving spiral of Archimedes as the graph of the function $r = f(\theta) = \theta$. An interesting family of spirals includes the graphs of polar functions $r = f(\theta)$ with the property that the angle $\gamma = \gamma(\theta)$ defined in Figure 12.9 is constant. Let $r = f(\theta)$ be such a function. Since $\gamma = \gamma(\theta)$ is constant, $c = \tan(\gamma(\theta) - \frac{\pi}{2})$ is a constant as well. The basic formula developed in Section 12.3 tells us that $f'(\theta) = f(\theta) \cdot \tan(\gamma(\theta) - \frac{\pi}{2})$. It follows that $r = f(\theta)$ satisfies

$$f'(\theta) = cf(\theta).$$

Since $r = f(\theta)$ is a solution of the differential equation $\frac{dr}{d\theta} = cr$, we know from the concluding comment of Section 7.10 that $f(\theta)$ has the form $f(\theta) = Ae^{c\theta}$, where A is a constant. Since $e^0 = 1$, we get $f(0) = A$, and therefore

$$f(\theta) = f(0)e^{c\theta} = f(0)e^{\tan(\gamma - \frac{\pi}{2})\theta}.$$

If $\gamma = \frac{\pi}{2}$, then $f(\theta) = f(0)e^0 = f(0)$, and the graph is a circle of radius $f(0)$. If $\gamma > \frac{\pi}{2}$, then the graph is a spiral that expands uniformly in a counterclockwise direction, as Figure 12.18a shows. If $\gamma < \frac{\pi}{2}$, then $f(\theta)$ is a decreasing function of θ, and the graph is a spiral that contracts uniformly in a counterclockwise

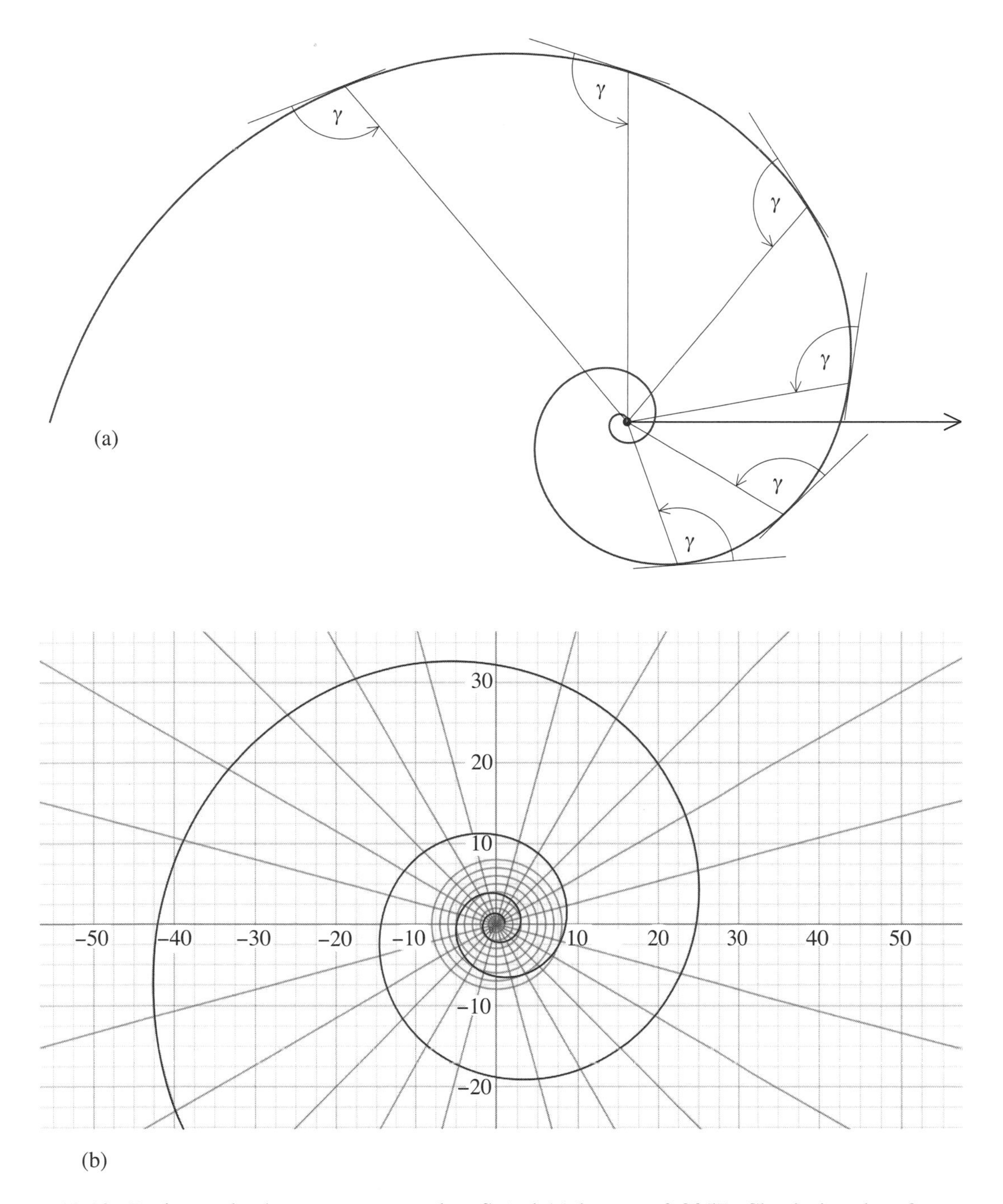

Fig. 12.18. Both spirals above are equiangular. Spiral (a) has $c = 0.3057$. Check that therefore $\gamma = 90° + \tan^{-1}c \approx 107°$. Spiral (b) has $c = 0.1700$, so that $\gamma = 90° + \tan^{-1}c \approx 99.65°$. The graphs of the spirals were produced with the graphing calculator of the website https://www.desmos.com/.

direction. Because γ remains constant as the spiral turns, such spirals are called *equiangular* (and also *logarithmic*). It follows from the data in the caption that the spiral of Figure 12.18a is the graph of $f(\theta) = e^{0.3057\theta}$ and that the spiral of Figure 2.18b is the graph of $f(\theta) = e^{0.1700\theta}$.

Spirals arise in the shell structure of a number of aquatic animals. The pearly nautilus is an example of an ocean animal with a smoothly spiraling shell. See Figure 12.19. The shell of a grown nautilus is about 25 cm (10 inches) in diameter and has about 30 chambers. The animal lives in the successive outermost chambers. The chambers are connected by a tube that lets the nautilus adjust the pressure of the gases

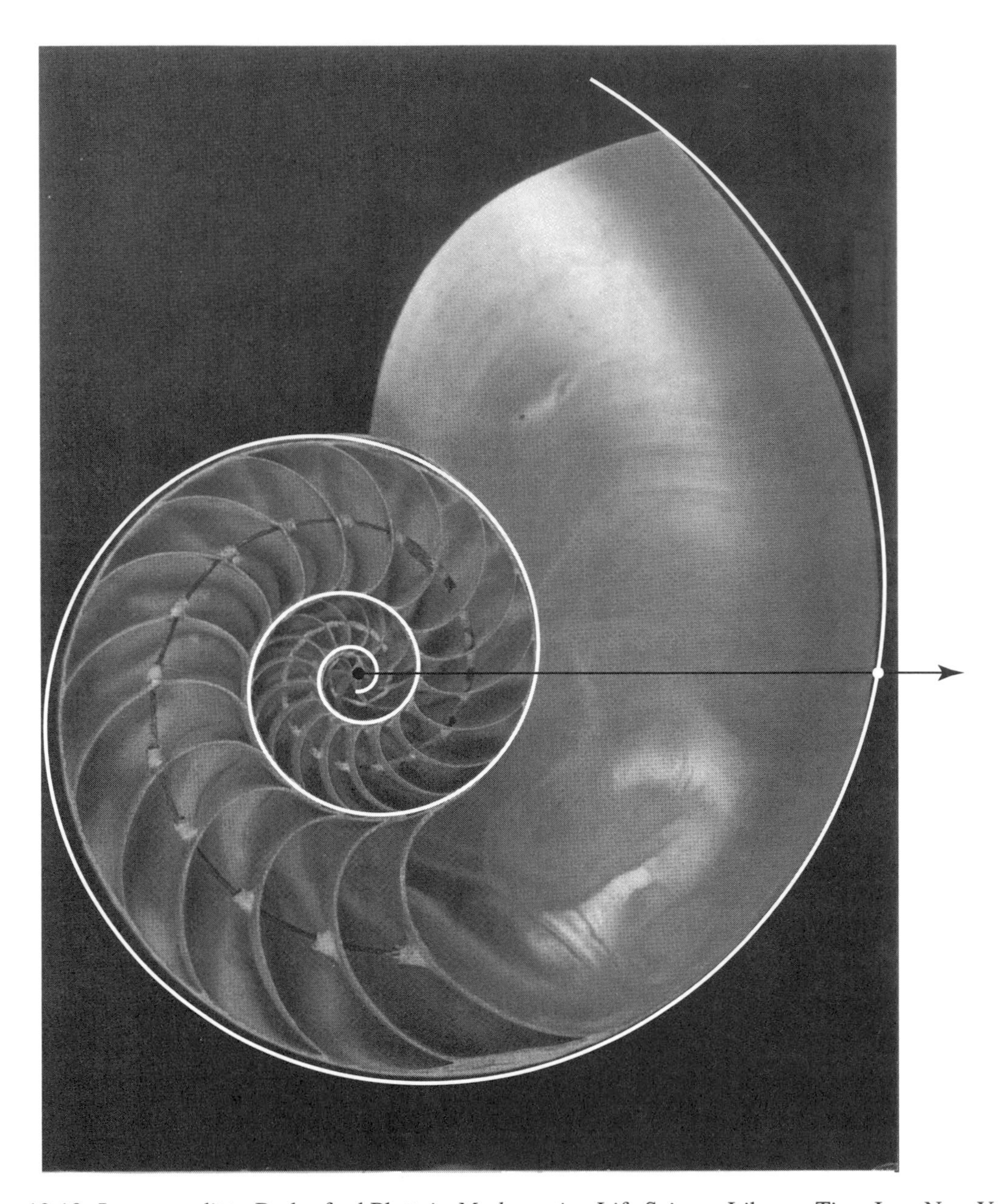

Fig. 12.19. Image credit to Rutherford Platt, in *Mathematics*, Life Science Library, Time Inc., New York, 1963. Permission of use granted by Time Inc.

within the shell. This ability allows it to change the depth at which it is swimming. A bottom feeder, it uses its tentacles (up to about 100 in number) for capturing prey. It lives at 300-m depths and is most commonly observed in its natural habitat through the television cameras of deep diving submersibles.

The spiral of the nautilus is essentially equiangular, as is illustrated in Figure 12.19. It shows the cross section of the shell of a nautilus together with an equiangular spiral that follows the contours of the cross section rather closely. It turns out that the constant angle γ of this spiral is $\gamma \approx 99.65°$. Therefore the spiral of the nautilus of the figure is in essence the spiral of Figure 12.18b.

Example 12.18. Turn to the image of the shell of the nautilus in Figure 12.19, and place a polar axis as shown. Let the rightmost point of intersection of the spiral and the polar axis correspond to $\theta = 0$. Let s be the distance from this point to the origin O at the eye of the spiral. Remarks already made tell us that the spiral of the image of the shell is closely approximated by the graph of the function $f(\theta) = f(0)\,e^{0.17\theta} = se^{0.17\theta}$. What aspects of the geometry of the shell do the two definite integrals

$$\int_{-\frac{13\pi}{2}}^{\frac{\pi}{3}} \sqrt{f(\theta)^2 + f'(\theta)^2}\, d\theta \quad \text{and} \quad \int_{-\frac{3\pi}{2}}^{\frac{\pi}{3}} \tfrac{1}{2} f(\theta)^2\, d\theta$$

represent at least approximately? Show that the values of these integrals are approximately equal to $6.94s$ and $2.40s^2$, respectively. Are these values consistent with your answers?

The combination of biological and mathematical factors by which the nautilus constructs one chamber of its shell after the other in such a way that an equiangular spiral emerges does not appear to be understood. What is the mechanism that regulates the shell's programmed sense to preserve its equiangular shape?

Nature generates much larger examples of equiangular spirals as well. A region of low atmospheric pressure will draw in air from surrounding areas of higher pressure. This pull of air toward the center of such a system combined with the rotation of the Earth results in a circulation of air around and toward this atmospheric low. In the Northern Hemisphere, such a circulation is counterclockwise and in the Southern

Fig. 12.20. Image by the Aqua MODIS instrument on September 4, 2003. Image credit to Jacques Descloitres, MODIS Rapid Response Team, NASA/GSFC.

Hemisphere, it is clockwise. Figure 12.20 shows a huge, beautifully formed low-pressure system swirling in the North Atlantic near Iceland. The changing geometry of such a rotating vortex is complicated, and many aspects of the dynamics are still unresolved. The trailing spiral bands arise because the angular velocity of the vortex decreases from the center outward, and this results in deformations of the bands into spirals that are generally equiangular. Near the vortex core, air can remain in the circulation near the center for many revolutions so that the bands are more circular. When conditions are right, often over warmer parts of oceans, such systems can develop into powerful cyclones and hurricanes.

We turn next to an important application of polar calculus. Newton's 1687 treatise *Principia Mathematica*, with its study of gravitational force and the shape of the trajectories of the bodies that it drives in our solar system, had been a miracle. It was a miracle for at least two reasons. First, for the synthesizing and penetrating way in which it combined basic physical laws and mathematical analysis to explain what Kepler had observed about the orbits of the planets. In doing this, Newton solved a problem that many of the best minds had puzzled over since the time of the Babylonians and Greeks 3000 years ago. A second miracle was the fact that Newton's "machine-gun-style" approach to a completely smoothly operating gravitational force was successful. This had been critical because by triangularizing the geometry and simplifying the mathematics, it became possible for Newton to draw out the connection between the force and the orbit. Refer to Sections 6.8 and 6.9.

Recall that a centripetal force is one that always acts in the direction of a fixed point. As in Newton's discussion, our focus will be on the abstract study of a centripetal force acting on a point-mass. The focus will again be on the relationship between the magnitude of the force and the geometry of the path—we'll refer to it as the orbit—of the point-mass that it propels. Polar coordinates, its connection with Cartesian coordinates, and polar calculus will play a decisive role. Instead of the assumption that the centripetal force acts "machine-gun-style" in intermittent bursts, the analysis relies on the realistic assumption that the action of this force is smooth.

12.7 CENTRIPETAL FORCE IN CARTESIAN COORDINATES

A point-mass P of mass m is in motion in a plane. It is acted on by a single force, a centripetal force with center in the same plane. Place an xy-coordinate system into the plane so that the center of force is at the origin O. Let x and y be the coordinates of P. The magnitude F of the force depends on the location (x, y) of P. Click your stopwatch at time $t = 0$, and suppose that P is in a typical position at an elapsed time $t > 0$ later. Both x and y vary with time t, so that both $x = x(t)$ and $y = y(t)$ are functions of time $t \geq 0$. The magnitude of the force is also a function $F = F(t)$ of t. The insertion of (t) simply emphasizes the

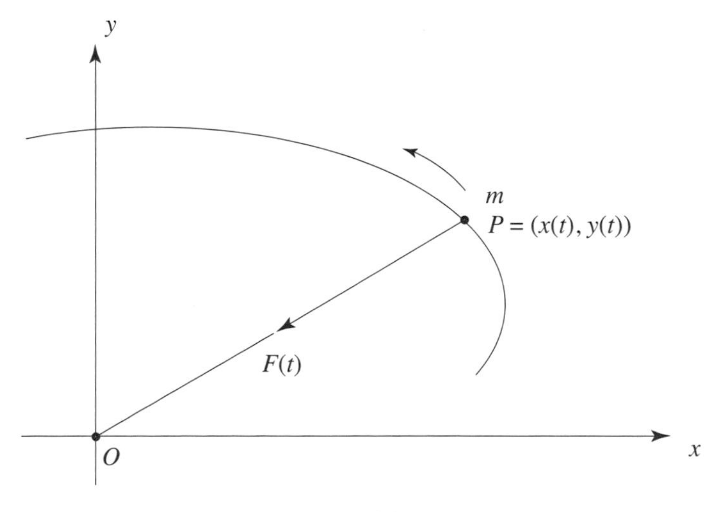

Fig. 12.21

fact that the quantities F, x, and y are all functions of time. Figure 12.21 captures what is happening.

We will assume that P moves smoothly—P does not zigzag sharply like a butterfly or a bat—and that F acts smoothly (and not in intermittent bursts). In particular, we assume that $F(t)$, $x(t)$, and $y(t)$ are differentiable functions of t and that the derivatives $\frac{dx}{dt} = x'(t)$ and $\frac{dy}{dt} = y'(t)$ are differentiable as well. Let r be the distance from P to O. Let F_x and F_y be the components of F in the x- and y-directions, respectively. The quantities r, F_x, and F_y are also differentiable functions of t. Figure 12.22 illustrates the

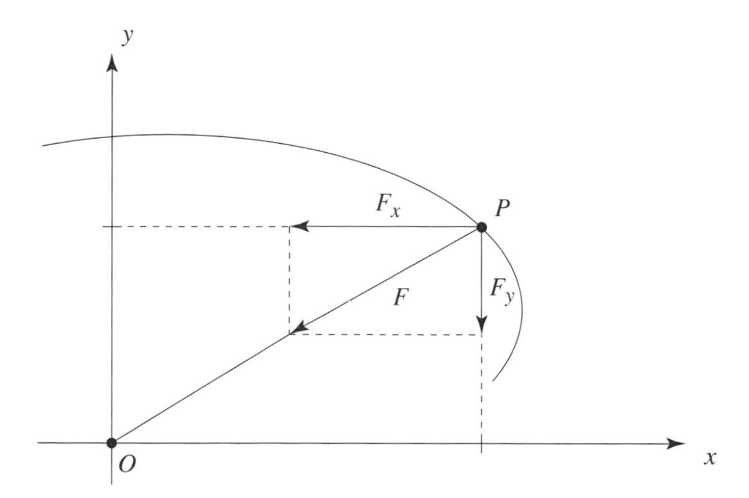

Fig. 12.22

relationship between F, F_x and F_y. By the parallelogram law of forces and similar triangles,

$$\frac{|F_x|}{F} = \frac{|x|}{r} \quad \text{and} \quad \frac{|F_y|}{F} = \frac{|y|}{r}.$$

Notice that when x is positive, the component F_x acts in the negative x-direction, and when y is positive, F_x acts in the negative y-direction. In fact, the sign of F_x is always opposite that of the x-coordinate of the position, and the same is true for F_y. This is why

$$\frac{F_x}{F} = \frac{-x}{r} \quad \text{and} \quad \frac{F_y}{F} = \frac{-y}{r}.$$

So $rF_x = -xF$ and $rF_y = -yF$. The derivatives $\frac{dx}{dt}$ and $\frac{dy}{dt}$ are the velocities of P in the x- and y-directions, respectively, and the second derivatives $\frac{d^2x}{dt^2}$ and $\frac{d^2y}{dt^2}$ are the respective accelerations in the x-and y-directions. As a consequence of Newton's second law, we get

$$F_x = m\frac{d^2x}{dt^2} \quad \text{and} \quad F_y = m\frac{d^2y}{dt^2}.$$

The physics of the matter—it involved the application of Newton's $F = ma$ in both the x- and y-directions—is now over. *The rest is mathematics!* By combining equations already derived, we get

$$mr\frac{d^2x}{dt^2} = -Fx \quad \text{and} \quad mr\frac{d^2y}{dt^2} = -Fy$$

and therefore that

$$mr\Big(x\frac{d^2y}{dt^2} - y\frac{d^2x}{dt^2}\Big) = mr\frac{d^2y}{dt^2}x - mr\frac{d^2x}{dt^2}y = -Fyx + Fxy = 0.$$

It follows that

$$y\frac{d^2x}{dt^2} = x\frac{d^2y}{dt^2}.$$

(Note that this also holds when $r = 0$, because then both $x = 0$ and $y = 0$.) Consider the difference

$x \cdot \frac{dy}{dt} - y \cdot \frac{dx}{dt}$. By the product rule, the derivative of this difference is

$$\begin{aligned}\frac{d}{dt}\Big(x \cdot \frac{dy}{dt} - y \cdot \frac{dx}{dt}\Big) &= \frac{d}{dt}\Big(x \cdot \frac{dy}{dt}\Big) - \frac{d}{dt}\Big(y \cdot \frac{dx}{dt}\Big)\\ &= \Big(\frac{dx}{dt} \cdot \frac{dy}{dt} + x \cdot \frac{d^2y}{dt^2}\Big) - \Big(\frac{dy}{dt} \cdot \frac{dx}{dt} + y \cdot \frac{d^2x}{dt^2}\Big)\\ &= \frac{dx}{dt} \cdot \frac{dy}{dt} - \frac{dy}{dt} \cdot \frac{dx}{dt} + x \cdot \frac{d^2y}{dt^2} - y \cdot \frac{d^2x}{dt^2}.\end{aligned}$$

In view of the equality $y\,\frac{d^2x}{dt^2} = x\,\frac{d^2y}{dt^2}$, it follows that $\frac{d}{dt}(y \cdot \frac{dx}{dt} - x \cdot \frac{dy}{dt}) = 0$. We can therefore conclude that

$$x \cdot \frac{dy}{dt} - y \cdot \frac{dx}{dt} = c,$$

where c is a constant. This fact contains essential information about the motion of P. To extract it, we will now "go polar."

12.8 GOING POLAR

The Cartesian part of our discussion is done. The task now will be to transfer the results that were obtained—in particular, the two force equations and the equality $x \cdot \frac{dy}{dt} - y \cdot \frac{dx}{dt} = c$—into the context of polar coordinates. So now regard the origin O and the positive x-axis to also be a polar coordinate system.

Let (r, θ), with $r > 0$ the distance from O to the point-mass P, be polar coordinates of P. The assumption $r > 0$ ensures that P does not crash into the center of force O. Let $r = f(\theta)$ be a function that has the orbit of the point P as its graph. Since $\theta = \theta(t)$ is a function of the elapsed time t, $r = r(t) = f(\theta(t))$ is a function of t. The magnitude of the centripetal force is determined by the location of P and hence by θ and $r = f(\theta)$. This magnitude is therefore also a function of θ. Combining this function with $\theta = \theta(t)$ provides its connection with the force function $F(t)$. Refer to Figure 12.23. Given our general smoothness assumptions, the function $r = f(\theta)$ of θ, as well as the functions $r(t)$ and $\theta(t)$ of t, all have first and second

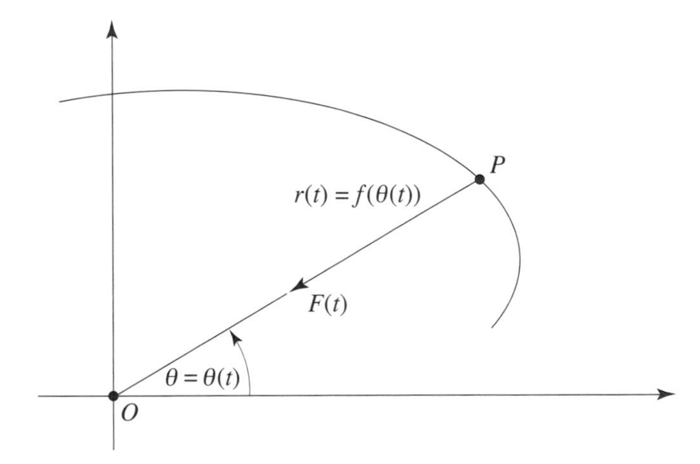

Fig. 12.23

derivatives. (The fact that θ is a differentiable and hence a continuous function of t means that as polar coordinate of the moving point P, it does not jump from θ to, say, $\theta + 2\pi$ from one moment to the next.)

Because we have P moving in a counterclockwise direction, it is now convenient to organize things as follows: let $t = 0$ be an instant at which P crosses the polar axis. Then $t > 0$ is the elapsed time thereafter, and $t < 0$ refers to the time before. (In televised launches of NASA space missions, it is common to hear announcements like "t equals minus 73 seconds and counting.") We will take $\theta(t) > 0$ for $t > 0$, $\theta(t) = 0$ for $t = 0$, and $\theta(t) < 0$ for $t < 0$.

Recall the relationships

$$x = r\cos\theta \quad \text{and} \quad y = r\sin\theta$$

between the polar and Cartesian coordinates (from Section 11.4). We will use them to rewrite the equality $x \cdot \frac{dy}{dt} - y \cdot \frac{dx}{dt} = c$ of Section 12.7 in terms of r and θ. By the product rule,

$$\frac{dx}{dt} = \frac{dr}{dt}\cos\theta + r\frac{d}{dt}(\cos\theta).$$

Since $\theta = \theta(t)$, we get by the chain rule that $\frac{d}{dt}(\cos\theta) = -\sin\theta \cdot \theta'(t)$ and hence that

$$\frac{dx}{dt} = \frac{dr}{dt}\cos\theta - r\sin\theta \cdot \frac{d\theta}{dt}.$$

In exactly the same way,

$$\frac{dy}{dt} = \frac{dr}{dt}\sin\theta + r\cos\theta \cdot \frac{d\theta}{dt}.$$

By combining the equations $x = r\cos\theta$ and $y = r\sin\theta$ with those just derived, we get

$$\begin{aligned} x \cdot \frac{dy}{dt} - y \cdot \frac{dx}{dt} &= (r\cos\theta)\Big(\frac{dr}{dt}\sin\theta + r\cos\theta \cdot \frac{d\theta}{dt}\Big) - (r\sin\theta)\Big(\frac{dr}{dt}\cos\theta - r\sin\theta \cdot \frac{d\theta}{dt}\Big) \\ &= (r^2\cos^2\theta)\frac{d\theta}{dt} + (r^2\sin^2\theta)\frac{d\theta}{dt} = r^2(\sin^2\theta + \cos^2\theta)\frac{d\theta}{dt} \\ &= r^2\frac{d\theta}{dt}. \end{aligned}$$

We have therefore verified that

$$r(t)^2\frac{d\theta}{dt} = r(t)^2\theta'(t) = c.$$

Example 12.19. Think for a moment about the equation $r(t)^2\theta'(t) = c$. What is the meaning of $\theta'(t)$? Consider P at two different times t_1 and t_2 in its orbit, and suppose that the distance from P to O is much greater at t_1 than at t_2. What can you say about the motion of P when you compare $\theta'(t_1)$ and $\theta'(t_2)$?

Continue to consider P at two different times t_1 and t_2 in its orbit. Suppose that $t_1 \le t_2$, and let A be the area that is swept out by the segment OP during the time interval $[t_1, t_2]$. Let $\theta(t_1) = a$ and $\theta(t_2) = b$.

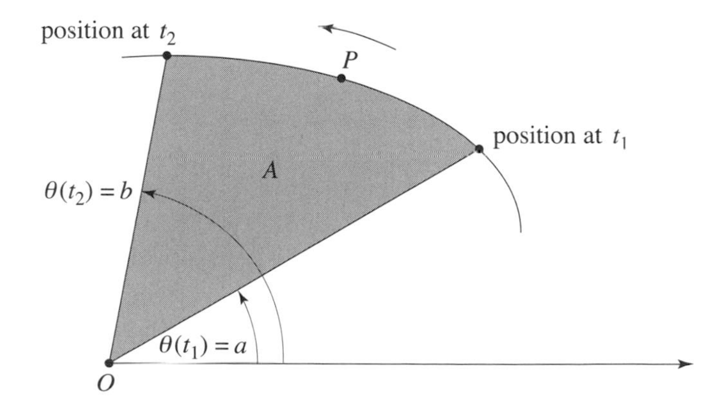

Fig. 12.24

Figure 12.24 illustrates what has been described. Now turn to Section 12.5. Note first that

$$A = \int_a^b \tfrac{1}{2} f(\theta)^2 \, d\theta \, .$$

Because $r(t) = f(\theta(t))$, the substitutions $\theta = \theta(t)$ and $d\theta = \theta'(t)dt$ and the discussion that concludes Section 12.5 tell us that

$$A = \int_a^b \tfrac{1}{2} f(\theta)^2 \, d\theta = \int_{t_1}^{t_2} \tfrac{1}{2} f(\theta(t))^2 \, \theta'(t)dt = \int_{t_1}^{t_2} \tfrac{1}{2} r(t)^2 \theta'(t)dt = \int_{t_1}^{t_2} \tfrac{1}{2} c \, dt = \tfrac{1}{2} ct \Big|_{t_1}^{t_2} = \tfrac{1}{2} c(t_2 - t_1).$$

Putting $\frac{1}{2}c = \kappa$, we see that the area A swept out by P is equal to κ times the time $t_2 - t_1$ that it takes to sweep it out. Taking $t_1 = 0$ and t_2 the elapsed time t establishes the equality

$$A = \kappa t$$

for the area A traced out by the segment PO from 0 to t. This equality is directly related to Kepler's second law (as discussed in Section 3.5). Newton verified it completely differently with his machine-gun-style argument (in Section 6.8). With a second bow to Kepler (the first came in Section 6.8), we'll refer to the constant κ as the Kepler constant of the orbit of P.

One more order of business is the conversion of the two Cartesian force equations

$$mr \frac{d^2x}{dt^2} = -Fx \quad \text{and} \quad mr \frac{d^2y}{dt^2} = -Fy$$

of Section 12.7 into a single force equation in polar form. Because $r(t)^2 \frac{d\theta}{dt} = c = 2\kappa$, we know that $0 = \frac{d}{dt}(r^2 \frac{d\theta}{dt}) = 2r\frac{dr}{dt} \cdot \frac{d\theta}{dt} + r^2 \cdot \frac{d^2\theta}{dt^2}$ and hence that

$$2 \frac{dr}{dt} \cdot \frac{d\theta}{dt} + r \cdot \frac{d^2\theta}{dt^2} = 0.$$

Recall that

$$\frac{dx}{dt} = \frac{dr}{dt} \cos\theta - r \sin\theta \cdot \frac{d\theta}{dt} \, .$$

By differentiating this equation (use the product and chain rules several times), we get that

$$\begin{aligned}
\frac{d^2x}{dt^2} &= \Big(\frac{d^2r}{dt^2} \cos\theta - \frac{dr}{dt} \sin\theta \cdot \frac{d\theta}{dt}\Big) - \frac{dr}{dt} \sin\theta \cdot \frac{d\theta}{dt} - r\Big(\cos\theta \cdot \frac{d\theta}{dt} \cdot \frac{d\theta}{dt} + \sin\theta \cdot \frac{d^2\theta}{dt^2}\Big) \\
&= \frac{d^2r}{dt^2} \cos\theta - 2\frac{dr}{dt} \sin\theta \cdot \frac{d\theta}{dt} - r\cos\theta \cdot \Big(\frac{d\theta}{dt}\Big)^2 - r\sin\theta \cdot \frac{d^2\theta}{dt^2} \\
&= \frac{d^2r}{dt^2} \cos\theta - r\cos\theta \cdot \Big(\frac{d\theta}{dt}\Big)^2 - \sin\theta\Big(2\frac{dr}{dt} \cdot \frac{d\theta}{dt} + r \cdot \frac{d^2\theta}{dt^2}\Big).
\end{aligned}$$

Since the last term is equal to 0,

$$\frac{d^2x}{dt^2} = \cos\theta \left[\frac{d^2r}{dt^2} - r \cdot \Big(\frac{d\theta}{dt}\Big)^2\right].$$

Doing the same thing with $\dfrac{dy}{dt} = \dfrac{dr}{dt} \sin\theta + r\cos\theta \cdot \dfrac{d\theta}{dt}$ gives us

$$\frac{d^2y}{dt^2} = \sin\theta \left[\frac{d^2r}{dt^2} - r \cdot \Big(\frac{d\theta}{dt}\Big)^2\right].$$

In view of the equations $-mr\frac{d^2x}{dt^2} = Fx = F(r\cos\theta)$ and $-mr\frac{d^2y}{dt^2} = Fy = F(r\sin\theta)$ established in

Section 12.7, we now get

$$F\cos\theta = m\cos\theta\left[r\cdot\left(\frac{d\theta}{dt}\right)^2 - \frac{d^2r}{dt^2}\right] \quad \text{and} \quad F\sin\theta = m\sin\theta\left[r\cdot\left(\frac{d\theta}{dt}\right)^2 - \frac{d^2r}{dt^2}\right].$$

A look at the graphs of $\sin\theta$ and $\cos\theta$ (see Section 4.6) tells us that they are never simultaneously equal to 0. Therefore

$$F = m\left[r\cdot\left(\frac{d\theta}{dt}\right)^2 - \frac{d^2r}{dt^2}\right].$$

One of our first important goals of our study of centripetal force has now been achieved. They are the equations

$$r^2\frac{d\theta}{dt} = 2\kappa \quad \text{and} \quad F(t) = m\left[\frac{4\kappa^2}{r(t)^3} - \frac{d^2r}{dt^2}\right].$$

The first is an expression of Kepler's second law and the second is the *centripetal force equation* obtained by inserting $\frac{d\theta}{dt} = \frac{2\kappa}{r^2}$ into the force equation just obtained.

Let's step back and summarize the results that were derived and the assumptions that were required. A point-mass P of mass m is regarded to be propelled by a centripetal force and to move smoothly in its orbit. A polar coordinate system is chosen with the origin O placed at the center of force. The polar function $r = f(\theta)$ expresses the distance r of P from O in terms of the polar angle θ of the changing position of P. As a consequence, the polar graph of this function describes the orbit of P. Let t be the elapsed time from some fixed instant. The fact that P moves means that both the positional angle $\theta = \theta(t)$ and the distance $r(t) = f(\theta(t))$ are functions of time t. The above description implies that the motion satisfies $r^2\frac{d\theta}{dt} = 2\kappa$ (a version of Kepler's second law), where κ is Kepler's constant of the orbit, and that the magnitude $F(t)$ of the centripetal force is related to $r(t)$ in an explicit way (the centripetal force equation).

It will be of advantage to have another version of the centripetal force equation. Let $g(\theta) = \frac{1}{f(\theta)}$. So $r(t) = f(\theta(t)) = g(\theta(t))^{-1}$. By the chain rule and the formula $\frac{d\theta}{dt} = \frac{2\kappa}{r(t)^2}$,

$$\begin{aligned}\frac{dr}{dt} &= -g(\theta(t))^{-2}\cdot\frac{d}{dt}g(\theta(t)) = -\frac{1}{g(\theta(t))^2}\,g'(\theta(t))\,\theta'(t)\\ &= -\frac{1}{g(\theta(t))^2}\,\frac{2\kappa}{r(t)^2}\,g'(\theta(t)) = -2\kappa g'(\theta(t)).\end{aligned}$$

By another application of the chain rule,

$$\frac{d^2r}{dt^2} = -2\kappa g''(\theta(t))\,\theta'(t) = -2\kappa g''(\theta(t))\frac{2\kappa}{r(t)^2} = -4\kappa^2 g''(\theta(t))\frac{1}{r(t)^2} = -4\kappa^2 g''(\theta(t))g(\theta(t))^2.$$

By substituting this and $r(t)^{-3} = g(\theta(t))^3$ into $F(t) = m\left[\frac{4\kappa^2}{r(t)^3} - \frac{d^2r}{dt^2}\right]$, we arrive at a second version

$$F(t) = 4m\kappa^2 g(\theta(t))^2[g(\theta(t)) + g''(\theta(t))]$$

of the centripetal force equation.

12.9 FROM CONIC SECTION TO INVERSE SQUARE LAW AND BACK AGAIN

We begin this section by assuming that the orbit of P (note that we have as yet made no assumptions about the geometry of the orbit) is a conic section—either an ellipse, a parabola, or a hyperbola—and that the center of force is at a focal point. A review of the analysis of the equation

$$r = \frac{d}{1+\varepsilon\cos\theta}$$

in Section 12.2 tells us that it is possible to place the polar axis so that the origin O is at the center of force and to take the orbit of P to be the graph of the polar function

$$r = f(\theta) = \frac{d}{1+\varepsilon\cos\theta}$$

with $r > 0$, $\varepsilon \geq 0$ the eccentricity of the conic section, and $d > 0$ a constant. If the orbit is an ellipse, then $0 \leq \varepsilon < 1$; if it is a parabola, then $\varepsilon = 1$; and if it is a hyperbola, then $\varepsilon > 1$.

Differentiate the equation

$$r(t) = d(1+\varepsilon\cos\theta(t))^{-1}$$

and use the fact that $\frac{d\theta}{dt} = \frac{2\kappa}{r(t)^2}$ to get

$$\frac{dr}{dt} = -d(1+\varepsilon\cos\theta(t))^{-2}\Big(-\varepsilon\sin\theta(t)\cdot\frac{d\theta}{dt}\Big) = \varepsilon d(1+\varepsilon\cos\theta(t))^{-2}(\sin\theta(t))2\kappa r(t)^{-2}.$$

After substituting $(1+\varepsilon\cos\theta(t))^{-2} = \frac{r(t)^2}{d^2}$, this becomes

$$\frac{dr}{dt} = \frac{2\varepsilon\kappa}{d}\sin\theta(t).$$

Differentiating once more, we get

$$\frac{d^2r}{dt^2} = \frac{2\varepsilon\kappa}{d}\Big(\cos\theta(t)\cdot\frac{d\theta}{dt}\Big) = \frac{2\varepsilon\kappa}{d}(\cos\theta(t))\frac{2\kappa}{r(t)^2} = \frac{4\kappa^2\varepsilon}{d}\cos\theta(t)\frac{1}{r(t)^2}.$$

Substituting this into the centripetal force equation of Section 12.8, we obtain

$$F(t) = m\left[\frac{4\kappa^2}{r(t)^3} - \frac{d^2r}{dt^2}\right] = m\left[\frac{4\kappa^2}{r(t)^3} - \frac{4\kappa^2\varepsilon}{d}\cos\theta(t)\frac{1}{r(t)^2}\right] = 4m\kappa^2\left[\frac{1}{r(t)} - \frac{\varepsilon}{d}\cos\theta(t)\right]\frac{1}{r(t)^2}.$$

Because $\frac{1}{r(t)} = \frac{1+\varepsilon\cos\theta(t)}{d} = \frac{1}{d} + \frac{\varepsilon}{d}\cos\theta(t)$, we find that $F(t) = \frac{4m\kappa^2}{d}\frac{1}{r(t)^2}$. We finally get

$$F(t) = \frac{8m\kappa^2}{L}\frac{1}{r(t)^2},$$

where m is the mass of the object, κ is Kepler's constant, $L = 2d$ is the latus rectum of the orbit (see Section 12.2), and $r(t)$ is the distance from the object to the point of origin of the force.

We have established that if a point-mass P is propelled by a centripetal force and *if it has an orbit that is either an ellipse, a parabola, or a hyperbola, with the center of force at a focal point,* then *the magnitude of the force is proportional to the inverse of the square of the distance between P and this focal point.*

This is the same result that Newton established with the argument described in Section 6.9. This time, however, the verification is complete, and it works for the ellipse, parabola, and hyperbola all at once.

If the orbit is an ellipse, with semimajor axis a and semiminor axis b, and if T is the period of the orbit, then $\kappa = \frac{ab\pi}{T}$ and $L = \frac{2b^2}{a}$ (see Problem 6.28), so that in this case,

$$F(t) = \frac{8m\kappa^2}{L}\frac{1}{r(t)^2} = 8m\frac{a^2b^2\pi^2}{T^2}\frac{a}{2b^2}\frac{1}{r(t)^2} = \frac{4\pi^2a^3m}{T^2}\frac{1}{r(t)^2}.$$

It has become tradition in the mathematical sciences to follow axiomatic approaches. Definitive versions of theories of mathematics or physics are often cast in the following form: certain basic underlying laws or principles, referred to as axioms or postulates, are taken as starting point, and all other relevant propositions are deduced from these by the force of logic and mathematics alone. The paradigm of such axiomatic approaches (apparently the first in the history of science) is the development of Plane Geometry in Euclid's

Elements. The theory of gravitation can be cast in this form. Taking Newton's three basic laws of motion (see Section 6.6) and his law of universal gravitation $F = G\frac{m_1 \cdot m_2}{r^2}$ (see Section 6.10) as axioms, Kepler's three laws of planetary motion can be derived mathematically. The second of Kepler's laws was already verified in Section 12.8. We now turn to a generalization of Kepler's first law, and show that the orbit of an object propelled by the single gravitational force of a much more massive body is an ellipse, a parabola, or a hyperbola, and that the center of the massive body is at a focal point.

Think of the situation of a planet of mass m being pulled along its orbit by the gravitational force of the Sun. With M the mass of the Sun, and r the distance between the planet and the Sun, the magnitude of this attractive gravitational force is—by Newton's law of universal gravitation—equal to $G\frac{Mm}{r^2}$. With $C = GM$, this magnitude is equal to $C\frac{m}{r^2}$.

So we will now assume that the magnitude $F(t)$ of the centripetal force on the point-mass P satisfies an inverse square law of the form

$$F(t) = C\frac{m}{r(t)^2},$$

where $C > 0$ is a constant. Combining $F(t) = C\frac{m}{r(t)^2} = Cmg(\theta(t))^2$ with the centripetal force equation, we get

$$4m\kappa^2 g(\theta(t))^2[g(\theta(t)) + g''(\theta(t))] = Cmg(\theta(t))^2.$$

Dividing through by $4m\kappa^2 \cdot g(\theta(t))^2$ gives us

$$g(\theta(t)) + g''(\theta(t)) = \frac{C}{4\kappa^2}.$$

We will see that this equation determines the shape of the orbit of P to be that of an ellipse, parabola, or hyperbola, with the center of force O at a focal point in all cases. Because we are interested in the shape of the orbit—namely, the precise form of the function $r = f(\theta)$—we now ignore the fact that θ is a function of t, and turn to the problem of solving the differential equation $g(\theta) + g''(\theta) = \frac{C}{4\kappa^2}$ for $g(\theta)$. Put $h(\theta) = g(\theta) - \frac{C}{4\kappa^2}$, and notice that

$$h''(\theta) + h(\theta) = 0.$$

But this means that the function $y = h(\theta)$ is a solution of the differential equation $y'' + y = 0$. It follows from the discussion toward the end of Section 11.6 that

$$h(\theta) = A\sin\theta + B\cos\theta$$

for some constants A and B (they are the constants D_1 and D_2, respectively, of this discussion). Therefore

$$g(\theta) = A\sin\theta + B\cos\theta + \frac{C}{4\kappa^2}.$$

We will now suppose that there is a point of "closest approach" for P, namely, an angle θ at which the distance $r = f(\theta)$ from P to O has a local minimum. (This is always so in any "real" situation.) Rotate the polar axis, while keeping the center of force at the origin O, so that that this local minimum occurs at $\theta = 0$. What we need next is some information about the constants A and B. The effort to get it is clarified by the Cartesian graph of $g(\theta) = A\sin\theta + B\cos\theta + \frac{C}{4\kappa^2}$. (What is meant by this is illustrated in Figure 12.3 for $\sin\theta$.) Because $g(\theta) = \frac{1}{f(\theta)}$, the function $g(\theta)$ has a local maximum at $\theta = 0$. So by the max-min theorem of Section 7.8, $g'(0) = 0$. Since $g'(\theta) = A\cos\theta - B\sin\theta$, it follows that $A = 0$. So $g'(\theta) = -B\sin\theta$ and $g''(\theta) = -B\cos\theta$. Assume for a moment that B is negative. Because $g''(0) = -B > 0$, the second derivative test of Section 7.8 would tell us that $g(\theta)$ has a local minimum at $\theta = 0$. But for a nonzero B, $g(\theta) = B\cos\theta + \frac{C}{4\kappa^2}$ cannot have both a local maximum and a local minimum at $\theta = 0$. Therefore $B < 0$ cannot be, and hence $B \geq 0$. In view of the fact that $f(\theta) = \frac{1}{g(\theta)} = \frac{1}{B\cos\theta + \frac{C}{4\kappa^2}}$, we now get

$$f(\theta) = \frac{1}{\frac{C}{4\kappa^2}(1 + \frac{4\kappa^2 B}{C}\cos\theta)} = \frac{\frac{4\kappa^2}{C}}{1 + \frac{4\kappa^2 B}{C}\cos\theta}.$$

Since this polar function has the form $f(\theta) = \frac{d}{1+\varepsilon\cos\theta}$ with $d = \frac{4\kappa^2}{C} > 0$ and $\varepsilon = \frac{4\kappa^2 B}{C} \geq 0$, we know from the study in Section 12.2 that its graph is a conic section with eccentricity ε and focal point the origin O. It is also shown in Section 12.2 that $\frac{4\kappa^2}{C} = d = \frac{L}{2}$, where L is the latus rectum of the orbit.

We have verified that *if the magnitude of the centripetal force acting on P satisfies an inverse square law,* then *the orbit of P is an ellipse, a parabola, or a hyperbola with the center of force at a focal point.*

From $\frac{4\kappa^2}{C} = \frac{L}{2}$, we get $C = \frac{8\kappa^2}{L}$, so that the equality $F(t) = C\frac{m}{r(t)^2}$ brings us back to the force formula

$$F(t) = \frac{8m\kappa^2}{L}\frac{1}{r(t)^2}.$$

In the case of an elliptical orbit with semimajor axis a, semiminor axis b, and period T, we know that $\kappa = \frac{ab\pi}{T}$ and $L = \frac{2b^2}{a}$, so that

$$F(t) = \frac{4\pi^2 a^3 m}{T^2}\frac{1}{r(t)^2}.$$

For an elliptical orbit, $C = \frac{8\kappa^2}{L} = 8(\frac{ab\pi}{T})^2 \cdot \frac{a}{2b^2} = \frac{4a^3\pi^2}{T^2}$. Combined with $C = GM$, this leads to the refined version $\frac{a^3}{T^2} = \frac{GM}{4\pi^2}$ of Kepler's third law.

Having demonstrated that *all three of Kepler's observations about the planetary orbits follow from Newton's basic laws of motion and his law of universal gravitation* (plus a lot of mathematical maneuvering!), we have completed what we set out to do.

There are two important issues that require further discussion. We have seen that the quantitative understanding of the orbit of P around O rests on the determination of the distance function $r = r(t)$. A look back tells us that $r = r(t)$ is the composite of the two functions $r = f(\theta)$ and $\theta = \theta(t)$. The polar function $r = f(\theta)$ describes the geometry of the orbit, and the function $\theta = \theta(t)$ tells us how this orbit is traced out over time. The function $r = f(\theta)$ was shown to be a conic section shaped by orbital constants. But what about $\theta = \theta(t)$? In the case of an elliptical orbit, this difficult question is answered by the analysis of the function $\alpha(t)$ in Section 10.4. (The parabolic and hyperbolic situations can be studied by similar approaches involving cubic polynomials and hyperbolic functions, respectively.) The second issue concerns the fact that our study has focused abstractly on a point-mass P moved along its orbit by a centripetal force. But does this study apply to our solar system? Does it apply to a planet in motion around the Sun? Can it be assumed that the gravitational pull by the Sun on a planet is directed from the center of mass of the planet toward the center of mass of the Sun? The next section will answer these questions—after some reasonable assumptions are made—in the affirmative.

12.10 GRAVITY AND GEOMETRY

Newton was aware that there was a fundamental unanswered problem that stood in the way of the application of his analysis of centripetal force for point-masses to the situation of the gravitational attraction of bodies in the solar system and beyond. Given that the gravitational attraction of any small particle of matter on any other satisfies the inverse square law, why should it be that large massive bodies—namely, huge collectives of such particles—should attract each other in the same way? Should the net force exerted by all the particles of matter of a massive sphere on all the particles of another massive sphere be directed from the center of one sphere to the center of the other and hence be inversely proportional to the square of the distance between these centers? Only if this is so does the study of gravitational force reduce to the situation of point-masses. This question presented a formidable challenge to Newton. Some scholars have in fact claimed that the matter was the cause (or at least one of them) for the 20-year delay between Newton's first thoughts about universal gravitation and the composition of the *Principia*.

The problem facing Newton was subtle. Consider the gravitational force F that a body B of mass M exerts on a point-mass m. (In the discussion of this section, m represents both the point-mass and its mass.) Then it is *not the case in general* that the magnitude F of this force is given by Newton's universal law of gravitation

$$F = G\frac{mM}{d^2},$$

where d is the distance from the point-mass to the center of mass of B. Both the shape of the body B and the distribution of the mass within it play a critical role. Both need to be configured symmetrically for Newton's law to hold. Fortunately, the formula is correct in the most important situations: those where the attracting body B is the Sun, the Earth, a planet, or, more generally, a sphere that has its mass distributed in a certain radially homogeneous way. The proof of this assertion makes use (more than once) of the essential strategy of integral calculus. Slice up the body B into a very large number of smaller pieces, and determine the force with which each of the pieces acts on m. To understand how B acts on m is a matter of computing the resultant of all the smaller forces, or in the language of integral calculus of adding them all up. This section presents an opportunity to have one last look at such addition strategies.

Part 1. To start, take B to be a thin homogeneous circle of matter of mass M and radius r. Think of a circular loop of wire. Homogeneous means that the matter of the circle is evenly distributed (in particular, there are no lumps). Suppose that the point-mass m lies on the perpendicular to the circle through its center O, at a distance c from the center. Divide the circle into a huge even number, say, $2k$, of small equal segments. Each segment is $\frac{2\pi r}{2k} = \frac{\pi r}{k}$ units long and has a mass of $M_{\text{seg}} = \frac{M}{2k}$. Refer to Figure 12.25. By

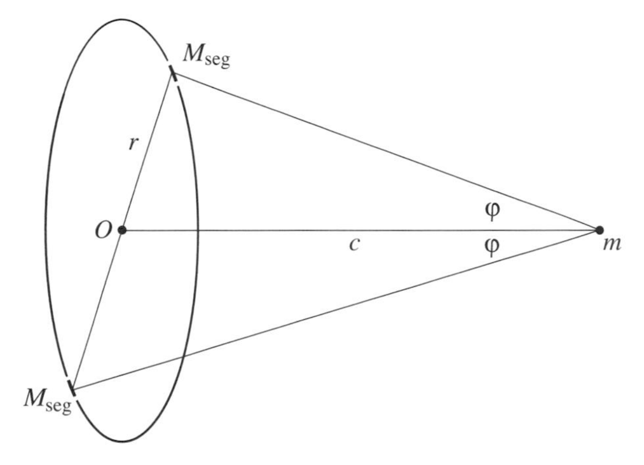

Fig. 12.25

the Pythagorean theorem, the distance between m and each of the segments is equal to $\sqrt{c^2 + r^2}$. Since the segments are very small, Newton's law of universal gravitation for point-masses tells us that the magnitude of the gravitational force with which each of them attracts m is equal to

$$\frac{GmM_{\text{seg}}}{(\sqrt{c^2 + r^2})^2} = \frac{GmM_{\text{seg}}}{c^2 + r^2}.$$

Since the number of identical segments is even, the segments can be paired, as Figure 12.25 indicates, each with the one on the opposite side of the circle. Figure 12.26 depicts as vectors both the forces of attraction that a matching pair of segments exerts on m as well as the indicated components of these forces. Notice that the two components along the line from m to O (only one of them is shown in the figure) are both equal to $F_{\text{seg}} = \frac{GmM_{\text{seg}}}{c^2+r^2}\cos\varphi$ and that the two components perpendicular to this line cancel each other out. By Figure 12.25, $\cos\varphi = \frac{c}{\sqrt{c^2+r^2}}$, so that

$$F_{\text{seg}} = \frac{GmM_{\text{seg}}}{c^2 + r^2} \cdot \frac{c}{\sqrt{c^2 + r^2}} = \frac{GmM_{\text{seg}}\, c}{(c^2 + r^2)^{\frac{3}{2}}}.$$

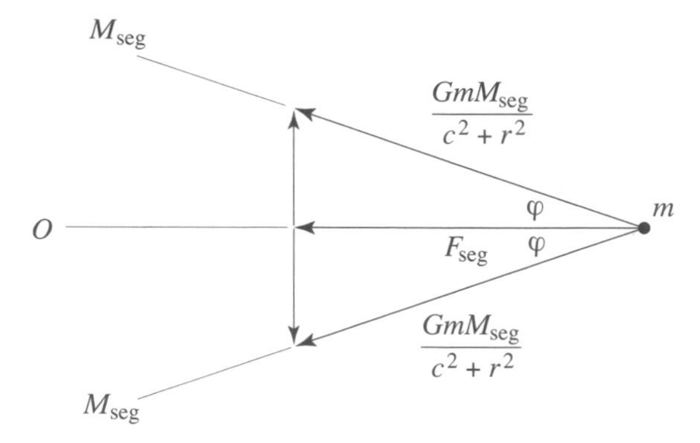

Fig. 12.26

The magnitudes of the two components along the line from m to O add to $2F_{\text{seg}} = 2\frac{GmM_{\text{seg}}c}{(c^2+r^2)^{\frac{3}{2}}}$. Considering the fact that the circular mass B is composed of k such opposite pairs, we find that the force of attraction of B on m points in the direction of O and has magnitude $2kF_{\text{seg}} = \frac{Gm(2kM_{\text{seg}})c}{(c^2+r^2)^{\frac{3}{2}}} = \frac{GmMc}{(c^2+r^2)^{\frac{3}{2}}}$.

The argument just completed informs us that the gravitational force with which the thin circular mass B of radius r attracts the point-mass m is directed to the center of the circle and has magnitude

$$F = \frac{GmMc}{(c^2+r^2)^{\frac{3}{2}}},$$

where M is the mass of B and c is the distance from the point-mass to the center of the circle. The various equalities in this discussion are in fact approximations. Only when k is pushed to infinity (in the style of the definite integral) do they become equalities. Because the center of mass of the thin homogeneous circle B is the center of the circle, the gravitational force of attraction does point to the center of mass of B. However, the fact that r is not zero means that the magnitude of this force is *not given by* Newton's law $F = \frac{GmM}{c^2}$.

Part 2. We now turn to consider a thin homogeneous spherical shell B of radius R and mass M. Since only the surface of the sphere is included, think of B as a ball with a thin spherical skin. Let O be the shell's center. Suppose that the point-mass m lies at a distance c from O. We will assume that $c \geq R$, so that the point-mass is on or outside the shell, and analyze the gravitational force that the shell exerts on m.

Begin by slicing up the spherical shell into a large number of very thin ring-like sections. All cuts are perpendicular to the axis—placed horizontally—that connects the center O with m. A typical ring is shown in gray in Figure 12.27. (It is not drawn "very thin" in the diagram, but thick enough so that the relevant mathematics can be explained.) The angle θ with $0 \leq \theta < \pi$ determines the point Q. The angle θ also determines the right boundary of the ring, and the sliver of an angle $d\theta$ determines its thickness and (along with θ) its left boundary. The points Q and Q' lie at distances $R\sin\theta$ and $R\sin(\theta + d\theta)$ from the horizontal axis, respectively. Applying the definition of radian measure to the angle $d\theta$, we see that the length of the arc QQ' is $Rd\theta$. An application of the formula in Section 9.4 for the area of a cone tells us that the surface area of the gray ring is tightly approximated by $\pi(R\sin(\theta + d\theta) + R\sin\theta)Rd\theta$. Let M_{ring} be the mass of the ring. Since the surface area of the entire spherical shell is $4\pi R^2$, the fact that mass and surface area are proportional (this is a consequence of the homogeneity) tells us that

$$\frac{M_{\text{ring}}}{M} \approx \frac{\pi R^2(\sin(\theta + d\theta) + \sin\theta)\,d\theta}{4\pi R^2} = \tfrac{1}{4}(\sin(\theta + d\theta) + \sin\theta)\,d\theta.$$

Therefore, $M_{\text{ring}} \approx \frac{1}{4}M(\sin(\theta + d\theta) + \sin\theta)\,d\theta$. Let s be the length of the segment connecting Q to the mass m. Let x be the base of the triangle with hypotenuse s. Since the distance from O to m is c, the remaining segment has length $c - x$. By applying Part 1 to the ring and the mass m, we get that the force

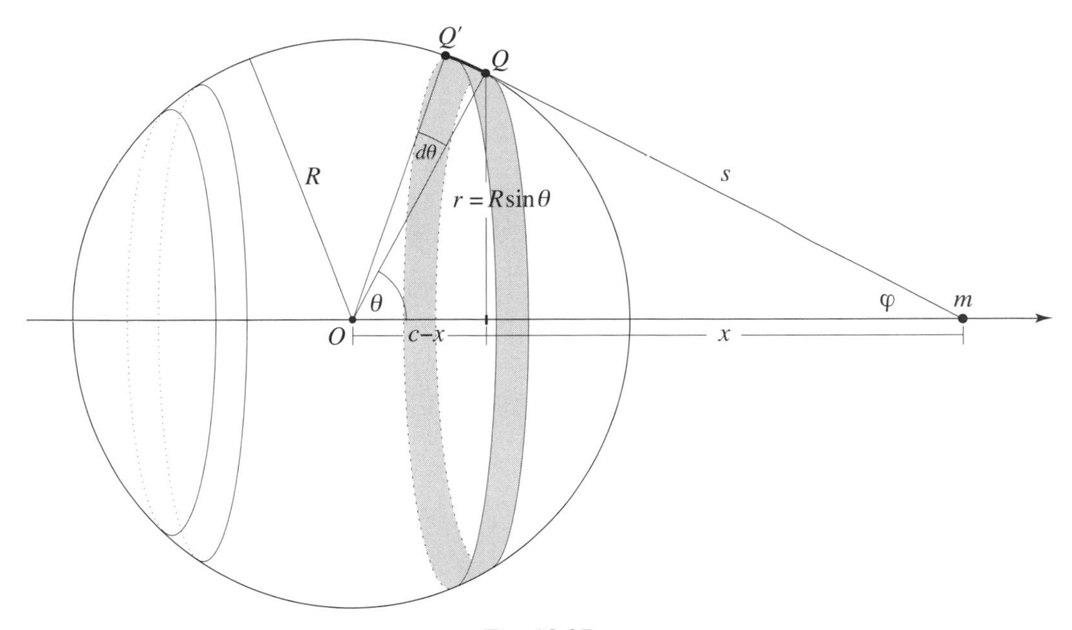

Fig. 12.27

of attraction of the ring on m is directed to the center O of the sphere and has approximate magnitude

$$F_{\text{ring}} \approx \frac{GmM_{\text{ring}}\,x}{(x^2+r^2)^{\frac{3}{2}}} \approx \frac{\frac{1}{4}GmMx(\sin(\theta+d\theta)+\sin\theta)\,d\theta}{(x^2+r^2)^{\frac{3}{2}}} \approx \frac{\frac{1}{2}GmMx\sin\theta}{(x^2+r^2)^{\frac{3}{2}}}\,d\theta.$$

Adding this approximation over all the ring-like sections that were cut (with θ varying from 0 to π) gives us an approximation of the gravitational force F_{shell} with which the entire shell pulls on m. It is the message of integral calculus that by slicing up the shell into sections that are thinner and thinner, these approximations of F_{shell} get tighter and tighter, and that in the limit,

$$F_{\text{shell}} = \int_0^{\pi} \frac{\frac{1}{2}GmMx\sin\theta}{(x^2+r^2)^{\frac{3}{2}}}\,d\theta.$$

Given that $r = R\sin\theta$ and x are both functions of θ, this would appear to be a complicated integral. However, it turns out that it can be solved rather quickly by making use of the substitution $s = \sqrt{x^2+r^2}$. Since $(x^2+r^2)^{\frac{3}{2}} = s^3$, it remains to express $\sin\theta\,d\theta$ and x in terms of s. Applying the law of cosines to the angle θ and the triangle ΔQOm, we get $s^2 = R^2+c^2-2Rc\cos\theta$. By differentiating, $2s\frac{ds}{d\theta} = -2Rc(-\sin\theta) = 2Rc\sin\theta$. Therefore $\sin\theta\,d\theta = \frac{1}{Rc}s\,ds$. To express x in terms of s, again use the law of cosines to get $R^2 = s^2+c^2-2sc\cos\varphi$. Since $\cos\varphi = \frac{x}{s}$, this implies that $R^2 = s^2+c^2-2cx$, so that $x = \frac{s^2+c^2-R^2}{2c}$. Feeding everything back into the integrand above, we get $\frac{1}{2}\frac{GmMx\sin\theta}{(x^2+r^2)^{\frac{3}{2}}}\,d\theta = \frac{1}{2}GmM\cdot\frac{s^2+c^2-R^2}{2c}\cdot\frac{1}{s^3}\cdot\frac{1}{Rc}s\,ds$. Notice that when $\theta = 0$, $s = c-R$, and when $\theta = \pi$, $s = c+R$. Therefore

$$F_{\text{shell}} = \frac{1}{4Rc^2}GmM\int_{c-R}^{c+R}\frac{s^2+c^2-R^2}{s^2}\,ds = \frac{1}{4Rc^2}GmM\int_{c-R}^{c+R}\left(1+\frac{c^2-R^2}{s^2}\right)ds.$$

Since $s-(c^2-R^2)s^{-1} = s-(c^2-R^2)\frac{1}{s}$ is an antiderivative of $1+\frac{c^2-R^2}{s^2}$,

$$\int_{c-R}^{c+R}\left(1+\frac{c^2-R^2}{s^2}\right)ds = s-(c^2-R^2)\frac{1}{s}\Big|_{c-R}^{c+R} = c+R-(c-R)-(c-R-(c+R)) = 4R.$$

We have therefore shown that

$$F_{\text{shell}} = \frac{GmM}{c^2}.$$

Since the spherical shell is homogeneous, O is its center of mass. The force of attraction of each ring on m points in the direction of O, so the same is true for the sum of all these forces. Since c is the distance between O and m, Newton's law of universal gravitation provides both the direction and magnitude of the force of attraction of a thin homogeneous spherical shell of mass M on a point-mass m.

Part 3. We have arrived at the important point of our discussion. Using what was already established, we will now show that the gravitational force F that a sphere of mass M exerts on a point-mass m a distance c from the center of the sphere is given by Newton's law of universal gravitation

$$F = \frac{GmM}{c^2},$$

provided that the matter within the sphere is distributed in a certain symmetric way. Some assumption about the way that the matter within the sphere is distributed is surely necessary. Why? Consider a sphere made of a light material that has embedded within it a small but dense and heavy kernel of matter. Since every particle attracts every other, surely the gravitational force that the larger sphere exerts on a point-mass depends decisively on the location of the small, heavy kernel within it.

Let's turn to a sphere B of matter with radius R and center O. The assumption that we will make is this: any two small particles of matter in B that are the same distance from the center of the sphere have the same density. This assumption is met by a sphere that is composed of concentric homogeneous layers, each in the shape of a spherical shell. Think of the way an onion is structured. Since each shell is homogeneous, the discussion of Part 2 applies to it. So each shell pulls on the point-mass m in accordance with Newton's law. Our intuition should tell us that therefore the entire sphere should pull on m in this way. Intuition is great. But a detailed argument is better.

The first thing to do is to define a density function for the sphere B. Let P be any point inside the sphere, and let S be a small sphere inside B with center P and radius r. Let m_S be the mass of S, V_S the volume of S, and consider the ratio $\frac{m_S}{V_S}$. This is the average density of the matter comprising the small sphere S. The density $\rho(P)$ at P is the limit

$$\rho(P) = \lim_{r \to 0} \frac{m_S}{V_S}.$$

If the point P is on the sphere, a similar definition works (with only a part of the sphere around P being relevant). The assumption that we will make about the matter within the sphere B is as follows.

If P and Q are any two points in B that are the same distance from O, then $\rho(P) = \rho(Q)$.

We can now define a density function $\rho(x)$ for the sphere. For any x with $0 \le x \le R$, choose a point P in the sphere that is distance x from O, and set $\rho(x) = \rho(P)$. We will assume that $\rho(x)$ is continuous function of x. Intuitively, a sphere that has such a density function is made up of shells—think of an onion—that are homogeneous.

Let's turn to our planet Earth for a moment. For Earth, the graph of the function $\rho(x)$ is sketched in Figure 12.28, with 1000 kg/m^3 the unit of density. What it tells us is that Earth has a very dense inner core with a radius of about 1250 km, a dense outer core about 2250 km thick, a less dense mantle about 2000 km thick, and that the crust, water, earth, and rock of Earth's surface are relatively light and thin.

We'll now verify Newton's law of universal gravitation for spheres with the density property that we have defined. We'll do so by revisiting a construction of Section 9.5. Let $\mathcal{P}$ be a partition

$$0 = x_0 < x_1 < x_2 < \cdots < x_{i-1} < x_i < \cdots < x_{n-2} < x_{n-1} < x_n = R$$

of the interval $[0, R]$ into n subintervals. Set $\Delta x_i = x_{i+1} - x_i$, for $i = 0, 1, \dots n-1$. The *norm* $\|\mathcal{P}\|$ of the partition $\mathcal{P}$ is the largest of the Δx_i. It is understood that n is large and that $\|\mathcal{P}\|$ is small, so that all Δx_i

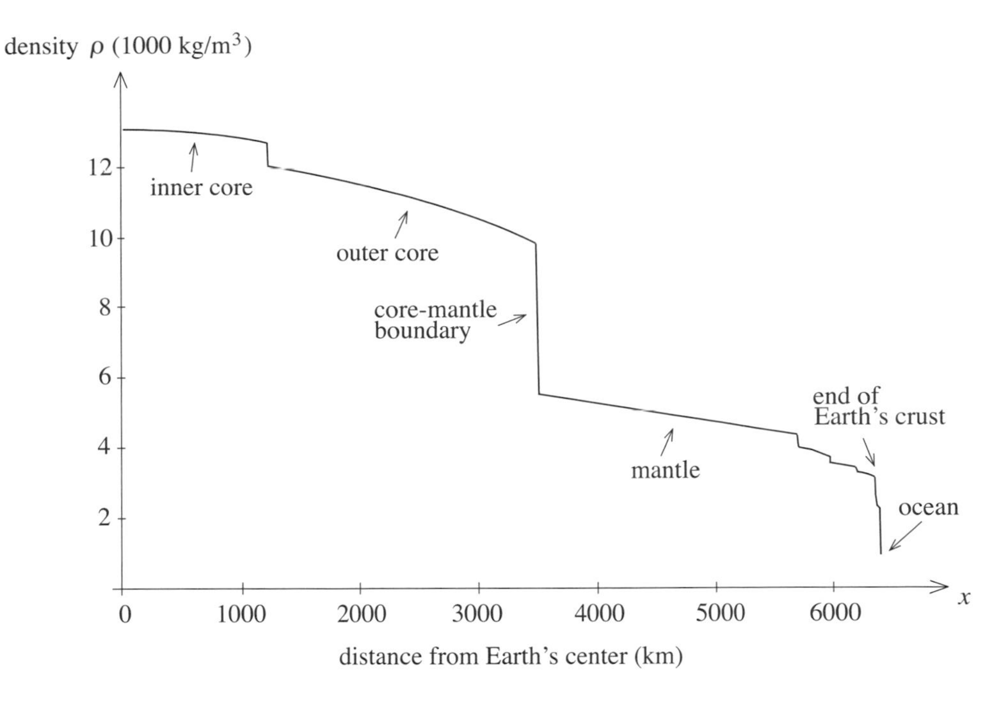

Fig. 12.28

are small. (Here, both large and small are relative to R.) Figure 12.29 depicts a cross section both of the sphere B and the thin spherical shell that the partition points x_i and x_{i+1} determine. (For purposes of "visibility," the segment $[x_i, x_{i+1}]$ and the spherical shell that it determines are both much thicker in the figure than in our description.) The volumes of the spheres of radii x_{i+1} and x_i are $\frac{4}{3}\pi x_{i+1}^3$ and $\frac{4}{3}\pi x_i^3$ respectively. Therefore the volume of the spherical shell is

$$\tfrac{4}{3}\pi x_{i+1}^3 - \tfrac{4}{3}\pi x_i^3 = \tfrac{4}{3}\pi((x_i + \Delta x_i)^3 - x_i^3) = \tfrac{4}{3}\pi(3x_i^2\Delta x_i + 3x_i(\Delta x_i)^2 + (\Delta x_i)^3) \approx 4\pi x_i^2 \Delta x_i.$$

Take c_0 to be a point in the subinterval $[x_0, x_1]$, c_1 a point in $[x_1, x_2]$, ..., c_i a point in $[x_i, x_{i+1}]$, and so on. Given that $\Delta x_i = x_{i+1} - x_i$ is small, this shell is thin. It follows that its density is nearly equal to $\rho(c_i)$ throughout, so that the shell is nearly homogeneous. It follows that the mass M_i of the shell is approximately

$$M_i \approx (4\pi c_i^2 \Delta x_i)\rho(c_i) = 4\pi c_i^2 \rho(c_i)\Delta x_i.$$

So by Part 2, the gravitational force F_i of this shell on the point-mass m is

$$F_i = G\frac{mM_i}{c^2} \approx \frac{Gm}{c^2} 4\pi c_i^2 \rho(c_i)\Delta x_i,$$

where c is the distance of the point-mass m from the shell's center, and hence the sphere's center O. By doing this for each of the n shells, adding the results, and noticing that the sum $\sum_{i=0}^{n-1} M_i$ of the masses of the n shells is the mass M of the sphere, we get that the force F of the entire sphere on m satisfies

$$F \approx \frac{Gm}{c^2} \sum_{i=0}^{n-1} 4\pi c_i^2 \rho(c_i)\,\Delta x_i \approx \frac{Gm}{c^2} \sum_{i=0}^{n-1} M_i = \frac{GmM}{c^2}.$$

Repeat this computation again and again for partitions $\mathcal{P}$ of smaller and smaller norm $||\mathcal{P}||$. Each time the Δx_i become smaller, in the process the various approximations derived above get tighter and tighter,

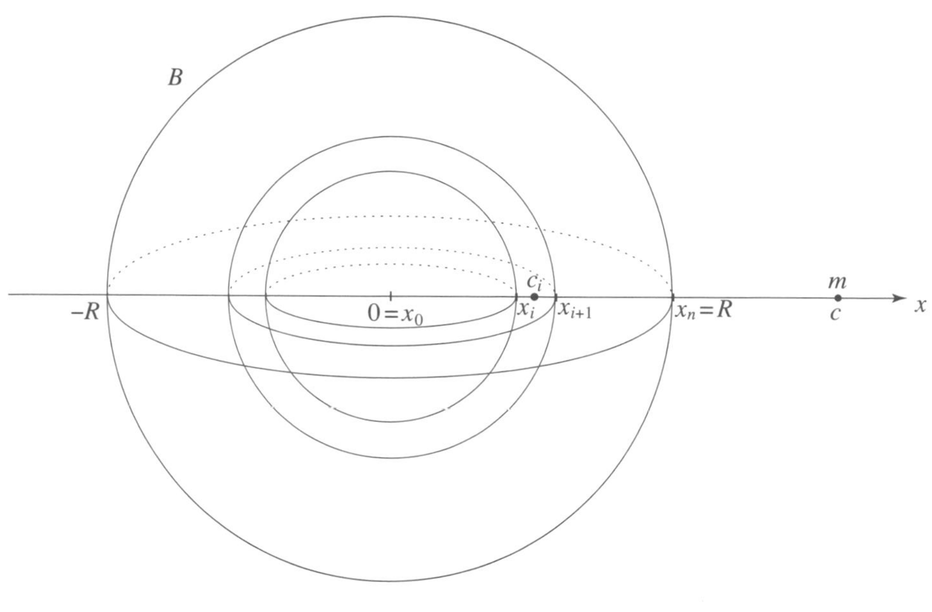

Fig. 12.29

and therefore in the limit,

$$F = \frac{Gm}{c^2} \lim_{\|\mathcal{P}\|\to 0} \sum_{i=0}^{n-1} 4\pi c_i^2 \rho(c_i)\,\Delta x_i \;=\; \frac{GmM}{c^2}.$$

We have established that if the density of a sphere of mass M satisfies the onion property described above, then its force of attraction on a point-mass m that is a distance c from its center satisfies Newton's universal law of gravitation.

Suppose, finally, that two spheres A and B of matter both satisfy the onion property. We know from Part 3 that every particle of A is attracted by B in accordance with Newton's formula, where c is the distance from the particle to the center of B. From the point of view of A, therefore, the entire mass of B can be considered to be concentrated at its center. In other words, B can be considered as a point-mass. But this point-mass is attracted by A in accordance with Newton's gravitational formula. Therefore Newton's formula holds for the two spheres A and B. The gravitational force with which they attract each other is equal to G times the product of their masses divided by the square of the distance between their centers.

We will close with some comments about the gravitational constants "little" g and "big" G. Since they both involve gravity, one would expect them to be related. In fact, we already saw in Section 6.10 that this is the case. For an object of mass m on or near Earth's surface, the gravitational force on the object is its weight mg. The universal law of gravitation applies as well. With the Earth's mass M concentrated at its center and r its radius, the gravitational force is equal to $G\frac{mM}{r^2}$. It follows that $mg = G\frac{mM}{r^2}$ and, after a cancellation,

$$g = \frac{GM}{r^2}.$$

Notice therefore that the gravitational acceleration g does not depend on the mass m of the object. Newton's theory of gravitation has provided the explanation for Galileo's observation that all objects falling near Earth's surface are accelerated in the same way (if they are heavy enough and small enough so that air resistance is not a factor). There are complications, however. Earth's massive mountain ranges and deep valleys tell us that its mass is not exactly distributed as the onion property requires. In addition, the fact that Earth is flattened at the poles and bulges at the equator means that the distance r from the surface of the Earth to its center varies. Therefore the gravitational "constant" g varies. The closer a location is to the equator, the farther it is from Earth's center, the smaller the value of g at that location. For instance, the value of g on the 50th parallel of the Northern Hemisphere—Vancouver, Winnipeg, Frankfurt, Prague, Krakow, and Kiev are near it—g is about 9.809 m/s^2. Farther south, on the 40th parallel—Denver, Philadelphia, Madrid, Ankara, and Beijing lie near this latitude—g is about 9.805 m/s^2. The standard metric value for g was specified to be 9.80665 m/s^2 in 1901 and is used to define the standard weight of an object. The value $g = 9.80665$ m/s^2 (this is equivalent to $g = 32.17405$ feet/second^2) corresponds roughly to the 45th parallel, the line through Minneapolis–St. Paul, Ottawa, Bordeaux, Turin, Belgrade, and the northern tip of the Japanese island of Hokkaido.

Modern experiments to measure G involve sophisticated versions of Cavendish's torsion balance. (See Section 6.10.) The puzzling thing is that recent advanced measurements have produced different values for this constant. The currently accepted value of

$$G = 6.67384 \times 10^{-11} \frac{\mathrm{m}^3}{\mathrm{kg} \cdot \mathrm{s}^2},$$

with an error estimate of 0.00080 m/s^2, is based on an experiment of 2010. However, a painstaking 10-year experiment to calculate G published in 2012 produced the value

$$6.67545 \times 10^{-11} \frac{\mathrm{m}^3}{\mathrm{kg} \cdot \mathrm{s}^2}.$$

Even though gravity seems to be a powerful force—when we jump up, we are quickly pulled back down—it is in fact very weak. The pull is the consequence of the enormous mass of the Earth. With regard to the magnitude of the force of gravity, notice that while the units meter, kilogram, and second are all on a human scale, the factor 10^{-11} means that G is off-the-charts small in this regard. This makes any attempt to calculate its strength such a difficult task. With regard to the fairly large discrepancy between the two measurements of G, most scientists think that the explanation is experimental error rather than changes in the value of G. However, these incompatible measurements may point to unknown subtleties of gravity. Perhaps its strength does depend on how it is measured or where on Earth the measurement is made. Or is the value of the measurement affected by the changing astronomical environment as the Earth moves around the Sun and as the solar system moves within the galaxy? Or are there inherent fluctuations in G? No doubt, further work is required to clarify the situation.

12.11 SPIRAL GALAXIES

We will now move from spirals on the scale of nautilus shells and tropical storms to spirals that are cosmically large. A spiral galaxy is a galaxy that has the shape of a flat disk that determines the "galactic plane." The disk of a galaxy consists of interstellar gases, young blue stars, and open star clusters. Everything in the disk, including billions of stars, rotates around the galactic center. Surrounding the galactic center is a bulge that consists of a huge, tightly packed group of small old yellow and red stars. Figure 12.30 shows two spiral galaxies on collision course. Billions of years from now, the larger galaxy on the left will have absorbed the smaller galaxy on the right. The two bright shining cosmic "eyes" are the galactic bulges. There is strong evidence that there is a supermassive *black hole* at the galactic center of such galaxies. This is a huge and tightly packed mass that exerts a gravitational force so strong that not even light can escape from it. Black holes therefore cannot be observed directly, but in some galaxies here are stars that are seen to move in rapid rotation around what appears to be empty space. This tells us that there exist massive invisible objects that exert powerful gravitational forces.

The interstellar gases and stars of a spiral galaxy orbit in elliptical trajectories around the galactic center, as Newton's gravitational theory predicts. See Figure 12.31a. As is the case in our solar system, objects closer to the center of mass move at greater speeds in their orbits than those farther away. This fact in combination with the powerful gravitational effects within the galaxy causes the ellipses to rotate so

Fig. 12.30. Image credit to Debra Meloy Elmegreen et al., and the Hubble Heritage Team.

that they change their orientations and align themselves as illustrated in Figure 12.31b. Observe from this figure that this rotated geometry creates spiral-shaped regions of matter of greater density. Hydrogen gas

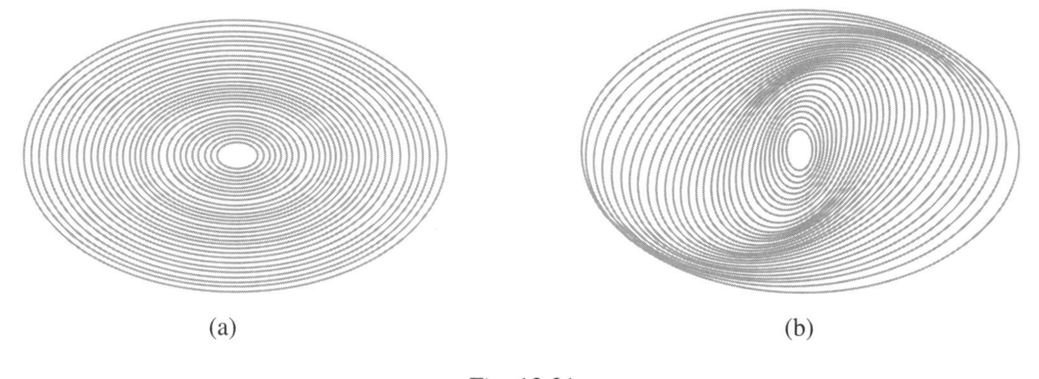

Fig. 12.31

is present throughout a galaxy, and new stars are created within huge hydrogen clouds of great density. Because new stars are brightest, such galaxies feature prominent spiraling threads of bright stars that emerge from the galactic bulge and wind outward.

Figure 12.32 shows an image of galaxy M51 and its spiral arms, highlighting the spirals, and showing that they are equiangular in a general way. Recall that light travels 300,000 km (or 186,000 miles) in one second, and that the light-year is the distance that light travels in one year. The galaxy M51 is an unimaginable 30 million light-years distant and in terms of size, a full 60,000 light-years across. In spite of this, it is close to the Milky Way in galactic terms. This "whirlpool galaxy" is one of the brightest and most beautiful galaxies in the sky.

Astronomers have known for some time that our own Milky Way galaxy is a spiral galaxy. The Milky Way's spiral structure is not easy to determine for two reasons. Not only is the Sun some 26,000 light-years from the galactic center of the Milky Way, but also from a vantage point within a galaxy, any galaxy can only be seen "edge on." The panoramic depiction of the Milky Way in Figure 12.33 shows the flatness of the disk of the Milky Way as seen from within it. A recent state-of-the-art survey deploying radio telescopes has looked for high-density regions of hydrogen in our galaxy. The analysis of shifts in the signals (like the "Doppler effect," in which the pitch of a horn of a car changes as the car's velocity changes relative to the listener) has made it possible to determine the distances of such high-density hydrogen

Fig. 12.32. This image of M51 is a digital combination of a telescope at Kitt Peak National Observatory in Arizona and the Hubble Space Telescope. Image credit: Hubble Heritage Team, NASA.

clouds. These findings have informed us that, like other spiral galaxies, the Milky Way has arms of bright new stars that closely approximate equiangular spirals. The outer arms are not as tightly wound as the inner arms, and the organization of the arms varies from precise and equiangular to imprecise and disorderly. The reason for these irregularities may well be the gravitational pull of other galaxies on ours, or earlier collisions of our galaxy with others. In any case, huge "grand design" equiangular spirals of stars are a common feature of our universe.

Late in the year 2013, the European Space Agency (ESA) launched the space telescope Gaia. It will chart our Milky Way Galaxy and capture it with a three-dimensional map by making accurate measurements of the positions and motions of 1 billion of its approximately 100 billion stars. In Greek mythology, Gaia

Fig. 12.33. Our own Milky Way galaxy rising skyward from the Salar de Atacama salt flat in northern Chile above the lights of a nearby town. Image credit to and copyright held by Alex Tudorica of the University of Bonn. His permission to use this image is gratefully acknowledged.

is the personification of Earth. There's irony here, given the spacecraft's focus on distant reaches and structures of the galaxy. Gaia is in orbit around a gravitationally stable point in space, where the gravity of the Sun and Earth balance each other out, some 1.5 million km beyond Earth as seen from the Sun. Its sunshield blocks heat and light from the Sun, providing the stable environment that its sophisticated instruments need to make their extraordinarily sensitive and precise census. Gaia carries twin telescopes

Fig. 12.34. Copyright/credit: European Space Agency, ESA/Gaia – CC BY-SA 3.0 IGO.

with an image sensitivity of more than 1 billion pixels. They are the highest-resolution telescopes ever sent into space. They will provide accurate positional and radial velocity measurements of the stars and reveal their temperature, luminosity, and physical composition. The three-dimensional map will tell us about the configuration of our region of the Milky Way and provide clarifying information about its spiral structure. Gaia will give insight "in the large" about the formation, evolution, and makeup of our galaxy. Figure 12.34 is an image of the Milky Way's galactic disk produced by Gaia.

The spacecraft is also expected to detect thousands of asteroids and comets in our solar system and a large number of distant "exoplanets" far beyond it. As planets orbit a star, gravitational effects cause the star to wobble in periodic and predictable patterns. By examining the star's movement and breaking down the different components of the wobble, astronomers can identify both the number and the mass of orbiting planets in a given planetary system. Studies have claimed that the Milky Way may be home to over 100 million planets capable of supporting complex life.

12.12 PROBLEMS AND PROJECTS

12A. Polar Equations and Polar Graphs. The first few problems study the conversion of some Cartesian equations to polar equations and the other way. They also engage some polar graphs.

12.1. Express each of the Cartesian equations below as a polar equation. Write the answer in the form $r = f(\theta)$ whenever possible.

i. $2x + 3y = 4$

ii. $x^2 + y^2 = 4y$

iii. $x^2 + y^2 = x(x^2 - 3y^2)$

12.2. Rewrite each of the polar equations as an equation in Cartesian coordinates.

i. $r = 5$

ii. $r = 3\cos\theta$

iii. $\tan\theta = 6$

iv. $r = 2\sin\theta\tan\theta$

12.3. In each case below, sketch the graph of the equation. For a complete understanding of the graph, it may be necessary to convert the equation to rectangular coordinates.

i. $r = -6$

ii. $\theta = -\frac{8\pi}{6}$

iii. $r = 4\sin\theta$

iv. $r(\sin\theta + \cos\theta) = 1$

12.4. Do we arrive at the complete graph of $f(\theta) = \sin\theta$ by considering only θ with $0 \le \theta \le \pi$? Or is there more? Continue graphing the function $f(\theta) = \sin\theta$ by using the information in columns 3 and 4 of Table 12.1. Then use columns 5 and 6, and finally, columns 7 and 8.

12.5. Consider Table 12.1, and produce a similar table for $\cos\theta$. Sketch a graph of $r = \cos\theta$. Confirm that the graph is a circle with center $(\frac{1}{2}, 0)$ (in both polar and Cartesian coordinates) and radius $\frac{1}{2}$.

12.6. Any line in the xy-plane has an equation of the form $ax + by + c = 0$, where a, b, and c are constants. Find a polar function $r = f(\theta)$ that has this line as its graph.

12.7. Sketch the graphs of the equations below. Before doing so, make use of the information in Section 12.2 to determine the Cartesian versions of the equations.

i. $r = \dfrac{4}{1+\cos\theta}$ and $r = \dfrac{8}{1+\cos\theta}$

ii. $r = \dfrac{2}{1+\frac{1}{5}\cos\theta}$ and $r = \dfrac{5}{1+\frac{1}{2}\cos\theta}$

iii. $r = \dfrac{3}{1+3\cos\theta}$ and $r = \dfrac{\frac{1}{2}}{1+5\cos\theta}$

12B. Conic Sections and Polar Coordinates

12.8. In each case, determine the equation $(*)$ (see Section 12.2) with the property that its graph has the required shape.

i. A parabola with distance between focal point and directrix equal to 7.

ii. An ellipse with semimajor axis $a = 6$ and semiminor axis $b = 4$.

iii. A hyperbola with semimajor axis $a = 6$ and semiminor axis $b = 4$.

Now let C be any conic section. A part of C is shown in Figure 12.35. Place a polar axis in such a way that O is at a focal point, the polar axis lies on the focal axis of the conic section, and the polar axis points outward as shown. Problems 12.9 to 12.13 have this figure as their starting point.

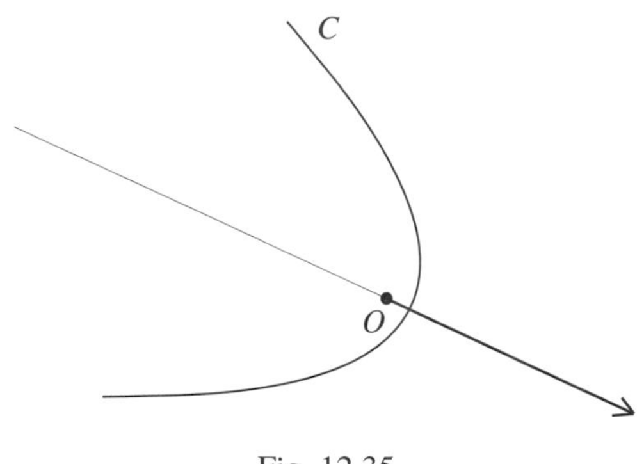

Fig. 12.35

12.9. Let C be a parabola. In equation $(*)$, set $\varepsilon = 1$ and take d to be the distance between the focal point and the directrix of C. Why does the graph of $(*)$ have the same shape as C?

12.10. Let C be a part of an ellipse with semimajor axis 7 and semiminor axis 4. Determine an ε and a d so that the graph of $(*)$ has the same shape as this ellipse.

12.11. Let C be a part of a hyperbola with semimajor axis $a = 5$ and semiminor axis $b = 3$. Determine an ε and a d so that graph of $(*)$ has the same shape as this hyperbola.

12.12. Let a and b be positive constants with $a \geq b$. Suppose that C is a part of an ellipse with semimajor axis a and semiminor axis b. For what d and ε does the graph of $(*)$ have the same shape as the ellipse?

12.13. Let a and b be positive constants. Suppose that C is a part of a hyperbola with semimajor axis a and semiminor axis b. For what d and ε does the graph of $(*)$ have the same shape as C?

12.14. Project: Let $\varepsilon \geq 0$ and $d > 0$ be constants. Modify the study of $r = \frac{d}{1+\varepsilon\cos\theta}$ in Section 12.2 to analyze the polar equation $r = \frac{d}{1+\varepsilon\sin\theta}$ and its graph. [Hint: The only difference turns out to be the orientation of the conic section.]

12C. Derivatives of Polar Functions. The next set of problems considers polar functions and the connection between their derivatives and their graphs. A number of examples of the formula $f'(\theta) = f(\theta)\cdot\tan(\gamma(\theta)-\frac{\pi}{2})$ are provided. Trigonometric identities such as $\cos(\theta-\frac{\pi}{2}) = \sin\theta$ and $\sin(\theta-\frac{\pi}{2}) = -\cos\theta$ and the observation that the tangent of a line at a point is that line are useful.

12.15. Consider the function $f(\theta) = \sin\theta$ and its graph as sketched in Figure 12.4. Study the graph of f over each of the intervals $0 \le \theta \le \frac{\pi}{2}$, $\frac{\pi}{2} \le \theta \le \pi$, $\pi \le \theta \le \frac{3\pi}{2}$, and $\frac{3\pi}{2} \le \theta \le 2\pi$ to confirm that $r = f(\theta)$ respectively, increases, decreases, decreases, and increases over these intervals. Then compare this to the behavior of $f'(\theta) = \cos\theta$ over each of the same intervals.

12.16. Consider Figure 12.10. Suppose that P is a point on the graph such that the graph of $r = f(\theta)$ lies below the circular arc to the left of P. Sketch such a situation, and go through the derivation of the equality $f'(\theta) = f(\theta)\cdot\tan(\gamma-\frac{\pi}{2})$ in this case.

12.17. Consider the function $r = \frac{2}{\sin\theta}$, and show that $y = 2$ is the corresponding Cartesian equation. Show that $\gamma = \theta$. Confirm the equality $f'(\theta) = f(\theta)\cdot\tan(\gamma(\theta)-\frac{\pi}{2})$.

12.18. Consider the function $f(\theta) = \cos\theta$. Its graph is described in Problem 12.5. Make use of a property of isosceles triangles to show that $\gamma+\theta = \frac{\pi}{2}$. Use this fact to confirm the identity $f'(\theta) = f(\theta)\cdot\tan(\gamma-\frac{\pi}{2})$.

12.19. Consider the function $r = f(\theta) = \frac{1}{\sin\theta-\cos\theta}$. Show that the graph is a line, and sketch its graph. Show for any point (r,θ) on the graph that $\gamma(\theta) = \theta-\frac{\pi}{4}$ and hence that $\tan(\gamma(\theta)-\frac{\pi}{2}) = \tan(\theta-\frac{3\pi}{4})$. Use the addition formulas for the sine and cosine (see Problem 1.25) to show that $\tan(\theta-\frac{3\pi}{4}) = -\frac{\sin\theta+\cos\theta}{\cos\theta-\sin\theta}$. Verify the formula $f'(\theta) = f(\theta)\cdot\tan(\gamma-\frac{\pi}{2})$.

12.20. Let a function $r = f(\theta)$ be given. Suppose that the polar coordinate systems of Figures 12.36a and 12.36b both have superimposed xy-coordinate systems. In each of the two cases of the figure, the point $P = (f(\theta),\theta)$ is a random point of the graph. The tangent line to the graph at P is depicted along with the positive angle φ that it makes with the horizontal. The slope of the tangent line is equal to $\tan\varphi$ in either case. The triangle ΔABP is placed so that its base AB is parallel to the pole.

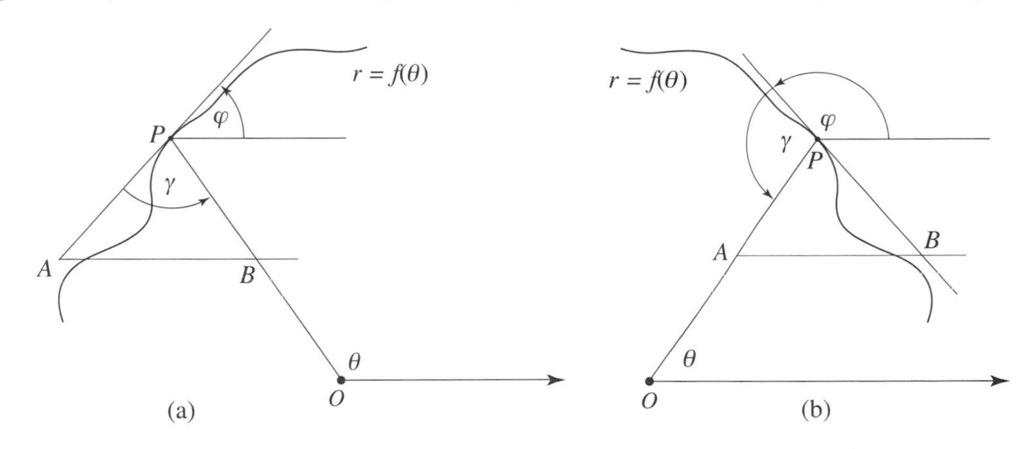

Fig. 12.36

i. Suppose that $r = f(\theta) \neq 0$. Draw in the angle γ and show that $\gamma = \tan^{-1}\frac{f'(\theta)}{f(\theta)} + \frac{\pi}{2}$. Adding up the angles at A, B, and P of the triangle ΔABP tells us that $\varphi + (\pi-\theta) + \gamma = \pi$ in case (a) and $\theta + (\pi-\varphi) + (\pi-\gamma) = \pi$ in case (b). So $\varphi = \theta-\gamma$ in case (a) and $\varphi = \theta-\gamma+\pi$ in case (b). Therefore the slope of the tangent line is

$$\tan\varphi = \tan(\theta-\gamma)$$

in either case.

ii. Recall the equations $x = r\cos\theta$ and $y = r\sin\theta$, and use them to derive the formula

$$\tan\varphi = \frac{dy}{dx} = \frac{\frac{dy}{d\theta}}{\frac{dx}{d\theta}} = \frac{\frac{dr}{d\theta}\cdot\sin\theta + r\cos\theta}{\frac{dr}{d\theta}\cdot\cos\theta - r\sin\theta}.$$

Suppose that $r = f(\theta) = 0$ but $\frac{dr}{d\theta} = f'(\theta) \neq 0$, and conclude that $\tan\varphi = \tan\theta$.

12D. Some Interesting Polar Graphs. We'll start with the graph of the polar function $r = f(\theta) = 1+\cos\theta$. To understand the relationship between r and θ, we begin by regarding $r = 1 + \cos\theta$ as an equation in Cartesian coordinates. Its graph is provided in Figure 12.37. Turning to polar coordinates, we see the

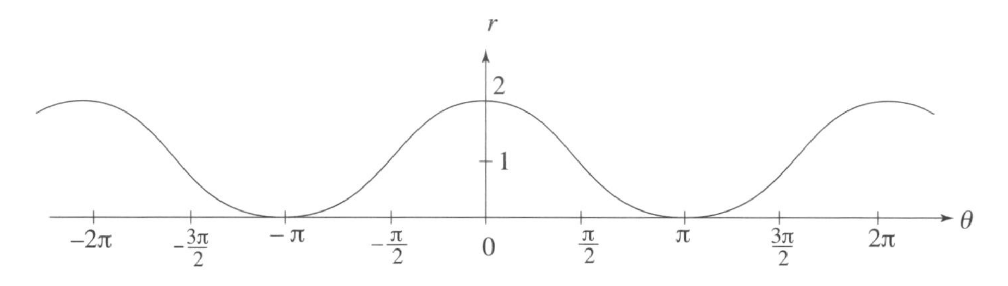

Fig. 12.37

following: as the ray determined by θ rotates from 0 to $\frac{\pi}{2}$, the r coordinate of the point (r, θ) slides from 2 to 1; as the ray rotates from $\frac{\pi}{2}$ to π, the r coordinate slides from 1 to 0; as θ rotates from π to $\frac{3\pi}{2}$, it slides from 0 to 1; and finally, as the ray given by θ rotates from $\frac{3\pi}{2}$ to 2π, r slides from 1 to 2. What happens for negative θ? Converting this information into a polar graph provides Figure 12.38. Since the graph resem-

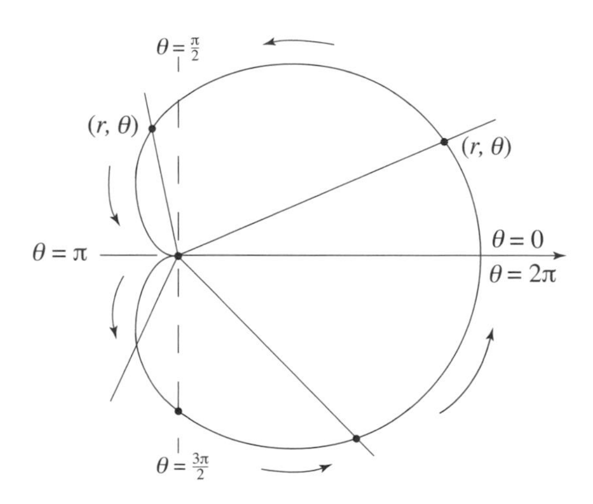

Fig. 12.38

bles a heart, it is called a *cardioid* (*kardia* is Greek for "heart").

12.21. Consider the cardioid $r = f(\theta) = 1 + \cos\theta$ over the interval $0 \leq \theta \leq \pi$. After referring to Problem 12.20, compute (use a calculator) the angle γ and then the slope $\tan(\theta - \gamma)$ of the tangent to the graph for $\theta = 0, \frac{\pi}{6}, \frac{\pi}{4}, \frac{\pi}{3}$, and $\frac{\pi}{2}$. What can you say about the location of the horizontal tangent? What about the slope of the cardioid at the point $(0, \pi)$? [Hint: For $(0, \pi)$, observe that $\lim\limits_{\theta\to\pi} \frac{f'(\theta}{f(\theta)} = \lim\limits_{\theta\to\pi} \frac{-\sin\theta}{1+\cos\theta}\cdot\frac{1-\cos\theta}{1-\cos\theta} = \lim\limits_{\theta\to\pi}\frac{\cos\theta-1}{\sin\theta} = -\infty$ and conclude that $\lim\limits_{\theta\to\pi}\gamma = \lim\limits_{\theta\to\pi}\tan^{-1}\frac{f'(\theta)}{f(\theta)} + \frac{\pi}{2} = 0$. Then analyze what this means geometrically.]

12.22. Sketch the graph of $r = 1 + 2\cos\theta$. (This is an example of the *limaçon of Pascal*. *Limaçon* is French for "snail.")

We'll continue with the polar function $r = \sin 2\theta$. Recall from Figure 12.3 and Table 12.1 that the Cartesian graph of $r = \sin\theta$ has a uniform description over each of the intervals $0 \le \theta \le \frac{\pi}{2}$, $\frac{\pi}{2} \le \theta \le \pi$, and so on as well as the analogous intervals for negative θ. In order to understand the graph of the polar function $r = \sin 2\theta$, we will therefore consider 2θ over the interval $0 \le 2\theta \le \frac{\pi}{2}$, then over $\frac{\pi}{2} \le 2\theta \le \pi$, then over $\pi \le 2\theta \le \frac{3\pi}{2}$, and so on. Hence θ will be taken over the intervals $0 \le \theta \le \frac{\pi}{4}$, then $\frac{\pi}{4} \le \theta \le \frac{\pi}{2}$, then $\frac{\pi}{2} \le \theta \le \pi$, and so on. The Cartesian graph of graph of $r = \sin 2\theta$ is obtained by contracting the

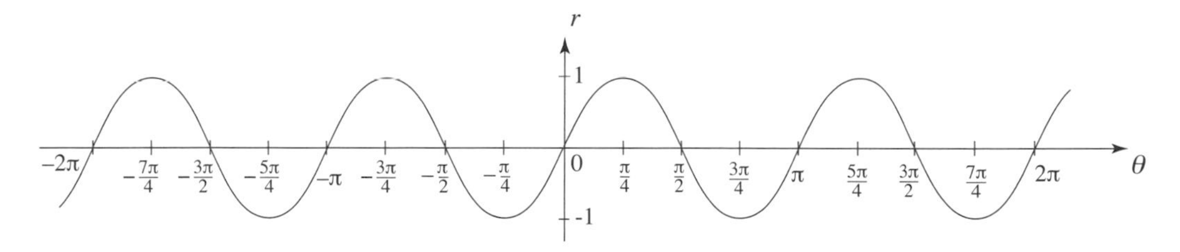

Fig. 12.39

graph of $r = \sin\theta$ in the horizontal direction by a factor of 2 as shown in Figure 12.39. (What is the slope of this graph at $\theta = 0$, $\frac{\pi}{4}$, and $\frac{\pi}{2}$?) The information in Table 12.2 can now be filled in. In turn, with the information in the table it is easy to graph $r = \sin 2\theta$. Notice that $(1, -\frac{\pi}{4})$ does not satisfy $r = \sin 2\theta$.

1	2	3	4	5	6	7	8
$0 \le \theta \le \frac{\pi}{4}$	$\frac{\pi}{4} \le \theta \le \frac{\pi}{2}$	$\frac{\pi}{2} \le \theta \le \frac{3\pi}{4}$	$\frac{3\pi}{4} \le \theta \le \pi$	$\pi \le \theta \le \frac{5\pi}{4}$	$\frac{5\pi}{4} \le \theta \le \frac{3\pi}{2}$	$\frac{3\pi}{2} \le \theta \le \frac{7\pi}{4}$	$\frac{7\pi}{4} \le \theta \le 2\pi$
$0 \le 2\theta \le \frac{\pi}{2}$	$\frac{\pi}{2} \le 2\theta \le \pi$	$\pi \le 2\theta \le \frac{3\pi}{2}$	$\frac{3\pi}{2} \le 2\theta \le 2\pi$	$2\pi \le 2\theta \le \frac{5\pi}{2}$	$\frac{5\pi}{2} \le 2\theta \le 3\pi$	$3\pi \le 2\theta \le \frac{7\pi}{2}$	$\frac{7\pi}{2} \le 2\theta \le 4\pi$
$0 \xrightarrow{\sin 2\theta} 1$	$1 \xrightarrow{\sin 2\theta} 0$	$0 \xrightarrow{\sin 2\theta} -1$	$-1 \xrightarrow{\sin 2\theta} 0$	$0 \xrightarrow{\sin 2\theta} 1$	$1 \xrightarrow{\sin 2\theta} 0$	$0 \xrightarrow{\sin 2\theta} -1$	$-1 \xrightarrow{\sin 2\theta} 0$

Table 12.2

Nonetheless, this point is on the graph of the function because $(-1, \frac{3\pi}{4})$ satisfies $r = \sin 2\theta$.

Going from one interval of the table to the next starting with $0 \le \theta \le \frac{\pi}{4}$, we get the polar graph of $r = \sin 2\theta$ as sketched in Figure 12.40. The graph is called a *four-leaf rose*. The numerical labels on the

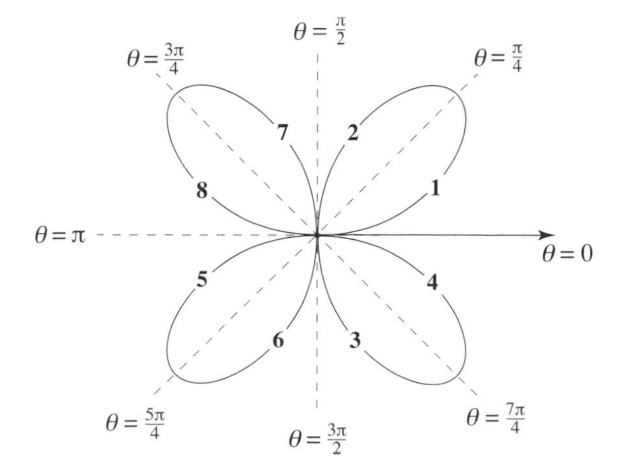

Fig. 12.40

loops correspond to those of the intervals listed in Table 12.2.

12.23. Consider the four-leaf rose $r = f(\theta) = \sin 2\theta$ over the interval $0 \le \theta \le \frac{\pi}{2}$. Review the conclusions of Problem 12.20. Then compute the angle γ and the slope $\tan(\theta - \gamma)$ of the tangent to the graph for $\theta = 0, \frac{\pi}{6}, \frac{\pi}{4}, \frac{\pi}{3}$, and $\frac{\pi}{2}$. What can you say about the location of the vertical and horizontal tangents?

12.24. Use the double-angle formula for the sine (see Problem 1.26i) to show that the Cartesian version of the equation $r = \sin 2\theta$ is $2xy = \pm(x^2 + y^2)^{\frac{3}{2}}$.

12.25. Sketch the graph of each of the equations below.

i. $r = 2\cos 2\theta$ (a *four-leaf rose*.)

ii. $r = -4\sin 3\theta$ (a *three-leaf rose*.)

iii. $r^2 = 9\sin 2\theta$ (a *lemniscate of Bernoulli*. The word *lemniscate* comes from the Greek *lemniscos*, meaning "ribbon.")

12.26. Use the formulas of Problems 1.25 or 1.26 to express each of the polar equations of the previous problem in Cartesian coordinates.

12E. Polar Spirals, Integrals, Lengths, and Areas. Consider a function $r = f(\theta)$ that is increasing and continuous. The fact that r increases as θ increases tells us that r increases as the ray determined by θ rotates counterclockwise. So the graph of $r = f(\theta)$ is a spiral that opens in a counterclockwise way. (The spiral may "wobble" in the sense that the increase in r may vary from smaller to larger, back to smaller, and so on.)

12.27. Consider the function $r = f(\theta) = 3\theta$. Its graph is an *Archimedean spiral*. Sketch it from $\theta = 0$ to $\theta = 2\pi$. Study the rate of expansion of this spiral by analyzing the equality $\frac{f'(\theta)}{f(\theta)} = \tan(\gamma(\theta) - \frac{\pi}{2})$. What can you say about γ for small positive θ? For large positive θ?

12.28. Consider the equiangular spiral $f(\theta) = 2e^{\frac{\theta}{\sqrt{3}}}$ over the interval $[0, 2\pi]$. What is the constant angle γ for this spiral? Determine the length of the spiral and the area that it encloses.

12.29. Use the fact that the Cartesian equation of the polar equation $r = \frac{1}{\sin\theta}$ is the line $y = 1$ to compute the integral $\displaystyle\int_{\frac{\pi}{4}}^{\frac{3\pi}{4}} \frac{1}{\sin^2\theta}\, d\theta$.

12.30. Show that the graph of the polar function $r = f(\theta) = \frac{4}{\cos\theta}$ is the line $x = 4$. Use this fact to evaluate the integral $\displaystyle\int_0^{\frac{\pi}{3}} \frac{8}{\cos^2\theta}\, d\theta$.

12.31. The graph of the function $r = f(\theta) = \frac{3}{\sin\theta + 2\cos\theta}$, $0 \le \theta < \pi$, lies on a line. Determine its Cartesian equation. Use the result to evaluate the integrals $\displaystyle\int_0^{\frac{\pi}{2}} \tfrac{1}{2} f(\theta)^2\, d\theta$ and $\displaystyle\int_0^{\frac{\pi}{2}} \sqrt{f(\theta)^2 + f'(\theta)^2}\, d\theta$.

12.32. Use definite integrals to determine the length of the graph of $r = \sin\theta$ from $\theta = \frac{\pi}{4}$ to $\theta = \frac{3\pi}{4}$, then from $\theta = 0$ to $\theta = \pi$, and finally from $\theta = 0$ to $\theta = 2\pi$. Explain your answers by referring to the graph of $r = \sin\theta$ in Figure 12.4.

12.33. Use the graph of the polar function $r = f(\theta) = \sin\theta$, $0 \le \theta \le \pi$, of Figure 12.4 to evaluate the integral $\displaystyle\int_0^{\pi} \tfrac{1}{2}\sin^2\theta\, d\theta$. Then evaluate the integral again, this time directly, by using the formula $\sin^2\theta = \frac{1-\cos 2\theta}{2}$ from Problem 1.26ii.

12.34. Study the graph of the function $r = f(\theta) = \frac{4}{1+\frac{2}{3}\cos\theta}$ by referring to Section 12.2. Then use what you know about ellipses to evaluate the integral $\int_0^{\pi} \frac{8}{(1+\frac{2}{3}\cos\theta)^2}\, d\theta$.

12.35. Let $d > 0$ and $\varepsilon \geq 0$. Show that $d \int_a^b \frac{\sqrt{1+\varepsilon^2+2\varepsilon\cos\theta}}{(1+\varepsilon\cos\theta)^2}\, d\theta$ is an expression for the length of the conic section given by $f(\theta) = \frac{d}{1+\varepsilon\cos\theta}$ between the rays $\theta = a$ to $\theta = b$. Show that in the parabolic case $\varepsilon = 1$, this reduces to $\sqrt{2}d \int_a^b \frac{1}{(1+\cos\theta)^{\frac{3}{2}}}\, d\theta$.

12.36. Study the solutions of Problems 12.30, 12.33, and 12.34, and then evaluate $\int_0^{\frac{\pi}{4}} \frac{3}{\cos^2 x}\, dx$, $\int_0^{\frac{\pi}{2}} \sin^2 x\, dx$, and $\int_0^{2\pi} \frac{2}{(1+\frac{2}{3}\cos x)^2}\, dx$.

12.37. Consider the circle $(x-1)^2+(y-1)^2 = 2$, and sketch its graph. Show that $r = f(\theta) = 2(\sin\theta+\cos\theta)$ is a polar function with graph this circle. Use the graph of the circle to evaluate the integrals $\int_0^{\frac{\pi}{2}} \frac{1}{2} f(\theta)^2\, d\theta$ and $\int_0^{\frac{\pi}{2}} \sqrt{f(\theta)^2 + f'(\theta)^2}\, d\theta$.

12.38. Consider the cardioid $r = 1 + \cos\theta$ of Figure 12.38.

i. Compute its area by making use of the formula $\cos^2\theta = \frac{1+\cos 2\theta}{2}$ from Problem 1.26ii.

ii. What is the connection between the integral $\sqrt{2}\int_0^{\pi} \sqrt{1+\cos\theta}\, d\theta$ and the length of the cardioid?

iii. Use the substitution $u = 1 + \cos\theta$ and a basic trig formula to show that the integral of (ii) is equal to $\sqrt{2}\int_0^2 \frac{1}{\sqrt{2-u}}\, du$.

iv. Use another substitution to show that the integral in (iii) is equal to 4.

v. What is the length of the cardioid?

12.39. Figure 12.41 shows the graph of a function $r = f(\theta)$ over the interval $a \leq \theta \leq b$ along with a partition $a = \theta_0 < \theta_1 < \cdots < \theta_i < \theta_{i+1} < \cdots < \theta_n = b$ of the interval. Let $d\theta = \theta_{i+1} - \theta_i$. The

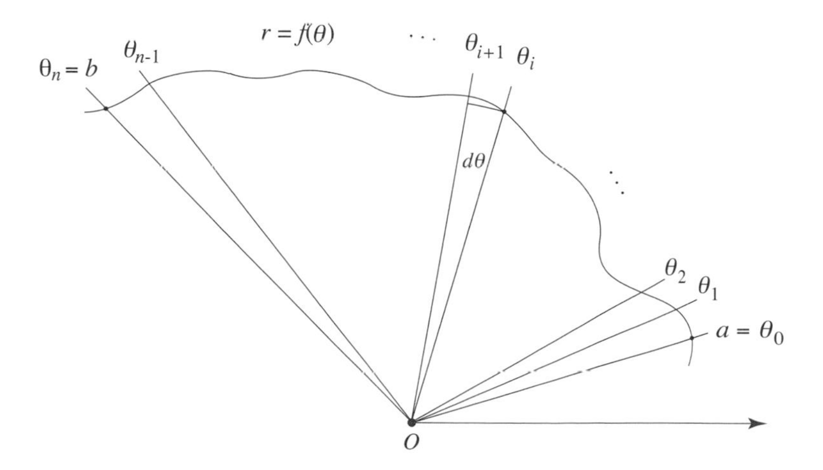

Fig. 12.41

circular arc between θ_i and θ_{i+1} shown in the figure lies on a circle of radius $f(\theta_i)$. Why does this arc have length $f(\theta_i)d\theta$? Since this approximates the length of the graph between the two rays, we get by the strategy of integral calculus that the length L of the graph of $r = f(\theta)$ between $\theta = a$ and $\theta = b$ is equal to

$$L = \int_a^b f(\theta)\, d\theta\,.$$

Why do we know that this formula is wrong? Give an example of a polar function $r = f(\theta)$ that verifies that this formula is wrong. Why does the derivation of this length formula fail.

12F. Centripetal Force, Inverse Cubes, and Cotes's Spirals. We'll begin with an example of an orbit of a point-mass P propelled by a centripetal force that satisfies the criteria of Section 12.7, but is unrelated to gravitational attraction. Let an xy-coordinate system be given and consider the ellipse $\frac{x^2}{a^2} + \frac{y^2}{b^2} = 1$, where a and b are the semimajor and semiminor axes (not necessarily respectively). Let $t \geq 0$ denote elapsed time, and let the location of P be given by the equations

$$x(t) = a\cos t \;\text{ and }\; y(t) = b\sin t.$$

Notice that P starts at the point $(a, 0)$ and moves counterclockwise around the ellipse with increasing t. Check that $x''(t) = -a\cos t$ and $y''(t) = -b\sin t$. Let m be the mass of P and let F_x and F_y be the forces on P in the x- and y-directions. Since force = mass × acceleration, their magnitudes are given by

$$F_x(t) = mx''(t) = -m(a\cos t) \;\text{ and }\; F_y(t) = my''(t) = -m(b\sin t).$$

The magnitude of the resultant of these forces is $F(t) = \sqrt{F_x(t)^2 + F_y(t)^2} = m\sqrt{a^2\cos^2 t + b^2\sin^2 t}$. Fig-

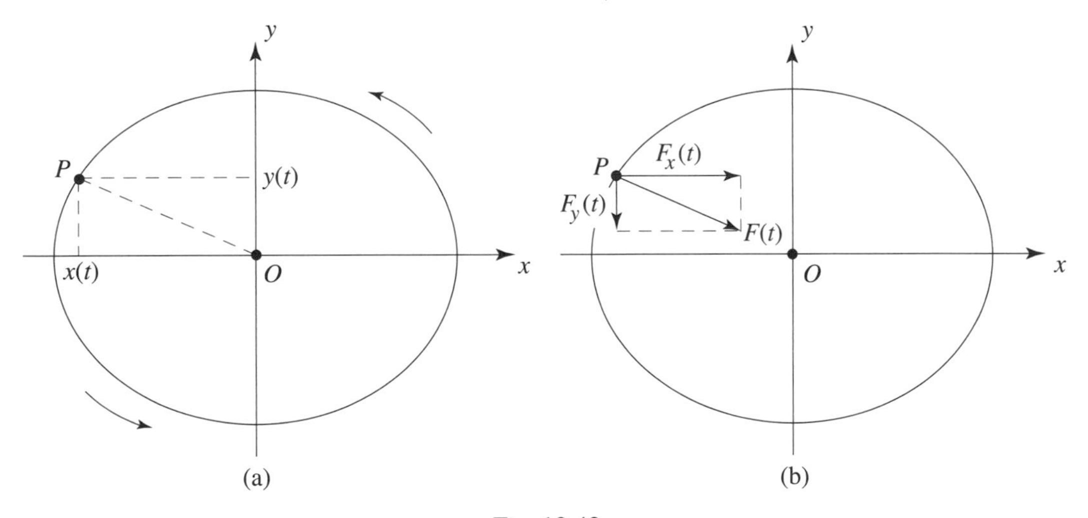

Fig. 12.42

ure 12.42 illustrates what has been described.

12.40. **i.** Check that when $x(t) > 0$, the force F_x acts to the left, and when $x(t) < 0$, the force F_x acts to the right. What similar relationship holds between $y(t)$ and F_y?

ii. Check that the slope of the slanting segment in Figure 12.42a that connects P and O is the same as the slope of the segment in Figure 12.42b that determines the direction of the resultant of the forces F_x and F_y.

iii. Conclude that the resultant of the two forces F_x and F_y is a centripetal force on P acting in the direction of the origin O.

iv. Show that $F(t) = mr(t)$, where $r(t)$ is the distance from P to O.

Since P completes its first revolution at time $t = 2\pi$, it follows that the Kepler constant of its orbit is $\kappa = \frac{ab\pi}{2\pi} = \frac{ab}{2}$. It is a consequence of Section 12.8 that $F(t)$ and $r(t)$ satisfy the centripetal force equation $F(t) = m\big[\frac{4\kappa^2}{r(t)^3} - \frac{d^2r}{dt^2}\big]$. This can be verified directly with a labor-intensive calculus exercise.

With regard to the relationship between $F(t)$ and $r(t)$, compare the equality $F(t) = mr(t)$ of the current situation with the inverse square law $F(t) = \frac{8m\kappa^2}{L}\frac{1}{r(t)^2}$ of Section 12.9 (in the elliptical case). In both situations, the point-mass P is propelled by a centripetal force and the orbit is elliptical. However, the different locations of the center of force (the center of the ellipse in one case, and a focal point in the other) results in very different force dynamics (except in the special case of a circle).

The context for the next several problems is given by Section 12.8. We are assuming that a point-mass P of mass m is moving smoothly in the polar plane driven by a single centripetal force directed to the origin O. The graph of the function $r = f(\theta)$ describes its orbit. The angle $\theta = \theta(t)$ is a function of time t, as are $r(t) = f(\theta(t))$, and the magnitude $F(t)$ of the force. The function $g(\theta)$ is defined by $g(\theta) = \frac{1}{f(\theta)} = f(\theta)^{-1}$. So $g(\theta(t)) = r(t)^{-1}$. It was shown—without any additional assumptions about the force or the geometry of the orbit—that $F(t)$ satisfies the centripetal force equation

$$F(t) = 4m\kappa^2 g(\theta(t))^2[g(\theta(t)) + g''(\theta(t))]$$

and that $r(t)^2\theta'(t) = 2\kappa$ or $\frac{d\theta}{dt} = 2\kappa g(\theta(t)^2$, where κ is Kepler's constant of the orbit of P.

12.41. Suppose that the angle $\theta(t)$ of Figure 12.23 increases at a constant rate. Show that $r(t)$ as well as $F(t)$ are constant and that the orbit is a circle.

12.42. Suppose that $r = f(\theta) = \frac{c}{a\sin\theta + b\cos\theta}$, where a, b, and c are constants with $c \neq 0$. Show that the graph of $r = f(\theta)$ is a line that does not go through the origin. Show that $g''(\theta) = -g(\theta)$, and conclude that $F(t) = 0$. Put another way, if a centripetal force acting on a point-mass P produces an orbit that is a straight line that does not go through the center of force, then its magnitude must be zero.

In his studies in the *Principia* of the orbital behavior of a point-mass P of mass m under the action of a centripetal force, Newton considered not only the situation where the magnitude $F(t)$ of the force satisfies an inverse square law, but also the case where it satisfies an inverse cube law of the form

$$F(t) = K\frac{m}{r(t)^3}$$

with K a constant. The centripetal force equation is key. Combining $F(t) = 4m\kappa^2 g(\theta(t))^2[g(\theta(t)) + g''(\theta(t))]$ with $F(t) = Kmg(\theta(t))^3$, we get $4\kappa^2[g(\theta(t)) + g''(\theta(t))] = Kg(\theta(t))$ and hence $g''(\theta(t)) + [1 - \frac{K}{4\kappa^2}]g(\theta(t)) = 0$. Ignoring the fact that θ is a function of t, we see that the function $g(\theta)$ satisfies

$$g''(\theta) + [1 - \tfrac{K}{4\kappa^2}]g(\theta) = 0.$$

So $y = g(\theta)$ is a solution of a second-order differential equation of the form studied in Section 11.6. With respect to the constants A, B, and C introduced there, observe that $A = 1, B = 0$, and $C = 1 - \frac{K}{4\kappa^2}$. Therefore $B^2 - 4AC = -4(1 - \frac{K}{4\kappa^2}) = \frac{K}{\kappa^2} - 4 = \frac{K - 4\kappa^2}{\kappa^2}$.

The solutions of the next two problem are consequences of the analysis of Section 11.6 in the first instance and the centripetal force equation in the second. For the initial problem express your solutions in terms of $g(\theta) = \frac{1}{f(\theta)}$ first. For the second problem, frame the assumption about $f(\theta)$ in terms of $g(\theta)$.

12.43. Let P be a point-mass propelled in a smooth orbit by a centripetal force. Let m be the mass of P and let κ be the Kepler constant of the orbit of P. The graph of the polar function $r = f(\theta)$ provides the orbit of P. Assume that the magnitude $F(t)$ of the centripetal force satisfies the inverse cube law $F(t) = K\frac{m}{r(t)^2}$ for some constant K. Then

Case 1. If $K > 4\kappa^2$, then $r = f(\theta) = \frac{1}{D_1 e^{\frac{\sqrt{K-4\kappa^2}}{2\kappa}\theta} + D_2 e^{-\frac{\sqrt{K-4\kappa^2}}{2\kappa}\theta}}$, for some real constants D_1 and D_2.

Case 2. If $K = 4\kappa^2$, then $r = f(\theta) = \frac{1}{D_1 + D_2\theta}$, for some real constants D_1 and D_2.

Case 3. If $K < 4\kappa^2$, then $r = f(\theta) = \frac{1}{D_1 \sin\left(\frac{\sqrt{4\kappa^2-K}}{2\kappa}\theta\right) + D_2 \cos\left(\frac{\sqrt{4\kappa^2-K}}{2\kappa}\theta\right)}$, where D_1 and D_2 are real constants.

12.44. Let P be a point-mass propelled in a smooth orbit by a centripetal force and assume that the function $r = f(\theta)$ describing the orbit has one of the following forms

i. $r = f(\theta) = \frac{1}{a\theta+c}$, where a and c are constants,

ii. $r = f(\theta) = \frac{1}{a\cos(b\theta+c)}$, where a, b, and c are constants, or

iii. $r = f(\theta) = \frac{1}{a\cosh(b\theta+c)}$, where a, b, and c are constants.

Then, with κ the Kepler constant of the orbit,

$$F(t) = K\frac{m}{r(t)^3},$$

where $K = 4\kappa^2$ in case (i), $K = 4\kappa^2(1 - b^2)$ in case (ii), and $K = 4\kappa^2(1 + b^2)$ in case (iii).

The three types of functions in the conclusion of Problem 12.43 and in the assumption of Problem 12.44 match up, but not as the numbering suggests. To see this, use the addition formulas for the cosine (Problem 1.25) and the hyperbolic cosine (Example 7.47). The graphs of these functions are spirals. They were investigated by Newton's contemporary Roger Cotes in his *Harmonia Mensurarum* of 1722. Today they are known as *Cotes's spirals*.

12G. Gravity and Geometry. The next two problems rely on the discussion of Section 12.10.

12.45. Show that the mass of a sphere of radius R satisfying the onion property with density function $\rho(x)$ is equal to $M = \int_0^R 4\pi x^2 \rho(x)\,dx$.

12.46. Consider a point-mass of mass m and a thin homogeneous circular disc D of radius R and mass M. Figure 12.43 depicts D along with an x-axis that lies in the plane of the disc. The origin O is the

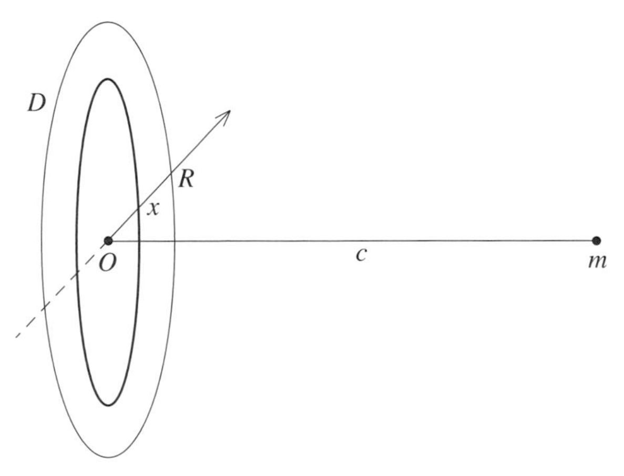

Fig. 12.43

center of D. The point-mass lies on the axis perpendicular to D through its center O at a distance c from O. The density of D is equal to $\rho = \frac{M}{\pi R^2}$. A typical circular ring of D is shown. It has radius

x and thickness dx. The circumference of the ring is $2\pi x$, so that it has mass $2\pi\rho x\, dx$. Show that the gravitational force of the disc on the point-mass is directed to the center of the disc and has magnitude $F = G\,\frac{2mM}{R^2}[1 - \frac{c}{\sqrt{R^2+c^2}}]$.

12.47. Consider a point-mass of mass m and a homogeneous cylinder C (include the interior of the cylinder as well as its surface) of radius R, height h, and mass M. Figure 12.44 depicts C along with its central axis. The cylinder extends from its circular base at $x = 0$ to its other circular boundary at $x = h$. The point-mass lies on the central axis at a distance $c \geq h$ from the base of the cylinder. The density of C is $\rho = \frac{M}{\pi R^2 h}$. A typical circular disc of C is shown in the figure. It is parallel to the base, intersects the axis at x, and has a thickness of dx. The disc has a mass of $\pi R^2 dx \cdot \rho$. Use the result of Problem 12.46 to determine the gravitational force of the disc on m. Set up an integral and then

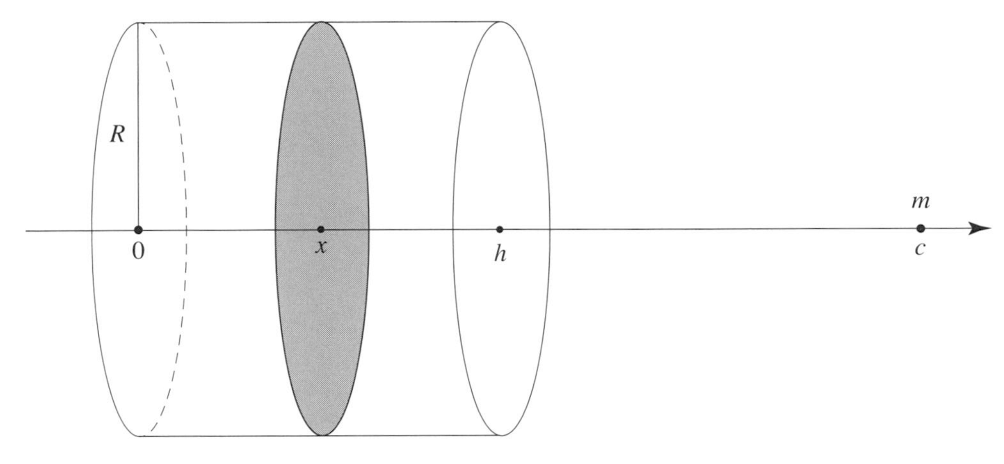

Fig. 12.44

solve it to show that the force of the cylinder on the point-mass is directed to the center of the cylinder and that its magnitude is $G\frac{2mM}{R^2h}[h - (R^2 + c^2)^{\frac{1}{2}} + (R^2 + (c - h)^2)^{\frac{1}{2}}]$. Observe therefore that Newton's inverse square law does not hold for a cylinder and a point-mass, even when the point-mass is symmetrically positioned on the cylinder's axis.

12H. Newton's Revolving Orbits. With regard to a precise description of the motion of the bodies of the solar system, in particular those of the planets, it is not enough to say that they move along elliptical paths. It has been known for some time that as each planet traces out its elliptical orbit, the ellipse itself moves. Its focal axis—fixed at the Sun—rotates like the hand of a clock. Unlike the hand of a clock, the focal axis rotates extremely slowly and carries its elliptical dial with it. Figure 12.45 illustrates this dual aspect of the

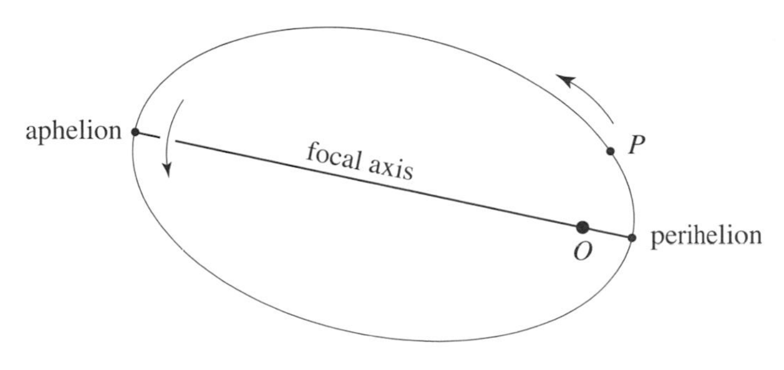

Fig. 12.45

motion of a planet P with the Sun at the point O. (The figure also exaggerates the flatness of the ellipse.) The two points of intersection of the ellipse and its focal axis—the perihelion and aphelion positions—are highlighted in the figure. Since the position of perihelion changes from year to year, this rotation of a planet's ellipse is known as its *precession of perihelion*. Careful observations have shown how many seconds of angle per year the focal axes of the planets rotate. Table 12.3 provides this information by

Mercury	Venus	Earth	Mars	Jupiter	Saturn	Uranus	Neptune
625	1756	314	222	550	183	1078	10,000

Table 12.3. The number of years it takes for a 1° precession of a planet's perihelion.

telling us how many years it takes for the focal axes of the planets to rotate through an angle of just one degree. Newton, thinking that such a precession influenced the orbit of the Moon around the Earth, attempted to understand this phenomenon mathematically. The *revolving orbit theorem* was his response.

Let's start Newton's analysis by considering a point-mass P in orbit in a polar plane with polar origin O. We'll label the polar angle by θ_1 and the polar function that describes the orbit by $r_1 = f_1(\theta_1)$. Figure 12.46a depicts a part of the orbit of P. No assumptions whatever are made about the force that drives P (even though

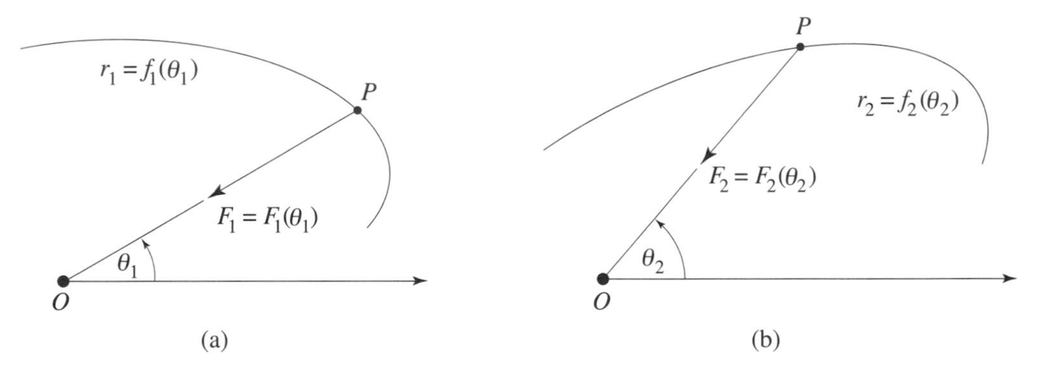

Fig. 12.46

it is depicted as a centripetal force in the figure). The focus for now is entirely on the orbit and not on the force.

Newton transforms the orbit of the point-mass P into a related orbit as follows. He introduces the polar angle θ_2 and polar function $r_2 = f_2(\theta_2)$ and considers this second orbit and motion of P to be connected to the first by the equalities

(1) $\theta_2 = s\theta_1$ for a constant s and all θ_1

(2) $f_2(\theta_2) = f_1(\theta_1)$ for all θ_2.

This second orbit is depicted in a generic way in Figure 12.46b.

We turn to an example that illustrates Newton's construction of the transformed orbit. Suppose that the original orbit of Figure 12.46a, in other words the graph of $r_1 = f_1(\theta_1)$, is an ellipse with O at a focus. It is shown in Figure 12.47a along with the progression of angles θ_1 in intervals of $\frac{\pi}{12}$ (or 15° degrees) from $\theta_1 = 0$ to π to 2π and beyond. The points on the orbit that correspond to these angles are labeled $0, 1, 2, 3, 4, \ldots, 24$, and 25 and 36 for the angles $2\pi + \frac{\pi}{12}$ and 3π. The points and their distances from O are determined by these angles and the function $r = f_1(\theta_1)$. For example, the point 11 corresponds to the angle $\theta_1 = \frac{11\pi}{12}$, and $f_1(\frac{11\pi}{12})$ is its distance from O. The point labeled both 0 and 24 corresponds to $\theta_1 = 0$ and $\theta_1 = 2\pi$, the point labeled 1 and 25 corresponds to $\theta_1 = \frac{\pi}{12}$ and $\theta_1 = 2\pi + \frac{\pi}{12}$, and similarly for the point labeled 12 and 36.

Take the constant s in Newton's equality (1) to be $s = 1.1$. So $\theta_2 = (1.1)\theta_1 = \theta_1 + (0.1)\theta_1$. Therefore θ_2 is obtained from θ_1 by adding 10% of θ_1. Now go to the Newton's equality (2). For each of the angles θ_1

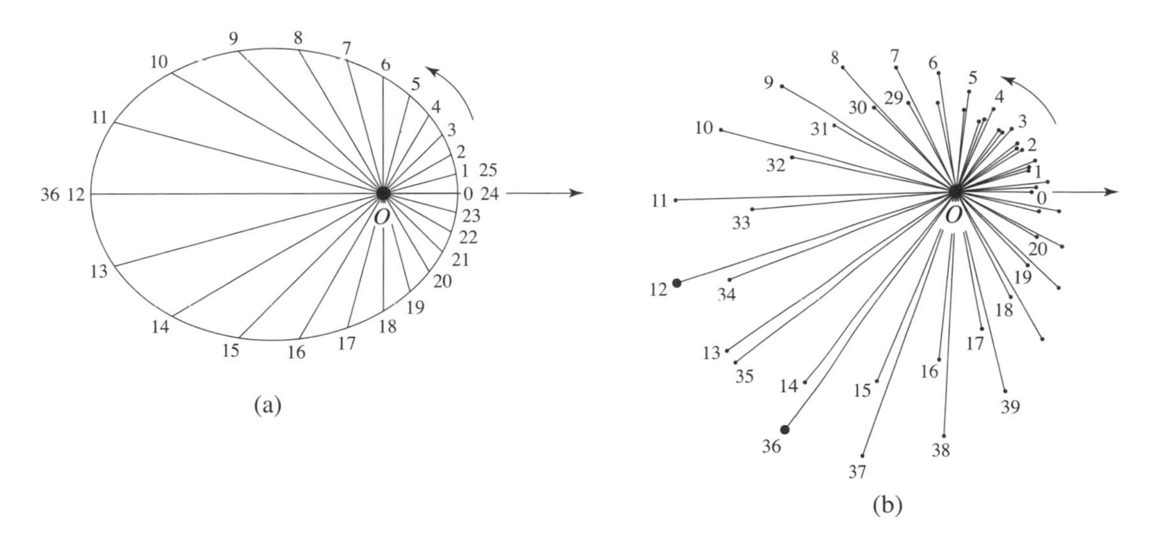

Fig. 12.47

singled out above, take the corresponding θ_2, and consider the point $(f_2(\theta_2), \theta_2)) = (f_1(\theta_1), \theta_2))$ on the graph of $r_2 = f_2(\theta_2)$ of Figure 12.46b. Figure 12.47b carries out the specifics. The point $(f_1(0), 0) = (f_2(0), 0)$ labeled 0 is the same in Figures 12.47a and 12.47b. The point labeled 1 is $(f_1(\frac{\pi}{12}), \frac{\pi}{12})$ in Figure 12.47a and $(f_1(\frac{\pi}{12}), \frac{\pi}{12} + \frac{\pi}{120})$ in Figure 12.47b. The point labeled 2 is $(f_1(\frac{\pi}{6}), \frac{\pi}{6})$ in Figure 12.47a and $(f_1(\frac{\pi}{6}), \frac{\pi}{6} + \frac{\pi}{60})$ in Figure 12.47b, and so on. In Figure 12.47a, aphelion occurs for the first time at $(f_1(\pi), \pi))$, the point 12 in the figure. The corresponding point 12 in Figure 12.47b is $(f_1(\pi), \pi + \frac{\pi}{10})$. It follows that the focal axis of the ellipse rotated 18° during the passage of the point-mass P through the first half of its orbit. After the point-mass has completed another orbit to the next aphelion, the focal axis has rotated another 36°. Compare the location of point 36 in Figure 12.47a with that of point 36 in Figure 12.47b.

Figure 12.48 has "connected the dots" that Figure 12.47b and the unfolding pattern provide into the continuous orbit that the graph of $r_2 = f_2(\theta_2)$ determines. This figure illustrates how Newton's construction

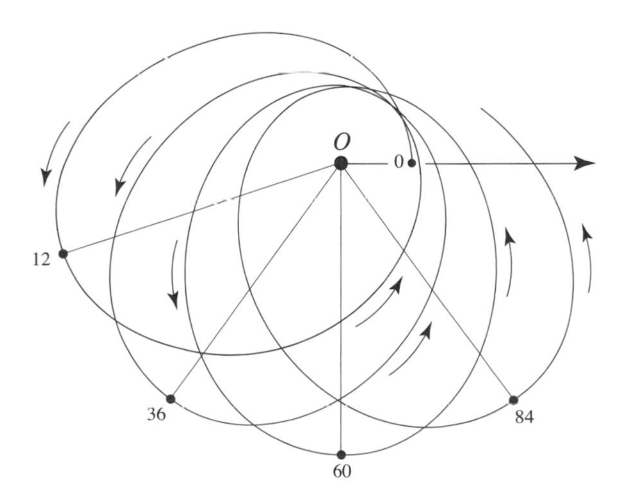

Fig. 12.48

is related to the phenomenon of the precession of the orbits of the planets. His procedure of transforming the orbit $r = f_1(\theta_1)$ into the orbit $r = f_2(\theta_2)$ supplies the original elliptical orbit with the kind of ongoing rotation that the orbits of the planets exhibit.

Not surprisingly, the question "what force on P gives rise to such a revolving orbit?" was of interest to Newton (in the context of his investigations of the orbit of the Moon). He asked, in particular, what the magnitude of such a force would be *if it were assumed to be a centripetal force*. He knew that such a force could not satisfy an inverse square law, because under this assumption the orbit would be a fixed ellipse. But what could be said?

We'll now assume that the forces driving the point-mass P both in its original orbit of Figure 12.46a and the transformed orbit of Figure 12.46b are centripetal in the direction of the respective polar origins O. We'll let m be the mass of P and study the two orbits simultaneously starting at some time t_0. The angle $\theta_1 = \theta_1(t)$ is now a function of time $t \geq t_0$. So $\theta_2(t) = s\theta_1(t)$ as well as $r_1(t) = f_1(\theta_1(t))$ and $r_2(t) = f_2(\theta_2(t)) = f_1(\theta_1(t)) = r_1(t)$ are also functions of t. The magnitudes $F_1(t)$ and $F_2(t)$ of the two centripetal forces are also functions of time. The question Newton considers is this. How does the magnitude of the centripetal force $F_2(t)$ that determines the motion of the point-mass illustrated in Figures 12.46b and 12.48 depend on the magnitude $F_1(t)$ that drives the original motion of the point-mass illustrated in Figure 12.46a? Following what was done in Section 12.8, we'll define the functions g_1 and g_2 by $g_1(\theta_1) = \frac{1}{f_1(\theta_1)} = f_1(\theta_1)^{-1}$ and $g_2(\theta_2) = \frac{1}{f_2(\theta_2)} = f_2(\theta_2)^{-1}$, respectively. In terms of g_1 and g_2, Newton's equality (2) becomes $g_2(\theta_2) = g_1(\theta_1)$.

Turn to the formulas derived in Section 12.8. With κ_1 and κ_2 the respective Kepler constants of the two orbits, we know that $\theta_1'(t) = 2\kappa_1 g_1(\theta_1(t))^2$ and $\theta_2'(t) = 2\kappa_2 g_2(\theta_2(t))^2$. By Newton's equality (1),

$$2\kappa_2 g_2(\theta_2(t))^2 = \theta_2'(t) = s\theta_1'(t) = 2s\kappa_1 g_1(\theta_1(t))^2 = 2s\kappa_1 g_2(\theta_2(t))^2,$$

so that $\kappa_2 = s\kappa_1$. Therefore s is the ratio $\frac{\kappa_2}{\kappa_1}$ of the two Kepler constants. Differentiating $g_2(\theta_2(t)) = g_1(\theta_1(t))$ using the chain rule, we get

$$g_2'(\theta_2(t)) \cdot \theta_2'(t) = g_1'(\theta_1(t)) \cdot \theta_1'(t)$$

and therefore that $g_2'(\theta_2(t)) = \frac{1}{s} g_1'(\theta_1(t))$. Doing this once more tells us that $g_2''(\theta_2(t)) = \frac{1}{s^2} g_1''(\theta_1(t))$. Check the algebra that confirms that

$$(3) \qquad g_2(\theta_2(t)) + g_2''(\theta_2(t)) = \tfrac{1}{s^2}\big(g_1(\theta_1(t)) + g_1''(\theta_1(t))\big) + \big(1 - \tfrac{1}{s^2}\big) g_1(\theta_1(t)).$$

The centripetal force equation for $F_1(t)$ tells us that

$$\tfrac{1}{s^2}\big(g_1(\theta_1(t)) + g_1''(\theta_1(t))\big) = \frac{F_1(t)}{4s^2 m\kappa_1^2 g_1(\theta_1(t))^2}.$$

Inserting it and the centripetal force equation for $F_2(t)$ into equation (3), we get

$$\frac{F_2(t)}{4m\kappa_2^2 g_2(\theta_2(t))^2} = \frac{F_1(t)}{4s^2 m\kappa_1^2 g_1(\theta_1(t))^2} + \big(1 - \tfrac{1}{s^2}\big) g_1(\theta_1(t)).$$

Multiplying through by $4m\kappa_2^2 g_2(\theta_2(t))^2$, using equalities derived above, and canceling and simplifying, provides the connection

$$F_2(t) = F_1(t) + 4m(\kappa_2^2 - \kappa_1^2)\frac{1}{r_1(t)^3}$$

between the magnitudes $F_1(t)$ and $F_2(t)$. This connection is Newton's revolving orbit theorem.

Newton has established that the magnitude $F_2(t)$ of the force that drives the point-mass P in the transformed orbit is obtained from the magnitude $F_1(t)$ of the force that drives it in its original orbit, by adding an inverse cube term. The most interesting aspect of this result was illustrated earlier. It is the assertion that a fixed elliptical orbit with Kepler constant κ_1 generated by a centripetal force satisfying an inverse square law can be transformed into a rotating elliptical orbit with prescribed Kepler constant

κ_2 by increasing or decreasing the magnitude of the centripetal force by such an inverse cube term. The revolving orbit theorem is also of historical interest. With the statement

> [the inverse-square law] is proved with great exactness from the fact that the aphelia are at rest. For the slightest departure from the ratio of the square would necessarily result in a noticeable motion of the apsides [in other words, in a precession] in a single revolution and an immense such motion in many revolutions

in the third book of the *Principia*, Newton relies on it to provide evidence for his central conclusion that gravitational force satisfies the inverse square law.

It is remarkable that the precession of the planets' perihelia appears to have been first observed (for Earth's orbit) by Islamic astronomers already in the 11th century. But among the astronomers of Western Europe there did not seem to be a general awareness of this phenomenon even as late as the 16th and 17th centuries. Newton's statement in the *Principia* tells us that he did not subscribe to this slight rotation of the focal axes of the planets, at least at the time he published his great opus. (Astronomers were well aware of the precession of the equinoxes, as this had already been documented by the ancient astronomers. But this is a consequence of the change in the tilt of Earth's north-south axis and not the precession of its perihelion.) After that time, the precession of a planet's perihelion became an observed fact and it was recognized that its cause is the collective gravitational pull of the other planets. (The pull of the massive Jupiter and Saturn is especially relevant). In the 19th century, a persistent discrepancy was discovered between the amount of angular precession predicted by Newtonian planetary theory and the amount that was actually observed. The difference—most pronounced in the case of Mercury, where it amounts to about 7.5%—was explained in the early part of the 20th century by Einstein's theory of general relativity. We have already seen the critical role played by the term $\frac{1}{r(t)^2}$ in Newton's inverse square formula for gravitational force. The relativistic correction of this result in the current context turns out to involve the term $\frac{1}{r(t)^4}$ (rather than the $\frac{1}{r(t)^3}$ of Newton's revolving orbit theorem). Einstein's definitive explanation of the precession of a planet's perihelion provided an early confirmation of his remarkable new point of view that gravity is the consequence of the curvature of space.

References

The principal references for this text, including the most relevant, are listed below.

The approach of *Calculus in Context*—to develop the essentials of differential and integral calculus and prerequisite subjects and to surround these with rich layers of historical background and relevant applications—follows the spirit of

1. Hahn, Alexander J., *Basic Calculus: From Archimedes to Newton to its Role in Science,* Springer-Verlag, New York, 1998.

Several chapters of *Calculus in Context* contain discussions that expand and modify material from this reference. This includes a number of the applications in Chapters 8 and 10. Some of the historical narratives draw on

2. Katz, Victor J., *A History of Mathematics: An Introduction,* Addison-Wesley, Boston, 2009,

and

3. Kline, Morris, *Mathematical Thought from Ancient to Modern Times,* Oxford University Press, Oxford, 1972.

Chapter 1

The story of the astronomy and geometry of the Greeks that this chapter presents includes insights from the following.

4. Pedersen, Olaf, *A Survey of the Almagest*, Odense University Press, Odense, 1974.

5. Heath, Thomas L., *The Thirteen Books of Euclid's Elements*, Dover, New York, 1956, originally published by Cambridge University Press, 1925, http://aleph0.clarku.edu/~djoyce/java/elements/toc.html.

6. Heath, Thomas L., *A History of Greek Mathematics*, vol. 1, *From Thales to Euclid,* Dover, New York, 1981.

7. Heath, Thomas L., *A History of Greek Mathematics*, vol. 2, *From Aristarchus to Diophantus,* Dover, New York, 1981, https://archive.org/stream/historyofgreekma029268mbp#page/n7/mode/2up.

8. Hartshorne, Robin, *Geometry: Euclid and Beyond,* Springer-Verlag, New York, 2000.

The maps of Hipparchus and Claudius Ptolemy are reproduced in the Historical Introduction of the volume

9. *Report on the scientific results of the voyage of H.M.S. Challenger during the years 1873–76 under the command of Captain George S. Nares and the late Captain Frank Tourle Thomson.* Prepared under the superintendence of Sir C. Wyville Thomson and John Murray, published by order of Her Majesty's government, http://www.biodiversitylibrary.org/item/194225#page/14/mode/1up.

The depiction of the Chinese diagram illustrating a proof of the Pythagorean theorem can be found in

10. Needham, Joseph, *Science and Civilization in China*, vol. 3, *Mathematics and the Sciences of the Heavens and the Earth*, Cave Books, Taipei, 1986.

Mariners from ancient to more recent times determined the latitude of a vessel at sea by measuring the angles of elevation of stars. The determination of longitude turned out to be a much more difficult problem. An interested reader should engage

11. Sobel, Dava, *Longitude: The True Story of a Lone Genius Who Solved the Greatest Scientific Problem of His Time,* Walker, New York, 1995.

Chapter 2

The stories about Archimedes in Syracuse are taken from the segment *Life of Marcellus* in

12. Plutarch, *Plutarch's Lives*. Dryden's translation as revised by A. H. Clough, Little Brown, Boston, 1906, http://oll.libertyfund.org/titles/plutarch-plutarchs-lives-dryden-trans-vol-2.

The mathematical studies of Archimedes make use of Reference 7 and

13. Heath, Thomas L., *The Works of Archimedes*, Dover, New York, 2002, https://archive.org/details/worksofarchimede029517mbp.

Lunes—derived from the Latin for "Moon"—are obtained as intersections of circular regions. Leonardo da Vinci, the polymath of the Renaissance, was fascinated by them. See

14. Coolidge, J. L., *The Mathematics of Great Amateurs*, Clarendon Press, Oxford, 1991.

Problem 2.18 is taken from

15. Shen, Alexander, ed., "Mathematical Entertainments," *Mathematical Intelligencer* 23, no. 2 (2001).

Chapter 3

Details about the scientific revolution incorporate information from

16. Butterfield, Herbert, *The Origins of Modern Science,* Revised Edition, The Free Press, New York, 1965,

and

17. Koyré, Alexandre, *From a Closed World to an Infinite Universe*, Johns Hopkins University Press, Baltimore, 1957.

Both

18. Nicolaus Copernicus, *On the Revolutions of the Heavenly Spheres,* trans. A. M. Duncan, Barnes & Noble, New York, 1976,

and

19. Johannes Kepler, *New Astronomy*, trans. William H. Donahue, Cambridge University Press, Cambridge, 1992 (esp. chapter 24, pp. 316–21),

are essential original sources. The article

20. Gingerich, Owen, "The Great Martian Catastrophe and How Kepler Fixed It," *Physics Today* 64 (2011), http://sites.apam.columbia.edu/courses/ap1601y/PhysToday-2011-Kepler.pdf,

describes the ingenious way in which Kepler went about "fixing" the orbits of Mars (and Earth). The references

21. Moreland, Parker, *The JJMO Mars Parallax Project* and *The 2003 JJMO Mars Parallax Project*, http://www.jjmo.org/astro/astmenu2.htm,

illustrate the method of the astronomer Flamsteed by computing the Earth-Mars distance at the end of August 2003 (a time when Mars was at perihelion and near opposition).

Chapter 4

The discussion about algebra and its advances makes use of Reference 7 and the following primary source,

22. René Descartes, *La Géométrie,* appendix in *Discourse de la Méthod,* trans. David Eugene Smith and Marcia L. Latham, Dover, New York, 1954, http://djm.cc/library/Geometry_of_Rene_Descartes_rev2.pdf,

and the discussion

23. Cox, David A., "Introduction to Fermat's Last Theorem," *American Mathematical Monthly,* 101 no. 1 (1994), 3-14, http://math.stanford.edu/~lekheng/flt/cox.pdf.

Chapter 5

The account of the calculus of Leibniz follows Reference 1. This incorporated conclusions from

24. Child, J. M., *The Early Mathematical Manuscripts of Leibniz,* Open Court, Chicago, 1920. New printing by Cosimo, 2008,

and

25. Hofmann, J. E., *Leibniz in Paris, 1672–1676,* Cambridge University Press, Cambridge, 1974.

The application to Kepler's wine barrels relies on Kepler's original account

26. Johannes Kepler, *Stereometria doliorum Vinariorum* (The Stereometry of Wine Barrels), 1621. Translated by R. Klug from Latin to German as *Neue Stereometrie der Fässer,* Verlag von Wilhelm Engel-mann, Leipzig, 1908, https://archive.org/stream/neuestereometri01kluggoog#page/n4/mode/2up.

The images of the wine barrels come from the two sources

27. "Liber maister fisirit mirs recht." Bestimmung des Spunddurchmessers, aus dem *Visierbüchlein von 1485* and *Visierbuch (Fassberechnung)*, Huldrichus Kern/1531/Sign.: II 8505.

Chapter 6

The presentation of Newton's calculus relies on Reference 1. This drew on Newton's early works and

28. Isaac Newton, *Mathematical Principles of Natural Philosophy*, trans. Robert Thorp, W. Strahan and T. Cadell, London, 1777. Reprinted by Dawsons of Pall Mall, London, 1969,

for Newton's planetary theory. The application to the record-breaking hammer throw uses data from

29. Otto, Ralph, *Hammer Throw World Record Photo Sequence—Yuriy Sedykh.* Photosequence by Gabriele Hommel (©Hommel AVS 1992). Translated from the original German by Jürgen Schiffer, reprinted with permission from New Studies in Athletics, http://www.hammercircle.com/help-and-advice/otto_sedykh_wr.pdf.

Yuriy Sedykh's dramatic record-setting performance is captured in the video https://www.youtube.com/watch?v=4qAE2PrCVhY.

The astronomer George Airy's estimate of the mass of the Moon is taken from

30. Hughes, David W., "Measuring the Moon's Mass (125th Anniversary Review)," *Observatory,* 122 (2002), 61-70.

Chapter 7

The topics of this chapter are standard matters of differential calculus, as these are developed in most texts. The arrangement of topics and the details of their presentation expand on the approach taken in Reference 1. The monograph

31. Grabiner, Judith V., *The Origins of Cauchy's Rigorous Calculus,* Dover, New York, 1981,

provides relevant historical insights.

The surprisingly delicate convergence of points required in Example 7.24 was proved by Laurence Taylor, Department of Mathematics, University of Notre Dame. It will be included on the website for the book.

Chapter 8

L'Hospital's pulley problem is taken from Reference 1. The account there was adapted from the original in

32. De L'Hospital, *Analyse des Infiniment Petits pour l'Intelligence des Lignes Courbes,* De L'Imprimerie Royale, Paris, 1696.

The analysis of the George Washington Bridge follows Reference 1 and draws on the record

33. "The George Washington Bridge across the Hudson River at New York, N.Y.," *Transactions of the American Society of Civil Engineers,* 97 (1933).

The study of the inclined plane experiment of Galileo is also taken from Reference 1. The book

34. Drake, Stillmann, *Galileo: Pioneer Scientist*, University of Toronto Press, Toronto, 1990,

provides additional information about Galileo's quest to understand motion and the role that his experiments played. The importance of Fermat's principle of least time to geometric optics is explained in Chapters 26 and 27 of the celebrated

35. Feynman, Richard P., Leighton, Robert B., Sands, Matthew, and Gottlieb, Michael A., *The Feynman Lectures on Physics*, the New Millenium Edition, Basic Books, New York, 2013.

The story of the world's largest refracting telescope is told in a series of articles. One of them is

36. Hale, G. E., "The Yerkes Observatory at the University of Chicago, IV. The Forty-Inch Dome, and Rising Floor," *Astrophysics Journal,* 6 (1897) 37.

For more information about the Giant Magellan Telescope with its 8 meter parabolic mirrors, refer to http://www.gmto.org/.

The analysis of the shape of the surface of a rotating liquid profits from

37. Sabatka, Z., and Dvorak, L., "Simple Verification of the Parabolic Shape of a Rotating Liquid and a Boat on Its Surface," *Physics Education* 45, no. 5 (2010), 462-468.

Chapter 9

This chapter develops the essentials of integral calculus as they can be found in most texts. The approach here focuses heavily on two aspects. The first is an evolving look—from simpler to more refined—at the summation process central to integral calculus in both theoretical and applied situations. The second is the study of the indefinite integral or antiderivative of a function in terms of both theoretical concerns and the development of a basic set of examples.

The historical development of some of the conceptual elements that underlie integral calculus is explored in

38. Boyer, Carl B., *The History of The Calculus and Its Conceptual Development,* Dover, New York, 1959.

Chapter 10

This chapter applies integral calculus to a diverse set of concerns. The study of the domes of the Hagia Sophia and the Pantheon is taken from

39. Hahn, Alexander J., *Mathematical Excursions to the World's Great Buildings,* Princeton University Press, Princeton, New Jersey, 2012,

and uses data from the article

40. Como, Mario, and Grimaldi, Antonio, "Large Structures Behaviour: The Past and the Future," in *Novel Approaches in Civil Engineering,* Springer-Verlag, Berlin, 2004.

The discussion of the main cables of the suspension bridge follows Reference 1. The story of the geometry of the tractrix and the pseudosphere applies the integral calculus already developed to Leibniz's observations in Paris and the consequences of these observations. The analysis of the motion of a planet in its elliptical orbit relies on Reference 1. The fact that basic concepts of our physical reality such as force, momentum, and impulse are one side of a coin that has calculus on its other side, is illustrated by the application to interior ballistics. This study also relies on Reference 1. This used information in

41. Webster, G. A., and Thompson, L. T. E., "A New Instrument for Measuring Pressures in a Gun," *Proceedings of the National Academy of Sciences,* 5, no. 7 (1919), 259-263,

and

42. Webster, A. G., "On the Springfield Rifle and the Le Duc Formula," *Proceedings of the National Academy of Sciences*, 6 (1920), 289.

Chapter 11

This chapter introduces standard aspects of first- and second-order differential equations and, along the way, the polar coordinate system, the complex numbers, and the fundamentals of power series. The self-contained derivation of the solution of the differential equation $y'' + y = 0$ was provided by Matt Dyer, Department of Mathematics, University of Notre Dame.

The application to free fall under air resistance is based on

43. Timmerman, Peter, and van der Weele, Jacobus P., "On the Rise and Fall of a Ball with Linear or Quadratic Drag," *American Journal of Physics*, 67, no. 6 (1999), 538-546.

The study of suspension systems of automobiles is streamlined and simplified. For a fuller sense of the complexities involved, refer to

44. Dixon, John C., *The Shock Absorber Handbook*, John Wiley & Sons, West Sussex, UK, 2007.

The analysis of the geometry of a cable without the assumption that the downward pull on it is constant throughout (the case of a cable that supports no load other than itself—such as an electric cable or a telephone wire—is an example) follows

45. Freeman, Ira, "A General Form of the Suspension Bridge Catenary," *Bulletin American Mathematical Society,* 31 (1925), 425-429.

Chapter 12

This final chapter begins with a development of the calculus of polar functions. It then returns to Newton's gravitational theory and develops its equations, including the inverse square law, by applying polar calculus. The chapter also studies equiangular spirals and considers their occurrences in the natural world.

The polar approach to Newton's planetary orbits was first published by Leonhard Euler in 1749 and appeared in textbooks a few decades later. It can be found in

46. Wright, J. M. F., *A Commentary on Newton's Principia*, T. T. & J. Tegg, London, 1833.

This book is "dedicated to the tutors of the several Colleges of Cambridge." What is apparent immediately is that this text uses the notation of Leibniz rather than that of Newton. One hundred years after Newton's death, the superior approach of Leibniz was firmly established even at Cambridge.

The study of Newton's revolving orbits theorem incorporates the insights of

47. Nguyen, Hieu D., "Rearing Its Ugly Head: The Cosmological Constant and Newton's Greatest Blunder," *American Mathematical Monthly,* 115 (2008).

Image Credits and Notes

The figures in this text—with the exception of those that have captions pointing to other sources—were either created by the author with Adobe Illustrator or are taken from Reference 1 (for which the author owns the copyright). Images that are not already recognized and provided with permissions of use in the captions of the text, are given credit and referenced with permissions of use below. Additional information about some of the images is also given.

Figure 1.8. Digital image reproduced by Denise Massa, Visual Resources Curator Hesburgh Library, University of Notre Dame, from Reference 9. The original image has landmasses in light ochre and seas in light blue. A special thank you goes to Denise Massa for this image and several others.

Figure 1.37. Digital image reproduced by Denise Massa, Visual Resources Curator Hesburgh Library, University of Notre Dame, from Reference 9. The original image has landmasses in light ochre and bodies of water in light blue.

Figure 1.41a. This image is in the public domain. Refer to Reference 10 and also to http://commons.wikimedia.org/wiki/File:Chinese_pythagoras.jpg.

Figure 2.57. Dr. Will Noel, who had been the curator of manuscripts and rare books at the Walters Museum and creator of the Digital Walters, made this high-resolution image available to the author. The left side (the side containing the religious text) of the original image is beige in color with the Greek script in brown. The original right side (the transcription of Archimedes's work) of the image is in grayscale—as reproduced here—but slightly darker.

Figure 5.19. Digital images reproduced by Denise Massa, Visual Resources Curator Hesburgh Library, University of Notre Dame, from Reference 27.

Figure 6.32. There are at least two versions of this banknote. One is in green tones (as that of author's note) with the depiction of the solar system in yellow. The other is in gray tones with the solar system in rose. A portrait of Queen Elizabeth II graces the reverse side in each case.

Figure 6.54. Image reproduced by Denise Massa, Visual Resources Curator Hesburgh Library, University of Notre Dame. Lieve Verschuier's original oil painting features a dramatic orange sky with the comet streaking in bright yellow.

Figure 6.57. This diagram of a typical configuration of the GPS satellites has been released into the public domain for the free reuse or publication for educational purposes by the U.S. government. See http://www.gps.gov/multimedia/images/.

Figure 8.25. A survey of http://www.imss.fi.it/ms72/INDEX.HTM shows that the folios of Codex 72 of Galileo's working papers are light ocher in color with Galileo's writing in brown ink.

Figures 8.37 and 8.60. These figures are adapted from the Wikipedia entry *Aspheric lens*, available at https://en.wikipedia.org/wiki/Aspheric_lens. Image credit to ArtMechanic. Used under the Creative Commons Attribution-ShareAlike License.

Figure 8.40. In the original image, Jupiter is depicted in yellow and light brown tones.

Figures 8.41 and **8.43.** These images are adapted from Wikipedia's entry *Reflecting telescope*, available at https://en.wikipedia.org/wiki/Reflecting_telescope. They were created by Krishnavedala with a text editor and published under the Creative Commons Attribution-Share Alike 4.0 International license.

Figure 8.44. The only essential difference with the original image depicting the manufacture of the mirror is that the outer boundary of the cylindrical furnace is red.

The author is grateful to Jerry Amerongen, creator of the nationally syndicated comic strip Ballard Street, for granting him permission to weave six of his cartoons into the sequence in Section 8.6 that makes light of the pulley problem of the Marquis de L'Hospital.

Figures 8.47 and **10.53.** Many thanks to Greg Young of Notre Dame's School of Architecture for creating the clown.

Figures 10.5 and 10.7. A warm thank-you to Adam Heet, digital projects specialist of the Architecture Library of the University of Notre Dame, for producing these clear, precise, high-resolution images from the originals of Desgodetz and Palladio, respectively.

Figure 11.22. The author thanks Professor Paul Aisopoulos of the Faculty of Applied Technologies in the Department of Automotive Engineering, Thessaloniki, Greece, for allowing the use of his high-resolution image.

Figure 11.23. This image—with the coil in bright orange—appeared in the August 2010 article "How to Replace Conventional Shock Absorbers in Your Automobile" in the online automobile news magazine *Motorward.com*. Another word of thanks to Mr. Bruno Silva, owner and CEO of Motorward.com (http://www.motorward.com/), for permitting its use.

Figure 11.25. This diagram is adapted from one created by Nuno Nogueira and featured on the site *Harmonic oscillator* at https://en.wikipedia.org/wiki/Damping. It is published under the Creative Commons Attribution-Share Alike 2.5 Generic License.

Figure 12.19. The large outer chamber of the nautilus in Rutherford Platt's original is rendered in variable, soft, pearly, purple and violet hues. The smaller inner chambers are in beiges and browns. The approval of Time Inc. comes by way of a "quitclaim" asserting that Time Inc. does not object to the onetime, nonassignable, nontransferable, nonexclusive reproduction of this image in the traditional print format depicted here.

Figures 12.30 and 12.33. These images were featured as Astronomy Picture of the Day for January 19th and September 16th, 2014, respectively, on the website http://apod.nasa.gov/apod/archivepix.html.

Index